METEOROLOGICAL MONOGRAPHS

VOLUME 28 | NOVEMBER 2001 | NUMBER 50

SEVERE CONVECTIVE STORMS

Edited by

Charles A. Doswell III

American Meteorological Society
45 Beacon Street, Boston, Massachusetts 02108

ISBN 1-878220-41-1
ISSN 0065-9401 CIP 98-074523

Published by the American Meteorological Society
45 Beacon St., Boston, MA 02108

Printed in the United States of America
by Cadmus Journal Services, Lancaster, PA

TABLE OF CONTENTS

PREFACE

This monograph owes its origins to a predecessor monograph published in 1963, entitled *Severe Local Storms,* also published by the American Meteorological Society (AMS), and edited by Dr. David Atlas (alas, it is now out of print). In his preface, Dr. Atlas notes that the 1963 monograph was written, in turn, as an update to an even earlier comprehensive review of the topic: *The Thunderstorm* (published by the U.S. government in 1949 as an outgrowth of a post–World War II study prompted by the importance of thunderstorms to aviation). The most recent comprehensive review of thunderstorms is the three-volume set edited by Dr. Edwin Kessler—*Thunderstorms: A Social, Scientific, & Technological Documentary*—published originally by the U.S. government in installments: volume 1 (*The Thunderstorm in Human Affairs*) in December 1981, volume 2 (*Thunderstorm Morphology and Dynamics*) in February 1982, and volume 3 (*Instruments and Techniques for Thunderstorm Observation and Analysis*) in April 1982. Of course, this three-volume set is addressed to thunderstorms in general, not specifically to the topic of *severe* thunderstorms.

As a student interested in severe convection, when I found out about the 1963 monograph, I quickly obtained a copy and became absorbed by the picture that emerged from reading the chapters. The material I read therein, including the interesting discussions appended to the chapters, became a major element in my understanding of severe convection. I have had many occasions during my professional career to reference material from that monograph. My colleagues have indicated that they, too, found the 1963 monograph to be useful.

Many of us have wished to see the subject reviewed once again, with the maturation of the new technologies of observation and analysis that were just beginning to emerge in 1963. The science of severe convection has undergone something of a renaissance in the time since the 1963 monograph. These scientific changes have been driven by exciting new observations (e.g., the implementation of a national network of Doppler radars, the development of a national lightning detection network, and the growth of scientific storm chasing, to name just a few) and by a revolution in computer capability that has allowed (a) the use of numerical simulation models for exploring the quantitative aspects of severe storm dynamics and (b) rapid quantitative analysis of the new observations. Hence, it was agreed that there was a need to gather another distinguished list of contributors to offer reviews of various aspects of severe convective storms.

The 1963 monograph was associated with an AMS conference on severe local storms in 1962 and included transcripts of the discussions that followed the papers presented by the chapter authors. For this updated review of the material, we chose to follow a different example of Dr. Atlas's fine editorial leadership: *Radar in Meteorology,* another AMS publication, wherein each chapter had a review panel assigned to it, with a panel chair to moderate the reviews of each chapter.

It was decided that we had to maintain as broad a context as possible for the word "severe" and therefore heavy precipitation is included in the list of topics, despite its usual separation from the traditional list of the hazards produced by severe convection in the United States: hail, damaging convective winds, and tornadoes. Mesoscale convective systems were not included within the 1963 monograph because they were not recognized at that time as a distinct phenomenon; clearly, this topic called for a comprehensive review. Whereas the 1963 monograph provided a forecasting review from the national perspective, the two forecasting chapters in this new monograph are written from a more local viewpoint. This was done deliberately, as new reviews of forecasting severe convection from the national perspective have appeared recently elsewhere. There was a need to have a chapter devoted to severe convection in the Tropics, since severe convection has generally and inaccurately been thought of almost exclusively as a midlatitude phenomenon. As with the 1963 monograph, material about hailstorms and the microphysical research associated with them was needed; in parallel, we have chapters describing wind- and rainstorms, and the physical process research tied to them, as well. The topic of storm electrification needed updating in light of many new ideas and observations since 1963. Moreover, a separate chapter reviewing numerical cloud modeling and its huge contributions seemed an obvious need. A chapter on tornadoes and tornadic storms was not included in the 1963 monograph, an omission noted by Dr. Atlas in his preface; it was agreed that such a review chapter would be needed for completeness. Finally, it was decided to separate the topics of mesoscale and synoptic-scale aspects of severe convection, since the content of these subdisciplines has grown so tremendously since 1963. Separate chapters about observation systems, like radars and satellites, were not included; not only have observing systems been reviewed extensively in recent years, but it was felt that authors should naturally refer to the tools of observation in the context of their chapters.

The content of each chapter has been structured to stand alone, and so there is some inevitable redundancy among the chapters. The authors have developed common, basic concepts to serve their own specific needs in developing the material covered in each chapter, and so this apparent redundancy is not without value. That is, it illustrates the variety of perspectives even within this fairly narrow subdiscipline of meteorology.

I hope you will share my belief that this monograph has been quite successful in several respects. The topic of severe convection has been given not only a wide-ranging review, but many of the authors have taken advantage of this opportunity to share some of their newest ideas as well. This set of chapter authors has more than fulfilled my expectations in not simply writing reviews consisting of a dry, third-person listing of existing publications. Rather, they have conveyed their passion for their subjects and written chapters that should stimulate readers to further study. Although the format of the 1963 monograph based upon presentations at a conference was not repeated, most of our chapters were presented as invited papers at the *19th AMS Conference on Severe Local Storms* in 1998. The feedback received about those presentations was that many of our colleagues were looking forward to this monograph's publication. Therefore, I expect that students of severe convection should find these reviews useful for many years to come. Moreover, in the long term, this monograph should serve our science in the same way that the 1963 monograph still does: it provides an encapsulated review of the science at the time the papers were written.

In rereading the 1963 monograph recently, I find it interesting in several respects from this historical view. First, several of the ideas contained in the 1963 monograph represent hints of directions that the science has followed since then. For example, in Dr. Frank Ludlam's chapter "Severe local storms: A review," he says,

> It is tempting to look for the spin of the tornado in the vorticity present in the general air stream as shear and tilted appropriately in the vicinity of the interface between the drafts as a consequence of the up- and down-motions.

This sentence is quite prescient when considered in light of subsequent research that has demonstrated the importance of vorticity tilting in supercells (although perhaps not directly in tornadogenesis), and yet it really amounts to little more than an offhand remark in Ludlam's chapter.

Another retrospective aspect of the 1963 monograph is the *lack* of clarification of some issues over the decades since that monograph appeared. In Mr. Don C. House's contribution, "Forecasting tornadoes and severe thunderstorms," for example, the author is obviously concerned with prefrontal lines of severe convection. This topic is a major focus in Dr. Chester Newton's chapter, "The dynamics of severe convective storms," as well. Nevertheless, our meteorological science has not yet provided a comprehensive consensus regarding the processes associated with prefrontal convective lines. In spite of a continuing recognition of the importance of prefrontal convection in the years since 1963, this remains one of many unresolved issues.

In contrast with these is the enduring value of the late Professor T. Theodore Fujita's contribution to the 1963 monograph, "Analytical mesometerology: A review." The passage of time has done little to diminish the widely recognized clarity and insight of this work. In fact, it was felt that this material needed no substantial updating or clarification.

Finally, it is interesting to note the shifting of paradigms with time. When reading historical scientific contributions, it is important to understand the context within which those works were originally written. For instance, in the 1963 monograph, Dr. Chester Newton's chapter, "The dynamics of severe convective storms," presents considerable insightful discussion of the interaction between deep convective storms and their environment. However, his paper was based primarily on linear theory and very limited observations. The development of numerical cloud models and improved observing tools has revolutionized our understanding of this interaction by allowing the quantitative diagnosis of how deep convection interacts with a sheared environment. Some might be tempted to regard Newton's views on the storm–environment interaction as "crude" or lacking in insight if the context in which the paper was written is ignored. That is, numerical cloud simulations in the early 1960s were still very much in their infancy and had not matured to the level where they could offer both qualitative and quantitative understanding of this complex topic. In fact, the relative immaturity of numerical simulation of deep convection can be seen in Professor Yoshimitsu Ogura's 1963 monograph chapter on "A review of numerical modeling research on small-scale convection in the atmosphere," when compared to what is now being done. Nevertheless, if our views have changed in consequence of new understanding that came from sophisticated numerical simulations and new observations since 1963, this does not detract in any meaningful way from Dr. Newton's valuable review. Quite the contrary is true: all of us benefit from the understanding that scientific ideas can change dramatically, as well as incrementally, as new tools are developed and new data are collected. Old, superceded ideas have lasting value as milestones on the path along which current ideas have evolved. The history of how our current understanding came about is important to know as we go about our research.

With this in mind, then, I believe it worthwhile to ponder just what severe convective storms meteorologists will think about the contents of *this* monograph in another 35 years or so. Which among the ideas contained herein will be seen as "crude" and which will be seen as "deep insight"? Which passing remarks will presage important new research directions and which paradigms will have been overthrown by new research? It is the nature of science that new and better ideas routinely supplant older ones, although usually not without considerable controversy. The health of our science is measured in a very real sense by the amount of ongoing controversy. To the extent that all scientists agree on some issue, that issue is scientifically dead, or at least dormant. It is only through continuing examination and revision of existing ideas that we make progress in any scientific field. Ideas can have intuitive appeal, have renowned scientists as advocates, and be cited in many publications and still turn out to be "wrong" in the sense of needing revision as new research is done. I'm pleased to say that my experience in the general domain of severe convection suggests that we are at present a *very* healthy science! Nothing in this monograph is sacred; readers of this monograph should keep this in mind as they plunge into the outstanding contributions herein. It is my hope that this material will serve to stimulate another generation of scientists in just exactly the way that the 1963 monograph stimulated my colleagues and me: to pursue research on various aspects of severe convective storms. May the contributions resulting from that research eventually require yet another update on this topic in the future.

It should not be a surprise that many people deserve acknowledgement for their efforts on behalf of this monograph. Apart from the chapter authors, the review panels and panel chairs deserve the lion's share of the credit for what I believe to be the successful completion of this monograph. Panel participants are listed prominently at the start of each chapter, and we owe them all a great deal of thanks for their commitment to excellence within their respective chapters.

I particularly want to thank Drs. Harry Orville and Jeff Trapp for their extraordinary contributions as the monograph neared completion. I benefited personally from valuable discussions with several of my current and former colleagues at the National Severe Storms Laboratory: Drs. Harold Brooks, Bob Maddox, Dave Schultz, Dave Stensrud, and Conrad Ziegler. Finally, Dr. Keith Seitter, Dr. Stuart Muench, and Mr. Ken Heideman of the AMS staff are gratefully acknowledged for their support and hard work, and for their patience in dealing with all our technical questions, as we completed this monograph.

Charles A. Doswell III
Editor
Norman, Oklahoma

CONTRIBUTORS

Dr. Gary M. Barnes
Department of Meteorology
University of Hawai'i at Mãnoa
2525 Correa Road
Honolulu, Hawai'i 96822-2219
E-mail: garyb@soest.hawaii.edu

Prof. Howard B. Bluestein
School of Meteorology
University of Oklahoma
100 E. Boyd, Rm. 1310
Norman, OK 73019
E-mail: hblue@ou.edu

Bruce A. Boe
ND Atmospheric Resource
900 E. Boulevard Ave.
Bismark, ND 58505-0850
E-mail: bboe@tic.bisman.com

Dr. Lance F. Bosart
Department of Earth and Atmospheric Sciences
The University of Albany/SUNY
1400 Washington Ave.
Albany, NY 12222
E-mail: bosart@atmos.albany.edu

Mr. Robert S. Davis
Pittsburgh National Weather Service Forecast Office
192 Shafer Road
Moon Township, PA 15108
E-mail: Robert.Davis@noaa.gov

Dr. Charles A. Doswell III
CIMMS, The University of Oklahoma
Sarkeys Energy Center
100 East Boyd Street, Room 1110
Norman, OK 73019-1011
E-mail: cdoswell@hoth.gcn.ou.edu

Dr. Gregory S. Forbes
Severe Weather Expert
The Weather Channel
300 Interstate North Parkway
Atlanta, GA 30339
E-mail: gforbes@weather.com

Mark A. Fresch
WSR-88D Radar Operations Center
1200 Westheimer Drive
Norman, OK 73069
Mark.A.Fresch@noaa.gov

Dr. J. Michael Fritsch
Department of Meteorology
Penn State University
University Park, PA 16802
E-mail: fritsch@ems.psu.edu

Bob Johns
Storm Prediction Center
1313 Halley Circle
Norman, OK 73069
E-mail: Robert.Johns@noaa.gov

Prof. Richard H. Johnson
Dept. of Atmospheric Science
Colorado State University
Fort Collins, CO 80523
E-mail: rhj@atmos.colostate.edu

Dr. Robert Davies-Jones
National Severe Storms Laboratory
1313 Halley Circle
Norman, OK 73069-8493
E-mail: bobdj@nssl.noaa.gov

Dr. Charles A. Knight
National Center for Atmospheric Research
P.O. Box 3000
Boulder, CO 80307-3000
E-mail: knightc@ucar.edu

Maj. Nancy C. Knight
National Center for Atmospheric Research
P.O. Box 3000
Boulder, CO 80307-3000
E-mail: knightn@ucar.edu

Prof. Dennis Lamb
Department of Meteorology
The Pennsylvania State University
503 Walker Bldg.
University Park, PA 16802
E-mail: lno@ems.psu.edu

Prof. Edward Lozowski
Department of Earth and Atmospheric Sciences
University of Alberta
Edmonton, Alberta, CANADA T6G 2E3
E-mail: Edward.Lozowski@ualberta.ca

Dr. Brian E. Mapes
NOAA-CIRES Climate Diagnostics Center
325 Broadway, mailcode R/CDC1
Boulder, CO 80305-3328
E-mail: bem@cdc.noaa.gov

Alan R. Moller
National Weather Service Forecast Office
3401 Northern Cross Boulevard
Fort Worth, TX 76137-3610
E-mail: al.moller@noaa.gov

Prof. Harold D. Orville
Institute of Atmospheric Sciences
South Dakota School of Mines and Technology
501 E. St. Joseph Street
Rapid City, SD 57701
E-mail: orville210@home.com

James H. Renick
Meteorological/Technical Consultant
TECH-KNOWLOGY Consulting Services
11 Warwick Drive
Red Deer, AB, Canada T4N 6L4
E-mail: renick@telusplanet.net

Dr. Paul Smith
South Dakota School of Mines and Technology
Rapid City, SD
E-mail: psmith@fire.ias.sdsmt.edu

Dr. R. Jeffrey Trapp
National Severe Storms Laboratory
1313 Halley Circle
Norman, OK 73069
E-mail: Jeff.Trapp@nssl.noaa.gov

Prof. Roger M. Wakimoto
Department of Atmospheric Science
UCLA
405 Hilgard Ave.
Los Angeles, CA 90095-1565
E-mail: roger@atmos.ucla.edu

Dr. Louis J. Wicker
National Severe Storms Laboratory
1313 Halley Circle, Norman, OK 73069
E-mail: Louis.Wicker@nssl.noaa.gov

Dr. Robert B. Wilhelmson
Department of Atmospheric Sciences
University of Illinois
105 S. Gregory Street
Urbana, IL 61801
E-mail: bw@ncsa.uiuc.edu

Dr. Earle R. Williams
Parsons Laboratory
Massachusetts Institute of Technology
MIT 48-211
Cambridge, MA 02139
E-mail: earlew@ll.mit.edu

Chapter 1

Severe Convective Storms—An Overview

CHARLES A. DOSWELL III

National Severe Storms Laboratory, Norman, Oklahoma

REVIEW PANEL: Frederick Sanders (Chair), Robert A. Maddox, Joseph Zehnder

1.1. Basic concepts of convection

a. What is convection?

In general, *convection* refers to the transport of some property by fluid movement, most often with reference to heat transport. As such, it is one of the three main processes by which heat is transported: radiation, conduction, and convection. Meteorologists typically use the term convection to refer to heat transport by the vertical component of the flow associated with buoyancy. Transport of heat (or any other property) by the nonbuoyant part of the atmospheric flow is usually called *advection* by meterologists; advection can be either horizontal or vertical.

Convection takes many forms in the atmosphere; a comprehensive treatment of the topic can be found in Emanuel (1994). This monograph's concern is severe convection—that is, the variety of hazardous events produced by deep, moist convection. Hazardous weather events (large hail, damaging wind gusts, tornadoes, and heavy rainfall) are generally the result of the energy released by phase changes of water. A circular convective cloud with a 5-km radius and 10-km depth contains, at any given instant, about 8×10^8 kg of condensed water, assuming an average condensed water content of 1 g m^{-3}. During the condensation of that water, roughly 10^{14} J of latent heat energy is released over a timescale of roughly 25 min (see below). For comparison purposes, a one-megaton bomb releases about 4×10^{15} J of heat (see Shortley and Williams 1961, p. 903), albeit in a tiny fraction of a second. Thus, at least in terms of released energy, this modest convective cloud is comparable to a 25-kiloton bomb. It is this released heat that powers convective storms. Most of the energy is expended against gravity, but some portion may also create hazardous weather.

The released heat contributes to buoyancy, B, an essential aspect of convective storms. Buoyancy is defined in somewhat simplified terms by

$$B \equiv g \frac{T - T'}{T'}, \tag{1.1}$$

where g is the acceleration due to gravity, T is the temperature of a parcel, and T' is the temperature of the surrounding environment. Buoyancy, of course, can be either negative or positive. If B is integrated from the level of free convection (LFC) to the equilibrium level (EL) above the LFC (see Fig. 1.1), the result is convective available potential energy (CAPE). It is this energy that is responsible for the convective updraft and for many of the hazards produced by the convection.

Downdrafts have their own source of energy, however. Downdrafts are driven by negative buoyancy, also derived from phase changes (mostly evaporation) and from the "loading" effect of precipitation. Whereas updrafts transport warm air upward, downdrafts transport cold air downward. Downdrafts are also responsible for some hazardous weather. In either updrafts or downdrafts, the net result is stabilization, but the mechanisms are distinct, which will be shown later.

b. Thunderstorms and deep, moist convection

I will use the phrase *deep, moist convection* (DMC) frequently in what follows, in lieu of the most common word associated with DMC: thunderstorms. This is because, in some instances, hazardous weather can be produced by nonthundering convection. Irrespective of this minor semantic distinction, DMC is the result of a type of instability. Consider the inviscid vertical momentum equation:

$$\frac{dw}{dt} = -\frac{1}{\rho}\frac{\partial p}{\partial z} - g, \tag{1.2}$$

where $w \equiv dz/dt$ is the vertical component of the flow, z is geometric height, ρ is the density, and p is the pressure. The vertical pressure gradient force is the

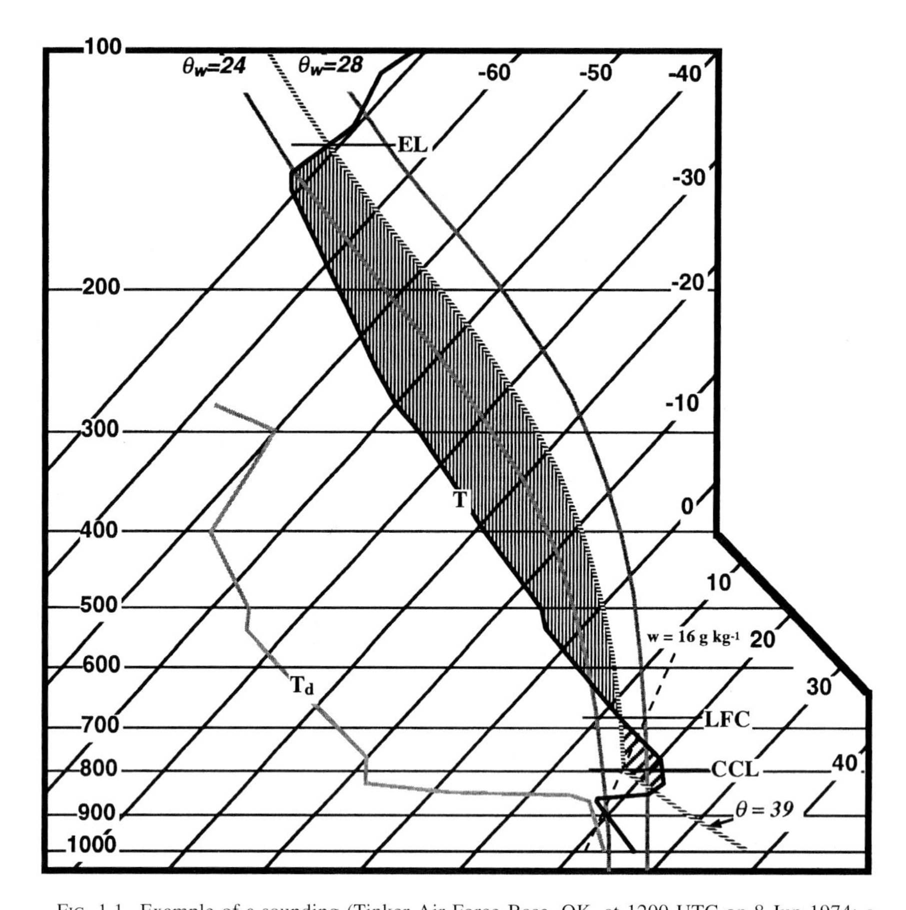

FIG. 1.1. Example of a sounding (Tinker Air Force Base, OK, at 1200 UTC on 8 Jun 1974; a day with significant tornadoes in the area) plotted on a skew T–log$p$ diagram; thin slanting solid lines are isotherms in °C, whereas the horizontal thin solid lines are isobars in hPa. The heavy solid line marked T is the temperature observation during balloon ascent and the heavy dashed line marked T_d is the observed dewpoint temperature. Two pseudoadiabats [thin curved lines labeled $\theta_w = 24$ (°C) and $\theta_w = 28$, where θ_w is the wet-bulb potential temperature] are shown, as well as one mixing ratio line (thin dashed line labeled $w = 16$ g kg^{-1}). The thin solid line labeled $\theta = 39$ is the dry adiabat through the forecast maximum surface temperature and the dash-dotted line labeled $\theta_w = 25.8$ is the pseudoadiabat associated with the lifted parcel ascent curve. The solid line labeled CCL is the condensation level for the forecast surface parcel; the solid line labeled LFC is the parcel's level of free convection; and the solid line labeled EL is the parcel's equilibrium level. The stippled area between the LFC and the EL represents the CAPE of the lifted parcel, and the hatched area below the LFC represents the CIN of the lifted parcel.

first term on the rhs of Eq. (1.2). Since the vertical acceleration is zero in a hydrostatic atmosphere, buoyancy is associated with an unbalanced pressure gradient force, caused by density perturbations. Equation (1.2) can be transformed using the definition of vertical motion and textbook linearization methods (e.g., Hess 1959, p. 95 ff.) to yield

$$\frac{dw}{dt} = \frac{d^2z}{dt^2} = B = -\frac{g}{T'}(\Gamma - \gamma)z, \qquad (1.3)$$

where T' is the environmental temperature, Γ is the parcel lapse rate $(-dT/dz)$, and γ is the environmental lapse rate $(-dT'/dz)$.[1] When the coefficients are constant, Eq. (1.3) is a simple, second-order differential equation that has a simple solution:

$$z(t) = z_o \exp[-iNt], \qquad (1.4)$$

where z_o is the initial height of the parcel and N is the so-called Brunt–Väisälä, or buoyancy, frequency

[1] The linearization leading to (1.3) ends up equating buoyancy with conditional instability, which is not quite correct. See Hess's (1959) textbook and Schultz et al. (2000) for a discussion of the approximations used to linearize the equation this way.

$$N^2 = \frac{g}{T'}(\Gamma - \gamma)z. \tag{1.5}$$

The solution (1.4) implies an instability (i.e., exponential growth of an infinitesimal upward displacement) whenever the square root of N^2 is imaginary; this occurs whenever the environmental lapse rate exceeds that of an ascending parcel (typically assumed to be adiabatic). Since γ is normally not greater than the dry adiabatic lapse rate, the context is clearly associated with conditional instability (i.e., $\gamma > \Gamma_m$, where Γ_m is the moist adiabatic lapse rate). The analysis of parcel instability leading to (1.4) is only appropriate when the textbook linearization assumptions are valid. Actual parcel instability leading to DMC is primarily associated with finite vertical displacements; hence, the key to the possibility for growth of convective storms is the presence of CAPE, not the environmental lapse rates alone (see Sherwood 2000; Schultz et al. 2000). Not all situations with conditional instability are characterized by parcels with CAPE.[2] Thus, the moisture content of the air is critical in knowing whether conditional instability actually contains the potential for parcels to become buoyant (i.e., to have CAPE). In most cases, energy must be supplied to lift the parcel through its condensation level to its LFC; the amount of this supplied energy is known as the convective inhibition (CIN). From the LFC to the EL, the parcel accelerates vertically, drawing the energy for this acceleration from the CAPE. Generally, parcels will overshoot the EL and then experience the stable version of (1.4), undergoing what are called Brunt–Väisälä oscillations that are ultimately damped by viscosity, so the rising parcels should (at least theoretically) accumulate in an anvil cloud at the height of the EL.

The origin of buoyant instability is heat, both latent and sensible, that is produced at low levels in the atmosphere as a result of solar heating and evapotranspiration of water vapor (also due to solar heating) into the lower troposphere.[3] The process of convection alleviates the instability created by the accumulation of heat at low levels. DMC takes the excess sensible heat and water vapor from low levels and expels it into the upper troposphere, and transports potentially cold, dry air downward, thereby alleviating the instability. If DMC begins, as long as there is instability of this special sort available and there is enough lift to realize it, DMC will continue until the instability is removed, whereupon the convection ceases.

According to Emanuel (1994, p. 479 ff.), in most of the Tropics, the atmosphere is approximately in a state of convective near-equilibrium because DMC stabilizes the environment much more quickly than the large-scale processes can act to destabilize the atmosphere. In other words, the instability is released soon after it is created, and the tropical atmosphere tends to stay rather close to neutral stability conditions with respect to deep convection. In midlatitude continental convection, however, the instability is not always realized as soon as it is created. Rather, the instability can be "stored" in the environment, owing to convective inhibition (or CIN), and can increase over a period of up to several days, finally to be released explosively in the form of severe convection.

Updraft instability is not necessarily equivalent to downdraft instability (see Johns and Doswell 1992). The processes associated with downdraft instability are discussed in considerable detail in chapter 7 of this monograph. The formation of precipitation inhibits updrafts owing to water loading, but it enhances downdrafts. However, without the precipitation caused by updrafts, no evaporation can take place to chill the air, creating the negative buoyancy that is a major factor in creating downdrafts. Hence, updrafts are necessary but not sufficient for downdraft instability. Destructive downdrafts can be associated with weak updrafts (Johns and Doswell 1992), so that parameters focused on updraft instability may not reveal such situations.

Although DMC is responsible for considerable destruction and perhaps thousands of casualties per year worldwide, there can be little doubt that most DMC is benign. The primary benefit of convection is rainfall, and much of the world's rainfall results from DMC, especially in the Tropics. There is also a contribution by lightning to the transformation of atmospheric nitrogen into forms useful to plants.

c. Storm electrification

The occurrence of electrification in DMC is common enough that most hazardous weather associated with convection is, indeed, produced by thunderstorms. Lightning arises in electrified clouds to alleviate the charge buildup associated with the DMC. Unfortunately, the charging mechanism continues to resist a definitive explanation (Vonnegut 1994). Various hypotheses exist (Vonnegut 1963; Williams 1985; Saunders 1993), and it might be that all these hypotheses are appropriate explanations in some cases. Even the basic conceptual models of charge distribution are under scrutiny at the moment (see Rust and Marshall 1996). Storm electrification in the context of severe convection is described in detail in chapter 13 of this monograph.

[2] When discussing CAPE in specific cases, it is critical to know which parcel is being lifted, as discussed by Doswell and Rasmussen (1994). For my purposes in this chapter, I am not specifying which parcel is being used, as this discussion is general.

[3] In some instances, the instability may be driven or maintained by sensible heat loss at upper levels, owing to radiative processes, rather than heat gain at lower levels. This is not likely to be at work in most forms of severe convection.

Lightning is arguably the most dramatic aspect of thunderstorms in general and it can be quite hazardous in its own right. Lightning casualties are as high as those associated with other DMC hazards, such as tornadoes, but tend to be singular rather than in groups and so may not be as well reported as tornado casualties (López et al. 1993). Because a thunderstorm need not be severe in any of the conventional ways for its lightning to pose a threat, no thunderstorm can be viewed as completely nonhazardous.

The distribution of thunderstorms worldwide has been known historically only in terms of human observers hearing thunder and reporting it, which creates biases in the perceived distribution of thunderstorms. Recently, however, remote sensing systems have been developed to detect lightning. The most widespread such systems, as of this writing, detect cloud-to-ground (CG) lightning strikes (Orville and Henderson 1986). In the near future, it is likely that intracloud (IC) lightning will also be detectable via sensitive optical systems to be carried aboard spacecraft (Christian et al. 1989). Thus, it may soon be possible to have a realistic picture of the worldwide distribution of thunderstorms.

In spite of the drama and importance of lightning, it appears that it is no more than a trivial component of a thunderstorm's energy budget. That is, the total energy expended in lightning flashes in a single thunderstorm cell is on the order of 10^9–10^{10} J (Williams and Lhermitte 1983), whereas we have seen that the energy released by condensation of the input water vapor is on the order of 10^{14} J. This should not be interpreted to mean that electrification is not important for processes within DMC. For example, it has long been known that electric field affects coalescence efficiency (Byers 1965, p. 158 ff.). Moreover, lightning activity should be considered a potentially useful indicator of DMC processes (MacGorman and Rust 1998), especially when combined with other data (e.g., radar, radiosondes, etc.).

d. The cell as a building block in organized convective systems

Modern concepts of DMC have their origins with the post–World War II Thunderstorm Project, culminating in the publication of the report by Byers and Braham (1949). This archetypal project set the pattern for many subsequent projects aimed at understanding more about convective storms [e.g., the mesonetwork used by Fujita (1955); the NSSL mesonetworks analyzed by Barnes (1978); the National Hail Research Experiment described by Foote and Knight (1979); or the Florida Area Cumulus Experiment described by Barnston et al. (1983)]. A key concept grew out of this project: the thunderstorm cell. As depicted in Byers and Braham (1949), the cell is the basic organizational structure of all thunderstorms (Fig. 1.2), and this notion became the fundamental paradigm for thunderstorms. Browning (1964) used it as a basis for developing his pioneering ideas associated with a special form of severe convection, the supercell. The cell was also the basis for a taxonomy of severe hailstorms developed by Marwitz (1972a,b,c).

The survival of the cell concept to this day is a tribute to the fundamental insights arising from the pioneering Thunderstorm Project studies. However, as with any conceptual model, it has limitations that tend to become more apparent with time. Although no conceptual model should be taken literally, a troubling aspect of the model is the depiction of convective cloud within regions of downdraft; generally, downdrafts are characterized by the disappearance of cloud particles, since they are small and tend to evaporate quickly (Kamburova and Ludlam 1962). This minor cosmetic modification was performed in Doswell (1985, Figs. 2.16, 2.18, 2.19).

Much controversy has centered on how entrainment occurs in thunderstorms (see Warner 1970, 1972; Simpson 1971, 1972; Newton 1963, his Fig. 12). Entrainment has been a troublesome issue related to DMC for a considerable time. As originally envisioned (Stommel 1947; Scorer and Ludlam 1953; Squires and Turner 1962), entrainment of environmental air occurs along the lateral boundaries of the cloud. A competing concept, discussed in Squires (1958), Paluch (1979), Deardorff (1980), and Emanuel (1981), has been the notion that penetrative downdrafts within the cloud might contribute substantially to entrainment.

Some time ago, Levine (1959) proposed entrainment based on a bubble theory. More recently, Blyth (1993) has revisited Levine's model, proposing that the major form of entrainment for convection is via the toroidal circulation (i.e., like a smoke ring) associated with the thermal (i.e., bubble) model of a convective cloud. Note that the thermal envisioned in this bubble theory of DMC is much larger than the small thermals, arising out of the convective boundary layer in shallow convection, that give rise to ordinary cumulus clouds.

Another interesting aspect of DMC is the steadiness of the updraft. DMC has been modeled as either plumes or thermals (bubbles), and some controversy has continued for years about which is most appropriate (see, e.g., Scorer and Ludlam 1953; Squires and Turner 1962). Real thunderstorms can have drafts ranging from quasi-steady plumelike to bubblelike in character. The steadiness of the drafts is used often in thunderstorm taxonomies like those developed by Marwitz. Unfortunately, real storms tend to resist being put into polarized categories (see Foote and Frank 1983).

For my purposes in this discussion, the prototypical thunderstorm cell is considered to be a single thermal/bubble: a collection of buoyant air parcels. I do not consider it to be an accident that the typical life cycle of the cells observed in the Thunderstorm Project was

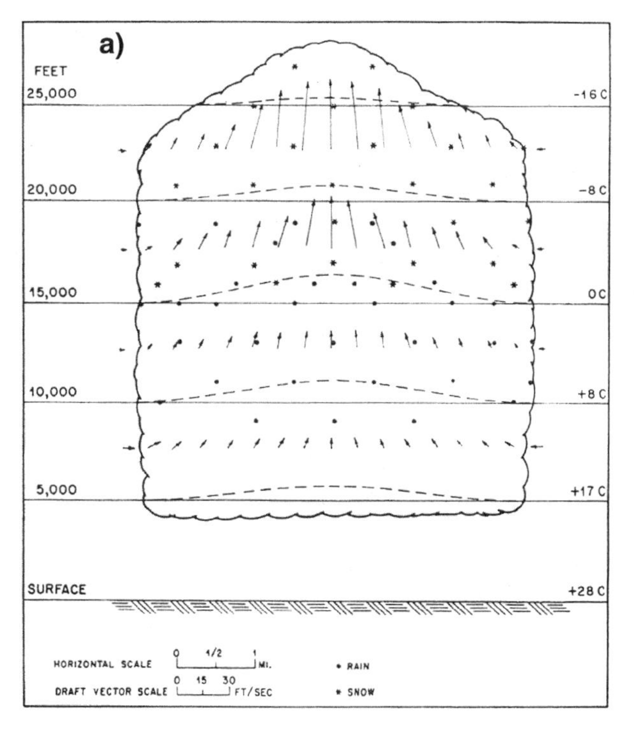

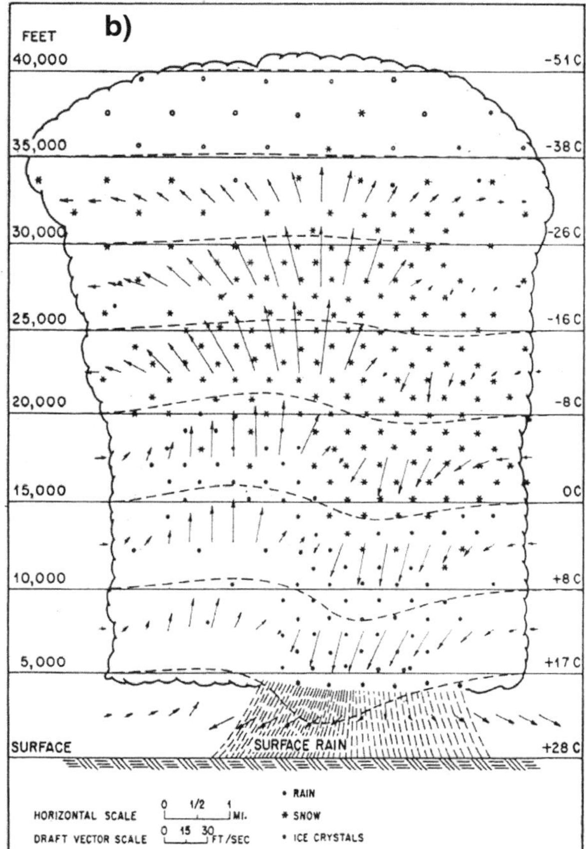

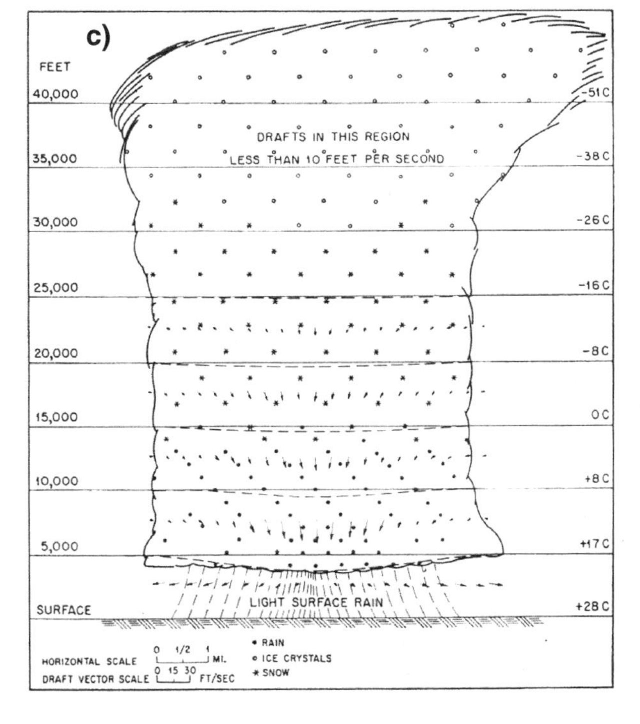

FIG. 1.2. Classic thunderstorm cell schematics from Byers and Braham (1949) showing (a) the cumulus stage, (b) the mature stage, and (c) the dissipating stage.

over a timescale corresponding closely to the time it takes for a lifted parcel to rise from its low-level origins to the EL; this is the *convective timescale,* τ_c. If a characteristic scale for vertical motion of 10 m s^{-1} is assumed, and the depth of the convection scales is assumed to be 10 km, then it takes on the order of $\tau_c = 10^3$ s (~25 min) for a parcel to rise through the depth of the storm.

Using this concept, a modified thunderstorm cell life cycle is illustrated in Fig. 1.3. This life cycle has stages corresponding to those depicted in Fig. 1.2, but show the cell in terms of a rising bubble of buoyant air. Note that the air motion on the exterior of the ascending bubble is downward (bubble-relative). Time-lapse images of convective clouds indicate clearly that this is the case. It is also noteworthy that

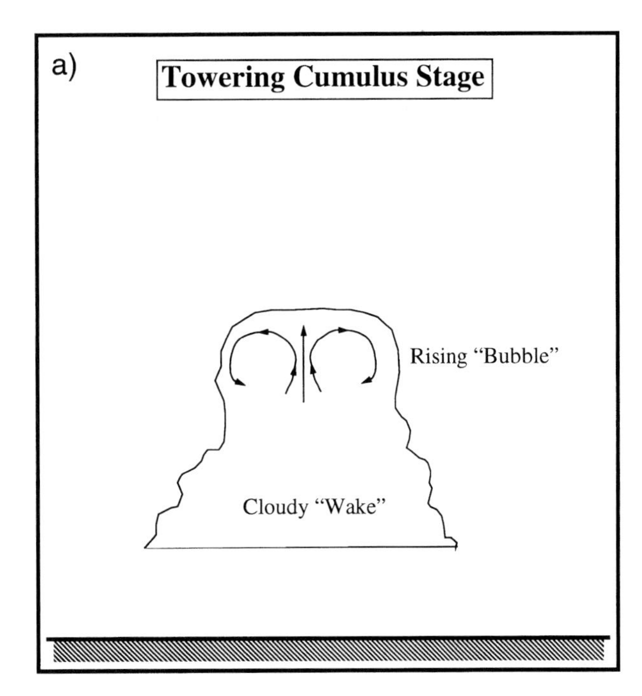

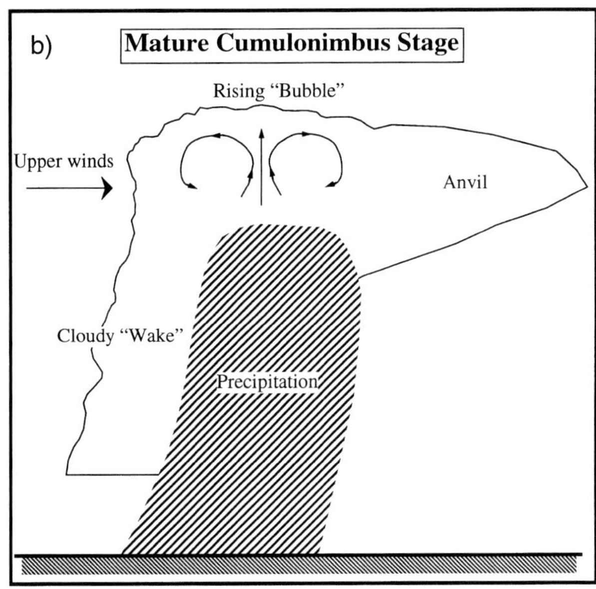

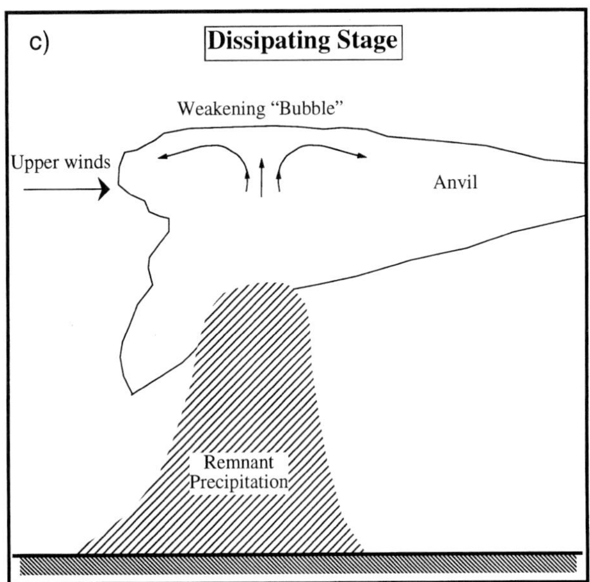

FIG. 1.3. A modified version of the life cycle of a single convective cell, showing the progression of a bubble of buoyant air. Features on the figure are labeled; (a) the towering cumulus stage, (b) the mature stage, and (c) the dissipating stage of Fig. 1.2.

the rise rate of the bubble is not necessarily equal to the vertical motion at the center of the spherical vortex circulation that defines the bubble (Turner 1973, p. 186 ff.). Figure 1.3 is very much in the spirit of that proposed by Scorer and Ludlam (1953). In Fig. 1.3a, the thermal has passed the LFC and is rising rapidly during the towering cumulus stage. This is followed by the mature stage, as precipitation has reached the surface (Fig. 1.3b). The downdraft in such a cell perhaps fits the "starting plume" model of Turner (1962) rather well, with a descending, negatively buoyant bubble leading a more or less continuing plume of downdraft in the precipitation cascade. When the rising bubble reaches its EL, the cell has reached its dissipating stage (Fig. 1.3c).

Such an isolated cell is unlikely in the real world; perhaps the initial developments of some DMC most resemble this prototypical single, isolated bubble (Fig. 1.4). However, the instability giving rise to the development of a cell will almost certainly involve the necessity for more such bubbles before that instability is alleviated. It is common for new cells to be initiated preferentially along the outflow boundary spreading out from the previous cells (Purdom 1976; Weisman and Klemp 1986; Fovell and Dailey 1995). In Fig. 1.3, unlike Fig. 1.2, the updraft is a bubblelike entity. This

FIG. 1.4. Example of so-called turkey towers (due to their resemblance to the heads of turkeys) on 20 May 1992 in western South Dakota (photo © 1992 C. Doswell).

means that the notion of a tilted updraft is not relevant; no plume of updraft is present to be tilted, although the thermals may follow a tilted path. A common notion is that the updraft can be "unloaded" of its precipitation only if it is tilted by shear (e.g., Ludlam 1963); otherwise, according to this concept, the cell's precipitation simply collapses downward upon the updraft. Within a prototypical cell in which the dynamics fits a buoyant bubble model rather than a plume, the tilted plume concept is simply irrelevant.

As first mentioned in Scorer and Ludlam (1953), the thermal model makes an interesting prediction. The buoyant bubble has a form drag associated with it, due to having to rise through still air (see Turner 1973, p. 188); energy is consumed in pushing aside the overlying air during the bubble's ascent. Recently, Renno and Williams (1995) have presented quasi-Lagrangian observations of convective boundary layer (CBL) thermals, noting that such thermals have roughly constant rise rates in spite of exhibiting positive buoyancy (thermal excesses ranging from 0.1°C to more than 1.0°C) through virtually all of their ascent. If buoyancy remains positive, the only way for vertical velocity to remain constant is for some equal but opposite force to exist that opposes the buoyancy. Renno and Williams postulated a mixing argument that would allow buoyancy to remain positive even as momentum mixing produces drag that opposes buoyancy. I believe that the opposing force should be viewed in terms of the form drag associated with the rising thermal; Young (1988) describes such an opposing force in terms of perturbation pressure gradients. Schumann and Moeng (1991) have shown that pressure gradients accelerate rising buoyant thermals low in the CBL, but oppose the buoyancy higher up in the CBL.

There are only limited observations of the rise rates of thunderstorm cloud tops[4] during their development (e.g., Scorer and Ludlam 1953), but it is quite possible that if the thermal/bubble model of cells is correct, the rise rate of the cloud tops would not be the same as, and would be likely to be slower than, the vertical velocities in the core of the thermals. At this point, it is not known if the approximately linear relationship between rise rate and buoyancy shown in Scorer and Ludlam (1953) extends into high CAPE, severe DMC ranges.

Other evidence suggesting a bubblelike character to DMC can be found in Hane and Ray (1985). Their Fig. 13 reveals considerable structure in the buoyancy field, even for a mesocyclonic thunderstorm. This implies that buoyant thermals pass through even quasi-steady supercell thunderstorm flows, typically in an episodic, rather than a steady, manner.

A persistent idea about DMC cells is that the outflow chokes off the updraft (e.g., Emanuel 1994, p. 233). This seems a rather unlikely general explanation for cell dissipation; an important problem for this hypothesis is that it cannot explain the finite life cycles of elevated DMC on the cold side of a surface front. Such thunderstorm cells are initiated above a cool, stable pool of surface-based air and apparently are not choked by it. Yet observations suggest that such elevated convective storms have cells that go through life cycles similar to DMC that does *not* develop above a cool pool. It is difficult to explain such events if the cessation of a plumelike cell updraft depends only on being cut off by outflow.

Our perceptions of convective structure are highly dependent on the observing system being employed. Classifications based on radar are not necessarily equivalent to those developed from satellite imagery, and so on. Although this topic will be expanded upon in subsequent chapters within this monograph, the likelihood of severe weather tends to increase with the degree of organization to the convection. The most intense severe convective weather is associated with *organized* DMC.

1) LINEAR ORGANIZATION

Linear organization is arguably the most common form of DMC organization. Since outflow is an effective mechanism for lifting near-surface parcels to their LFCs, once outflow develops it can have a dominant role in the development of subsequent cells. The horizontal convergence along outflow boundaries can be on the order of 10^{-3} s^{-1} or larger. If a value of 10^{-2} s^{-1} is sustained through a layer as deep as 1 km, the resulting upward motions at a height of 1 km are of order 10 m s^{-1}, quite likely to be capable of initiating deep convection. As convection continues, the merger of new outflows with old ones results in an expanding pool of cold, stable air at low levels, often with new convection on its leading edge, as the outflow pushes into untapped, potentially buoyant air

[4] Numerous observations of the storm's summit as depicted by radar are available, but this is not necessarily equivalent to its cloud top.

ahead of the outflow. It is common to refer to DMC organized linearly as a *squall line,* although the term *instability line* was in common use for a time (see House 1959).

How might the term "squall line" be defined? The term first arose in the context of synoptic analysis by the Norwegian school; it was applied first to what we now call a cold front (see Friedmann 1999, p. 31). The word "squall" is derived from what is now a largely archaic description of a gusty wind with a specific character: a gust of 16 knots ($\sim$8 m s^{-1}) or higher that is sustained for at least two minutes. According to this criterion, a squall line presumably would involve "squally" winds meeting this criterion, observed at two or more locations along the line. In practice, no one requires this criterion to be met in using the term "squall line" as applied to DMC, nor do I believe it should be. Perhaps it was the association between DMC and cold fronts so often seen in the conceptual models of the Norwegian school that led eventually to the use of the term in connection with lines of convection. A minimal definition of squall line in the context of DMC could begin with as few as two isolated cells; two points define a line, after all. However, if the linear organization is to be the dominant characteristic, a more reasonable definition should require that the cells be close enough together that the perturbation flows they generate are interacting. An outflow boundary with gusty postboundary winds (virtually all such outflows are gusty and probably meet the "squall" criteria most of the time) should suffice for a linear organization of convective cells to constitute a squall line.

A related factor in developing a linear structure is the nature of the process responsible for the first convective cell initiation. Often the lifting mechanism is a front, a dryline, or a prefrontal trough (perhaps associated with an upper-level front; see Stoelinger et al. 2000). There are typically along-line variations in the lift associated with such processes, as well as variability in the thermodynamic characteristics of the air being lifted. Therefore, the first developments occur at relatively isolated points as individual cells, but the overall linear nature of the initiating mechanism for DMC often results in a rapid filling-in of convective elements along the line. Subsequent development of cold outflows serves to reinforce this evolution, hence the high frequency of this sort of organization to convective systems.

Severe weather may or may not be closely associated with linear structures. If the constituent cells are competing with each other on a nearly equal footing, as when the squall line exhibits a quasi-two-dimensional (or Q2D) character (Fig. 1.5), no cell is particularly favored. Whereas marginal severe events (at or just above the arbitrary limits of "severe" winds and hail) are not uncommon and may indeed be widespread, the extremes of severe convective weather are unlikely in such situations, although not impossible. Note that whereas Fig. 1.5a shows a nearly constant band of reflectivity at low levels, even such a Q2D example breaks into separate convective cells aloft (Fig. 1.5b). Another form of squall lines is a line of supercells (Fig. 1.6). Although the cells are aligned laterally, they are not necessarily competing with one another (Fujita 1975). Finally, parts of a Q2D line may not move equally fast, creating the so-called line echo wave pattern (LEWP; see Nolen 1959) structure, or bow echoes (Przybylinski 1995). Bow echoes are often associated with damaging convective wind gusts.

A Q2D structure is a very efficient way of overturning an unstable stratification in comparison with isolated convective cells, even when the updrafts are not especially intense. The development of new convection at the leading edge of the outflow with a front-to-rear ascending inflow and embedded convective cells can "process" the warm moist air from low levels quite readily. The precipitation falling out of the cells maturing rearward of the leading edge promotes the downdraft that maintains the outflow boundary. Often, a rear-to-front flow develops in the wake of the leading edge that promotes the advancement of the outflow. This overall structure, illustrated schematically in Fig. 1.7, can stabilize the stratification over a large area when a laterally extensive squall line moves rapidly.

2) MESOSCALE CONVECTIVE SYSTEMS

The term *mesoscale convective system* (MCS) is virtually uniquely associated with the satellite perspective. Maddox (1980) developed the term *mesoscale convective complex* (MCC) with a specific set of criteria (Table 1.1) related to the infrared (IR) imagery views of convective systems. MCC criteria were designed in Maddox's study to limit the initial study to the largest, most circular, and longest-lasting examples within a large class of such systems. Subsequently, the entire class of convective systems observed via IR imagery has been dubbed MCSs (Zipser 1982); the MCC criteria are essentially arbitrary and have no particular physical significance, other than that they refer to particularly evident forms in the IR images. MCSs are covered in detail in chapter 9 within this monograph, and in chapter 3 as well.

The typical life cycle of an MCS is illustrated in Fig. 1.8, as seen in a series of IR satellite images. The convective system begins as a number of relatively isolated convective cells, usually during the afternoon. By late evening, the anvils of the individual cells merge; the characteristic cold cloud shield develops toward maturity sometime after midnight, local time. Dissipation then occurs, typically sometime in the morning. Although by no means restricted to the nocturnal hours, MCSs most frequently reach maturity after sunset.

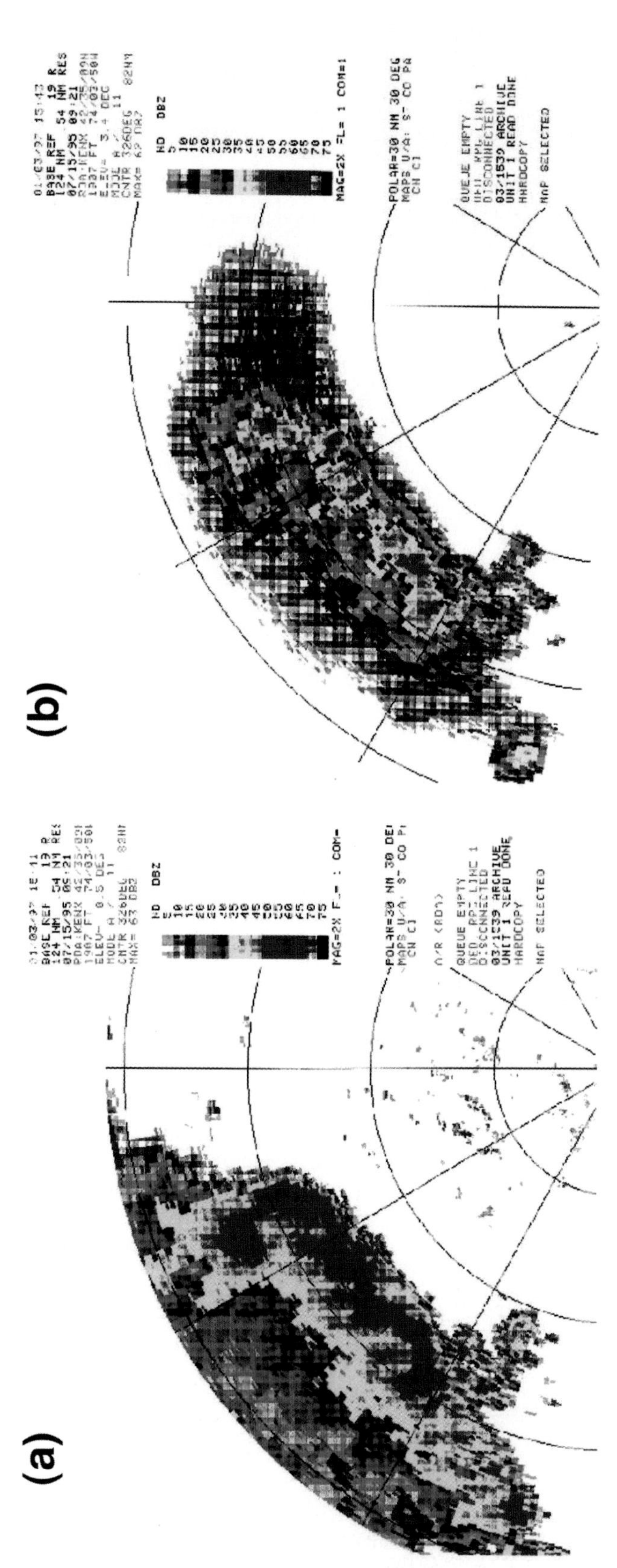

FIG. 1.5. Example of a squall line, as seen from the Albany, NY, WSR-88D radar, which (a) shows relatively little along-line variation in intensity at the lowest elevation (0.5°) but which (b) breaks down into individual convective cells at higher levels (3.4°), on 15 Jul 1995 at 0921 UTC.

FIG. 1.6. A line of isolated tornadic supercell convective storms in eastern Oklahoma on 26 May 1973 (photo taken from an aircraft by P. Sinclair).

MCS circular cloud tops often mask a linear structure of the convective cells when viewed on radar. The deep, intense convection in MCSs is usually organized in a linear structure on the leading edge of the cold outflow, with weaker stratiform precipitation trailing behind, over the outflow. Other organizations are possible, including having the deep convection on the trailing edge of the outflow, but are not as frequent as cases where the deep convection is on the forward side of the MCS. Severe weather events tend to occur during the early stages of MCS development, when the individual convective cells have not yet agglomerated into the characteristic MCS structure. However, widespread severe windstorms known as *derechos* (Johns and Hirt 1987) are often associated with mature MCSs. When MCSs generate heavy precipitation, they can be well into the mature MCS phase.

Occasionally, MCSs can produce a persistent mesoscale circulation (Bosart and Sanders 1981; Bartels and Maddox 1991; Davis and Weisman 1994), apparently above the cool, stable pool of outflow, that can persist well after the convection dissipates. These circulations have been observed to be associated with redevelopment of another MCS (e.g., Bosart and Sanders 1981; Menard and Fritsch 1989), so that the system as a whole can live longer than 24 h. In fact, some examples of long-lived MCSs have persisted as deep convective systems continuously for more than 24 h, defying the trend for dissipation by late morning.

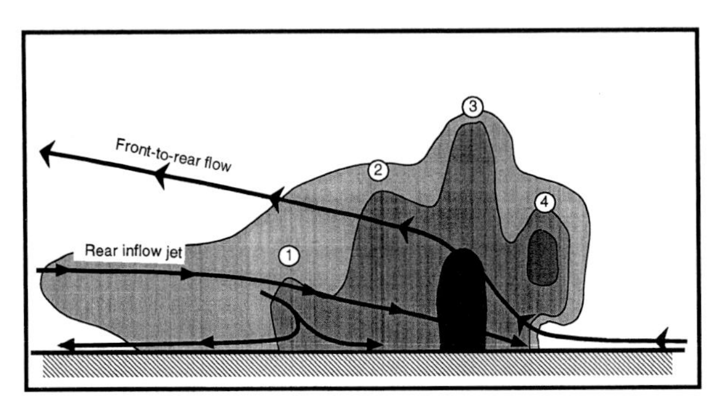

FIG. 1.7. Schematic cross section of the airflow in an MCS; the stippling denotes low, moderate, and high radar reflectivities by progressively darker stippling; the front-to-rear flow and rear inflow jet are indicated; and the circled numbers 1–4 represent individual convective cells in the order in which they have developed. Since the MCS movement includes a component due to development of new convective cells on its leading edge, the cells tend to move rearward through the MCS as they mature and dissipate. See Smull and Houze (1987).

TABLE 1.1. Physical characteristics of a mesoscale convective complex (MCC), based on an analysis of the enhanced IR satellite imagery; criteria developed by Maddox (1980). The intent was to select the largest, most persistent, most nearly circular mesoscale convective systems (MCSs) and to define their initiation, maximum extent, and termination.

Size:	A—cloud shield with continuously low IR temperature ≤ −32°C, must have an area ≤ 100 000 km^2 B—Interior cold cloud region with temperature ≤ −52°C, must have an area ≥ 50 000 km^2
Initiate:	Size definitions A and B are first satisfied.
Duration:	Size definitions A and B must be met for a period ≥ 6 h.
Maximum extent:	Contiguous cold cloud shield (IR temperature ≤ −32°C) reaches maximum extent.
Shape:	Eccentricity (minor axis/major axis) > 0.7 at time of maximum extent.
Terminate:	Size definitions A and B are no longer satisfied.

Although most case studies of MCSs have been located in the United States, they are by no means confined to there, or even to North America. MCSs occur in virtually all convective-prone areas of the world (see, e.g., Velasco and Fritsch 1987; Smull 1995; Laing and Fritsch 1997), notably including the Tropics. Within the Tropics (see chapter 10), convection tends to develop in loosely organized clusters that contain individual elements that might or might not meet MCS criteria individually, depending on how the essentially arbitrary MCS criteria are chosen. Except in tropical cyclones, tropical convection generally is not as persistent as midlatitude convective systems.

Not all MCSs are nearly circular. Included within the category of MCS is a linearly organized band of cold cloud tops (Fig. 1.9). Such a structure is nearly always associated with a frontal boundary and/or with a cold front aloft (Locatelli et al. 1998). The individual convective elements in such a line may or may not be made up of radar-observed linear convective structures.

3) SUPERCELLS

The most dramatic form of organized convection is arguably the supercell. First proposed by Browning (1964), the supercell model was initially conceived as a quasi-steady form of an ordinary cell. Browning himself (1977) later developed a new definition of a

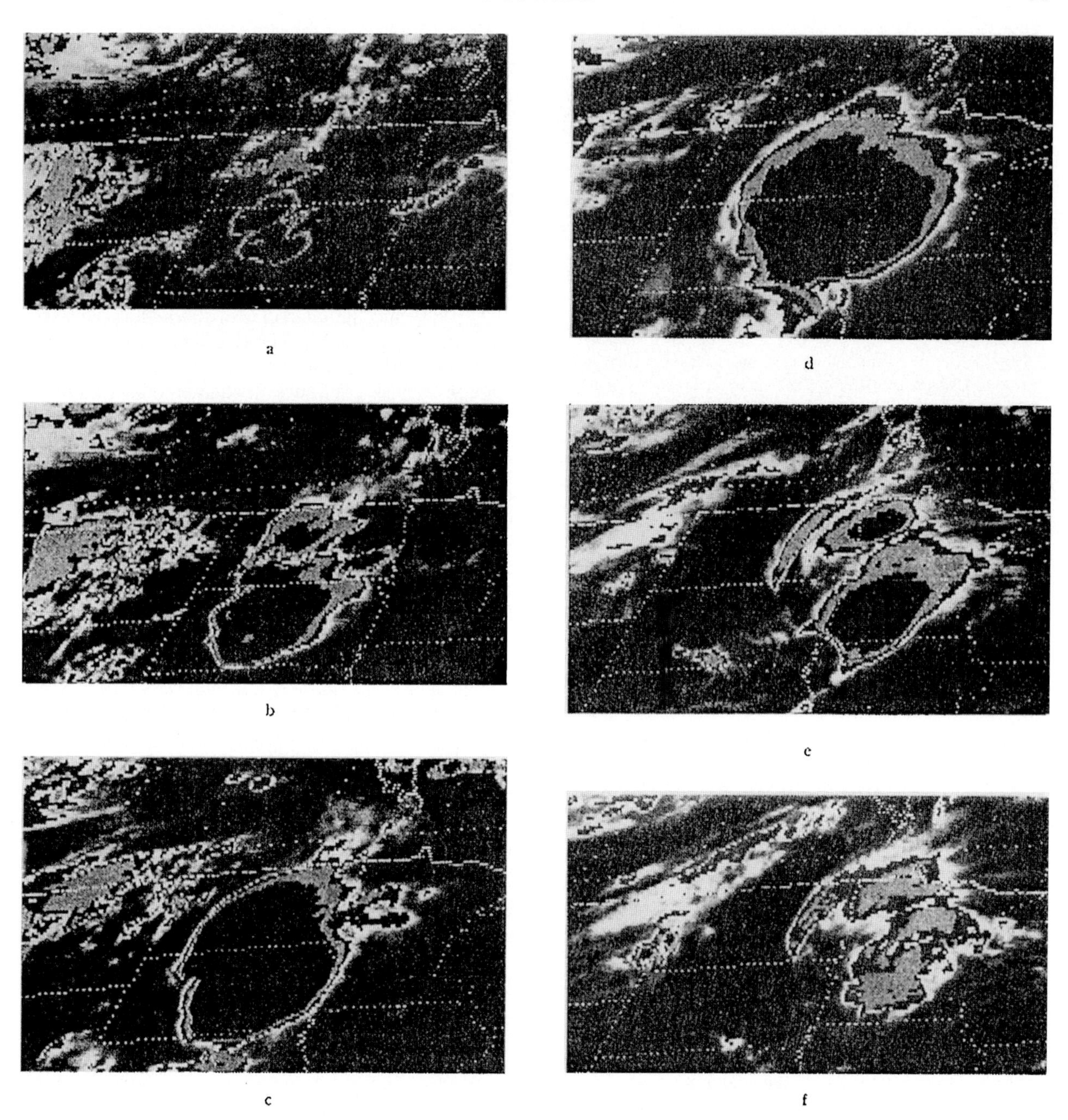

FIG. 1.8. Life cycle of an MCS as seen in IR satellite imagery for an MCC on 12 Jul 1979 at (a) 0030 UTC, (b) 0300 UTC, (c) 0600 UTC, (d) 0900 UTC, (e) 1430 UTC, and (f) 1630 UTC. From Maddox (1980).

supercell as a convective storm having a mesocyclonic circulation. Weisman and Klemp (1982) adopted this definition as the means by which they could distinguish supercells from "ordinary" cells. Doswell and Burgess (1993) suggested a slightly modified version of Browning's (1977) definition of a supercell, incorporating somewhat arbitrary criteria. Although debate continues about the details of the definition, Browning's basic notions remain as the heart of whatever consensus exists. Supercells are discussed in detail in chapter 5 of this monograph.

A supercell is made distinguishable from nonsupercells by the presence of its mesocyclone. The mesocyclone creates the radar reflectivity morphology [distinctive features (Forbes 1981) such as hook echo structures and LEWPs] typically associated with supercells. With the pioneering numerical simulations by Schlesinger (1975), Klemp and Wilhelmson (1978),

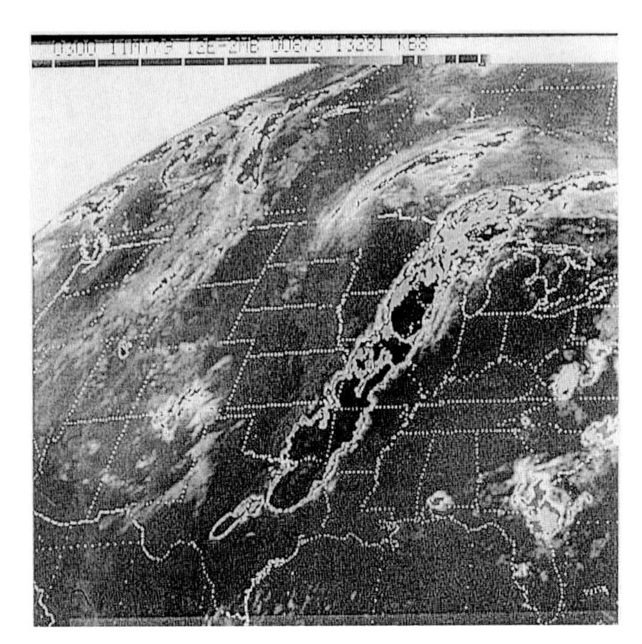

FIG. 1.9. An infrared satellite image (using a standard MB enhancement) depicting a linearly organized MCS, on 11 May 1973 at 0300 UTC.

and Weisman and Klemp (1982, 1984), it has also become clear that the supercell storm involves dynamical processes that do not arise in ordinary DMC. The interaction of the updraft with an environment characterized by strong vertical shear of the horizontal wind permits some storms to develop nonhydrostatic vertical pressure gradients that can be as influential in developing updrafts as the buoyancy effects (Weisman and Klemp 1984). Therefore, such storms can have strong updrafts even when the static instability, as measured by CAPE, is modest (McCaul 1993).

Storms with large, strong, and persistent updrafts process substantial amounts of mass; if the average updraft over a 5-km radius is 10 m s^{-1} and the average density of the inflowing moist air is 1 kg m^{-3}, the mass flux is roughly 10^9 kg s^{-1}. For a water vapor mixing ratio of 10 g kg^{-1}, the water mass flux in the updraft is 10^7 kg s^{-1}.[5] The mass flux is most sensitive to the size of the updraft (it increases as the square of the radius), and supercells generally have larger updrafts than ordinary storms. It appears that supercells tend to dominate their near environments, outcompeting nonsupercellular neighbors for the low-level warm, moist air that sustains them. Mass continuity requires the development of compensating subsidence around DMC storms, since the convective downdrafts typically do not process as much mass as the updrafts (Fritsch 1975; Wetzel et al. 1983). Thus, the most intense updrafts will virtually always be isolated (cf. Fig. 1.6). In clear contradistinction to the Q2D squall line, with its numerous competing cells, supercells are relatively isolated storms that, nevertheless, can also process large quantities of air and water vapor. A dominant factor in the development of supercells versus Q2D squall lines is the vertical wind shear structure in the environment of the developing convection. The magnitude of the shear (Weisman and Klemp 1982, 1984) and the character of the hodograph associated with that shear (Davies-Jones et al. 1990; Brooks et al. 1994) appear to be quite pertinent issues, although debate continues about the relative importance of various measures of the vertical wind shear.

Supercells are relatively rare events and are predominantly a midlatitude phenomenon. Tropical environments usually do not have adequate shear to develop deep, persistent convective mesocyclones (as noted in chapter 10). Given the wind shear of midlatitudes associated with near-geostrophic wind balance, coupled with sufficient instability from the thermal and moisture stratification, it is obvious that midlatitudes are the normal spawning grounds for supercellular DMC. There is growing evidence that supercells are more common around the world than formerly thought (e.g., Houze et al. 1993; Dessens and Snow 1993; Colquhoun 1995); the perception that supercells are primarily a phenomenon confined to the central plains of North America is simply false. Even within the United States, data are accumulating from the newly implemented WSR-88D Doppler radars to indicate that supercells might form a larger fraction of the overall thunderstorm population than once believed. Nevertheless, it seems clear that supercells are by no means frequent compared to other forms of organized DMC; the ratio of supercells to nonsupercells is probably on the order of 10 percent. Comprehensive studies to determine this ratio are still lacking, unfortunately.[6]

By virtue of their large, intense vertical drafts, supercells create a disproportionate share of the most intense forms of convective severe weather, excluding heavy precipitation. Nonsupercells probably account for the majority of convective severe weather overall, but nonsupercellular severe weather events do not attain the extremes most typically associated with supercells. Suppose that meteorologically significant[7]

[5] If 100% of that condensing water were to fall out as precipitation over a circle 5 km in radius, it would represent a rainfall rate of about 250 mm h^{-1} (roughly 10 in h^{-1}). Precipitation efficiencies are typically much less than 100%, of course.

[6] It is far from obvious just how such a study might be done, since it is not clear just how a "thunderstorm" is defined. How many thunderstorms does a convective line or a convective complex contain at a given instant?

[7] Meteorological significance is tied to the intensity of the event, rather than to the effect of the event on humans, which is contingent on the meteorological event's interaction with human habitation and use of a location.

severe convective hailfall events are defined as those producing hailstones with diameters of 5 cm (~2 in.) or larger, significant severe convective wind gust events are defined as those with speeds 33 m s^{-1} (~65 knots) or greater, and significant tornadoes are defined as those with intensities F3 or greater (using the Fujita damage/intensity scale, Fujita 1971). Then a large fraction of meteorologically significant events is produced by supercells, with perhaps a majority of significant hailfalls and tornadoes being supercell associated. Recently, it has been recognized that supercell storms can also produce prodigious rainfall rates at times (Doswell 1994, 1999), simply by virtue of their intense updrafts. Nevertheless, supercell storms are not associated with the majority of heavy rainfall events.

1.2. Severe convection

a. Definitions

There is considerable arbitrariness in defining severe forms of hail, wind, and precipitation (Doswell 1985). That is, the typical method for deciding whether an event is severe involves some threshold criteria. In the United States, for a hailfall to be considered severe, the hailstone diameters must be ~2 cm (¾ in.) or larger, and severe convective wind gusts must be ~25 m s^{-1} (50 knots) or greater. Is there a meaningful physical distinction between a storm that produces hailstones that are 1.9 cm in diameter and one producing hailstones 2.1 cm in diameter? Is a convective storm that produces a wind gust of 24 m s^{-1} physically different in some unambiguous way from a storm that can muster a 26 m s^{-1} gust? The answers to both these questions are obviously negative, but some sort of threshold is required for classification purposes. The basic principle behind the existing thresholds for defining severe convective weather in the United States is that the probability of damage increases substantially as hailstone diameters and wind gust speeds increase beyond the current thresholds.

There are other issues, however. Substantial accumulations of subthreshold hail might well cause significant damage, especially to crops when wind driven. If there is a population of hailstone sizes (which is typical), with only one or two isolated examples meeting or exceeding the threshold to be considered "severe," is the storm considered severe? In the United States, the answer is "yes." Real hailfalls are associated with swaths of hail, and it is quite possible that somewhere within an extensive swath of subthreshold hailstones, a stone or two equaling or exceeding the threshold could be found. Will someone see and report the isolated examples that exceed the thresholds? Should we be reporting the average hailstone size rather than the largest observed?

The same sorts of problems arise with respect to convective wind gusts. Subthreshold wind speed might well be capable of creating damage; for example, aircraft in the process of taking off and landing are very vulnerable to convective winds (see Fujita and Caracena 1977). As with hailfalls, strong convective winds arise in swaths of varying length and width; it is basically just an accident when what might be only a momentary peak within the swath either ends up being measured or results in damage, leading to a report of the event.

Damaging convective winds, hailfalls, and tornadoes all occur over some area (typically a swath), but severe reports are in terms of points, with the exception of tornadoes; for most tornadoes, a path length and width are reported. Although the width usually varies along the track, only the widest portion is recorded. In any case, with the exception of tornadoes, there is a clear mismatch between the events and the reports of the events. This mismatch leads to a reduction in the fidelity with which our climatological records of severe convective events portray the reality of the events themselves. At the moment, there appears to be little we can do to change this situation.

In the United States there is no officially defined threshold for precipitation, beyond which the rainfall is considered to be severe. Officially, heavy rain is not considered to be severe weather. Generally speaking, the highest sampling frequency for precipitation rates is hourly, but the majority of our precipitation measurements are 24 h totals. Hourly rates of ~25 mm (1 in.) or more are generally considered minimally intense, but it is unlikely that such a rate would produce an important event if sustained for only one hour. Heavy precipitation's greatest threat to life is associated with flash floods (chapter 12 of this monograph). Flash floods are the result of heavy rainfalls (chapter 8 of this monograph) in combination with various types of hydrological situations (e.g., terrain relief, antecedent precipitation, drainage basin soil and usage characteristics, etc.). It is possible for heavy rainfalls to produce little or no danger, owing to the hydrological factors (Doswell et al. 1996). By the same token, in some hydrologic circumstances, rainfalls of only moderate intensities can result in flash flood disasters (as with the Shadyside, Ohio, event of 14 June 1990; see National Weather Service 1991 and chapter 2 of this monograph).

All tornadoes are considered severe and are the focus of chapter 5 in this monograph. However, even in this category, there are gray areas. For instance, roughly two-thirds of all tornadoes are relatively weak, being rated F0 or F1 on the Fujita scale (Kelly et al. 1978). Further, although waterspouts are simply tornadoes over water, in the United States they are not entered in the climatological record of severe events unless they move onshore as tornadoes. As we learn more about tornadoes, it seems that there are many aspects of the tornado identification problem that we are only beginning to recognize (see Forbes and Wakimoto 1983; Doswell and Burgess 1993).

There is no easy way around these problems with defining convective event severity using essentially arbitrary thresholds. It is clear that virtually any set of thresholds will miss some important events and might include events that arguably might not deserve to be considered severe. Hales (1993) has proposed a two-tiered system that retains the current United States thresholds but adds a set of higher thresholds to define "significant" severe weather, similar to the definitions described above. Hales presents some evidence that the observed frequency of significant events (by his criteria) in the United States has been much more nearly constant over a long period than the frequency of events that are marginally severe (i.e., that meet the current criteria but do not attain the thresholds he has defined as significant).

Worldwide, there has been even less agreement over the definitions of severe convective weather events than in the United States. Therefore, the climatological record of severe convective weather worldwide is correspondingly uncertain. It has been my experience that the reporting of severe weather worldwide is pretty erratic, with only major disasters having much chance of being recorded. The situation is not unlike a good part of the record of severe weather in the United States prior to the institution of the severe weather watch/warning program in the 1950s (see Galway 1989). In effect, the chances of severe weather being reported are in direct proportion to the effort expended in trying to mitigate the effects of severe forms of DMC. When the awareness of the threat posed by severe DMC grows, the probability that a given event will be reported also grows. Hence, a significant contribution to the climatological record is associated with perceptions, both within the public and within the local meterological community, of how important severe forms of DMC are in a given region. I shall return to this point later.

b. The distinction between severe and nonsevere convection

The potential impacts of severe forms of DMC on society virtually demand that forecasters make the attempt to distinguish between severe and nonsevere storms, irrespective of the scientific problems with arbitrary thresholds. The only logical basis for making the distinction between severe and nonsevere DMC is to determine the likelihood of severe convective events, given the available information. The subject of severe convective weather forecasting is covered in detail in chapter 11 of this monograph, from the perspective of the local forecaster.

This distinction is closely connected to the association between DMC storm types (i.e., DMC taxonomies) and the weather events they produce. For example, if it can be established that a DMC storm is a supercell, then the probability of significant severe convective weather increases substantially over nonsupercellular types. To the extent that a taxonomy allows one to make this sort of distinction among the categories, that taxonomy has value, either in a practical (forecasting) sense, or in a scientific classification. Storm classification is not always given a great deal of respect in comparison to, for example, studies of storm dynamics. Nevertheless, as I have noted elsewhere (Doswell 1991), taxonomy inevitably influences our scientific perceptions to the extent that we may overlook obvious real aspects of storm structure and behavior because of the way we see such storms.

1.3. Observations of severe convection

a. Observations of processes leading to severe convection

1) LARGE SCALE

The primary observational tool for assessing the large-scale structure of the atmosphere remains the rawinsonde. New observations are becoming an important component worldwide, notably satellite images, quantitative satellite data (e.g., Chesters et al. 1982; Spencer et al. 1989), observations from commercial aircraft (Schwartz and Benjamin 1995), etc. However, much of what we know about synoptic-scale structure and evolution today has been derived from balloon soundings. The network of such observations has some large gaps, notably over the oceans (especially in the Southern Hemisphere), in sparsely populated regions, and within economically disadvantaged nations. Therefore, our understanding of the synoptic-scale structures associated with severe convection tends to be dominated by continental, Northern Hemispheric, American, and western European systems. Chapter 2 gives a summary of that understanding.

To a considerable extent, the synoptic-scale situations chosen for detailed analysis have tended to be large "outbreak"-type events [e.g., 11 April 1965 (the "Palm Sunday" outbreak) or 3–4 April 1974 (the "Super" outbreak)]. As noted in Johns and Doswell (1992), there are relatively few such cases in any given year. In their summary, Barnes and Newton (1983) present this outbreak-centered view (Fig. 1.10), noting that their composite chart [i.e., a depiction of features at different levels on a single chart, pioneered by R. C. Miller (see, e.g., Miller 1972)] is "especially favorable" for severe weather. Whereas outbreak days are relatively uncommon during the year, when they do happen, they contribute a significant fraction of the total number of severe events (see Galway 1977). Moreover, the severe convective events during outbreaks often achieve high impact because the intensity of the events tends to be high. For example, Galway (1977) notes that in the period 1950–75, outbreak

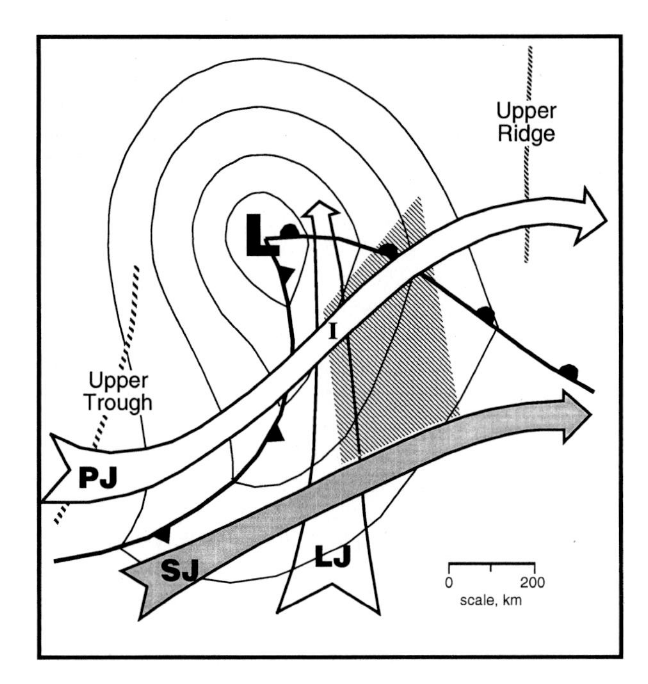

FIG. 1.10. Idealized example of important synoptic-scale features in an outbreak of severe convective storms; thin lines are sea level isobars, surface feature symbols are conventional, the broad arrow labeled LJ is the low-level jetstream, that labeled PJ denotes the polar jetstream aloft, the shaded broad arrow labeled SJ depicts the subtropical jetstream aloft (which may not always be present). The hatched area shows where severe convective storms are most likely during the ensuing 6–12 h; the severe storms are considered most likely to begin near the point labeled I where the LJ and PJ intersect. Nonsevere convection can occur outside the hatched region. After Barnes and Newton (1983).

tornadoes contributed more than 20% of the total number of tornadoes for the period and accounted for more than half the tornado fatalities.

Nevertheless, most severe convective event days are *not* outbreaks (e.g., Maddox and Doswell 1982), and so any description of the "typical" synoptic situation for severe convection might prove to be an elusive goal. Doswell et al. (1996) have suggested that an ingredients-based approach to understanding weather events is preferable to trying to define characteristic synoptic-scale patterns. If the ingredients for a particular event can be brought together in an uncharacteristic pattern, the event still ensues (as illustrated by Bosart and Lackmann 1995, albeit in a different context).

Poleward of the Tropics, synoptic-scale processes are dominated by quasigeostrophic (QG) processes. Although there are non-QG aspects to the large-scale weather systems that affect severe convection, most of the basic elements of those systems are described remarkably well by the relatively simple notions of QG theory.

Within the Tropics, however, the relative unimportance of geostrophic balance means that higher-order balances (see, e.g., McWilliams and Gent 1980; Davies-Jones 1991) must be considered at synoptic scales. Convection plays a large role in tropical meteorology, and synoptic-scale systems are much less apparent, whereas mesoscale systems are clearly important. It does not help that so much of the Tropics is oceanic and so in situ observations tend to be sparse; nevertheless, considerable areas on several continents are within the Tropics. Notably, tropical regions in Africa, Latin America, Asia, and Australia all have monsoon-dominated convection (see chapters 2 and 10 of this monograph).

2) MESOSCALE

Mesoscale observations have until recently tended to be associated with research-driven special observing networks. The routine surface observations can only arguably be considered mesoscale; no one would suggest that the routine rawinsonde observations offer any mesoscale resolution. Of course, meteorological satellite images have offered considerable mesoscale detail (see Purdom 1976; Zehr et al. 1988). However, the images are primarily qualitative rather than quantitative. Efforts to obtain quantitative information on the mesoscale from satellite platforms are both useful and difficult to achieve (e.g., Hilger and Purdom 1990). Radar could also be considered to be a source for some mesoscale information, but its line-of-sight geometry makes it marginal as a source for mesoscale information and, like satellites, it does not collect quantitative information about common meteorological variables (temperature, pressure, humidity). Doppler radars can collect wind information in a "clear air" mode (see Matejka and Srivastrava 1991; Boccippio 1995), and with the implementation of a network of Doppler radars in the United States, those wind data may become an important part of a mesoscale analysis. Fujita (1963) provides an excellent summary of the state of mesoscale analysis prior to the introduction of the new observing technologies. In chapter 3, a discussion of mesoscale aspects of severe convection is given.

There is no completely satisfactory way to delineate scale boundaries (for different perspectives, see Orlanski 1975; Fujita 1981; Doswell 1987), so some arbitrariness is inevitable. In spite of this difficulty, there does not seem to be much dispute about the mesoscale issues relevant to severe convection. Certainly, an important mesoscale aspect associated with convection is *convective outflow*. Precipitation-cooled air from DMC processes tends to spread out, with outflows from nearby convection interacting and merging to form regions that can approach 10^6 km^2. The larger and stronger the initial airmass contrast with the undisturbed environment, the longer such outflow structures can exist beyond the end of the DMC that produced them. Note that convective outflow also occurs at the storm top, where the updraft spreads out at its level of neutral buoyancy. Thus, convective outflow can have an important role in subsequent convection by altering the environment in which new convection develops (see, e.g., Ninomiya 1971; Maddox et al. 1980) and by serving to initiate new convective cells.

Mesoscale processes other than those created by DMC can be thought of in several groups: free "internal" instabilities, forced "external" processes, fronts, and gravity waves. Not many free internal instabilities have been identified that have maximum growth rates (from linear, normal mode analysis) in the mesoscale; "symmetric" instabilities (see Emanuel 1985; Xu 1987) are perhaps the best known. There may be other, fundamentally nonlinear processes that result in mesoscale instabilities, but nonlinearity inhibits the development of clear physical understanding.

Forced external mesoscale processes, many of which owe their existence to various topographical characteristics, are relatively common. Mountain–valley and sea–land breeze circulations result from diurnal heating differences; they are driven by mesoscale baroclinity. Topographic effects include strong downslope winds (Klemp and Lilly 1975; Peltier and Clark 1979; Durran 1986), horizontal mesoscale vortices (i.e., with vertical vorticity) created as wind interacts with terrain (Wilczak and Glendening 1988; Mass and Albright 1989), flow changes associated with differences in surface roughness, and such subtle concepts as mesoscale variations in evapotranspiration (Lanicci et al. 1987; Segal et al. 1995; Pan et al. 1996; Lynn et al. 1998).

Fronts form a special class of mesoscale processes.[8] Generally, their mesoscale aspects are in the direction normal to the boundary, whereas tangent to the boundary, their spatial scales can fit readily into the domain of "synoptic scale." It has been shown that QG processes are unable to describe fronts satisfactorily (see, e.g., Williams and Plotkin 1968; Hoskins and Bretherton 1972). To the extent that QG dynamics can be equated with midlatitude synoptic-scale processes, then fronts should not be considered synoptic scale. Although fronts per se have been under study for many decades, the dynamics of nontraditional boundaries (e.g., the dryline, prefrontal windshifts and troughs, etc.) are relatively poorly known. During the 1950s and early 1960s, there was considerable attention to prefrontal instability lines (e.g., Newton 1950; House 1959, 1963). However, this work has not been pursued to the extent that a clear understanding of such boundaries is available. For example, the dryline has long been known to be a locus for convective initiation (Rhea 1966) in spite of weak (or nonexistent) baroclinity at the time convection commences. Schaefer (1975) suggested one explanation (nonlinear biconstituent diffusion) to create ascent along the dryline in the absence of baroclinity, but Lilly and Gal-Chen (1990) showed that Schaefer's proposed mechanism is an unlikely solution. To date, no completely satisfactory explanation for ascent along the dryline exists, although that ascent's existence is undeniable (see Ziegler et al. 1995).

In general, since lapse rates are almost always less than dry adiabatic through most of the troposphere, gravity waves are possible and so common as to be nearly ubiquitous (see Hooke 1986). However, only occasionally do their amplitudes and size make them obviously important to DMC (see Eom 1975; Uccellini 1975). If a gravity wave is to initiate convection, it must exist at low levels where potentially buoyant parcels exist. However, gravity wave energy tends to "leak" upward with time, so a waveguide or duct must exist to trap that energy, allowing the wave to retain its amplitude over time and distance (Lindzen and Tung 1976). It appears that the conditions favorable for large-amplitude gravity waves to propagate for significant time and distance and also to initiate DMC are relatively infrequent, occurring perhaps a few times per year in the United States (see Hoffman et al. 1995; Koppel et al. 2000).

1.4. Prediction of severe convection

In chapters 11 and 12, discussions of the severe convection forecasting problem (including flash floods) from the local forecast office viewpoint are given. It is not my intent here to repeat that content. Further, there have been recent reviews of various aspects of forecasting severe convection (including heavy precipitation) at a national level (Johns and Doswell 1992; Doswell et al. 1993; Olson et al. 1995). My intentions here are 1) to provide a brief summary of current forecasting accuracy for severe convective events, and 2) to suggest some goals for any implementation of severe convective storm forecasting aimed at mitigating the threats posed by such storms.

a. Current levels of accuracy

The forecasting situation is presently a mixture of clear progress and frustration over a lack of understanding of essential issues. In the United States, the hazardous convection problem is pervasive and important, so progress has been made in forecasting of severe thunderstorms (and, especially, tornadoes) since the effort began in earnest in the early 1950s (when

[8] Within the term "front," I am including boundaries that may or may not have substantial baroclinity. Thus, for purposes of this discussion, I am specifically including drylines and other forms of nonfrontal boundaries. According to Sanders and Doswell (1995), many such boundaries are not true fronts, but they have in common with fronts that some atmospheric variable has strong variability in the direction normal to the boundary.

the U.S. public weather service began to issue severe thunderstorm and tornado forecasts). See Galway (1989) for some historical perspectives on the forecasting effort.

Grazulis (1993, p. 14 ff.) notes several items about the tornado casualty figures in the United States. A trend toward decreasing annual fatalities seems to have begun in the late 1930s and has continued to the present. Whereas, early in the century, U.S. annual tornado fatalities often exceeded 200, that figure has not been reached since 1974, in spite of a continued upward trend in the number of reported tornadoes and an increasing population. Over the last 80 years, the majority of the increase in tornadoes reported annually has been in the "weak" category (F0 and F1 on the Fujita scale) of events, and these meteorologically minor events only account for only about 2% of the casualties (Kelly et al. 1978). As shown in Fig. 1.11, the increase in reporting of weaker tornadoes is indicative of an exponential distribution of tornadoes with intensity. The number of "violent" events (F4 and F5) per year has not changed much with time, and these strongest tornadoes typically account for a disproportionate share (more than 50%) of the annual casualties. Some of the decrease in fatalities is simply related to enhanced public awareness and communication (notably radio and television). Nevertheless, there is evidence that indicates increasing accuracy in severe thunderstorm and tornado prediction over the period from the 1950s to the present. Doswell et al. (1990) have suggested that the skill of forecasting severe convection has increased by roughly a factor of 2 (Fig. 1.12), after accounting for the inflation in the number of severe thunderstorm reports over the decades. It appears that the infrastructure created by the public weather service in the 1950s for dealing with the severe thunderstorm and tornado hazard has helped reduce the tornadic fatality figures. Major tornado events continue to cause extensive damage, but the fatality rate relative to the inflation-adjusted damage figures has decreased since 1953 (Doswell et al. 1999).

Nontornadic severe thunderstorm forecasting has not received as much attention as tornadic storm forecasting in the United States. Casualties from nontornadic severe convection are relatively rare, but only because the official U.S. definition of severe convection excludes heavy precipitation. Flash flood–related fatality figures on the order of 100 persons annually are an indication that heavy precipitation forecasting can still be improved. Flash flood forecasting is a combination of meteorological and hydrological factors, and as noted by several authors (e.g., Larson et al. 1995; Doswell et al. 1996), there is considerable difficulty associated with the quantitative aspects of precipitation forecasting. A factor in the continued relatively high death tolls from flash floods is the challenge to convey an appropriate sense of urgency in flash flood situations. Whereas there is little difficulty recognizing a tornado as a threat to life, even observations of heavy rainfall may not be frightening enough to trigger appropriate responses in some situations (see, e.g., the anecdotes compiled by Anderson and Wamsley 1996).

Hail formation (chapter 6) is difficult to predict in detail, especially in terms that are accessible and useful to operational forecasters. Moreover, short-term mitigation of hail damage[9] is difficult; not much can be done to reduce hail damage to crops and buildings,

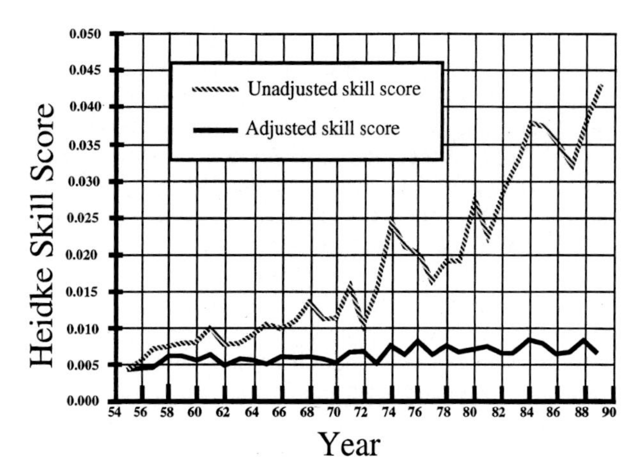

FIG. 1.12. Yearly values of Heidke skill score for tornado and severe thunderstorm watches, showing the effect of adjusting for the inflation of severe weather reports as described in Doswell et al. (1990).

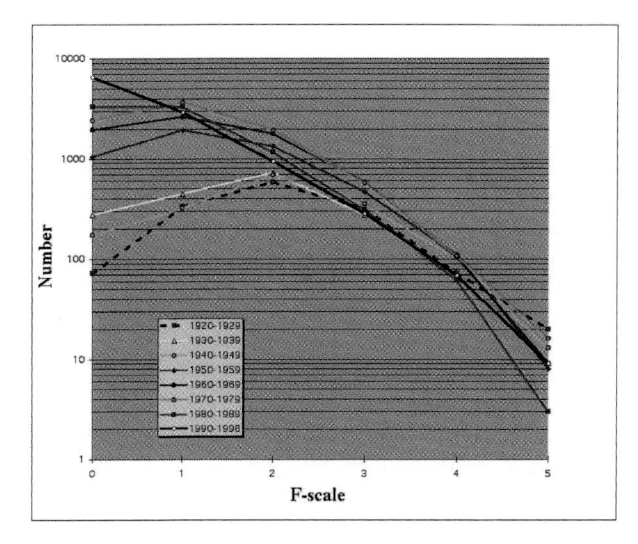

FIG. 1.11. Plot of the distribution of tornadoes in the United States as a function of F-scale, by decade. The number of events is shown on a logarithmic scale.

[9] This ignores the issue of the ongoing controversial attempts to prevent hail, or reduce its size, by seeding (Foote and Frank 1979).

even when a reasonably accurate forecast of an impending event is available on the day of the event. Although forecasting of severe hail is done as part of the severe weather watch and warning process in the United States, a separate verification of event forecasts is not done, perhaps because results of such a verification have not been very encouraging (Doswell et al. 1982). This lack of verification may be at least partially explained by the inadequate observations of hailfalls (see section 1.2a above). Physically, most hail forecasting methods depend on prediction of a strong updraft, with the assumptions being 1) that large hail requires strong updrafts, and 2) that hail size is positively correlated with updraft strength. Although these assumptions may well be valid, there seems to be much more to hailstone size prediction than updraft strength (including, possibly, microphysical factors).

Forecasts for damaging convective wind events are relatively easy to do in principle but not so easy to do in practice (as noted in Doswell et al. 1982). Convective wind gusts are driven by downdrafts, and the physical processes associated with downdrafts are reasonably well understood. Like hailfalls, convective wind gust damage is difficult to mitigate with short-range forecasts and warnings. Convective storm-associated wind gust casualty statistics have been compiled only for a short time in the United States, so it is not possible to assess how the figures may have changed over the decades. It appears that, with the important exception of aviation accidents associated with microbursts, convective wind gusts are not particularly threatening to human life; recreational boaters are at risk, and there are currently a few fatalities per year attributed to falling trees or tree limbs.

Brooks and Doswell (1993) have indicated that, a few times per year in North America, a particularly dangerous form of high wind–producing convective storm can occur. Such storms can produce swaths of intense winds more than 20 km wide, affecting areas as large as 2000 km^2, with peak winds exceeding 35 m s^{-1} (perhaps also including large hail within the winds). These events are apparently the result of supercell convective storms that can at times be embedded within a larger-scale damaging wind event (i.e., a derecho; Johns and Hirt 1987). Such events are not specifically identified in operational forecasts.

An important facet to nontornadic convective wind events is the microburst that affects aviation (Fujita and Caracena 1977). Downdrafts (including microbursts) need not be officially severe to be hazardous to aviation, notably during takeoffs and landings. There is some informal literature (e.g., Caracena et al. 1983) devoted to forecasting microbursts, but formal papers on the topic (e.g., Wakimoto 1985) are not numerous. A possible explanation for this dearth of microburst forecasting studies is that the historical record of convective wind events does not distinguish the meteorological character of severe convective winds in the United States; that is, the fact that a reported convective wind event was caused by a microburst is not recorded. This situation is certain closely tied to the absence in the public forecasts of operational convective wind forecasts formally identified as due to microbursts.

b. As applied to disaster mitigation

If a nation's forecasting service is going to take on the task of predicting the likelihood of severe convection, the perception that the threat posed by DMC is significant must already exist within the political sphere. It has been my experience that in many nations outside North America, residents have little or no appreciation of the threats associated with DMC storms. In such countries, there is no systematic reporting of severe convective events, so when residents experience severe convective storms, no mechanism exists for including the events in a climatological database. Without such a database, there is no factual information useful for informing citizens about the threat, so this ignorance tends to perpetuate itself. Those unfortunates affected are on their own to deal with the events when they occur, and the occurrence is often unforecast.

When it is perceived that severe convective weather is rare in a particular nation or region, it is unlikely that planning for such an event has been done. It is precisely the relative rarity of such events, in regions where severe convection remains possible (albeit unlikely), that creates a potential for disastrous events when the occasional significant event does occur. If it is decided, for whatever reason, that some sort of forecasting process for severe convective weather is needed, some basic elements must be considered in designing a system for mitigating the effects of severe convection through forecasting.

1) Basic forecast verification and climatology

An infrastructure for obtaining and saving basic information about the occurrence of severe convection has several important uses, including hazard planning and developing public awareness. Moreover, it is an essential component of any substantive forecast verification program. Any forecast that is not verified does not receive, nor does it deserve, much credibility. Of course, the verification must be done properly (see Murphy and Winkler 1987; Brooks and Doswell 1996) if it is to have much value. A critical point to establish in doing verification is that the primary objective is forecast improvement. It is difficult to imagine any systematic approach to forecast improvement that is not based on verification and, in turn, verification is based on knowledge of what weather events actually occurred. Hence, a serious severe convective storms forecast program must begin with as thorough a knowledge of the occurrence of severe convective weather events as is economically feasible.

2) SCIENTIFIC FORECASTING APPROACHES

Forecasting, per se, can use any of several different bases: empiricism, numerical prediction, statistical modeling, conceptual models (pattern recognition), and combinations of these. Forecasting severe convection has its roots in empirical checklists and subjective decision trees (see Miller 1972; Schaefer 1986; Colquhoun 1987), with parameters often chosen on the basis of perceived utility and subjective arguments. If the parameters have been chosen objectively, the approach tends to be statistical (see, e.g., Reap and Foster 1979; Charba 1979). Because statistical methods focus on statistical correlations between predictors and predictands, they can make no direct statement about physical cause and effect. However, arguments can be raised (e.g., Doswell et al. 1996) that with the growth of research into severe convective storms, more physically based concepts can be used to develop forecasting methods. Purely associative methods, wherein weather events are associated with parameters, either statistically or via such approaches as checklists, are difficult to adapt to new concepts arising from research. Furthermore, it is not easy to predict when such associative methods will succeed and when they will fail. Statistical techniques tend to have difficulty with events that are not typical in a statistical sense, but the most important severe convective storms are inevitably rare events. Therefore, statistical approaches often fail to be useful in situations where a forecaster needs guidance the most. Using methods having minimal physical bases means it will be difficult to predict when the method will fail because the understanding of how it works is not available.

3) EVENT MITIGATION PLANS

Even when forecasters do their job perfectly and identify the severe convective weather threat accurately in their forecast products, the effort can have no value if there is no plan in place to use those forecasts. People and businesses need to be aware not only of the forthcoming severe convective weather, but also of how to act to mitigate its effects. In many cases, relatively little can be done on short notice to reduce the damage from severe convective storms. The timescale of such events is usually too short to permit hazard reduction efforts like evacuations and efforts to reinforce construction (e.g., boarding up windows). Any efforts at damage mitigation from severe convective storms, typically involving building construction practice (see FEMA 1999), have to take place long before the event is under way or even forecast to be under way. With short-range forecasts and warnings, the main object is to reduce casualties, not damage.

Casualty reduction, in turn, requires considerable effort at making residents aware of what to do. Historically, especially with regard to tornadoes but also with other severe convective weather, we have made errors based on ignorance and misconceptions in telling citizens what to do. The case with regard to opening windows as tornadoes approach is typical of this sort of error. In times past, it was felt that tornadoes had such low pressure in their cores that buildings would simply explode from the rapid drop in external pressure. Residents were instructed to open windows in an effort to minimize the likelihood of their homes exploding if struck by a tornado. This notion has been thoroughly studied by structural engineers and is now discredited (Marshall 1993). Unfortunately, this means we have to "unteach" the public something that we had been teaching them for years. This is by no means the only such mistake we have made in the past in our well-intentioned efforts to educate the public. We need to make residents aware that our ideas can change and to be tolerant of, and receptive to, such changes.

In the case of severe convective storms, damage mitigation is possible but, as already noted, not on short notice. Engineering studies (Mehta 1976; Minor and Mehta 1979; McDonald 1993; FEMA 1999) have shown that a large part of structural wind damage (including that from tornadoes) can be prevented or minimized through sound construction practices and building code enforcement. Even in a violent tornado, the strongest winds occur only in a small fraction of the total damage area. Homes that are not hit by the strongest winds can survive on the periphery of a violent tornado with minimal effects if they have been constructed properly (FEMA 1999). In the case of tornadoes especially, damage mitigation can even reduce casualties by reducing airborne debris, a major cause of tornado casualties.

To a considerable extent, the public needs to be made aware that they have a share of the responsibility for their own safety and property in severe convective weather. Making the public aware of the threat on the day of the event is just part of the story. Residents need to be well informed about the risks they face from such actions as building a home on a flood plain, living in a mobile home in tornado-prone parts of the country, driving in hazardous weather situations, and so forth. Public officials have a responsibility to make information available, but it remains the duty of the inhabitants to avail themselves of the opportunity to read and heed the available information. Local communities need to develop an infrastructure for dealing with weather-related hazards (tornado spotters, emergency operations centers, school and workplace safety training, etc.), even if those hazards are relatively rare. Preparatory activities require resources, and many communities only take serious steps to mitigate a disaster in the immediate aftermath of the disaster itself. It is all too easy to dismiss the hazard as too infrequent to justify spending time and resources to develop a mitigation infrastructure.

1.5. Prospects and unsolved problems

a. Forecasting

Tornado forecasting in the United States is the subject of continuing research (see Rasmussen et al. 1994). It is likely that tornado forecasting skill is going to continue to improve as more is learned from the newly implemented WSR-88D radar system, as well as other new observing systems (see section 1.5c below). Some early results related to tornado warnings (e.g., Polger et al. 1994) seem to indicate that this is already happening. These early findings, however, can be misleading, since the effects of other aspects of the warning system (including human factors) are convolved with the effects of installing the new radars, and no effort has been made as yet to account for these other effects on warning performance (see Maddox and Forsyth 1994; Polger 1994). Dramatic further gains, especially in differentiating tornadic from nontornadic supercells and in forecasting nonsupercell tornado situations, may be difficult to achieve. Furthermore, the fatality figures from tornadoes in the United States may well be approaching some nearly irreducible minimum. I hasten to add that there are still things we can do to improve upon our tornado warnings and forecasts, so I am not saying that we have attained that irreducible minimum. Nevertheless, it is unrealistic to believe that there will ever be a time when tornadoes will cause zero fatalities owing to the perfection of the forecasts and warnings (and to citizen response).

I believe the time is ripe to develop improvements in the total system for mitigating flash floods. With mesoscale and cloud-scale numerical simulation models, it is quite likely that we will develop new and useful understanding upon which to base improved precipitation forecasts. I hope to see the implementation of probabilistic approaches (e.g., Krzysztofowicz et al. 1993) that will offer improvements not only in forecasting skill, but also in forecasting *value* for heavy precipitation events (Murphy 1993). At the time of this writing, the research investments aimed at improvement of quantitative precipitation forecasting are just beginning. Moreover, relatively little is as yet being done to couple precipitation forecast models with hydrological models so as to produce flood and flash flood forecasts directly. Thus, flash flood forecasting is open to considerable experimentation, with burgeoning new techniques and technologies. New methods for quantitative rainfall estimation (e.g., Ryzhkov and Zrnić 1996) and automated streamflow measurements are promising better flash flood detection and warning capability.

Because of the importance of aviation, convective wind gust forecasting and warning should continue to see investments in infrastructure and research. This should lead to some continued improvement in forecasting convective wind events, notably microbursts.

Although by no means unimportant in terms of damage, hailstorms remain difficult to mitigate. Unless it can be demonstrated that short-term mitigation is possible, it is going to be a challenge to find resources for sustained hailstorm research. Perhaps the best source for support will come from the insurance industry, which is already concerned about losses due to severe convective storms and is thus interested in refined hail climatology information. Improved climatological databases do nothing for understanding the processes by which hail is formed, but knowing more accurately than at present where and when hailstorms are most common can be a good starting point for both basic and applied research.

b. Modification

The idea of weather modification appeals to many citizens, especially in the wake of an especially damaging event. It seems logical to ask why a society that can develop hydrogen bombs, travel to the moon, and accomplish other technological feats cannot prevent weather-related disasters. Moreover, weather modification offers an apparent cure for all the ills produced by the weather. It appears that we are engaged in substantial weather modification, both planned and inadvertent, at the present, so why not do something about severe convective storms? Unless some mechanism can be exploited, like the metastability of clouds containing supercooled water, the titanic energies involved in convective storms make it pretty daunting to imagine making beneficial modifications. Furthermore, with our current lack of understanding of many of the processes associated with severe convective weather, it seems potentially dangerous to make a concerted effort to alter the weather. The results could be negative in ways we might not be able to anticipate. In spite of several ongoing private weather modification programs, it is by no means certain that they can validate claims of a purely beneficial effect. Several federally sponsored convective weather modification programs—for example, NHRE (Foote and Knight 1979) and FACE (Barnston et al. 1983)—were terminated by the early 1980s, with the realization that any substantial effects were at best only marginally detectable. Thus, I have little optimism for future benefits associated with modification of convective severe storms.

c. New observations

The prospects for new observations continue to be relatively bright. The implementation of a network of

Doppler radars is not likely to be the end of improvements to radar; as noted earlier, the potential value of polarimetric capability for rainfall measurement means that the retrofitting of the WSR-88D radars with the means for dual polarization observations is possible. This can have a beneficial impact on hail detection (Ryzhkov and Zrnić 1994) as well as improving rainfall estimates. Algorithm development for utilizing the capabilities of the new radars for automated detection of a variety of important convective weather events will certainly continue (Eilts et al. 1996). New strategies for faster and more thorough volumetric scanning by radars, including considering the employment of phased array radars, will be an important contribution to the value of radar in severe convective storms of all sorts.

Improvements to the overall sampling of atmospheric processes are also on the near horizon. Meteorological measurements by commercial aircraft (Schwartz and Benjamin 1995) are likely to make a large impact on the global observing system. Although the vertical wind profilers using Doppler radar (Gage and Balsley 1978) have encountered a number of problems in their implementation, they clearly demonstrate the value of such observations (e.g., Neiman and Shapiro 1989; Spencer et al. 1996).

Other technologies—such as radio acoustic sounding systems (RASS; see Moran et al. 1991; Neiman et al. 1992) and systems using the global positioning satellites (e.g., Ware et al. 2000)—promise additional remote measurement of profiles of thermodynamic variables. Although wind observations are important by themselves, the importance of thermodynamic structure (including details of the moisture distribution) for DMC processes cannot be underestimated. Perhaps pure wind observations might be very useful on larger scales, where balance relationships can be exploited to infer a lot about the thermodynamic structure from wind observations alone, but it is obvious that we should not be satisfied with wind-only enhancements to the observing system for severe convection-related problems.

Satellite-borne remote sensors continue to improve, and the multispectral character of geostationary platforms, along with the high temporal resolution capability afforded by the current and future generations of meteorological satellites, offer much promise. The difficulties with inversion of the radiative transfer equation (associated with nonuniqueness when solving an integro-differential equation) are not likely to disappear, but there is reason to believe that methods for dealing with the problem will evolve to fit a new mix of observational capabilities.

The picture is not entirely rosy, of course. Economic difficulties are driving national observing capabilities toward cheaper alternatives. In addition to reduced spatial density and perhaps even the temporal frequency of rawinsonde observations, this also means an increasing dependence on automated surface observations. The latter is a mixed blessing, offering greater density and consistency of observations at the price of not being able to provide the full suite of observations performed by a human observer. The implementation of small mesonetworks funded locally rather than federally (e.g., the Oklahoma Mesonetwork described by Brock et al. 1995) is likely to continue simply because the data are useful and the technology is affordable. Of course, the affordable instruments may not be state-of-the-art sensors; for example, a dewcel (a humidity instrument involving a temperature-controlled mirror for sensing the formation of dew or frost) is typically a better humidity sensor than a humicap (an electronic sensor whose capacitance is a function of humidity), but the cost and power requirement differentials make the humicap a logical choice for inexpensive networks of automated surface observations.

d. Scale interactions and chaos

Although synoptic-scale meteorology has been quite successful in describing the general features of weather systems, the majority of the weather events that affect people substantially are associated with mesoscale and smaller processes. To a great extent, this "sensible" weather (i.e., that which most people "sense") is dominated by the presence of ascending moist air: clouds and precipitation. Mesoscale processes and DMC are notoriously intermittent, that is, they are not ubiquitous and omnipresent. Whereas it would be extremely unusual to look at a hemispheric weather map and not see one or more synoptic-scale extratropical weather systems in midlatitudes, mesoscale weather and deep convection are not present every day within large regions. Intermittency is characteristic of the nonlinear dynamics associated with chaotic systems. The sensitive dependence on initial conditions that is central to the development of chaos (Lorenz 1993) is almost certainly at work in situations of DMC. The difference between a major outbreak of severe convective weather and no convection at all might be associated with a very small difference in, for example, the initial convective inhibition. Even forecasts of the meteorological environment for DMC that we now consider to be relatively good might be inadequate, as noted by Brooks et al. (1993).

It is the nonlinear dynamics that make both understanding the physical processes and the forecasting so difficult. Scale interaction is virtually by definition a product of this nonlinearity, so a treatment of scale interactions must necessarily include a nonlinear treatment of the physical processes. Since analytical solutions to nonlinear problems are confined to special cases that tend to be of only marginal application to real events, it is inevitable that numerical simulations are going to be essential both to gaining understanding and to forecasting of the real events. To date, numerical simulations have indeed been of considerable value in gaining insights into severe convective storms. Chapter 4 in this volume addresses numerical modeling in the context of DMC. However, simulations alone are not going to solve all our problems. We need observations to validate the simulations, and it is quite likely that there are many physically important processes that we have yet to observe. The history of our science suggests that theory and modeling in the absence of observations have not been very productive.

e. Important unobservables

It is an unfortunate reality that potentially important observations are presently unavailable and are not likely to be obtained in the near future. Perhaps the most frustrating class of unobservables is the suite of measurements associated with the microphysics of convection. It is possible to infer various aspects of the distribution of water substance in clouds with radar (see Jameson and Johnson 1990). Unfortunately, short of flying an aircraft into a cloud and attempting to gather up the various forms of condensed water (not an easy task to accomplish successfully), there is no simple way to validate those inferred measurements. Moreover, measurements of in-cloud water distributions are not sufficient; in a sense, such knowledge is only part of a successful understanding that would lead to improved forecasts. For instance, we have virtually nothing available in terms of condensation and/or freezing nuclei counts as a function of space and time in and near DMC, and prior to DMC onset. These unobserved variables might have a large impact on how water substance is distributed in the ensuing convection. In turn, the distribution of water substance in convective clouds plays an important role in cloud dynamics (Kessler 1969). Regrettably, there is virtually no information available operationally that could be used to take advantage of any new scientific understanding based on microphysical aspects of DMC. As of this writing, I see no feasible methods even being proposed to provide a set of microphysical observations on a routine basis.

The occurrence of severe convective events continues to be beyond the capabilities of our observing system; recall the mismatch between the real events and the mostly pointwise reports of the events discussed earlier. In order to observe severe convective weather properly, we need to be able to detect and record the actual distributions of tornadoes, hail, wind, and precipitation on scales well below the size of a convective cloud. Anything less means that we are not obtaining an accurate picture of the events, and it is obvious we are indeed not able to assume an accurate climatology; see Kelly et al. (1985) or Doswell and Burgess (1988) for some discussion. The consequences of this lack of knowledge have already been described in section 1.4a. Any progress beyond the capacity of existing systems is going to require resources at a time when most economic indicators suggest retrenchment, not expansion. Therefore, it is difficult to be optimistic that the future holds much promise for increasing the quality of our event observations.

The VORTEX project (Rasmussen et al. 1994) has made it clear that there is considerable structure in atmospheric variables present on scales not much larger than that of large convective systems. A few examples have appeared (e.g., Brooks et al. 1995) that offer hints of structures that might be important. Doppler radars operating in "clear air" mode and high-resolution satellite images present us with a complex picture of things happening in the lower troposphere, especially apparent lines of discontinuity. In many cases, we currently know relatively little about these features. When we have access to high-resolution data throughout the atmosphere, we invariably find structures of which we had not previously been aware (e.g., profiler data as in Carr et al. 1995). The extent to which these structures influence DMC and the associated severe weather events remains problematic because we are not at present able to observe these features reliably. It takes time to integrate new observations into our scientific understanding. Thus, as long as there are unobserved phenomena and unobserved physically relevant variables, there will always be gaps in our understanding and in our corresponding ability to forecast. This is the paradox of all science: We are both excited and frustrated by what we do not know, even as we create new understanding.

Acknowledgments. I am grateful to the members of my review panel for their helpful comments and suggestions, and I appreciate encouragement from Roger Wakimoto early in the process of writing this paper. Help with finding Fig. 1.5 was provided by Don Burgess. I have benefited from numerous discussions with colleagues on this topic, notably George Young, Fred Sanders, Bob Maddox, Mike Fritsch, Harold Brooks, Dave Stensrud, Bob Davies-Jones, and others too numerous to mention.

REFERENCES

Anderson, B., and S. Wamsley, 1996: Reflection of the hearts. *The Big Thompson Canyon Flood of July 31, 1976*, G. Wamsley and L. Roesener, Eds., C&M Press, 272 pp.

Barnes, S. L., 1978: Oklahoma thunderstorms on 29–30 April 1970. Part I: Morphology of a tornadic storm. *Mon. Wea. Rev.*, **106,** 673–684.

———, and C. W. Newton, 1983: Thunderstorms in the synoptic setting. *Thunderstorm Morphology and Dynamics,* E. Kessler, Ed., University of Oklahoma, 75–112.

Barnston, A. G., W. L. Woodley, J. A. Flueck, and M. H. Brown, 1983: The Florida Area Cumulus Experiment's second phase

(FACE-2). Part I: The experimental design, implementation and basic data. *J. Climate Appl. Meteor.,* **22,** 1504–1528.

Bartels, D. L., and R. A. Maddox, 1991: Midlevel cyclonic vortices generated by mesoscale convective systems. *Mon. Wea. Rev.,* **119,** 104–118.

Blyth, A. M., 1993: Entrainment in cumulus clouds. *J. Appl. Meteor.,* **32,** 626–641.

Boccippio, D. J., 1995: A diagnostic analysis of the VVP single-Doppler retrieval technique. *J. Atmos. Oceanic Technol.,* **12,** 230–248.

Bosart, L. F., and F. Sanders, 1981: The Johnstown flood of July 1977: A long-lived convective system. *J. Atmos. Sci.,* **38,** 1616–1642.

———, and G. M. Lackmann, 1995: Postlandfall tropical cyclone reintensification in a weakly baroclinic environment: A case study of Hurricane David (September 1979). *Mon. Wea. Rev.,* **123,** 3268–3291.

Brock, F. V., K. C. Crawford, R. L. Elliott, G. W. Cuperus, S. J. Stadler, H. L. Johnson, and M. D. Eilts, 1995: The Oklahoma mesonet: A technical overview. *J. Atmos. Oceanic Technol.,* **12,** 5–19.

Brooks, H. E., and C. A. Doswell III, 1993: Extreme winds in high-precipitation supercells. Preprints, *17th Conf. on Severe Local Storms,* St. Louis, MO, Amer. Meteor. Soc., 173–177.

———, and ———, 1996: A comparison between measures-oriented and distributions-oriented approaches to forecast verification. *Wea. Forecasting,* **11,** 288–303.

———, ———, and L. J. Wicker, 1993: STORMTIPE: A forecasting experiment using a three-dimensional cloud model. *Wea. Forecasting,* **8,** 352–362.

———, ———, and R. B. Wilhelmson, 1994: The role of midtropospheric winds in the evolution and maintenance of low-level mesocyclones. *Mon. Wea. Rev.,* **122,** 126–136.

———, ———, E. N. Rasmussen, and S. Lasher-Trapp, 1995: Detailed observations of complex dryline structure in Oklahoma on 14 April, 1994. Preprints, *14th Conf. on Weather Analysis and Forecasting,* Vienna, VA, Amer. Meteor. Soc., 62–67.

Browning, K. A., 1964: Airflow and precipitation trajectories within severe local storms which travel to the right of the winds. *J. Atmos. Sci.,* **21,** 634–639.

———, 1977: The structure and mechanism of hailstorms. *Hail: A Review of Hail Science and Hail Suppression, Meteor. Monogr.,* No. 38, Amer. Meteor. Soc., 1–39.

Byers, H. R., 1965: *Elements of Cloud Physics.* University of Chicago Press, 191 pp.

———, and R. R. Braham Jr., 1949: *The Thunderstorm.* U.S. Government Printing Office, Washington, D.C., 287 pp.

Caracena, F., J. McCarthy, and J. A. Flueck, 1983: Forecasting the likelihood of microbursts along the front range of Colorado. Preprints, *13th Conf. on Severe Local Storms,* Tulsa, OK, Amer. Meteor. Soc., 261–264.

Carr, F. H., P. L. Spencer, C. A. Doswell III, and J. D. Powell, 1995: A comparison of two objective analysis techniques for profiler time-height data. *Mon. Wea. Rev.,* **123,** 2165–2180.

Charba, J. P., 1979: Two to six hour severe local storm probabilities: An operational forecasting system. *Mon. Wea. Rev.,* **107,** 268–282.

Chesters, D. L., L. W. Uccellini, and A. Mostek, 1982: VISSR Atmospheric Sounder (VAS) simulation experiment for a severe storm environment. *Mon. Wea. Rev.,* **110,** 198–216.

Christian, H. J., R. J. Blakeslee, and S. L. Goodman, 1989: The detection of lightning from geostationary orbit. *J. Geophys. Res.,* **94,** 13 329–13 337.

Colquhoun, J. R., 1987: A decision tree method of forecasting thunderstorms, severe thunderstorms, and tornadoes. *Wea. Forecasting,* **2,** 337–345.

———, 1995: The Sydney supercell thunderstorm? *Weather,* **50,** 15–18.

Davies-Jones, R., 1991: The frontogenetical forcing of secondary circulations. Part I: The duality and generalization of the Q vector. *J. Atmos. Sci.,* **48,** 497–509.

———, D. Burgess, and M. Foster, 1990: Test of helicity as a forecast parameter. Preprints, *16th Conf. on Severe Local Storms,* Kananaskis Park, AB, Canada, Amer. Meteor. Soc., 588–592.

Davis, C. A., and M. L. Weisman, 1994: Balanced dynamics of mesoscale vortices produced in simulated convective systems. *J. Atmos. Sci.,* **51,** 2005–2030.

Deardorff, J. W., 1980: Cloud top entrainment instability. *J. Atmos. Sci.,* **37,** 131–147.

Dessens, J., and J. T. Snow, 1993: Comparative description of tornadoes in France and the United States. *The Tornado: Its Structure, Dynamics, Prediction, and Hazards, Geophys. Monogr.,* No. 79, Amer. Geophys. Union, 427–434.

Doswell, C. A., III, 1985: The operational meteorology of convective weather. Vol. II: Storm scale analysis. NOAA Tech. Memo. ERL ESG-15, NTIS PB85–226959, 240 pp.

———, 1987: The distinction between large-scale and mesoscale contribution to severe convection: A case study example. *Wea. Forecasting,* **2,** 3–16.

———, 1991: Comments on "Mesoscale convective patterns of the southern High Plains." *Bull. Amer. Meteor. Soc.,* **72,** 389–390.

———, 1994: Flash flood-producing convective storms: Current understanding and research. *Proc. U.S.-Spain Workshop on Natural Hazards,* Barcelona, Spain, National Science Foundation, 97–107.

———, 1999: Seeing supercells as heavy rain producers. Preprints, *13th Conf. on Hydrology,* Dallas, TX, Amer. Meteor. Soc., 73–76.

———, and D. W. Burgess, 1988: On some issues of United States tornado climatology. *Mon. Wea. Rev.,* **116,** 495–501.

———, and ———, 1993: Tornadoes and tornadic storms: A review of conceptual models. *The Tornado: Its Structure, Dynamics, Prediction, and Hazards, Geophys. Monogr.,* No. 79, Amer. Geophys. Union, 161–172.

———, and E. N. Rasmussen, 1994: The effect of neglecting the virtual temperature correction on CAPE calculations. *Wea. Forecasting,* **9,** 625–629.

———, J. T. Schaefer, D. W. McCann, T. W. Schlatter, and H. B. Wobus, 1982: Thermodynamic analysis procedures at the National Severe Storms Forecast Center. Preprints, *Ninth Conf. on Weather Forecasting and Analysis,* Seattle, WA, Amer. Meteor. Soc., 304–309.

———, D. L. Keller, and S. J. Weiss, 1990: An analysis of the temporal and spatial variation of tornado and severe thunderstorm watch verification. Preprints, *16th Conf. on Severe Local Storms,* Kananaskis Park, AB, Canada, Amer. Meteor. Soc., 294–299.

———, R. H. Johns, and S. J. Weiss, 1993: Tornado forecasting: A review. *The Tornado: Its Structure, Dynamics, Hazards, and Prediction, Geophys. Monogr.,* No. 79, Amer. Geophys. Union, 557–571.

———, H. E. Brooks, and R. A. Maddox, 1996: Flash flood forecasting: An ingredients-based methodology. *Wea. Forecasting,* **11,** 560–581.

———, A. R. Moller, and H. E. Brooks, 1999: Storm spotting and public awareness since the first tornado forecasts of 1948. *Wea. Forecasting,* **14,** 544–557.

Durran, D. R., 1986: Mountain waves. *Mesoscale Meteorology and Forecasting,* P. S. Ray, Ed., Amer. Meteor. Soc., 472–492.

Eilts, M. D., and Coauthors, 1996: Severe weather warning decision support system. Preprints, *18th Conf. on Severe Local Storms,* San Francisco, CA, Amer. Meteor. Soc., 536–540.

Emanuel, K. A., 1981: A similarity theory for unsaturated downdrafts within clouds. *J. Atmos. Sci.,* **38,** 1541–1557.

———, 1985: Frontal circulations in the presence of small moist symmetric instability. *J. Atmos. Sci.,* **42,** 1062–1071.

———, 1994: *Atmospheric Convection.* Oxford University Press, 580 pp.

Eom, J., 1975: Analysis of the internal gravity wave occurrence of April 19, 1970 in the Midwest. *Mon. Wea. Rev.,* **103,** 217–226.

FEMA, 1999: Midwest tornadoes of May 3, 1999: Observations, recommendations and technical guidance. Building Performance Assessment Rep. FEMA-342, Federal Emergency Management Agency, Washington, D.C., 189 pp.

Foote, G. B., and C. A. Knight, 1979: Results of a randomized hail suppression experiment in northeast Colorado. Part I: Design and conduct of the experiment. *J. Appl. Meteor.,* **18,** 1526–1537.

———, and H. W. Frank, 1983: Case study of a hailstorm in Colorado. Part III: Airflow from triple-Doppler measurements. *J. Atmos. Sci.,* **40,** 686–707.

Forbes, G. S., 1981: On the reliability of hook echoes as tornado indicators. *Mon. Wea. Rev.,* **109,** 1457–1466.

———, and R. Wakimoto, 1983: A concentrated outbreak of tornadoes, downbursts and microbursts, and implications regarding vortex classification. *Mon. Wea. Rev.,* **111,** 220–235.

Fovell, R. G., and P. S. Dailey, 1995: The temporal behavior of numerically simulated multicell-type storms. Part I: Modes of behavior. *J. Atmos. Sci.,* **52,** 2073–2095.

Friedman, R. M., 1999: Constituting the polar front, 1919–1920. *The Life Cycle of Extratropical Cyclones,* M. A. Shapiro and S. Grønas, Eds., Amer. Meteor. Soc., 29–39.

Fritsch, J. M., 1975: Cumulus dynamics: Local compensating subsidence and its implications for cumulus parameterization. *Pure Appl. Geophys.,* **113,** 851–867.

Fujita, T. T., 1955: Results of detailed synoptic studies of squall lines. *Tellus,* **4,** 405–436.

———, 1963: Analytical mesometeorology: A review. *Severe Local Storms, Meteor. Monogr.,* No. 27, Amer. Meteor. Soc., 77–125.

———, 1971: Proposed characterization of tornadoes and hurricanes by area and intensity. SMRP Research Paper 91, University of Chicago, Chicago, IL, 42 pp.

———, 1975: New evidence from April 3–4, 1974 tornadoes. Preprints, *Ninth Conf. on Severe Local Storms,* Norman, OK, Amer. Meteor. Soc., 248–255.

———, 1981: Tornadoes and downbursts in the context of generalized planetary scales. *J. Atmos. Sci.,* **38,** 1511–1534.

———, and F. Caracena, 1977: An analysis of three weather-related aircraft accidents. *Bull. Amer. Meteor. Soc.,* **58,** 1164–1181.

Gage, K. S., and B. B. Balsley, 1978: Doppler radar probing of the clear atmosphere. *Bull. Amer. Meteor. Soc.,* **59,** 1074–1093.

Galway, J. G., 1977: Some climatological aspects of tornado outbreaks. *Mon. Wea. Rev.,* **105,** 477–484.

———, 1989: The evolution of severe thunderstorm criteria within the Weather Service. *Wea. Forecasting,* **4,** 585–592.

Grazulis, T. P., 1993: *Significant Tornadoes 1680–1991.* Environmental Films, St. Johnsbury, VT, 1326 pp.

Hales, J. E., Jr., 1993: Biases in the severe thunderstorm data base: Ramifications and solutions. Preprints, *13th Conf. on Weather Analysis and Forecasting,* Vienna, VA, Amer. Meteor. Soc., 504–507.

Hane, C. E., and P. S. Ray, 1985: Pressure and buoyancy fields derived from Doppler radar data in a tornadic thunderstorm. *J. Atmos. Sci.,* **42,** 18–35.

Hess, S. L., 1959: *Introduction to Theoretical Meteorology.* Holt, Rhinehart and Winston, 362 pp.

Hilger, D. W., and J. F. W. Purdom, 1990: Clustering of satellite sounding radiances to enhance mesoscale meteorological retrievals. *J. Appl. Meteor.,* **29,** 1344–1351.

Hoffman, E. G., L. F. Bosart, and D. Keyser, 1995: Large-amplitude inertia-gravity wave environments: Vertical structure and evolution. Preprints, *15th Conf. on Weather Analysis and Forecasting,* Norfolk, VA, Amer. Meteor. Soc., 245–248.

Hooke, W. H., 1986: Gravity waves. *Mesoscale Meteorology and Forecasting,* P. Ray, Ed., Amer. Meteor. Soc., 272–288.

Hoskins, B. J., and F. P. Bretherton, 1972: Atmospheric frontogenesis models: Mathematical formulation and solution. *J. Atmos. Sci.,* **29,** 11–37.

House, D. C., 1959: The mechanics of instability line formation. *J. Meteor.,* **16,** 108–120.

———, 1963: Forecasting tornadoes and severe thunderstorms. *Severe Local Storms, Meteor. Monogr.,* No. 27, Amer. Meteor. Soc., 141–155.

Houze, R. A., Jr., W. Schmid, R. G. Fovell, and H. H. Schiesser, 1993: Hailstorms in Switzerland: Left movers, right movers, and false hooks. *Mon. Wea. Rev.,* **121,** 3345–3370.

Jameson, A. R., and D. B. Johnson, 1990: Cloud microphysics and radar. *Radar in Meterology,* D. Atlas, Ed., Amer. Meteor. Soc., 323–340.

Johns, R. H., and W. D. Hirt, 1987: Derechos: Widespread convectively induced windstorms. *Wea. Forecasting,* **2,** 32–49.

———, and C. A. Doswell III, 1992: Severe local storms forecasting. *Wea. Forecasting,* **7,** 588–612.

Kamburova, P. L., and F. H. Ludlam, 1962: Rainfall evaporation in thunderstorm downdraughts. *Quart. J. Roy. Meteor. Soc.,* **88,** 510–518.

Kelly, D. L., J. T. Schaefer, R. P. McNulty, C. A. Doswell III, and R. F. Abbey Jr., 1978: An augmented tornado climatology. *Mon. Wea. Rev.,* **106,** 1172–1183.

———, ———, and C. A. Doswell III, 1985: Climatology of nontornadic severe thunderstorm events in the United States. *Mon. Wea. Rev.,* **113,** 1997–2014.

Kessler, E., 1969: *On the Distribution and Continuity of Water Substance in Atmospheric Circulations. Meteor. Monogr.,* No. 32, Amer. Meteor. Soc., 84 pp.

Klemp, J. B., and D. K. Lilly, 1975: The dynamics of wave-induced downslope winds. *J. Atmos. Sci.,* **32,** 320–339.

———, and R. B. Wilhelmson, 1978: The simulation of three-dimensional convective storm dynamics. *J. Atmos. Sci.,* **35,** 1070–1096.

Koppel, L. L., L. F. Bosart, and D. Keyser, 2000: A 25-yr climatology of large-amplitude hourly surface pressure changes over the conterminous United States. *Mon. Wea. Rev.,* **128,** 51–68.

Krzysztofowicz, R., W. J. Drzal, T. R. Drake, J. C. Weyman, and L. A. Giordano, 1993: Probabilistic quantitative precipitation forecasts for river basins. *Wea. Forecasting,* **8,** 424–439.

Laing, A. G., and J. M. Fritsch, 1997: The global population of mesoscale convective complexes. *Quart. J. Roy. Meteor. Soc.,* **123,** 389–405.

Lanicci, J. M., T. N. Carlson, and T. T. Warner, 1987: Sensitivity of the Great Plains severe-storm environment to soil-moisture distribution. *Mon. Wea. Rev.,* **115,** 2660–2673.

Larson, L. W., and Coauthors, 1995: Operational responsibilities of the National Weather Service river and flood program. *Wea. Forecasting,* **10,** 465–476.

Levine, J., 1959: Spherical vortex theory of bubble-like motion in cumulus clouds. *J. Meteor.,* **16,** 653–662.

Lilly, D. K., and T. Gal-Chen, 1990: Can dryline mixing create buoyancy? *J. Atmos. Sci.,* **47,** 1170–1171.

Lindzen, R. S., and K. K. Tung, 1976: Banded convective activity and ducted gravity waves. *Mon. Wea. Rev.,* **104,** 1602–1617.

Locatelli, J. D., M. T. Stoelinga, and P. V. Hobbs, 1998: Structure and evolution of winter cyclones in the central United States and their effects on the distribution of precipitation. Part V: Thermodynamic and dual-Doppler radar analysis of a squall line associated with a cold front aloft. *Mon. Wea. Rev.,* **126,** 860–875.

López, R. E., R. L. Holle, T. A. Heitkamp, M. Boyson, M. Cherington, and K. Langford, 1993: The underreporting of lighting injuries and deaths. Preprints, *17th Conf. on Severe Local Storms/Conf. on Atmospheric Electricity,* St. Louis, MO, Amer. Meteor. Soc., 775–778.

Lorenz, E. N., 1993: *The Essence of Chaos.* University of Washington, 227 pp.

Ludlam, F. H., 1963: Severe local storms: A review. *Severe Local Storms, Meteor. Monogr.,* No. 27, Amer. Meteor. Soc., 1–30.
Lynn, B. H., W.-K. Tao, and P. J. Wetzel, 1998: A study of landscape-generated deep moist convection. *Mon. Wea. Rev.,* **126,** 928–942.
MacGorman, D. R., and W. D. Rust, 1998: *The Electrical Nature of Storms.* Oxford University Press, 422 pp.
Maddox, R. A., 1980: Mesoscale convective complexes. *Bull. Amer. Meteor. Soc.,* **61,** 1374–1387.
———, and C. A. Doswell III, 1982: An examination of jet stream configurations, 500 mb vorticity advection and low-level thermal advection patterns during extended periods of intense convection. *Mon. Wea. Rev.,* **110,** 184–197.
———, and D. E. Forsyth, 1994: Comments on "National Weather Service warning performance based on the WSR-88D." *Bull. Amer. Meteor. Soc.,* **75,** 2175.
———, L. R. Hoxit, and C. F. Chappell, 1980: A study of tornadic thunderstorms interactions with thermal boundaries. *Mon. Wea. Rev.,* **108,** 1866–1877.
Marshall, T. P., 1993: Lessons learned from analyzing tornado damage. *The Tornado: Its Structure, Dynamics, Prediction, and Hazards, Geophys. Monogr.,* No. 79, Amer. Geophys. Union, 495–499.
Marwitz, J. D., 1972a: The structure and motion of severe hailstorms. Part I: Supercell storms. *J. Appl. Meteor.,* **11,** 166–179.
———, 1972b: The structure and motion of severe hailstorms. Part II: Multi-cell storms. *J. Appl. Meteor.,* **11,** 180–188.
———, 1972c: The structure and motion of severe hailstorms. Part III: Severely sheared storms. *J. Appl. Meteor.,* **11,** 189–201.
Mass, C. F., and M. D. Albright, 1989: Origin of the Catalina eddy. *Mon. Wea. Rev.,* **117,** 2406–2436.
Matejka, T., and R. C. Srivastrava, 1991: An improved version of the extended VAD analysis of single-Doppler radar data. *J. Atmos. Oceanic Technol.,* **8,** 435–466.
McCaul, E. W., Jr., 1993: Observations and simulation of hurricane-spawned tornadic storms. *The Tornado: Its Structure, Dynamics, Prediction, and Hazards, Geophys. Monogr.,* No. 79, Amer. Geophys. Union, 119–142.
McDonald, J. R., 1993: Damage mitigation and occupant safety. *The Tornado: Its Structure, Dynamics, Prediction, and Hazards, Geophys. Monogr.,* No. 79, Amer. Geophys. Union, 523–528.
McWilliams, J. C., and P. R. Gent, 1980: Intermediate models of planetary circulations in the atmosphere and ocean. *J. Atmos. Sci.,* **37,** 1657–1678.
Mehta, K. C., 1976: Windspeed estimates: Engineering analyses. *Proc. Symp. on Tornadoes,* Lubbock, TX, Texas Tech. University, 89–103.
Menard, R. D., and J. M. Fritsch, 1989: A mesoscale convective complex-generated inertial stable warm core vortex. *Mon. Wea. Rev.,* **117,** 1237–1261.
Miller, R. C., 1972: Notes on the analysis and severe-storm forecasting procedures of the Air Force Global Weather Central. AWS Tech. Rep. 200 (rev.), Headquarters, Air Weather Service, Scott Air Force Base, IL, 190 pp.
Minor, J. E., and K. C. Mehta, 1979: Wind damage observations and implications. *J. Struc. Div. Amer. Soc. Civ. Eng.,* **105,** 2279–2291.
Moran, K. P., D. B. Wuertz, R. G. Strauch, N. L. Abshire, and D. C. Law, 1991: Temperature sounding with wind profiler radars. *J. Atmos. Oceanic Technol.,* **8,** 606–611.
Murphy, A. H., 1993: What is a good forecast? An essay on the nature of goodness in forecasting. *Wea. Forecasting,* **8,** 281–293.
———, and R. L. Winkler, 1987: A general framework for forecast verification. *Mon. Wea. Rev.,* **115,** 1330–1338.
National Weather Service, 1991: Shadyside, Ohio, flash floods—June 14, 1990. Natural Disaster Survey Rep., 124 pp. [Available from NOAA/National Weather Service, 1325 East-West Highway, Silver Spring, MD 20910.]
Neiman, P. J., and M. A. Shapiro, 1989: Retrieving horizontal temperature gradients and advections from single-station wind profiler observations. *Wea. Forecasting,* **4,** 222–233.
———, P. T. May, and M. A. Shapiro, 1992: Radio acoustic sounding system (RASS) and wind profiler observations of lower- and midtropospheric weather systems. *Mon. Wea. Rev.,* **120,** 2298–2313.
Newton, C. W., 1950: Structure and mechanism of the pre-frontal squall line. *J. Meteor.,* **7,** 210–222.
———, 1963: Dynamics of severe convective storms. *Severe Local Storms, Meteor. Monogr.,* No. 27, Amer. Meteor. Soc., 33–58.
Ninomiya, K., 1971: Mesoscale modification of synoptic situations from thunderstorm development as revealed by ATS III and aerological data. *J. Appl. Meteor.,* **10,** 1103–1121.
Nolen, R. H., 1959: A radar pattern associated with tornadoes. *Bull. Amer. Meteor. Soc.,* **40,** 277–279.
Olson, D. A., N. W. Junker, and B. Korty, 1995: Evaluation of 33 years of quantitative precipitation forecasting at the NMC. *Wea. Forecasting,* **10,** 498–511.
Orlanski, I., 1975: A rational subdivision of scales for atmospheric processes. *Bull. Amer. Meteor. Soc.,* **56,** 527–530.
Orville, R. E., and R. W. Henderson, 1986: Global distribution of midnight lightning: September 1977 to August 1978. *Mon. Wea. Rev.,* **114,** 2640–2653.
Paluch, I. R., 1979: The entrainment mechanism in Colorado cumuli. *J. Atmos. Sci.,* **36,** 2467–2478.
Pan, Z., E. Takle, M. Segal, and R. Turner, 1996: Influences of model parameterization schemes on the response of rainfall to soil moisture in the central United States. *Mon. Wea. Rev.,* **124,** 1786–1802.
Peltier, W. R., and T. L. Clark, 1979: The evolution and stability of finite-amplitude mountain waves. Part II: Surface wave drag and severe downslope windstorms. *J. Atmos. Sci.,* **36,** 1498–1529.
Polger, P. D., 1994: Reply to "Comments." *Bull. Amer. Meteor. Soc.,* **75,** 2175–2176.
———, B. S. Goldsmith, R. C. Przywarty, and J. R. Bocchieri, 1994: National Weather Service warning performance based on the WSR-88D. *Bull. Amer. Meteor. Soc.,* **75,** 203–214.
Przybylinski, R. W., 1995: The bow echo: Observations, numerical simulations, and severe weather detection methods. *Wea. Forecasting,* **10,** 203–217.
Purdom, J. F. W., 1976: Some uses of high-resolution GOES imagery in the mesoscale forecasting of convection and its behavior. *Mon. Wea. Rev.,* **104,** 1474–1483.
Rasmussen, E. N., J. M. Straka, R. Davies-Jones, C. A. Doswell III, F. H. Carr, M. D. Eilts, and D. R. MacGorman, 1994: Verification of the Origins of Rotation in Tornadoes Experiment: VORTEX. *Bull. Amer. Meteor. Soc.,* **75,** 995–1006.
Reap, R. M., and D. S. Foster, 1979: Automated 12–36 hour probability forecasts of thunderstorms and severe local storms. *J. Appl. Meteor.,* **18,** 1304–1315.
Renno, N. O., and E. R. Williams, 1995: Quasi-Lagrangian measurements in convective boundary layer plumes and their implications for the calculation of CAPE. *Mon. Wea. Rev.,* **123,** 2733–2742.
Rhea, J. O., 1966: A study of thunderstorm formation along dry lines. *J. Appl. Meteor.,* **5,** 59–63.
Rust, W. D., and T. C. Marshall, 1996: On abandoning the thunderstorm tripole charge paradigm. *J. Geophys. Res.,* **101,** 23 499–23 504.
Ryzhkov, A., and D. Zrnić, 1994: Precipitation observed in Oklahoma mesoscale convective systems with a polarimetric radar. *J. Appl. Meteor.,* **33,** 455–464.
———, and ———, 1996: Assessment of rainfall measurement that uses specific differential phase. *J. Appl. Meteor.,* **35,** 2080–2090.
Sanders, F., and C. A. Doswell III, 1995: A case for detailed surface analysis. *Bull. Amer. Meteor. Soc.,* **76,** 505–521.

Saunders, C. P. R., 1993: A review of thunderstorm electrification processes. *J. Appl. Meteor.,* **32,** 642–655.

Schaefer, J. T., 1975: Nonlinear biconstituent diffusion: A possible trigger of convection. *J. Atmos. Sci.,* **32,** 2278–2284.

———, 1986: Severe thunderstorm forecasting: A historical perspective. *Wea. Forecasting,* **1,** 164–189.

Schlesinger, R. E., 1975: A three-dimensional numerical model of an isolated deep convective cloud: Preliminary results. *J. Atmos. Sci.,* **32,** 934–957.

Schultz, D. M., P. N. Schumacher, and C. A. Doswell III, 2000: The intracacies of instabilities. *Mon. Wea. Rev.,* **128,** 4143–4148.

Schumann, U., and C.-H. Moeng, 1991: Plume budgets in clear and cloudy convective boundary layers. *J. Atmos. Sci.,* **48,** 1758–1770.

Schwartz, B., and S. G. Benjamin, 1995: A comparison of temperature and wind measurements from ACARS-equipped aircraft and rawinsondes. *Wea. Forecasting,* **10,** 528–544.

Scorer, R. S., and F. H. Ludlam, 1953: Bubble theory of penetrative convection. *Quart. J. Roy. Meteor. Soc.,* **79,** 94–103.

Segal, M., R. W. Arritt, C. Clark, R. Rabin, and J. M. Brown, 1995: Scaling evaluation of the effect of surface characteristics on potential for deep convection over uniform terrain. *Mon. Wea. Rev.,* **123,** 383–400.

Sherwood, S. C., 2000: On moist instability. *Mon. Wea. Rev.,* **128,** 4139–4142.

Shortley, G., and D. Williams, 1961: *Elements of Physics.* Prentice-Hall, 928 pp.

Simpson, J., 1971: On cumulus entrainment and one-dimensional models. *J. Atmos. Sci.,* **28,** 449–455.

———, 1972: Reply to comments. *J. Atmos. Sci.,* **29,** 220–225.

Smull, B. F., 1995: Convectively induced mesoscale weather systems in the tropical and warm-season midlatitude atmosphere. *Rev. Geophys.,* **33** (Suppl.), 897–906.

———, and R. A. Houze Jr., 1987: Rear inflow in squall lines with trailing stratiform precipitation. *Mon. Wea. Rev.,* **115,** 2869–2889.

Spencer, P. L., F. H. Carr, and C. A. Doswell III, 1996: Investigation of an amplifying and a decaying wave using a network of wind profilers. *Mon. Wea. Rev.,* **124,** 209–223.

Spencer, R. W., H. M. Goodman, and R. E. Hood, 1989: Precipitation retrieval over land and ocean with the SSM/I: Identification and characteristics of the scattering signal. *J. Atmos. Oceanic Technol.,* **6,** 254–273.

Squires, P., 1958: Penetrative downdraughts in cumuli. *Tellus,* **10,** 381–393.

———, and J. S. Turner, 1962: An entraining jet model for cumulonimbus updraughts. *Tellus,* **14,** 422–434.

Stoelinga, M. T., J. D. Locatelli, and P. V. Hobbs, 2000: Structure and evolution of winter cyclones in the central United States and their effects on the distribution of precipitation. Part IV: A mesoscale modeling study of the initiation of convective rainbands. *Mon. Wea. Rev.,* **128,** 3481–3500.

Stommel, H., 1947: Entrainment of air into a cumulus cloud. *J. Meteor.,* **4,** 91–94.

Turner, J. S., 1962: The starting plume in neutral surroundings. *J. Fluid Mech.,* **13,** 356–368.

———, 1973: *Buoyancy Effects in Fluids.* Cambridge University Press, 367 pp.

Uccellini, L. W., 1975: A case study of apparent gravity wave initiation of severe convective storms. *Mon. Wea. Rev.,* **103,** 497–513.

Velasco, I., and J. M. Fritsch, 1987: Mesoscale convective complexes in the Americas. *J. Geophys. Res.,* **92,** 9591–9613.

Vonnegut, B., 1963: Some facts and speculations concerning the origin and role of thunderstorm electricity. *Severe Local Storms, Meteor. Monogr.,* No. 27, Amer. Meteor. Soc., 224–241.

———, 1994: The atmospheric electricity paradigm. *Bull. Amer. Meteor. Soc.,* **75,** 53–61.

Wakimoto, R. M., 1985: Forecasting dry microburst activity over the High Plains. *Mon. Wea. Rev.,* **113,** 1131–1143.

Ware, R. H., and Coauthors, 2000: SuomiNet: A real-time national GPS network for atmospheric research and education. *Bull. Amer. Meteor. Soc.,* **81,** 677–694.

Warner, J., 1970: On steady-state one-dimensional models of cumulus convection. *J. Atmos. Sci.,* **27,** 1035–1040.

———, 1972: Comments on "On cumulus entrainment and one-dimensional models." *J. Atmos. Sci.,* **29,** 218–219.

Weisman, M. L., and J. B. Klemp, 1982: The dependence of numerically simulated convective storms on vertical wind shear and buoyancy. *Mon. Wea. Rev.,* **110,** 504–520.

———, and ———, 1984: The structure and classification of numerically simulated convective storms in directionally-varying wind shears. *Mon. Wea. Rev.,* **112,** 2479–2498.

———, and ———, 1986: Characteristics of isolated convective storms. *Mesoscale Meteorology and Forecasting,* P. S. Ray, Ed., Amer. Meteor. Soc., 331–358.

Wetzel, P. J., W. R. Cotton, and R. L. McAnelly, 1983: A long-lived mesoscale convective complex. Part II: Evolution and structure of the mature complex. *Mon. Wea. Rev.,* **111,** 1919–1937.

Wilczak, J. M., and J. W. Glendening, 1988: Observations and mixed-layer modeling of a terrain-induced mesoscale gyre: The Denver cyclone. *Mon. Wea. Rev.,* **116,** 2688–2711.

Williams, E. R., 1985: Large-scale charge separation in thunderclouds. *J. Geophys. Res.,* **90,** 6013–6025.

———, and R. M. Lhermitte, 1983: Radar test of the precipitation hypothesis for thunderstorm electrification. *J. Geophys. Res.,* **88,** 10 984–10 992.

Williams, R. T., and J. Plotkin, 1968: Quasi-geostrophic frontogenesis. *J. Atmos. Sci.,* **25,** 201–206.

Xu, Q., 1987: The existence and stability of steady circulations in a conditionally symmetrically unstable basic flow. *J. Atmos. Sci.,* **44,** 3020–3029.

Young, G., 1988: Convection in the atmospheric boundary layer. *Earth-Sci. Rev.,* **25,** 179–188.

Zehr, R. M., J. F. W. Purdom, J. F. Weaver, and R. N. Green, 1988: Use of VAS data to diagnose the mesoscale environment of convective storms. *Wea. Forecasting,* **3,** 33–49.

Ziegler, C. L., W. J. Martin, R. A. Pielke, and R. L. Walko, 1995: A modeling study of the dryline. *J. Atmos. Sci.,* **52,** 263–285.

Zipser, E. J., 1982: Use of a conceptual model of the life-cycle of mesoscale convective systems to improve very short-range forecasts. *Nowcasting,* K. A. Browning, Ed., Academic Press, 191–204.

Chapter 2

Extratropical Synoptic-Scale Processes and Severe Convection

CHARLES A. DOSWELL III

National Severe Storms Laboratory, Norman, Oklahoma

LANCE F. BOSART

Department of Earth and Atmospheric Sciences, The University at Albany, State University of New York, Albany, New York

REVIEW PANEL: Howard B. Bluestein (Chair), John M. Brown, Bradley Colman, and Christopher Davis

2.1. Introduction

Within this chapter, we intend to give a broad perspective on the interaction between severe convection and extratropical synoptic-scale processes. A traditional view of this interaction is that the synoptic-scale processes simply provide a setting in which severe convection develops (see, e.g., Newton 1963; Barnes and Newton 1983; Johns and Doswell 1992). This view could be interpreted as implying that convection has little or no direct impact on synoptic scales. However, there have been many recent developments in mesoscale meteorology as it relates to severe convection (as described in chapter 3 of this volume), wherein upscale effects of convection are seen most clearly. Mesoscale processes often act as a sort of intermediary between convective and synoptic scales. We take the view that, in spite of the intermediation by mesoscale processes, it is still possible to take a synoptic-scale view of the impacts of deep, moist convection, especially in its most severe manifestations.

The subject of scale separation is always a thorny one. Orlanski's (1975) scale divisions are essentially arbitrary, based on powers of 10 in space and time. There are also dynamically motivated ways to divide scales (Emanuel 1986; Doswell 1987), but there is no universally accepted way to separate scales of motion. For the purposes of this review, we are concerned with the processes associated with extratropical weather systems in midlatitudes; tropical synoptic-scale processes are considered in chapter 7 of this volume.

Quasigeostrophic (QG) theory is arguably the simplest statement of what it means to be synoptic-scale (Doswell 1987), at least outside the Tropics. In section 2.2, we provide brief overviews of QG principles, potential vorticity thinking, and basic jet streak-related processes. Section 2.3 presents a discussion of planetary boundary layer processes, focusing on how these relate to both synoptic scales and severe convection. section 2.4 provides some basic climatological distributions of convection, both in space and in time. These observed climatological distributions provide important clues as to the interaction between synoptic-scale processes and convection. The climatology of severe forms of deep, moist convection is the topic of section 2.5. In section 2.6, brief overviews of a number of cases are presented, in part to illustrate the principles developed, but also to show the variety of synoptic-scale structures in which severe convection can develop. Section 2.7 presents some perspectives on the synoptic contributions to severe convection, and section 2.8 provides a discussion of the reverse feedbacks of convection to the synoptic scale. Finally, section 2.9 provides some discussion and conclusions.

2.2. Brief overviews

Deep, moist convection (DMC)[1] is associated with a triad of necessary and sufficient ingredients: moisture, low static stability, and ascent of parcels to their level of free convection (LFC) by some lifting mechanism (Doswell 1987).[2] Synoptic-scale processes, notably the extratropical cyclone (ETC), play a reasonably well-understood role in moistening and destabilization, but the relatively weak vertical motions (on the order of a few centimeters per second) of synoptic-scale systems are usually too slow to lift potentially buoyant parcels to their LFCs in less than about a day. On the other

[1] As in chapter 1 of this volume, we use the term "deep moist convection" rather than "thunderstorm," since not all severe forms of deep, moist convection produce lightning and, hence, thunder.

[2] We assume that the presence of the LFC implies that the convection will, indeed, be "deep," although rare exceptions to this might be found.

hand, ETCs provide an environment that favors processes operating on smaller scales, such as drylines (see Schaefer 1986; Ziegler and Rasmussen 1998) and fronts, and vertical motions substantially larger than those of the synoptic scale can be created by those subsynoptic processes. Thus, there is a strong association between the development of DMC and ETCs, even though DMC is not confined exclusively to the ETC environment. Severe convection also exhibits this association, perhaps even more strongly than ordinary convection. To understand this connection, we begin with consideration of the simplest model of synoptic-scale processes in midlatitudes, quasigeostropic theory. Then we move to consider potential vorticity, a modern perspective on synoptic-scale processes. Jet streaks are reviewed briefly in this same context.

a. QG principles

Quasigeostrophic (QG) theory is not a concept that springs from a single, brilliant exposition; instead, it is a child of many parents. This is manifest in the excellent historical treatments of QG theory's development by Phillips (1990) or Bosart (1999).[3] Many respected scientists in the history of modern meteorology have made contributions to QG theory. The quasigeostrophic system is contained in the two equations (following Holton 1992, p. 158 ff.)

$$\left[\nabla^2 + \frac{\partial}{\partial p}\left(\frac{f_0^2}{\sigma}\frac{\partial}{\partial p}\right)\right]\chi = -f_0\mathbf{V}_g\cdot\nabla\left(\frac{1}{f_0}\nabla^2\Phi + f\right) - \frac{\partial}{\partial p}\left[-\frac{f_0^2}{\sigma}\mathbf{V}_g\cdot\nabla\left(-\frac{\partial\Phi}{\partial p}\right)\right], \quad (2.1)$$

and

$$\left(\nabla^2 + \frac{f_0^2}{\sigma}\frac{\partial^2}{\partial p^2}\right)\omega = \frac{f_0}{\sigma}\frac{\partial}{\partial p}\left[\mathbf{V}_g\cdot\nabla\left(\frac{1}{f_0}\nabla^2\Phi + f\right)\right] + \frac{1}{\sigma}\nabla^2\left[\mathbf{V}_g\cdot\nabla\left(-\frac{\partial\Phi}{\partial p}\right)\right], \quad (2.2)$$

where ∇ is the horizontal gradient operator on a p surface, Φ is geopotential height, the height tendency (χ) is defined by $\chi \equiv \partial\Phi/\partial t$, $\omega \equiv dp/dt$, and the static stability (σ) is defined by $\sigma \equiv -(\alpha_0/\theta_0)(d\theta_0/dp)$. Static stability and the basic-state variables α_0 (specific volume, or inverse density) and θ_0 (potential temperature) are assumed to be functions of p alone, and f_0 denotes a constant reference value of the Coriolis parameter. Equations (2.1) and (2.2) exhibit a near-symmetry that is apparent in the physical discussions presented by Holton (1992, p. 177 ff.) and Bluestein (1992, chapter 5). Together, (2.1) and (2.2) constitute the QG forecasting system. It is possible to use this system, along with a mass continuity equation, to make forecasts; this system was used operationally at the National Meteorological Center (now the National Centers for Environmental Prediction, or NCEP) in the 1960s.

However, the real value in the QG system today is not in prediction, but in qualitative understanding of midlatitude, synoptic-scale processes (Durran and Snellman 1987). Equations (2.1) and (2.2) are derived from the primitive equations by making a number of assumptions about the flow they describe: namely, that the flow is adiabatic and hydrostatic and the ageostrophic part of the flow makes no contribution to advection, etc. Textbook discussions (e.g., Holton 1992, p. 166 ff.; Bluestein 1992, chapter 5) point out that vertical motion in the QG system is simply a theoretical response to the disruptions of geostrophic and thermal wind balance caused by thermal and differential vorticity advection, with the QG response acting to restore hydrostatic and geostrophic balance. Disturbances in the height field (that are reflected in the relative vorticity) move by vorticity advection, weakening or intensifying as a consequence of differential thermal advection.

Generally speaking, QG theory predicts rising motion ahead of cyclonic disturbances and descending motion behind. In QG theory, however, σ is assumed to be, at most, a function of pressure, whereas in reality, σ varies in space and time. Rising motion favors a decrease in stability below the level of peak ascent, usually somewhere in midtroposphere, whereas sinking motion increases the stability below the level of maximum descent.[4] Therefore, ETCs should exhibit some asymmetry in their vertical motion patterns beyond that derived from QG theory: ascent should be more intense than descent. Further, Emanuel et al. (1987) have shown (in a non-QG context) that if ascent is saturated and descent is unsaturated, the difference between moist ascent and dry descent can be parameterized by using a dry static stability for descent and a weaker, moist static stability for ascent. Doing so results in ascent being localized and relatively intense, whereas descent is weaker and more widespread (Whitaker and Barcilon 1992; Fantini 1995). These concepts are generally consistent with the observed behavior in midlatitude cyclones (e.g., Whitaker et al. 1988), which certainly include non-QG processes; upward motion is typically stronger than

[3] Phillips attributes the first use of the term "quasi-geostrophic" to Sutcliffe (1939). However, Sutcliffe (1938) includes the following passage: "It is suggested that the term 'quasi-geostrophic' would be a better description in that the motion is geostrophic only to a first approximation."

[4] This can also be inferred from a static stability tendency equation; e.g., Panfosky (1964, p. 105 ff.).

downward motion and more localized (at times, of course, ascent is concentrated in narrow bands associated with fronts).

ETC development (e.g., as described in Palmén and Newton 1969, chapter 11; Uccellini 1990; Bosart 1999) is strongly dependent on the vertical motions. Vertical motion, in turn, depends on σ (even in the QG system considered above), so the environmental static stability is an important factor in cyclogenesis and its associated frontogenesis (Roebber 1993). Moreover, convection acts to stabilize the troposphere, in general, so convection can modify the environment seen by cyclones, as well as the other way around. ETCs are constantly advecting and changing the static stability, as well, representing an important nonlinear interaction. Besides their impacts on (and modification by) the static stability field, ETCs are also quite important in transporting heat and moisture (Lorenz 1965, chapter IV). Thus, the modulation of convection through modification of its environment by ETCs is quite substantial, and the interaction with DMC is complex and likely to be hard to forecast accurately.

Most (but not all) DMC involves parcels originating from near the surface. Although the observed three-dimensional distributions of heat and moisture are complex in detail, it is generally the case that both heat and moisture tend to increase equatorward and decrease with height. Poleward flows bringing what is generally warm air poleward often include low-level moisture as well.[5] Thus, strongly meridional low-level flows, in general, transport moisture for convection and significantly modify the global sensible and latent heat distributions.

The environmental vertical wind shear structure has been shown to be a very important factor in determining the severity of any resulting convection (Newton 1963; Klemp and Wilhelmson 1978; Weisman and Klemp 1982, 1984; Brooks et al. 1994b). The geostrophic vertical wind shear is associated with thermal advection [an important factor in both Eqs. (2.1) and (2.2)]. In the Northern Hemisphere, warm advection favors a veering wind profile (at least to the extent that the vertical wind shear is dominated by the thermal wind contribution), as well as being associated with synoptic-scale ascent. It has been shown (e.g., Browning 1964; Rotunno 1993) that strong vertical wind shear (of order 10^{-3} s^{-1} and larger) is a major factor in promoting the supercellular forms of convection that are the most prolific producers of severe weather. Some shear profiles can also favor the organization of convection into more intense and persistent forms (Rotunno and Klemp 1982; Rotunno 1993).

On the other hand, strong vertical wind shear has been viewed as an inhibiting factor for convection, in the sense that it tends to reduce the intensity of updrafts (Asai 1970; Brooks and Wilhelmson 1993). That strong vertical wind shear can interact with convection to enhance it in some circumstances was recognized in Newton's (1963) review paper, but this concept has been extensively refined by means of numerical cloud model simulations (Weisman and Klemp 1982; 1984). Not only does vertical wind shear promote new convective cell development via the interaction between existing updrafts and the shear (Rotunno and Klemp 1982), it also affects the local distribution of hydrometeors produced by convection (Brooks et al. 1994a). It has been shown that the interaction between wind shear and convection can result in a contribution to the updraft as large as that of buoyancy (Weisman and Klemp 1984).

For synoptically evident,[6] major outbreaks of tornadic severe weather (Doswell et al. 1993), favorable wind shear is widespread. In fact, baroclinic instability (linked to the development of ETCs) is closely related to the vertical wind shear through the thermal wind relationship. If this synoptic-scale vertical wind shear occurs in combination with potential instability, the stage is set for severe convection. Unfortunately, vertical wind shear-derived parameters can vary substantially within the synoptic-scale and mesoscale environments (Davies-Jones 1993). Although the synoptic-scale environment can be characterized by some average vertical wind shear parameter, this is virtually certain not be representative of a particular convective storm's local environment. The mesoscale structure of the wind field can be understood in a QG context only to a limited extent, that is, only on the largest scales that might be considered to be "mesoscale."

b. Potential vorticity (PV) perspectives

The seminal paper describing potential vorticity thinking is that by Hoskins et al. (1985). Detailed descriptions of potential vorticity (PV) can be found in textbooks (e.g., Holton 1992; Bluestein 1992). The intensification of cyclones is associated with cyclonic

[5] Obviously, there are geographical circumstances where this may not be the case. Notable exceptions include Europe and Australia, where large, arid subtropical and tropical landmasses are present equatorward of midlatitudes. Meridional flows in such places typically transport dry air poleward; moisture is brought into midlatitude regions by predominantly easterly or westerly low-level flows that have a long fetch over warm, open waters. Such flows are also strongly modulated by ETCs, of course.

[6] The term "synoptically evident" was used in Doswell et al. (1993) to imply a synoptic pattern associated with tornado outbreaks. These involve strong synoptic-scale advections and the presence of substantial CAPE in the presence of strong flow and vertical wind shear. It is by no means "evident" in advance which days are going to produce tornado outbreaks, but it is often possible to say retrospectively that there was a strong synoptic-scale signal suggesting the possibility of such an outbreak.

PV maxima that generally have ascent (descent) on their downshear (upshear) sides, even in weakly baroclinic environments, as discussed by Bosart and Lackmann (1995). Raymond and Jiang (1990), Raymond (1992), and Montgomery and Farrell (1993) have shown this theoretically, as well as observationally (e.g., Davis and Emanuel 1991). By being conserved for adiabatic, inviscid flow, an immediate benefit of PV diagnostics is that when PV is not conserved following the flow, it indicates important diabatic or viscous processes, with deep, moist convection often being an important contributor (Dickinson et al. 1997).

Standard texts (e.g., Holton 1992; Bluestein 1992) have shown that the QG form of PV (QGPV, or q) is a linearized form of the full Ertel PV for a frictionless, adiabatic flow and is defined by

$$q \equiv \frac{1}{f_0}\nabla^2\Phi + f + \frac{\partial}{\partial p}\left(\frac{f_0}{\sigma}\frac{\partial \Phi}{\partial p}\right). \tag{2.3}$$

This quantity is conserved following the geostrophic flow, such that

$$\frac{D_g q}{Dt} \equiv \frac{\partial q}{\partial t} + \mathbf{V}_g \cdot \nabla q = 0. \tag{2.4}$$

Bluestein (1992, p. 372) has observed that (2.4) is equivalent to (2.1). QGPV is a good approximation to the "true" PV.[7] As can be seen easily in (2.3), q is simply the sum of the geostrophic absolute vorticity and a static stability term.

In its QG form, the PV perspective does not contain any physics beyond traditional QG approaches, but thinking in PV terms can be useful, as discussed in Hoskins et al. (1985), Davis and Emanuel (1991), or Morgan and Neilsen-Gammon (1998), among others. Mobile upper-atmospheric troughs can be interpreted as PV anomalies that have their own induced flows extending vertically above and below the level(s) containing the PV anomaly. The induced flows act to redistribute PV, creating new anomalies by the advections they develop. Many aspects of midlatitude weather systems can be seen as a consequence of interactions between two (or more) PV anomalies, laterally and/or vertically. The induced flows can be derived by using the invertibility principle associated with PV thinking; see Charney (1962) or Raymond (1992), as well as other references in this section on this topic.

Traveling upper-level systems can also be diagnosed using the concept of the dynamic tropopause (DT) described in Hoskins et al. (1985) or Morgan and Nielsen-Gammon (1998). The DT is defined as a surface of constant PV (typically 1–2 $\times$ 10^{-6} K m^2 kg^{-1} s^{-1}, where the factor of 10^{-6} K m^2 kg^{-1} s^{-1} is often referred to as a PV unit or PVU). On the DT, troughs are depicted as regions of high pressure (a low tropopause) and relatively low values of potential temperature. Jets and jet streaks are manifested as ribbons of closely packed potential temperature and pressure contours, indicative of a sloping DT (see Morgan and Neilsen-Gammon 1998). Moving with a positive PV anomaly, then, air parcels from the lower part of the troposphere are being overtaken by the PV anomaly and ascending along sloping isentropic surfaces.

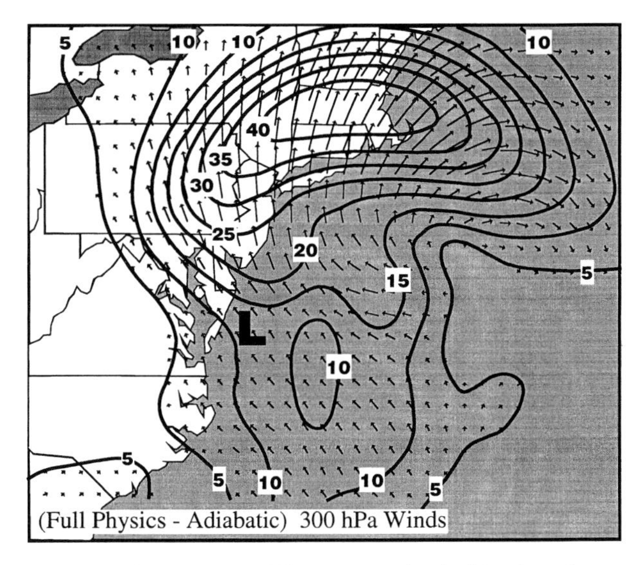

FIG. 2.1. 300-hPa wind differences associated with subtracting an adiabatic numerical model run from a full-physics version that includes parameterized convection, valid at 1200 UTC on 19 Feb 1979. Contours show the isotachs of the wind differences in m s^{-1}, while the arrows show the direction and the speed; the boldface "L" depicts the full-physics run surface low pressure center location. The figure is based on data supplied by L. Uccellini for the case described in Whitaker et al. (1988).

However, when air parcels participate in DMC, they are no longer subject to the balance constraints of large-scale processes. Rather than rising along gently sloping surfaces, buoyant parcels have a large vertical component of motion. This can have an important impact on the structures aloft ahead of a PV anomaly, since parcels ascending in DMC are carried rapidly to high levels, disturbing the dynamic balances near the tropopause, at least locally. That is, they create regions of high gradients of potential temperature, implying the creation of unbalanced flow, not unlike jet streaks. Diabatic effects will be seen as nonconservation of PV, of course, as noted in Whitaker et al. (1988) and illustrated in Fig. 2.1. Above an elevated maximum of diabatic heating (of which DMC is clearly an example), PV decreases and anticyclonic outflow increases, which is quite evident in the figure.[8] Beneath that

[7] For some minor technical reasons, however, the QG form of PV is not a "proper" vorticity, as mentioned in Hoskins et al. (1985), Davies-Jones (1991), and Hakim et al. (1996).

[8] Note that there are two maxima in the difference field: One is north and northeast of the surface low in a region of strong,

diabatic heating maximum, PV and cyclonic inflow increase. The result (at least in the upper levels) is analogous to the structures created by mesoscale convective complexes (MCCs; Maddox 1980; Fritsch and Maddox 1981), but on the synoptic scale.

The 13–14 March 1993 "Superstorm" over eastern North America provided an excellent example of the importance of diabatic processes associated with deep convection on explosive cyclogenesis. Dickinson et al. (1997, their Fig. 5) show that at the time of incipient cyclogenesis over the northwestern Gulf of Mexico, surface θ_e values exceeded 340 K along the lower Texas coast and adjacent northwest Gulf of Mexico. This high θ_e air at the surface was overridden by air on the DT with a potential temperature $<$330 K, indicative of exceptionally deep potential instability in the precyclogenetic environment over the northwestern Gulf of Mexico (Dickinson et al. 1997, their Fig. 3). The net result was the rapid onset of widespread deep convection in the vicinity of the cyclone, as seen in the satellite imagery and cloud-to-ground lightning distribution (Dickinson et al. 1997, their Figs. 6–8). It is possible that this convection was a factor in the ensuing rapid cyclogenesis (see section 2.7, below).

Bosart et al. (1999, their Fig. 2) showed that derechos (Johns and Hirt 1987) could be associated with long-lived, upper-tropospheric mesoscale disturbances moving eastward in a relatively strong flow aloft on the poleward flank of an intense continental anticyclone. The existence of these disturbances aloft during 13–15 July 1995 was revealed by an analysis of pressure and potential temperature on the DT. The disturbances aloft could be tracked for several days on the basis of an area of higher pressure and lower potential temperature on the DT. Deep convection began where ascent associated with these disturbances was superimposed on very high θ_e values at the surface ($>$360 K), producing extreme CAPE values (exceeding 3000 J kg^{-1}) along weak surface baroclinic zones.

c. Upper-level jet streaks

Sutcliffe (1939) and Sutcliffe and Forsdyke (1950) recognized that upper-level divergence and the associated vertical motions were closely connected to synoptic-scale cyclogenesis. These ideas have been confirmed repeatedly (e.g., Palmén and Holopainen 1962) and have evolved into what is now known as Sutcliffe–Petterssen development theory (see Uccellini 1990; Bosart 1999).

When observations of the flow in the upper troposphere became available, high-speed westerly flow aloft was discovered, and it was recognized that this high-speed flow was organized into long ribbons that came to be known as *jet streams* (see the textbook by Reiter 1961). Bjerknes and Holmboe (1944) and Bjerknes (1951) developed an explanation for at least some of the along-stream variations in upper-level divergence in terms of gradient wind imbalance. However, the observed structures in jet streams include features that have considerable along-flow variation in wind speed—known as *jet streaks*—that cannot be attributed solely to curvature effects. It has been proposed (Newton 1959, 1981; Newton and Persson 1962) that inertial oscillations play an important role in the along-flow wind speed variations represented by jet streaks. Although inertial oscillations may be a factor in developing jet streaks, it is likely that other factors (such as persistent convection) also create along-stream changes of the flow speed in the jet streams.

A conceptual model (Fig. 2.2) of the vertical motions and ageostrophic circulations associated with jet streaks has been developed by numerous authors (Riehl et al. 1952; Beebe and Bates 1955; Shapiro

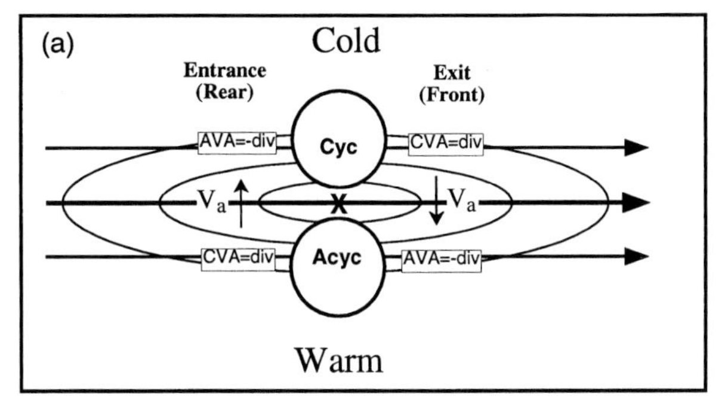

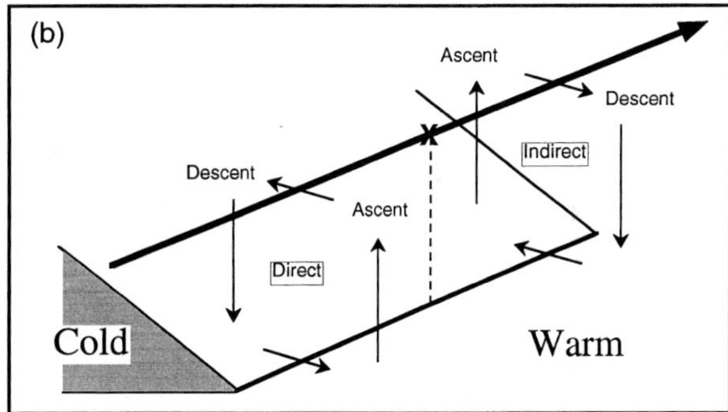

FIG. 2.2. Schematic model of a Northern Hemispheric jet streak in straight flow; cold air is indicated to the left side of the jet streak and warm air to the right. In (a), the isotachs (thin solid lines) depict the jet streak; the thick solid lines with arrows indicate the flow, an "X" marks the wind speed maximum. The circle marked "Cyc" shows a cyclonic vorticity maximum and that marked "Acyc" is a comparable anticyclonic vorticity maximum. Cyclonic vorticity advection is located by "CVA" and "AVA" locates anticyclonic vorticity advection, while "div" stands for divergence and "−div" for convergence. The vectors labeled $\mathbf{V}_a$ are the transverse ageostrophic wind vectors at the height of the jet streak, implied by the along-stream variations in the wind speed. In (b), the conceptual circulations are depicted in three dimensions. The vertical circulation [including the ageostrophic cross-jet flows, as in (a)], in the exit region is indirect (warm air sinking and cold air rising), while that in the entrance region is direct (warm air rising and cold air sinking). The "wedge" of cold air is indicated schematically, with its leading edge under the jet axis.

saturated ascent over the warm front, while the other is southeast of the surface low, associated with parameterized convection.

1981; Bluestein 1993, p. 405; Cunningham and Keyser 1997) and used extensively to diagnose regions of ascent. Ascent, of course, is favorable for cyclogenesis, widespread precipitation, and organized deep, moist convection. Thus, it has been suggested that jet streaks are associated with severe convection via their vertical motions and coupled ageostrophic flows (Uccellini and Johnson 1979). The accelerations associated with curved flow and along-flow thermal advection can modify this simple conceptual model significantly (see, e.g., Shapiro 1982; Bluestein 1986; Keyser 1986). At the level of QG theory, the intensity of the cross-jet secondary circulation is proportional to the shear of the ageostrophic wind; the ageostrophic wind is enhanced for compact jet maxima with large along- and cross-flow speed variations.

Major cool season cyclogenesis events involving severe convection seem to show a preference for poleward jet exit regions (e.g., Uccellini 1990). In some instances, overlapping divergence regions in the upper troposphere between the poleward exit region of one jet streak and the equatorward entrance region of another, roughly parallel, jet streak can interact to support intense and deep synoptic-scale ascent. This has been associated with explosive cyclogenesis, as in the so-called Presidents' Day storm of 18–19 February 1979 (see, e.g., Bosart 1981; Uccellini et al. 1984; Uccellini and Kocin 1987), that, in turn, is seen as setting the stage for severe convection.

Severe forms of DMC can develop within the warm sector of a rapidly deepening cyclone during the cool season, where moist, unstably stratified air is being advected poleward in low-level jet streams that are coupled to the upper-level jet streaks (Uccellini and Johnson 1979). In the warm season, severe convection can spread well poleward of the warm sector in cyclones. An association between jet streaks and severe convection can often be found during the warm season, as well. During the early spring, intense (or intensifying) extratropical cyclones can bring moist, unstably stratified air into their warm sectors, just as in the winter. Later in the spring, as hemispheric baroclinity weakens, the extratropical cyclones and jet streams are correspondingly less intense, but static stability may be relatively low, such that a given amount of geostrophic advection [i.e., the rhs of Eqn. (2.2)] produces stronger vertical motions than in the more stable wintertime.

In the summer, the polar jet stream weakens still more and continues to retreat poleward. Jet streams and the jet streaks embedded within them remain tied to severe convection, however. The retreat of the polar jet stream is typically accompanied by poleward penetration of moisture and instability, and the subtropical jet stream can become important in providing vertical wind shear. Over much of the United States in summer, for example, a subtropical anticyclone dominates the continental plains equatorward of, say, 40° latitude, with active extratropical weather systems and jet streams dominating the northern continental plains and the Great Lakes region. The extratropical synoptic-scale systems of summer are not as intense as their cool-season counterparts but can have easy access to abundant moisture and instability, with a correspondingly large frequency of deep, moist convective storms. Such events can become derechos (Johns and Hirt 1987) embedded in northwesterly flow (Johns 1982) that form on the equatorward sides of jet entrance regions, within a deep layer of warm air advection, with ample moisture and low static stability.

2.3. Planetary boundary layer processes

The planetary boundary layer (PBL) is generally defined to be the tropospheric layer within which the effects of the surface are important; the depth of the PBL varies considerably in both space and time, but is generally on the order of 1 km. In one sense, the PBL might not be considered a component of synoptic-scale processes. As noted by Stull (1988, p. 2), the timescales of PBL processes can be as small as an hour or less. Obviously, processes on that short a timescale are not appropriate for a synoptic-scale discussion, but the PBL clearly interacts with synoptic-scale systems.

What makes the PBL so relevant in this chapter (which attempts to connect extratropical synoptic-scale processes with severe convection) are the surface sensible and latent heat fluxes. The extratropical cyclone is the dominant process outside the Tropics for mitigating the meridional gradients in temperature caused by differential solar heating (Palmén and Newton 1969, section 10.6). Since most of the retained heat is absorbed at the surface, it is the heat flux from the surface to the atmosphere that dominates the heat transfers governed by ETCs. Some fraction of that incoming heat is used to evaporate liquid water, which thereby becomes latent heat available for transport by the synoptic-scale flow.[9] The majority of the diurnal temperature wave is within the PBL, so PBL processes are relevant in understanding the heat and moisture transports in ETCs and their relationship to DMC, which also typically has a strong diurnal dependence.

a. Simple Ekman theory

A very simple model of the effect of friction on the flow in the PBL can be developed using so-called Ekman theory (e.g., Holton 1992, p. 129 ff.). This simple theory indicates that the effect of friction is to cause air to flow across the isobars toward lower pressure at and near the surface, but the cross-isobaric

[9] The ratio of the sensible heat flux to the latent heat flux is known as the *Bowen ratio* (see Stull 1988, p. 274 ff.). The surface heat balance requires that the net incoming solar radiation minus any upward heat flux from the soil must equal the sum of the sensible and latent heat fluxes.

flow decreases with height such that the effect of friction becomes nearly negligible at the top of the PBL. Based on Ekman theory, a profile of the horizontal wind as a function of height can be derived, such that the wind approaches geostrophic at the top of the PBL. This profile, incidentally, is unstable (see Faller and Kaylor 1966; Lilly 1966). The net result is called *Ekman pumping,* with rising motion associated with cyclones (which should exhibit low-level convergence via Ekman pumping) and sinking motion with anticyclones (which should be characterized by low-level divergence from Ekman pumping). Using this theory, it can be shown that the resulting vertical motion at the top of an Ekman layer is proportional to the low-level geostrophic relative vorticity, and inversely proportional to the so-called Ekman parameter: $\gamma \equiv \sqrt{f/2K}$, where f is the Coriolis parameter and K is the eddy viscosity coefficient (assumed constant in this simple theory). Of course, Ekman pumping is only one of several processes that create vertical motion. At the level of QG theory, Ekman pumping can be incorporated as a bottom boundary condition on ω for the solution of Eq. (2.2). Bannon (1998) has considered Ekman pumping in situations where the free atmospheric flow is not assumed to be steady state.

The eddy viscosity coefficient K is not really constant with height; in fact, it has been found that K typically increases through a relatively shallow surface-based layer (~100 m in depth) to a maximum somewhat above that, and then decreases rapidly to relatively small values at the top of the PBL (see, e.g., Stull 1988, p. 210). Thus, even in uncomplicated conditions, the actual PBL may not fit the purely theoretical Ekman profile very accurately. Furthermore, the eddy viscosity is not constant with time, which can have some important impacts on the low-level flow. If it is assumed, only for the sake of simplicity, that the major physical mechanism by which the free atmosphere "feels" the surface is via mixing from convection in the PBL,[10] then the eddy viscosity should be highest at the time of maximum heating during the afternoon. Thus, the isobaric crossing angle (from high to low pressure) associated with an Ekman-like force balance looks something like Fig. 2.3, where the friction-induced departure from pure geostrophic balance should be a maximum at this time of the day. To the limited extent that the real atmosphere fits this simple model, Ekman pumping should vary during the day, reaching a peak at the time of maximum mixing in the PBL.

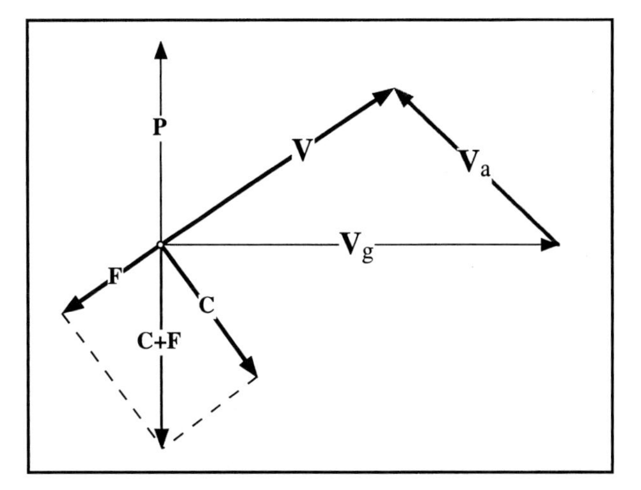

FIG. 2.3. Force balance for a wind **V,** including the Coriolis force **C,** pressure gradient force **P,** and a simple friction force (opposite to the wind direction) **F** vector. $\mathbf{V}_g$ is the geostrophic wind and $\mathbf{V}_a$ is the ageostrophic wind.

b. Diurnal variations and residual layers

Under typical conditions involving solar heating during a sunny day, the PBL has an inversion at its top that ascends as the PBL deepens. During this process, the inversion weakens as it ascends, because potential temperature typically increases with height. This stable layer can act as an inhibitor of deep convection, and its erosion by diurnal heating is one reason why deep convection often begins during the afternoon. However, this simple picture of the diurnal modulation of convection has numerous exceptions.

As noted first by Carlson and Ludlam (1968) and developed in some detail by Lanicci and Warner (1991) and Stensrud (1993), surface-based layers can be carried away in the flow from their original locations, resulting in *elevated* residual layers, some of which may be "well-mixed" layers. When such a residual layer is created over dry regions, it tends to be a relatively dry, high-lapse rate mixed layer, the superposition of which over another, potentially cooler surface boundary layer can result in a very strong inversion, sometimes referred to as a *capping inversion* (often referred to as a *cap* or *lid*) that can prevent the release of potential instability. Graziano and Carlson (1987) have shown that when the "lid strength index"[11]

[10] It is not obvious that viscosity in the PBL is always dominated by the presence or absence of dry convection; the presence of vertical wind shear in the PBL also contributes to eddy mixing. The diurnal growth of a neutrally stratified boundary layer promotes mixing and so in a sheared environment produces vertical momentum transports. The momentum from higher levels mixed downward tends to reflect the flow aloft and so the isobaric crossing angles observed need not fit those expected from simple Ekman pumping. Thermal advection in baroclinic PBLs also complicates the simple picture of a convectively dominated PBL (Hoxit 1974).

[11] Graziano and Carlson (1987) defined this index as $(\theta_{sw1} - \bar{\theta}_w)$, where θ_{sw1} is the saturation wet-bulb potential temperature at the warmest point in the inversion, and $\bar{\theta}_w$ is the vertically averaged value of wet-bulb potential temperature between 30 and 80 mb above the ground.

is more than about 2°C, deep convection is likely to be suppressed.

The presence of a capping inversion can inhibit convection at some locations, but it can also promote it elsewhere through the seeming paradox of preventing the release of potential instability. Convective inhibition results in "storage" of PBL parcels with high CAPE. When the unstable parcels at low levels flow out from under the margins of an elevated mixed layer, thereby permitting DMC development, they are said to be *underrunning* the cap. When some process erodes the cap locally, perhaps by forced ascent along some boundary, DMC ensues. Irrespective of what process eventually enables DMC to occur, its intensity is associated with parcels whose CAPE was created and maintained by PBL processes that perhaps can be far removed from where the convection actually occurs. The challenge of knowing when, where, or even if the cap will be overcome is one of several factors that make forecasting severe convection so difficult (see chapter 11 in this volume).

As the solar heating begins to decrease in the afternoon, at some point the insolation falls below the outgoing longwave radiation level, and so the net radiation becomes negative. This radiation loss near the surface promotes surface cooling. The resulting surface-based stable layer effectively decouples the surface from the air above it, thereby reducing the mixing within the stable layer. A residual well-mixed layer (see Carlson and Ludlam 1968; Stensrud 1993) will remain in place above the stable layer, even as continued cooling at the surface slowly deepens the surface-based stable layer (see Stull 1988, his Fig. 1.7, for a schematic of this process). By preventing the free atmosphere from "seeing" the surface, development of this surface-based stable layer can cause a substantial reduction in eddy viscosity.

During the day, when mixing is large, it is possible to imagine how the friction force (directed opposite to the wind) and the resulting Coriolis force combine to balance the horizontal pressure gradient force, as shown in Fig. 2.3. Imagine that the mixing is "turned off" impulsively, such that the friction force vanishes.[12] Then, as shown in Fig. 2.4, the loss of friction means an unbalanced pressure gradient force, a component of which is along the wind. Part of the pressure gradient force temporarily balances the Coriolis force at the instant the friction disappears. The unbalanced part of the pressure gradient force, however, increases the wind, which, in turn, increases the Coriolis force. Therefore, the wind increases in speed and turns to the right (in the Northern Hemisphere), producing a rotation of the resultant wind about the geostrophic, the

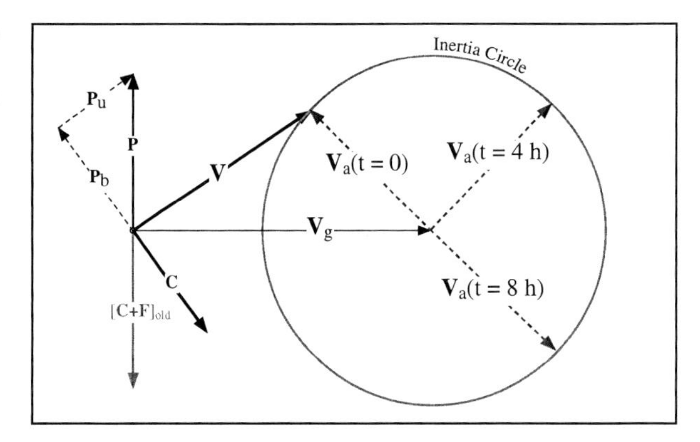

FIG. 2.4. The force balance immediately after "switching off" the friction force, showing that the pressure gradient force **P** includes a component $\mathbf{P}_b$ in balance with the Coriolis force **C** and a component $\mathbf{P}_u$ that is unbalanced; **F** is the friction force vector. Also shown are the "old" vector sum of **C** and **F** (as in Fig. 2.3) that has vanished when the friction was turned off, as well as the inertia circle, showing how the ageostrophic wind $\mathbf{V}_a$ rotates anticyclonically about the geostrophic wind $\mathbf{V}_g$ at selected times (labeled).

so-called *inertia circle*. The rotation period ($2\pi/f$) is known as the *half-pendulum day;* at 45° latitude, this period is about 16 h.[13] Of course, long before a complete rotation has been made, the heating cycle again begins to recouple the surface and the air above it. About 4–8 h (depending on the latitude) after the decoupling begins, the geostrophic departure vector is closely aligned with the geostrophic wind, producing a supergeostrophic wind maximum at that time. This process results in the so-called *nocturnal boundary layer wind maximum* (NBLWM), first elucidated by Blackadar (1957) and since discussed in many papers (e.g., Bonner 1966; Holton 1967; Thorpe and Guymer 1977; Stensrud 1996b). This process operates whenever the diurnal cycle of surface heating is strong and is suppressed when the diurnal heating cycle is weak, as on cloudy days.

c. Baroclinic planetary boundary layers

The foregoing is still an argument that assumes barotropy. There is an additional baroclinic process that can be superimposed upon the NBLWM (Holton 1967). Differential heating is often observed over sloping terrain; for example, the dry elevated high plains of the United States typically heat up faster during the day than regions to the east and cool off faster at night. This causes a diurnal cycle in the horizontal temperature gradient superimposed on the synoptic-scale gradient. The result is a diurnal cycle in the thermal wind vector, such that the poleward flow

[12] The reduction of eddy viscosity to negligible levels does not occur instantaneously, of course; this simplification is simply for illustrative purposes.

[13] At 30° latitude, the period is very close to exactly 24 h and so the inertial oscillation is in phase with the heating cycle at that latitude.

should increase with height at night, at least within the baroclinic zone created by this differential heating. Thus, this mechanism is roughly synchronous with the NBLWM and can create significant horizontal variations in the NBLWM flow, resulting in a low-level jet stream (LLJS). LLJSs are herein distinguished from NBLWMs and the ambiguous term "low-level jet" is avoided specifically to make this distinction (as did Stensrud 1996b). It appears that relatively narrow ribbons of strong low-level flow (LLJSs) arise in a variety of contexts (see, e.g., Browning and Pardoe 1973; Uccellini and Johnson 1979; Nagata and Ogura 1991; Doyle and Warner 1993; Chen et al. 1998).

Under baroclinic conditions, furthermore, thermal advection within the boundary layer can create significant departures from simple Ekman theory (see, e.g., Hoxit 1974; Bannon and Salem 1995). In the Northern Hemisphere, the thermal wind relation indicates that the geostrophic wind veers (backs) with height under conditions of warm (cold) thermal advection. As noted in Maddox (1993), diurnal changes in the PBL wind profile can have a large impact on the hodograph (especially in the critical part near the surface) and, hence, on the potential for severe convection.

Further, given the progression of synoptic-scale cyclones from west to east across midlatitudes, the low-level flow switches repeatedly from poleward to equatorward and back again to poleward (see Lanicci and Warner 1991) with the passage of each cyclone and the approach of the next. The tendency to develop a narrow ribbon of strong poleward flow ahead of approaching ETCs is enhanced by leeside troughing processes (Uccellini 1980; Lanicci and Warner 1991) in the plains region of the United States. Therefore, the LLJS is often tied dynamically to processes operating in the upper troposphere (Uccellini 1980), as well as being modulated by PBL-associated processes that have diurnal cycles.

The LLJS and NBLWM have long been recognized as major factors in promoting nocturnal convection (Means 1944, 1952; Pitchford and London 1962; Bonner 1966; Maddox 1983) during the summer in the plains of the United States. Typically, they prolong the life of convective systems that were initiated during the day, extending the threat of severe weather well into the night. They also play a role in enhancing poleward heat transport (both sensible and latent heat) in ETCs. For a review of the factors in low-level winds, see Stensrud (1996b); more information about the relevant processes can be found in Bluestein (1993, p. 391 ff.). We have already observed that meridional heat transport for purposes of mitigating the unequal solar heating can be viewed as a major reason for the existence of extratropical cyclones; to this extent, then, PBL processes must be considered an important element of synoptic-scale processes. As we have only been able to describe briefly here, the wide range of processes that govern the PBL's structure and evolution (e.g., solar heating cycles, cloud cover, antecedent precipitation, fronts, drylines, the character of the underlying surface, etc.) are critical in determining when and where stored CAPE is released in its severe manifestations.

2.4. Climatology of deep, moist convection

It is of interest to develop a synoptic-scale sense of the climatological occurrence of DMC. A way of viewing the existence of convection within midlatitudes is that convection develops whenever the redistribution of heat by synoptic-scale processes is not sufficient to mitigate the imbalances resulting from differential heating at the surface. ETCs produce prodigious meridional and vertical heat transports; Palmén and Newton (1969, p. 301 ff.) estimate that the poleward heat transport per cyclone across 45°N on an April day is about 10^{12} kJ s^{-1}; they note that about six such disturbances around the globe at any one time would suffice to compensate for the associated meridional imbalance in radiative heating. They also estimate the upward heat transport across 500 hPa per cyclone to be about 2×10^{11} kJ s^{-1}. Again, they note that this amount is roughly comparable to that needed to balance the excess heat input at low levels.

Whereas the typical convective cell transports but a miniscule fraction of the amount of heat transported by an ETC, the large vertical motions of DMC can transport heat much more rapidly than the relatively weak vertical motions of ETCs. The temporal and spatial distribution of convective storms can be viewed as depicting those times and locations when nonconvective heat transport processes (especially in the vertical) simply have not been adequate to mitigate the excess heat (both sensible and latent, of course) accumulating at low levels. Thus, as convective cells become more numerous and intense, the implication is that more such rapid vertical transport is needed.

The concatenation of moisture and instability necessary for DMC is quite obviously tied to this process of excessive accumulation of sensible and latent heat in the lower troposphere. Clearly, convective transports are primarily vertical, but they also affect horizontal transports by carrying heat upward to upper-tropospheric levels where it can be redistributed horizontally by the relatively strong winds aloft. The coupling of convection to synoptic-scale processes is still being explored (see, e.g., Gutowski and Jiang 1998).

The release of the excess sensible and latent heat may not be exactly where it was created because the associated CAPE was "stored" in the atmosphere (Emanuel 1994) until it could be released. Convective inhibition is the main reason for this storage, and so synoptic-scale processes can transport the excess heat to a location well away from where it entered the atmosphere. Release of the instability is intermittent and may not occur at the time and location of the destabilization processes that created it. This makes parameterizations of convection difficult, since its occurrence is tied to a number of ingredients that can interact in nonlinear ways, rather than to a single parameter.

a. The global spatial distribution of convective precipitation

On a worldwide basis, an average of about 1 m of precipitation falls annually. This precipitation is actually distributed quite unevenly, of course, with the bulk of it falling in the Tropics along the so-called Intertropical convergence zone (ITCZ), the South Atlantic and South Pacific convergence zones (Vincent 1994), the western Pacific oceanic warm pool (Ramage 1975), Amazonia and Latin America, central Africa, across the oceanic storm tracks of midlatitudes, and on the windward slopes of mountains. These regional maxima also appear in the global climatology presented by Huffman et al. (1997), based on satellite image datasets obtained in conjunction with the Global Precipitation Climatology Project. Indirect measures of precipitation (e.g., based on satellite imagery) can also be compared with other recent global climatologies (Jaeger 1976; Legates 1987; Legates and Wilmont 1990). Agreement is generally good among these different climatological distributions, although uncertainties remain in the absence of in situ measurements over the vast oceanic areas.

Hsu and Wallace (1976) have summarized the annual variation of global precipitation. They have shown that precipitation over the continents peaks during the warm season in mid- and low latitudes, and tends to follow the sun in the Tropics, except in the deep Tropics, where there is little annual variation. Other exceptions are associated with tropical monsoon rainfall, where rainfall is heavily concentrated in the warm season.

Although these global precipitation climatologies do not reveal the distribution of rainfall from DMC directly, the vast majority of the rainfall within the Tropics and during the warm season outside of the Tropics is dominated by the DMC contribution. Orville and Henderson (1986) mapped the global distribution of midnight lightning (i.e., local midnight) from September 1977 to August 1978, showing that midnight lightning exhibits a strong preference for the continents in low and middle latitudes. This suggests that maritime tropical DMC is only weakly electrified, a hypothesis that has recently been confirmed by Zipser (1994).

Laing and Fritsch (1997) have mapped the global distribution of MCCs (a subset of mesoscale convective systems; Maddox 1980; Smull 1995). Their distribution is shown in Fig. 2.5. Figure 2.5 shows that MCCs in North and South America occur preferentially to the lee of mountain ranges in midlatitudes, where LLJSs are common (see Fig. 4 in Laing and Fritsch 1997). The distribution in the Old World is predominantly in the Tropics and subtropics. Given that considerable DMC occurs in regions other than those favored by MCCs (e.g., Amazonia), organized mesoscale convective systems seem to be associated with some particular combination of thermodynamic and synoptic-dynamic processes, perhaps modulated by topographic effects (see chapter 9 of this volume).

Wallace (1975) examined the diurnal variation of precipitation and thunderstorm frequency across the United States. His results confirmed the presence of a warm-season rainfall maximum over much of the

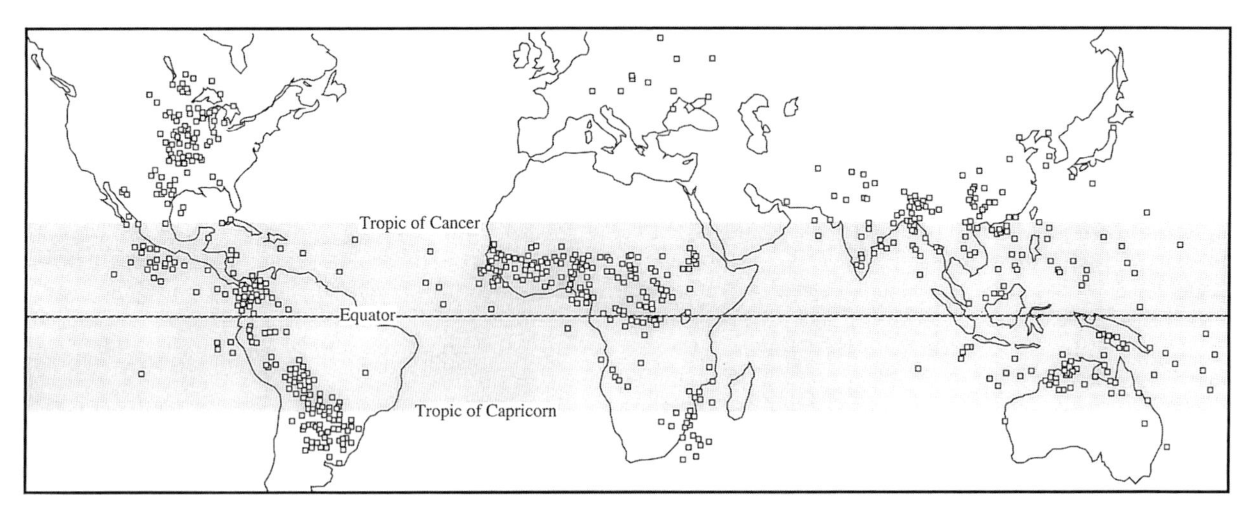

FIG. 2.5. The global distribution of MCCs, based on satellite imagery as described in Laing and Fritsch (1997). Small squares indicate location of MCC at time of maximum extent. Adapted from Laing and Fritsch (1997).

United States, with the exception of coastal New England and much of the coastal west. The High Plains region to the lee of the Rocky Mountains has a notable warm-season precipitation peak, with much of the rainfall associated with nocturnal DMC activity. Thunderstorms over the Rocky Mountains, Florida, and the East Coast tend to occur more in the late afternoon and early evening. As shown in Riley et al. (1987), the nocturnal storms over the Plains tend to have two sources: 1) afternoon storms that develop initially over and near the Rocky Mountains and then move eastward, and 2) storms that develop locally over the High Plains in associations with a variety of mesoscale weather systems. These findings have recently been verified in a comprehensive investigation of summertime precipitation over the United States by Higgins et al. (1997).

Globally, DMC is not uniformly observed owing to little or no data over the oceans and the sparsely populated regions around the globe. The recent availability of various remote sensing systems provides views that approach being truly global in scope (especially satellite-borne sensors). In fact, this is a major aspect of the ongoing Tropical Rainfall Measuring Mission (TRMM; see Simpson et al. 1988). From these observations, DMC occurs most frequently over land, with the important exception of the ITCZ band near the equator. The predominance of DMC over land surfaces could be the result of the lower heat capacity of soil versus water. A given amount of input radiation produces a larger near-surface temperature increase over land than over water, increasing the amount of instability. The result is that the land typically develops larger potential instability than the oceans (and large lakes). When sufficient low-level moisture is present, this means more frequent DMC over land (assuming that some lifting mechanism is present). However, it is not entirely obvious why continental updrafts in the Tropics are typically stronger than their maritime counterparts, since according to Lucas et al. (1994), large differences in CAPE have not been found.

The equatorial band of convection is usually found on the "summer" side of the equator. Since a major proportion of the surface in the tropical band (between the Tropics of Cancer and Capricorn at 23.5°N and S, respectively) is maritime, much of the excess insolation goes into warming tropical seas. Tropical synoptic weather systems do not usually involve much vertical wind shear (as noted in chapter 7 of this volume) and a typical maritime sounding in the Tropics develops a near neutrality to DMC through much of the troposphere (e.g., Fig. 2.6), as discussed by Xu and Emanuel (1989) or Williams and Renno (1993), among others. Tropical maritime convection is therefore relatively uninhibited but is also not characterized by much CAPE. Tropical continental regions are famous for the presence of rain forests, with a predominance

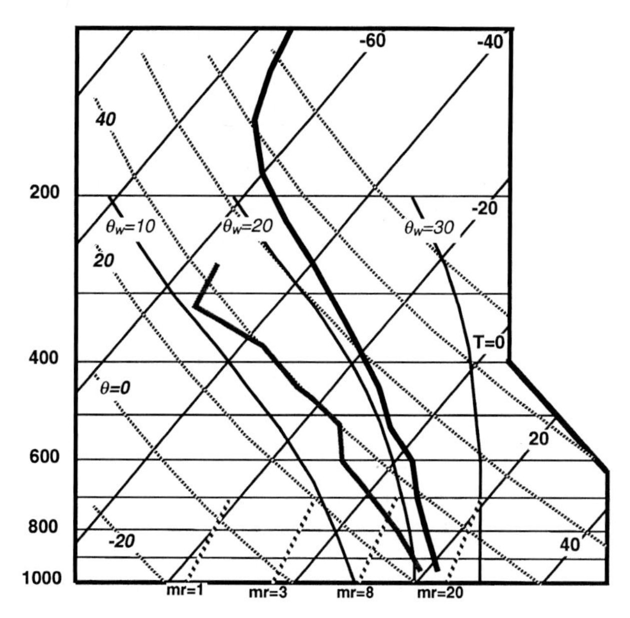

FIG. 2.6. C. L. Jordan's (1958) mean West Indies tropical sounding plotted on a skew T–log p diagram.

of convective rains. DMC plays a very large role in the synoptic-scale tropical heat transport, which can be viewed successfully as a Hadley cell on average (Lorenz 1965).

Another exception to the dominance of continents over oceans in the distribution of DMC is over the surface currents of warm water flowing poleward on the eastern boundaries of continents (e.g., the Gulf Stream or the Kurioshio current). As sources of both low-level sensible and latent heat, these currents also show up as regions of enhanced frequency of DMC, notably in the cool season, at a time when the continents tend to be dominated by conditions not conducive to DMC. Much of this convection is associated with strong synoptic-scale cyclones.

Regions of complex terrain often reveal local regions favoring the frequent development of DMC, even when relatively moisture-poor. Acting as an elevated heat source, especially in the warm season, complex terrain can be associated with considerable convective weather, although the resulting DMC does not necessarily produce much precipitation. For example, the Rocky Mountains of North America display a pronounced peak in summer convective frequency (López and Holle 1986), despite being relatively arid. High-based thunderstorms that produce little or no precipitation often initiate fires that can be a significant societal problem even if they are not usually considered to be "severe" thunderstorms. In Australia, for example, thunderstorms that produce little precipitation represent a significant fire hazard. Australian operational severe thunderstorm forecasters regularly review fire danger and issue fire weather forecasts (Love and Downey 1986; Williams and Karoly 1999).

In subtropical areas, or in regions with extensive regions of warm oceans (e.g., the Mediterranean Sea), the land–sea breeze circulations can be important in developing DMC even in synoptically unfavorable regimes. This role for land–sea breeze circulations can be enhanced for islands and peninsulas (Pielke 1974). Thus, for example, subtropical peninsular Florida is noted for its frequency of warm-season thunderstorms even though the synoptic flows in the region remain generally weak. However, relatively minor synoptic-scale fluctuations can still modulate the frequency and location of DMC (see Blanchard and López 1985).

b. The global temporal variation

As with the spatial distribution, the temporal distribution of DMC tends to be associated with processes that modulate conditional instability, notably, the diurnal and seasonal heating cycles. In the regions where most DMC is associated with complex terrain or sea-breeze circulations, the greatest DMC frequency tends to be coupled to, but lagging somewhat behind, the peak solar heating. However, there are at least three notable departures from this average behavior. First, strong synoptic-scale processes can become the dominant factor in developing DMC, and these may or may not occur in phase with the daily heating cycle. Thus, in the cool season in the southern United States around the Gulf of Mexico, for example, DMC is not so dominated by the diurnal cycle because DMC can be tied to intense synoptic-scale processes.

Second, at the terminus of an LLJS, the moisture-bearing poleward flows typically impinge on a thermal boundary (either a front or an outflow boundary from previous DMC), favoring ascent and convection initiation. Since this is enhanced by the typical NBLWM process, such convection can develop after sunset. Although the initial convection often develops late in the afternoon, it can be sustained well after dark in places where the LLJS (enhanced by the NBLWM) encounters a thermal boundary (e.g., Trier and Parsons 1993). This favors nocturnal convective systems (Laing and Fritsch 1997), a process that occurs in many places around the world (recall Fig. 2.5). Severe weather with these convective systems tends to occur before or shortly after dark, whereas the systems' precipitation may continue for many hours after sunset (Maddox 1983).

Third, recent evidence (e.g., Gray and Jacobson 1977; Janowiak et al. 1994) indicates that, over the tropical oceans, convective activity tends to peak during the night and early morning hours. This cycle is out of phase with the diurnal cycle, but its amplitude is not as large as that over land areas (which, with some exceptions as noted, is much closer to being in phase with the diurnal cycle). Nevertheless, the evidence for this modest diurnal variation in oceanic tropical convection is compelling. Brier and Simpson (1969) have suggested an important role for the semidiurnal tide in modulating tropical rainfall, which is predominantly convective. Chen and Houze (1997) have argued that the nocturnal peak in convective activity in the tropical oceans is due to a complex interaction among the diurnal heating cycle, large-scale processes in the Tropics, radiative effects, and the typical life cycles of convective systems.

The seasonal cycle of DMC also tends to follow the seasonal cycle of conditional instability. Naturally, synoptic-scale processes modulate this significantly. DMC is a feature near the cores of many explosively deepening maritime cyclones (e.g., Bosart 1981; Reed and Albright 1986; Nieman and Shapiro 1993; Dickinson et al. 1997), even though such cyclones are infrequent outside of the cold season. The impact of warm ocean currents on developing DMC in the cold season has already been mentioned. Maritime and subtropical regions have a much less pronounced seasonal cycle, and so may experience DMC events occasionally outside of their respective warm-season maxima; synoptic-scale control on such events is typically high.

Occasionally, thunderstorms develop in winter weather systems. These arise in at least three distinct ways. The first way involves the development of a warm sector that has near warm-season characteristics, and so the DMC is virtually the same as that developing in the warm season. A second process is when high lapse rates (in relatively dry air) are carried over a cold and stable but relatively moist surface-based air mass, producing "elevated" DMC that simply produces snow (or freezing rain) rather than simple rain. The third process is associated with slantwise ascent and, possibly, derives its energy from conditional symmetric instability (CSI; see Emmanuel 1994, chapter 12; Schultz and Schumacher 1999). The release of CSI is confined to saturated environments but can produce rain, as well as either freezing or frozen forms of precipitation. Thus, two of the three processes producing thunderstorms in association with winter weather systems involve "elevated" DMC; that is, the parcels participating in the DMC are not surface based because the surface-based air mass is markedly stable (Colman 1990a,b). Winter thunderstorms are never common and so represent a mostly negligible element in the climatology of DMC, although they certainly can be important when they occur; snowfall rates associated with snow thunderstorms can reach magnitudes of 10–30 cm h^{-1}. Virtually no severe weather other than heavy precipitation is associated with the second and third forms of winter DMC.

2.5. Climatology of severe convection

A global climatology of severe convection is not generally available, owing to the absence of international commitment to the development of such a database. Even in the United States, which is unquestionably the single nation with the highest frequency of severe convection, a number of deficiencies exist in the climatological record of severe storms. Further, the definition of what constitutes a severe convective event is typically arbitrary, whereby some phenomenon is called severe when it meets or exceeds some specified criterion, and the criteria vary around the world. For convective wind gusts, a speed threshold is used; for hail, hailstone diameter is the variable. Tornadoes are an exception, in that a convective storm producing any sort tornado is usually considered to be severe. However, even this can become somewhat arbitrary. When a tornado occurs over the water, thereby being called a waterspout, the associated deep convective storms that produce the waterspouts are usually not designated as severe. In the United States, heavy rainfall is not considered severe, although many other countries around the world do have some sort of threshold rainfall (or rainfall rate) that is deemed severe.

Nevertheless, we shall attempt to give at least some picture of the global distribution of severe weather, but the reader should be aware of its deficiencies. Our main interest in the climatology of severe convective weather is that this gives some clues about the synoptic-scale patterns that favor severe weather, as implied by Ludlam (1963). Our presentation focuses on hail and tornadoes. Convective windstorms are quite difficult to classify into severe and nonsevere categories, since subjective estimation of wind speed is so poor and the actual wind speed measurements are so sparse relative to the events. Heavy rainfalls are probably the most global in scope of all DMC-associated events, but the lack of consistent criteria and high-resolution rain gauge networks worldwide make it problematic to depict heavy rainfall climatology in detail. Even in the United States, there is no systematic record kept of important convective rainfall events (such as flash floods), other than the routine precipitation climatological data (see Brooks and Stensrud 2000).

Owing to their economic impact worldwide, hailstorms are sufficiently important that some records are kept in most nations that have hailfalls at all regularly. Figure 2.7 shows one picture of the global annual hail day (i.e., a day with one or more hail events) frequency distribution, indicating where the frequency is at least one hail day per year. The size of the hail is not included in this record, which apparently is mostly associated with crop losses. The apparent infrequency of hail days over Australia is probably a direct reflection of the extremely low population density (and, hence, observations) away from the coasts in that continent. Nevertheless, Sydney, Australia, has been hit recently by two devastating hail events (in 1990 and 1999), in spite of what might well be a truly low annual frequency.

Ludlam (1963) presented a global climatology of "severe squall-thunderstorms" (his Fig. 5), which bears some resemblance to Fig. 2.7. He indicates that

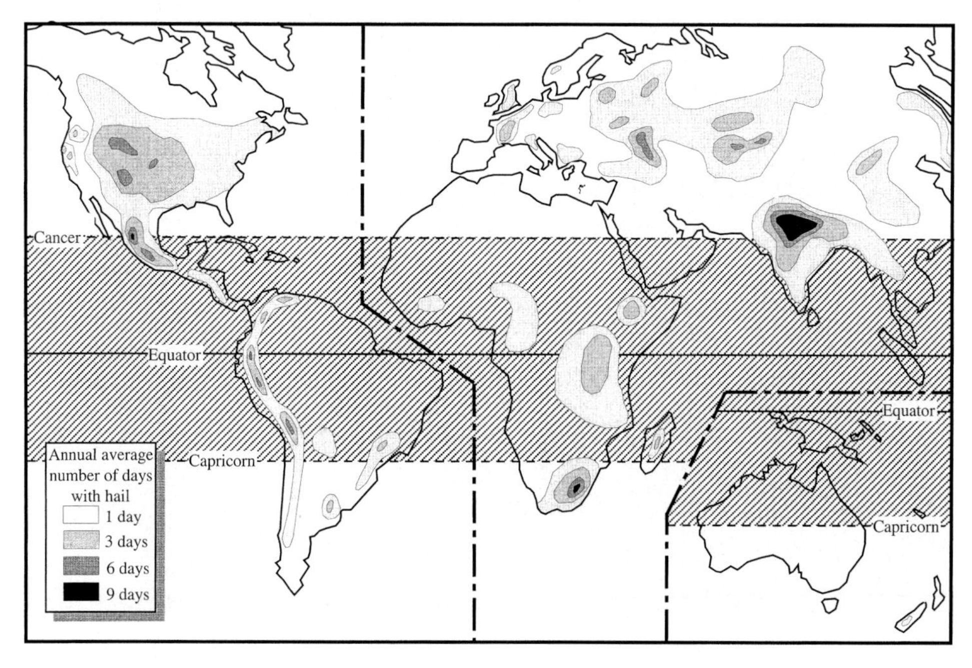

FIG. 2.7. Depiction of the global frequency of hail. After Frisby and Sansom (1967).

the distribution of midlatitude events appears to be associated with the jet stream axis and its associated vertical windshear. Given what we know about hailstorms, this seems like a plausible hypothesis. As hailstone diameter increases, the likelihood of its occurrence from a supercell also increases (since supercells typically have the strongest updrafts), and supercells are known to be associated with the presence of significant vertical windshear (Weisman and Klemp 1982, 1984).

Over the United States, hailstone size is estimated (albeit imperfectly; see Sammler 1993), and so it becomes possible to depict the climatological frequency as a function of hailstone size. As shown in Fig. 2.8, the distribution of large hail [≥2 cm (3/4 in)] in the United States is concentrated in the central and high plains, to the lee of the Rocky Mountains. This is even more evident when considering the distribution of giant hail [≥5 cm (2 in.); Fig. 2.9]. Heavy falls of hail <2 cm in diameter can have significant economic impact, especially on crops. As already noted, giant hailstones (≥5 cm) are mostly associated with supercell thunderstorms (Nelson and Young 1979; Ludlam 1980, p. 256 ff.; Burgess and Lemon 1990). Supercells are capable of producing the full range of severe convective phenomena (damaging convective wind gusts, tornadoes, and heavy rainfalls), in addition to giant hail.

Since the association between severe convective weather events and supercells is quite high, it is likely that the reason for the dominance of the United States in the global climatology of severe forms of convection is its topography, which favors the occurrence of supercells (Ludlam 1963; Doswell 1993) on its west-central plains. This does not mean that supercells are unheard of outside the United States (see, e.g., Browning and Ludlam 1962; Ludlam 1963; Doswell and Brooks 1993b; Schmid et al. 1997).

In places where supercells are possible, the occurrence of tornadoes becomes much more likely, although by no means guaranteed. In the United States, prior to the 1980s, it was believed that thunderstorms in Colorado were prolific hail producers and it was known widely that hail events in Colorado were sometimes associated with supercells (see Marwitz 1972), but tornadoes were considered rare. Since the 1980s, there has been a rapid growth of tornado reports in Colorado, to the point where the state is often among the national leaders in reported tornadoes each year. A belief in the infrequency of tornadoes can be a self-fulfilling prophecy; if tornadoes are considered unlikely, there can be little motivation for keeping track of the number of tornadoes, especially in sparsely populated regions where it is difficult to get any information at all about what might have happened. For example, Snow and Wyatt (1997) found that no

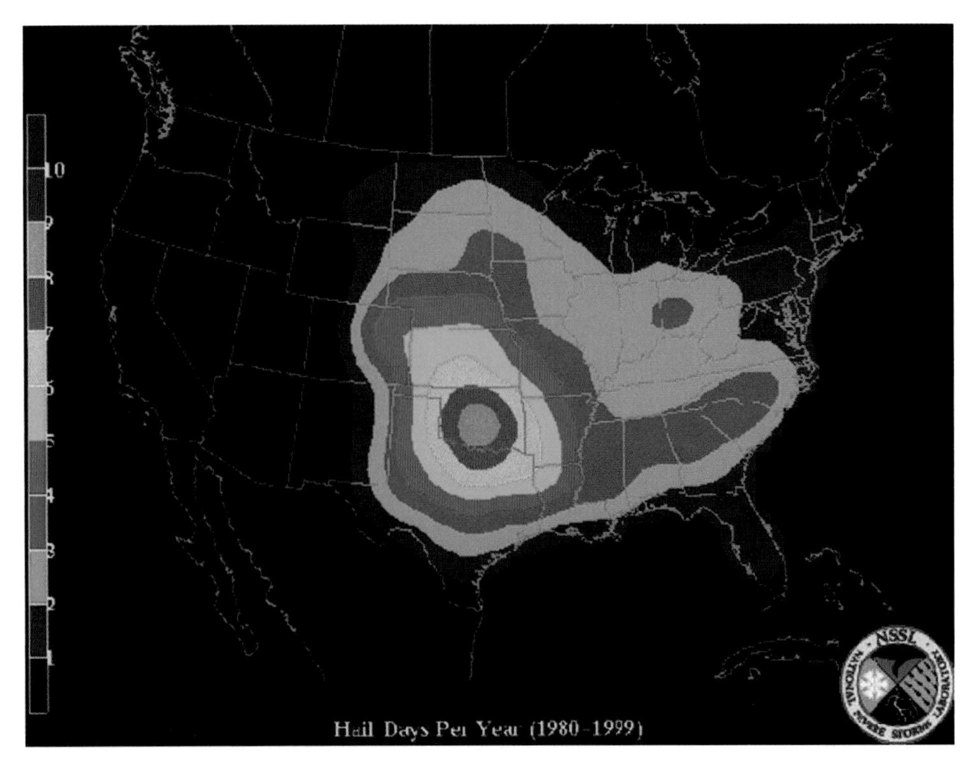

FIG. 2.8. Annual frequency (see the key) of hail reports with diameters ≥¾ in (2 cm) within an 80 km grid square, based on data for the period 1980–99.

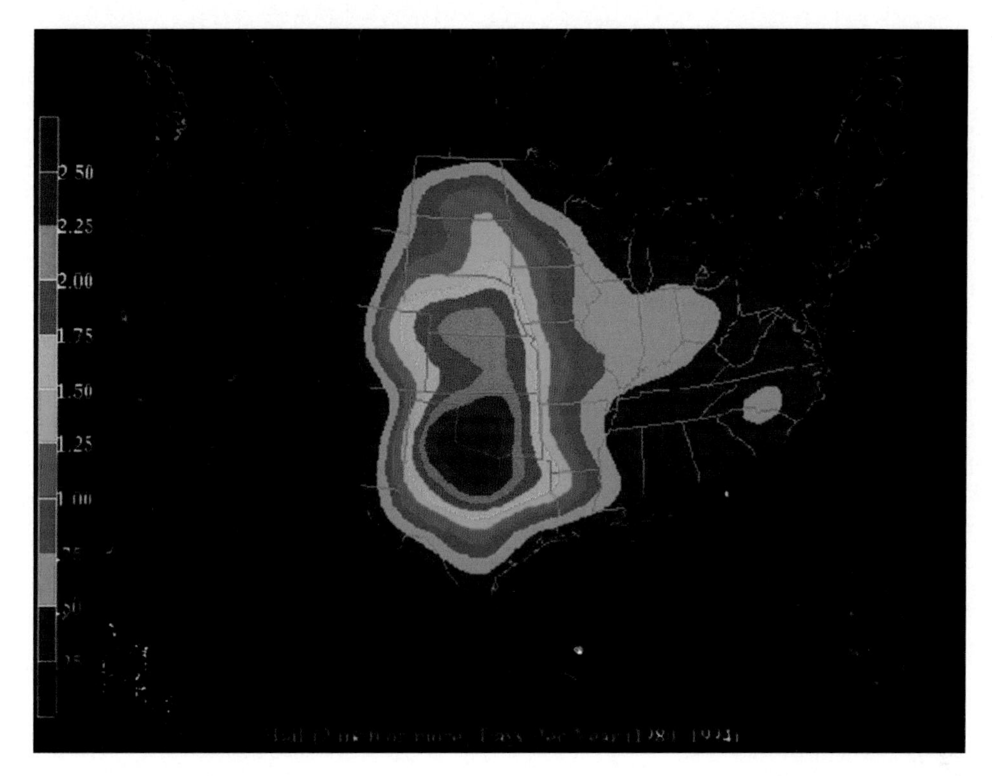

FIG. 2.9. As in Fig. 2.8, except for diameters ≥2 in. (5 cm) and period 1980–94.

tornadoes have been documented in Spain since World War I. This is perhaps nominally true but it does not correspond with the known facts; for example, tornadoes in Spain are reported by Ramis et al. (1995; 1997); they are also reported in the Balearic Islands, a part of Spain (Gayá et al. 1997); there are even widely circulated videotapes of Spanish tornadoes. Nevertheless, there is as yet no formal, national program to document tornadoes in Spain and so none have been "recorded."

Fujita (1973) attempted to discuss the worldwide climatological distribution of tornadoes, but his presentation acknowledges the same lack of consistent international documentation. What is known about tornadoes outside the United States tends to be anecdotal rather than systematic (e.g., Dotzek et al. 1998; Peterson 1998; Xu et al. 1993). Only in isolated nations has there been any attempt to be comprehensive [e.g., Dessens and Snow (1993) reviewed tornadoes in France] and there remain problems with relatively short records of erratic quality. Taking into account all the known factors associated with underreporting of tornadoes, it is probable that the worldwide pattern of tornado occurrences is not all that dissimilar from that of hail, although at a lower frequency than hail events. This appears to be the case in the United States (Fig. 2.10; cf. Fig. 2.8). The distribution of violent tornadoes [F4–F5 on the Fujita (1971) scale; Fig. 2.11] shows the same west-central plains peak as that of giant hail.

Curiously, in the United States, there is considerable difference between the distribution of all severe convective wind reports [including wind damage reports without an estimate of the wind speed, as well as estimated wind speeds exceeding 25 m s^{-1} (50 kt)] and that of hail (Fig. 2.12; cf. Fig. 2.8). It appears that convective wind events arise in a large variety of ways, in situations with both weak and strong synoptic-scale processes under way (e.g., Johns and Hirt 1987; Johns 1993), and so are more widely distributed in the United States, with a rather different climatological distribution from tornadoes and hail. Worldwide, unfortunately, there is virtually no documentation of severe convective wind events. The distribution of extreme convective wind gusts [≥32 m s^{-1} (65 kt); Fig. 2.13] is again similar to that of giant hail, with a peak in the west-central plains.

The superposition of high lapse rate air in the lower midtroposphere flowing off the high terrain to the west of the plains above the low-level poleward flow of moisture creates the extreme CAPE values seen in classic severe weather soundings (Doswell et al. 1985). Combining this CAPE with sufficient baroclinity to enhance the vertical wind shear means that the U.S. west-central plains region experiences the highest frequency of extreme forms of severe convection. The decrease in giant hail, extreme convective wind gusts, and significant tornadoes to the east of the Mississippi River suggests that, in the United States, CAPE is typically released before synoptic-scale processes can

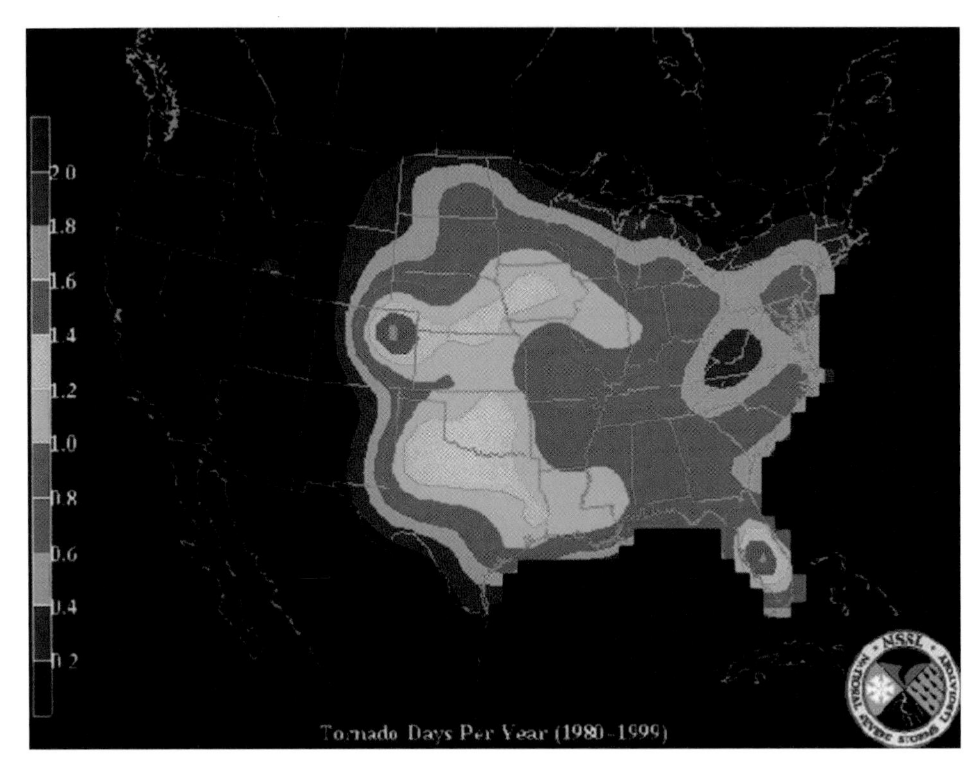

FIG. 2.10. As in Fig. 2.8, except for any tornado report.

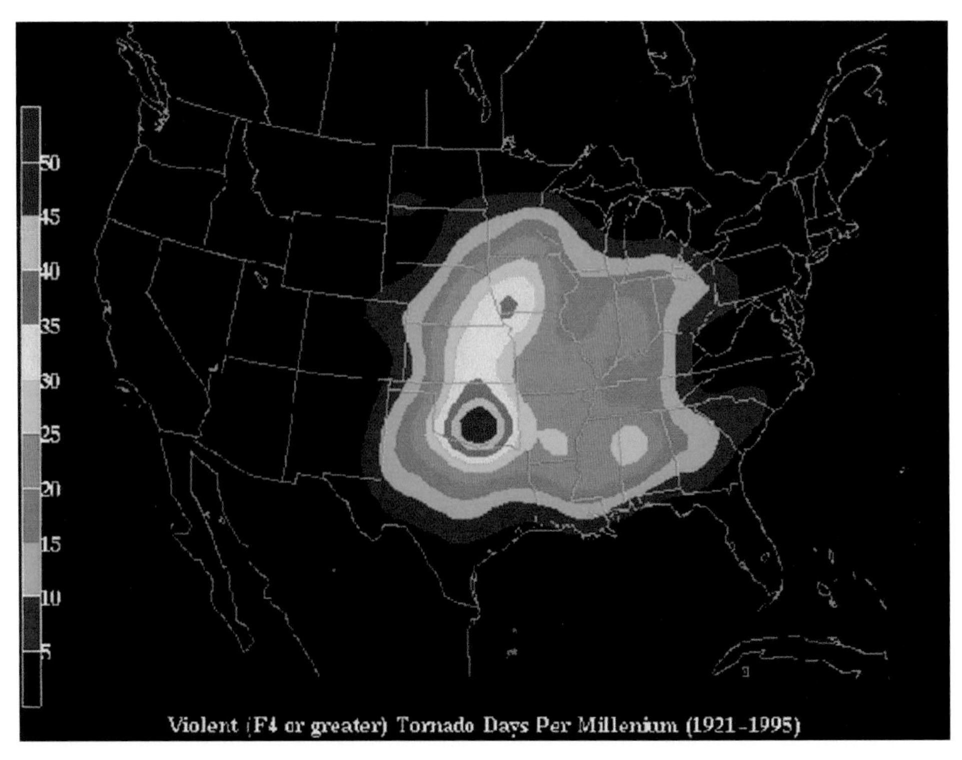

FIG. 2.11. Frequency of violent tornadoes [F4–F5 on the Fujita (1971) scale] normalized to per thousand years, based on data for the period 1921–95.

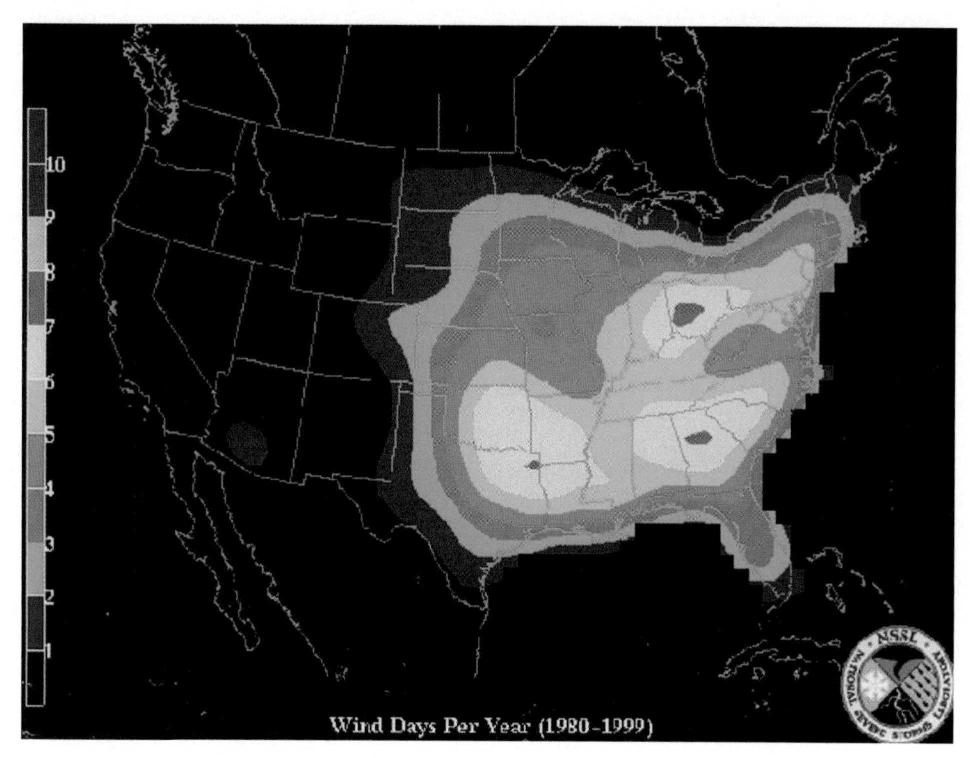

FIG. 2.12. As in Fig. 2.8, except for any convective wind gust report.

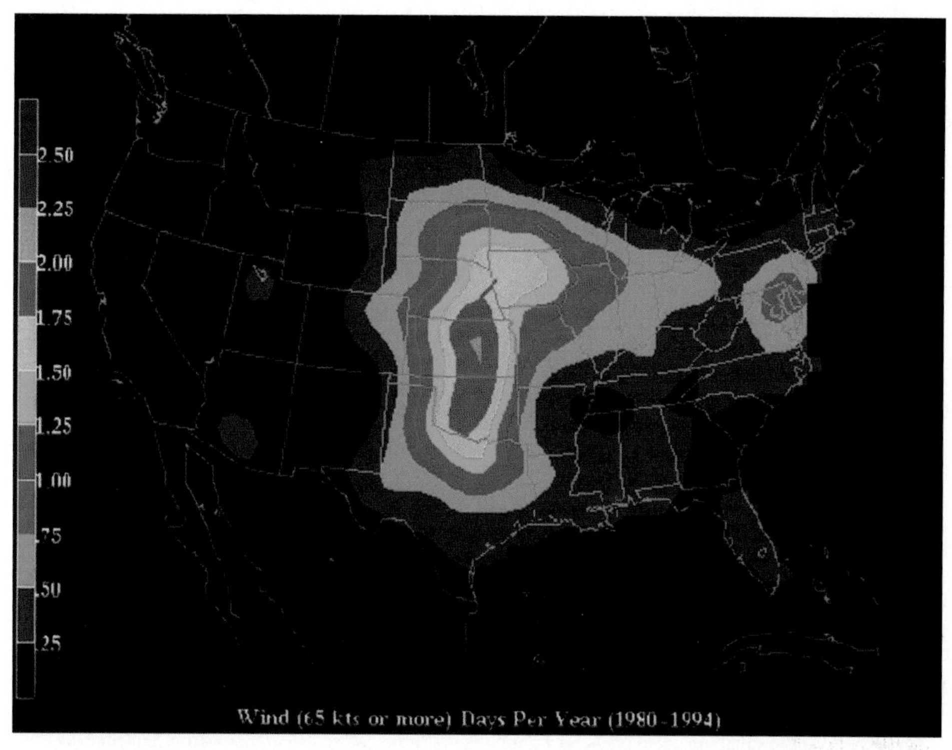

FIG. 2.13. As in Fig. 2.8, except for wind gusts exceeding 65 kt (32 m s^{-1}).

carry it very far. When DMC is inhibited by a cap, however, it can be transported eastward, resulting in the occasional outbreaks of extreme severe convection east of the Mississippi.

2.6. Synoptic factors associated with severe convection

As already noted, DMC requires three basic ingredients: moisture, instability (high lapse rates), and lift. Assuming that sufficient moisture and conditional instability are already present, a frequently important forecast issue is the sufficiency of the lift to overcome convective inhibition (CIN). Some uncertainty remains about the precise details of how DMC is initiated. Certainly if there is forced ascent of parcels that have negative buoyancy during the lifting process, then mass continuity requires that there must be mass convergence beneath the ascending currents, often manifested as "boundaries" exhibiting this convergence (e.g., fronts, drylines, and nonfrontal windshift lines). Deep convection usually begins as isolated convective cells along boundaries exhibiting low-level convergence or in groups of relatively isolated cells (Bluestein and Jain 1985). As we will discuss below, topographic processes are also important focusing mechanisms for the initiation of DMC.

The vertical motions in convective storms are of order 10 m s^{-1}, whereas synoptic-scale vertical motions are typically of order 1–10 cm s^{-1}, a value two to three orders of magnitude smaller than that of the convective drafts. It can be argued that on the scale of a convective cloud (~10 km horizontal length), synoptic-scale vertical motion is quite negligible. Nevertheless, as mentioned earlier, the occurrence of DMC in association with synoptic-scale weather systems is well known. How does this association arise?

Consider Fig. 2.14; in both schematic examples, the parcel must be lifted 150 hPa (corresponding to about 1.5 km) to attain the LFC, but in the one case, there is a CIN of 10 J kg^{-1}, and in the other the CIN is 100 J kg^{-1}. Obviously, more energy must be supplied to lift the parcel that distance in the second example. In either case, if the parcel is rising at a synoptic-scale rate of 3×10^{-3} hPa s^{-1}, such an ascent will take on the order of 5×10^{4} s, which is more than half a day. Clearly, this is too long to represent an important contribution to convection

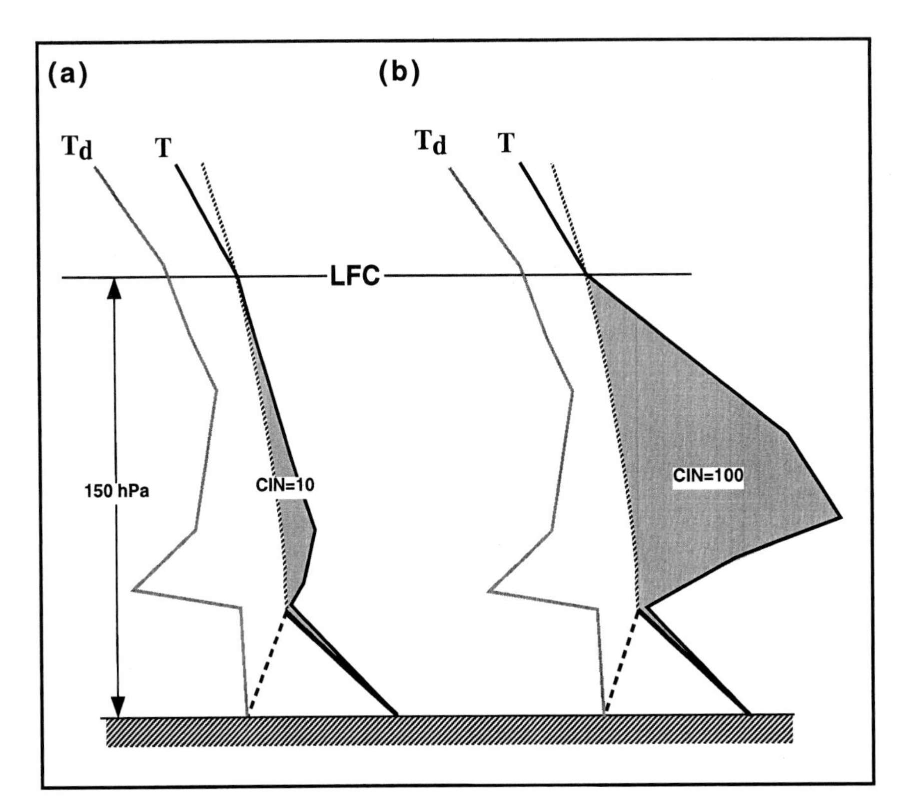

FIG. 2.14. Schematic comparing two surroundings, each with an LFC 150 hPa above the surface. The trajectory followed by the surface parcel is suggested with the solid, dashed, and hatched lines that show the dry adiabatic ascent, the moist adiabatic ascent, and the mixing ratio of the surface parcel, respectively. The environmental temperature (T) and dewpoint temperature (T_d) soundings are shown, and the associated CIN is indicated by the stippling.

initiation in many cases.[14] Presumably, some process operating on subsynoptic scales (e.g., fronts, drylines, gravity waves, etc.) must supply the energy needed to overcome the CIN and lift potentially buoyant parcels to their LFCs (e.g., see Wilson and Schrieber 1986). However, the synoptic-scale motions condition the environment by this slow ascent, causing a reduction in CIN by lifting and weakening any convection-inhibiting stable layers.

This also suggests that convection can be initiated in the absence of synoptic-scale ascent (or even in regions of synoptic-scale descent), provided that some subsynoptic lifting process can initiate DMC, whenever the moisture and instability are already sufficient to create the potential for DMC. Subsynoptic-scale processes (like upslope flows, fronts, etc.) often operate in association with synoptic-scale systems, so the association between cyclones and DMC is not coincidental, but there is no unique relationship, either.

a. Topographic influences

Topographic features play an important role in the evolution of synoptic-scale weather systems and, therefore, in the development of convection. Note that "topography" is a general term for the physical features of a region (including bodies of water, vegetative cover, snow and ice, soil type, and so on) and is not limited to features of the surface height (orography). There are several quite different types of topographic influences; these are discussed in the following sections.

1) OROGRAPHIC EFFECTS

An important factor in the evolution of a synoptic weather system is its interaction with large-scale orographic features. Synoptic-scale cyclogenesis does not occur completely randomly about the midlatitudes but rather is concentrated in certain key geographic areas, often in the lee of major mountainous regions (Roebber 1984; Tibaldi et al. 1990). Although lee cyclogenesis is not completely understood even now (e.g., Mattocks and Bleck 1986; Egger 1988), the characteristic structures that develop as a cyclone intensifies are dependent on the topography of the region. For example, in North America, there are three main regions of cyclogenesis: to the lee of the Canadian Rocky Mountains in Alberta, to the lee of the American Rocky Mountains in Colorado, and over the eastern coast of the United States near the Gulf Stream. Although the East Coast cyclogenesis region also is to the lee of the Appalachian Mountains, it is not quite so obviously dominated by the lee cyclogenetic effect, since the Gulf Stream makes the East Coast the location of a quasipermanent baroclinic zone, notably during the cool season (Bell and Bosart 1988).

The Alberta and Colorado cyclogenesis regions are most active when the main belt of zonal flow is in their vicinity. Hence, both can be active in the cool season, but the Colorado region is not so active in the warm season, since the westerlies in the warm season can migrate well poleward of Colorado for extended periods (Atallah and Bosart 1996). Whereas cyclones developing in Colorado have reasonably consistent access to the high low-level moisture of the Gulf of Mexico, even in the cool season, the Alberta cyclones can develop without much absolute moisture content. During the warm season, with the poleward progression of the subtropical anticyclones into North America, however, cyclones developing in the northern plains of the United States and the southern parts of the prairie provinces of Canada can have rich moisture nearby. Clearly, the ability of a cyclone to import substantial low-level moisture into a region is a key contributor to DMC activity.

Another aspect of topographic effects is the development of thermal contrasts. Midlatitudes are the focus for the main thermal contrast between the equator and the poles, of course, but east–west-oriented mountains can inhibit the meridional flow of contrasting air masses. The plains of North America are ideal in this regard for the development of conditions conducive to severe weather; there is no mountain barrier between the Tropics and the pole. Cyclones that affect North America can easily bring together the ingredients for severe forms of DMC simply because of the unique geography of the region. That is, moisture and instability can be brought together readily by these ETCs, with the baroclinic zone favoring the presence of vertical wind shear. Further, the lee slopes of the Rocky Mountain barrier promote the development of poleward flows of sensible and latent heat and, at the same time, the high terrain of the west acts as an elevated heat source, as described already.

2) OTHER TOPOGRAPHIC EFFECTS

Some nonorographic topography can play a substantial role in modulating the behavior of cyclones. For instance, the Great Lakes of the United States have been shown to favor cyclonic circulations in their vicinity during the winter (e.g., Petterssen and Calabrese 1959; Sousounis and Fritsch 1994). On the other hand, Harman (1987) has shown that anticyclones avoid the Great Lakes in the winter and tend to pass over them frequently in the summer. Thus, if a lake, grouping of lakes, or an inland sea is large enough, it can alter the dynamics of extratropical weather systems, as well as the more obvious effect of evaporation of moisture into the airstream passing over the

[14] This ignores the effect of synoptic-scale ascent on the sounding itself, of course. That ascent would tend to reduce any CIN and make initiation of convection more likely.

water. The diagnoses by Alpert et al. (1996; see especially their Fig. 2) suggest strongly that cyclogenesis can be the result of a variety of distinct processes and interactions. Topographic effects can interact synergistically with other physical processes (e.g., convection) in complex ways. As a result, the mixture of processes promoting cyclogenesis at any particular time can vary during a particular cyclogenetic event.

In different parts of the world, the ingredients for severe DMC are brought together in different ways. In the western Mediterranean, for example, a long fetch of poleward flow ahead of an approaching cyclone brings in low-level air from the deserts of northern Africa, which is not conducive to severe forms of convection. When the low-level flow has a long easterly fetch over the warm waters of the Mediterranean, however, it can be quite rich in moisture (see Doswell et al. 1998). The ingredients remain the same, but the topography changes the details by which the ingredients are concatenated as synoptic-scale systems interact with that topography.

In regions of complex terrain, subsynoptic-scale, solenoidally driven flows are quite common; Whiteman (1990) and Egger (1990) provide recent summaries of these. Banta (1990) reviews the impact of complex terrain on clouds, and on convective rain in particular. In essence, complex terrain offers a number of different processes that can provide sufficient lift to make DMC possible. Therefore, when sufficient moisture is present, mountainous terrain often supports vigorous convective activity even in the absence of favorable synoptic-scale processes. This is reflected in the climatological frequency of thunderstorms in favored locations in complex terrain noted earlier, wherein thunderstorms become a nearly daily occurrence in selected mountain areas at the height of the warm season.

3) AIR MASS CHARACTERISTICS

The vertical structure of the atmosphere depends on, among other factors, the character of the various airstreams at different levels moving over a point. Given the existence of substantial vertical shear of the horizontal wind in midlatitudes, the patterns of vertical thermodynamic structure associated with severe forms of DMC arise in part as a consequence of the superposition of air masses. Fawbush and Miller (1954) presented examples of the "typical" structures. As noted by Carlson and Ludlam (1968) and Doswell et al. (1985), these structures can arise from superposing airstreams with different properties at different levels. The classic "loaded gun" sounding typical of tornadic situations arises from the superposition of a flow of moist air at low levels (see below) and a midtropospheric air mass with high lapse rates that originates elsewhere. In the plains of the United States, midtropospheric high lapse rates originate on the elevated terrain of the Rocky Mountains and are advected as an elevated residual mixed layer (see Carlson and Ludlam 1968; Lanicci and Warner 1991; Stensrud 1993) over the low-level poleward flow of moist air in the southwesterly airstream ahead of an approaching ETC. The weak static stability of this air mass enhances the upward motion ahead of the advancing cyclone, as we have discussed.

Note that the low-level moisture is also a critical ingredient. For the plains of the United States, this moisture is typically brought poleward first at or near the surface in the airstreams ahead of the migrating synoptic-scale cyclones. The presence of adequate moisture in that low-level flow, however, is by no means assured, especially early in the spring (Crisp and Lewis 1992; Lewis and Crisp 1992). Many issues influence the structure of the low-level moisture field; hence, its forecasting can be a challenge (Thompson et al. 1994).

b. Solenoidal circulations over simple terrain

Many flows associated with solenoidal circulations are subsynoptic in scale, including classic fronts (not adequately described by QG theory; e.g., see Williams and Plotkin 1968). Naturally, fronts are an important aspect of ETCs. That fronts are associated with vertical circulations is of course well known (e.g., Eliassen 1962; Bluestein 1993, p. 297 ff.). As already noted, vertical motions produced along and near fronts are a ready source of subsynoptic-scale lift (of order 10 cm s^{-1} or larger) for initiating DMC. Classical fronts that develop in ETCs are not, however, the only solenoidally associated flows that can provide initiation mechanisms for DMC. The majority of these other thermally driven circulations probably fall into the time and space scales known as *mesoscale* [reviewed in Atkinson (1981, p. 123 ff.) and discussed in chapter 3 of this monograph], and so are only mentioned briefly here.

Another solenoidally driven flow that has a large impact on DMC is the land–sea breeze. This process has been recognized as an important contributor to initiating DMC for some time (see, e.g., Pielke and Segal 1986). As with the comparable thermally driven circulations in complex terrain, the local topography has an important impact on where and when DMC is likely to be initiated. There is a modulation of the process by its interaction with the synoptic-scale flow. The movement of the land–sea breeze is retarded when it is moving against that synoptic flow and enhanced when moving with it. Typical diurnal movement of the boundary defining the land–sea breeze "front" is on the order of 10 km, so the favored zones for DMC initiation are usually confined to the vicinity of the coastline.

Convectively generated outflow boundaries are yet another solenoidally driven circulation that can initiate convection. In situations where DMC evolves into

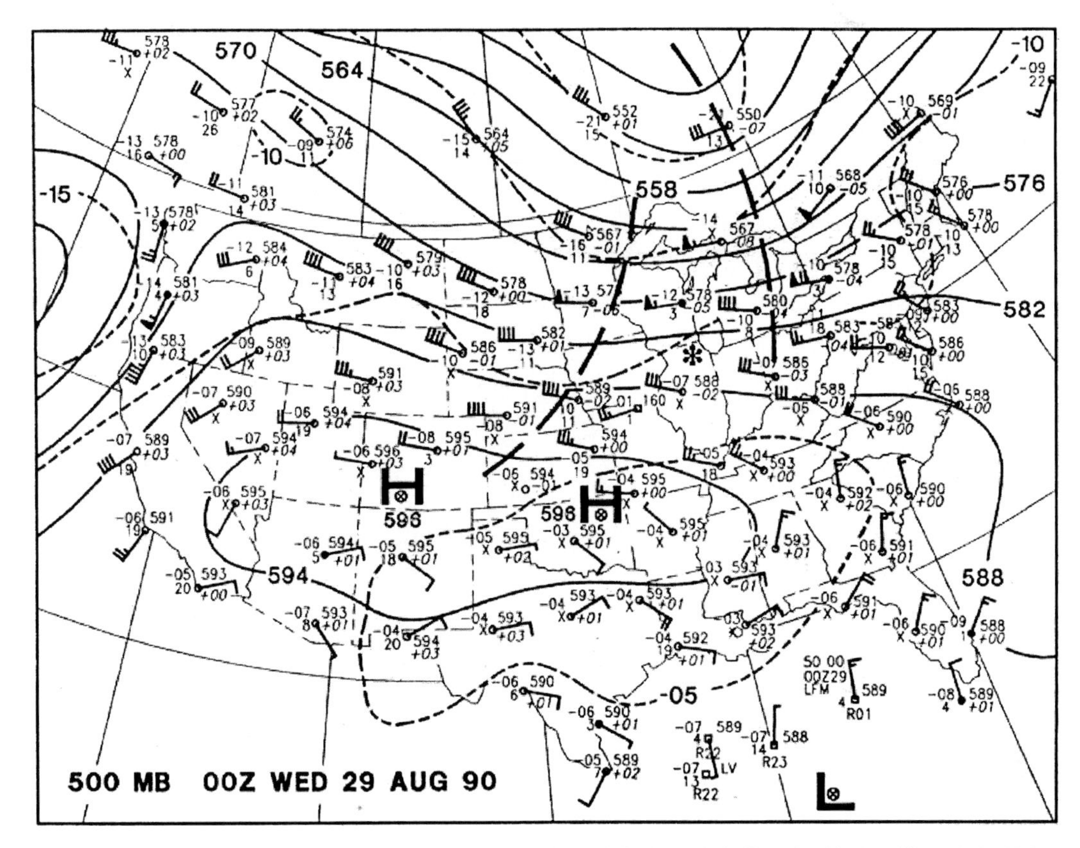

FIG. 2.15. 500-hPa chart at 0000 UTC on 29 August 1990. Plainfield, IL, is indicated with the "*" symbol; thick dashed lines denote shortwave trough axes.

large mesoscale convective systems (MCSs), outflow regions grow to meso-β scale (Orlanski 1975) or larger when numerous convective cells contribute to the pool of precipitation-cooled outflow. These large outflows can be detected and followed readily in many surface observation networks, although the details cannot be said to be well resolved outside of a true mesonetwork.

There is growing evidence for a variety of solenoidally generated flows associated with inhomogeneities in the underlying topography (e.g., Sun and Ogura 1979; Segal and Arritt 1992; Hane et al. 1997). These might be associated with albedo differences, vegetated versus bare soil or irrigated versus nonirrigated regions, soil moisture availability differences due to antecedent precipitation (Ookouchi et al. 1984), variations in snow cover (Johnson et al. 1984), and so on. At times, they have been described as an "inland sea breeze" (Sun and Ogura 1979) in analogy with ordinary sea breezes. As discussed in chapter 3, evidence is accumulating that these nontraditional sources of solenoidally driven flows can be an important factor in DMC initiation (e.g., Colby 1984).

Finally, the dryline (Schaefer 1986) is a special case where solenoidal circulations may or may not be present. The thermal gradient across the dryline in the United States undergoes a diurnal oscillation, chang-

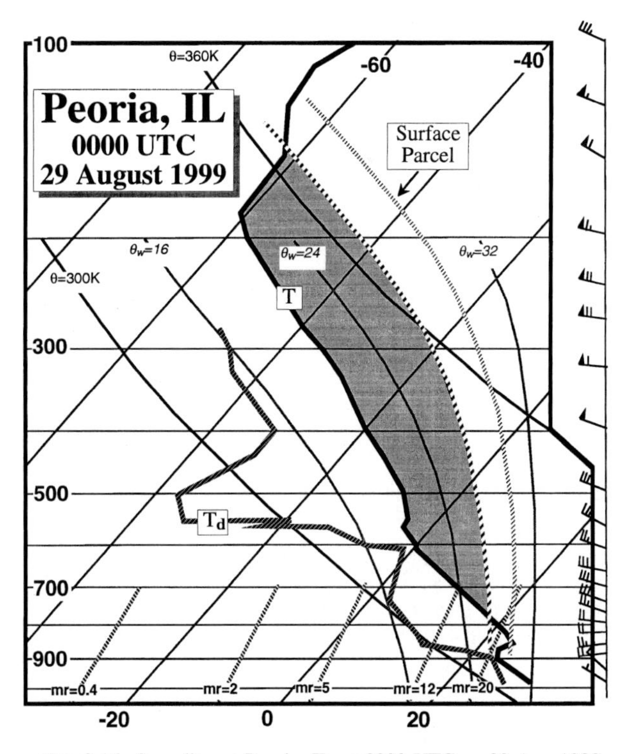

FIG. 2.16. Sounding at Peoria, IL, at 0000 UTC on 29 Aug 1990, including plotted wind profile. The positive area (CAPE) is stippled, for a surface-based mixed layer parcel (hatched curve). Also shown is the ascent curve for a surface parcel (fine-hatched curve).

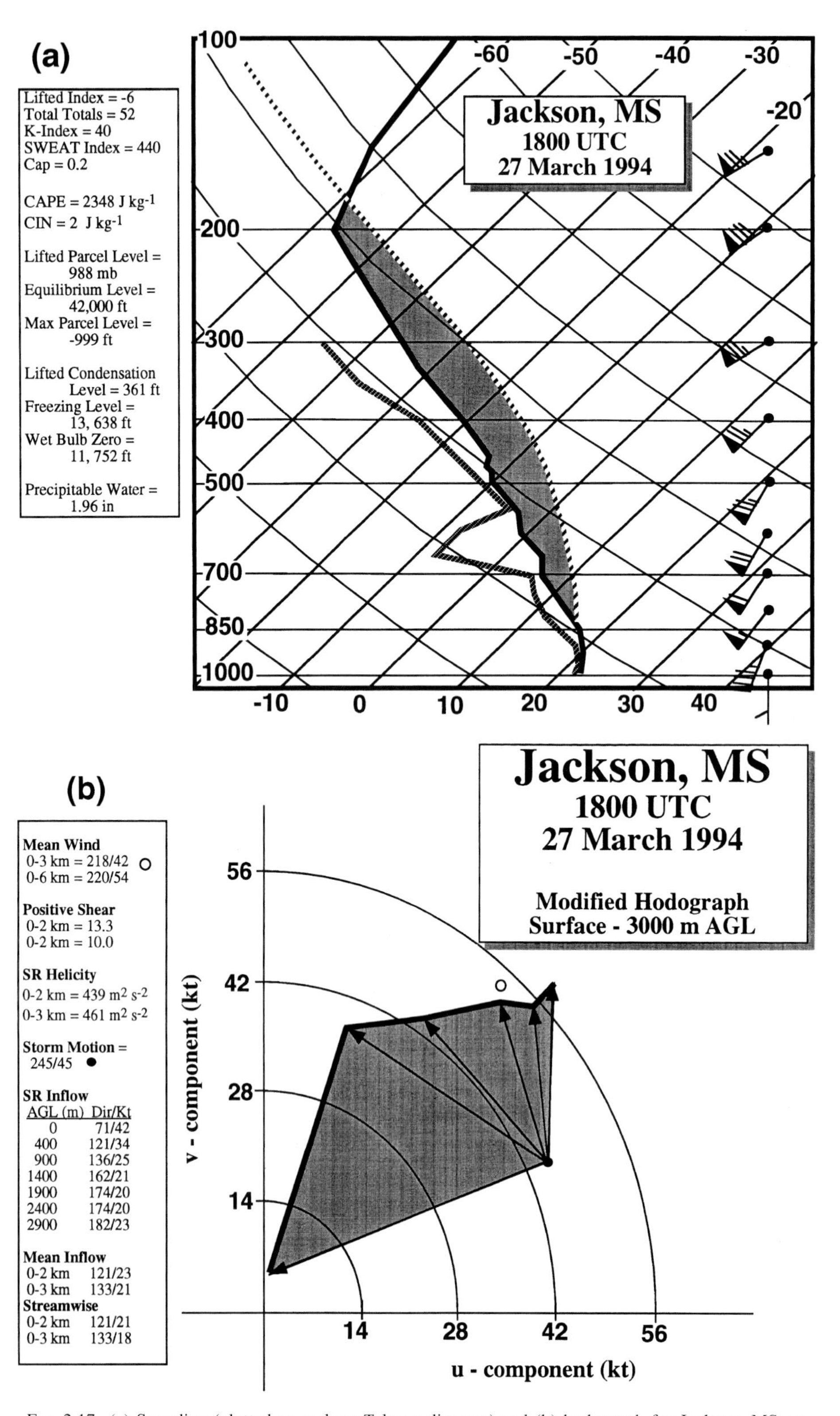

FIG. 2.17. (a) Sounding (plotted on a skew T–log p diagram) and (b) hodograph for Jackson, MS, at 1800 UTC on 27 Mar 1994. The hodograph was modified as in National Weather Service (1994).

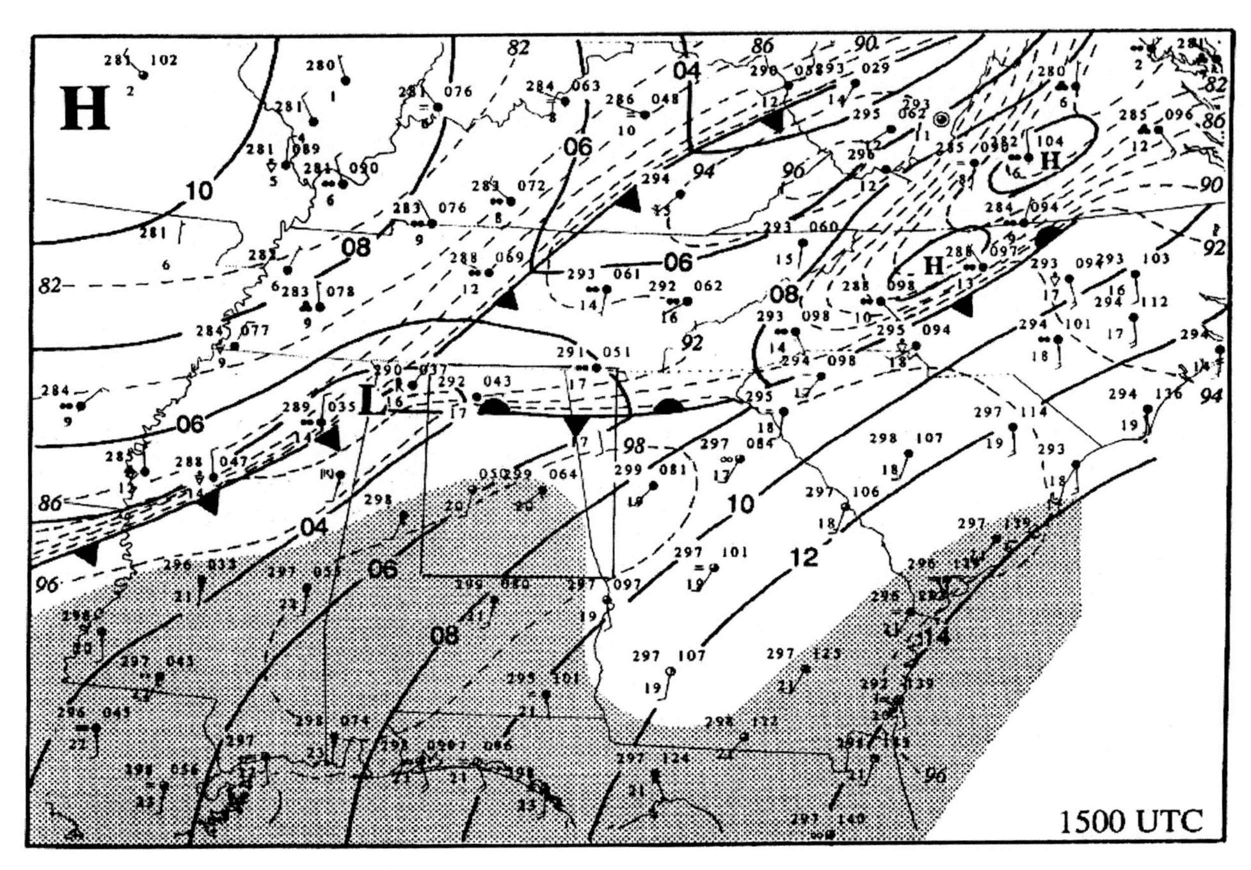

FIG. 2.18. Surface analysis for 1500 UTC on 27 Mar 1994 (from Langmaid and Riordan 1998); isobars and frontal symbols are conventional, isotherms are dashed lines (°F), and shading denotes areas with dewpoints ≥20°C (68°F).

ing directions during the heating cycle. In the morning, on the dry air side, temperatures are colder than on the moist side; by afternoon, this reverses. What makes it challenging is that the moisture content in the afternoon contributes to density in the opposite way to the temperature, apparently resulting in a near-cancellation of the density gradient. Given natural variability, limited sampling, and finite measurement accuracy, it has not yet been shown definitively that the ascent at the dryline during the afternoon is driven by solenoids. For instance, Ziegler and Hane (1993) present a case study where a virtual temperature (and, hence, density) gradient still exists across the dryline in the afternoon, implying that solenoidal circulation might explain the ascent. However, their sample was a single case, and their capacity to sample the structure and evolution of the dryline was limited. Much remains to be shown about how the dryline acts to initiate DMC (Ziegler and Rasmussen 1998).

c. Monsoons

Monsoons have been defined as a seasonal shift in the direction of the surface winds (Huschke 1959). Monsoons have been reviewed several times in the literature (e.g., Ramage 1975; Boyle and Chen 1987; Webster et al. 1998). Although it is certainly an oversimplification, monsoons can be thought of as a seasonal analog to the solenoidally driven diurnal flows. Basically, the seasonal shifts in surface flow

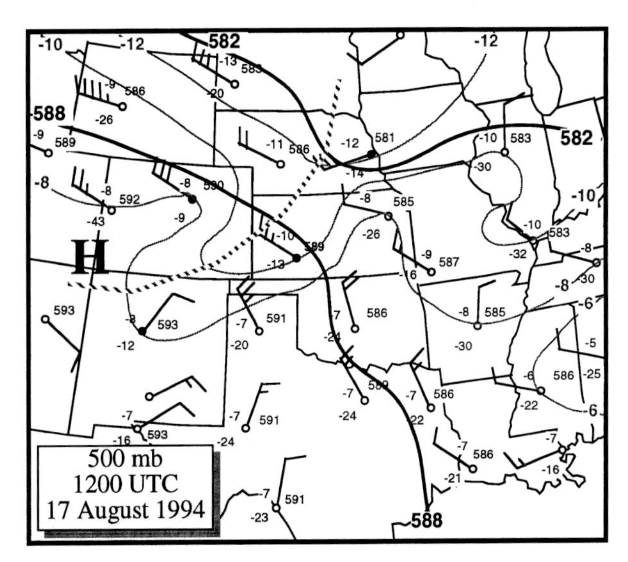

FIG. 2.19. Analysis at 500 hPa at 1200 UTC on 17 Aug 1994; thick solid lines are isohypses (dam), thin gray lines are isotherms (°C), the thick hatched line denotes a thermal trough.

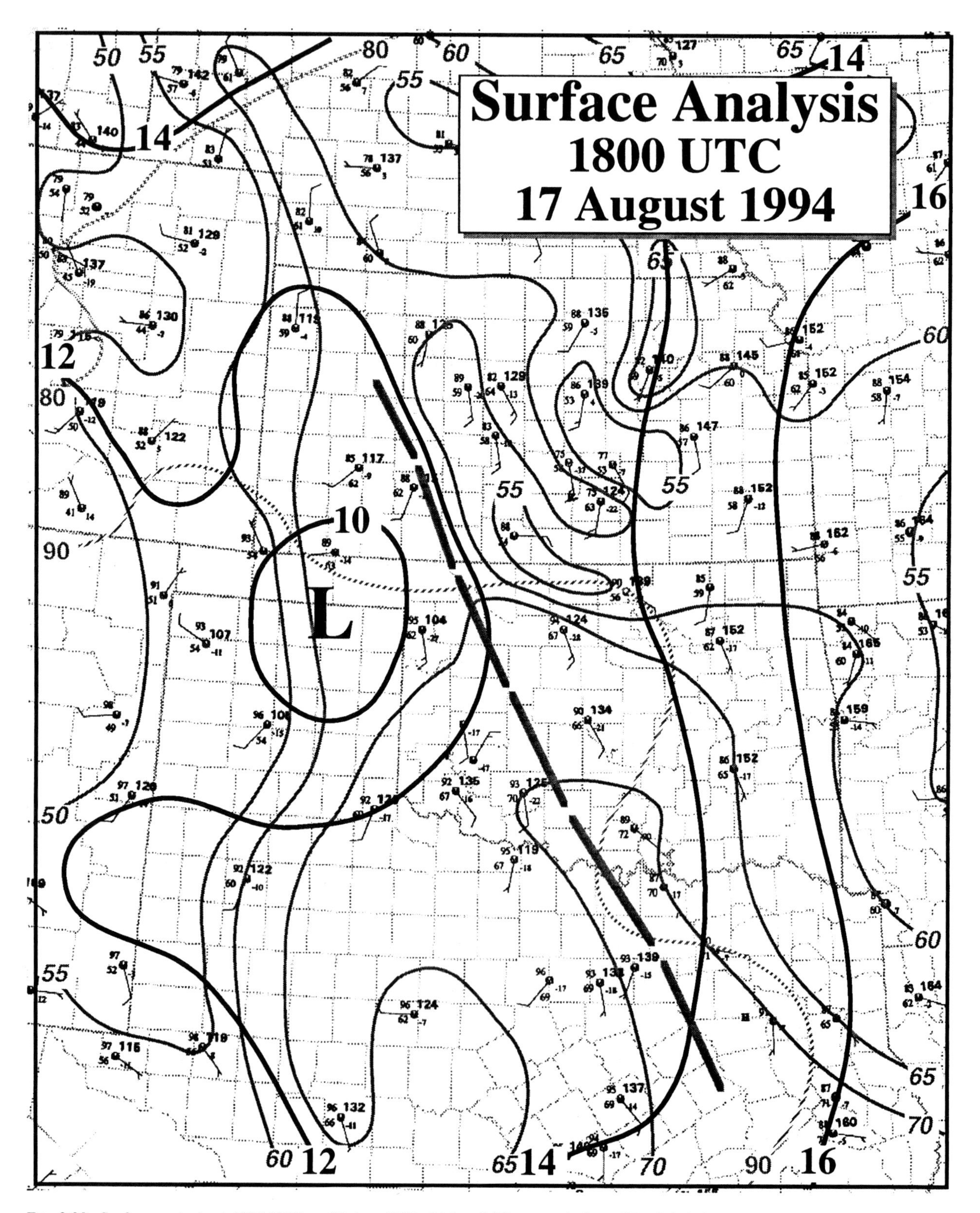

FIG. 2.20. Surface analysis at 1800 UTC on 17 Aug 1994; thick solid lines are isobars (hPa, labeled conventionally), thin hatched lines are isotherms (10°F interval), thin gray lines are isodrosotherms (5°C interval), and the moisture axis is depicted by the thick dashed line.

are being driven by continental-scale land–sea temperature differences. A tendency for ascent over the warm continents in the warm season drives a flow toward the continent from the cooler oceans, whereas descent over the continents in the cool season drives a flow outward from the continent. The monsoons do not initiate convection, per se, but they are associated directly with processes that *do* provide subsynoptic-scale ascent, such as upslope flows, monsoon-associated weather systems, and embedded solenoidally driven flows. The most well-known monsoons globally tend to be associated with the Tropics and subtropics in Asia (e.g., Ramage 1975; Chang and Krishnamurti 1987), but the midlatitudes have monsoon-like flows as well (see, e.g., Lau and Li 1984; Douglas et al. 1993). These seasonal tendencies can enhance diurnal flows that are also thermally driven and so can play a role in the development or suppression of DMC (e.g., Doswell 1980). Moreover, they can import low-level moisture from over the oceans to participate in DMC over land, and so can be a factor in creating a suitable environment for DMC as well as in the initiation process.

d. Vertical wind shear

Our understanding of the processes by which severe convective events arise has come to include an awareness of the importance of the vertical wind shear structure in the convective storm environment. DMC arising in certain circumstances can produce quasi-stationary convective systems (Chappell 1986) that are common in heavy rainfall events. Supercell storms are a product of the interaction of the convective updrafts within a sheared environment (see, e.g., Rotunno and Klemp 1982; Weisman and Klemp 1986; Rotunno 1993; Brooks et al. 1994a,b). The vertical wind shear is not the only measure of character of the wind profile. Recently, some theoretical work (Davies-Jones 1984; Lilly 1986) has suggested a relatively new measure; the helicity of the wind profile. The ground-relative helicity, H, is defined by

$$H(z) = -\int_{z_0}^{z} \mathbf{k} \cdot \left(\mathbf{V} \times \frac{\partial \mathbf{V}}{\partial z}\right) dz, \quad (2.5)$$

and is useful as a measure of the curvature of the hodograph. In a storm-relative framework, (2.5) becomes

$$H_{\mathrm{SR}}(z) = -\int_{z_0}^{z} \mathbf{k} \cdot \left((\mathbf{V} - \mathbf{C}) \times \frac{\partial \mathbf{V}}{\partial z}\right) dz, \quad (2.6)$$

where **C** is the observed (or forecast) storm motion. Numerous studies (Davies-Jones and Brooks 1993; Davies and Johns 1993; Moller et al. 1994) have

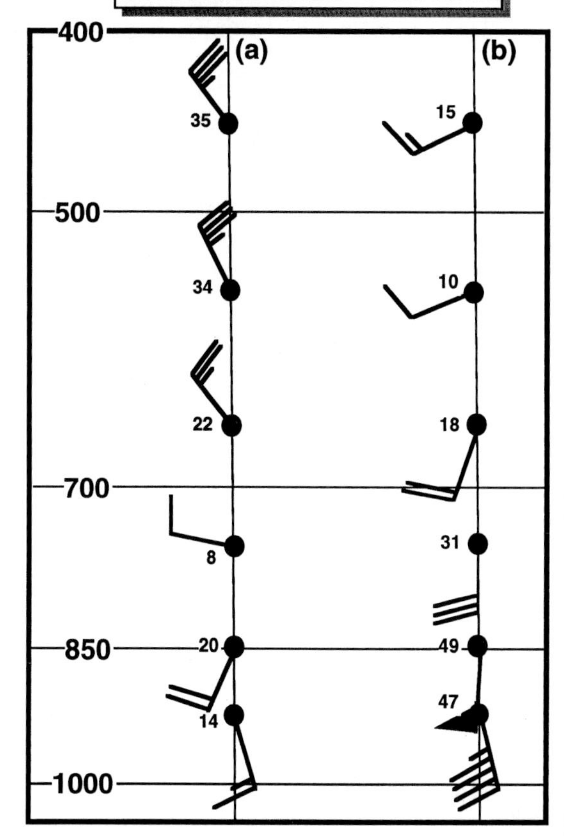

FIG. 2.21. (a) Ground-relative and (b) storm-relative wind profiles from the Vici, OK, profiler (in northwestern OK) at 1800 UTC on 17 Aug 1994; wind barbs are conventional.

suggested that this parameter can be useful in assessing the potential for severe thunderstorms.

It should be noted that helicity is associated with winds changing direction with height, which is especially important when the framework is shifted to a storm-relative one. The Galilean coordinate transformation from one uniform velocity to another (as in going from a ground-relative to a storm-relative viewpoint) is of significance, because the physics of the atmosphere are invariant to such a transformation. Vector wind shear is indeed a Galilean invariant, whereas helicity is not. However, the use of diagnostic variables is not necessarily restricted to those that are Galilean invariant. The value of evaluating the threat of severe weather using storm-relative helicity is well established (see Davies-Jones et al. 1990; Moller et al.

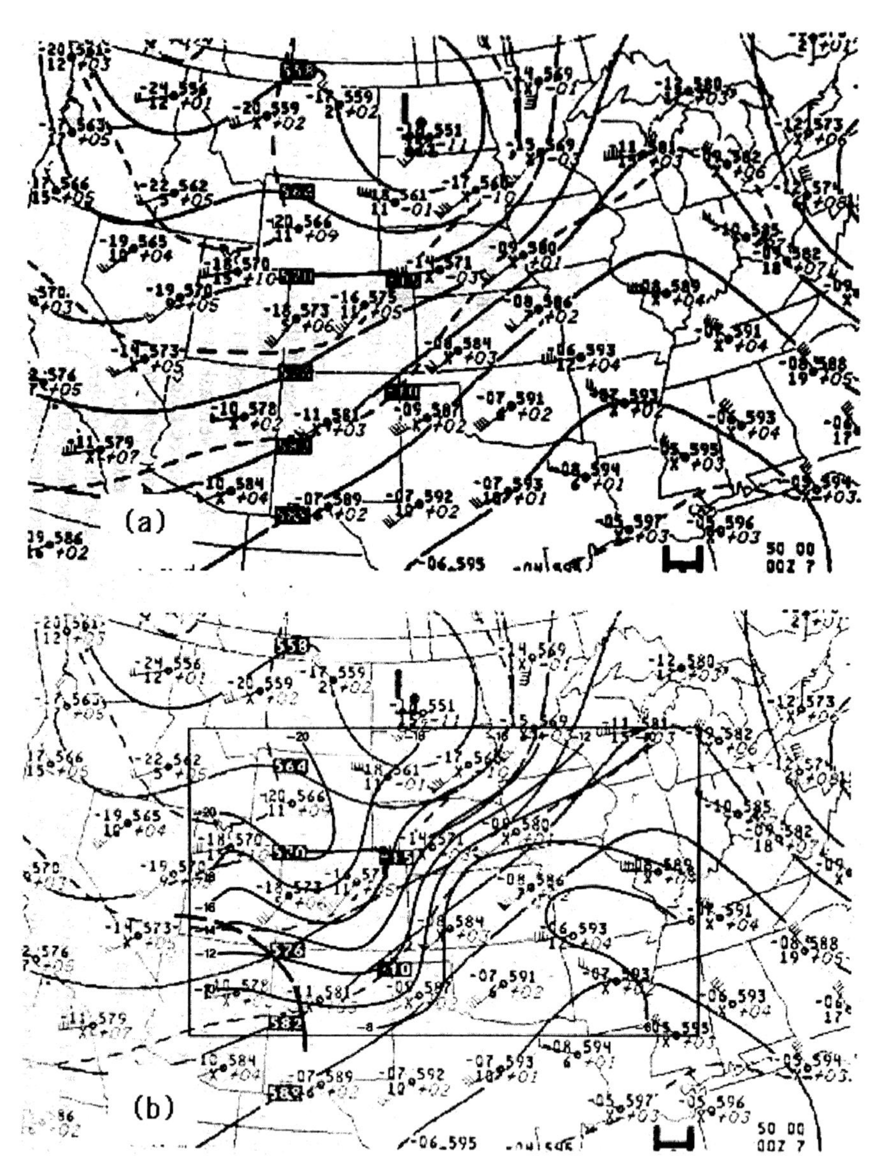

FIG. 2.22. A comparison of (a) basic National Meteorological Center analysis at 850 hPa for 0000 UTC on 7 Jun 1982, and (b) as reanalyzed by Rockwood and Maddox (1988).

1994; Rasmussen and Blanchard 1998) in spite of whatever drawbacks might accrue from its violation of Galilean invariance. Since the storm motion is not known in advance of storm formation, this makes evaluation of storm-relative helicity problematic. A number of approximations can be used (see Bunkers et al. 2000) based on existing observations. From a purely theoretical viewpoint, however, storm motion can be viewed as implicit in the governing equations as a sort of eigenvalue (Davies-Jones 1998); it is not something entirely external to the problem of DMC. However difficult it may be to forecast, it is still physically relevant.

No matter how the character of the wind profile is evaluated, ETCs impose both temporal and spatial complexity on that vertical wind profile. In addition to processes in the PBL, various processes in the free atmosphere (both geostrophic and ageostrophic) in association with ETCs alter the winds aloft. Notably, thermal advection imposes backing and veering of the wind with height, and isallobaric accelerations associated with moving and evolving ETCs are important in modifying the wind profile. If the thermal wind ($\partial \mathbf{V}_g/\partial z$) is substituted into the integrand of (2.5) and its definition in terms of the thermal gradient is used, along with some vector identities, it can be shown that

$$\mathbf{k} \cdot \left(\mathbf{V} \times \frac{\partial \mathbf{V}}{\partial z} \right) = \frac{g}{fT} \mathbf{V} \cdot \nabla_z T. \qquad (2.7)$$

Thus, H is roughly proportional to the integrated thermal advection in the layer z_0 to z in midlatitudes

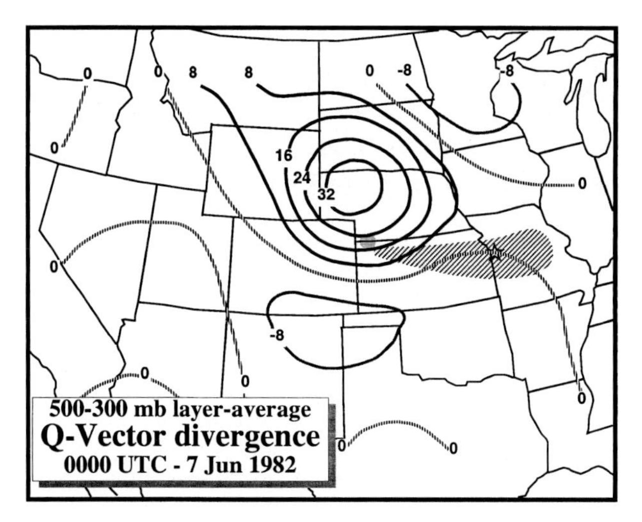

FIG. 2.23. 500–300-hPa layer averaged Q-vector divergence field at 0000 UTC on 7 Jun 1982, showing forcing for vertical motion; positive values imply downward motion, isopleths are in units of 10^{-17} s^{-3} hPa^{-1} (after Rockwood and Maddox 1988). The calculation follows described in Barnes (1985). Shading indicates location of first storm development, hatched area shows the region of reported severe weather, and the Kansas City, MO, area is indicated by a star.

where QG approximations are reasonably good (as first demonstrated by Tudurí and Ramis 1997). Warm advection favors veering of the wind with height in the Northern Hemisphere, corresponding to positive ground-relative helicity, which is considered to be a favorable wind profile for the development of severe convection, as already noted. However, it is not the case that veering wind profiles are confined to regions of warm advection.

The importance of the low-level wind shear on severe forms of convection is beyond doubt (e.g., Weisman and Klemp 1982, 1984; Brooks et al. 1994a). However, it can be challenging to forecast the winds accurately. To the extent that an accurate wind profile forecast can be made, it becomes possible to anticipate the influence of synoptic systems on the character of any DMC that develops, especially the likelihood of supercells. Maddox (1993) has noted the potential for interesting diurnal variations of helicity due to PBL effects, but this is only a small part of the picture.

e. Organized convective systems and the low-level jet stream

Mesoscale convective systems are a particularly large and persistent class of DMC. As described by Maddox (1983), the largest and most persistent of these (MCCs) have some characteristics that appear to be tied to the synoptic structures in which they tend to occur. Rather than being associated with vigorous ETCs, these systems often arise in relatively quiescent conditions in the middle and upper troposphere but with considerable warm thermal advection at low levels. This thermal advection is enhanced, at least in the plains of the United States, by diurnal variations in the LLJS, as noted in Maddox and Doswell (1982) and discussed earlier.

The structure and evolution of the LLJS in the United States does not take place independently of the synoptic structure, unlike the NBLWM. This has been demonstrated by, among others, Lanicci and Warner (1991) or Djuric and Ladwig (1983), who have shown that the LLJS undergoes a quasi-regular cycle associated with the passage of midlatitude cyclones and anticyclones. Further, Uccellini (1980) has shown that the LLJS is often coupled to upper-level jet streaks. Thus, the synoptic-scale structure is quite pertinent in the distribution and intensity of MCSs, and perhaps other aspects of severe DMC. The topic of MCSs is discussed in considerable detail in chapter 9.

2.7. Feedbacks to the synoptic scale

Most of the content of this chapter has been directed at revealing how the synoptic structure organizes the ingredients for DMC and modulates convective events. However, this addresses only half of the interaction. How does convection feed back to synoptic-scale processes? Uccellini (1990) concludes that "The role of convection in enhancing cyclogenesis . . . remains unresolved." We concur with this assessment, but observe that evidence of upscale feedbacks can certainly be observed when DMC is widespread and persistent (see, e.g., Stensrud 1996a; Dickinson et al.

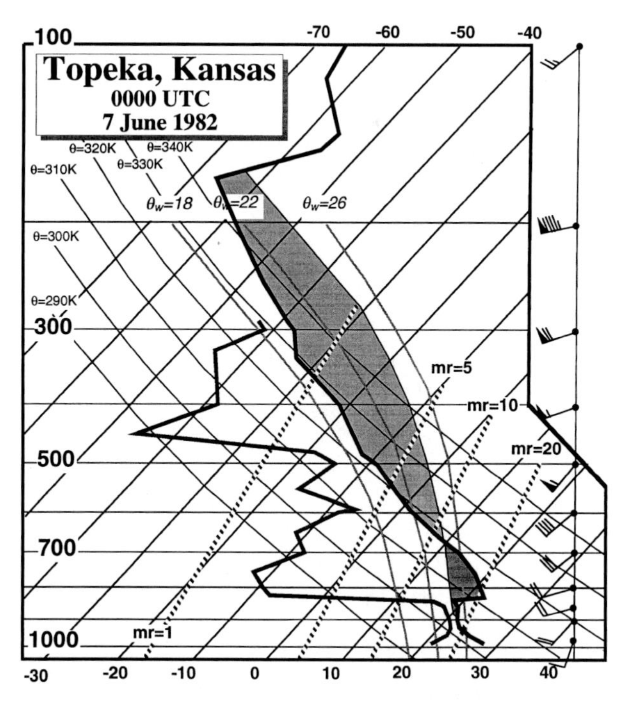

FIG. 2.24. Topeka, KS, sounding at 0000 UTC on 7 Jun 1982, plotted on a skew T–log p diagram, also showing plotted wind profile. The dark shaded area is the negative area, and the light shaded area is the positive area, for a parcel lifted from the surface layer.

1997). The climatological minimum of DMC in the winter (in midlatitudes) can be attributed to the enhanced vigor of cool-season synoptic systems and the reduced heat input at low levels. In the warm season, synoptic systems have reduced baroclinity and, therefore, are correspondingly less intense. At the same time, the heat input at low levels is considerably larger in the warm season, so convection is needed more often than in the cool season. Convection and ETCs operate together to maintain the global heat balance.

In this context, we note that static stability is a physically relevant variable at virtually all scales in meteorology. Notably, if the atmosphere is too stable, cyclogenesis is inhibited, since static stability measures the "resistance" of the atmosphere to forcing for vertical motions (Gates 1961) and, at the same time, DMC is inhibited. It is widely accepted that ETCs provide the means of alleviating baroclinity by poleward heat transport; their upward heat transport stabilizes the atmosphere as well. Both heat transports serve to dampen cyclogenesis. Tracton (1973), among others, has suggested that when deep convection is near the cyclone center, it can enhance cyclogenesis. As noted already, the evidence supporting a positive impact on ETCs is inconclusive.

The 13–14 March 1993 Superstorm over eastern North America mentioned in section 2.2b also provides an excellent example of how convection feeds back to the synoptic scale (Bosart et al. 1996; Dickinson et al. 1997; Bosart 1999). A dynamical tropopause (DT) analysis revealed the presence of highly nonconservative PV behavior on the basis of an abrupt poleward shift of potential temperature contours on the DT over the southeastern United States in the 24 h ending 0000 UTC 13 March 1993. This poleward shift, occurring downstream of a massive convection outbreak over the northwestern Gulf of Mexico (as mentioned earlier in section 2.2b), could not be explained by simple advection as would be the case if PV were conserved. This PV nonconservation strengthened the potential temperature gradient on the DT downstream of the deepening trough, indicative of a near-doubling of the jet strength in the upper troposphere over a 24-h period (see Dickinson et al. 1997, their Figs. 3 and 4; Bosart 1999, his Fig. 9). The inability of the operational forecast models to simulate the explosive convection and its impact on the strength of the downstream jet resulted in significant errors in modeling the incipient cyclogenesis (Dickinson et al. 1997, their Figs. 14–22; Bosart 1999, his Fig. 9).

Convective modification of the larger-scale environment has also been noted in conjunction with relatively weak forcing aloft. Bosart and Lackmann (1995) documented the reintensification of Tropical Storm David (September 1979) over land while it traversed the eastern slopes of the central and northern Appala-

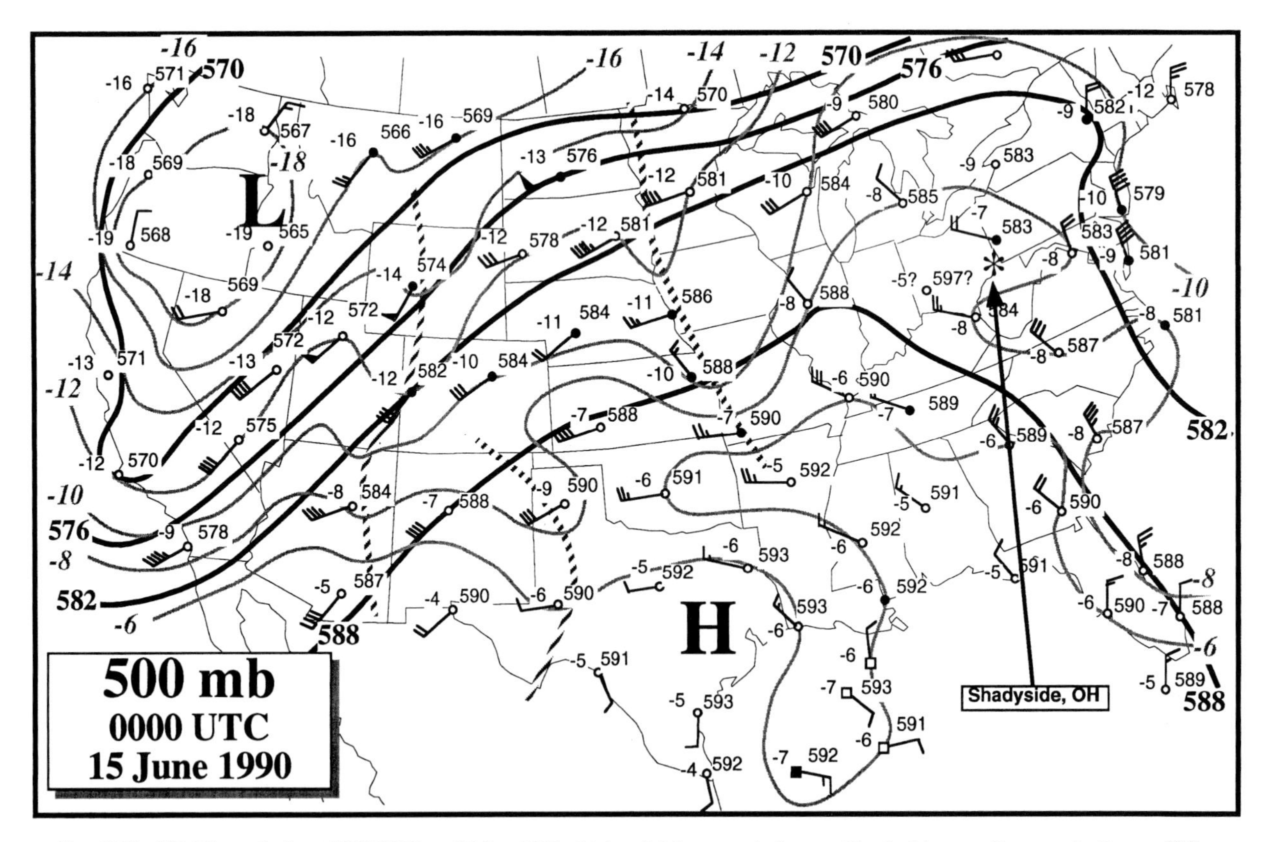

FIG. 2.25. 500-hPa analysis at 0000 UTC on 15 Jun 1990; thick solid lines are isohypses (dam), thin grey lines are isotherms (°C), thick hatched lines denote thermal troughs; Shadyside, OH, is indicated by the "*" symbol.

chians. They showed that diabatic heating associated with deep convection on the eastern side of David resulted in amplification of a ridge aloft just to the east of the storm on the basis of a DT analysis. This ridge amplification created an area of positive PV advection over David ahead of a weak trough to the west that culminated in the reintensification of the tropical storm. The slope of the DT across David increased, with the DT higher to the northeast (Bosart and Lackmann 1995, their Figs. 11–14). In effect, the ability of the upstream trough to contribute to cyclogenesis by means of positive PV advection was enhanced by downstream ridging associated with widespread latent heat release from DMC associated with the tropical storm.

In the United States, the existence of strong synoptic-scale systems in the spring, combined with increasing conditional instability created by surface heating while the middle and upper troposphere is still cool, produces the most intense and widespread severe convection. In the autumn, after a warm season's worth of DMC, the middle and upper troposphere are relatively warm; the asymmetry in conditional instability makes autumn much less likely to produce the intense convective events. However, the autumnal DMC can be very effective at producing heavy precipitation. In midlatitude summer, synoptic-scale processes are weak and the heat input must be balanced primarily with convective transports (as in the Tropics).

As an interesting example of geographical variations on this theme, a relatively cool sea dominates the Mediterranean basin in the spring, so its most vigorous convective season tends to be in the autumn. After a summer's heating, the warm, moist boundary layer over the sea basin is the source of most of the region's severe convection. As it turns out, the predominant form of such severe convection in the Mediterranean basin is also heavy precipitation (as in North America during the autumn).

Another variation can be found in Australia. A considerable portion of the continent is tropical or subtropical, but no barrier exists between the continent and Antarctica. The Southern Hemispheric flow is predominantly zonal, perhaps as a result of fewer mountain barriers and much less landmass. A large,

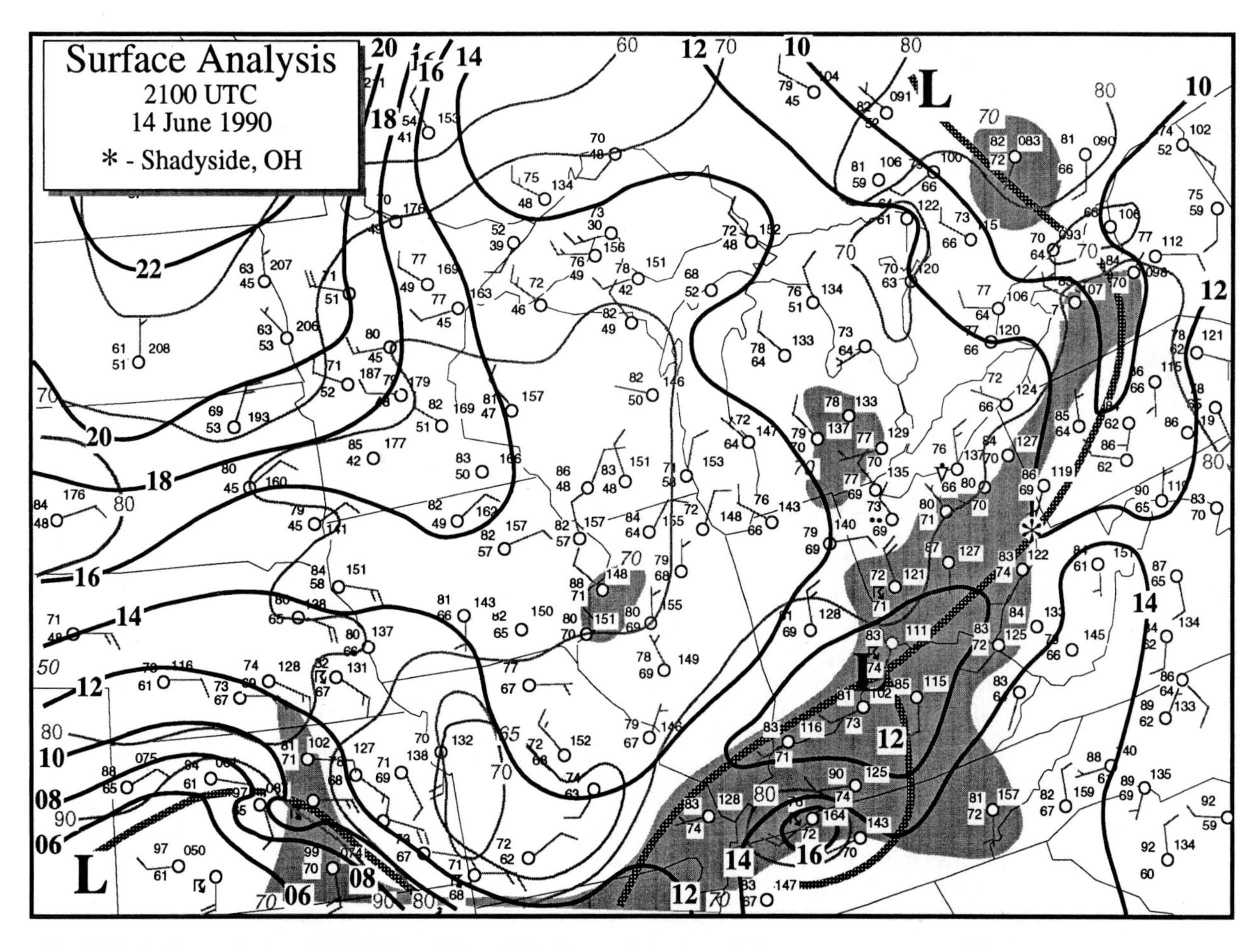

FIG. 2.26. Surface analysis at 2100 UTC on 14 Jun 1990; thick solid lines are isobars (2-hPa interval), thin gray lines are isotherms (10°F interval), and shading denotes areas with dewpoints ≥70°F; Shadyside, OH, is indicated by the "*" symbol. Thick stippled lines denote surface boundaries; station model is conventional.

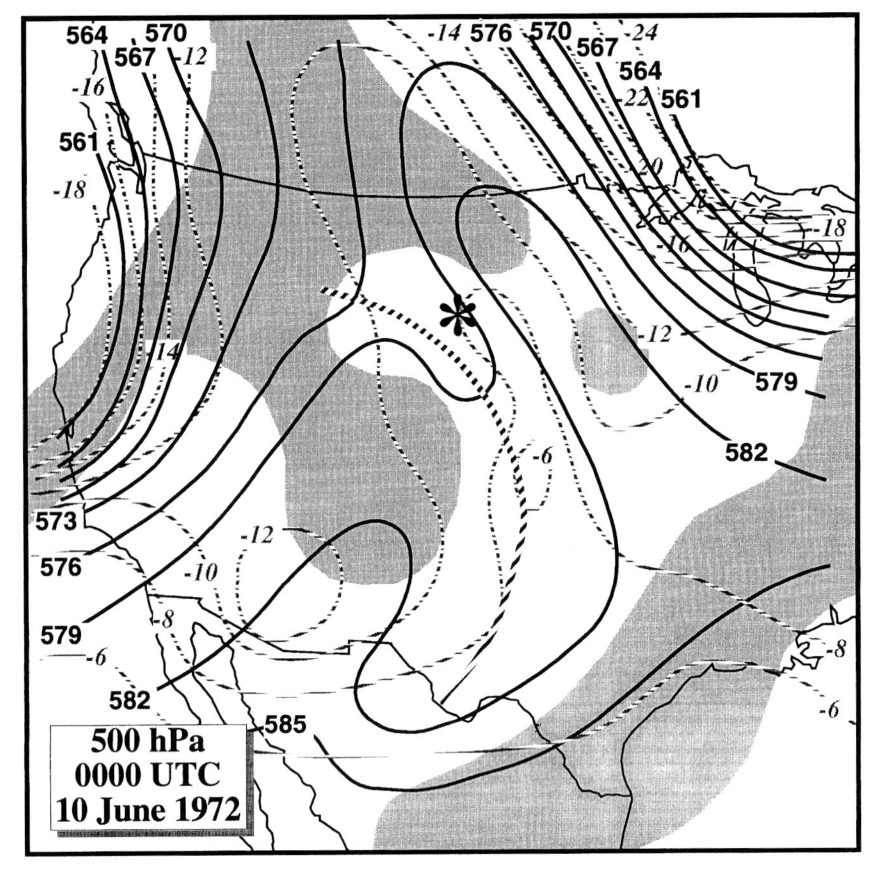

FIG. 2.27. 500-hPa analysis at 0000 UTC on 10 Jun 1972 (after Maddox et al. 1978), with heavy solid lines showing isohypses (dam), gray hatched lines showing isotherms (°C) and the heavy hatched line depicting a shortwave trough axis; shading denotes regions with dewpoint depressions of 6°C or less. The location of Rapid City, SD, is indicated by the "*" symbol.

predominantly arid interior promotes deep boundary layers and steep lapse rates, so poleward flow into nontropical Australia is often unstable but dry. A mountain range close to the east coast limits the penetration of moisture into the interior in the prevailing low-level easterly flow. Thus, moisture can be the major limiting factor on severe convection during the spring and autumn, when extratropical cyclones brush across the poleward half of the continent. Low-level moisture is available to synoptic-scale systems typically during landfall on Australia's west coast and perhaps again as they exit the east coast.

In qualitative terms, DMC must have an important role to play in the global heat balance. The well-recognized importance of DMC in the Tropics is a direct consequence of the need for a near-equilibrium between convection and those processes leading to destabilization (Emanuel 1994; p. 479 ff.). As noted in chapter 1 of this monograph, midlatitudes are not so well characterized by a sort of quasi-equilibrium, at least on timescales of a day or less. DMC does not always develop in midlatitudes as soon as destabilization begins; rather, it can be "stored" for hours or even days before DMC commences.

Although an individual convective cloud is too small and represents too brief a disturbance to have much direct impact on synoptic-scale processes, it is not uncommon for convection to develop into large mesoscale convective systems (MCSs), and in some situations, a succession of MCSs can exist for several days (Bosart and Sanders 1981; Fritsch et al. 1986). Stensrud (1996a) has shown that the accumulated effects of several days of persistent convection can have a significant impact on the synoptic structure. Whereas the long-term effect of convection must inevitably be an increase of static stability and a decrease in the vertical wind shear, the short-term effect on the environment can sometimes be to maintain conditions for DMC. In Stensrud's case study, an increased flow of low-level moisture (latent heat) into the convective area as a result of the convection allowed DMC to persist in spite of the slow stabilization of the lapse rates.

An interesting issue regarding feedback of convection to synoptic scales is the production of perturbations in the upper troposphere (and perhaps the lower stratosphere) from convection. Several studies (e.g., Ninomiya 1971; Fritsch and Maddox 1981; Maddox

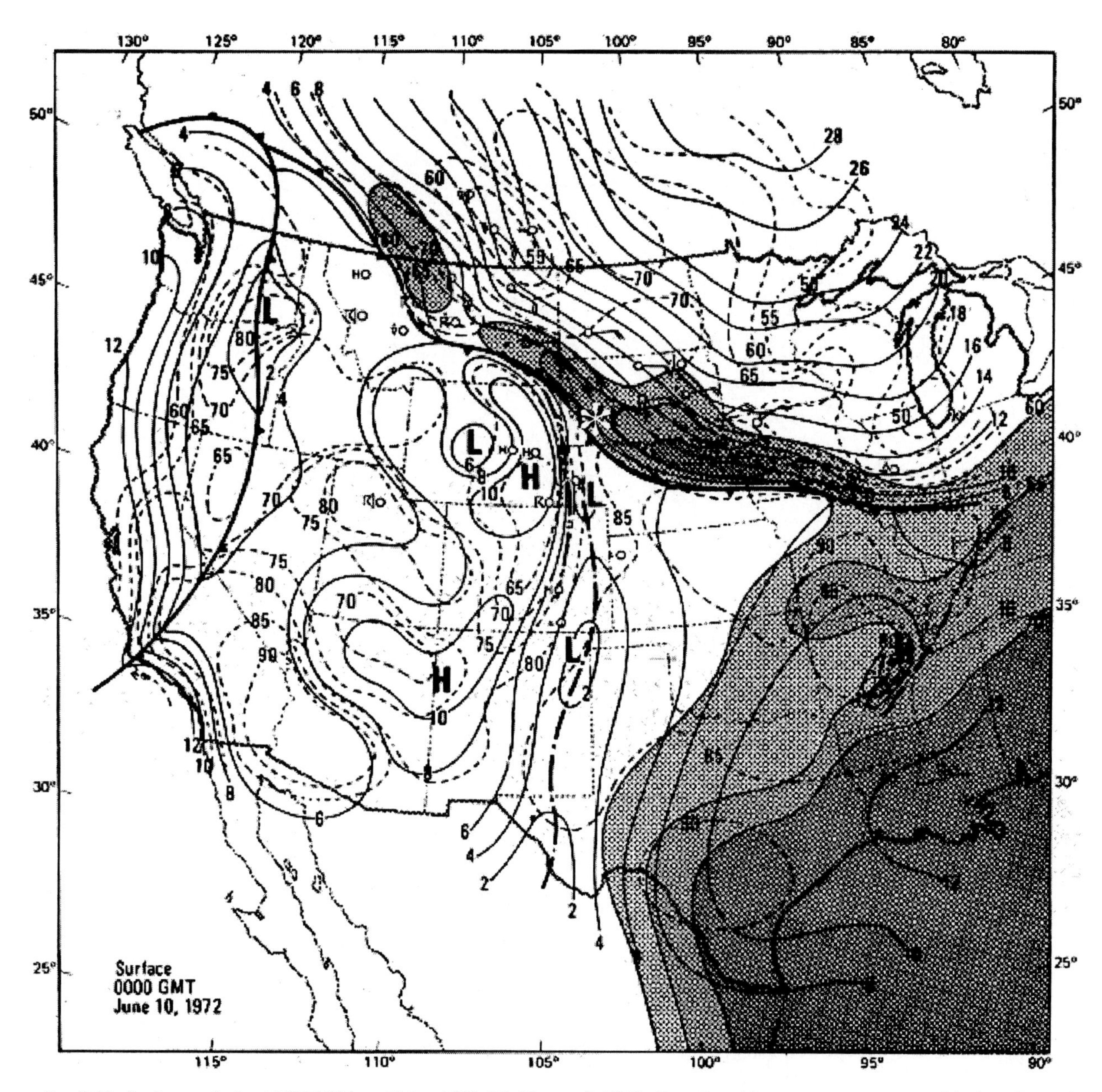

FIG. 2.28. Surface analysis at 0000 UTC on 10 Jun 1972 (Maddox et al. 1978). Frontal positions, pressure centers, and isobars (interval of 2 mb) are solid lines. Isotherms (interval of 5°F) are dashed lines; isodrosotherms ≥60°F are shaded, with solid lines (interval of 5°F). The location of Rapid City is indicated by the white "*" symbol.

et al. 1981; Maddox 1983) reveal that MCSs can produce significant perturbations in the meso-α-scale wind and height field aloft. Such structures do not usually persist for long beyond the demise of the convective system that produced them, but they do so occasionally. It is not known, in general, to what extent these perturbations influence the synoptic evolution downstream, subsequent to their development.

It is clear that, at times, convectively produced mesoscale features, notably the so-called mesoscale vorticity centers (MVCs; see, e.g., Bartels and Maddox 1991), occasionally persist well beyond the demise of the convection that produced them. As with perturbations in the upper troposphere, these can persist and even be associated with the redevelopment of a convective system (e.g., Fritsch et al. 1994). MVCs appear to be a product of the preferential enhancement by the Coriolis effect of the cyclonic member of a couplet of vortices produced at the ends of convective lines (Davis and Weisman 1994). Jiang and Raymond (1995) have used a nonlinear balance model to produce a simulation of a convective system that develops an MVC, allowing exploration of such topics as the energy budget of the system.

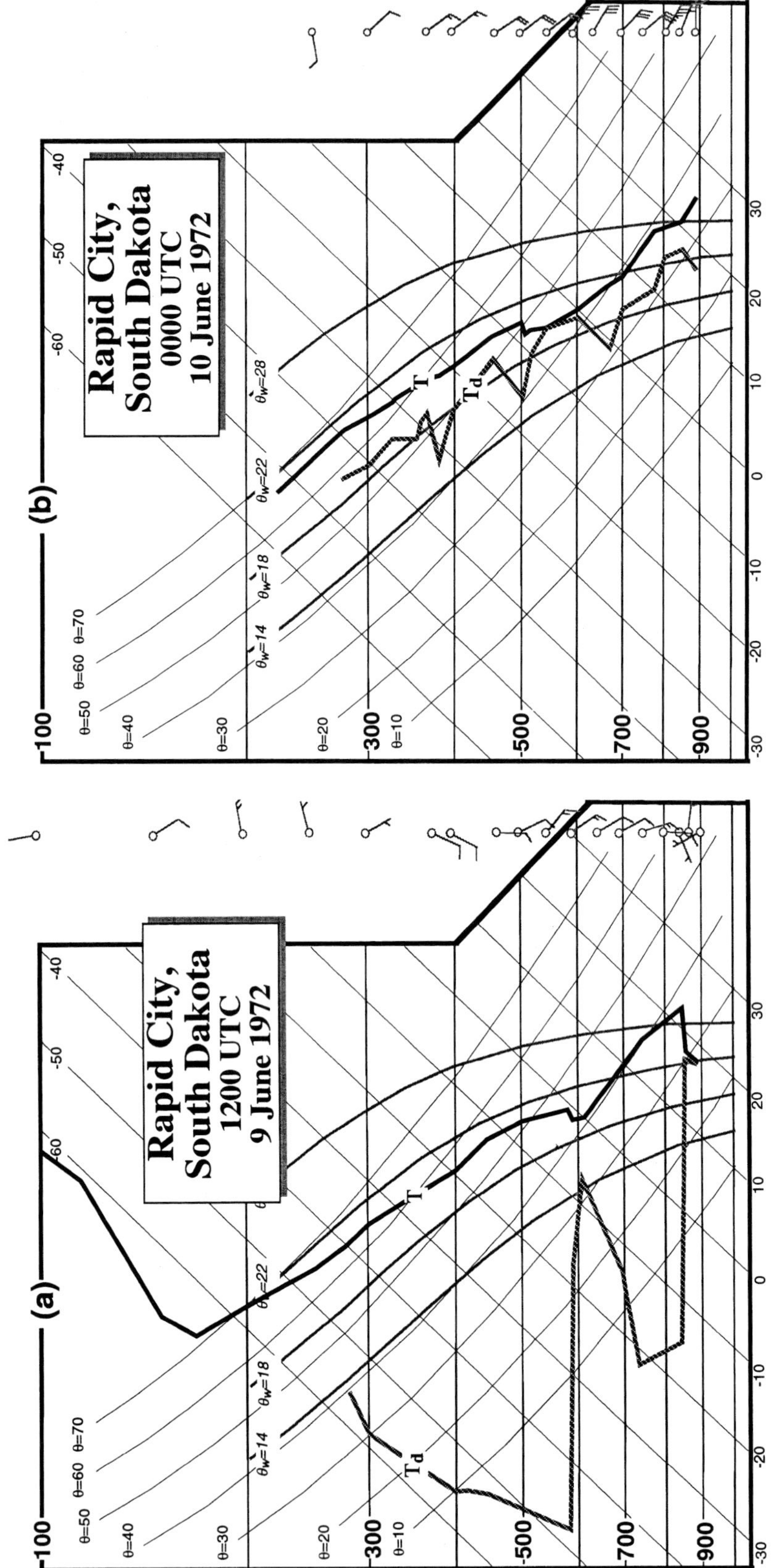

FIG. 2.29. Rapid City, SD, soundings at (a) 1200 UTC 9 Jun 1972 and (b) 0000 UTC 10 Jun 1972, plotted on a skew T–log$p$ diagram.

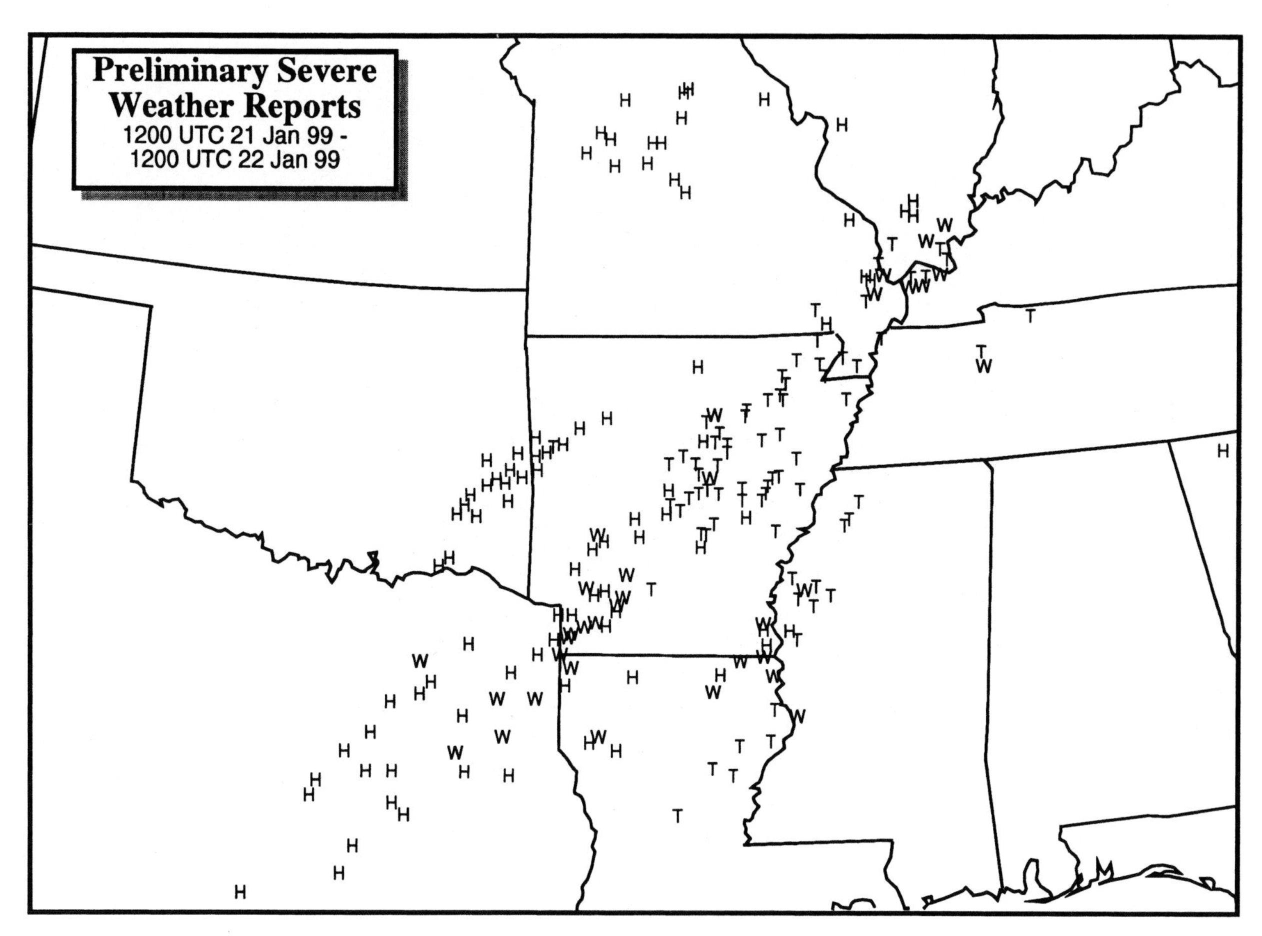

FIG. 2.30. Plot of the preliminary severe weather reports from 1200 UTC 21 Jan to 1200 UTC 22 Jan 1999. Hail, convective wind gust, and tornado reports are denoted by "H," "W," and "T," respectively.

2.8. Examples of severe convective events from a synoptic perspective

Rather than using only major outbreaks of severe weather to illustrate the synoptic-scale viewpoint, the idea here is to present as examples some events that are rather more subtle and certainly have proven to be more challenging to forecast. Severe convection in major outbreak situations during the climatological frequency maxima for severe convection (see Miller 1972; Doswell et al. 1993) is relatively easy to forecast and understand.[15] Many studies involving severe weather outbreaks (typically tornadic) of varying magnitude can be found (e.g., Fujita et al. 1970; Galway 1977; Benjamin and Carlson 1986). However, as noted in chapter 1 of this monograph, presenting only cases that involve major outbreaks of severe DMC gives an inappropriate bias to the implied relationship between the synoptic weather pattern and severe DMC.

Severe convection can develop in many different synoptic environments. All such environments have in common the potential for DMC (see Johns and Doswell 1992), but specific severe weather events have different basic ingredients. For example, the conditions favoring strong downdrafts and, hence, potentially damaging convective outflow winds, need not be the same as the conditions favoring strong updrafts, which can favor, for example, large hail. Thus, hazardous convection can be found in what might seem to be a bewildering array of synoptic-scale conditions. This makes generalizations difficult and leads to some concern for the value of individual event case studies, since to some extent every case is unique. Our approach here must necessarily be limited. Case studies are presented not because they exemplify the entire range of severe convection events, but because they illustrate the variety of synoptic structures that can be found with such events. The examples we have provided by no means exhaust that variety.

The format used for each short case study will not be identical. Space does not permit an extravagant presentation of the details for the case. Rather, the

[15] However, not every synoptically evident situation produces significant outbreaks of severe weather. Although major outbreaks of severe convection are often associated with vigorous ETCs, not every vigorous ETC results in severe convection.

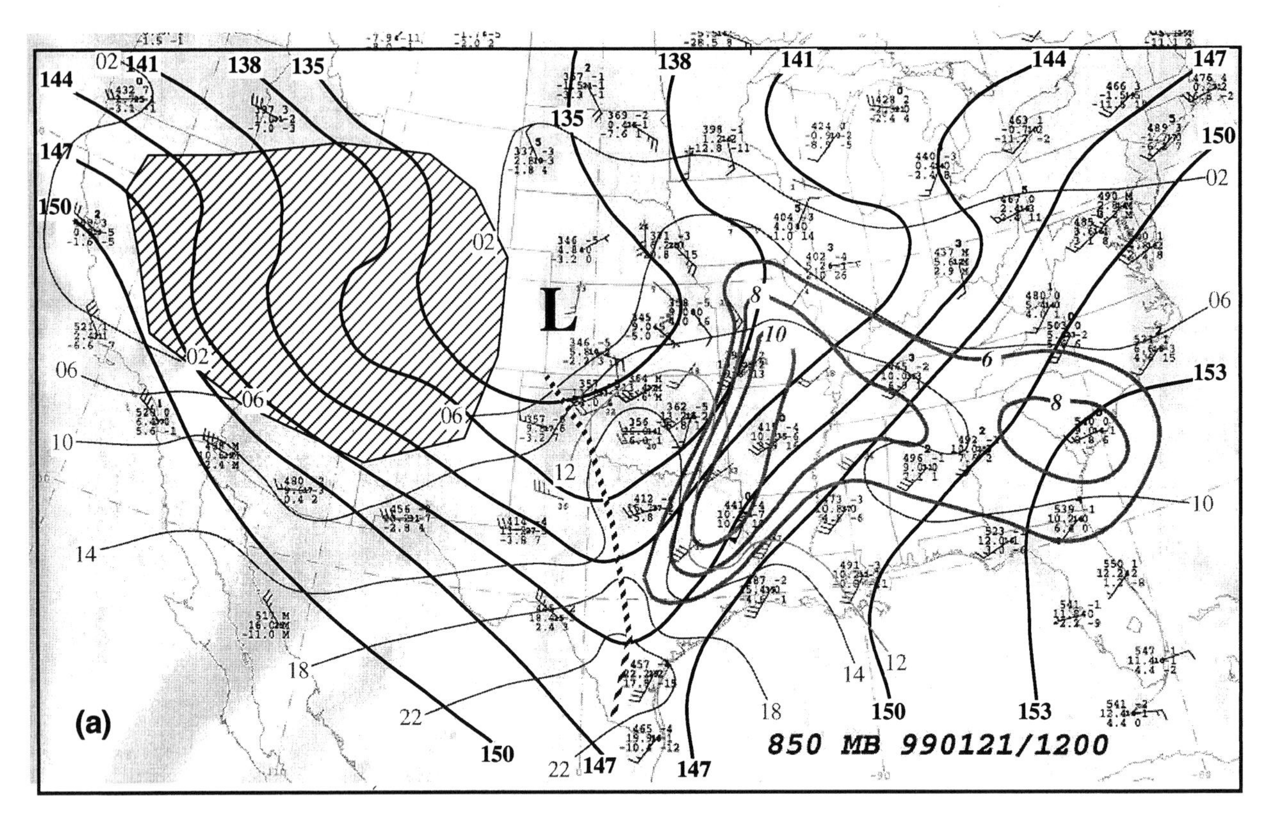

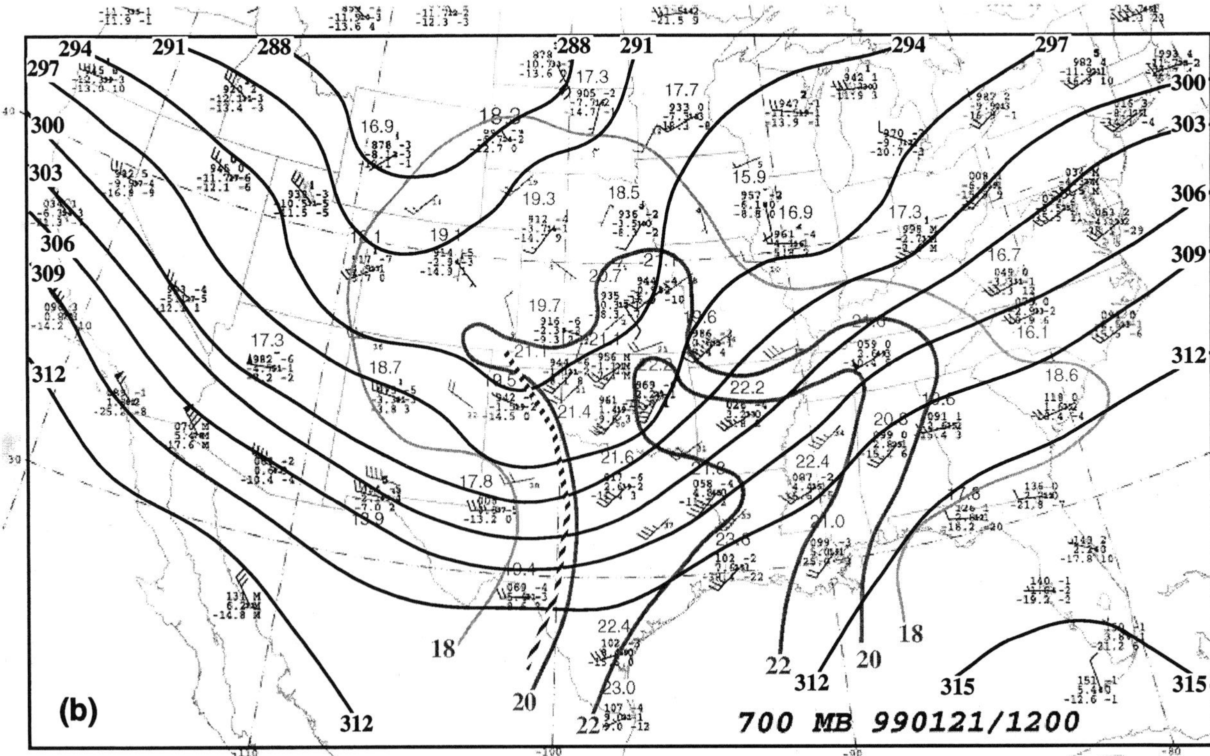

FIG. 2.31. Analyses at (a) 850, (b) 700, and (c) 500 hPa. At 850 hPa, thick solid lines depict isohypes (in dam), thin solid lines are isotherms (°C), and thick gray lines depict isodrosotherms (°C), with a trough line depicted by a thick hatched line; at 700 hPa, the convention is the same, except the thick gray lines depict isopleths of the temperature difference (°C) between 700 and 500 hPa; at 850 hPa, the convention is the same, except the thick gray lines are isopleths of geopotential height changes (m). The hatched area on the 850 hPa is excluded from any thermal analysis since the associated stations are generally below the earth's surface. The station model for all charts is conventional, except that the heights are plotted on the upper left side and dewpoint temperatures are used instead of dewpoint depressions; plotted numbers on the right side of the station model are 12-h changes.

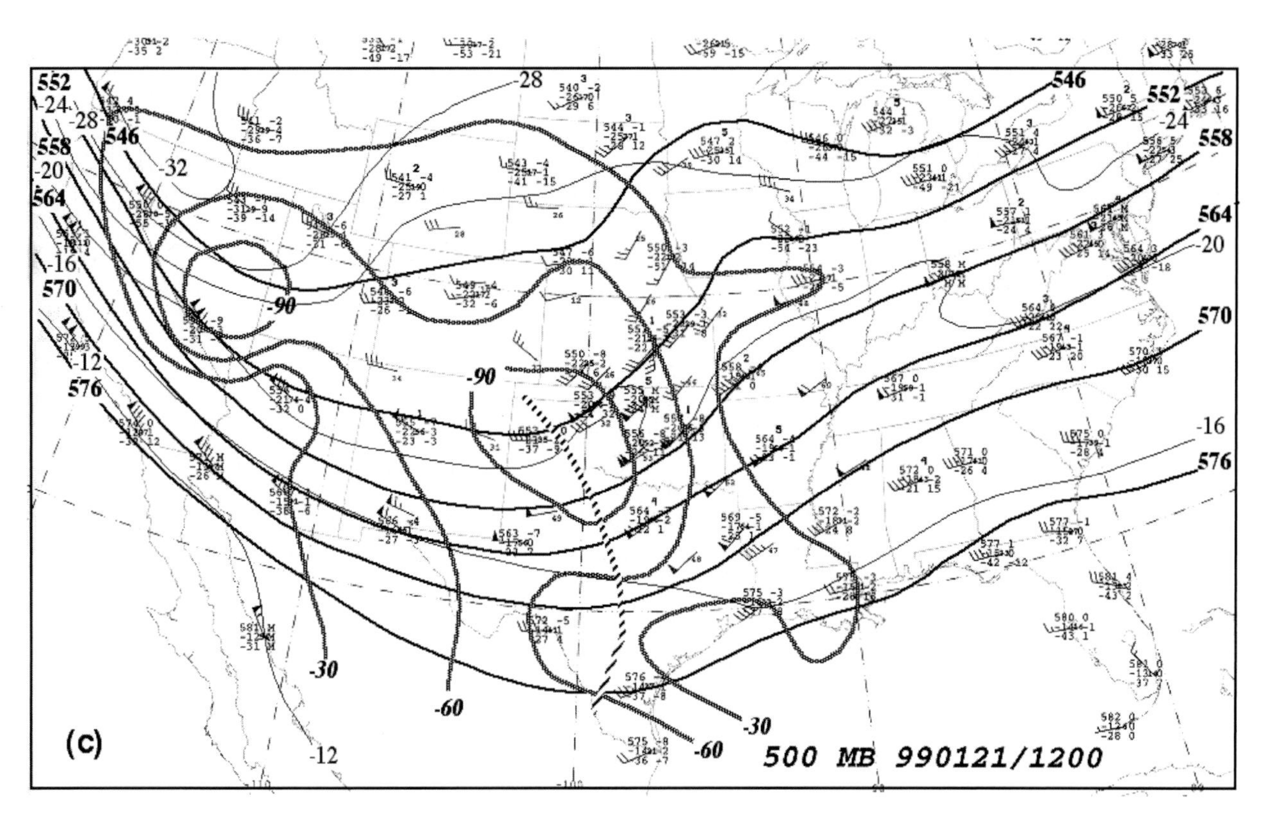

FIG. 2.31. (*Continued*)

intent is to provide a few key figures that illustrate the main points associated with each case. All but the last of our cases have been documented to a greater or lesser extent in publications, either in the refereed literature or in NOAA Disaster Surveys. Interested readers can consult the documentation for more details.

a. Tornadic systems

1) 28 AUG 1990—PLAINFIELD, IL, AN ISOLATED TORNADO EVENT

Johns (1982) has documented many aspects of severe convection associated with midlevel flows exhibiting an equatorward component, rather than the better-recognized situations involving flows having a poleward component in midtroposphere. This devastating tornado developed in such a case, striking Plainfield at about 1815 UTC, and lasted about 30 min; it was rated F5 on the Fujita scale. A survey of the event can be found in the National Weather Service (1991b) and was discussed by Fujita (1993) and Seimon (1993). As shown in Fig. 2.15, the event occurs on the margins of a synoptic-scale ridge in midtroposphere, just ahead of a weak short-wave trough. However, the region is characterized by extreme lapse rates in the lower midtroposphere, giving rise to very large CAPE (Fig. 2.16), with estimates ranging from nearly 6000 J kg^{-1} to around 8000 J kg^{-1}, depending on which parcel is chosen for ascent (Doswell and Brooks 1993a). CAPE values exceeding 3000 J kg^{-1} are uncommon, as documented by Doswell and Rasmussen (1994). This CAPE has clearly arisen because of convective inhibition, preventing the destruction of the high lapse rates and allowing them to be carried relatively far east of the west-central plains severe weather maximum.

The Plainfield event represents a challenge to our current concepts of environmental influences on tornadogenesis, in that the tornado was of violent intensity but occurred in what appears to be an environment with relatively weak wind shear and helicity (see Korotky et al. 1993; Doswell and Brooks 1993a). As noted by Brooks et al. (1994a), the actual storm environment in any given situation is always open to question, especially in cases that are not synoptically evident. Further, it is clear that we have so few examples of extreme CAPE-associated supercells that our understanding of convective storm dynamics is not adequate to explain every event.

2) 27 MAR 1994—PALM SUNDAY 1994 TORNADO OUTBREAK

This event was the subject of an entire session at the American Meteorological Society's 18th Severe Local Storms Conference (e.g., see Hales and Vescio 1996).

It has been characterized as not exhibiting the "classic" features of an outbreak; notably, a prominent midtropospheric cyclone was not observed. Nevertheless, the National Weather Service (1994) concludes that synoptic-scale processes were important in creating an environment favorable for the outbreak. During the time leading up to this event, the synoptic evolution created a widespread region with considerable CAPE and substantial vertical wind shear (Fig. 2.17). This evolution took place without involving strong surface cyclogenesis (Fig. 2.18). Within the broad region of favorable instability and wind shear, mesoscale processes played a large role in developing and organizing convection (see Kaplan et al. 1998; Langmaid and Riordan 1998). The presence of a jet stream with embedded subsynoptic-scale wind speed maxima certainly may have been an important factor in initiating the outbreak of tornadic storms via transverse circulations (Kaplan et al. 1998).

Two tornadic storms were responsible for most of the damage and casualties. One supercell developed out of thunderstorms that began in eastern Mississippi before 1500 UTC, and the storm tracked across Alabama, producing four tornadoes, dissipating as it moved into Georgia after 1800 UTC. The second supercell evolved from thunderstorms in west-central Alabama before 1500 UTC, remaining south of the other major supercell. This second storm also produced at least four long-track tornadoes as it moved across Alabama and on into Georgia, dissipating in northwestern South Carolina after 1800 UTC.

Although it is not uncommon for mesoscale processes to be important, outbreaks are associated classically with major cyclones (e.g., Fujita et al. 1970). The absence of such a prominent cyclone in this case makes it clear that the ingredients for a significant outbreak of severe weather can be assembled in many different synoptic-scale structures. The fact that it is typical for outbreaks to occur in association with cyclogenesis does not mean that the absence of cyclogenesis precludes major outbreaks of tornadic severe convection.

b. Nontornadic systems

1) 17 AUG 1994—LAHOMA, OK, AN ISOLATED INTENSE WIND AND HAIL STORM

This storm was also the topic for most of an entire (poster) session at the American Meteorological Society's 18th Severe Local Storms Conference. Janish et al. (1996) note that it was not well anticipated, perhaps owing to the relatively quiescent appearance of the synoptic environment (Figs. 2.19, 2.20). The most obvious feature in midtroposphere is the approaching thermal trough (Fig. 2.19), which is exaggerated by using an isotherm interval of 2°C. It was apparently produced by a relatively rare form of supercell storm first documented by Brooks and Doswell (1993) that is notable for the intensity and duration of strong convective outflows over a broad swath. Based on a limited sample, Brooks and Doswell conclude that these events are characterized by extreme CAPE, low midtropospheric relative humidity, and significant storm-relative helicity in the 0–3-km layer, factors shared with many other supercells. However, they surmise that the factor leading to the dominance of strong outflow is relatively weak midtropospheric storm-relative flow. In their conceptual model, this weakness in storm-relative flow suggests that midtropospheric mesocyclones would tend to be filled with precipitation, rather than having most of the precipitation carried downstream to fall in the storm's forward flank precipitation cascade. This case seems to fit the conceptual model, with high CAPE (not shown) and weak storm-relative windflow in midtroposphere (Fig. 2.21).[16] It also provides a compelling example of the potentially devastating character of such nontornadic severe convective wind events, in spite of a seemingly innocuous synoptic-scale setting.

This event, which became severe soon after 1900 UTC and ended at about 2100 UTC, produced a swath of large hail (up to 12 cm) and strong winds (gusts exceeding 50 m s^{-1} were reported) that was 6–12 km wide over a path about 90 km long. This represents an affected area approaching 1000 km^2, which means that, should such an event strike an urban area, the potential for substantial nontornadic damage exists. Apparently, events of this magnitude are relatively rare; over all of North America, Brooks and Doswell (1993) estimate an occurrence rate of perhaps a few times per year, at most.

2) 7 JUNE 1982—KANSAS CITY, MO, A DERECHO

Johns and Hirt (1987) were the first to document the widespread windstorm event now known as a *derecho*. This case, in which severe weather began (in northwestern Kansas) at about 0530 UTC on 7 June and continued until after 1400 UTC, has been presented in detail by Rockwood and Maddox (1988) as an example of the complex interactions that lead to convective events. It also illustrates the importance of a detailed analysis in some situations, since many features of significance (e.g., as illustrated in Fig. 2.22) are not easily detected in a smoothed analysis, typical of conventional synoptic-scale analyses done at a national center.

[16] The veering of the wind through a deep layer also suggests deep warm thermal advection, which is associated with ascent, quasi-geostrophically.

The synoptic setting cannot be described as benign, but neither can it be said to be clearly portentous of the event to come. The features of importance are not readily diagnosed using the operational objective analyses. Although the Q-vector divergence (Fig. 2.23) implied synoptic-scale descent in the area where storms first developed (in extreme southwestern Nebraska),[17] the thermodynamic conditions indicated considerable potential for strong storms, with CAPE of about 3000 J kg^{-1} (Fig. 2.24) if convection could be initiated. In this case, Rockwood and Maddox (1988) show that the initiation of convection required a complex sequence of mesoscale events that would be difficult to anticipate. Once initiated, however, an important convective event ensued; the widespread damage in Kansas City associated with the storm disrupted activities for several days, and there were numerous injuries.

c. Heavy precipitation systems

1) 14 JUN 1990—SHADYSIDE, OH, A CHALLENGE

As a measure of the importance of heavy rainfall as a severe convective event, this flash flood caused more than 25 fatalities (National Weather Service 1991a). Although the National Weather Service had issued a flash flood watch for the area at 1941 EDT (2341 UTC), no flash flood warning was issued for the event. It is typical of many flash flood events in that it occurred with the passage of a weak short-wave trough in midtroposphere (Fig. 2.25) on the margins of a large anticyclone. Very high surface moisture was evident on the poleward side of a nonfrontal convergence boundary (Fig. 2.26), which originated as an outflow from previous convection. Considerable rainfall preceded the Shadyside flash flood; indeed, the preceding May had well above average precipitation across the state of Ohio. This sort of situation is described reasonably as a "meso-high" event (using the conceptual model developed by Maddox et al. 1979), but it is certainly not a perfect fit to that pattern.

Rainfall began over the basins that flooded at about 2345 UTC and lasted for about 2 h. The convection was quasi-stationary over two small catchments with steep, rocky terrain (although there was no history of flash flooding in the catchments). The peak rainfalls were not measured by any rain gauge, but estimated peak amounts were in the range of 75–100 mm, most of which fell in about 1 h. Such peak rainfalls are by no means remarkable in themselves; the hydrological setting (steep terrain, antecedent rainfalls) clearly played a large role in this event's societal impact, as it often does. The available radar data (the WSR-88D radars were not operational for this case) in real time proved an unreliable tool for detecting the event and, in spite of favorable synoptic conditions, the magnitude of the event went undetected. In spite of having relatively low flash flood guidance values of about 50 mm in 3 h, there was no obvious indication from the available data that such an event was under way. Regrettably, it took about 4 h for information about the flooding at Shadyside to reach the National Weather Service office.

2) 9 JUNE 1972—RAPID CITY, SD, A WESTERN EVENT

The circumstances of a dominating midtropospheric anticyclone with a weak short-wave trough (Fig. 2.27) are found again for this event. Dennis et al. (1973), Maddox et al. (1978), and Doswell et al. (1996) have documented the event, which killed more than 230 people. In this case, the orography of the region seems to have been an important factor in focusing the convective development. At low levels, easterly flow behind a surface front (Fig. 2.28) was bringing high moisture values upslope, in the presence of strong lower midtropospheric lapse rates (Fig. 2.29a) at Rapid City. By the evening of the event, the arrival of the moist easterly flow at low levels and the increase in tropospheric humidity are quite evident (Fig. 2.29b). The arrival of this moisture, combined with orographic lifting and the approach of a slow-moving short-wave trough in midtroposphere, resulted in strong thunderstorms, beginning about 0000 UTC. The thunderstorms were moving relatively slowly, apparently owing to the weak winds throughout the middle and upper troposphere on this day, as well as to the importance of orographic lifting. The presence of a weak, negatively tilted midtropospheric ridge apparently favored the slow movement of the storms and the northward track of the weak short-wave trough. Peak 24-h rainfall amounts are estimated to be about 380 mm, most of which fell within about 4 h on the steep slopes of the Black Hills to the immediate west of Rapid City.

Maddox et al. (1978) specifically noted many similarities of the Rapid City case to another flash flood event, in the Big Thompson Canyon of Colorado four years later. The Big Thompson flash flood killed more than 140 people, thus having the dubious distinction of being the most recent (as of this writing) flash flood in the United States to produce more than 100 fatalities.

d. 21–22 January 1999—A cold-season severe weather case

On 21–22 January 1999, a substantial outbreak of severe thunderstorms and tornadoes (Fig. 2.30) devel-

[17] Note that inferring vertical motion from the rhs of the Omega equation [Eq. (2.2)] can be perilous. In this case, the pattern had considerable vertical coherence and is correspondingly likely to represent the vertical motion pattern reasonably well.

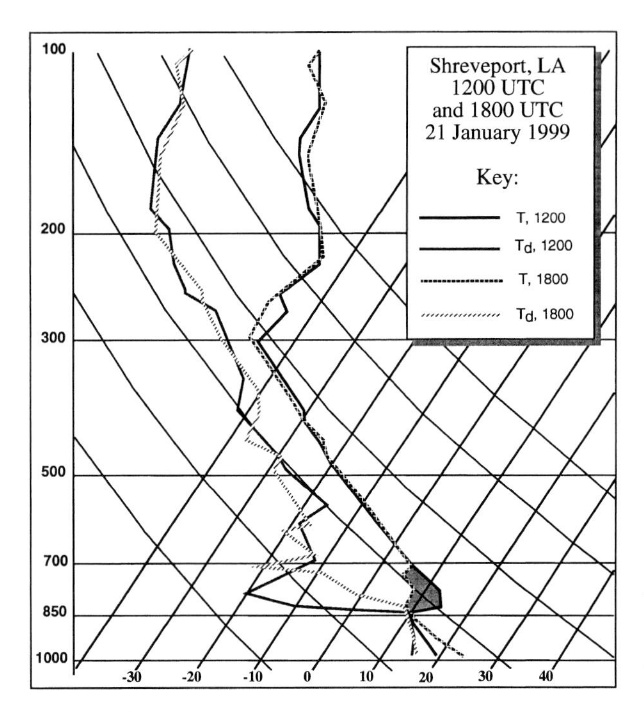

FIG. 2.32. Soundings for Shreveport, LA, at 1200 and 1800 UTC on 21 Jan 1999 plotted on a skew *T*–log*p* diagram. Shaded area depicts the temperature change in the 700 and 850 hPa from 1200 to 1800 UTC.

oped, including several tornadoes rated F3 on the Fujita scale. The first severe weather began about 2000 UTC on 21 January, diminishing after 0430 UTC on 22 January. What makes this event interesting is the season in which it occurred; tornado outbreaks in the depths of winter are certainly not common. Although vigorous extratropical cyclones (Fig. 2.31c) occur routinely during the winter, they do not typically include both relatively high midtropospheric lapse rates (Fig. 2.31b) and substantial low-level moisture (Fig. 2.31a). An indication of the effect of the synoptic-scale ascent can be seen in the 6-h evolution of the sounding at Shreveport, Louisiana (Fig. 2.32). Clearly, there was considerable erosion of what was originally a strong capping inversion, resulting in widespread severe convection.

The pattern shown in Fig. 2.31 includes a negatively tilted trough in the midtroposphere, with a strong jet stream upstream of the trough axis and diffluent contours ahead of the trough. This is by no means a necessary structure for severe weather outbreaks, but it may not be coincidental that this pattern is similar to that of 3–4 April 1974, the infamous "superoutbreak" described by Fujita (1974). Such a pattern favors ascent on the cyclonic shear side of the jet exit region and can put strong flow aloft over the warm sector of the developing surface cyclone, producing favorable wind shears. When the right lapse rate and moisture conditions can be created within a major winter storm, outbreaks of severe convection (see Galway and Pearson 1979; Kocin et al. 1995) can occur.

Acknowledgments. The authors are grateful for assistance by Dr. David Schultz in obtaining references and for many helpful discussions and comments. We also appreciate the many valuable suggestions from the review panel, chaired by Prof. Howie Bluestein. Numerous people provided assistance in developing figures, including Roger Edwards, Paul Janish, Dr. Louis Uccellini, Dr. Arlene Laing, and Dr. J. Michael Fritsch. Dr. Harold E. Brooks did the calculations for the severe weather climatology figures.

REFERENCES

Alpert, P., M. Tsidulko, S. Kricha, and U. Stein, 1996: A multi-stage evolution of an ALPEX cyclone. *Tellus,* **48A,** 209–220.

Asai, T., 1970: Stability of a plane parallel flow with a variable vertical shear and unstable stratification. *J. Meteor. Soc. Japan,* **48,** 129–138.

Atallah, E. H., and L. F. Bosart, 1996: Preferential storm pathways through western North America: A climatology and case studies. Preprints, *15th Conf. Weather Analysis and Forecasting,* Norfolk, VA, Amer. Meteor. Soc., 610–611.

Atkinson, B. W., 1981: *Meso-scale Atmospheric Circulations.* Academic Press, 495 pp.

Bannon, P. R., 1998: A comparison of Ekman pumping in approximate models of the accelerating planetary boundary layer. *J. Atmos. Sci.,* **55,** 1446–1451.

———, and T. L. Salem Jr., 1995: Aspects of the baroclinic boundary layer. *J. Atmos. Sci.,* **52,** 574–596.

Banta, R. M., 1990: The role of mountain flows in making clouds. *Atmospheric Processes over Complex Terrain, Meteor. Monogr.,* No. 45, Amer. Meteor. Soc., 229–283.

Barnes, S. L., 1985: Omega diagnostics as a supplement to LFM/MOS guidance in weakly forced convective situations. *Mon. Wea. Rev.,* **113,** 2122–2141.

———, and C. W. Newton, 1983: Thunderstorms in the synoptic setting. *Thunderstorm Morphology and Dynamics,* E. Kessler, Ed., University of Oklahoma, 75–112.

Bartels, D. L., and R. A. Maddox, 1991: Midlevel cyclonic vortices generated by mesoscale convective systems. *Mon. Wea. Rev.,* **119,** 104–118.

Beebe, R. G., and F. C. Bates, 1955: A mechanism for assisting in the release of convective instability. *Mon. Wea. Rev.,* **83,** 1–10.

Bell, G. P., and L. F. Bosart, 1988: Appalachian cold-air damming. *Mon. Wea. Rev.,* **116,** 137–161.

Benjamin, S. G., and T. N. Carlson, 1986: Some effects of surface heating and topography on the regional severe storm environment. Part I: Three-dimensional simulations. *Mon. Wea. Rev.,* **114,** 307–329.

Bjerknes, J., 1951: Extratropical cyclones. *Compendium of Meteorology,* T. F. Malone, Ed., Amer. Meteor. Soc., 577–598.

———, and J. Holmboe, 1944: On theory of cyclones. *J. Meteor.,* **1,** 1–22.

Blackadar, A. K., 1957: Boundary layer wind maxima and their significance for the growth of nocturnal inversions. *Bull. Amer. Meteor. Soc.,* **38,** 283–290.

Blanchard, D. O., and R. López, 1985: Spatial patterns of convection in south Florida. *Mon. Wea. Rev.,* **113,** 1282–1299.

Bluestein, H. B., 1986: Fronts and jet streaks: A theoretical perspective. *Mesoscale Meteorology and Forecasting,* P. S. Ray, Ed., Amer. Meteor. Soc., 173–215.

———, 1992: *Synoptic-Dynamic Meteorology in Midlatitudes.* Vol. I, *Kinematics and Dynamics.* Oxford University Press, 431 pp.

———, 1993: *Synoptic-Dynamic Meteorology in Midlatitudes.* Vol. II, *Observations and Theory of Weather Systems.* Oxford University Press, 594 pp.

———, and M. H. Jain, 1985: Formation of mesoscale lines of precipitation: Severe squall lines in Oklahoma during the spring. *J. Atmos. Sci.,* **42,** 1711–1732.

Bonner, W. D., 1966: Case study of thunderstorm activity in relation to the low level jet. *Mon. Wea. Rev.,* **94,** 167–178.

Bosart, L. F., 1981: The President's Day snowstorm of 18–19 February 1979: A subsynoptic-scale event. *Mon. Wea. Rev.,* **109,** 1542–1566.

———, 1999: Observed cyclone life cycles. *The Life Cycles of Extratropical Cyclones,* M. A. Shapiro and S. Grønås, Eds., Amer. Meteor. Soc., 187–213.

———, and F. Sanders, 1981: The Johnstown flood of July 1977: A long-lived convective system. *J. Atmos. Sci.,* **38,** 1616–1642.

———, and G. M. Lackmann, 1995: Postlandfall tropical cyclone reintensification in a weakly baroclinic environment: A case study of Hurricane David (September 1979). *Mon. Wea. Rev.,* **123,** 3268–3291.

———, G. J. Hakim, K. R. Tyle, M. A. Bedrick, W. E. Bracken, M. J. Dickinson, and D. M. Schultz, 1996: Large-scale antecedent conditions associated with the 12–14 March 1993 cyclone ("Superstorm") over eastern North America. *Mon. Wea. Rev.,* **124,** 1865–1891.

———, W. E. Bracken, and A. Seimon, 1999: Evolution of warm season continental anticyclones and their contribution to derecho environments. Preprints, *17th Conf. on Weather Analysis and Forecasting,* Denver, CO, Amer. Meteor. Soc., 100–103.

Boyle, J. S., and T.-J. Chen, 1987: Synoptic aspects of the wintertime East Asian monsoon. *Monsoon Meteorology,* C. P. Chang and T. N. Krishnamurti, Eds., Oxford University Press, 125–160.

Brier, G. W., and J. Simpson, 1969: Tropical cloudiness and rainfall related to pressure and tidal variations. *Quart. J. Roy. Meteor. Soc.,* **95,** 120–147.

Brooks, H. E., and C. A. Doswell III, 1993: Extreme winds in high precipitation supercells. Preprints, *17th Conf. on Severe Local Storms,* St. Louis, MO, Amer. Meteor. Soc., 173–177.

———, and D. J. Stensrud, 2000: Climatology of heavy rain events in the United States from hourly precipitation observations. *Mon. Wea. Rev.,* **128,** 1194–1201.

———, and R. B. Wilhelmson, 1993: Hodograph curvature and updraft intensity in numerically modeled supercells. *J. Atmos. Sci.,* **50,** 1824–1833.

———, C. A. Doswell III, and J. Cooper, 1994a: On the environments of tornadic and non-tornadic mesocyclones. *Wea. Forecasting,* **9,** 606–618.

———, ———, and R. B. Wilhelmson, 1994b: The role of midtropospheric winds in the evolution and maintenance of low-level mesocyclones. *Mon. Wea. Rev.,* **122,** 126–136.

Browning, K. A., 1964: Airflow and precipitation trajectories within severe local storms which travel to the right of the winds. *J. Atmos. Sci.,* **21,** 634–639.

———, and F. H. Ludlam, 1962: Radar analysis of a hailstorm. Tech. Note 5, Dept. Meteor., Imperial College, London.

———, and C. W. Pardoe, 1973: Structure of low-level jet streams ahead of mid-latitude cold fronts. *Quart. J. Roy. Meteor. Soc.,* **99,** 619–638.

Bunkers, M. J., B. A. Klimowski, J. W. Zeitler, R. L. Thompson, and M. L. Weisman, 2000: Predicting supercell motion using a new hodograph technique. *Wea. Forecasting,* **15,** 61–79.

Burgess, D. W., and L. R. Lemon, 1990: Severe thunderstorm detection by radar. *Radar in Meteorology,* D. Atlas, Ed., Amer. Meteor. Soc., 619–647.

Carlson, T. N., and F. H. Ludlam, 1968: Conditions for the occurrence of severe local storms. *Tellus,* **20,** 203–226.

Chang, C.-P., and T. N. Krishnamurti, Eds., 1987: *Monsoon Meteorology.* Oxford University Press, 544 pp.

Chappell, C. F., 1986: Quasi-stationary convective events. *Mesoscale Meteorology and Forecasting,* P. S. Ray, Ed., Amer. Meteor. Soc., 289–310.

Charney, J. G., 1962: Integration of the primitive and balance equations. *Proc. Int. Symp. on Numerical Weather Prediction,* Tokyo, Japan, Meteor. Soc. Japan, 131–151.

Chen, S.-J., Y.-H. Kuo, W. Wang, Z.-Y. Tao, and B. Cui, 1998: A modeling case study of heavy rainstorms along the Mei-Yu front. *Mon. Wea. Rev.,* **126,** 2330–2351.

Chen, S. S., and R. A. Houze Jr., 1997: Diurnal variation and life-cycle of deep convective systems over the tropical Pacific warm pool. *Quart. J. Roy. Meteor. Soc.,* **123,** 357–388.

Colby, F. P., Jr., 1984: Convective inhibition as a predictor of convection during AVE-SESAME II. *Mon. Wea. Rev.,* **112,** 2239–2252.

Colman, B. R., 1990a: Thunderstorms above frontal surfaces in environments without positive CAPE. Part I: A climatology. *Mon. Wea. Rev.,* **118,** 1103–1121.

———, 1990b: Thunderstorms above frontal surfaces in environments without positive CAPE. Part II: Organization and instability mechanisms. *Mon. Wea. Rev.,* **118,** 1123–1144.

Crisp, C. A., and J. M. Lewis, 1992: Return flow in the Gulf of Mexico. Part I: A classificatory approach with a global historical perspective. *J. Appl. Meteor.,* **31,** 868–881.

Cunningham, P., and D. Keyser, 1997: Analytical and numerical modeling of jet-streak dynamics. Preprints, *11th Conf. Atmospheric and Oceanic Fluid Dynamics,* Tacoma, WA, Amer. Meteor. Soc., 106–110.

Davies, J. M., and R. H. Johns, 1993: Some wind and instability parameters associated with strong and violent tornadoes. *The Tornado: Its Structure, Dynamics, Prediction, and Hazards, Geophys. Monogr.,* No. 79, Amer. Geophys. Union, 573–582.

Davies-Jones, R., 1984: Streamwise vorticity: The origins of updraft rotation in supercell storms. *J. Atmos. Sci.,* **41,** 2991–3006.

———, 1991: The frontogenetical forcing of secondary circulations. Part I: The duality and generalization of the Q vector. *J. Atmos. Sci.,* **48,** 497–509.

———, 1993: Helicity trends in tornado outbreaks. Preprints, *17th Conf. on Severe Local Storms,* St. Louis, MO, Amer. Meteor. Soc., 56–60.

———, 1998: Tornadoes and tornadic storms. Preprints, *19th Conf. Severe Local Storms,* Minneapolis, MN, Amer. Meteor. Soc., 185.

———, and H. Brooks, 1993: Mesocyclogenesis from a theoretical perspective. *The Tornado: Its Structure, Dynamics, Prediction, and Hazards. Geophys. Monogr.,* No. 79, Amer. Geophys. Union, 105–114.

———, D. Burgess, and M. Foster, 1990: Test of helicity as a tornado forecast parameter. Preprints, *16th Conf. Severe Local Storms,* Kananaskis Park, AB, Canada, Amer. Meteor. Soc., 588–592.

Davis, C. A., and K. Emanuel, 1991: Potential vorticity diagnostics of cyclogenesis. *Mon. Wea. Rev.,* **119,** 1929–1953.

———, and M. L. Weisman, 1994: Balanced dynamics of mesoscale vortices produced in simulated convective systems. *J. Atmos. Sci.,* **51,** 2005–2030.

Dennis, A. S., R. A. Schlensener, J. H. Hirsch, and A. Koscielski, 1973: Meteorology of the Black Hills flood of 1972. Inst. Atmos. Sci. Rep. 73–4, South Dakota School of Mines and Technology, 41 pp.

Dessens, J., and J. T. Snow, 1993: Comparative description of tornadoes in France and the United States. *The Tornado: Its Structure, Dynamics, Prediction, and Hazards, Geophys. Monogr.,* No. 79, Amer. Geophys. Union, 110–132.

Dickinson, M. J., L. F. Bosart, W. E. Bracken, G. J. Hakim, D. M. Schultz, M. A. Bedrick, and K. R. Tyle, 1997: The March 1993 superstorm cyclogenesis. Incipient phase synoptic- and convective-scale interaction and model performance. *Mon. Wea. Rev.,* **125,** 3041–3072.

Djuric, D., and D. S. Ladwig, 1983: Southerly low-level jet in the winter cyclones of the southwestern United States. *Mon. Wea. Rev.,* **111,** 2275–2281.

Doswell, C. A., III, 1980: Synoptic scale environments associated with High Plains severe thunderstorms. *Bull. Amer. Meteor. Soc.,* **60,** 1388–1400.

———, 1987: The distinction between large-scale and mesoscale contribution to severe convection: A case study example. *Wea. Forecasting,* **2,** 3–16.

———, 1993: Severe convective weather and associated disasters in North America. *Proc. Int. Workshop on Observations/Forecasting of Meso-scale Severe Weather and Technology of Reduction of Relevant Disasters,* Tokyo, Japan, Science and Technology Agency, 21–28.

———, and H. E. Brooks, 1993a: Comments on "Anomalous cloud-to-ground lightning in an F5 tornado-producing supercell thunderstorm on 28 August 1990." *Bull. Amer. Meteor. Soc.,* **74,** 2213–2218.

———, and ———, 1993b: Supercell thunderstorms. *Weather,* **48,** 209–210.

———, and E. N. Rasmussen, 1994: The effect of neglecting the virtual temperature correction on CAPE calculations. *Wea. Forecasting,* **9,** 625–629.

———, F. Caracena, and M. Magnano, 1985: Temporal evolution of 700–500 mb lapse rates as a forecasting tool—A case study. Preprints, *14th Conf. on Severe Local Storms,* Indianapolis, IN, Amer. Meteor. Soc., 398–401.

———, R. H. Johns, and S. J. Weiss, 1993: Tornado forecasting: A review. *The Tornado: Its Structure, Dynamics, Prediction, and Hazards. Geophys. Monogr.,* No. 79, Amer. Geophys. Union, 557–571.

———, C. Ramis, R. Romero, and S. Alonso, 1998: A diagnostic study of three heavy precipitation episodes in the western Mediterranean. *Wea. Forecasting,* **13,** 102–124.

Dotzek, N., R. Hannesen, and R. E. Peterson, 1998: Tornadoes in Germany, Austria, and Switzerland. Preprints, *19th Conf. on Severe Local Storms,* Minneapolis, MN, Amer. Meteor. Soc., 93–96.

Douglas, M. W., R. A. Maddox, K. W. Howard, and S. Reyes, 1993: The Mexican monsoon. *J. Climate,* **6,** 1665–1677.

Doyle, J. D., and T. T. Warner, 1993: A three-dimensional numerical investigation of a Carolina coastal low-level jet during GALE IOP 2. *Mon. Wea. Rev.,* **121,** 1030–1047.

Durran, D. R., and L. W. Snellman, 1987: The diagnosis of synoptic-scale vertical motion in an operational environment. *Wea. Forecasting,* **2,** 17–31.

Egger, J., 1988: Alpine lee cyclogenesis: Verification of theories. *J. Atmos. Sci.,* **42,** 2176–2186.

———, 1990: Thermally forced flows: Theory. *Atmospheric Processes over Complex Terrain, Meteor. Monogr.,* No. 45, Amer. Meteor. Soc., 43–58.

Eliassen, A., 1962: On the vertical circulation in frontal zones. *Geofys. Publ.,* **24,** 147–160.

Emanuel, K. A., 1986: Overview and definition of mesoscale meteorology. *Mesoscale Meteorology and Forecasting,* P. S. Ray, Ed., Amer. Meteor. Soc., 1–17.

———, 1994: *Atmospheric Convection.* Oxford University Press, 580 pp.

———, M. Fantini, and A. J. Thorpe, 1987: Baroclinic instability in an environment of small stability to slantwise moist convection. *J. Atmos. Sci.,* **43,** 585–604.

Faller, A. J., and R. E. Kaylor, 1966: A numerical study of the instability of the laminar Ekman boundary layer. *J. Atmos. Sci.,* **23,** 466–480.

Fantini, M., 1995: Moist Eady waves in a quasigeostrophic three-dimensional model. *J. Atmos. Sci.,* **52,** 2473–2485.

Fawbush, E. J., and R. C. Miller, 1954: The types of air masses in which North American tornadoes form. *Bull. Amer. Meteor. Soc.,* **35,** 154–165.

Frisby, E. M., and H. W. Sansom, 1967: Hail incidence in the Tropics. *J. Appl. Meteor.,* **6,** 339–354.

Fritsch, J. M., and R. A. Maddox, 1981: Convectively driven mesoscale weather systems aloft. Part I: Observations. *J. Appl. Meteor.,* **20,** 9–19.

———, R. J. Kane, and C. R. Chelius, 1986: The contribution of mesoscale convective weather systems to the warm-season precipitation in the United States. *J. Climate Appl. Meteor.,* **25,** 1333–1345.

———, J. D. Murphy, and J. S. Kain, 1994: Warm core vortex amplification over land. *J. Atmos. Sci.,* **51,** 1780–1807.

Fujita, T. T., 1971: Proposed characterization of tornadoes and hurricanes by area and intensity. SMRP Research Paper 91, University of Chicago, Chicago, IL, 42 pp.

———, 1973: Tornadoes around the world. *Weatherwise,* **26,** 56–62.

———, 1974: Jumbo tornado outbreak of 3 April 1974. *Weatherwise,* **27,** 116–119, 122–126.

———, 1993: Plainfield tornado of August 28, 1990. *The Tornado: Its Structure, Dynamics, Prediction, and Hazards, Geophys. Monogr.,* No. 79, Amer. Geophys. Union, 1–17.

———, D. L. Bradbury, and C. F. van Thullenar, 1970: Palm Sunday tornadoes of April 11, 1965. *Mon. Wea. Rev.,* **98,** 29–69.

Galway, J. G., 1977: Some climatological aspects of tornado outbreaks. *Mon. Wea. Rev.,* **105,** 477–484.

———, and A. D. Pearson, 1979: Winter tornado outbreaks. Preprints, *11th Conf. on Severe Local Storms,* Kansas City, MO, Amer. Meteor. Soc., 1–6.

Gates, W. L., 1961: Static stability measures in the atmosphere. *J. Meteor.,* **18,** 526–533.

Gayá, M., C. Ramis, R. Romero, and C. A. Doswell III, 1997: Tornadoes in the Balearic Islands (Spain): Meteorological setting. *Proc. INM/MO Int. Symp. on Cyclones and Hazardous Weather in the Mediterranean,* Palma de Mallorca, Spain, Universitat de les Isles Baleares, 525–534.

Gray, W., and R. W. Jacobson, 1977: Diurnal variation of deep cumulus convection. *Mon. Wea. Rev.,* **105,** 1171–1188.

Graziano, T. M., and T. N. Carlson, 1987: A statistical evaluation of lid strength on deep convection. *Wea. Forecasting,* **2,** 127–139.

Gutowski, W. J., Jr., and W. Jiang, 1998: Surface-flux regulation of the coupling between cumulus convection and baroclinic waves. *J. Atmos. Sci.,* **55,** 940–953.

Hakim, G. J., D. Keyser, and L. F. Bosart, 1996: The Ohio Valley wave-merger cyclogenesis event of 25–26 January 1978. Part II: Diagnosis using quasigeostrophic potential vorticity inversion. *Mon. Wea. Rev.,* **124,** 2176–2205.

Hales, J. E., and M. D. Vescio, 1996: The March 1994 tornado outbreak in the southeast U.S. The forecast process from an SPC perspective. Preprints, *18th Conf. on Severe Local Storms,* San Francisco, CA, Amer. Meteor. Soc., 32–36.

Hane, C. E., H. B. Bluestein, T. M. Crawford, M. E. Baldwin, and R. M. Rabin, 1997: Severe thunderstorm development in relation to along-dryline variability: A case study. *Mon. Wea. Rev.,* **125,** 231–251.

Harman, J. R., 1987: Mean monthly North American anticyclone frequencies, 1950–1979. *Mon. Wea. Rev.,* **115,** 2840–2848.

Higgins, R. W., Y.-P. Yao, E. S. Yarosh, J. E. Janowiak, and K. C. Mo, 1997: Influence of the Great Plains low-level jet on summertime precipitation and moisture transport over the central United States. *J. Climate,* **10,** 481–507.

Holton, J. R., 1967: The diurnal boundary layer wind oscillation above sloping terrain. *Tellus,* **19,** 199–205.

———, 1992: *An Introduction to Dynamic Meteorology.* 3d ed. Academic Press, 511 pp.

Hoskins, B. J., M. E. McIntyre, and A. W. Robertson, 1985: One the use and significance of isentropic potential vorticity maps. *Quart. J. Roy. Meteor. Soc.,* **111,** 877–946.

Hoxit, L. R., 1974: Planetary boundary layer winds in baroclinic conditions. *J. Atmos. Sci.,* **31,** 1003–1020.

Hsu, C.-P., and J. M. Wallace, 1976: The global distribution of annual and semiannual cycles in precipitation. *Mon. Wea. Rev.,* **104,** 1093–1101.

Huffman, G. J., and Coauthors, 1997: The Global Precipitation Climatology Project (GPCP) combined precipitation dataset. *Bull. Amer. Meteor. Soc.,* **78,** 5–20.

Huschke, R. E., 1959: *Glossary of Meteorology.* Amer. Meteor. Soc., 638 pp.

Jaeger, L., 1976: Monatskarten des Niedershlags für die ganze Erde. *Ber. Dtsch. Wetterdienstes,* **139,** 33 pp. plus plates.

Janish, P. R., R. H. Johns, and K. C. Crawford, 1996: An evaluation of the 17 August 1994—Lahoma, Oklahoma supercell/MCS event using conventional and non-conventional analyses and forecasting techniques. Preprints, *18th Conf. on Severe Local Storms,* San Francisco, CA, Amer. Meteor. Soc., 76–80.

Janowiak, J. E., P. A. Arkin, and M. Morrissey, 1994: An examination of the diurnal cycle in oceanic tropical rainfall using satellite and in situ data. *Mon. Wea. Rev.,* **122,** 2296–2311.

Jiang, H., and D. J. Raymond, 1995: Simulation of a mature mesoscale convective system using a nonlinear balance model. *J. Atmos. Sci.,* **52,** 161–175.

Johns, R. H., 1982: A synoptic climatology of northwest flow severe weather outbreaks. Part I: Nature and significance. *Mon. Wea. Rev.,* **110,** 1653–1663.

———, 1993: Meteorological conditions associated with bow echo development in convective storms. *Wea. Forecasting,* **8,** 294–299.

———, and W. D. Hirt, 1987: Derechos: Widespread convectively induced windstorms. *Wea. Forecasting,* **2,** 32–49.

———, and C. A. Doswell III, 1992: Severe local storms forecasting. *Wea. Forecasting,* **7,** 588–612.

Johnson, R. H., G. S. Young, J. J. Toth, and R. M. Zehr, 1984: Mesoscale weather effects of variable snow cover over northeast Colorado. *Mon. Wea. Rev.,* **112,** 1141–1152.

Jordan, C. L., 1958: Mean soundings for the West Indies area. *J. Meteor.,* **15,** 91–97.

Kaplan, M. L., Y.-L. Lin, D. W. Hamilton, and R. A. Rozumalski, 1998: The numerical simulation of an unbalanced jetlet and its role in the Palm Sunday 1994 tornado outbreak in Alabama and Georgia. *Mon. Wea. Rev.,* **126,** 2133–2165.

Keyser, D., 1986: Atmospheric fronts: An observational perspective. *Mesoscale Meteorology and Forecasting,* P. S. Ray, Ed., Amer. Meteor. Soc., 216–258.

Klemp, J. B., and R. B. Wilhelmson, 1978: Simulations of right- and left-moving storms produced through storm splitting. *J. Atmos. Sci.,* **38,** 1558–1580.

Kocin, P. J., P. N. Schumacher, R. F. Morales, and L. W. Uccellini, 1995: Overview of the 12–14 March 1993 superstorm. *Bull. Amer. Meteor. Soc.,* **76,** 165–182.

Korotky, W. R., R. W. Przybylinski, and J. A. Hart, 1993: The Plainfield Illinois tornado of 28 August 1990: The evolution of synoptic and mesoscale environments. *The Tornado: Its Structure, Dynamics, Prediction, and Hazards, Geophys. Monogr.,* No. 79, Amer. Geophys. Union, 611–624.

Laing, A. G., and J. M. Fritsch, 1997: The global population of mesoscale convective complexes. *Quart. J. Roy. Meteor. Soc.,* **123,** 389–405.

Langmaid, A. H., and A. J. Riordan, 1998: Surface mesoscale process during the 1994 Palm Sunday tornado outbreak. *Mon. Wea. Rev.,* **126,** 2117–2132.

Lanicci, J. M., and T. T. Warner, 1991: A synoptic climatology of the elevated mixed-layer inversion over the southern Great Plains in spring. Part II: The life cycle of the lid. *Wea. Forecasting,* **6,** 198–213.

Lau, K.-M.,and M.-T. Li, 1984: The monsoon of East Asia and its global associations—A survey. *Bull. Amer. Meteor. Soc.,* **65,** 114–125.

Legates, D. R., 1987: A climatology of global precipitation. *Publ. Climatol.,* **40** (1), 85 pp.

———, and C. J. Wilmott, 1990: Mean seasonal and spatial variability in gauge-corrected global precipitation. *Int. J. Climatol.,* **10,** 111–127.

Lewis, J. M., and C. A. Crisp, 1992: Return flow in the Gulf of Mexico. Part II: Variability of return-flow thermodynamics inferred from trajectories over the Gulf. *J. Appl. Meteor.,* **31,** 868–881.

Lilly, D. K., 1966: On the instability of Ekman boundary flow. *J. Atmos. Sci.,* **23,** 481–494.

———, 1986: The structure, energetics and propagation of rotating convective storms, II: Helicity and storm stabilization. *J. Atmos. Sci.,* **43,** 126–140.

López, R. E., and R. L. Holle, 1986: Diurnal and spatial variability of lightning activity in northeastern Colorado and central Florida during the summer. *Mon. Wea. Rev.,* **114,** 1288–1312.

Lorenz, E. N., 1965: *The Nature and Theory of the General Circulation of the Atmosphere.* World Meteor. Org., 161 pp.

Love, G., and A. Downey, 1986: The prediction of bushfires in central Australia. *Austr. Meteor. Mag.,* **34,** 93–101.

Lucas, C., E. J. Zipser, and M. A. LeMone, 1994: Convective available potential energy in the environment of oceanic and continental clouds: Correction and comments. *J. Atmos. Sci.,* **51,** 3829–3830.

Ludlam, F. H., 1963: Severe local storms: A review. *Severe Local Storms, Meteor. Monogr.,* No. 27, Amer. Meteor. Soc., 1–30.

———, 1980. *Clouds and Storms.* The Pennsylvania State University Press, 405 pp.

Maddox, R. A., 1980: Mesoscale convective complexes. *Bull. Amer. Meteor. Soc.,* **61,** 1374–1387.

———, 1984: Large-scale meteorological conditions associated with midlatitude, mesoscale convective complexes. *Mon. Wea. Rev.,* **111,** 1475–1493.

———, 1993: Diurnal low-level wind oscillations and storm-relative helicity. *The Tornado: Its Structure, Dynamics, Prediction, and Hazards, Geophys. Monogr.,* No. 79, Amer. Geophys. Union, 591–598.

———, and C. A. Doswell III, 1982: An examination of jet stream configurations, 500 mb vorticity advection and low-level thermal advection patterns during extended periods of intense convection. *Mon. Wea. Rev.,* **110,** 184–197.

———, L. R. Hoxit, C. F. Chappell, and F. Caracena, 1978: Comparison of meteorological aspects of the Big Thompson and Rapid City flash floods. *Mon. Wea. Rev.,* **106,** 375–389.

———, C. F. Chappell, and L. R. Hoxit, 1979: Synoptic and meso-α scale aspects of flash flood events. *Bull. Amer. Meteor. Soc.,* **60,** 115–123.

———, D. J. Perky, and J. M. Fritsch, 1981: Evolution of upper tropospheric features during the development of a mesoscale convective complex. *J. Atmos. Sci.,* **38,** 1664–1674.

Marwitz, J. D., 1972: The structure and motion of severe hailstorms. Part I: Supercell storms. *J. Appl. Meteor.,* **11,** 166–179.

Mattocks, C., and R. Bleck, 1986: Jet streak dynamics and geostrophic adjustment processes during the initial stages of lee cyclogenesis. *Mon. Wea. Rev.,* **114,** 2033–2056.

Means, L. L., 1944: The nocturnal maximum occurrence of thunderstorms in the Midwestern states. Misc. Rep. 16, University of Chicago, 38 pp.

———, 1952: On thunderstorms forecasting in the central United States. *Mon. Wea. Rev.,* **80,** 165–189.

Miller, R. C., 1972: Notes on analysis and severe-storm forecasting procedures of the Air Force Global Weather Central. Air Weather Service Tech. Rep. 200 (Rev.), Air Weather Service, Scott Air Force Base, IL, 190 pp.

Moller, A. R., C. A. Doswell III, M. P. Foster, and G. R. Woodall, 1994: The operational recognition of supercell thunderstorm environments and storm structures. *Wea. Forecasting,* **9,** 327–347.

Montgomery, M. T., and B. Farrell, 1993: Tropical cyclone formation. *J. Atmos. Sci.,* **50,** 285–310.

Morgan, M. C., and J. W. Neilsen-Gammon, 1998: Using tropopause maps to diagnose midlatitude weather systems. *Mon. Wea. Rev.,* **126,** 2555–2579.

Nagata, M., and Y. Ogura, 1991: A modeling case study of interaction between heavy precipitation and a low-level jet over Japan in the Baiu season. *Mon. Wea. Rev.,* **119,** 1309–1336.

National Weather Service, 1991a: Shadyside, Ohio, flash floods: June 14, 1990. U.S. Dept. of Commerce, 43 pp. [Available from NOAA/National Weather Service, Warming and Forecast Branch, 1325 East-West Highway, Silver Spring, MD 20910.]

———, 1991b: The Plainfield/Crest Hill Tornado. Northern Illinois, August 28, 1990. U.S. Dept. of Commerce, 43 pp. [Available from NOAA/National Weather Service, Warning and Forecast Branch, 1325 East-West Highway, Silver Spring, MD 20910.]

———, 1994: Southeastern United States Palm Sunday Tornado Outbreak of March 27, 1994. U.S. Dept. of Commerce, 54 pp. [Available from NOAA/National Weather Service, Warning and Forecast Branch, 1325 East-West Highway, Silver Spring, MD 20910.]

Neiman, P. J., and M. A. Shapiro, 1993: The life cycle of an extratropical marine cyclone. Part I: Frontal-cyclone evolution and thermodynamic air-sea interaction. *Mon. Wea. Rev.,* **121,** 2153–2176.

Nelson, S. P., and S. K. Young, 1979: Characteristics of Oklahoma hailfalls and hailstorms. *J. Appl. Meteor.,* **18,** 339–347.

Newton, C. W., 1959: Axial velocity streaks in the jet stream: Ageostrophic "inertial" oscillations. *J. Meteor.,* **16,** 638–645.

———, 1963: Dynamics of severe convective storms. *Severe Local Storms, Meteor. Monogr.,* No. 27, Amer. Meteor. Soc., 33–58.

———, 1981: Lagrangian partial-inertial oscillations and subtropical and low-level monsoon jet streaks. *Mon. Wea. Rev.,* **109,** 2474–2486.

———, and A. V. Persson, 1962: Structural characteristics of the subtropical jet stream and certain lower stratospheric wind systems. *Tellus,* **14,** 221–241.

Ninomiya, K., 1971: Mesoscale modification of synoptic situations from thunderstorm development as revealed by ATS III and aerological data. *J. Appl. Meteor.,* **10,** 1103–1121.

Ookouchi, Y., M. Segal, R. C. Kessler, and R. A. Pielke, 1984: Evaluation of soil moisture effects on the generation and modification of mesoscale circulations. *Mon. Wea. Rev.,* **112,** 2281–2292.

Orlanski, I., 1975: A rational subdivision of scales for atmospheric processes. *Bull. Amer. Meteor. Soc.,* **56,** 527–530.

Orville, R. E., and R. W. Henderson, 1986: Global distribution of midnight lightning: September 1977 to August 1978. *Mon. Wea. Rev.,* **114,** 2640–2653.

Palmén, E., and E. O. Holopainen, 1962: Divergence, vertical velocity and conversion between potential and kinetic energy in an extratropical disturbance. *Geophysica,* **8,** 89–113.

———, and C. W. Newton, 1969: *Atmospheric Circulation Systems.* Academic Press, 603 pp.

Panofsky, H. A., 1964: *Introduction to Dynamic Meteorology.* The Pennsylvania State University, 243 pp.

Peterson, R. E., 1998: Tornadoes in Sweden. Preprints, *19th Conf. on Severe Local Storms,* Minneapolis, MN, Amer. Meteor. Soc., 89–92.

Petterssen, S., and P. A. Calabrese, 1959: On some weather influences due to warming of the air by the Great Lakes in winter. *J. Meteor.,* **16,** 646–652.

Pielke, R. A., 1974: A three-dimensional numerical model of the sea breezes over south Florida. *Mon. Wea. Rev.,* **102,** 115–139.

———, and M. Segal, 1986: Mesoscale circulations forced by differential terrain heating. *Mesoscale Meteorology and Forecasting,* P. S. Ray, Ed., Amer. Meteor. Soc., 516–548.

Phillips, N. A., 1990: The emergence of quasi-geostrophic theory. *The Atmosphere—A Challenge: The Science of Jule Gregory Charney,* R. Lindzen et al., Eds., Amer. Meteor. Soc., 177–206.

Pitchford, K., and J. London, 1962: The low-level jet as related to nocturnal thunderstorms over Midwest United States. *J. Appl. Meteor.,* **1,** 43–47.

Ramage, C. S., 1975: *Monsoon Meteorology.* Academic Press, 296 pp.

Ramis, C., J. Arús, J. M. López, and A. Mestre, 1995: Two cases of severe weather in Catalonia (Spain). An observational study. *Meteor. Appl.,* **2,** 207–217.

———, J. M. López, and J. Arús, 1997: Two cases of severe weather in Catalonia (Spain). An observational study. *Meteor. Appl.,* **6,** 11–28.

Rasmussen, E. N., and D. O. Blanchard, 1998: A baseline climatology of sounding-derived supercell and tornado forecast parameters. *Wea. Forecasting,* **13,** 1148–1164.

Raymond, D. J., 1992: Nonlinear balance, and potential-vorticity thinking at large Rossby number. *Quart. J. Roy. Meteor. Soc.,* **118,** 987–1015.

———, and H. Jiang, 1990: A theory for long-lived mesoscale convective system. *J. Atmos. Sci.,* **47,** 3067–3077.

Reed, R. J., and M. D. Albright, 1986: A case study of explosive cyclogenesis in the eastern Pacific. *Mon. Wea. Rev.,* **114,** 2297–2319.

Reiter, E. R., 1961: *Jet Stream Meteorology.* University of Chicago, 515 pp.

Riehl, H., and Coauthors, 1952: *Forecasting in Middle Latitudes. Meteor. Monogr.,* No. 5, Amer. Meteor. Soc., 1–80.

Riley, G. T., M. G. Landin, and L. F. Bosart, 1987: The diurnal variability of precipitation across the Central Rockies and adjacent Great Plains. *Mon. Wea. Rev.,* **115,** 1161–1172.

Rockwood, A. A., and R. A. Maddox, 1988: Mesoscale and synoptic scale interactions leading to intense convection: The case of 7 June 1982. *Wea. Forecasting,* **3,** 51–68.

Roebber, P. J., 1984: Statistical analysis and updated climatology of explosive cyclones. *Mon. Wea. Rev.,* **112,** 1577–1589.

———, 1993: A diagnostic case study of self-development as an antecedent conditioning process in explosive cyclogenesis. *Mon. Wea. Rev.,* **121,** 976–1006.

Rotunno, R., 1993: Supercell thunderstorm modeling and theory. *The Tornado: Its Structure, Dynamics, Prediction, and Hazards, Geophys. Monogr.,* No. 79, Amer. Geophys. Union, 57–73.

———, and J. B. Klemp, 1982: The influence of the shear-induced pressure gradient on thunderstorm motion. *Mon. Wea. Rev.,* **110,** 136–151.

Sammler, W. R., 1993: An updated climatology of large hail based on 1970–1990 data. Preprints, *17th Conf. on Severe Local Storms,* St. Louis, MO, Amer. Meteor. Soc., 32–35.

Schaefer, J. T., 1986: The dryline. *Mesoscale Meteorology and Forecasting,* P. S. Ray, Ed., Amer. Meteor. Soc., 549–572.

Schmid, W., H.-H., Schiesser, and B. Bauer-Messmer, 1997: Supercell storms in Switzerland: Case studies and implications for nowcasting. *Meteor. Appl.,* **4,** 49–68.

Schultz, D. M., and P. N. Schumacher, 1999: The use and misuse of conditional symmetric instability. *Mon. Wea. Rev.,* **127,** 2709–2732.

Segal, M., and R. W. Arritt, 1992: Nonclassical mesoscale circulations caused by surface sensible heat-flux gradients. *Bull. Amer. Meteor. Soc.,* **73,** 1593–1604.

Seimon, A., 1993: Anomalous cloud-to-ground lightning in an F5-tornado-producing supercell thunderstorm on 28 August 1990. *Bull. Amer. Meteor. Soc.,* **74,** 189–203.

Shapiro, 1981: Frontogenesis and geostrophically forced secondary circulations in the vicinity of jet stream-frontal zone systems. *J. Atmos. Sci.,* **38,** 954–973.

———, 1982: Mesoscale weather systems of the central United States. CIRES, University of Colorado/NOAA, 78 pp.

Simpson, J., R. F. Adler, and G. R. North, 1988: A proposed Tropical Rainfall Measuring Mission (TRMM) satellite. *Bull. Amer. Meteor. Soc.,* **69,** 278–295.

Smull, B. F., 1995: Convectively induced mesoscale weather systems in the tropical and warm-season midlatitude atmosphere. *Rev. Geophys.,* **33** (Suppl.), 897–906.

Snow, J. T., and A. L. Wyatt, 1997: The tornado, nature's most violent wind: Part I—World-wide occurrence and categorisation. *Weather,* **52,** 298–304.

Sousounis, P. J., and J. M. Fritsch, 1994: Lake-aggregated mesoscale disturbances. Part II: A case study of the effect on regional and synoptic-scale weather systems. *Bull. Amer. Meteor. Soc.,* **75,** 1793–1811.

Stensrud, D. J., 1993: Elevated residual layers and their influence on surface boundary-layer evolution. *J. Atmos. Sci.,* **50,** 2284–2293.

———, 1996a: Effects of persistent, midlatitude mesoscale regions of convection on the large-scale environment during the warm season. *J. Atmos. Sci.,* **53,** 3503–3527.

———, 1996b: Importance of low-level jets to climate: A review. *J. Climate,* **9,** 1698–1711.

Stull, R. B., 1988: *An Introduction to Boundary Layer Meteorology.* Kluwer Academic Publishers, 666 pp.

Sun, W.-Y., and Y. Ogura, 1979: Boundary-layer forcing as a possible trigger to a squall-line formation. *J. Atmos. Sci.,* **36,** 235–254.

Sutcliffe, R. C., 1938: On development in the field of barometric pressure. *Quart. J. Roy. Meteor. Soc.,* **64,** 495–509.

———, 1939: Cyclonic and anticyclonic development. *Quart. J. Roy. Meteor. Soc.,* **65,** 518–524.

———, and A. G. Forsdyke, 1950: The theory and use of upper air thickness patterns in forecasting. *Quart. J. Roy. Meteor. Soc.,* **76,** 189–217.

Thompson, R. L., J. M. Lewis, and R. A. Maddox, 1994: Autumnal return of tropical air to the Gulf of Mexico's coastal plain. *Wea. Forecasting,* **9,** 348–360.

Thorpe, A. J., and T. H. Guymer, 1977: The nocturnal jet. *Quart. J. Roy. Meteor. Soc.,* **103,** 633–653.

Tibaldi, S., A. Buzzi, and A. Speranza, 1990: Orographic cyclogenesis. *Extratropical Cyclones,* C. Newton and E. O. Holopainen, Eds., Amer. Meteor. Soc., 107–127.

Tracton, M. S., 1973: The role of cumulus convection in the development of extratropical cyclones. *Mon. Wea. Rev.,* **101,** 573–592.

Trier, S. B., and D. B. Parsons, 1993: Evolution of environmental conditions preceding the development of a nocturnal mesoscale convective complex. *Mon. Wea. Rev.,* **121,** 1078–1098.

Tudurí, E., and C. Ramis, 1997: On the environments of significant convective events in the western Mediterranean. *Wea. Forecasting,* **12,** 294–306.

Uccellini, L. W., 1980: On the role of upper tropospheric jet streaks and leeside cyclogenesis in the development of low-level jets in the Great Plains. *Mon. Wea. Rev.,* **108,** 1689–1696.

———, 1990: Processes contributing to the rapid development of extratropical cyclones. *Extratropical Cyclones,* C. Newton and E. O. Holopainen, Eds., Amer. Meteor. Soc., 81–105.

———, and D. R. Johnson, 1979: The coupling of upper and low tropospheric jet streaks and implications for the development of severe convective storms. *Mon. Wea. Rev.,* **107,** 682–703.

———, and P. J. Kocin, 1987: The interaction of jet streak circulations during heavy snow events along the East Coast of the United States. *Wea. Forecasting,* **2,** 289–308.

———, ———, R. A. Petersen, C. H. Wash, and K. F. Brill, 1984: The President's Day cyclone of 18–19 February 1979: Synoptic overview and analysis of the subtropical jet streak influencing the pre-cyclogenetic period. *Mon. Wea. Rev.,* **112,** 31–55.

Vincent, D. G., 1994: The South Pacific Convergence Zone (SPCZ): A review. *Mon. Wea. Rev.,* **122,** 1949–1970.

Wallace, J. M., 1975: Diurnal variations in precipitation and thunderstorm frequency over the conterminous United States. *Mon. Wea. Rev.,* **103,** 406–419.

Webster, P. J., V. O. Magaña, T. N. Palmer, J. Shukla, R. A. Tomas, M. Yanai, and T. Yasunari, 1998: Monsoons: Processes, predictability, and the prospects for prediction. *J. Geophys. Res.,* **103,** 14 541–14 510.

Weisman, M. L., and J. B. Klemp, 1982: The dependence of numerically simulated convective storms on vertical wind shear and buoyancy. *Mon. Wea. Rev.,* **110,** 504–520.

———, and ———, 1984: The structure and classification of numerically simulated convective storms in directionally varying wind shears. *Mon. Wea. Rev.,* **112,** 2479–2498.

———, and ———, 1986: Characteristics of isolated convective storms. *Mesoscale Meteorology and Forecasting,* P. S. Ray, Ed., Amer. Meteor. Soc., 331–358.

Whitaker, J. S., and A. Barcilon, 1992: Type B cyclogenesis in a zonally varying flow. *J. Atmos. Sci.,* **49,** 1877–1892.

———, L. W. Uccellini, and K. F. Brill, 1988: A model-based diagnostic study of the rapid development phase of the President's Day cyclone. *Mon. Wea. Rev.,* **116,** 2337–2365.

Whiteman, C. D., 1990: Observations of thermally developed wind systems in mountainous terrain. *Atmospheric Processes over Complex Terrain, Meteor. Monogr.,* No. 45, 5–42.

Williams, A. A. J., and D. J. Karoly, 1999: Extreme fire weather in Australia and the impact of the El Niño-Southern Oscillation. *Austr. Meteor. Mag.,* **48,** 15–22.

Williams, E. R., and N. Renno, 1993: An analysis of the conditional instability of the tropical atmosphere. *Mon. Wea. Rev.,* **121,** 21–36.

Williams, R. T., and J. Plotkin, 1968: Quasi-geostrophic frontogenesis. *J. Atmos. Sci.,* **25,** 201–206.

Wilson, J. W., and W. E. Schreiber, 1986: Initiation of convective storms at radar-observed boundary-layer convergence lines. *Mon. Wea. Rev.,* **114,** 2516–2536.

Xu, K., and K. A. Emanuel, 1989: Is the tropical atmosphere conditionally unstable? *Mon. Wea. Rev.,* **117,** 1471–1479.

Xu, Z., P. Wang, and X. Lin, 1993: Tornadoes of China. *The Tornado: Its Structure, Dynamics, Prediction, and Hazards, Geophys. Monogr.,* No. 79, Amer. Geophys. Union, 435–444.

Ziegler, C. L., and C. E. Hane, 1993: An observational study of the dryline. *Mon. Wea. Rev.,* **121,** 1134–1151.

———, and E. N. Rasmussen, 1998: The initiation of moist convection at the dryline: Forecasting issues from a case study perspective. *Wea. Forecasting,* **13,** 1106–1131.

Zipser, E. J., 1994: Deep cumulonimbus cloud systems in the Tropics with and without lightning. *Mon. Wea. Rev.,* **122,** 1837–1851.

Chapter 3

Mesoscale Processes and Severe Convective Weather

RICHARD H. JOHNSON

Department of Atmospheric Science, Colorado State University, Fort Collins, Colorado

BRIAN E. MAPES

CIRES/CDC, University of Colorado, Boulder, Colorado

REVIEW PANEL: David B. Parsons (Chair), K. Emanuel, J. M. Fritsch, M. Weisman, D.-L. Zhang

3.1. Introduction

Severe convective weather events—tornadoes, hailstorms, high winds, flash floods—are inherently mesoscale phenomena. While the large-scale flow establishes environmental conditions favorable for severe weather, processes on the mesoscale initiate such storms, affect their evolution, and influence their environment. A rich variety of mesocale processes are involved in severe weather, ranging from environmental preconditioning to storm initiation to feedback of convection on the environment. In the space available, it is not possible to treat all of these processes in detail. Rather, we will introduce several general classifications of mesoscale processes relating to severe weather and give illustrative examples. Although processes on the mesoscale are often intimately linked with those on smaller and larger scales, we will exclude from discussion those that obviously lie outside the mesoscale domain (e.g., baroclinic waves on the synoptic scale or charge separation in clouds on the microscale).

a. Definition of mesoscale

There are several definitions of "mesoscale," and even "scale," in common currency. Some use fixed geometrical scales (Fujita 1963, 1981; Ogura 1963; Orlanski 1975) while others are based on dynamical considerations (Ooyama 1982; Emanuel 1986; Doswell 1987).

Ooyama (1982) defines mesoscale flows as those having a horizontal scale between the scale height H of the atmosphere and the Rossby radius of deformation $\lambda_R \approx NH/f$, where N is the Brunt–Väisälä frequency and f the Coriolis parameter.[1] By this definition, mesoscale phenomena occur on horizontal scales between ten and several hundred kilometers. This range generally encompasses motions for which both ageostrophic advections and Coriolis effects are important (Emanuel 1986). In general, we apply such a definition here; however, strict application is difficult since so many mesoscale phenomena are "multiscale." For example, a ~100-km-long gust front can be less than ~1 km across. The triggering of a storm by the collision of gust fronts can actually occur on a ~100-m scale (the microscale). Nevertheless, we will treat this overall process (and others similar to it) as mesoscale since gust fronts are generally regarded as mesoscale phenomena.

b. Scope of paper

The range of mesoscale processes associated with severe weather is enormous. Therefore, to provide focus, we present a division of mesoscale processes according to whether they help to generate severe weather (termed *preconditioning* and *triggering*) or arise from the convection itself. Moreover, we will draw a distinction between preconditioning and triggering, much in the same way as Newton (1963) has done. Newton distinguished factors that precondition (destabilize) the environment, for example, an approaching upper-level trough, differential horizontal advection, low-level jets, from those that release the instability, such as rapid lifting by fronts, cold domes from thunderstorms, drylines, and topography (although slow quasigeostrophic lifting was also suggested as a possibility).

A list of common mesoscale preconditioning processes is provided in Table 3.1. In most instances, these mechanisms serve to gradually destabilize the environment and modify the wind shear profile, thereby setting the stage for severe weather. However,

[1] Ooyama (1982) notes that if the relative rotation and vorticity are increased in an area, f should be replaced by the geometric mean of the absolute vorticity and absolute angular speed.

TABLE 3.1. Mesoscale preconditioning processes for severe weather.

Local	Advective	Dynamical
Boundary layer processes • deepening the mixed layer • deepening the moist layer • convergence along dryline • nocturnal inversion, low-level jet formation Terrain effects • creation of convergence zones • development of slope flows • modification of hodograph Surface effects • evaporation, heating • surface, discontinuities —soil moisture —roughness	Differential advection • creation of capping inversion • destabilization • formation of deep, dry PBL (leading to microbursts) Convergence lines • fronts • drylines • sea/land/lake breezes • mountain/valley breezes Moisture advection • increase CAPE, lower LFC • local cumulus moistening	Secondary circulations • geostrophic adjustment • jets Gravity currents, waves • cold pool lifting • localized reduction of CIN • modification of vertical shear Mesoscale instabilities Boundary layer processes • horizontal convective rolls • inertial oscillation (low-level jets)

if the destabilization occurs rapidly enough, some of these processes may actually trigger convection, thus blurring the distinction between preconditioning and triggering.

The mesoscale processes in Table 3.1 have been subdivided into local, advective, and dynamical. Local preconditioning processes include boundary layer mixing and interactions of the atmosphere with geographically fixed features such as topography or gradients of surface properties. Advective processes involve the physical transport of air masses. Advection acts as an important preconditioning process in the prestorm environment (e.g., differential advection of cold air over warm, or the development and convergence of humid air masses). Mesoscale dynamical processes are harder to observe, as they often involve rapidly evolving motions in clear regions of the atmosphere. Events in one location can affect events in another location through the propagation of gravity waves that may travel faster than the wind at any level. Much of the unsaturated fluid dynamics of the atmosphere for which horizontal advection and rotation processes are secondary can be described as gravity wave processes, not just the rare cases of coherent propagating phenomena with a single, well-defined frequency and wavelength. These processes are important both in prestorm environments and inside severe storms. In the former situation, they can include secondary circulations associated with geostrophic adjustment, upper-level jets, and low-level jets. In addition, the atmosphere can be subject to mesoscale dynamical instabilities that may cause convective preconditioning. In this case, there is no "upstream precursor" for a weather development, as there is for the advective phenomena discussed above. Discussion of individual phenomena listed in Table 3.1 is given in section 3.3.

Specific processes involved in triggering convection are identified in Table 3.2. As mentioned earlier, some triggering and preconditioning processes are the same. For example, some cold fronts trigger convection everywhere along their leading edge, whereas others precondition the atmosphere by providing mesoscale lifting and moisture convergence. However, in general, the lifting required for triggering is much stronger than that for preconditioning, particularly

TABLE 3.2. Mesoscale triggering processes for severe weather.

Local	Advective	Dynamical
Boundary layer circulations • thermals Terrain effects • orographic lifting • thermal forcing • obstacle effects Surface effects • sensible/latent heat flux discontinuities	Convergence lines • cold fronts • gust fronts • sea/lake breezes • drylines Boundary intersections • triple point • colliding fronts, sea breezes	Gravity currents, waves Boundary layer horizontal convective rolls

Combined lifting processes:
thermals along fronts
fronts intersecting terrain
boundary layer rolls intersecting fronts, sea breezes, drylines
gravity waves interacting with fronts, drylines

when convective inhibition (CIN) is present. Probably the most common triggering mechanisms are advective in nature: convergence lines (gust fronts, cold fronts, sea/lake breezes, drylines) or boundary intersections (e.g., triple points, colliding gust fronts). Mesoscale dynamical processes are less common, but there are some instances where gravity waves or boundary layer rolls trigger convection. Superpositions of triggering processes are particularly effective at initiating storms (e.g., thermals along fronts, fronts intersecting terrain, rolls intersecting boundaries, etc.). These processes will be discussed in detail in section 3.4.

Once initiated, severe storms generate mesoscale phenomena that impact storm evolution as well as the growth of neighboring storms (Table 3.3). On the local scale, radiation and microphysics are two such processes. Microphysical effects acting on the mesoscale are key in downdraft and cold pool production, generation of microbursts and other high-wind events, generation of midlevel convergence due to melting, and lightning production.

Advective effects include particle advection, fall, and phase changes, which influence downdraft development and upscale growth of convection. Cold pool advective processes lead to cell regeneration and mesoscale convective system (MCS) evolution. Momentum transport and sloping flows are important factors in severe surface winds. Vortex tilting/stretching can generate vertical vorticity, leading to supercell development or MCS mesovortices.

Severe storms have a number of important dynamical effects. Convectively generated gravity waves influence storm evolution and the development of neighboring storms. Mesoscale pressure fields produced by buoyancy and dynamic effects (e.g., shear/updraft interactions) impact supercell evolution and propagation. Baroclinic vorticity generation at gust fronts may play a role in tornadogenesis. These processes are discussed in detail in section 3.5.

3.2. Instability of the atmosphere to mesoscale convection: General considerations

a. Elementary deep convective instability (buoyancy only)

Severe convective weather owes its existence to the buoyant ascent of cloudy air whose source is in the lowest layers of the troposphere. This convective ascent occurs in regions with a temperature and humidity stratification that is typically stable toward small-amplitude vertical displacements, but unstable toward large-amplitude upward displacements of low-level air. Here we review some definitions, measures, and indices of deep convective instability, as tools for considering the effects of mesoscale processes on storm potential and organization.

Buoyancy is defined as the acceleration of gravity times the fractional density difference between a parcel of air and its environment. The density of air + water mixtures at a given pressure level depends on temperature, humidity, and condensed water content. If thermodynamic equilibrium (or a specific disequilibrium) among water phases is assumed, the latter two quantites can be calculated from the total water content.

The density of the parcel and its environment depend quite strongly on the definitions of "parcel" and "environment." Defining these from sounding data can be difficult, because air properties vary considerably within the distances and time intervals between soundings. More problematically, convection modifies its own environment in several ways. Section 3.5 will highlight some of these mechanisms, in light of the products and indices introduced here.

1) PARCELS, SOUNDINGS, AND DEEP CONVECTIVE INSTABILITY

Parcel-oriented sounding diagnostics are useful conceptual and forecasting tools for convective weather situations. Regions of absolute stability can be identi-

TABLE 3.3. Storm-generated mesoscale effects.

Local	Advective	Dynamical
Radiation	Particle advection, fall, and phase changes	Gravity currents, waves
Microphysics	• downdraft generation	• impact on cell/MCS growth
• downdraft, cold pool production	• upscale growth	• influence on neighboring convection
• microburst generation	Cold pool processes	Mesoscale pressure fields
• melting-generated midlevel convergence	• cell regeneration	• buoyancy contribution
• lightning production	• MCS evolution	• dynamic contribution (storm splitting, propagation)
	Momentum transport/sloping flows	• surface mesohighs/wake lows
	• severe surface winds	Baroclinic vorticity generation
	Vortex tilting/stretching	• horizontal vorticity at gust front
	• vertical vorticity generation (supercells, MCS mesovortices)	Vortex breakdown
		• mesocyclone
		• tornado

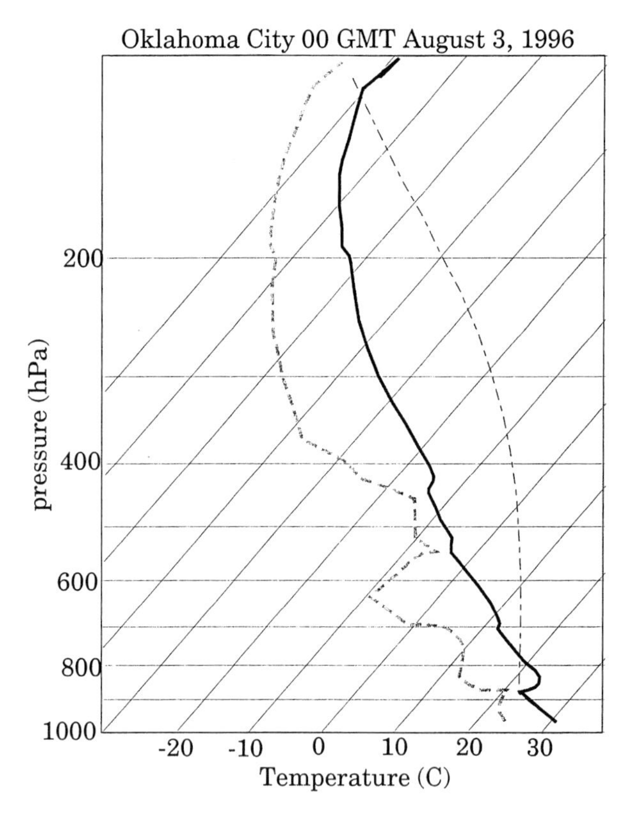

FIG. 3.1. Skew T–log$p$ of Oklahoma City sounding for 0000 UTC 2 August 1996. The temperature of a pseudoadiabatically ascending parcel from the surface (970 mb) is indicated. This parcel temperature is much warmer than the prevailing midtropospheric temperature, indicating a tremendous amount of potential buoyancy. However, this air experiences some negative buoyancy in the capping inversion layer, from 790 to 890 mb, which prevents the atmosphere from simply overturning everywhere.

fied, while relative values of some indices may delineate relative risks of various types of severe weather. Choosing the parcel, environment, and processes is an ambiguous exercise, best illustrated with an analysis of a particular severe storm sounding. Consider the unstable sounding of Fig. 3.1.

A more thorough diagram indicating the possible instability of this sounding is shown in Fig. 3.2. Here the buoyancy of every parcel below the 600-mb level is shown, as a function of pressure on a log scale. When total buoyancy (including water vapor and liquid water density effects) is expressed in temperature units (as a cloud virtual temperature T_{cv}), and a log p coordinate is used in the vertical, integrals of the contoured values in the vertical direction are proportional to work done by the buoyancy force. For example, the convective available potential energy (CAPE, units J kg^{-1}) for air originating at the surface is the integral of the positive values of buoyancy along the left-hand edge of Fig. 3.2:

$$\mathrm{CAPE} = \int_{T'_{cv}>0} RT'_{cv}\, d(\ln p), \tag{3.1}$$

where prime refers to parcel-environment differences.

Likewise, the convective inhibition energy (CIN; Colby 1984) for surface air is the integral of the negative values along the left-hand edge. The integrated CAPE over the entire depth of the unstable layer (ICAPE, units J m^{-2}) is proportional to the volume of the "mountain" delineated by the positive contours in Fig. 3.2. ICAPE is a property of an atmospheric column rather than of an arbitrarily cho-

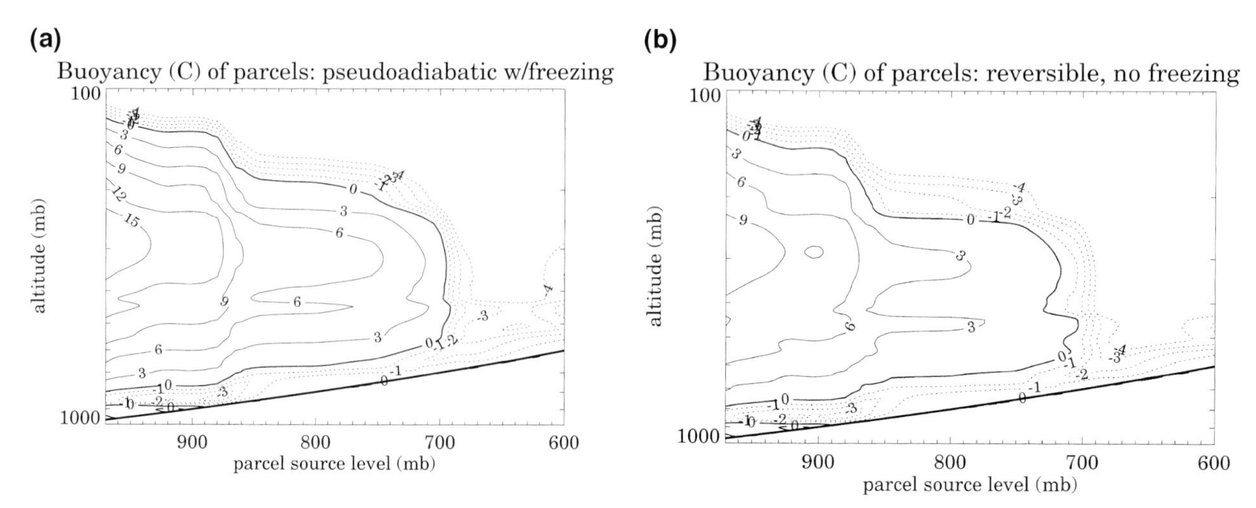

FIG. 3.2. Contours of buoyancy (total density difference, including water vapor and condensate loading effects, but expressed as equivalent degrees Celsius) for all parcels originating between the surface (970 mb) and 600 mb. The altitude scale is logarithmic in pressure, so that area is proportional to work done by the buoyancy force, as on a skew T–log$p$ diagram. Calculations are based on (left panel) pseudoadiabatic ascent of saturated air, with the latent heat of fusion included for vapor that condenses below 0°C; and (right panel) reversible ascent, with no freezing processes. Contours are 0 (heavy), negative (dashed, interval 1°C), and positive (solid, interval 3°C).

sen parcel (Mapes 1993; see also "cumulative CAPE" in Emanuel 1994). Thus, it has the advantage of assessing instability in cases where the CAPE of near-surface air is zero, for example, in the morning (when a convective temperature has to be forecast to compute CAPE) or in frontal overrunning situations.

Figure 3.2 indicates that parcels of air originating anywhere below the 700-mb level can attain positive buoyancy if lifted through the ambient environment for a distance of about 100 mb. The surface air has the highest CAPE, though the entire-mixed layer (below the 880-mb level) has similarly large values. This mixed-layer air has considerable CIN, with negative buoyancies of $-2°$ to $-3°C$ over a $\sim$1-km layer. A second, elevated mixed layer between about 700 and 870 mb also makes a significant contribution to ICAPE. This air only experiences negative buoyancy of $-1°C$ when lifted through its negative area. Interestingly, the elevated mixed layer serves both as the cap on the extremely unstable air at the surface, and as a possible source layer for convection in its own right. This air is capable of achieving respectable buoyancies of 5°–6°C under pseudoadiabatic ascent, with a smaller CIN than the mixed-layer air.

The CAPE, CIN, and ICAPE indices discussed above are tabulated in Table 3.4 for various assumptions about the precipitation, freezing, and mixing processes in the parcel. The parenthetical values in the entraining cases are for the hypothetical case where the relative humidity of the entrained air is artificially constrained to be no less than 80%.

Large amounts of energy are available to any process that can muster enough energy to overcome the much smaller CIN. If a convective structure can harness just a tiny fraction of the available energy toward overcoming CIN, it can become extremely vigorous. In essence, the mean state is set to amplify convective and mesoscale variability, rendering the prediction problem quite difficult.

Severe convection depends not only on the availability of large CAPE but also on convective initiation processes (overcoming CIN). For example, Graziano and Carlson (1987) note that for a given value of buoyancy, the probability of severe convection increases with increasing lid strength (up to some cutoff value), although the total probability of convection diminishes. In some instances, the actual form convection takes is influenced by details of the initiation process. For example, Fritsch and Rodgers (1981) report on a large hail-bearing storm in Colorado that had radically different structure and movement from other storms in the same large-scale environment, presumably due to a gust front that propagated southward along a north–south barrier. This situation is not uncommon and points to a continuing problem of predicting the form of convection, even though 12–24-h forecasts of the large-scale environment have become quite good.

TABLE 3.4. Instability indices for the sounding of Fig. 3.1 under various assumptions about the parcel process. The last three entries are for an entraining parcel, which continuously mixes with 10%, 20%, or 40% of its original mass per 100 mb traveled.

Parcel processes	$CAPE_{970}$ (J kg^{-1})	CIN_{970} (J kg^{-1})	ICAPE (10^6 J m^{-2})
Pseudoadiabatic + freezing	5986	25	8.14
Pseudoadiabatic	4909	25	6.32
Reversible	3847	34	4.67
Reversible, 10%/100 mb (RH > 80)	2131 (2375)	41 (38)	2.94 (3.46)
Reversible, 20%/100 mb (RH > 80)	1496 (1827)	48 (42)	2.27 (3.00)
Reversible, 40%/100 mb (RH > 80)	993 (1363)	59 (48)	1.68 (2.68)

2) DRY AIR ALOFT

The existence of dry air aloft, for example, from subsynoptic-scale dry intrusions (Carr and Millard 1985), can enhance the evaporation of precipitation and hence the strength of downdrafts and cold outflows from convection. Elevated dry layers, especially with a large storm-relative wind velocity, might enhance the severity and longevity of squall lines, bow echoes, and microbursts by producing vigorous downdraft circulations. Johns and Doswell (1992) note that forecasters consider that the presence of at least some dry air in the downdraft entrainment region is necessary for both bow echo-induced damaging winds and supercell tornado development.

A diagram similar to Fig. 3.2, but with downdraft potential buoyancy contoured below the line of unit slope, may be seen in Emanuel (1994). The downdraft buoyancy can be integrated over pressure to yield DCAPE, a measure of maximum possible kinetic energy production by downdrafts. However, in nature, the thermodynamics of evaporatively driven downdrafts lies much farther from equilibrium among water phases than does updraft thermodynamics. For example, a very dry layer aloft will have a very large potential negative buoyancy, under the assumption that it is brought to saturation isobarically (to its wet-bulb temperature), and then maintained at saturation while it descends. However, this process is unlikely to be realized in this ideal configuration in the real atmosphere. As a result, DCAPE is a difficult quantity to interpret (also see Gilmore and Wicker 1998). In addition, this potentially dense quality of dry air is realized not only as a force to accelerate energetic downdrafts but also as a drag on updrafts that entrain the dry air (Table 3.4). Without a specific idea of the form of convection one expects—highly entraining small cells versus lines, quasi-steady versus intermittent, sloped versus upright, etc.—the overall effects of dry air aloft cannot be easily foreseen.

b. Effects of wind shear

Wind shear significantly influences what form convection is likely to take. Wind shear can be incorporated into indices that may be better predictors of severe weather than buoyant instability alone (Miller 1972; Moncrieff and Green 1972; Moncrieff and Miller 1976; Weisman and Klemp 1982). In an effort to classify the mode of convection, Weisman and Klemp (1982) introduced a bulk Richardson number R similar to the one proposed by Moncrieff and Green (1972) that combines the effects of buoyant energy and shear:

$$R = \frac{\text{CAPE}}{\frac{1}{2}\bar{u}^2}, \tag{3.2}$$

where $\bar{u}$ is defined as the difference between the density-weighted mean wind speed taken over the lowest 6 km and an average surface wind speed taken over the lowest 500 m. Numerical modeling results for storms having CAPEs in the range ~1000–3500 m^2 s^{-2} and calculations of R for a series of documented storms led Weisman and Klemp (1982, 1984, 1986) to conclude that multicell storm growth occurs most readily for $R > 30$ and supercell storm growth for $10 < R < 40$.

However, when applying these or other parameters to real situations, there is inherently scatter. Part of the reason lies in the difficulty in obtaining a representative environmental sounding. Additionally, there is the problem of knowing how much CAPE will actually be realized for a particular storm owing to uncertainties about parcel properties at cloud base or dilution by entrainment, water loading, and ice loading. While it has been well established that strong, deep shear layers are supportive of supercellular convection (e.g., Weisman and Klemp 1986), mesoscale terrain effects, outflow boundaries, or other mesoscale phenomena can modify the mean shear profile and create mesoscale variability in severe-storm potential. Last, there is considerable natural variability in storm evolution owing to the history of convection itself, mesoscale forcing mechanisms, and the impact of neighboring cells.

All of these factors not only serve to make forecasting difficult but also suggest that slight changes in environmental conditions—say, by perturbations on the mesoscale—can dramatically affect storm development. For example, local orography or low-level jets can modify the storm environmental hodograph, thereby yielding different storm types in a region having the same synoptic-scale flow. Similarly, mesoscale processes can locally weaken CIN, allowing storms to develop in some locations and not others.

Some types of severe storms (e.g., hailstorms, tornadoes) occur in moderate-to-strong shear environments, whereas others (e.g., microbursts, flash floods) often occur with weak-to-moderate midtropospheric shear. Damaging microburst winds are more dependent on the thermodynamic profile than shear, as in the case of dry microbursts where deep dry adiabatic layers lead to intense downdrafts (Wakimoto 1985). While flash floods also typically have weak shear throughout the cloud depth (Maddox et al. 1980a), they often occur in the presence of strong low-level flow, which can contribute to repeated storm formation and motion over the same area (Chappell 1986). Since it is the low-level flow that provides the lifting, flash floods are particularly sensitive to mesoscale effects such as topography or outflow boundaries, which may determine the areas of maximum ascent.

Mesoscale organization of convection is also sensitive to wind shear, but the relationships are complex. For example, when CAPE and shear are large, bow echoes can form (e.g., Weisman 1993). Fujita (1978) defined the bow echo as a bowed convective line (25–150 km long in the cases he presented) with a cyclonic circulation at the northern end and an anticyclonic circulation at the southern end. Bow echoes may occur in isolation or as multiple features (e.g., line echo wave patterns) along squall lines. A phenomenon often related to bow echoes is the *derecho* (Johns and Hirt 1987). Derechos are convective systems that produce straight-line wind gusts > 26 m s^{-1} within a concentrated area with a major axis of at least 400 km. Several wind events are common, frequently only a few hours apart. Some have lasted 18 h or longer. Several studies have shown that many derechos are associated with bow echo structure (Przybylinski and DeCaire 1985; Johns and Hirt 1987; Johns 1993; Przybylinski 1995), but not all. For other combinations of CAPE and shear, other types of mesoscale organization can occur, for example, squall lines, mesoscale convective complexes (MCCs; Maddox 1980). However, the occurrence of these phenomena is also related to the existence of synoptic or mesoscale features such as jets, fronts, and convergence lines.

3.3. Mesoscale mechanisms for environment preconditioning

Before severe storms can develop, synoptic and/or mesoscale processes must act to provide adequate moisture and instability for convection to initiate. Once initiated, the interaction of convection with shear produces a pattern of storm evolution that can lead to severe weather. In this section, we consider processes that produce the instability and shear, specifically, those that occur on the mesoscale. As a framework for discussion, we will consider preconditioning processes according to whether they act locally, or are advective or dynamic in nature, as outlined in Table 3.1.

a. Local processes

1) VERTICAL MIXING IN THE BOUNDARY LAYER

Daytime heating of the convective boundary layer (CBL) is probably the most common preconditioning process for convection over land. Typically, once the nocturnal inversion is burned off, clouds can form as boundary layer thermals reach their LCL. The rate at which this occurs depends on the morning inversion depth, sky cover, and surface wetness. Entrainment at the CBL top acts to dry out the boundary layer and reduce the potential for deep convection, but if moisture is sufficient, clouds can still form. If there is a strong capping inversion, the growth of clouds may be restricted or even suppressed. However, moderate capping inversions can enhance the potential for deep convection by allowing shallow cumulus to be suppressed, but then later in the day, as the inversion has been weakened by heating, a lifting mechanism can release the instability quickly (e.g., Carlson et al. 1983).

To illustrate how daytime CBL heating and cumulus cloud development ensue, a time series of reflectivity, CBL height, and cloud base derived from a 915-MHz wind profiler in Illinois on 16 August 1995 is shown in Fig. 3.3 (Angevine et al. 1998). The series of virtual temperature soundings shows a fairly well-defined convective boundary layer developing through the morning hours, with a weak capping inversion above. During the course of the day, the water vapor mixing ratio increases as the CBL deepens, presumably due to surface evaporation and advection. Winds in the CBL (not shown) were 6–8 m s^{-1} from the south, with southwesterly winds above. The top of the CBL is marked by the peak in reflectivity, which is a result of strong gradients of temperature and humidity. The boundary layer top can be seen rising from approximately 500 m AGL at 0900 CST to about 1000 m at 1300 CST. Fluctuations in the entrainment zone atop the CBL during the morning are due to up- and downdrafts associated with thermals or CBL rolls (to be discussed later).

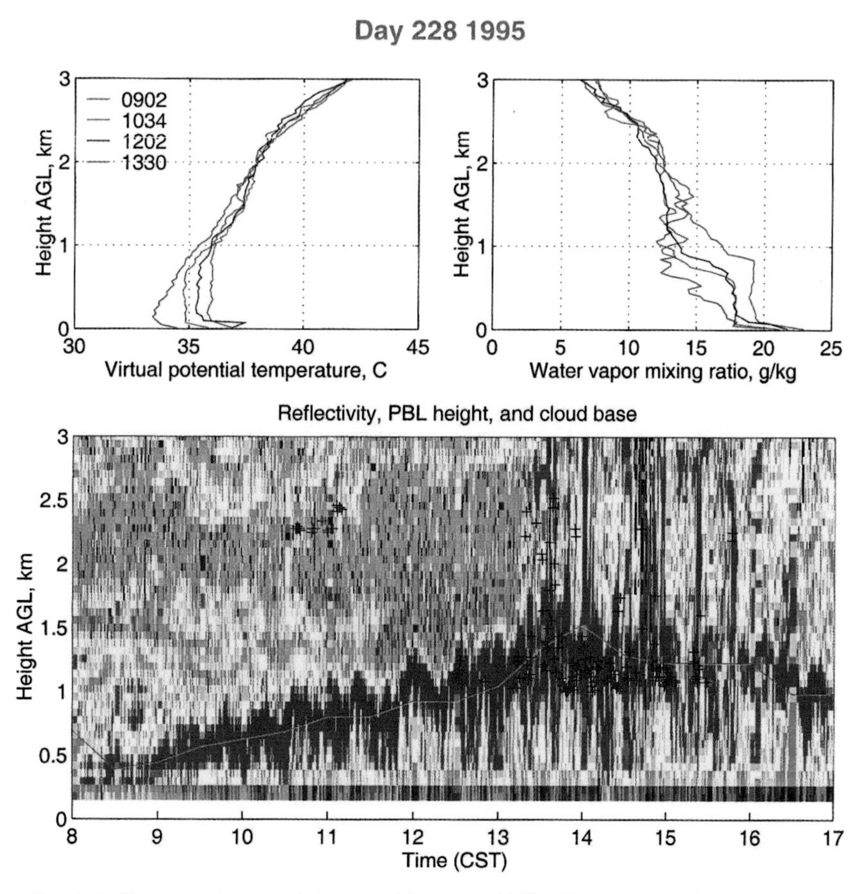

FIG. 3.3. Boundary layer evolution on 16 August 1995. Virtual potential temperature and water vapor mixing ratio from four soundings at the times shown are plotted in the upper panels. The lower panel shows the profiler reflectivity (arbitrary scale) in pseudocolor. The solid green line is the automatically determined boundary layer height, and the blue crosses are cloud-base heights detected by the ceilometer. From Angevine et al. (1998).

Cloud development was apparently triggered by these perturbations once the CBL became deep enough for the LCL of CBL air to be reached (cloud bases shown by crosses).

The upward mixing of moisture in the CBL, acting in combination with low-level convergence, can precondition the atmosphere for deep convection (e.g., Wilson et al. 1992). Vertical mixing can also lead to enhanced convergence along drylines. For example, to the west of drylines, westerly momentum is efficiently transported down to the surface in the deep, dry CBL there, whereas vertical mixing on the moist east side is weaker, thereby enhancing convergence along the dryline axis (Danielsen 1974; Ogura and Chen 1977; McCarthy and Koch 1982). The vertical structures of water vapor mixing ratio, wind, potential, and virtual potential temperature across the dryline of 24 May 1989 (Fig. 3.4) is illustrative of this process (Ziegler and Hane 1983).

2) TERRAIN EFFECTS

Surface relief, whether it be small hills, ridges, escarpments, or mountain ranges, can have profound effects on convection. An excellent review of this topic is given by Banta (1990). He identified three classifications of topographic effects on convection: mechanical lifting to the level of free convection (LFC), thermally generated circulations, and aerodynamic effects (e.g., blocking, flow deflection, gravity waves). The first is directly related to triggering, so examples of these will be given later.

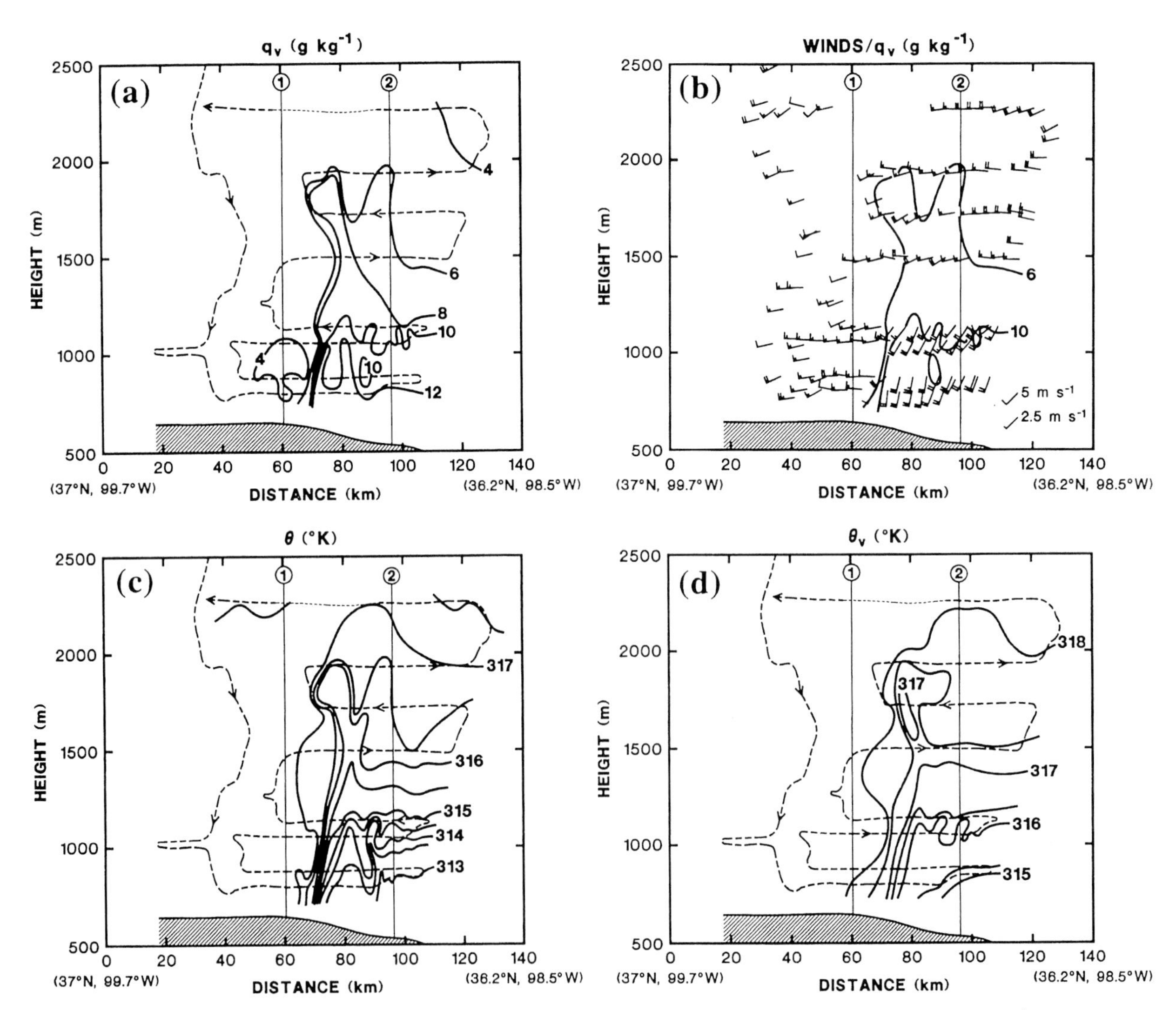

FIG. 3.4. Subjective analyses of P-3 measurements during series of vertically stepped aircraft passes through dryline; (a) q_v (g kg^{-1}); (b) horizontal, ground-relative wind vectors (full barb equals 5 m s^{-1}); (c) θ (K); (d) θ_v (K). Dashed curve in (a), (c), and (d) represents the P-3 flight track. The vertical lines are the locations of M-CLASS soundings projected onto the cross section. The hatching denotes terrain elevation subjectively smoothed as determined from the radar altimeter on the P-3. From Ziegler and Hane (1993).

Thermally generated flows such as upslope and upvalley wind systems can play a prominent role in the initiation and development of hailstorms, tornadoes, flash floods, and high winds associated with dry microbursts. Large mountain barriers like the Rocky Mountains, Tibetan Plateau, Andes, and so forth, generate large-scale, diurnally varying circulation features that are instrumental in establishing the thermodynamic and wind profiles conducive to these types of severe weather. On the mesoscale, smaller topographic features generate thermally forced flows that provide focus areas for convective initiation. As an example, consider the vector-mean surface flow over the plains of northeastern Colorado at 1100 LST for July 1981 (Toth and Johnson 1985) superimposed upon the radar climatology for the summers of 1971 and 1972 (Wetzel 1973; Fig. 3.5). This figure indicates that the preferred regions for convective development coincide in most instances with zones of maximum confluence (and convergence) during the late morning along the two prominent east–west ridges north and south of Denver, Colorado: the Cheyenne Ridge and the Palmer Lake Divide, respectively. That such ridges are a focal point for intense hailstorms was documented in the 1973 National Hail Research Experiment (e.g., Browning et al. 1976). The specific conditions that favor hail or other severe weather occurrence on a particular day consist of the superposition of flows on both the synoptic scale and the mesoscale. For example, Modahl (1979) finds that on hail days in northeastern Colorado the afternoon easterly (upslope) component of the flow is much stronger than on no-hail days,

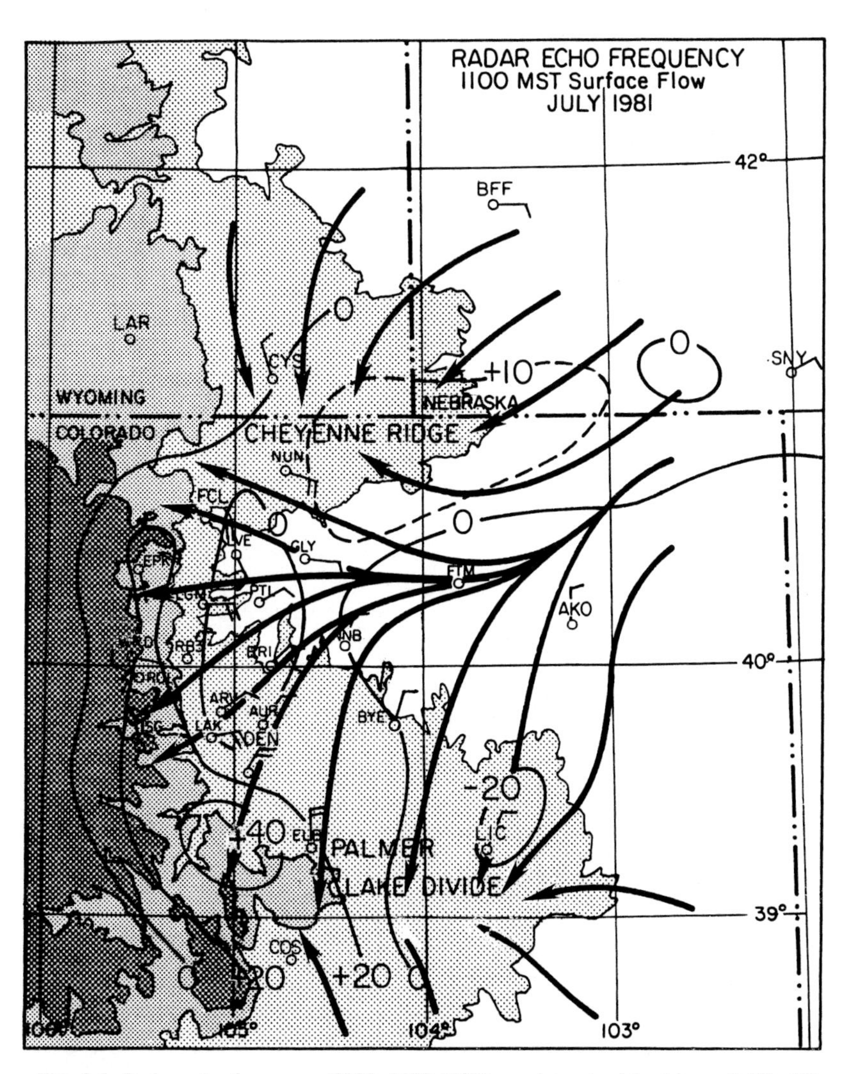

FIG. 3.5. Radar echo frequency (0900–2100 MST), as determined by Limon (LIC), CO, WSR-57 radar, given as percent deviation from azimuthally averaged mean at each radius from Limon (reproduced from Wetzel 1973). Surface streamlines and vector-averaged winds (one full barb = 1 m s^{-1}) are shown for 1100 MST. Dashed line is intermediate contour. From Toth and Johnson (1985).

which leads to enhanced moisture and wind shear favorable for severe storms. However, viewed in a larger-scale context, this finding is consistent with the well-known, favorable Front Range severe weather synoptic pattern characterized by an east–west-oriented front to the south of the threat area and surface high over the northern Great Plains (Doswell 1980).

Flash flood environments are another example of preconditioning by mechanically or thermally forced upslope flows, particularly in the western United States where the time of onset of heavy rains is in the afternoon (Maddox et al. 1980a). Striking examples are the Black Hills and Rapid City, South Dakota, flood of 1972 (Maddox et al. 1978); the Big Thompson River, Colorado, flood of 1976 (Maddox et al. 1978; Caracena et al. 1979); the Johnstown, Pennsylvania flood of 1977 (Bosart and Sanders 1981; Zhang and Fritsch 1986); and the many floods throughout Asia on the windward slopes of mountain ranges during the summer monsoon (Ramage 1971). But there are also cases of floods associated with more subtle topographic features, such as the Balcones Escarpment of Texas, which played a role in the Texas Hill Country flash floods of 1978 (Caracena and Fritsch 1983). Typically, these floods occur in association with low-level jets, weak flow at midlevels, moderate-to-large CAPE, and a low-level inversion (Maddox et al. 1978). The precise locations of the storms producing the flash floods are often determined by complex triggering processes involving interaction of outflow boundaries with terrain, direct orographic lift, and other mesoscale features.

Thermally induced topographic flows also influence the development of dry microbursts (Wakimoto 1985) and their associated high surface winds, which are a common occurrence during the summer along the Front Range of the Rocky Mountains. Typical morning and evening soundings for dry microburst days over the High Plains are shown in Fig. 3.6 (Wakimoto 1985). Characteristic of these soundings is a nearly dry adiabatic lapse rate from just above the surface to near 500 mb, the approximate level of cloud base. Wakimoto (1985) notes that it is a deep dry adiabatic layer that, following precipitation sublimation and evaporation, permits extreme downward vertical velocities to be attained (Brown et al. 1982; Proctor 1989). The mountains play an important role in the microburst process in at least two ways: 1) by providing a deep dry adiabatic layer whose upper portions are in part composed of mixed layers advected from the mountains to the west (Carlson et al. 1983; Wilczak and Christian 1990), and 2) by generating the showers that are the sources of the initial downdrafts.

Flow deflection and blocking by terrain often influence the location and development of convection. Details of flow deflection and blocking are often

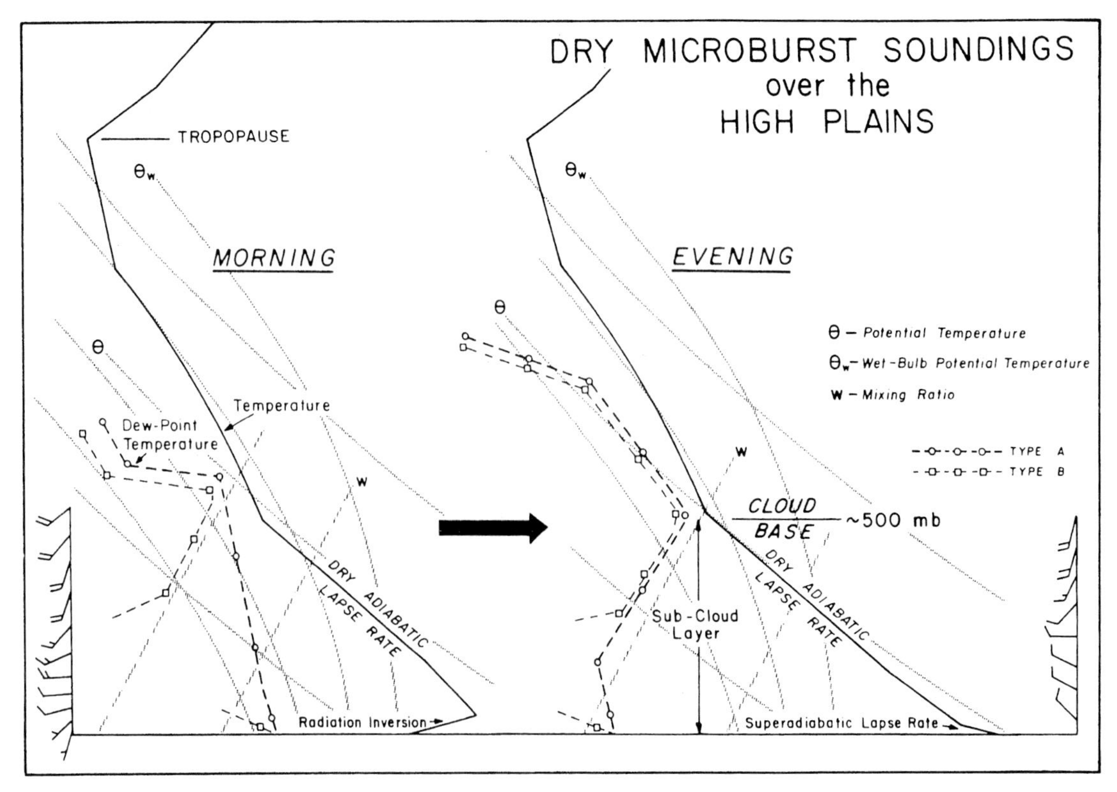

FIG. 3.6. Model of the characteristics of the morning and evening soundings favorable for dry-microburst activity over the High Plains. From Wakimoto (1985).

complex, but they generally depend upon the incident flow speed U and stratification represented by Brunt-Väisälä frequency N through the Froude number $F = U/NH$, where H is the height of the barrier (e.g., Carruthers and Hunt 1990). When $F \lesssim 1$, for example, in stably stratified situations, the flow is typically blocked and goes around a barrier, whereas when $F > 1$, the flow can go over the barrier, except when a very stable layer exists just above it (Fritsch et al. 1992).[2] There are countless examples of these effects worldwide, the most notable of which are in the vicinity of isolated mountains or hills, mountain ranges, and mountainous islands. Banta (1990) provides an extensive review of effects of flow deflection and blocking on convection in the United States and other parts of the world.

Perturbations to the flow downstream of both large-scale and small-scale barriers can affect the occurrence of severe convective weather. A prominent example of the former situation is the southwest vortex to the lee of the Tibetan Plateau, where low-level shear lines and midlevel vortices frequently develop that are often linked to the production of heavy rains (Tao and Ding 1981). The southwest vortex was partly responsible for the 1981 Sichuan flood, which claimed more than 1000 lives (Kuo et al. 1984). In a modeling study, Kuo et al. (1988) showed that the formation of the southwest vortex actually represents a coupling between large- and smaller-scale topographic effects. The Tibetan Plateau is instrumental in setting up a large-scale latitudinal temperature gradient that drives the moist, southwest monsoon flow (e.g., Yanai et al. 1992). This monsoon flow then impinges upon the mesoscale Yun-Gui Plateau that extends from the southeastern corner of the main Tibetan Plateau, and the low-level flow is blocked. The flow then descends into the Sichuan basin on the lee side of the mesoscale plateau, creating cyclonic relative vorticity over the basin by vortex stretching. Kuo et al. (1988) found that latent heat release intensified the southwest vortex and that the interaction between convection and the vortex was crucial for the production of heavy rainfall.

There are also instances where mesoscale topographic features generate downstream eddies and convergence zones that are the sites for severe weather such as hailstorms and tornadoes. Such a convergence zone often exists to the lee of the Olympic Mountains in Washington State (Mass 1981; Mass and Dempsey 1985), although the weather is typically not severe. Another well-known example of this phenomenon is the *Denver cyclone*, which occurs downstream (for south-to-southeasterly flow) of the Palmer Lake Divide on the eastern plains of Colorado (Szoke et al.

[2] More precisely, flow blocking characteristics depend upon two vertical scales: the height H of the barrier and the depth h of the fluid or stable layer (Simpson 1987). The condition described here is for $h \lesssim H$.

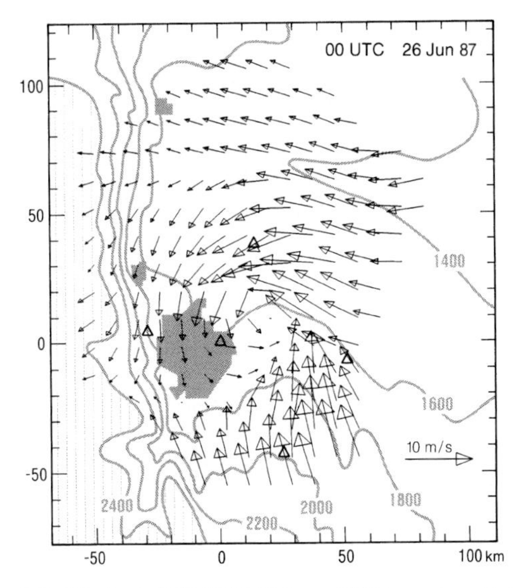

FIG. 3.7. Winds at 0000 UTC 26 June 1987 in eastern Colorado. Terrain contours (plotted every 200 m) and cities are shown in light gray. Elevations above 2400 m are hatched. Triangles and the diamond indicate the locations of the 5 CLASS sites and the BAO tower, respectively, and the origin is Stapleton International Airport where the CP3 radar, the wind profiler, and the Denver CLASS sounding site were located. Axes coordinates are kilometers from CP3. From Wilczak and Christian (1990).

1984; Blanchard and Howard 1986; Brady and Szoke 1989). One case of the Denver cyclone is shown in Fig. 3.7 (from Wilczak and Christian 1990). A strong, surface cyclonic circulation center exists just east of the Denver metropolitan area at 0000 UTC 26 June 1987. Wilczak and Christian found that in the late afternoon, as the surface heat flux began to decrease, flow at the vortex center that had earlier been divergent became convergent (Fig. 3.7). The vortex contracted, vorticity increased rapidly, and shortly thereafter intense thunderstorms (up to 60–70 dBZ, producing 4.5-cm diameter hail) developed along the convergence zone on the east side of the vortex center. Although no tornadoes were observed, several small cyclonic vortices developed along the convergence zone. The size and azimuthal shears of these vortices were similar to those of tornado-associated misocyclones observed on other Denver cyclone days (Wakimoto and Wilson 1989). Brady and Szoke (1989) and Wakimoto and Wilson (1989) have proposed that such shear-induced low-level misocyclones, when positioned underneath strong cumulus updrafts, can, through vortex stretching, generate nonsupercell tornadoes (Fig. 3.8). These findings are supported by other studies that have shown that the Denver cyclone and its associated convergence zone are active sites for nonsupercell tornadoes (Szoke et al. 1984; Szoke and Brady 1989). Because the Denver cyclone is a topographically quasi-locked feature, it appears to produce a local maximum in tornado frequency just east and

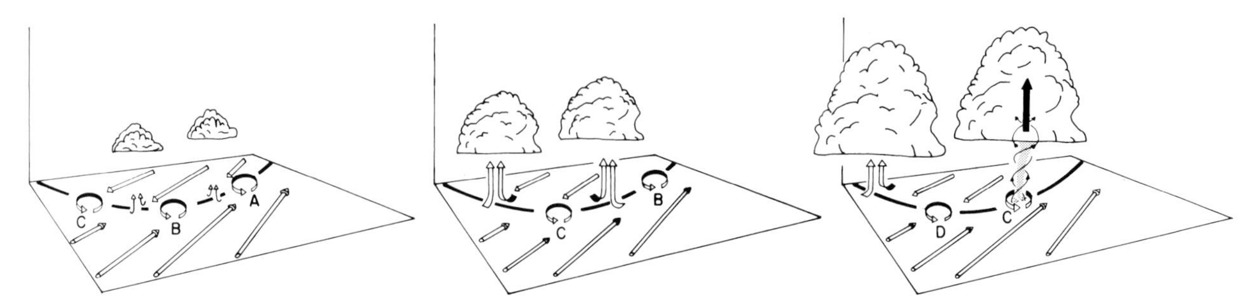

FIG. 3.8. Schematic model of the life cycle of the nonsupercell tornado. The black line is the radar detectable convergence boundary. Low-level vortices are labeled with letters. From Wakimoto and Wilson (1989).

northeast of the Denver metropolitan area (Golden 1978; Wakimoto and Wilson 1989).

Lee vortex phenomena such as the Denver cyclone have been attributed to a variety of mechanisms, operating singly or in combination: tilting of frictionally created horizontal vorticity (Thorpe et al. 1993), baroclinic vorticity tilting (Smolarkiewicz and Rotunno 1989), potential vorticity generation by surface friction or wave breaking (Smith 1989), horizontal variations in the Reynolds stress divergence over sloping terrain (Wilczak and Glendening 1988; Dempsey and Rotunno 1988; Davis 1997), mountain waves (Smith 1982), and Coriolis turning of the decelerated (subgeostrophic) southeasterly flow as it approaches the Continental Divide (Crook et al. 1990a).

Flow deflection by orography can also affect severe weather development by modifying the environmental wind profile. An example of such a situation pertains to hailstorms in Switzerland (Houze et al. 1993). In their study of eight years of data, Houze et al. (1993) found that hailstorms are nearly equally divided between left- and right-moving storms. They obtain realistic numerical simulations of the storms by using observed thermodynamic and wind profiles, while assuming a flat lower boundary. While this finding suggests that the basic characteristics of the storms are not directly a function of flow over complex terrain, the fact that the same thermodynamic sounding and wind hodograph can support a multiplicity of storm structures (left-moving, right-moving, supercell; see Weisman and Klemp 1982, 1984, 1986), indicates that, through modification of the wind profile, local orography may actually determine which type of storm would be favored at a specific time and place.

In another example, Carbone (1982) found the low-level horizontal shear in a prefrontal jet in the Central Valley of California (Fig. 3.9a) to be instrumental in tornado formation along the front. Radar reflectivity and velocity signatures of the tornado are shown in Figs. 3.9c and 3.9d. The prefrontal jet in this case may have been enhanced by barrier winds upstream of the Sierra Nevada, as indicated in aircraft observations (Fig. 3.9b) of yet another case by Parish (1982). Flow deflection by topography has also been observed to generate bulges in drylines (Atkins et al. 1998). In their case, a river valley along the Caprock Escarpment in the Texas panhandle contributed to the bulge, at which point there was enhanced convergence and density contrast across the dryline. Such bulges may become favored locations for severe storm development.

Topographically generated gravity or mountain waves can also affect deep convection. Booker (1963) found that mountain wave activity can affect the distribution of summertime convective rainfall in the mountains of Pennsylvania. Thunderstorm echoes were observed to dissipate in the subsiding air immediately downwind of a mountain barrier only to reform in (presumed) rising air several kilometers farther downwind. Tripoli and Cotton (1989a,b) found support for this concept in a numerical modeling study of convective development to the lee of the Rocky Mountains. On a typical afternoon with westerly flow, convection develops on the west and east slopes of the barrier (Fig. 3.10). The cells on the east slope arise from convergence between mountain wave flow and the developing upslope flow. Several hours later, the west-slope storms merge with those in the east and an explosive growth of convection takes place. The overall convective system then propagates eastward, experiencing an initial weakening owing to the strong change in topographic slope from the Rockies to the plains, but then rebounding in intensity as midlevel warming and growth to larger scales occurs (Fig. 3.11). Pulsing of the intensity of the storm on a timescale of several hours occurs due to oppositely propagating transient gravity wave circulations, but further intensification occurs at sunset as destabilization of the upper-anvil cloud inhibits gravity wave propagation into the stratosphere. The trapping of gravity waves in the upper troposphere contributes to continued growth to larger scales and eventually a quasi-balanced system that moves out onto the plains.

3) SURFACE EFFECTS

Two aspects of surface properties can affect environmental preconditioning: 1) the actual state of the

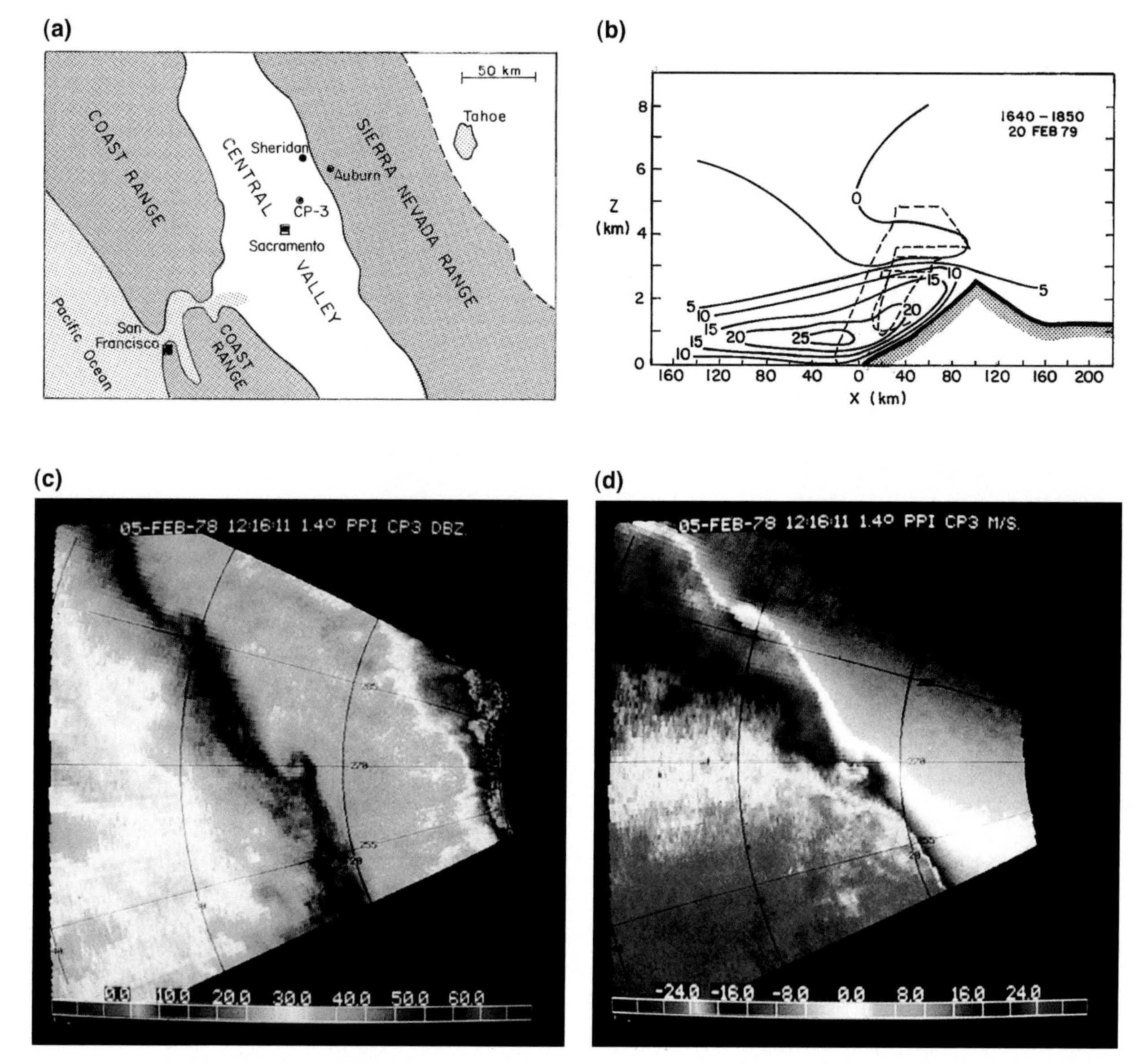

FIG. 3.9. (a) Map of northern California. Three Doppler radar sites are indicated by black dots. Sierra Crest is shown by dashed line and corresponds to 2.0–2.5 km height (above MSL). Central Valley is near sea level. (b) Mountain-parallel wind components (m s^{-1}) derived from rawinsonde and aircraft data for 20 February 1979. Flight track shown by dashed line; flight time listed at top. (c) Radar reflectivity and (d) velocity data from CP-3 radar two minutes prior to tornado damage on ground on 5 February 1978. Range marks are at 10-km intervals. Note two vortex signatures along the line 13 km apart. Panels (a), (c), and (d) are from Carbone (1982). Panel (b) is from Parish (1982).

surface (e.g., dry vs wet soil) and 2) heterogeneities in surface conditions (e.g., dry land adjacent to wet land). In the first instance, soil moisture is one of the most important factors since it affects the partitioning between latent and sensible heat fluxes. As noted by Segal et al. (1995), wet surfaces under clear-sky conditions are generally more conducive to deep convection than dry surfaces. When the surface is wet, large latent heat fluxes can increase the CBL specific humidity in the afternoon, thereby enhancing CAPE. If the capping inversion is weak or absent, convection can readily break out. However, in cases of stronger capping inversions, dry surfaces may be more conducive to deep convection even if there is less moisture since the larger sensible heat flux in those cases can erode the capping layer (e.g., McGinley 1986). Prolonged evaporation and transpiration from vegetation can in some instances lead to a "pooling" of higher moisture in convergence zones (Segal et al. 1989; Chang and Wetzel 1991). The enhanced moisture can provide additional fuel for severe storms, as reported in the derecho study by Johns and Hirt (1985).

Land surface heterogeneities in roughness, wetness, albedo, vegetation cover, snow cover, urbanization,

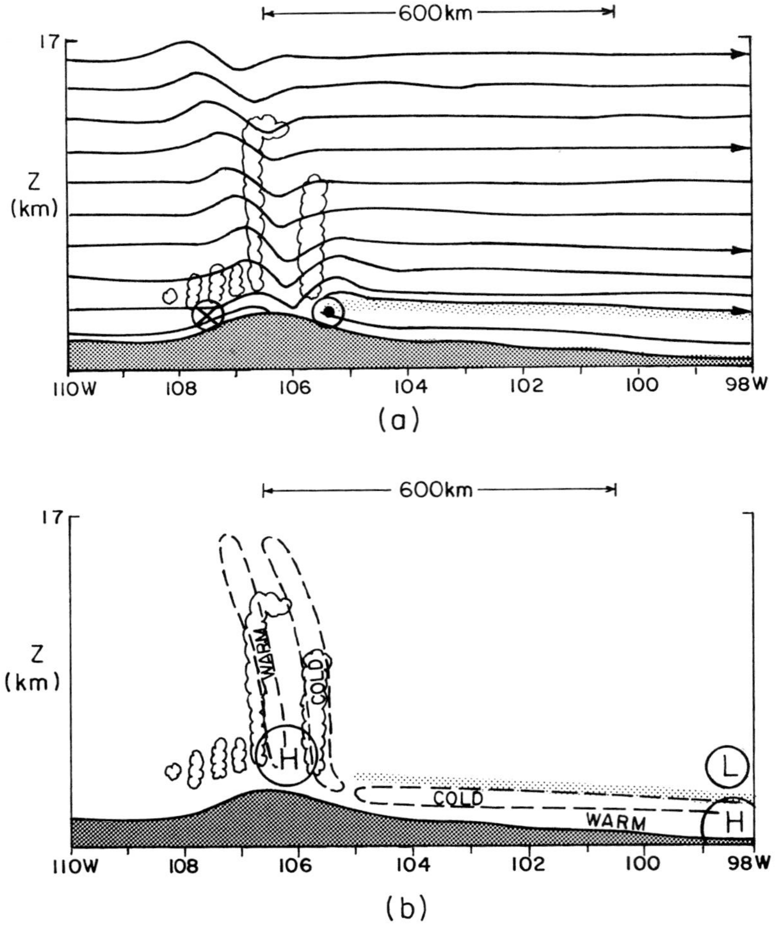

Fig. 3.10. Conceptual model showing flow field and position of convective elements at the time when deep convection forms over the Rocky Mountains. The stippled line represents the position of the plains inversion. Regions of cloud are indicated. (a) depicts the flow field with ground-relative mesoβ-scale streamlines. Circles depict flow perturbation normal to plane. (b) depicts the pressure and temperature response. Pressure centers are depicted by solid closed contours and temperature by dashed contours. The length scale of 600 km ($2 \times$ the Rossby radius) is indicated. From Tripoli and Cotton (1989a).

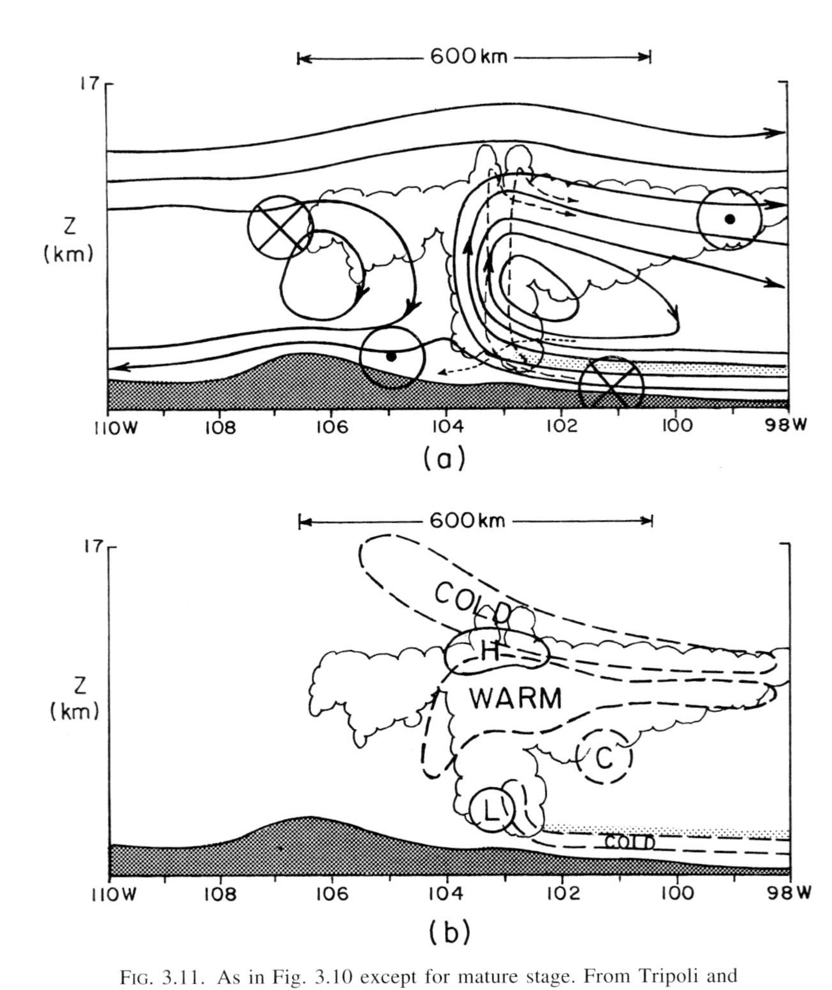

Fig. 3.11. As in Fig. 3.10 except for mature stage. From Tripoli and Cotton (1989b).

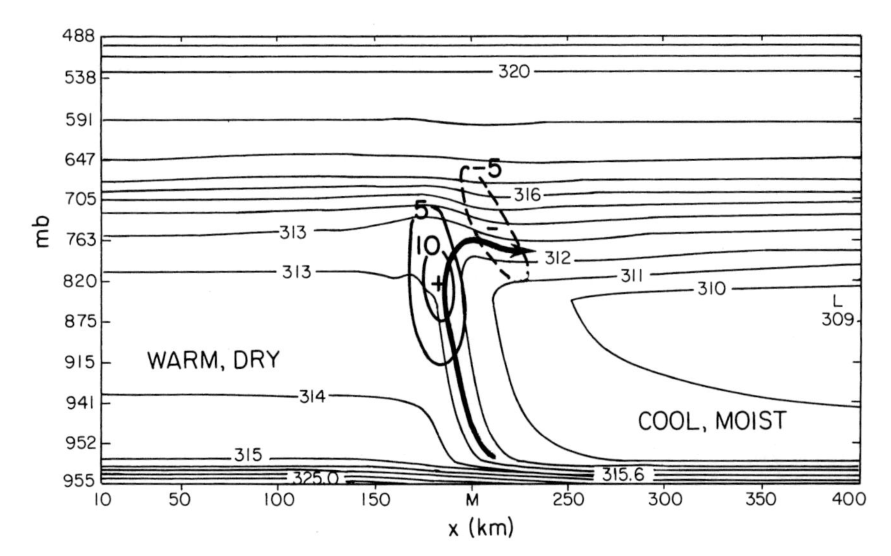

FIG. 3.12. Distributions of potential temperature (*K*, light solid), vertical motion (cm s^{-1}, heavy solid and dashed), and circulation (arrow) within a developing dryline at 8 h. From Sun and Ogura (1979).

and other factors can produce mesoscale circulations that may lead to convection (e.g., Anthes 1984; Ookouchi et al. 1984; Pielke and Segal 1986; Yan and Anthes 1988; Segal and Arritt 1992). Lanicci et al. (1987) showed that regional variations in soil moisture over the Texas–Oklahoma–Mexico area impact the structure and evolution of the elevated mixed layer (lid), the dryline, easterly ageostrophic flow over east Texas and the Gulf of Mexico, the potential instability of air under the lid, and the location and intensity of precipitation. Collins and Avissar (1994) identified those land surface characteristics that are most important in redistributing energy into turbulent sensible and latent heat fluxes, namely, land surface wetness, surface roughness, albedo, and, when vegetation covers the ground, leaf area index and plant stomatal conductance. Mesoscale circulations induced by surface moisture variability (whether a result of wetness or vegetation heterogeneity) appear to have a marked impact on the atmosphere on timescales ranging from hours (affecting cloud formation, e.g., Rabin et al. 1990) to weeks (contributing to heavy rains such as the 1993 Midwest floods, e.g., Paegle et al. 1996) to seasons (affecting the development of drought conditions, as in the Sahel, e.g., Charney 1975).

To illustrate the possible effects of moisture discontinuities on convection, consider the hypothetical dryline simulations of Sun and Ogura (1979). Since dry surfaces exhibit a much larger diurnal variation of temperature than wet surfaces, the daily march of the temperature contrast across a wet–dry boundary can resemble that across a dryline. Sun and Ogura used a two-dimensional model of the planetary boundary layer to simulate the diurnal variation in temperature across a dryline by specifying the cross-line gradient in surface potential temperature. The potential temperature and vertical velocity after about eight hours, starting from a weak westerly flow and horizontally uniform atmospheric conditions, are shown in Fig. 3.12. As a result of greater vertical mixing due to the warm surface conditions on the left (to the west) than on the right, a deeper mixed layer develops on the left[3] and a horizontal temperature gradient develops in the middle of the domain (the dryline). An up–down vertical motion couplet develops near the top of the mixed layer at the dryline position. The air flow (schematically shown in Fig. 3.12) indicates that the potential for convection to develop along the dryline is enhanced as moist air is drawn westward and upward to the top of the mixed layer. This type of circulation is referred to as an *inland sea breeze* by Ogura and Chen (1977). Sun and Ogura (1979) hypothesized that the localized lifting along the dryline by this mechanism may have accounted for the initiation of the 8 June 1966 squall line. While the results of Sun and Ogura are useful in illustrating circulations that can develop in association with moisture contrasts, additional processes normally occur in real dryline situations, for example, differential vertical mixing of momentum, along-dryline variability, and so forth.

In addition to moisture and vegetation gradients, other factors contribute to mesoscale variability of severe weather and cloudiness. Some studies have suggested an increase of rain, thunderstorms, and hailstorms within or downwind of urban areas due to increased cloud buoyancy, mechanical or thermodynamic effects that produce confluence zones, and en-

[3] The somewhat exaggerated superadiabatic layers may be an artifact of the turbulence parameterization scheme used.

hancement of the coalescence process due to giant nuclei from industrial activity (e.g., Changnon et al. 1976).

b. Advective processes

1) MOISTURE ADVECTION

Direct advection of moisture into a region can increase CAPE and lower the LFC, thereby increasing the potential for deep convection. Strong moisture advection is essential for extreme-precipitation events since rainout far exceeds local evaporation. In such cases low-level jets provide the required moisture transport, for example, for the development of MCCs (Maddox 1983), widespread floods such as the 1993 Mississippi Valley flood (Paegle et al. 1996; Arritt et al. 1997), and localized flash floods (Caracena et al. 1979). Moisture advection on the cloud scale may also be important for local moistening promoting new cloud growth. For example, Perry and Hobbs (1996) find significant humidity enhancements on the downshear and cross-shear sides of cumulus clouds.

2) DIFFERENTIAL ADVECTION

Much of the environmental preconditioning for severe weather arises from differential advection on the synoptic scale, for example, in destabilizing the atmosphere (Newton 1963), in providing the vertical wind shear (Ludlam 1963), or in establishing capping inversions (Carlson and Ludlam 1968). There are a number of examples of similar processes that occur on the mesoscale.

- Low-level jets with high-θ_e air overrunning fronts and cold pools leading to long-lived bow echoes (Johns 1984) and MCCs or mesoscale vorticity centers (MVCs; Maddox 1983; Fritsch et al. 1994).
- The flow of moist boundary layer air out from beneath an inversion or lid (a process called *underrunning*) due to ageostrophic circulations about a mesoscale jet streak leading to intense convection (Carlson et al. 1983).
- Transport of clouds and moisture aloft downstream of mountain barriers producing conditions conducive to dry microbursts (Wakimoto 1985; Wilczak and Christian 1990).
- Differential advection associated with jet streak circulations and boundary layer heating changing markedly over very short periods of time and acting to destabilize relatively small regions immediately prior to convective outbreaks (Kocin et al. 1986).

3) CONVERGENCE LINES

Convergence lines can serve as both preconditioning and triggering mechanisms. For example, lifting by a convergence line may in some cases be adequate to generate convection all along it; however, in other cases, enhanced lifting—say, by collisions or intersections with other boundaries—may be required to lift air to the LFC. The latter subject will be treated in section 3.4d.

Although cold fronts are synoptic scale in the along-front dimension, cross-front circulations are of mesoscale dimension. Localized convergence and lifting at the front destabilize the environment and reduce CIN, thereby making the atmosphere susceptible to deep convection. In some cases, convergence can be so strong that deep or severe convection can occur even in the absence of CAPE (Browning and Harrold 1970; Carbone 1982). In such instances, the front serves as a convective trigger, but in some aspects may still be a preconditioning mechanism. For example, in the case of a strong California cold front (Carbone 1982), the surface frontal zone provided strong cyclonic horizontal shear, which led to perturbations (and eventually tornadoes) along the front. These perturbations arose apparently from a horizontal shearing instability. Lifting can also occur out ahead of fronts, leading to prefrontal squall line formation as a result of cold fronts aloft (Browning and Monk 1982; Locatelli et al. 1995, 1997), jet streak secondary circulations (Browning and Pardoe 1973; Shapiro 1982), or prefrontal wind shifts associated with lee troughs (Hutchinson and Bluestein 1998).

Stationary or warm fronts can also be instrumental in preconditioning the environment for severe weather. Nocturnal MCC development to the north of quasi-stationary surface fronts has been attributed to low-level warm advection (Maddox and Doswell 1982) and to destabilization by diurnally modulated low-level jets, mesoscale ascent produced by the fronts, and convergence near the terminus of jets (Trier and Parsons 1993). Quasi-stationary east–west fronts have also been implicated in bow echo and derecho development (Johns 1984; Johns and Hirt 1987).

Precipitation-driven convective downdrafts, with their associated cold, spreading gravity currents or "gust fronts," probably represent the most common mechanism for generating localized surface convergence in regions where convection already exists. Results from the Thunderstorm Project clearly showed that the spreading cold air was instrumental in generating new convective elements in squall-line or multicell-type storms (Byers and Braham 1949; Newton and Newton 1959). Satellite observations provide convincing evidence that thunderstorm outflows contribute to remote effects of thunderstorms, with new cell growth at distances up to several hundred km from preexisting cells (e.g., Purdom 1973; Gurka 1976; Purdom 1982). On smaller scales, downdraft outflows in supercells (the forward-flank and rear-flank downdrafts described by Lemon and Doswell 1979) have been hypothesized to play varying roles in tornadogenesis (Barnes 1978;

Lemon and Doswell 1979; Klemp and Rotunno 1983; Brandes 1984a,b; Rotunno and Klemp 1985; Klemp 1987). Although lifting at the leading edge of gust fronts can directly trigger new convection, there are many situations where it can be considered a preconditioning mechanism, with convection initiation occurring as a result of collisions or intersections with other low-level perturbations to the flow (e.g., other gust fronts, cold fronts, drylines, terrain features, horizontal convective rolls in the CBL). The localized enhancement of vertical wind shear and convergence along preexisting boundaries can lead to increases in the occurrence of severe weather there (Maddox et al. 1980b).

Drylines also provide a focus for convection through localized convergence (Fujita 1958; Rhea 1966). The two most likely mechanisms contributing to the convergence are 1) solenoidally forced, frontogenetic circulations (Ogura and Chen 1977; Sun and Ogura 1979; Parsons et al. 1991; Ziegler and Hane 1993) and 2) vertical momentum mixing (Ogura and Chen 1977; McCarthy and Koch 1982). However, storms typically do not form everywhere along drylines. Therefore, mechanisms in addition to the above are required to trigger convection.

Low-level convergence also occurs at the leading edge of sea and land breezes. Summer sea breezes often produce vigorous afternoon showers over land as a result of zones of convergence between the synoptic-scale and onshore flow, for example, over Florida (Byers and Rodebush 1948; Gentry and Moore 1954; Atlas 1960; Pielke 1974) or the Texas coast (Hsu 1969). Even lake breezes can contribute to the formation of severe thunderstorms and tornadoes (Lyons and Chandik 1971; King and Sills 1998).

Land breezes are also known to produce nocturnal thunderstorms offshore (e.g., Neumann 1951; Preston-Whyte 1970; Williams and Houze 1987) and contribute to the formation of MCCs in low-latitude locations such as the southern South China Sea off the north coast of Borneo (Houze et al. 1981) and the Gulf of Panama (Danielsen 1982; Velasco and Fritsch 1987).

A number of observational studies indicate that land breezes and their associated cloud lines may contribute to waterspout formation at coastal locations. Wakimoto and Lew (1993) documented the development of a waterspout from a relatively small cumulus cloud (photograph in Fig. 3.13a) that occurred in the early morning along a line of cumuli just off the east coast of Florida. A satellite image taken just after the waterspout occurrence (Fig. 3.13b), when the parent cloud moved ashore, shows several offshore cloud bands. Some of these bands could be shoreward-propagating remnants of land-breeze convergence zones or, as Wakimoto and Lew suggest, mesoscale frontal zones produced by small variations in SST. While it is not known if the other bands produced waterspouts, the tendency of waterspouts to form along cloud lines has been well documented (e.g., Dinwiddie 1959; Golden 1973; Simpson et al. 1986). In fact, Golden (1974) argues that less than 5% of waterspouts develop from isolated cumulus. Considering the findings of Wakimoto and Wilson (1989) that surface convergence zones (and associated small-scale vorticity centers) are important for nonsupercell tornadoes, it may be that the offshore cloud lines are a reflection of similar convergence/vorticity zones that provide a low-level vorticity source upon which stretching by cumulus growth can act.

In some cases, the low-level lifting provided by sea, land, or lake breezes is adequate by itself to generate convection along the boundary. However, as in the case of fronts, drylines, and convective outflows, the actual triggering of convection is more often controlled by the intersection of such fronts with other phenomena that accentuate the lift.

Mountain/valley breezes are also important for preparing the atmosphere for a number of types of severe weather. For example, the preference for afternoon onset of flash floods in the western United States (Maddox et al. 1980a) indicates the importance of upslope flow in providing convergence and lifting to prepare the atmosphere for such storms. Schmid and Lehre (1998) found drainage flows in the Swiss Alps to be an important factor in providing a wind profile conducive to severe storms in eastern Switzerland.

c. *Dynamical processes*

1) SECONDARY CIRCULATIONS

Upper-level wind maxima (jet streaks) and low-level jets have long been associated with severe convective weather. In the case of upper-level jets, transverse ageostrophic circulations about the jet axis are argued to help initiate some severe storms (Uccellini and Johnson 1979; Bluestein and Thomas 1984). An example of such circulations within an idealized straight jet streak (which propagates more slowly than the maximum wind in the jet itself) is shown in Fig. 3.14. In the entrance region of the jet, wind speeds are subgeostrophic, resulting in a cross-stream ageostrophic component of the flow toward the cyclonic side of the jet (Fig. 3.14a). This component is the upper branch of a thermally direct circulation cell (Fig. 3.14b). In the exit region, winds are supergeostrophic and the transverse circulations are reversed. These circulations arise from geostrophic confluence and diffluence forcing, but can be modulated in many situations by the effects of horizontal shearing deformation (Shapiro 1981; Keyser and Shapiro 1986), flow curvature (Beebe and Bates 1955; Shapiro and Kennedy 1981; Moore and VanKnowe 1992), and transience (Ziv and Paldor 1999). The pattern of vertical motion induced by the confluence/diffluence forcing supports the concept that clouds and precipita-

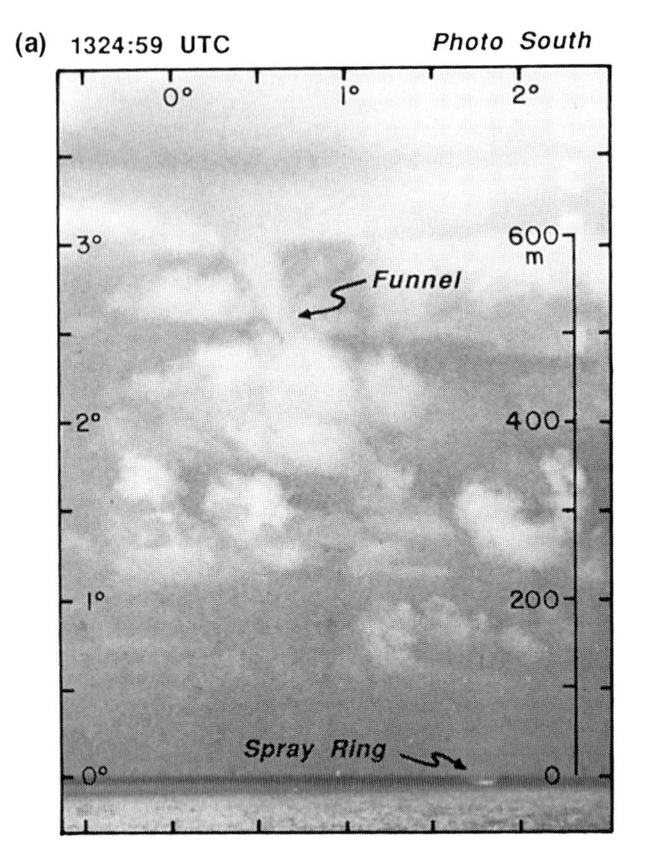

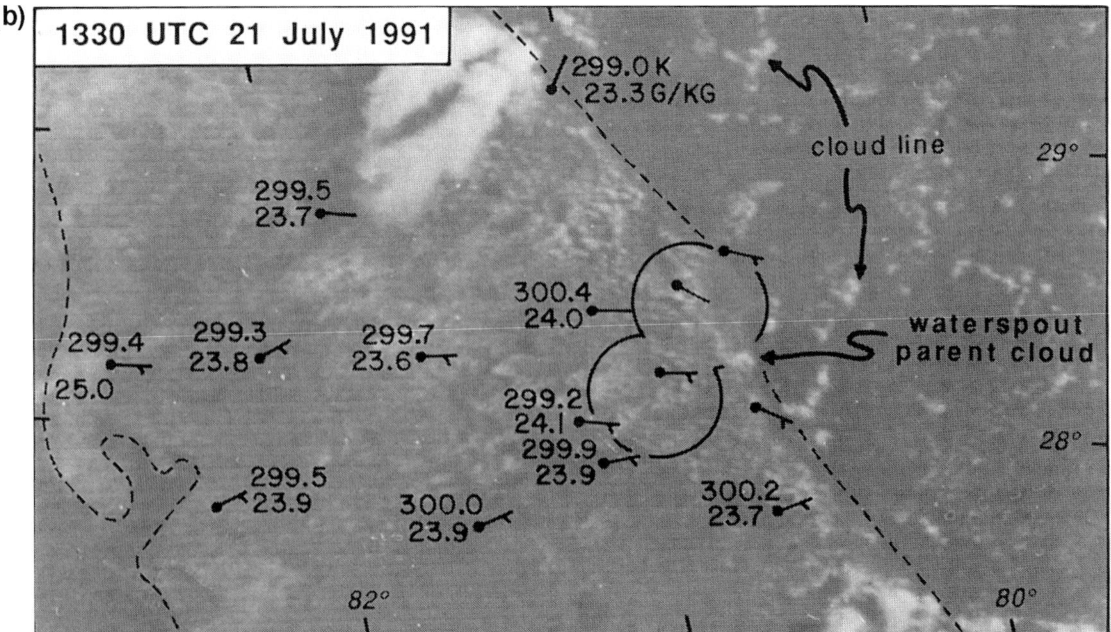

FIG. 3.13. (a) Photograph of cloud base and the waterspout taken from the Photo-South site at 1324:59 UTC. The height scale is valid at the distance of the waterspout. Azimuth and elevation angle grids from the photo sites are also indicated in the figure. (b) Visual satellite image at 1330 UTC on 21 July 1991. Potential temperature, mixing ratio, wind speed, and direction for select PAM stations are shown in the figure. Wind vectors are drawn with one barb and half-barb representing 5 and 2.5 m s^{-1}, respectively. The dual-Doppler lobes for the CP-3 and CP-4 baseline are also indicated in the figure. From Wakimoto and Lew (1993).

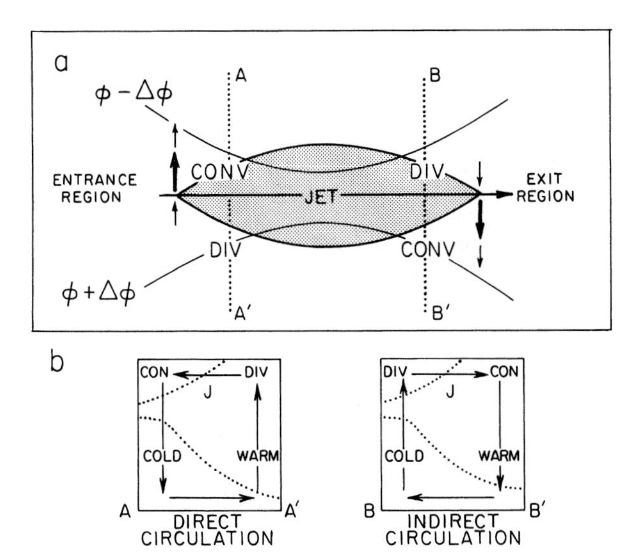

FIG. 3.14. (a) Schematic of transverse ageostrophic wind components and patterns of divergence (DIV) and convergence (CON) associated with the entrance and exit regions of a straight jet streak. ϕ refers to geopotential height. (b) Vertical cross section illustrates vertical motions and direct and indirect circulations in the entrance region (line A–A′) and exit region (line B–B′) of a jet streak. Cross sections include two representative isentropes (dotted), upper-level jet (J) location, upper-level divergence, and horizontal ageostrophic wind components within the plane of each cross section. Adapted from Uccellini (1990).

tion should be most prevalent in the right entrance and left exit region of straight jet streaks, a pattern that is often observed (e.g., Namias and Clapp 1949; Uccellini and Johnson 1979).

Low-level jets (LLJs) have also been linked with the generation of severe weather, presumably through enhancement of moisture and temperature advection, localized increase in low-level convergence, and an increase in the vertical wind shear (Means 1952; Bonner 1966; Wallace 1975; Maddox 1983; Trier and Parsons 1993). Such jets are most common at night, which has led to a theory for their development based on nocturnal boundary layer decoupling (Blackadar 1957). Recent observations (Mitchell et al. 1995; Arritt et al. 1997; Whiteman et al. 1997) support Blackadar's (1957) inertial oscillation theory for the LLJ, but other factors are probably also involved. In particular, diurnal oscillations associated with sloping terrain over the plains appear to contribute to a nocturnal maximum in the LLJ (Holton 1967; Paegle 1978; McNider and Pielke 1981).

The coupling between upper- and lower-level jets has been studied by Uccellini and Johnson (1979), Kocin et al. (1986), and others. They find that the mass adjustment associated with the upper-level jet ageostrophic flow forces an isallobaric wind that represents a return branch of the indirect circulation in the jet exit region (Fig. 3.14b). This return branch is argued to be an important factor in the development of LLJs in active synoptic situations. The LLJ is shown to be enhanced by diabatic heating (e.g., CBL heating or convection).

These processes have been studied in detail by Sortais et al. (1993) using data from the FRONTS-87 experiment. They found a coupling between the indirect circulation in the exit region of an upper-level jet and a low-level jet, and a cold front and its diabatic heating (Fig. 3.15). The lower branch of the upper-level jet transverse ageostrophic circulation came into phase with the low-level jet, which advected warm, moist air toward the ascending branch where deep convection occurred. In addition, mesoscale transverse ageostrophic circulations associated with the LLJ assisted in the formation of convection ahead of the cold front.

Other cases have been documented where secondary circulations (producing rising motion) ahead of fronts (Shapiro 1982) have interacted with boundary layer horizontal convective rolls (Trier et al. 1991) or drylines (Nieman and Wakimoto 1999) to generate severe convection.

2) GRAVITY CURRENTS, WAVES, BORES, AND SOLITARY WAVES

Many convergence-producing phenomena leading to convection possess the characteristics of density or gravity currents: fronts (Shapiro 1985), gust fronts (Charba 1974), sea and land breezes (Simpson 1969; Wakimoto and Atkins 1994), and drylines (Schaefer 1974b; Parsons et al. 1991). The movement of these features is described fairly well by gravity current theory (Simpson 1987), although in the case of drylines, vertical mixing may significantly affect dryline propagation (Schaefer 1974a).

Thunderstorm initiation depends sensitively on the vertical structure (depth) as well as strength of lifting by gravity currents. Lower-tropospheric wind shear plays a significant role in determining the depth and uprightness of lift at the leading edge of gravity currents (Droegemeier and Wilhelmson 1985; Rotunno et al. 1988; Crook 1996). Specifically, the role of the low-level shear in lifting at the leading edge of a thunderstorm cold pool is illustrated in Fig. 3.16 (from Rotunno et al. 1988). Horizontal vorticity produced by horizontal buoyancy gradients in a cloud updraft (Fig. 3.16a), when combined with the negative vorticity of an underlying cold pool in a no-shear environment, can cause the updraft to lean upshear (Fig. 3.16b). When shear exists in the absence of a cold pool, the updraft leans downshear (Fig. 3.16c). When the vorticity of the cold pool and shear are in balance, the updraft is erect (Fig. 3.16d). Rotunno et al. define this as an *optimal state* where the full CAPE can be realized without being inhibited by the cold pool or shear. Such situations should be characterized by a narrow leading convective line, such as the cases

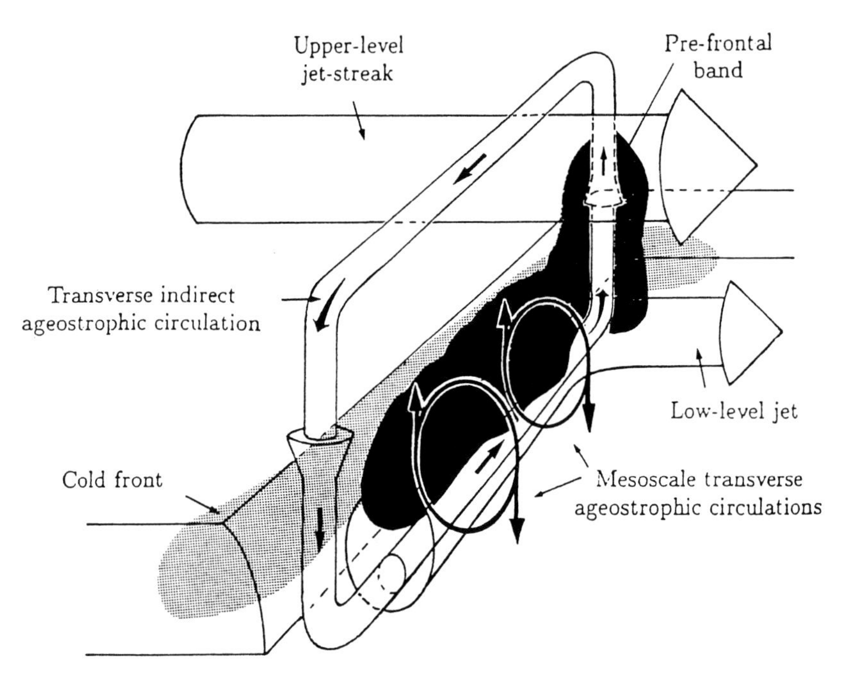

FIG. 3.15. Conceptual model showing the structure and orientation of the ageostrophic circulations associated with low- and upper-level jets in the vicinity of a cold front. From Sortais et al. (1993).

studied by Smull and Houze (1987), Ogura and Liou (1980), and Carbone (1982).

There are a number of boundary layer phenomena resembling or often associated with gravity currents—gravity waves, internal undular bores, and solitary waves—that at times are linked with the initiation of severe convective weather. Disturbances of this type often occur in connection with surface-based stable layers, such as nocturnal inversions, thunderstorm outflows, or marine inversions. Gravity waves represent a periodic oscillation of the upper surface of the stable layer. They may be generated by an impulsive forcing such as a downdraft impinging upon a stable layer, in which case there is no change in the mean depth as the oscillations pass by, or by a gravity current (e.g., a thunderstorm outflow) intruding on the stable layer, in which case an *undular bore* can be generated. A bore consists of an increase in depth of a fluid advancing with a series of waves on its surface that typically separate from the gravity current and move ahead of it. Figure 3.17 (from Simpson 1987) illustrates the production of such a disturbance by this process. Whether or not an undular bore develops depends upon the density current speed U normalized by $(g'h)^{1/2}$ (the internal Froude number F, where g' is the reduced gravity $g\Delta\rho/\rho$, $\Delta\rho$ being the density difference between the two fluids) and the ratio of the gravity current depth d to the depth of the undisturbed stable layer h (Fig. 3.17, lower diagram). If the gravity current is shallow relative to the stable-layer depth (d/h small) and F is >1, then the flow is *supercritical* and the gravity current moves faster than any disturbance can move forward along the top of the dense layer; the interface between the fluids rises smoothly over the head as the gravity current advances. If d/h is small and $F < 1$, the flow is *subcritical* and the only distur-

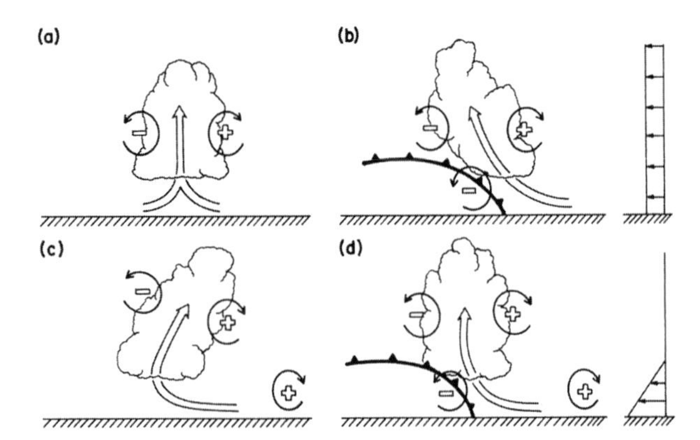

FIG. 3.16. Schematic diagram showing how a buoyant updraft may be influenced by wind shear and/or a cold pool. (a) With no shear and no cold pool, the axis of the updraft produced by the thermally created, symmetric vorticity distribution is vertical. (b) With a cold pool, the distribution is biased by the negative vorticity of the underlying cold pool and causes the updraft to tilt upshear. (c) With shear, the distribution is biased toward positive vorticity and this causes the updraft to lean back over the cold pool. (d) With both a cold pool and shear, the two effects may negate each other, and allow an erect updraft. From Rotunno et al. (1988).

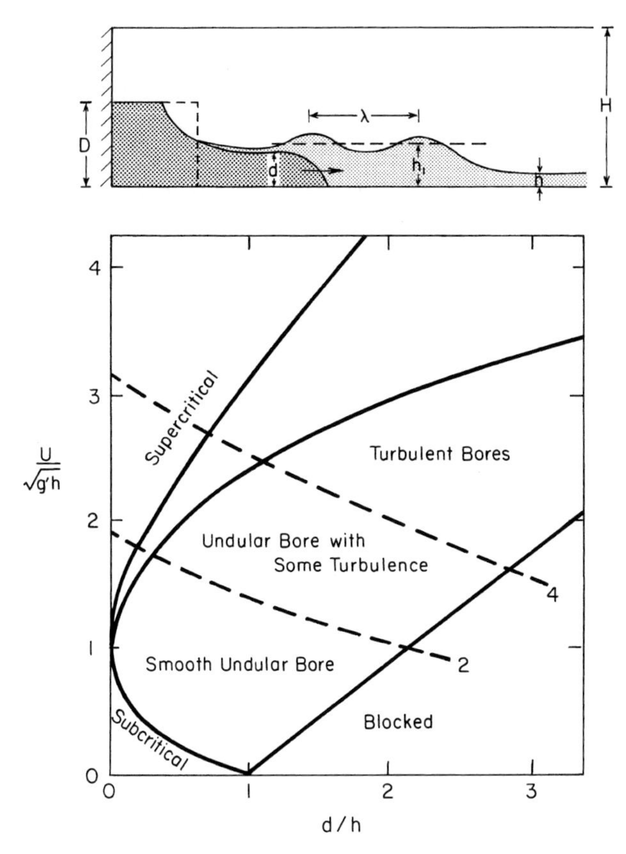

FIG. 3.17. (Upper) Depiction of internal bore generated by gravity current in the laboratory. (Lower) Flow regime diagram for gravity current of depth *d* impinging on a stable layer of depth *h*. Ordinate is internal Froude number $U/(g'h)^{0.5}$, where U is speed of the gravity current and abscissa is d/h. Dashed lines refer to undular bore magnitude (h_1/h). D is the depth of the reservoir containing denser fluid (a dense salt solution) on the left (dark shading) that is released into the less-dense fluid (water) on the right (light shading) by opening a gate (vertical dashed line). Redrafted from Simpson (1987).

bance is a small depression in the layer that moves along above the advancing gravity current head. Undular bores typically occur if the density current depth is comparable to or greater than the stable-layer depth and if F is not too small (otherwise the gravity current will be blocked or rise on top of the stable layer).

Probably the most dramatic atmospheric example of an undular bore is the morning glory of northeastern Australia (Clarke 1972; Smith 1988; Christie 1992). The morning glory is characterized by a spectacular low-level roll cloud or series of roll clouds often extending over several hundred kilometers in length. Although the precise mechanisms for its origin are uncertain, it is thought to develop as a result of colliding sea breezes over the Cape York peninsula (Clarke 1984). There have been numerous studies based on surface, tower, and Doppler radar data of similar atmospheric bore-like phenomena associated with thunderstorm outflows (Pothecary 1954; Shreffler and Binkowski 1981; Doviak and Ge 1984; Haase and Smith 1984; Carbone et al. 1990; Fulton et al. 1990). In some of these cases, the bores were observed to evolve into solitary waves as the density current weakened and slowed down. Undular bores or solitary waves may also be the explanation for the surface pressure jumps and oscillations reported in papers by Tepper (1950, 1951), Curry and Murty (1974), Uccellini (1975), and Balachandran (1980), particularly since many of the cases occurred at night, but these authors did not describe them as such. The gravity waves in these cases were argued to have originated by convection or cold fronts and evidence was presented for them initiating thunderstorms downstream by lifting air to the LFC.

However, there are problems in extending the laboratory two-fluid analog of Fig. 3.17 to the atmosphere, since atmospheric stratification above bores or solitary waves should allow energy to radiate vertically and limit the amplitude of the disturbances (Lindzen and Tung 1976). But Lindzen and Tung showed that a stable layer capped by an unstable layer can, depending on flow conditions, reflect wave energy and form a duct that allows wave propagation over great distances with little loss of amplitude. Similar trapping has been found to occur when there is a curvature in the wind profile (or a wind reversal) aloft (Crook 1986), a jet in the lower layer that opposes the wave motion, or an inversion at a certain height above the stable layer (Crook 1988). For "morning glories," all of the above three trapping conditions appear to be met (Crook 1988), whereas low-level opposing flow appears to be the trapping mechanism in several Midwest borelike disturbances (Shreffler and Binkowski 1981; Fulton et al. 1990). On the other hand, Karyampudi et al. (1995) found curvature in the wind profile of the low-level jet to be important in sustaining a prefrontal bore that developed downstream of the Rocky Mountains and played a role in the generation of severe weather over Kansas and Nebraska.

A unique set of observations of a mesoscale ducted gravity wave was obtained in the FRONTS-84 field experiment (Ralph et al. 1993). Surface, wind profiler, and rawinsonde data were used to determine the vertical structure of such a wave ducted between the ground and a critical level (where the phase velocity of the wave matches the wave-parallel background wind speed) (Fig. 3.18). The vertical velocity is a maximum (and the horizontal perturbation velocities are zero) at the antinode of the oscillation, which corresponds to the top of the layer of strong static stability, and which is at an altitude of one-quarter of a vertical wavelength. The temperature and vertical velocity perturbations at midlevels are in quadrature, and the horizontal velocity perturbations are in phase with the pressure perturbations at the surface, which allows detection of such waves in surface data (Koch and Golus 1988). Ralph et al. (1993) could not determine whether the wave was initiated by convection or induced convection. However, Koch et al. (1988)

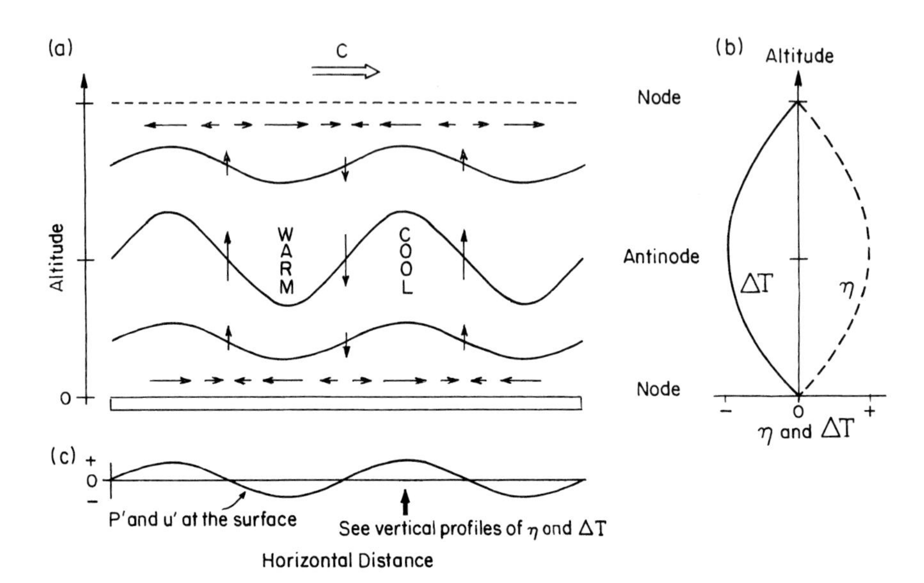

FIG. 3.18. Schematic representation of a ducted mesoscale gravity wave with one-half of a vertical wavelength contained between the ground and a critical level. (a) Horizontal cross section perpendicular to wave phase lines, showing wave-induced vertical and horizontal motions (arrows), streamlines or isentropes (solid lines), the ground (shaded), the critical level (dashed), and the direction of wave motion (labeled C). Regions of cool and warm air created by the vertical displacements are also shown. (b) Vertical profiles of vertical displacement (η) and temperature change (ΔT) for the phase of the wave marked in (c). (c) Wave-induced surface pressure perturbations (P') and wave-parallel wind perturbations (u') drawn for the same wave segment shown in (a). Notice that u' and P' are in phase, and that they lag behind the phase of the vertical motion within the duct by 90°. From Ralph et al. (1993).

argued for a close linkage between the waves and convection, and found deep convection and precipitation near the pressure maximum and low pressure ahead of and behind the squall line. The coupling of convection with gravity waves was argued by Koch et al. (1988) to be in qualitative agreement with predictions from wave-CISK (Lindzen 1974; Raymond 1975), where it is proposed that the gravity wave provides moisture convergence into the storm and the heating/cooling distributions from the storm in turn provide the energy to drive the wave disturbance.

Sources of convection-affecting mesocale gravity waves might all fit under the umbrella of "geostrophic adjustment processes" associated with (a) wind imbalances, for example, in the right-exit region of a jet streak approaching a ridge (Koch and Dorian 1988; Koch and Golus 1988; Koch et al. 1988, 1993; Koch and Siedlarz 1999) and (b) mass imbalances, for example, as a response to heating (Miller and Sanders 1980; Ulanski and Heymsfield 1986; Crook 1987; and Crook et al. 1990b's 7-h mesoscale oscillation). Mesoscale instability was also cited by Koch and Dorian (1988) as a possible explanation for their gravity waves.

3) MESOSCALE INSTABILITIES

Numerous instability mechanisms have been proposed to explain various mesoscale phenomena. Conditional instability is associated with the development of cumulus clouds, which occurs on small scales, but growth to the mesoscale can occur, especially when wind shear exists. Other instability mechanisms involving latent heat release have been proposed: CISK (conditional instability of the second kind; Charney and Eliassen 1964), wave-CISK (Lindzen 1974), CSI (conditional symmetric instability; Bennetts and Hoskins 1979), and others. Some instabilities do not require the release of latent heat: inertial instability, (dry) symmetric instability and Kelvin–Helmholtz instability (although the latter is typically associated with submesoscale phenomena, e.g., billow clouds).

On the synoptic scale, the condition for inertial instability ($\zeta + f < 0$) is rather rare, although it can occur on the anticyclonic side of strong upper-level jets (Knox 1997). One example of such a situation is found in Ciesielski et al. (1989), where the instability appeared to be manifested by a series of disturbances in the cirrus canopy on the anticyclonic side of the jet axis. On smaller scales, inertial instability can develop within the upper-level outflow jets of MCSs and severe storms (Maddox 1983; Blanchard et al. 1998).

When horizontal shears become very large, even if $\zeta + f$ is everywhere positive, a hydrodynamic instability can arise, sometimes called *Rayleigh* or *shearing instability*. From linear theory, disturbances that develop have a wavelength approximately 7.5 times the width of the shear zone (Miles and Howard 1964). This instability has been used to explain the formation of small-scale vortices—the precursors to dust devils—along low-level shear zones (Barcilon and Drazin 1972). Other possible examples of mesoscale phenomena arising from this instability are vortices along cold fronts (Carbone 1982, 1983) or along outflow boundaries (Mueller and Carbone 1987; Wakimoto and Wilson 1989) that may be subsequently related to tornadogenesis. Lee and Wilhelmson (1997) successfully simulated the development of vortices (misocyclones) arising from shearing instability (caused by an outflow boundary advancing into boundary parallel flow) and found updraft maxima to develop adjacent to the misocyclones, which may provide an explanation for the colocation of cumulus updrafts and surface vortices illustrated in Fig. 3.8 (from Wakimoto and Wilson 1989).

The concept of CSI considers an atmosphere stable with respect to vertical (buoyancy) and horizontal (inertial) displacements (hence no CAPE), but unstable with respect to displacement along slant paths (Bennetts and Hoskins 1979; Emanuel 1979; see Schultz and Schumacher 1999 for a review). If CAPE is present and the LFC is reached by a displacement, then it seems logical that conditional gravitational instability (CGI) should be realized as opposed to CSI since the former has the fastest growth rate. However, a coupling between convective and mesoscale motions has been hypothesized to occur in environments of weak symmetric stability where CGI exists (Emanuel 1980; Xu 1986; Jascourt et al. 1988; Seman 1994). Emanuel (1980) showed that if the secondary circulations are hydrostatic and isentropic, the amount of work done by subsidence in the environment of upward fluid displacements decreases as the symmetric stability is reduced, at some point becoming less than the kinetic energy generated by the convective updraft. When this point is reached, the mesoscale circulation in the environment enhances the convection.

This idea has been extended by Seman (1994) to the nonlinear, nonhydrostatic case with CAPE, where vertical momentum transports in deep convection are found to produce inertial instability aloft (strictly speaking, negative isentropic absolute vorticity, the condition for CSI, is generated aloft). It is suggested that the instability enhances horizontal mass transport in the outflow branch, which ventilates the upper levels of the system, thereby promoting further convection. Blanchard et al. (1998) have provided some observational evidence that supports this positive feedback process, which they and Seman (1994) argue leads to the upscale growth of convection. Such coupled convective–mesoscale motions may have occurred in the early stages of the development of parallel, deep convective bands of precipitation over the south-central United States (Jascourt et al. 1988). Satellite and conventional data analysis suggested that a layer of weak symmetric stability modified the atmosphere's response to free convective instability, contributing to highly organized banded convective structure.

Zhang and Cho (1992) have shown that in the stratiform region of a squall line,[4] moist potential vorticity (MPV; potential vorticity defined using θ_e) can become negative as a result of modification of the stability and absolute vorticity fields by upward and rearward transport in the front-to-rear flow. The resulting symmetric instability considerably enhances the vertical motion and precipitation rate in the stratiform clouds. They also suggest that the generation of negative MPV contributes to inertial instability aloft that enhances the anticyclonic outflow at upper levels.

CSI has been argued by Colman (1990) to also play a possible role in the development of elevated thunderstorms in frontal overrunning situations. The cases studied exhibited negligible CAPE, yet strong thunderstorms occurred. He found that the storms developed in a strongly baroclinic environment and, in general, were aligned along the geostrophic shear. These observations are consistent with the theory of moist symmetric instability (Emanuel 1979, 1983). However, frontogenetical forcing in a symmetrically neutral environment was also suggested as a possible initiation mechanism (Emanuel 1985). In addition, it may be that while coarse sounding data indicate negligible CAPE in overrunning situations, significant CAPE may exist on smaller scales not sampled by the sounding network.

Mesoscale cloud bands have also been attributed to parallel instability, an instability of the Ekman boundary flow (Lilly 1966; Raymond 1978). CSI and parallel instability are analogous in the sense that both instabilities arise in vertically sheared flows; however, the shear in the case of CSI depends on the thermal wind (i.e., horizontal temperature gradient and the earth's rotation), whereas in the case of parallel instability the shear is due to the Ekman wind profile (i.e., boundary layer friction and the earth's rotation). Raymond developed a theory for parallel instability within the shear layer below a low-level jet and applied this idea to the development of three powerful squall lines and the massive tornado outbreak of 3 April 1974. The rolls connected with this instability (wavelength ~100 km) were hypothesized to concentrate low-level vorticity into narrow shear lines that generated banded cloud structure prior to the onset of deep convection.

The wave-CISK approach to the problem of thunderstorm generation considers the storm as a forced

[4] Here we use conventional terminology "stratiform" for the trailing light-rain region of a squall line, even though this region may contain some embedded convective elements aloft and it owes its existence largely to hydrometeor transport from the leading convective line (Houze 1997).

gravity wave and uses linear theory to predict the modes of maximum growth rate (Lindzen 1974; Raymond 1975, 1976). Wave-CISK as applied to severe storms has been reviewed by Lilly (1979). Ooyama (1982) has argued against the usage of the term "wave-CISK" for squall lines, etc., since CISK generally refers to a process introduced by Charney and Eliassen (1964) to explain hurricane genesis as a cooperative instability involving deep convection and large-scale moisture convergence. Raymond (1987) recommends that a more appropriate terminology for the wave-CISK-type process is *forced gravity wave mechanism*. This concept has been used by a number of authors to explain the development and propagation of squall lines, but results are highly parameterization dependent. According to Raymond (1987), "In spite of the difficulties with existing forced gravity wave models, the *idea* that gravity waves interact constructively with convection to produce propagating convective systems remains an attractive one. However, a more accurate treatment of convection is needed before the idea can be seriously tested against observation."

4) ORGANIZED BOUNDARY LAYER CIRCULATIONS

When the atmospheric boundary layer is heated from below, organized circulations often develop having a vertical scale equivalent to the depth h of the boundary layer and horizontal scales ~1 to 50 times h. For winds ~5–10 m s^{-1} or greater, boundary layer plumes or thermals tend to become approximately aligned with the wind, where they constitute the updraft portions of horizontal rolls or helical circulations (LeMone 1973). Rolls are generally thought to be formed by the along-roll wind shear and thermal instability (Kuo 1963; Asai 1970) or wind-shear curvature (Kuettner 1971), which are argued to organize convection into wind-parallel bands. Linear theory of inflection-point or convective instability predicts roll circulations with aspect (width to depth) ratios of 2–4 (Brown 1980). However, rolls with aspect ratios much larger than these (up to 15) have been observed, which are attributed by Etling and Brown (1993) to vortex pairing or merger and interactions with gravity waves in the free troposphere (Balaji and Clark 1988). When adequate moisture is available, the visible evidence of the rolls is cloud streets, where the individual clouds are initiated by periodic along-roll vertical velocity maxima (Kuettner 1959; Christian and Wakimoto 1989). There are instances over the ocean where complex, but organized, patterns of boundary layer circulations (e.g., lines, hexagonal cells, etc.) develop whose horizontal dimension is much greater than h (~30–50 km) and whose boundaries are foci for showers and thunderstorms (e.g., Agee 1984).

3.4. Triggering of convection

Convective-triggering mechanisms can operate in isolation (e.g., storms along a gust front) or in combination (e.g., gust fronts colliding with each other or with topography). In this section, we first consider these processes in isolation. Then combinations of these processes are treated.

a. Local processes

Buoyancy-driven circulations in the CBL are often sufficient to trigger convection. They can take the form of thermals or, in the case of a flat surface and moderate winds, horizontal convective rolls (HCRs). The growth of many isolated cells arises from thermals. In some cases the growth of cells is restricted by inversions or entrainment. Large numbers of fair-weather cumulus clouds arise simply from overshooting buoyant eddies in the CBL, referred to as *forced cumulus* (Esbensen 1978; Stull 1985). Others achieve sufficient buoyancy from latent heat release to rise to great heights, referred to as *active cumulus*. HCRs often lead to cloud rows or streets (Kuettner 1959), but seldom to severe weather in isolation. However, from a forecasting standpoint, it is important to recognize that HCRs lead to substantial inhomogeneities in the boundary layer moisture field (the updraft portions of HCRs are relatively moist; Weckwerth et al. 1996) so that an individual rawinsonde observation may not properly represent the true convective potential of the boundary layer.

Terrain forcing is a common trigger of convection. Houze (1993) discusses various ways in which this can occur (see his Fig. 12.24). Cloud initiation often arises from leeside convergence (Banta 1990) as the cold pool "burns off" in the morning and upslope flow develops (Fig. 3.19). Such storms do not usually

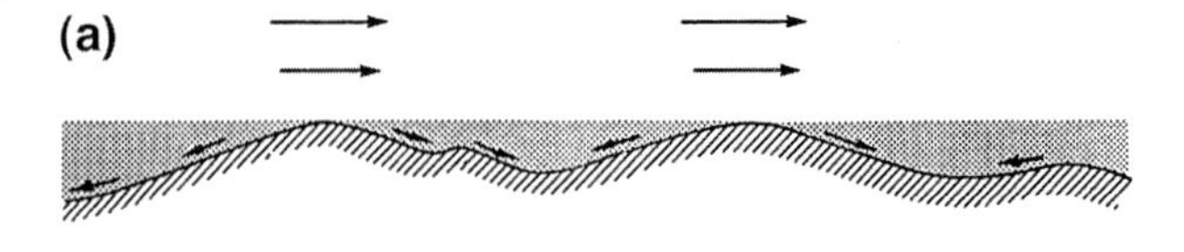

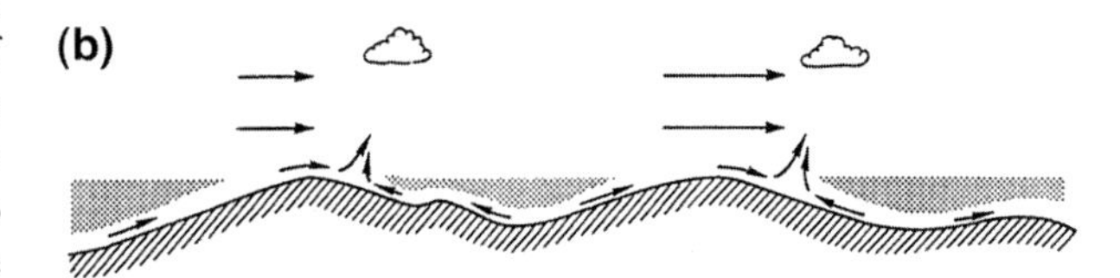

FIG. 3.19. Schematic cross section of (a) the nocturnal inversion layer, or "cold pool" buildup at night, and (b) the appearance of the shallow mixed layer the next morning. The inversion layer is shaded. The convergence zone forms at the uphill edge of the cold pools and leads to cloud initiation. From Banta (1990).

become severe; however, as they propagate downstream, they may develop into severe storms or organize into mesoscale convective systems (e.g., Tripoli and Cotton 1989a,b). Small-scale topographic features such as the "Caprock" escarpment in the Texas panhandle (Newton 1963), the Ozark Mountains (Hagemeyer 1984), and even the small ~100-m high Wichita Mountains in Oklahoma (Bluestein and Woodall 1990; Bluestein and Hutchinson 1996) can contribute directly to the initiation of severe convective weather through localized thermally induced lifting or earlier attainment of the convective temperature. Flow deflection by topography can also trigger severe weather by generating low-level convergence, as demonstrated by Watanabe and Ogura (1987) in a study of a flash flood in western Japan, or by orographic lift, as in the 1997 Fort Collins flood (Petersen et al. 1999).

Surface inhomogeneities in soil moisture or vegetation type can lead to both preconditioning and direct triggering of convection (Anthes 1984). This effect has been demonstrated in observational studies showing cumulus clouds to form first in Oklahoma over harvested wheat where the ground was warmer than adjoining areas dominated by growing vegetation (Rabin et al. 1990) and in Brazil where cumulus clouds developed preferentially over deforested land (Cutrim et al. 1995). This effect has been modeled by Lynn et al. (1998), who found that the total accumulated rainfall from mesoscale circulations generated by adjacent wet/dry patches was a maximum for patch sizes comparable to the radius of deformation (~130 km).

b. Advective processes

Convection often initiates along sharp boundary layer convergence lines of some type. Sometimes these lines are associated with density contrasts, but not always (Wilson and Schreiber 1986). Often they are outflows from previous convection. These convection-initiating convergence lines are typically very narrow, with horizontal dimensions of just a few km; really they are convective scale in the cross-line dimension, almost by definition (Byers and Braham 1949; Purdom 1982; Wilson and Schrieber 1986; Carbone et al. 1990; Lee et al. 1991; Wilson et al. 1992; Kingsmill 1995; Fankhauser et al. 1995). Shallow convergence lines, inadequate to trigger convection, may deepen the moist layer such that subsequent or additional lifting is more effective at initiating convection. These narrow convergence lines are visible as radar echo lines, apparently because insect concentrations are increased (Wilson and Schreiber 1986).

Fankhauser et al. (1995, their Figs. 20 and 21) and Crook and Tuttle (1994) show how initially broad convergence zones rapidly collapse to become fine lines of convergence in a flow model (the same process as in frontal collapse). The implication is that many diagnosed "mesoscale" convergence zones may actually be narrow convective-scale convergence lines, undersampled by sparse wind measurements. Convergence features above the surface can also affect convective development, especially in cases in which the convectively unstable air lies above the surface (e.g., Rochette and Moore 1996).

Several examples of direct triggering of deep convection by cold-frontal lifting were cited earlier (Browning and Harrold 1970; Carbone 1982). In another case, Ogura and Portis (1982) found a direct vertical circulation with moist warm air ascending directly above the surface front that apparently triggered and sustained severe storms. In a numerical modeling study of the Carbone (1982) California squall line, Parsons (1992) found that the intense updrafts (up to 20 m s^{-1}) at the front could be explained in terms of the gravity current–shear interaction concept of Rotunno et al. (1988). Parsons's results are illustrated in Fig. 3.20, which shows the most vigorous updrafts occurring when there is an optimal balance between the horizontal vorticity associated with the low-level wind shear and that associated with the cold air behind the front. An important distinction between the California cold front and the optimal balance for squall lines described by Rotunno et al. (1988) is the absence of CAPE and therefore appreciable buoyancy effects in the approaching flow. Intense upward motion and a narrow band and heavy precipitation were achieved solely by a strong upward-directed pressure force from convergence at the front.

While advective phenomena such as gust fronts, sea/lake breezes, and drylines represent loci for convection, deep convective cells typically do not form everywhere along them. However, intersections of these features can trigger storms. For example, the intersection of a dryline with a front (sometimes referred to as a *triple point* since it separates three different air masses) can lead to convective storm formation in its proximity (e.g., Bluestein 1993). The collision of gust fronts (Mahoney 1988) or sea breezes can lead to vigorous convection along the axes of intersection. Sea breeze collision often occurs over relatively narrow peninsulas or flat islands. A notable example is the daily occurrence of an intense thunderstorm "Hector" over Bathhurst and Melville Islands just north of Darwin, Australia, during the summer monsoon (Keenan et al. 1990, 1993; Simpson et al. 1993).

Whatever contributes to inhomogeneities along fronts, drylines, etc., may serve as a trigger for convection. CBL thermals may be sufficient in some cases. In other cases, horizontal convective rolls may provide the enhanced lift. It is the collision or intersection of these phenomena with each other, or with other features such as terrain or horizontal convective rolls, that provides the most energetic triggering of convection (Wilson and Schreiber 1986). The sub-

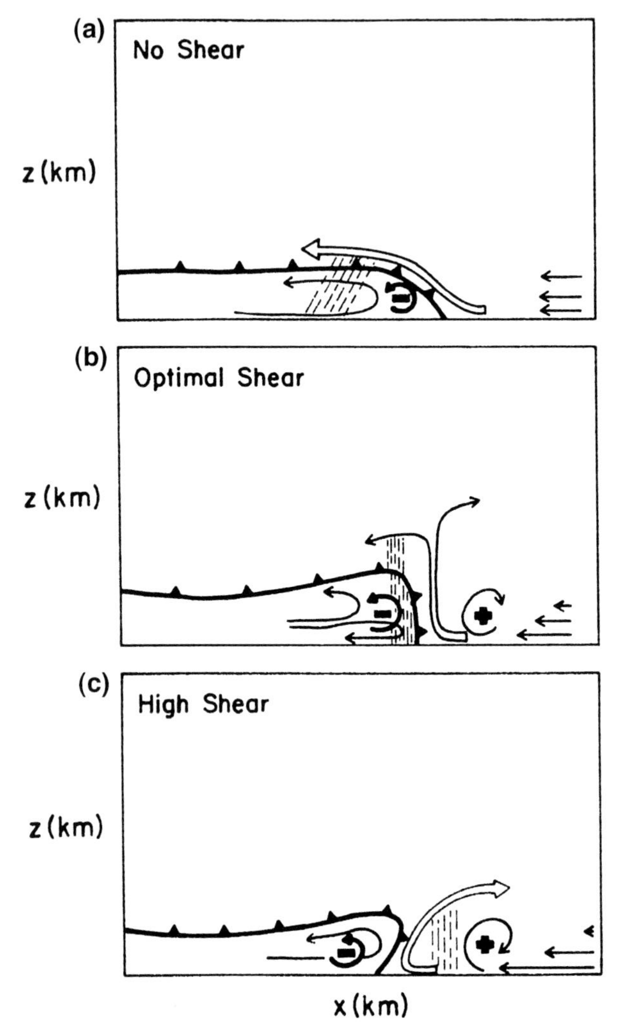

FIG. 3.20. A schematic of the dependence of the circulations at a cold front upon vertical shear. The horizontal vorticity due to the low-level vertical shear ahead of the front and that due to the horizontal gradient of buoyancy associated with the leading edge of the cold air mass are indicated. (a) Low-shear simulations with a sloped updraft and a broad area of precipitation that trails the front. The flow within the cold air mass is in the same sense as in a classical dry gravity current. (b) Optimal vertical shear with a deep and intense vertically oriented updraft and a narrow band of heavy rainfall. (c) Higher-than-optimal vertical shear with an updraft sloping into the warm air. The flow is unsteady due to the effect of precipitation interrupting the inflow and the influence of potential instability created by the overrunning of cold air. The airflow within the cold air mass is different from a classical gravity current due to an initial flow being prescribed within the cold air mass but is without the flow reversal evident in the optimal-shear simulations. From Parsons (1992).

ject of combined lifting processes will be treated in section 3.4d.

c. Dynamical processes

Horizontal convective rolls are thought to represent some form of a dynamical instability (e.g., Brown 1980), but as mentioned before, they generally do not trigger severe storms in isolation from other lifting mechanisms. In some instances, gravity waves have been linked to the triggering of severe convective weather (Uccellini 1975), but more commonly the triggering of convection by them occurs in conjunction with other processes, as described below. In the Tropics, where CIN is relatively weak, gravity waves or bores have been identified as direct triggers for squall lines, for example, morning glory-generated squall lines over northern Australia (Smith and Page 1985; Drosdowsky and Holland 1987).

Certain mesoscale instabilities may be related to the triggering of severe weather. For example, horizontal shearing instabilities have been linked with tornadogenesis (Carbone 1982; Wakimoto and Wilson 1989), although the parent clouds were presumably initiated by other processes. Similar instabilities may have initiated the tornado-like vortices in Hurricane Andrew (Wakimoto and Black 1994; Schubert et al. 1999). A mesoscale instability was implicated in triggering convection in a localized outbreak of severe storms in the Oklahoma–Kansas area (Sanders and Blanchard 1993). A very strong inversion existed in the region of the outbreak. While transverse circulations associated with a jet streak were not adequate to locally break the lid in the area of convection, they did produce an environment within which a shearing instability developed characterized by strong vertical motions on a 200-km scale. It was this oscillation that apparently triggered the convection.

d. Combined lifting processes

The genesis of severe storms can often be traced back to combinations of local, advective, and dynamical lifting processes.

Recent work with radar data has shown that when boundary layer rolls intersect convergence lines, they can locally amplify vertical motion and trigger deep convection. This process is illustrated in Fig. 3.21 (from Wilson et al. 1992). On 17 July 1987, a Denver cyclone was present during the afternoon (1500 MDT or Mountain Daylight Time) in the Denver area (Fig. 3.21a). Note the prominent north–south convergence line just east of Denver (DEN). An enlargement of an area along the convergence zone (thick line) centered near DEN at 1645 MDT is shown in Fig. 3.21b. Several showers were located 5–10 km east of the convergence line at this time. Also indicated are the updraft axes of horizontal rolls (thin lines). The rolls were subjectively identified using "clear-air" reflectivity and velocity data from a multiple-Doppler radar network. The analysis shows a clear correspondence between the roll updrafts and the clouds, suggesting that a superposition of lifting mechanisms—the convergence line and the rolls—contributed to cloud development. The primary effect of the convergence

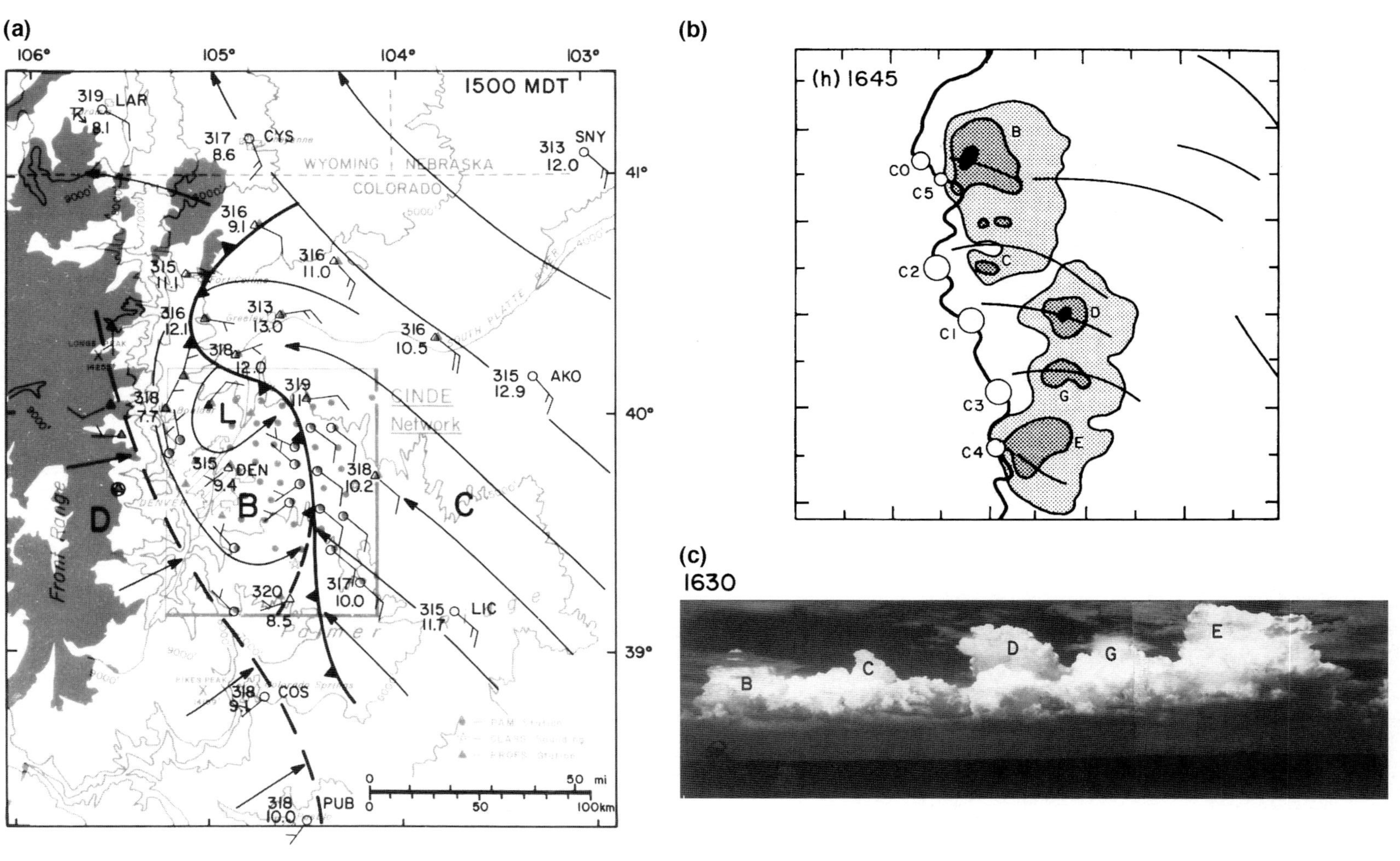

FIG. 3.21. (a) Surface map at 1500 MDT on 17 July 1987. The indicated winds and streamlines are based on NWS, PROFS, and selected PAM stations. A full barb is equivalent to 5 m s^{-1}, and a half-barb, 2.5 m s^{-1}. Potential temperatures and mixing ratios are shown next to the selected stations. The low is the Denver cyclone, and the cold-front type boundary is the Denver convergence line. The letters represent different air masses, and the inner box delineates the CINDE network and boundaries of (b). (b) Relative location of convergence line (heavy solid line), horizontal rolls (light solid lines), misocyclones (labeled open circles beginning with letter C), and precipitation echo at 6.5 km MSL (contours shown are 10, 30, and 50 dBZ_e) at 1645 MDT. (c) Photograph of clouds B, C, D, E, and G in (b). From Wilson et al. (1992).

line was to deepen the moist layer locally and provide a region potentially favorable to deep convection. A photograph of clouds B, C, D, E, and G from Fig. 3.21b is shown in Fig. 3.21c. Also indicated in Fig. 3.21b are positions of small-scale vortices or misocyclones along the convergence line. Wilson et al. (1992) found that these misocyclones formed where the rolls intersected the convergence line and that they were important in the initiation phase of a number of the radar echoes. When misocyclones became colocated with convective updrafts, they produced, in several instances, nonsupercell tornadoes (Wakimoto and Wilson 1989).

The above example illustrates convective triggering associated with the intersection of horizontal rolls and a terrain-induced convergence zone. However, similar triggering has recently been documented in connection with sea breeze front and roll intersections over south Florida (Wakimoto and Atkins 1994; Atkins et al. 1995; Fankhauser et al. 1995; Kingsmill 1995). Atkins et al. found that close to the front the roll axes were tilted upward and lifted by the frontal updrafts, leading to a deeper updraft and an additional impetus for cloud development (Fig. 3.22). Also, the same mechanism for cloud formation has recently been reported to occur along a dryline (Atkins et al. 1998). Based on these results, convective triggering may be rather commonplace as a result of intersections of rolls with a wide variety of convergence zones, for example, fronts, outflow boundaries, and so forth.

Observations suggest that dryline convection tends to be isolated, exhibiting a variety of modes of development (Bluestein and Parker 1993). The inhibiting factor is typically a strong capping inversion, which is broken only by enhanced lifting (Hane et al. 1997). A number of mechanisms have been suggested that could contribute to localized, enhanced lifting along drylines leading to severe storms: mesoscale low pressure areas (Bluestein et al. 1988), intersection of fronts with drylines (Shapiro 1982; Schaefer 1986; Parsons et al. 1991), intersection of boundary layer rolls or cloud lines with drylines (Hane et al. 1997; Atkins et al. 1998), and mesoscale gravity waves interacting with drylines (Koch and McCarthy 1982; McCarthy and Koch 1982).

To illustrate the complexity of processes associated with severe storm initiation along a dryline, consider the dryline of 26 May 1991 investigated by Hane et al. (1997) with aircraft, sounding, and mesonetwork data. In Fig. 3.23 results are shown from aircraft sawtooth traverses of the dryline at 860 mb (~1 km AGL). A cloud line was observed to intersect the dryline near the center of the domain, at which location the first

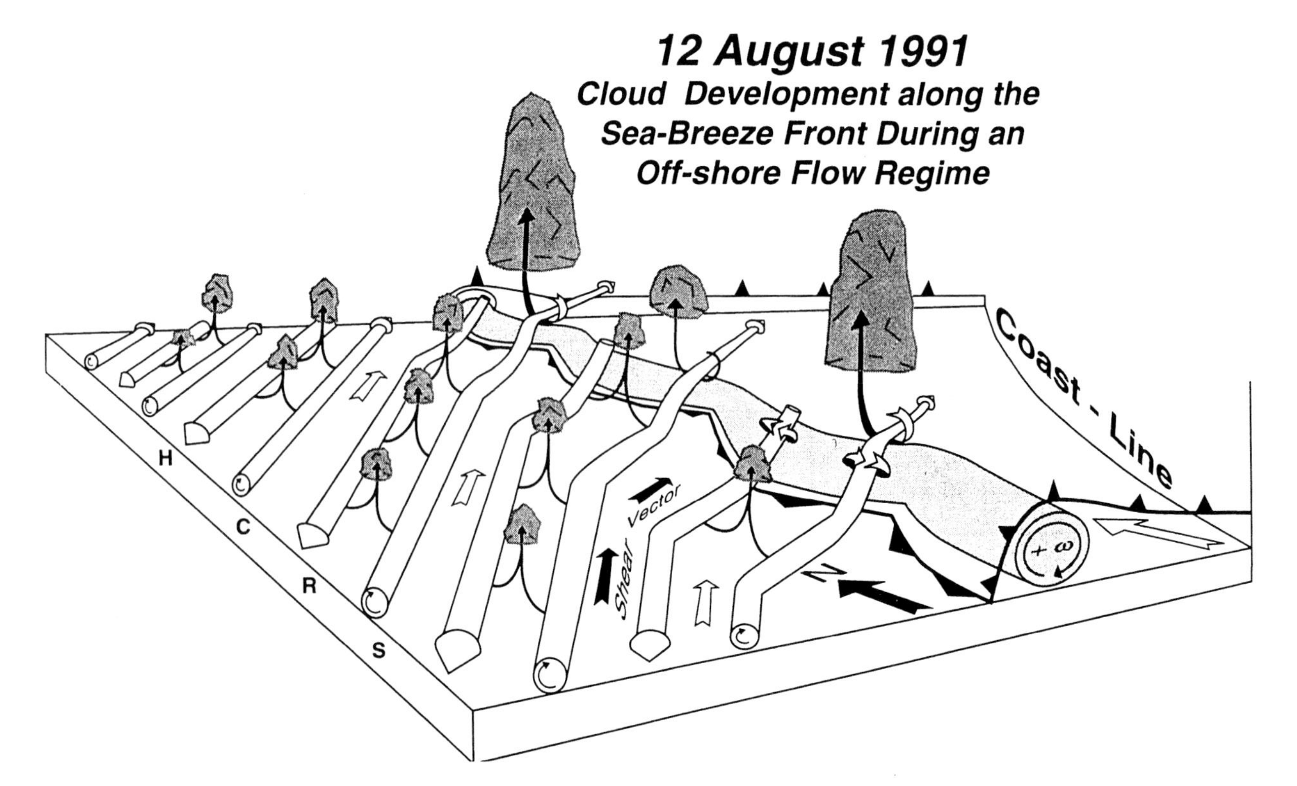

FIG. 3.22. Schematic diagram showing the interaction between the sea-breeze front and horizontal convective rolls (HCRs) and how it relates to cloud development on 12 August 1991. The sea-breeze front is delineated by the heavy, barbed line. The head circulation is lightly shaded. The horizontal vorticity vectors associated with the counterrotating roll circulations are shown. Clouds along the HCRs and at the intersection points along the front are shaded gray. The shear vector (solid, 2D arrow) and low-level winds (white, 2D arrow) are also shown. From Atkins et al. (1995).

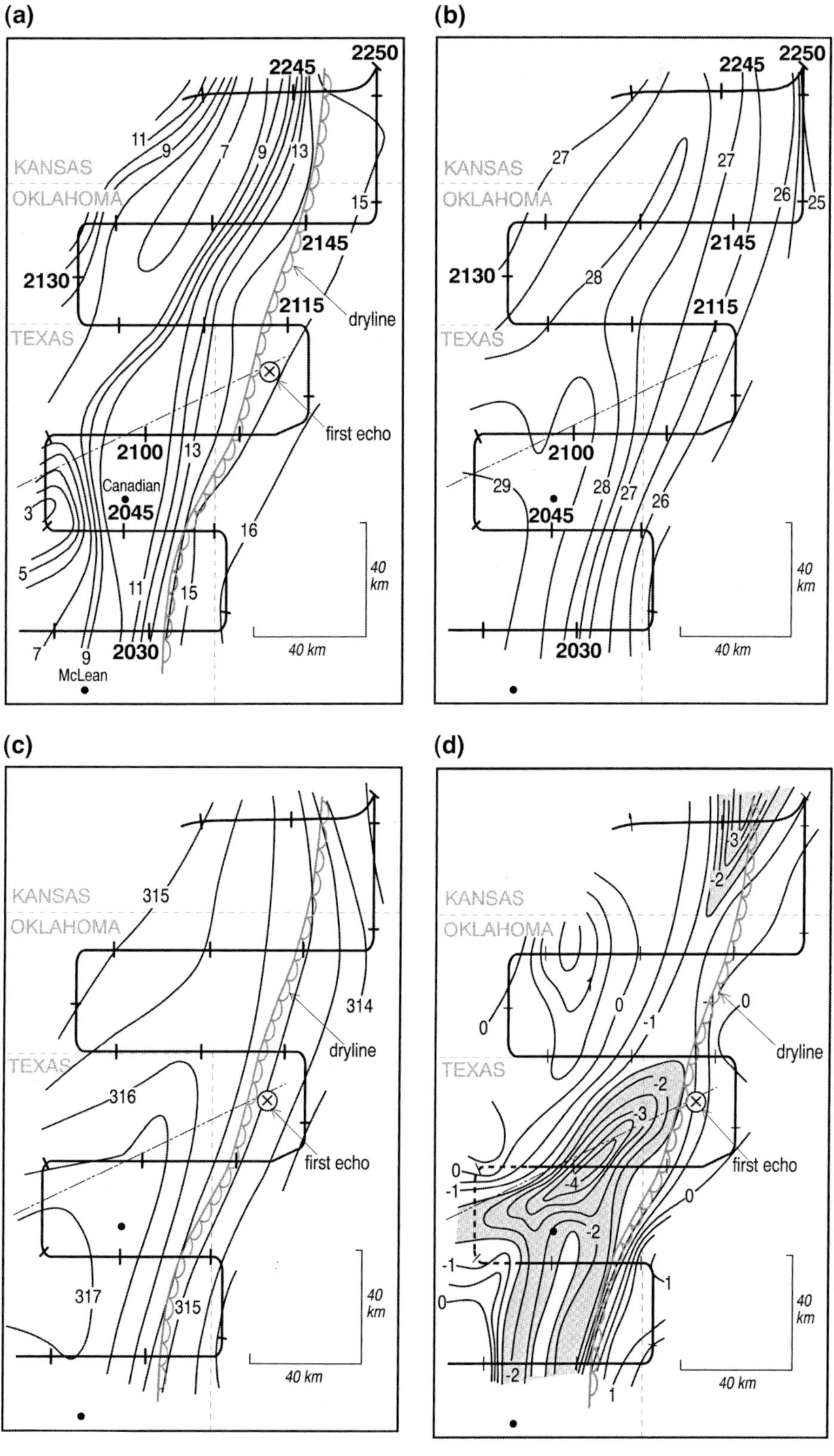

FIG. 3.23. (a) Dewpoint distribution in the dryline region based on aircraft observations (sawtooth pattern) at 860 mb. Dewpoint temperature (°C) shown by solid contours is derived from aircraft measurements along track shown. Dryline is indicated by scalloped line and cloud line by dash-dot-dot line; location of first echo also noted. (b) Temperature field (°C), (c) virtual potential temperature (K), and (d) horizontal divergence (10^{-4} s^{-1}). From Hane et al. (1997).

echo formed (at 2115 UTC), which later became one of several tornadic storms that developed along the dryline on this day. The rather sharp moisture discontinuity across the dryline can be seen in Fig. 3.23a, although the moisture drop is not everywhere uniform behind the line. Temperatures at this altitude were 2°–3°C warmer to the west of the dryline (Fig. 3.23b) as a result of the stronger surface sensible heat fluxes and a deeper mixed layer there. A 1°–2°C gradient in θ_v existed across the dryline (Fig. 3.23c), strongest in the southern part, supporting the existence of a dryline solenoidal circulation (Parsons et al. 1991; Ziegler and Hane 1993; Ziegler et al. 1995). Axes of convergence were detected at flight level both along and just behind the dryline and along the cloud line (Fig. 3.23d). Hane et al. argue that convective clouds developed along the cloud line in dry air and were advected across the dryline zone through a strongly convergent region. They encountered increasing low-level moisture as they moved east-northeastward, and upon reaching the deep moisture east of the dryline, they grew vigorously. A key question, though, is what produced the cloud line in the first place? An explanation for this feature turned out to be elusive, although satellite data suggest that surface vegetative inhomogeneities may have played some role in its generation. This example points to the difficulty in short-term (several hours) forecasting of convection along drylines but some potential for reliable nowcasting (0–30 min) of storm initiation with frequent, high-resolution visible satellite imagery.

There are other examples of combined lifting processes: gust fronts intersecting terrain (e.g., the 1979 Fort Collins hailstorm with softball-sized hail; Fritsch and Rodgers 1981); undular bores interacting with the dryline (e.g., the 14 April 1986 high plains severe weather outbreak; Karyampudi et al. 1995); interactions among gravity currents, bores, drylines, and low-level jets (e.g., the explosive convective development of 26–27 May 1985 in Kansas; Carbone et al. 1990); interactions between gust-front-generated internal gravity waves and Kelvin–Helmholtz waves to produce new convective cells atop a thunderstorm outflow in Alabama (Weckworth and Wakimoto 1992); and a combination of cloud forcing from thermally direct boundary layer circulations and gravity waves (Balaji and Clark 1988). The complexity of such processes represents a serious challenge for short-term forecasting.

3.5. Storm-generated mesoscale processes

Severe storms generate a host of mesoscale effects. Some act to promote storm development, severity, and longevity, for example, cold pool lifting, vortex tilting/stretching, generation of mesoscale pressure fields by dynamic and buoyant effects, vortex breakdown. Others act to weaken storms, for example, rapidly spreading cold pools, cloud shading, changed shear profiles due to gravity waves. A list of such processes is given in Table 3.3. In this section, we discuss these processes according to whether they are local, advective, or dynamical in nature.

a. Local effects

An example of a local process inside an existing cloud system is radiation. Radiative transfer is not local in the vertical direction, but it tends to transfer energy within a single vertical column of the atmosphere, in contrast to the horizontal transports effected by advective and dynamical processes. On the ~1–2-h timescale of most severe storms, cloud–radiative effects are not important in storm evolution. However, they may be important in the development of new storms, for example, from cloud shading of cirrus anvils generating inhomogeneities in boundary layer properties and subsequent convergence zones (McNider et al. 1995; Markowski et al. 1998). Also, on the longer (~6–12 h) timescale of MCSs, cloud–radiative interactions may be important in promoting the longevity of storms by enhancing the mass circulation within them (Gray and Jacobson 1977; Chen and Cotton 1988; Dudhia 1989) or by trapping storm-generated gravity waves in the upper troposphere through cloud-top radiative destabilization (Tripoli and Cotton 1989a,b). They may also increase storm total precipitation through longwave radiative cooling at cloud top and heating at cloud base (Tao et al. 1993).

Likewise, microphysical processes such as phase changes or particle spectrum evolution tend to have local effects, although these effects can later be advected to other regions. Production of the surface cold pool occurs principally through evaporation (e.g., Sawyer 1946) and melting (Atlas et al. 1969). The strength of the cold pool is important to supercell behavior and longevity (Weisman and Klemp 1982), to squall-line intensity and longevity (Thorpe et al. 1982; Rotunno et al. 1988), and to baroclinic vorticity generation in tornadic storms (Klemp 1987). In MCSs, sublimation, evaporation, or both, is important in causing the rear-to-front flow aloft (the *rear-inflow jet* after Smull and Houze 1987) to descend to the lower troposphere (Zhang and Gao 1989; Stensrud et al. 1991; Braun and Houze 1997). In microbursts, particle sizes are important in determining downdraft intensity, with smaller hydrometeors being most conducive to strong downdrafts (Brown et al. 1982; Srivastava 1985, 1987; Proctor 1989). In frontal rainbands, evaporation, sublimation, and melting can have the effect of enhancing the thermal contrast across cold fronts, thereby increasing their intensity and longevity (Parsons et al. 1987; Rutledge 1989; Barth and Parsons 1996).

b. Advective effects

A very common effect of convection on subsequent convection involves the outflow of cold air from convective downdrafts in a density current along the earth's surface. Cold pools are responsible for cell regeneration in multicell storms (Browning et al. 1976). As the cold pool spreads out over a large area and becomes shallow, its lifting may no longer be sufficient to cause air parcels to reach their LFC, and new cell growth will cease. Downdraft outflows are also important in supercells, where forward-flank downdrafts (FFD, produced by the downstream advection and evaporation of condensate) and rear-flank downdrafts (RFD, dynamically induced as strong low-level rotation lowers the pressure locally and draws air from above; Klemp and Rotunno 1983) produce storm-scale "fronts" (Fig. 3.24) that intersect at the main center of supercell rotation (the mesocyclone center; Lemon and Doswell 1979). Tornadoes generally form in the updraft ahead of the RFD (Fig. 3.24). As RFD advances, cold air is ingested into the updraft at the point of occlusion of the "fronts," thereby weakening the mesocyclone. However, as shown by Burgess et al. (1982), on some occasions new mesocyclones may form at the occlusion, leading to a succession of tornadoes with near-parallel tracks (Fig. 3.25).

Advection of condensate is also an important factor in the development of trailing stratiform regions of squall lines. Falling snow particles generated at upper levels in the leading convective line are advected rearward by the storm-relative front-to-rear flow aloft (Fig. 3.26), where they eventually fall out in a secondary band of precipitation (Biggerstaff and Houze 1991). The precipitation in the secondary band is also produced in part by in situ generation of condensate by a mesoscale updraft (e.g., Rutledge and Houze 1987). Houze et al. (1990), in a study of six years of springtime squall lines in Oklahoma, found that nearly

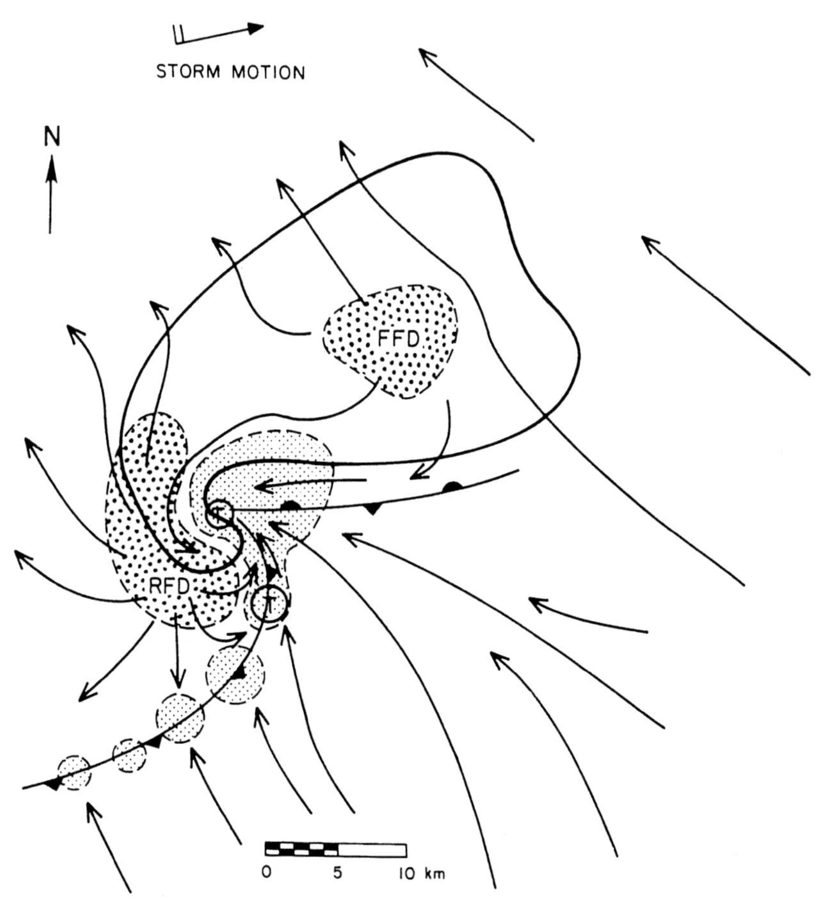

FIG. 3.24. Schematic plan view of a tornadic thunderstorm near the surface. The thick line encompasses the radar echo. The cold-front symbol denotes the boundary between the warm inflow and cold outflow and illustrates the occluding gust front. Low-level position of the updraft is finely stippled, while the forward-flank (FFD) and rear-flank (RFD) downdrafts are coarsely stippled. Storm-relative surface flow is shown along with the likely location of tornadoes (encircled T's). From Lemon and Doswell (1979).

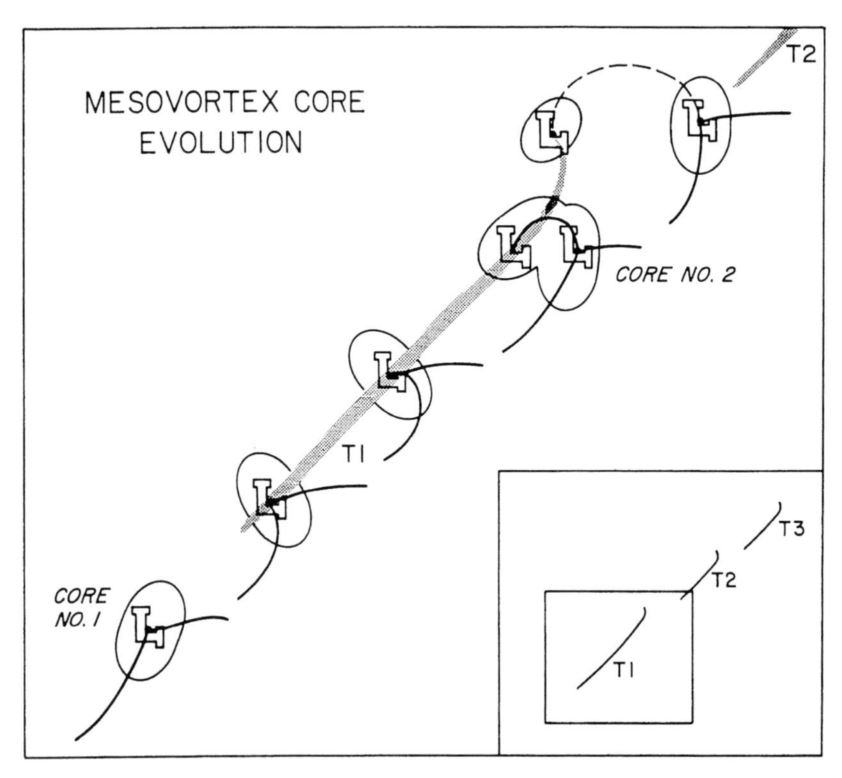

FIG. 3.25. Conceptual model of mesocyclone core evolution. Low-level wind discontinuities (thick lines) and tornado tracks (shaded). Inset shows the tracks of the tornado family and the small square is the expanded region in the figure. From Burgess et al. (1982).

two-thirds of the lines exhibited a leading-line/trailing-stratiform-region structure.

Momentum transport also represents an advective process in severe storms. Newton (1950) has shown that in squall-line systems, vertical transport of horizontal momentum helps to generate convergence at the leading convective line. A similar process may be operating in bow echoes associated with derechos where rear-inflow jets descend to the surface (Weisman 1993; Przybylinski 1995), although extreme surface winds in derechos may also be a direct consequence of downburst winds impacting the surface (Fujita and Caracena 1977) or intense surface pressure gradients that develop within the storm (Schmidt and Cotton 1989).

Another advective effect is vortex tilting, which in vertically sheared environments leads to the development of vertical vorticity couplets in storm updrafts (Rotunno 1981; Davies-Jones 1984; Klemp 1987). This process is illustrated in Fig. 3.27a, where a tube of low-level horizontal vorticity associated with unidirectional shear is tilted vertically, generating positive and negative vorticity centers within the updraft. Later, as the downdraft develops (Fig. 3.27b), the vortex tube is tilted downward, producing two vortex pairs. Given sufficient vertical shear, the storm can split into right- and left-moving cells as a result of dynamical effects, namely, lifting pressure gradients that reinforce new updraft growth on the southern and northern flanks of the central updraft (Schlesinger 1980; Rotunno and Klemp 1982).

Tilting can also produce vertical vorticity in MCSs. An example is the bow echo (Fig. 3.28), which exhibits a vortex couplet (cyclonic on the north end and anticyclonic on the south; Fujita 1978). Bow echoes are often associated with severe surface winds. Weisman (1993) argues that the primary mechanism for the generation of the vortex couplet or "bookend vortices" is tilting of horizontal vorticity associated with the ambient vertical shear by precipitation downdrafts (note cyclonic and anticyclonic eddies at northern and southern ends of the downdraft in Fig. 3.27b). In a numerical modeling study of the bow echo, Weisman (1993) found that the bookend vortex pair associated with the downdraft persisted throughout the storm simulation, while the anticyclonic and cyclonic eddies associated with the updrafts on the northern and southern ends of the line (Fig. 3.27b) did not persist as new, shorter-lived updrafts were generated at the leading gust front.

The development of mesovortices in MCSs such as bow echoes has been studied from a potential vorticity perspective by Davis and Weisman (1994) and Weisman and Davis (1998). They find that, in addition to the mechanism described by Weisman (1993) involving an interaction between the downdraft and the

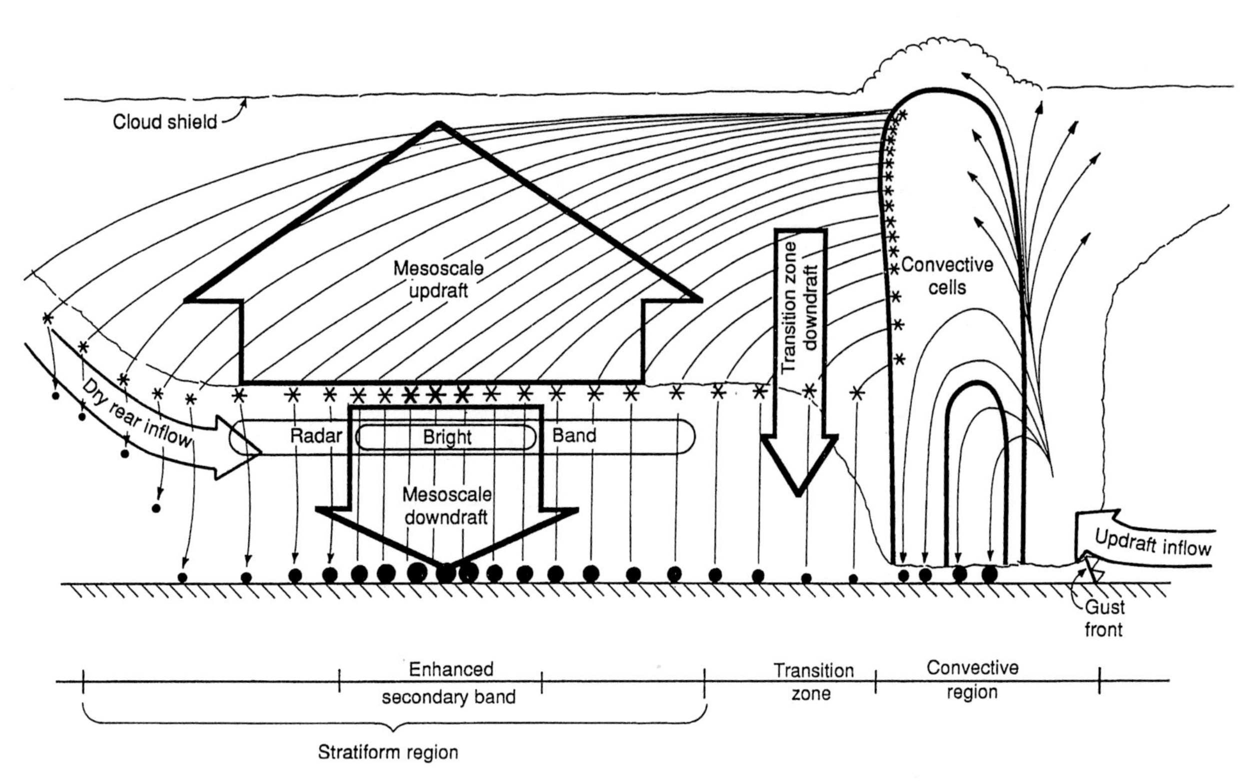

FIG. 3.26. Conceptual model of the two-dimensional hydrometeor trajectories through the stratiform region of a squall line with trailing stratiform precipitation. From Biggerstaff and Houze (1991).

ambient westerly shear (Fig. 3.29b), another tilting mechanism is operative involving perturbation shears associated with the storm itself. Buoyancy forces act to generate front-to-rear (FTR) and rear-to-front (RTF) flow, that is, perturbation shears and a horizontal vortex tube pointed toward the south (Fig. 3.29a). The tilting of this vortex tube by the ascending FTR flow at the leading edge of the storm serves to strengthen the vortex couplet. The numerical simulations by Weisman and Davis (1998) indicate that tilting of perturbation shears generated by the cold pool is important in the production of line-end vortex pairs in environments with weak-to-moderate shear, whereas tilting of ambient vorticity is operative in environments with stronger and deeper shear. The cyclonic circulation at the north end of a bow echo often resembles the mesovortex observed within the trailing stratiform region of mature squall lines (Zhang and Fritsch 1986, 1988; Houze et al. 1989; Skamarock et al. 1994; Loehrer and Johnson 1995); however, the latter are frequently of much larger scale (several hundred km across; Bartels and Maddox 1991). As squall lines mature, Coriolis effects appear to be important in the development of mesovortices on their north ends (Zhang 1992; Skamarock et al. 1994; Weisman and Davis 1998), leading to the eventual evolution of many squall lines to an asymmetric precipitation pattern (Houze et al. 1990; Loehrer and Johnson 1995).

Advective changes also include the effects of the plume of heat and moisture, both vapor and ice, lofted by primary convection. Weisman (1992) argues that the convective/advective warming of the midtroposphere is the primary mechanism for the generation of the midlevel rear-inflow jets in squall lines. Moistening can increase the buoyancy of secondary convection that entrains the air, compared to the buoyancy it would have if it entrained drier unmodified environmental air. For example, Table 3.2 shows the (mostly modest) effects on instability indices if the relative humidity above the 900-mb level is raised to 80%, and this air is entrained by the updraft parcel. Moistening of the environment, particularly at low levels, by the detrainment of moist cloudy air might play a larger role in encouraging the development of deep convection than is indicated by the small continuous entrainment rate calculations of Table 3.2.

c. *Dynamical effects*

Dynamical effects of convection are numerous, occurring on scales ranging from individual cells up to the scale of MCSs and even the synoptic scale. One important process is the generation of mesoscale pressure fields through buoyancy and dynamic effects. On the storm scale, dynamical effects (lifting pressure gradients on the flanks of updrafts) contrib-

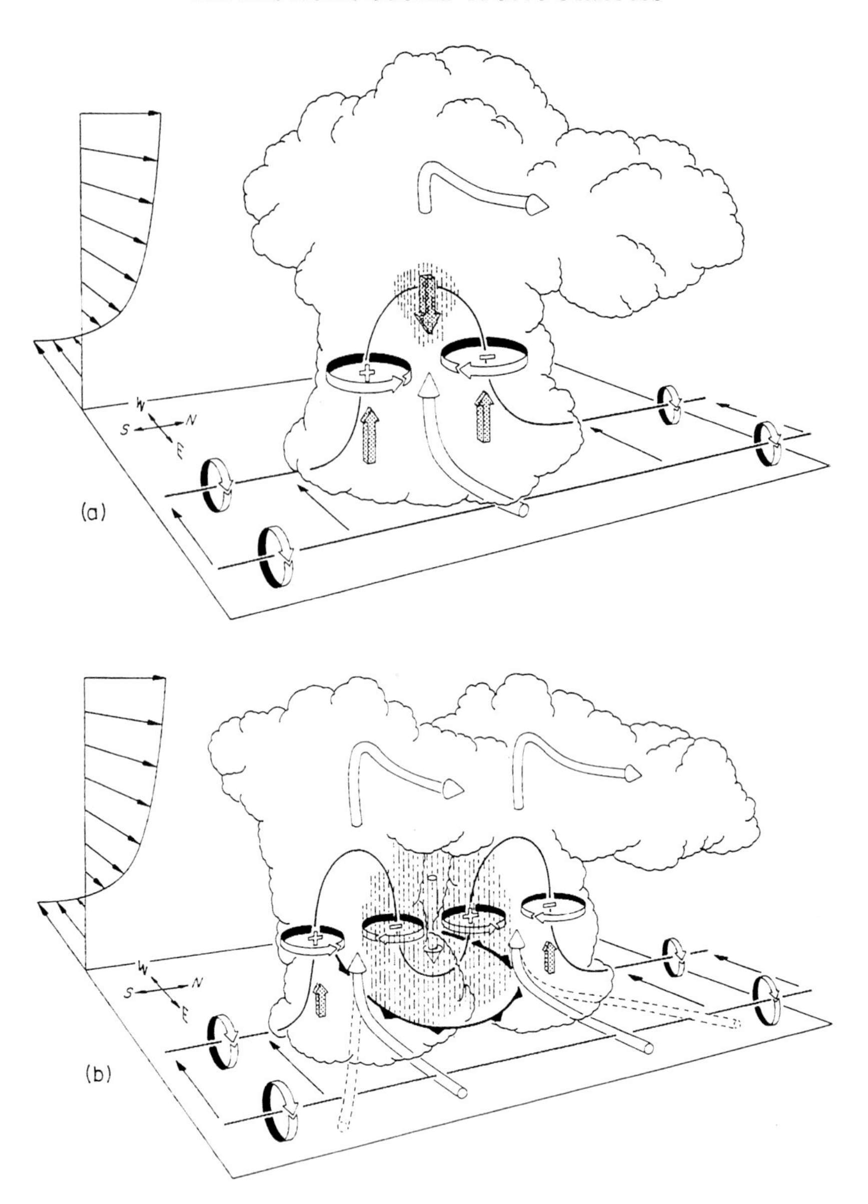

FIG. 3.27. Schematic depicting how a typical vortex tube contained within (westerly) environmental shear is deformed as it interacts with a convective cell (viewed from the southeast). Cylindrical arrows show the direction of cloud-relative airflow, and heavy solid lines represent vortex lines with the sense of rotation indicated by circular arrows. Shaded arrows represent the forcing influences that promote new updraft and downdraft growth. Vertical dashed lines denote regions of precipitation. (a) Initial stage: Vortex tube loops into the vertical as it is swept into the updraft. (b) Splitting stage: Downdraft forming between the splitting updraft cells tilts vortex tubes downward, producing two vortex pairs. The barbed line at the surface marks the boundary of the cold air spreading out beneath the storm. From Klemp (1987); adapted from Rotunno (1981).

ute to the splitting of supercells (Rotunno and Klemp 1982). The lifting arises from the dynamic lowering of pressure at midlevels within vorticity centers of both sign along the flanks of the updraft (Fig. 3.27a). Updraft growth is thereby induced on these flanks, leading to storm splitting. Environments with unidirectional shear yield an equal preference for right- and left-moving storms. However, in the central United States, the shear vector in severe storm situations normally turns clockwise with height (Maddox 1976), which leads to a favoring of right-moving supercells. Rotunno and Klemp (1982) explain this behavior by showing that high (low) pressure perturbations develop when the shear vector points toward

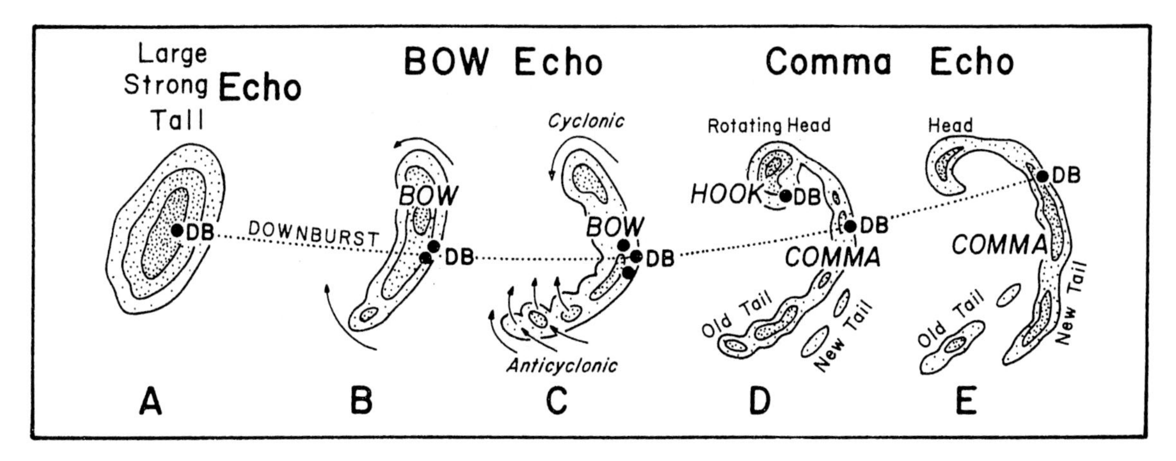

FIG. 3.28. A typical morphology of radar echoes associated with bow echoes accompanied by strong and extensive downbursts. Some bow echoes disintegrate before turning into comma echoes. During the period of strongest downbursts, the echo often takes the shape of a spearhead or a kink pointing toward the direction of motion. From Fujita (1978).

(away from) the updraft. In the case of unidirectional shear (Fig. 3.30a), this leads to an upward-directed pressure gradient on the leading edge of the storm and a downward-directed gradient to the rear. This configuration does not contribute to preferential growth on either of the flanks transverse to the shear. However, when the shear vector turns clockwise with height (Fig. 3.30b), the interaction of the shear with updrafts leads to an upward-directed pressure gradient on the right flank of the storm and a downward-directed gradient on the left, thus favoring new cell growth on the right.

The evolution of supercells to the tornadic phase involves a number of mesoscale dynamical processes that are not completely understood. Among others, they include the ingestion of streamwise vorticity into updrafts (Davies-Jones 1984), baroclinic vorticity generation along supercell forward-flank downdrafts (Klemp and Rotunno 1983; Rotunno and Klemp 1985), and mesocyclone vortex breakdown (Brandes 1978; Wakimoto and Liu 1998). The tornado itself can be subject to vortex breakdown, leading to smaller-scale suction vortices (Rotunno 1984; Davies-Jones 1986). These topics are treated in detail elsewhere in this volume.

Another dynamical process involves cold pool–shear interactions (Fig. 3.16), which influence the generation of new convective cells. It has been argued that continual regeneration is favored if there is an optimal balance between horizontal vorticity generated by the cold pool and that associated with the ambient low-level shear (Fig. 3.16d), yielding deeper, stronger, and more erect updrafts (Thorpe et al. 1982; Rotunno et al. 1988). However, the situation is often complicated by the existence of additional sources of vorticity when storms evolve to larger scales (Lafore and Moncrieff 1989; Weisman 1992). Moreover, the "optimal" state is not a requirement for long-lived

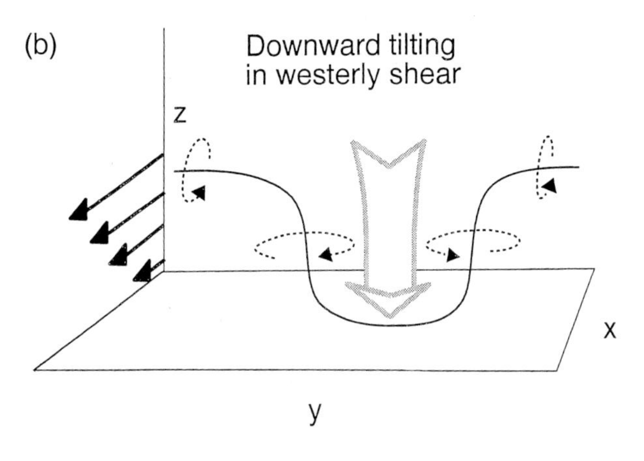

FIG. 3.29. Schematic of vertical vorticity generation through vortex tilting. For easterly shear (a), ascending motion tilts the vortex lines, resulting in cyclonic rotation on the north end and anticyclonic rotation on the south end. Localized descent in westerly shear (b) also produces the same pattern of vertical vorticity through tilting. From Weisman and Davis (1998).

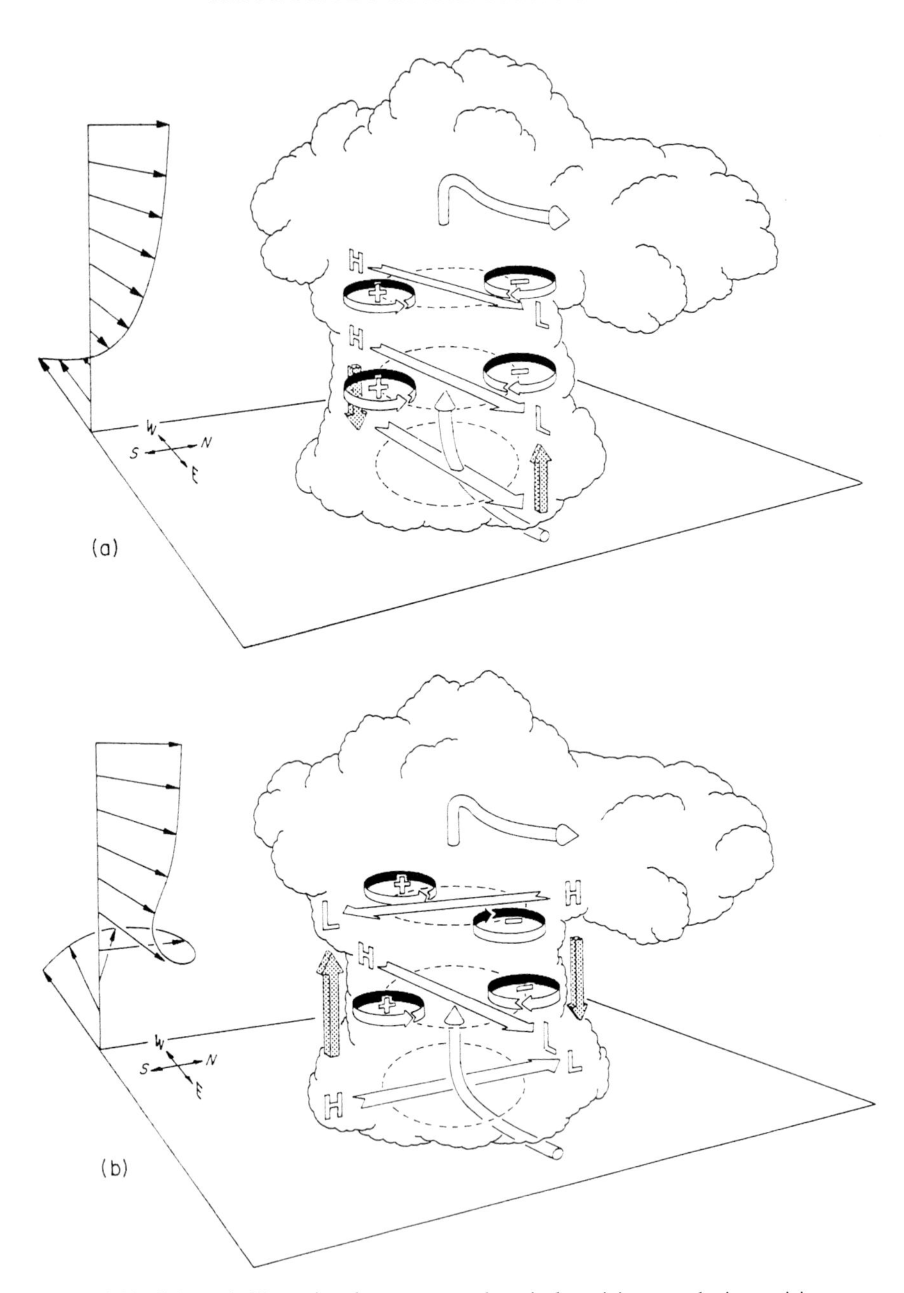

FIG. 3.30. Schematic illustrating the pressure and vertical vorticity perturbations arising as an updraft interacts with an environmental wind shear that (a) does not change direction with height and (b) turns clockwise with height. The high (H) to low (L) horizontal pressure gradients parallel to the shear vectors (flat arrows) are labeled along with the preferred location of cyclonic (+) and anticyclonic (−) vorticity. The shaded arrows depict the orientation of the resulting vertical pressure gradients. From Klemp (1987); adapted from Rotunno and Klemp (1982).

squall lines in all numerical simulations (Fovell and Ogura 1988, 1989; Grady and Verlinde 1997). Fovell and Ogura find the upshear sloping phase (Fig. 3.16b) to possess a quasi-equilibrium state.

Lafore and Moncrieff (1989) and Weisman (1992) propose that mesoscale influences come about through the following sequence of events.

1) An initial updraft leans downshear in response to the ambient vertical shear (Fig. 3.31a).
2) The circulation generated by the storm-induced cold pool balances the ambient shear, and the system becomes upright (Fig. 3.31b).
3) The cold pool overwhelms the ambient shear and the system tilts upshear. A rear-inflow jet (Smull

and Houze 1987) develops as horizontal buoyancy gradients along the rear edge of the buoyant plume aloft and the cold pool near the surface generate horizontal vorticity favoring rear inflow (Fig. 3.31c).

The development of rear-inflow jets can be important in producing intense surface winds (Weisman 1992; Przybylinski 1995). Rear-inflow jets can also

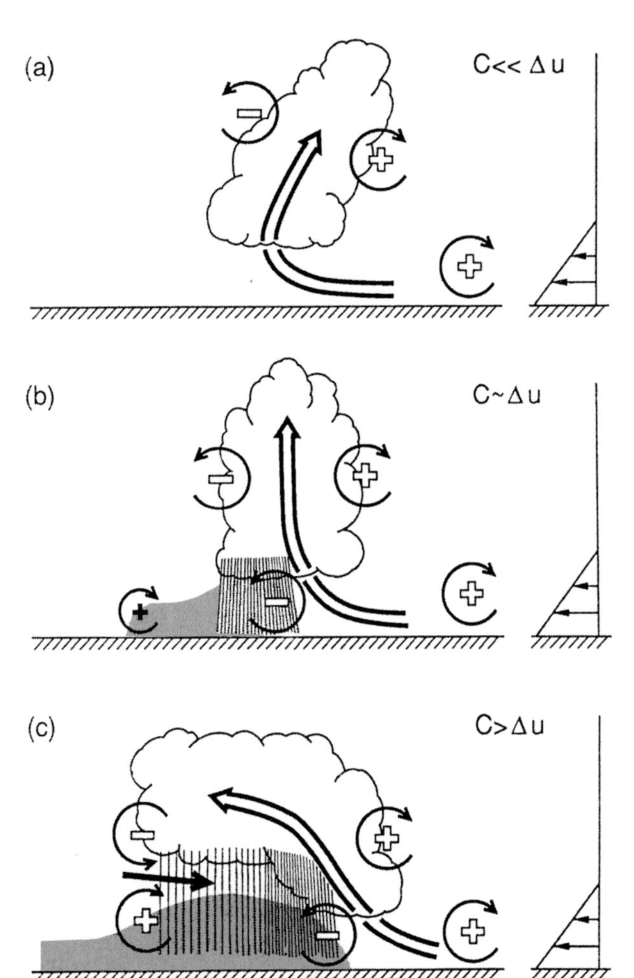

FIG. 3.31. Three stages in the evolution of a convective system. (a) An initial updraft leans downshear in response to the ambient vertical wind shear, which is shown on the right. (b) The circulation generated by the storm-induced cold pool balances the ambient shear, and the system becomes upright. (c) The cold pool circulation overwhelms the ambient shear and the system tilts upshear, producing a rear-inflow jet. The updraft current is denoted by the thick, double-lined flow vector, with the rear-inflow current in (c) denoted by the thick, solid vector. The shading denotes the surface cold pool. The thin, circular arrows depict the most significant sources of horizontal vorticity, which are either associated with the ambient shear or are generated within the convective system. Regions of lighter or heavier rainfall are indicated by the more sparsely or densely packed vertical lines, respectively. The scalloped line denotes the outline of the cloud; C represents the strength of the cold pool, while Δu represents the strength of the ambient low-level vertical wind shear. From Weisman (1992).

influence storm evolution. On the scale of MCSs, buoyancy effects can be used to explain the existence of mesoscale pressure fields and storm circulations. Lafore and Moncrieff (1989) show specifically how a rear-inflow jet develops in response to horizontal and vertical buoyancy gradients aloft (Fig. 3.32). The vertical buoyancy gradient influences the perturbation pressure p' field, which can be seen by taking the divergence of the horizontal momentum equations:

$$\frac{1}{\rho_0}\nabla^2 p' = -\nabla\cdot(\mathbf{v}\cdot\nabla\mathbf{v}) + \frac{\partial B}{\partial z}, \tag{3.3}$$

where $B = g\theta'_v/\bar{\theta}_v$. A positive buoyancy anomaly in the upper troposphere and negative anomaly in the lower troposphere produce a positive $\partial B/\partial z$ in the midtroposphere, thereby creating a mesolow ($\nabla^2 p' > 0$). The horizontal gradient in B aloft (Fig. 3.32) produces a horizontal pressure gradient that contributes to both the rear inflow and the front-to-rear flow in the convective line. This mechanism does not require an extensive stratiform region and, indeed, there is evidence to indicate that rear-inflow jets can form independently of the existence of the stratiform region (Klimowski 1994), although they are enhanced as the stratiform region develops. Equation (3.3) can also be used to explain the common occurrence of mesohighs aloft in MCSs (Fritsch and Maddox 1981; Maddox et al. 1981), since $\partial B/\partial z < 0$ in the upper troposphere.

Another interpretation of rear-inflow jets is that they can form as a gravity wave response to convective heating (Pandya and Durran 1996). The heating function specified by Pandya and Durran represents the heating associated with the sloping front-to-rear ascending and rear-to-front descending branches of a mature squall line, which are produced in large part by advective effects within the squall line. The gravity wave response that is then generated explains the far-field generation of a rear-inflow jet as a response to an "advectively" produced nearer-field component heating described by Weisman (1992) and Lafore and Moncrieff (1989). Additional complications in the interpretation of rear-inflow jets are the roles of processes on large scales—for example, upper-level jets and shortwave troughs—and on the storm scale—for example, mesovortices—in their development (Zhang and Gao 1989; Belair and Zhang 1997).

Convection can also change the ambient density field, and hence convective stability, in distant regions through dynamical effects. These dynamics have received considerable theoretical attention in recent years, and gravity wave motions are a central mechanism. Convection has at least three effects on the gravity wave field. The temporal transience of convection and gust front circulations excites abundant high-frequency waves, essentially by mechanical agitation. In addition, wind blowing across the tops of convective clouds can generate the equivalent of mountain

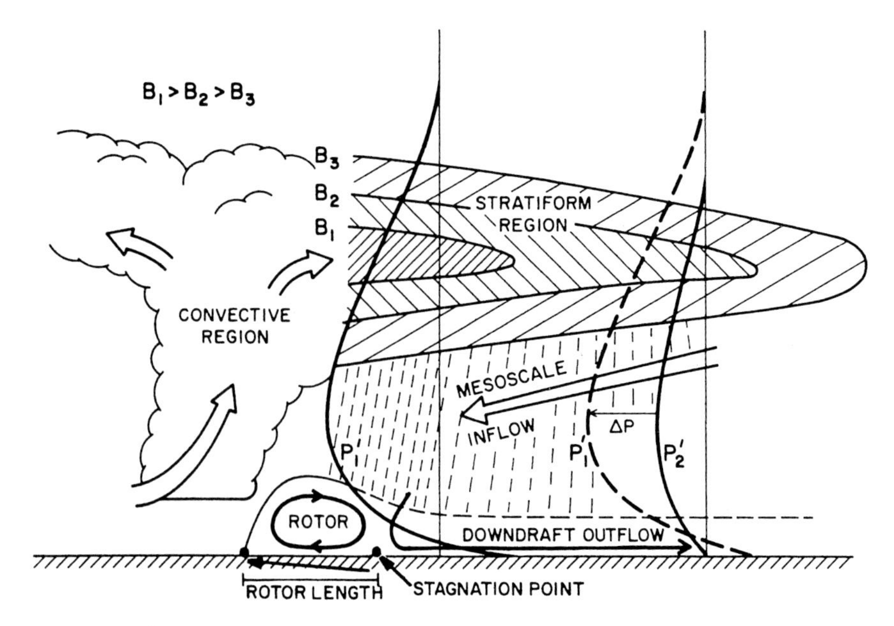

FIG. 3.32. Schema of the relation between buoyancy and the pressure perturbation fields in the mesoscale region of the squall line. Here B_1, B_2, and B_3 represent the perturbation buoyancy fields. From Lafore and Moncrieff (1989).

waves. These waves occur in the troposphere as winds blow across shallow boundary layer cumulus (Hauf 1993), and in the stratosphere above deep convective systems (Fovell et al. 1992). These waves may also be responsible for V-shaped structures in the cloud-top temperature field (Heymsfield and Fulton 1994). Both of these mechanically forced types of waves tend to propagate upward out of the troposphere within a relatively short distance of their source, because they have high frequency. However, for certain configurations of static stability, wind shear, and wavelengths, these oscillatory waves can be trapped or "ducted" and so propagate long distances horizontally without appreciable loss of energy to vertical propagation. Such ducted waves can have important effects on future convective developments, especially if they are generated and trapped at low levels (section 3.3c).

A third gravity wave effect involves the response of the atmosphere to the irreversible, diabatic rearrangement of mass in deep convection (i.e., the net heating associated with precipitation). Because the heating is irreversible, the response includes a very low frequency component that is fairly independent of the details of the convection. This response constitutes the warming caused by convective heating, and it remains in the troposphere for all time, at least in a linear approximation (Bretherton and Smolarkiewicz 1989; Nicholls et al. 1991; Pandya et al. 1993; Bretherton 1993; Mapes 1998). Because this part of the response is less dependent on the details of the convection, we examine it in some detail (but in extreme idealization).

Figure 3.33a illustrates how gravity waves can redistribute heat added to one region of a nonrotating stratified fluid, as by a precipitating cloud. The region of elevated temperature (contours), which contains the heat added to a narrow zone near the origin some time before the time of Fig. 3.33a, propagates continuously to the right. A downwelling wavefront adiabatically warms air at the leading edge, while a trailing wavefront of ascent restores air to its original temperature.

The passage of this structure at a point causes net horizontal displacements of air, toward the origin at low levels and away from the origin aloft. In a rotating fluid, such horizontal displacements generate geostrophic currents, which require thermal perturbations for balance. As a result, heat added to a rotating stratified fluid is exponentially trapped within a Rossby deformation radius given by $R = c/f$, where c is the gravity wave speed and f is the Coriolis parameter. This trapping becomes apparent when the gravity wave structure has traveled for a time f^{-1}. Figure 3.33b shows a situation equivalent to Fig. 3.33a, but with rotation. The warming (contours) is now trapped in a mesoscale region near the origin, with an e-folding distance of $R = c/f \sim 300$ km, while the wavefront structures have become ordinary oscillatory inertio-gravity waves, with no warming between them. In an unbounded atmosphere, these waves would propagate up out of the troposphere.

Figure 3.33 depicts the geostrophic adjustment process in response to localized heating for a single vertical mode with gravity wave speed c, valid only for the case of a special heating profile in a stratified

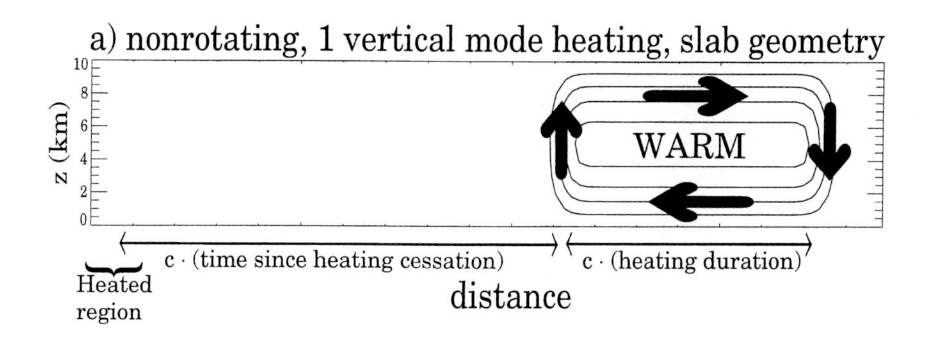

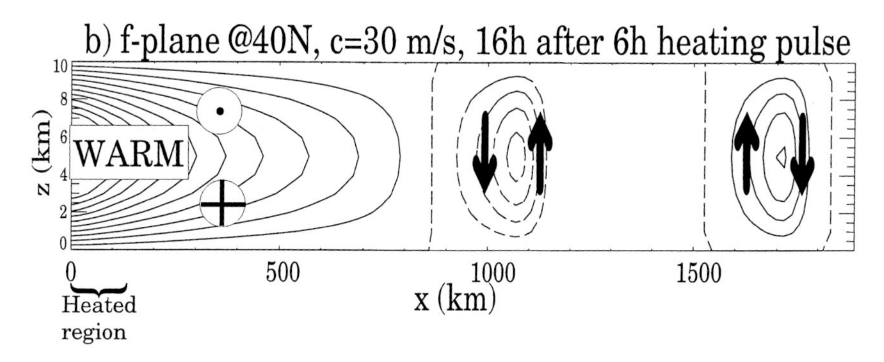

FIG. 3.33. Temperature (contours) and schematically indicated winds (arrows) that occur in a bounded linear stratified fluid in response to a transient slab-symmetric heating at the origin some time ago. (a) In the nonrotating case, the warming propagates away from the heated region as a roll circulation traveling at gravity wave speed, leaving no trace. See Nicholls et al. (1991) for details. (b) In a rotating fluid, the warming is trapped within a Rossby deformation radius of the heated area by geostrophic currents (point: out of page; cross: into page). The high-frequency inertio-gravity waves at x = 1000 and 1700 km in (b) would propagate vertically in an unbounded atmosphere.

fluid bounded above by a lid.[5] In the unbounded atmosphere, tropospheric heating generates gravity waves with a spectrum of values of c (vertical modes). This spectrum of waves separates out in space as the waves dispersively propagate away from the source, with the deepest spectral components traveling faster than shallower components. In a rotating fluid, these different vertical modes have different Rossby radii c/f. This dispersion can have some surprising consequences.

If the mass convergence into a convecting region has its peak value above the surface, then the response of the atmosphere to this convection can include low-level upward displacements, and hence adiabatic cooling, in the near vicinity of the convection. Surprisingly, this cooling can occur even for heating profiles that are positive at all levels (as overall convective system heating profiles tend to be). If these low-level upward displacements are more important to convection than the concomitant upper-level subsidence, then this effect could be destabilizing, rendering convection "gregarious," as hypothesized by Mapes (1993). On the other hand, if the subsidence aloft is strong enough to prevent deep convection, then the result might be merely a burst of cumulus congestus capped by a midlevel inversion, as studied by Stensrud and Maddox (1988) and Johnson et al. (1995). A more complex sequence of events was envisioned for MCSs by McAnelly et al. (1997). They proposed that the initial gravity wave response to an intensifying convective system, involving the gavest vertical mode generated by convective heating, leads to brief MCS weakening, followed by a resurgence of growth due to a second vertical mode associated with condensational heating aloft and evaporative cooling at low levels.

[5] Figure 3.33 shows the special case of a single-mode heating profile in a fluid trapped beneath a reflecting upper boundary. While atmospheric conditions can sometimes reflect gravity waves, in general they propagate vertically, dispersing along sloping ray paths. Since larger horizontal wavenumbers propagate preferentially upward, the tropospheric parts of the solution become spatially smoothed with time (e.g., Appendix 1 of Mapes 1998). The tight spatial gradients in Fig. 3.33 are therefore artifacts of the rigid lid approximation.

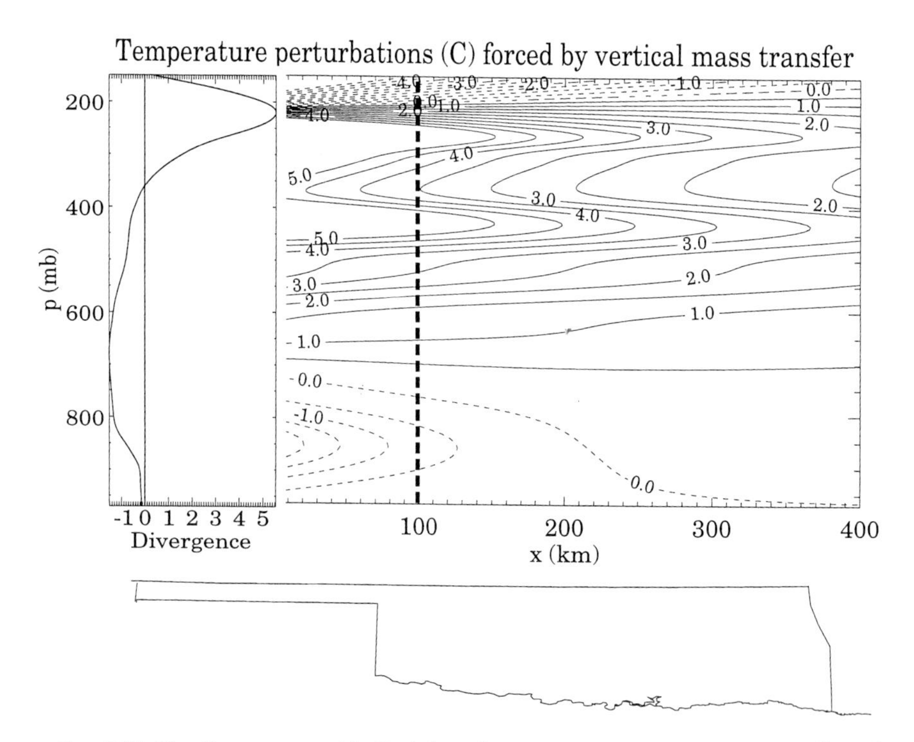

FIG. 3.34. The linear, geostrophically balanced temperature response to a narrow line of heating at the origin, in a uniform resting atmosphere with basic state density stratification taken from the sounding of Fig. 3.1. The heating profile is a cross-isentropic (vertical) mass transfer, characterized by the horizontal wind divergence profile, as plotted here. This profile is adapted from divergence observations around mesoscale convective complexes (Fig. 7 of Maddox 1983; units 10^{-5} s^{-1}). Vertical shear of geostrophic currents balances the temperature gradients. The magnitude of the heating is equivalent to 3 in. of rain in a 100-km-wide zone. The outline of Oklahoma is shown for scale.

Midlatitude organized convection tends to have its maximum convergence near 700 mb (e.g., Fig. 7 of Maddox 1983; data replotted in Fig. 3.34 here). This translates into a net heating profile that is positive at all levels but top-heavy, with its peak in the upper troposphere. The elevated convergence is largely inflow into downdrafts driven by the evaporation of precipitation.

Figure 3.34 shows the geostrophically balanced temperature response to a line of convection-like heating at the origin, in a linear resting atmosphere on an f plane at 45° latitude. The model is forced with the heating profile implied by the composite divergence profile of mature MCCs from Maddox (1983), acting in a static stability profile derived from a smoothed version of Fig. 3.1. The calculation procedure is described and discussed in Mapes and Houze (1995) and Mapes (1998). For each vertical wavelength in the spectrum excited by this heating profile the associated temperature perturbation falls off with distance as $e^{-x/R}$, where $R = c/f$ is the Rossby radius of deformation as discussed above (see chapter 7 of Gill 1982).

At a distance of 100 km from the zone of heating (heavy dashed line), the temperature is ~1°C cooler at low levels, and 5°C or more warmer aloft. Rerunning the parcel instability diagnostics of Table 3.4 with this change to the temperature sounding indicates that the pseudoadiabatic ICAPE including freezing is decreased by 33%, to 5.5 × 10^6 J m^{-2}, by the warming aloft. However, the CIN of surface (970 mb) air has decreased from 24 J kg^{-1} to 10 J kg^{-1}. CIN for air originating at other levels undergoes a similar numerical decrease. Such an alteration to the sounding might actually favor new convective development, if the CIN change outweighs the CAPE change. If the line of heating were not infinitely long, for example, if a circular patch of heating were used, the cool core at low levels would be even more pronounced relative to the deep warming that decreases the CAPE and ICAPE.

The temperature changes shown in Fig. 3.34 are initially caused by adiabatic vertical displacements in gravity wave fronts (as in Fig. 3.33a), and are rendered permanent by the development of geostrophic currents (as in Fig. 3.33b). Although the cooling processes in downdrafts are responsible for the elevated convergence peak, which in turn causes the cool core at low levels in Fig. 3.34, it is worth noting that there is no actual diabatic cooling in this model—only a top-heavy diabatic heating process, representing mesoscale convective complex heating in its totality, with its peak in the upper troposphere. Of course, horizontal advection by mean storm-relative wind and wind shear can be expected to deform and redistribute the warmed and cooled regions seen in Fig. 3.34, so this calculation is only a rough approximation to the true situation in any real case. Similar calculations of the balanced response to squall line heating events have been calculated using nonlinear potential vorticity methods by Schubert et al. (1989) and Hertenstein and Schubert (1991).

As noted earlier, gravity waves occurring in environments of variable stability can have some interesting effects. Wave ducting or trapping can occur if a near-adiabatic layer exists above a stable layer. Since stable layers are often generated near the ground in thunderstorm outflows, wave trapping conditions may exist within some storms. Schmidt and Cotton (1990) have proposed that such conditions may occur in squall lines and that the interaction of high-amplitude gravity waves with the stable layer can produce severe surface winds. Since derechos often occur along and to the north of stationary frontal boundaries (Johns 1984; Johns and Hirt 1987), this interaction may explain extreme surface winds accompanying these phenomena.

When a stratiform region exists, rear-to-front flow aloft in an MCS typically descends to lower levels as a result of sublimation or evaporation, or both, along the lower boundary of the stratiform cloud (Zhang and Gao 1989; Stensrud et al. 1991; Braun and Houze 1997). Rear-inflow jets can descend all the way to the surface and play a significant role in the triggering of new convection by the gust front circulations with which they eventually link up (Lafore and Moncrieff 1989; Weisman 1992) or in the dissipation of the MCS (Zhang and Gao 1989). Weisman (1992) further elucidates these relationships, noting, in particular, that the degree to which a rear-inflow jet may help trigger new convection depends, among other things, on whether the jet descends to the surface well behind the leading edge of the system or remains elevated to near the leading edge of the system.

Upon encountering trailing stratiform precipitation regions, rear-inflow jets have been observed to descend very rapidly. Just behind the stratiform region, strong warming in the lower troposphere is observed. As a result, strong surface pressure falls are often produced locally in the region of warming, leading to the development of wake lows (Fujita 1955, 1963; Pedgley 1962; Williams 1963; Zipser 1977; Johnson and Hamilton 1988). These features are distinct from presquall mesolows, which occur ahead of squall lines (Hoxit et al. 1976) and are typically of smaller amplitude (1–2 mb). However, both may be a gravity wave response to convective heating and/or cooling in the squall line (e.g., Koch et al. 1988; Koch and Siedlarz 1999).

In support of this idea, Haertel and Johnson (2000) simulated MCS mesohighs and wake lows using a linear dynamical system in which the only forcing was the lower-tropospheric cooling associated with stratiform precipitation. The response consisted entirely of gravity waves, whose amplitudes were enhanced in the direction of the cool source motion. When the moving cool source was defined to have a three-dimensional structure, both a mesohigh and mesolow developed having characteristics of squall-line mesohighs and wake lows (Fig. 3.35). The simulated evolution closely resembles that described by Pedgley (1962), Fujita (1963), and Johnson and Hamilton (1988). When an upper boundary was introduced directly above the cooling, the response approached a steady state in which a mesohigh–mesolow couplet was centered on the cooling. An analytic solution showed that the large-amplitude response to stratiform cooling in squall lines is a unique consequence of the fact that the stratiform region's forward speed of motion typically approaches the gravity wave speed associated with the vertical wavelength of the stratiform cooling. The modeled wake low intensified when the stratiform precipitation terminated (after 4 h in Fig. 3.35), consistent with Fujita's (1963) observations of wake low

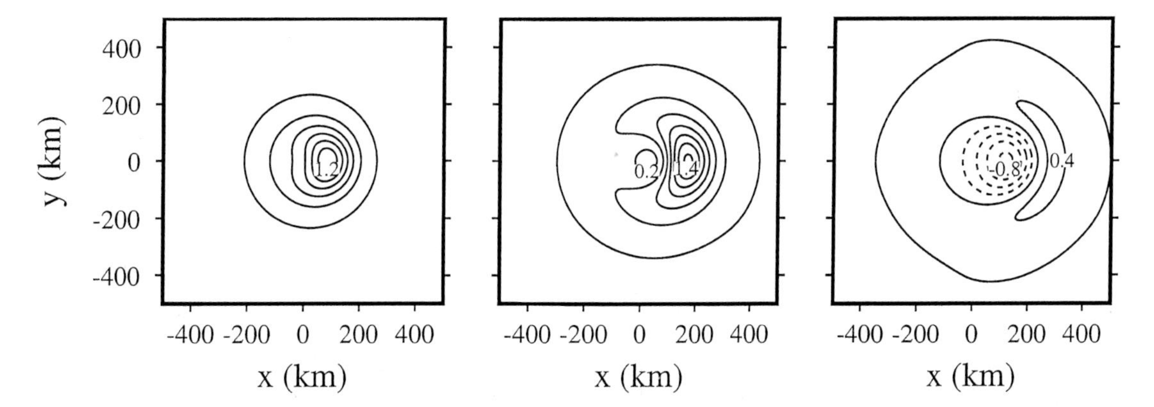

FIG. 3.35. The surface pressure response to a moving, axisymmetric cooling at 2, 4, and 6 h. The forcing is chosen to represent lower-tropospheric cooling associated with the stratiform precipitation region of a squall line (width = 150 km; height = 4 km; lifetime = 4 h). The contour interval is 0.2 mb. From Haertel and Johnson (2000).

intensification during the latter stages of the life cycle of squall lines. These findings are also consistent with observations showing that wake lows tend to occur only when trailing stratiform regions exist.

In rare instances, the rapidly descending flow in an MCS reaches the surface as a hot blast of air, referred to as a *heat burst*. The early work by Williams (1963) actually suggested an association between rear inflow and heat bursts; however, the details of the relationship between the rear-inflow jet and cloud and precipitation structures were limited due to the sparsity of radar and sounding data. Johnson (1983) proposed that heat bursts are a consequence of strong downdrafts penetrating a shallow layer of cool air near the surface. This idea is supported by the modeling study of Proctor (1989) and the observational study of the 23–24 June 1985 heat burst by Johnson et al. (1989). The lower-tropospheric thermodynamic structure in the 23–24 June case closely resembled a dry microburst environment (Wakimoto 1985), except that a shallow, ~500 m deep stable layer existed near the surface. If an evaporating parcel were introduced from cloud base (near 500 mb), then sufficiently cooled, it could descend all the way to the surface if it had enough downward momentum to penetrate the surface stable layer. Johnson et al. (1989) found that on 23–24 June downdrafts of 6–8 m s^{-1} were sufficient to reach the surface. Later dual-Doppler analyses by Johnson and Bartels (1992) and Bernstein and Johnson (1994) confirmed that downdrafts of this magnitude existed.

In some instances, the pressure gradients to the rear of the stratiform region are intense (Bosart and Seimon 1988; Loehrer and Johnson 1995; Johnson et al. 1996). One such case occurred at 0210 UTC on 6 May 1995 (Fig. 3.36, from Johnson et al. 1996). In this situation a mesoscale convective system was moving through east-central Oklahoma with a wake low at the back edge of the northern stratiform region (the southern portion of the system was not captured by the Oklahoma surface mesonetwork). A mesohigh was within the region of heaviest rainfall with a wake low immediately to the rear of the precipitation band. The most intense pressure gradient appeared to "hug" the back edge of the stratiform rain area, consistent with the findings of Johnson and Hamilton (1988), Stumpf et al. (1991), and Loehrer and Johnson (1995). The most intense wake low was at the far southern boundary of the surface network and therefore could not be fully resolved. Nevertheless, the portion that was resolved revealed a pressure gradient exceeding 5 mb $(20\ \mathrm{km})^{-1}$ (a corresponding surface pressure fall of 10 mb in 20 min!). Five-minute average surface winds within this intense gradient were 18.4 m s^{-1} from 092°, with a peak gust of 23.7 m s^{-1}. In some cases, as a result of reduced friction, strong and damaging winds may develop in the vicinity of wake lows that pass over open water areas or smooth terrain (Ely 1982). In addition, the low-level wind shear in the wake low region represents a significant aviation hazard. Recently, Meuse et al. (1996) reported that strong low-level wind shear near the back edge of the trailing stratiform region of a squall line nearly caused an airline crash on 12 April 1996 at the Dallas–Fort Worth International Airport. While the linear theory of Haertel and Johnson (2000) can explain the general structure of the mesohigh–wake low couplets, extreme pressure gradients such as those depicted in Fig. 3.36 are clearly influenced by nonlinear effects (e.g. rapidly descending rear inflow jets).

3.6. Conclusions and outlook

Thirty-five years ago, Byers and Atlas (1963) noted that "The last decade has seen the birth of mesometeorology and with it a vast improvement in visualization of the small-scale circulations which are both the cause and effect of the severe local storm." In the intervening period, there has been considerable progress in identifying many of these processes, principally due to major advances in observing systems (radars, aircraft, sounding systems, mesonets, satellites) and numerical modeling. Although the sheer number of mesoscale processes has made it difficult to treat all topics thoroughly in this review, an attempt has been made to identify the main ones.

However, many questions remain.

- To what extent is the development of severe storms dependent on mesoscale conditions or initiation characteristics? How well can we observe these conditions? How well can we forecast them?

  Over the past two decades, there have been major advances in understanding the basic mechanisms for severe storms through observational studies and idealized numerical simulations. Despite this progress, forecasting severe storms remains a major challenge. Much of the difficulty stems from mesoscale processes inadequately observed and not fully understood. Rapid advances in observing technologies in recent years—for example, wind profilers, surface and airborne Doppler radars, and Raman lidars (measuring humidity profiles)—have greatly enhanced our capability of observing the environment of severe storms. For practical reasons, most past field experiments have focused on developing or mature storms. Perhaps now we have the resources to tackle the convective initiation problem in a major way. Moreover, since mesoscale prediction models have advanced considerably, research resources could be targeted by the use of forecasts. Of course, resources will always be limited, so decisions need to be made about the balance between case

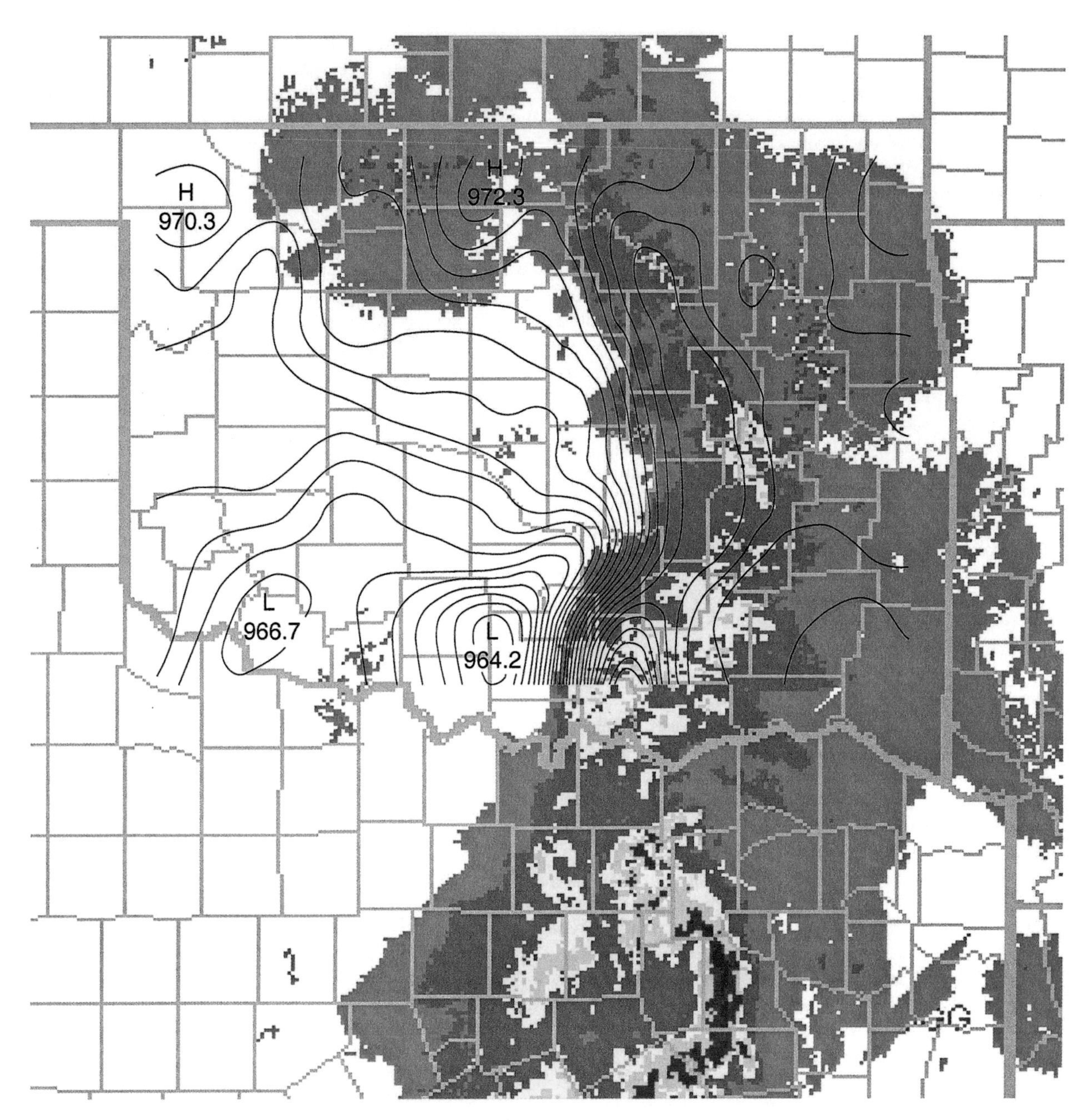

FIG. 3.36. Example of intense mesohigh–wake low couplet. Base-scan radar reflectivity at 0210 UTC 6 May 1995 over Oklahoma (thick state outline). Colors correspond to reflectivity thresholds of 18, 30, 41, 46, and 50 dBZ. Pressure field at 0.5-mb intervals is analyzed at 390 m (the mean station elevation) and is based on time-to-space conversion of 5-min Oklahoma mesonet data. From Johnson et al. (1996).

study and statistical approaches. And an optimum balance between precision and coverage of measurement resources must be determined.

- What are the important triggering mechanisms for different types of severe convective weather and what is their frequency distribution?

Climatologies of triggering mechanisms for severe weather are virtually nonexistent, largely as a result of inadequate measurements. However, with the recently deployed NWS WSR-88D Doppler radar network, and with other research radars and profilers having sufficient sensitivity to measure boundary layer properties, perhaps some headway could be made on this problem. One approach would be to extend the climatologies of mature-phase severe storms back to the initial stages of development. Also, it would be valuable to survey climatologies of all convection for clues as to why only a small fraction becomes severe.

- How do storms modify their local environment? What processes contribute to the enhancement of convection? To its decay?

  Here again, new measurement technologies will help, but an emphasis on modeling is also suggested. What types of modeling are right or wrong for answering these questions? For example, do storm models need to be run for a longer time, to be less initial condition dependent? Can more sophisticated decompositions of model data, such as into balanced versus unbalanced components, teach us more about the mechanisms at work? Or is complexity of analysis already too great, so that simple robust analyses applied to more cases would be more enlightening?
- In what ways do mesoscale processes (jets, topography, etc.) control the geographical distribution and diurnal variability of severe weather?

  With the advent of the NWS WSR-88D and wind profiler networks, it will be possible within several years to construct meaningful climatologies of severe weather in relation to mesoscale features such as jets, drylines, topography, boundary intersections, and so forth. The mechanisms by which such features control severe storm development will inevitably be exposed by field campaigns and modeling studies.
- What are the main factors in storm severity, propagation, and longevity?

  Many severe storms are unanticipated by forecasters. In addition, there are numerous instances where severe and nonsevere storms develop in the same large-scale environment. How does this happen? Some of the processes are already understood. Numerical simulations by Weisman, Klemp, and others have shown that convective evolution for a given shear profile naturally results in a variety of convective structures, ranging from symmetric splitting storms, preferred right- or left-movers, multicellular development, and so forth. However, in many instances variability can also be attributed to mesoscale effects.

  CAPE and wind shear are obviously first-order factors in storm severity; however, these parameters are ordinarily determined from the synoptic-scale sounding network. A serious challenge is that the growth of storms to severity is often influenced by mesoscale effects, which are not sampled by the operational large-scale network. For example, water vapor, surface conditions, the low-level wind field, and so forth, are known to vary significantly in the mesoscale environment of storms. Cases of differing storm evolution in the same large-scale environment are deserving of thorough study, both from observational and modeling perspectives. The real-time mesoscale modeling now under way in many U.S. universities and research laboratories may be of great help in this matter since in many cases those models are designed to capture regional/mesoscale datasets that otherwise would be lost. Data rejection experiments may be useful in determining the most important factors in storm severity. Of course, in addition to severity, the track, time of formation, and longevity of storms are all important issues and the above approaches should be applied to them too.
- What are the mechanisms for heavy rainfall and flash floods and how can they be better forecast?

  Heavy rainfall is the leading cause of convective weather-related fatalities, yet skillful forecasting of floods remains elusive. Floods generally occur when storms repeatedly form or move over the same area and new cell development is often influenced by complex triggering processes involving interaction of outflow boundaries with terrain or other mesoscale features. Mesoscale models hold some promise for improving flash flood forecasting, but limitations in initialization and data assimilation on the mesoscale will undoubtedly make progress slow. Significant improvements in measuring heavy rainfall from the WSR-88D radar network will necessarily involve the implementation of multiparameter measurements from those radars.
- What factors control the upscale growth of convection? Why do some MCSs develop bow echoes, mesovortices, rear inflows, and so forth, and others not?

  These questions are currently being addressed by modeling studies. As capabilities advance, further modeling studies are inevitable; however, it should be emphasized that many of the processes are sufficiently complex to demand rather sophisticated diagnostic analysis of model results. Major efforts will have to be put into diagnostic analyses and interpretation in light of theoretical considerations.

In summary, while we are beginning to get answers to these questions, much work is yet to be done. The NWS modernization (WSR-88D radars, wind profilers, etc.) will help, but focused field campaigns, such as those anticipated in connection with the U.S. Weather Research Program (USWRP), will be necessary to provide adequate data to study remaining key issues. Advances in theory and numerical modeling will also be essential to make significant strides on these problems.

Acknowledgments. This research has been supported by the National Science Foundation under Grant ATM-9618684 (RJ) and by the National Oceanographic and Atmospheric Administration's Office of Global Programs (BM). The comments of Kerry Emanuel, Mike Fritsch, Dave Parsons, Morris Weisman, and Da-Lin Zhang have significantly improved the manuscript.

REFERENCES

Agee, E. M., 1984: Observations from space and thermal convection: A historical perspective. *Bull. Amer. Meteor. Soc.,* **65,** 938–949.

Angevine, W. M., A. W. Grimsdell, L. M. Hartten, and A. C. Delany, 1998: The Flatland boundary layer experiments. *Bull. Amer. Meteor. Soc.,* **79,** 419–431.

Anthes, R. A., 1984: Enhancement of convective precipitation by mesoscale variations in vegetative covering in semiarid regions. *J. Climate Appl. Meteor.,* **23,** 541–554.

Arritt, R. W., T. D. Rink, M. Segal, D. P. Todey, C. A. Clark, M. J. Mitchell, and K. M. Labas, 1997: The Great Plains low-level jet during the warm season of 1993. *Mon. Wea. Rev.,* **125,** 2176–2192.

Asai, T., 1970: Thermal instability of a plane parallel flow with variable vertical shear and unstable stratification. *J. Meteor. Soc. Japan,* **48,** 129–139.

Atkins, N. T., R. M. Wakimoto, and T. M. Weckwerth, 1995: Observations of the sea-breeze front during CaPE. Part II: Dual-Doppler and aircraft analysis. *Mon. Wea. Rev.,* **123,** 944–969.

———, ———, and C. L. Ziegler, 1998: Observations of the finescale structure of a dryline during VORTEX 95. *Mon. Wea. Rev.,* **126,** 525–550.

Atlas, D., 1960: Radar detection of the sea breeze. *J. Meteor.,* **17,** 244–258.

———, R. Tatehira, R. C. Srivastava, W. Marker, and R. E. Carbone, 1969: Precipitation-induced mesoscale wind perturbations in the melting layer. *Quart. J. Roy. Meteor. Soc.,* **95,** 544–560.

Balachandran, N. K., 1980: Gravity waves from thunderstorms. *Mon. Wea. Rev.,* **108,** 804–816.

Balaji, V., and T. L. Clark, 1988: Scale selection in locally forced convective fields and the initiation of deep cumulus. *J. Atmos. Sci.,* **45,** 3188–3211.

Banta, R. M., 1990: The role of mountain flows in making clouds. *Atmospheric Processes over Complex Terrain, Meteor. Monogr.,* No. 45, Amer. Meteor. Soc., 229–283.

Barcilon, A. I., and P. G. Drazin, 1972: Dust devil formation. *Geophys. Fluid Dyn.,* **4,** 147–158.

Barnes, S. L., 1978: Oklahoma thunderstorms on 29–30 April 1970. Part I: Morphology of a tornadic storm. *Mon. Wea. Rev.,* **106,** 673–684.

Bartels, D. L., and R. A. Maddox, 1991: Midlevel cyclonic vortices generated by mesoscale convective systems. *Mon. Wea. Rev.,* **119,** 104–118.

Barth, M. C., and D. B. Parsons, 1996: Microphysical processes associated with intense frontal rainbands and the effect of evaporation and melting on frontal dynamics. *J. Atmos. Sci.,* **53,** 1569–1586.

Beebe, R. G., and F. C. Bates, 1955: A mechanism for assisting in the release of convective instability. *Mon. Wea. Rev.,* **83,** 1–10.

Belair, S., and D.-L. Zhang, 1997: A numerical study of the along-line variability of a frontal squall line during PRE-STORM. *Mon. Wea. Rev.,* **125,** 2544–2561.

Bennetts, D. A., and B. J. Hoskins, 1979: Conditional symmetric instability—A possible explanation for frontal rainbands. *Quart. J. Roy. Meteor. Soc.,* **105,** 945–962.

Bernstein, B. C., and R. H. Johnson, 1994: A dual-Doppler radar study of an OK PRE-STORM heat burst event. *Mon. Wea. Rev.,* **122,** 259–273.

Biggerstaff, M. I., and R. A. Houze Jr., 1991: Kinematic and precipitation structure of the 10–11 June 1985 squall line. *Mon. Wea. Rev.,* **119,** 3034–3065.

Blackadar, A. K., 1957: Boundary-layer wind maxima and their significance for the growth of nocturnal inversions. *Bull. Amer. Meteor. Soc.,* **38,** 283–290.

Blanchard, D. O., and K. W. Howard, 1986: The Denver hailstorm of 13 June 1984. *Bull. Amer. Meteor. Soc.,* **67,** 1123–1131.

———, W. R. Cotton, and J. M. Brown, 1998: Mesoscale circulation growth under conditions of weak inertial instability. *Mon. Wea. Rev.,* **126,** 118–140.

Bluestein, H. B., 1993: *Synoptic-Dynamic Meteorology in Midlatitudes.* Vol. II, Oxford Press, 594 pp.

———, and K. W. Thomas, 1984: Diagnosis of a jet streak in the vicinity of a severe weather outbreak in the Texas panhandle. *Mon. Wea. Rev.,* **112,** 2499–2520.

———, and G. R. Woodall, 1990: Doppler-radar analysis of a low-precipitation (LP) severe storm. *Mon. Wea. Rev.,* **118,** 1640–1664.

———, and S. Parker, 1993: Modes of isolated convective storm formation along the dryline. *Mon. Wea. Rev.,* **121,** 1354–1372.

———, and T. Hutchinson, 1996: Convective-storm initiation and organization near the intersection of a cold front and the dryline. Preprints, *Seventh Conf. on Mesoscale Processes,* Reading, UK, Amer. Meteor. Soc., 41–43.

———, E. W. McCaul Jr., G. P. Byrd, and G. R. Woodall, 1988: Mobile sounding observations of a tornadic storm near the dryline: The Canadian, Texas storm of 7 May 1986. *Mon. Wea. Rev.,* **116,** 1790–1804.

Bonner, W. D., 1966: Case study of thunderstorm activity in relation to the low-level jet. *Mon. Wea. Rev.,* **94,** 167–178.

Booker, D. R., 1963: Modification of convective storms by lee waves. *Severe Local Storms, Meteor. Monogr.,* No. 27, Amer. Meteor. Soc., 129–140.

Bosart, L. F., and F. Sanders, 1981: Johnstown flood of July 1977: A long-lived convective system. *J. Atmos. Sci.,* **38,** 1616–1642.

———, and A. Seimon, 1988: Case study of an unusually intense atmospheric gravity wave. *Mon. Wea. Rev.,* **116,** 1857–1886.

Brady, R. H., and E. J. Szoke, 1989: A case study of nonmesocyclone tornado development in northeast Colorado: Similarities to waterspout formation. *Mon. Wea. Rev.,* **117,** 843–856.

Brandes, E. A., 1978: Mesocyclone evolution and tornadogenesis: Some observations. *Mon. Wea. Rev.,* **106,** 995–1011.

———, 1984a: Relationships between radar-derived thermodynamic variables and tornadogenesis. *Mon. Wea. Rev.,* **112,** 1033–1052.

———, 1984b: Vertical vorticity generation and mesocyclone sustenance in tornadic thunderstorms: The observational evidence. *Mon. Wea. Rev.,* **112,** 2253–2269.

Braun, S. A., and R. A. Houze Jr., 1997: The evolution of the 10–11 June 1985 PRE-STORM squall line: Initiation, development of rear inflow, and dissipation. *Mon. Wea. Rev.,* **125,** 478–504.

Bretherton, C. S., 1993: The nature of adjustment in cumulus cloud fields. *The Representation of Cumulus Convection in Numerical Models,* K. A. Emanuel and D. J. Raymond, Eds., Amer. Meteor. Soc., 63–74.

———, and P. K. Smolarkiewicz, 1989: Gravity waves, compensating subsidence and detrainment around cumulus clouds. *J. Atmos. Sci.,* **46,** 740–759.

Brown, J. M., K. R. Knupp, and F. Caracena, 1982: Destructive winds from shallow, high-based cumulonimbus. Preprints, *12th Conf. on Severe Local Storms,* San Antonio, TX, Amer. Meteor. Soc., 272–275.

Brown, R. A., 1980: Longitudinal instabilities and secondary flows in the planetary boundary layer: A review. *Rev. Geophys. Space Phys.,* **18,** 683–697.

Browning, K. A., and T. W. Harrold, 1970: Air motion and precipitation growth at a cold front. *Quart. J. Roy. Meteor. Soc.,* **96,** 369–389.

———, and C. W. Pardoe, 1973: Structure of low-level jet streams ahead of mid-latitude cold fronts. *Quart. J. Roy. Meteor. Soc.,* **99,** 619–638.

———, and G. A. Monk, 1982: Simple model for the synoptic analysis of cloud fronts. *Quart. J. Roy. Meteor. Soc.*, **108**, 435–452.

———, and Coauthors, 1976: Structure of an evolving hailstorm, Part V: Synthesis and implications for hail growth and hail suppression. *Mon. Wea. Rev.*, **104**, 603–610.

Burgess, D., V. Wood, and R. Brown, 1982: Mesocyclone evolution statistics. Preprints, *12th Conf. on Severe Local Storms*, San Antonio, TX, Amer. Meteor. Soc., 422–424.

Byers, H. R., and H. R. Rodebush, 1948: Causes of thunderstorms of the Florida peninsula. *J. Meteor.*, **5**, 275–280.

———, and R. R. Braham Jr., 1949: *The Thunderstorm.* U.S. Government Printing Office, 287 pp.

———, and D. Atlas, 1963: Severe local storms: In retrospect. *Severe Local Storms, Meteor. Monogr.*, No. 27, Amer. Meteor. Soc., 242–247.

Caracena, F., and J. M. Fritsch, 1983: Focusing mechanisms in the Texas Hill Country flash floods of 1978. *Mon. Wea. Rev.*, **111**, 2319–2332.

———, R. A. Maddox, L. R. Hoxit, and C. F. Chappell, 1979: Mesoanalysis of the Big Thompson storm. *Mon. Wea. Rev.*, **107**, 1–17.

Carbone, R. E., 1982: A severe frontal rainband. Part I: Stormwide hydrodynamic structure. *J. Atmos. Sci.*, **39**, 258–279.

———, 1983: A severe frontal rainband. Part II: Tornado parent vortex circulation. *J. Atmos. Sci.*, **40**, 2639–2654.

———, J. W. Conway, N. A. Crook, and M. W. Moncrieff, 1990: The generation and propagation of a nocturnal squall line. Part I: Observations and implications for mesoscale predictability. *Mon. Wea. Rev.*, **118**, 26–49.

Carlson, T. N., and F. H. Ludlam, 1968: Conditions for the occurrence of severe local storms. *Tellus*, **20**, 203–226.

———, S. G. Benjamin, G. S. Forbes, and Y.-F. Li, 1983: Elevated mixed layers in the regional severe storm environment: Conceptual model and case studies. *Mon. Wea. Rev.*, **111**, 1453–1473.

Carr, F. H., and J. P. Millard, 1985: Composite study of comma clouds and their association with severe weather over the Great Plains. *Mon. Wea. Rev.*, **113**, 370–387.

Carruthers, D. J., and J. C. R. Hunt, 1990: Fluid mechanics of airflow over hills: Turbulence, fluxes, and waves in the boundary layer. *Atmospheric Processes over Complex Terrain, Meteor. Monogr.*, No. 45, Amer. Meteor. Soc., 83–103.

Chang, J.-T., and P. J. Wetzel, 1991: Effects of spatial variations of soil moisture and vegetation on the evolution of a prestorm environment: A numerical case study. *Mon. Wea. Rev.*, **119**, 1368–1390.

Changnon, S. A., R. G. Semonin, and F. A. Huff, 1976: A hypothesis for urban rainfall anomalies. *J. Appl. Meteor.*, **15**, 544–560.

Chappell, C. F., 1986: Quasi-stationary convective events. *Mesoscale Meteorology and Forecasting*, P. S. Ray, Ed., Amer. Meteor. Soc., 289–310.

Charba, J., 1974: Application of the gravity current model to analysis of squall-line gust fronts. *Mon. Wea. Rev.*, **102**, 140–156.

Charney, J. G., 1975: Dynamics of deserts and drought in the Sahel. *Quart. J. Roy. Meteor. Soc.*, **101**, 193–202.

———, and A. Eliassen, 1964: On the growth of the hurricane depression. *J. Atmos. Sci.*, **21**, 68–75.

Chen, S., and W. R. Cotton, 1988: The sensitivity of a simulated extratropical mesoscale convective system to longwave radiation and ice-phase microphysics. *J. Atmos. Sci.*, **45**, 3897–3910.

Christian, T. W., and R. M. Wakimoto, 1989: The relationship between radar reflectivities and clouds associated with horizontal roll convection on 8 August 1982. *Mon. Wea. Rev.*, **117**, 1530–1544.

Christie, D. R., 1992: The morning glory of the Gulf of Carpenteria: A paradigm for nonlinear waves in the lower atmosphere. *Austr. Meteor. Mag.*, **41**, 21–60.

Ciesielski, P. E., D. E. Stevens, R. H. Johnson, and K. R. Dean, 1989: Observational evidence for asymmetric inertial instability. *J. Atmos. Sci.*, **46**, 817–831.

Clarke, R. H., 1972: The morning glory: an atmospheric hydraulic jump. *J. Appl. Meteor.*, **11**, 304–311.

———, 1984: Colliding sea breezes and the creation of internal atmospheric bore waves: two dimensional numerical studies. *Austr. Meteor. Mag.*, **32**, 207–226.

Colby, F. P., Jr., 1984: Convective inhibition as a predictor of convection during AVE-SESAME II. *Mon. Wea. Rev.*, **112**, 2239–2252.

Collins, D. C., and R. Avissar, 1994: An evaluation with the Fourier Amplitude Sensitivity Test (FAST) of which land-surface parameters are of greatest importance in atmospheric modeling. *J. Climate*, **7**, 681–703.

Colman, B. R., 1990: Thunderstorms above frontal surfaces in environments without positive CAPE. Part II: Organization and instability mechanisms. *Mon. Wea. Rev.*, **118**, 1123–1144.

Crook, N. A., 1986: The effect of ambient stratification and moisture on the motion of atmospheric undular bores. *J. Atmos. Sci.*, **43**, 171–181.

———, 1987: Moist convection at a surface cold front. *J. Atmos. Sci.*, **44**, 3469–3494.

———, 1988: Trapping of low-level internal gravity waves. *J. Atmos. Sci.*, **45**, 1533–1541.

———, 1996: Sensitivity of moist convection forced by boundary layer processes to low-level thermodynamic fields. *Mon. Wea. Rev.*, **124**, 1767–1785.

———, and J. D. Tuttle, 1994: Numerical simulations initialized with radar-derived winds. Part II: Forecasts of three gust-front cases. *Mon. Wea. Rev.*, **122**, 1204–1217.

———, T. L. Clark, and M. W. Moncrieff, 1990a: The Denver Cyclone. Part I: Generation in low Froude number flow. *J. Atmos. Sci.*, **47**, 2725–2742.

———, R. E. Carbone, M. W. Moncrieff, and J. W. Conway, 1990b: The generation and propagation of a nocturnal squall line. Part II: Numerical simulations. *Mon. Wea. Rev.*, **118**, 50–65.

Curry, M. J., and R. C. Murty, 1974: Thunderstorm-generated gravity waves. *J. Atmos. Sci.*, **31**, 1402–1408.

Cutrim, E., D. W. Martin, and R. Rabin, 1995: Enhancement of cumulus clouds over deforested lands in Amazonia. *Bull. Amer. Meteor. Soc.*, **76**, 1801–1805.

Danielsen, E. F., 1974: The relationship between severe weather, major duststorms, and rapid large-scale cyclogenesis. Part I. *Subsynoptic Extratropical Weather Systems*, M. Shapiro, Ed., National Center for Atmospheric Research, 215–241.

———, 1982: Statistics of cold cumulonimbus anvils based on enhanced infrared photographs. *Geophys. Res. Lett.*, **9**, 601–604.

Davies-Jones, R. P., 1984: Streamwise vorticity: The origin of updraft rotation in supercell storms. *J. Atmos. Sci.*, **41**, 2991–3006.

———, 1986: Tornado dynamics. *Thunderstorms: A Social and Technological Documentary*, 2d ed., Vol. II, E. Kessler, Ed. University of Oklahoma Press, 197–236.

Davis, C. A., 1997: Mesoscale anticyclonic circulations in the lee of the central Rocky Mountains. *Mon. Wea. Rev.*, **125**, 2838–2855.

———, and M. L. Weisman, 1994: Balanced dynamics of mesoscale vortices produced in simulated convective systems. *J. Atmos. Sci.*, **51**, 2005–2030.

Dempsey, D. P., and R. Rotunno, 1988: Topographic generation of mesoscale vortices in mixed layer models. *J. Atmos. Sci.*, **45**, 2961–2978.

Dinwiddie, F. B., 1959: Waterspout-tornado structure and behavior at Nags Head, N.C., August 12, 1952. *Mon. Wea. Rev.,* **87,** 239–250.
Doswell, C. A., III, 1980: Synoptic-scale environments associated with High Plains severe thunderstorms. *Bull. Amer. Meteor. Soc.,* **61,** 1388–1400.
———, 1987: The distinction between large-scale and mesoscale contribution to severe convection: A case study example. *Wea. Forecasting,* **2,** 3–16.
Doviak, R. J., and R. Ge, 1984: An atmospheric solitary gust observed with a Doppler radar, a tall tower and a surface network. *J. Atmos. Sci.,* **41,** 2559–2573.
Droegemeier, K. K., and R. B. Wilhelmson, 1985: Three-dimensional numerical modeling of convection produced by interacting outflows. Part II: Variations in vertical wind shear. *J. Atmos. Sci.,* **42,** 2404–2414.
Drosdowsky, W., and G. J. Holland, 1987: North Australian cloud lines. *Mon. Wea. Rev.,* **115,** 2645–2659.
Dudhia, J., 1989: Numerical study of convection observed during the Winter Monsoon Experiment using a mesoscale two-dimensional model. *J. Atmos. Sci.,* **46,** 3077–3107.
Ely, G. F., 1982: Case study of a significant thunderstorm wake depression along the Texas coast: May 29–30 1981. NOAA Tech. Memo NWS SR-105, 64 pp.
Emanuel, K. A., 1979: Inertial instability and mesoscale convective systems. Part I: Linear theory of inertial instability in rotating viscous fluid. *J. Atmos. Sci.,* **36,** 2425–2449.
———, 1980: Forced and free mesoscale motions in the atmosphere. *CIMMS Symp. Collection of Lecture Notes on Dynamics of Mesometeorological Disturbances*, Norman, OK, University of Oklahoma/NOAA, 189–259.
———, 1983: The Lagrangian parcel dynamics of moist symmetric instability. *J. Atmos. Sci.,* **40,** 2368–2376.
———, 1985: Frontal circulations in the presence of small moist symmetric stability. *J. Atmos. Sci.,* **42,** 1062–1071.
———, 1986: Overview and definition of mesoscale meteorology. *Mesoscale Meteorology and Forecasting,* P. S. Ray, Ed., Amer. Meteor. Soc., 1–17.
———, 1994: *Atmospheric Convection.* Oxford Press, 580 pp.
Esbensen, S., 1978: Bulk thermodynamic effects and properties of small tropical cumuli. *J. Atmos. Sci.,* **35,** 826–837.
Etling, D., and R. A. Brown, 1993: Roll vortices in the planetary boundary layer: a review. *Bound.-Layer Meteor.,* **65,** 215–248.
Fankhauser, J. C., N. A. Crook, J. Tuttle, and C. G. Wade, 1995: Initiation of deep convection along boundary-layer convergence lines in a semitropical environment. *Mon. Wea. Rev.,* **123,** 291–313.
Fovell, R., and Y. Ogura, 1988: Numerical simulation of a midlatitude squall line in two dimensions. *J. Atmos. Sci.,* **45,** 3846–3879.
———, and ———, 1989: Effect of vertical wind shear on numerically simulated multicell storm structure. *J. Atmos. Sci.,* **46,** 3144–3176.
———, D. Durran, and J. R. Holton, 1992: Numerical simulations of convectively generated stratospheric gravity waves. *J. Atmos. Sci.,* **49,** 1427–1442.
Fritsch, J. M., and R. A. Maddox, 1981: Convectively driven mesoscale pressure systems aloft. Part I: Observations. *J. Appl. Meteor.,* **20,** 9–19.
———, and D. M. Rodgers, 1981: Ft. Collins hailstorm: An example of the short-term forecast enigma. *Bull. Amer. Meteor. Soc.,* **62,** 1560–1569.
———, J. Kapolka, and P. A. Hirschberg, 1992: The effects of subcloud-layer diabatic processes on cold air damming. *J. Atmos. Sci.,* **49,** 49–70.
———, J. D. Murphy, and J. S. Kain, 1994: Warm core vortex amplication over land. *J. Atmos. Sci.,* **51,** 1780–1807.
Fujita, T. T., 1955: Results of detailed synoptic studies of squall lines. *Tellus,* **7,** 405–436.
———, 1958: Structure and movement of a dry front. *Bull. Amer. Meteor. Soc.,* **39,** 574–582.
———, 1963: Analytical mesometeorology: A review. *Severe Local Storms, Meteor. Monogr.,* No. 27, Amer. Meteor. Soc., 77–125.
———, 1978: Manual of downburst identification for project NIMROD. SMRP Research Paper 156, University of Chicago, 42 pp.
———, 1981: Tornadoes and downbursts in the context of generalized planetary scales. *J. Atmos. Sci.,* **38,** 1511–1534.
———, and F. Caracena, 1977: Analysis of three weather-related aircraft accidents. *Bull. Amer. Meteor. Soc.,* **58,** 1164–1181.
Fulton, R., D. S. Zrnić, and R. J. Doviak, 1990: Initiation of a solitary wave family in the demise of a nocturnal thunderstorm density current. *J. Atmos. Sci.,* **47,** 319–337.
Gentry, R. C., and P. L. Moore, 1954: Relation of local and general wind interaction near the sea coast to time and location of air-mass showers. *J. Meteor.,* **11,** 507–511.
Gill, A. E., 1982: *Atmosphere-Ocean Dynamics.* Academic Press, 662 pp.
Gilmore, M. S., and L. J. Wicker, 1998: The influence of midtropospheric dryness on supercell morphology and evolution. *Mon. Wea. Rev.,* **126,** 943–958.
Golden, J. H., 1973: Some statistical aspects of waterspout formation. *Weatherwise,* **26,** 108–117.
———, 1974: Scale-interaction implications for the waterspout life cycle. II. *J. Appl. Meteor.,* **13,** 693–709.
———, 1978: Picture of the month: Jet aircraft flying through a Denver tornado? *Mon. Wea. Rev.,* **106,** 574–578.
Grady, R.L., and J. Verlinde, 1997: Triple-Doppler analysis of a discretely propagating, long-lived, High Plains squall line. *J. Atmos. Sci.,* **54,** 2729–2748.
Gray, W. M., and R. W. Jacobson, Jr., 1977: Diurnal variation of deep cumulus convection. *Mon. Wea. Rev.,* **105,** 1171–1188.
Graziano, T. M., and T. N. Carlson, 1987: A statistical evaluation of lid strength on deep convection. *Wea. Forecasting,* **2,** 127–139.
Gurka, J. J., 1976: Satellite and surface observations of strong wind zones accompanying thunderstorms. *Mon. Wea. Rev.,* **104,** 1484–1493.
Haase, S. P., and R. K. Smith, 1984: Morning glory wave clouds in Oklahoma: A case study. *Mon. Wea. Rev.,* **112,** 2078–2089.
Haertel, P. T., and R. H. Johnson, 2000: The linear dynamics of squall-line mesohighs and wake lows. *J. Atmos. Sci.,* **57,** 93–107.
Hagemeyer, B. C., 1984: An investigation of summertime convection over the Upper Current River Valley of Southwest Missouri. NOAA Tech. Memo. NWS CR-71, 35 pp.
Hane, C. E., H. B. Bluestein, T. M. Crawford, M. E. Baldwin, and R. M. Rabin, 1997: Severe thunderstorm development in relation to along-dryline variability: A case study. *Mon. Wea. Rev.,* **125,** 231–251.
Hauf, T., 1993: Aircraft observation of convection waves over southern Germany—A case study. *Mon. Wea. Rev.,* **121,** 3282–3290.
Hertenstein, R. F. A., and W. H. Schubert, 1991: Potential vorticity anomalies associated with squall lines. *Mon. Wea. Rev.,* **116,** 1663–1672.
Heymsfield, G. M., and R. Fulton, 1994: Passive microwave and visible/infrared structure of mesoscale precipitation systems. *Meteor. Atmos. Phys.,* **54,** 123–139.
Holton, J. R., 1967: The diurnal boundary-layer wind oscillation above sloping terrain. *Tellus,* **19,** 199–205.
Houze, R. A., Jr., 1993: *Cloud Dynamics.* Academic Press, 573 pp.
———, 1997: Stratiform precipitation in regions of convection: A meteorological paradox? *Bull. Amer. Meteor. Soc.,* **78,** 2179–2196.
———, S. G. Geotis, F. D. Marks, and A. K. West, 1981: Winter monsoon convection in the vicinity of north Borneo. Part I:

Structure and time variation of the clouds and precipitation. *Mon. Wea. Rev.,* **108,** 1595–1614.
———, S. A. Rutledge, M. I. Biggerstaff, and B. F. Smull, 1989: Interpretation of Doppler weather radar displays of midlatitude mesoscale convective systems. *Bull. Amer. Meteor. Soc.,* **70,** 608–619.
———, B. F. Smull, and P. Dodge, 1990: Mesoscale organization of springtime rainstorms in Oklahoma. *Mon. Wea. Rev.,* **118,** 613–654.
———, W. Schmid, R. G. Fovell, and H.-H. Schiesser, 1993: Hailstorms in Switzerland: Left movers, right movers, and false hooks. *Mon. Wea. Rev.,* **121,** 3345–3370.
Hoxit, L. R., C. F. Chappell, and J. M. Fritsch, 1976: Formation of mesolows or pressure troughs in advance of cumulonimbus clouds. *Mon. Wea. Rev.,* **104,** 1419–1428.
Hsu, S. A., 1969: Mesoscale structure of the Texas coast sea breeze. Rep. 16, Atmospheric Science Group, University of Texas, 237 pp.
Hutchinson, T. A., and H. B. Bluestein, 1998: Prefrontal wind-shift lines in the Plains of the United States. *Mon. Wea. Rev.,* **126,** 141–166.
Jascourt, S. D., S. S. Lindstrom, C. J. Seman, and D. D. Houghton, 1988: Observation of banded convective development in the presence of weak symmetric stability. *Mon. Wea. Rev.,* **116,** 175–191.
Johns, R. H., 1984: Synoptic climatology of northwest flow severe weather outbreaks. Part II: Meteorological parameters and synoptic patterns. *Mon. Wea. Rev.,* **112,** 449–464.
———, 1993: Meteorological conditions associated with bow echo development in convective storms. *Wea. Forecasting,* **8,** 294–299.
———, and W. D. Hirt, 1985: The derecho of 19–20 July 1983: A case study. *Natl. Wea. Dig.,* **10,** 17–32.
———, and ———, 1987: Derechos: Widespread convectively induced windstorms. *Wea. Forecasting,* **2,** 32–49.
———, and C. A. Doswell III, 1992: Severe local storms forecasting. *Wea. Forecasting,* **7,** 588–612.
Johnson, B. C., 1983: The heat burst of 29 May 1976. *Mon. Wea. Rev.,* **111,** 1776–1792.
Johnson, R. H., and P. J. Hamilton, 1988: The relationship of surface pressure features to the precipitation and air flow structure of an intense midlatitude squall line. *Mon. Wea. Rev.,* **116,** 1444–1472.
———, and D. L. Bartels, 1992: Circulations associated with a mature-to-decaying midlatitude mesoscale convective system. Part II: Upper-level features. *Mon. Wea. Rev.,* **120,** 1301–1320.
———, S. Chen, and J. J. Toth, 1989: Circulations associated with a mature-to-decaying midlatitude mesoscale convective system. Part I: Surface features—Heat bursts and mesolow development. *Mon. Wea. Rev.,* **117,** 942–959.
———, B. D. Miner, and P. E. Ciesielski, 1995: Circulations between mesoscale convective systems along a cold front. *Mon. Wea. Rev.,* **123,** 585–599.
———, E. R. Hilgendorf, and K. C. Crawford, 1996: Study of midwestern mesoscale convective systems using new operational networks. Preprints, *18th Conf. on Severe Local Storms,* San Francisco, CA, Amer. Meteor. Soc., 308–312.
Karyampudi, V. M., S. E. Koch, C. Chen, J. W. Rottman, and M. L. Kaplan, 1995: The influence of the Rocky Mountains on the 13–14 April 1986 severe weather outbreak. Part II: Evolution of a prefrontal bore and its role in triggering a squall line. *Mon. Wea. Rev.,* **123,** 1423–1446.
Keenan, T. D., B. R. Morton, Y. S. Zhang, and K. Nyguen, 1990: Some characteristics of thunderstorms over Bathurst and Melville Islands near Darwin, Australia. *Quart. J. Roy. Meteor. Soc.,* **116,** 1153–1172.
———, B. Ferrier, and J. Simpson, 1993: Development and structure of a maritime continent thunderstorm. *Meteor. Atmos. Phys.,* **53,** 185–222.
Keyser, D., and M. A. Shapiro, 1986: A review of the structure and dynamics of upper-level frontal zones. *Mon. Wea. Rev.,* **114,** 452–499.
King, P., and D. M. L. Sills, 1998: The 1987 ELBOW Project: An experiment to study the effects of lake breezes on weather in southern Ontario. Preprints, *19th Conf. on Severe Local Storms,* Minneapolis, MN, Amer. Meteor. Soc., 317–320.
Kingsmill, D. E., 1995: Convection initiation associated with a sea-breeze front, a gust front, and their collision. *Mon. Wea. Rev.,* **123,** 2913–2933.
Klemp, J. B., 1987: Dynamics of tornadic thunderstorms. *Ann. Rev. Fluid Mech.,* **19,** 369–402.
———, and R. Rotunno, 1983: A study of the tornadic region within a supercell thunderstorm. *J. Atmos. Sci.,* **40,** 359–377.
Klimowski, B. A., 1994: Initiation and development of rear inflow within the 28–29 June 1989 North Dakota mesoconvective system. *Mon. Wea. Rev.,* **122,** 765–779.
Knox, J. A., 1997: Possible mechanisms of clear-air turbulence in strongly anticyclonic flows. *Mon. Wea. Rev.,* **125,** 1251–1259.
Koch, S. E., and J. McCarthy, 1982: The evolution of a Oklahoma dryline. Part II: Boundary-layer forcing of mesoconvective systems. *J. Atmos. Sci.,* **39,** 237–257.
———, and P. B. Dorian, 1988: A mesoscale gravity wave event observed during CCOPE. Part III: Wave environment and probable source mechanisms. *Mon. Wea. Rev.,* **116,** 2570–2592.
———, and R. E. Golus, 1988: A mesoscale gravity wave event observed during CCOPE. Part I: Multiscale statistical analysis of wave characteristics. *Mon. Wea. Rev.,* **116,** 2527–2544.
———, and L. M. Siedlarz, 1999: Mesoscale gravity waves and their environment in the central United States during STORM-FEST. *Mon. Wea. Rev.,* **127,** 2854–2879.
———, R. E. Golus, and P. B. Dorian, 1988: A mesoscale gravity wave event observed during CCOPE. Part II: Interactions between mesoscale convective systems and antecedent waves. *Mon. Wea. Rev.,* **116,** 2545–2569.
———, F. Einaudi, P. B. Dorian, S. Lang, and G. M. Heymsfield, 1993: Mesoscale gravity-wave event observed during CCOPE. Part IV: Stability analysis and Doppler-derived wave vertical structure. *Mon. Wea. Rev.,* **121,** 2483–2510.
Kocin, P. J., L. W. Uccellini, and R. A. Petersen, 1986: Rapid evolution of a jet streak circulation in a pre-convective environment. *Meteor. Atmos. Phys.,* **35,** 103–138.
Kuettner, J. P., 1959: The band structure of the atmosphere. *Tellus,* **2,** 267–294.
Kuettner, J. P., 1971: Cloud bands in the earth's atmosphere: Observations and theory. *Tellus,* **23,** 404–426.
Kuo, H. L., 1963: Perturbations of plane Couette flow in stratified fluid and origin of cloud streets. *Phys. Fluids,* **6,** 195–211.
———, L. Cheng, and R. A. Anthes, 1984: Mesoscale analyses of the Sichuan Flood catastrophe, 11–15 July 1981. *Mon. Wea. Rev.,* **114,** 1984–2003.
———, ———, and J.-W. Bao, 1988: Numerical simulation of the 1981 Sichuan Flood. Part I: Evolution of a mesoscale southwest vortex. *Mon. Wea. Rev.,* **116,** 2481–2504.
Lafore, J.-P., and M. W. Moncrieff, 1989: A numerical investigation of the organization and interaction of the convective and stratiform regions of tropical squall lines. *J. Atmos. Sci.,* **46,** 521–544.
Lanicci, J. M., T. N. Carlson, and T. T. Warner, 1987: Sensibility of the Great Plains severe storm environment to soil-moisture distribution. *Mon. Wea. Rev.,* **115,** 2660–2673.
Lee, B. D., and R. B. Wilhelmson, 1997: The numerical simulation of non-supercell tornadogenesis. Part I: Initiation and evolution of pretornadic misocyclone circulations along a dry outflow boundary. *J. Atmos. Sci.,* **54,** 32–60.
———, R. D. Farley, and M. R. Hjelmfelt, 1991: A numerical case study of convection initiation along colliding convergence boundaries in northeast Colorado. *J. Atmos. Sci.,* **48,** 2350–2366.

Lemon, L. R., and C. A. Doswell III, 1979: Severe thunderstorm evolution and meso-cyclone structure as related to tornadogenesis. *Mon. Wea. Rev.,* **107,** 1184–1197.

LeMone, M. A., 1973: The structure and dynamics of horizontal roll vortices in the planetary boundary layer. *J. Atmos. Sci.,* **30,** 1077–1091.

Lilly, D. K., 1966: On the instability of Ekman boundary flow. *J. Atmos. Sci.,* **23,** 481–494.

———, 1979: The dynamical structure and evolution of thunderstorms and squall lines. *Ann. Rev. Earth Planet. Sci.,* **7,** 117–161.

Lindzen, R. S., 1974: Wave-CISK in the tropics. *J. Atmos. Sci.,* **31,** 156–179.

———, and K.-K. Tung, 1976: Banded convective activity and ducted gravity waves. *Mon. Wea. Rev.,* **104,** 1602–1617.

Locatelli, J. D., J. E. Martin, J. A. Castle, and P. V. Hobbs, 1995: Structure and evolution of winter cyclones in the central United States and their effects on the distribution of precipitation. Part III: The development of a squall line associated with weak cold frontogenesis aloft. *J. Atmos. Sci.,* **123,** 2641–2662.

———, M. T. Stoelinga, R. D. Schwartz, and P. V. Hobbs, 1997: Surface convergence induced by cold fronts aloft and prefrontal surges. *Mon. Wea. Rev.,* **125,** 2808–2820.

Loehrer, S. M., and R. H. Johnson, 1995: Surface pressure and precipitation life cycle characteristics of PRE-STORM mesoscale convective systems. *Mon. Wea. Rev.,* **123,** 600–621.

Ludlam, F. H., 1963: Severe local storms—a review. *Severe Local Storms, Meteor. Monogr.,* No. 27, Amer. Meteor. Soc., 1–30.

Lynn, B. H., W.-K. Tao, and P. J. Wetzel, 1998: A study of landscape-generated deep moist convection. *Mon. Wea. Rev.,* **126,** 928–942.

Lyons, W. A., and J. F. Chandik, 1971: Thunderstorms and the lake breeze front. Preprints, *Seventh Conf. on Severe Local Storms,* Kansas City, MO, Amer. Meteor. Soc., 46–54.

Maddox, R. A., 1976: An evaluation of tornado proximity wind and stability data. *Mon. Wea. Rev.,* **104,** 133–142.

———, 1980: Mesoscale convective complexes. *Bull. Amer. Meteor. Soc.,* **61,** 1374–1387.

———, 1983: Large-scale meteorological conditions associated with midlatitude, mesoscale convective complexes. *Mon. Wea. Rev.,* **111,** 1475–1493.

———, and C. A. Doswell, 1982: Examination of jet stream configurations, 500-mb vorticity advection, and low-level thermal advection patterns during extended periods of intense convection. *Mon. Wea. Rev.,* **110,** 184–197.

———, L. R. Hoxit, C. F. Chappell, and F. Caracena, 1978: Comparison of meteorological aspects of the Big Thompson and Rapid City flash floods. *Mon. Wea. Rev.,* **106,** 375–389.

———, F. Canova, and L. R. Hoxit, 1980a: Meteorological characteristics of flash flood events over the western United States. *Mon. Wea. Rev.,* **108,** 1866–1877.

———, L. R. Hoxit, and C. F. Chappell, 1980b: A study of tornadic thunderstorm interactions with thermal boundaries. *Mon. Wea. Rev.,* **108,** 322–336.

———, D. J. Perkey, and J. M. Fritsch, 1981: Evolution of upper tropospheric features during the development of a mesoscale convective complex. *J. Atmos. Sci.,* **38,** 1664–1674.

Mahoney, W. P., 1988: Gust front characteristics and the kinematics associated with interacting thunderstorm outflows. *Mon. Wea. Rev.,* **116,** 1474–1491.

Mapes, B. E., 1993: Gregarious tropical convection. *J. Atmos. Sci.,* **50,** 2026–2037.

———, 1998: The large-scale part of a mesoscale convective system circulations: A linear vertical spectral band model. *J. Meteor. Soc. Japan,* **76,** 29–55.

———, and R. A. Houze Jr., 1995: Diabatic divergence profiles in western Pacific mesoscale convective systems. *J. Atmos. Sci.,* **52,** 1807–1828.

Markowski, P. M., E. N. Rasmussen, J. M. Straka, and D. C. Dowell, 1998: Observations of low-level baroclinity generated by anvil shadows. *Mon. Wea. Rev.,* **126,** 2942–2958.

Mass, C. F., 1981: Topographically forced convergence in western Washington State. *Mon. Wea. Rev.,* **109,** 1335–1347.

———, and D. P. Dempsey, 1985: One-level, mesoscale model for diagnosing surface winds in mountainous and coastal regions. *Mon. Wea. Rev.,* **113,** 1211–1227.

McAnelly, R. L., J. E. Nachamkin, W. R. Cotton, and M. E. Nicholls, 1997: Upscale evolution of MCSs: Doppler radar analysis and analytical investigation. *Mon. Wea. Rev.,* **125,** 1083–1110.

McCarthy, J., and S. E. Koch, 1982: The evolution of an Oklahoma dryline. Part I: A mesoscale and subsynoptic-scale analysis. *J. Atmos. Sci.,* **39,** 225–236.

McGinley, J. A., 1986: Nowcasting mesoscale phenomena. *Mesoscale Meteorology and Forecasting,* P. S. Ray, Ed., Amer. Meteor. Soc., 657–688.

McNider, R. T., and R. A. Pielke, 1981: Diurnal boundary layer development over sloping terrain. *J. Atmos. Sci.,* **38,** 2198–2212.

———, J. A. Song, and S. Q. Kidder, 1995: Assimilation of GOES-derived solar insolation into a mesoscale model for studies of cloud shading effects. *Int. J. Remote Sensing,* **16,** 2207–2231.

Means, L. L., 1952: On thunderstorm forecasting in the central United States. *Mon. Wea. Rev.,* **80,** 165–189.

Meuse, C., L. Galusha, M. Isaminger, M. Moore, D. Rhoda, F. Robasky, and M. Wolfson, 1996: Analysis of the 12 April 1996 wind shear incident at DFW airport. Preprints, *Workshop on Wind Shear and Wind Shear Alert Systems,* Oklahoma City, OK, Amer. Meteor. Soc., 23–33.

Miles, J. W., and L. N. Howard, 1964: Note on a heterogeneous shear flow. *J. Fluid Mech.,* **20,** 331–336.

Miller, D. A., and F. Sanders, 1980: Mesoscale conditions for the severe convection of 3 April 1974 in the east-central United States. *J. Atmos. Sci.,* **37,** 1041–1055.

Miller, R. C., 1972: Notes on Analysis and Severe-Storm Forecasting Procedures of the Air Force Global Weather Central. Tech. Rep. 200 (rev. 1975), U.S. Air Force, Air Weather Service, 190 pp.

Mitchell, M. J., R. W. Arritt, and K. Labas, 1995: A climatology of the warm season Great Plains low-level jet using wind profiler observations. *Wea. Forecasting,* **10,** 576–591.

Modahl, A. C., 1979: Low-level wind and moisture variations preceding and following hailstorms in northeast Colorado. *Mon. Wea. Rev.,* **107,** 442–450.

Moncrieff, M. W., and J. S. A. Green, 1972: The propagation and transfer properties of steady convective overturning in shear. *Quart. J. Roy. Meteor. Soc.,* **98,** 336–352.

———, and M. J. Miller, 1976: The dynamics and simulation of tropical cumulonimbus and squall lines. *Quart. J. Roy. Meteor. Soc.,* **102,** 373–394.

Moore, J. T., and G. E. VanKnowe, 1992: The effect of jet-streak curvature on kinematic fields. *Mon. Wea. Rev.,* **120,** 2429–2441.

Mueller, C. K., and R. E. Carbone, 1987: Dynamics of a thunderstorm outflow. *J. Atmos. Sci.,* **44,** 1879–1898.

Namias, J., and P. F. Clapp, 1949: Confluence theory of the high tropospheric jet stream. *J. Meteor.,* **6,** 330–336.

Neumann, J., 1951: Land breezes and nocturnal thunderstorms. *J. Meteor.,* **8,** 60–67.

Newton, C. W., 1950: Structure and mechanisms of the prefrontal squall line. *J. Meteor.,* **7,** 210–222.

———, 1963: Dynamics of severe convective storms. *Severe Local Storms, Meteor. Monogr.,* No. 27, Amer. Meteor. Soc., 33–58.

———, and H. R. Newton, 1959: Dynamical interactions between large convective clouds and environment with vertical shear. *J. Meteor.,* **16,** 483–496.

Nicholls, M. E., R. A. Pielke, and W. R. Cotton, 1991: Thermally forced gravity waves in an atmosphere at rest. *J. Atmos. Sci.,* **48,** 1869–1884.

Nieman, P. J., and R. M. Wakimoto, 1999: The interaction of a Pacific cold front with shallow air masses east of the Rocky Mountains. *Mon. Wea. Rev.,* **127,** 2102–2127.

Ogura, Y., 1963: A review of numerical modeling research on small-scale convection in the atmosphere. *Severe Local Storms, Meteor. Monogr.,* No. 27, Amer. Meteor. Soc., 65–76.

———, and Y. L. Chen, 1977: A life history of an intense mesoscale convective storm in Oklahoma. *J. Atmos. Sci.,* **34,** 1458–1476.

———, and M.-T. Liou, 1980: The structure of a midlatitude squall line: A case study. *J. Atmos. Sci.,* **37,** 553–567.

———, and D. Portis, 1982: Structure of the cold front observed in SESAME-AVE III and its comparison with the Hoskins-Bretherton frontogenesis model. *J. Atmos. Sci.,* **39,** 2773–2792.

Ookouchi, Y., M. Segal, R. C. Kessler, and R. A. Pielke, 1984: Evaluation of soil moisture effects on the generation and modification of mesoscale circulations. *Mon. Wea. Rev.,* **112,** 2281–2292.

Ooyama, K. V., 1982: Conceptual evolution of the theory and modeling of the tropical cyclone. *J. Meteor. Soc. Japan,* **60,** 369–380.

Orlanski, I., 1975: A rational subdivision of scales for atmospheric processes. *Bull. Amer. Meteor. Soc.,* **56,** 527–530.

Paegle, J., 1978: A linearized analysis of diurnal boundary layer convergence over the topography of the United States. *Mon. Wea. Rev.,* **106,** 492–502.

———, K. C. Mo, and J. Nogues-Paegle, 1996: Dependence of simulated precipitation on surface evaporation during the 1993 United States summer floods. *Mon. Wea. Rev.,* **124,** 345–361.

Pandya, R. E., and D. R. Durran, 1996: The influence of convectively generated thermal forcing on the mesoscale circulation around squall lines. *J. Atmos. Sci.,* **53,** 2924–2951.

———, ———, and C. Bretherton, 1993: Comments on "Thermally forced gravity waves in an atmosphere at rest." *J. Atmos. Sci.,* **50,** 4097–4101.

Parish, T. R., 1982: Barrier winds along the Sierra Nevada Mountains. *J. Appl. Meteor.,* **21,** 925–930.

Parsons, D. B., 1992: An explanation for intense frontal updrafts and narrow cold-frontal rainbands. *J. Atmos. Sci.,* **49,** 1810–1825.

———, C. G. Mohr, and T. Gal-Chen, 1987: Severe frontal rainband. Part III: Derived thermodynamic structure. *J. Atmos. Sci.,* **44,** 1615–1631.

———, M. A. Shapiro, R. M. Hardesty, R. J. Zamora, and J. M. Intreri, 1991: The finescale structure of a west Texas dryline. *Mon. Wea. Rev.,* **119,** 1242–1258.

Pedgley, D. E., 1962: A meso-synoptic analysis of the thunderstorms on 28 August 1958. U.K. Met Office, Geophys. Mem. 106, 74 pp.

Perry, K. D., and P. V. Hobbs, 1996: Influences of isolated cumulus clouds on the humidity of their surroundings. *J. Atmos. Sci.,* **53,** 159–174.

Petersen, W. A., L. D. Carey, S. A. Rutledge, J. C. Knievel, N. J. Doesken, R. H. Johnson, T. B. McKee, T. H. Vonder Haar, and J. F. Weaver, 1999: Mesoscale and radar observations of the Fort Collins flash flood of 28 July 1997. *Bull. Amer. Meteor. Soc.,* **80,** 191–216.

Pielke, R. A., 1974: A three-dimensional numerical model of the sea breezes over south Florida. *Mon. Wea. Rev.,* **102,** 115–139.

———, and M. Segal, 1986: Mesoscale circulations forced by differential terrain heating. *Mesoscale Meteorology and Forecasting,* P. S. Ray, Ed., Amer. Meteor. Soc., 516–548.

Pothecary, I. J. W., 1954: Short-period variations in surface pressure and wind. *Quart. J. Roy. Meteor. Soc.,* **80,** 395–401.

Preston-Whyte, R. A., 1970: Land breezes and rainfall on the Natal coast. *S. Afr. Geogr. J.,* **52,** 38–43.

Proctor, F. H., 1989: Numerical simulation of an isolated microburst. Part II: Sensitivity experiments. *J. Atmos. Sci.,* **46,** 2143–2165.

Przybylinski, R. W., 1995: The bow echo: Observations, numerical simulations, and severe weather detection methods. *Wea. Forecasting,* **10,** 203–218.

———, and D. M. DeCaire, 1985: Radar signatures associated with the derecho, a type of mesoscale convective system. Preprints, *13th Conf. on Severe Local Storms,* Tulsa, OK, Amer. Meteor. Soc., 270–273.

Purdom, J. F. W., 1973: Meso-highs and satellite imagery. *Mon. Wea. Rev.,* **101,** 180–181.

———, 1982: Subjective interpretations of geostationary satellite data for nowcasting. *Nowcasting,* K. Browning, Ed., Academic Press, 149–166.

Rabin, R. M., S. Stadler, P. J. Wetzel, D. J. Stensrud, and M. Gregory, 1990: Observed effects of landscape variability on convective clouds. *Bull. Amer. Meteor. Soc.,* **71,** 272–280.

Ralph, F. M., M. Crochet, and S. V. Venkateswaran, 1993: Observations of a mesoscale ducted gravity wave. *J. Atmos. Sci.,* **50,** 3277–3291.

Ramage, C. S., 1971: *Monsoon Meteorology.* Academic Press, New York, 296 pp.

Raymond, D. J., 1975: A model for predicting the movement of continuously propagating convective storms. *J. Atmos. Sci.,* **32,** 1308–1317.

———, 1976: Wave-CISK and convective mesosystems. *J. Atmos. Sci.,* **33,** 2392–2398.

———, 1978: Instability of the low-level jet and severe storm formation. *J. Atmos. Sci.,* **35,** 2274–2280.

———, 1987: A forced gravity wave model of self-organizing convection. *J. Atmos. Sci.,* **44,** 3528–3543.

Rhea, J. O., 1966: A study of thunderstorm formation along dry lines. *J. Appl. Meteor.,* **5,** 58–63.

Rochette, S. M., and J. T. Moore, 1996: Initiation of an elevated mesoscale convective system associated with heavy rainfall. *Wea. Forecasting,* **11,** 443–457.

Rotunno, R., 1981: On the evolution of thunderstorm rotation. *Mon. Wea. Rev.* **109,** 577–586.

———, 1984: Investigation of a three-dimensional asymmetric vortex. *J. Atmos. Sci.,* **41,** 283–298.

———, and J. B. Klemp, 1982: The influence of shear-induced pressure gradient on thunderstorm motion. *Mon. Wea. Rev.,* **110,** 136–151.

———, and ———, 1985: On the rotation and propagation of simulated supercell thunderstorms. *J. Atmos. Sci.,* **42,** 271–292.

———, ———, and M. L. Weisman, 1988: A theory for strong, long-lived squall lines. *J. Atmos. Sci.,* **45,** 463–485.

Rutledge, S. A., 1989: A severe frontal rainband. Part IV: Precipitation mechanisms, diabatic processes and rainband maintenance. *J. Atmos. Sci.,* **46,** 3570–3594.

———, and R. A. Houze Jr., 1987: Diagnostic modeling study of the trailing stratiform region of a midlatitude squall line. *J. Atmos. Sci.,* **44,** 2640–2656.

Sanders, F., and D. O. Blanchard, 1993: The origin of a severe thunderstorm in Kansas on 10 May 1985. *Mon. Wea. Rev.,* **121,** 133–149.

Sawyer, J. S., 1946: Cooling by rain as the cause of the pressure rise in convective squalls. *Quart. J. Roy. Meteor. Soc.,* **72,** 168.

Schaefer, J. T., 1974a: A simulative model of dryline motion. *J. Atmos. Sci.,* **31,** 956–964.

———, 1974b: Life cycle of the dryline. *J. Appl. Meteor.,* **13,** 444–449.

———, 1986: The dryline. *Mesoscale Meteorology and Forecasting,* P. S. Ray, Ed., Amer. Meteor. Soc., 549–572.

Schlesinger, R. E., 1980: A three-dimensional numerical model of an isolated deep thunderstorm. Part II: Dynamics of updraft splitting and mesovortex couplet evolution. *J. Atmos. Sci.,* **37,** 395–420.

Schmid, W., and M. Lehre, 1998: Drainage flow: A key factor for prediction of severe storms near mountain chains? Preprints, *19th Conf. on Severe Local Storms,* Minneapolis, MN, Amer. Meteor. Soc., 22–25.
Schmidt, J. M., and W. R. Cotton, 1989: High Plains squall line associated with severe surface winds. *J. Atmos. Sci.,* **46,** 281–302.
———, and ———, 1990: Interactions between upper and lower tropospheric gravity waves on squall line structure and maintenance. *J. Atmos. Sci.,* **47,** 1205–1222.
Schubert, W. H., S. R. Fulton, and R. F. A. Hertenstein, 1989: Balanced atmospheric response to squall lines. *J. Atmos. Sci.,* **46,** 2478–2483.
———, M. T. Montgomery, R. K. Taft, T. A. Guinn, S. R. Fulton, J. P. Kossin, and J. P. Edwards, 1999: Polygonal eyewalls, asymmetric eye contraction, and potential vorticity mixing in hurricanes. *J. Atmos. Sci.,* **56,** 1197–1223.
Schultz, D. M., and P. N. Schumacher, 1999: The use and misuse of conditional symmetric instability. *Mon. Wea. Rev.,* **127,** 2709–2732.
Segal, M., and R. W. Arritt, 1992: Nonclassical mesoscale circulations caused by surface sensible heat-flux gradients. *Bull. Amer. Meteor. Soc.,* **73,** 1593–1604.
———, W. E. Schreiber, G. Kallos, J. R. Garratt, A. Rodi, J. Weaver, and R. A. Pielke, 1989: The impact of crop areas in northeast Colorado on midsummer mesoscale thermal circulations. *Mon. Wea. Rev.,* **117,** 809–825.
———, R. W. Arritt, C. Clark, R. Rabin, and J. Brown, 1995: Scaling evaluation of the effect of surface characteristics on potential for deep convection over uniform terrain. *Mon. Wea. Rev.,* **123,** 383–400.
Seman, C. J., 1994: A numerical study of nonlinear nonhydrostatic conditional symmetric instability in a convectively unstable atmosphere. *J. Atmos. Sci.,* **51,** 1352–1371.
Shapiro, M. A., 1981: Frontogenesis and geostrophically forced secondary circulations in the vicinity of jet stream-frontal zone systems. *J. Atmos. Sci.,* **38,** 954–973.
———, 1982: Mesoscale weather systems of the central United States. *The National STORM Program,* UCAR, Boulder, CO, 3-1–3-77.
———, 1985: Frontal hydraulic head: A microscale (approximately 1-km) triggering mechanism for mesoconvective weather systems. *Mon. Wea. Rev.,* **113,** 1166–1183.
———, and P. J. Kennedy, 1981: Research aircraft measurements of jet stream geostrophic and ageostrophic winds. *J. Atmos. Sci.,* **38,** 2642–2652.
Shreffler, J. H., and F. S. Binkowski, 1981: Observations of pressure jump lines in the Midwest, 10–12 August 1976. *Mon. Wea. Rev.,* **109,** 1713–1725.
Simpson, J. E., 1969: A comparison between laboratory and atmospheric density currents. *Quart. J. Roy. Meteor. Soc.,* **95,** 758–765.
Simpson, J., B. R. Morton, M. C. McCumber, and R. S. Penc, 1986: Observations and mechanisms of GATE waterspouts. *J. Atmos. Sci.,* **43,** 753–782.
———, T. D. Keenan, B. Ferrier, R. H. Simpson, and G. J. Holland, 1993: Cumulus mergers in the Maritime Continent region. *Meteor. Atmos. Phys.,* **51,** 73–99.
———, 1987: *Gravity Currents.* Ellis Horwood Limited, 244 pp.
Skamarock, W. C., M. L. Weisman, and J. B. Klemp, 1994: Three-dimensional evolution of simulated long-lived squall lines. *J. Atmos. Sci.,* **51,** 2563–2584.
Smith, R. B., 1982: Synoptic observations and theory of orographically disturbed wind and pressure. *J. Atmos. Sci.,* **39,** 60–70.
———, 1989: Comments on "Low Froude number flow past three-dimensional obstacles. Part I: Baroclinically generated lee vortices." *J. Atmos. Sci.,* **46,** 3611–3613.
Smith, R. K., 1988: Travelling waves and bores in the lower atmosphere: The "Morning Glory" and related phenomena. *Earth-Sci. Rev.,* **25,** 267–290.
———, and M. A. Page, 1985: Morning glory wind surges and the Gulf of Carpentaria cloud line of 25–26 Oct. 1984. *Austr. Met. Mag.,* **33,** 185–194.
Smolarkiewicz, P. K., and R. Rotunno, 1989: Low Froude number flow past three-dimensional obstacles. Part I: Baroclinically generated lee vortices. *J. Atmos. Sci.,* **46,** 1154–1164.
Smull, B. F., and R. A. Houze Jr., 1987: Rear inflow in squall lines with trailing stratiform precipitation. *Mon. Wea. Rev.,* **115,** 2869–2889.
Sortais, J.-L., J.-P. Cammas, X. D. Yu, E. Richard, and R. Rosset, 1993: A case study of coupling between low- and upper-level jet-front systems: Investigation of dynamical and diabatic processes. *Mon. Wea. Rev.,* **121,** 2239–2253.
Srivastava, R. C., 1985: Simple model of evaporatively driven downdraft: application to microburst downdraft. *J. Atmos. Sci.,* **42,** 1004–1023.
———, 1987: Model of intense downdrafts driven by the melting and evaporation of precipitation. *J. Atmos. Sci.,* **44,** 1752–1773.
Stensrud, D. J., and R. A. Maddox, 1988: Opposing mesoscale circulations: A case study. *Wea. Forecasting,* **3,** 189–204.
———, ———, and C. L. Ziegler, 1991: A sublimation-initiated mesoscale downdraft and its relation to the wind field below a precipitating anvil cloud. *Mon. Wea. Rev.,* **119,** 2124–2139.
Stull, R. B., 1985: Fair-weather cumulus cloud classification scheme for mixed-layer studies. *J. Climate Appl. Meteor.,* **24,** 49–56.
Stumpf, G. J., R. H. Johnson, and B. F. Smull, 1991: The wake low in a midlatitude mesoscale convective system having complex organization. *Mon. Wea. Rev.,* **119,** 134–158.
Sun, W.-Y., and Y. Ogura, 1979: Boundary-layer forcing as a possible trigger to a squall-line formation. *J. Atmos. Sci.,* **36,** 235–253.
Szoke, E. J., and R. Brady, 1989: Forecasting implications of the 26 July 1985 northeastern Colorado thunderstorm case. *Mon. Wea. Rev.,* **117,** 1834–1860.
———, M. L. Weisman, J. M. Brown, F. Caracena, and T. W. Schlatter, 1984: A subsynoptic analysis of the Denver tornadoes of 3 June 1981. *Mon. Wea. Rev.,* **112,** 790–808.
Tao, S., and Y.-H. Ding, 1981: Observational evidence of the influence of the Qinghai-Xizang (Tibet) Plateau on the occurrence of heavy rain and severe convective storms in China. *Bull. Amer. Meteor. Soc.,* **62,** 23–30.
Tao, W.-K., J. Simpson, C.-H. Sui, B. Ferrier, S. Lang, J. Scala, M.-D. Chou, and K. Pickering, 1993: Heating, moisture, and water budgets of tropical and midlatitude squall lines: Comparisons and sensitivity to longwave radiation. *J. Atmos. Sci.,* **50,** 673–690.
Tepper, M., 1950: A proposed mechanism of squall lines: The pressure jump line. *J. Meteor.,* **7,** 21–29.
———, 1951: On the desiccation of a cloud bank by a propagated pressure wave. *Mon. Wea. Rev.,* **79,** 61–70.
Thorpe, A. J., M. J. Miller, and M. W. Moncrieff, 1982: Two-dimensional convection in nonconstant shear: a model of midlatitude squall lines. *Quart. J. Roy. Meteor. Soc.,* **108,** 739–762.
———, H. Volkert, D. Heimann, 1993: Potential vorticity of flow along the Alps. *J. Atmos. Sci.,* **50,** 1573–1591.
Toth, J. J., and R. H. Johnson, 1985: Summer surface flow characteristics over northeast Colorado. *Mon. Wea. Rev.,* **113,** 1458–1469.
Trier, S. B., and D. B. Parsons, 1993: Evolution of environmental conditions preceding the development of a nocturnal mesoscale convective complex. *Mon. Wea. Rev.,* **121,** 1078–1098.
———, ———, and J. H. E. Clark, 1991: Environment and evolution of a cold-frontal mesoscale convective system. *Mon. Wea. Rev.,* **119,** 2429–2455.
Tripoli, G. J., and W. R. Cotton, 1989a: Numerical study of an observed orogenic mesoscale convective system. Part I: Simulated genesis and comparison with observations. *Mon. Wea. Rev.,* **117,** 283–304.

———, and ———, 1989b: Numerical study of an observed orogenic mesoscale convective system. Part II: Analysis of governing dynamics. *Mon. Wea. Rev.,* **117,** 305–328.

Uccellini, L. W., 1975: A case study of apparent gravity wave initiation of severe convective storms. *Mon. Wea. Rev.,* **103,** 497–513.

———, 1990: Processes contributing to the rapid development of extratropical cyclones. *Extratropical Cyclones: The Erik Palmén Memorial Volume,* C. W. Newton and E. O. Holopainen, Eds., Amer. Meteor. Soc., 81–105.

———, and D. R. Johnson, 1979: The coupling of upper- and lower-tropospheric jet streaks and implications for the development of severe convective storms. *Mon. Wea. Rev.,* **107,** 682–703.

Ulanski, S. L., and G. M. Heymsfield, 1986: Meso-β perturbations of the wind field by thunderstorm cells. *Mon. Wea. Rev.,* **114,** 780–787.

Velasco, I., and J. M. Fritsch, 1987: Mesoscale convective complexes in the Americas. *J. Geophys. Res.,* **92,** 9591–9613.

Wakimoto, R. M., 1985: Forecasting dry microburst activity over the High Plains. *Mon. Wea. Rev.,* **113,** 1131–1143.

———, and J. W. Wilson, 1989: Non-supercell tornadoes. *Mon. Wea. Rev.,* **117,** 1113–1140.

———, and J. K. Lew, 1993: Observations of a Florida waterspout during CaPE. *Wea. Forecasting,* **8,** 412–423.

———, and N. T. Atkins, 1994: Observations of the sea-breeze front during CaPE. Part I: Single-Doppler, satellite, and cloud-photogrammetry analysis. *Mon. Wea. Rev.,* **122,** 1092–1114.

———, and P. G. Black, 1994: Damage survey of Hurricane Andrew and its relationship to the eyewall. *Bull. Amer. Meteor. Soc.,* **75,** 189–200.

———, and C. Liu, 1998: The Garden City, Kansas, storm during VORTEX 95. Part II: The wall cloud and tornado. *Mon. Wea. Rev.,* **126,** 393–408.

Wallace, J. M., 1975: Diurnal variations in precipitation and thunderstorm frequency over the conterminous United States. *Mon. Wea. Rev.,* **103,** 406–419.

Watanabe, H., and Y. Ogura, 1987: Effects of orographically forced upstream lifting on mesoscale heavy precipitation: A case study. *J. Atmos. Sci.,* **44,** 661–675.

Weckwerth, T. M., and R. M. Wakimoto, 1992: The initiation and organization of convective cells atop a cold-air outflow boundary. *Mon. Wea. Rev.,* **120,** 2169–2187.

———, J. W. Wilson, and R. M. Wakimoto, 1996: Thermodynamic variability within the convective boundary layer due to horizontal convective rolls. *Mon. Wea. Rev.,* **124,** 769–784.

Weisman, M. L., 1992: The role of convectively generated rear-inflow jets in the evolution of long-lived mesoconvective systems. *J. Atmos. Sci.,* **49,** 1826–1847.

———, 1993: The genesis of severe, long-lived bow echoes. *J. Atmos. Sci.,* **50,** 645–670.

———, and J. B. Klemp, 1982: The dependence of numerically simulated convective storms on vertical wind shear and buoyancy. *Mon. Wea. Rev.,* **110,** 504–520.

———, and ———, 1984: The structure and classification of numerically simulated convective storms in directionally varying wind shears. *Mon. Wea. Rev.,* **112,** 2479–2498.

———, and ———, 1986: Characteristics of isolated convective storms. *Mesoscale Meteorology and Forecasting,* P. S. Ray, Ed., Amer. Meteor. Soc., 331–358.

———, and C. A. Davis, 1998: Mechanisms for the generation of mesoscale vortices within quasi-linear convective systems. *J. Atmos. Sci.,* **55,** 2603–2622.

Wetzel, P. J., 1973: Moisture sources and flow patterns during the northeast Colorado hail season. Master's thesis, Dept. of Atmos. Sci., Colorado State University, 90 pp.

Whiteman, C. D., B. Xindi, and S. Zhong, 1997: Low-level jet climatology from enhanced rawinsonde observations at a site in the southern Great Plains. *J. Appl. Meteor.,* **36,** 1363–1376.

Wilczak, J. M., and J. W. Glendening, 1988: Observations and mixed-layer modeling of a terrain-induced mesoscale gyre: The Denver cyclone. *Mon. Wea. Rev.,* **116,** 2688–2711.

———, and T. W. Christian, 1990: Case study of an orographically induced mesoscale vortex (Denver cyclone). *Mon. Wea. Rev.,* **118,** 1082–1102.

Williams, D. T., 1963: The thunderstorm wake of May 4, 1961. Natl. Severe Storms Project Rep. 18, U.S. Dept. of Commerce, Washington, DC, 23 pp. [NTIS PB-168223.]

Williams, M., and R. A. Houze Jr., 1987: Satellite-observed characteristics of winter monsoon cloud clusters. *Mon. Wea. Rev.,* **115,** 505–519.

Wilson, J. W., and W. E. Schreiber, 1986: Initiation of convective storms at radar-observed boundary-layer convergence lines. *Mon. Wea. Rev.,* **114,** 2516–2536.

———, G. B. Foote, N. A. Crook, J. C. Fankhauser, C. G. Wade, J. D. Tuttle, C. K. Mueller, and S. K. Krueger, 1992: The role of boundary-layer convergence zones and horizontal rolls in the initiation of thunderstorms: A case study. *Mon. Wea. Rev.,* **120,** 1785–1815.

Xu, Q., 1986: Conditional symmetric instability and mesoscale rainbands. *Quart. J. Roy. Meteor. Soc.,* **112,** 315–334.

Yan, H., and R. A. Anthes, 1988: The effect of variations in surface moisture on mesoscale circulations. *Mon. Wea. Rev.,* **116,** 192–208.

Yanai, M., C. Li, and Z. Song, 1992: Seasonal heating of the Tibetan Plateau and its effects on the evolution of the Asian summer monsoon. *J. Meteor. Soc. Japan,* **70,** 189–221.

Zhang, D.-L., 1992: The formation of a cooling-induced mesovortex in the trailing stratiform region of a midlatitude squall line. *Mon. Wea. Rev.,* **120,** 2763–2785.

———, and J. M. Fritsch, 1986: Numerical simulation of the meso-beta-scale structure and evolution of the 1977 Johnstown flood. Part I: Model description and verification. *J. Atmos. Sci.,* **43,** 1913–1943.

———, and ———, 1988: Numerical investigation of a convectively generated, inertially stable, extratropical warm-core mesovortex over land. Part I: Structure and evolution. *Mon. Wea. Rev.,* **116,** 2660–2687.

———, and K. Gao, 1989: Numerical simulation of an intense squall line during 10–11 June 1985 PRE-STORM. Part II: Rear inflow, surface pressure perturbations and stratiform precipitation. *Mon. Wea. Rev.,* **117,** 2067–2094.

———, and H.-R. Cho, 1992: The development of negative moist potential vorticity in the stratiform region of a simulated squall line. *Mon. Wea. Rev.,* **120,** 1322–1341.

Ziegler, C. L., and C. E. Hane, 1993: An observational study of the dryline. *Mon. Wea. Rev.,* **121,** 1134–1151.

———, W. J. Martin, R. A. Pielke, and R. L. Walko, 1995: A modeling study of the dryline. *J. Atmos. Sci.,* **52,** 263–285.

Zipser, E. J., 1977: Mesoscale and convective-scale downdrafts as distinct components of squall-line circulation. *Mon. Wea. Rev.,* **105,** 1568–1589.

Ziv, B., and N. Paldor, 1999: The divergence fields associated with time-dependent jet streams. *J. Atmos. Sci.,* **56,** 1843–1857.

Chapter 4

Numerical Modeling of Severe Local Storms

ROBERT B. WILHELMSON

Department of Atmospheric Sciences and National Center for Supercomputer Applications, University of Illinois at Urbana–Champaign, Champaign, Illinois

LOUIS J. WICKER

NOAA/National Severe Storms Laboratory, Norman, Oklahoma

REVIEW PANEL: Joe Klemp (Chair), Dale Durran, Greg Tripoli, and Morris Weisman

---

4.1. Introduction

Numerical modeling of clouds is as old as computers capable of solving discrete versions of the fundamental dynamical equations. With the limited memories and computer power available in the 1960s and 1970s, most modelers employed two-dimensional slab or axisymmetric approximations to study convective dynamics (e.g., Lilly 1962; Ogura and Charney 1962; Orville 1968; Takeda 1971; Wilhelmson and Ogura 1972; Hane 1973; Soong and Ogura 1973; Schlesinger 1973; Soong 1974). The slab models represented the convective growth of infinitely long convective bands forming in environments with or without vertical wind shear, while the axisymmetric simulations were constrained to shearless environments. However, as computer power grew and vector computers were developed (e.g., the CRAY 7600 and the succeeding CRAY computers; Kaufmann and Smarr 1993), it became possible to solve the three-dimensional equations of motion on relatively coarse-mesh grids (e.g., Steiner 1973; Deardorff 1972; Wilhelmson 1974; Schlesinger 1975; Klemp and Wilhelmson 1978b).

Over the years the use of convective models has led to better understanding and improved forecasting of severe storms and their associated high winds, tornadoes, and hail. Researchers have been able to augment observational studies through the provision of consistent datasets over regular spatial grids. In practice, convective storm models are used to carry out case studies of observed storm events, to depict the behavior of particular storm features such as outflows or tornadoes, and to study the response of a storm cloud or system to changes in the storm environment. Furthermore, they are used to explore the important dynamical or physical processes that influence a particular storm's behavior. Computational power has now increased sufficiently so that these models are beginning to be used in forecasting severe weather events.

Somewhat remarkably, numerical simulations of convective storms have been able to reproduce most of the salient features associated with atmospheric convection. Three-dimensional simulations of severe storms have reproduced, sometimes with uncanny accuracy, such phenomena as supercells, their mesocyclones, and tornadoes (both mesocyclonic and nonmesocyclonic), as well as severe weather associated with convective lines including bow echoes, severe squall lines, mesoscale convective vortices, downbursts, and microbursts. Recent experiments have even begun to include detailed representations of cloud hydrometeors, their complex interactions, and the effect these processes have on the storm's electric field. Numerical weather prediction of individual convective storms in the two- to six-hour time frame has also been demonstrated. All of this has been accomplished despite the glaring limitations in model physics, grid resolution, and initial conditions.

The discussion of severe storm modeling in this chapter focuses primarily on the understanding of storm structure and evolution enabled by the ability to represent three-dimensional spatial features of storms and storm systems (e.g., supercells, finite squall lines, tornadic development, and structure). The use of three-dimensional models also removes some of the artifacts of lower-dimensional models that include the upscale growth of small turbulent eddies. For example, two-dimensional density current simulations reported by Droegemeier and Wilhelmson (1987) show a well-defined Kelvin–Helmholtz roll well behind the gust front, while laboratory studies do not show this feature (Droegemeier and Wilhelmson 1986). Three-dimensional simulations involving density currents investigated by Lee and Wilhelmson (1997a) are consistent with the laboratory studies. The focus here is therefore directed on models and modeling results that represent three-dimensional flows associated with atmospheric convection. Even with this limitation, it is impossible to adequately discuss all of the three-dimensional models and studies that have contributed to our knowledge of these storms. The intent here is to provide the reader with a broad overview of the major modeling efforts and the new fundamental scientific knowledge generated through the use of numerical modeling of severe convective storms during the last two decades.

This chapter is divided into two main parts. In the first part, the history and framework for severe storm models is reviewed. The intent is to provide an overview of key issues faced by the modeler and how some of these have been addressed over the past 40 years. In the second part, examples from the hundreds of papers published during this time are used to discuss the impact of modeling on efforts to understand and predict severe storm behavior. Several areas have been selected in which modeling has contributed significantly to our understanding and predictive capabilities. These include the numerical simulation of supercell and multicell storms, severe storm lines, bow echoes, gust fronts, and tornado vortices and tornadogenesis. Some of these topics have been covered elsewhere in this monograph. Their brief inclusion here is designed to illustrate the impact that modeling can have on our understanding of these phenomena and how modeling can be used in conjunction with observations and laboratory studies to understand some of the flow features associated with severe convection. Past reviews provide additional material in book form (e.g., Cotton and Anthes 1989; Houze 1993; Emanuel 1994) and in review articles (e.g., Lilly 1975, 1979; Houze and Hobbs 1982; Klemp 1987; Rotunno 1993). The chapter ends with some comments on key issues that remain to be addressed.

4.2. Nonhydrostatic model systems for local severe storm modeling

a. The dynamic equations

The numerical modeling of storms is based on the ability to solve a system of equations that describes the range of atmospheric flows on horizontal scales ranging from tens of meters to hundreds of kilometers. This system can be illustrated by the following equations expressed in Cartesian coordinates and in advective form[1]:

$$\frac{\partial u}{\partial t} + c_p\theta_v \frac{\partial \pi'}{\partial x} = -u\frac{\partial u}{\partial x} - v\frac{\partial u}{\partial y} - w\frac{\partial u}{\partial z} + fv + D_u = f_u \tag{4.1}$$

$$\frac{\partial v}{\partial t} + c_p\theta_v \frac{\partial \pi'}{\partial y} = -u\frac{\partial v}{\partial x} - v\frac{\partial v}{\partial y} - w\frac{\partial v}{\partial z} - fu + D_v = f_v \tag{4.2}$$

$$\frac{\partial w}{\partial t} + c_p\theta_v \frac{\partial \pi'}{\partial z} = g\left[\frac{\theta - \bar{\theta}}{\bar{\theta}} + 0.61(q_v - \bar{q}_v) - q_T\right] - u\frac{\partial w}{\partial x} - v\frac{\partial w}{\partial y} - w\frac{\partial w}{\partial z} + D_w = f_w \tag{4.3}$$

$$\frac{\partial \theta}{\partial t} = -u\frac{\partial \theta}{\partial x} - v\frac{\partial \theta}{\partial y} - w\frac{\partial \theta}{\partial z} + M_\theta + D_\theta \tag{4.4}$$

$$\frac{\partial \rho}{\partial t} = -\frac{\partial \rho u}{\partial x} - \frac{\partial \rho v}{\partial y} - \frac{\partial \rho w}{\partial z}. \tag{4.5}$$

Using (4.4), (4.5), and the equation of state, a nondimensional pressure equation can be derived:

$$\frac{\partial \pi'}{\partial t} + \frac{\bar{c}^2}{c_p\bar{\rho}\bar{\theta}_v^2}\left[\frac{\partial(\bar{\rho}\bar{\theta}_v u)}{\partial x} + \frac{\partial(\bar{\rho}\bar{\theta}_v v)}{\partial y} + \frac{\partial(\bar{\rho}\bar{\theta}_v w)}{\partial z}\right] = -u\frac{\partial \pi'}{\partial x} - v\frac{\partial \pi'}{\partial y} - w\frac{\partial \pi'}{\partial z} - \frac{R_d\pi}{c_v}\left(\frac{\partial u}{\partial x} + \frac{\partial v}{\partial y} + \frac{\partial w}{\partial z}\right) + \frac{c^2}{c_p\theta_v^2}\frac{d\theta_v}{dt} + D_\pi = f_\pi . \tag{4.6}$$

This equation then replaces (4.5) as the prognostic mass equation for the system. The equations for the remaining dependent variables, such as water mixing ratios or chemical concentrations, can be expressed as

$$\frac{dq_i}{dt} = \frac{\partial q_i}{\partial t} + u\frac{\partial q_i}{\partial x} + v\frac{\partial q_i}{\partial y} + w\frac{\partial q_i}{\partial z} = M_{q_i} + D_{q_i}. \tag{4.7}$$

In the above equations, $\pi' = \pi - \bar{\pi}(x, y, z) = (R_d\rho\theta_v/p_o)^{R_d/c_v} - (R_d\bar{\rho}\bar{\theta}_v/p_o)^{R_d/c_v}$ is the nondimensional pressure deviation from a motionless base state defined by the barred quantities where $\rho = \bar{\rho}(x, y, z) + \rho'$ and $\theta = \bar{\theta}(x, y, z) + \theta'$. The base state pressure at the ground is represented by p_o, the virtual potential temperature is defined by $\theta_v = \theta(1 + 0.61q_v)$, and $\bar{c}^2 = c_pR\bar{\pi}\bar{\theta}_v/c_v$ is

[1] The only difference between these equations and those presented by Klemp and Wilhelmson (1978b) are in the coefficients of the pressure derivative terms and the sign correction in the pressure equation.

the square of the speed of sound for the base state. Here q_i represents the mixing ratio for the water and ice species represented in the model (q_T is the sum of these categories but does not include water vapor). The term D represents the subgrid turbulence terms and approximates the subgrid fluxes that result from Reynold's averaging. If instead they are considered to represent viscous dissipation before any averaging, the equations when applied to dry or moist (no condensation) motion involve no approximations beyond the standard ones used in deriving the general Navier–Stokes equations. The parameter M refers to the appropriate microphysical adjustments including condensation, freezing, evaporation, and sublimation, as well as interactions between the various water species. The resulting system is a fully prognostic system and includes both gravity and sound waves.

Various simplifications to this nonhydrostatic system have been proposed and used in convective models through the use of scale analysis that depended on the particular phenomena being studied (e.g., small cumulus or cumulonimbus, deep storms lasting several hours or several days). Historically, simplification was motivated by the desire to focus on "key" dynamical processes and by limitations imposed by computer speed, memory, input/output (I/O) performance, and data storage capacity. Some of the first convective storm modeling results were reviewed by Ogura (1963) in the first severe local storms monograph. The nonhydrostatic model equations used were based on the work of Ogura and Phillips (1962), who removed both acoustic (sound) and external gravity waves. In this system the mass continuity equation took on the form $\partial \bar{\rho}u/\partial x + \partial \bar{\rho}v/\partial y + \partial \bar{\rho}w/\partial z = 0$ and the base-state potential temperature, θ_0, was considered to be constant. This constant then replaced θ_v in the pressure derivative terms in (4.1)–(4.3). Ogura and Phillips referred to their equation set as *anelastic* because the associated kinetic–potential energy equation did not contain an elastic energy term related to the square of the pressure fluctuation and the presence of sound waves. It differed from the traditional Boussinesq system in that $\bar{\rho}(z)$ was substituted for total density in the continuity equation. In their version they also used entropy and the total water mixing ratio as Eq. (4.7) variables and determined the portion of the condensate by subtracting the saturation mixing ratio from the predicted total water mixing ratio.

Since 1963 many different forms of the fluid equations have been used to simulate convective flow. The systems fall into two classes, those that are incompressible or anelastic and those that are compressible (sound waves explicitly represented).[2] The original anelastic scheme derived by Ogura and Phillips (1962) was based on the assumption that $\theta' \ll \theta_0$, where, θ_0 was taken to be a constant base-state potential temperature. This is appropriate for three-dimensional simulations of shallow convection (e.g., Steiner 1973) but not for deep convection that extends into the stably stratified stratosphere. Thus, Wilhelmson and Ogura (1972) introduced the modified anelastic system in which θ_0 was replaced by an initially specified mean $\bar{\theta}$ that varied with height.

Then, Klemp and Wilhelmson (1978b) introduced an elastic system referred to as KW78 and demonstrated that it could be solved efficiently despite the presence of sound waves. The pressure equation (4.6) was simplified in their model code from that given in (4.6) to

$$\frac{\partial \pi'}{\partial t} + \frac{\bar{c}^2}{c_p \bar{\rho} \bar{\theta}^2}\left[\frac{\partial(\bar{\rho}\bar{\theta}u)}{\partial x} + \frac{\partial(\bar{\rho}\bar{\theta}v)}{\partial y} + \frac{\partial(\bar{\rho}\bar{\theta}w)}{\partial z}\right] = 0, \tag{4.8}$$

where $\bar{c}^2 = c_p R \bar{\pi} \bar{\theta} / c_v$ is the square of the speed of sound. Durran (1989, 1990) noted that the bracketed term in (4.6) must approach zero for the pressure perturbation to remain finite. The resulting non-time-dependent equation differs from the modified anelastic continuity equation, $\partial \bar{\rho}u/\partial x + \partial \bar{\rho}v/\partial y + \partial \bar{\rho}w/\partial z = 0$, by the inclusion of $\bar{\theta}$ inside the derivatives. This and other forms of the anelastic continuity equation and energy conservation are reviewed by Durran (1999).

Three-dimensional convective and storm models formulated with the anelastic approach and reported on in the 1970s include Clark (1977), Fox (1972), Schlesinger (1975), Steiner (1973), and Wilhelmson (1974). In these systems the mass continuity equation was combined with the momentum equations to form an elliptic partial differential equation for pressure. Many researchers have also chosen to solve some form of the compressible (elastic) model system. The most straightforward solution technique is to use explicit finite difference approximations to predict the prognostic variables as done by Lilly (1962) in early two-dimensional simulations of buoyant convection. The time step is then constrained by the more restrictive sound wave speed rather than by the significantly slower advective or gravity wave speeds. Since that time, approaches to solving systems in which sound waves are present, but not of significant interest, have been developed that are computationally competitive with anelastic solutions. The use of compressible nonhydrostatic systems is motivated by several factors including the basic fact that no dynamic approximations to the Euler equations are necessary (although they are often made). Further, sound waves do have

[2] Several books have appeared in the past decade that provide more in-depth information on convective modeling systems than can be presented in this chapter. They include Cotton and Anthes (1989), Houze (1993), and Emanuel (1994).

some influence on gravity wave speeds that could be important in situations where accurate predictions are desired [Ogura and Phillips 1962; Durran 1999, (Eq. 7.46)].[3] Because all the equations are prognostic (none are elliptic), different coordinate systems and boundary condition specifications can be more easily included when semi-implicit numerical approximations that lead to an elliptic Helmholtz equation are not used. It is also easier to implement adaptive mesh refinement (AMR) that may include rotated fine grids because exact mass conservation is not crucial (Skamarock and Klemp 1993). Further, there has been a growing movement over the past 10 years aimed at developing compressible nonhydrostatic models that can be used on convective to mesoscale to global scales. This is motivated in part by the rapid growth in computer power that has made it possible to use increasingly higher spatial resolution. Eventually gravity wave speeds in large-scale hydrostatic models begin deviating significantly from nonhydrostatic models as the horizontal wavenumber increases (Tapp and White 1976). With nonhydrostatic models it is possible to successfully model trapped gravity waves.

b. Numerical solutions to the anelastic and compressible system

A wide variety of finite difference methods have been used in local storm models to approximate the set of partial differential equations used, whether they are in anelastic or compressible form. Numerical cloud models have typically used finite difference methods to approximate the partial differential equations. The momentum equations have traditionally been handled with forward or centered time approximations, centered in space pressure differences, and upstream-biased (with forward time) or centered (with centered time) advection approximations (e.g., Schlesinger 1975; Clark 1977; Klemp and Wilhelmson 1978b). The scalar variables (temperature, water vapor, and hydrometeor mixing ratios) have also been treated with similar approximations. A staggered grid is typically used to increase the accuracy of the pressure gradient and divergence calculations as well as to simplify the solution of the elliptic pressure equation in anelastic systems. Further, the computational dispersion properties of staggered grids are usually better (Fox-Rabinovitz 1996) than those of nonstaggered grids, although they usually require smaller time steps. The elliptic equation in anelastic and in some semi-implicit compressible approximations has been solved using direct methods such as fast Fourier transforms or cyclic reduction or using iterative methods (e.g., Wilhelmson 1974). The latter require stopping criteria that have most recently been addressed by Smolarkiewicz et al. (1997). Today's choices also include multigrid methods (Adams et al. 1992) and preconditioned conjugate-residual solvers for the Helmholtz equations (Skamarock et al. 1997).

More specifically and for brevity, a brief history of the Klemp–Wilhelmson model is reviewed as an example of model evolution typical of the history of other storm models. In the original KW78 Cartesian model, the finite differences for time and space were centered and second order (e.g., the classic leapfrog scheme). Wilhelmson and Chen (1982) introduced a stretched vertical grid to more accurately resolve low-level features while limiting computational cost. They also introduced fourth-order flux approximations to the advective terms in order to reduce phase speed errors. This form of the model was used until Wicker and Wilhelmson (1995) began advecting the scalar variables using a monotonic forward-in-time and upwind-in-space scheme. Also, the model code was substantially rewritten in order to take advantage of the new large in-core memory computers that were becoming available. It then became known as Collaborative Model for Multiscale Atmospheric Simulation (COMMAS). COMMAS initially included a nested-grid capability in which nonrotating finer grid meshes could be introduced into the parent grid mesh by user intervention. In the early 1990s this capability was replaced by the more generalized automatic strategy based on flow evolution discussed by Skamarock and Klemp (1993). More recently, the time differencing of the momentum equations was changed from a second-order centered time (i.e., the leapfrog time scheme) to a second- and third- order Runge–Kutta time scheme (Wicker and Skamarock 1998). The Runge–Kutta schemes enabled the use of third- and fifth-order upwind horizontal differences for the advective terms in the momentum equations, resulting in a significant reduction in phase speed errors. The actual implementation, when coupled with the forward–backward method for high-frequency modes, reduced the total number of small time steps required from the KW78 (see below); however, two or three evaluations of the advection terms are required for each large time step.

Compressible systems can be solved using time-splitting methods, semi-implicit methods, or other compressible solution techniques.[4] Klemp and Wilhelmson (1978b) introduced the time-splitting approach to the cloud modeling community. This approach separates low-frequency processes, such as advection, mixing, and buoyancy, from the high-frequency motions associated with sound wave propagation. The scalar variables are integrated in time using a single large time step. For the integration of the momentum and pressure equations, the low-frequency

[3] The sound wave speed in anelastic systems is infinity, rather than typical atmospheric values of 350 m s^{-1}.

[4] The first two approaches for handling systems of equations that include physically insignificant fast waves are discussed in chapter 7 of Durran (1999).

terms are computed once during a large time step and stored as f_u, f_w, and f_p (e.g., see appendix A in Wicker and Wilhelmson 1995). The high-frequency terms (e.g., the pressure gradient force and the divergence) are then computed using these f's and a smaller time step, $\Delta t_{\text{small}} = \Delta t/N$, where N is the number of small time steps. The small time step is chosen such that the sound waves are propagated by the numerical scheme in a stable manner. This is accomplished using a forward–backward scheme (Mesinger 1977) for the pressure gradient and divergence terms in the momentum and pressure equations. If the large time step scheme is a leapfrog scheme, then the following two-dimensional finite difference equations represent the small time step integration procedure:

$$\left[\frac{u^{t-\Delta t+n\Delta\tau} - u^{t-\Delta t+(n-1)\Delta\tau}}{\Delta\tau}\right] + c_p\bar{\theta}_v\delta_x\pi^{t-\Delta t+(n-1)\Delta\tau} + K_{\text{div}}\delta_x D^{t-\Delta t+(n-1)\Delta\tau} = f_u^t \quad (4.9)$$

$$\left[\frac{w^{t-\Delta t+n\Delta\tau} - w^{t-\Delta t+(n-1)\Delta\tau}}{\Delta\tau}\right] + c_p\bar{\theta}_v\delta_z(\alpha\pi^{t-\Delta t+n\Delta\tau} + \beta\pi^{t-\Delta t+(n-1)\Delta\tau}) + K_{\text{div}}\delta_z D^{t-\Delta t+(n-1)\Delta\tau} = f_w^t \quad (4.10)$$

$$\left[\frac{\pi^{t-\Delta t+n\Delta\tau} - \pi^{t-\Delta t+(n-1)\Delta\tau}}{\Delta\tau}\right] + \frac{\bar{c}_s^2}{c_p\bar{\rho}\bar{\theta}_v^2}\{\delta_x(\bar{\rho}\bar{\theta}_v u)^{t-\Delta t+n\Delta\tau} + \delta_z[\bar{\rho}\bar{\theta}_v(\alpha w^{t-\Delta t+n\Delta\tau} + \beta w^{t-\Delta t+(n-1)\Delta\tau})]\} = f_p^t, \quad (4.11)$$

where $\frac{1}{2} \lesssim \alpha \lesssim 1$ and $\alpha + \beta = 1$. The finite differencing is applied at the u, w, and π staggered grid locations (Arakawa C grid), respectively; $\delta_x u = [u(x + \Delta x/2) - u(x - \Delta x/2)]/\Delta x$ and similarly for other derivatives; $\Delta\tau$ is the small time step; and $D = \delta_x(\bar{\rho}u) + \delta_z(\bar{\rho}w)$. Using the leapfrog scheme for the large time step, (4.9)–(4.11) are stepped from $t - \Delta t$ to $t + \Delta t$ using 2N small time steps. Here the sound wave terms are treated implicitly in the vertical and require the solution of a tridiagonal system for every column in the integration domain. With the vertical implicit treatment, the small time step size is determined by the sound wave speed and horizontal grid spacing rather than the vertical grid spacing, which is often much smaller. Also, when $\alpha > 0.5$, vertical damping is introduced and helps control instability that can develop with the time splitting technique (see section 7.3.2 of Durran 1999). The K_{div} terms introduce divergence damping that is useful in suppressing these small grid-scale sound wave disturbances. A complete analysis of the time-splitting approach was presented by Skamarock and Klemp (1992) and was partially based on an uncompleted analysis by Klemp in 1978 when the original Klemp and Wilhelmson papers were written.

Skamarock and Klemp (1992) also note that for mesoscale and larger-scale application of the nonhydrostatic equations, it is computationally advantageous to include the buoyancy term in the vertical momentum equations and the vertical advection of θ on the small time step. This was done so that gravity wave speeds do not constrain the large time step. Cullen (1990) employed a similar method in an operational mesoscale model. This allows the use of a greater large time step than would otherwise be possible if high-stability regions such as the stratosphere are present. Overall, Skamarock and Klemp (1994) note that the KW method using time-centered differencing is efficient and accurate for the integration of the nonhydrostatic compressible equations for grids in which $\Delta z/\Delta x \approx 1$ or less and thus can additionally be used for large-scale hydrostatic flows. More recently Wicker and Skamarock (1998) have shown that stable integration using a time-splitting method can also be obtained in a similar way using a second-order Runge–Kutta time integrator for the large time step. As shown by Skamarock and Klemp (1994), pure forward-in-time approximations for the advection terms (such as those presented by Tremback et al. 1987) cannot be combined with the forward–backward scheme to produce a formally stable time-split scheme. However, with the use of filtering or reduced large time steps, several investigators have successfully combined these two methods to study convective and mesoscale systems (Proctor 1987; Tripoli 1992; Purser and Leslie 1991).

It is also possible to solve variations of (4.1)–(4.7) using semi-implicit techniques such as the ones presented by Tapp and White (1976), Cullen (1990), Tanguay et al. (1990), and Benoit et al. (1997). In one form (Skamarock et al. 1997), the sound wave terms are differenced implicitly while the remaining terms are explicit. The implicit differences for the same system used in creating (4.9)–(4.11) using a two time-level scheme are

$$\Delta u + \Delta t\sigma c_p\bar{\theta}\delta_x(\Delta\pi') = \Delta t[-c_p\bar{\theta}\delta_x(\pi')] + f_u^t \quad (4.12)$$

$$\Delta w + \Delta t\sigma c_p\bar{\theta}\delta_z(\Delta\pi') = \Delta t[-c_p\bar{\theta}\delta_z(\pi')] + f_w^t \quad (4.13)$$

$$\Delta\pi + \Delta t\sigma\frac{\bar{c}^2}{c_p\bar{\rho}\bar{\theta}^2}[\delta_x\bar{\rho}\bar{\theta}\Delta u^t + \delta_z\bar{\rho}\bar{\theta}\Delta w^t] = -\Delta t\frac{\bar{c}^2}{c_p\bar{\rho}\bar{\theta}^2}[\delta_x\bar{\rho}\bar{\theta}u^t + \delta_z\bar{\rho}\bar{\theta}w^t]. \quad (4.14)$$

Here $\Delta u = u^{n+1} - u^n$, $\Delta w = w^{n+1} - w^n$, and $\Delta\pi = \pi^{n+1} - \pi^n$. The time-averaging parameter, σ, sets the implicit character of the scheme. If $\sigma = ½$ the scheme weights the pressure derivatives equally at time n and $n + 1$. A three-dimensional Helmholtz equation for pressure can be derived by substituting (4.12) and (4.13) into (4.14), yielding

$$\frac{\bar{c}^2}{c_p\bar{\rho}\bar{\theta}^2}\sigma^2\Delta t^2[\delta_x(c_p\bar{\rho}\bar{\theta}^2\delta_x\Delta\pi) + \delta_z(c_p\bar{\rho}^2\delta_z\Delta\pi)] - \Delta\pi = R^n, \quad (4.15)$$

where R is computed from known variables at time n. The solution can then be obtained by solving (4.15) for pressure and then calculating (4.12) and (4.13). With the recent improvement in efficient techniques for solving Helmholtz equations with variable coefficients (Skamarock et al. 1997), the use of semi-implicit schemes becomes competitive with split-explicit schemes. Further, the improvements are of particular value in solving model systems in which a semi-Lagrangian integration is used since stable split-explicit techniques have not been found.

It is also possible to solve the compressible equations using explicit methods without employing time-split or semi-implicit techniques. For example, Carpenter et al. (1990) solved the equations using piecewise parabolic approximations for momentum and scalar advection. In this method, the low- and high-frequency terms are integrated using the same time step. While the time step in these calculations was about ten times smaller than in time-split models, Carpenter et al. argue that their solutions, for a given accuracy, are comparable to the centered time and space time-split schemes and cost about the same computationally. Mendez-Nunez and Carroll (1994) suggested the use of the predictor-corrector MacCormack scheme rather than centered time and space schemes to solve the compressible equations. In their convective model tests, the time step was four times larger for similar accuracy as the centered scheme. To date, these fully compressible solution techniques have not been widely implemented and research continues on the best way to approximate and efficiently solve the equations on parallel computers. It has also been demonstrated that compressible systems can be integrated with some accuracy using a reduced sound speed, as long as the Mach number remains less than one-half and the sound speed remains at least twice as fast as the fastest gravity wave modes (Droegemeier and Davies-Jones 1987). This supercompressible system has proven very useful in increasing the computational efficiency for integrating the compressible equations.

Another important choice severe storm modelers must take into consideration is the thermodynamic variable used for model integration. Most local severe storm modelers have utilized a thermodynamic formulation involving a potential temperature that is conserved during either dry, moist, or water and ice conditions (e.g., Wilhelmson 1977; Tripoli and Cotton 1981). The use of conservative prognostic variables can simplify the formulation of subgrid-scale flux approximations (Redelsperger and Sommeria 1986). Alternatively, Ogura and Phillips (1962) employed entropy as a thermodynamic variable within their Boussinesq warm-water system. The potential temperature and water vapor mixing ratios were then directly diagnosed from entropy. The extension of this approach to anelastic or fully compressible equations was discussed by Ooyama (1990), who argues for its use in universal nonhydrostatic models simulating atmospheric motions from cloud to planetary scales.

c. Model parameterizations

Storm modelers must not only select a dynamical formulation but also develop or choose parameterizations for subgrid processes, microphysics, boundary layer physics, radiation, and electrification that fit with their objectives. To illustrate the type of choices made historically, a brief overview is presented in this section. More detail can be found in Cotton and Anthes (1989), Emanuel (1994), and Houze (1993), as well as in the published literature.

Ogura mentioned in the 1963 monograph, "A point to be remarked on here is that motions whose scales are comparable with or larger than the prescribed spatial grid size are presumably well expressed by a set of finite-difference versions of the basic equations. What we must consider therefore is whether or not motions whose size is comparable with or smaller than the grid mesh contribute appreciably to the gross features of convective evolution and, if they do, to formulate mathematically the contribution from these fine-scale 'eddies.' No satisfactory treatments of the mixing process have so far been presented in the numerical experiments."

Many subgrid parameterizations are formulated through perturbation analysis in which variables are represented by grid volume averages and perturbations to these averages. Subgrid-scale fluxes are then approximated. Approximations can be as simple as K theory (Lilly 1962; Schlesinger 1978; Clark 1979; Wicker and Wilhelmson 1995) with diffusion coefficients that depend on the buoyancy and deformation (first-order closure). They can also involve the solution of one or more time-dependent subgrid-scale equations (e.g., Klemp and Wilhelmson 1978b; Redelsperger and Sommeria 1986). Many cloud models include an equation for the subgrid-scale turbulent kinetic energy E. Changes in this energy are related to resolvable changes in buoyancy, shear, diffusion, and dissipation. The diffusion coefficient for momentum is then taken to be proportional to a grid-based length scale such as $\ell = (\Delta x \Delta y \Delta z)^{1/3}$ times $E^{1/2}$. If Δx and Δy are substantially different from Δz, ℓ may be defined differently in the horizontal and vertical directions, or if nested grids are used, care must be taken to prevent undesirable gradients at nested boundaries. The successful use of these relatively simple parameterizations indicates that many of the dynamical features of severe convection can be adequately represented using sufficient grid resolution coupled with a qualitative approximation to the transfer of energy into subgrid scales. However, there are times when a better representation of subgrid processes is needed. For example, simulations of convective clouds when surface insolation is included may require the turbulence scheme to adequately treat the subgrid transport of heat and moisture in a convective boundary layer. Typical turbulence schemes used in cloud models often underpredict the transport by these eddies (Holt and Raman 1988). Further, simulations of tornado-producing convective storms (Wicker and Wilhelmson 1995; Lee and Wilhelmson 1997b) demonstrate the need for accurate treatment of frictional and mixing processes near the ground. Computational limitations generally require that these simulations resolve the vortex flow with a minimal number of grid points. Therefore, the treatment of subgrid turbulent fluxes significantly impacts the solution. This is demonstrated by Lewellen and Lewellen (1997), who developed a new turbulence parameterization near the ground that attempts to represent the highly anisotropic effects of mixing in tornado-like vortices.

Storm modelers are well aware of the impact of and complexities involved in introducing microphysics within their models, even when simple representations are used. Ogura noted in the original monograph that "A great deal of complexity will be brought into the computation, however, if one attempts to combine the evolution of water-droplet spectra with dynamical models." Although there have been some warm convective modeling studies where many water-drop categories have been included (e.g., Clark 1973; Takahashi 1988), most storm modelers have used some form of the Kessler (1969) bulk water parameterization in which cloud water and rainwater within each grid zone are assumed to be exponentially distributed by mass. Cloud water, a drop-size category to represent small cloud drops, forms instantaneously when air becomes saturated and evaporates instantaneously up to the saturation level in unsaturated air. Rainwater is generated from cloud water through simple expressions for autoconversion and collection, while cloud water increases through rainwater evaporation at a rate determined from a bulk evaporation formula (e.g., see Klemp and Wilhelmson 1978a). An approximate terminal velocity for rain is also needed, and is an important factor that differentiates rainwater that can fall to the ground from cloud water that moves with the air.

Bulk categories for representing frozen water have also been used in studies of local severe storms. Parameterizations include those of Hsie et al. (1980), Lin et al. (1983), Rutledge and Hobbs (1983), Lord et al. (1984), Ferrier (1994), Krueger et al. (1995), and Straka (see Johnson et al. 1994). Bulk categories can include ice, snow, graupel, and hail. One of the major concerns in developing bulk parameterizations is the use of exponential distributions for the different categories and the specification of the parameters associated with these distributions that are assumed to be applicable throughout a storm system. Despite this concern, modelers have been able to proceed because the parameterizations have been sufficient for representing the bulk effects of microphysics on overall storm structure. This is possible even when frozen categories are not included or are represented very crudely. For example, Wilhelmson et al. (1990) assumed that cloud water below $-20°C$ was ice and should not melt instantaneously in subsaturated regions of the anvil. If it did, holes would appear in the storm anvil. More sophisticated approaches were developed by Cotton et al. (1982) and Meyers et al. (1992). These schemes removed some of the assumptions made about particle size distributions as well as predicting cloud and ice nuclei number concentrations and can be used in a wider variety of convective cloud environments. Recently, Grabowski (1998) approximated the presence of ice by changing the growth and terminal velocities of "rainwater" to those appropriate for ice when the air was well below freezing. In this way, reasonable anvils formed in tropical convective clouds without the explicit presence of ice categories. The presence of these anvils in the evolution of convective systems over periods longer than a few hours is important for radiation parameterizations to be effective. Despite some success in using these bulk parameterizations, their impact on storm evolution and structure remains an important area of investigation.

Radiation parameterizations have been primarily developed over the past decade for use in studying long-lived squall lines and tropical convection (e.g., Chin et al. 1995; Tao et al. 1996; Petch 1998). Nearly all of the severe storm simulations reported during the 1970s and 1980s were carried out for less than four hours, a time period over which radiation would not significantly influence the solution. Radiation parameterizations are also available in mesoscale models that have been downsized to study local severe convection (e.g., MM5 and RAMS). Their impact on severe storm evolution remains to be carefully evaluated. An idealized study by Crook and Weisman (1998) showed that stabilization of the atmospheric boundary layer via surface radiational cooling could help organize low-level storm rotation. Other studies of these effects are likely to be forthcoming as more sophisticated physics are used in cloud models.

Only a few studies of the electrical behavior of convective storms have been undertaken due to the computational constraints of adding more variables to the model and the need to link electrical parameterizations to appropriate microphysical parameterizations (e.g., Takahashi 1984, 1987; Helsdon and Farley 1987; Mansell et al. 2000). The latter depends in part on the type of electrification processes being studied (e.g., Saunders and Peck 1998). Despite the complexity, recent simulations by Mansell (2000) demonstrate that the electrical behavior of clouds, including lightening discharge, can be reasonably simulated in a variety of storms and environments. Improvements in lightning detection networks will enable researchers to qualitatively verify these and other new modeling results.

d. Other model choices

Other model choices include integration domain and grid size, boundary conditions, and initialization mechanisms. These choices are in part limited by the computational power available to the modeler. In the 1960s this limitation meant that three-dimensional simulations were unfeasible. In the 1970s this meant that many three-dimensional simulations were carried out in modest domains (e.g., $60 \times 60 \times 15$ km) using a small number of grid points (e.g., $30 \times 30 \times 30$) with coarse-grid resolution (e.g., $2 \times 2 \times 0.5$ km). It is now common to use horizontal domain sizes extending beyond 200 km with 1-km resolution or to use horizontal domain sizes of 20 km with grid resolutions from 50 to 100 m. This increased capability has enabled modelers to simulate severe convective systems—including finite length squall lines composed of many individual cells, as well as individual storms—with very high resolution.

In order to make more efficient use of existing computer power to study multiscale severe storm phenomena, nested models have been developed and used extensively during the last decade. Most major modeling systems employ some form of horizontal nesting (e.g., Clark and Farley 1984; MM5, Zhang et al. 1986; RAMS, Pielke et al. 1992; COMMAS, Wicker and Wilhelmson 1995). In these models, multiple grid systems are used with finer and finer resolution. These grids can be horizontally "nested" inside one another or the new grids can overlap in a fully AMR approach as shown in Fig. 4.1. In either case, higher resolution is used only where needed for the phenomena being investigated. For squall lines, this means that the highest resolution grids are placed over regions of active convection. In supercell simulations, high resolution grids are placed near the storm's low-level mesocyclone in order to capture potential tornado development. The placement of the new grids can be manual (Wicker and Wilhelmson 1995; Grasso and Cotton 1995; Finley et al. 1998a, b) or automatic (Skamarock and Klemp 1993).

The use of nested grids allows the modeler to push the boundaries of the outer grid far away from active convection. These outer boundary conditions are often handled through simple formulations. For example, in models using Cartesian coordinates, the vertical velocity at the lower and upper boundary is usually set to zero. This means that upward traveling gravity waves are reflected and can contaminate the solution in the interior of the domain. This situation is handled through the use of a wave-absorbing layer, and usually requires extra grid points in the vertical and/or the use of a radiation upper boundary condition (e.g., Klemp and Durran 1983).

The lateral boundaries are artificial and in the ideal world should allow disturbances to pass through them from the interior of the domain. To handle this, lateral boundary conditions that utilize the sum of the horizontal advection velocity normal to the boundary and a characteristic gravity wave velocity are often employed. If the net flow is outward, advection at a rate determined by this net sum is used together with an upstream approximation to the normal derivative. If it is inward, there is no contribution to the time rate of change to model variables from terms with derivatives in the normal direction (Klemp and Wilhelmson 1978b) unless some external forcing is applied. A general discussion of this approximate radiation condition and the handling of artificial lateral boundaries has been given by Durran (1999, pp. 395–397).

The severe storm modeler also needs to choose a method of initializing convection consistent with the objectives of a particular study. This can include using a horizontally homogeneous sounding in which convection is triggered through the use of one or more warm/moist bubbles (e.g., Klemp and Wilhelmson 1978a) or some form of lifting. For example, localized lifting can be provided through the specification of a density current (e.g., Rotunno et al. 1988; Lee and Wilhelmson 1997a) or the placement of a hydrometeor-filled parcel above the surface to emulate a water- and ice-loaded downdraft (Proctor 1988; Droegemeier and Davies-Jones 1987). In fact, the study of density currents themselves has been given substantial attention during the past decade (e.g., Droegemeier and Wilhelmson 1987; Chen 1995; Liu and Moncrieff 1996; Klemp et al. 1997; Xue et al. 1997). Mesoscale or large-scale convergence can also be used either by specifying them in the initial conditions and/or through the use of external forcing terms (e.g., Chen and Orville 1980; Crook and Moncrieff 1988).

In recent years horizontal variability in the base state has been implemented in local storm models through specification of a balanced analytic state (Richardson et al. 1998), through specification of several soundings with interpolation between them (e.g., Kay and Wicker 1998), through specification of a time-varying base state (Richardson 1999), and through data assimilation. The latter is usually carried out using nested grid models where the outer grid domain

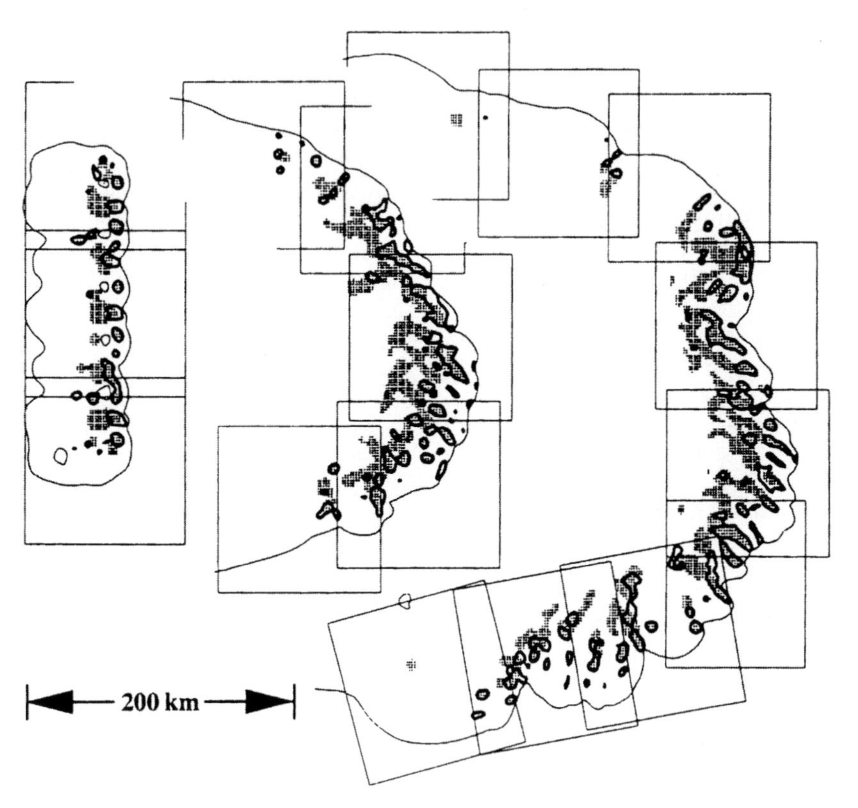

FIG. 4.1. This figure illustrates the use of adaptive mesh refinement with increased horizontal resolution for a simulation of a long-lived squall line. Horizontal cross sections in ground-relative coordinates at a height of 3 km are shown at 2, 6, and 10 h. The surface gust front ($\theta' = -1$ K at $z = 350$ m) is the thin contour, $w = 2$ m s^{-1} is the thick contour, and the rainwater field greater than 0.5 g kg^{-1} is stippled. The finest resolution grids are included at the three times. From Skamarock et al. (1994).

is large enough to usefully incorporate observational data on the mesoscale. This approach has become more popular in case studies and in forecast prediction where the assimilation of Doppler radar data on the storm scale is being explored (e.g., Droegemeier et al. 1999; Sun and Crook 1997).

e. The modeling enterprise

1) MODEL VERIFICATION/VALIDATION

The development and execution of models is done with the intent of capturing the dynamics, physics, and/or statistics of severe storms. When it is done in the context of exploration, it is possible to create a model and study its behavior based on the set of discrete equations derived from approximating a set of differential equations. Some initial and boundary conditions are needed for the model integration to proceed. Sets of experiments are often carried out to explore how storm dynamics and physics respond to varying initial conditions or to varying parameterizations. At times this is done to explore the impact of the changes in order to better understand the model. However, the usual intent is to better interpret what is happening in the real world, or to predict it. Thus, it is necessary to talk about validation/verification against observational data and about forecast model calibration (or tuning) in the context of improving agreement with observations (Randall and Wielicki 1997).[5] It is important to understand that when a scientist says he or she is simulating the 3 April 1964 splitting storms (Wilhelmson and Klemp 1981), he or she is in fact simulating a facsimile or virtual event in which some observed features may be adequately represented while others may not.

The history of severe storm modeling can be viewed as a process of discovering what can be simulated and validated. It was not apparent in the early three-

[5] The word "validation" has been suggested as more appropriate than "verification" because the latter is reserved in the philosophical arena to refer to "an assertion or establishment of truth," whereas validation refers to "establishment of legitimacy" (Oreskes et al. 1994). Thus, the latter leaves open the possibility that the model is not a replica of reality. However, the terms are used interchangeably here.

dimensional modeling days that a coarse-grid dynamic model with crude representation of the microphysics and subgrid-scale processes would be sufficient to successfully simulate a severe storm or the now known splitting process.[6] It turned out that better representation of the microphysics and subgrid-scale mixing was unnecessary for simulating splitting and the basic dynamics of observed supercell storms because their character is strongly linked to the interaction between the storm's updraft and the environmental vertical shear. However, there are many instances when microphysics and subgrid-scale processes may be important, not only quantitatively but also qualitatively. For example, in situations where control of storm dynamics is not as tightly coupled to environmental conditions, direct comparison against detailed observations of storm evolution is not practical.

Another unknown during the early modeling days was how long a simulation needed to be integrated to simulate splitting. The few observations at the time indicated an hour or so from the first appearance on radar. But when the updraft strengths within the initial modeled cloud would decay substantially from their maximum value, there was no indication that convection would continue and reamplify. It was only when a simulation was made for well over an hour that it became clear that splitting could be simulated. A specific example of this was the 3 April 1964 simulation (Wilhelmson and Klemp 1981) that has been widely used to argue that it might be possible to use models to accurately predict storm evolution (discussed in more depth later). In this simulation, the model updraft fell to approximately 3 m s^{-1} before the left- and right-moving storms developed. There was great excitement when the Klemp–Wilhelmson model, initialized with a "representative" sounding and a small thermal embedded in an otherwise horizontally uniform environment, yielded splitting storms. Further, as observed, the left-moving storm in the model split again. The reported simulation, however, was only one of a handful of simulations of this case that resulted in a second split. If one were to use the convective model in a forecasting mode at the time, the natural question would have been, Which model simulation to believe? So in this case only one model simulation was validated against observations.[7]

Another form of validation is related to the solution itself. One of the fundamental questions asked in most courses on numerical methods in fluid dynamics is whether the solution to the system of equations converges to the analytic solution as the grid spacing and time step approach zero. According to the Lax theorem, consistent methods that can be shown to be stable for linear problems are convergent. When solving fully nonlinear problems, however, convergence will depend on how the parameterizations (e.g., turbulent mixing) are formulated. For example, most turbulence mixing parameterizations are functions of the grid spacing. Thus, as the grid spacing becomes very small, the influence of the parameterization approaches zero in some manner. Ideally, the parameterization at very small scales should represent the effects of molecular viscosity, but these effects are rarely included in turbulent mixing schemes. Therefore the way in which the parameterization behaves as the grid spacing gets small will determine whether, or at what rate, convergence takes place. Generally, all computer power available is exhausted before the solution is converged.

An alternative is to examine the convergence of the model's numerical solution when a constant subgrid length scale is assumed (e.g., a constant eddy diffusivity). For example, Straka et al. (1993) reported on a two-dimensional model comparison study that was undertaken with the assumption of constant eddy diffusivity to address the accuracy of nonlinear density current simulations. Two-dimensional simulations were used to enable researchers to reach a converged solution at high resolution while limiting the computational cost. For both compressible and incompressible equation sets, a reference model was simulated at ultra-high resolution in order to provide a quantitative error measure for the other model runs at lower resolutions. The results for all the models tested were surprising consistent. When the grid spacing was fine enough to adequately resolve the flow, all of the schemes and systems (compressible and incompressible) captured the basic dynamics, although some timing and location differences were noted in the development of Kelvin–Helmholtz instabilities that developed. In marginally resolved cases, upstream and monotonic models had some difficulties in adequately simulating the Kelvin–Helmholtz rotors, while higher-order models did noticeably better. It was also discovered that some quantitative differences existed between compressible and incompressible models. One conclusion drawn from these results is that modelers must be careful in interpreting details from simulations of phenomena that are marginally resolved.

Tests of this kind are usually restricted to two-dimensional models for computational reasons. Another alternative is to run a model at several resolutions without holding the subgrid-scale mixing length fixed. The goal is not to determine convergence but

[6] Before the simulation of splitting storm cells in models, there were only a few documented cases of observed splitting.

[7] Houze (1993) reported another simulation in which a left mover split when the model was initialized in a different environment. This suggested that a second split is a reproducable dynamic feature of some storms.

to determine whether the results change significantly for the specific scientific issues being addressed. For example, the three-dimensional simulations of Weisman et al. (1997) showed that, while the detailed evolution of individual convective elements changed considerably as the resolution decreased, the 4-km horizontal resolution simulation captured much of the mesoscale structure and evolution of squall systems simulated using higher 1-km resolution.

Finally, validation can be thought of in more general terms as the ability of a model to represent the spectrum of observed storms under appropriate environmental conditions. For example, the Klemp–Wilhelmson and COMMAS models have been used successfully to simulate supercells, multicells, squall lines, and bow echoes under representative conditions appropriate for these different storm types, as determined from observational studies. These conditions are most easily related to the distribution of environmental buoyancy and shear with height in the absence of mesoscale and/or synoptic-scale forcing. Examples of this are presented in later sections of this chapter.

2) MODEL IMPLEMENTATION

In 1963, when the first severe storms monograph appeared, computers were scarce in comparison to today's computer-driven world, in which "information technology" is a common phrase (Kaufman and Smarr 1993). Since the early 1960s, computing throughput has increased from thousands to billions of floating point operations per second (flops). At the beginning of the twenty-first century, some high-performance parallel computers will provide trillions of flops (teraflops). Computers familiar to severe storm modelers include the CDC 7600 and the Cray series shown among others in Fig. 4.2. In the past decade, memory size has increased dramatically, from megabytes to many gigabytes, while disk capacity in high-performance systems has grown to hundreds of gigabytes and mass store systems to multiple terabytes. Kaufmann and Smarr (1993), in their review of the growth in computational capability through the early 1990s, summarized the computational efforts that span a variety of scientific disciplines, including the atmospheric sciences (severe storms and climate).

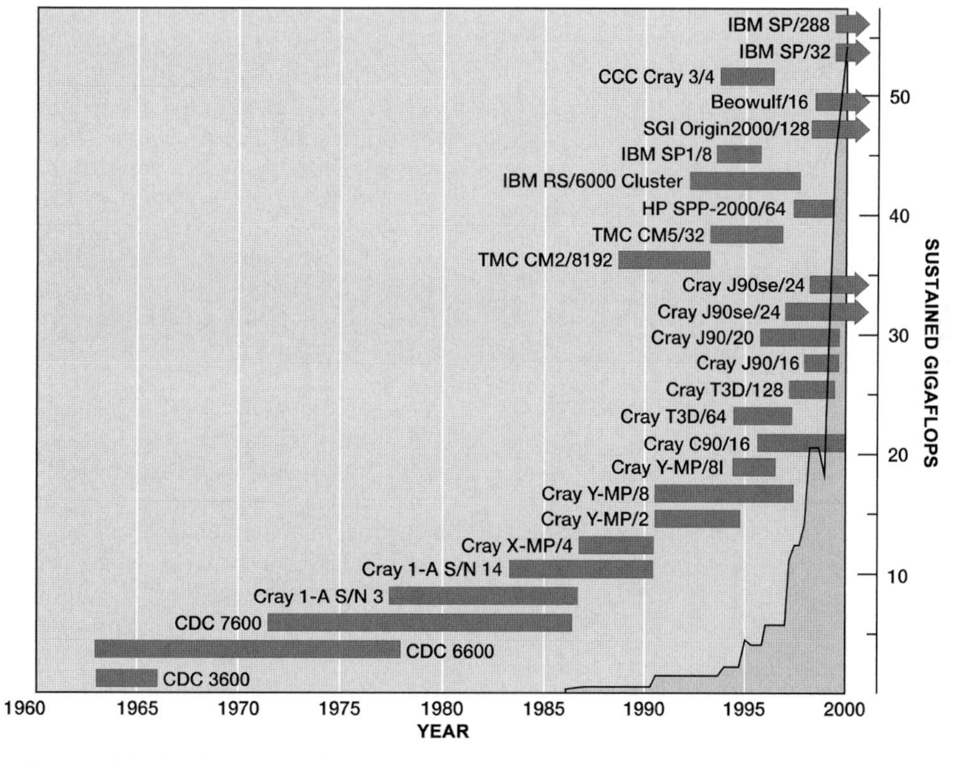

FIG. 4.2. Scientific Computing Division (SCD) major computers and sustained performance. The gray bars show the life spans of these computers at the National Center for Atmospheric Research (NCAR) where many of the early storm simulations were carried out. Arrows at the right indicate machines that were in operation at the start of fiscal year 2000. On the right-hand side of the chart, the shaded gray area shows sustained gigaflops (billions of floating-point calculations per second) attained by all SCD machines from 1986 to the end of FY99. The Cray 1-A is now on display at the Smithsonian Institute in Washington, D.C. The chart is adapted from one used by SCD. NCAR is sponsored by the National Science Foundation. Special credit goes to T. Engel and M. Shibao at NCAR.

TABLE 4.1. Exemplary list of models used to study severe convection.

Model name	Reference	HTTP address
Advanced Regional Prediction System (ARPS)	Xue et al. (1995, 2000, 2001)	www.caps.ou.edu
Clark Model	Clark et al. (1996)	None
Goddard Cumulus Ensemble Model (GCE)	Tao and Simpson (1993)	rsd.gsfc.nasa.gov/912/code912/model.html
Klemp and Wilhelmson/COMMAS	Klemp and Wilhelmson (1978a); Wicker and Wilhelmson (1995)	www.nssl.noaa.gov/~1wicker
MM5	Dudhia (1993)	www.mmm.ucar.edu/mm5
Regional Atmospheric Modeling System (RAMS)	Pielke et al. (1992)	www.aster.com/rams.shtml
UW-NMS	Tripoli (1992)	mocha.meteor.wisc.edu

The greater than 10^7 increase in raw computing power in high-performance machines since the 1960s[8] has not been matched by a similar increase in the number of model grid cells used in everyday simulations. This is partly because the percentage of peak power that fluid dynamic codes can realize has dropped due to the use of different processor architectures. Also, the scalability of many model codes for a fixed number of grid cells drops dramatically because the workload for each processor drops while the overall interprocessor communication remains the same at best. Although some parallel machines have had a common memory for all processors (e.g., Cray T90), most current machines do not (e.g., distributed memory machines like the IBM SP2 and distributed shared memory machines such as the Cray/Origin 2000). Advances in compilers (Fortran 90, OpenMP[9]) and Fortran source translators[10] help to alleviate the need for modelers to understand the details of computer architecture; however, high-performance codes require attention to code structure and layout. For parallel machines, this often means that parallelism is achieved by dividing the model integration domain into many subdomains surrounded by ghost zones. The calculation associated with each subdomain is then distributed to an available processor during run time. However, the user must now explicitly move data from each subdomain into neighboring domain ghost zones using communication software such as the Message Passage Interface (MPI; see chapter 2 in Foster 1995). This significantly complicates the code and can make modifications to a code difficult.

3) MODEL DATA VISUALIZATION

Data stored from a single severe storm simulation now can easily reach 100 gigabytes. Fortunately, this volume of data was not produced in the 1960s because visualization of it was typically handled by printing out pages of numbers and then hand contouring the printed fields. The first animations of three-dimensional storm model data were recorded in the early 1980s on 16-mm movie film in the form of three-dimensional wire frames obtained using NCAR graphics. In the latter half of the 1980s, computer animation became a reality, first through the use of motion picture and commercial quality visualization software and then through workstation and personal computer animation software (e.g., Droegemeier and Wilhelmson 1986; Hibbard and Santek 1990; Wilhelmson et al. 1989; Schiavone and Papathomas 1990; Wilhelmson et al. 1990; Kaufmann and Smarr 1992; Hibbard et al. 1994; Wilhelmson et al. 1995, 1996; Tufte 1997; Sherman et al. 1997). A recent still from a 1990s animation created for the Omnimax movie *Stormchasers* is shown in Fig. 4.3. A nested grid simulation of a supercell and the tornado it spawned were visualized using surface rendering and specially developed software for displaying the continual release of weightless tracer particles that exceeded 10 000 at any given time (Wilhelmson et al. 1996).

Perhaps the best-known visualization of a severe local storm, called *Study of a Numerically Modeled Storm,* was scientifically led by one of the authors (Wilhelmson). "The resulting video animation became an international sensation, awakening many people's imaginations about what could be accomplished with supercomputers," according to Kaufmann and Smarr (1993). It was also nominated for an Academy Award in the animation category. The six-minute model animation was created directly from the model output of a 3 April 1964 supercell storm simulation through the use of high-end commercial visualization software, coupled with software developed at the National Center for Supercomputing Applications (NCSA). Over an 11-month period, from the first meeting to the completion of the final version of the color video, NCSA staff, including four scientific animators, worked on the project for approximately one man-year. In addition, scriptwriters, artistic consultants, and post-pro-

[8] In the 1980s and for the first half of the 1990s these were called *supercomputers*. Today the term *high-performance computer* is often used.

[9] http://www.openmp.org

[10] http://www-unix.mcs.anl.gov/~michalak/B/mpmm_index.html

FIG. 4.3. A still adapted from a 1990s movie sequence created for the Omnimax film *Stormchasers* from model simulation data (Wilhelmson et al. 1996). The sequence depicts the growth of a cumulus into a supercell that produces a tornado. The tornado is revealed through the use of thousands of tracer particles introduced into the high-vorticity region within the simulated flow.

duction personnel were involved. The final version took about 200 hours of computer time on a high-performance graphics computer.[11] A black-and-white version of a frame from the color video is shown in Fig. 4.4a; another frame, from a reworked gray-shade animation with a time line on the bottom, is shown in Fig. 4.4b.

f. Historical models that have survived the test of time

Many models have been written over the years to solve convective and local severe storm problems. Most of these models have undergone significant changes during their existence and many are now used with a wide variety of microphysical, turbulence, and radiation parameterizations on parallel computers. Some began as local convective models; others began as mesoscale models. Table 4.1 contains an exemplary list of models that have a relatively long history and are actively in use today to study severe convection. Some are used to make daily local storm, mesoscale, and regional forecasts. The papers given are starting points for understanding the models; they do not necessarily provide the latest information, some of which can be found on the Internet as indicated.

Many of these models have parallel implementations (e.g., for ARPS, Sathye et al. 1996, 1997; for MM5, Michalakes 2000; for RAMS, C. J. Tremback and R. L. Walko[12]). Some, like MM5 and ARPS, are community models, while others, like COMMAS and GCE, are in use by the developers and close collaborators.

4.3. Modeling contributions to the understanding of isolated severe storms

Storm models provide a framework for understanding the initiation, evolution, structure, and decay of severe storms. Early modeling in three dimensions focused primarily on the dynamics associated with individual supercell or multicell storm elements and infinitely long lines composed of these elements. With increased computer power/memory and new numerical algorithms (e.g., AMR), simulations supporting greater

[11] Stills from this animation have appeared on many covers including the July 1990 issue of the *Bulletin of the American Meteorological Society*. Stills and animations can also be found at http://redrock.ncsa.uiuc.edu/AOS/home_images.html.

[12] http://www.aster.com/papers.shtml

(a)

(b)

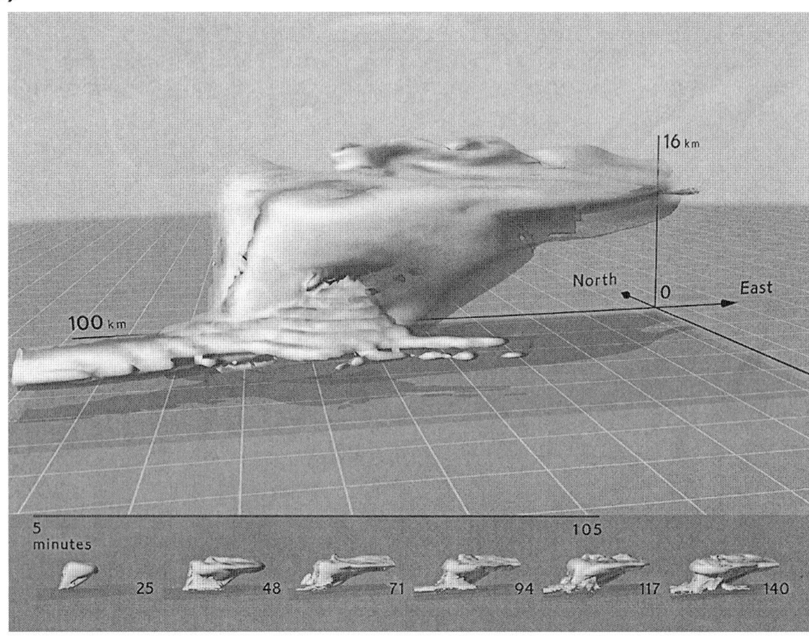

FIG. 4.4. (a) A video snapshot showing the rain field structure at 2 h in a simulated supercell, along with a two-dimensional slice taken at z-4.75 km above the ground. The pointer on the color scale on the right indicates the maximum water content (g kg^{-1}) for the inset. The black region around the rain surface indicates the outer location of the inset. Adapted from the video "A Study of the Evolution of a Numerically Modeled Severe Storm" and further described in Wilhelmson et al. (1990). (b) An adaptation of a redesigned frame from this video that originally used gray shades and minimal color. It includes a time line of storm evolution in minutes along with selected storm surfaces at the bottom. The original redesign by E. Tufte and C. B. Bushell. For further discussion see Tufte (1997).

detail have been enabled and larger integration domains have been used to focus on the dynamics of finite line behavior (e.g., see Fig. 4.1b) and interactions within clusters of cells. Further, severe storm models are now being used in numerical weather prediction. In this section several areas are highlighted as exemplary illustrations of the role that modeling has played in understanding severe storms.

a. Supercells[13]

Supercells produce tornadoes, high winds, and large hail. In their simplest form they have a distinctive structure in which an updraft and downdraft exist cooperatively, as first documented by Browning and Donaldson (1963). Numerical modeling has played a vital role in improving our understanding of their structure and evolution, and in forecasting their development. Early numerical simulations of supercells in horizontally homogeneous environments in which the vertical wind shear was predominately east–west and confined to the lowest 4 km were reported by Klemp and Wilhelmson (1978a,b) and Wilhelmson and Klemp (1978). In these papers they demonstrated that initial cloud splitting and long-lived supercells could be simulated using coarse-grid resolution in an integration domain that traveled along with the storms.

Splitting of the initial convective cloud occurred as a rainy downdraft formed and "split" the low-level updraft. At the time, the authors thought that rain was necessary for splitting to occur. However, Rotunno and Klemp (1982, 1985) demonstrated that splitting occurred even in the absence of precipitation. Low-level right- and left-flank updrafts developed in response to pressure-driven vertical accelerations (noted by Schlesinger 1980) that resulted from midlevel rotation (counterclockwise when viewed from the top in the southern storm and clockwise in the northern storm). This midlevel rotation results from the tilting of horizontal vorticity present in the vertically sheared storm environment. The continued separation of the initially split storms was a continued response to an upward vertical pressure gradient on the flanks of these storms induced by the midlevel rotation and the fact that the precipitation within these storms did not fall back into these favorable regions. Thus, the rise of new parcels from the boundary layer to the level of free convection was continually supported by the storm's interaction with its environment.

Klemp and Wilhelmson also demonstrated that, even for the short period of two hours, the Coriolis effect was noticeable in that the precipitation structure in the right mover (southern storm) was more representative of the classic observed supercell than that in the left mover. Finally, they clearly demonstrated that the clockwise (counterclockwise) turning of the environment's wind shear vector was a key factor in favoring southern (northern) storm development. This explains why most supercells in the Great Plains of the United States have cyclonically rotating updrafts and tornadoes. This fundamental insight was enabled by their ability to change the storm environment, a control that observationalists do not possess. A more in-depth review of the dynamics of early rotational development, storm splitting, preferential enhancement of cyclonically rotating storms, storm propagation, and transition to the tornadic phase has been provided by Klemp (1987).

It is noteworthy that the Klemp and Wilhelmson results were obtained using 2-km horizontal resolution within a domain that was only 10 km high. Further, only warm rain microphysics was used, neglecting ice processes within the storm. The simulations demonstrated that overall supercell dynamics could be explored without representation of many storm features (e.g., flanking lines, tail clouds, cloud turrets, and hail). In addition, these simulations helped explain the few documented cases of splitting storms available at the time (Fujita and Grandoso 1968; Achtemeier 1969; Charba and Sasaki 1971; Brown 1976). It also prompted researchers and forecasters to look more carefully for splitting storms in radar data.

Further analysis of splitting storms by Weisman and Klemp (1982) and others revealed the gamut of structures that could evolve from splitting storms. For example, in linear vertical shear environments (without Coriolis force), northern and southern supercells possess horizontal mirror-image symmetry. When the wind shear is positive, the southern storm rotates cyclonically and the northern storm anticyclonically. However, multicell development may replace supercell structure in the northern storm when there is significant clockwise turning of the environmental wind vector with increasing height (Weisman and Klemp 1984). This results from a reduction in the vertical pressure gradient force associated with the left split storm and increased dependence for its existence on low-level convergence along the storm gust front. Further, as the storms continue to separate, the cold pool they leave behind can interact with the low-level environmental winds to produce multicell behavior between the original split storms.

The analysis of modeled supercell evolution and mature structure as reviewed above has been aided through the use of three-dimensional visualizations derived directly from model data. For example, both static and animated visualizations of model-generated supercell structure have substantiated the nondestructive relation between main updraft and downdraft that was originally derived by Browning (1964). Klemp et al. (1981) documented this structure by releasing weightless particles in a fixed-flow field taken from the mature stage of a supercell simulation. More recently, Wilhelmson et al. (1990) augmented this picture using higher resolution and time-dependent data.[14] The updraft tracer ribbons were obtained by placing weightless particles in a line within the middle of the updraft and integrating forward and backward in time. The downdraft trajectories were obtained by placing a line of particles at a height of 2 km to the south of the storm 1 h into the 2.3-h simulation, yielding the structure shown in Fig. 4.5.

Animated artist schematics made indirectly from model data have also been created to communicate the

[13] A supercell is defined as a thunderstorm containing a persistent rotating updraft (Johns and Doswell 1992).

[14] http://redrock.ncsa.uiuc.edu/AOS/image_89video.html

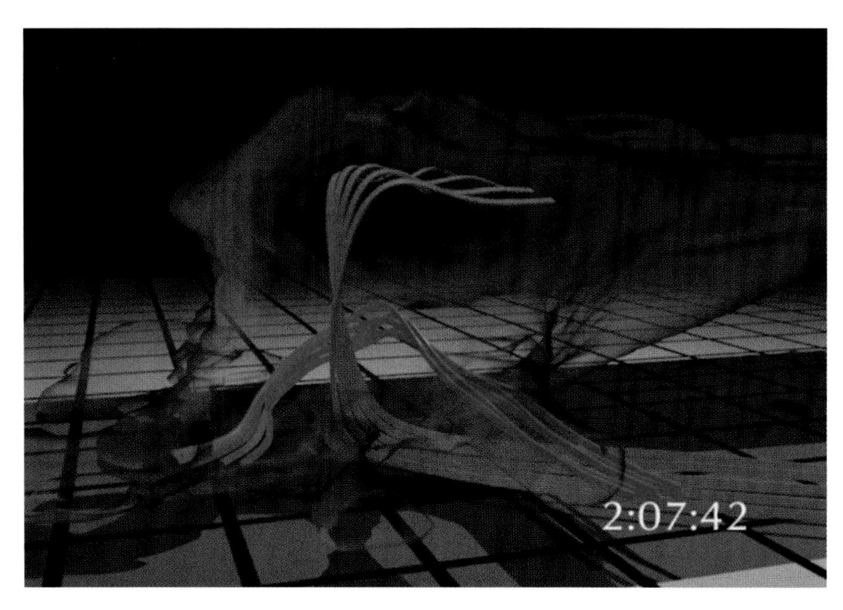

FIG. 4.5. A snapshot from a video in which weightless tracer particles are used to depict the flow of air through time in a simulated supercell. The orange/red ribbons represent air tracers rising in the storm, while the blue ribbons represent air tracers that eventually go around the updraft and then sink in the downdraft. In the video, the ribbons grow in length as the air tracers move through the storm. (See Wilhelmson et al. 1990 for further discussion.)

findings of severe storm modelers. For example, qualitative descriptions of supercell evolution in terms of the tilting of vortex lines, splitting, and subsequent separation of cells have been painstakenly created through collaboration of computer artists and scientists. Such animated schematics, whether derived directly from the data or created by an artist depicting current scientific understanding, are becoming more widely available and provide excellent educational material accessible from CD-ROM or the World Wide Web (WWW).[15]

b. Supercell case studies

The success in the late 1970s in simulating right- and left-moving supercells using idealized soundings and small integration domains prompted numerical modelers to experiment with the use of observed data associated with supercell development. Numerical simulations at this time were made with relatively small integration domains; therefore, the inclusion of mesoscale variations in the initialization was not practical. Each column in a model would be initialized using a single point sounding, resulting in a horizontally homogeneous/uniform environment. In this environment, storms had to be artificially triggered. This was typically accomplished by placing a warm buoyant thermal at low levels. If the thermal were sufficiently buoyant to rise to the level of free convection in spite of the model's mixing processes, it would potentially trigger a severe storm. Whether storms initialized in this way would be representative of some of those observed was open to serious question, particularly when also considering the very qualitative representation of cloud physics and subgrid-scale mixing.

Nevertheless, some attempts at simulating the general dynamic behavior of observed severe storms were successful. One of the first cases was reported by Wilhelmson and Klemp (1981). This study simulated portions of the complex evolution associated with a severe storm outbreak on 3 April 1964. Figure 4.6 shows observed and simulated echo evolution for this case. Both the real and observed cases initially began with a single storm. The initial storm splitting was followed by a second split of the left-moving storm. Overall, a line of storms developed in response to the splits and other new cell development along the low-level outflows between the original left mover and right mover. Lilly (1990) argued that this simulation is an example where supercell storms may be more predictable because the large-scale environment determines much of the overall storm structure and evolution.

[15] A schematic animation can be found in the Physical Processes: Curved Hodograph Dynamic Processes section (page 6) of the COMET Training CD entitled *Anticipating Convective Storm Structure and Evolution*. This CD, with the *A Convective Storm Matrix: Buoyancy/Shear Dependencies* CD and the *Mesoscale Convective Systems: Squall Lines and Bow Echoes* WWW-based instructional module, provide excellent sources for learning about severe local storms and the key role modeling has played in the development of our understanding of these phenomena. The WWW-based module is at http://www.comet.ucar.edu/modules/MCS.htm.

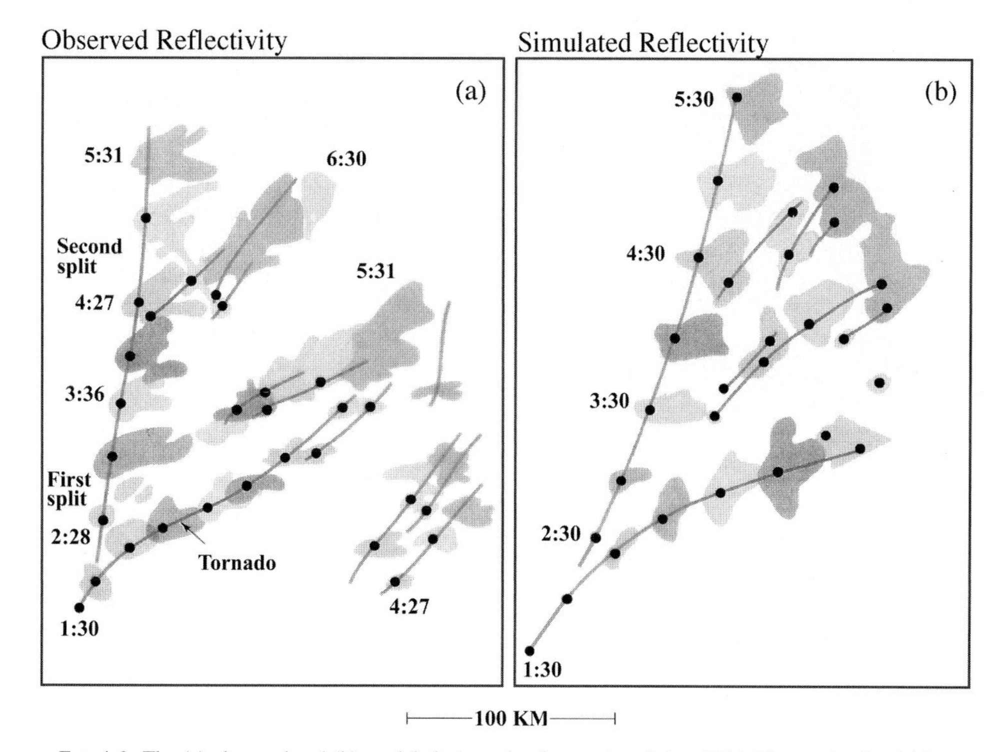

FIG. 4.6. The (a) observed and (b) modeled storm development on 3 Apr 1964. Observed reflectivities > 12 dBZ at the 0° elevation and modeled rainwater contents > 0.5 g kg-1 at $z = 0.4$ km are enclosed by alternating solid and dashed contours about every 30 min. Maxima in these fields are connected by solid lines. The storms are labeled at several times and the contoured regions are stippled for better visualization of the storm development. Labels for the modeled storms are at the same corresponding observed times except for the inclusion of "M". The scale shown in (a) applies to (b). Adapted from Wilhelmson and Klemp (1981).

The use of multiple Doppler radar to construct the three-dimensional flow field within severe storms was emerging at the same time as three-dimensional storm modeling in the late 1970s. The first study to combine both tools to better understand supercell structure was reported by Klemp et al. (1981). A detailed comparison between a dual Doppler analysis of the 20 May 1977 Del City, Oklahoma, tornadic thunderstorm and a numerical simulation initialized with a representative homogeneous environment on that day was presented. Figure 4.7 shows the observed and modeled three-dimensional precipitation surfaces using wire-frame 3D graphics—which, compared to today's rendering capabilities, seem relatively primitive. Despite significant horizontal inhomogeneities that were present in the atmosphere, the observed and numerically generated structures were qualitatively similar. Even weightless air parcel trajectories computed from the three-dimensional Doppler, or model winds and smaller-scale features, such as the location of the maximum vertical vorticity, were remarkably similar. The simulation also developed a horseshoe-shaped updraft at low levels, a structure that has often been observed within evolving tornadic storms, including the 20 May storm. This feature is associated with the hook echo in the storm's low-level reflectivity field. Both the appearance of the horseshoe updraft and the behavior of the hook echo are consistent with the development of strong low-level rotation within the storm and with the tornadic supercell paradigm presented by Lemon and Doswell (1979).

Over the past two decades, the Del City case has become a benchmark on which cloud modelers test their simulation codes. It is interesting to note that the numerically simulated storm structure is qualitatively insensitive to variations in initial conditions, model physics, and model numerics. Further, tests using a wide range of models produce similar evolution and structure. One possible explanation for this apparent robustness is that the input sounding is moist through a deep layer.[16] This implies that the impact of horizontal mixing and entrainment between the updraft

[16] There can be a significant impact on storm size and evolution when the moisture above the boundary layer is substantially reduced. In a recent non–Del City supercell simulation carried out by Wilhelmson, there was a more than 75% reduction in updraft area at 4 km above the ground when the moisture above the boundary layer was reduced from relatively high values to near zero.

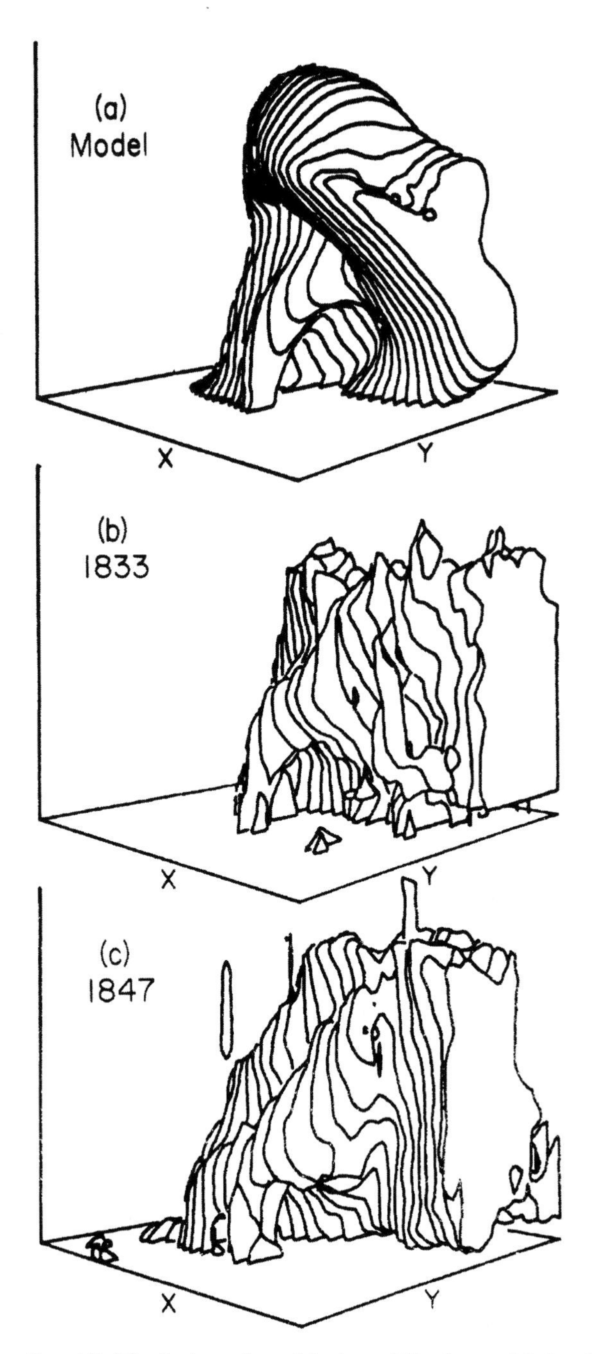

FIG. 4.7. Distribution of precipitation within the modeled and observed storms as viewed from the southeast. The contoured wire-frame surfaces represent the 0.5 g kg^{-1} rainwater surface in the model and the 35-dB*Z* reflectivity surfaces in the modeled storm. The x–y base plane corresponds to a 29 × 29 km domain and the vertical axis extends to 14 km. From Klemp et al. (1981).

and ambient environment from cloud base to midlevels is minimized and not strongly sensitive to the particular type of subgrid turbulent mixing scheme used. Note that Klemp et al. (1981, see their Fig. 13) showed that a significant portion of storm updraft air at $z = 7$ km does not originate within the boundary layer. Also, inspection of the sounding shows that the vertical gradients of temperature, humidity, and horizontal wind are relatively smooth. Therefore interpolation of the sounding data to the differing vertical levels of a particular model grid will generally produce similar profiles. Experience has shown that this case is one of the few where the properties of the initial data produce similar numerical solutions for a wide range of cloud models and their numerical parameters.

During the past ten years, increased attention has been given to the role that microphysics plays in determining storm structure and evolution. Unfortunately, few observational data exist about the microphysical structure of deep convective clouds, especially severe storms. The Johnson et al. (1994) case study focused on the microphysical properties of an observed and simulated supercell storm. Using a three-class ice microphysical parameterization, the evolution of an intense supercell that passed through the Cooperative Convective Precipitation Experiment (CCOPE) in August 1981 was studied. The numerical simulation produced a storm where the potential temperature in the updraft was nearly identical to that observed by aircraft penetration. Further, cloud water mixing ratios in the simulation updraft and in the observed storm were both approximately 0.5 g kg^{-1}. Similar structures in the hail shaft were also identified in the observed and simulated storm.

c. Environmental parameter studies

These case studies clearly demonstrate that severe storm modeling can be a useful tool for studying the basic dynamics of supercell storms, even when homogeneous initial environments are used. This is underscored by the many parameter studies of these storms that focus on how storm evolution and structure change as the initial state is systematically varied. These studies addressed Ogura's statement in the 1963 monograph where he noted that "There have appeared some indications that, while vertical shear of the horizontal wind is inimical to the development of shower clouds, squall-lines and large thunderstorms show a preference for the jet-stream region where this shear is pronounced. All of the numerical experiments of convection carried out so far have ignored the interaction of convective elements with a prevailing wind field. In view of these facts it would be interesting to make numerical experiments which include the vertical variation of the environmental wind field . . .."

Weisman and Klemp (1982) produced the seminal parameter study defining the impact of CAPE and vertical shear on storm type and morphology. They showed that, by systematically varying unidirectional hodographs (the vertical shear vector was unidirectional) and CAPE, convection changed from short-lived, single cells (pulse cells) to multicells and then

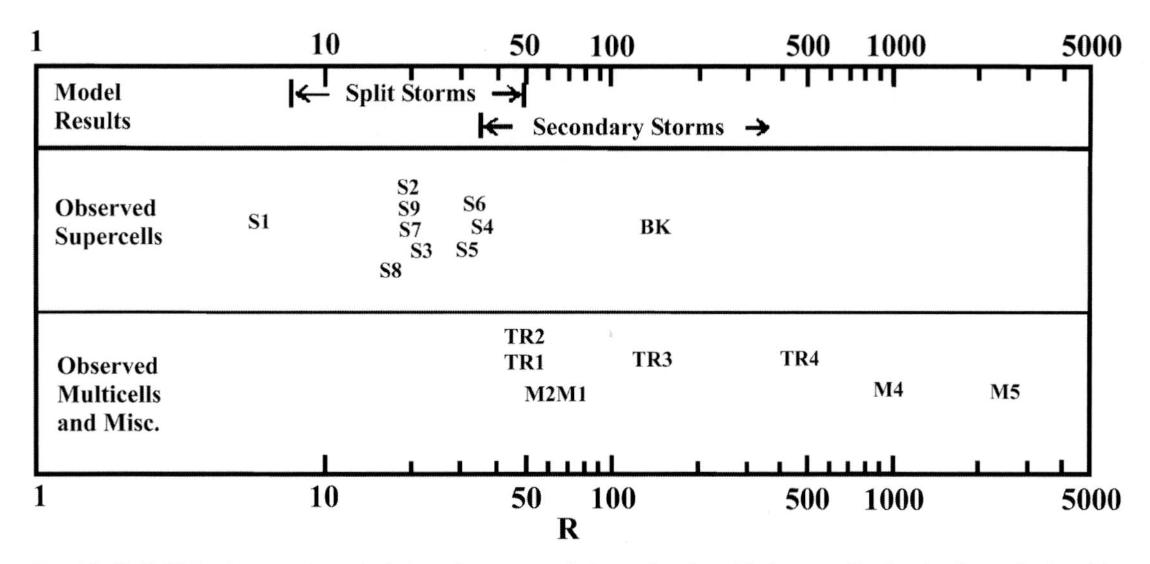

FIG. 4.8. Bulk Richardson number calculations for a range of observed and modeled storms. Further details can be found in Table 1 of Weisman and Klemp (1982). From Weisman and Klemp (1982).

to supercells. The relationship among CAPE, shear, and storm type was then quantified in a single parameter called the bulk Richardson number,

$$R_{ib} = \frac{\text{CAPE}}{0.5(\Delta u^2 + \Delta v^2)}, \tag{4.16}$$

where Δu and Δv represent the difference in wind components between the density-weighted mean wind at 6 km and 500 m. This parameter, originally proposed by Moncrieff and Green (1972), is a fair predictor of storm type. Figure 4.8 relates the bulk Richardson number to the modeled and observed storm types. Low values ($R_{ib} < 35$) are associated with supercells, while higher values ($R_{ib} > 35$) are associated with multicell storms. It should be noted that supercells are observed over a more varied range of R_{ib} than the modeling study. This difference is believed to be primarily due to the simple initial conditions and physics contained within the model versus the complexity found in nature in actual storms. Forecasters had known for many years that significant vertical shear was necessary for severe storms and tornadoes. However, the bulk Richardson number provided a quantitative measure of the potential storm type. Weisman and Klemp (1984) extended this conclusion to environments in which the wind hodograph had significant curvature. Various combinations of shear and buoyancy led to a spectrum of storm types that could be discriminated from one another using the bulk Richardson number. The use of the bulk Richardson number in forecasting represented one of the first widespread applications of three-dimensional cloud modeling results to severe storm prediction.

Other researchers probed deeper into the details of supercell behavior. Brooks et al. (1994) studied the relationship between midlevel shear and precipitation structure and how it influenced the generation of low-level mesocyclones. In weak-shear regimes, midlevel storm-relative flow was weak, and precipitation fell out west of a storm's main updraft. This produced more evaporative cooling at the surface in this region and generated a stronger gust front at low levels. Often this gust front moved eastward faster than the midlevel updraft and cut off the inflow into the storm. Increasing the midlevel shear increased the storm-relative flow and transported the precipitation further north and east of the midlevel updraft. The cold air near the gust front in this region supported the low-level solenoidal development of horizontal vorticity that was favorable, when tilted, for the development of persistent low-level rotation. As midlevel shear was further increased, the storm-relative flow transported precipitation farther east of the updraft. The depth and strength of the cold pool near the low-level updraft was significantly reduced, and low-level rotation did not develop. The results suggest that there may be an optimal midlevel flow regime in which the precipitation fallout and associated cooling are optimal for the development and maintenance of low-level mesocyclones. Thus, microphysics could play an important role in determining low-level mesocyclone development.

In another study, Droegemeier et al. (1993) focused on the parameter space associated with storm-relative helicity (hereafter SRH; Davies-Jones et al. 1990) in supercell and multicell environments. Overall, the results indicated that the bulk Richardson number is a better predictor of storm type, while SRH is a better predictor of storm rotation. The simulations also showed that low-level storm-relative inflow must be at least 10 m s^{-1} for supercell generation, and that supercell storms are able to extract helicity from the mean flow while multicells are not.

Most parameter studies (e.g., Wickers 1998), including those already mentioned, have focused primarily on the impact of vertical variations in wind shear. Few studies, outside those by Weisman and Klemp (1982, 1984), discuss variations in vertical temperature and moisture structure. Recently, Gilmore and Wicker (1998) studied the role that midlevel dry air can have on supercell structure and evolution. Midlevel dry air, through the process of entrainment, can significantly alter the thermodynamic properties of storm downdrafts. This, in turn, influences the depth and strength of the cold pool underneath the storm. For a sounding having identical CAPE and vertical shear, the storm morphology was significantly impacted by changing the relative humidity between the levels of 800 and 700 hPa. The control run generated a classic supercell similar to the one studied by Rotunno and Klemp (1985). Strong midlevel and low-level rotation were present when the storm matured. By progressively decreasing the relative humidity in the prescribed layer, stronger downdrafts were created. A relatively moderate change in the relative humidity changed the downdraft strength sufficiently to create an outflow-dominated storm with little low-level rotation. This study also concluded that downdraft convective available potential energy[17] (DCAPE; Emanuel 1994) is not a good indicator of potential downdraft strength. This is due to the fact that downdrafts are often subsaturated and experience significant mixing with the environment, a violation of parcel assumptions used to compute DCAPE. This apparent sensitivity to midtropospheric dryness may provide one explanation for observed tornado failure modes when the environment otherwise appears favorable. It also indicates the potential importance of microphysics through the evaporation and melting in downdrafts.

d. Simulation of various classes of supercells

Observations over the last 20 years by storm intercept teams and Doppler radar indicate that several subclasses of supercell storms exist. Moller et al. (1994) summarize these observations and describe three basic subclasses: the low-precipitation, classic, and high-precipitation supercell. Low-precipitation supercells have little or no visible precipitation shaft near the updraft or low-level mesocyclone associated with the storm. High-precipitation supercells are characterized by substantial precipitation within the mesocyclone. The classic storm is somewhere in between and is probably the type of supercell initially described by Browning (1964).

Only a few simulations of low-precipitation (LP) supercells have been attempted because almost all model microphysical parameterizations produce significant precipitation. Weisman and Bluestein (1985) compared simulations in which precipitation was allowed to fall with others in which precipitation simply moved with the air (the precipitation's terminal velocity was set to zero). They noted that both storms developed some classic supercell characteristics such as midlevel rotation and propagation to the right of the mean winds. Brooks and Wilhelmson (1992) were able to simulate an LP storm in a model with water-only microphysics. The initial storm grew slowly due to the use of an initial weakly buoyant bubble to trigger the convection. This storm produced little precipitation and only weak downdrafts for over 1.5 h. Eventually the simulated storm became a classic supercell, a transformation that has been noted in nature by Bluestein and Woodall (1990). They also noted that a given storm could have low-precipitation, classic, and high-precipitation characteristics as it evolves. The dependence on the initial conditions may be only one factor that influences low-precipitation storm development. Microphysics probably plays a crucial but yet unidentified role.

Kulie and Lin (1998) recently reported on a high-precipitation storm simulation. As noted by Moller et al. (1994), these storms are often associated with complicated reflectivity evolution and structure. The 28 November 1988 Raleigh tornadic thunderstorm complex simulated by Kulie and Lin evolved from a multicell-type storm to a multiple-updraft supercell. Interestingly, the supercellular updraft developed significant rotation as it moved along a boundary created by older simulated cells. Because the high-precipitation supercell is the dominant form of supercell (Johns and Doswell 1992), further studies and analyses are needed to understand the interplay of updrafts and downdrafts and the dynamic meaning of updraft mergers noted by Kulie and Lin. Since updrafts emanate over time from the boundary layer, a merger may simply mean that an old updraft dies and is replaced by a neighboring new one, perhaps in a more favorable (moist) environment for producing a stronger updraft.

During the past decade, it has become apparent that supercells like those discussed above are occasionally much smaller in both the horizontal and vertical dimensions than the majority of supercell storms. Miniature supercells have been discussed in the literature as early as 1979 (Burgess and Davies-Jones 1979). McCaul (1987) documented a tornado outbreak generated

[17] Using the environmental sounding, DCAPE is computed by determining the θ_w of a potential downdraft parcel from above the boundary layer. The parcel is assumed to decend moist-adiabatically to the surface. The negative area swept out by this moist adiabat and the environmental temperature profile is the DCAPE. See Emanuel (1994) for details.

by miniature supercells associated with a land-falling tropical storm along the Alabama coast. Tornado outbreaks have long been observed in association with land-falling tropical storms and hurricanes. McCaul (1987) showed that miniature supercells formed in an environment that had little CAPE and very strong vertical shear. Numerical simulations using a composite sounding (McCaul and Weisman 1996) demonstrated that shallow supercells could form in environments whose lapse rates are close to moist adiabatic. Storm updrafts reach peak intensity at low levels ($z = 2$ km) in comparison to $z = 7$–8 km in Great Plains supercells. These intense low-level updrafts approaching 10 m s^{-1} were driven primarily by vertical accelerations associated with the dynamic pressure much like the low-level updrafts in larger Great Plains supercells.

More recent modeling studies have attempted to understand the complex processes that influence supercell behavior. A recent study by Adlerman et al. (1999) studied the physical processes associated with cyclic mesocyclogenesis. The 20 May 1997 sounding from the Del City storm often produces multiple mesocyclone cores during the model simulations. Adlerman's simulations indicated that the rear-flank downdraft was not only important in developing the initial low-level rotation, but that the occlusion process that eventually dissipates the initial low-level mesocyclone also creates a favorable storm structure for the generation of additional mesocyclones. Therefore, while the initial mesocyclone required nearly two hours of simulation before occurring, subsequent mesocyclone spinup occurred much more quickly. This type of evolution is very consistent with observed storms having multiple mesocyclone cores (Burgess et al. 1982). Another study by Atkins et al. (1999) focused on the influence of low-level boundaries on low-level rotation in supercell storms. A tornadic supercell on 16 May 1995 near Garden City, Kansas, was extensively observed by the Verifications of the Origins of Rotation in Tornadoes Experiment (VORTEX; Rasmussen et al. 1994). These observations indicated that the storm developed and interacted with a weak low-level thermal boundary during much of its lifetime. Numerical experiments were conducted that explored the differences between storms that are initiated in a domain, using the observed sounding from VORTEX, with and without a low-level thermal boundary present. The simulations show that low-level mesocyclogenesis occurs sooner, is stronger, and is more persistent in the simulations that contain a boundary. This is due to the fact that the boundary's solenoids, not the storm's forward-flank solenoids, now provide the significant source of streamwise vorticity in the storm's inflow. Therefore the development and strength of the storm-generated forward-flank cold pool is less important to low-level vorticity generation, as the initial storm almost immediately begins to ingest significant low-level streamwise vorticity from its mesoscale environment.

4.4. Modeling contributions directed at explaining convective storm lines

Squall lines and severe windstorms have been discussed in some detail elsewhere in this monograph; here the discussion will focus on 3D modeling studies. Idealized simulations of these storms were initially employed to study the fundamental dynamics of links, similar to early idealized supercell studies. Using this methodology led to an increased understanding of how the large-scale environment controls storm evolution and structure. Most of these studies occurred in the late 1980s, after the initial supercell studies of Wilhelmson, Klemp, and Weisman, because the three-dimensional simulations of storm lines required larger domains and longer integration periods than initial supercell studies. Computational power needed for these simulations was not routinely available until then.

Rotunno et al. (1988) and Weisman et al. (1988) carried out parameter studies that focused on the relationship between CAPE and shear in squall line development. Similar to the earlier studies by Weisman and Klemp (1982, 1984), these papers generated new conceptual models relating the large-scale environment to the observed structure of squall line systems. The vertical wind shear in the parameter space was shallower than the vertical wind shear typical of supercell environments. Variations in magnitude of the vertical shear seemed to explain the observed upshear and downshear tilting behavior of squall line updrafts. The CAPE in these experiments was held constant. Both Rotunno et al. and Weisman et al. define a squall line as "optimal" when the individual convective updrafts are nearly vertical. This optimal state is considered to be a state where the squall line is most intense, consistent with observed squall lines. Their results showed that environments with shallow and weak shear tend to produce squall lines that quickly evolve into an upshear tilted system that contains weak and broad updraft cells. As the shear is increased, the individual convective updrafts are more upright and maintain this configuration for longer periods of time before potentially tilting upshear. Strong and deep shear produces a line of quasi-steady supercells that have midlevel mesocyclones. Their results showed that the relationship between the cold pool strength and the low-level shear appears to be the primary factor controlling squall line intensity. They cast this relationship as a balance between horizontal vorticity generated by the cold pool's solenoids versus the horizontal vorticity present in the low-level shear. If these two vorticities balance, then the system will have an upright orientation; if one dominates, then the system will tilt upshear or downshear.

Weisman (1993) extended this work further into examining the mechanisms that generate severe, low-level bow echo storms. This work showed that when the environmental shear has shallow and strong low-level shear, the squall line system generates a rear-inflow jet. This inflow jet is oriented normal to the line and remains above the surface until it reaches the convective line. Here the jet is forced downward by the convective-scale downdrafts that transfer horizontal momentum to the surface, creating severe surface winds and the formation of the bow-shaped convective line. At each end of the squall line, bookend vortices are created via the tilting and stretching of the horizontal vorticity in the inflow. These circulations also accelerate the rear flow normal to the squall line, intensifying the rear-inflow jet and surface winds.

Skamarock et al. (1994) explored the development of asymmetric squall lines and mesoscale convective vortices by combining an adaptive grid technique with the Klemp and Wilhelmson cloud model (see their Fig. 2). This enhancement of the more traditional nested-grid technique allowed Skamarock to simulate a long-lived large squall line in a domain that was 1000 km by 1000 km in horizontal size, yet still be able to resolve the convection using 2-km grid spacing with reasonable computational resources. The research study depended on the ability of the model to simulate finite lines of cells without using periodic lateral boundaries or allowing the convection to travel out of the domain. Skamarock et al.'s simulation showed that the development of the mesoscale convective vortex behind the squall line is crucial to the development of asymmetric systems. Further, by turning the Coriolis force on and off in the model simulations, it was determined that the mesoscale convergence of the Coriolis rotation was responsible for the development of the mesoscale vortex.

Another modeling study of squall lines by Pandya and Durran (1996) revealed that the principal flow and thermodynamic features associated with the mesoscale region (e.g., ascending front-to-rear flow and descending rear inflow) are fundamentally dynamic in character and can be viewed as a gravity wave. This was verified by carrying out a model simulation with microphysics that produced the mesoscale features. Simulations were then carried out using a dry model in which the spatial and temporal distribution of the thermodynamic heating is taken from the convective model simulation near the convection. The dry model's simulation produced similar mesoscale features to the moist simulation. This lends support to a hypothesis proposed by Yang and Houze (1995) that these features could be qualitatively reproduced without ice-phase microphysics. Again, modeling was instrumental in determining the fundamental dynamic behavior of squall lines.

Multicells occur in most line simulations; they also occur in splitting storm simulations, usually to the north of the southern supercell if one is present. Models have also been useful in studying the active convective cell life cycle and cell regeneration in these storms, as exemplified by Fovell and Tan (1998). Using both two- and three-dimensional models, they demonstrated that new cells that form near the squall gust front induce local circulations that eventually cut off the flow of unstable air into them, resulting in their decay. At the same time, these circulations indirectly support new cell development. Sensitivity of the results to model parameters is also discussed.

Idealized studies have not been the only modeling methodology used to investigate severe convective lines and systems. Early modeling case studies by numerous authors using two-dimensional models (e.g., Tripoli and Cotton 1989) led the way to three-dimensional studies where a wide range of motions, from the synoptic to the meso-γ scale, are represented using nested grid models. Using movable nested grid models, Bernardet and Cotton (1998) studied the development of a severe mesoscale convective system during a 24-h integration beginning on 12 May 1985. A swath of severe wind events (surface winds were reported in excess of 35 m s^{-1}) was generated from this system in northwest Kansas during the nighttime hours. Synoptic and mesoscale forcing in the simulation initiated convection in southeast Colorado (i.e., no warm bubbles or other mechanisms were used) and the subsequent modeled convective system propagated northeast into Kansas. Consistent with the observations, the convective system produced strong surface winds during the late night hours. Bernardet and Cotton found that the development of the nocturnal boundary layer, combined with the mesoscale downdraft, generated an enhanced region of high pressure in the system's cold pool that accelerated the low-level flow to severe levels. They point out that the interaction between the nocturnal boundary layer thermodynamic structure and the storm's downdrafts was key to the development of severe winds. Previous idealized studies, using afternoon soundings and homogeneous initial environments, had never created severe winds via this mechanism.[18] Therefore both idealized and case study simulations are needed in order to capture and explain the wide variety of convective phenomena observed.

4.5. Modeling contributions to storm forecasting

Numerical simulations can be used to provide information useful to forecasters both indirectly and directly. The former includes simulation-driven schematics of different types of storm events that can be

[18] In fact, it probably would not be possible to model nocturnal convection in idealized homogeneous environments since these these systems probably owe their existence to nonhomogeneous mesoscale forcing.

used in the interpretation of current radar data and the identification and usefulness of various bulk parameters in distinguishing possible storm evolution. The latter includes the real-time prediction of possible storm types based on simulations using projected horizontally homogeneous storm soundings or forecast simulations where the storm scale is resolved and assimilation of Doppler data is used in creating two- to four-hour forecasts.

The results from early modeling parameter studies by Weisman and Klemp (1982, 1984) were rapidly applied to operational forecasting of convective storm type with some success in the mid- to late 1980s. Weisman and Klemp's classic diagram depicting the relationship between the bulk Richardson number and storm type was shown in Fig. 4.8. However, experience indicated that the bulk Richardson number was a poor discriminator between tornadic and nontornadic storms. Davies-Jones et al. (1990) introduced the parameter SRH as a measure of the storm's inflow streamwise vorticity, which is a measure of potential updraft rotation. Early application of this parameter indicated that it was a better discriminator between tornadic and nontornadic supercells than the bulk Richardson number. The modeling study of storm-relative helicity by Droegemeier et al. (1993) showed that this parameter was the best indicator of storm rotation when the low-level storm-relative inflow was at least 10 m s^{-1}. The modest storm-relative flow was needed to keep storms from being "undercut" by their outflow. These results were very consistent with observed storm behavior. Another modeling study by Wicker (1996) indicated that the orientation of the low-level environmental horizontal vorticity vector could be an even better discriminator of potential storm rotation than SRH. This study showed that large values of SRH do not always generate significant near-surface rotation in the supercell. However, this method requires knowing with some accuracy the vertical profile of wind in the lowest 500 m, which makes it difficult to test operationally, due to the limited availability of wind profiles combined with the significant mesoscale variability of this parameter.

The usefulness of the bulk Richardson number, SRH, and other parameters to determine storm type and rotation strongly depends on the availability of a representative observational sounding from which they can be computed. Because they are not typically available, they must be estimated from early morning sounding change projections or inferred from mesoscale model predictions. Even then, it may not be possible to use these parameters to determine tornadic potential. The value of storm-relative helicity depends on the cell motion and is sensitive to modest changes in the low-level environmental wind profile. Markowski et al. (1998) found that the storm-relative helicity was highly variable in four observational case studies, implying that the use of storm-relative helicity in determining the tornadic potential of a storm is difficult. Another important factor in applying these parameters operationally is the relative vertical scale of the convection, which can be quite different from the vertical scale of the simulated storms. Normally, the bulk Richardson number is computed over a 6-km depth; however, some supercell environments produce rather shallow convection. Nevertheless, Trier and Parsons (1995) noted that the bulk Richardson number is also applicable in shallow shear environments and environments characterized by jet features (Chang and Yoshizaki 1993; McCaul 1993; McCaul and Weisman 1996; Chin and Wilhelmson 1998). It may also be useful in weak to moderate buoyancy environments (McCaul 1993; Kennedy et al. 1993).

With the rapid increase in computer speed and availability in the early 1990s, several investigators attempted to apply cloud models more directly to operational forecasts. These efforts used a simple methodology for generating "storm-type" forecasts. By obtaining a single environmental sounding from the region of interest, a horizontally homogeneous domain was initialized using the sounding as if one were doing an idealized simulation experiment. The model was integrated for several hours after triggering a storm with a conventional warm bubble. The first work to use this methodology in an operational setting was Brooks et al. (1993b). In this experiment, forecasters at the National Weather Service Forecast Office in Norman, Oklahoma, generated an afternoon sounding that was subsequently sent to the researchers at the National Severe Storms Laboratory and University of Illinois, where it was used to initiate the COMMAS cloud model for a two-hour forecast. The results from the simulation were interpreted by the modelers and sent back to the forecasters in Norman. The interpretation included storm type (no storm, pulse, multicell, or supercell), rotational character, potential for hail, and a movie of the low-level storm reflectivity. An important aspect of this work is that the information transfer, both from the forecasters to the modelers and vice versa, would not have been possible without the Internet being available to all the participants. A more extensive use of this particular methodology was shown in Janish et al. (1995). Here the input sounding was obtained from the Eta model, altered by Norman forecasters, and then employed in initiating the ARPS model. Brooks et al. (1994) used output soundings from MM5 to initialize the cloud model with little or no human intervention. Wicker et al. (1997) performed a similar type of experiment with forecasters at the Fort Worth National Weather Service Office using the predicted 0000 UTC Eta soundings for that day. Data from the cloud model simulation were made available to Fort Worth forecasters via the World Wide Web. Overall, forecasters found output from the cloud model to have useful information about storm type and potential rotation. The storm-type prediction methodologies appeared to work best in situations where significant low-level mesoscale forcing was not present (e.g., where the model setup is closest to the atmospheric situation). In these cases, many of the simulations had remarkable skill, at least subjectively, in generating the correct storm type and rotational characteristics. Obviously, the use of a horizontally homogeneous initial state in the cloud model severely limits the general application of this methodology to real-time operational forecasting.

During the past decade the horizontal grid sizes used in mesoscale models have decreased to the point where the finest grid in the mesoscale model is sufficient to resolve individual convective clouds. The horizontal grid resolution can now be obtained in the 2–3-km range within domains of up to 600 km. This is similar to the resolution used in early cloud modeling studies. Therefore initial work has begun to explicitly predict the initiation and evolution of convective storms within the mesoscale model on a 3–6-h timescale. There are several significant challenges to be overcome in order to develop storm-scale numerical weather prediction. These include obtaining good initial states for the model and understanding the associated error characteristics associated with this initial state and its potential impact on forecast error. The initial conditions at the convective scale must be either internally generated by the mesoscale model or assimilated into the model from observed data, which are typically Doppler radar data. For example, simulations have been reported using radar-derived winds to study gust fronts (Crook and Tuttle 1994). Sun and Crook (1996) compare two methods for retrieving three-dimensional wind and thermodynamic fields from dual Doppler radar. They conclude that adjoint retrieval is more accurate and robust than more traditional retrieval methods (Gal-Chen 1978). However, the use of a model adjoint is expensive and not practical computationally for the operational forecasting environment. Therefore much recent work has focused on using three- and four-dimensional variational techniques in order to retrieve the wind and temperature fields in a robust and efficient manner (Sun and Crook 1997; Gao et al. 1999).

Despite the difficulty in obtaining an accurate initial condition on the storm-scale in these models, several investigators have successfully simulated both mesoscale and convective-scale phenomena, if not in forecast mode, at least in research mode, using a cloud resolving mesoscale model. Ziegler et al. (1997) showed—using the CSU RAMS model (Pielke et al. 1992) and several levels of nested grids—that the simulations could predict fairly well the development of convective clouds and their subsequent organization. The model also predicted that no storms would form on days where no convection was observed. Another study using the RAMS model (Grasso and Cotton 1998) demonstrated the development of a tornado-like vortex within a supercell storm generated from a fully three-dimensional regional mesoscale simulation. Here six levels of grid nesting were used, with the finest grid resolution being 111 m and centered on the storm's mesocyclone. Both of these cases were simulated in hindcast mode, after the fact, for research purposes. During the past decade, the Center for the Analysis and Prediction of Storms (CAPS[19]) has conducted several quasi-operational storm prediction experiments in order to investigate and develop the scientific methodologies needed for storm-scale numerical weather prediction (Droegemeier et al. 1996a). For example, during the spring of 1995, CAPS ran the ARPS model in support of the VORTEX-95 field program (Droegemeier et al. 1996b; Xue et al. 1996a,b; Wang et al. 1996). The results from this experiment are similar to the results found in the simpler "storm-type" experiments previously discussed. At times the model predicts fairly well the development, position, movement, and even rotational character of convection. Another example of a "good" forecast occurred during the 3 May 1999 Oklahoma tornado outbreak, one of the largest outbreaks of significant tornadoes in Oklahoma and Kansas. The ARPS forecast predicted the development of two areas of convection, as well as their movement, fairly accurately.[20] However, significant forecast errors are often observed in these simulations. These forecasts errors are not only associated with the convection, as the prediction of mesoscale features is often equally in error and results in incorrect initiation or development of convection. Therefore storm-scale numerical weather prediction is closely tied to accurate mesoscale prediction. In order to reduce these errors, ensemble forecast methods are now being investigated as a method of reducing forecast uncertainty. SAMEX, a Storm and Mesoscale Ensemble Experiment, is one such effort (Hou et al. 2001).[21]

4.6. Modeling contributions aimed at understanding tornado genesis and structure

During the last two decades research on tornadoes and tornadogenesis using numerical models has progressed steadily. Lewellen's (1976) schematic documented the important flow regimes associated with a tornado. These are shown in Fig. 4.9, where tornadic flows have been divided into five principal regions: the vortex core, the outer flow, the upper flow (storm updrafts and downdrafts), the surface inflow, and the

[19] http://www.caps.ou.edu

[20] A special issue of *Weather and Forecasting* will be devoted to the 3 May 1999 outbreak.

[21] http://www.caps.ou.edu/CAPS/samex.html

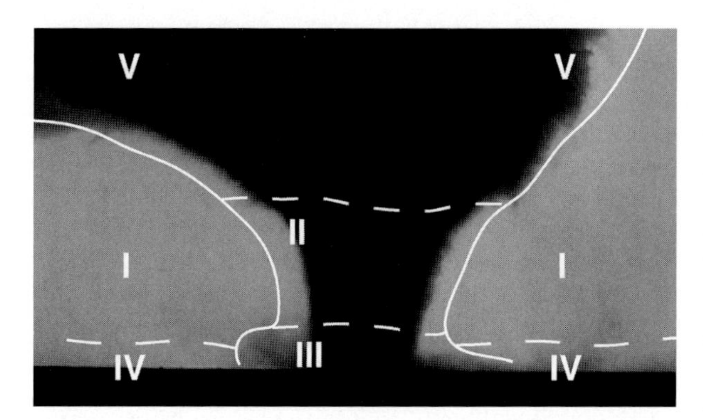

FIG. 4.9. Seymour, TX, tornado from 10 Apr 1979. Overlay is adapted from Lewellen's (1976) schematic showing the five regions of tornadic flow: region I, outer flow; region II, core; region III, corner flow; region IV, inflow layer; and region V, convective plume. Delineation between regions is approximate and is denoted by dashed lines. Photograph courtesy of National Severe Storms Laboratory.

corner region. Each of the five regions requires representation of different physical processes by a numerical model. Since the flows in the surface inflow and core regions require very high spatial and temporal resolution, initially axisymmetric and later three-dimensional vortex models were used to study this region. These simulations were generally carried out without explicit representation of the parent thunderstorm (the upper flow). Storm rotation and updraft parameters were therefore specified through the initial and boundary conditions of the model. This permitted the study of the corner, surface inflow, and core dynamic regions with a reasonable amount of computational effort. The significant limitation, even within three-dimensional vortex simulations, is the lack of interaction of the vortex with the thunderstorm cell itself. Understanding the storm processes that generate and destroy tornadoes from these vortex simulations is therefore limited.

The second approach, which attempts to rectify this limitation, emerged from three-dimensional cloud simulations of supercell thunderstorms. By the early 1980s, high-resolution three-dimensional supercell simulations produced many storm features believed to be associated with tornado generation. As computational power increased during the decade (in both memory and speed), it was natural to increase the grid resolution in an attempt to resolve smaller-scale storm features, including the tornado vortex itself. However, inclusion of the thunderstorm required domain sizes to be an order of magnitude larger than in vortex models. This forced a corresponding reduction in grid resolution, thereby eliminating the representation of fine-scale features of the vortex flow, especially in the surface inflow and corner region. On the other hand, these simulations more accurately simulated the updraft, downdrafts, and rotational structures of the storm, so that the processes responsible for generating tornadoes in the inhomogeneous storm flow could be accurately studied. Even today, computational limitations hamper the adequate representation of the corner and surface inflow regions in three-dimensional storm and vortex simulations, so that both modeling methodologies continue to be used in order to further our understanding of tornadoes and tornadogenesis.

a. Axisymmetric tornado models

Early studies of tornado-like flows were done using laboratory vortex chambers (Ward 1972). These experiments were crucial in defining the relevant physical parameters that influence the resulting vortex structure and dynamics. Aside from the physical constraints of the laboratory instrument itself (i.e., the height and width of the chamber), these studies indicated that the swirl parameter was the primary predictor of the resultant vortex flow. The swirl parameter is defined as the ratio of the imposed vertical vorticity to the low-level convergence within the vortex chamber, or equivalently, the ratio of the imposed mean tangential inflow and the outflow vertical velocity at the top of the domain. Low values of the swirl ratio generated narrow columnar vortices with an updraft along the axis of rotation (single-celled vortices), while larger values generated larger vortices having a downdraft along the axis of rotation surrounded by a rotating updraft (two-celled vortices).

Whether such vortices in the form of tornadoes are one- or two-celled, it is important to determine the maximum wind speeds and pressure falls that accompany them. Knowledge of this information near the ground is required for designing tornado-proof buildings such as nuclear reactors. Early estimates of tornadic wind speeds approached the speed of sound (Flora 1954). Later studies of tornado damage and photogrammetry reduced estimates of maximum wind speeds to less than 200 m s^{-1} (Keller and Vonnegut 1976). However, vortex core hydrostatic pressure deficits, estimated from sounding data (Snow and Pauley 1984; Lilly 1969), only permit maximum vortex wind speeds of 60–70 m s^{-1}. Maximum wind speeds estimated from hydrostatic pressure considerations are called the *thermodynamic speed limit.* The apparent ability of tornadoes to generate wind speeds in excess of those estimated from hydrostatic considerations indicates that another mechanism must be responsible for generating the estimated velocities.

Fiedler and Rotunno (1986) proposed a theory to explain this. Frictional forces at the base of the tornado generate a secondary circulation that, in certain circumstances, allows inflowing parcels to penetrate close to the axis of rotation while retaining most of their initial angular momentum. This somewhat paradoxical process generates large tangential velocities as well as a significant axial jet near the axis as the inward radial flow is forced upward. Their theory predicted that this "end-wall" vortex configuration could exceed the thermodynamic speed limit by 70% for the tangential flow and by more than 100% for the vertical velocity. Subsequent numerical studies support this theory (e.g., Fiedler 1994).

The first systematic numerical simulation studies of tornado-like flows were done using axisymmetric models in the mid-1970s. Initial experiments were constrained to studying the dynamics of the basic vortex flow near the core (Fig. 4.9), given computer speed and memory at the time. These studies reproduced the primary vortex circulation, and the numerical solutions were similar to laboratory results. As computer power increased, more sophisticated models with smaller grid spacing were used. The background storm flow in which the tornado formed was prescribed through the initial and boundary conditions. The use of axisymmetric models precluded the use of any type of realistic specification of the three-dimensional thunderstorm flow.

Howells et al. (1988) presented an overview of many of the important aspects and issues associated with simulations of tornado-like flows. This study demonstrated that the secondary circulation, generated by surface friction, was clearly responsible for generating large radial, tangential, and vertical velocities near the axis of rotation. The strength of the secondary circulation was also found to be proportional to the magnitude of the swirl ratio. The simulations also showed that the amplitude of the velocity field in the corner region is especially sensitive to the specification of the eddy diffusion coefficient near the surface. This suggests that turbulent eddies and their associated momentum fluxes are an important component in determining the dynamical character of the flow. Unfortunately, little is known about the nature of turbulence in strongly rotating flows. Therefore, Howells et al. suggested careful consideration when modeling these processes.

One of the first significant efforts to improve the parameterization or turbulent flux was presented by Lewellen and Sheng (1980). Their model included prognostic equations for the second-order turbulent fluxes in an effort to more accurately model the turbulent flow in the corner and surface inflow regions. They systematically studied the effects of the swirl ratio and surface roughness parameters on the strength of the tornado circulation. Maximum wind speeds were obtained when the swirl ratio was approximately 1.0. The vortex strength was also sensitive to the specified surface roughness, with rougher surfaces reducing maximum wind speeds. The maximum wind speeds were approximately 40–50 m above the surface while the wind speeds at a height of 10 m were approximately 90% of the maximum wind speeds aloft. The turbulent velocity fluctuations (for the given grid spacing) had the same magnitude as the mean flow near the surface. Therefore, the turbulent eddies within the tornado circulation represented a significant portion of the kinetic energy of the flow. Scaling the model parameters to the atmosphere, Lewellen and Sheng indicated that maximum wind speeds near the surface should not greatly exceed 125 m s^{-1}.

Fiedler's (1994) modeling study focused on the processes and parameters that allow velocities in the vortex to exceed the thermodynamic speed limit. His study showed that many vortices do not exceed this limit because their steady-state configuration does not represent an end-wall vortex (e.g., the flow is subcritical). Steady end-wall vortices (i.e., supercritical vortices) have wind speeds about twice the thermodynamic speed limit. When nonsteady flow is considered, transient supercritical vortices were produced having winds speeds six times the thermodynamic speed limit. Fiedler also demonstrated that subsidence warming in the vortex core, which had been hypothesized by others as an important mechanism for increasing wind speeds (Walco 1988), can also produce a vortex whose winds exceed the thermodynamic limit. Therefore, this study validated and extended the earlier theoretical results of Fiedler and Rotunno (1986).

Fiedler (1995) pointed out a significant shortcoming associated with axisymmetric vortex model simulations. The upper boundary conditions are usually held fixed during simulation. A series of experiments demonstrated that changing the upper boundary condition altered the maximum tangential velocity within the vortex by a factor of 2. Therefore, care is needed in extrapolating results from axisymmetric simulations to tornadoes without accounting for changes associated with the "boundary condition" imposed by the parent storm.

b. Three-dimensional vortex simulations

As computational power increased during the 1980s, three-dimensional vortex simulations became possible. One of the first studies was presented in Rotunno (1984). He used a three-dimensional cylindrical numerical model of the Ward laboratory chamber to examine the generation of subvortices within the parent vortex. For large swirl ratios, vortex chamber experiments showed the development of smaller-scale vortices that revolve around the axis of the principal vortex. These *suction vortices* (Fujita 1971) appear to develop from instability in the three-dimensional vortex sheet located between the outer-core updraft and inner-core downdraft in the large-scale vortex. The numerical solution also developed small-scale vortices within the parent vortex (Fig. 4.10). The asymmetric flow shown in Fig. 4.10a became quasi-steady with two subvortices rotating around the parent vortex. The subvortices tilt clockwise with height (Fig. 4.10b), consistent with both laboratory experiments and atmospheric observations of tornado suction vortices. Ro-

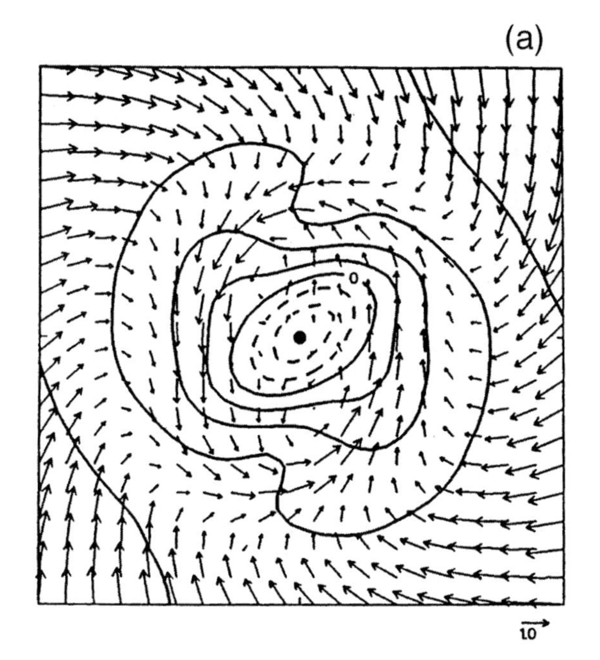

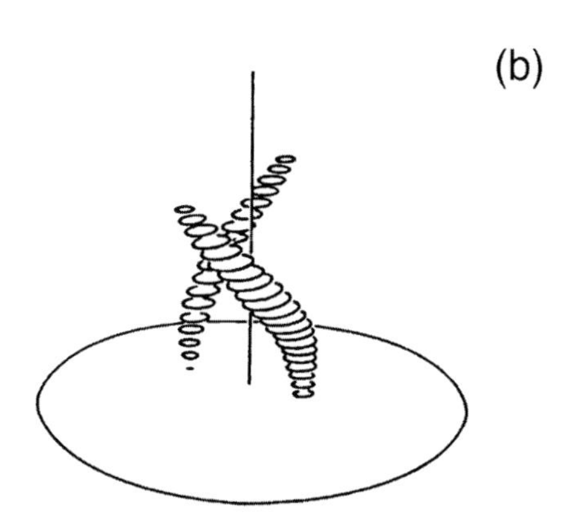

FIG. 4.10. Three-dimensional vortex structure from Rotunno (1984). (a) Low-level flow field shown with the vortex propagation velocity (1.9 radian per unit time) subtracted out. Superimposed is the field of vertical motion. Note that the centers of rotation are near the location where $w = 0$. (b) Three-dimensional contour representation of the pressure field at $t = 40$; the contour value is -1.3. This illustrates that the axes of the multiple vortices tilt clockwise with height, although they are embedded within a counterclockwise flow.

tunno also argued that the dynamics of this rotating flow could be representative of mesocyclonic flow when the rear-flank downdraft "divides" the mesocyclone to create a two-celled vortex (Lemon and Doswell 1979). At this scale, a subvortex within the mesocyclone represents an incipient tornado cyclone or tornado vortex, and therefore tornadogenesis could be closely tied to this three-dimensional vortex sheet instability.

Fiedler (1998) extended his earlier axisymmetric studies to three-dimensional vortices to again study what parameters and resulting vortex configurationsgenerate wind velocities greater than the thermodynamic speed limit. No-slip lower boundary conditions were used on the horizontal velocities, thereby including the effects of friction into the model physics. These simulations were carried out using a three-dimensional adaptive grid to accurately represent the details near the surface and in the corner region of the vortex. Figure 4.11 shows the results from a comparison between axisymmetric and three-dimensional simulations using identical flow and viscosity parameters. Each axisymmetric solution generated a vortex where the flow exceeded the thermodynamic speed limit for a short period of time (Figs. 4.11a–c). After this time, vortex breakdown occurred and the flow transitioned to a weaker state. The three-dimensional simulations also revealed an initial strong spinup close to the same time as the axisymmetric simulations. However, after the initial vortex flow transitioned to a two-celled state, suction vortices developed within the three-dimensional vortex sheet and these entities consistently produced wind speeds that exceeded the thermodynamic limit (Figs. 4.11d–f). Typical suction vortex wind maxima were twice that of the thermodynamic speed limit and did not require an end-wall (supercritical) state. These results were consistent with Fujita (1971), who estimated that suction vortices could have wind speeds twice that of the parent vortex. Similar to the solution discussed in Rotunno (1984), the vortex sheet developed two suction vortices.

Most vortex models (including the ones previously discussed) have been built with a fixed of simply varying viscosity coefficient. However, Lewellen (1993) used a large-eddy simulation to study the dynamics and the turbulent properties of a fully developed three-dimensional vortex. With a three-dimensional stretched mesh, the grid resolution was about 10 m over a significant portion of the domain and 5 m in the core of the tornado. Figure 4.12 is a simulation from Lewellen et al. (2000) where the tangential and vertical velocities were greater than 75 m s^{-1} in the corner region of the flow. Similar to earlier studies, the inward radial momentum flux due to the secondary circulation increased the tangential flow by about 50% from the downstream rotational flow farther aloft. Several suction vortices were located along the inner edge of the core updraft and can be seen as local updrafts and downdraft maxima in Fig. 4.12a. These suction vortices were highly transient and lasted only a few seconds. The time-averaged surface pressure deficit is consistent with the applied upper boundary conditions, about −40 hPa. Transient surface pressure deficits associated with suction vortices were large, occasionally amplifying the time-averaged pressure deficit by a factor of 2 or more. The rapid translation of these subvortices produces pressure variations on a timescale of tenths of a second. The three-dimensional structure of these subvortices is shown in

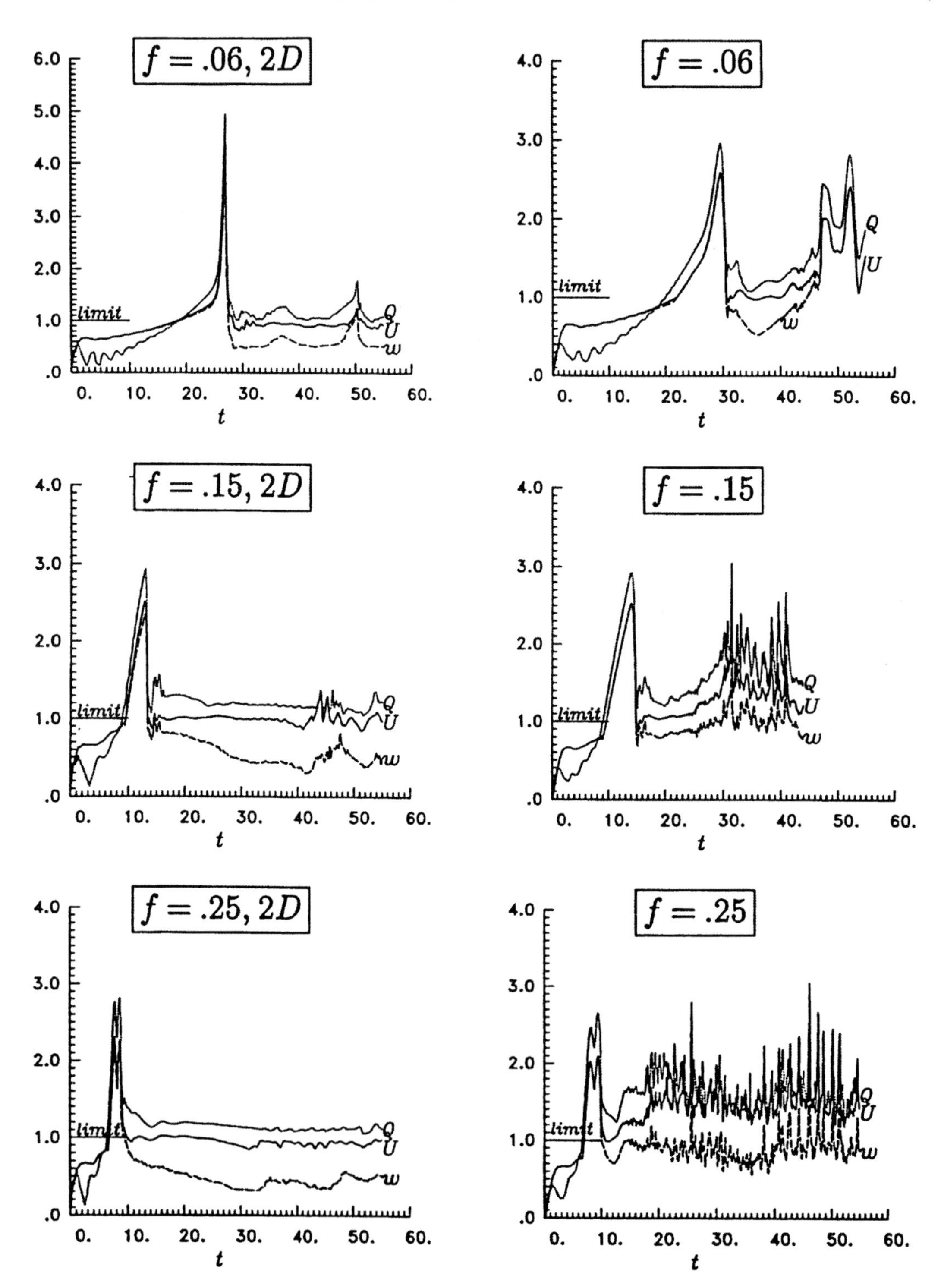

FIG. 4.11. Time histories of the maxima in the domain for both axisymmetric (2D) and three-dimensional vortex simulations. One unit of speed is marked limit, and is the thermodynamic speed limit of the model. Curve w is the vertical velocity, U is the maximum wind speed, and Q represents a scaled pressure. Adapted from Fiedler (1998).

Fig. 4.12b. Careful study of these simulations demonstrated that the numerical grid used to resolve the tornado flow was sufficiently small that the solution was relatively insensitive to changes in the turbulent mixing parameters. This is a significant improvement from other numerical vortex studies, where the flow has often been shown to be very sensitive to the eddy viscosity specification.

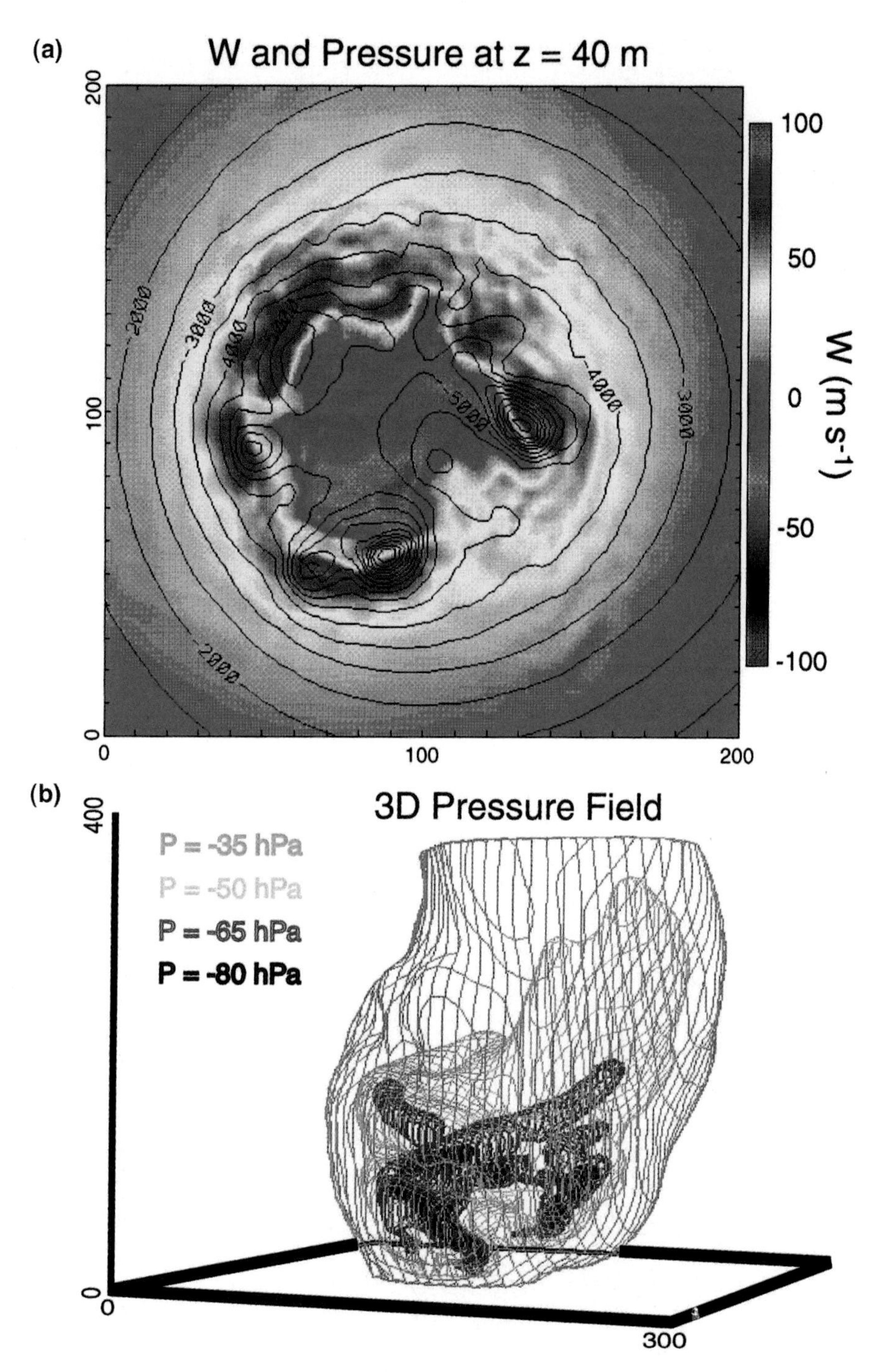

FIG. 4.12. (a) Horizontal cross section of pressure field and vertical velocities from an LES tornado simulation using medium swirl. Pressure is contoured, vertical velocity is shown as the color image. Shown is a horizontal domain 200 m by 200 m. (b) A three-dimensional plot of the instantaneous pressure field within the tornado. A domain 300 m square by 400 m deep is shown. Four pressure isosurfaces are contoured. Images courtesy of S. Lewellen and D. Lewellen (2000).

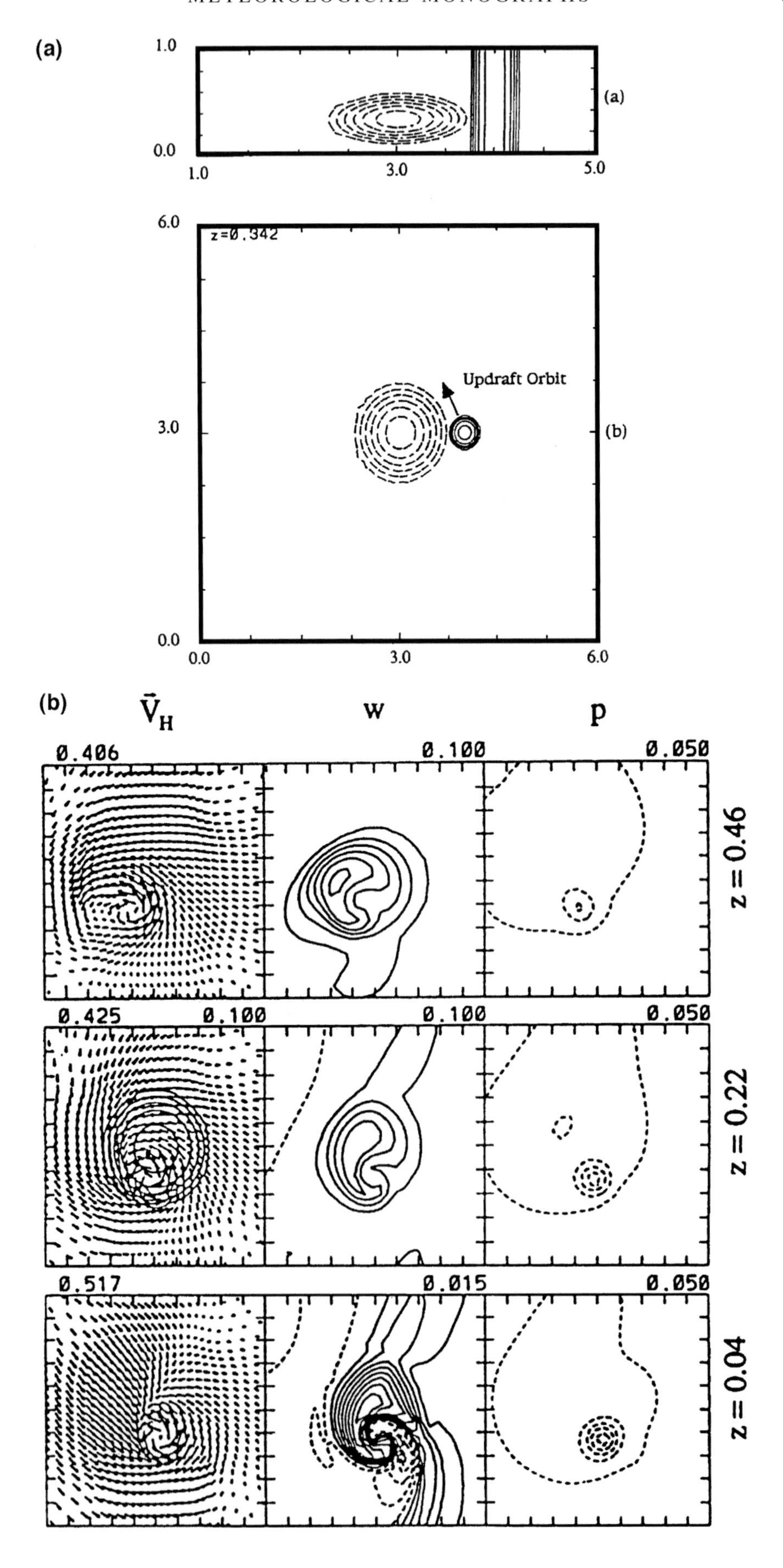
(a)
1.0
0.0
1.0
3.0
5.0
(a)
6.0
z=0.342
Updraft Orbit
3.0
(b)
0.0
0.0
3.0
6.0
(b)
$\vec{V}_H$
w
p
0.406
0.100
0.050
z = 0.46
0.425
0.100
0.100
0.050
z = 0.22
0.517
0.015
0.050
z = 0.04

c. Pseudostorm tornado simulations

The study of tornadogenesis in a fully three-dimensional model with moist physics and turbulence is a complex task. Some investigators believed that even if a tornado were generated in a three-dimensional supercell simulation, the flow would be so turbulent and complex that one would not be able to understand the physics of the flow.[22] In an effort to reduce the problem's complexity, several investigators designed numerical studies with "simplified" storm representations.

Walko (1993) reported the first study using this approach with the aid of the Colorado State University RAMS model in which an idealized supercell was created. This was accomplished by specifying a cylindrical updraft region using an analytical buoyancy function. Several wind distributions were then explored. The simulations clearly showed that combinations of vertical wind shear and updraft were not sufficient for the generation of low-level rotation. Tornado formation required the presence of some form of boundary layer vertical vorticity and an updraft. The boundary layer vertical vorticity could be generated by a downdraft transporting higher momentum air to the surface when vertical wind shear is present (as in supercell storms) or simply available through a preexisting wind shift line (similar to those observed near landspout tornadoes). This is one of the first studies to indicate that downdrafts could play an important role in the generation of low-level circulations. It also revealed that vertical vorticity for the tornado is generated on a horizontal scale ten times larger than the tornado vortex itself. The rotation at this radius is then converged by the updraft and generates the tornado vortex.

Trapp and Fiedler (1995) developed an even more realistic "pseudostorm" to study tornadogenesis. Their simulations contained both warm and cold buoyancy fields representing updrafts and downdrafts, as well as a surface cold pool. Figure 4.13a shows the initial distribution of positive and negative buoyancy in the domain. This distribution of buoyancy was used to generate the necessary vertical motion in the storm simulation. The cylindrical positive buoyancy field was translated around the outer edge of the circular cold pool and an updraft–downdraft couplet generated and centered on the low-level baroclinic zone. This pseudostorm contained most of the physics believed necessary for tornado formation within an actual supercell thunderstorm. Using this methodology, a wide variety of parameters, such as downdraft intensity, storm propagation, and viscosity, can then be studied. Figure 4.13b shows the results from a free-slip simulation. At low levels, a strong vortex is present in the flow field and the vertical velocity pattern has a horseshoe-shaped updraft. The surface pressure field has a circular region of low pressure induced cyclostrophically by the rotational flow. Vertical vorticity, generated by tilting and stretching near the surface, has been stretched farther and transported upward into midlevels of the pseudostorm. Overall, the low-level and midlevel vorticity, flow, and updraft structures are remarkably similar to observed and modeled supercell storms.

Extensive parameter tests by Trapp and Fiedler showed that vortex intensity was weakly modulated by downdraft intensity. Weaker downdrafts were generated in the model by weaker negative buoyancy. This weakened the vortex in two ways. First, a weaker cold pool implied that less baroclinically generated horizontal vorticity was available to be tilted into the updraft. Weaker downdrafts also implied weaker horizontal gradients of vertical velocity, which also reduced the tilting. Vortex intensity was strongly influenced by the residency time of parcels within the baroclinic zone prior to entering the updraft. This was controlled by the translational speed of the updraft around the cold pool. If the translation speed was slow, there was insufficient inflow to generate positively correlated updraft and vertical vorticity. If the translation speed was large, any vertical vorticity generated from tilting was immediately transported upward and out of the boundary layer. Further, several simulations with no-slip lower boundary conditions were found to be sensitive to the amount of viscosity. Increased viscosity strongly reduced both the low-level rotation and low-level convergence. Reduction in low-level vertical vorticity and convergence significantly reduced vortex strength or inhibited vortex formation. Overall, Trapp and Fiedler found that vortex formation was confined to a limited region of the parameter space. This is consistent with the fact that only a limited number of supercells actually produce tornadoes and that the optimal parameters for pseudostorm vortex formation appear to be consistent with the observed parameters associated with actual tornadoes.

[22] This statement was made by a prominent senior research scientist to the second author (LJW) while he was pursuing his Ph.D. research on tornadogenesis!

FIG. 4.13. (a) Initial buoyancy field in a vertical cross section (top) and a horizontal slice (bottom) for the pseudostorm experiments. The contour interval for positive buoyancy is 0.1; for negative buoyancy it is 0.01. (b) Horizontal velocity vectors (left), vertical velocity contours (middle), and pressure contours (right) shown at three different heights above the surface at $t = 70$. Velocity vectors are plotted at every other grid point. Magnitude of vector shown in upper left corner of plot. The position of the storm at $t = 70$ is shown in the small window on top. Adapted from Trapp and Fiedler (1995).

d. Modeled supercells with tornado-like vortices

With the advent of three-dimensional models in the late 1970s, a second line of modeling work began with Klemp and Wilhelmson's initial work on supercells, and subsequentially progressed toward understanding tornadogenesis starting at the storm scale and progressing downscale to the tornado vortex. Early simulations of supercell storms displayed many of the observed radar characteristics of supercells (bounded weak echo regions, rotation at both mid- and low levels, hook echoes, etc.). In the early 1980s, the primary factor limiting the study of tornadogenesis in storm models was computing memory and speed. In order to circumvent these limitations, Klemp and Rotunno (1983) used a simple horizontal grid-nesting scheme to achieve 250-m grid resolution within the storm over a limited domain. Figure 4.14 is a comparison of the low-level solutions from the 1-km and 250-m horizontal grid resolution simulations. An important observation tied to this methodology is that the solution on the fine grid primarily represents an adjustment of the flow to the newly resolved features and subgrid mixing effects, rather than representing evolu-

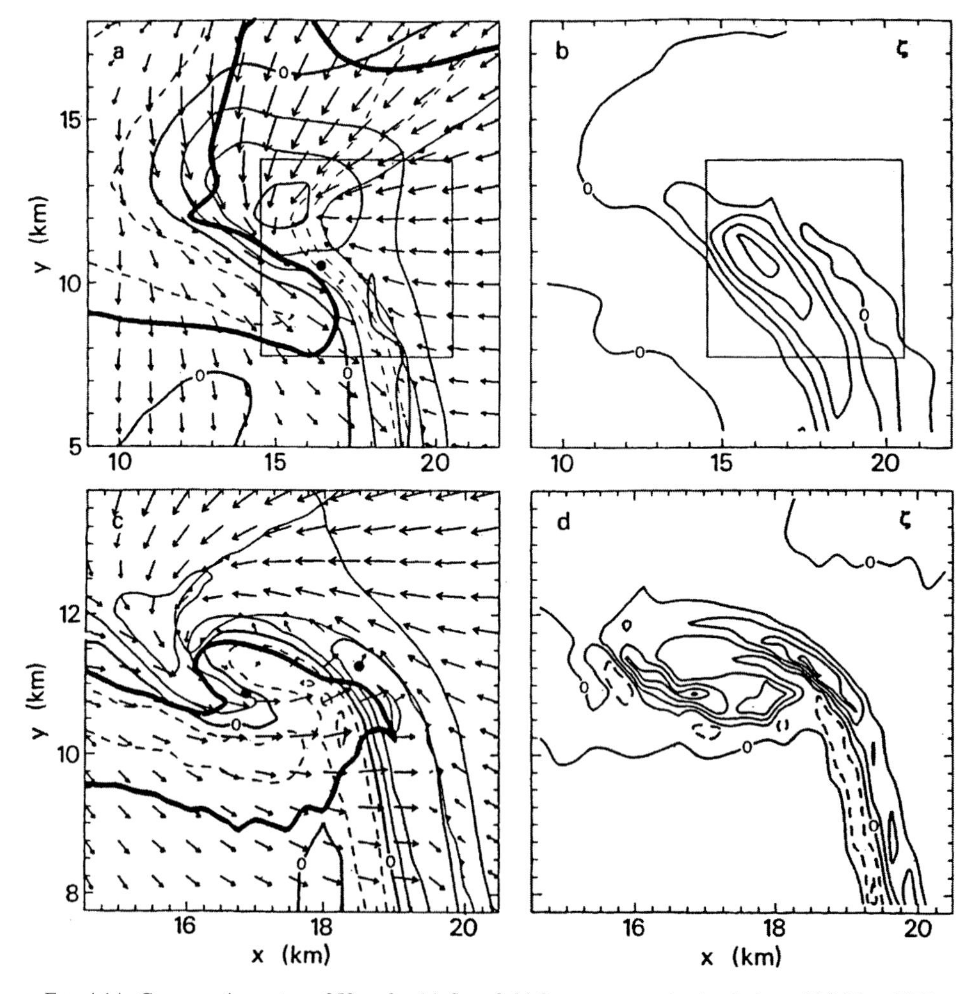

FIG. 4.14. Cross sections at $z = 250$ m for (a) flow field from storm-scale simulation of 20 May 1977 at T = 120 min. (b) Vertical vorticity associated with the flow field in (a) contoured using 0.005 s^{-1} intervals; (c) flow field at 6 min in the high-resolution simulation displayed in the region indicated by the box in (a); and (d) vertical vorticity at 6 min displayed in the same region as shown in (c) with a 0.02 s^{-1} contour interval. In (a), the vertical velocity is contoured in 1 m s^{-1} intervals and the heavy line represents the 0.5 g kg^{-1} rainwater contour. One grid interval indicates 20 m s^{-1} for the horizontal flow vectors. The -1°C perturbation potential temperature is denoted by the cold frontal boundary, while the two thin dashed lines behind this boundary indicate the -2° and -3°C isotherms. The location of the maximum vertical vorticity is marked with a black circle. In (c), the notation is as in (a), except that w is contoured using a 2 m s^{-1} intervals and one grid point represents 10 m s^{-1} for horizontal flow vectors, which are plotted at every other grid point. From Klemp and Rotunno (1983).

tion from one solution to another. Figure 4.14a is the 1-km solution and has large hook echo with a strong low-level updraft. The low-level vorticity field is elongated along the storm's rear flank gust front (Fig. 4.14b). After six minutes of integration (Figs. 4.14c, d), the solution had adjusted to the increased resolution and had captured many details of the low-level mesocyclonic flow. The mesocyclone developed an occlusion downdraft (Fig. 4.14c) as discussed by Lemon and Doswell (1979). This two-celled structure was very similar to the structure simulated in the Rotunno (1984) three-dimensional vortex simulations. The vertical vorticity field also contained several localized maxima (Fig. 4.14d) similar to the Rotunno (1984) results, and these were hypothesized to possibly represent tornado cyclones.

Using trajectory analysis, Klemp and Rotunno also showed that vertical vorticity at low levels is first generated from tilting of horizontal vorticity and then stretched in the storm's low-level updraft. They concluded that the occlusion downdraft is generated by the intense near-surface rotation and associated downward-directed vertical pressure gradient accelerations. Rotunno and Klemp (1985) showed that baroclinic generation of horizontal vorticity along the storm's low-level cold pool was crucial to the development of low-level rotation. Further, using Bjerknes's first circulation theorem, Rotunno and Klemp demonstrated that low-level rotation is initially horizontal and subsequently reoriented to produce vertical circulation. This was demonstrated by the evolution of a material line placed around the low-level mesocyclone (Fig. 4.15a). Using trajectories, the material line was integrated backward in time for 15 minutes. During this period the loop expanded and a portion became vertical. The vertical portion of the material line experienced a net cyclonic torque from the negative buoyancy field shown in Fig. 4.15b. Figure 4.15c shows

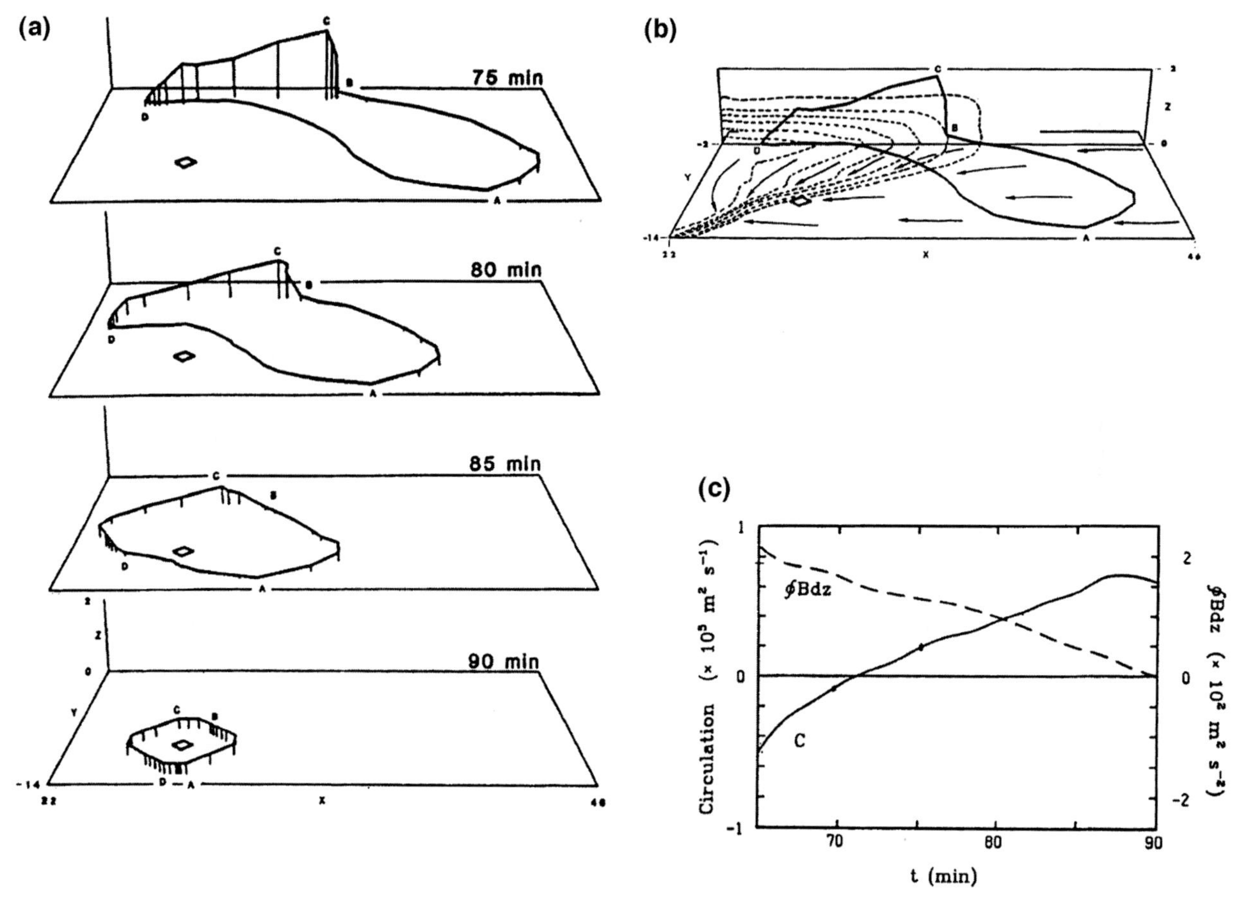

FIG. 4.15. (a) Time sequence of perspective plots (looking to the north) that show the three-dimensional evolution, in 5-min intervals, of the material curve on a 24 × 24 × 2 km portion of the computational domain. The locations of points A–D are shown at each time for reference. (b) An illustration of how the baroclinic generation is positive as the curve evolves. The surface buoyancy field and buoyancy fields in an x–z plane are displayed together with the material curve at 75 min. This display is chosen because the curve has two primary branches: DAB, which lies close to the $z = 0$, and BCD, which stands nearly vertical in the x–z plane. The surface flow is indicated by the arrows. (c) The circulation and generation of circulation computed around the material curve shown in (a) and (b). From Klemp and Rotunno (1983).

that the material line's circulation is negative. During the next 15 minutes, the buoyancy force torques the material line such that a significant positive circulation is generated. This circulation is then converged by the storm's low-level updraft, forming the mesocyclone. Using Stokes' theorem, one can show that this process is equivalent to the tilting of horizontal vorticity generated by the cold pool's solenoids.

These studies suggest that tornadogenesis could be studied using a storm model at very high spatial and temporal resolution. Using high-resolution two-way nested grids, both Wicker and Wilhelmson (1995) and

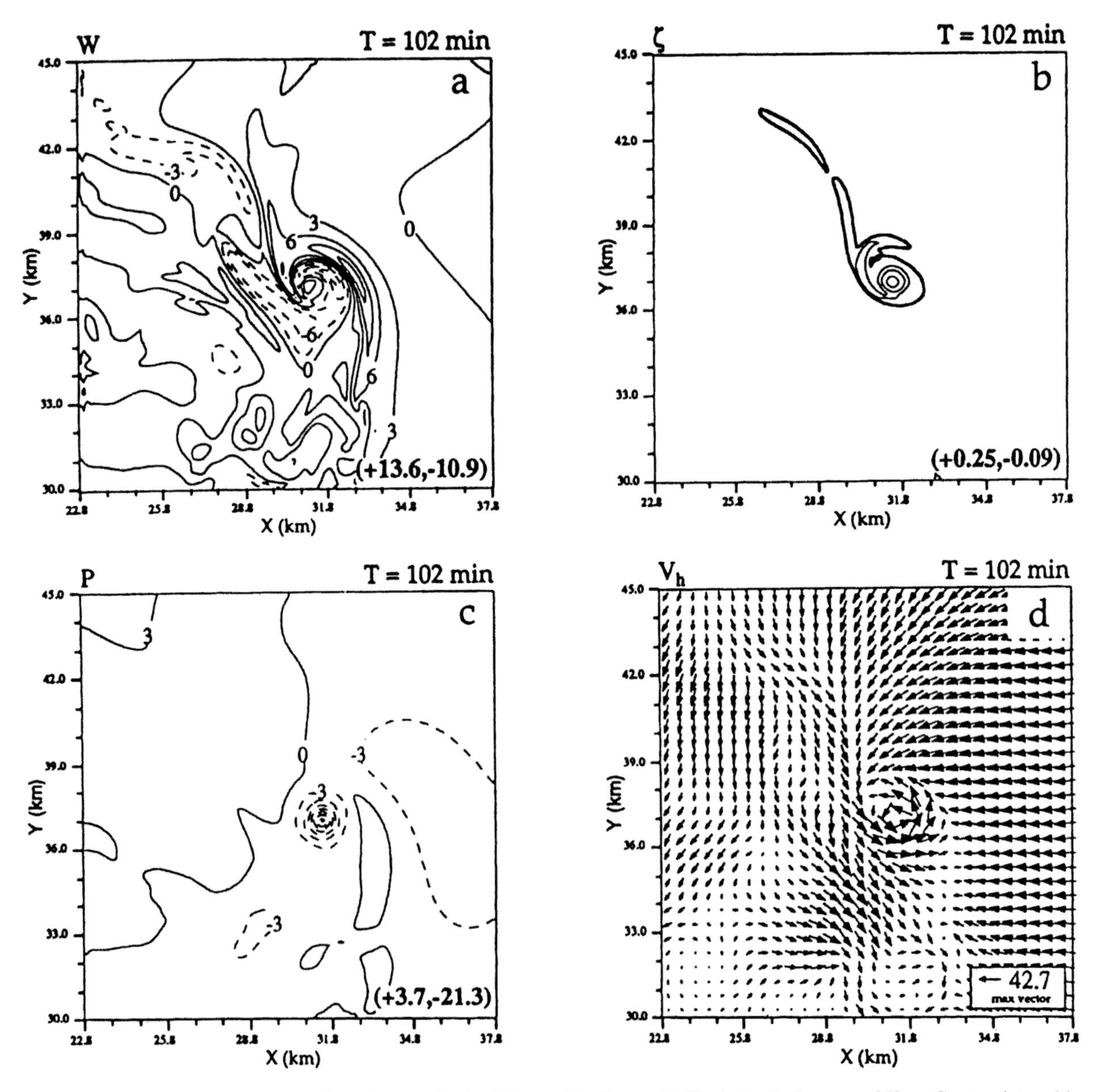

FIG. 4.16. Horizontal cross section from the tornado simulation at 80 minutes. (a) Vertical velocity at $z = 250$ m. Contour interval is 3 m s^{-1}. Maximum and minimum values for the field are shown at top right. (b) Vertical vorticity at $z = 100$ m. Contour intervals are 0.01, 0.05, 0.1, 0.2, 0.3 s^{-1}. The 0.01 s^{-1} contour is the thick dark line and indicates the location of the low-level mesocyclone. Maximum and minimum values for the field are shown at top right. (c) Pressure field at $z = 100$ m. Contour interval is 3 hPa. Maximum and minimum values for the field are shown at top right. (d) Horizontal wind vectors for $z = 100$ m. Maximum wind vector magnitude is shown at bottom right. Every third vector is plotted. (e) Time–height cross section of variables from fine-grid simulation showing maximum updraft. Contour interval is 5 m s^{-1}. Bold dashed line indicates updraft pulses that occur near the mesocyclone. Shaded areas are regions where vorticity exceeds 0.21 s^{-1}. (f) Maximum vertical vorticity. Contour interval is 0.03 s^{-1}. Bold dashed line indicates updraft pulses that occur near the mesocyclone. Regions where shear is greater than 0.21 s^{-1} are shaded. From Wicker and Wilhelmson (1995).

Grasso and Cotton (1995) generated high-resolution simulations of supercells containing tornado-like vortices. Grasso and Cotton (1995) simulated an intense tornado-like vortex using an observed sounding from 20 May 1977. With three two-way interactive grids, a vortex was resolved using a 100-m horizontal and 25-m vertical grid spacing. Surface friction was also included. Their simulation generated a tornado-like vortex with 50 m s^{-1} tangential winds and updraft speeds of over 50 m s^{-1}. Analysis showed that low pressure at the base of the updraft generated vertical motion near the surface. A vortex then formed in the gradient of the vertical velocity near the low pressure. The tilting and stretching of additional vorticity lowered the subcloud pressure field further, and the vortex built down to the surface via this process.

Another tornado-scale simulation is presented in Wicker and Wilhelmson (1995). The life cycles of two tornado-like vortices were simulated on a high-resolution grid during a 40-minute period. Figure 4.16 shows the model solution when the second vortex was at its maximum strength. Maximum ground-relative wind speeds exceeded 55 m s^{-1} for over five minutes and exceeded 60 m s^{-1} at this time. Vertical vorticity near the surface also exceeded 0.25 s^{-1} at this time. Both vortices appeared to have been generated from strong vertical motion that periodically intensified at the base of the storm. Figures 4.16e and 4.16f are a time–height plot of the maximum updraft within the storm over the 40-minute period. Two of the updraft pulses, at 83 and 96 minutes, appeared to develop first at low levels and then build upward into the storm (Fig. 4.16e). Updrafts in the storm at a height of 3 km accelerated over 10 m s^{-1} within a few minutes, immediately preceding the development of the tornado-like vortices in the boundary layer. An analysis showed that updraft accelerations were driven by rapid changes in the vertical pressure gradient aloft due to increased midlevel rotation in the storm's mesocyclone.

Analysis of the tornado vortex itself showed that the sources of vorticity were very similar to those originally presented in Klemp and Rotunno (1983). Trajectories integrated backward from the vortex (Fig. 4.17) show that parcels came from three locations around the storm. Parcels traveling within the storm's downdraft northwest of the tornado and in the inflow east of the gust front originated close to the surface and were the principal contributors to the genesis of tornado rotation. The parcels from the northwest of the vortex obtained significant horizontal vorticity along the storm's forward-flank solenoids. Parcels entering from the east also obtained horizontal vorticity along the path toward the storm's updraft. As the updraft aloft intensified, this horizontal vorticity subsequently tilted and stretched into the tornado vortex. Figure 4.17b also shows that some air descends rapidly to a height of 1 km prior to entering the vortex. However, the analysis indicated that these parcels do not significantly contribute to the vorticity field; rather, the

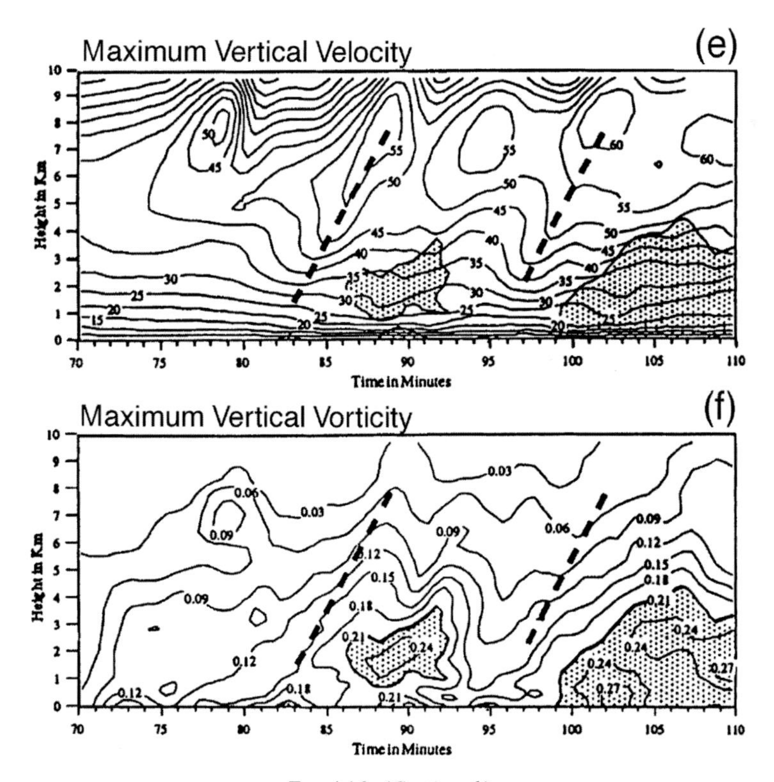

FIG. 4.16. (*Continued.*)

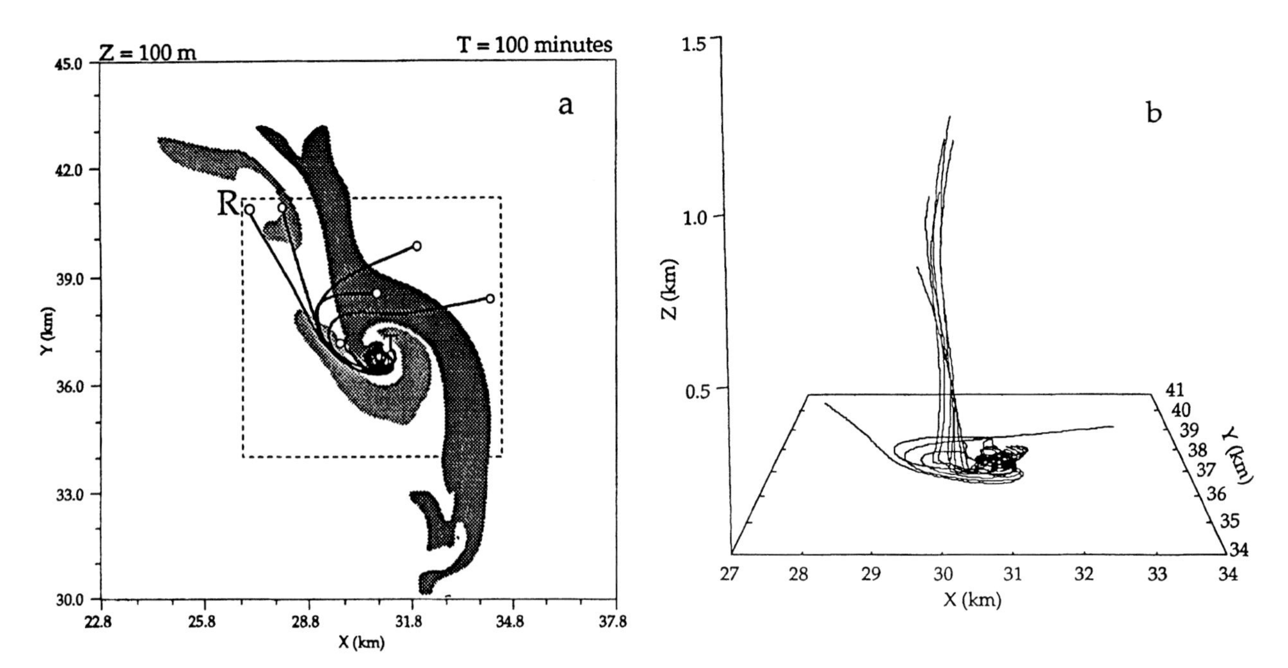

FIG. 4.17. (a) Plot of vertical velocity at 100 min for z = 100 m. Darkly shaded regions are updrafts greater than 2 m s^{-1}. Lightly shaded regions are downdrafts less than -2 m s^{-1}. Overlaid are backward trajectories beginning at 100 min (filled-in circles) and ending at 92.5 min (open circles). The position of the tornado at 100 min is marked "T." The dashed square indicates the region used for the three-dimensional plot in (b). (b) Three-dimensional trajectory plot of trajectories from 100 min backward to 92.5 min. From Wicker and Wilhelmson (1995).

parcels are accelerated downward by the adverse vertical pressure gradients associated with the vortex circulation. Tornado demise occurs when the updraft over the tonardo weakens significantly. At this time, the occlusion downdraft surrounds the tornado and associated low-level divergence. Cut off from its source of positive vorticity, the vortex then dissipates.

More recent work by Finley et al. (1998a) demonstrates the potentially wide range of situations in which tornadoes form. Their simulations produced two tornado-like vortices within a high-precipitation supercell. These vortices were very different from those simulated by Wicker and Wilhelmson (1995), as two weak vortices developed along the flanking line of the high-precipitation supercell and were not associated with the storm's main mesocyclone. Both tornado-like vortices developed in the boundary layer and then extended upward into the storm. This mode of development is consistent with observations that show that nearly 50% of all tornado vortex signatures observed on Doppler radar develop near the surface (Trapp et al. 1999). Finley et al. also note that tornado decay in their simulation, as well as in Wicker and Wilhelmson (1995) and Grasso and Cotton (1995), does not appear realistic, as all of the vortices appear to weaken and broaden during decay. This is in contrast to many observed tornadoes, which appear to shrink rapidly.

e. Simulations of nonsupercell tornadoes

During the late 1980s, Wakimoto and Wilson (1989) and Brady and Szoke (1989) showed that not all tornadoes are associated with supercell thunderstorms. Frequently observed along the Front Range of Colorado, these nonsupercell tornadoes (NSTs) occur in convection that has no significant midlevel rotation. These vortices appear to grow in association with the development of convection over low-level convergence boundaries.

Initial dry simulation studies by Lee and Wilhelmson (1997a) demonstrated that shearing instability along a low-level convergence boundary led to the development of small circulations called *misocyclones,* and that vortex–vortex interactions led to coalescence of misocyclones into larger ones. Development of misocyclones occurred when the outflow encountered low-level winds parallel to it. Lee and Wilhelmson (1997b) showed numerically that these misocyclones contract to form NSTs by including moisture and cloud physics into their earlier model (Fig. 4.18a). Early in the simulation, shearing instability created seven misocyclones along the low-level outflow. The merging and intensification of the misocyclones created vertical motion generated by the secondary circulation due to surface friction. This intensified the convection over the misocyclone, further intensifying the low-level convergence. As the convection intensified, vertical vorticity was transported upward from the boundary layer misocyclone by the cloud's updraft. The stretching associated with this process was sufficient to generate the NST. The simulated tornado-like vortices had F1 winds lasting over six minutes. Figure 4.18b presents the complete evolution associated with NST development. Stage V is noteworthy because it shows that, as the NST matures, the cold pool created by the storm's precipitation downdraft intensifies the NST through generation of horizontal vorticity that is subsequently tilted and stretched. The cold outflow also increases the low-level convergence. While the formation of the downdraft and cold pool acted to initially intensify the NST, it ultimately led to its demise as the outflow eventually moved the base of the NST away from the updraft into the downdraft.

4.7. Simulating storms in the new millennium

At the turn of the century, the computational infrastructure needed to advance our understanding of severe local storms and to simulate their behavior for forecasting purposes continues to evolve rapidly. Talk of supercomputers has now been replaced by talk of computational grids interconnected by high-speed networks that allow access to computers through global operating systems (Foster and Kesselman 1999). Computer scientists are hard at work developing the underlying computational and grid infrastructure, while computational scientists are working on the infrastructure necessary to couple models whose execution can be distributed on the grid. One such collaborative team effort, focusing on environmental hydrology, is being pursued within the NCSA Alliance.[23] It involves an interdisciplinary team of researchers with specialties in fluid flow, the atmospheric and hydrologic sciences, ecology, geographical information systems, and visualization. The goal of this team is to create a computational infrastructure using new and emerging technologies, including collaborative visualization tools, to support scientists and the public in understanding the complex interactions between geophysical and ecologic components of the environment.[24] One of the focus problems relevant to severe storms is flash flooding.

In addition, individual models are being rewritten to perform billion grid zone calculations on today's highly parallel computers and tomorrow's petaflop machines. Such efforts will open new doors of exploration where the microphysical and electrical structure of local storms can be pursued aggressively, where a U.S. local storm forecast model with 1-km horizontal resolution is feasible, and where multiple-scale resolving simulations are possible. Storms will no longer be treated in isolation from mesoscale variability (many simulations are still being carried out using a single sounding for the initial conditions).

The incorporation of mesoscale variability can be handled through introduction of more complex initial states into convective models or through use of adaptive mesh refinement or nesting strategies to span regional to convective scales. An example of the former approach was discussed earlier in this chapter (Atkins et al. 1999), where storms crossing initially specified boundaries were studied. The latter approach is being pursued in the forecasting arena by CAPS as already indicated and in the research community by a growing number of researchers. For example, one of the authors (Wilhelmson) is involved in several case studies aimed at representing convection within regional and mesoscale environments including nested-grid simulations of Hurricane Opal (inner grid 1.1 km), a tornadic outbreak in Illinois in 1996, and a study of the 1996 heavy rain event in northern Illinois.[25] These efforts exemplify the challenges faced by modelers in the twenty-first century, which include the need for ample and good data for initialization and data assimilation coupled with high-performance modeling systems. The former is needed in order to provide the stage for model storm evolution from hours to days. It is important in the tornadic outbreak and heavy flooding simulations noted above to capture the overall system evolution accurately in order to study the particular convective events of interest. For example, in the latter case study, heavy rain was observed over northern Illinois associated with a larger-scale system in which storms formed repeatedly in and moved over the same area (i.e., training).

Improvements in model dynamics and physics are also needed in multiscale simulations as well as in studies that are directed primarily at the convective scale. New investigations directed at microphysical and electrical parameterizations are needed, as well as future investigation into subgrid parameterizations appropriate for storm environments. One such effort in its early stages is the Weather Research and Forecast (WRF) model project.[26] This project is developing the next-generation mesoscale forecast model and assimilation system that will advance both the understanding and prediction of important mesoscale weather, and will promote closer ties between the research and operational forecasting communities. The model is being developed as a collaborative effort among the National Center for Atmospheric Research (NCAR), the Forecast Systems Laboratory and National Centers

[23] http://alliance.ncsa.uiuc.edu/

[24] http://www.ncsa.uiuc.edu/alliance/partners/ApplicationTechnologies/EnvironmentalHydrology.html

[25] http://redrock.ncsa.uiuc.edu/AOL/home.html

[26] http://www.wrf-model.org

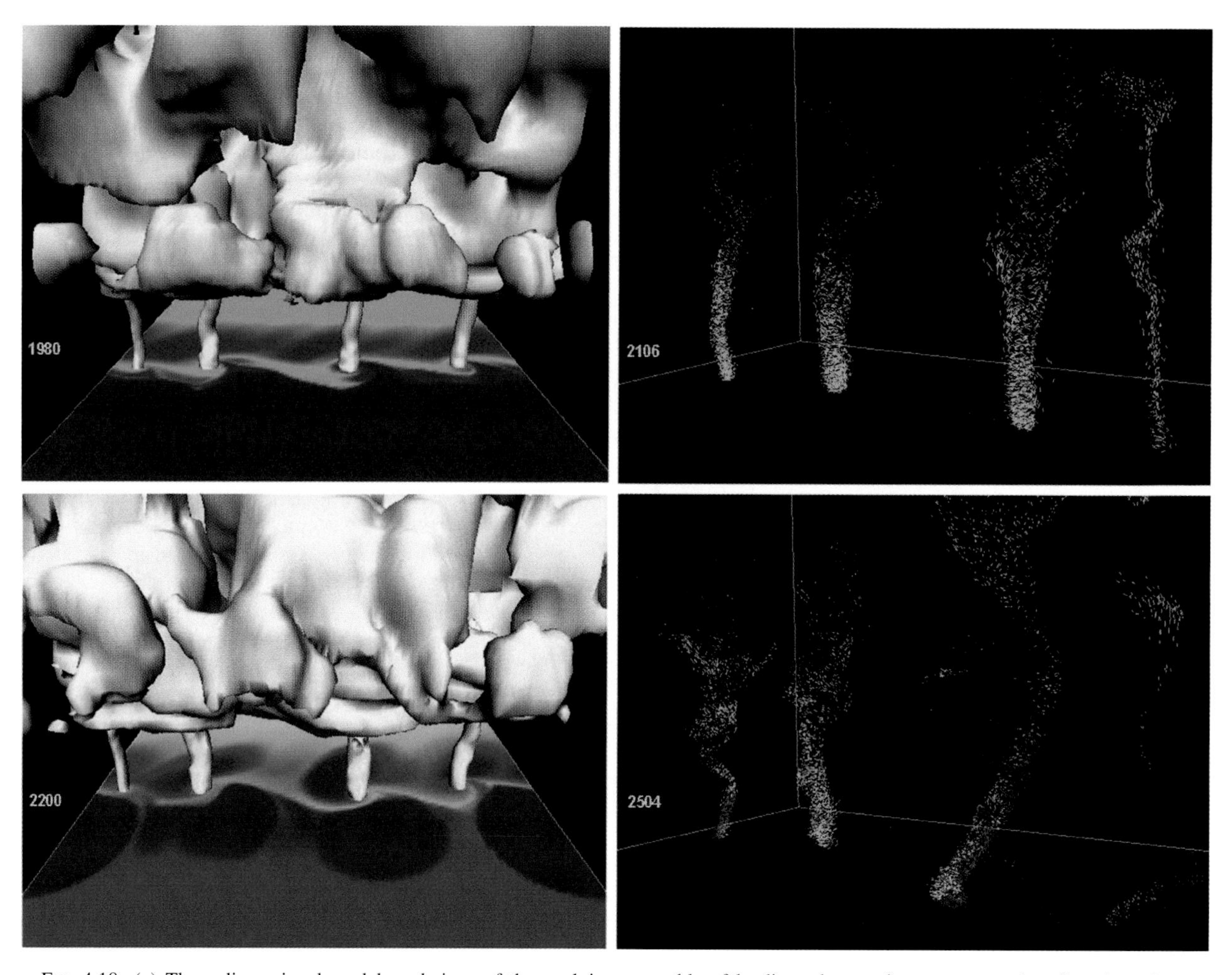

FIG. 4.18. (a) Three-dimensional model renderings of the evolving ensemble of leading edge vortices, storm, and outflow boundary at 1980 and 2200 s for four nonsupercell tornadoes (NSTs). At the domain base the blue color spectrum delineates the outflow boundary (darker = colder) in the left two panels. The yellow vertical vortex tubes are representative of vertical vorticity greater than 0.1 s^{-1}. The gray-scaled cloud isosurface represents cloud water greater than 0.2 g kg^{-1}. In the right two panels, massive particle trajectory visualization is used to display the flow at 2106 and 2504 s. The green and red color coding for the 6500 weightless particles indicates whether the particles are moving upward or downward, respectively. The viewing perspective in the left two panels is from an elevated position looking east while the viewing perspective for the right two panels is from an elevated position looking southwest.

for Environmental Prediction of the National Oceanic and Atmospheric Administration (FSL, NCEP/NOAA), the Center for Analysis and Prediction of Storms (CAPS) at the University of Oklahoma, and the Air Force Weather Agency (AFWA). Scientists from a number of other universities and organizations are also collaborating in this project. Research efforts directed at the building of WRF include exploration of various dynamic formulations solved with time-split and implicit solution techniques. The model is being constructed to run on single to hundreds of computer processors. The final community model will incorporate advanced numerical methods and data assimilation techniques, multiple relocatable nesting capability, and improved physics, particularly for treatment of convection and mesoscale precipitation systems.

The time and effort spent in model development and subsequent research are rewarding to serious modelers who recognize that the virtual world they create and use has many similarities to the real world that they live in. This virtual world can be changed by inclusion/exclusion of different physical processes and continues to afford a nonthreatening methodology for exploring and simplifying the complexity of severe local storms. As high-performance computing power continues to increase, the links between various scales of motion will continue to unfold. Coupled with assimilation of new data sources, numerically aided forecasts should continue to improve. In the new millennium it will become feasible to use high resolution (1 km or less) over the United States as needed. The virtual modeling world of the twenty-first century

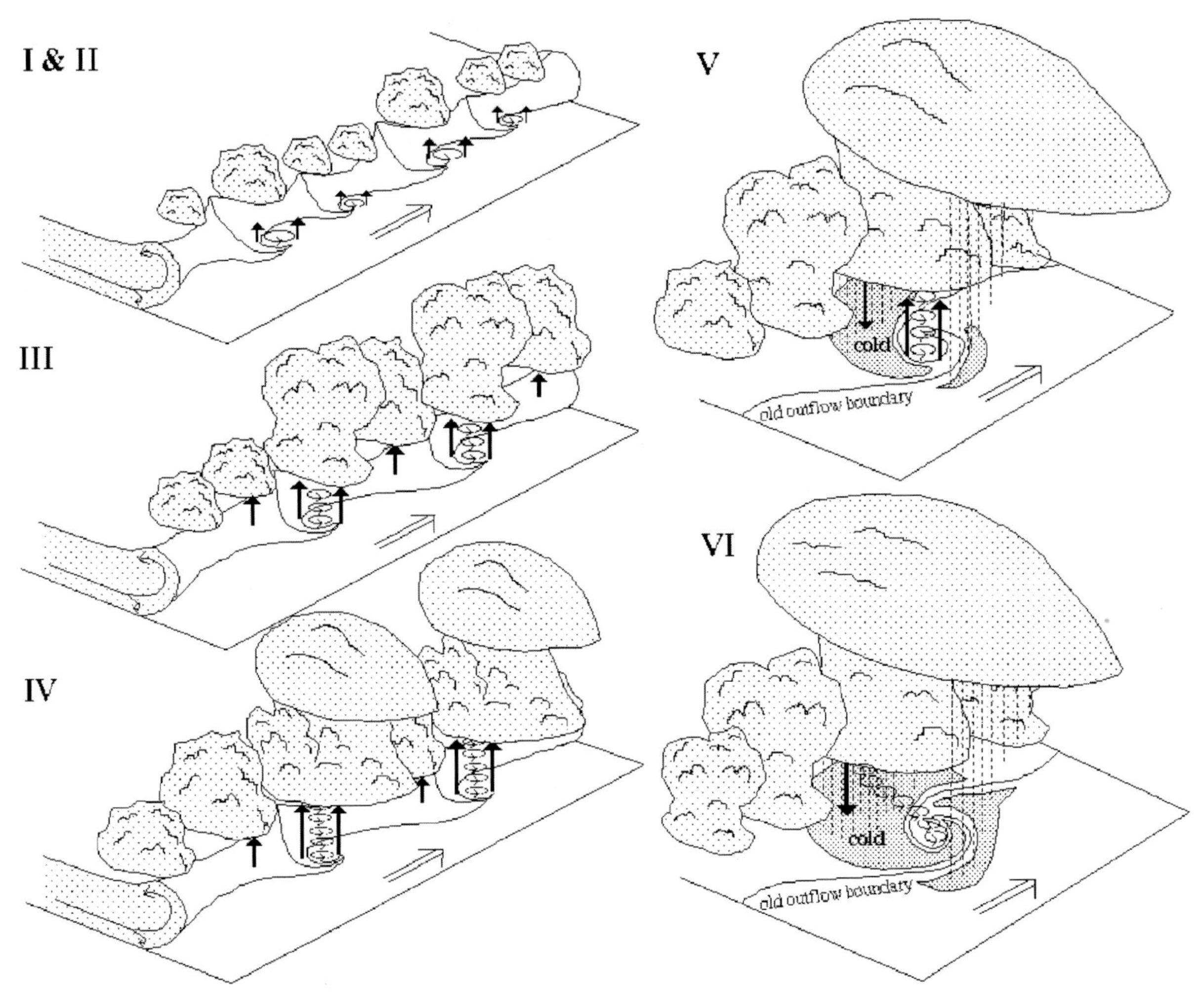

FIG. 4.18. (*Continued.*) (b) Schematic presentation of the life cycle stages of NST evolution. Stage I, vortex sheet development; stage II, vortex sheet rollup; stage III, misocyclone interaction and merger; stage IV, early mature NST; stage V, late morning NST; and stage VI, dissipation. The diagrams in stages V and VI focus on just one member of the NST family. The viewing perspective is from an elevated position looking northwest. From Lee and Wilhelmson (1997b).

will bring us new understanding as we continue to explore the world around us and how the negative impact of natural hazards can be reduced.

Acknowledgments. This review was supported by the National Science Foundation under Grants ATM 9633228, ATM-9318914, ATM-9707815, and ATM-0000412. The review panel (Dale Durran, Greg Tripoli, and Morris Weisman) headed by Joseph Klemp provided many useful suggestions. Thanks also to Brian Jewett and Steve Peckham for proofreading the review and to Crystal Shaw for her help with the figures.

REFERENCES

Achtemeier, G. L., 1969: Some observations of splitting thunderstorms over Iowa on August 25–26, 1965. Preprints, *Sixth Conf. on Severe Local Storms,* Chicago, IL, Amer. Meteor. Soc., 89–94.

Adams, J., R. Garcia, B. Gross, J. Hack, D. Haidvogel, and V. Pizzo, 1992: Applications of multigrid software in the atmospheric sciences. *Mon. Wea. Rev.,* **120,** 1447–1458.

Adlerman, E. J., K. K. Droegemeier, and R. Davies-Jones, 1999: A numerical simulation of cyclic mesocyclogenesis. *J. Atmos. Sci.,* **56,** 2045–2069.

Atkins, N. T., M. L. Weisman, and L. J. Wicker, 1999: The influence of preexisting boundaries on supercell evolution. *Mon. Wea. Rev.,* **127,** 2910–2927.

Benoit, R., M. Desgagne, P. Pellerin, S. Pellerin, and Y. Chartier, 1997: The Canadian MC2: A semi-Lagrangian, semi-implicit wideband atmospheric model suited for finescape process studies and simulation. *Mon. Wea. Rev.,* **125,** 2382–2415.

Bernardet, L. R., and W. R. Cotton, 1998: Multiscale evolution of a derecho-producing mesoscale convective system. *Mon. Wea. Rev.,* **126,** 2991–3015.

Bluestein, H. B., and G. R. Woodall, 1990: Doppler radar analysis of a low-precipitation severe storm. *Mon. Wea. Rev.,* **118,** 1640–1664.

Brady, R. H., and E. J. Szoke, 1989: A case study of non-mesocyclone tornado development in northeast Colorado: Similarities to waterspout formation. *Mon. Wea. Rev.,* **117,** 843–856.

Brooks, H. E., and R. B. Wilhelmson, 1992: Numerical simulation of a low-precipitation supercell thunderstorm. *Meteor. Atmos. Phys.,* **49,** 3–17.

———, D. J. Stensrud, and J. V. Cortinas Jr., 1993a: The use of mesoscale models to initialize cloud-scale models for convective forecasting. Preprints, *13th Conf. on Weather Analysis and Forecasting,* Vienna, VA, Amer. Meteor. Soc., 301–304.

———, L. J. Wicker, and C. A. Doswell III, 1993b: STORMTIPE: A forecasting experiment using a three-dimensional cloud model. *Wea. Forecasting,* **8,** 352–362.

———, C. A. Doswell III, and J. Cooper, 1994: On the environments of tornadic and nontornadic mesocyclones. *Wea. Forecasting,* **9,** 606–618.

Brown, R. A., Ed., 1976: The Union City, Oklahoma tornado of 24 May 19973. NOAA Tech. Memo. ERL NSSL-80, National Severe Storms Laboratory, Norman, OK.

Browning, K. A., 1964: Airflow and precipitation trajectories within severe local storms which travel to the right of the winds. *J. Atmos. Sci.,* **21,** 634–639.

———, and R. J. Donaldson Jr., 1963: Airflow structure of a tornadic storm. *J. Atmos. Sci.,* **20,** 533–545.

Burgess, D. W., and R. P. Davies-Jones, 1979: Unusual tornadic storms in eastern Oklahoma on 5 December 1975. *Mon. Wea. Rev.,* **107,** 451–457.

———, V. T. Wood, and R. A. Brown, 1982: Mesocyclone evolution statistics. Preprints, *12th Conf. on Severe Local Storms,* San Antonio, TX, Amer. Meteor. Soc., 422–424.

Carpenter, R. L., Jr., K. K. Droegemeier, P. R. Woodward, and C. E. Hane, 1990: Application of the Piecewise Parabolic Method (PPM) to meteorological modeling. *Mon. Wea. Rev.,* **118,** 586–612.

Chang, C.-Y., and M. Yoshizaki, 1993: Three-dimensional modeling study of squall lines observed in COPT81. *J. Atmos. Sci.,* **50,** 161–183.

Charba, J., and Y. Sasaki, 1971: Structure and movement of the severe thunderstorms of 3 April 1964 as revealed from radar and surface mesonetwork data analysis. *J. Meteor. Soc. Japan,* **49,** 191–213.

Chen, C., 1995: Numerical simulations of gravity currents in uniform shear flows. *Mon. Wea. Rev.,* **123,** 3240–3253.

Chen, C.-H., and H. D. Orville, 1980: Effects of mesoscale convergence on cloud convection. *J. Appl. Meteor.,* **19,** 256–274.

Chin, H.-N. S., and R. B. Wilhelmson, 1998: Evolution and structure of tropical squall line elements within a moderate CAPE and strong low-level jet environment. *J. Atmos. Sci.,* **55,** 3089–3113.

———, Q. Fu, M. M. Bradley, and C. R. Molenkamp, 1995: Modeling of a tropical squall line in two dimensions: Sensitivity to radiation and comparison with a midlatitude case. *J. Atmos. Sci.,* **52,** 3172–3193.

Clark, T. L., 1973: Numerical modeling of the dynamics and microphysics of warm cumulus convection. *J. Atmos. Sci.,* **30,** 857–878.

———, 1977: A small-scale dynamic model using a terrain-following coordinate transformation. *J. Comput. Phys.,* **24,** 186–215.

———, 1979: Numerical simulations with a three-dimensional cloud model: Lateral boundary condition experiments and multicellular severe storm simulations. *J. Atmos. Sci.,* **36,** 2191–2215.

———, and R. D. Farley, 1984: Severe downslope windstorm calculations in two and three spatial dimensions using anelastic interactive grid nesting: A possible mechanism for gustiness. *J. Atmos. Sci.,* **41,** 329–350.

———, W. D. Hall, and J. L. Coen, 1996: Source code documentation for the Clark-Hall cloud-scale model: Code version G3CH01. NCAR Tech. Note NCAR/TN-426+STR, 137 pp. [Available from NCAR Information Service, P.O. Box 3000, Boulder, CO 80307.]

Cotton, W. R., and R. A. Anthes, 1989: *Storm and Cloud Dynamics.* Academic Press, 883 pp.

———, M. A. Stephens, T. Nehrkorn, and G. J. Tripoli, 1982: The Colorado State University three-dimensional mesoscale model. Part II: An ice phase parameterization. *J. Rech. Atmos.,* **16,** 295–320.

Crook, N. A., and M. W. Moncrieff, 1988: The effect of large-scale convergence on the generation and maintenance of deep moist convection. *J. Atmos. Sci.,* **45,** 3606–3624.

———, and J. D. Tuttle, 1994: Numerical simulations initialized with radar-derived winds. Part II: Forecasts of three gust-front cases. *Mon. Wea. Rev.,* **122,** 1204–1217.

———, and M. L. Weisman, 1998: Comparison of supercell behavior in a convective boundary layer with that in horizontally-homogeneous environment. Preprints, *19th Conf. on Severe Local Storms,* Minneapolis, MN, Amer. Meteor. Soc., 253–256.

Cullen, M. J. P., 1990: A test of a semi-implicit integration technique for a fully compressible non-hydrostatic model. *Quart. J. Roy. Meteor. Soc.,* **116,** 1253–1258.

Davies-Jones, R. P., D. W. Burgess, and M. P. Foster, 1990: Test of helicity as a tornado forecast parameter. Preprints, *16th Conf. on Severe Local Storms,* Kananaskis Park, AB, Canada, Amer. Meteor. Soc., 588–592.

Deardorff, J. W., 1972: Numerical investigation of neutral and unstable planetary boundary layers. *J. Atmos. Sci.,* **29,** 91–115.

Droegemeier, K. K., and R. B. Wilhelmson, 1986: Kelvin-Helmholtz instability in a numerically simulated thunderstorm outflow. *Bull. Amer. Meteor. Soc.,* **67,** 416–417.

———, and ———, 1987: Numerical simulation of thunderstorm outflow dynamics. Part I: Outflow sensitivity experiments and turbulence dynamics. *J. Atmos. Sci.,* **44,** 1180–1210.

———, and R. P. Davies-Jones, 1987: Simulation of thunderstorm microbursts with a supercompressible numerical model. Preprints, *Fifth Int. Conf. on Numerical Methods in Laminar and Turbulent Flow,* Montreal, PQ, Canada, Amer. Meteor. Soc.

———, S. M. Lazarus, and R. Davies-Jones, 1993: The influence of helicity on numerically simulated convective storms. *Mon. Wea. Rev.,* **121,** 2005–2029.

———, and Coauthors, 1996a: Realtime numerical prediction of storm-scale weather during VORTEX '95, Part I: Goals and methodology. Preprints, *18th Conf. on Severe Local Storms,* San Francisco, CA, Amer. Meteor. Soc., 6–10.

———, and Coauthors, 1996b: The 1996 CAPS spring operational forecasting period—Realtime storm-scale NWP, Part I: Goals and methodology. Preprints, *11th Conf. on Numerical Weather Prediction,* Norfolk, VA, Amer. Meteor. Soc., 294–296.

———, and Coauthors, 1999: The explicit numerical prediction of an intense hailstorm using WSR-88D observations: The need for realtime access to Level II data and plans for a prototype acquisition system. Preprints, *15th Int. Conf. on Interactive Information Processing Systems (IIPS) for Meteorology, Oceanography, and Hydrology,* Dallas, TX, Amer. Meteor. Soc., 295–299.

Dudhia, J., 1993: A nonhydrostatic version of the Penn State-NCAR Mesoscale Model: Validation tests and simulations of an Atlantic cyclone and cold front. *Mon. Wea. Rev.,* **121,** 1493–1513.

Durran, D. R., 1989: Improving the anelastic approximation. *J. Atmos. Sci.,* **46,** 1453–1461.

———, 1990: Reply. *J. Atmos. Sci.,* **47,** 1819–1820.

———, 1999: *Numerical Methods for Wave Equations in Geophysical Fluid Dynamics.* Springer-Verlag, 468 pp.

Emanuel, K. A., 1994: *Atmospheric Convection.* Oxford University Press, 580 pp.

Ferrier, B. S., 1994: A double-moment multiple-phase four-class bulk ice scheme. Part I: Description. *J. Atmos. Sci.,* **51,** 249–280.

Fiedler, B. H., 1994: The thermodynamic speed limit and its violation in axisymmetric numerical simulations of tornado-like vortices. *Atmos.–Ocean,* **32,** 335–339.

———, 1995: On modeling tornadoes in isolation from the parent storm. *Atmos.–Ocean,* **33,** 501–512.

———, 1998: Windspeed limits in numerically-simulated tornadoes with suction vortices. *Quart. J. Roy. Meteor. Soc.,* **124,** 2377–2392.

———, and R. Rotunno, 1986: A theory for maximum wind speeds in tornado-like vortices. *J. Atmos. Sci.,* **43,** 2328–2340.

Finley, C. A., W. R. Cotton, and R. A. Pielke, 1998a: Numerical simulation of two tornadoes produced by a high-precipitation supercell. Preprints, *19th Conf. on Severe Local Storms,* Minneapolis, MN, Amer. Meteor. Soc., 206–209.

———, ———, and ———, 1998b: Secondary vortex development in a tornado vortex produced by a simulated supercell thunderstorm. Preprints, *19th Conf. on Severe Local Storms,* Minneapolis, MN, Amer. Meteor. Soc., 359–362.

Flora, S. D., 1954: *Tornadoes of the United States.* University of Oklahoma Press, 194 pp.

Foster, I., 1995: *Designing and Building Parallel Programs.* Addison-Wesley, 381 pp.

———, and C. Kesselman, 1999: Computational Grids. *The Grid: Blueprint for a New Computing Intrastructure,* J. Foster and C. Kesselman, Eds., Morgan Kaufmann Publishers, 15–50.

Fovell, R. G., and P.-H. Tan, 1998: The temporal behavior of numerically simulated multicell-type storms. Part II: The convective cell life cycle and cell regeneration. *Mon. Wea. Rev.,* **126,** 551–577.

Fox, D. G., 1972: Numerical simulation of three-dimensional, shape-preserving convective elements. *J. Atmos. Sci.,* **29,** 322–341.

Fox-Rabinovitz, M. S., 1996: Computational dispersion properties of 3D staggered grids for a nonhydrostatic anelastic system. *Mon. Wea. Rev.,* **124,** 498–510.

Fujita, T. T., 1971: Proposed mechanisms of suction spots accompanied by tornadoes. Preprints, *Seventh Conf. on Severe Local Storms,* Kansas City, MO, Amer. Meteor. Soc., 208–213.

———, and H. Grandoso, 1968: Split of a thunderstorm into anticyclonic and cyclonic storms and their motion as determined from numerical model experiments. *J. Atmos. Sci.,* **25,** 416–439.

Gal-Chen, T., 1978: A method for the initialization of the anelastic equations: Implications for matching models with observations. *Mon. Wea. Rev.,* **106,** 587–606.

Gao, J., M. Xue, A. Shapiro, and K. K. Droegemeier, 1999: A variational method for the analysis of three-dimensional wind fields from two Doppler radars. *Mon. Wea. Rev.,* **127,** 2128–2142.

Gilmore, M., and L. J. Wicker, 1998: The influence of mid-tropospheric dryness on supercell morphology and evolution. *Mon. Wea. Rev.,* **126,** 943–958.

Grabowski, W. W., 1998: Toward cloud resolving modeling of large-scale tropical circulations: A simple cloud microphysics parameterization. *J. Atmos. Sci.,* **55,** 3283–3298.

Grasso, L. D., and W. R. Cotton, 1995: Numerical simulation of a tornado vortex. *J. Atmos. Sci.,* **52,** 1192–1203.

———, and ———, 1998: Numerical simulation of the May 15, 1991 Laverne, Oklahoma tornado. Preprints, *19th Conf. on Severe Local Storms,* Minneapolis, MN, Amer. Meteor. Soc., 278–282.

———, and E. R. Hilgendorf, 2001: Observations of a severe left moving thunderstorm. *Wea. Forecasting,* **16,** 500–511.

Hane, C. E., 1973: The squall line thunderstorm: Numerical experimentation. *J. Atmos. Sci.,* **30,** 1672–1690.

Helsdon, J., Jr., and R. D. Farley, 1987: A numerical modeling study of a Montana thunderstorm: 2. Model results versus observations involving electrical aspects. *J. Geophys. Res.,* **92** (D5), 5661–5675.

Hibbard, W., and D. Santek, 1990: The Vis5D system for easy interactive visualization. *Proc. Visualization '90,* San Francisco, CA, IEEE, 28–35.

———, B. E. Paul, D. A. Santek, C. R. Dyer, A. L. Battaiola, and M.-F. Voidrot-Martinez, 1994: Interactive visualization of earth and space science computations. *Computer,* **27,** 65–72.

Holt, T., and S. Raman, 1988: A review and comparative evaluation of multilevel boundary layer parameterizations for first-order and turbulent kinetic energy closure schemes. *Rev. Geophys.,* **26,** 761–780.

Hou, D., E. Kalnay, and K. K. Droegemeier, 2001: Objective verification of the SAMEX '98 ensemble forecasts. *Mon. Wea. Rev.,* **129,** 73–91.

Houze, R. A., Jr., 1993: *Cloud Dynamics.* Academic Press, 573 pp.

———, and P. V. Hobbs, 1982: Organization and structure of precipitating cloud systems. *Advances in Geophysics,* Vol. 24, Academic Press, 225–315.

———, W. Schmid, R. G. Fovell, and H.-H. Schiesser, 1993: Hailstorms in Switzerland: Left movers, right movers, and false hooks. *Mon. Wea. Rev.,* **121,** 3345–3370.

Howells, P., R. Rotunno, and R. K. Smith, 1988: A comparative study of atmospheric and laboratory-analogue numerical tornado-vortex models. *Quart. J. Roy. Meteor. Soc.,* **114,** 801–822.

Hsie, E. Y., R. D. Farley, and H. D. Orville, 1980: Numerical simulation of ice phase convective cloud seeding. *J. Appl. Meteor.,* **19,** 950–977.

Janish, P. R., K. K. Droegemeier, M. Xue, K. Brewster, and J. Levit, 1995: Evaluation of the advanced regional prediction system (ARPS) for storm-scale modeling applications in operational forecasting. *Proc. 14th Conf. on Weather Analysis and Forecasting,* Dallas, TX, Amer. Meteor. Soc., 224–229.

Johns, R. H., and C. A. Doswell, 1992: Severe local storms forecasting. *Wea. Forecasting,* **7,** 588–612.

Johnson, D. E., P. K. Wang, and J. M. Straka, 1994: A study of microphysical processes in the 2 August 1981 CCOPE supercell storm. *Atmos. Res.,* **33,** 93–123.

Kaufmann, W. J., III, and L. L. Smarr, 1993: *Supercomputing and the Transformation of Science.* Scientific American Library, 238 pp.

Kay, M. P., and L. J. Wicker, 1998: Numerical simulations of supercell interactions with thermal boundaries. Preprints, *19th Conf. on Severe Local Storms,* Minneapolis, MN, Amer. Meteor. Soc., 246–248.

Keller, D., and B. Vonnegut, 1976: Wind speeds required to drive straws and splinters into wood. *J. Appl. Meteor.,* **59,** 899–901.

Kennedy, P. C., N. E. Westcott, and R. W. Scott, 1993: Single-Doppler radar observations of a mini supercell tornadic thunderstorm. *Mon. Wea. Rev.,* **121,** 1860–1870.

Kessler, E., 1969: *On the Distribution and Continuity of Water Substance in Atmospheric Circulation. Meteor. Monogr.,* No. 32, Amer. Meteor. Soc., 84 pp.

Klemp, J. B., 1987: Dynamics of tornadic thunderstorms. *Ann. Rev. Fluid Mech.,* **19,** 369–402.

———, and R. Wilhelmson, 1978a: The simulation of three-dimensional convective storm dynamics. *J. Atmos. Sci.,* **35,** 1070–1096.

———, and ———, 1978b: Simulations of right- and left-moving storms produced through storm splitting. *J. Atmos. Soc.,* **35,** 1097–1110.

———, and D. R. Durran, 1983: An upper boundary condition permitting internal gravity wave radiation in numerical mesoscale models. *Mon. Wea. Rev.,* **111,** 430–444.

———, and R. Rotunno, 1983: A study of the tornadic region within a supercell thunderstorm. *J. Atmos. Sci.,* **40,** 359–377.

———, R. B. Wilhelmson, and P. S. Ray, 1981: Observed and numerically simulated structure of a mature supercell thunderstorm. *J. Atmos. Sci.,* **38,** 1558–1580.

———, R. Rotunno, and W. C. Skamarock, 1997: On the propagation of internal bores. *J. Fluid Mech.,* **331,** 81–106.

Krueger, S. K., Q. Fu, K. N. Liou, and H.-N. S. Chin, 1995: Improvements of an ice-phase microphysics parameterization for use in numerical simulations of tropical convection. *J. Appl. Meteor.,* **34,** 281–287.

Kulie, M. S., and Y.-L. Lin, 1998: The structure and evolution of a numerically simulated high-precipitation supercell thunderstorm. *Mon. Wea. Rev.,* **126,** 2090–2116.

Lee, B. D., and R. B. Wilhelmson, 1997a: The numerical simulation of nonsupercell tornadogenesis. Part I: Initiation and evolution of pre-tornadic misocyclone circulations along a dry outflow boundary. *J. Atmos. Sci.,* **54,** 32–60.

———, and ———, 1997b: The numerical simulation of nonsupercell tornadogenesis. Part II: Evolution of a family of tornadoes along a weak outflow boundary. *J. Atmos. Sci.,* **54,** 2387–2415.

Lemon, L. R., and C. A. Doswell III, 1979: Severe thunderstorm evolution and mesocyclone structure as related to tornadogenesis. *Mon. Wea. Rev.,* **107,** 1184–1197.

Lewellen, D. C., W. S. Lewellen, and J. Xia, 2000: The influence of a local swirl ratio on tornado intensification near the surface. *J. Atmos. Sci.,* **57,** 527–544.

Lewellen, W. S., 1976: Theoretical models of the tornado vortex. *Symp. on Tornadoes: Assessment of Knowledge and Implications for Man,* Lubbock, TX, Texas Tech. University, 107–143.

———, 1993: Tornado vortex theory. *The Tornado: Its Structure, Dynamics, Predictions, and Hazards, Geophys. Monogr.,* No. 79, Amer. Geophys. Union, 19–40.

———, and Y. P. Sheng, 1980: *Modeling Tornado Dynamics.* U.S. Nuclear Regulatory Commission, NTIS NUREG/CR-2585.

———, and D. C. Lewellen, 1997: Large-eddy simulation of a tornado's interaction with the surface. *J. Atmos. Sci.,* **54,** 581–605.

Lilly, D. K., 1962: On the numerical simulation of buoyant convection. *Tellus,* **XIV,** 148–172.

———, 1969: Tornado dynamics. NCAR Manuscript 69–117, 39 pp. [Available from NCAR, P.O. Box 3000, Boulder, CO 80307.]

———, 1975: Severe storms and storm systems: Scientific background, methods, and critical questions. *Pure Appl. Geophys.,* **113,** 713–734.

———, 1979: The dynamical structure and evolution of thunderstorms and squall lines. *Ann. Rev. Earth Planet. Sci.,* **7,** 117–161.

———, 1990: Numerical prediction of thunderstorms—Has its time come? *Quart. J. Roy. Meteor. Soc.,* **116,** 779–797.

Lin, Y.-L., R. D. Farley, and H. D. Orville, 1983: Bulk parameterization of the snow field in a cloud model. *J. Climate Appl. Meteor.,* **22,** 1065–1092.

Liu, C. L., and M. W. Moncrieff, 1996: A numerical study of the effects of ambient flow and shear on density currents. *Mon. Wea. Rev.,* **124,** 2282–2303.

Lord, S. J., H. E. Willoughby, and J. M. Piotrowicz, 1984: Role of a parameterized ice-phase microphysics in an axisymmetric, nonhydrostatic tropical cyclone model. *J. Atmos. Sci.,* **41,** 2836–2848.

Mansell, E. R., 2000: Electrification and lightening in simulated supercell and non-supercell thunderstorms. Ph.D. dissertation, Dept. of Physics, University of Oklahoma, 211 pp.

Markowski, P. M., J. M. Straka, E. N. Rasmussen, and D. O. Blanchard, 1998: Variability of storm-relative helicity during VORTEX. *Mon. Wea. Rev.,* **126,** 2959–2971.

McCaul, E. W., Jr., 1987: Observations of the Hurricane Danny tornado outbreak of 16 August 1985. *Mon. Wea. Rev.,* **115,** 1206–1223.

———, 1993: Observations and simulations of hurricane-spawned tornadic storms. *The Tornado: Its Structure, Dynamics, Prediction, and Hazards, Geophys. Monogr.,* No. 79, Amer. Geophys. Union, 119–142.

———, and M. L. Weisman, 1996: Simulations of shallow supercell storms in landfalling hurricane environments. *Mon. Wea. Rev.,* **124,** 408–429.

Mendez-Nunez, L. R., and J. J. Carroll, 1994: Application of the MacCormack scheme to atmospheric nonhydrostatic models. *Mon. Wea. Rev.,* **122,** 984–1000.

Mesinger, F. M., 1977: The forward-backward scheme and its use in a limited-area model. *Contrib. Atmos. Phys.,* **50,** 200–210.

Meyers, M. P., P. J. DeMott, and W. R. Cotton, 1992: New primary ice nucleation parameterizations in an explicit cloud model. *J. Appl. Meteor.,* **31,** 26–50.

Michalakes, J., 2000: The same-source parallel implementation of MM5. *J. Sci. Computing,* **8,** (1), 5–12.

Moller, A. R., C. A. Doswell, III, M. P. Foster, and G. R. Woodall, 1994: The operational recognition of supercell thunderstorm environments and storm structures. *Wea. Forecasting,* **9,** 327–347.

Moncrieff, M. W., and J. S. A. Green, 1972: The propagation and transfer properties of steady convective overturning in shear. *Quart. J. Roy. Meteor. Soc.,* **98,** 336–352.

Ogura, Y., 1963: A review of numerical modeling research on small scale convection in the atmosphere. *Severe Local Storms, Meteor. Monogr.,* No. 27, Amer. Meteor. Soc., 65–75.

———, and J. C. Charney, 1962: A numerical model of thermal convection in the atmosphere. *Proc. Int. Symp. Numerical Weather Prediction,* Tokyo, Japan, Meteor. Soc. Japan, 431–451.

———, and N. A. Phillips, 1962: Scale analysis of deep and shallow convection in the atmosphere. *J. Atmos. Sci.,* **19,** 173–179.

Ooyama, K. V., 1990: A thermodynamic foundation for modeling the moist atmosphere. *J. Atmos. Sci.,* **47,** 2580–2593.

Oreskes, N., K. Shrader-Frechette, and K. Belitz, 1994: Verification, validation, and confirmation of numerical models in the earth sciences. *Science,* **263,** 641–646.

Orville, H. D., 1968: Ambient wind effects on the initiation and development of cumulus clouds over mountains. *J. Atmos. Sci.,* **25,** 385–403.

Pandya, R. E., and D. R. Durran, 1996: The influence of convectively generated thermal forcing on the mesoscale circulation around squall lines. *J. Atmos. Sci.,* **53,** 2924–2951.

Petch, J. C., 1998: Improved radiative transfer calculations from information provided by bulk microphysical schemes. *J. Atmos. Sci.,* **55,** 1846–1858.

Pielke, R. A., and Coauthors, 1992: A comprehensive meteorological modeling system—RAMS. *Meteor. Atmos. Phys.,* **9,** 69–91.

Proctor, F. H., 1987: The terminal area simulation system. Volume I: Theoretical formulation. NASA Contractor Rep. 4046, NASA, Washington, DC, 176 pp.

———, 1988: Numerical simulations of an isolated microburst: Part I: Dynamics and structure. *J. Atmos. Sci.,* **45,** 3137–3160.

Purser, R. J., and L. M. Leslie, 1991: Reducing the error in a time-split finite-difference scheme using an incremental technique. *Mon. Wea. Rev.,* **119,** 578–585.

Randall, D. A., and B. A. Wielicki, 1997: Measurements, models, and hypotheses in the atmospheric sciences. *Bull. Amer. Meteor. Soc.,* **78,** 399–406.

Rasmussen, E. N., J. M. Straka, R. Davies-Jones, C. A. Doswell III, F. H. Carr, M. D. Eilts, and D. R. MacGorman, 1994: Verifications of the Origins of Rotation in Tornadoes Experiment: VORTEX. *Bull. Amer. Meteor. Soc.,* **75,** 995–1006.

Redelsperger, J. L., and G. Sommeria, 1986: Three-dimensional simulation of a convective storm: Sensitivity studies on subgrid parameterization and spatial resolution. *J. Atmos. Sci.,* **22,** 2619–2635.

Richardson, Y. P., 1999: The influence of horizontal variations in vertical shear and low-level moisture on numerically simulated convective storms. Ph.D. dissertation, University of Oklahoma, 236 pp.

———, K. K. Drogemeier, and R. P. Davies-Jones, 1998: A study of the horizontally-varying vertical shear and CAPE on numerical simulated convective storms. Preprints, *19th Conf. on Severe Local Storms,* Minneapolis, MN, Amer. Meteor. Soc., 249–251.

Rotunno, R., 1984: An investigation of a three-dimensional asymmetric vortex. *J. Atmos. Sci.,* **41,** 283–298.

———, 1993: Supercell thunderstorm modeling and theory. *The Tornado: Its Structure, Dynamics, Prediction, and Hazards, Geophys. Monogr.,* No. 79, Amer. Geophys. Union, 57–73.

———, and J. B. Klemp, 1982: The influence of the shear-induced pressure gradient on thunderstorm motion. *J. Atmos. Sci.,* **42,** 271–292.

———, and ———, 1985: On the rotation and propagation of simulated supercell thunderstorms. *Mon. Wea. Rev.,* **110,** 136–151.

———, ———, and M. L. Weisman, 1988: A theory for strong, long-lived squall lines. *J. Atmos. Sci.,* **45,** 463–485.

Rutledge, S. A., and P. V. Hobbs, 1983: The mesoscale and microscale structure and organization of clouds and precipitation in midlatitude cyclones. Part VIII: A model for the "seeder-feeder" process in warm-frontal rainbands. *J. Atmos. Sci.,* **40,** 1185–1206.

Sathye, A., G. Bassett, K. Droegemeier, M. Xue, and K. Brewster, 1996: Experiences using high performance computing for operational storm scale weather prediction. *Concurrency: Practice and Experience,* **8,** 731–740.

———, M. Xue, G. Bassett, and K. Droegemeier, 1997: Parallel weather modeling with the advanced regional prediction system. *Parallel Computing,* **23,** 2243–2256.

Saunders, C. P., and S. L. Peck, 1998: Laboratory studies of the influence of the rime accretion rate on charge transfer during crystal/graupel collisions. *J. Geophys. Res.,* **103,** 13 949–13 956.

Schiavone, J. A., and T. V. Papathomas, 1990: Visualizing meteorological data. *Bull. Amer. Meteor. Soc.,* **71,** 1012–1020.

Schlesinger, R. E., 1973: A numerical model of deep moist convection. Part I: Comparative experiments for variable ambient moisture and wind shear. *J. Atmos. Sci.,* **30,** 835–856.

———, 1975: A three-dimensional numerical model of an isolated deep convective cloud: Preliminary results. *J. Atmos. Sci.,* **35,** 2268–2273.

———, 1978: Nonlinear eddy-viscosity turbulence parameterization in anelastic three-dimensional flow: Some mathematical aspects. *J. Atmos. Sci.,* **35,** 2268–2273.

———, 1980: A three-dimensional numerical model of an isolated deep thunderstorm. Part II: Dynamics of updraft splitting and mesovortex couplet evolution. *J. Atmos. Sci.,* **37,** 395–490.

Sherman, W. R., A. B. Craig, M. P. Baker, and C. Bushell, 1997: Scientific visualization, *The Computer Science and Engineering Handbook,* A. B. Tucker Jr., Ed., CRC Press, 820–846.

Skamarock, W. C., and J. B. Klemp, 1992: The stability of time-split numerical methods for the hydrostatic and the nonhydrostatic elastic equations. *Mon. Wea. Rev.,* **120,** 2109–2127.

———, and ———, 1993: Adaptive grid refinement for two-dimensional and three-dimensional nonhydrostatic atmospheric flow. *Mon. Wea. Rev.,* **121,** 788–804.

———, and ———, 1994: Efficiency and accuracy of the Klemp-Wilhelmson time-splitting technique. *Mon. Wea. Rev.,* **122,** 2623–2630.

———, M. L. Weisman, and J. B. Klemp, 1994: Three-dimensional evolution of simulated long-lived squall lines. *J. Atmos. Sci.,* **51,** 2563–2584.

———, P. K. Smolarkiewicz, and J. B. Klemp, 1997: Preconditioned conjugate-residual solvers for Helmholtz equations in nonhydrostatic models. *Mon. Wea. Rev.,* **125,** 587–599.

Smolarkiewicz, P. K., V. Grubisic, and L. G. Margolin, 1997: On forward-in-time differencing for fluids: Stopping criteria for iterative solutions of anelastic pressure equations. *Mon. Wea. Rev.,* **125,** 647–654.

Snow, J. T., and R. L. Pauley, 1984: On a thermodynamic method for estimating maximum tornado wind speeds. *J. Climate Appl. Meteor.,* **23,** 1465–1468.

Soong, S.-T., 1974: Numerical simulation of warm rain development in an axisymmetric cloud model. *J. Atmos. Sci.,* **31,** 1262–1285.

———, and Y. Ogura, 1973: A comparison between axi-symmetric and slab-symmetric cumulus cloud models. *J. Atmos. Sci.,* **30,** 879–883.

Steiner, J. T., 1973: A three-dimensional model of cumulus cloud development. *J. Atmos. Sci.,* **30,** 414–434.

Straka, J. M., R. B. Wilhelmson, L. J. Wicker, J. R. Anderson, and K. K. Droegemeier, 1993: Numerical solutions of a non-linear density current: A benchmark solution and comparisons. *Int. J. Num. Meth. Fluids,* **17,** 1–22.

Sun, J., and N. A. Crook, 1996: Comparison of thermodynamic retrieval by the adjoint method with the traditional retrieval method. *Mon. Wea. Rev.,* **124,** 308–324.

———, and ———, 1997: Dynamical and microphysical retrieval from Doppler radar observations using a cloud model and its adjoint: Model development and simulated data experiments. *Mon. Wea. Rev.,* **125,** 1642–1661.

Takahashi, T., 1984: Thunderstorm electrification—A numerical study. *J. Atmos. Sci.,* **41,** 2541–2558.

———, 1987: Determination of lightning origins in a thunderstorm model. *J. Meteor. Soc. Japan,* **65,** 777–794.

———, 1988: Long-lasting trade-wind showers in a three-dimensional model. *J. Atmos. Sci.,* **45,** 3333–3353.

Takeda, T., 1971: Numerical simulation of a precipitating convective cloud: The formation of a "long-lasting" cloud. *J. Atmos. Sci.,* **28,** 350–376.

Tanguay, M., A. Robert, and R. Laprise, 1990: A semi-implicit semi-Lagrangian fully compressible regional forecast model. *Mon. Wea. Rev.,* **118,** 1970–1980.

Tao, W.-K., and J. Simpson, 1993: Goddard cumulus ensemble model. Part I: Model Description. *TAO,* **4,** 35–72.

———, S. Lang, J. Simpson, C.-H. Sui, B. Ferrier, and M.-D. Chou, 1996: Mechanisms of cloud-radiation interaction in the tropics and midlatitudes. *J. Atmos. Sci.,* **53,** 2624–2651.

Tapp, M. C., and P. W. White, 1976: A non-hydrostatic mesoscale model. *Quart. J. Roy. Meteor. Soc.,* **102,** 277–296.

Trapp, R. J., and B. H. Fiedler, 1995: Tornado-like vortexgenesis in a simplified numerical model. *J. Atmos. Sci.,* **52,** 3757–3778.

———, E. D. Mitchell, G. A. Tipton, D. W. Effertz, A. I. Watson, D. L. Andra Jr., and M. A. Magsig, 1999: Descending and nondescending tornadic vortex signatures detected by WSR-88Ds. *Wea. Forecasting,* **14,** 625–639.

Tremback, C. J., J. Powell, W. R. Cotton, and R. A. Pielke, 1987: The forward-in-time upstream advection scheme: Extension to higher orders. *Mon. Wea. Rev.,* **115,** 540–555.

Trier, S. B., and D. B. Parsons, 1995: Updraft dynamics within a numerically simulated subtropical rainband. *Mon. Wea. Rev.,* **123,** 39–58.

Tripoli, G. J., 1992: A nonhydrostatic mesoscale model designed to simulate scale interaction. *Mon. Wea. Rev.,* **120,** 1342–1359.

———, and W. R. Cotton, 1981: The use of ice-liquid water potential temperature as a thermodynamic variable in deep atmospheric models. *Mon. Wea. Rev.,* **109,** 1094–1102.

———, and ———, 1989: Numerical study of an observed mesoscale convective system. Part I: Simulated genesis and comparison with observations. *Mon. Wea. Rev.,* **117,** 273–304.

Tufte, E. R., 1997: *Visual Explanations.* Graphics Press, 157 pp.

Wakimoto, R. M., and J. W. Wilson, 1989: Non-supercell tornadoes. *Mon. Wea. Rev.,* **117,** 1113–1140.

Walko, R. L., 1988: Plausibility of substantial dry adiabatic subsidence in a tornado core. *J. Atmos. Sci.,* **45,** 2251–2267.

———, 1993: Tornado spin-up beneath a convective cell: Required basic structure of the near-field boundary layer winds. *The Tornado: Its Structure, Dynamics, Prediction, and Hazards, Geophys. Monogr.,* No. 79, Amer. Geophys. Union, 89–95.

Wang, D., M. Xue, V. C. Wong, and K. K. Droegemeier, 1996: Prediction and simulation of convective storms during VORTEX '95. Preprints, *11th Conf. on Numerical Weather Prediction,* Norfolk, VA, Amer. Meteor. Soc., 301–303.

Ward, N. B., 1972: The exploration of certain laboratory features of tornado dynamics using a laboratory model. *J. Atmos. Sci.,* **29,** 1194–1204.

Weisman, M. L., 1993: The genesis of severe, long-lived bow echos. *J. Atmos. Sci.,* **50,** 645–670.

———, and J. B. Klemp, 1982: The dependence of numerically simulated convective storms on vertical wind shear and buoyancy. *Mon. Wea. Rev.,* **110,** 504–520.

———, and ———, 1984: The structure and classification of numerically simulated convective storms in directionally varying wind shears. *Mon. Wea. Rev.,* **112,** 2479–2498.

———, and H. B. Bluestein, 1985: Dynamics of numerically simulated LP storms. Preprints, *14th Conf. on Severe Local Storms,* Indianapolis, IN, Amer. Meteor. Soc., 267–270.

———, J. B. Klemp, and R. Rotunno, 1988: Structure and evolution of numerically simulated squall lines. *J. Atmos. Sci.,* **45,** 1990–2013.

———, W. C. Skamarock, and J. B. Klemp, 1997: The resolution dependence of explicitly modeled convective systems. *Mon. Wea. Rev.,* **125,** 527–548.

Wicker, L. J., 1996: The role of near surface wind shear on low-level mesocyclone generation and tornadoes. Preprints, *18th Conf. on Severe Local Storms,* San Francisco, CA, Amer. Meteor. Soc., 115–119.

———, 1998: The role of low-level shear, mid-level shear, and CAPE in low-level mesocyclone generation. Preprints, *19th Conf. on Severe Local Storms,* Minneapolis, MN, Amer. Meteor. Soc., 222–225.

———, and R. B. Wilhelmson, 1995: Simulation and analysis of tornado development and decay within a three-dimensional supercell thunderstorm. *J. Atmos. Sci.,* **52,** 2675–2703.

———, and W. C. Skamarock, 1998: A time-splitting scheme for the elastic equations incorporating second-order Runge–Kutta time differencing. *Mon. Wea. Rev.,* **126,** 1992–1999.

———, M. P. Kay, and M. P. Foster, 1997: STORMTIPE-95: A convective storm forecast experiment. *Wea. Forecasting,* **12,** 427–436.

Wilhelmson, R. B., 1974: The life cycle of a thunderstorm in three dimensions. *J. Atmos. Sci.,* **31,** 1629–1651.

———, 1977: On the thermodynamic equation for deep convection. *Mon. Wea. Rev.,* **105,** 545–549.

———, and Y. Ogura, 1972: The pressure perturbation and the numerical modeling of a cloud. *J. Atmos. Sci.,* **29,** 1295–1307.

———, and J. B. Klemp, 1978: A three-dimensional numerical simulation of splitting that leads to long-lived storms. *J. Atmos. Sci.,* **35,** 1974–1986.

———, and ———, 1981: A three-dimensional numerical simulation of splitting severe storms on 3 April 1964. *J. Atmos. Sci.,* **38,** 1581–1600.

———, and C.-S. Chen, 1982: A simulation of the development of successive cells along a cold outflow boundary. *J. Atmos. Sci.,* **39,** 1466–1483.

———, L. J. Wicker, H. E. Brooks, and C. Shaw, 1989: The display of modeled storms. Preprints, *Fifth Int. Conf. on Interactive and Information Processing Systems for Meteorology, Oceanography, and Hydrology,* Anaheim, CA, Amer. Meteor. Soc., 166–171.

———, and Coauthors, 1990: A study of the evolution of a numerically modeled severe storm. *Int. J. Supercomputing Appl.,* **4,** 20–36.

———, D. P. Wojtowicz, C. Shaw, J. Hagedorn, and S. Koch, 1995: NCSA PATHFINDER: Probing ATmospHeric Flows in an INtegrated and Distributed EnviRonment. *Visualization Techniques in Space and Atmospheric Sciences,* E. P. Szuszczewicz and J. H. Bredekamp, Eds., NASA, 289–296.

———, and Coauthors, 1996: Visualization of storm and tornado development for an OMNIMAX film and for the CAVE. Preprints, *12th Int. Conf. on Interactive Information and Processing Systems for Meteorology, Oceanography, and Hydrology,* Atlanta, GA, Amer. Meteor. Soc., 135–138.

Xue, M., K. K. Droegemeier, V. Wong, A. Shapiro, and K. Brewster, 1995: ARPS User's Guide. CAPS, 375 pp. [Available from Center for Analysis and Prediction of Storms, University of Oklahoma, Sarkeys Energy Center, Room 1110, 100 East Boyd Street, Norman, OK 73019 or online at http://wwwcaps.ou.edu/ARPS/.]

———, and Coauthors, 1996a: Realtime numerical prediction of storm-scale weather during VORTEX '95, Part II: Operations summary and example predictions. Preprints, *18th Conf. on Severe Local Storms,* San Francisco, CA, Amer. Meteor. Soc., 178–182.

———, and Coauthors, 1996b: The 1996 CAPS spring operational forecasting period—Realtime storm-scale NWP, Part II: Operational summary and sample cases. Preprints, *11th Conf. on Numerical Weather Prediction,* Norfolk, VA, Amer. Meteor. Soc., 297–300.

———, Q. Xu, and K. K. Droegemeier, 1997: A theoretical and numerical study of density currents in non-constant shear flows. *J. Atmos. Sci.,* **54,** 1998–2019.

———, K. K. Droegemeier, and V. Wong, 2000: The Advanced Regional Prediction System (ARPS)—A multiscale nonhydrostatic atmospheric simulation and prediction tool. Part I: Model dynamics and verification. *Meteor. Atmos. Phys.,* **75,** 161–193.

———, ———, ———, A. Shapiro, K. Brewster, F. Carr, D. Weber, Y. Liu, and D.-H. Wang, 2001: The Advanced Regional Prediction System (ARPS)—A multiscale nonhydrostatic atmospheric simulation and prediction tool. Part II: Model physics and applications. *Meteor. Atmos. Phys.,* **76,** 134–165.

Yang, M.-J., and R. A. Houze Jr., 1995: Sensitivity of squall line rear inflow to ice microphysics and environmental humidity. *Mon. Wea. Rev.,* **123,** 3175–3193.

Zhang, D.-L., H.-R. Chang, N. L. Seaman, T. T. Warner, and J. M. Fritsch, 1986: A two-way interactive nesting procedure with variable terrain resolution. *Mon. Wea. Rev.,* **114,** 1330–1339.

Ziegler, C. L., T. J. Lee, and R. A. Pielke Sr., 1997: Convective initiation at the dryline: A modeling study. *Mon. Wea. Rev.,* **125,** 1001–1026.

Chapter 5

Tornadoes and Tornadic Storms

ROBERT DAVIES-JONES

National Severe Storms Laboratory, NOAA, Norman, Oklahoma

R. JEFFREY TRAPP

National Severe Storms Laboratory, NOAA, Norman, Oklahoma, and Cooperative Institute for Mesoscale Meteorological Studies, University of Oklahoma, Norman, Oklahoma

HOWARD B. BLUESTEIN

School of Meteorology, University of Oklahoma, Norman, Oklahoma

REVIEW PANEL: John T. Snow (Chair), Eugene W. McCaul Jr., Joseph H. Golden, Jean Dessens, Richard Peterson, and W. S. Lewellen

5.1. Introduction

a. Definitions

Tornadoes, with measured wind speeds of 125 m s^{-1} to perhaps 140 m s^{-1}, are the most violent of atmospheric storms (Fig. 5.1). A tornado is defined here as a violently rotating, narrow column of air, averaging about 100 m in diameter, that extends to the ground from the interior of a cumulonimbus (or occasionally a cumulus congestus) cloud and appears as a condensation funnel pendant from cloud base and/or as a swirling cloud of dust and debris rising from the ground. Significant damage can occur at the ground even when the condensation funnel does not reach the surface. A condensation funnel associated with a tornadic vortex that fails to contact the ground is called a *funnel cloud.* A waterspout is a tornado over a body of water.

Tornadoes can be divided into two types. A *type I* tornado forms within a mesocyclone, a larger-scale parent circulation. To a first approximation, the tangential winds in a mesocyclone may be modeled as a Rankine combined vortex, which consists of a core in solid-body rotation surrounded by a potential vortex where the tangential wind is inversely proportional to distance from the center of circulation. Core diameters vary from 3 to 9 km, with an average value around 5 km. The parent storm of a type I tornado can be an isolated supercell storm (Fig. 5.2), a supercell in a line of thunderstorms (Browning 1986), or a miniature supercell containing a small mesocyclone (Davies 1993b; Kennedy et al. 1993). Large and violent tornadoes almost invariably fall into this class. Tornadoes within minisupercells in rainbands of landfalling tropical cyclones (e.g., McCaul 1993; Novlan and Gray 1974) are also type I.

For the purposes of this review, a *supercell* is defined as a long-lived (> 1 h) thunderstorm with a high degree of spatial correlation between its mesocyclone and updraft. All supercells produce their overall most significant tornadoes under or very near the wall cloud (Moller 1978) (Fig. 5.3); other less significant tornadoes form along the rear-flank downdraft gust front. Supercell storms may be visually (and also by radar) generalized as follows, according to the position and extent of heavy rain relative to the storm's main updraft: low precipitation (LP), "classic" (or moderate precipitation), and high (or heavy) precipitation (HP) (Doswell and Burgess 1993; see also Rasmussen and Straka 1998). As can be inferred from the discussions in sections 5.2 and 5.3, the relatively higher propensity of a classic supercell to become tornadic is determined somewhat by the storm organization implied by this classification. Nevertheless, even LP supercells (Burgess and Davies-Jones 1979; Bluestein and Parks 1983; Bluestein and Woodall 1990) produce tornadoes (Bluestein and MacGorman 1998), even though it is not expected that there will be any significant evaporatively cooled pool of air near the ground (Fig. 5.4).

A *type II* tornado is not associated with a mesocirculation. It is generally a small and weak vortex that forms along a stationary or slowly moving windshift line, from the rolling-up of the associated vortex sheet into individual vortices (Barcilon and Drazin 1972; Davies-Jones and Kessler 1974). In the case of a "landspout" (Bluestein 1985a) for example, the wind-

FIG. 5.1. Tornado at Osnabrock, North Dakota, on 24 July 1978 as seen looking NE. The eastward-moving tornado passed one-half mile to the north of the photographer. Note the cloud band spiralling inward and cyclonically around the tornado from the east. Copyright photo by Edi Ann Otto, permission granted for scientific use.

shift line precedes the tornado's parent cloud. A "gustnado" (Bluestein 1980) is a type II vortex that forms along the storm's internally generated outflow boundary well away from any mesocyclone. Lack of a vigorous parent updraft superimposed over the outflow boundary usually precludes it from intensifying into a strong tornado. It is visible near the ground as a dust whirl and sometimes at cloud base as a small rotating eddy or a short condensation funnel. Cold-air funnels are spawned during daytime by high-based, low-topped, moderate thunderstorms that form in deep, cold-core synoptic-scale lows (Cooley 1978). These funnel clouds are long, slender, and ropelike. Occasionally they reach the ground and cause light damage. High-based funnels (e.g., Bluestein 1994) and vortices embedded within synoptic-scale fronts (Carbone 1983) can probably also be categorized as type II tornadoes.

b. Climatological distribution of tornadoes

Tornadoes occur worldwide, but are most prevalent in the Great Plains of the United States and in northeast India–Bangladesh, which are both areas that lie to the east of a mountain range and poleward of a warm ocean (see Fujita 1973). Based on data in the United States from 1921 to 1995 (see Grazulis 1993), the maximum in the mean number of days per century of F2 intensity (based on the Fujita scale; Fujita 1981) or greater tornado occurrence is located in southcentral Oklahoma (Fig. 5.5). This spatial distribution, in conjunction with an annual consistency in the seasonal distribution of such significant tornado occurrence, suggests a "Tornado Alley" that extends from north Texas northward into western Iowa (Concannon et al. 2000).

Over the past 50 years, the annual number of all tornadoes reported in the United States has increased from approximately 200 to 1200. This increase is unlikely to be physical; rather, it reflects an overall increase in population density, improved reporting procedures, and organized networks of "storm spotters," etc. (e.g., Doswell et al. 1999); there is currently no evidence to suggest that the threat from significant tornadoes is increasing due to global climate change (e.g., Concannon et al. 2000). Contemporary tornado verification data can still be problematic, though. Tornado formation times, in particular, may contain errors of as much as 1 h, for a variety of nonmeteorological reasons (Witt et al. 1998). The reported tornado inten-

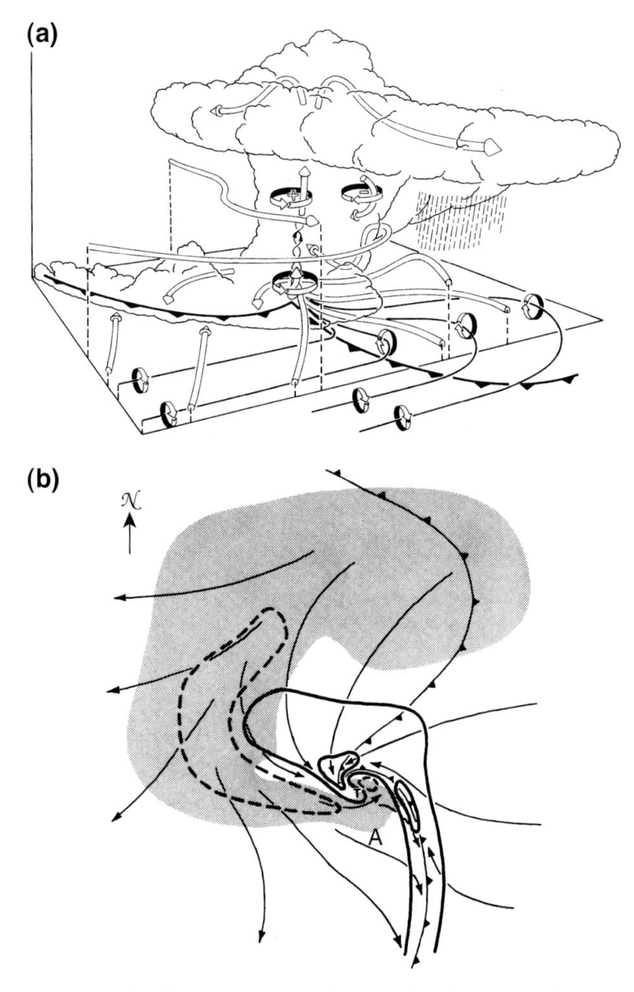

FIG. 5.2. (a) Three-dimensional schematic view from the southeast of a supercell thunderstorm in environmental winds that veer with height. The cylindrical arrows depict the flow in and around the storm. The thick lines show some of the vortex lines, with the sense of rotation indicated by the circular-ribbon arrows. The heavy barbed line marks the boundary of the cool-air outflow beneath the storm. From Klemp (1987; reprinted with permission). (b) Schematic flow field at 250 m above the ground in a tornadic supercell. The radar echo is indicated by the stippling. Vertical velocity is contoured at 2 m s^{-1} intervals with zero contour omitted and negative contour dashed. Note the storm-scale rear-flank downdraft west of the updraft and the occlusion downdraft near the vorticity maximum. The pseudo–cold front is drawn where the temperature is 1 K cooler than the environment. The flow arrows depict storm-relative surface streamlines. The T marks the location of the vertical vorticity maximum and cyclonic tornado, and A indicates where a rare anticyclonic tornado might occur. In time the secondary updraft maximum on the bulge in the gust front becomes dominant and a new mesocyclone forms in its vicinity as the storm-relative flow pattern shifts southward. From Klemp and Rotunno (1983).

sity is based primarily on a description of damage intensity as it appears to non-engineers. Each category (F0 to F5 in ascending order of severity) has been assigned a range of wind speeds without rigorous justification. Engineers can reliably estimate the minimum wind speeds required to account for observed damage only if structures of known structural integrity are damaged.

In contrast to the tornado-reporting trend, Fig. 5.6 shows that the annual, population-normalized death toll due to tornadoes in the United States has declined steadily since 1925 (Doswell et al. 1999). The decline since 1950 can be attributed to improved public awareness and education, the issuance of severe-weather warnings and watches, vast improvements in forecasting and dissemination of information to the public, and the identification of dangerous storms on radar and by trained spotters in the field. Operational use of the most current operational radar, the Weather Surveillance Radar-1988 Doppler (WSR-88D), can be linked to the current mean tornado-warning lead time of 13 min, a 5-min improvement over warnings issued prior to the installation of the WSR-88D network (Bieringer and Ray 1996). Better warnings and watches are of course also related to a greater scientific understanding of tornadoes and tornadic storms, the focus of the remainder of this chapter.

Since tornadoes and tornadic storms have been well reviewed in the past 30 years, we will not reiterate excellent descriptions that are available elsewhere. For background material the reader is referred to individual review articles by Morton (1966), Davies-Jones and Kessler (1974), Snow (1982, 1984), Rotunno (1986), Klemp (1987), and Davies-Jones (1995), and to many papers in the books edited by Peterson (1976), Kessler (1986), and Church et al. (1993). Other chapters in this monograph also touch on various aspects of tornadoes, particularly the one by Wilhelmson and Wicker on numerical simulations of severe storms.

Morton (1966) states that the "study of tornado vortices may be separated into two parts: mechanisms for the generation of an enhanced level of vorticity in some neighborhood of a thunderstorm, and the development of an actual vortex from this background." We believe that this statement is still generally valid and so reflect it in the organization of sections 5.2 and 5.3. We then examine, in sections 5.4–5.6, the observed and modeled flow structure of tornadoes to show how these vortices, once well established, are stable and intense. The chapter concludes with comments on some of the unsolved problems/unanswered questions and technological needs of this field of study. Our discussions hereafter apply to Northern Hemisphere storms, but carry over to Southern Hemisphere ones with the appropriate modifications (interchange left and right, north and south, clockwise and counterclockwise).

5.2. Midlevel and near-ground mesocyclogenesis

We separate our discussions of midlevel (5-km altitude, nominally) and near-ground mesocyclogenesis according to the respective processes by which vertical vorticity on the mesocyclone scale is generated. At midlevels, the mesocyclone develops initially

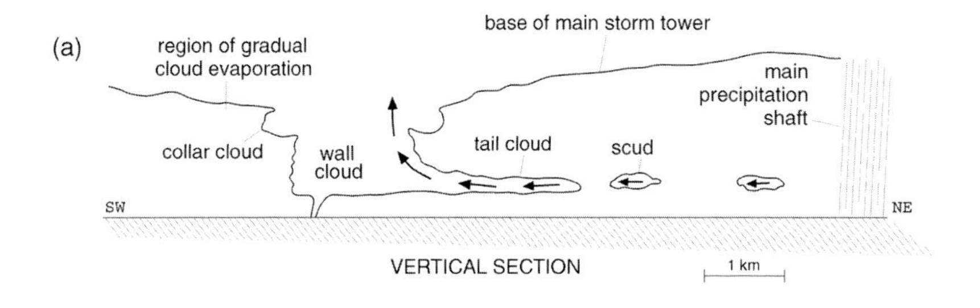

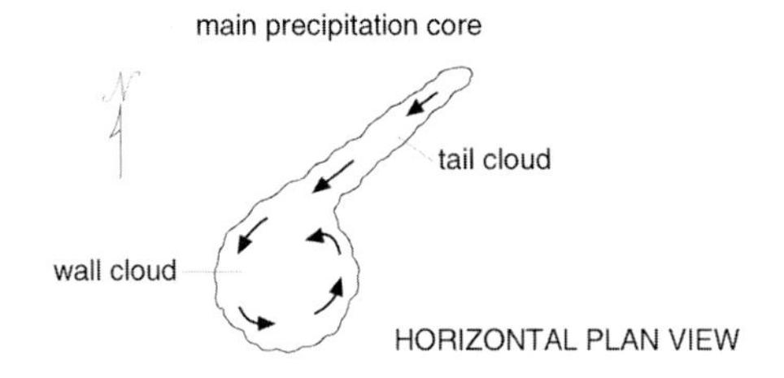

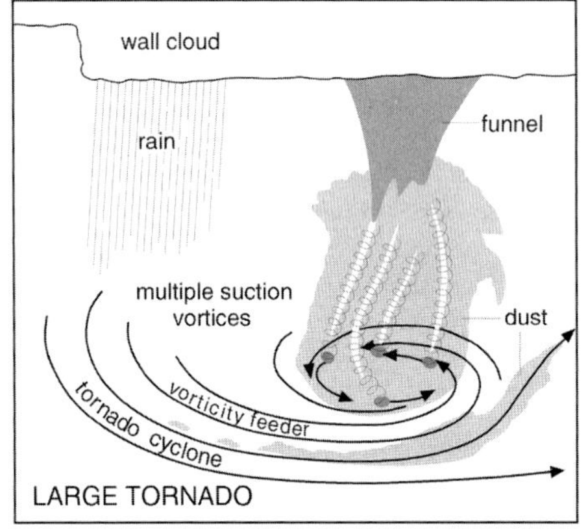

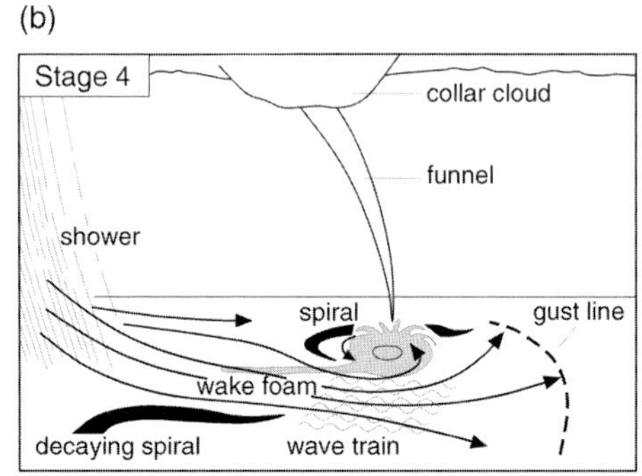

FIG. 5.3. Diagrams showing relationship of nearby rain shaft and cloud features to (a) tornado (based on Fujita 1959 and from Fujita et al. 1976), and (b) waterspout (based on Golden 1974a).

from the tilting of the horizontal vorticity associated with the vertical shear of the environmental winds. The theoretical basis for this statement is derived below from the linearized and then fully nonlinear equations of motion. Near the ground, the mesocyclone may form due to the tilting of horizontal vorticity generated within low-level density gradients. Barotropic processes may also play a role in near-ground mesocyclogenesis, and hence they are also treated below. Before venturing into this and other explanations, however, we will first review the observations of mesocyclones, which motivate the theory.

a. Mesocyclone observations

The mesocyclone observations presented below result in some way from informal and formally organized storm-intercept endeavors (e.g., Browning 1964; Donaldson and Lamkin 1964; Agee 1969, 1970; Golden and Morgan 1972; Moller et al. 1974; Golden and Purcell 1978b; Moller 1978; Bluestein 1980; Davies-Jones 1983; Bluestein 1984; Bluestein 1985a; Bluestein 1986; Bluestein and Golden 1993; Rasmussen et al. 1994; Bluestein 1996; Bluestein 1999a,b).

Visual evidence of rotation in the updraft portion (labeled "main storm tower" in Fig. 5.4) of cumulonimbus clouds is given particularly well by time-lapse photography. Complementary evidence is provided by Doppler radar data, which represent mesocyclones as range-constant and azimuthally adjacent regions of inbound and outbound Doppler velocities (in the direction of the radar beam, the speed of precipitation particles and other radar targets), whose peaks are azimuthally separated. Objective mesocyclone identification criteria, first established by Donaldson (1970) (who is also responsible for the first observation of a mesocyclone by Doppler radar; see Donaldson 1990), consist now in U.S. National Weather Service (NWS) operations of Doppler velocity shear ≥ 6 m s^{-1} km^{-1} and a differential velocity ≥ 30 m s^{-1} (both applicable within ranges ≤ 100 km, and thereafter reduced by

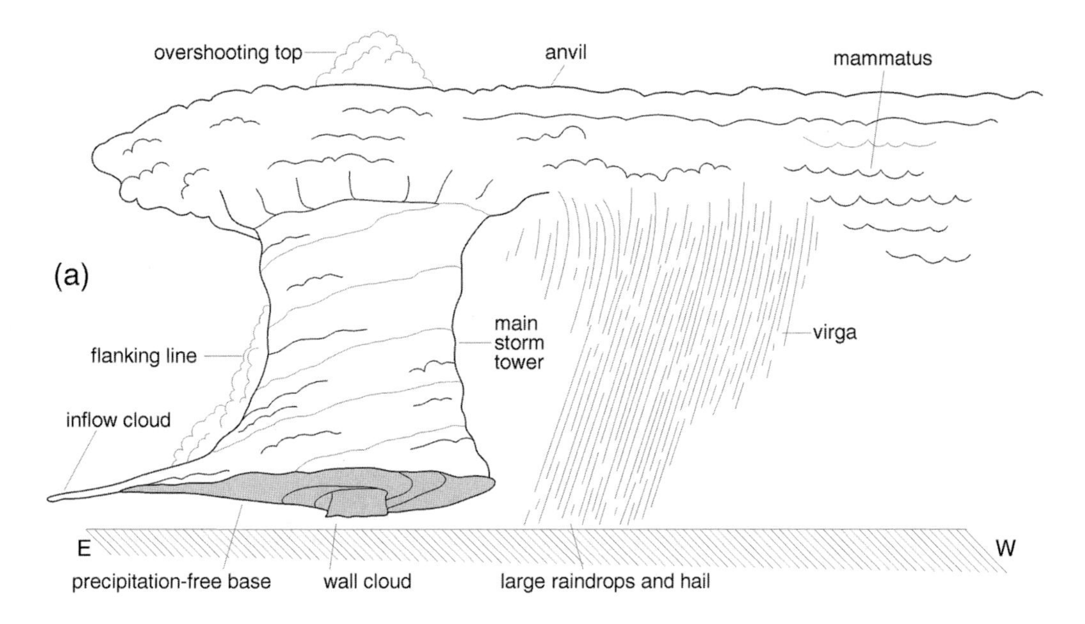

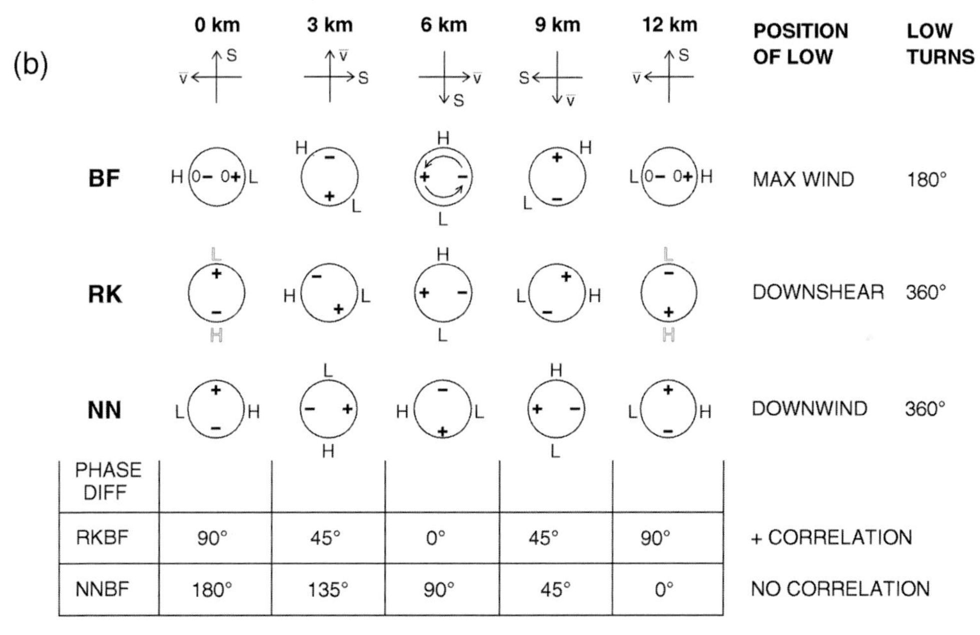

PHASE DIFF						
RKBF	90°	45°	0°	45°	90°	+ CORRELATION
NNBF	180°	135°	90°	45°	0°	NO CORRELATION

BF - Beltrami Flow
RK - Rotunno and Klemp (1982)
NN - Newton and Newton (1959)

FIG. 5.4. (a) Depiction of how a low-precipitation supercell storm might look in an environment with a circle hodograph. The entire updraft rotates in this case. (b) Schematic showing the positions of the high (H) and lows (L) of the linear pressure and of the associated maximum upward (+) and downward (−) VPPGF, as functions of height from the surface (0 km) to storm top (12 km) according to the exact Beltrami flow solution (BF) and to the Rotunno and Klemp (RK) and Newton and Newton (NN) models. The circle depicts axisymmetric updraft, which is rotating as a whole at midlevels (curved arrows). Outlined H and L (in RK) and 0+ and 0− (in BF) indicate positions of extrema just inside the domain in cases where the quantity vanishes on the boundary. Arrows give environmental wind and shear at each level. Also tabulated for each model are the phase errors, position of the low relative to wind features, and the angle through which the low turns. The nonlinear pressure has only one low in this case and it is at the midpoint of the updraft. From Davies-Jones (1985, 1996b).

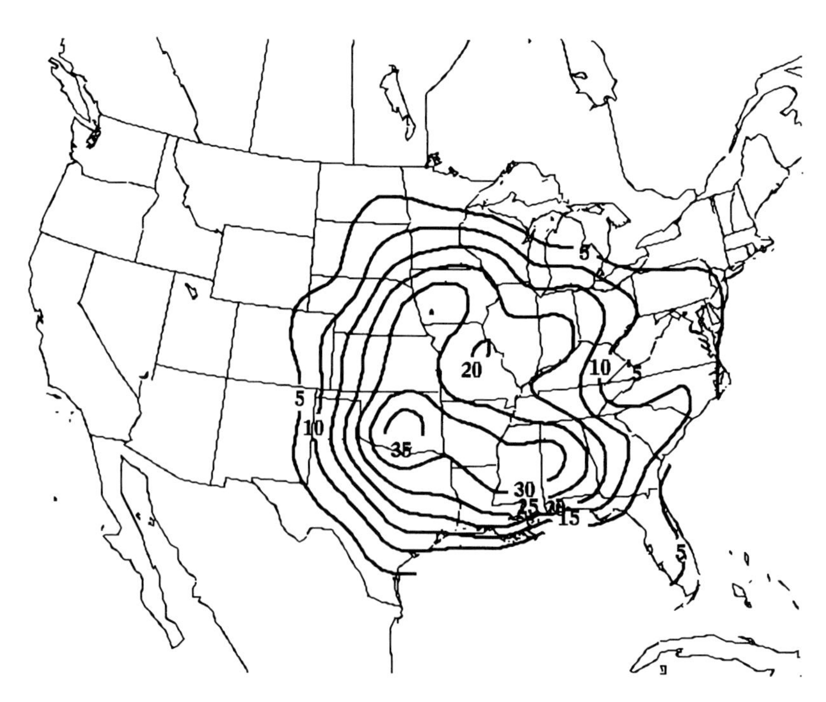

FIG. 5.5. Based on data from 1921 to 1995, the mean number of days per century of F2 intensity or greater tornado occurrence within 40km of a point in the United States (Concannon et al. 2000). Contour interval is 5 days, with minimum level equal to 5.

some percentage), met over a depth $\geq$ 3 km and with a base at altitudes $\leq$ 5 km above radar level, and persisting longer than 5–6 min (more than one radar volume scan). Some form of these criteria—with complementary visual observations and poststorm surveys—has been used to establish the oft-cited statistic that only 50% (or 25%, based on more recent work) of all objectively defined mesocyclones are associated with tornadoes (e.g., Burgess and Lemon 1990). The implication here—that mesocyclone detection leads to only a modest probability of tornado detection—serves as the impetus for the continued work on automated algorithmic identification of specific attributes of those mesocyclones most likely to spawn tornadoes (e.g., Stumpf et al. 1998).

Dual- or multiple-Doppler radar observations, from which the 3D wind may be retrieved, confirm the vortical structure of mesocyclones (the existence of a vortex may only be inferred from single Doppler radar). Analyses of such observations (Brandes 1977; Heymsfield 1978; Brandes 1978; Ray et al. 1980, 1981; Brandes 1981; 1984a,b; Hane and Ray 1985; Johnson et al. 1987; Brandes et al. 1988; Ray and Stephenson 1990; Wakimoto et al. 1996; Dowell et al. 1997; Dowell and Bluestein 1997; Bluestein et al. 1997; Wakimoto et al. 1998; Wakimoto and Liu 1998; Trapp 1999) additionally confirm/expand on kinematical relationships among the updraft, downdraft, and mesocyclone that are described below. Computed from the analyses is the mesocyclonic vertical vorticity scale of 0.01 s^{-1}, a benchmark value in many regards. The mesocyclonic vertical vorticity exhibits a vertical variability that depends on the stage in the supercell storm life cycle: Early (late) in the life of a supercell, the vertical vorticity is greatest aloft (near the ground). Indeed, we note the particularly relevant observation by Brandes (1984b) of a "low-level multiplicative growth of vertical vorticity" that coincides with the tornadic-stage transition; this characteristic motivates in part the consideration of different mechanisms for midlevel and near-ground mesocyclogenesis, which we are now ready to pursue.

b. Linear theory of initial midlevel rotation

The origins of the initial rotation of supercell updrafts at low to midlevels can be deduced from the linear Boussinesq equations for dry inviscid motion applied to an isolated growing convective cell in a sheared unstably stratified environment (Lilly 1982; Davies-Jones 1984). The cell is modeled as an exponentially amplifying disturbance with an axisymmetric buoyancy field or "thermal plume"; the linearity assumption in the governing equations is valid only when the disturbance amplitude is small. Perturbations owing to the disturbance are relative to a basic state that represents the undisturbed environment. A perturbation variable of key importance in the approach due to Davies-Jones (1984) is the vertical displacement $h'(x, y, z, t)$ of air parcels in the disturbance, which also describes the perturbed height of isentropic surfaces and vortex lines, which are horizontal initially at

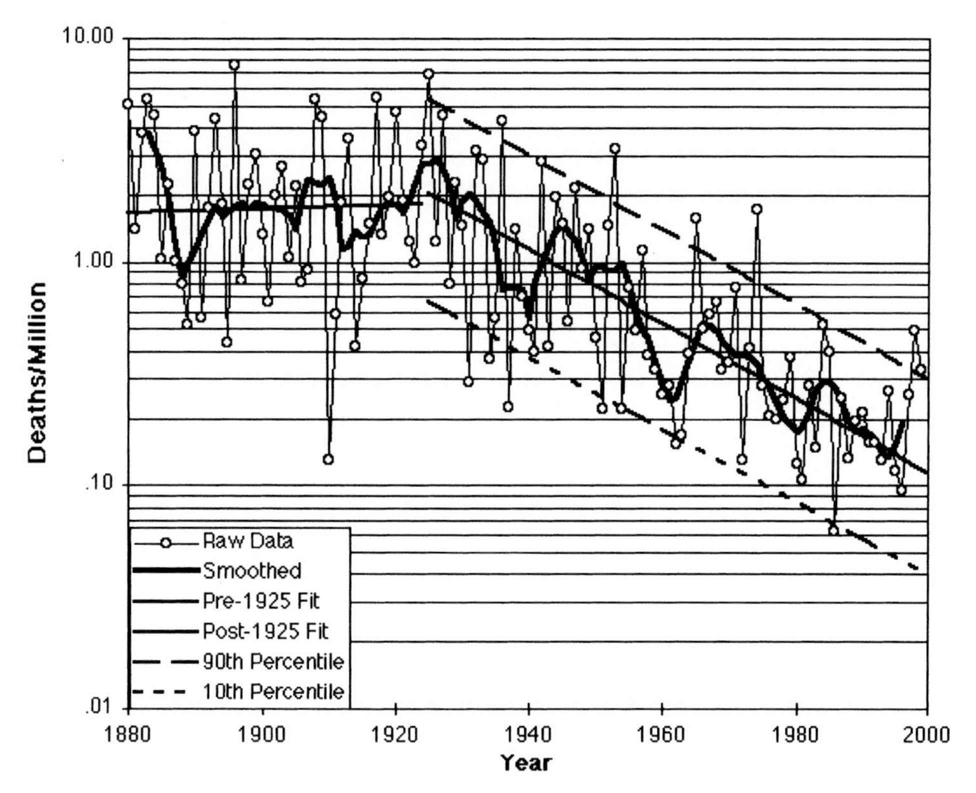

FIG. 5.6. Population-normalized annual death toll due to tornadoes in the United States (Doswell et al. 1999).

all levels (see below). The vertical displacement is important because it is proportional to buoyancy and is related to vertical velocity and vertical vorticity via simple expressions.

Since the horizontal vorticity associated with vertical shear is often 50 times greater than the large-scale background vertical vorticity associated with the earth's rotation and midlatitude cyclones, Coriolis terms can be safely omitted, and we can assume that the velocity and potential temperature of the environment (basic state) are of the form $\bar{\mathbf{v}}(z) \equiv (\bar{u}(z), \bar{v}(z), 0)$ and $\bar{\theta}(z)$ (where z is height). The environmental shear and vorticity are given by $\bar{\mathbf{S}}(z) \equiv (\bar{S}_1, \bar{S}_2) \equiv (\bar{u}_z, \bar{v}_z, 0)$ and $\bar{\boldsymbol{\omega}}(z) \equiv \nabla \times \bar{\mathbf{v}}(z) = (-\bar{v}_z, \bar{u}_z, 0)$. Note that the environmental vorticity has the same magnitude as the shear and is directed 90° to the left of it. Furthermore, the potential vorticity (PV), $\alpha\boldsymbol{\omega} \cdot \nabla\theta$, is zero in the environment. Since potential vorticity is conserved following a parcel in the absence of turbulent mixing and diabatic heating or cooling, the potential vorticity is constrained to remain zero throughout the flow.

According to the linear theory, vertical vorticity is generated solely via the upward tilting of environmental vorticity. Our interest here is in applying the theory to explain initiation of cyclonically rotating updrafts in sheared environmental flow, or, specifically, how perturbation vertical velocity (w') and vertical vorticity (ζ') become positively correlated through the tilting of environmental vortex lines. The concept of *streamwise* versus *crosswise* vorticity is indispensable for this purpose: The streamwise (crosswise) vorticity component is parallel (normal) to the local storm-relative velocity vector, and accounts for the storm-relative directional (speed) shear.

The derivation by Davies-Jones (1984) yields the following approximate formula for the theoretical correlation coefficient r between w' and ζ':

$$r = \frac{\langle \zeta' w' \rangle}{[\langle \zeta'^2 \rangle \langle w'^2 \rangle]^{1/2}} \approx \frac{|\bar{\mathbf{v}} - \mathbf{c}|}{\sqrt{\sigma^2 D^2 + |\bar{\mathbf{v}} - \mathbf{c}|^2}} \frac{\bar{\omega}_s}{|\bar{\boldsymbol{\omega}}|} \tag{5.1}$$

where angular brackets denote a surface integral, D ($\sim$3 km) is the length scale of the horizontal gradients associated with the disturbance, σ is the exponential growth rate of the disturbance ($\sim 3 \times 10^{-3}$ s^{-1}), $\mathbf{c}$ is the disturbance- or "storm"-motion vector, and $\bar{\omega}_s$ denotes the streamwise component of the environmental vorticity vector. Equation (5.1) demonstrates that the correlation is high when the environmental vorticity is predominantly streamwise and storm-relative winds are of order σD ($\sim$10 m s^{-1}) or greater so that the advective timescale $D/|\bar{\mathbf{v}} - \mathbf{c}|$ is comparable to the growth timescale $1/\sigma$.

A qualitative explanation for the development of this correlation can be provided in terms of isentropic surfaces (Fig. 5.7). Since the potential vorticity remains zero, the vortex lines are embedded in these

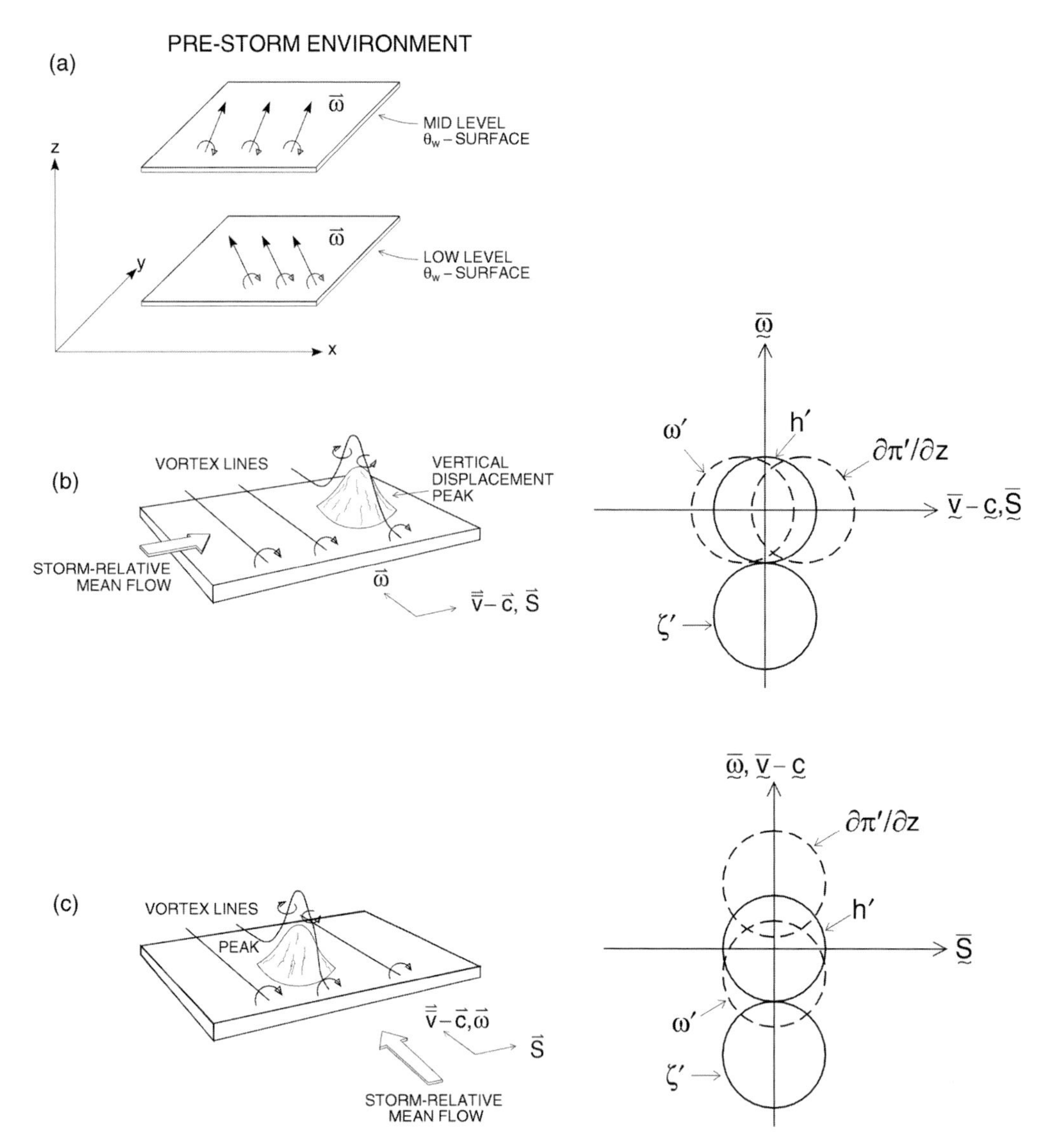

FIG. 5.7. Linear barotropic effect on vortex lines of a small, amplifying, isolated, axisymmetric peak (vertical displacement maximum) in an isentropic surface, as observed in a storm-relative reference frame (in which the peak does not translate and all winds are storm relative). In the horizontally homogeneous prestorm environment, depicted in (a), the isentropic surfaces and vorticity vectors are horizontal. The peak pulls up loops of vortex lines (shown slightly above, instead of in, the surface for clarity), giving rise to cyclonic (anticyclonic) vorticity on the right (left) side of the shear vector $\bar{\mathbf{S}}$ drawn through the peak. The environmental flow over the peak displaces the maximum updraft to the upstream side of the peak owing to the upslope flow there. In (b) the winds $\bar{\mathbf{v}} - \mathbf{c}$ increase with height without veering or backing so that the environmental vorticity $\bar{\boldsymbol{\omega}}$ is purely crosswise, i.e., perpendicular to the flow. In this case the cyclonic (anticyclonic) vortex is on the right (left) side of the flow, resulting in no net updraft rotation. In (c) the winds $\bar{\mathbf{v}} - \mathbf{c}$ veer with height without changing speed so that $\bar{\boldsymbol{\omega}}$ is purely streamwise, i.e., aligned with the flow. In this case, the cyclonic vortex is on the upslope side of the peak, resulting in positive correlation between vertical velocity and vertical vorticity. The schematic diagrams on the right illustrate the linear theory relationships among vertical displacement h', vertical velocity w', vertical vorticity ζ', and vertical perturbation pressure gradient $\partial\pi'/\partial z$ in the two extreme cases. The centers of the circles indicate the relative positions of the maxima. In all cases, the maximum updraft is on the upstream side of the peak in h', the maximum $\partial\pi'/\partial z$ is on the downstream side, and the cyclonic vorticity maximum is on the right side (relative to $\bar{\mathbf{S}}$). Overlapping, touching, or disjointed circles signify positive, zero, or negative correlation between the two variables involved. Note, however, that the correlation between w' and $\partial\pi'/\partial z$ becomes negative at high storm-relative wind speed. Adapted from Davies-Jones (1984).

surfaces. Let us consider a particular surface that is deformed from its initial flat shape by the convection. A developing isolated convective cell raises a growing peak in this surface. The storm-relative wind advects the vortex lines over the peak, and the resulting tilting of vortex lines produces a cyclonic (anticyclonic) vortex to the right (left) of the peak, looking downshear. If the vorticity is crosswise, the flow is upslope on the upshear side of the peak, so the updraft does not rotate as a whole. If the vorticity is streamwise, the relative flow is from right to left and the upslope side is the cyclonic side. The updraft rotates cyclonically as a whole because the vorticity is along the streamlines and is tilted upward as the air enters the updraft. The anticyclonic vortex on the downstream side of the peak is either in downdraft or in less intense updraft.

The relative locations of the maxima of h', w', ζ', and the vertical perturbation pressure-gradient force (VPPGF; $\partial\pi'/\partial z$) in a given level for a growing convective cell can be deduced from linear theory (Fig. 5.7). Recall that the h' field is coincident with the buoyancy field b'. Looking downshear, the cyclonic and anticyclonic vortices lie on the right and left side of the buoyancy peak, within buoyancy gradients (Figs. 5.7b,c), so that ζ' is uncorrelated with b' (Davies-Jones 1984; Kanak and Lilly 1996). In the storm's reference frame the maximum updraft is upstream of this peak because of the "upslope" flow along an isentropic surface; this effect has been observed in numerical simulations by Brooks and Wilhelmson (1993). Furthermore, the maximum VPPGF is upstream of the maximum updraft: Physically, there has to be an upward pressure-gradient force in this location to account for the upstream displacement of the updraft relative to the peak buoyancy, since the buoyancy force by itself would give rise to a downstream displacement.

Without using all the equations in the linearized set we have successfully identified streamwise vorticity as the origin of updraft rotation, as first suggested by Browning and Landry (1963) and Barnes (1970). Our task is not complete because the streamwise direction is a function of the as yet unknown storm motion. Storm motion can be determined as an eigenvalue of the solution of the governing equation in h' (see appendix of Davies-Jones 1984), or deduced qualitatively from the inferredVPPGF around an axisymmetric updraft (Rotunno and Klemp 1982; Davies-Jones 1996b). According to linear theory, $\mathbf{c}$ for an axisymmetric thermal plume lies on the concave side of a simple curved hodograph (one that does not intersect itself), and on the hodograph if it is straight. This implies that updrafts will rotate cyclonically (anticyclonically) if the hodograph curves clockwise (anticlockwise), and will not rotate if the hodograph is straight.

At this point, we have exhausted linear theory. It has served us well by revealing the basic mechanism that causes updrafts to rotate at midlevels. It fails to explain storm splitting and nonlinear updraft propagation (a component to the movement arising from nonlinear forcing of new updraft on one side and suppression of updraft on the opposite side). Thus, it does not account for the development of mirror-image right- and left-moving supercells when the hodograph is straight, and it underestimates the intensity of updraft rotation for storms that deviate to the right of the mean wind when the hodograph turns clockwise with height. These are nonlinear effects that we describe next.

c. Nonlinear theory of midlevel rotation

1) GENERAL REMARKS

Two important processes that generate vertical vorticity are missing in the above linear theory. The first is amplification of vertical vorticity by vertical stretching of vortex tubes. This is a second-order effect because vertical vorticity is generated by tilting of the horizontal environmental vorticity in the linear solution and subsequently stretched in a second-order solution. The second is the tilting (at second order) of horizontal vorticity that has been generated (at first order) by horizontal buoyancy gradients, a process that is often important for the generation of rotation near the ground (section 5.2d). We also have to consider the nonlinear forcing terms in the diagnostic pressure equation because the associated VPPGF may play important dynamical roles in storm propagation and splitting.

Nonlinear theory is further complicated by moist processes such as evaporative cooling and latent heat released by condensation, which can no longer be ignored. For moist adiabatic processes, moist specific entropy S_E and equivalent potential temperature θ_E tend to be conserved in lieu of the dry specific entropy S and θ. Even in inviscid and isentropic flow, moist potential vorticity MPV $\equiv \alpha\boldsymbol{\omega}\cdot\nabla S_E$ is not conserved exactly because S_E depends on vapor mixing ratio q in addition to two independent-state variables (say T and p). But MPV is conserved for processes for which $q = q(T, p)$, for example, dry/moist adiabatic ascent of a parcel in a horizontally homogeneous environment.

We first investigate how strong the shear has to be to organize thunderstorm structure and then to induce splitting. A rough answer to this problem is supplied by a scale analysis of the forcing terms in the diagnostic pressure equation (see, e.g., appendix of Rotunno and Klemp 1982). Since the vertical advection of vertical velocity is comparable to buoyancy, the scale of the vertical velocity $W \sim \sqrt{\mathrm{CAPE}}$, where CAPE $\equiv g\int_{LFC}^{EL}\{[\rho_p(z) - \bar{\rho}(z)]/[\bar{\rho}(z)]\}dz$ is the buoyant or convective available potential energy (Weisman and Klemp 1982). Here $\rho_p(z)$ is the density of a representative near-surface parcel that is lifted dry and then

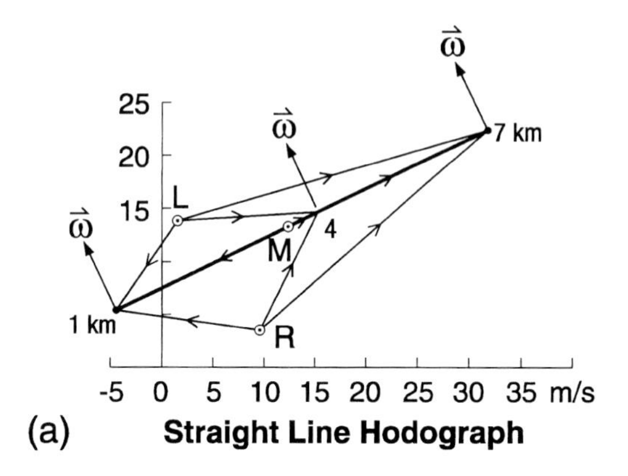

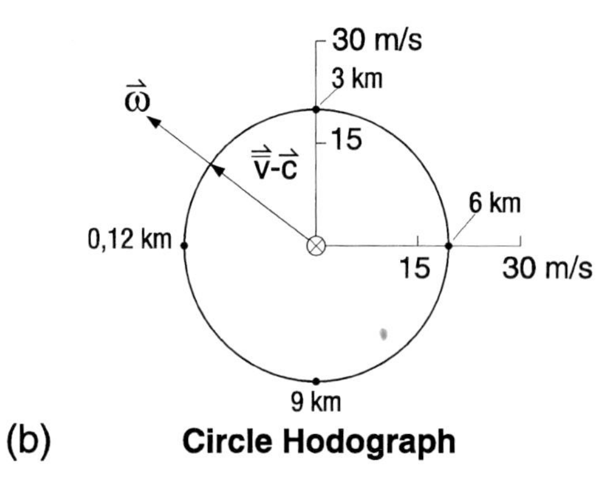

FIG. 5.8. Idealized (a) straight and (b) circle hodographs.

moist adiabatically. This parcel is buoyant from its level of free convection (LFC) to its equilibrium level (EL). From the conservation of (zero) moist potential vorticity, the vertical vorticity of a parcel is given by

$$\zeta = \boldsymbol{\omega} \cdot \nabla h, \tag{5.2}$$

where h is the finite vertical displacement of the parcel (Davies-Jones 1984; Rotunno and Klemp 1985). Thus, if S_0 is the scale of the low- to midlevel shear magnitude, D and H are the length scales for horizontal and vertical gradients, respectively, and $U_S \equiv S_0 H$, $\zeta \sim U_S/D$. Note that U_S can be interpreted as the length of the lower portion of a smooth hodograph (Weisman and Klemp 1982, 1984). For $D/H \sim 1/2$, the ratios of the forcing functions in the diagnostic pressure equation are

$$F_B{:}F_L{:}F_{NL} \sim \mathrm{Ri}{:}\sqrt{\mathrm{Ri}}{:}1, \tag{5.3}$$

where $\mathrm{Ri} \equiv \mathrm{CAPE}/U_S^2$ is a Richardson number, and F_B, F_L, F_{NL} are the forcing functions that involve buoyancy, linear dynamical terms, and nonlinear dynamical terms, respectively (Rotunno and Klemp 1982). Clearly, buoyancy effects dominate those of shear for $\sqrt{\mathrm{Ri}} \equiv W/U_S \gg 1$. As $\mathrm{Ri} \rightarrow 1$ from above, first the linear and then the nonlinear dynamic forcing become important, and storms should become more organized. Indeed, $\sqrt{\mathrm{Ri}}$ less than 2–2.5 is a general condition for supercells to form, at least in numerical simulations. Consistently, storm splitting in unidirectional shear occurs below a critical value of $\sqrt{\mathrm{Ri}}$ around 2–2.5. In mathematics, such a complete change in behavior at a critical value of a parameter is called a *bifurcation*.

Storms are affected by the shape of the hodograph as well as by its length. We now describe the behavior of supercell storms in two extreme environments, characterized by hodographs that are a straight line and a complete circle (Fig. 5.8). Nature occasionally produces hodographs that approximate one or other of these cases (Fig. 5.9). More importantly, by going from one extreme to the other we pass through a spectrum of different storm behaviors.

2) THE STRAIGHT-LINE HODOGRAPH

The rotational characteristics of convective storms growing in a unidirectionally sheared environment have been modeled analytically (Rotunno 1981) and numerically (e.g., Schlesinger 1978; Wilhelmson and

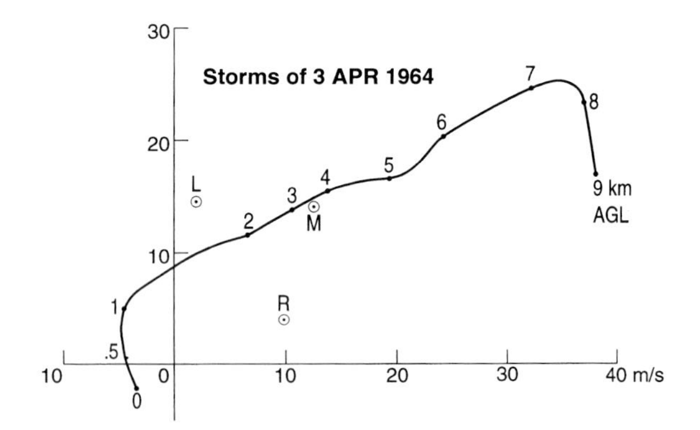

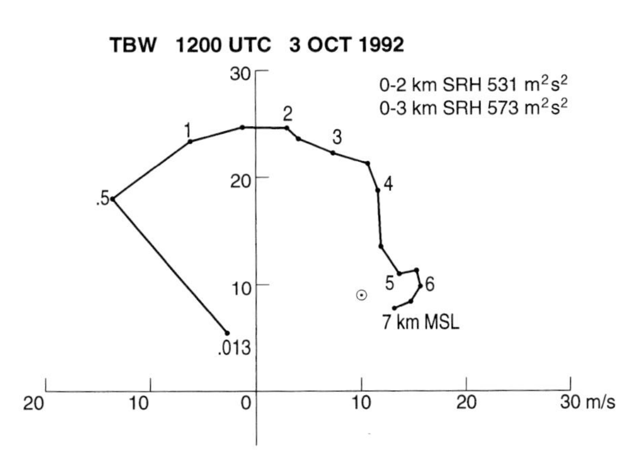

FIG. 5.9. Examples of actual hodographs that approximate (a) a straight hodograph, (b) a circle hodograph.

Klemp 1978). The results of the three-dimensional supercell simulations suffice to illustrate the nonlinear effects that lead to storm splitting, off-hodograph storm motion, and consequential updraft rotation.

For purposes of discussion, assume that the shear is westerly. If the earth's rotation is switched off, the numerical fields possess north–south symmetry or antisymmetry (depending on the variable) about a west–east vertical plane (the plane of symmetry). In agreement with linear theory, the storm moves initially with a velocity that lies on the hodograph and the updraft tilts the northward environmental vorticity, producing a vortex pair that is most intense at midlevels (e.g., Wilhelmson and Klemp 1978). Thus, the southern (northern) half of the updraft rotates cyclonically (anticyclonically), but the updraft does not rotate as a whole (Fig. 5.10).

At around 40 min into a simulation with strong shear, the initial storm splits along the plane of symmetry into two storms—a severe left-moving (SL) supercell and a severe right-moving (SR) supercell—that are mirror images of each other. The southern (northern) storm moves southeast (northeast), that is, off the hodograph to the right (left) of the shear vector. The initial updraft splits as a result of two nonlinear effects. First, precipitation accumulates in the center of the updraft and loads it down to such an extent that the updraft turns into a downdraft in the vicinity of the symmetry plane. Second, the midlevel vortices are centers of low pressure that induce new updraft growth on the flank. This new growth is enhanced by low-level convergence along the storm's gust front where cool, outflowing downdraft air converges with warm, moist inflow air. The southern (northern) storm propa-

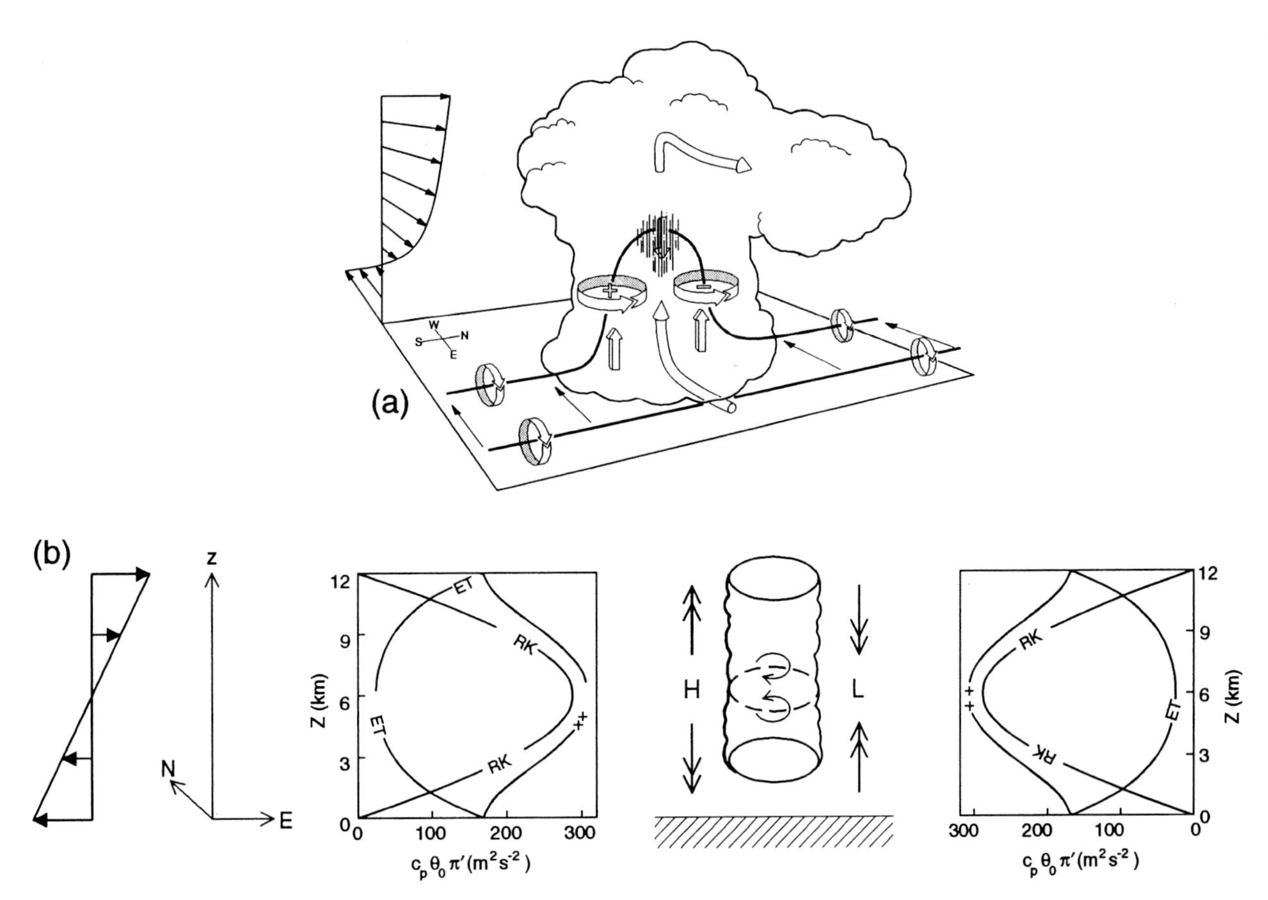

FIG. 5.10. (a) A storm in westerly shear prior to the onset of splitting. The cylindrical arrows depict the flow in the storm. The thick lines are vortex lines with the sense of rotation indicated by the circular-ribbon arrows. The shaded arrows give the direction of the VPPGF. From Klemp (1987; reprinted with permission). The updraft draws up loops of environmental vortex tubes, producing a cyclonic vortex on its right side and an anticyclonic vortex on its left side. (b) The linear pressure as a function of height on the upshear and downshear side of the updraft. The locations of the high and low are marked by H and L. The double arrows denote the direction of the VPPGF. The RK curves show the pressure according to Rotunno and Klemp's (1981) heuristic theory. The ++ curves include an extra term derived by Davies-Jones (1996b) to enable satisfaction of boundary conditions on pressure. ET is the extra term that has to be added to RK in order to satisfy the boundary conditions. The curved arrows depict the midlatitude vortices and associated lows of nonlinear pressure on the flanks of the updraft. From Davies-Jones (1996b).

gates to the right (left) of the shear vector as a result of updraft growth on its right (left) flank and decay on the opposite flank. By virtue of this propagation, the inflow into the right (left) mover has storm-relative streamwise (antistreamwise) vorticity. Thus, the right- (left-) moving updraft acquires overall cyclonic (anticyclonic) rotation as predicted by linear theory, given the "nonlinear storm motion," and the storms quickly become supercells. After the storms become widely separated, they may continue their leftward and rightward movement for several hours. The initial deviate motion from the split initiates overall updraft rotation, and the updraft rotation maintains the deviate motion that it needs for its own sustenance by the following mechanism. In the right- (left-) moving storm, the cyclonic (anticyclonic) vortex is the "leading vortex" because it is located close the the updraft maximum on its forward side, and the anticyclonic (cyclonic) vortex, which is in lesser updraft or even in downdraft above the cold pool on the storm's rear flank, is the "trailing vortex." Both midlevel vortices exert suction on the low-level air beneath them. However, the leading vortex has far more influence on updraft propagation because it is closer to the updraft maximum and because it lifts the more unstable air ahead of the updraft. Thus the updraft continues its deviate motion toward the leading vortex.

In nature perfect symmetry of SL and SR supercells is never realized because of Coriolis effects. First, amplification of the background vertical vorticity of the earth's rotation makes the southern updraft rotate slightly faster than the northern updraft. Second, frictional effects near the ground cause the hodograph to turn clockwise with height in the lowest 1 km (the Ekman layer), and this turning is enhanced by low-level warm advection. As evident in the veering of the near-surface winds relative to the left-moving storm in Fig. 5.9a, this can result in the "wrong" (i.e., streamwise) vorticity entering the left mover's updraft at very low levels. This may be the reason why left movers seldom produce tornadoes. The statistic that mesocyclones are about 50 times more frequent than mesoanticyclones (Davies-Jones 1986) indicates that nearly straight hodographs are comparatively rare.

3) THE CIRCLE HODOGRAPH

It is convenient to consider also the case of a circle hodograph, since in this case there is an exact steady-state solution of the inviscid, Boussinesq equations of motion[1] (Lilly 1982, 1983; Davies-Jones 1985). The solution represents a dry updraft in a neutrally stable environment. The governing equations are

$$\frac{\partial \mathbf{v}}{\partial t} + \nabla\left(\frac{\mathbf{v}\cdot\mathbf{v}}{2}\right) - \mathbf{v}\times\boldsymbol{\omega} = -\nabla\Pi - \nabla(gz), \quad (5.4)$$

$$\nabla\cdot(\mathbf{v}) = 0, \quad (5.5)$$

where $\Pi \equiv c_p\theta_0(p/p_0)^\kappa$, $\kappa \equiv R/c_p$, and θ_0 is the constant potential temperature of the dry adiabatic atmosphere. The price that one has to pay for the exact solution is the unrealistic absence of buoyancy. However, the exact solution is still useful because storms in environments with fairly low CAPE and quasi-circular hodographs with strong shear will strive toward this steady state without ever quite reaching it.

The exact solution belongs to the Beltrami class of flows. In a Beltrami flow, vorticity is everywhere parallel to velocity, that is, the Lamb vector $\mathbf{v}\times\boldsymbol{\omega}$ vanishes and

$$\boldsymbol{\omega} \equiv \nabla\times\mathbf{v} = \lambda\mathbf{v}, \quad (5.6)$$

where λ is a scalar called the abnormality (Truesdell 1954). The superposition of two Beltrami flows is also a Beltrami flow if—and only if—the abnormalities are the same. In our case we will superpose one Beltrami flow, which represents the "perturbed flow" associated with a rotating updraft, on another Beltrami flow, which represents a steady-state, horizontally homogeneous environment. The abnormality becomes the rate at which the storm-relative winds veer with height. Vorticity is divergence-free, so λ must satisfy the constraint

$$\nabla\cdot(\lambda\mathbf{v}) = 0. \quad (5.7)$$

Hence when (5.5) holds, λ must be a constant. The vector vorticity equation reduces simply to

$$\partial\boldsymbol{\omega}/\partial t = 0, \quad (5.8)$$

which implies that the solution we are seeking is steady. The vorticity tendency is zero because in the vector vorticity equation the solenoid and diffusion terms are zero and the advection term is exactly canceled by the stretching and tilting term.

As shown by Davies-Jones (1985), a Beltrami solution exists for an axisymmetric, rotating updraft within a Beltrami-flow-modeled environment that is characterized by an arc-shaped hodograph that turns more than 180°. A hodograph in the form of a clockwise-turning complete circle with center at the origin and a radius of 20 m s^{-1} is chosen here because Lilly (1982) found reasonable agreement in this case between the

[1] An exact solution also exists if the anelastic approximation is made in the continuity equation. We wish only to reveal the basic properties of the flow. Hence, the Boussinesq approximation is made, which simplifies the solutions. Strictly, the Boussinesq approximation is valid only for shallow convection. However, the solutions for deep convection are qualitatively similar.

nonbuoyant Beltrami solution and a numerically simulated storm in a moderately unstable environment. For instance, the simulated storm was quasi-steady and nearly stationary, and it had a strongly rotating updraft. Furthermore, the ratio of maximum vertical vorticity to maximum vertical velocity in the simulation agreed with the theoretical value to within 20%.

For a circle hodograph and values of W_0 (30 m s^{-1}) and H (12 km) characteristic of severe storms, where W_0 is a constant equal to the maximum vertical velocity and H is the depth of the updraft, the central flow consists of a rotating updraft that resembles a midlevel mesocyclone in structure, diameter, and intensity. Since the flow is Beltrami and steady, the streamlines, trajectories, and vortex lines all coincide. The trajectory through the midpoint of the updraft $(r, z) = (0, H/2)$ turns anticyclonically, even though the updraft is rotating cyclonically, because the effect of environmental wind veering overcomes that of cyclonic circulation (Klemp et al. 1981; Lilly 1983). Such trajectories have been observed in Doppler radar analyses and numerical simulations of rotating storms. The trajectories and vortex lines near the ground are almost horizontal; thus parcels entering the updraft do not develop significant cyclonic spin until they ascend above $H/6$.

For a steady Beltrami flow, Eq. (5.4) dictates that the pressure obeys the universal Bernoulli relationship

$$\Pi \quad + \quad gz$$
specific [enthalpy + potential energy
$$+ \quad \mathbf{v}\cdot\mathbf{v}/2 \quad = \quad C,$$
+ kinetic energy] = constant , (5.9)

a consequence of the Lamb vector being zero. We now use the perturbed and environmental wind solutions in this relationship to obtain the pressure. The solution for pressure can be decomposed into a hydrostatic environmental part $\bar{\Pi}$, a linear part due to the storm-environment interaction Π'_L, and a nonlinear part due to intrastorm interactions Π'_{NL} where

$$\bar{\Pi} = -gz + C - M^2/2,$$
$$\Pi'_L = -\bar{\mathbf{v}}\cdot\mathbf{v}',$$
$$\Pi'_{NL} = -\mathbf{v}'\cdot\mathbf{v}'/2. \qquad (5.10)$$

The pressure field has some surprising features. The flow is not in cyclostrophic balance. The nonlinear pressure is axisymmetric and so does not contribute to storm propagation. On the axis, it is proportional to W_0^2 with a spin-forced deficit of $H/2$ of 3 mb when $W_0 = 30$ m s^{-1}. There is no axial pressure deficit at the ground and no axial pressure excess at the top of the updraft. In contrast to the straight-hodograph case, the nonlinear pressure does not feature two lows on opposite flanks of the updraft, because the cyclonic vortex is now coincident with the updraft and the anticyclonic vorticity is in the ring of downdraft that surrounds the updraft. Consequently, the storm-splitting mechanism is absent here. Instead, the low at the updraft's midpoint exerts a suction on air beneath it, thus enhancing the updraft at low levels (Lilly 1986). Brooks and Wilhelmson (1993) demonstrated this effect by showing that updraft speed increases with hodograph curvature in simulated supercells.

The linear pressure is asymmetric and is proportional to MW_0, where M is the environmental wind speed. For $M = 20$ m s^{-1} it is comparable in magnitude to the axisymmetric component. At each level the linear pressure is low on the flank of the updraft where the deviation wind is in the same direction as the environmental wind, and high on the opposite flank (Fig. 5.4). Consequently, the highs and lows twist 180° around the rotating updraft from the ground to the top. Only at the midlevel is the linear horizontal pressure gradient in the direction of the shear vector (Rotunno and Klemp 1982). At the top, the pressure field resembles that predicted by the "flow around an obstacle" model of Newton and Newton (1959). At other levels the updraft is rather porous. It is evident from the asymmetric part of the vertical equation of motion,

$$-\frac{\partial \Pi'}{\partial z} = (\bar{U} + U')\frac{\partial w'}{\partial r} + w'\frac{\partial w'}{\partial z}, \qquad (5.11)$$

that the maximum VPPGF is located on the upwind side of the updraft, as in linear theory. The whole updraft does not propagate toward this side because, in the absence of buoyancy, the asymmetric part of the VPPGF is exactly balanced by the advection of vertical velocity by the environmental wind, $-\bar{U}\partial w'/\partial r$. This simple analytical solution, and a companion numerical simulation of a supercell in a moderately unstable environment with the same hodograph (Lilly 1982, 1983), demonstrate that updrafts do not have to propagate relative to the mean wind in order to rotate, and that rotating updrafts do not necessarily propagate.

The Beltrami model offers an explanation of why, in surface observations of mesocyclones, the mesolow is often several kilometers ahead of the circulation center (Barnes 1978). This phenomenon is also apparent in many tornado pressure traces where the mesolow precedes the abrupt V-shaped fall associated with the tornado (which, based on Doppler radar data, is usually near the circulation center) by several minutes. In the storm's inflow the environmental and storm-induced winds are aligned, giving rise to an "inflow low" with a pressure deficit of a few millibars through the Bernoulli effect (Davies-Jones 1985; Brooks et al. 1993). The presence of the inflow low, along with lower hydrostatic pressure in the inflow, separates the mesolow center from the circulation center. The inflow

low is the location of the severe weather in our example, surface winds of 30 m s^{-1}. Stronger winds are present at the midlevel owing to the tangential and vertical components of the perturbation wind being greater than the radial component.

4) Intermediate cases

Traditionally, supercells in general environments have been viewed as modifications of ones in unidirectional shear (e.g., Klemp 1987). The effect of clockwise turning of the shear vector with height has been viewed as enhancing (suppressing) the right- (left-) moving storm. This is a valid viewpoint for slightly curved hodographs, but does not seem appropriate for environments with strong hodograph curvature in which initial updrafts rotate and storm splitting either does not occur or is dynamically insignificant (Lilly 1983; Klemp 1987). An alternative viewpoint consists of recognizing the two idealized extremes (straight-line and circle hodographs). With increasing clockwise hodograph curvature, splitting disappears, the cyclonic vortex moves closer to the center of the updraft, linear effects gain in importance, and updraft rotation becomes increasingly disassociated from deviate updraft motion (Table 5.1).

d. Theory of rotation near the ground

As mentioned, rotation within the storm in roughly the lowest kilometer above the ground seems to develop from a separate mechanism than the one described in sections 5.2b and 5.2c. Although meso- or synoptic-scale vertical vorticity that preexists the storm has apparently been linked to near-ground mesocyclogenesis (e.g., Wakimoto et al. 1998), we limit the following discussion to near-ground mesocyclogenesis mechanisms involving storm-scale processes, since these have drawn most of the attention in the recent literature.

Significant near-ground rotation does not develop in the numerical simulations of Rotunno and Klemp (1985) and Davies-Jones and Brooks (1993) when the production of hydrometeors is turned off. This result was predicted by Davies-Jones (1982), who argued that, in a sheared environment with negligible background vertical vorticity, an "in, up, and out" type circulation driven by forces primarily aloft would fail to produce rotation close to the ground because vertical vorticity is generated in rising air. A strong updraft that is rotating only aloft may draw strong inflow from below, but its strongest winds are still high in the clouds; it seldom produces damaging winds at the ground without an accompanying downdraft. Rotation can be produced near the ground if the streamlines and vortex lines turned upward abruptly, instead of gradually, as in the Beltrami model. One way this could occur is by the steep uplifting of streamwise vortex lines by a gust front. The effect is not important in published numerical simulations, however, since it is not resolved by most numerical models and yet the models reproduce low-level mesocyclogenesis. The only other obvious way requires intense low pressure at the ground beneath the updraft, which in turn requires near-ground rotation! The dynamic pipe effect (DPE), which builds some tornadoes down to the ground via a type of bootstrap process, could lead to such a situation. However, the Beltrami mesocyclone is not in cyclostrophic balance, so that the DPE probably does not apply to mesocyclones in nature. Thus, streamwise vorticity in air flowing along the ground cannot cause a mesocyclone to make contact with the ground although it could, in principle, help to maintain an already established near-ground mesocyclone.

TABLE 5.1. Comparison of the properties of supercell storms in environments with nearly straight hodographs and with strongly curved hodographs that turn clockwise with height.

Property	(Nearly) straight hodograph	Strongly curved hodograph
Left/right symmetry	For straight hodograph & $\Omega = 0$	None
Net updraft rotation in initial storm	No	Yes
Cyclonic vortex in initial storm	On right side of updraft	In strong updraft
Anticyclonic vortex in initial storm	On left side of updraft	In downdraft or weak updraft
Storm splitting	Highly significant	Insignificant or absent
Deviate motion	Essential for mesocyclone (meso.)	Not needed for meso. formation
Time to first mesocyclone (meso.)	Slower	Faster
Low and midlevel meso. intensity	Generally less intense	Generally more intense
Mesoanticyclone	In left mover	Generally absent
Updraft strength in right mover (RM)	Weaker	Stronger, esp. at low $\sqrt{Ri}$
RM strength vs $\sqrt{Ri}$	Maximum at $\sqrt{Ri} \approx 1.7$	Increases with decreasing $\sqrt{Ri}$
Near-ground ζ vs $\sqrt{Ri}$	Maximum at $\sqrt{Ri} \approx 1.7$	Increases with decreasing $\sqrt{Ri}$
Barotropic and baroclinic vorticity	In opposite directions	In same direction
Violent tornado possible?	Yes	Yes
Tornado outbreak	Rare	More likely
Rotational dynamics	Highly nonlinear	More linear

Consistent with the numerical simulations, field observations confirm that tornadoes do not generally occur in the absence of rainy downdrafts. Water-loaded and/or rain-cooled downdrafts can cause strong surface winds and/or generate vertical vorticity by several mechanisms. First, in contrast to a nonbuoyant jet that slows down considerably as it approaches a boundary, a negatively buoyant downdraft can impact the ground with considerable force and spread out rapidly (Fujita 1985). Second, a downdraft may transport high-momentum air down to the surface. This may take the form of downward transport of horizontal momentum, which is associated with tilting of existing horizontal vorticity toward the vertical (Wiin-Nielsen 1973; Walko 1993) or of transport, within a mesocyclone, of angular momentum downward and inward toward the mesocyclone center. Finally, a cool downdraft affords the baroclinic generation of horizontal vorticity in the form of a toroidal circulation around the downdraft perimeter. This vorticity may also be tilted by differential vertical velocity.

1) BAROCLINIC MECHANISM

Klemp and Rotunno (1983) and Rotunno and Klemp (1985) showed that air entering the mesocyclone near the ground had traveled in the forward flank region of their simulated storm, along a strong baroclinic zone with warmer (cooler) air to its left (right), and acquired a large amount of streamwise vorticity generated by the associated buoyancy torque. As this advancing, moderately chilled air is ingested into the updraft, its spin is tilted upward and amplified by vertical stretching, giving rise to slightly elevated rotation because the vertical spin develops as the air is rising. In some fortuitous encounters, the cool-air outflow may come from a neighboring storm instead of the storm's own downdraft. Davies-Jones and Brooks (1993) and Davies-Jones (1996a, 2000a) demonstrated that near-ground cyclonic vorticity develops first in the downdraft on the left side of its centerline (looking downwind). The passage of cyclonically spinning air from the downdraft into the updraft and subsequent vertical stretching of this air completes a cyclonic vortex column from very close to the ground to midaltitudes or higher. We now review the concepts of baroclinic and barotropic vorticity, and then apply these to theory that explains the baroclinic generation of vorticity in a downdraft.

Dutton (1976) showed from an integral of the vector vorticity equation that absolute vorticity in dry, inviscid, isentropic flows is the sum of barotropic and baroclinic components, that is,

$$(\xi, \eta, \zeta + f) \equiv \boldsymbol{\omega}(x, y, z, t) = \boldsymbol{\omega}_{\mathrm{BT}}(x, y, z, t) + \boldsymbol{\omega}_{\mathrm{BC}}(x, y, z, t), \quad (5.12)$$

where the components satisfy the initial conditions

$$\boldsymbol{\omega}_{\mathrm{BT}}(x, y, z, 0) = \boldsymbol{\omega}(x, y, z, 0), \quad \boldsymbol{\omega}_{\mathrm{BC}}(x, y, z, 0) = 0 \quad (5.13)$$

at some arbitrarily chosen initial time $t = 0$; $\boldsymbol{\omega} = \boldsymbol{\omega}_{\mathrm{BC}}$ is a solution of the inviscid vector vorticity equation for isentropic flow and $\boldsymbol{\omega} = \boldsymbol{\omega}_{\mathrm{BT}}$ satisfies the inviscid barotropic version of the same equation (without the solenoidal term), which is the necessary and sufficient condition for frozen vortex lines (Borisenko and Tarpov 1979).

Physically, the barotropic vorticity at time t is the vorticity that would develop from the amplification and reorientation of initial relative vorticity and the earth's vorticity if there were no baroclinic generation of vorticity. Barotropic vortex lines are "frozen" into the fluid and behave like elastic strings that the flow moves, stretches, and reorients. Thus, the barotropic vorticity of a parcel depends on initial vorticity, and the initial and current positions of the parcel and its neighbors along the vortex line through the parcel, but not on the parcel positions at intermediate times. The baroclinic vorticity of a parcel, given by

$$\boldsymbol{\omega}_{\mathrm{BC}} = \nabla\Lambda \times \nabla S, \quad (5.14)$$

where $\Lambda \equiv \int_0^t T d\tau$ following a parcel, T is temperature, and S is entropy, depends on its cumulative temperature since $t = 0$ (Dutton 1976); the vorticity is subsequently affected by vortex-tube stretching and tilting. The baroclinic vortex lines lie in isentropic surfaces; hence the baroclinic component does not contribute to potential vorticity.

A modification of Eq. (5.14) is used in a Lagrangian model by Davies-Jones (2000a) to deduce analytically that cyclonic (anticyclonic) vertical vorticity forms from baroclinic vorticity on the left (right) side of a downdraft. He showed that the baroclinic vorticity can be expressed as

$$\boldsymbol{\omega}_{\mathrm{BC}} = (\xi, \eta, \zeta)_{\mathrm{BC}} = N^2\left(-\frac{\partial H}{\partial y}, \frac{\partial H}{\partial x}, \frac{\partial H}{\partial x}\frac{\partial h}{\partial y} - \frac{\partial H}{\partial y}\frac{\partial h}{\partial x}\right), \quad (5.15a)$$

where $N^2 \equiv (g/c_p)\partial S/\partial z$ is the local static stability or the square of the Brunt–Väisälä frequency ($N^2 < 0$ for unstable stratification); $h(x, y, S, t)$ is the height of an isentropic surface of entropy S, which has height $h_{-\infty}$ at upstream infinity; $h'(x, y, s, t) \equiv h(x, y, S, t) - h_{-\infty}$ is the "perturbation" height of a point on an isentropic surface, relative to the height of the surface at upstream infinity; and

$$H(x, y, S, t) \equiv \int_0^t h'(X, Y, S, \tau)d\tau$$

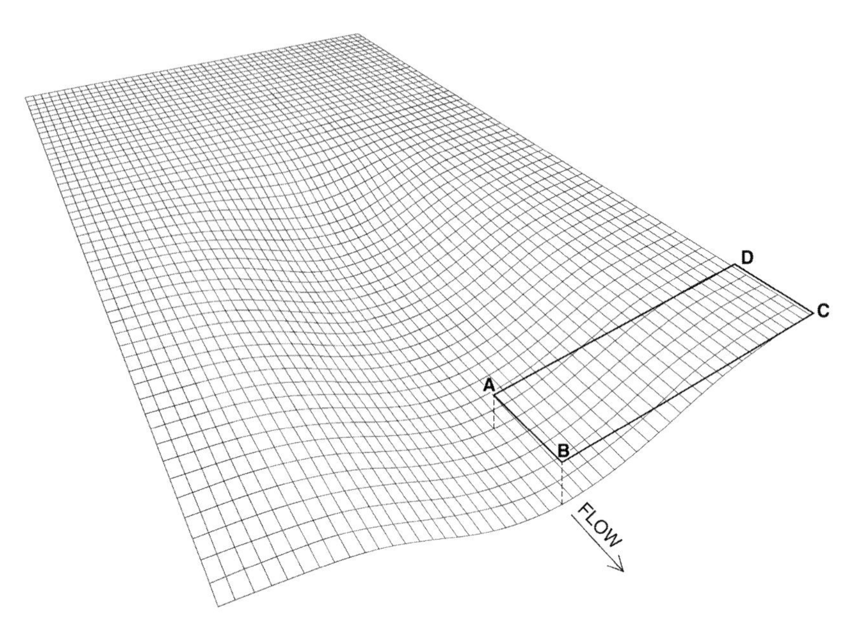

FIG. 5.11. Example of an isentropic surface in a horizontally homogeneous upstream environment. The isentropic surface has the shape of a semi-infinite valley extending downstream. ABCD is a material area, which far downstream is rectangular, and level at time t. Side CD is in the surface, but AB, which lies along the x axis, is in a higher isentropic surface. The baroclinic term in the vorticity equation continually generates vorticity within the surface in the shape of hairpins.

is the cumulative height perturbation, computed following the parcel initially located in isotropic coordinates at (X, Y, S). Note from Eq. (5.15a) that

$$\zeta_{\mathrm{BC}} = \xi_{\mathrm{BC}}\frac{\partial h}{\partial x} + \eta_{\mathrm{BC}}\frac{\partial h}{\partial y}. \tag{5.15b}$$

The vorticity field described by (5.15) has the following properties. The quantity N^2H is the "streamfunction" of the horizontal vorticity field (Figs. 5.11, 5.12). In other words, the horizontal projections of the vortex lines are also contours of H, the magnitude of horizontal vorticity is proportional to the spacing of the H contours and the local static stability, and the direction is given by the rule that historically warmer (cooler) air is located to the left (right) side of an observer looking along the local horizontal vorticity vector. Because the vortex lines lie in the isentropic surface (zero potential vorticity), the vertical vorticity is cyclonic (anticyclonic) where the horizontal vorticity vector points across the height contours toward higher (lower) values (Fig. 5.12). Height contours coincide with contours of baroclinic generation of vorticity and also with isotherms. The magnitude of the vertical vorticity is proportional to N^2 and the number of solenoids of H and h per unit area. If $|\nabla h| \ll 1$, the vertical vorticity is much smaller than the horizontal vorticity.

As with the linear theory of midlevel rotation, we again illustrate the vortex dynamics using isentropic surfaces that are deformed in some simple way. The isentropic surface height field can be provided as in Davies-Jones (2000a) through a three-dimensional potential-flow solution of the inviscid momentum and continuity equations in the absence of stratification and shear; the solution constitutes the primary flow, in a primary-flow–secondary-flow approach (Taylor 1972; Scorer 1978). The secondary flow is the first-order correction arising from the inclusion of stratification and vertical shear as functions of the height ($h_{-\infty}$) in a horizontally homogeneous upstream environment. The secondary vorticity is determined via Eq. (5.15) by the environmental stratification and shear and by the primary flow; it does not modify the primary flow and, hence, acts as a passive vector, an obvious limitation of the model.

Consider the flow under an axisymmetric fairing, placed in a uniform stream $U\mathbf{i}$ (Batchelor 1967) (Figs. 5.11, 5.13). This primary-flow case of Davies-Jones (2000a) illustrates the danger that may be latent in narrow currents of descending, rain-cooled air. Assuming unstable stratification, the air that sinks adiabatically is cooler than its surroundings and the baroclinic term continually generates vorticity in the isentropic surfaces (Fig. 5.11). Mesocyclonic-scale cyclonic (anticyclonic) vertical vorticity originates from cumulative tilting of baroclinically generated crosswise vorticity [the $\eta\partial h/\partial y$ term in (5.15b)] on the north (south) side of the depression in the isentropic

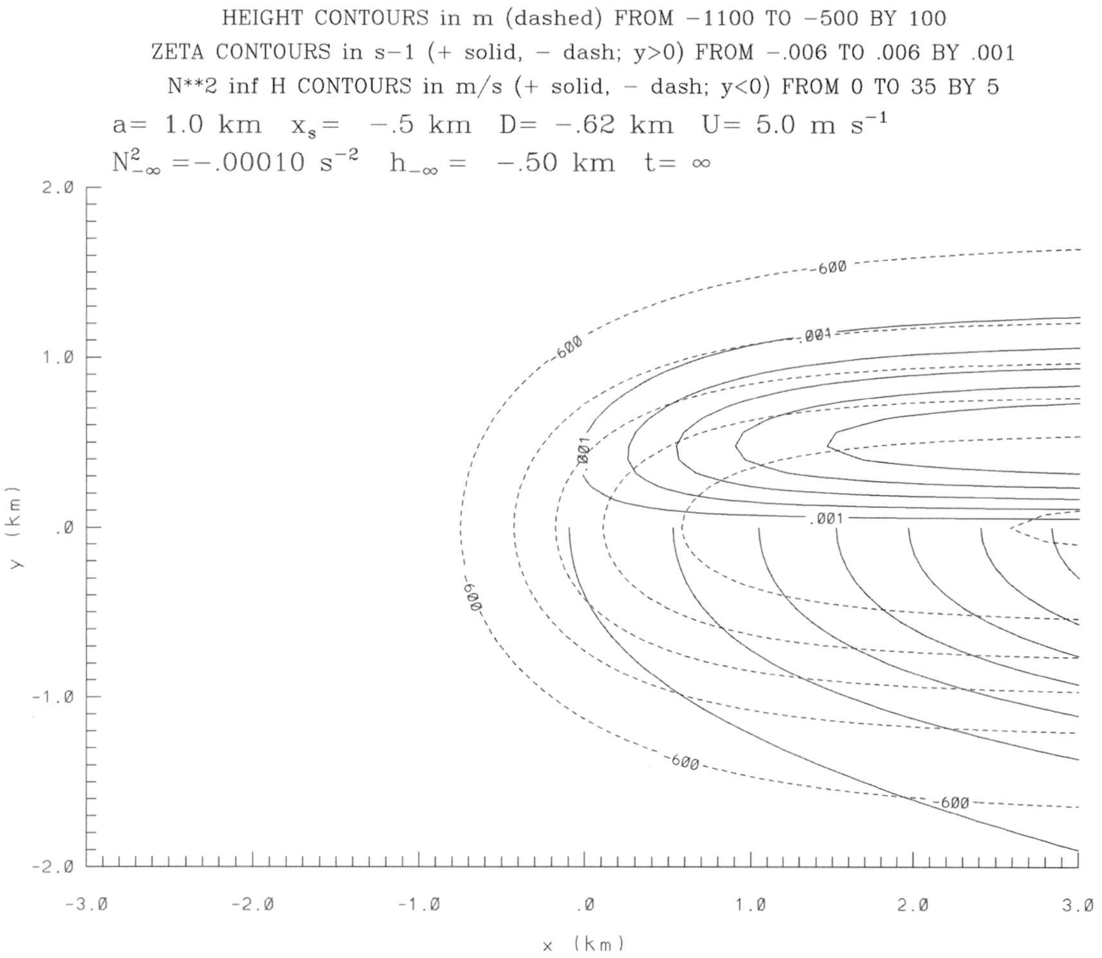

HEIGHT CONTOURS in m (dashed) FROM −1100 TO −500 BY 100
ZETA CONTOURS in s−1 (+ solid, − dash; y>0) FROM −.006 TO .006 BY .001
N**2 inf H CONTOURS in m/s (+ solid, − dash; y<0) FROM 0 TO 35 BY 5
a= 1.0 km x_s= −.5 km D= −.62 km U= 5.0 m s^{-1}
$N^2_{-\infty}$ =−.00010 s^{-2} $h_{-\infty}$ = −.50 km t= ∞

FIG. 5.12. Height contours (dashed) of the isentropic surface depicted in Fig. 5.11, contours of cumulative height (solid, drawn only for $y \leq 0$), and contours of vertical vorticity (solid, drawn only for $y > 0$). The accumulated height and vertical vorticity fields are symmetric and antisymmetric, respectively, about $y = 0$. In the surface, temperature contours and contours of baroclinic generation of vorticity both coincide with the height contours, and cumulative temperature and vortex lines both coincide with the cumulative height contours.

surfaces (Fig. 5.12). Cumulative tilting of baroclinically generated streamwise vorticity, $\xi \partial h/\partial x$, produces vertical vorticity of the opposite sign during the descent and vanishes at downstream infinity.

We can also explain the development of rotation about a vertical axis in terms of circulation Γ instead of vorticity. The origin of the circulation can be determined by advecting the material area ABCD (Figs. 5.11, 5.14)—which is far downstream, rectangular, and level at time t—backward in time. The side AB lies on the axis and lies in the trough of a surface of constant entropy, say, S_1. The side CD on the left side of the flow lies in a surface of higher constant entropy, say, $S_2 > S_1$ because the stratification is unstable. At an earlier time the back edge DA is higher than the leading edge BC so that the mean temperature along side DA, $\bar{T}_{DA}$, is lower than $\bar{T}_{BC}$. From Bjerknes's circulation theorem (Dutton 1976),

$$\frac{d\Gamma}{dt} = \oint_{ABCD} T dS$$

$$\approx (S_2 - S_1)\bar{T}_{BC} + (S_1 - S_2)\bar{T}_{DA} > 0, \quad (5.16)$$

so positive circulation around the material circuit is generated in the downdraft. The contour and its circulation evolve in a similar way to those in diagnostic studies of the origins of near-ground rotation in numerical simulations (Rotunno and Klemp 1985; Davies-Jones and Brooks 1993; Trapp and Fiedler 1995). Thus, the essential physics of rotation near the ground seems to be contained in the analytical model.

A scale analysis of Eq. (5.15) (Davies-Jones 2000a) indicates that long, narrow, deep downdrafts with moderate flow through them in highly unstable envi-

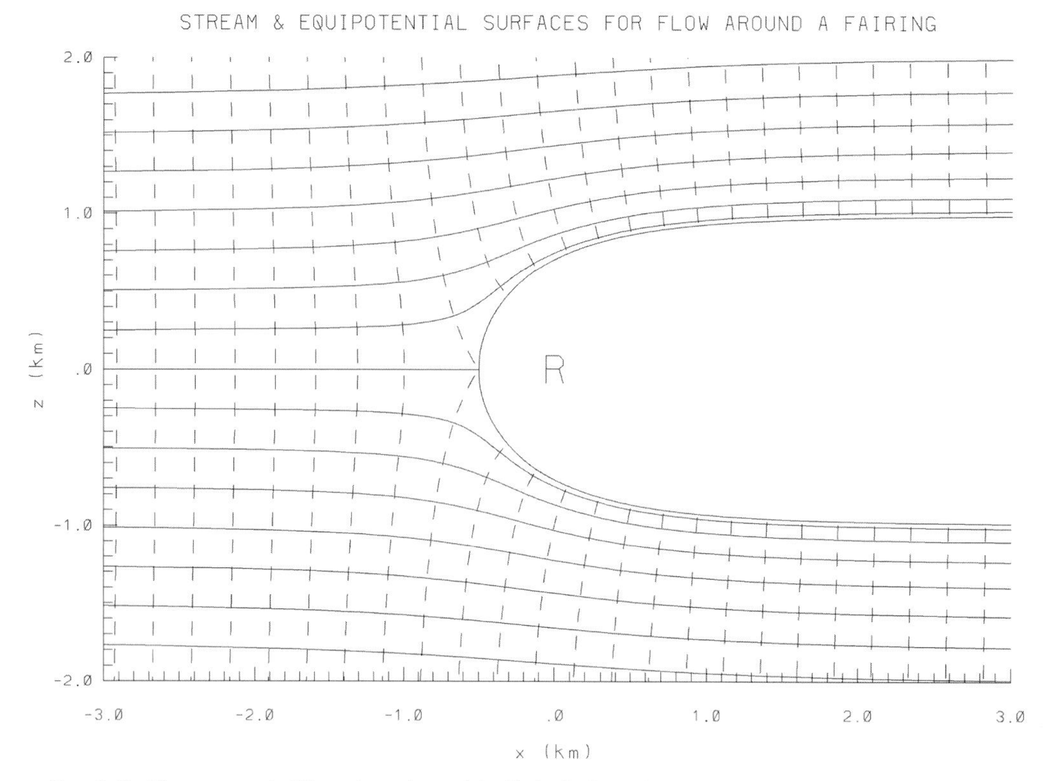

FIG. 5.13. The stream (solid) and equipotential (dashed) lines in the $y = 0$ plane for three-dimensional irrotational flow around an axisymmetric semi-infinite fairing of 2-km maximum width. The flow is in the positive x direction and is uniform at upstream infinity. The stream surfaces are the surfaces of revolution obtained by revolving the streamlines about the x axis. R is the location of the source that when placed in the uniform stream gives rise to the flow as shown. The isentropic surface in Fig. 5.11 is the material surface of this flow that consists of the union of all streamlines with the same height, −0.5 km, at upstream infinity. The thickly drawn streamline is the intersection of the isentropic surface with the $y = 0$ plane.

ronments produce the most vorticity in a moderate time (the vorticity forms on the advective timescale). This scale analysis, and also the analysis of Klemp and Rotunno (1983, p. 365), indicates additionally that the baroclinic vorticity is proportional to U^{-1}, where U is an advective velocity scale. Hence if the flow is too strong the parcels pass through the baroclinic zone too quickly to acquire appreciable baroclinic vorticity. On the other hand, if the flow is too weak, the parcels take too long to pass through the zone and the process becomes vulnerable to disruption by other events such as the cool air spreading out and undercutting the main updraft.

We note that the ultimate result of a near-ground mesocyclone may be influenced positively or negatively by barotropic vorticity (which, again, depends on the initial vorticity), as demonstrated by the model simulations of Davies-Jones and Brooks (1993) and Walko (1993). Consider, for example, the case when the initial vorticity is purely positive (negative) crosswise: Tilting of barotropic vorticity produces cyclonic (anticyclonic) vertical vorticity on the left side of the downdraft and the opposite sign of vorticity on the right side of the downdraft. When the initial vorticity is purely streamwise (antistreamwise), tilting gives rise to anticyclonic (cyclonic) vorticity during descent but the vertical component vanishes where the streamlines bottom out. As the air reascends, tilting of barotropic vorticity imparts cyclonic (anticyclonic) vorticity to the updraft.

Synthesizing the results of Davies-Jones (2000a), Davies-Jones and Brooks (1993), and Rotunno and Klemp (1985), among others, we can conceptualize an answer to the question of how near-ground rotation forms baroclinically in a supercell: Initially, the supercell's main updraft has strong cyclonic rotation aloft as a result of tilting of streamwise vorticity associated with the storm-relative environmental winds veering with height. There is little rotation near the ground at this stage because the cyclonic vorticity is being generated in rising air. The mesocyclonic rotation draws a thin curtain of rain around the rear side of the mesocyclone, where it falls into dry air that is overtaking the storm and is diverted around the updraft. As it enters and flows through the precipitation region, the dry air is cooled and moistened by evaporating rain, becomes negatively buoyant, and descends to near the ground. If the air flow is cyclonically curved instead of

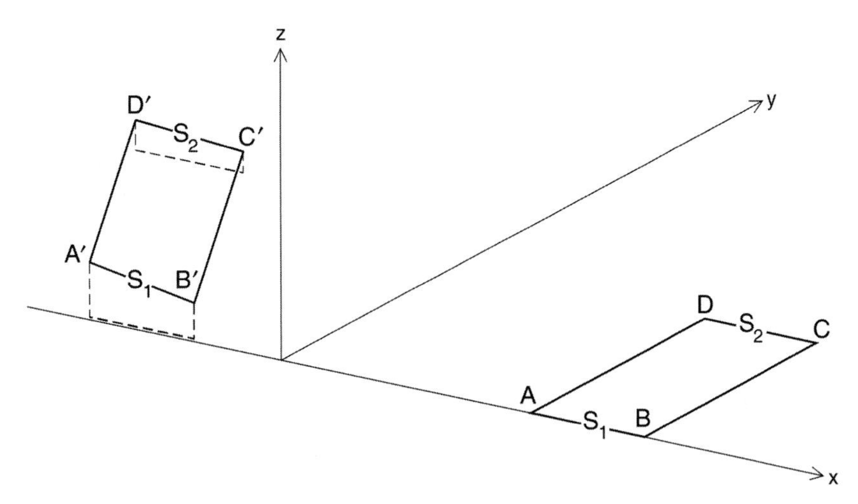

FIG. 5.14. Schematic showing the material circuit in Fig. 5.11 at time t (ABCD) and at an earlier time when it was farther upstream (A′B′C′D′). The side A′D′ is higher and cooler on average than the side B′C′. The material lines AB and CD lie in the surfaces of constant entropy S_1 and S_2, respectively. In unstable stratification $S_1 < S_2$ and the circuit acquires positive circulation according to Bjerknes's circulation theorem.

straight as in the analytical model, crosswise vorticity will be converted into streamwise vorticity by the "river-bend effect" (Fig. 5.15; Shapiro 1972; Scorer 1978; Adlerman et al. 1999). Horizontal vorticity is generated solenoidally at the entrance and along the sides of the downdraft (Fig. 5.12). The vortex lines are tilted upward in the downdraft as described above, producing cyclonic (anticyclonic) vorticity on the left (right) side of the downdraft. The air then exits the downdraft and flows along the ground. The storm's main updraft, which is rotating strongly aloft, lies

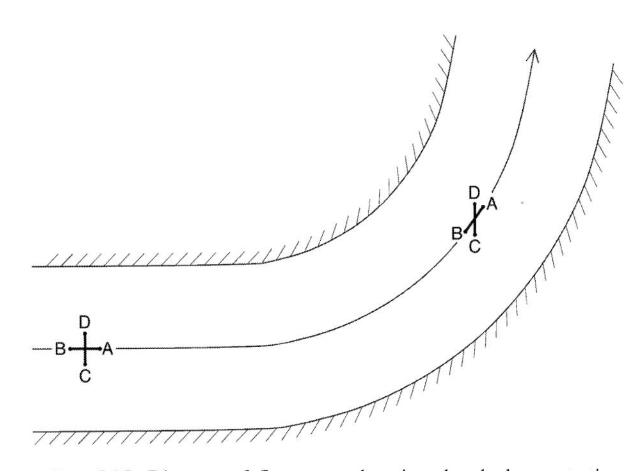

FIG. 5.15. Diagram of flow around a river bend, demonstrating the development of streamwise vorticity. Upstream of the bend the flow is parallel with speed shear (crosswise vorticity) owing to friction at the river bottom. Consider the fluid cross ABCD with arm AB along a streamline and CD along a vortex line. Since flow around the bend generates no vertical vorticity to a first approximation, the arms of the cross must rotate in opposite directions. Thus the vortex line CD turns toward the streamwise direction AB.

ahead and to the left of the advancing air (Fig. 5.3), a location that breaks the left–right symmetry that the analytical model is forced to have for the sake of an easy solution. The main rotating updraft sucks up cyclonically rotating, almost saturated, rain-cooled air (through the suction effect described in section 5.3) and stretches it vertically, thus amplifying the vertical vorticity close to the ground and creating a rapidly rotating wall cloud in the process (Rotunno and Klemp 1985). As this air rises, its streamwise vorticity is tilted upward and the resulting vertical vorticity is stretched, a process that adds to its already cyclonic spin. Although the mesocyclone consists simply of a cyclonically rotating updraft at midlevels, near the ground it is composed of both rotating updraft and downdraft (hence the term *divided mesocyclone;* Lemon and Doswell 1979) with the vertical vorticity maximum lying just on the updraft side of the interface between the updraft and a downdraft on the storm's rear flank. The ingestion into the right rear side of the updraft of cyclonically spinning air that has passed through the left side of the downdraft completes a huge vortex column that extends from the ground to 12 km in extreme cases. The rain curtain is seen on a nearby radar as a hook-shaped appendage (henceforth called a hook) extending from the storm's echo (Fig. 5.16). Surface observations (Fujita 1958; Rasmussen and Straka 1996) reveal a zone of high pressure and divergent flow at the surface beneath the hook. In the above scenario, Davies-Jones has theorized that the rain curtain/hook is not simply a passive indication of near-ground mesocyclonic rotation, but the instigator of it, and the link between rotation aloft and near the ground.

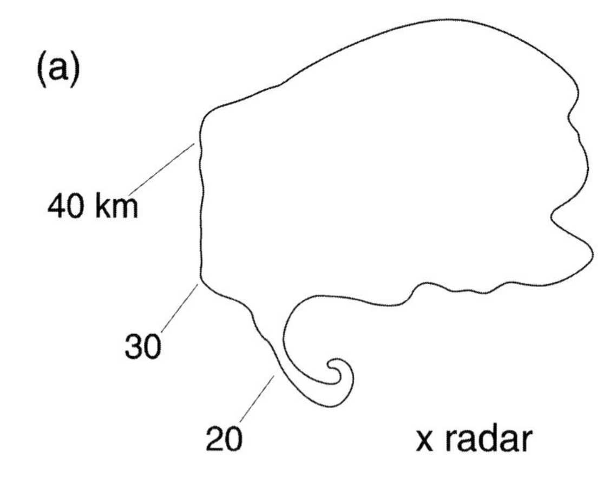

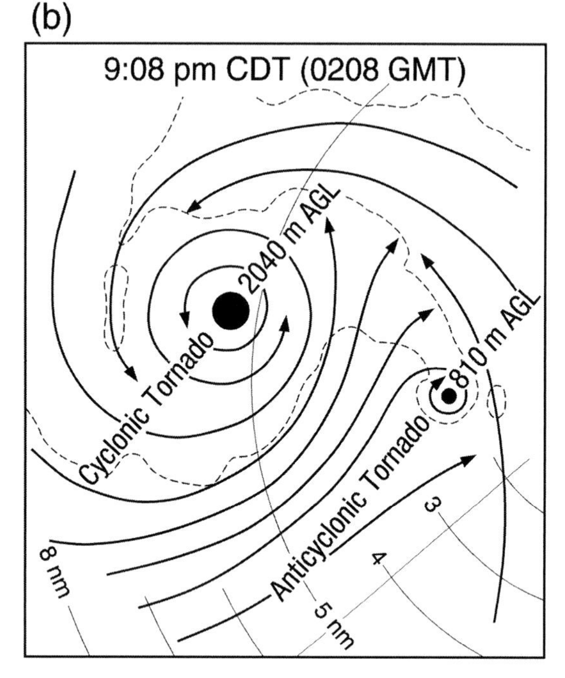

FIG. 5.16. (a) A cyclonic hook (from Garrett and Rockney 1962). (b) Surface flow (arrows) and radar echo (dashed) associated with simultaneous cyclonic and anticyclonic tornadoes at Grand Island, Nebraska, 3 June 1980. (From Fujita and Wakimoto 1982.) Note the wrapped-up cyclonic hook with a central eye and the anticyclonic hook to its ESE.

What happens to the anticyclonic vortex on the right side of the downdraft? In some cases it may be entrained into an updraft in the flanking line of convective towers that extend outward from the right rear of the main updraft. This gives rise to a low-level mesoanticyclone and occasionally anticyclonic tornadoes (Fig. 5.16; Brown and Knupp 1980; Fujita and Wakimoto 1982). The mesoanticyclone is generally weaker than the mesocyclone because the baroclinic streamwise vorticity is antistreamwise, that is, in the opposite direction to the streamwise vorticity generally present in the environment. This makes an intense, deep, mesoanticyclone unlikely. Instead, the weaker mesoanticyclone will tend to revolve partly around the stronger mesocyclone to its north.

The near-ground mesocyclone should form more slowly and be less intense if the barotropic vorticity opposes the baroclinic vorticity and hence retards and weakens the developing baroclinic circulation (e.g., Wicker 1996). This seems to be the case when the shear is unidirectional, as in Rotunno and Klemp's (1985) simulation where the circulation around a fluid circuit encompassing the near-surface vorticity maximum becomes negative when the circuit is traced back 20 min in time. As it is taken backward in time, the circuit starts climbing the plane of symmetry. It is evident in their Figs. 11 and 13 that around the elevated portion of the circuit at the earlier time the circulation around the circuit associated with the mean flow is opposed to the thermal circulation. In Walko's (1993) simulation, the vortex formed as a result of tilting of barotropic vorticity in the downdraft with the solenoidal effects acting to weaken the circulation. This finding may be a result of the imbalance in his initial conditions. The cold pool is present initially without any of the vorticity that would be generated naturally by buoyancy torques as the cold pool formed.

2) BAROTROPIC MECHANISMS

Brandes (1984a) retrieved buoyancy from dual-Doppler wind fields of a tornadic supercell to investigate the hypothesized role in mesocyclogenesis of horizontal buoyancy gradients and attendant baroclinic vorticity generation. His analyses suggested little contribution of horizontal baroclinic vorticity to mesocyclone intensification and subsequent tornadogenesis, although cautions were raised about uncertainties in the retrieved buoyancy. Recent surface measurements are consistent, however, with this apparent lack of baroclinity: During the Verification of the Origins of Rotation in Tornadoes Experiment (VORTEX; Rasmussen et al. 1994) and its follow-on experiments, the "mobile mesonet" (see Straka et al. 1996) detected cool temperatures and moist entropies at 3 m AGL beneath a few of the tornadic hook echoes observed at close range on (airborne and/or mobile) Doppler radars, but found negligible surface temperature and entropy perturbations in hook echoes that were associated with the stronger tornadoes (Markowski 2000). In the clear slot south of the hook, temperatures were warmer and moist entropies lower than ambient, consistent with forced unsaturated descent of air in a potentially unstable environment. One must conclude that there is (i) baroclinity of the correct sense near the ground that was not sampled by the mobile mesonet (which is limited by existence of adequate roads and by the relatively small number of sensors that comprise the mesonet), (ii) baroclinity of the correct sense aloft that is not revealed by observations at 3 m AGL, or (iii) a barotropic mechanism for near-ground mesocyclogenesis in addition to the baroclinic one. Parcels descending in a counterclockwise direction around the updraft would travel along an appropriate baroclinic zone aloft if the updraft were warmer than the environment at low levels (E. N. Rasmussen and K. A. Browning 1999, personal communication). This would require a low level of free convection. Some support for this hypothesis is found in the extensive analyses by Rasmussen of the VORTEX data collected near the Dimmitt, Texas, tornado on 2 June 1995. However, the analyzed fields are consistent also with a barotropic mechanism. We now explore the possibility of such a barotropic mechanism.

As already discussed, the vortex in Walko's (1993) idealized simulation formed from barotropic vorticity. Rothfusz and Lilly (1989) showed how near-ground rotation and even a tornado might be generated from a helical environment without baroclinic effects via theoretical analysis and a modified Ward tornado simulator (section 5.4). The simulator was changed so as to provide an inflow that veered with height without net circulation. Their theoretical analysis and that of Brooks et al. (1993) show that, as the inflow accelerated into the center of the apparatus, the streamwise vortex lines were stretched, generating a roll-type circulation with ascent on the left side and descent on the right side of the inflow at each level (Fig. 5.17). The vortex is produced by vertical eddy transport of angular momentum in the secondary helical flow. The positive angular momentum on the right side of the inflow remains at low levels until it reaches small radii, while the negative angular momentum on the left side is transported upward at large radii. A vortex forms near the center as a result of this process.

Davies-Jones (2000b) has constructed a simplified axisymmetric numerical model to demonstrate how rotation can be lowered to the ground by a barotropic mechanism. The model's initial condition consists of a central axisymmetric nonbuoyant updraft that rotates cyclonically at midlevels (a midlevel mesocyclone) surrounded by a concentric anticyclonic downdraft in which the angular momentum is still positive. The initial state is perturbed by adding potential energy through the introduction of a prescribed distribution of hydrometeors with constant fall velocity at the top boundary, a procedure used by Eskridge and Das (1976) and Proctor (1988). The rain falls in a curtain near the updraft–downdraft interface and most of it reaches the ground without being recirculated in the updraft. The model excludes conventional buoyancy forces associated with temperature differences to avoid the occurrence in the closed domain of unrealistic buoyancy recirculations or oscillations. The descent of the liquid water represents a conversion of potential energy into meridional kinetic energy.

Preliminary results indicate that the associated drag force intensifies the downdraft and causes a downward transport of angular momentum. Part of the outflow from the downdraft is directed inward, thus transporting high-angular-momentum air and increasing convergence into the central updraft's lower regions. From a vorticity perspective the low-level spinup is a result of upward tilting of inward radial barotropic vorticity and stretching of cyclonic vertical barotropic

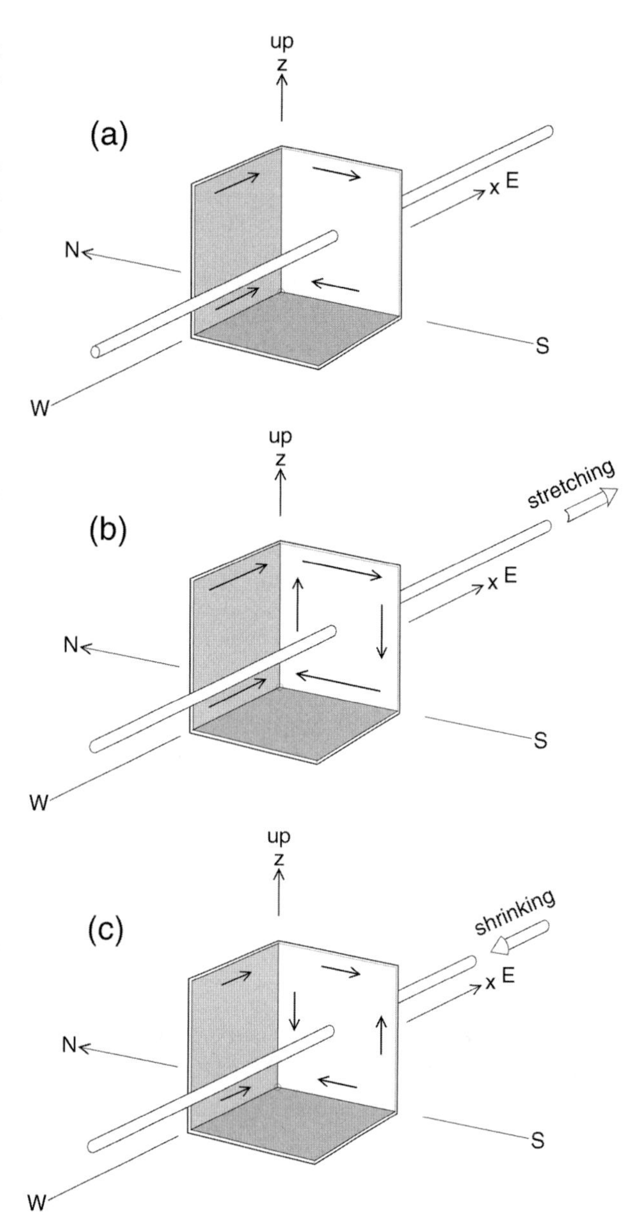

FIG. 5.17. Generation of vertical motion by longitudinal stretching or shrinking of streamwise vorticity. The length of the flow arrows is proportional to wind speed. (a) Conditions in the far upstream environment. Here there is vertical shear transverse to the wind associated with the streamwise vorticity, but no vertical motion. (b) Local stretching of the vortex lines in accelerating flow. The transverse shear intensifies and upward (downward) motion develops on the left (right) side of the flow. (c) Local shrinking of the vortex lines in decelerating flow. The transverse shear weakens and downward (upward) motion develops on the left (right) side of the flow.

vorticity. Baroclinic vorticity is generated as azimuthal vorticity by horizontal torques associated with radial gradients of the drag force, but it remains in this component because the axisymmetry prohibits tilting of azimuthal vorticity. Since angular momentum is nearly conserved owing to relatively slow diffusion at large Reynolds number, the maximum tangential velocity intensifies as parcels move downward and inward. Simultaneously the pressure low deepens and travels down the axis from its initial midlevel position. The associated pressure-gradient forces cause the low-level streamlines to slope downward (analogous to a rear-flank downdraft/clear slot) before turning sharply upward into the contracting central vortex. This penetration of the anticyclonic downdraft toward the axis surrounds the vortex with anticyclonic vorticity, resulting in a surface pressure profile $p(r)$ with a steeper than expected slope near the radius of maximum winds and a flatter than expected slope at larger radii (as in the tornadic pressure trace obtained by Winn et al. 1999). Far aloft, the top of the updraft turns to downdraft (a collapsing top) in response to a downward axial pressure-gradient force. The near-ground rotation clearly originates from inward and downward transport of angular momentum associated with the initial mesocyclone aloft, not from a baroclinic mechanism.

e. Forecast parameters

We can now interpret various forecast parameters in the light of the above paradigms of mesocyclone formation. The properties of three modern parameters, each of which contain some, but not all, of the important physics, are summarized in Table 5.2.

The dimensionless bulk Richardson number (BRN; Weisman and Klemp 1982, 1984) has been quite successful in predicting the conditions under which storms will split and supercells will form. It is defined by BRN $\equiv$ CAPE/BRNKE. Here CAPE is the buoyant energy defined in section 5.2c and BRNKE $\equiv (\Delta\bar{U})^2/2$, where $\Delta\bar{U} \equiv |\bar{\mathbf{v}}_{0-6} - \bar{\mathbf{v}}_{0-.5}|$ and $\bar{\mathbf{v}}_{0-6}$ and $\bar{\mathbf{v}}_{0-.5}$ are the density-weighted mean winds in the lowest 6 and 0.5 km, respectively. The maximum updraft velocity is $\sqrt{2\,\mathrm{CAPE}}$ according to parcel theory. The denominator, BRNKE, is the kinetic energy of the mean wind in the lowest 500 m in a reference frame moving with the mean wind in the lowest 6 km. If the latter is viewed as an estimate of storm motion, then BRNKE is simply the storm-relative kinetic energy of the inflow in the lowest 500 m. Alternatively, one can interpret $\Delta\bar{U}$ as a measure of the environmental shear in the lowest 6 km, with large $\Delta\bar{U}$ being the condition for the formation of strong midlevel vortices. In theory, supercell storms will not develop at large BRN

TABLE 5.2. Comparison of the properties of three forecast parameters (ignoring dependence of BRN on CAPE because other two parameters may be nondimensionalized by CAPE and $\sqrt{\mathrm{CAPE}}$ if desired).

Property	SRH	BRN	Length of HODO.
Predictor of	Net updraft rotation (including sign)	Supercells, splitting storms	Tornadic storms, mesocyclones
Physics contained	Potential for midlevel updraft rotation	Buoyant energy/inflow kinetic energy	Mean magnitude of shear in 0–4-km layer
Geometric interpretation on hodograph diagram	$-2 \times$ signed area swept out by storm-relative wind between 0 and 3 km	BRNKE $\equiv (\Delta\bar{U})^2/2$ $\Delta\bar{U} \equiv |\bar{\mathbf{v}}_{0-6} - \bar{\mathbf{v}}_{0-.5}|$	Arc length
Dependence on hodograph shape	Highly dependent	Less dependent	Independent
Function of storm motion **c**?	Linear function of observed (or forecast) **c**	No	No
Sensitivity to **c** (nearly straight hodographs)	Highly sensitive	None	None
Sensitivity to **c** (highly curved hodographs)	Insensitive. Substituting mean wind for **c** gives good estimate of helicity	None	None
Differentiates storms in same environment by their motion?	Yes, may warn on storm that deviates to right (e.g., along a boundary)	No	No
Mirror image storms	Opposite signs	Same value	Same value
Sensitivity to surface-inflow wind	Moderate	High (important for supercell vs multicell)	Low
Volatility with respect to hodograph changes	More volatile	Less volatile	Less volatile
Galilean invariant?	Yes (no if estimate of **c** is not invariant)	Yes	Yes
Winds included	0–2 or 3 km explicitly but higher winds affect **c**	0–6 km	0–4 km
Sensitive to fractal hodographs	No	No	Yes, needs prescription for smoothing hodographs

(greater than 50) because the storm-relative inflow kinetic energy is too small to prevent the cold-air outflow from undercutting the updraft and surging far ahead of the storm, the dynamic pressure-gradient forces associated with the midlevel vortices are too weak to promote storm splitting, and the updraft will not rotate as a whole because the storm-relative environmental winds are too weak in comparison to storm-induced winds.

Another parameter for forecasting storm type is the mean shear (Rasmussen and Wilhelmson 1983) or, equivalently, the arc length U_{S4} (m s^{-1}) of the hodograph curve between 0 and 4 km. (The mean shear is simply $U_{S4}/4000$ m.) Except for the 0–4-km restriction, U_{S4} is the parameter U_S introduced by Weisman and Klemp (1982, 1984) to characterize the shear magnitudes of the analytical hodographs, which they used in their sets of comparative simulations. As shown in section 5.2c, effects of shear are weak when $U_S \ll \sqrt{\mathrm{CAPE}}$. Storms are forecast to be tornadic if the CAPE and mean shear are large. The mean shear suffers in practice because it is ill-defined for real-life (fractal) hodographs, it is independent of hodograph shape, and it can be large when the storm-relative surface wind is weak (Droegemeier et al. 1993).

Given observed or predicted storm motions, storm-relative helicity (SRH; Davies-Jones et al. 1990; Droegemeier et al. 1993) is a useful parameter for predicting whether updrafts will rotate cyclonically or anticyclonically as a whole. The helicity hierarchy (in a column of unit cross-sectional area) of the inflow layer of a storm translating with velocity $\mathbf{c}$ in a horizontally homogeneous nonrotating environment can be defined as $\bar{H}_{mn}(\mathbf{c}, h) \equiv \int_0^h [\nabla \times]^m[\bar{\mathbf{v}}(z) - \mathbf{c}] \cdot [\nabla\times]^n[\bar{\mathbf{v}}(z) - \mathbf{c}]dz$, where h is the nominal depth of the layer (usually chosen arbitrarily to be 3 or 2 km). Note that $\bar{H}_{01}(\mathbf{c}, h)$ is the storm-relative helicity; $\bar{H}_{00}(\mathbf{c}, h)/2h$ is the mean storm-relative kinetic energy per unit mass of the environment in the inflow layer and is related to BRNKE by BRNKE $= \frac{1}{2}\bar{H}_{00}(\bar{\mathbf{v}}_{0-6}, 0.5\ \mathrm{km})$. For a straight hodograph, $\bar{H}_{00}$ is symmetric, that is, it is the same for the severe left- and right-moving products of the split, while $\bar{H}_{01}$ is antisymmetric (the SRH of the SL storm is the negative of that of the SR storm). A mean-wind-relative helicity (MWRH) can be defined by substituting $\bar{\mathbf{v}}_{0-6}$ for $\mathbf{c}$. For a straight hodograph, MWRH is zero, which applies to the initial storm (no net updraft rotation), but not to the SR and SL storms individually. For highly curved hodographs, MWRH underestimates SRH only slightly. SRH is a linear function of both $\mathbf{c}$ and h. For fixed h, the contours of SRH as a function of $\mathbf{c}$ on a hodograph diagram are straight lines parallel to the shear vector from 0 to h. Thus the effect of different storm motions on SRH are readily apparent to a forecaster. For given $\mathbf{c}$, the variation of SRH from layer to layer can be found from a plot of $\bar{H}_{01}(\mathbf{c}, h)$ versus h. The biggest threat for tornadoes is associated with large environmental helicities in the lowest 1 km (Markowski et al. 1998a) because such helicities are converted into large vertical helicity densities at quite low altitudes within storm updrafts.

A nondimensional number similar to BRN can be obtained by dividing CAPE by SRH. High values of this number imply that the storm-relative environmental winds are negligible compared to the storm-induced winds. CAPE/SRH is not used in practice because CAPE can vary widely in proximity to storms, and violent tornadoes occur occasionally in warm-season, high-CAPE, low-shear (high BRN) environments. In fact, a threshold value of the energy-helicity index (EHI $\propto$ CAPE $\times$ SRH) fits the data (scatter diagrams of proximity values of CAPE and SRH for strong and violent tornadoes) far better than a threshold value of CAPE/SRH (Davies 1993a). How warm-season tornadic storms develop is not well understood because they are rarely observed in field projects and they have not yet been simulated successfully by computer models. It would seem that the large amounts of precipitable water in the highly unstable environment would result in heavy precipitation, excessive water loading, substantial evaporative cooling, and outflow-dominated storms. This often occurs, but not always. High CAPE could favor supercell formation in some cases. For instance, low-level inflow winds induced by intense updrafts may be sufficient to prevent the outflow from propagating ahead of the storm. Tilting and stretching of relatively weak environmental horizontal vorticity by a high-speed updraft may still result in a strong mesocyclone aloft. The high instability may give rise to an intense baroclinic zone just behind the gust front. The updraft can remain buoyant even after ingesting a large amount of rain-cooled air, allowing occluded mesocyclones to live longer than usual and to have more time to become tornadic.

All of the predictors that we have considered so far only predict the formation of supercells or mesocyclones aloft. Predicting a persistent low-level mesocyclone, the next step toward tornadogenesis, is more difficult because it involves forecasting the distribution of precipitation and rain-cooled air within the storm (Brooks et al. 1994b). If the mid- and upper-level storm-relative winds are weak, precipitation falls too near the updraft and rain-cooled air rapidly undercuts the updraft and occludes the mesocyclone. If, on the other hand, the mid- and upper-level winds are very strong, the precipitation falls too far downstream of the rotating updraft to be drawn around to the rear of the mesocyclone. Brooks et al. (1994a) went a step forward in considering a balance among midtropospheric storm-relative winds, SRH, and low-level mixing ratio. Another factor that also affects the inflow-outflow balance is the dryness of midlevel air, which is included only in the DCAPE (downdraft CAPE; Emanuel 1994, p. 172; Gilmore and Wicker 1998).

According to parcel theory, $\sqrt{2\,\mathrm{DCAPE}}$ is the maximum downward velocity that can occur in a given environment via evaporative cooling at constant pressure of relatively dry midlevel air to its wet-bulb temperature and then descent along a pseudoadiabat with just enough evaporation to keep the air saturated.

5.3. Tornadogenesis, maintenance, and decay

a. Mesocyclonic or type I tornadoes

We now are ready to address tornadogenesis, the stage that follows the development of near-ground rotation on the mesocyclone scale. Given now a complete vortex column in the updraft and strong low-level convergence that persists without interruption for, say, 10–15 min, a concentrated vortex (i.e., a tornado) should form as in a tornado simulator. The convergence associated with the updraft is enhanced by convergence in the swirling boundary layer of the mesocyclone; the boundary layer also provides an important feed of vorticity for vortex sustenance (section 5.4). Since "seed" vertical vorticity is already present near the ground, analysis of the vertical-vorticity equation always shows the dominance of the convergence term in a bull's-eye around the incipient tornado. Sometimes this is misinterpreted as evidence that the origin of the tornado's rotation is simply preexisting vertical vorticity.

The significance of strong low-level convergence to tornado formation is apparent from the following simple problem. Assume that uniform horizontal convergence $C = 5 \times 10^{-3}\ \mathrm{s}^{-1}$ stretches an initial vortex of radius $R_0 = 1$ km with maximum tangential wind $V_0 = 10\ \mathrm{m\ s}^{-1}$ into a tornado of radius $R_1 = 100$ m with maximum tangential wind $V_1 = 100\ \mathrm{m\ s}^{-1}$ without any loss of angular momentum. What is the timescale τ for tornado formation? From the kinematic formula for divergence, $2d\ \ln R/dt = -C$, which has the solution $\tau = 2/C\ \ln R_0/R_1 = 920$ s = 15 min. Doubling the convergence to $10^{-2}\ \mathrm{s}^{-1}$ reduces the time to 8 min.

The embryonic or fully developed tornadic vortex is detected on Doppler radar as a tornadic vortex signature (TVS; Brown et al. 1978), a region of large cyclonic shear between azimuthally adjacent sampling volumes in the field of Doppler velocities. The first TVS observed was that of an embryonic tornado initially detected at 3–4 km above the ground, near Union City, Oklahoma, on 24 May 1973 (see also Lemon et al. 1978). Over the course of 30 min, the TVS slowly extended downward, reaching near the ground coincident with tornado touchdown, and upward, reaching a height of 12 km. With only a few cases from research radars available to them, observationalists believed for several years that almost all large tornadoes formed first in the clouds in this way—labeled mode I tornadogenesis (Figs. 5.18a,b) by Trapp and Davies-Jones (1997)—enabling warnings to be issued roughly 20 min prior to their touchdown. Theoreticians, on the other hand, thought that tornadoes should build from the ground up (mode II tornadogenesis of Trapp and Davies-Jones 1997; Fig. 5.18c) because air with high angular momentum would approach the axis of rotation faster in the mesocyclone's boundary layer than above it (Rotunno 1986). Since the installation of the WSR-88D network across the United States, it is now realized that ~50% of tornadoes apparently form either very close to the ground or almost simultaneously in a column in the lowest 2 km, within 5–10 min (Trapp et al. 1999). Since the WSR-88Ds take 5 min to complete a volume scan, these mode II resultant tornadoes often develop with little advance warning. There is little difference between the mean intensities of mode I and mode II tornadoes.

Trapp and Davies-Jones utilized numerical and analytical models of an idealized mesocyclone to provide simple explanations for both modes of tornadogenesis. A time-dependent version of the Burgers–Rott vortex due to Rott (1958), in which the circulation at infinity and the convergence are both constant with height, provides an example of the mode II genesis. High-angular-momentum air approaches the axis at the ground and aloft simultaneously. The vortex in this case forms as a cylindrical column that is independent of height. If neither the radial inflow nor the circulation increases with height, and one or both of them has a maximum at the ground, the vortex will form from the ground up. If the radial inflow and the circulation are both nondecreasing, and one of these parameters increases with height, the high-angular-momentum air arrives near the axis first aloft and mode I genesis ensues. If the low-level rotation is insufficient, the vortex remains aloft (Smith and Leslie 1978). Otherwise the vortex develops downward via a bootstrap process called the dynamic pipe effect (Leslie 1971; Smith and Leslie 1979). As demonstrated in section 5.4d, the vortex core is in cyclostrophic balance and is stable to radial displacements so that air is prevented from entering the vortex through its sides. In this respect the vortex acts like a suction tube with solid walls. As air is drawn into the lower end of the vortex, it spins faster, centrifugal forces increase and balance the inward pressure-gradient force. This air then becomes part of the dynamic pipe. The vortex builds itself down by this process until it makes contact with the ground.

The mechanisms just described do not explain those tornadoes that develop outside the mesocyclone's central axis, that is, tornadogenesis within two-celled mesocyclones. A two-celled vortex is characterized by downflow along the central axis, terminating in low-level radial outflow that turns vertical in an annular updraft at an outer radius (see section 5.4). Radar observations of two-celled mesocyclones have been

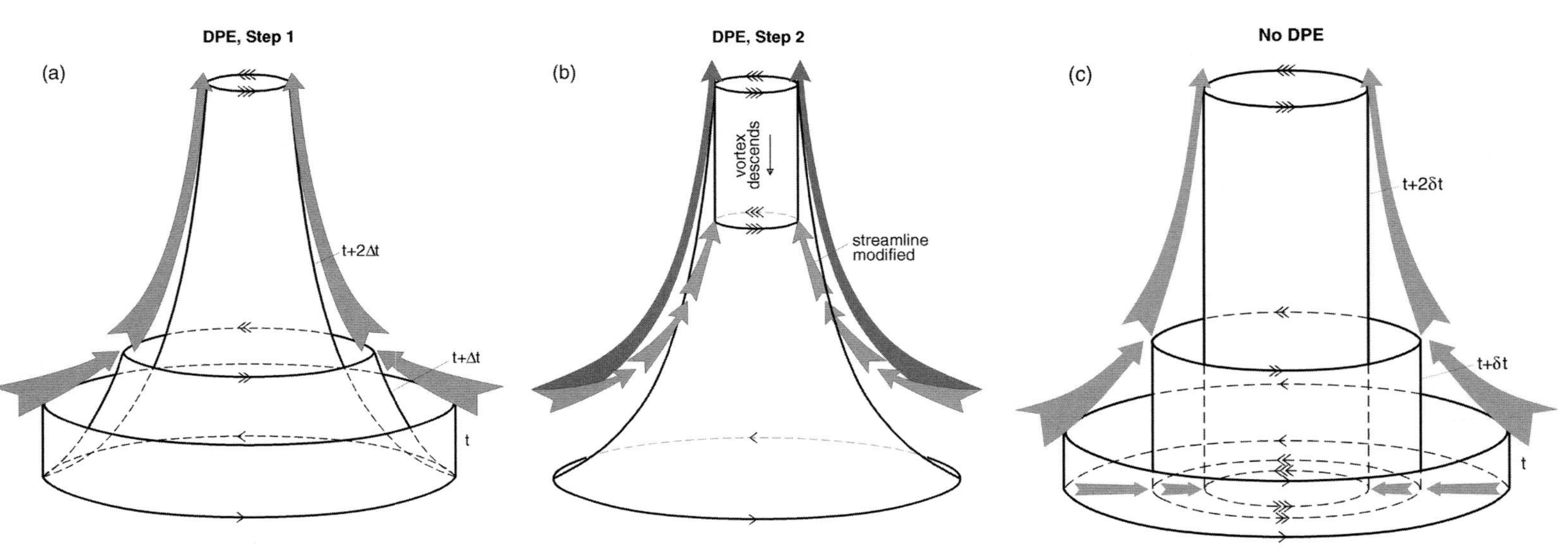

FIG. 5.18. Idealization of two modes of vortex formation within a rotating updraft (Trapp and Davies-Jones 1997). (a) and (b) illustrate the dynamic pipe effect. In (a) the radial inflow increases with height so an initial cylindrical vortex tube is stretched into a cone and a funnel cloud forms aloft first. In (b) the vortex pipe builds downward by increasing the radial inflow into its lower end through a suction effect as described in the text. The vortex descends to the ground relatively slowly. In (c) the radial inflow is constant with height so the DPE is absent. The initial cylindrical vortex tube remains cylindrical as it is stretched vertically so that the vortex spins up simultaneously at all heights, resulting in rapid tornado formation.

reported by Brandes (1978) and more recently by Wakimoto et al. (1998) and Trapp (1999); visual evidence is provided in Bluestein (1985b). Brandes (1984a) confirmed observationally that the VPPGF may become negative in the vicinity of tornadic mesocyclones as the vertical vorticity in the low-level mesocyclone exceeds that in the midlevel mesocyclone. As also demonstrated in a numerically simulated storm by Klemp and Rotunno [(1983); and in subsequent storm observations with airborne Doppler radar, by Wakimoto et al. (1998)], adverse pressure-gradient forces cause downward motion in what has become known as an "occlusion downdraft," which often (though not always; Wakimoto and Cai 2000) precedes tornado formation.

The two-celled vortex is susceptible to a cylindrical vortex-sheet instability (Rotunno 1984). In analogy with the formation of satellite vortices around large tornadoes, the release of this instability in the parent mesocyclonic vortex leads to the formation of submesocyclone-scale vortices outside the central downdraft, in an annular region where both vertical velocity and its radial gradient are positive (Fig. 11 of Rotunno 1984). Tornadogenesis occurs if one or more of these smaller-scale vortices interacts with the ground and consequently contracts into tornadoes (Rotunno 1986). It is unclear how often such a mechanism may be active, however: The percentage of all tornadoes that develop out of two-celled mesocyclones is unknown, as is the percentage of all mesocyclones that become two-celled some time during their life cycle (Trapp 1999).

The genesis of type I tornadoes is often associated with the interaction of the parent supercell with a preexisting thermal boundary such as a storm outflow boundary or a warm or stationary front. A recent study suggests that 66% of the significant (F2 or greater) tornadoes in the VORTEX-95 domain occurred near some type of low-level boundary not associated with the parent storm itself (Markowski et al. 1998b; see also Rasmussen et al. 2000). Storm–boundary interactions may affect low-level rotation (and presumably tornadogenesis) as follows: (i) horizontal streamwise vorticity, baroclinically generated within the thermal boundary, is ingested by storms with inflows along the boundary, and subsequently is vertically tilted in the downdraft and updraft and stretched in the updraft (Markowski et al. 1998b; Atkins et al. 2000); and (ii) vertical vorticity owing to cross-boundary horizontal variations in the boundary layer wind profile is vertically stretched as the storm's updraft encounters the boundary (Maddox et al. 1980). Note that convergence along a boundary may also cause a deviant storm motion that in turn affects the updraft's rotation.

Before leaving the discussion of type I tornadogenesis we note that, according to airborne Doppler radar data collected during VORTEX, not all instances of a low-level mesocyclone with persistent, low-level convergence result in a tornado. Such lack of tornado formation, subjectively classified as "tornadogenesis failure" by Trapp (1999), may among other reasons have been due to a thin (tens of meters), surface-based, yet unobserved layer of low angular-momentum air imparted to the mesocyclone (Lewellen et al. 2000). Surface-layer inflow of low angular momentum air increases, for example, with an increase in effective surface roughness, or an increase in the translation speed of the mesocyclone. When a "local corner flow swirl ratio" (see section 5.4)—which is inversely proportional to such depletion of angular momentum air in the surface layer—is prescribed below some critical value in Lewellen et al.'s numerical model, only very little vortex intensification, well off the lower surface, occurs.

Long-lived tornadoes typically evolve through the following five stages. The tornado is first visible as dust swirling upward from the surface and a short funnel pendant from cloud base (the *dust whirl stage*). It then goes through an *organizing stage* where its funnel descends and it intensifies. In its *mature stage* it reaches maximum intensity and its funnel reaches its greatest width and is almost vertical. The baroclinic or barotropic mechanisms provide a constant feed of vorticity to the tornado, as does horizontal vorticity generation above the no-slip ground (e.g., Trapp and Fiedler 1995), which may explain how some tornadoes can remain in the mature stage for a considerable time. During its *shrinking stage,* it decreases in width and becomes more tilted. It may still be extremely damaging. In its *decay stage* the tornado dies as a result of its base being overtaken by a divergent cool downdraft and its circulation weakening. It is no longer able to resist the shear of the surrounding flow, and so becomes tilted over and greatly stretched into a rope shape as its top travels with the updraft and its base is moved in a different direction by the low-level outflow. Although its circulation has decreased significantly, its winds can still be high in a small area because its core radius is also much reduced. Thus the tornado can still be very destructive over a narrow path. The funnel finally undergoes wavelike contortions, with parts of it disappearing and reappearing, before it finally dissipates. This behavior suggests that the tornado decays by becoming unstable to sinusoidal long-wave displacement of its axis. A few intense tornadoes have tracks that widen at the end, suggesting that they decayed by spreading out and becoming diffuse as their parent updraft weakened abruptly (Agee et al. 1976).

A cyclic repetition of this tornado evolution within a single supercell has been documented on numerous occasions. Indeed, Fujita (1963) and Darkow (1971) noticed that roughly 20% of tornadic supercell storms spawned tornadoes periodically, thus producing "tornado families." The median interval between tornadoes was 45 min. Ward and Arnett (1963) observed

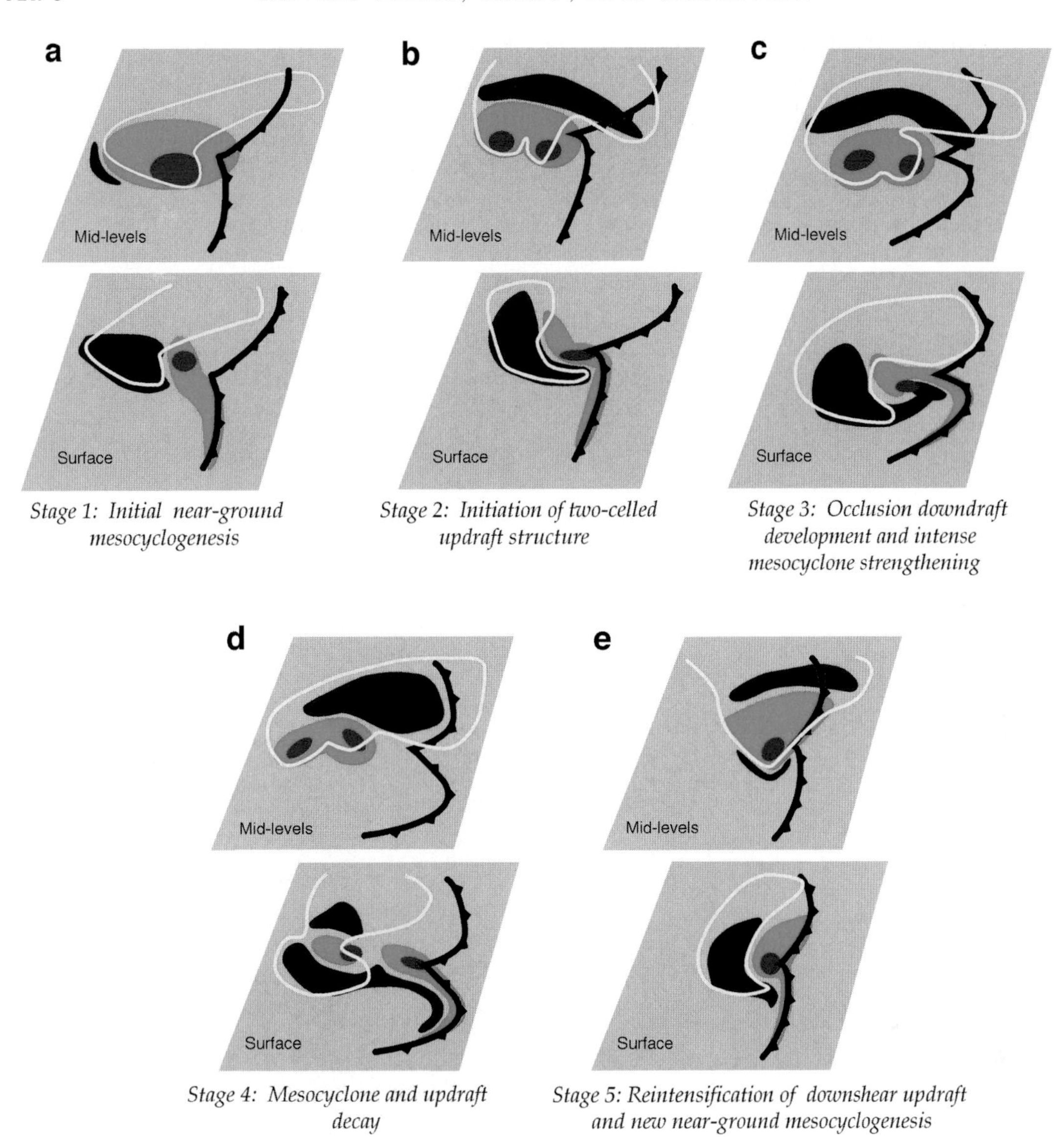

FIG. 5.19. Five-stage conceptual model of the cyclic genesis of mesocyclones. The surface cold-pool boundary is shown by the scalloped black line. Red marks vorticity maxima. Updraft and downdraft are shown in light and dark blue, respectively. The boundary of the rain area is outlined by the yellow contour (Adlerman et al. 1999).

that supercell storms generated periodic surges in outflow with an average cycle of 50 min. Forbes (1978) studied the hook echoes and tornado tracks associated with cyclic supercells during the massive tornado outbreak on 3 April 1974. Most of the tornadoes turned to the left before dissipating. Burgess et al. (1982) showed, from Doppler radar observations, that about one-quarter of supercells produced more than one mesocyclone with a roughly 40-min interval between successive formations.

The initial mesocyclone in a cyclic supercell is typically the longest lived (90 min). In numerical simulations (Fig. 5.19), this mesocyclone forms first at midlevels owing to the tilting of streamwise environmental vorticity, and then ~30 min later at low levels owing to tilting and stretching of baroclinic vorticity in the rear-flank downdraft and in the updraft (Adlerman et al. 1999; see section 5.2). The mature stage of the first cycle is marked by strong rotation through a large depth (from 0 to 7 km or higher), tangential velocities ~25 m s^{-1}, circulation ~5×10^5 m^2 s^{-1}, and low-level convergence ~5×10^{-3} s^{-1}, and lasts around 40 min. Subsequent mesocyclones are shorter lived (45 min) and have mature stages that last only 20 min. However, the near-ground rotation develops within 9 min of the beginning of the new cycle. According to Adlerman et al., the mechanism for subsequent near-ground mesocyclogenesis is identical to that of the initial near-ground mesocyclogenesis; it simply proceeds more rapidly because cold air residual from the previous cycle lends to a buoyancy gradient orientation that is ideal for horizontal baroclinic vorticity generation. These mesocyclones typically decay in around 15 min as their height, radius, tangential winds, and circulation decrease, and the flow around them becomes divergent. Tornadoes touch down midway through the mature stage at a time when the downdraft is intensifying, and dissipate during the decay stage as the updraft is undercut near the surface by subsiding air. Burgess et al. proposed (and Adlerman et al. verified) that, as the previous mesocyclone and tornado are being occluded by cold outflow, a new mesocyclone and tornado develop on the gust front ahead of and to the right of the old mesocyclone. The old mesocyclone is weakened aloft since its associated updraft now draws air with lesser moist entropy, but it may still contain a tornado that is displaced by the new outflow to the rear side (instead of the base) of the main storm tower. Jensen et al. (1983) reported visual observations of a storm that corroborates this model.

A somewhat different evolution is revealed by a recent analysis of VORTEX data. Dowell and Bluestein (2000) used airborne Doppler radar to characterize as "inflow dominated" the cyclic supercell that produced a series of tornadoes (one of which was a long-lived F4 tornado) that formed in the Texas panhandle on 8 June 1995 (see Fig. 5.20). They concluded that individual tornadoes were not kinemat-

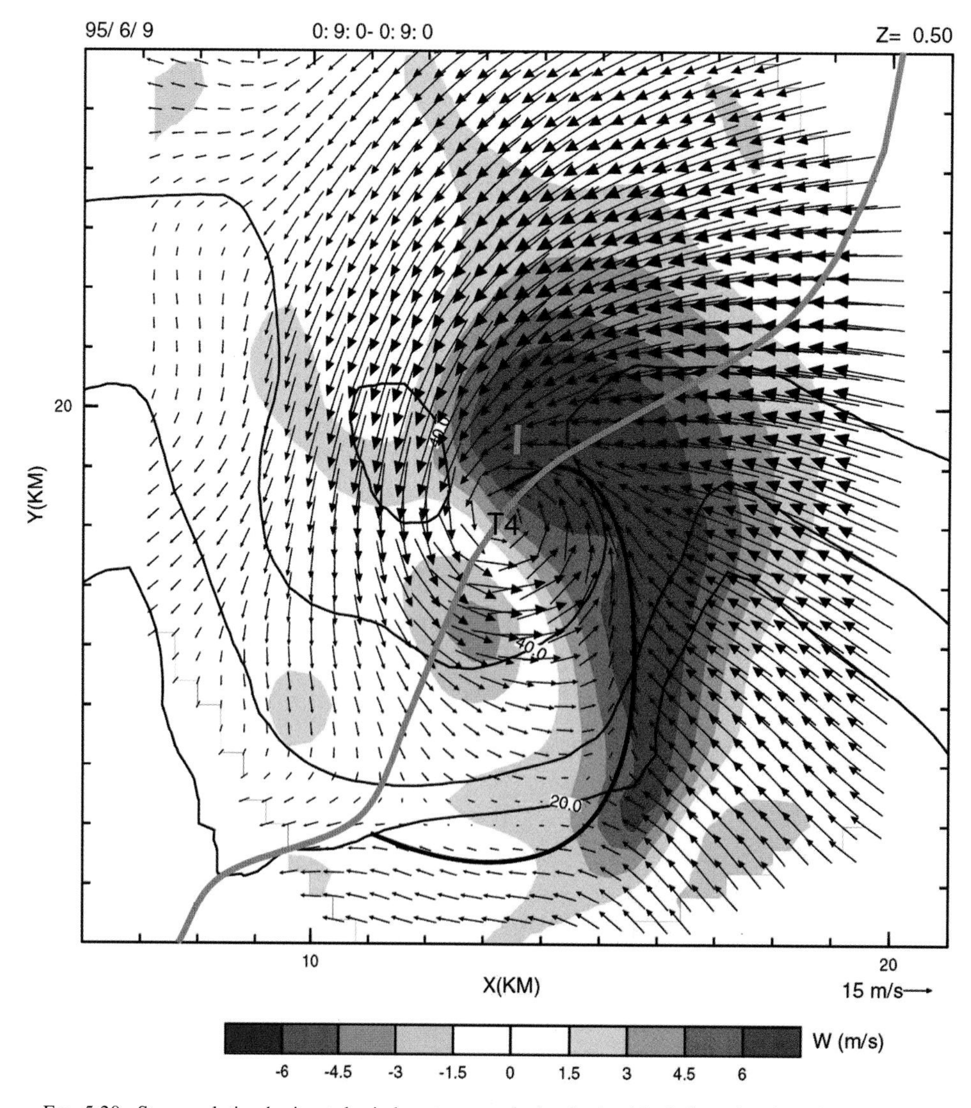

FIG. 5.20. Storm-relative horizontal wind vectors, vertical velocity (shaded), and radar reflectivity factor (contoured), derived from airborne Doppler radar data of the tornadic storm near McLean, Texas, at 0009 UTC 9 June 1995. The analysis level is 500 m AGL, and the domain is relative (in km) to McLean, Texas. The black line denotes the rear-flank gust front position, and the curvy line, the tornado track. "T4" is the location of the fourth tornado in the family of tornadoes observed on this date. Courtesy of D. Dowell.

ically occluded, as in the aforementioned "outflow-dominated" model, but rather were advected rearward, away from the updraft, by strong, low-level, front-to-rear storm-relative inflow.

These conceptual models explain why the damage tracks of tornadoes in a family are often a series of parallel arcs that are staggered to the right as one track ends and the next one in the series begins. The cyclic process does not always conform to these models, however. For instance, the old and new mesocyclones revolve partly around each other if they are close together, because of the velocity induced on each vortex by the other one (Brown et al. 1973). A film of the Hesston, Kansas, tornadic storm of 13 March 1990 showed the Hesston tornado in its rope stage moving into the condensation funnel of a newly formed large tornado.

b. Nonmesocyclonic or type II tornadoes

As stated in the introduction, a type II tornado forms along a stationary or slowly moving front or windshift line apparently from the rolling-up of the associated quasi-vertical vortex sheet into individual vortices; these vortices, which at times may be classified as "misocyclones"—for our purposes tornado-parent vortices that by definition have diameters < 4 km (Fujita 1981)—then are vertically stretched by convective updrafts under which the vortices move or over which they develop (Fig. 5.21). The horizontal shearing instability mechanism has been proposed to explain the formation of lines of dust devils, by Barcilon and Drazin (1972); waterspouts in cloud lines along the sea-breeze front and weak tornadoes in flanking lines, by Davies-Jones and Kessler (1974) and Davies-Jones (1986); and landspouts, by Brady and Szoke (1989) and Wakimoto and Wilson (1989). According to Lee and Wilhelmson (1997), the release of the horizontal shearing instability leads to first-generation vortices that interact and merge to create the misocyclones within which landspouts ultimately develop. We note that the presence of a spiral rain curtain (or hook echo on radar) close to waterspouts (Golden 1974a,b) and landspouts (Wakimoto and Wilson 1989; Wilczak et al. 1992; Wakimoto and Martner 1992) suggests that the vorticity of type II tornadoes may be augmented by the baroclinic process that plays an important role in type I tornadogenesis (see section 5.2). Indeed, Wilczak et al. (1992) provided evidence that the vertical tilting of baroclinically generated, streamwise horizontal vorticity was important in the generation of the low-level rotation of a nonmesocyclone tornado on 2 July 1987 in northeastern Colorado. In this regard, the proximity of a nontornadic storm to an apparent type II tornado may muddle the mechanisms responsible for the tornado as well as the characterization of the tornado type itself. This statement is illustrated in the dissimilar discussions by Wakimoto and Atkins (1996) and later by Ziegler et al. (2001) of the origin and nature of the F3 tornado observed during VORTEX on 29 May 1994.

Waterspouts and landspouts are generally cyclonic because the vorticity of the preexisting windshift line is usually cyclonic owing to the earth's rotation. They typically have circulations of 2×10^3 to 10^4 m^2 s^{-1}. Consider, for example, a straight boundary or front within which resides an infinite number of vertical vortices that are spaced 3 km apart. If the vortices are associated with a cross-front difference in the along-front wind component of $\Delta u = 6$ m s^{-1}, the circulation available to each vortex is $\Gamma \equiv \oint_c \mathbf{u} \cdot \mathbf{ds} \cong \Delta u \Delta s \approx 2 \times 10^4$ m^2 s^{-1}, where $\Delta s = 3$ km is the along-front length of the circuit around one of the vortices; the cross-front wind component (relative to the vortex sheet) does not contribute to circulation. This value is small compared to the circulation of the average mesocyclone (say, 5×10^4 m^2 s^{-1}), but the merger of several misovortices could lead to a sizable circulation.

In some instances the tornado forms from a cumulonimbus that develops rapidly above a convergence zone associated with two colliding boundaries (such as an outflow boundary emanating from another storm and a sea-breeze front; Holle and Maier 1980). Note,

FIG. 5.21. Diagram showing genesis of a nonmesocyclone tornado (Wakimoto and Wilson 1989). Misoscale vortices (identified by letters) form along a windshift line (black line) as a result of shearing instability. A tornado results from one of these vortices (C) being stretched by overhead convection.

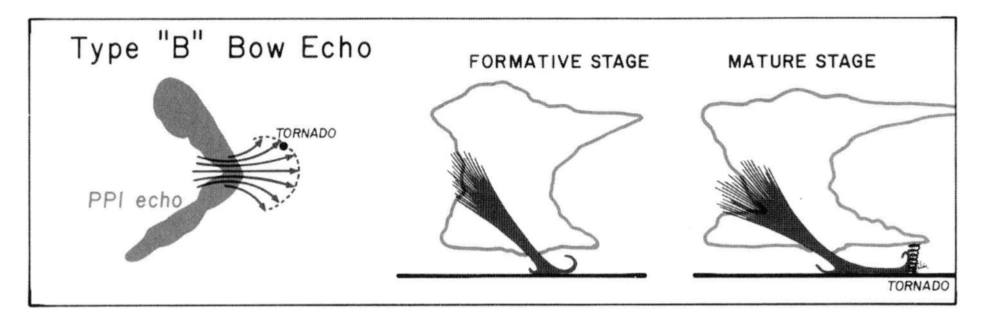

FIG. 5.22. Tornado formation within a bow echo according to Fujita (1985). Note the location of the tornado on the left side of the strong outflow.

however, that colliding outflows are far more common than tornadoes, implying that type II tornadoes require either a fortuitous or a physically intentional coincidence of maxima in vertical velocity and vertical vorticity. Regarding the latter, incipient type II tornadoes may form at the intersections of the windshift line and the ascending branches of the circulations of boundary layer rolls. In this case, coincidence of vortex and convective updraft is established immediately since convective initiation tends also to be favored at line–roll intersections (Wilson et al. 1992). Alternatively, the "locational relationship" between the incipient tornado and the updraft may be due to the influence of the vortex itself on the low-level convective forcing, which, therefore, precludes the need for a line–roll intersection (Lee and Wilhelmson 1997).

Squall-line tornadoes occur in lines devoid of supercells as well as in supercells within lines. The nonsupercell variety quite frequently form in conjunction with high straight-line winds in short lines or line segments with bow-shaped radar echoes. As indicated in Fig. 5.22, cyclonic tornadoes may occur on the cyclonic side of the jet of strong outflow winds. Assuming that the environmental shear is westerly, the cyclonic (anticyclonic) vorticity on the north (south) side of the jet probably originates from a combination of the baroclinic mechanism described in section 5.2 and downward momentum transport (or, equivalently, tilting of horizontal barotropic vorticity). Tornadoes may also occur anywhere along a still quasi-linear convective system, apparently via the vertical vortex sheet roll-up mechanism, and within the system's cyclonic "bookend" vortex (Pfost and Gerard 1997; Trapp et al. 1999).

5.4. Tornado structure

Investigations of the tornadic vortex itself—particularly via laboratory simulators or numerical models thereof—typically idealize it as a nontranslating vortex in an axisymmetric near environment extending out to about 1 km from the axis (Lewellen 1993); analyses of tornado observations often employ azimuthal averaging techniques yielding comparable idealizations. Within this axisymmetric setting, the tornado may be asymmetric in three-dimensional models if it has a multivortex structure. In actuality, even a single-vortex tornado is asymmetric because it moves across the ground in response to a steering current, because its inflow—when visible in blowing dust/patterns on a water surface or observed by radar—appears to be concentrated in a spiral band (Fig. 5.23; see also Bluestein and Pazmany 2000), and because it tilts with height. Even though a mature tornado may be quite erect near the ground, the TVS on radar generally has an overall tilt of about 25° toward the left forward side of the storm (Brown et al. 1978).

The idealized axisymmetric tornado is conveniently divided into five strongly interacting regions (Fig. 5.24): the vortex core (Ia), the outer flow at radii outside the core (Ib), the boundary layer (II), the corner region where inflow in the boundary layer turns the corner and enters the core from below (III), and the flow aloft that caps the vortex within the parent cloud (IV) (Lewellen 1976, 1993; Snow 1982; Davies-Jones 1986). Prior to describing these regions and their inherent dynamics, we provide critical comments on a representative sample of the different means with which tornado structure has been deduced: by laboratory simulations, numerical simulations, and direct and remotely sensed observations. Thereafter, the dependence of tornadic vortex behavior and structure on flow parameters is discussed, as are some of the more significant measurements made in tornado simulators.

a. *Comments on laboratory simulations of tornadoes*

Although their use has declined significantly in recent years due mostly to cost (see Doswell and Grazulis 1998), tornado simulators have provided a wealth of information about tornado structure (see reviews by Davies-Jones 1976, 1986; Maxworthy 1982; Wilkins 1988; Church and Snow 1993). The Ward-type vortex generator (Fig. 5.25) replicates a wide range of tornadic vortices and reproduces observed surface pressure profiles of tornadoes (Ward

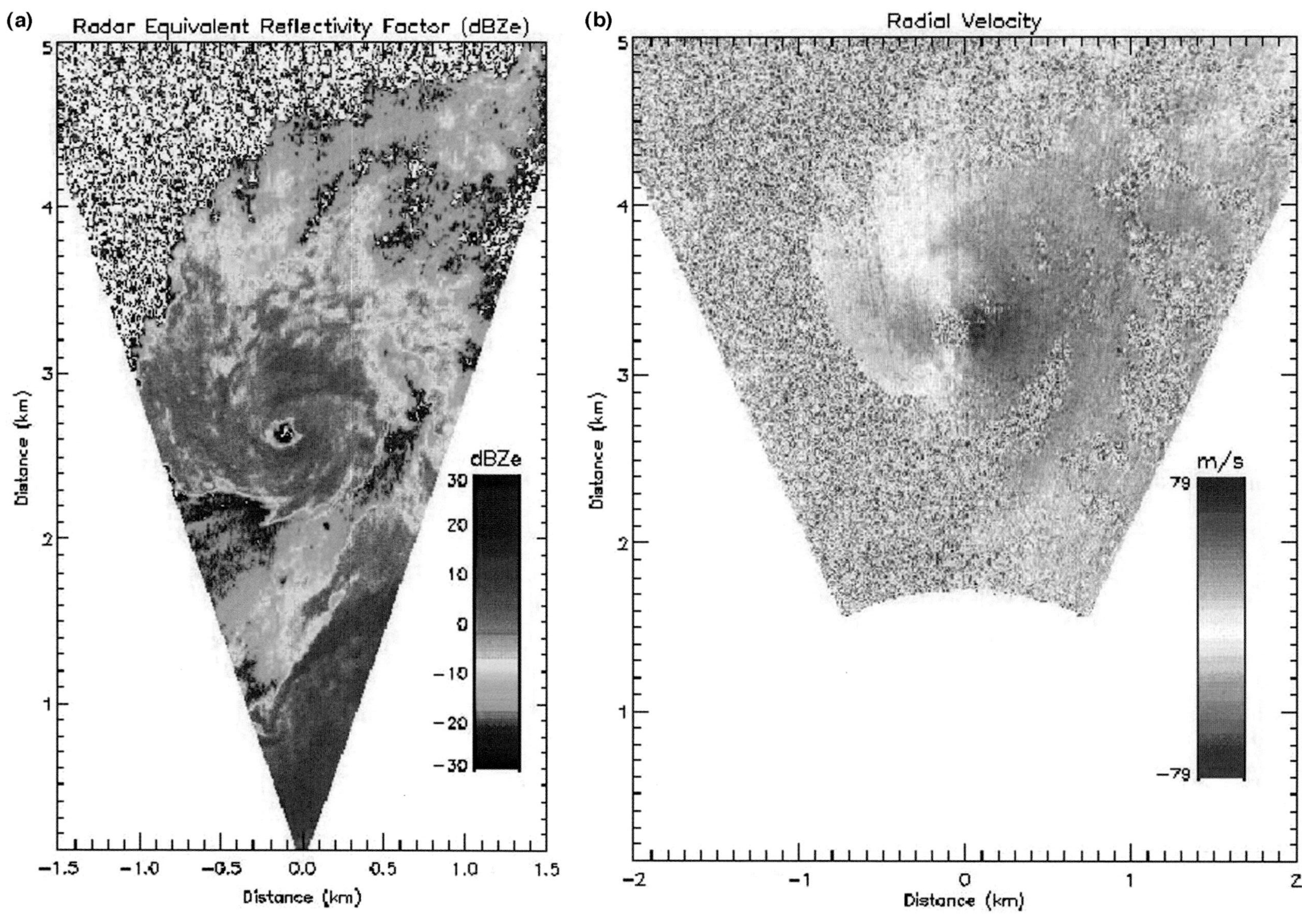

FIG. 5.23. Data collected by the University of Massachusetts, Amherst, 3-mm wavelength radar of an F-3 tornado near Verden, Oklahoma, on 3 May 1999. (a) Radar reflectivity factor, at 2254:57 UTC. (b) Doppler velocity, at 2255:53 UTC. Courtesy of A. Pazmany and H. Bluestein.

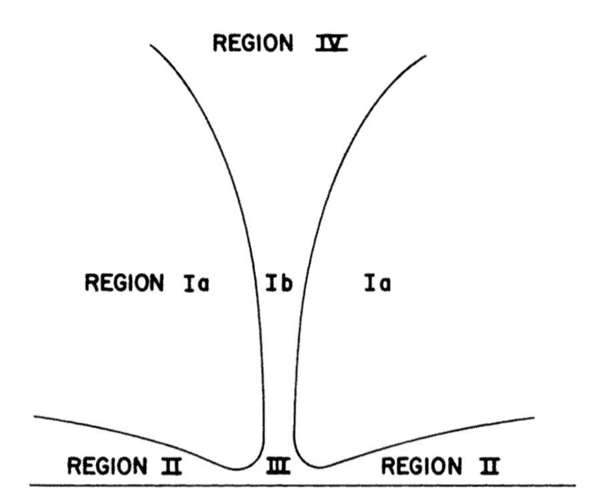

FIG. 5.24. Different flow regions of a tornado (adapted from Lewellen 1976). Region Ib is the core, II is the boundary layer, Ia is the region above the boundary layer outside the core, III is the corner region, and IV is the termination region within the parent storm at mid- or high altitudes.

1972; Church et al. 1977; Church and Snow 1993). The Ward model is more relevant to tornadoes than most other vortex chambers because it allows descent of initially nonrotating fluid in the core and because it is geometrically and dynamically similar to tornado-producing flows. In particular it has an aspect ratio, $a \equiv h/r_0$, where r_0 is the radius of the updraft hole and h is the height of the inflow to the updraft, that is of order 1, which is characteristic of tornadic thunderstorms; many other simulators have high aspect ratios. The flow in the apparatus is governed to a considerable extent by a single parameter, the swirl ratio, defined by $S \equiv r_0M/2Q$ where $2\pi M$ is the circulation at the edge of the updraft ($M = vr$) and $2\pi Q$ is the volume flow rate of the updraft ($Q = \int_0^r wr'dr'$) (Davies-Jones 1973, 1976). The swirl ratio can also be expressed by $S \equiv v_0/\bar{w}$, where v_0 is the tangential velocity at the edge of the updraft and $\bar{w}$ is the mean vertical velocity through the updraft hole. It is tempting to scale up measurements made in the Ward-type simulator to obtain estimates of tornado wind speeds and pressure drops. These estimates must be treated with some skepticism, however, because the updraft radius in the simulator is artificially constrained instead of being able to adjust to increases in circulation (Maxworthy 1982); the vortex is terminated aloft artificially by the honeycomb baffle, which can have a large effect on vortex size and structure (Smith 1987); the stagnation pressure of the air that descends along

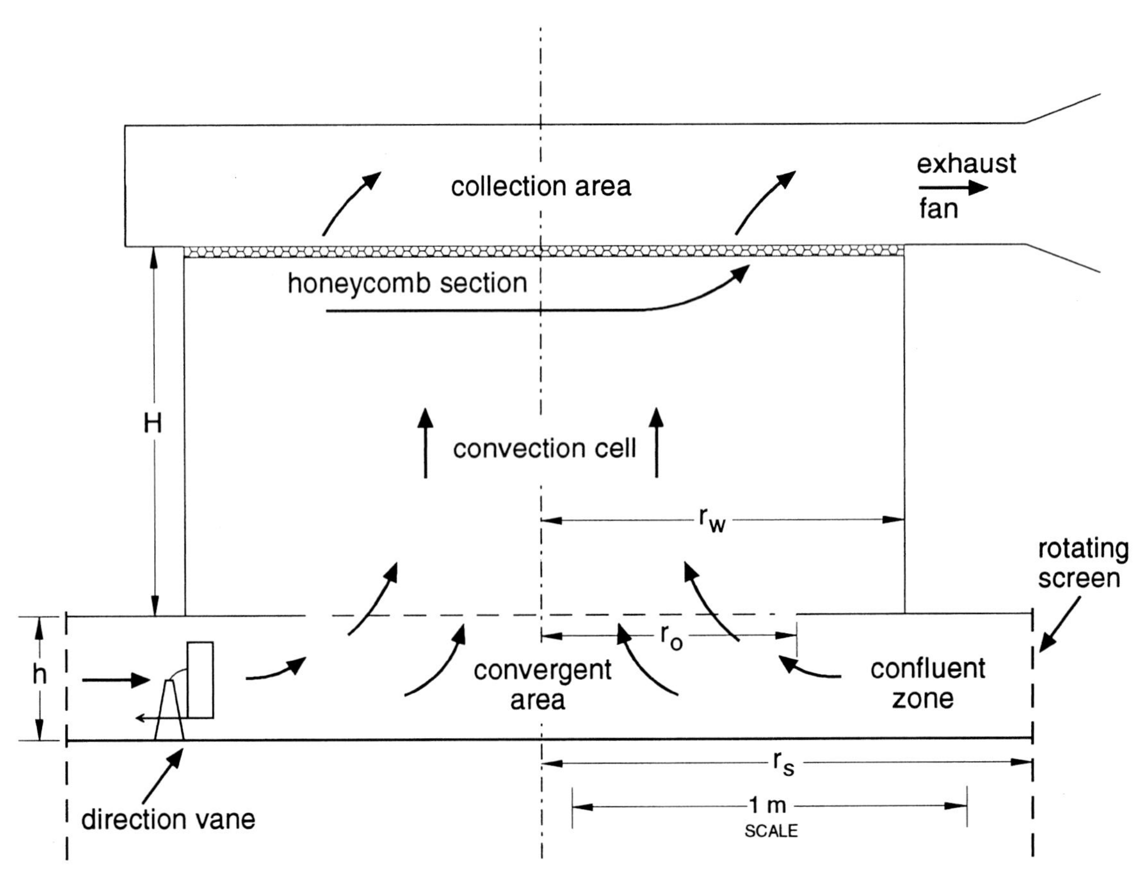

FIG. 5.25. Ward's tornado simulator.

the axis from above the baffle is unknown; the flow is driven mechanically instead of by buoyancy; there is no warming in the core when subsidence occurs there; and the Reynolds number is low compared to atmospheric values, resulting in flows that are partly laminar at some settings of the external parameters.

b. Comments on numerical simulations of tornadoes

Although three-dimensional cloud models have not yet been used successfully to routinely simulate actual storms, they do reproduce many of the features of tornadic storms and even generate "tornado-like vortices" (e.g., Wicker and Wilhelmson 1995; Grasso and Cotton 1995). The origins of these vortices can be revealed by diagnoses of the numerical results. To produce a bona fide tornado, the cloud-model characteristics must include an innermost grid with a grid-point spacing of 10 m or less in the horizontal and the same in the vertical near the ground; realistic two-way interaction between the grids; lower boundary conditions that are representative of the interaction of the turbulent flow with the ground; and a realistic turbulence parameterization that includes effects of rotational damping. Because these requirements are rather daunting (see Wilhelmson and Wicker's chapter in this volume), many investigators have turned their attention to numerical models either of the tornado simulators or of a tornado in an axisymmetric setting; our subsequent discussion focuses on such models.

With the exception of the Lewellen and Sheng (1980) turbulent model of the Purdue University simulator, all known models of tornado simulators have assumed laminar flow and simplified the geometry of the simulator to a cylinder. The main difficulty with modeling the tornado simulator lies in formulating an appropriate top boundary condition for representing the influence of the honeycomb baffle on the flow; the other boundary conditions are well established. The obvious choices for the lower boundary are either a no-slip or a free-slip solid surface (e.g., Rotunno 1977, 1979). Symmetry conditions are imposed at the axis if the flow is axisymmetric; for asymmetric flow in a three-dimensional model it is convenient to place around the axis a narrow impervious free-slip cylindrical inner boundary (Rotunno 1984). The outer cylindrical boundary is open as an inlet up to the height h of the top of the inflow layer, above which the outer boundary is impermeable and free slip. The radial and tangential velocities of the swirling inflow entering through this boundary are typically assumed to be constant with height. Rotunno (1977) realized that the largely irrotational flow observed in the simulator was reproduced better when the third inlet condition was zero azimuthal vorticity, $\eta \equiv \partial u/\partial z - \partial w/\partial r$, rather than zero vertical velocity as imposed by Harlow and Stein (1974).

As mentioned, the top boundary is problematic. In none of the known models is the baffle explicitly incorporated into the computer model domain. Therefore, the top boundary must correspond to the lower surface of the baffle, which we define to be at $z = H$. Assuming that it completely removes the radial and swirling components from the flow and has negligible resistance to vertical flow, the baffle has two effects. First, fluid reenters the apparatus from above with no initial rotation so an appropriate inflow boundary condition at the top is $v = 0$. Second, the radial distribution of pressure on the upper side of the baffle is quite uniform over the outflow portion because of the removal of the horizontal components from the flow. Making the simulator's convection cell deep simplifies the formulation of the top boundary conditions because the throughflow is practically vertical by the time it reaches the top and the baffle. The baffle's role of eliminating the radial flow is then redundant and we can make the first boundary condition $u = 0$ at the top. This implies from continuity that $\partial w/\partial z = 0$ at $z = H$, which is not an independent condition. In some simulations, $v = 0$ or $p =$ constant has been applied across the entire top. These boundary conditions clearly do not apply to the underside of the baffle. Since $u = \partial w/\partial z = 0$ at $z = H$, imposing zero radial viscous force, that is, $\partial \eta/\partial z = 0$, seems a reasonable choice for the second condition, especially since it is equivalent to imposing cyclostrophic balance there. It is also equivalent to $\partial^2 u/\partial z^2 = 0$ since $\partial^2 w/\partial r \partial z = 0$ from the first condition. A third independent condition is $v = 0$ for inflow, as reasoned above, and $\partial v/\partial z = 0$ for outflow, which is consistent with vertical throughflow and near conservation of angular momentum along streamlines below the top. These are close to the conditions used by Rotunno (1977, 1979), who did not distinguish inflow from outflow and so did not impose zero rotation on inflow through the top. Therefore his model is probably the most relevant to flow in the simulator.

In open boundary, limited-domain models of the tornado itself (e.g., Smith and Leslie 1979; Howells et al. 1988; Walko 1988; Lewellen et al. 1997), the effects of the larger-scale flow are represented through the lateral boundary conditions, where swirling inflow is imposed, and also through the top boundary conditions if the model height does not extend into the stratosphere. Apparently, all such models have at least one of the following artificial features: porous lids at low heights with dubious boundary conditions imposed, prescribed eddy viscosity, homogeneous fluids with fixed body forces in lieu of buoyancy forces, parameterized subsidence warming in lieu of a reservoir of overlying potentially warm air that the vortex can draw downward, and the anelastic assumption. Eliminating all these constraints even in an axisymmetric model would require a large amount of computing resources.

Lewellen et al. (1997) have developed a sophisticated three-dimensional large-eddy simulation model (the subgrid model of which now includes the effects of rotational damping; see Lewellen et al. 2000) with open boundaries to investigate the turbulent interaction of the tornado vortex with the surface. The Lewellen et al. model has been successful in improving the quantification of the processes in the corner region that yield the largest tangential velocities in the flow (Lewellen et al. 2000). It has a rather limited domain size of 1 km × 1 km × 2 km in the x, y, z directions, respectively, to allow resolution of the large turbulent eddies. Gridpoint spacings of 1.5 m in the vertical and 2.5 m in the horizontal are attained by grid stretching; Lewellen et al. (1997) conclude that such resolution in their model is sufficient so that the time-averaged velocity and pressure distributions in the turbulent corner region show little sensitivity to either finer resolution or a modified subgrid turbulence model. A 1-km diameter, uniform upward vertical velocity (one of a number of possible conditions) is prescribed at 2 km above the ground. Hence, although axial downdrafts can still occur, they are severely limited in extent.

Lewellen et al.'s model design restrains the interaction of the vortex with the "storm" to be one way. Fiedler's (1995) valid criticism of such open boundary, limited-domain modeling is as follows: The vortex is not allowed to adjust as a result of axial downdrafts developing in response to the surface low. For example, a vortex breakdown might enter the domain from above, descend to the surface, and eliminate the supercritical vortex (see section 5.4d) altogether. Fiedler's (1993, 1994) remedy to this problem has been a closed-domain modeling approach. This approach is not without its own limitations, however. He imposes in his model a body force that has maximum strength at the midpoint of the axis and is supposed to simulate buoyancy in the parent updraft, in lieu of or in addition to an entropy equation with advection and diffusion terms. Since the body force is unrealistically immutable, a mutual interaction between buoyancy field and the vortex is precluded. Also unrealistic is the instant loss of parcel "warmth" as a parcel leaves the fixed region of the body force. The domain is a closed rotating rank of radius r_0 that is large enough so that it is effectively infinite (i.e., the results have become insensitive to r_0). Since the flow is closed, the kinetic energy would increase without bound in the nonrotating, inviscid limit owing to production of kinetic energy each time parcels cycled through the body force region. The viscosity near the top of the tank is increased by an order of magnitude so that the vertical velocity at steady state is less than $(2\ \mathrm{CAPE})^{1/2}$, which is the vertical velocity a parcel would acquire, if there were no VPPGF, in proceeding upward along the axis once through the body force region. In the inviscid limit with rotation, there is an exact solution consisting of a vortex in hydrostatic and gradient-wind balance with no meridional motion. For a body force like Fiedler's that for all practical purposes does not extend to the boundaries, this solution satisfies free-slip boundary conditions at top and bottom. Hence, the free-slip viscous solution tends to a steady state consisting of a wide vortex in gradient-wind and hydrostatic balance and an outer region of weak meridional flow that largely avoids the region of the body force altogether.

In both types of models, a critical effect of the lower boundary condition on the vortex structure is exhibited, as has been demonstrated by Rotunno (1977), Howells et al. (1988), Fiedler (1993), and others. Indeed, the use of a frictional lower boundary affords the development of a rotating boundary layer, the existence of which (at sufficiently high Reynolds number and swirl ratio) has implications on high tornadic wind speeds (see section 5.5).

As mentioned, a limited-area model of a tornado should include a realistic updraft driven by buoyancy forces associated with latent heat release in a conditionally unstable atmosphere. It should also extend into the stratosphere so that the vortex can terminate naturally in radially divergent flow and is exposed to overlying potentially warm air, which could be drawn down into the core and undergo compressional warming. One such model exists (Walko 1988), but the size of the domain is too large to permit a large-eddy simulation. He used a prescribed eddy viscosity instead. He showed that subsidence in the core of an intense tornado can extend practically to the ground. Unfortunately the model is anelastic, meaning that variations in density, ρ', from ambient density, $\rho_a(z)$, are neglected in all terms except when multiplied by the gravitational acceleration (i.e., $\rho' g$ is retained). In the region of high tangential winds in a tornado, the centrifugal acceleration, v^2/r, is many times g. For example, $v^2/rg \approx 10$ for $v = 100\ \mathrm{m\ s^{-1}}$, $r = 100$ m. For a potential vortex the ratio of accelerations decreases as r^{-3} so the "radial buoyancy acceleration" $(\rho'/\rho_a)v^2/r$ is significant compared to the normal vertical acceleration due to buoyancy, $-\rho' g/\rho_a$, out to a distance of 350 m from the axis in the above example. The "radial buoyancy" effect, which enhances the stability of the vortex, is not included in an anelastic model, which therefore may not simulate accurately the trajectories of air parcels in and near the core.

c. *Comments on tornado observations*

Photogrammetry, aerial surveys of tornado damage, in situ measuring devices, and remote-sensing techniques comprise the means by which vortex structure has been deduced from tornadoes observed in nature. Consider aerial damage surveys, a practice pioneered by Fujita. These have been used to determine if the damaging winds were straight-line or vortical, and if multiple vortices were present (Fujita and Smith 1993; Fujita 1993); the existence of suction vortices that attend some tornadoes was suggested on the basis of such surveys. As inherent also in F-scale determination (see section 5.1), information from aerial surveys is necessarily limited by the type of ground cover and structures in the tornado's path.

Portable devices have been designed to make in situ measurements of tornado properties typically near the ground only. These devices are placed in the projected path of a tornado; their success is bound by the potentially complicated and hazardous deployment logistics. TOTO (totable tornado observatory; Bedard and Ramzy 1983; Bluestein 1983a,b; Burgess et al. 1985), equipped with wind and thermodynamic property sensors, was used only during the early and mid-1980s. Data were collected in mesocyclones, but unfortunately never in a mature tornado: Deployment logistics were complicated further when it was determined that unless TOTO were staked down, it could tip over at F1 wind speeds. "Turtles" (Brock et al. 1987) consist only of temperature and pressure sensors encapsulated in heavy concave shells; as the instrument name implies, the aerodynamic design of turtles allows them to withstand higher wind speeds. The strategy during VORTEX was to deploy an array of several turtles across the projected tornado path, thus increasing the likelihood of a direct measurement (Rasmussen 1995). Successful measurements were made by two turtlelike instruments (also known as "E-turtles," since these also measured electric field) in an F4 tornado on 8 June 1995 near Allison, Texas (Fig. 5.26; Winn et al. 1999): Pressure deficits as high as 55–60 mb were found at a distance of approximately 660 m from the center of the tornado.

Variously configured Doppler radars can remotely sense tornado airflow characteristics, distribution of airborne debris and precipitation scatterers in and around the tornado core (using radar reflectivity factor), and turbulence (with certain assumptions, using spectrum width data; see Istok and Doviak 1986). Consider the recent development of mobile radars. Obviously, unlike their fixed-site counterparts, mobile radars can be brought very close (within a distance of a few kilometers) to the tornado, providing for the high spatial resolution and also for radar scans at very low altitudes above radar level. The mobility, however, restricts the radar dish size, which in turn limits the antenna beamwidth (hence, effective resolution), given a specific wavelength choice. Moreover, the wavelength is inversely related to the signal attenuation, limiting the type of weather conditions (i.e., the storm-relative tornado-viewing position) under which the radar can effectively operate. Such attenuation is comparably more severe with the 3-mm wavelength, pulsed, mobile Doppler radar, designed and built by A. Pazmany and colleagues at the University of Massachusetts, Amherst (see Bluestein et al. 1995). How-

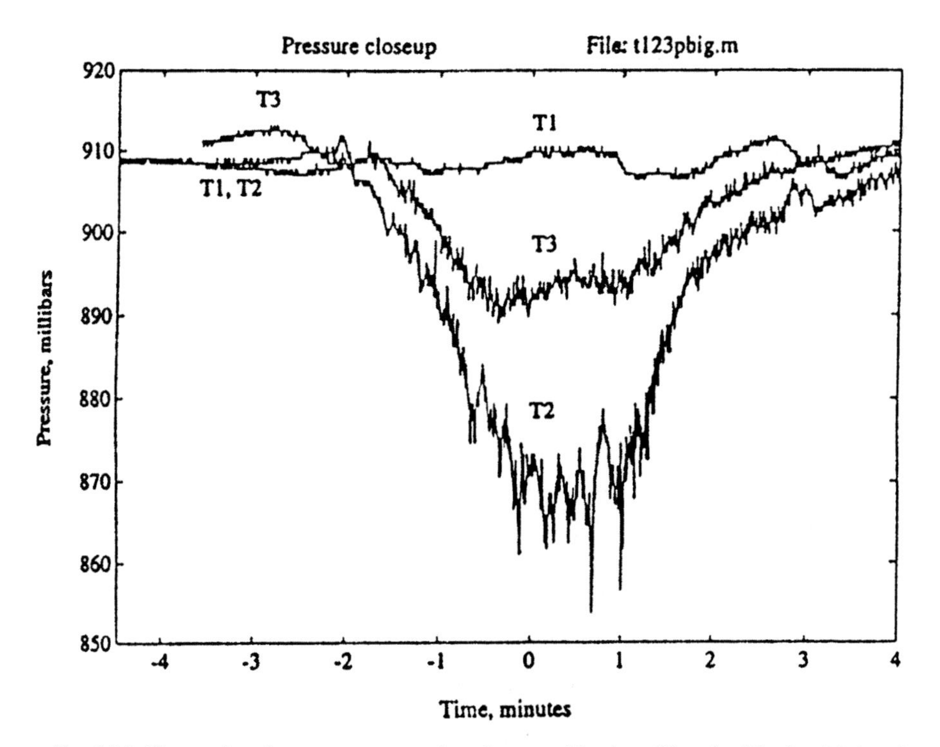

FIG. 5.26. Time series of pressure, measured on the ground by three "E-turtles," in the vicinity of an F-4 tornado near Allison, Texas, on 8 June 1995. Time $t = 0$ is 0100 UTC. From Winn et al. (1999).

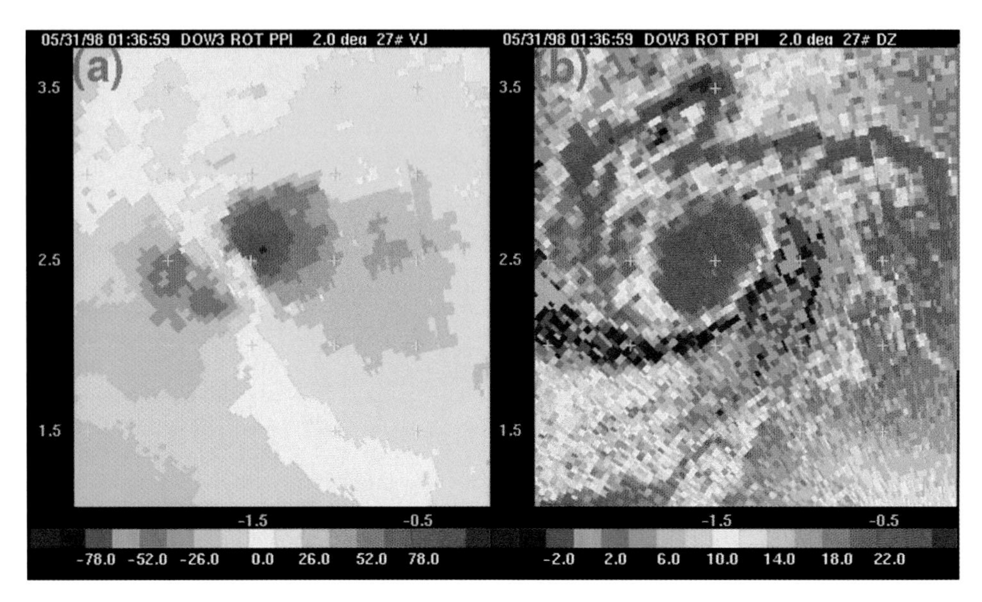

FIG. 5.27. Data collected by the Doppler on Wheels radar of an F-4 tornado near Spencer, South Dakota, on 31 May 1998. (a) Radar reflectivity factor and (b) Doppler velocity, at 0136:59 UTC. Courtesy of J. Wurman.

ever, this radar affords very high spatial resolution: The along-beam resolution is 12 m and, owing to the 0.18° antenna beamwidth, the cross-beam resolution (which also depends on the range to the tornado) can be as low as 5–10 m, as was attained in the 3 May 1999 tornado near Oklahoma City, Oklahoma. As a consequence, this radar is capable of resolving radar reflectivity structures, such as within an apparent tornado "eyewall," suggestive of asymmetries in the tornado vortex in the form of waves or multiple vortices (Fig. 5.23; see also Bluestein and Pazmany 2000). In 1995, J. Wurman and colleagues developed a pulsed, 3-cm wavelength, mobile Doppler radar (Wurman et al. 1997), known as the DOW (Doppler on Wheels). The antenna beamwidth is 0.93°; (tens of meters)^3 resolution volumes and 30–60-s volume scans are possible. Structures resolved by the DOW include a weak-echo hole like that documented by Fujita (1981; see Wurman et al. 1996) and also spiral bands, like those observed in hurricanes (Fig. 5.27). DOW data have also proved useful for the evaluation of azimuthal velocity models (such as the Rankine vortex; see Wurman and Gill 2000). We note that a two-DOW deployment has been attempted with some success since 1997. However, these operations have been focused on gathering data on tornadogenesis rather than tornado structure.

d. Tornado flow regions

1) CORE AND OUTER FLOWS

The outer flow (region Ia) extends outward from the core at least 1 km. It consists of air that approaches and rises around the core, while approximately conserving its angular momentum. Consequently the air spins faster as it gets closer to the axis.

The core (region Ib) is the region surrounding the axis and extending outward to roughly the radius of maximum tangential winds. Core radii vary from tens to hundreds of meters. Narrow cores are approximately in solid-body rotation, owing to the importance of turbulent diffusion. In contrast, the angular velocity in wide cores may increase outward from the axis to close to the core wall. The axial flow in the core may be either upward or downward and there may be axial stagnation points aloft separating axial upflow and downflow.

Flows of the type $u = 0$, $v = v(r)$, $w = 0$ are inertially stable (unstable) to axisymmetric radial displacements according to the criterion $\partial M^2/\partial r > 0$ everywhere ($\partial M^2/\partial r < 0$ somewhere), where M is angular momentum (Rayleigh 1916). For small displacements δ a parcel's radial acceleration is governed by

$$\frac{d^2\delta}{dt^2} \approx -\frac{1}{r_0^3}\frac{\partial M^2}{\partial r}\bigg|_{r_0}\delta. \qquad (5.17)$$

If M^2 decreases with radius, the parcel experiences a force that acts to move it farther away from its original radius r_0 and the flow is unstable. If M^2 increases with radius, the parcel experiences a restoring force and oscillates about its original position with a frequency $[(1/r_0^3)(\partial M^2/\partial r)|_{r_0}]^{1/2}$. These oscillations are called *inertial* or *centrifugal waves.*

Since both narrow and wide cores are inertially stable, perturbations can generate inertial waves that may propagate vertically along the core and there is virtually no entrainment of air into the core from region Ia. The core must consist of air that has entered the vortex from its lower end through the boundary layer and corner region and/or from its upper end.

More general flows of the type $u = 0$, $v = v(r)$, $w = w(r)$ are stable to axisymmetric disturbances if $[(1/r^3)(\partial M^2/\partial r)]/(\partial w/\partial r)^2 > 1/4$ everywhere (Howard and Gupta 1962), a stability criterion analogous to the Richardson number criterion for stability of stratified flows with the local inertial frequency replacing the buoyancy frequency and $\partial w/\partial r$ replacing $\partial u/\partial z$. Thus large radial shears in vertical velocity can overcome the rotational stability of the core flow, perhaps explaining why some cores appear highly turbulent, although breaking waves may be another cause (Rotunno 1979). If the flow is heterogeneous, the stability criterion becomes roughly $(1/r^3)(\partial M^2/\partial r) - (M^2/r^3) \times (1/\theta)(\partial \theta/\partial r) > 1/4(\partial w/\partial r)^2$ (Lewellen 1993). Cold air in the core of the vortex ($\partial \theta/\partial r > 0$) is another destabilizing factor because the potentially cool air may be centrifuged outward and replaced by potentially warmer air. Such replacement would also lower the central pressure because warm air is less dense. These considerations set limits on how potentially cold the air in the core can be compared to surrounding air. The reduction of entrainment in strongly rotating flows explains why the introduction of swirl into the inflow intensifies warm-core updrafts and flames in fire whirls. In a subsiding core, air can be potentially very warm if it descends dry adiabatically from a considerable height. The presence of buoyant air in the core again dampens turbulence and lowers the central pressure hydrostatically.

Exact analytical solutions of the governing equations for the core and outer flow in a real fluid are reviewed in Lewellen (1976) and Davies-Jones (1986). These solutions are not totally realistic because they fail to satisfy realistic conditions at the ground and at radial and/or vertical infinity. They are informative, however; for example, they reveal that the radius of the core of steady laminar vortices r_m is determined by the balance between the inward advection and outward diffusion of angular momentum, resulting in core radius varying as $(\upsilon/C)^{1/2}$ where C is the applied horizontal convergence (e.g., Burgers 1948; Rott 1958; Sullivan 1959; Kuo 1966, 1967). (In the simulator r_m varies as $[\upsilon/M]^{1/2}$ for reasons explained below.)

However, the above balance does not apply to turbulent vortices because Lilly (1969) calculated from photogrammetric data that the turbulent diffusion of angular momentum in a tornado is inward (a negative eddy viscosity phenomenon) and because Davies-Jones (1973) and Baker and Church (1979) found that the core radius of turbulent vortices r_c in the simulator depends primarily on the swirl ratio and increases with it, and for fixed volume flow rate is nearly independent of the depth of the confluent layer zone feeding the updraft. Baker and Church refined Ward's core radii measurements and extended them to large S, where multiple vortices develop in the simulator (see below). In the multivortex regime, the core radius pertains to the azimuthally averaged flow. Originally it had been thought that the core radius was a function of the tangent of the inflow angle M/Qh in the confluent zone (Ward 1972). In the limit of inviscid flow ($\upsilon \to 0$), angular momentum M is conserved and the outer flow cannot penetrate to the axis because this would require infinite inward pressure-gradient force to counteract the infinite centrifugal force M^2/r^3 at $r = 0$. There has to be a nonrotating core composed either of air that was already present prior to the onset of rotation or of nonrotating air that has descended through the top boundary of the simulator. The simplest inviscid axisymmetric model of this flow consists of a stagnant core separated from irrotational outer flow by a vortex sheet. Because the core is stagnant, the core pressure p_c is constant. The radius of the core varies from r_1 at the ground to r_2 at the top of the apparatus. The updraft above the inflow is assumed to be cylindrical with the meridional flow becoming increasingly vertical with height (i.e., $u \to 0$). Since the outer flow is irrotational, $\partial w/\partial r \to 0$ and $w \to 2Q/(r_0^2 - r_2^2)$ with height in the outer flow. Just outside the intersection of the core wall with the ground, $u = w = 0$ and $v = M/r_1$. Since pressure is continuous across the core wall, Bernoulli's equation applied to the streamline just outside the core yields the relationships

$$p_\infty = p_c + \frac{\rho}{2}\left[\frac{M^2}{r_2^2} + \frac{4Q^2}{(r_0^2 - r_2^2)^2}\right]$$
$$= p_c + \frac{\rho}{2}\left[\frac{M^2}{r_1^2}\right], \tag{5.18}$$

where p_∞ is the pressure at infinity and the stagnation pressure of the outer flow. In terms of the pressure drop driving the vortex $\Delta p \equiv p_\infty - p_c$, the core radius at the ground is given by

$$r_1 = (2\Delta p/\rho)^{-1/2}M. \tag{5.19}$$

Also from (5.18), the core radii at the ground and near the top are related by

$$\frac{1}{x_1} = \frac{1}{x_2} + \frac{1}{S^2}\frac{1}{(1 - x_2)^2}, \tag{5.20}$$

where $x_1 \equiv (r_1/r_0)^2$ and $x_2 \equiv (r_2/r_0)^2$. In the simulator, the core is composed of initially nonrotating air that descends through the honeycomb baffle at the top of the updraft. The stagnation pressure of this air is lower than that of the inflow air by an unknown amount so that the core radius cannot be obtained directly from

(5.19). Davies-Jones (1973) postulated that the core radius can be determined by minimizing Δp for given Q, M, and r_0. This is equivalent to minimizing the maximum wind speed in the flow, which occurs at the core wall, or minimizing $1/x_1$ in (5.20). According to Binnie and Hookings (1948), the resulting flow is critical in the sense that the exit vertical velocity is equal to the phase speed of a long small-amplitude inertial wave. The minimum strength vortex occurs when

$$S^2(1 - x_2)^3 = 2x_2^2. \qquad (5.21)$$

An optimum core radius for minimizing extreme wind speed exists for fixed Q, M, and r_0 because tangential velocity becomes large as $x_2 \rightarrow 0$ and vertical velocity becomes large as $x_2 \rightarrow 1$ owing to the cross-sectional area of the updraft $\pi r_0^2(1 - x_2) \rightarrow 0$. The theory can account for the core radius increasing from 0 at $S = 0$ to r_0 as $S \rightarrow \infty$. The method predicts the core radius fairly well at large S, but is too large at small S, probably owing to the neglect of turbulent stresses along the core wall, the effects of which should increase with the length of the vortex (Baker and Church 1979). Wilkins and Diamond (1987) demonstrated that for fixed updraft radius the core radius decreases as the height of the baffle is increased. Other theories may yield better fits, but they have at least one tuning parameter and some assume a radius that does not vary with height. Note that although the radius of a rotating thunderstorm updraft is probably an optimum value according to some variational principle, it is not obvious that this value has to be independent of swirl ratio. This casts doubt on the straightforward applicability of the laboratory core radius results to tornadoes.

Unlike the simple inviscid solution, the similarity solutions of the full equations presented by Burgers (1948), Rott (1958), Sullivan (1959), and Kuo (1966) all suffer from decoupling of the swirling and meridional flows because the radial pressure gradient is independent of height. This condition cannot be satisfied in atmospheric vortices because they are not infinitely tall. Long's (1958) similarity solution and Morton's (1966) scale analysis for a tall thin axisymmetric vortex that spreads slowly with height showed that 1) the core is in cyclostrophic balance, 2) vertical and tangential velocities in the core are of the same order of magnitude, while radial velocities are much smaller, and 3) swirling and updraft flows interact strongly. Of the above solutions, Kuo's are the only

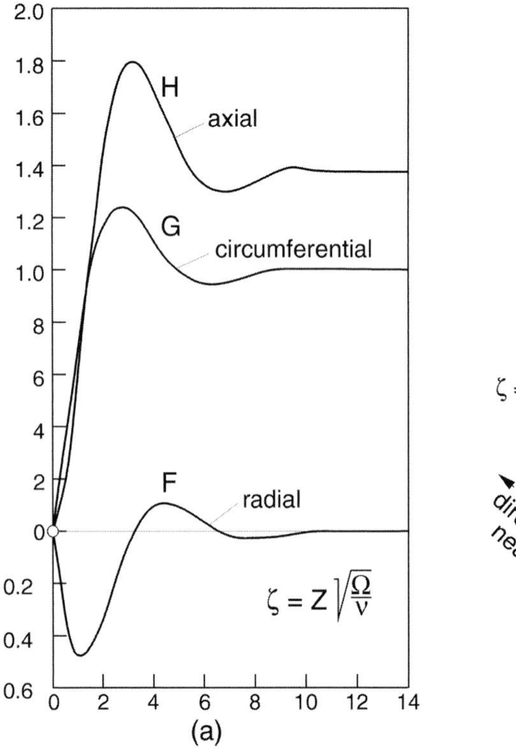

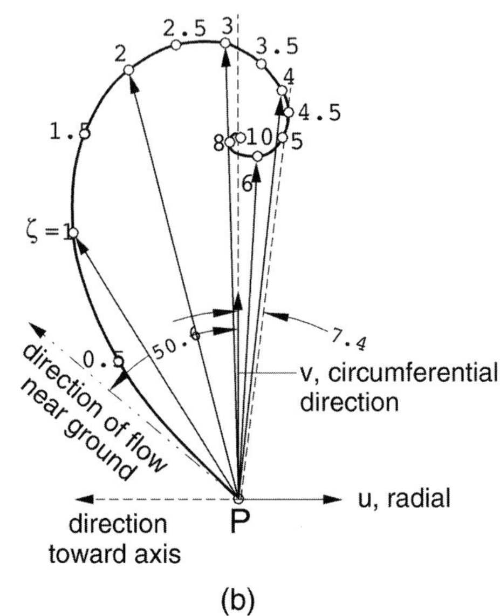

FIG. 5.28. The boundary layer beneath a fluid in solid-body rotation. (a) The functions $F(\zeta)$, $G(\zeta)$ $H(\zeta)$ where $\zeta \equiv z(\Omega/v)^{1/2}$ is the nondimensional height. The radial, tangential and vertical velocities are given by $u = r\Omega F(\zeta)$, $v = r\Omega G(\zeta)$ and $w = (v\Omega)^{1/2}H(\zeta)$. (b) Hodograph of horizontal wind. From Schlichting (1960; reprinted with permission).

ones for convective vortices in a (conditionally) unstable atmosphere, albeit a saturated one. Both the one- and two-cell solutions have cold cores as a result of assuming that saturation is maintained everywhere, an unlikely circumstance in the core of a two-cell vortex. Lewellen (1993) concludes that buoyancy forces are unimportant in the corner region except for their role in stabilizing the flow, but become quite important in the core at higher levels.

2) THE BOUNDARY LAYER

Frictional interaction of a "primary" rotating flow with the ground reduces tangential velocities near the surface. Pressure does not vary much across the boundary layer and therefore the imbalance between the radial pressure-gradient force and the reduced centrifugal force drives a strong radial inflow within the lower part of the boundary layer; the boundary layer of a tornado is roughly 100 m deep. Bödewadt (1940) found an exact similarity solution describing the laminar interaction of a flow rotating as a solid body at the rate Ω with a stationary solid rigid surface (Fig. 5.28). The boundary layer thickness $\delta \approx 8(\upsilon/\Omega)^{1/2}$, where υ is the kinematic viscosity. The maximum radial inflow velocity of $0.5\Omega r$ occurs at a height of around $\delta/8$. The vertical velocity, equaling $1.4(\upsilon/\Omega)^{1/2}$ far from the ground, is independent of radius and is upward at all heights. All three velocity components overshoot as they tend to their values at great height. The radial inflow forces the overshoot in tangential velocity, resulting in rotation in the midboundary layer that exceeds the rotation aloft by 25%. Paradoxically, friction can increase extreme wind speeds!

Of more relevance to tornadoes and hence to region II is the case in which the primary swirling flow is a potential vortex, wherein $\upsilon \propto 1/r$ (rather than a solid-body rotation, wherein $\upsilon \propto r$). Burggraf et al. (1971) matched a laminar boundary layer on a surface of radius r_s to a potential vortex. They found that the boundary layer has a double structure with an inner viscous layer next to the surface with variable thickness of order $(\upsilon/M)^{1/2}r$, in which the flow is primarily radial, and an outer inertial layer of order $(\upsilon/M)^{1/2}r_s$ thick, in which the flow recovers to the external potential flow; the head $B \equiv (p/\rho) + \frac{1}{2}\mathbf{v}\cdot\mathbf{v} + gz$ is nearly conserved in the outer layer. In contrast to the Bödewadt solution, the vertical velocity is negative and thus the inward mass flux in the boundary layer cannot be compensated for by flux out of the top of the boundary layer: The fluid has no escape until it reaches the core radius where M is no longer constant; then it turns violently upward inside a narrow core. The laminar boundary layer in the simulator has a similar double-layer structure; the radius r_s at which the swirling boundary layer effectively begins is said to correspond to the radius of the rotating screen or the vanes (Fiedler and Rotunno 1986). Note, however, that the flow above the boundary layer in the confluent zone and in the outer part of the convergence zone has a strong radial component. Hence, the effective beginning radius of the potential vortex boundary layer thickness may only be $0.6r_0$ (Wilson and Rotunno 1986).

The nature of the flow in the boundary layer is Reynolds number dependent. The boundary layer of flow past a flat plate (for which the pressure gradient is neutral) becomes turbulent at a Reynolds number of around 3×10^5 for a smooth surface and at lower values for rough surfaces. The radial Reynolds number $\mathrm{Re}_r \equiv Q/\upsilon h$ in the simulator is typically close to this critical value. Given that the radial pressure-gradient force is inward and hence favorable for the damping of perturbations in the inflow, we conclude that the flow in the boundary layer may be either laminar or turbulent, depending on the settings of the external parameters. In the atmosphere, Re_r is much higher, $\vartheta(10^9)$, and the ground is rough, so it is generally accepted that the ~100-m boundary layer of a tornado is turbulent rather than laminar, although the flow may be relatively smooth because of the favorable pressure gradient. For reviews of turbulent boundary layers pertinent to tornadoes, we defer to Lewellen's (1976, 1993) articles. Since the boundary layer is rotating, it is susceptible to Ekman instability and consequent formation of horizontal roll vortices resembling the spiral bands in hurricanes (Faller 1963). The boundary layer rolls are oriented about 15° to the left of the flow above the boundary layer.

3) THE CORNER REGION

The boundary layer erupts upward in region III, the corner region. It is here that missiles are generated and debris is lofted into the air. Light debris such as bank checks drawn up into the storm through this region may be transported hundreds of kilometers downstream (Snow et al. 1995; Magsig and Snow 1998). The vertical pressure gradient, which is small in boundary layers, is large here. The mass flux of virtually nonrotating air that flows into the vortex through the lower part of the boundary layer and the corner region is insufficient to "fill" or eliminate the vortex from below. Lewellen (1976) suggested that the maximum tangential velocity in the tornado occurs in the upper half of the corner region: Parcels flowing into region III from the upper part of region II penetrate closer to the axis than parcels in region Ia and consequently gain more tangential velocity (the loss of angular momentum is not large enough to annul this gain).

Vortex breakdown is most important to high tornadic wind speeds when it occurs in the corner region, and therefore this phenomenon is introduced here. Note that the following discussion is relevant to the tornado but not to its mesocyclonic parent circulation, despite some recent observations that have been interpreted otherwise (Trapp 2000). The vortex breakdown

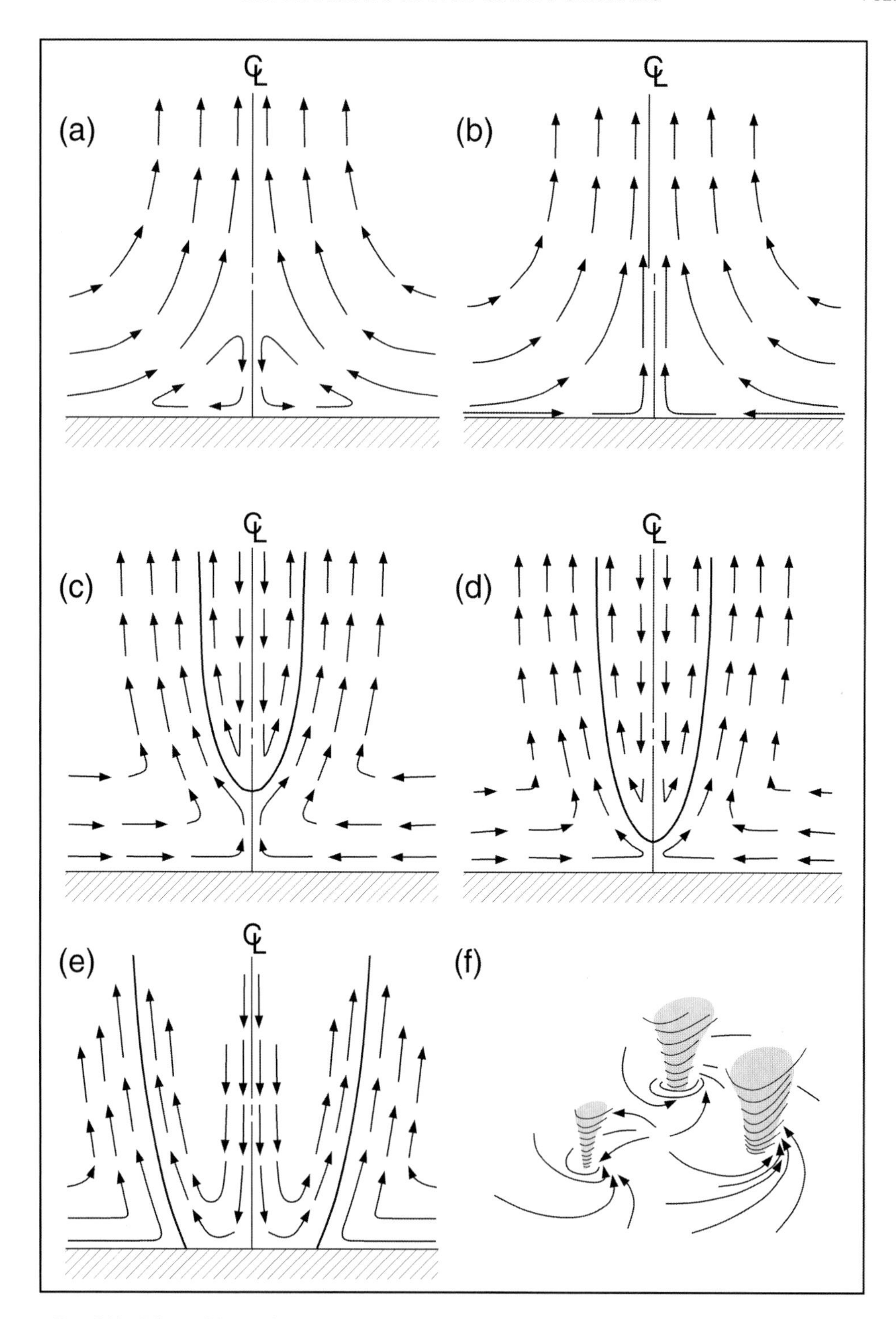

FIG. 5.29. Effect of increasing swirl ratio, S, on tornadic vortex flows. (a) Very weak swirl—flow in boundary layer separates and passes around corner region and there is no tornado; (b) Low S—smooth-flowing one-cell weak tornado; (c) Moderate S—end-wall boundary layer erupting upward into strong smooth-flowing end-wall vortex that breaks down near the top of the boundary layer into a turbulent two-cell vortex aloft: (d) Slightly higher S—drowned vortex jump (DVJ) with the defining characteristic of the vortex-breakdown stagnation point very close to the ground; (e) Turbulent two-cell tornado at higher S—the central downdraft now impinges on the ground, eliminating the stagnation point aloft, inflow in the boundary layer erupts upward in a now-annular corner region, and the core radius increases rapidly with S; (f) Large S—tornado "splits" into multiple vortices (2, 3, . . . , 6 as S increases). (Modified from Davies-Jones 1986.)

is the rotating-flow counterpart of the hydraulic jump in stratified flows. It is a transition between two conjugate vortex states (Fig. 5.29c) that occurs when the swirl ratio increases beyond a critical value of order one. For tornado-like vortices, the upstream state (below the breakdown level) is a smooth, high-speed swirling jet that arises from the eruption of the boundary layer near the axis. It is also referred to as an end-wall vortex. In the downstream state above the breakdown, the vortex "bursts" into a much wider two-celled structure with highly reduced axial flow (often downdraft) surrounded by updraft, much higher levels of turbulence, lower vertical and tangential velocities, and reduced pressure deficit. In some cases the upper flow is asymmetric and assumes the form of a precessing single or double helix. With vortex breakdown, it is possible to have a tornado that appears as a condensation funnel below a bowl-shaped lowering of cloud base with a cloud-free gap in between (see Pauley and Snow 1988).

The upstream state is supercritical, meaning that the upward flow is faster than the speed of the fastest inertial waves and information of the downstream conditions cannot be carried upstream (i.e., downward) by the waves. The flow above the breakdown is subcritical so that waves are able to progress upstream as far as the breakdown and "impart knowledge" of the downstream conditions or the top boundary condition throughout the subcritical vortex. The conjugate states arise because the inviscid swirl equation (Wilson and Rotunno 1986) has both a supercritical and a subcritical solution for the same distribution of angular momentum across streamlines $M(\psi)$. Like the hydraulic jump, the vortex breakdown is a transition between conjugate states. Maxworthy (1972) views vortex breakdown as a "buffer zone" through which incompatible upstream and downstream boundary conditions can be matched. The downstream flow can become supercritical again with the subcritical flow confined to a bubble-shaped enlargement of the core. In these cases a second breakdown can occur farther downstream.

Fiedler (1989) has argued that the flow in supercritical end-wall vortices might be laminar if the boundary layer is turbulent. This would account for the smooth appearance of some slender condensation funnels. Lewellen (1993) disagrees since the advective timescale in region III is too small for the total elimination of turbulent eddies. Condensation funnels may appear laminar simply as a result of rotational damping of turbulent eddies. Furthermore, the balance between inward advection and outward viscous diffusion of angular momentum yields an unrealistically small core radius of less than 1 m in Kuo's (1966) vortex solution.

4) THE UPPER FLOW

The upper-flow region of the tornado is uncertain because it is embedded in the parent storm and has yet to be observed in detail with present instruments (Lewellen 1993). The core of a moderate tornado that has a TVS only to midaltitudes (perhaps only up to 3 km; Brandes 1981) may become unstable at some height and change into a turbulent buoyant nonrotating plume with the angular momentum at this level transported outward by turbulent eddies (Lewellen 1993), as apparently occurs in the experiment of Mullen and Maxworthy (1977). The strong buoyancy of the air above the vortex acts like a "buoyant cork" (Fiedler 1995) by preventing the core low pressure from being filled from above.

A large tornado that reaches to near the tropopause (Brown et al. 1978; Johnson and Ziegler 1984) probably terminates in surrounding upflow that is strongly divergent. As pointed out by Lilly (1969), the dynamics of rotating and nonrotating updrafts differ near the tropopause. The anvil outflow from a nonrotating updraft is driven by an outward pressure-gradient force that arises as follows. The updraft penetrates above the equilibrium level (EL) into the stratosphere creating an overshooting storm top and, relative to the environment, a cold dome of air that gives rise hydrostatically to high pressure beneath it. The cold dome cancels the effect on the surface pressure of the warmth of the updraft column below the EL, so there is little hydrostatic pressure deficit at the ground. Using the inviscid swirl equation, Lilly showed that a rotating updraft can terminate without overshooting the EL. In this case the radial pressure gradient goes to zero and unbalanced centrifugal forces can drive the outflow. Without a cold dome, there is low hydrostatic pressure at the surface, which can be maintained in a vortex because it has "dynamic walls" owing to its cyclostrophically balanced state. Even though the vortex is tilted so that its top does not overlie its base, its nonporous core acts like the glass walls of a tilted mercury barometer in transmitting the hydrostatic pressure deficit down the axis. Some of the environmental potential energy can now be realized in the form of violent surface winds (see section 5.5).

e. Different flow regimes

As the swirl ratio in the simulator is increased, vortex structure changes as shown in Fig. 5.29. The critical swirl ratio values given below are approximate because they depend slightly on Reynolds number and they are machine dependent because of differences in the height of the baffle H compared to r_0 (Wilkins and Diamond 1987). At $S = 0$ high pressure at the stagnation point at the foot of the axis causes the boundary layer to separate in the region of adverse pressure gradient, and the outer flow circumvents the corner region that consists of a recirculating eddy. A very small amount of swirl does not change this pattern of meridional circulation so that even if a weak vortex forms it will not be in contact with the ground. Low swirl is sufficient to cause the boundary layer to reattach itself, resulting in a one-cell vortex, which by definition has a central updraft (except near the top baffle where the vortex breaks down). In the one-cell vortex, the central pressure deficit at the surface is surrounded by a ring of high pressure. This occurs at a radius where the wind speed just above the boundary layer is a minimum, owing to a rapid decrease of tangential velocity with respect to radius outside the core, and an increase in radial inflow with respect to radius from the axis to $r = r_0$ (Ward 1972). Addition of more swirl keeps the boundary layer attached to the ground, eliminates the adverse pressure gradient and high pressure ring, and lowers the vortex breakdown. The subcritical flow consists of a bubble or eddy of recirculating air at the base of an enlarged turbulent wakelike vortex core, which is capped at a downstream axial stagnation point by either stagnant flow or downflow near the axis. With increasing swirl, the flow between the two stagnation points evolves into first a single helix and then a double helix (Church and Snow 1979).

At moderate swirl ratio, the boundary layer erupts into a smooth, high-speed vortex jet that ends at a low-level vortex breakdown, above which exists a much larger turbulent two-celled vortex. At a critical swirl ratio $S^* \approx 0.45$, the breakdown point descends into the surface boundary layer. This case is called a *drowned vortex jump* (DVJ; Maxworthy 1973; Snow 1982). Note that the critical swirl ratio for this and other transitions decreases to an asymptotic value as the radial Reynolds number (Re_r) is increased (Church et al. 1979). As S increases beyond S^*, the breakdown point is eliminated altogether as the central downdraft impinges on the ground and the boundary layer erupts in a now annular corner region surrounding the inner cell of the vortex. With increasing S, the core radius increases and the vortex resembles a wide single-vortex tornado.

At large swirl ratio, $S \approx 0.8$, the single vortex becomes unstable and "splits" instantaneously into two secondary vortices that form near the radius of maximum tangential wind of the primary vortex. These vortices revolve around the axis under the influence of each other and their parent circulation at the rate of about half the maximum tangential wind speed of the primary vortex. As S is increased to 3 or beyond (depending on Re_r and the version of the simulator), further transitions to three, four, five, and six vortices occur. A system of seven vortices has not been observed, perhaps because seven equal-strength line vortices spaced regularly around a circle is an unstable configuration (Saffman 1992, p. 119).

Increasing the swirl ratio with a vortex pair initially present sends the flow into a disordered state before the swirl ratio becomes large enough for a triad of vortices. The higher transitions also occur following episodes of chaotic flow. This behavior is typical of nonlinear dynamical systems and flows in transition to turbulence (Sreenivasan 1985). As S is decreased, the reverse transition—from two vortices to one vortex—occurs at a higher critical swirl ratio. This hysteresis effect is characteristic of a subcritical bifurcation in the theory of dynamical systems (Bergé et al. 1984, pp. 40–42).

We note here that it is possible to define a local corner flow swirl ratio, $S_c \equiv r_c M_\infty^2/\Upsilon$, where r_c is the radius of maximum swirl velocity in the upper-core region, M_∞ is angular momentum outside the core and boundary layer, and Υ is the total depleted M flux flowing through the corner flow region (Lewellen et al. 2000). Whereas S characterizes the swirling, converging flow within which the tornado vortex forms, S_c characterizes the surface-layer core flow embedded within this larger-scale swirling converging flow; both affect tornado structure. Hence, as Lewellen et al. state, S_c does not replace or redefine S, yet S_c can more completely determine the corner-flow structure. Indeed, since only S_c can compensate for the existence of/variations in surface-layer inflow of low angular momentum fluid, flows with the same S can exhibit very different vortex structure in the corner flow region.

Ward (1972), Davies-Jones and Kessler (1974), and Davies-Jones (1976) idealized the two-cell vortex as a cylindrical vortex sheet in an otherwise irrotational flow and suggested, without proof, that the multiple vortices formed as a result of this vortex sheet becoming barotropically unstable and rolling up into individual vortices. Rotunno (1978) proved this conjecture by demonstrating that a cylindrical vortex sheet with an inner uniform downdraft and outer uniform updraft is unstable to three-dimensional perturbations. The most unstable perturbation is a helical one with negative pitch (i.e., turns anticyclonically with height in a cyclonic vortex), in agreement with the observed negative pitch of secondary vortices in tornadoes and in laboratory and numerically simulated vortices. The perturbations also transported angular momentum radially inward, consistent with Lilly's (1969) analysis of Hoecker's (1960) photogrammetric wind measurements in the 1957 Dallas tornado and with the pressure profile observed by Winn et al. (1999). Extreme winds are located in the secondary vortices. In the simulator the instantaneous winds in a secondary vortex may be almost double the maximum winds in the azimuthally averaged flow (Leslie 1977).

Increasing the surface roughness generally increases the core radius, radial and vertical velocities, turbulence levels, and the critical swirl ratios for transitions to higher numbers of vortices, while decreasing tangential velocities. Effects of surface roughness on tornadoes and laboratory vortices are discussed further in Davies-Jones (1986) and Church and Snow (1993).

f. Measurements in the simulator

Extensive pressure measurements, both at the surface (Snow et al. 1980; Pauley et al. 1982) and aloft (Church and Snow 1985; Pauley 1989) have been made in the Purdue University version (TVC I) of the Ward-type simulator. Church and Snow (1985) found that the nondimensional axial pressure deficit $\Delta p^* \equiv \Delta p/\rho\bar{w}^2$ is a function of S and has a maximum value of 60 at S just less than $S^* \equiv 0.45$. The minimum occurred in the laminar supercritical vortex at a height of several core radii above the surface. The elevated pressure minimum is a common feature of the laminar vortices. To avoid confusion we denote the core radius and maximum velocities in laminar supercritical and turbulent subcritical vortices by subscripts m and c, respectively. Church and Snow also found that the core radius of the supercritical vortex $r_m \approx \delta \equiv r_s(v/M)^{1/2}$ where δ is the depth of the laminar boundary layer. This indicates that the end-wall vortex is simply a continuation of the boundary layer. In terms of $\beta_1 \equiv \Delta p/\rho v_m^2$, where $v_m \approx M/2\delta$ is the maximum tangential wind speed (allowing for loss of angular momentum), $\Delta p^* \sim \beta_1 (r_0^2/r_s^2)(M/v)S^2$, so the nondimensional pressure deficit increases rapidly with S until S^* is reached. Since head is nearly conserved from the rotating screen to the axis in the upper boundary layer, the maximum vertical velocity w_m in the end-wall vortex at $S \approx S^*$ is given by $w_m^2/2 \approx \Delta p/\rho = 60\bar{w}^2$ or $w_m \approx 11\bar{w}$ (Fig. 5.30). According to velocity measurements in laminar vortices $v_m \approx 0.54w_m$, so $v_m \approx 6\bar{w}$ and $\beta_1 \approx 1.7$. Baker and Church (1979) showed that the maximum tangential velocity in turbulent cores v_c was given by $v_c \approx 2.6\bar{w}$ over a wide range of S. Assuming that $r_c \approx M/2v_c$, this is consistent with $r_c/r_0 \propto S$. For these vortices the nondimensional pressure deficit is much smaller, $\Delta p^* \approx 6.8\beta_2$, where $\beta_2 \equiv \Delta p/\rho v_c^2$ is probably around 1 based on Kuo's two-cell vortex. The maximum tangential velocity is 2.3 times larger below the breakdown than above it. Fiedler and Rotunno (1986) have shown from a solution of the inviscid swirl equation that this is roughly commensurate with conjugate states with flow force (Morton 1966, 1969) invariant across the breakdown, but with a loss of head owing to turbulence at the top of the jet. Their solution also predicted the value of S^*.

According to the Purdue measurements, the azimuthally averaged surface pressure field has the following characteristics. The lowest pressure is at the center for one-celled vortices, and maximizes at $S \approx S^*$. For $S > S^*$ the vortices become two-celled at the surface, the lowest surface pressure becomes located in a ring off center with a slight pressure maximum on the axis, and Δp^* is much less than in the end-wall vortex because the tangential velocity is much reduced. Although Δp^* increases once more at swirls beyond the transition to multiple vortices, it never reaches the maximum value attained at $S \approx S^*$. The pressure deficits in individual secondary vortices were also measured; these were 2–3 times greater than the Δp^* in the azimuthally averaged flow, but were still less than Δp^* at S^*.

The variation of central pressure with height was also measured. For the laminar vortices, core pressure

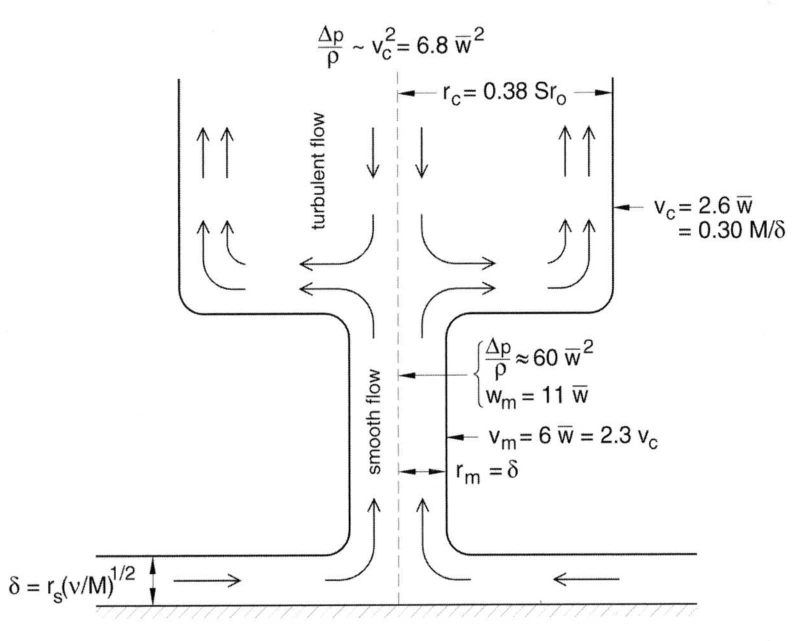

FIG. 5.30. The geometry of and the velocities in the supercritical end-wall vortex and the subcritical turbulent vortex aloft at S slightly less than $S^* = 8.7\delta/r_0$.

decreases rapidly with height, then becomes constant. As shown above, Δp^* aloft in laminar vortices increases as the cube of the circulation so that the largest pressure deficit occurs at $S \approx S^*$. The strong upward pressure-gradient force near the surface provides the upward acceleration for air in the corner region. The pressure deficit at the surface is less than the maximum deficit aloft by a factor that ranges up to 7. The pressure deficit declines abruptly with height across the vortex breakdown. The largest surface pressure deficit is associated with the DVJ. In this case the pressure deficit declines rapidly with height. In the turbulent two-celled vortices that occur at $S > S^*$, the axial pressure deficit is much less than in the DVJ, but declines little with height. Over the region of parameter space that has been investigated, 60 appears to be an upper limit for the maximum nondimensional pressure deficit that can occur in the simulator.

5.5. Tornado intensity

The issue of tornado intensity can be addressed using measurements of maximum tornado wind speed, through theoretical considerations of the kinetic energy of the tornado, and/or by atmospheric scaling of atmosphere laboratory or numerical model results, the caveats of which are discussed in section 5.4. In the spirit of the preceding sections, we begin with a discussion of relevant tornado observations, which, as mentioned above, indicate wind speeds as high as 125 m s^{-1} to perhaps 140 m s^{-1}.

a. Measurements of maximum tornado wind speeds

Photogrammetric analyses of tornado-debris movies (e.g., Hoecker 1960; Golden 1976; Golden and Purcell 1978a; and many others) have yielded maximum wind speed estimates ranging from 56 to 95 m s^{-1}, at 15–200 m above the ground; aloft, in a suction vortex, Forbes (1978) estimated wind speeds as high as 125 m s^{-1}. Analyses additionally suggest upward vertical velocities in tornadoes as high as 25–60 m s^{-1} between 25 and 60 m AGL. The photogrammetric techniques require accurate knowledge of the distance of visible landmarks (in the film) and of the tornado to the camera site. Air tracers (tornado debris, cloud tags) can then be scaled on each movie loop and subsequently tracked on a grid (Golden and Purcell 1978a). The techniques may underestimate maximum wind speed owing to the following reasons: (i) the tracer motion may be less than the actual wind speed, (ii) the location of the maximum wind speeds may be hidden by an opaque dust column or condensation funnel, and (iii) the line-of-sight motions cannot be determined. In addition, debris tracers may be absent in some parts of the vortex.

With Doppler radar, the maximum wind speed estimation proceeds as follows. Facilitated by the use of a relatively high pulse repetition frequency, unfolded Doppler spectra can be computed; maximum wind speeds in the radar volume are then determined by locating the highest frequency in the spectrum above the noise floor (see also Doviak and Zrnic 1993, pp. 344–350). Such a Doppler spectrum of a tornado was first reported by Smith and Holmes (1961): Using a fixed-site CW (continuous wave) Doppler radar in Kansas in 1958, they found maximum wind speeds in a tornado of 92 m s^{-1}. Comparable estimates (65–92 m s^{-1}) from a number of tornadoes that passed within a 100-km range of NSSL's 10-cm wavelength, fixed-site, pulsed Doppler radar have been reported by Zrnic and Doviak (1975), Zrnic et al. (1977), Zrnic and Istok (1980), and Zrnic et al. (1985).

A portable, 3-cm wavelength, CW Doppler radar, designed and built at the Los Alamos National Laboratory (LANL) (and later modified to have the capability to collect Doppler spectra in range bins of 78-m spacing, using FM-CW signal processing), has been utilized to increase the probability of determining the Doppler velocity spectrum at and below cloud base in tornadoes (Bluestein and Unruh 1989; Bluestein and Unruh 1993; Bluestein et al. 1993, 1997). At close range (i.e., within sight, or less than 5–10 km in range), visual documentation by a boresighted video camera mounted on the radar antenna was used to determine the location of the maximum wind speeds relative to the condensation funnel, if present. Most of the portable radar–based estimates (collected during 1987, 1990, 1991, and 1994) of maximum tornadic wind speed were between 50 and 100 m s^{-1}, at some distance (as much as 100 m) above the ground (mea-

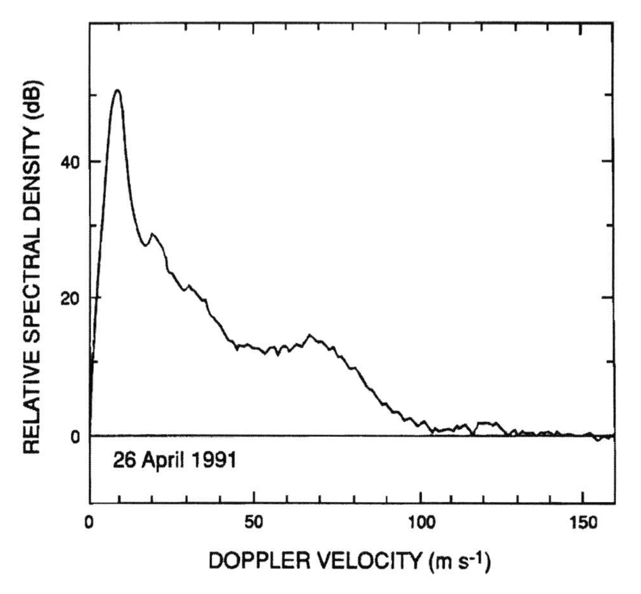

FIG. 5.31. Receding portion of the Doppler velocity spectrum of an F-5 tornado, as measured by a portable, 3-cm wavelength CW Doppler radar near Ceres and Red Rock, Oklahoma, on 26 April 1991. From Bluestein et al. (1993).

surements confined to volumes near the ground were not possible owing to the 5° beamwidth of the antennas). For example, maximum wind speeds in the tornado near Northfield, Texas, on 25 May 1994 reached 65 m s^{-1} and possibly 75 m s^{-1}; the 2-km diameter parent mesocyclone had wind speeds of 45–50 m s^{-1} (Bluestein et al. 1997). Higher wind speeds—up to 120–125 m s^{-1}—were estimated in a tornado on 26 April 1991 near Ceres and Red Rock, Oklahoma (Fig. 5.31; see Bluestein et al. 1993). These measurements were the first confirmation of F5 wind speeds in a tornado based on Doppler radar measurements. [On 3 May 1999, DOW data were collected on a tornado that was associated with F5 damage near Oklahoma City, Oklahoma. Wind speeds possibly as high as 140 m s^{-1}—the upper end of the F5 scale—were measured over a very short duration (~50 ms) at about 50 m AGL; J. Wurman 1999, personal communication.] Not surprisingly, however, wind speeds corresponding to the F-scale ratings for other tornadoes were less than those indicated by the Doppler spectra. Hence, as alluded to in section 5.1, estimates of tornado intensity based on damage surveys alone must be viewed with caution (see also Doswell and Burgess 1988).

A comparison by Bluestein et al. (1993) of simulated wind spectra with actual spectra suggests that the highest radar reflectivities are located well outside the core of the tornadoes, presumably in the debris cloud. This inference is consistent with other radar observations of weak-echo holes. It is therefore possible that there are even higher wind speeds closer to the center of the tornado where radar reflectivities may be too weak to detect a signal and near the ground where wind speeds could not be measured owing to the 5° beam width of the antennas.

b. Theoretical limits

The parent thunderstorm contains far more energy than does the tornado simply because of the thunderstorm's far greater volume. The kinetic energy of the thunderstorm circulation is derived largely from the latent heat released in the updraft and evaporative cooling in the downdraft, and possibly from some conversion from environmental kinetic energy. The high density of kinetic energy in the tornado is associated with a large local drop in pressure, which occurs by mechanisms described below. Investigators have attacked the question of the maximum wind speed that can occur in a tornado by addressing the equivalent question of how large a pressure deficit can be supported in the atmosphere. Since nature abhors even a partial vacuum, a mechanism is needed that prevents fluid from rushing in to fill the void.

Various attempts have been made to place upper bounds on tornado wind speeds by calculating a maximum hydrostatic pressure drop (e.g., Lilly 1969, 1976). The conceptual model assumes a warm axisymmetric updraft column at $r = 0$ in an environment that is in hydrostatic equilibrium at $r = \infty$. The vertical momentum equation

$$\frac{\partial w}{\partial t} + \frac{1}{r}\frac{\partial}{\partial r}(ruw) + \frac{\partial}{\partial z}(w^2) = -c_p\theta(z)\frac{\partial \pi}{\partial z} - g + v\left[\frac{1}{r}\frac{\partial}{\partial r}\left(r\frac{\partial w}{\partial r}\right) + \frac{\partial^2 w}{\partial z^2}\right]$$

reduces at radial infinity to the hydrostatic equation, which can be written as

$$0 = -c_p\theta_\infty(z)\partial\pi_\infty/\partial z - g, \qquad (5.22)$$

where the subscript ∞ denotes $r = \infty$. On the axis, denoted by subscript 0, the steady-state vertical momentum equation can be approximated by

$$\frac{1}{2}\frac{\partial w_0^2}{\partial z} = -c_p\theta_0(z)\frac{\partial \pi_0}{\partial z} - g + 2v\left.\frac{\partial^2 w}{\partial r^2}\right|_{r=0} - \left.\frac{1}{r}\frac{\partial}{\partial r}\overline{(ru'w')}\right|_{r=0}, \qquad (5.23)$$

where l'Hôpital's rule has been used, and the last term is a turbulent stress term for quasi-cylindrical flow (Pauley 1989). Using (5.22), (5.23) may be written

$$\frac{1}{2}\frac{\partial w_0^2}{\partial z} = c_p\theta_0\frac{\partial(\pi_\infty - \pi_0)}{\partial z} + g\frac{\theta_0 - \theta_\infty}{\theta_\infty} + 2v\left.\frac{\partial^2 w}{\partial r^2}\right|_{r=0} - \left.\frac{1}{r}\frac{\partial}{\partial r}\overline{(ru'w')}\right|_{r=0}. \qquad (5.24)$$

Let the updraft terminate at the equilibrium level $z = H$ without overshooting so that $\pi_0(H) = \pi_\infty(H)$ and $w_0(H) = 0$. Then integrating (24) from the top of the corner region $z = \delta$ to $z = H$ yields a formula for the pressure drop near the surface,

$$c_p\tilde{\theta}_S[\pi_\infty(\delta) - \pi_0(\delta)] \approx \text{CAPE} + \frac{1}{2}w_0^2(\delta) + 2\int_\delta^H v\left.\frac{\partial^2 w}{\partial r^2}\right|_0 dz - \int_\delta^H \left.\frac{1}{r}\frac{\partial}{\partial r}\overline{(ru'w')}\right|_{r=0} dz, \qquad (5.25)$$

where $\tilde{\theta}_S$ is an intermediate value of $\theta|_{z=0}$ and $\text{CAPE} \equiv g\int_\delta^H [\theta_0(z) - \theta_\infty(z)]/[\theta_\infty(z)]dz$. The pressure drop is related to a maximum tangential wind speed v_M by assuming cyclostrophic balance and some tangential-wind profile $v(r)$. For several different types of vortices this results in the relationship $c_p\tilde{\theta}_S[\pi_\infty(\delta) - \pi_0(\delta)] = \beta v_M^2$, where the constant $\beta = 0.5$ for a stagnant-core vortex, 1 for a Rankine combined vortex, 1.14 for Kuo's two-cell vortex, 1.74 for his one-cell vortex, and 1.68 for the Burgers–Rott vortex. For a steady nondiffusive flow, $\delta = 0$, $v = 0$, and $w_0(\delta) = u' = w' = 0$, and the maximum tangential velocity on the ground is given by the "thermodynamic speed limit" $v_M = (\text{CAPE}/\beta)^{1/2}$, where $\theta_0(z)$ is the potential temperature of a surface parcel lifted adiabatically to cloud base and pseudoadiabatically above cloud base. This limit pertains only to inviscid tornadoes with axial updrafts.

Based on an environmental sounding, Bluestein et al. (1993) calculated a CAPE of 3900 m^2 s^{-2} for the Red Rock, Oklahoma, tornado and a thermodynamic speed limit of 62 m s^{-1} based on a Rankine combined vortex ($\beta = 1$). As discussed above, this is roughly one-half the observed wind speed measured somewhere below cloud base. Since this tornado was large, the inner part of the core may have been rotating slowly compared to the core wall, hence β may have been closer to 0.5, which would increase the speed limit to 88 m s^{-1}. Since the tornado was moving to the east-northeast at about 15 m s^{-1} and the measurement was obtained when the tornado was northeast of the radar, the maximum tornado-relative wind speed may be only 105 m s^{-1}. By this reckoning the speed limit is broken not by a factor of 2, but by only 20%. To explain away the remaining 20% without invoking subsidence warming in the core, we could postulate that the measurements were made near the top of the tornado's boundary layer where inflowing parcels overshot their equilibrium radius by 20%. Or we might point out that the CAPE is relative to a pristine horizontally homogeneous environment, instead of the actual inhomogeneous surroundings that include the storm's cold pool (Lewellen 1976, 1993; Trapp and Davies-Jones 1997). These remarks are intended only to raise uncertainty, not to lend support to the thermodynamic speed limit that is based on constant equivalent potential temperature along the axis.

The simplest mechanism for breaking the thermodynamic speed limit is via subsidence in the core forced by a downward axial pressure-gradient force (Lilly 1969, 1976). The CAPE in (5.25) is now based on the temperature difference between a depressed (instead of lifted) parcel and the environment, resulting in a much higher speed limit based on constant potential temperature along the axis, which is labeled here the *upper thermodynamic speed limit.* Subsidence in the core can result in a much larger CAPE if the core is filled almost down to the ground with parcels that have entered a large tornado through its top near the equilibrium level. The core of such a tornado would be very stable (since $\partial\theta/\partial r < 0$; see section 5.4) and hence very warm because the air would descend dry adiabatically in the absence of turbulent mixing with air outside the core. There is limited observational evidence for the existence of tornadoes with hurricane-like eyes from Doppler radar observations of an echo-weak hole associated with a TVS (Johnson and Ziegler 1984; see also Wakimoto et al. 1996) and from mobile Doppler radar observation of a suspected strong downdraft near cloud base in a tornado (Wurman et al. 1996; see also Figs. 5.23 and 5.27).

For $\beta = 0.5$ both limits are commensurate with assuming a stagnant-core vortex in hydrostatic and cyclostrophic balance. The balanced state can be illustrated in the case of core subsidence with a two-layer atmosphere consisting of a troposphere of constant potential temperature θ_1 and a stratosphere of constant potential temperature θ_2, where $\theta_2 > \theta_1$ (Fig. 5.32a). The unperturbed tropopause is at $z = H$. Assume that a thunderstorm generates a vortex that draws θ_2 air down to the surface as shown in Fig. 5.32b. The new

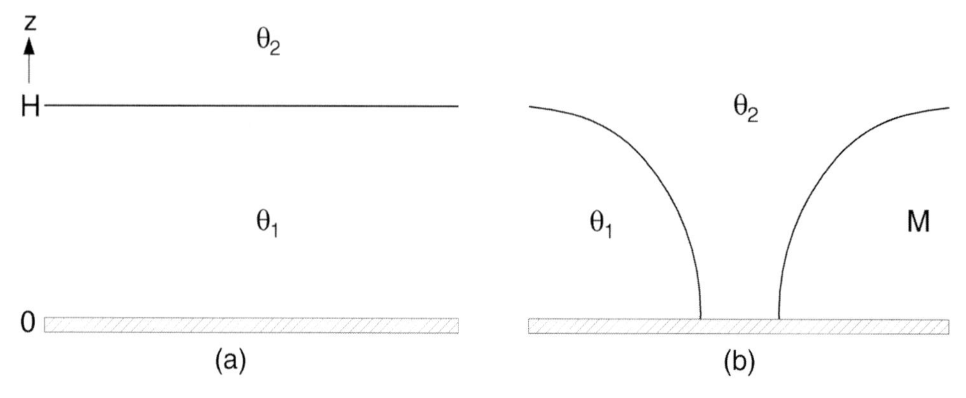

FIG. 5.32. (a) Equilibrium state in a two-layer atmosphere in the absence of rotation. The troposphere of constant potential temperature θ_1 is overlaid by a stratosphere of constant potential temperature θ_2 ($>\theta_1$). (b) Balanced inviscid vortex in the same two-layer atmosphere. There is no flow in the radial and vertical directions. An outer potential vortex of tropospheric air with constant angular momentum M surrounds a stagnant core of air drawn down from the stratosphere.

state has higher potential energy than did the original one so the transition requires (i) the addition of energy, which has to be supplied by the parent storm, and (ii) the prior presence of strong low-level rotation, probably the tornadic vortex in its formative stages, to provide the downward pressure-gradient force needed to draw down the θ_2 air. Figure 5.32b may represent an ideal state for tornadoes.

The core pressure deficit and maximum wind can be derived for the two-layer model as follows. The nondimensional pressure at height $z \leq H$ in the environment, characterized by constant potential temperature θ_1, is found by integrating (5.22) upward from the surface and assuming $\pi(\infty, 0) = 1$, and is given by $\pi(\infty, z) = 1 - gz/c_p\theta_1$. The nondimensional pressure at $z = H$ is independent of r and hence is $\pi(r, H) = 1 - gH/c_p\theta_1$. The pressure along the axis, found by integrating the hydrostatic equation downward from the tropopause, is $\pi(0, z) = \pi(0, H) + g(H - z)/c_p\theta_2$. The axial pressure deficit is

$$\Delta\pi(z) \equiv \pi(\infty, z) - \pi(0, z)$$

$$= \frac{g}{c_p}\left(\frac{1}{\theta_1} - \frac{1}{\theta_2}\right)(H - z), \qquad (5.26)$$

which also applies to the entire stagnant core. The Froude number based on the total height of the tornado, defined as $\mathrm{Fr} \equiv \{(c_p\theta/gH)/[\Delta\pi(0)/\Delta\theta/\theta)]\}^{1/2}$ (Lewellen 1993), is 1 for our two-layer model. In the outer potential part of the vortex centrifugal forces balance the inward pressure-gradient force. The maximum tangential velocity and the core radius at each level are given by

$$v_c(z) = [2c_p\theta_1\Delta\pi(z)]$$

$$= \left[2g\left(1 - \frac{\theta_1}{\theta_2}\right)(H - z)\right]^{1/2} \qquad (5.27)$$

and

$$r_c(z) = M/v_c(z). \qquad (5.28)$$

These formulas for maximum pressure drop and tangential velocity can be generalized easily to include a general sounding and a general vortex core with potential temperatures $\theta_\infty(z)$ and $\theta_0(z)$, respectively. Now H becomes the height of the equilibrium level [where $\theta_0(H) = \theta_\infty(H)$]. Integrating the hydrostatic equation yields

$$\Delta\pi(z) = \frac{g}{c_p}[\overline{(1/\theta_\infty)} - \overline{(1/\theta_0)}](H - z), \qquad (5.29)$$

where the overbar denotes average values between $H - z$ and H. The maximum wind is

$$v_c \approx \left\{\frac{gH}{\beta}\,\theta_\infty(0)[\overline{(1/\theta_\infty)} - \overline{(1/\theta_0)}]\right\}^{1/2}. \qquad (5.30)$$

Adopting the sounding shown in Lilly (1976), we can use the formulas to make estimates of tornado parameters. Assuming that the source height of the subsiding air is 180 mb, $H = 13\,000$ m and θ_2=340 K. Here $\bar{\theta}_1$=315 K is representative of a mean tropospheric value of θ and $M = 2 \times 10^4$ m^2 s^{-1} is typical for a strong tornado. For a dry adiabatic core, these values yield a central pressure drop of 100 mb, a maximum tangential velocity of 137 m s^{-1}, a core radius of 146 m, and a ground temperature in the tornado of 57°C if the subsiding air reaches the ground unmodified (although such warm air has never been observed). The pressure drop and speed limit (with the translation speed added) seem to be reasonable *upper* bounds in the light of existing measurements. Higher values can be obtained by assuming that the air descends from 160 mb, the parcel equilibrium level, instead of 180 mb. With pseudoadiabatic descent, the surface pressure deficit and the maximum tangential wind speeds are only about 30% and 55%, respectively, of the above values.

In addition to the hydrostatic lowering of pressure, Eq. (5.25) offers two dynamical mechanisms for maintaining a large pressure deficit: the turbulent stress term (Ward 1972; Pauley 1989) and large vertical kinetic energy of parcels at the top of the corner region (Davies-Jones and Kessler 1974; Lewellen 1976; Snow and Pauley 1984; Fiedler and Rotunno 1986). These dynamical effects allow the lower thermodynamic speed limit to be broken without subsidence warming in the core (Lewellen 1976; Snow and Pauley 1984; Fiedler and Rotunno 1986). The dynamical mechanisms may also be responsible for the relatively larger pressure drops and associated higher wind speeds that occur in secondary vortices: Photogrammetric measurements in tornadoes (Forbes 1976) and laboratory vortices (Leslie 1977) and three-dimensional numerical modeling results (Rotunno 1984; Lewellen et al. 1997) show that tornado-relative winds may be 30%–50% higher in secondary vortices than in the azimuthally averaged flow.

Ward (1972) proposed that the low axial pressure near the surface could be maintained by turbulent stresses preventing filling of the vortex from above. Pauley's (1989) measurements support this conclusion, but the turbulent stress term does not seem to be important in the axial momentum budget in some axisymmetric numerical simulations (Walko 1988; Fiedler 1994).

The most complete explanation of the mechanism involving the vertical kinetic energy term is that due to Fiedler and Rotunno (1986). They suggest that, in some instances, intense tornadoes could be supercritical end-wall vortices with axial jets in their cores, arising from the frictional interaction of a vortex with the ground; thus, as already explained, the vertical kinetic energy term is large for end-wall vortices and drowned vortex jumps. An example of a tornado of this type may be the narrow, intense, and well videographed Pampa, Texas, tornado of 8 June 1995, which lofted a flattened pickup truck 30 m in the air. Indications of vortex breakdowns in tornadoes have been reported by Hoecker (1960), Ward (1972), Burggraf and Foster (1977), and Pauley and Snow (1988). Because end-wall vortices are supercritical flows that do not allow upstream (i.e., downward) wave propagation, they can contain intense dynamical pressure drops that cannot be reduced by filling from above. With increasing circulation M, the end-wall vortex intensifies because $r_m \propto M^{-1/2}$, and $(\Delta p)_m \propto v_m^2 \propto M^3$. There is still a limit, however, on the maximum pressure drop that can be realized in a given parent updraft because adding more swirl to the flow eventually leads to the less intense subcritical vortex reaching the surface and preventing the more intense end-wall vortex from forming. Conversely, reducing the swirl leads to a less intense end-wall vortex and less disparity between upstream and downstream conditions that can be matched by a vortex breakdown further aloft (Fiedler and Rotunno 1986).

Fiedler and Rotunno estimated the intensity of a supercritical end-wall tornado by assuming that the maximum tangential velocity in the subcritical vortex aloft was $v_c = (\text{CAPE})^{1/2}$, the thermodynamic speed limit for a Rankine combined vortex. Applying their theoretical results that $v_m \approx 1.7v_c$, $w_m \approx 2v_m$, and $(\Delta p)_m \approx \rho w_m^2/2$ yields for $v_c = 65$ m s^{-1}, $v_m \sim 110$ m s^{-1}, $w_m \sim 220$ m s^{-1}, and $(\Delta p)_m \sim 242$ mb. The maximum inward velocity is about half of v_m according to Fiedler's (1994) model. These extreme values are aloft where measurements of axial pressure and vertical velocity are difficult to obtain. Can the simulator results and the theory be applied to tornadoes where the flow in the boundary layer and corner region is turbulent, the earth's surface is rougher than plywood, the flow force has a buoyancy term (Morton 1966, 1969), the termination of the tornado aloft may be radically different in the absence of an atmospheric counterpart to the baffle, the flow is no longer approximately incompressible at the estimated wind speeds, and the downstream conditions probably are not simply $v_c \approx 2.6\bar{w}$? The latter condition appears inconsistent because it makes $\bar{w} = (\text{CAPE})^{1/2}/2.6$, whereas $\bar{w} = (\text{CAPE})^{1/2}$ would seem more in keeping with observations. Regarding the first condition, however, even if the boundary layer and end-wall vortex are turbulent, the ratio v_m/v_c may still be around 2 (Chi 1977) or even 2.5 (Lewellen et al. 2000), although w_m/v_m may be reduced (to 1.4, as given by Lewellen et al. 2000; see also Fiedler and Rotunno 1986; Lewellen 1993).

The supercritical end-wall vortex tornado, although high velocity, is arguably not the most damaging at the ground (aside from hits from large falling objects ejected from the tornado) because its high winds and low pressure are located some 50 m above the ground in the axial jet very close to the axis. The wind damage in this type of tornado should be worst at treetop level, although the recent modeling study by Lewellen et al. (2000) suggests that the most significant damage might be much lower; the location of the extreme winds and pressure deficit makes observational verification (using current observing tools and strategies) of the theory and model almost impossible. The most damaging types of tornado should be the drowned vortex jump and large two-celled vortices with (or even sometimes without) secondary vortices. The majority of intense tornadoes should be of the latter type since it occurs over a much wider range of swirl ratios than the end-wall vortex and DVJ. A damage survey of the violent Union City tornado (Davies-Jones et al. 1978) gave indications that it was equally damaging to houses in its path when it was two-celled at the surface (damaging winds circulatory) as it was when it later became a smaller vortex with signs in the lowest few meters of strong low-level convergence into its center (and with damage and debris dispersal caused predominantly by winds that were in the same direction as tornado translation).

5.6. Concluding remarks

A buoyant updraft, rainy downdrafts, and a deep, mesocyclonic vortex (preexisting vertical vorticity) are tornadic supercell (nonsupercell) storm "ingredients." These are, to varying degrees, mutually dependent processes and interact in some way to yield a tornado. It is unlikely that a single such "recipe" can be generalized to all instances of tornadogenesis. For example, the exact sequence of events leading to tornado formation will depend, as alluded to in section 5.2, on whether or not the mechanism for the genesis of the near-ground mesocyclone is effectively baroclinic or barotropic.

Observationally, the basic mechanism for midlevel mesocyclogenesis (section 5.2) has been confirmed. Refuting or verifying other model- or theoretical-based mechanisms such as those involving near-ground mesocyclogenesis and the modes of tornadogenesis (section 5.3) may be hard to verify or refute with present data of tornadic storms. Similarly, much about the tornado structure deduced from models and theory, particularly that which governs tornado intensity (sections 5.4 and 5.5), awaits observational confirmation.

Current and future researchers are faced with additional yet related unresolved scientific issues too numerous to list here. One such issue regards the current inability among meteorologists to reliably forecast or even "nowcast" tornado intensity and longevity. Another that has critical operational implications is the current lack of a means to discriminate tornadic storms from those storms that appear quantitatively and qualitatively identical to tornadic storms yet do not produce tornadoes. The exact nature of tornado demise remains speculative, as does the physics of the storm–boundary interactions that sometimes precede tornadogenesis. And, of course, we have yet to solve the mysteries of whether the cores of large tornadoes are warm and how the vortex terminates aloft. Solutions to these problems will require cooperation and complementary efforts among observationalists, modelers, and theoreticians.

Apart from limited temporal and spatial resolution in inadequately small temporal and spatial domains, datasets currently available for mesocyclone and tornadogenesis study also suffer from (i) lack of wind information in the lowest few hundred meters between the ground and the lowest elevation angle of a typical Doppler radar scan, where horizontal convergence and baroclinic generation of vorticity are concentrated; and (ii) lack of temperature and humidity measurements in the lowest 1–2 km in the vicinity of the wall cloud, tail cloud, and clear slot, and within the rear- and forward-flank downdraft regions. Data appropriate for investigations of tornado dynamics necessarily must have even higher temporal and spatial resolution. On both the mesocyclone and the tornado scales, a sufficient number of observations must be gathered to allow for generalization of the results.

In addition to fixed-site and mobile rawinsondes and surface mesonets that provide in situ measurements, collecting the needed data in the future will probably require remotely controlled instruments (perhaps aboard a remotely piloted aircraft; e.g., see Bluth et al. 1996) and/or other remote-sensing techniques, and yet will still be very challenging owing to the logistics of intercepting hazardous and quickly moving and evolving storms. Clearly, single- and coordinated multiple-Doppler radars will continue to play a prominent role in tornado and tornadic storm research. We emphasize the future importance of radars installed on mobile ground-based and airborne platforms (see sections 5.2 and 5.3), which can ease (albeit with their unique set of limitations) the primary shortcomings of ground-based, fixed-site radars. For example, owing to the tens of kilometers typical distance between the storm and the radar(s), the fixed-site data suffer from relatively coarse vertical resolution. Such degraded resolution and the related lack of data close to the ground diminish the accuracy of horizontal convergence calculations within the convective boundary layer and therefore of vertical velocity and subsequent retrieved variables (the computations of which are sensitive to vertical velocity). Another shortcoming is that the probability is low that a storm will pass through a fixed-site Doppler-radar network such that data can be collected to document tornado formation, maintenance, and decay. In contrast, an aircraft equipped with Doppler radar[2] can fly close to a storm at low levels and can follow it during its entire life cycle. An analogous statement can be made about ground-based mobile radars, with the caveat that the deployment and data collection are subject to the availability of a suitable road network.

These shortcomings aside, fixed-site radars will continue to provide invaluable information. Data from operational radars such as the WSR-88D and the Terminal Doppler Weather Radar can be used to compile large-sample statistics on observable storm attributes, to help focus research endeavors, and to improve warning operations. In addition, retrieval of the 3D wind in thunderstorms using single Doppler radar data has become possible recently using techniques based on assumptions of conservation of radar reflectivity or velocity stationarity (e.g., Tuttle and Foot 1990; Qiu and Xu 1992; Laroche and Zawadzki 1994; Shapiro et al. 1995; Zhang and Gal-Chen 1996; Sun and Crook 1997). In addition to their potential for storm diagnostic purposes (e.g. Sun and Crook 1998), the retrieved winds and subsequently (or concurrently) retrieved thermodynamic variables can be assimilated into numerical models, for the purpose of producing very short-term prognoses of the future state of the storm. Weygandt et al. (1998) have reported some success recently with such predictions of the 17 May 1981 Arcadia, Oklahoma, tornadic supercell. Storm-scale predictions (and also diagnoses) will be limited by the characteristics of the data sampling, as mentioned above, and additionally by the assumptions that underlie the single-Doppler retrieval techniques.

It is appropriate to close our review as did Davies-Jones (1986) with comments on tornado "modification." Numerous suggestions on tornado destruction/wind modification have been offered[3] (by meteorologists and nonmeteorologists alike) that purport to disrupt the flow in and around the tornado. An explosive device administered to the tornado by missiles is an example of one of the more popular ideas. Such an approach carries with it the grave consequences associated with a number of unintended results, the most obvious of which is damage to life and property by explosives themselves. Similarly, cloud seeding, if effective in enhancing rainfall, would have uncertain effects on tornado modification. For example, in storms with too little natural precipitation, it is plausible that an increase in rainfall would then allow the storm to support

[2] Details of radar scanning strategies, and how these radar scans comprise "pseudo-dual-Doppler" radar data, are given by Jorgensen et al. (1996). A discussion of airborne Doppler radar sampling limitations and of error sources in pseudo-dual-Doppler analyses can be found in Ray and Stephenson (1990) and elsewhere.

[3] It has been our experience that these tend to follow highly publicized tornado outbreaks with numerous fatalities such as the event on 3 May 1999 near Oklahoma City, Oklahoma.

the processes that lead to tornadogenesis! Other proposals are outlined by Davies-Jones (1986).

To our knowledge, none of these techniques have been advanced. Thus, as stated by Davies-Jones (1986), we are "probably many years away" from practical implementation of any such techniques, which will continue to be limited by the gaps in our understanding of tornadoes and tornadic storms. In the meantime, we recommend that steps be taken toward continued mitigation of loss of life, with appropriate structural engineering and building practices, installation of concrete-reinforced "safe rooms," public outreach and education on storm and tornado safety, and basic research on tornadoes and tornadic storms, with rapid transfer of these research results to the operational sector.

Acknowledgments. The authors wish to thank the members of the review panel, and in particular S. Lewellen and J. Snow, for their constructive comments on various versions of the manuscript. We are grateful to Joan O'Bannon for her time and effort in drafting many of the figures. The second author contributed to this chapter while he was a visiting scientist with the Mesoscale and Microscale Meteorology Division of the National Center for Atmospheric Research. The National Center for Atmospheric Research is sponsored by the National Science Foundation.

REFERENCES

Adlerman, E. J., K. K. Droegemeier, and R. Davies-Jones, 1999: A numerical simulation of cyclic mesocyclogenesis. *J. Atmos. Sci.,* **56,** 2045–2069.

Agee, E. M., 1969: Tornado Project activities at Purdue University. *Bull. Amer. Meteor. Soc.,* **50,** 806–807.

———, 1970: Purdue tornado project activities—Part II. *Bull. Amer. Meteor. Soc.,* **51,** 951.

———, J. T. Snow, and P. R. Clare, 1976: Multiple vortex features in the tornado cyclone and the occurrence of tornado families. *Mon. Wea. Rev.,* **104,** 552–563.

Atkins, N. T., M. L. Weisman, and L. J. Wicker, 1999: The influence of preexisting boundaries on supercell evolution. *Mon. Wea. Rev.,* **127,** 2910–2927.

Baker, G. L., and C. R. Church, 1979: Measurements of core radii and peak velocities in modeled atmospheric vortices. *J. Atmos. Sci.,* **36,** 2413–2424.

Barcilon, A., and P. G. Drazin, 1972: Dust devil formation. *Geophys. Fluid Dyn.,* **4,** 147–158.

Barnes, S. L., 1970: Some aspects of a severe, right-moving thunderstorm deduced from mesonet rawinsonde observations. *J. Atmos. Sci.,* **27,** 634–648.

———, 1978: Oklahoma thunderstorms on 29–30 April 1970. Part I: Morphology of a tornadic storm. *Mon. Wea. Rev.,* **106,** 673–684.

Batchelor, G. K., 1967: *An Introduction to Fluid Dynamics.* Cambridge University Press, 615 pp.

Bedard, A. J., Jr., and C. Ramzy, 1983: Surface meteorological observations in severe thunderstorms. Part I: Design details of TOTO. *J. Climate Appl. Meteor.,* **22,** 911–918.

Bergé, P., Y. Pomeau, and C. Vidal, 1984: *Order Within Chaos.* Wiley, 329 pp.

Bieringer, P., and P. S. Ray, 1996: A comparison of tornado warning lead times with and without NEXRAD Doppler radar. *Wea. Forecasting,* **11,** 47–52.

Binnie, A. M., and G. A. Hookings, 1948: Laboratory experiments on whirlpools. *Proc. Roy. Soc. London A,* **194,** 348–415.

Bluestein, H. B., 1980: The University of Oklahoma Severe Storms Intercept Project—1979. *Bull. Amer. Meteor. Soc.,* **61,** 560–567.

———, 1983a: Measurements in the vicinity of severe thunderstorms and tornadoes with TOTO: 1982–1983 results. Preprints, *13th Conf. on Severe Local Storms,* Tulsa, OK, Amer. Meteor. Soc., 89–92.

———, 1983b: Surface meteorological observations in severe thunderstorms. Part II: Field experiments with TOTO. *J. Climate Appl. Meteor.,* **22,** 919–930.

———, 1984: Photographs of the Canyon, TX storm on 26 May 1978. *Mon. Wea. Rev.,* **112,** 2521–2523.

———, 1985a: The formation of a "landspout" in a "broken-line" squall line in Oklahoma. Preprints, *14th Conf. on Severe Local Storms,* Indianapolis, IN, Amer. Meteor. Soc., 267–270.

———, 1985b: Wall clouds with eyes. *Mon. Wea. Rev.,* **113,** 1081–1085.

———, 1986: Visual aspects of the flanking line in severe thunderstorms. *Mon. Wea. Rev.,* **114,** 788–795.

———, 1994: High-based funnel clouds in the Southern Plains. *Mon. Wea. Rev.,* **122,** 2631–2638.

———, 1996: 3-D stereo photography of supercell tornadoes. Preprints, *18th Conf. on Severe Local Storms,* San Francisco, CA, Amer. Meteor. Soc., 469–470.

———, 1999a: A history of severe-storm intercept field programs. *Wea. Forecasting,* **14,** 558–577.

———, 1999b: *Tornado Alley: Monster Storms of the Great Plains.* Oxford University Press, 180 pp.

———, and C. Parks, 1983: A synoptic and photographic climatology of low-precipitation severe thunderstorms in the Southern Plains. *Mon. Wea. Rev.,* **111,** 2034–2046.

———, and W. P. Unruh, 1989: Observations of the wind field in tornadoes, funnel clouds, and wall clouds with a portable Doppler radar. *Bull. Amer. Meteor. Soc.,* **70,** 1514–1525.

———, and G. R. Woodall, 1990: Doppler-radar analysis of a low-precipitation severe storm. *Mon. Wea. Rev.,* **118,** 1640–1664.

———, and J. H. Golden, 1993: A review of tornado observations. *The Tornado: Its Structure, Dynamics, Prediction, and Hazards, Geophys. Monogr.,* No. 79, Amer. Geophys. Union, 319–352.

———, and W. P. Unruh, 1993: On the use of a portable FM-CW Doppler radar for tornado research. *The Tornado: Its Structure, Dynamics, Prediction, and Hazards, Geophys. Monogr.,* No. 79, Amer. Geophys. Union, 367–376.

———, and D. R. MacGorman, 1998: Evolution of cloud-to-ground lightning characteristics and storm structure in the Spearman, Texas, tornadic supercells of 31 May 1990. *Mon. Wea. Rev.,* **126,** 1451–1467.

———, and A. Pazmany, 2000: Observations of tornadoes and other convective phenomena with a mobile 3-mm wavelength, Doppler radar: The spring 1999 field experiment. *Bull. Amer. Meteor. Soc.,* **81,** 2939–2951.

———, W. P. Unruh, J. LaDue, H. Stein, and D. Speheger, 1993: Doppler-radar wind spectra of supercell tornadoes. *Mon. Wea. Rev.,* **121,** 2200–2221.

———, A. L. Pazmany, J. C. Galloway, and R. E. McIntosh, 1995: Studies of the substructure of severe convective storms using a mobile 3-mm wavelength Doppler radar. *Bull. Amer. Meteor. Soc.,* **76,** 2155–2169.

———, W. P. Unruh, D. C. Dowell, T. A. Hutchinson, T. M. Crawford, A. C. Wood, and H. Stein, 1997: Doppler radar analysis of the Northfield, Texas tornado of 25 May 1994. *Mon. Wea. Rev.,* **125,** 212–230.

Bluth, R. T., P. A. Durkee, J. H. Seinfeld, R. C. Flagan, L. M. Russell, P. A. Crowley, and P. Finn, 1996: Center for interdisciplinary remotely-piloted aircraft studies (CIRPAS). *Bull. Amer. Meteor. Soc.,* **77,** 2691–2699.

Bödewadt, U. T., 1940: Die Drehströmung über festem Grund. *Z. Angew Math. Mech.,* **20,** 241–253.

Borisenko, A. I., and I. E. Tarapov, 1979: *Vector and Tensor Analysis with Applications.* Dover, 257 pp.

Brady, R. H., and E. J. Szoke, 1989: A case study of nonmesocyclone tornado development in northeast Colorado: Similarities to waterspout formation. *Mon. Wea. Rev.,* **117,** 843–856.

Brandes, E. A., 1977: Gust front evolution and tornadogenesis as viewed by Doppler radar. *J. Appl. Meteor.,* **16,** 333–338.

———, 1978: Mesocyclone evolution and tornadogenesis: Some observations. *Mon. Wea. Rev.,* **106,** 995–1011.

———, 1981: Fine-structure of the Del City-Edmond tornado mesocirculation. *Mon. Wea. Rev.,* **109,** 635–647.

———, 1984a: Relationships between radar-derived thermodynamic variables and tornadogenesis. *Mon. Wea. Rev.,* **112,** 1033–1052.

———, 1984b: Vertical vorticity generation and mesocyclone sustenance in tornadic thunderstorms: The observational evidence. *Mon. Wea. Rev.,* **112,** 2253–2269.

———, R. P. Davies-Jones, and B. C. Johnson, 1988: Streamwise vorticity effects on supercell morphology and persistence. *J. Atmos. Sci.,* **45,** 947–963.

Brock, F. V., G. Lesins, and R. Walko, 1987: Measurement of pressure and air temperature near severe thunderstorms: An inexpensive and portable instrument. *Extended Abstracts, Sixth Symp. Meteorological Observations and Instrumentation,* New Orleans, LA, Amer. Meteor. Soc., 320–323.

Brooks, H. E., and R. B. Wilhelmson, 1993: Hodograph curvature and updraft intensity in numerically modeled supercells. *J. Atmos. Sci.,* **50,** 1824–1833.

———, C. A. Doswell III, and R. Davies-Jones, 1993: Environmental helicity and the maintenance and evolution of low-level mesocyclones. *The Tornado: Its Structure, Dynamics, Prediction, and Hazards, Geophys. Monogr.,* No. 79, Amer. Geophys. Union, 97–104.

———, ———, and J. Cooper, 1994a: On the environments of tornadic and nontornadic mesocyclones. *Wea. Forecasting,* **9,** 606–618.

———, ———, and R. B. Wilhelmson, 1994b: The role of midtropospheric winds in the evolution and maintenance of low-level mesocyclones. *Mon. Wea. Rev.,* **122,** 126–136.

Brown, J. M., and K. R. Knupp, 1980: The Iowa cyclonic-anticyclonic tornado pair and its parent thunderstorm. *Mon. Wea. Rev.,* **108,** 1626–1646.

Brown, R. A., D. W. Burgess, and K. C. Crawford, 1973: Twin tornado cyclones within a severe thunderstorm: Single Doppler radar observations. *Weatherwise,* **26,** 63–71.

———, L. R. Lemon, and D. W. Burgess, 1978: Tornado detection by pulsed Doppler radar. *Mon. Wea. Rev.,* **106,** 29–39.

Browning, K. A., 1964: Airflow and precipitation trajectories within severe local storms which travel to the right of the winds. *J. Atmos. Sci.,* **21,** 634–639.

———, 1986: Morphology and classification of middle-latitude thunderstorms. *Thunderstorm Morphology and Dynamics,* 2d ed., E. Kessler, Ed., University of Oklahoma Press, 133–152.

———, and C. R. Landry, 1963: Airflow within a tornadic thunderstorm. Preprints, *10th Weather Radar Conf.,* Washington, DC, Amer. Meteor. Soc., 116–122.

Burgers, J. M., 1948: A mathematical model illustrating the theory of turbulence. *Adv. Appl. Mech.,* **1,** 197–199.

Burgess, D. W., and R. P. Davies-Jones, 1979: Unusual tornadic storms in eastern Oklahoma on 5 December 1975. *Mon. Wea. Rev.,* **107,** 451–457.

———, and L. R. Lemon, 1990: Severe thunderstorm detection by radar. *Radar in Meteorology,* D. Atlas, Ed., Amer. Meteor. Soc., 619–647.

———, V. T. Wood, and R. A. Brown, 1982: Mesocyclone evolution statistics. Preprints, *12th Conf. on Severe Local Storms,* San Antonio, TX, Amer. Meteor. Soc., 84–89.

———, S. V. Vasiloff, R. P. Davies-Jones, D. S. Zrnic, and S. E. Frederickson, 1985: Recent NSSL work on windspeed measurement in tornadoes. *Proc. Fifth U.S. National Conf. of Wind Engineering,* Lubbock, TX, Texas Tech University, 1A-53–1A-60.

Burggraf, O. R., and M. R. Foster, 1977: Continuation or breakdown in tornado-like vortices. *J. Fluid Mech.,* **80,** 685–703.

———, K. Stewartson, and R. Belcher, 1971: Boundary layer induced by a potential vortex. *Phys. Fluids,* **14,** 1821–1833.

Carbone, R. E., 1983: A severe frontal rainband. Part II: Tornado parent vortex circulation. *J. Atmos. Sci.,* **40,** 2639–2654.

Chi, J., 1977: Numerical analysis of turbulent end-wall boundary layers of intense vortices. *J. Fluid Mech.,* **82,** 209–222.

Church, C. R., and J. T. Snow, 1979: The dynamics of natural tornadoes as inferred from laboratory simulations. *J. Rech. Atmos.,* **12,** 111–133.

———, and ———, 1985: Measurements of axial pressure in tornado-like vortices. *J. Atmos. Sci.,* **42,** 576–582.

———, and ———, 1993: Laboratory models of tornadoes. *The Tornado: Its Structure, Dynamics, Prediction, and Hazards, Geophys. Monogr.,* No. 79, Amer. Geophys. Union, 277–295.

———, ———, and E. M. Agee, 1977: Tornado vortex simulation at Purdue University. *Bull. Amer. Meteor. Soc.,* **58,** 900–908.

———, ———, G. L. Baker, and E. M. Agee, 1979: Characteristics of tornado-like vortices as a function of swirl ratio: A laboratory investigation. *J. Atmos. Sci.,* **36,** 1755–1776.

———, D. Burgess, C. Doswell, and R. Davies-Jones, Eds., 1993: *The Tornado: Its Structure, Dynamics, Prediction, and Hazards, Geophys. Monogr.,* No. 79, Amer. Geophys. Union, 637 pp.

Concannon, P. R., H. E. Brooks, and C. A. Doswell III, 2000: Climatological risk of strong and violent tornadoes in the United States. Preprints, *Second Conf. on Environmental Applications,* Long Beach, CA, Amer. Meteor. Soc., 212–219.

Cooley, J. R., 1978: Cold air funnel clouds. *Mon. Wea. Rev.,* **106,** 1368–1372.

Darkow, G. L., 1971: Periodic tornado production by long-lived parent thunderstorms. Preprints, *Seventh Conf. on Severe Local Storms,* Kansas City, MO, Amer. Meteor. Soc., 214–217.

Davies, J. M., 1993a: Hourly helicity, instability and EHI in forecasting supercell tornadoes. Preprints, *17th Conf. on Severe Local Storms,* St. Louis, MO, Amer. Meteor. Soc., 56–60.

———, 1993b: Small tornadic supercells in the central Plains. Preprints, *17th Conf. Severe Local Storms,* St. Louis, MO, Amer. Meteor. Soc., 305–309.

Davies-Jones, R. P., 1973: The dependence of core radius on swirl ratio in a tornado simulator. *J. Atmos. Sci.,* **30,** 1427–1430.

———, 1976: Laboratory simulations of tornadoes. *Proc. Symp. on Tornadoes: Assessment of Knowledge and Implications for Man,* Lubbock, TX, Texas Tech University, 151–174.

———, 1982: Observational and theoretical aspects of tornadogenesis. *Intense Atmospheric Vortices,* L. Bengtsson and J. Lighthill, Eds., Springer-Verlag, 175–189.

———, 1983: Tornado interception with mobile teams. *Instruments and Techniques for Thunderstorm Observation and Analysis. Vol. III, Thunderstorms: A Social, Scientific, and Technological Documentary,* E. Kessler, Ed., University of Oklahoma Press, 23–32.

———, 1984: Streamwise vorticity: The origin of updraft rotation in supercell storms. *J. Atmos. Sci.,* **41,** 2991–3006.

———, 1985: Dynamical interaction between an isolated convective cell and a veering environmental wind. Preprints, *14th Conf. on Severe Local Storms,* Indianapolis, IN, Amer. Meteor. Soc., 216–219.

———, 1986: Tornado dynamics. *Thunderstorm Morphology and Dynamics,* 2d ed., E. Kessler, Ed., University of Oklahoma Press, 197–236.

———, 1995: Tornadoes. *Sci. Amer.,* **273,** (2), 34–41.

———, 1996a: Formulas for the barotropic and baroclinic components of vorticity with applications to vortex formation near the ground. Preprints, *Seventh Conf. on Mesoscale Processes,* Reading, UK, Amer. Meteor. Soc., 14–16.

———, 1996b: Inclusion of boundary conditions on pressure in conceptual models of updraft-environment interaction. Preprints, *18th Conf. Severe Local Storms,* San Francisco, CA, Amer. Meteor. Soc., 713–717.

———, 2000a: A Lagrangian model for baroclinic genesis of mesoscale vortices. Part I: Theory. *J. Atmos. Sci.,* **57,** 715–736.

———, 2000b: Can the hook echo instigate tornadogenesis barotropically? Preprints, *20th Conf. on Severe Local Storms,* Orlando, FL, Amer. Meteor. Soc., 269–272.

———, and E. Kessler, 1974: Tornadoes. *Weather and Climate Modification,* W. N. Hess, Ed., Wiley, 552–595.

———, and H. E. Brooks, 1993: Mesocyclogenesis from a theoretical perspective. *The Tornado: Its Structure, Dynamics, Prediction, and Hazards, Geophys. Monogr.,* No. 79, Amer. Geophys. Union, 105–114.

———, D. W. Burgess, L. R. Lemon, and D. Purcell, 1978: Interpretation of surface marks and debris patterns from the 24 May 1973 Union City, Oklahoma, tornado. *Mon. Wea. Rev.,* **106,** 12–21.

———, D. Burgess, and M. Foster, 1990: Test of helicity as a tornado forecast parameter. Preprints, *16th Conf. on Severe Local Storms,* Kananaskis Park, AB, Canada, Amer. Meteor. Soc., 588–592.

Donaldson, R. J., and W. E. Lamkin, 1964: Visual observations beneath a developing tornado. *Mon. Wea. Rev.,* **92,** 326–328.

Donaldson, R. J., Jr., 1970: Vortex signature recognition by a Doppler radar. *J. Appl. Meteor.,* **9,** 661–670.

———, 1990: Foundations of severe storm detection by radar. *Radar in Meteorology,* D. Atlas, Ed., Amer. Meteor. Soc., 115–121.

Doswell, C. A., III, and D. W. Burgess, 1988: On some issues of United States tornado climatology. *Mon. Wea. Rev.,* **116,** 495–501.

———, and D. W. Burgess, 1993: Tornadoes and tornadic storms: A review of conceptual models. *The Tornado: Its Structure, Dynamics, Prediction, and Hazards, Geophys. Monogr.,* No. 79, Amer. Geophys. Union, 161–172.

———, and T. P. Grazulis, 1998: A demonstration of vortex configurations in an inexpensive tornado simulator. Preprints, *19th Conf. on Severe Local Storms,* Minneapolis, MN, Amer. Meteor. Soc., 85–88.

———, A. R. Moller, and H. E. Brooks, 1999: Storm spotting and public awareness since the first tornado forecasts of 1948. *Wea. Forecasting,* **14,** 544–557.

Doviak, R. J., and D. S. Zrnić, 1993: *Doppler Radar and Weather Observations.* Academic Press, 562 pp.

Dowell, D. C., and H. B. Bluestein, 1997: The Arcadia, Oklahoma, storm of 17 May 1981: Analysis of a supercell during tornadogenesis. *Mon. Wea. Rev.,* **125,** 2562–2582.

———, and ———, 2000: Conceptual models of cyclic supercell tornadogenesis. Preprints, *20th Conf. on Severe Local Storms,* Orlando, FL, Amer. Meteor. Soc., 259–262.

———, ———, and D. P. Jorgensen, 1997: Airborne Doppler radar analysis of supercells during COPS-91. *Mon. Wea. Rev.,* **125,** 365–383.

Droegemeier, K. K., S. M. Lazarus, and R. Davies-Jones, 1993: The influence of helicity on numerically simulated convective storms. *Mon. Wea. Rev.,* **121,** 2005–2029.

Dutton, J. A., 1976: *The Ceaseless Wind.* McGraw-Hill, 579 pp.

Emanuel, K. A., 1994: *Atmospheric Convection.* Oxford University Press, 580 pp.

Eskridge, R. E., and P. Das, 1976: Effect of a precipitation-driven downdraft on a rotating wind field: A possible trigger mechanism for tornadoes? *J. Atmos. Sci.,* **33,** 70–84.

Faller, A. J., 1963: An experimental study of the instability of the laminar Ekman layer. *J. Fluid Mech.,* **15,** 560–576.

Fiedler, B. H., 1989: Conditions for laminar flow in geophysical vortices. *J. Atmos. Sci.,* **46,** 252–259.

———, 1993: Numerical simulations of axisymmetric tornadogenesis in forced convection. *The Tornado: Its Structure, Dynamics, Prediction, and Hazards, Geophys. Monogr.,* No. 79, Amer. Geophys. Union, 41–48.

———, 1994: The thermodynamic speed limit and its violation in axisymmetric numerical simulations of tornado-like vortices. *Atmos.–Ocean,* **32,** 335–359.

———, 1995: On modeling tornadoes in isolation from the parent storm. *Atmos.–Ocean,* **33,** 501–512.

———, and R. Rotunno, 1986: A theory for the maximum windspeeds in tornado-like vortices. *J. Atmos. Sci.,* **43,** 2328–2340.

Forbes, G. S., 1976: Photogrammetric characteristics of the Parker tornado of April 3, 1974. *Proc. Symp. on Tornadoes: Assessment of Knowledge and Implications for Man,* Lubbock, TX, Texas Tech University, 58–77.

———, 1978: Three scales of motion associated with tornadoes. U.S. Nuclear Regulatory Commission Contract Rep. NUREG/CR-0363, 359 pp.

Fujita, T. T., 1958: Tornado cyclone: bearing system of tornadoes. *Proc. Seventh Weather Radar Conf.,* Miami Beach, FL, Amer. Meteor. Soc., K31–K38.

———, 1959: Detailed analysis of the Fargo tornadoes of June 20, 1957. U.S. Weather Bureau Tech. Rep. 5, Severe Local Storms Project, University of Chicago, 29 pp. plus figures.

———, 1963: Analytical mesometeorology: A review. *Severe Local Storms, Meteor. Monogr.* No. 27, Amer. Meteor. Soc., 77–125.

———, 1973: Tornadoes around the world. *Weatherwise,* **26,** 56–62 78–83.

———, 1981: Tornadoes and downbursts in the context of generalized planetary scales. *J. Atmos. Sci.,* **38,** 1511–1534.

———, 1985: The downburst. Satellite and Mesometeorology Research Project (SMRP), Dept. of Geophysical Sciences, University of Chicago, 122 pp.

———, 1993: Plainfield tornado of August 28, 1990. *The Tornado: Its Structure, Dynamics, Prediction, and Hazards, Geophys. Monogr.* No. 79, Amer. Geophys. Union, 1–17.

———, and R. M. Wakimoto, 1982: Anticyclonic tornadoes in 1980 and 1981. Preprints, *12th Conf. on Severe Local Storms,* San Antonio, TX, Amer. Meteor. Soc., 401–404.

———, and B. E. Smith, 1993: Aerial surveys and photography of tornado and microburst damage. *The Tornado: Its Structure, Dynamics, Prediction, and Hazards, Geophys. Monogr.,* No. 79, Amer. Geophys. Union, 479–493.

———, G. S. Forbes, and T. A. Umenhofer, 1976: Close-up view of 20 March 1976 tornadoes: Sinking cloud tops to suction vortices. *Weatherwise,* **29,** 116–131, 145.

Garrett, R. A., and V. D. Rockney, 1962: Tornadoes in northeastern Kansas, 19 May 1960. *Mon. Wea. Rev.,* **90,** 231–240.

Gilmore, M. S., and L. J. Wicker, 1998: The influence of midtropospheric dryness on supercell morphology and evolution. *Mon. Wea. Rev.,* **126,** 943–958.

Golden, J. H., 1974a: The life-cycle of Florida Keys' waterspouts, I. *J. Appl. Meteor.,* **13,** 676–692.

———, 1974b: Scale-interaction implications for the waterspout life cycle. II. *J. Appl. Meteor.,* **13,** 693–709.

———, 1976: An assessment of wind speeds in tornadoes. *Proc. Symp. on Tornadoes,* Lubbock, TX, Texas Tech University, 5–42.

———, and B. J. Morgan, 1972: The NSSL/Notre Dame tornado intercept program, spring 1972. *Bull. Amer. Meteor. Soc.,* **53,** 1178–1180.

———, and D. Purcell, 1978a: Airflow characteristics around the Union City tornado. *Mon. Wea. Rev.,* **106,** 22–28.

———, and ———, 1978b: Life-cycle of the Union City, OK tornado and comparison with waterspouts. *Mon. Wea. Rev.,* **106,** 3–11.

Grasso, L. D., and W. R. Cotton, 1995: Numerical simulation of a tornado vortex. *J. Atmos. Sci.,* **52,** 1192–1203.

Grazulis, T. P., 1993: *Significant Tornadoes 1680–1991.* Environmental Films, 1326 pp.

Hane, C. E., and P. S. Ray, 1985: Pressure and buoyancy fields derived from Doppler radar in a tornadic thunderstorm. *J. Atmos. Sci.,* **42,** 18–35.

Harlow, F. H., and L. R. Stein, 1974: Structural analysis of tornado-like vortices. *J. Atmos. Sci.,* **31,** 2081–2098.

Heymsfield, G. M., 1978: Kinematic and dynamic aspects of the Harrah tornadic storm analyzed from dual-Doppler radar data. *Mon. Wea. Rev.,* **106,** 233–254.

Hoecker, W. H., 1960: Wind speed and airflow patterns in the Dallas tornado of April 2 1957. *Mon. Wea. Rev.,* **88,** 167–180.

Holle, R. L., and M. W. Maier, 1980: Tornado formation from downdraft interaction in the FACE network. *Mon. Wea. Rev.,* **108,** 1010–1028.

Howard, L. N., and A. S. Gupta, 1962: On the hydrodynamic and hydromagnetic stability of swirling flows. *J. Fluid Mech.,* **14,** 463–476.

Howells, P. A. C., R. Rotunno, and R. K. Smith, 1988: A comparative study of atmospheric and laboratory-analogue numerical tornado-vortex models. *Quart. J. Roy. Meteor. Soc.,* **114,** 801–822.

Istok, M. J., and R. J. Doviak, 1986: Analysis of the relation between Doppler spectral width and thunderstorm turbulence. *J. Atmos. Sci.,* **43,** 2199–2214.

Jensen, B., T. P. Marshall, M. A. Mabey, and E. N. Rasmussen, 1983: Storm scale structure of the Pampa storm. Preprints, *13th Conf. on Severe Local Storms,* Tulsa, OK, Amer. Meteor. Soc., 85–88.

Johnson, B. C., and C. L. Ziegler, 1984: Doppler observations and retrieved thermal and microphysical variables for the Binger tornadic storm. Preprints, *22d Weather Radar Conf.,* Zurich, Switzerland, Amer. Meteor. Soc., 31–36.

Johnson, K. W., P. S. Ray, B. C. Johnson, and R. P. Davies-Jones, 1987: Observations related to the rotational dynamics of the 20 May 1977 tornadic storms. *Mon. Wea. Rev.,* **115,** 2463–2478.

Jorgensen, D. P., T. Matejka, and J. D. DuGranrut, 1996: Multibeam techniques for deriving wind fields from airborne Doppler radars. *J. Meteor. Atmos. Phys.,* **59,** 83–104.

Kanak, K. M., and D. K. Lilly, 1996: The linear stability and structure of convection in a mean circular shear. *J. Atmos. Sci.,* **53,** 2578–2593.

Kennedy, P. C., N. E. Westcott, and R. W. Scott, 1993: Single-Doppler radar observations of a mini-supercell tornadic thunderstorm. *Mon. Wea. Rev.,* **121,** 1860–1870.

Kessler, E., Ed., 1986: *Thunderstorm Morphology and Dynamics.* 2d ed. University of Oklahoma Press, 411 pp.

Klemp, J. B., 1987: Dynamics of tornadic thunderstorms. *Ann. Rev. Fluid Mech.,* **19,** 369–402.

———, and R. Rotunno, 1983: A study of the tornadic region within a supercell thunderstorm. *J. Atmos. Sci.,* **40,** 359–377.

———, R. B. Wilhelmson, and P. S. Ray, 1981: Observed and numerically simulated structure of a mature supercell thunderstorm. *J. Atmos. Sci.,* **38,** 1558–1580.

Kuo, H. L., 1966: On the dynamics of convective atmospheric vortices. *J. Atmos. Sci.,* **23,** 25–42.

———, 1967: Note on the similarity solutions of the vortex equations in an unstably stratified atmosphere. *J. Atmos. Sci.,* **24,** 95–97.

Laroche, S., and I. Zawadsky, 1994: A variational analysis method for the retrieval of three-dimensional wind field from single-Doppler data. *J. Atmos. Sci.,* **51,** 2664–2682.

Lee, B. D., and R. B. Wilhelmson, 1997: The numerical simulation of nonsupercell tornadogenesis. Part II: Evolution of a family of tornadoes along a weak outflow boundary. *J. Atmos. Sci.,* **54,** 2387–2415.

Lemon, L. R., and C. A. Doswell III, 1979: Severe thunderstorm evolution and mesocyclone structure as related to tornadogenesis. *Mon. Wea. Rev.,* **107,** 1184–1197.

———, D. W. Burgess, and R. A. Brown, 1978: Tornadic storm airflow and morphology derived from single-Doppler radar measurements. *Mon. Wea. Rev.,* **106,** 48–61.

Leslie, F. W., 1977: Surface roughness effects on suction vortex formation: A laboratory simulation. *J. Atmos. Sci.,* **34,** 1022–1027.

Leslie, L. M., 1971: The development of concentrated vortices: A numerical study. *J. Fluid Mech.,* **48,** 1–21.

Lewellen, D. C., W. S. Lewellen, and J. Xia, 2000: The influence of a local swirl ratio on tornado intensification near the surface. *J. Atmos. Sci.,* **57,** 527–544.

Lewellen, W. S., 1976: Theoretical models of the tornado vortex. *Proc. Symp. on Tornadoes: Assessment of Knowledge and Implications for Man,* Lubbock, TX, Texas Tech University, 107–143.

———, 1993: Tornado vortex theory. *The Tornado: Its Structure, Dynamics, Prediction, and Hazards, Geophys. Monogr.,* No. 79, Amer. Geophys. Union, 19–39.

———, and Y. P. Sheng, 1980: Modeling tornado dynamics. U.S. Nuclear Regulation Committee Rep. NUREG/CR-2585, Washington, DC.

———, D. C. Lewellen, and R. I. Sykes, 1997: Large-eddy simulation of a tornado's interaction with the surface. *J. Atmos. Sci.,* **54,** 581–605.

Lilly, D. K., 1969: Tornado dynamics. NCAR Manuscript 69-117. [Available from National Center for Atmospheric Research, P.O. Box 3000 Boulder, CO 80307-3000.]

———, 1976: Sources of rotation and energy in the tornado. *Proc. Symp. on Tornadoes: Assessment of Knowledge and Implications for Man,* Lubbock, TX, Texas Tech University, 145–150.

———, 1982: The development and maintenance of rotation in convective storms. *Intense Atmospheric Vortices,* L. Bengtsson and J. Lighthill, Eds., Springer-Verlag, 149–160.

———, 1983: Dynamics of rotating thunderstorms. *Mesoscale Meteorology—Theories, Observations and Models,* D. K. Lilly and T. Gal-Chen, Eds., Reidel, 531–543.

———, 1986: The structure, energetics and propagation of rotating convective storms. Part I: Energy exchange with the mean flow. *J. Atmos. Sci.,* **43,** 113–125.

Long, R. R., 1958: Vortex motion in a viscous fluid. *J. Meteor.,* **15,** 108–112.

Maddox, R. A., J. R. Hoxit, and C. F. Chappell, 1980: Study of tornadic thunderstorm interactions with thermal boundaries. *Mon. Wea. Rev.,* **108,** 322–336.

Magsig, M. A., and J. T. Snow, 1998: Long-distance debris transport by tornadic thunderstorms. Part I: The 7 May 1995 supercell thunderstorm. *Mon. Wea. Rev.,* **126,** 1430–1449.

Markowski, P. M., 2000: Surface thermodynamic characteristics of RFDs as measured by a mobile mesonet. Preprints, *20th Conf. on Severe Local Storms,* Orlando, FL, Amer. Meteor. Soc., 251–254.

———, J. M. Straka, and E. N. Rasmussen, 1998a: A preliminary investigation of the importance of helicity location in the hodograph. Preprints, *19th Conf. on Severe Local Storms,* Minneapolis, MN, Amer. Meteor. Soc., 230–233.

———, E. N. Rasmussen, and J. M. Straka, 1998b: The occurrence of tornadoes in supercells interacting with boundaries during VORTEX-95. *Wea. Forecasting,* **13,** 852–859.

Maxworthy, T., 1972: On the structure of concentrated columnar vortices. *Astronaut. Acta,* **17,** 363–374.

———, 1973: Vorticity source for large scale dust devils and other comments on naturally occurring vortices. *J. Atmos. Sci.,* **30,** 1717–1720.

———, 1982: The laboratory modelling of atmospheric vortices: A critical review. *Intense Atmospheric Vortices,* L. Bengtsson and J. Lighthill, Eds., Springer-Verlag, 229–246.

McCaul, E. W., 1993: Observations and simulations of hurricane-spawned tornadic storms. *The Tornado: Its Structure, Dynam-*

ics, Prediction, and Hazards, Geophys. Monogr. No. 79, Amer. Geophys. Union, 119–142.
Moller, A., 1978: The improved NWS storm spotters' training program at Ft. Worth, Tex. *Bull. Amer. Meteor. Soc.*, **59,** 1574–1582.
———, C. Doswell, J. McGinley, S. Tegtmeier, and R. Zipser, 1974: Field observations of the Union City tornado in Oklahoma. *Weatherwise,* **27,** 68–77.
Morton, B. R., 1966: Geophysical vortices. *Progr. Aeronaut. Sci.*, **7,** 145–193.
———, 1969: The strength of vortex and swirling core flows. *J. Fluid Mech.*, **38,** 315–333.
Mullen, J. B., and T. Maxworthy, 1977: A laboratory model of dust devil vortices. *Dyn. Atmos. Oceans,* **1,** 181–214.
Newton, C. W., and H. R. Newton, 1959: Dynamical interactions between large convective clouds and environment with vertical shear. *J. Meteor.*, **16,** 483–496.
Novlan, D. J., and W. M. Gray, 1974: Hurricane-spawned tornadoes. *Mon. Wea. Rev.*, **102,** 476–488.
Pauley, R. L., 1989: Laboratory measurements of axial pressure in two-celled tornado-like vortices. *J. Atmos. Sci.*, **46,** 3392–3399.
———, and J. T. Snow, 1988: On the kinematics and dynamics of the 18 June 1986 Minneapolis tornado. *Mon. Wea. Rev.*, **116,** 2731–2736.
———, C. R. Church, and J. T. Snow, 1982: Measurements of maximum surface pressure deficits in modeled atmospheric vortices. *J. Atmos. Sci.*, **39,** 369–377.
Peterson, R. E., Ed., 1976: *Proc. Symp. on Tornadoes: Assessment of Knowledge and Implications for Man,* Lubbock, TX, Texas Tech University, 696 pp.
Pfost, R. L., and A. E. Gerard, 1997: "Bookend vortex" induced tornadoes along the Natchez Trace. *Wea. Forecasting,* **12,** 572–580.
Proctor, F. H., 1988: Numerical solution of an isolated microburst. Part I: Dynamics and structure. *J. Atmos. Sci.*, **45,** 3137–3160.
Qui, C.-J., and Q. Xu, 1992: A simple adjoint method of wind analysis for single-Doppler radar. *J. Atmos. Oceanic Technol.*, **9,** 588–598.
Rasmussen, E. N., 1995: VORTEX operations plan. 141 pp. [Available from the National Severe Storms Laboratory, 1313 Halley Circle, Norman, OK 73069.]
———, and R. B. Wilhelmson, 1983: Relationships between storm characteristics and 1200 GMT hodographs. low-level shear, and stability. Preprints, *13th Conf. on Severe Local Storms,* Tulsa, OK, Amer. Meteor. Soc., J5–J8.
———, and J. M. Straka, 1996: Mobile mesonet observations of tornadoes during VORTEX-95. Preprints, *18th Conf. on Severe Local Storms,* San Francisco, CA, Amer. Meteor. Soc., 1–5.
———, and ———, 1998: Variations in supercell morphology. Part I: Observations of the role of upper-level storm-relative flow. *Mon. Wea. Rev.*, **126,** 2406–2421.
———, ———, R. Davies-Jones, C. A. Doswell, F. H. Carr, M. D. Eilts, and D. R. MacGorman, 1994: Verification of the origins of rotation in tornadoes experiment: VORTEX. *Bull. Amer. Meteor. Soc.*, **75,** 995–1006.
———, S. Richardson, J. M. Straka, P. M. Markowski, and D. O. Blanchard, 2000: The association of significant tornadoes with a baroclinic boundary on 2 June 1995. *Mon. Wea. Rev.*, **128,** 174–191.
Ray, P. S., and M. Stephenson, 1990: Assessment of the geometric and temporal errors associated with airborne Doppler radar measurements of a convective storm. *J. Atmos. Oceanic Technol.*, **7,** 206–217.
———, C. L. Ziegler, W. Bumgarner, and R. J. Serafin, 1980: Single- and multiple-Doppler radar observations of tornadic storms. *Mon. Wea. Rev.*, **108,** 1607–1625.
———, B. C. Johnson, K. W. Johnson, J. S. Bradberry, J. J. Stephens, K. K. Wagner, R. B. Wilhelmson, and J. B. Klemp, 1981: The morphology of several tornadic storms on 20 May 1977. *J. Atmos. Sci.*, **38,** 1643–1663.
Rayleigh, Lord, 1916: On the dynamics of revolving fluids. *Proc. Roy. Soc. London A,* **93,** 148–154.
Rothfusz, L. P., and D. K. Lilly, 1989: Quantitative and theoretical analysis of an experimental helical vortex. *J. Atmos. Sci.*, **46,** 2265–2279.
Rott, N., 1958: On the viscous core of a line vortex. *Z. Angew. Math. Physik,* **96,** 543–553.
Rotunno, R., 1977: Numerical simulation of a laboratory vortex. *J. Atmos. Sci.*, **34,** 1942–1956.
———, 1978: A note on the stability of a cylindrical vortex sheet. *J. Fluid Mech.*, **87,** 761–771.
———, 1979: A study in tornado-like vortex dynamics. *J. Atmos. Sci.*, **36,** 140–155.
———, 1981: On the evolution of thunderstorm rotation. *Mon. Wea. Rev.*, **109,** 577–586.
———, 1984: An investigation of a three-dimensional asymmetric vortex. *J. Atmos. Sci.*, **41,** 283–298.
———, 1986: Tornadoes and tornadogenesis. *Mesoscale Meteorology and Forecasting,* P. S. Ray, Ed., Amer. Meteor. Soc., 414–436.
———, and J. B. Klemp, 1982: The influence of the shear-induced pressure gradient on thunderstorm motion. *Mon. Wea. Rev.*, **110,** 136–151.
———, and ———, 1985: On the rotation and propagation of simulated supercell thunderstorms. *J. Atmos. Sci.*, **42,** 271–292.
Saffman, P. G., 1992: *Vortex Dynamics.* Cambridge University Press, 311 pp.
Schlesinger, R. E., 1978: A three-dimensional model of an isolated thunderstorm: Part I. Comparative experiments for variable ambient wind shear. *J. Atmos. Sci.*, **35,** 690–713.
Schlichting, H., 1960: *Boundary Layer Theory.* 4th ed. McGraw-Hill, 647 pp.
Scorer, R. S., 1978: *Environmental Aerodynamics.* Ellis Horwood, 488 pp.
Shapiro, A. H., 1972: Vorticity. *Illustrated Experiments in Fluid Mechanics, The National Committee for Fluid Mechanics Films Book of Film Notes,* The MIT Press, 63–74.
Shapiro, A., S. Ellis, and J. Shaw, 1995: Single-Doppler retrievals with Phoenix II data: Clear air and microburst wind retrievals in the planetary boundary layer. *J. Atmos. Sci.*, **52,** 1265–1287.
Smith, D. R., 1987: Effect of boundary conditions on numerically simulated vortices. *J. Atmos. Sci.*, **44,** 648–656.
Smith, R. K., and L. M. Leslie, 1978: Tornadogenesis. *Quart. J. Roy. Meteor. Soc.*, **104,** 189–199.
———, and ———, 1979: A numerical study of tornadogenesis in a rotating thunderstorm. *Quart. J. Roy. Meteor. Soc.*, **105,** 107–127.
Smith, R. L., and D. W. Holmes, 1961: Use of Doppler radar in meteorological observations. *Mon. Wea. Rev.*, **89,** 1–7.
Snow, J. T., 1982: A review of recent advances in tornado vortex dynamics. *Rev. Geophys. Space Phys.*, **20,** 953–964.
———, 1984: The Tornado. *Sci. Amer.*, **250** (4), 56–66.
———, and R. L. Pauley, 1984: On the thermodynamic method for estimating tornado windspeeds. *J. Climate Appl. Meteor.*, **23,** 1465–1468.
———, C. R. Church, and B. J. Barnhart, 1980: An investigation of the surface pressure fields beneath simulated tornado cyclones. *J. Atmos. Sci.*, **37,** 1013–1026.
———, A. L. Wyatt, A. K. McCarthy, and E. K. Bishop, 1995: Fallout of debris from tornadic thunderstorms: An historical perspective and two examples from VORTEX. *Bull. Amer. Meteor. Soc.*, **76,** 1777–1790.
Sreenivasan, K. R., 1985: Transition and turbulence in fluid flows and low-dimensional chaos. *Frontiers in Fluid Mechanics,* S. H. Davis and J. L. Lumley, Eds., Springer-Verlag. 41–67.
Straka, J. M., E. N. Rasmussen, and S. E. Fredrickson, 1996: A mobile mesonet for finescale meteorological observations. *J. Atmos. Oceanic Technol.*, **13,** 921–936.

Stumpf, G. J., A. Witt, E. D. Mitchell, P. L. Spencer, J. T. Johnson, M. D. Eilts, K. W. Thomas, and D. W. Burgess, 1998: The National Severe Storms Laboratory mesocyclone detection algorithm for the WSR-88D. *Wea. Forecasting,* **13,** 304–326.

Sullivan, R. D., 1959: A two-cell vortex solution of the Navier-Stokes equations. *J. Aerospace Sci.,* **26,** 767–768.

Sun, J., and N. A. Crook, 1997: Dynamical and microphysical retrieval from Doppler radar observations using a cloud model and its adjoint. Part I: Model development and simulated data experiments. *J. Atmos. Sci.,* **54,** 1642–1661.

———, and ———, 1998: Dynamical and microphysical retrieval from Doppler radar observations using a cloud model and its adjoint. Part II: Retrieval experiments of an observed Florida convective storm. *J. Atmos. Sci.,* **55,** 835–852.

Taylor, E. S., 1972: Secondary flow. *Illustrated Experiments in Fluid Mechanics, The National Committee for Fluid Mechanics Films Book of Film Notes,* The MIT Press, 97–104.

Trapp, R. J., 1999: Observations of nontornadic low-level mesocyclones and attendant tornadogenesis failure during VORTEX. *Mon. Wea. Rev.,* **127,** 1693–1705.

———, 2000: A clarification of vortex breakdown and tornadogenesis. *Mon. Wea. Rev.,* **128,** 888–895.

———, and B. H. Fiedler, 1995: Tornado-like vortexgenesis in a simplified numerical model. *J. Atmos. Sci.,* **52,** 3757–3778.

———, and R. Davies-Jones, 1997: Tornadogenesis with and without a dynamic pipe effect. *J. Atmos. Sci.,* **54,** 113–133.

———, E. D. Mitchell, G. A. Tipton, D. A. Effertz, A. I. Watson, D. L. Andra, and M. A. Magsig, 1999: Descending and nondescending tornadic vortex signatures detected by WSR-88D's. *Wea. Forecasting,* **14,** 625–639.

Truesdell, C., 1954: *The Kinematics of Vorticity.* Indiana University Press, 232 pp.

Tuttle, J. D., and G. B. Foote, 1990: Determination of the boundary-layer airflow from a single Doppler radar. *J. Atmos. Oceanic Technol.,* **7,** 218–232.

Wakimoto, R. M., and J. W. Wilson, 1989: Non-supercell tornadoes. *Mon. Wea. Rev.,* **117,** 1113–1140.

———, and B. E. Martner, 1992: Observations of a Colorado tornado. Part II: Combined photogrammetric and Doppler radar analysis. *Mon. Wea. Rev.,* **120,** 522–543.

———, and N. T. Atkins, 1996: Observations on the origins of rotation: The Newcastle tornado during VORTEX 94. *Mon. Wea. Rev.,* **124,** 384–407.

———, and C. Liu, 1998: The Garden City, Kansas, storm during VORTEX 95. Part II: The wall cloud and tornado. *Mon. Wea. Rev.,* **126,** 393–408.

———, and H. Cai, 2000: Analysis of a nontornadic storm during VORTEX 95. *Mon. Wea. Rev.,* **128,** 565–592.

———, W.-C. Lee, H. B. Bluestein, C.-H. Liu, and P. H. Hildebrand, 1996: ELDORA observations during VORTEX 95. *Bull. Amer. Meteor. Soc.,* **77,** 1465–1481.

———, C. Liu, and H. Cai, 1998: The Garden City, Kansas, storm during VORTEX 95. Part I: Overview of the storm's life cycle and mesocyclogenesis. *Mon. Wea. Rev.,* **126,** 372–392.

Walko, R. L., 1988: Plausibility of substantial dry adiabatic subsidence in a tornado core. *J. Atmos. Sci.,* **45,** 2251–2267.

———, 1993: Tornado spin-up beneath a convective cell: Required basic structure of the near-field boundary layer winds. *The Tornado: Its Structure, Dynamics, Prediction, and Hazards, Geophys. Monogr.,* No. 79, Amer. Geophys. Union, 89–95.

Ward, N. B., 1972: The exploration of certain features of tornado dynamics using a laboratory model. *J. Atmos. Sci.,* **29,** 1194–1204.

———, and A. B. Arnett Jr., 1963: Some relations between surface wind fields and radar echoes. *Conf. Review, Third Conf. on Severe Local Storms,* Urbana, IL, Amer. Meteor. Soc.

Weisman, M. L., and J. B. Klemp, 1982: The dependence of numerically simulated convective storms on vertical wind shear and buoyancy. *Mon. Wea. Rev.,* **110,** 504–520.

———, and ———, 1984: The structure and classification of numerically simulated convective storms in directionally varying wind shears. *Mon. Wea. Rev.,* **112,** 2479–2498.

Weygandt, S. S., A. Shapiro, and K. K. Droegemeier, 1998: The use of wind and thermodynamic retrievals to create initial forecast fields from single-Doppler observations of a supercell thunderstorm. Preprints, *16th Conf. on Weather Analysis and Forecasting,* Phoenix, AZ, Amer. Meteor. Soc., 286–288.

Wicker, L. J., 1996: The role of near surface wind shear on low-level mesocyclone generation and tornadoes. Preprints, *18th Conf. on Severe Local Storms,* San Francisco, CA, Amer. Meteor. Soc., 115–119.

———, and R. B. Wilhelmson, 1995: Simulation and analysis of tornado development and decay within a three-dimensional supercell thunderstorm. *J. Atmos. Sci.,* **52,** 2675–2703.

Wiin-Nielsen, A., 1973: *Compendium of Meteorology for Class I and II Personnel.* Vol. 1, Part 1, *Dynamic Meteorology,* World Meteorological Organization, Geneva, 334 pp.

Wilczak, J. M., T. W. Christian, D. E. Wolfe, R. J. Zamora, and B. Stankov, 1992: Observations of a Colorado tornado. Part I: Mesoscale environment and tornadogenesis. *Mon. Wea. Rev.,* **120,** 497–520.

Wilhelmson, R. B., and J. B. Klemp, 1978: A numerical study of storm splitting that leads to long-lived storms. *J. Atmos. Sci.,* **35,** 1974–1986.

Wilkins, E. M., 1988: Influence of Neil Ward's simulator on tornado research. CIMMS Rep. 87, Cooperative Institute for Mesoscale Meteorological Studies, 64 pp.

———, and C. J. Diamond, 1987: Effects of convection cell geometry on simulated tornadoes. *J. Atmos. Sci.,* **44,** 140–147.

Wilson, J. W., G. B. Foote, N. A. Crook, J. C. Fankhauser, C. G. Wade, J. D. Tuttle, and D. K. Mueller, 1992: The role of boundary-layer convergence zones and horizontal rolls in the initiation of thunderstorms: A case study. *Mon. Wea. Rev.,* **120,** 1785–1815.

Wilson, T., and R. Rotunno, 1986: Numerical simulation of a laminar end-wall vortex and boundary layer. *Phys. Fluids,* **29,** 3993–4005.

Winn, W. P., S. J. Hunyady, and G. D. Aulich, 1999: Pressure at the ground within and near a large tornado. *J. Geophys. Res.,* **104** (D18), 22 067–22 082.

Witt, A., M. D. Eilts, G. J. Stumpf, E. D. Mitchell, J. T. Johnson, and K. W. Thomas, 1998: Evaluating the performance of the WSR-88D severe storm detection algorithms. *Wea. Forecasting,* **13,** 513–518.

Wurman, J., and S. Gill, 2000: Finescale radar observations of the Dimmitt, Texas (2 June 1995), tornado. *Mon. Wea. Rev.,* **128,** 2135–2164.

———, J. M. Straka, and E. N. Rasmussen, 1996: Fine-scale Doppler radar observations of tornadoes. *Science,* **272,** 1774–1777.

———, ———, ———, M. Randall, and A. Zahrai, 1997: Design and deployment of a portable, pencil-beam, pulsed, 3-cm Doppler radar. *J. Atmos. Oceanic Technol.,* **14,** 1502–1512.

Zhang, J., and T. Gal-Chen, 1996: Single-Doppler wind retrieval in the moving frame of reference. *J. Atmos. Sci.,* **53,** 2609–2623.

Ziegler, C. L., E. N. Rasmussen, T. R. Shepherd, A. I. Watson, and J. M. Straka, 2001: Evolution of low-level rotation in the 29 May 1994 Newcastle–Graham, Texas, storm complex during VORTEX. *Mon. Wea. Rev.,* **129,** 1339–1368.

Zrnić, D. S., and R. J. Doviak, 1975: Velocity spectra of vortices scanned with a pulse-Doppler radar. *J. Appl. Meteor.,* **14,** 1531–1539.

———, and M. Istok, 1980: Wind speeds in two tornadic storms and a tornado, deduced from Doppler spectra. *J. Appl. Meteor.,* **19,** 1405–1415.

———, R. J. Doviak, and D. W. Burgess, 1977: Probing tornadoes with a pulse Doppler radar. *Quart. J. Roy. Meteor. Soc.,* **103,** 707–720.

———, D. W. Burgess, and L. Hennington, 1985: Doppler spectra and estimated windspeed of a violent tornado. *J. Climate Appl. Meteor.,* **24,** 1068–1081.

Chapter 6

Hailstorms

CHARLES A. KNIGHT AND NANCY C. KNIGHT
National Center for Atmospheric Research, Boulder, Colorado*

REVIEW PANEL: Paul L. Smith (Chair), Bruce Boe, Mark Fresch, Robert Johns, Edward Lozowski, James Renick, and Harold Orville

6.1. Introduction

The purposes in this chapter are to provide a critical summary of current thinking about the distinction between hailstorms and other severe storms, and to discuss the outstanding problems of hailstorms and approaches toward their solution. The last reviews of this subject were by Browning (1977), Macklin (1977), and other authors in the AMS monograph *Hail: A Review of Hail Science and Hail Suppression* (Foote and Knight 1977), and articles by List (1985), Morgan and Summers (1985), and Foote (1985). Some repetition here of material in these references is unavoidable. They provide discussions that are still current in most ways, and are useful entries into the literature prior to their publication dates. The review by Browning (1977) in particular is more complete in discussing some aspects of hailstorms than is the present chapter. We have purposely taken an organizational approach that is different from his, because the setting of this article is a volume on severe storms instead of one on hail, and because our viewpoint is somewhat different.

Because of the existence of *Meteorological and Geophysical Abstracts*, we do not attempt to provide exhaustive referencing. That would be a nearly overwhelming task [1495 references keyed to hail, hailstorm(s), or hailstone(s) are listed, starting about 1976 and ending about August 1996], and the result would be unreadable. There are so many references in a topic this broad that the present selection is certain to be incomplete, and we apologize for the inevitable neglect of some that should have been included. We hope not to have missed too many.

There is no sharp and meaningful distinction between storms that do and do not produce hailstones. The AMS *Glossary* definition of a hailstone (Huschke 1959) can be paraphrased as "a ball or irregular lump of ice ranging in size from that of a pea to that of a grapefruit." "Ice" is clearly meant to imply nearly solid ice, to distinguish hailstones from large graupel (snow pellets) and large snowflakes. Strictly speaking, nearly all severe convective storms are probably hailstorms, in that they produce some hailstones aloft even if not at the ground. Thus it may be better to state the main concern of this chapter as, "Why do some storms produce more and/or larger hail than others?"

The plan is to work from the small scale to the large. Since the distinguishing feature of hailstorms is hailstones, the conditions of their formation are discussed first, followed by the storm organizations relevant to providing those conditions, and finally a discussion of major problems. Severe hailstorms constitute a somewhat special class of severe storms, but the distinguishing features of that class are not yet firmly established.

If one believes that progress in understanding necessarily implies simplification of a subject, then there has been little progress since the previous surveys. The present authors feel that, while simplification is a goal, in the meantime, better understanding often means gaining a more realistic appreciation of the true complexities of a subject. Progress has been made in this sense, and also in the developing of new observational tools and analytical approaches.

6.2. How hailstones form

The principles of hailstone growth have been well understood for many years. The review by Macklin (1977) covers the fundamentals, and that by List (1985) emphasizes some of the more complex aspects. The principles involved are basic cloud physics (e.g., Rogers and Yau 1989), as follows.

The updrafts of convective storms are the location of condensation of water vapor into cloud droplets. Most droplets, if not collected by larger hydrometeors

* The National Center for Atmospheric Research is funded by the National Science Foundation.

or evaporated (because of entrainment or descent), remain liquid—supercooled—in the height interval that corresponds to temperatures between the freezing point and the homogeneous ice nucleation temperature, about −40°C.

The terminal fall velocity of a hydrometeor is the speed at which its downward acceleration due to gravity balances the upward acceleration due to the drag of the air and is, of course, relative to the air itself. The updrafts in storms range up to several tens of meters per second, enough to elevate large hailstones. The terminal velocity of the cloud water droplets is small, centimeters to tens of centimeters per second. They can be considered to be traveling with the airflow, but a larger piece of ice in this environment falls much faster with respect to the air, one to tens of meters per second, and collides with droplets, which freeze upon impact. The ice grows thereby, its fall velocity increases, and it continues collecting and freezing yet more droplets as long as it remains within supercooled cloud. It becomes a hailstone at some threshold size defined by convention as pea-size (Huschke 1959), about 5 mm diameter. Hydrometeors this large have terminal velocities of about 10 m s^{-1}, but large hailstones may fall as fast as 50 m s^{-1}. As hailstones grow they may be moving upward or downward relative to the earth, depending upon the vertical velocity of the air. Their trajectories deviate greatly from the airflow trajectories.

In subsequent discussions it will be necessary to refer to classes of hydrometeors by size and by growth process, and Fig. 6.1 illustrates the meanings of terms to be used. The word *hydrometeor* refers to all liquid- and solid-phase water particles, and is therefore not useful in making distinctions. Cloud droplets (liquid), grown entirely by condensation, are the source of nearly all precipitation mass from severe storms, but they do not grow by condensation to more than about 50 μm diameter. They grow larger by collection of other droplets, or they may freeze, grow from the vapor for a while, and then collect and freeze supercooled droplets. Different kinds of collection can be called *riming, coalescence,* and *accretion,* but the major process distinction is between condensation and collection, and we use the word *collectors* to refer to all hydrometeors growing by collecting cloud droplets. Hailstones are the large ice collectors (diameter above

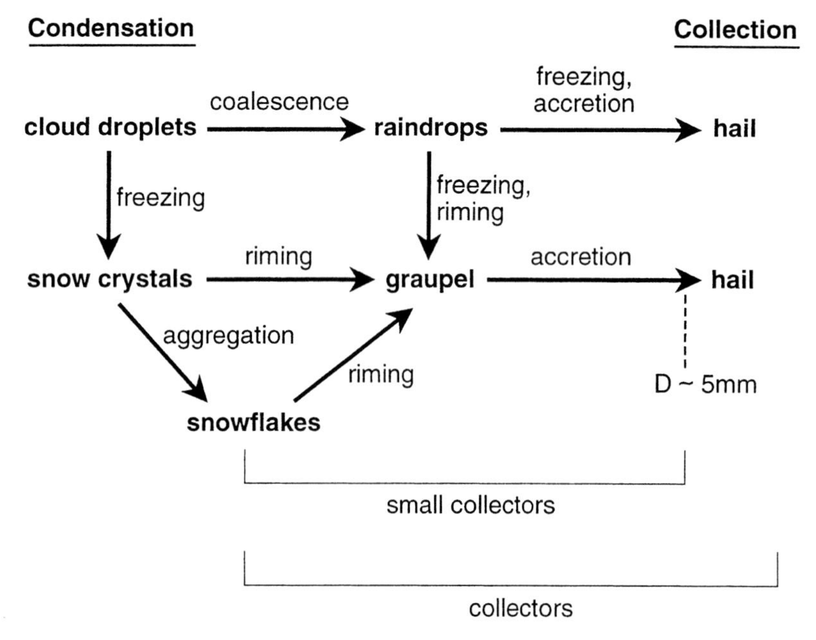

FIG. 6.1. Rough classification of the primary hydrometeor growth processes. Condensation of vapor (into either liquid or ice) and collection are the two basic growth processes. "Coalescence" refers to the collision and merging of two water drops, "accretion" generally refers to a sizable ice particle collecting either supercooled water droplets or small ice crystals (the latter being a less significant process), and "aggregation" generally refers to clumping of snow in the air to form snowflakes. "Riming" is usually used to mean the accretion of supercooled cloud droplets onto ice in a low-density deposit. The figure defines what is meant by collectors and small collectors, as used in the text.

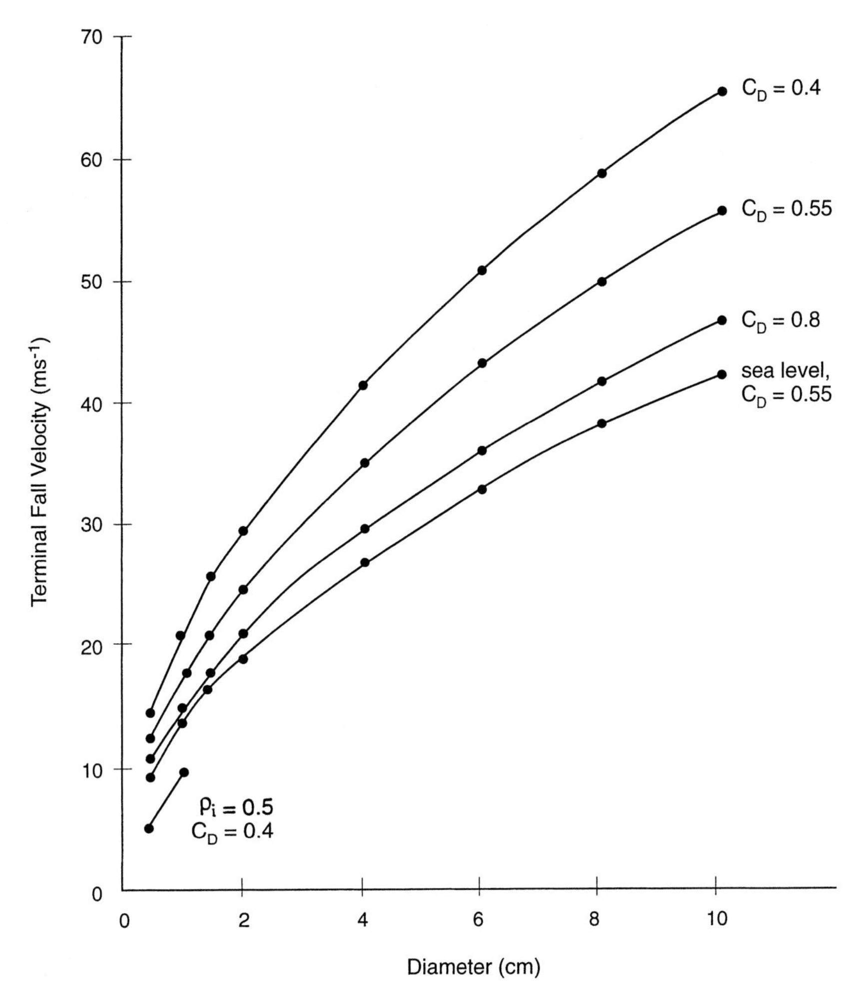

FIG. 6.2. Terminal fall velocities of hailstones according to Eq. (6.1), with different prescriptions.

about 5 mm, density[1] usually above about 0.8 g cm^{-3}); the term *small collectors* will refer to everything else. Figure 6.1 is restricted to primary growth and is not strictly complete, so as to avoid the denser forest of words and arrows that otherwise would be present, but it is adequate for this chapter. Secondary formation of small collectors by the shedding of water drops from hailstones is discussed later.

It is useful to provide some quantitative information about the parameters important for hailstone growth: terminal velocities and growth rates, and the important variables upon which they depend. Figure 6.2 presents terminal velocities, V_T, as a function of diameter, according to the formulation for a falling sphere in Eq. (6.1),

$$V_T = \left\{ \frac{4g\rho_i D}{3C_D\rho_a} \right\}^{1/2}, \tag{6.1}$$

where g is the acceleration due to gravity, ρ_i and ρ_a the densities of the ice deposit and the air, C_D is the drag coefficient, and D is the diameter of the hailstone. The top three curves in Fig. 6.2 represent fall velocities with C_D values of 0.4, 0.55, and 0.8; ice density 0.9 g cm^{-3}; and air density at 500 mb and $-20°C$. Drag coefficients vary because of nonspherical shapes, variable surface roughness, and tumbling during fall, and are only roughly independent of D, which is often defined as the equivalent spherical diameter. A value of 0.55 is a reasonable approximate drag coefficient for hail, but there is considerable variation. The bottom full curve represents fall velocities using a drag coefficient of 0.55 but at sea level air density. The short curve from $D = 0.5$ to 1.0 cm represents a drag coefficient of 1 and a density of 0.5 g cm^{-3}, which can often be more appropriate at these smaller sizes.

[1] Here the rather high density is added, qualified by "usually," because "stone" in the word hailstone implies solidity to most people. Word usage in this area is neither consistent nor precise.

Figure 6.3 shows growth rates for hail with the terminal velocity specifications of Fig. 6.2, in terms of growth of the diameter in centimeters per minute, for an effective liquid water content (actual cloud water content multiplied by the collection efficiency) of 2.5 g m^{-3} (2.5 cm^3 m^{-3}): Liquid water contents where most significant hail growth occurs probably vary between about 1 and 5 g m^{-3}. The equation for growth rate is simply

$$dD/dt = V_T \times \text{LWC}_{\text{eff}} \times \rho_w/2\rho_i, \qquad (6.2)$$

where LWC_{eff} is the effective liquid water content of the supercooled cloud droplets expressed as the volume ratio instead of the usual mass per volume. LWC_{eff} is usually less than LWC because both the collection and the coalescence efficiencies are usually less than 1. LWC_{eff} is by far the most critical factor in Fig. 6.3, and the growth rates of hailstones may vary from zero to about twice those of Fig. 6.3. The growth of giant hail, 10 cm in diameter, must take at least 15 minutes, and probably longer, starting from a diameter of 0.5 cm.

Figure 6.4 is a representation of adiabatic liquid water content (the maximum possible amount of condensate) within a cloud whose base is 900 mb, as a function of cloud-base temperature, with the adiabatic temperature within the cloud also indicated.

Figures 6.2–6.4 provide the information needed to gain a feeling for hailstone growth in the hailstorm environment, and to appreciate in general terms what the important factors are. Sizable hailstones can grow quite quickly from small hail: 0.5 to 3 cm diameter in about 10 minutes at an extreme value of 5 g m^{-3} LWC_{eff}, or 20 minutes at half that value. Over these times the hailstone has an average terminal velocity of roughly 20 m s^{-1}. It has therefore fallen 12–24 km with respect to the air, so it must grow within updraft. It appears likely that most hailstone growth occurs between about $-10°$ and $-30°$C (discussed below), which gives a height interval of about 3 km within which the growth occurs. Thus, an updraft averaging at least 15 m s^{-1} is required.

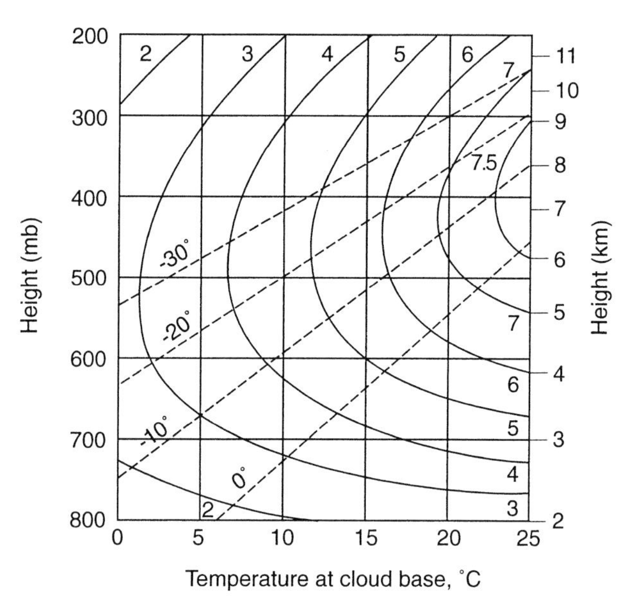

FIG. 6.4. The solid curves represent adiabatic liquid water content as a function of height within a cloud whose base is at 900 mb and the temperature shown on the abscissa. Temperatures within cloud are indicated by the dashed lines. Redrawn from Browning (1967).

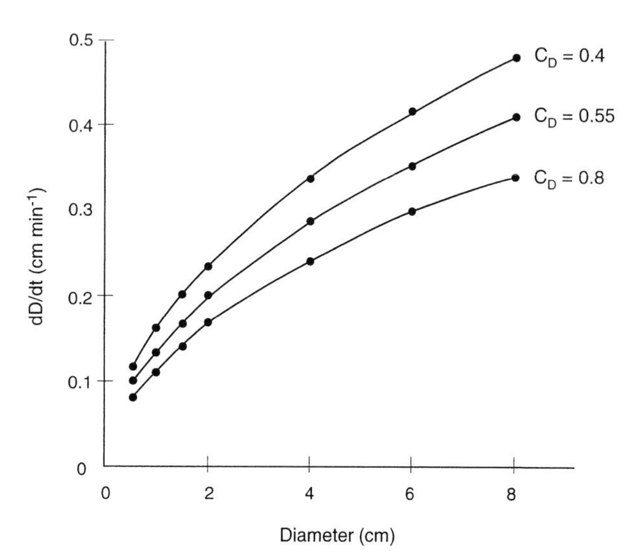

FIG. 6.3. The growth rates of hailstones, for the cases in Fig. 6.2 (except for the sea level case) according to Eq. (6.2), with LWC_{eff} = 2.5 g m^{-3}.

This illustrates well the important conditions for hailstone formation: adequate updraft to keep the hailstone aloft long enough, supercooled water content to enable it to grow fast enough before falling out, and a piece of ice for it to grow upon. The updraft and the liquid water are correlated in convective storms because the upward motion causes the condensation, but the correlation is imperfect because of depletion of the cloud water by the growing precipitation, mixing of the cloud with dry air from outside the cloud, and the fact that updrafts are influenced by dynamic effects as well as by buoyancy. The piece of ice upon which the hailstone grows represents a very important part of the hail problem, and its many possible histories will be a major preoccupation herein.

6.3. Uses of hailstones in the hailstorm problem

The major use of hailstones in research on hailstorms is deducing the conditions of their growth environments. Hailstones provide a rich variety of shapes, growth layering, internal air bubble assemblages, crystal textures and orientations, and chemical composition, all of which provide some evidence about the conditions in which they grew. However, the physics of hailstone growth is complicated and still incompletely understood, perhaps especially the tumbling motions of hailstones in free-fall and their effects upon the growth parameters.

The environmental variables that influence the internal structures of hailstone growth layers are primarily the cloud water content, which determines the rate of accretion, and the cloud temperature, which influences the heat economy of the growing hail. The effects of these variables are strongly modulated by hailstone size (fall velocity) and the tumbling motions. According to laboratory experiments, the crystal size within a hailstone growth layer provides semiquantitative information about the cloud temperature, and hence the height at which the layer formed (Levi and Aufdermaur 1970). Recrystallization often clouds the evidence from natural hailstones, however, to a degree that reduces trust in the interpretations (Knight et al. 1978).

Another method that has been used to deduce the height of formation of growth layers uses the deuterium content (or O^{18} content) of the ice. This method (described by Macklin 1977) uses the isotopic fractionation that occurs in the condensation process in rising parcels of cloud air, and the assumption that during accretion the isotopic composition of the droplets is preserved in the ice. Deductions from this method are only semiquantitative, since they rely upon uncertain assumptions about entrainment within the cloud and the isotopic composition of the cloud inflow. Both of these methods of estimating growth heights from hailstones have been discussed in technical detail in the literature, and there probably is not complete agreement on their reliability. We do not pursue these complex topics here, but refer the interested reader to references and to the previous reviews. Results especially from the isotope method will be noted and referenced later.

Two problems of hailstone structure do deserve special mention, however, as matters of possible general import to the understanding of hailstone formation. They concern two extremes of hailstone growth. One is spongy growth, in which a hailstone collects supercooled water so fast that the heat exchange is not adequate to freeze it all.[2] The hail may then grow as a "spongy" mixture of ice and water at 0°C even when the environment is substantially colder than that. The other is low-density growth, which can occur when the liquid water content of the cloud is so low that the ice deposit grows as a low-density latticework of frozen drops. Such deposits are permeable and can soak up liquid water either in subsequent growth or below the melting level, but the liquid must freeze later if the hailstone is to fall to the ground as solid ice. Low-density riming followed by soaking and freezing has been called *two-stage growth* (Prodi 1986).

Laboratory simulations of hail growth first showed that spongy growth might be very important. It proved easy to grow deposits in conditions that might be encountered in nature that had upward of 50% liquid water included within a finescale ice framework (List 1959). Since the conditions for spongy growth involve collecting too much water to freeze, it is more likely to occur at warmer subfreezing cloud temperatures and higher cloud water contents. Macklin's review (1977) includes a diagram of the limits for spongy growth as a function of temperature and water content, showing that 4 or more g m^{-3} at -20°C, or 2 or more g m^{-3} at -10°C, should produce this kind of growth. The problem has been that large, natural hailstones very rarely exhibit growth that appears to have been appreciably spongy, whereas the conditions for spongy growth would be expected to be fairly common. More recent experimental results that include the effects of tumbling motions of the hail have shown that centrifugal forces can cause shedding of water from the growing hail, lessening the sponginess of the deposit (Lesins and List 1986; Garcia-Garcia and List 1992; Levi and Lubart 1998). Since the character of the ice formed in these experiments has not been documented, comparisons with natural hail cannot be made, but the tumbling effects may explain the scarcity of observations of spongy, large hailstones in nature.

Low-density growth does occur for graupel hail embryos, which often soak up liquid water and then freeze solid at some later time (within their hailstones) before descending to earth. The original suggestion that low-density growth may be an important factor for considerably larger hailstones was by Pflaum (1984). However, at diameters larger than 2 cm, the fall speed is about 20 m s^{-1} or more (Fig. 6.2), and impacts at these speeds flatten the droplets on the ice surface as they freeze (Brownscombe and Hallett 1967). The resulting ice is unlikely to have low density, and indeed simulation experiments by Prodi et al. (1986) showed this to be the case. Thus it is unlikely that two-stage growth is important for hailstones larger than 2 cm, beyond being one among many microphysical complications that probably have little importance in the overall context of this chapter.

[2] Since the heat of fusion of water is about 80 cal gm^{-1} and its heat capacity 1 cal gm^{-1} $C^{\circ -1}$, it is easy to see that when water is supercooled to (say) -20°C and nucleated, only a quarter of it freezes spontaneously. In spongy growth the accretion rate of supercooled water outstrips the ability of heat transfer to the air to freeze it all. Some of it remains trapped within the ice and some may be shed.

6.4. The concept of the hailstone embryo

The subject of hailstone embryos is one of the central topics in the thinking about hail, but is difficult to write about because the embryo has no exact definition. It is a concept, and hail researchers find it meaningful and useful beyond the truism that hailstones must have been small before they became large. Since the graphs of fall velocity and growth rate for hail in Figs. 6.2 and 6.3 start at a diameter of 5 mm, the discussion of hailstone embryos can start with about 5 mm as their size.

Much of cloud microphysics would have to be summarized in order to inquire in detail into the mechanisms and rates of hailstone embryo formation. The discussion here must be cursory in the extreme, and a reader interested in more detail is referred to cloud physics texts (e.g., Rogers and Yau 1989; Young 1993). The two most direct routes to a 5-mm ice particle are the ice process and the freezing of a large raindrop formed by coalescence. In the ice process, typically, a cloud droplet forms, freezes, grows into a snow crystal, starts collecting supercooled water droplets, and becomes a graupel that constitutes the hailstone embryo. This process takes about 20 to 30 minutes, and the fall velocity averaged over that time is roughly 1 m s^{-1}. Coalescence to form a large raindrop also takes about 20 to 30 minutes, with a comparable fall velocity averaged over its formation time.

Thus the formation of the embryos for the first hailstones a storm produces requires a few tens of minutes and the average fall velocity during that time is low. This makes it impossible for embryos to form within the 15–20 m s^{-1} updrafts needed to grow the final hailstones. With a 1 m s^{-1} fall speed, a growing embryo would rise at an average velocity of 14–19 m s^{-1}, and a 20-minute formation time would therefore require about 20 km of ascent relative to the earth. The nascent embryo would be prematurely ejected into the anvil.

From this kind of elementary reasoning, it was realized early in the study of hail (Ludlam 1958) that the formation of hail embryos is likely to be distinct from that of hailstones and to occur in a different dynamical environment from that in which the hailstone grew. This is the embryo concept. It defines the concept of the embryo but does not set a well-defined size or fall velocity at which an embryo becomes a hailstone.

Hailstone embryos have received a lot of attention because their formation in a source region and subsequent delivery into a strong, wet updraft emerge as necessary conditions for hailstone formation. Also, embryo-sized ice particles within a strong updraft may limit each other's growth by depleting the cloud water. This process is sometimes called *beneficial competition* because of the idea that high enough concentrations of potential embryos could prevent hail formation. It becomes critical for understanding hail (as well as for examining the possibility of hail suppression) to discover the embryo sources and means of delivery into the strong updrafts by examining growth trajectories within the complex and evolving wind fields of the hailstorms.

Although the topic of hail embryos will pervade the rest of this chapter, it should be kept in mind that it is a fundamentally qualitative concept and that it carries different meanings. In the context of storm dynamics, the *hail embryo* refers to that hydrometeor that enters the strong updraft and grows into a hailstone. This is generally an extremely small fraction of the total population of embryo-sized hydrometeors a storm produces. There are no discontinuities here, no fixed updraft strength that is strong enough, and no specific size requirement. The point is just that there need to be two distinguishable stages of growth to form a hailstone, and the first one, the embryo stage, produces a hydrometeor roughly 1–10 mm in diameter with a fall velocity of several meters per second.

Another context of the word *embryo* involves its growth process. This can be as simple as just droplet coalescence or the ice process, or any of a host of multistage processes, as, for example, ice nucleation followed by vapor growth, then aggregation, riming, perhaps partial or complete melting, and finally partial or complete freezing.

Yet another context of the word *embryo* is its application to the smallest identifiable growth unit within a hailstone. This does have a specific size (though in some cases the identification is subjective), but that size may not correspond to the embryo size that one would select from the dynamical context if one knew the complete history of the hailstone.

The embryo concept appears to be necessary for sensible discussion of hail formation within hailstorms, but it requires some vigilance to keep its contexts sorted out.

6.5. Embryos as seen within hailstones

Hailstones virtually always exhibit an early growth unit that is visually distinct from the rest of the stone and that can usually be identified as a frozen drop or a graupel. This is the embryo by definition in this context. [A recurring topic in notes on meteorological curiosities is foreign objects within hailstones. One example of hail containing coal and gravel up to 50 g is noteworthy (Meaden 1995), but in this chapter the embryos under discussion are all ice.] Figure 6.5 gives some examples. While detailed classifications of embryo types exist (e.g., Federer et al. 1978), the present writers have employed a simpler one: frozen drop, conical graupel, spherical bubbly, and "other," for cases that do not fit fairly clearly (there is appreciable subjectivity here) into one of the first three. *Frozen drop embryos* consist of clear ice with a little, chaotic group of bubbles in the center and usually a crack from expansion during the inward freezing, with a bubbly ring around the outside. While they look like frozen water drops, they are sometimes too big, show-

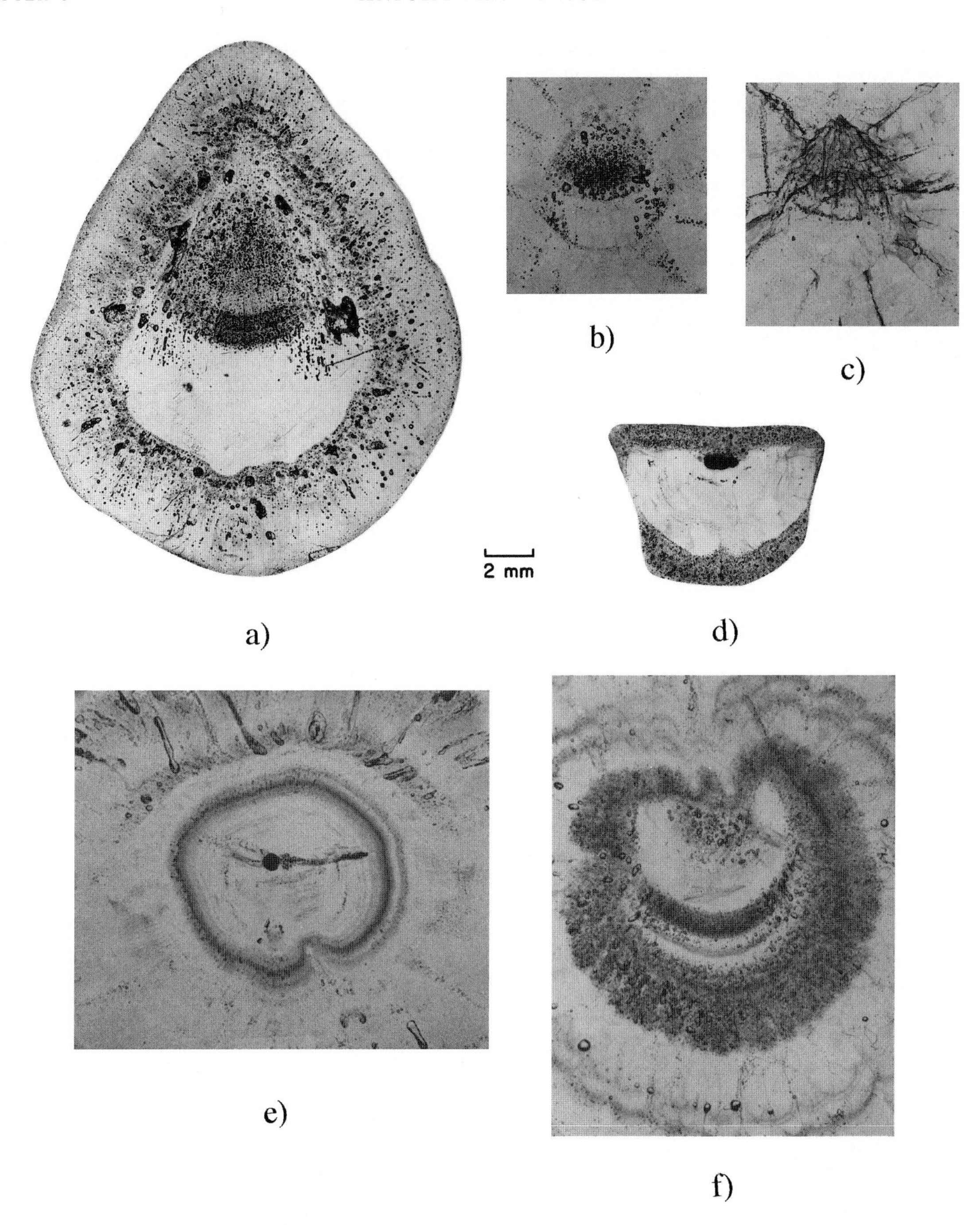

FIG. 6.5. Some examples of the inner growth units, the embryos, of hailstones, in cross section seen by diffuse transmitted light, so that air bubbles of different sizes provide the dark contrast. (a) A complete conical hailstone with a graupel embryo, showing a final layer of uniform thickness after it started tumbling. (b) and (c) Conical graupel embryos from within hailstones; (c) was quenched to reveal the presence of liquid water in the embryo, which was originally low-density rime but became permeated with water that was still liquid when the hailstone reached the ground, but (b) was not quenched. (d) A section through a very small hailstone, the embryo of which is a broken half of a frozen drop. The final layer of bubbly ice (dry growth) is probably missing at the sides because of melting. (e) Frozen-drop embryo with a prominent crack, from expansion during the final freezing just before the thin, bubbly layer was deposited. One can never determine the size of the initial drop when it starts freezing, because wet growth proceeds after nucleation, before the final freezing. (f) This embryo is also characterized as a broken frozen drop. Broken frozen drops typically fall at a constant attitude with their rounded sides down. Scales are all the same, as indicated. These are exceptionally clear examples.

ing that substantial accretion of liquid water continued after ice nucleation but before final freezing and crack formation (Knight and Knight 1974). *Conical graupel embryos* look like loose rime (actually two-stage growth, as discussed above) followed by denser rime while falling in a fairly constant orientation, accreting droplets at the bottom to produce the conical shape. As a rule, we have not found evidence in hailstones of the nature of the original ice particle that started the riming. It may be a small frozen drop or a snow particle of some kind. *Spherical bubbly embryos* might once have been snow aggregates or graupel that grew while tumbling, or partially melted and refrozen snow or graupel.

Examination of natural hail embryo types has yielded some correlations. Knight and Knight (1979) systematically collected hail on the high plains of northeastern Colorado during the National Hail Research Experiment and found a very great predominance of graupel embryos. This fit with the conclusion that the precipitation process in the cumulus of this continental area was nearly exclusively the ice process (Dye et al. 1974). While many of the hail collections were all graupel, some had some frozen drop embryos as well. One explanation for these results was that the embryos recycled into the updrafts at low levels and grew into hailstones while ascending. The clouds of northeastern Colorado have relatively cold bases, and the results can be rationalized by supposing that sometimes all the recycling was above the freezing level, producing hail with all graupel embryos, while in other cases the recycling straddled the freezing level, resulting in mixtures of types because some of the graupel melted.

Similar subsequent studies in other regions have shown higher percentages of frozen drop embryos in hailstones from clouds with warmer cloud bases (Knight 1981; Federer and Waldvogel 1978). This fits with the recycling explanation, but the drop embryos may also form by primary coalescence. Another correlation that appeared was a higher percentage of drop embryos (and a lower percentage of conical graupel embryos) in the larger hailstones.

Perhaps the most remarkable fact about hailstone embryo types is that so many analyses of hailstone collections from significant hailstorms reveal mixtures of types. The all graupel collections from northeastern Colorado were mostly very small hail from weak hailstorms, and the few collections that the authors have studied with all drop embryos have also been small hail, often with odd shapes indicating warm and very wet growth (e.g., Briggs 1968). Either embryo delivery into strong updrafts usually straddles the freezing level or embryo sources are typically diverse within storms that produce large hail—or, probably, both of these statements are true.

6.6. Studies of hailstorms

The primary observational methods of studying single hailstorms employ radar, aircraft, and hailstone examination, either singly to try to establish general characteristics of storms or in combinations to try to gain more complete pictures of individual storms using case-study analysis. Brief introductions to the methods and to some common features of the radar echo structures of hailstorms are required before discussing the case studies, which form the core of this section.

Radar is the essential observing system, since it alone provides an overall view of the storms that can be spatially and temporally complete. Radar data are the framework within which all other measurements are interpreted. The radar echo—the map of equivalent reflectivity factor Z_e (expressed as dBZ)—is a general measure of hydrometeor size, but it also depends upon concentration, phase (water, ice, or a mixture), and shape, so that it alone is never an unequivocal indicator of hail. Intense echoes can always be produced by nearly any combination of heavy rain, great quantities of small ice particles, and/or hail. An intense radar echo at high enough altitude (low enough temperature) is certainly from ice hydrometeors, but the size/concentration ambiguity remains. Polarization techniques are being used to remove some of this ambiguity, and are discussed later.

Doppler techniques allow radar to acquire radial velocity information, and if two or more radars are used in different locations to scan the same storm, three-dimensional wind fields within storms can be retrieved from the data. The wind fields are used for calculating growth trajectories of embryos and hailstones within the storms.

The only direct measures of the characteristics of hailstone growth environments that are available come from aircraft penetrating the storms, and the armored T-28 (e.g., Musil et al. 1991) is the only well-instrumented aircraft that has been used systematically for hailstorm studies.

Before commencing the case-study summaries, several concepts need to be introduced. The most important term in storm description is the *convective cell,* but it can also be the term most loaded with potentially misleading implications. Dating at least from the Thunderstorm Project (Byers and Braham 1949), the observation is that storms can be single cell or multicell, where a cell is defined as a single convective impulse: an updraft that starts, reaches maturity, and dies. Unfortunately, detailed modern storm data reveal difficulties in delineating just what a cell is. The distinctions are vague between steady and pulsing updrafts and between pulsing and successive updrafts. Thunderstorms are turbulent and embody temporal and spatial scales of variability down to very small sizes, with no clear separation of scales. Thus "cell" is a loaded term in very much the same way that "embryo" is. It is a useful and meaningful concept, but it can become a misleading oversimplification.

Echo-free vaults, weak-echo regions (WERs), bounded weak-echo regions (BWERs), and radar overhangs are observationally much more straightforward, and often appear in descriptions of the radar structure of hailstorms. Echo-free vault and BWER are synonyms, referring to a region of weak radar echo projecting upward into the stronger echo of the storm. These are interpreted (correctly) as the location of major updrafts, and the echo is weaker than the surroundings because the updraft is so strong that large collectors can neither fall into its highest-velocity central portions nor grow there because their residence time would be too short. A BWER is considered to be a sign of a particularly severe storm, and all storms of adequate size and intensity that possess BWERs probably produce some hail. Direct measurements within BWERs have been obtained with the armored T-28 aircraft, and one showed a 14-km-wide updraft with a maximum exceeding 50 m s^{-1} and cloud water contents, probably adiabatic, above 6 g m^{-3} (Musil et al. 1986).

BWERs are generally bounded on one side by a major precipitation shaft, shown by intense echo extending down to the ground. A BWER is not completely surrounded by echo at the ground, but only at some distance above. When it is not bounded by echo at any level, it is termed an echo overhang, or simply a WER: still the site of the main storm updraft and for the same reason, and usually fairly easy to interpret. Figure 6.6 shows simple schematics of these features, and Browning (1977) contains a more complete exposition.

A special interpretation of the WER radar structure formed the basis for extensive hail suppression operations in the Soviet Union (Sulakvelidze et al. 1974), that were reportedly highly successful and led to the National Hail Research Experiment in the United States. In their model, the radar echo overhanging the WER was thought to contain a very high liquid water content composed of large, supercooled water drops balanced in the updraft but undergoing breakup. In this scenario the smaller drops resulting from breakup rise in the updraft but descend back to the same level as they grow by accreting cloud droplets. Thus a few tens of grams per cubic meter of liquid water accumulate, providing an environment in which hail can grow rapidly. The hail suppression, accordingly, involved seeding and glaciating these big drop zones. Direct observations within clouds by aircraft and examination

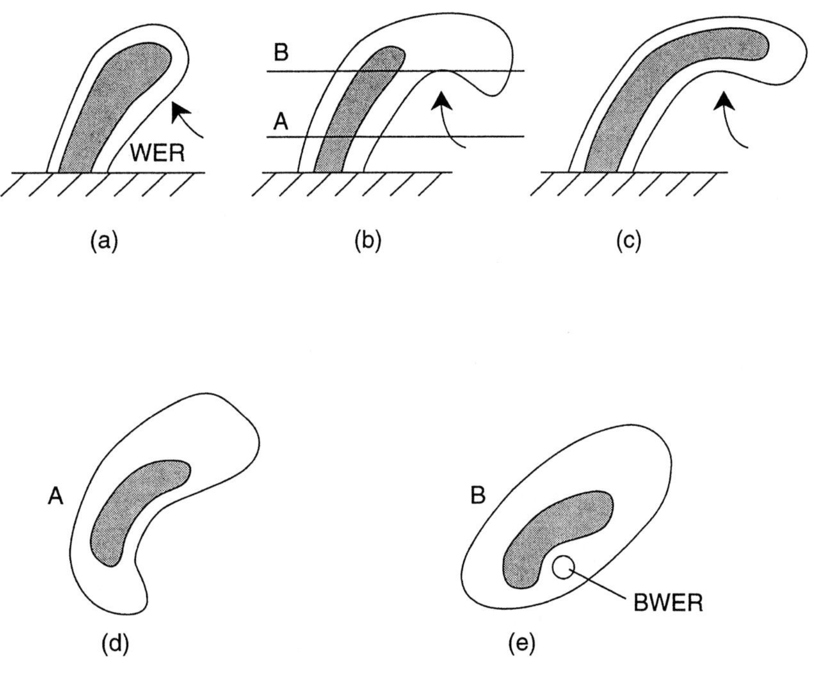

FIG. 6.6. Sketches to illustrate some descriptive terms of radar echo features in hailstorms. A simple echo overhang (a) shows higher reflectivity values aloft in the updraft, which is indicated by the arrow. The region beneath the overhang is sometimes called the weak-echo region (WER). In (b), the echo overhangs so much that the WER is bounded (BWER). (d) A low-level, horizontal cross section showing the hook echo [level A from (b)], and a higher-level cross section shown in (e) shows the BWER. (c) In some cases the overhang can extend out far over the inflow. In fact, the distinction between this overhang and a series of cells is sometimes obscure.

of hailstone embryos in the National Hail Research Experiment showed conclusively that this model did not apply to the storms of the Colorado high plains (see summary and references in Knight et al. 1979), and aircraft results have indicated the same for Swiss hailstorms (Musil and Smith 1986; Blackmore 1989).

The steadiness of a storm refers to how much it changes with time and relates somewhat to its cellularity. For the purposes of hail formation, a storm is considered steady if it does not evolve much (a subjective judgment) in a period of about 40 minutes to 1 hour: roughly, the complete time for the formation of significant hail, including the embryo. A steady-state storm is necessarily a single, long-lived cell, but a single cell need not be steady at any time. Of course no storm is ever perfectly steady, but the judgment in using steady state in the context of hail is that if the storm really were steady, it would produce about the same amount and size of hail.

Supercell is a term used very widely but in different ways. Within this volume it generally implies a mesocyclone associated with the storm, though the storm itself can be multicellular, resulting in a mixture of meanings for "cell" that seems unfortunate. In the context of hail, the word "supercell" has been used without the "mesocyclone" as part of the definition, simply to denote a long-lived, intense cell with a relatively steady, usually bounded, weak-echo region.

The main issue in this chapter is the storm organizations that provide embryos and growth environments that lead to hailstones. Accordingly, hailstorm case studies that contain discussions of hail embryo sources and/or hailstone growth trajectories are provided with very brief summaries in the following list, presented chronologically. These summaries have to be so brief here that they are necessarily oversimplified. Note also that statements about embryo sources and growth trajectories are all interpretations, resting upon substantial frameworks of assumptions.

- Browning (1963). Recycling of hail in connection with a steady-state BWER. Embryo sources might be from satellite cumulus, falling into the main updraft, and/or the freezing of drops shed from hailstones falling at the edge of the updraft. This is one of the two "archetypal," conceptual storm models that include explicit consideration of hail formation. The other is the next entry.
- Browning et al. (1976). A multicell storm, in which there is an orderly succession of cells moving through their life cycles, each supplanting its predecessor. Hailstone embryos form in the early stage of each cell, when the updraft is weak, and grow into hail in the mature stage within the same cell.
- Browning and Foote (1976). Elaboration of the interpretation of Browning (1963), with a much more complete consideration of the precipitation. A steady-state storm, with embryos growing on the right flank of the main updraft, then recycling into the updraft and growing while traveling over or around the BWER.
- Heymsfield (1983). Embryo formation largely by aggregation of snow crystals followed by riming, in the region of the forward stagnation point of the environmental flow around the storm, in weak and intermittent updrafts. Growth in feeder cells and then in almost horizontal trajectories across the main updraft.
- Foote (1984). Same storm as above. Hail size limited by horizontal extent of main updraft. Diverse, interweaving growth trajectories, but conform in general to Browning and Foote (1976).
- Knight et al. (1982); Miller and Fankhauser (1983). Stationary, multicell storm but with steady overall airflow structure. Embryos form in cells ahead of the main updraft (out in the extensive echo overhang), move into and through the main updraft, and then fall out. Random recycling, some graupel, and some drop embryos.
- Ziegler et al. (1983). Hail grows while traversing main updraft horizontally at 7–8 km (roughly $-10°$ to $-20°C$), from embryos introduced into upwind side of updraft from feeder cells ("embryo corridor") upwind.
- Dye et al. (1983); Miller et al. (1983). Documents systematic recycling of hydrometeors into the updraft of a weak storm to form graupel and small hail in 10–20 m s^{-1} updrafts.
- Nelson (1983). Hail growth in one pass across strong updraft at warmer than $-25°C$. Embryos from recycling and from broad, moderate-updraft region.
- Grenier et al. (1983). Most postembryonic growth in a narrow height range, $-11°$ to $-19°C$.
- Kraus and Marwitz (1984). Feeder clouds supply embryos for hail into a supercell.
- Knight (1984); Knight et al. (1981). Severe hailstorm with one slow-moving BWER lasting about 40 minutes. The heavy hailfall was associated with the formation of BWER and its early stages, but stopped before the BWER collapsed. The embryos may have been present in the region where the strong updraft penetrated. Hail growth largely between $-13°$ and $-28°C$ according to deuterium analyses.
- Knight and Knupp (1986); Knight (1987). Small, steady storm, trajectory analysis in Doppler wind field, embryo source regions and trajectories diverse.
- Nelson and Knight (1987); Nelson (1987). Case study and generalization of a storm type called *hybrid multicellular–supercellular,* a large, severe hailstorm with long-lived but evolving BWER. Very broad updraft with unsteady BWER structure, more than one kind of growth trajectory indicated.
- Rasmussen and Heymsfield (1987); Kubesh et al. (1988); Miller et al. (1990). The 1 August 1981 CCOPE storm, a supercell with no BWER, because

the updraft was filled with large hail. Embryos were graupel from the updraft fringes and drops shed from hailstones. Growth trajectories were through the updraft core or around it on either side, generally simple up-and-down paths.

- Musil et al. (1986); Miller et al. (1988). The 2 August 1981 CCOPE storm was another severe, well-documented hailstorm with supercell features. Embryos tended to originate as frozen drops near the forward stagnation zone or from overhanging echo ahead (upwind) of the updraft. Growth trajectories were all simple up-and-down paths but were diverse in their relation to the main updraft. Some embryos may have come from shedding.
- Cheng and Rogers (1988). Discrete streaks of hailfall at the ground are related to *feeder cells*—radar reflectivity maxima moving into the storm—suggesting that each feeder cell supplies a pulse of embryos.
- Raymond and Blyth (1989). Growth trajectories of small hail and graupel are traced back to the top of a growing, but weak, thunderstorm at the 7–8-km level.
- Conway and Zrnić (1993). Polarization radar shows a column of drops and wet hydrometeors extending 2 km above the freezing level in updraft, formed from melted ice fallen from the back-sheared anvil. Some of these would become drop hail embryos. Embryo sources from shedding are also possible, as well as graupel that enter updraft above the freezing level. Hail growth trajectories quite diverse, with some recirculation.
- Höller et al. (1994). Application of multiparameter radar techniques to a hybrid storm. Embryos mainly graupel from the edge of the updraft, with some melting and recirculating.
- Kennedy and Rutledge (1995). Hail embryos from cells ahead of the main updraft.
- Brandes et al. (1995). Embryos mostly from "feeder band" within storm inflow. Large drops from melted ice are carried in updraft 2 km above freezing level but probably do not become hail embryos. Hail production complex.
- Bringi et al. (1996). Two multiparameter radars and T-28 penetrations allowed the deduction of the existence of supercooled raindrops rising in updraft, becoming hail at about the $-15°C$ level. The authors suggest a primary coalescence process, but no trajectory calculations are presented.
- Hubbert et al. (1998). A severe hailstorm was analyzed with multiparameter and dual-Doppler radar and precipitation observations including hailstone interpretation. As in the above case, raindrops rise in the updraft and freeze; 30%–40% of the large hailstones have frozen drop embryos. Origin of drops inferred to be shedding from wet hailstones.

A lot of effort has been spent on these detailed case studies aimed at understanding how hail forms within single storms. Perhaps the major shortcomings in the growth trajectory calculations are the smoothing of the airflow that is inherent in the analyses, potential errors in the deduced mean airflow patterns, and the prescription of the supercooled liquid water contents within the storms, upon which the hail growth rate so largely depends. These all introduce important uncertainties into the results, so that the trajectory calculations should not be taken too literally. On the other hand, comparisons of calculated hailfall locations with the low-level radar echo and with hailfall data when available look generally reasonable. The writers' attitude is that the calculated growth trajectories probably represent kinds of hail formation trajectories that do exist in the storms examined. They are more likely to miss some, or underestimate maximum hailstone size because of the simplifications, than to add impossible ones or overestimate hail size. Another shortcoming of trajectory analysis is that it provides no estimate of hail amount.

Attempting to maintain the right balance between skepticism and credulity, then, one might glean the following conclusions from these case studies.

1. Embryos. There are usually smaller "cells" upwind of the strong, hail-producing updraft (within or upwind of the storm, depending upon how "the storm" is defined) that are probable sources of hail embryos. This source may be very close to the main updraft or it may extend 10–20 km or more upwind. Growth within the upwind edges of the main updraft[3] is also cited as a viable embryo source. Another source of embryos is shedding from hailstones experiencing wet growth above the freezing level, or melting below it. (We will call embryos formed in these ways *secondary,* because their formation requires hail already in the storm; other embryos are *primary.*) All of these sources may operate in a single storm. The conclusion that primary embryos generally must cycle into a strong updraft from elsewhere was already shown from fundamental considerations in section 6.4.
2. Growth trajectories. The studies typically show a variety, often a rich variety, of growth trajectories capable of producing hail within a storm. In general, the larger the storm, the more detailed the data and the analysis, and the more complicated the flow field, the greater the variety that is found. The trajectories themselves, however, are usually quite simple, given embryos to start with: single, up-and-down paths through or around the main updraft.

[3] The wind and the storm motion vectors generally do not coincide. This is upwind relative to the storm and can be very different directions at different altitudes. See section 6.7.

Recycling paths are found, but not very often, and when recycling trajectories are found the decision of what part belongs to the embryo stage can be quite arbitrary. (The trajectory modeling usually starts by initializing "embryos" of different sizes at different locations. This is the way the word "embryo" is used in these studies. If, for instance, a 1-mm embryo grows to 5 mm while rising and then falling, and thence into a large hailstone while rising in the main updraft, it is a largely semantic issue whether the embryo is the 5-mm or the 1-mm particle.) Some cases show most of the hailstone growth confined to a fairly narrow temperature range, roughly $-10°$ to $-25°C$, a range of about 2.5 km in altitude.

Past these brief summaries, the next two sections contain further discussions of hailstone growth trajectories and embryos.

6.7. More on hailstone growth trajectories

Subjects to be discussed in this section are growth trajectories according to hailstone studies, trajectories in idealized flow fields, trajectories in idealized storms, and small-scale influences.

a. General conclusions on growth trajectories from evidence provided by hailstones

While many of the case studies did not examine the altitude (temperature) ranges in which most of the hail growth occurred, three did note that most of it was in a relatively narrow interval. Grenier et al. (1983), relying upon isotope analyses of the growth layers of six hailstones, reported that most growth occurred between $-11°$ and $-19°C$; Nelson (1983) reported that "most hailstone mass is acquired in one pass across an updraft at a nearly constant level and at temperatures warmer than $-19°C$"; and Foote (1984) reported that most growth occurs between $-10°$ and $-25°C$. The latter two studies used growth computations within storm wind fields from Doppler radar. Many of the trajectory studies show examples of bigger overall altitude excursions than this, but usually without noting where most of the growth occurred.

Growth altitudes deduced from isotopic analysis of hailstone growth layers (the deuterium to hydrogen ratios in particular) from several storms have shown a similar result (Knight et al. 1981), and it looks as if it is fairly general. Federer et al. (1982) reached the same conclusion and added that "trajectories of the large hailstones are surprisingly flat," in conformity with Nelson's analysis quoted above. Exceptions are expected and are found in both the isotope studies and the trajectory computations in the case studies, but the nature of the hail problem is the search for general truths amidst a welter of variability, and this may be one.

Hailstone structural layering cannot be used to test whether big height excursions exist or not. Crystal size changes from large to small with decreasing cloud temperature at $-15°$ to $-20°C$ (Levi and Aufdermaur 1970), but recrystallization makes detailed deductions difficult (Knight et al. 1978). Air bubble content reflects hailstone temperature more than cloud temperature and is not useful for growth height deduction. In the past, the onionlike layering that hailstone structure is supposed to resemble (but rarely does in reality) was used as evidence that hail necessarily performs multiple, extreme up-and-down excursions (see, e.g., Huschke 1959, the description of the multiple incursion theory, in the entry for hailstone). Hailstones do often have a distinct layering, shown especially by air bubble content in the ice, but in large hailstones in particular it is usually much coarser than the layering of an onion, numbering two to four prominent layers past an embryo stage. Hailstones with a single layer past the embryo stage are not uncommon—probably more common, in fact, than stones with more than four prominent layers.

While hailstone structural layering gives only limited information on growth trajectories, it provides important qualitative information about the uniformity of the hail growth histories within single storms and between different storms. The picture that one gains from most of the case study trajectory deductions from using the internal wind fields of the storms—especially the most detailed ones—is of a variety of viable growth trajectories operating at once, and this is already a simplified view because of the smoothed flow-fields. Without the smoothing, the variety of trajectories would be greater. The qualitative aspects of hailstone structural layering within single-storm collections are nearly always variable, with different numbers and thicknesses of growth layers from stone to stone. The exceptions to this rule tend to be cases in which all or nearly all of the growth was warm, so that most of the stones had no bubbly layers and no small crystals at all.

Superimposed upon the variability of hailstone structures within single collections and single storms, there are often consistent differences between the hailstones of different storms. Thus, at this qualitative level, the hailstones themselves carry a message of great variability in detail but at the same time potentially informative differences between different storms. They do not carry a message of drastic, repeated vertical excursions, but if anything the opposite: of relatively simple growth trajectories, often with most of the growth within a fairly narrow altitude and temperature range.

b. Growth trajectories in idealized flow fields

The purpose here is to discuss principles, starting simple and adding complexities as long as new principles are involved. A lot of complications will be left out. One is that the terminal velocity of a growing hailstone increases with time. The reasoning below assumes that that rate of increase is slow compared with any changes in updraft speed, which is not necessarily true. Also ignored are manifold microphysical complexities involving terminal velocity changes caused by drag coefficient changes. These can be significant when surface roughness changes quickly, because the growth mode changes from wet to dry or vice versa, or as melting starts, or if the tumbling mode changes. In the context of this chapter it is sufficient to note that these complexities exist.

The simplest case is an updraft constant in space and time. If a growing hailstone is balanced, it immediately starts to fall relative to the ground, and the rate of fall increases as it grows further. If the updraft exceeds its terminal velocity, it will rise until it grows large enough to be balanced, and then it will fall.

If the updraft is steady and uniform in the horizontal but increases with height, a balanced hailstone is likely to be unstable: If it were a little lower, it would fall; a little higher, ascend. If the updraft decreases with height, however, the balanced hailstone is stable, and as it grows it descends just enough to maintain that balance. Since the updraft is zero at both top and bottom of a storm, there is necessarily a maximum somewhere in the middle. Thus a growing hailstone balanced above that updraft maximum would descend as it grew until it reached the level of the maximum. With any further growth it would descend to the ground. For this reason, in this idealization, the hailstone size that balances the maximum updraft would be about the maximum possible in a storm. Forecasting maximum updrafts is a common method of forecasting maximum hailstone size.

The next step of complexity is to consider the variation in the horizontal. If there is no wind shear, the updraft may be vertical at its axis, and as shown in Fig. 6.7 there is horizontal convergence of the winds below the level of maximum updraft, and horizontal divergence above it. The growing hailstones move with the horizontal component of the flow and vertically at the vertical velocity of the air minus their terminal velocity, so they cross the airflow streamlines in the senses shown in Fig. 6.7. If they are balanced in tilted airflow they move horizontally, but if unbalanced, they may move up or down in the ranges indicated. The updraft strength also decreases away from the axis at all levels in this model, adding more complications to the overall trajectories. The situation in Fig. 6.7 obviously provides an opportunity for recycling: for hydrometeors to diverge from the flow axis at high elevations, fall down into the converging flow lower down, and be swept back up in the updraft, providing more growth. Hydrometeor growth in recycling trajectories like this was analyzed by Woodward

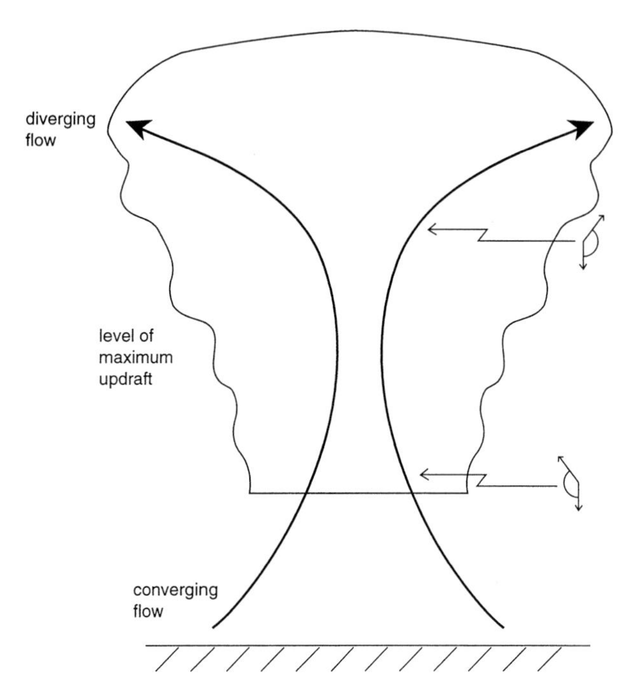

FIG. 6.7. A stylized, steady-state stationary thunderstorm in a zero-wind environment, illustrating convergent airflow beneath and divergence above. Hydrometeors on the right-hand airflow streamline move toward the axis in the convergent flow and away from it in the divergent flow, travelling within the sectors indicated: the vector sum of their fall velocity (vertical) and the air velocity.

(1959) in a thermal—a ring vortex—in which the air recycled as well.

Adding an environmental wind with uniform one-directional shear adds major new elements to the picture: storm motion and overall updraft tilt. The assumption of steady-state conditions is retained, and we first assume that the storm is stationary with respect to the ground. (The meaningful coordinate system for hydrometeor trajectories is always fixed to the storm, so in this case it is fixed to the ground too. Storm motion always tracks its inflow, and a storm may be motionless if, for instance, its inflow is anchored to a topographic feature.) As shown in Fig. 6.8a, the updraft is tilted downshear. There are opportunities for recycling on both sides, but perhaps more on the upshear side—the left—if the shear is not so strong as to completely overwhelm the upshear branch of the divergence at the upper levels. The horizontal momentum of the rising air tends to maintain itself against the environmental flow, which must then divert around it, as in Fig. 6.8b. Turbulent eddies may exist downshear of the storm, as shown. Now there is also horizontal flow through the updraft. An object that responds only to the horizontal component of the wind—a hailstone, if it stays balanced in the updraft—can move right through it at constant altitude. Thus a growing hailstone balanced in the updraft and approaching from the left in Fig. 6.8b moves through the

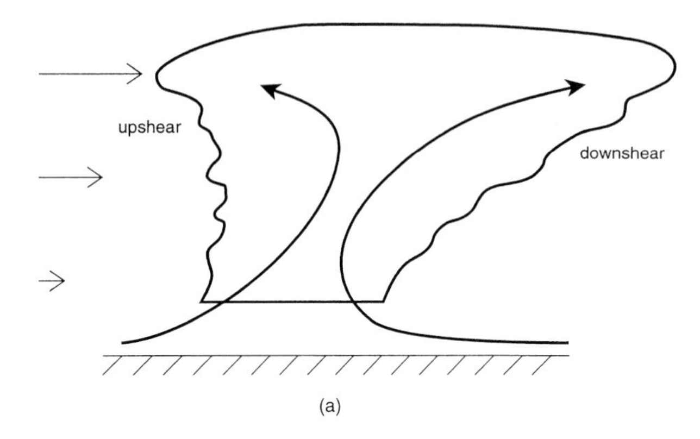

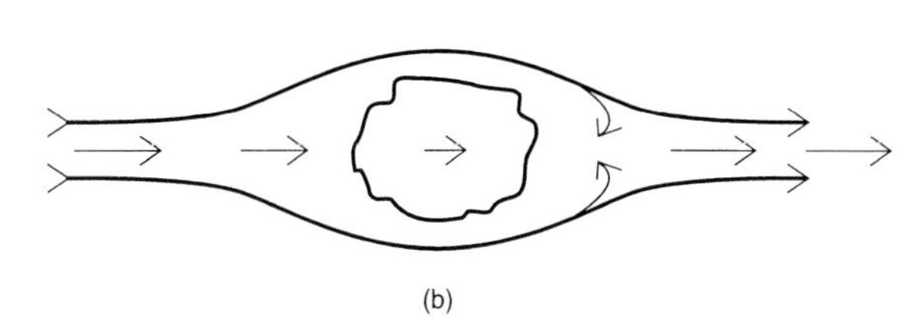

FIG. 6.8. (a) Like Fig. 6.7, storm again stationary, but now in an environment with wind shear indicated at left. (b) Midlevel airflow around the storm: The updraft behaves like an obstacle to the environmental flow.

center of the updraft, falling out on the downshear side of the cloud. Several researchers have emphasized that the residence time of growing hail in an updraft, and hence its maximum size, can be limited by the horizontal extent of updraft and the storm-relative, horizontal wind through it (related to the updraft tilt), not by the maximum updraft velocity.

Usually the surface convergence that feeds the storm is not fixed with respect to the ground. In principle it, and therefore the storm itself, can move at any speed in any direction. As is discussed elsewhere in this volume, storm motion is often strongly influenced by the storm's own outflow (when the low-level convergence occurs at the outflow boundary) and/or by its modulation of the low-level pressure field. Suppose the storm of Fig. 6.8a moves with the midlevel winds, in the same shear environment. The airflow relative to the storm is now different, producing a new storm shape as suggested in Fig. 6.9a, in which the storm-relative wind shear is indicated. The storm's low-level inflow may now be exclusively from the right side, and the hydrometeor recycling potential is also on the right side. The storm is an obstacle to the environmental winds in opposite senses at high and low altitudes, with a level in the middle at which its motion coincides with the environmental flow. Figure 6.9b is a sketch of the horizontal airflow in horizontal cross section, above the level at which the storm moves with the wind. Here the horizontal airflow within the updraft is opposite to the environmental winds, and in this ideal picture the horizontal flow must have two stagnation points, or regions. Since a hydrometeor in either of these regions moves with the storm in the horizontal, these stagnation points may have special importance for hail if they are within an updraft.

These kinds of idealizations are useful in understanding some of the principles of hail and embryo growth trajectories, but they are still a rather long way from the reality. Three-dimensional asymmetries, smaller-scale variabilities, and especially changes of the airflow with time can, separately or in combination, vitiate the direct application of models like that in Fig. 6.9 to real storms, but the principles illustrated by models remain valid. The next step after Fig. 6.9 might be, for instance, for the storm to split and the right-hand one of the new pair of storms to become a supercell with embedded mesocyclone. The environmental streamline that deviates around the storm at the bottom in Fig. 6.9b (the south, in the Northern Hemisphere) could then curve around into the easterly flow within the updraft. This kind of storm circulation is discussed elsewhere within this monograph.

The study of hail growth trajectories within modeled storm flow fields has been pursued very little, though the flow fields have been systematically modeled as a function of environmental shear and buoyancy, as

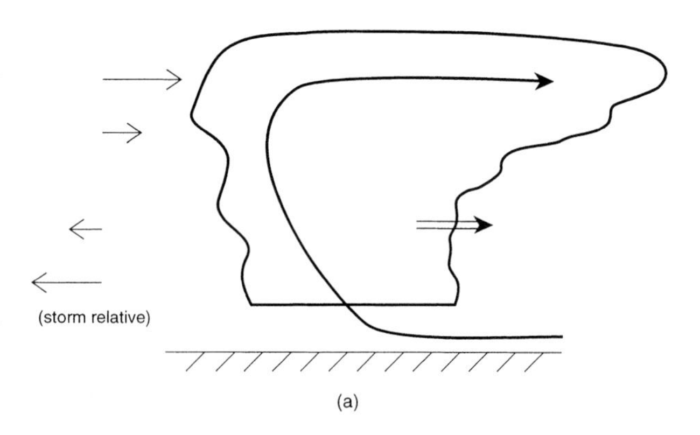

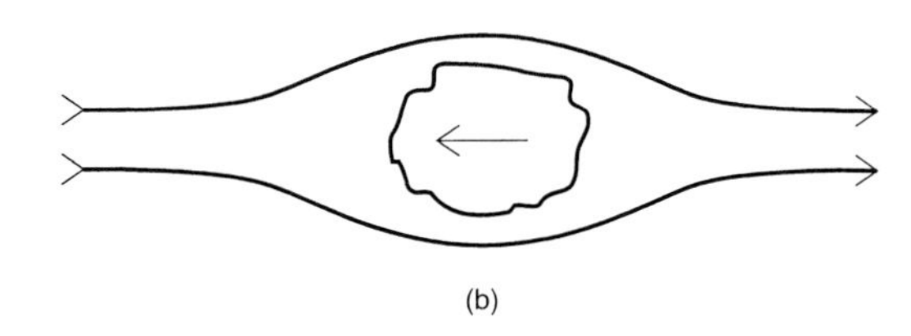

FIG. 6.9. Like Fig. 6.8, except that the storm is moving with the midlevel winds. With the same environmental wind shown in Fig. 6.8, the storm-relative winds are now as shown in (a), with a sketch of the storm shape and its main internal airflow. Its motion relative to the ground is shown by the double-barred arrow. (b) A horizontal cross section above the level of zero storm-relative environmental flow, showing opposing flow within the updraft.

indicated by the convective available potential energy, or CAPE (Weisman and Klemp 1986), and the results look quite realistic. This might be worth doing, though the results would still be highly dependent upon unsubstantiated microphysical assumptions.

c. Growth trajectories in idealized storms

The prominent example here is Browning and Foote's (1976) conceptual model of hailstone growth in a steady-state supercell with a BWER. Figure 6.10 gives their interpretation of the midlevel airflow and its relation to the low-level inflow and the midlevel radar reflectivity pattern in that storm. There is obstacle flow around the storm in the storm-relative environmental winds from the west, flow toward the north in the updraft air within and to the north of the BWER (the vault), inherited from the storm inflow from the south, and some turning of that flow in the vicinity of the upwind stagnation point, carrying some of it around the south side of the storm and off to the east. The hailstone growth envisioned within this flow has embryos forming along rising trajectories starting in the vicinity of A at low levels, reaching the vicinity of the stagnation point fairly high in the storm at B. By this time they have acquired a significant terminal velocity, and as they travel around the south side of the vault they descend into the storm inflow once more at C and then ascend once again, northward over the vault, growing to large size and finally falling out to the north. The airflow in this storm was deduced qualitatively from aircraft measurements outside the storm and interpretation of the radar echo pattern. Wind fields from Doppler radar measurements were not available for this case.

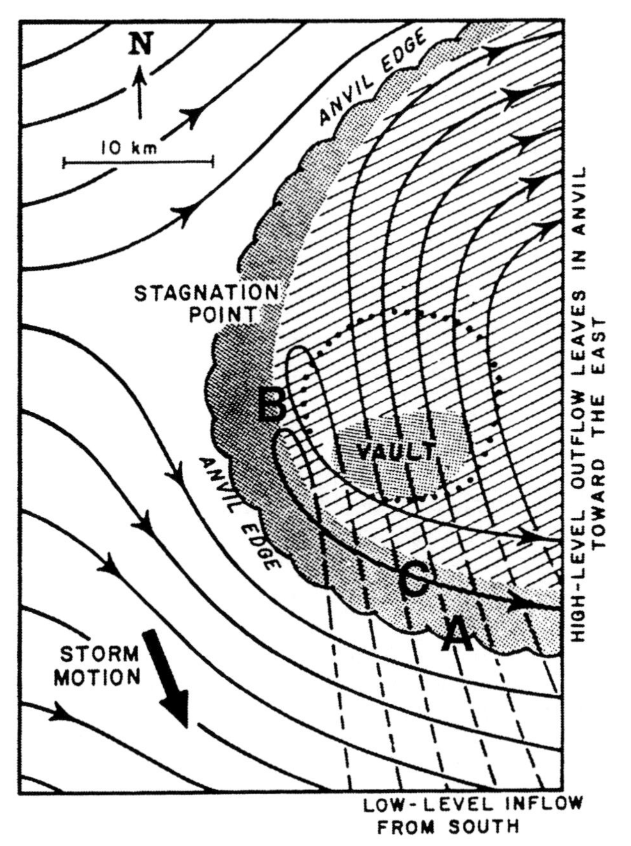

FIG. 6.10. Conceptualization of airflow within a supercell storm, from Browning and Foote (1976), accompanying discussion of hail growth trajectories in the text. The radar echo is shown hatched. The environmental flow is at mid levels, and the airflow within the storm is at levels as indicated. The main updraft region is the dotted circle.

A general feature of this kind of recycling hail growth trajectory is that the terminal velocity history (the rate of growth history) needs to match the airflow pattern. In this example, if the potential embryo is too small—falls too slowly—when it reaches B, it will move outward in the diverging flow aloft and eventually evaporate as it falls, or fall too far from the storm to be caught back up into the updraft. If it is too big—falls too fast—it can remain in the northward flow and fall as smaller hail or as rain. There may be no size in between "too big" and "too small," so that at present it is a conjecture that growth trajectories like this one actually exist in some supercell storms. The conjecture does seem a reasonable one, however, and there is some support from evidence of the location of hailfall with respect to radar echo patterns.

The other hail growth idealization that provides time for embryo formation and also strong updraft for hail growth is simply a growing storm cell. Updraft velocity may increase fairly slowly in a growing cell. Browning et al. (1976) interpreted hail formation in an organized, multicellular storm as if the hail formation were confined within each cell and simply followed the cell's life cycle. By the time the updraft reached hailstone-forming velocity (e.g., 20 m s^{-1}) the embryos were already within it.[4] Each cell in the succession followed the same history, so that the multicell storm produced a series of discrete but orderly hailfall events.

Browning (1977) drew a major and sharp distinction between supercell and multicell storms, a distinction that merits further scrutiny now that a number of detailed hailstorm case studies have been performed. The present writers find that the distinction does not hold up very well in the hail context. The "archetypal" supercells of the Wokingham and Fleming storms (Browning 1963; Browning and Foote 1976) were not observed in great detail, and perhaps partly

[4] This statement sounds like a good explanation until one tries to add specifics of just how it happens. The increasing updraft in a growing cell does not have a clear model. Is it a plume with slowly increasing inflow? A series of successively stronger thermals? Some hybrid between the two? This is the kind of microphysical and dynamical issue that can be examined with numerical models.

because of that they lent themselves rather well to the assumption of having steady-state airflow. The more detailed case studies of storms close to radar networks include several storms with notably supercell-like overall airflow patterns but a cellularity superimposed: of vaults (BWERS) with cells (meaning trackable radar reflectivity maxima) moving into and around them. Almost all of the detailed case studies postulate embryo sources in radar echo that is somehow distinct from the main storm: "feeder cells" is a phrase often used, "embryo corridor" one used occasionally. Nelson (1987) emphasized cellularity within an overall supercell-like organization as especially important for intense hail formation, and Kraus and Marwitz (1984) and Cheng and Rogers (1988) emphasized feeder cells supplying pulses of embryos accounting for pulses of hailstone formation. The very severe 1 and 2 August 1981 CCOPE hailstorms (Miller et al. 1990; Miller et al. 1988) do not fit the Browning and Foote (1976) archetype well. Almost all of the case study examples in section 6.6 are more noteworthy as exceptions to the proposed rules than as examples of them. Thus it appears at this point that the multicell/supercell dichotomy is not a comfortable one for real hailstorms.

Foote (1985) discussed more of the factors relating to classifying storms in ways relevant to hailfall, including degrees of cellularity in the organization of the airflow. Looking though the case studies suggests, to the present authors, that the very concept of storm, cellularity may become misleading in the context of hail formation. As was noted above, there is no agreed upon, objective definition of what a cell is, and detailed radar data do not seem to reveal useful scale separations. It seems almost heresy to suggest downplaying the cell concept because it is so deeply ingrained in thought about severe storms, yet it is possible that a useful classification of storms in terms of hail production might be found more quickly by ignoring it.

d. Smaller-scale effects

Smaller than what? The ready answer would be "smaller than a cell," which leaves it suitably vague. Perhaps the best answer here would be "smaller than the scales accounted for in Figs. 6.6 through 6.10," which are idealizations that are smoothed in space, and even more so in time, since they and their discussion assume that these patterns are virtually steady state. Real clouds and storms include major, smaller-scale variations (turbulence), one effect of which is to disperse air and particle trajectories. The ultimate effect of turbulence is homogenization through complete mixing, but the intermediate stages are just the opposite, to increase variability on these smaller scales. Observing methods that are sensitive to smaller scales find marked variations within thunderstorms. Very small scales of variation are emphasized in detailed looks at hailfall distributions at the ground (e.g., Changnon 1977), in radar data taken at close range (e.g., Battan 1975), and in all aircraft data from storms, as reported in many of the case-study papers.

The larger-scale, systematic organizations are real, too, and these smaller-scale features are superimposed onto them. The radar echo patterns of WERs, overhangs, and even the larger of what are called cells represent these, and it is true that they are often systematic. It is meaningful to think in terms of the larger-scale features, but it is necessary to deal somehow with the smaller scales, and one of the major, largely unanswered questions is the role of the smaller-scale variability in hailstone formation.

Numerical models have not faced this issue. It would not be difficult to add random dispersion into trajectories, but doing this in a way that bears some quantitative resemblance to reality would probably be next to impossible at this time. Another source of trajectory dispersion mentioned above is microphysical—from the spread of terminal velocities even at constant hailstone size, from the variability of surface roughness and, from the shape. It is hard to say which source of dispersion is greater. Saying anything more about this is speculative, but that represents the status of knowledge. The impression of the present writers, based on extensive experience in northeastern Colorado, is that, in that area, almost all thunderstorms produce a very small amount of small hail along with rain. This suggests that, in weaker storms, the mean conditions might not produce any hail at all: Their little bit of hail might result entirely from smaller-scale variations from mean conditions. Indeed, that may be the case for the great majority of hailstorms in general. For the big, organized, severe hail producers, however, it certainly is not the case. Such storms can produce large hail in continuous swaths both long and wide (see Changnon 1977; Nelson and Young 1979). The very heavy hail swath closely correlated to a BWER documented by Knight (1984) is another clear example.

6.8. More on hailstone embryo origins

The main issue about hailstone embryo origins is how embryos arrive within the strong updraft that is capable of growing them to large hailstones. It seems doubtful that closely delineated source regions exist for, say, the ice nucleation events that eventually lead to hail embryos. Inasmuch as there exists no real distinction between an embryo and a small hailstone, some of the principles involved in that transport are covered in the previous section in reference to hailstone growth trajectories.

In general, introduction of embryos above the level of maximum updraft must be less likely than below, because of the divergence of the airflow above and convergence below. Transport of potential embryos around the updraft in the midlevel, storm-relative airflow may be important to deliver them to the location from which they can be drawn into a tilted updraft at low levels. Several authors have appealed to this mechanism in case studies, and Browning (1977) systematized it by considering the angle between the inflow direction and the midlevel winds (all relative to the storm) reproduced in Fig. 6.11. For instance, if they are nearly parallel (L_3 in Fig. 6.11), there is little opportunity for this to happen except close to the upwind stagnation point. It is also clear that the variation with height of the storm-relative, midlevel flow—discussed in the previous section—can be important here.

The stagnation regions ("points") in the airflow around the storm may also have special importance in embryo production, by providing locations within weaker updraft at the edges of the main updraft in which the embryos can stay with the storm long enough to grow.

There is some promise in learning more about embryo introduction from hailstone structure studies of time resolved and/or spatially resolved collections, correlated with detailed radar observations of the storms that produced the hail. Roos et al. (1977) found that embryos generally formed between −10° and −18°C, whereas the bulk of the hail growth occurred between −20° and −25°C, with fairly simple trajectories, but not uniform, even in stones from the same collection. Federer and Waldvogel (1978) found that graupel embryos generally appeared earlier in a hailfall than frozen-drop embryos, but there was no size difference between stones with the two types. Knight and English (1980), however, found that stones with frozen-drop embryos fell closer to the updraft than those with graupel embryos, and the drop-embryo stones tended to be bigger. These kinds of results can be rationalized in terms of embryo formation mechanism, locations of embryo "delivery" into the main updraft, and the kinds of growth trajectories taken, but it is difficult to get complete enough data to be conclusive, and even more difficult to get enough cases to generalize.

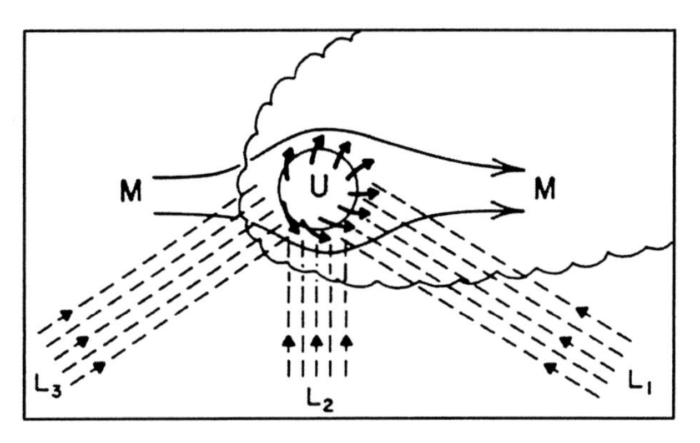

FIG. 6.11. Schematic of the midlevel flow (M) around a storm updraft (U), with three possible low-level inflow directions labeled L_1, L_2, and L_3. From Browning (1977).

6.9. Hail and rain: Depletion of cloud water, precipitation efficiency; wind shear and storm type; hail climatology

The most thorough comparison of amounts of hail and rain was done to evaluate a hail suppression test, and unfortunately was in terms of "hail days" and a fixed area rather than single storms (Morgan 1982). It showed that the hail mass averaged only 2.2% that of the rain, with a maximum of 10.7% for the 33 days of the sample. In one major storm that hailed over the same area and was analyzed in detail (Knight et al. 1982) the hail was less than 3% of the total precipitation. Quantitative data on hail/rain ratios for single storms are rare, but it is probable that the values are nearly always a few percent or less. Since most thunderstorm rain is melted ice, much of the distinction here is simply whether or not the ice particles grow large enough not to melt completely before reaching the ground: In all respects the hail- and rain-formation processes form a continuum.

a. Precipitation efficiency, depletion, and competition (see Browning 1977)

Hail (meaning hail at the ground) is a minor factor in the precipitation efficiency of storms, but one might ask whether hail occurrence increases as precipitation efficiency decreases, and if so, whether that correlation might reflect cause and effect. There are reasons to expect the answers to be qualified affirmatives. The connection between hail formation and precipitation efficiency in deep convective storms is depletion of the cloud water by the accretors. Both hailstones and the smaller accretors that end up as rain grow by gravitational sweep-out, collecting multitudes of the cloud water droplets in their paths. If there were no accretors in the cloud, these cloud droplets would all evaporate, some as the cloudy air mixes with drier air in the surroundings, and the rest in the outflow near the tropopause, as tiny ice particles. No precipitation would reach the ground, and the cloud's precipitation efficiency would be zero. If there were just a few accretors in the updraft but so widely spaced that they had negligible influence on the population of cloud droplets, they might grow to hailstone size or not, but the precipitation efficiency would still be very small. If there were enough accretors that they collected nearly all of the cloud water droplets, the precipitation efficiency could be very high, but now this depletion itself must influence the maximum size to which they grow. This feedback has been called *competition,* because the accretors can limit each other's growth rates, and therefore their final sizes, by depleting the cloud water. In this simplified view one would expect little hail when the precipitation efficiency is high and more and larger hail when it is low, other things being equal.

Depletion must increase with precipitation efficiency, and this depletion might reasonably be expected to decrease the size and amount of hail.[5] This is called *beneficial competition,* and is where cloud seeding enters in. Enough beneficial competition, and only rain could be produced. Another widely mentioned concept is *unfair competition,* which occurs because of size sorting. Bigger accretors fall faster than the smaller ones and may separate from them. Thus, starting with a region in an updraft containing accretors of different sizes, lying above a region with fewer accretors but containing cloud water, the bigger ones descend faster and get unfair access to undepleted cloud water. They grow bigger yet, while depleting the cloud water before their erstwhile companions, the smaller ones, get access to it.[6] Unfair competition is especially anticipated in radar overhangs and BWERs, where there is radar echo—presumably representing hail—overlying a weak-echo updraft (Browning and Foote 1976), and in fact large hail often (though by no means always) is produced in storms with these features. The size sorting that produces unfair competition would also be expected to produce large hail falling by itself, unaccompanied by the rain or small hail that would reach the ground later or in a different location, depending upon storm motion. This has also been observed. The extreme and abrupt echo gradient on one side of the vault is thought often to comprise large hail alone and has been called the *hail cascade.*

Depletion and competition were stressed in the analysis of data from the National Hail Research Experiment. Depletion times (the calculated time to deplete the cloud water to about one-third of its value), calculated from the sizes and concentrations of precipitation measured from the T-28 aircraft, ranged from one to tens of minutes (Knight et al. 1982). This ranges from important to essentially negligible in the strong updrafts. A more general conclusion was that the size spectra of the collectors were usually such that most of the depletion was by those about 1 to 2 mm in diameter, not by the large graupel or the hail. The larger collectors were rarely found without much greater numbers of these smaller ones, so the aircraft data showed little evidence of unfair competition. However, the data were biased toward weaker storms both because of aircraft safety and because the weaker storms are so much more frequent. Big, severe hailstorms with well-developed, long-lived BWERs were not sampled in the flights.

Two large, severe hailstorms were sampled by the T-28 in CCOPE, in eastern Montana on 1 and 2 August 1981. Both had broad updrafts with maxima greater than 40 m s^{-1} and both had long depletion times (little depletion) in the updraft cores, according to the aircraft data (Kubesh et al. 1988; Musil et al. 1986). Thunderstorm data from the T-28 in Oklahoma (Heymsfield and Hjelmfelt 1984), Alabama (Musil and Smith 1988), and Switzerland (Blackmore et al. 1989) were not analyzed in terms of depletion rates.

b. *Storm type, wind shear, precipitation efficiency, and hailfall*

Classifications of storms into multicellular and supercellular organizations and their correlation with wind shear (Marwitz 1972b,c; Chisholm and Renick 1972; summarized in Browning 1977) were based upon observations, largely by radar but also by aircraft and time-lapse photography. More recently, storm dynamical structure as a function of wind shear and buoyant instability (CAPE) has been explored systematically with numerical models (e.g., Weisman and Klemp 1986). The results have a lot of realism and represent a major improvement in the conceptual understanding of storm dynamics. However, the modeling approach is not yet advanced enough to treat precipitation formation (especially hail) systematically and in realistic detail, and so the main approach to hail study has been observational and deductive: the case studies of section 6.6 above, and the supercell and multicell hail formation concepts summarized by Browning (1977) and briefly described in section 6.8.

Wind helps to produce organized, long-lived, traveling storms in many instances, and these storms can have unusual severity, both in terms of their wind and in terms of hail. The role of wind shear in the precipitation efficiency of storms was examined by Marwitz (1972a) and discussed by Foote and Fankhauser (1973), finding an inverse correlation: the more shear, the lower the precipitation efficiency. This kind of result is susceptible to simple explanation: Shear tilts the updraft, tending to prevent recycling of collectors because most of them are blown off downwind while still at small sizes, and they mostly evaporate before reaching the ground. Browning and Foote (1976) and Browning (1977) argue (discussed briefly

[5] This larger-scale reasoning represents a step backward from thoughts of storm organization and hydrometeor growth trajectories, in that it considers the storm as a whole and assumes a certain uniformity of process. It is of course possible for a storm to be organized so as to be quite precipitation efficient overall, and yet produce significant hail from a local region in which the precipitation process is inefficient.

[6] There is a cloud model attached to this argument. In this case, for the argument to hold, the updraft needs to be neither too tilted nor too temporary. The competition/size-sorting relations can get quite complex when considered in detail, as by Young (1977) and Paluch (1978).

above) that the shear can also organize storms into producers of large hail by providing systematic recycling paths for the larger accretors to function as hail embryos and grow into large hailstones in the largely undepleted updrafts of BWERs. These idealizations fit well with the general idea of inefficient precipitation formation correlating with a tendency toward hail formation.

Recently the observations of so-called *LP (low precipitation) storms* (Bluestein and Parks 1983; Bluestein and Woodall 1990) seem to provide an extreme example of this. The extreme of LP storms is probably the "highly sheared storms" of Marwitz (1972d), in which the weak-echo vault extends entirely through the storm's radar echo: a vault without a ceiling. In a climatological study, Rasmussen and Straka (1998) have correlated LP supercells with upper-level wind speed relative to the storm. There seems little doubt that LP storms are inefficient precipitation producers because of insufficient delivery of accretors into the updraft. LP storms are characterized by very weak or invisible precipitation shafts below cloud base (Fig. 6.12), but they evidently do produce some hail: often, according to storm chase reports, large hail. There are no quantitative data on precipitation from them, however, and no detailed studies have yet aimed at understanding them in terms of their precipitation production. Doswell and Burgess (1993) added a category of supercell called *HP (high precipitation),* in which very heavy precipitation falls through the cloud base.

A major stumbling block in research on storm precipitation is still the difficulty of obtaining good and complete enough observations to provide adequate estimates of the important facts. The most complete examination of precipitation efficiency, still plagued by large uncertainties in all the terms—the amount of precipitation itself as well as the water vapor inflow to the storm—failed to reveal a meaningful correlation between precipitation efficiency and either shear or any other tested potential correlate (Fankhauser 1988). Likewise, the most complete case studies of hail formation from severe hailstorms—the 1 and 2 August cases from CCOPE, both "supercells" (Miller et al. 1990, 1988)—fail to conform to a simple picture. There is a long way to go yet before adequate idealizations of hail formation in storms are arrived at, and one must wonder about the degree to which simple idealizations will ever be adequate.

c. Hail climatology

Another approach toward understanding what it is about hailstorms that distinguishes them from others is to look at the climatology of hail. Hail occurs in many regions of the world and has been studied fairly systematically in a few, though even in the best-studied areas the data fall far short of what one would wish. It is perhaps unfortunate that the definition of hail has reached to such small sizes, because pea-sized ice particles are normal forms of convective precipitation that ordinarily melt before descending to the ground unless the ground is exceptionally close to the cloud base. Thus the fact that there is a lot of hail in high mountainous regions (Court and Griffiths 1985) is not informative about hailstorms in the context of this

FIG. 6.12. An LP storm in northeastern Colorado, 1996, photographed from the air. There is no visible precipitation whatever below the cloud base. Photo by A. Seimon.

chapter, and the high hail incidence in northeastern Colorado, for example, may merely reflect the thunderstorm frequency and the relatively high terrain, since most of the hail is very small.

A reliable hail climatology that included information about hail size and amount, and that included information on storm frequency, measures of intensity, atmospheric soundings, and so forth would be most valuable, but the attempts in these directions are inevitably short of what one would like for research purposes because of the shortcomings of the data. Attempts in these directions, such as a climatology of large hail for the United States by Kelly et al. (1985) and a correlation of large hail occurrences with storm radar characteristics by Amburn and Wolf (1997), show some promising relationships, but the data shortcomings noted by the authors are rather discouraging.

One intriguing possible climatological difference between two areas with relatively frequent, severe hailstorms is that between the South African high veld and the central plains of the United States and Canada. Carte (1979) has noted the scarcity of supercell storms on the South African high veld and suggested that this may relate to the fact that the wind shear there is usually weak. (The high veld is one of the few high-hail areas that is not in the lee of a mountain range, which may be a relevant factor.) Yet the Pretoria–Johannesburg area has evidently been subjected to severe hailstorms with large and giant hailstones as often as any other highly populated region. This would not be expected if supercellular organizations were in any sense needed for the growth of large hail and is at least a caution against assuming a necessary connection between supercells and severe hail.

There may be a relative scarcity of hail in the Tropics and in severe maritime storms. Takahashi (1987) described a hailstorm on Hawaii that was most notable because it was so unusual. However, lacking good data on the incidence of large hail anywhere, it is difficult to assess the extent to which there really is a relative scarcity of it in the Tropics or over the ocean. If there is, it cannot be explained just by melting in descent beneath the freezing level. The most direct explanation would be generally weaker updrafts in the Tropics, which is documented for convection in general (Lucas et al. 1994b), but whether this holds for the rarer, severe events that might give rise to hail is not as certain. Also, it is not clear why the convective updrafts are weaker: It does not appear to be because of deficient CAPE (Lucas et al. 1994a; Michaud 1996; Lucas et al. 1996). Other explanations for less large hail in the Tropics could be generally weaker shear in the environment, a much more active depletion of cloud water by rain formed through a primary coalescence process, and a very active microphysical production of secondary ice particles to produce more depletors above the freezing level (Hallett et al. 1978). Black and Hallett (1986) found that "hurricane convection is almost completely glaciated at the $-5°C$ level," which would certainly discourage hail formation.

The correlation between hail and areas in the lee of mountain ranges has also been noted, but here again, the reason may simply be high thunderstorm frequency with or without elevated terrain, producing many small-hail events.

A recent preliminary study has shown that nearly all tornadic storms in the central United States are accompanied by large hail ($D > 2$ cm), whereas along the Gulf and Atlantic coasts and in the southeastern states, some severe tornadic storms are not accompanied by large hail (Johns and Hart 1998). This is consistent with the general thought that the more efficient precipitation processes in warmer-based and/or more maritime clouds (clouds with smaller cloud droplet concentrations and therefore larger droplets, so that coalescence processes start more rapidly) leads to increased precipitation and decreased hailfall. However, it may be more likely that the explanation of this correlation lies in consistent differences in tornadic-storm dynamics between the two regions, resulting from consistently differing buoyancy and/or shear profiles. Questions involving hail nearly always combine dynamics and microphysics, leading to the quandary of which is the more important.

With the advent of radar techniques for detecting hail in storms (see below) and other remote sensing methods for detecting and locating lightning, distinguishing between intracloud and cloud-to-ground strokes and distinguishing the polarity of the latter, interesting correlations between hail, other precipitation characteristics, and lightning are emerging (e.g., MacGorman and Burgess 1994; Carey and Rutledge 1996, 1998; Lopez and Aubaguac 1997). While this research may be more likely to lead to better understanding of lightning than of hail, the establishment of correlations between lightning and severe hail or supercell hail may show that lightning data can be used indirectly as a useful source of hail climatological information.

6.10. Microphysics and storm dynamics: Is microphysical variability important?

Hailstorms involve storm dynamics and microphysical processes acting together in ways that are difficult to sort out. The two aspects are usually separated for the sake of simplicity, and it is interesting that the two major practical aims of hailstorm research—hail forecasting and hail suppression—have entirely separate rationales. Hail forecasting is based upon predicting gross storm properties such as updrafts and, for very short-term forecasting, radar echo parameters; hail suppression is based upon microphysical intervention by seeding.

Present hail forecasting schemes take the approach of predicting maximum hailstone size by predicting storm severity in terms of updraft velocity (Moore and Pino 1990), sometimes with additional factors that are largely empirical (Johns and Doswell 1992; Billet et al. 1997). This ignores all the convoluted considerations of embryo formation and delivery, the depletion of cloud water by accretion, and the likely size limitations due to hailstone growth rate versus residence time crossing an updraft. Short-term forecasting of hail uses empirical radar echo methods (e.g., Amburn and Wolf 1997). Hail amount has not been considered in forecasting schemes at all.

Hail suppression is also a practical endeavor for which the complications emphasized by the hail researcher are either ignored or bypassed by means of strong assumptions. Its basis is the supposition that hailstorms are organized so that more ice nuclei will produce more potential embryos that will act one way or another to reduce hailstone size.

Hail forecasting and hail suppression as presently practiced may produce results that are useful in practice, but in either case, the results are hard to evaluate. It is certain that strong updrafts are necessary to produce large hail, and the issues for forecasting are the degree to which they are sufficient for forecasting hail size and the problem of forecasting hail amount. It is well known that concentrations of both ice nuclei (IN) and cloud condensation nuclei (CCN) are highly variable in nature (e.g., Vali et al. 1982), and the unresolved, fundamental issue that is crucial for hail suppression (and of interest for forecasting as well) concerns the relation that hailfall bears to this variability, both within the natural range and that accessible through seeding. Knowledge of this relation would clarify the prospects of hail suppression and would aid in appreciating the predictability of hail.

It is remarkable how little can be said about this. Neither the size of the effect nor its sign is known with any degree of confidence, nor is it confidently expected that the size and sign of the effect would be consistent from storm to storm. It can be argued, however, that the effect is probably significant.

Since homogeneous ice nucleation probably only contributes small ice particles that are carried into storm anvils, there would be a one-to-one correspondence between ice nuclei and ice hydrometeors in storms if it were not for ice multiplication and aggregation (see Lamb 2001). Without these two processes, doubling IN would double the number of ice hydrometeors. This would surely be significant for the precipitation in a storm in which the ice process dominates, as on the high plains. If depletion was insignificant, doubling IN might approximately double the hail (and rain) amount; if depletion was already significant, doubling IN might decrease hail size. The T-28 data reveal great local variability in the depletion time constants in the middle levels of thunderstorms (discussed above), so the actual results could fluctuate widely.

There also appear to be neither evidence nor strong reasoning to support an argument that ice multiplication and aggregation systematically counteract the results of varying the IN population. Aggregation could do this if there was a great deal of ice already, but that is not the case in many of the updrafts. Multiplication might be expected to amplify IN variations, not compensate for them. Thus, while the clouds are complex and interactions between the microphysical processes and the cloud kinematics make all arguments of this kind rather tenuous, the weight of expectation favors the position that ice nucleus concentration is an important factor in the hail production in storms dominated by the ice process, as on the U.S. high plains. The argument goes no further than that, though. There is no general argument of this kind to suggest what the effect is, or even whether it is consistent from storm to storm.

The picture is less clear in maritime severe storms with warmer bases, where primary liquid processes are much more involved in the formation of accretors. Giant or ultragiant CCN (see the review by Beard and Ochs 1993) might fill some of the role of IN in initiating the collectors, and secondary ice formation processes are likely to be more active. As noted above, the relative rarity of hail in maritime severe storms may result in part from these factors, but the whole picture is very complex.

Microphysical variability is probably important for hail, but the specifics of that importance are poorly understood.

6.11. Approaches

The rarity and small scale of hailfall are major obstacles to research. In particular, it can be next to impossible in practice to obtain good, coordinated datasets on the truly severe hailstorms (severe in size and/or amount of the hailstones) using radars on the ground and aircraft for direct probing, because they are simply too infrequent. The experience in the National Hail Research Experiment is instructive, in that of the 32 hailstorms sampled in three years of intensive effort in a relatively small area, no hail bigger than 2.5 cm in diameter was found (Knight and Knight 1979). The only storms that did produce larger hail were outside the network of hail-sensing instruments and were relatively poorly documented.

Unfortunately, there are sound reasons to concentrate research on the severest storms. They are far more important than average hailstorms in economic terms,[7] and there

[7] With increasing population and urbanization, economically disastrous hailstorms are becoming more frequent. Recent storms in Edmonton and Denver caused damage running into hundreds of millions of dollars, mainly to roofs and automobiles (Charlton et al. 1995). These are storms with frequencies per any single metropoli-

is every expectation that they might be more systematic in their production of hail as well. BWERs are one suggestion of this, and big storms with broad, strong updrafts would be expected in general to be less chaotic than ordinary airmass thunderstorms. This is especially true, of course, for organized supercells with their own mesocyclonic circulations.

What realistic approaches are there to acquire useful data on the formation of hail, especially in truly severe hailstorms? The detailed case study approach with coordinated radars and aircraft, employed in NHRE (northeastern Colorado), CCOPE (easternern Montana), and to some extent at NSSL in Oklahoma (the emphasis there has traditionally been tornadoes, not hail), can still be valuable and can be improved with new radar techniques (see below), but a lot of waiting is involved to get much information.

Three present approaches may produce valuable new insights into the unsolved problems in the near future: airborne radar, multiparameter radar, and numerical modeling. These all have strong potential for future advances of knowledge.

1. Airborne Doppler radar has the advantage of being mobile, allowing the severest storms over a wide area to be observed at close range. In this way one can obtain very good spatial resolution to examine details of the reflectivity patterns (Wakimoto et al. 1996). Some of the time resolution is lost, and also some of the ability to deduce three-dimensional wind fields that is available from two or more Doppler radars at the ground, but only for storms that cooperate by being within a small area. Multiparameter techniques (discussed below), to the extent that they enable remote hail detection and some measure of hail size, may provide new insights when they are used with airborne radar.
2. Multiparameter radar. Advanced radar techniques to characterize the forms of precipitation at a distance and to quantify even roughly the amounts and sizes of hailstones within the storms can have great value for hailstorm research. This is a rapidly developing, highly technical subject with many recent references, and no semblance of a review is offered here. The reader may refer to papers in the review edited by Atlas (1990), to Zrnić et al. (1993), to the more recent of the case studies referenced above, to Vivekanandan et al. (1999), and to recent AMS Radar Conference Proceedings volumes to enter into the literature on these methods.

The polarization techniques provide measures of the shapes and orientations of the hydrometeors responsible for dominating the backscattered radar signal. Z_{DR} is a comparison of the average horizontal to the average vertical dimensions of the hydrometeors. Echoes from raindrops bigger than about 1 mm diameter have positive Z_{DR} because raindrops are deformed by the drag of the air into flattened (oblate) shapes, longer in the horizontal than in the vertical. Hail, however, is usually found to have negative Z_{DR}, so the stones must fall with the vertical dimension greater on the average than the horizontal, though the situation is complicated because large hailstones can have various shapes and their tumbling motions are complex. LDR is the depolarization of the backscattered radiation from a polarized, transmitted beam, and this reveals information on both hydrometeor asymmetry and randomness of orientation in space. These and other radar quantities are becoming more available, particularly with advances in the ability to record and process the immense amounts of data. They all act to lessen the ambiguities of interpretation of the radar echo: to enable some hail detection, to start to enable some deductions about hailstone size, and to start to make better measurements of precipitation amounts.

The observational potential here is considerable for the hail problem. In the detailed case studies, if hail within storms can be detected with two- or three-minute, several-hundred-meter resolutions in time and space and with some size information, it would have great value in testing ideas about hail formation.

Hail climatology would also offer valuable information about general characteristics of hailstorms and rainstorms, if only one had routine, reliable data on hailfall that included measures of hail amount and size. The multiparameter radar techniques offer hope in this direction, using research radars and perhaps ultimately operational radar networks like the NEXRAD system. Likewise, if new radar techniques for estimating rainfall as distinct from hail (differential phase shift—K_{DP}—is one promising approach) are developed and prove reliable enough, issues of hail fall versus rainfall will be much more accessible to study. Remote-sensing techniques applied to LP storms might be used to quantify their rain to see just how low their precipitation amount really is, and they could be applied to storms in general, to see if there are storms with very strong updrafts that produce great amounts of rain but little or no hail. These kinds of issues are very difficult to address with present means of documenting precipitation.

Interesting results are already being obtained. One is the finding of positive Z_{DR} shafts within updrafts of storms, extending above the freezing level (e.g., Conway and Zrnić 1993; Brandes et al. 1995; Bringi et al. 1997; and several of the case studies cited above). Clearly raindrops are being carried aloft, to freeze and (potentially) become frozen-drop hailstone embryos.

---

tan area of probably one in many tens of years, if not hundreds of years, making them exceedingly difficult to study in any detail, yet they are the storms that it would be most worthwhile to forecast and (if possible) to mitigate. These storms are becoming more significant for insurance losses than the much more widespread agricultural damage caused largely by the numerous, more moderate hailstorms.

Smith et al. (1999) point out that the upper parts of these Z_{DR} shafts represent partly frozen drops. The multiparameter techniques are under active development at this time and are beginning to be applied fruitfully to hailstorm study.

3. Numerical modeling. We have commented that the results of hail suppression operations or tests do not provide information useful to hailstorm research because they have all been inconclusive. That a particular suppression technique does not produce detectable results is a piece of information with little or no content of knowledge about the hail process. The limits of detectability are too wide: The statistics are so unfavorable that only huge effects, positive or negative, can be detected. This comment is reiterated here because identical considerations apply to the large numerical models of hailstorms that include interactions between hydrodynamical and microphysical processes (e.g., Farley and Orville 1986; Farley 1987; Kubesh et al. 1988; Wang and Chang 1993; Johnson et al. 1994). At the present time, these models are necessarily rather crude in their representation of microphysical processes, and perhaps especially in treating the large-diameter tail of the hydrometeor size spectrum, the hail. They typically use large amounts of computer resources, and they typically compare fairly well with observation when intended to simulate observed hailstorms. However, testing a storm modeling scheme would entail much the same kind of objective, statistical comparison of model results with a number of different cases as does testing a hail suppression technique. This has not been attempted, and it probably cannot be accomplished usefully at this stage of either the modeling itself or the ability to get the observations needed for the test. Thus it seems impractical to use these models to explore questions like the relation between ice nuclei and hail, until they reach the point that their results can be trusted in considerable detail, without any test. That point clearly has not yet arrived.

If this sounds pessimistic, it also needs to be said that the numerical modeling approach is probably the only way of treating the most fundamental questions about hailfall from storms. The model must be cheap enough to run many times, varying microphysical inputs—CCN, giant nuclei, IN, secondary ice production schemes, and so forth—for determining the importance of variability in the nuclei as well as the sensitivity of hail and rain formation to the wide array of microphysical factors that may be important. Model studies published to date are promising steps in this direction, but it is still a goal to use them to learn more about how hailstorms work and to discover simplifying and unifying relationships.

6.12. Storm updrafts and weak-echo regions; conclusions

A valuable device in problem-solving research is to return repeatedly and in different ways to the fundamentals of the problem. Two of the most fundamental phenomena for hail production in storms are the updraft, where the hailstones must grow, and weak-echo regions, which are evidently portions of the updrafts within cloud that are deficient in collectors, and therefore regions where more hailstones would grow if appropriate embryos were supplied. Graphic representations of updrafts and their relations to weak-echo regions at cloud base were presented by Fankhauser (1988) and are reproduced in Fig. 6.13. The updraft is represented here by the upward flux of water vapor and the radar echo by dBZ, and one sees that at cloud base the updrafts of these six storms are mostly but not completely echo-free (meaning less than 25 dBZ). These figures correspond well with the visual aspect of many storms viewed from the ground: a flat, dark cloud base bordered on one side by a solid-looking gray wall of precipitation extending to the ground. Doswell and Burgess (1993) present schematics of low-precipitation, classic, and high-precipitation supercells that are distinguished primarily by the relative locations of the precipitation and the storm inflow in the region at and below cloud base.

Figure 6.13 can be used as a device to illustrate most of the things one would like to know about hail production within storms, all of which concern the interrelations between the precipitation and the updraft. One would like to know the size and phases of the accretors within the updraft (of which the radar reflectivity alone gives an inadequate estimate); the distribution of cloud water within the updraft (largely invisible to the radar); all of this along with updraft velocity distribution; the same information in three spatial dimensions, not just one slice; and how it changes with time. Hail formation in storms is poorly understood because a woefully small part of this information is known or even knowable using present observational techniques. Multiparameter radar developments may help a lot in supplying some parts of this picture.

The height at which the updraft fills in with the radar echo is probably critical in hailstorms, and a radar method is available for this but has not been utilized. Bragg scattering makes cloud boundaries visible in the range of 0–10 dBZ at wavelengths of 5 and 10 cm, so the profile of the outer cloud edge on the side away from the strong echo is observable, even when the cloud does not contain precipitation (Knight and Miller 1993). The cloud boundary itself shows as an echo maximum. This effect was not appreciated at the time of most of the case studies, and an apparent example appears in Miller et al. (1990, Fig. 3c).

Storms may lack WERs because of hail so large that it can fall throughout the updraft (the 1 August 1981 CCOPE storm appears to be an example) and it might

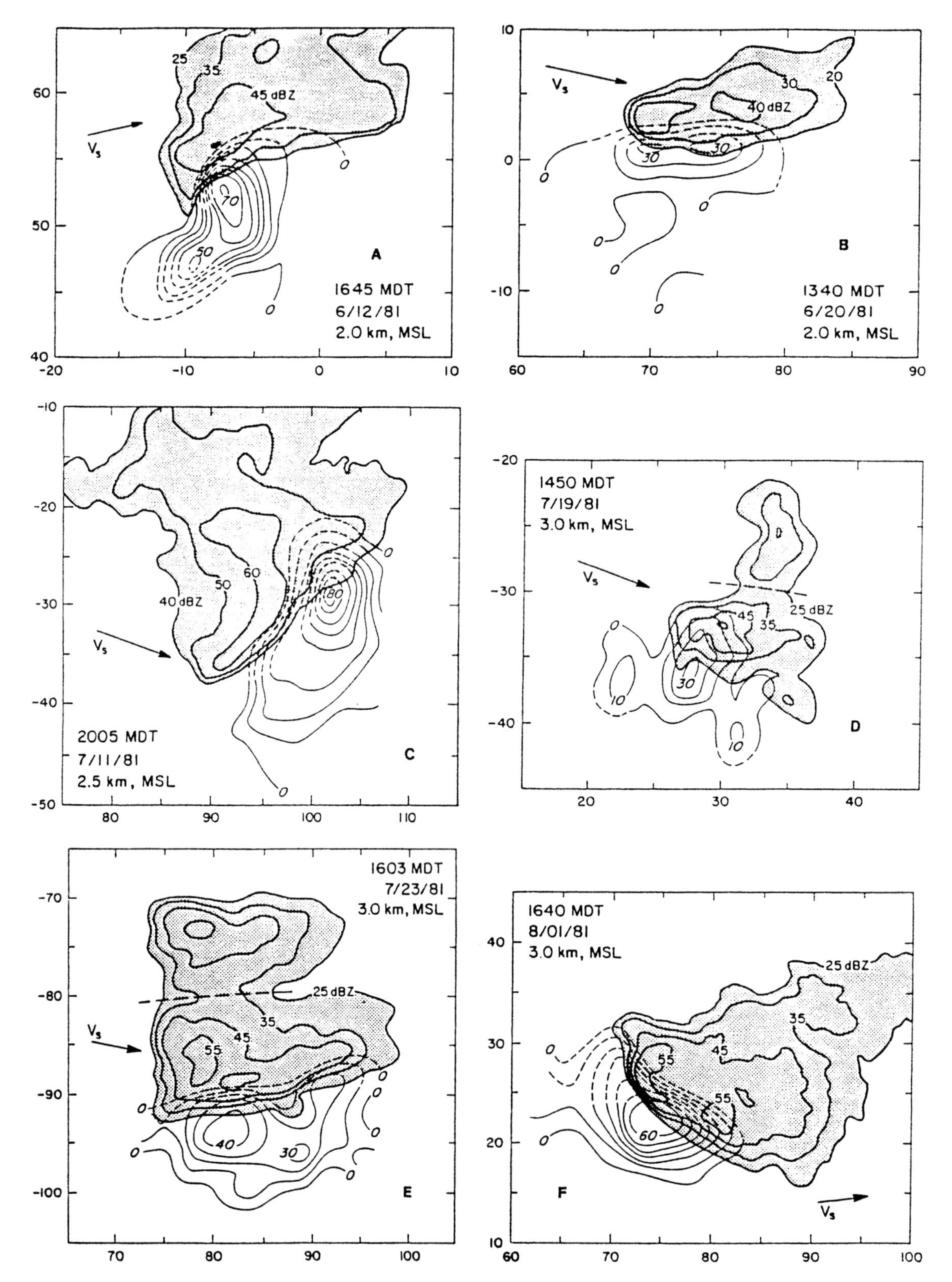

FIG. 6.13. Cloud-base radar reflectivity patterns (shaded area with heavy contours) and contours of vertical water vapor flux in increments of 10 g m^{-2} s^{-1} for six storms in CCOPE, eastern Montana. Map scale in km, and storm motion vectors (V_S) have a scale of 5 km = 10 m s^{-1}. From Fankhauser (1988).

be possible in some cases for great amounts of rain to be ingested throughout an updraft at low levels, in quantities that would provide echo throughout and might even deplete the cloud water so effectively that hail could not form at all. It would be interesting if storms with very powerful updrafts exist that produce copious rain but no hail at all. They would be the high-precipitation counterpart to the LP storms, which presumably produce hail but little rain: They would not possess the dark, rain-free cloud bases at all.

All of the extreme cases—storms with very large hail, great quantities of hail, hail but no rain, or rain and strong updrafts but little or no hail—are worthy of intense study as extreme forms of the kinds of storm organization that distinguish storms with different precipitation behavior. Their study could add much to knowledge of the factors that encourage or discourage hail formation in storms.

REFERENCES

Atlas, D., Ed., 1990: *Radar in Meteorology.* Amer. Meteor. Soc., 806 pp.

Amburn, S. A., and P. L. Wolf, 1997: VIL density as a hail indicator. *Wea. Forecasting,* **12,** 473–478.

Battan, L. J., 1975: Doppler radar observations of a hailstorm. *J. Appl. Meteor.,* **14,** 98–108.

Beard, K. V., and H. T. Ochs III, 1993: Warm-rain initiation: An overview of microphysical mechanisms. *J. Appl. Meteor.,* **32,** 608–625.

Billet, J., M. DeLisi, and B. G. Smith, 1997: Use of regression techniques to predict hail size and the probability of large hail. *Wea. Forecasting,* **12,** 154–164.

Black, R. A., and J. Hallett, 1986: Observations of the distribution of ice in hurricanes. *J. Atmos. Sci.,* **43,** 802–822.

Blackmore, W. H., III, D. J. Musil, P. L. Smith, and A. Waldvogel, 1989: Spatial and temporal variations of the interior characteristics of Swiss thunderstorms. *Atmos. Res.,* **23,** 135–161.

Bluestein, H. B., and C. R. Parks, 1983: A synoptic and photographic climatology of low-precipitation severe thunderstorms in the southern plains. *Mon. Wea. Rev.,* **111,** 2034–2046.

——, and G. R. Woodall, 1990: Doppler-radar analysis of a low-precipitation severe storm. *Mon. Wea. Rev.,* **118,** 1640–1664.

Brandes, E. A., J. Vivekanandan, J. D. Tuttle, and C. J. Kessinger, 1995: A study of thunderstorm microphysics with multiparameter radar and aircraft observations. *Mon. Wea. Rev.,* **123,** 3129–3143.

Briggs, G. A., 1968: Hailstones, starfish and daggers—Spiked hail falls in Oak Ridge, Tennessee. *Mon. Wea. Rev.,* **96,** 744–745.

Bringi, V. N., L. Liu, P. C. Kennedy, V. Chandrasekar, and S. A. Rutledge, 1996: Dual multiparameter radar observations of intense convective storms: The 24 June 1992 case study. *Meteor. Atmos. Phys.,* **59,** 3–31.

———, K. Knupp, A. Detwiler, L. Liu, I. J. Caylor, and R. A. Black, 1997: Evolution of a Florida thunderstorm during the Convection and Precipitation/Electrification Experiment: the case of 9 August 1991. *Mon. Wea. Rev.,* **125,** 2131–2160.

Browning, K. A., 1963: The growth of large hail within a steady updraught. *Quart. J. Roy. Meteor. Soc.,* **89,** 490–506.

———, 1967: The growth environment of hailstones. *Meteor. Mag.,* **96,** 202–211.

———, 1977: The structure and mechanisms of hailstorms. *Hail: A Review of Hail Science and Hail Suppression, Meteor. Monogr.,* No. 38, Amer. Meteor. Soc., 1–43.

———, and G. B. Foote, 1976: Airflow and hail growth in supercell storms and some implications for hail suppression. *Quart. J. Roy. Meteor. Soc.,* **102,** 499–533.

———, J-P. Chalon, P. J. Eccles, R. C. Strauch, F. H. Merrem, D. J. Musil, E. L. May and W. R. Sand, 1976: Structure of an evolving hailstorm. Part V: Synthesis and implications for hail growth and hail suppression. *Mon. Wea. Rev.,* **104,** 603–610.

Brownscombe, J. L., and J. Hallett, 1967: Experimental and field studies of precipitation particles formed by the freezing of supercooled water. *Quart. J. Roy. Meteor. Soc.,* **93,** 455–473.

Byers, H. R., and R. R. Braham, Jr., 1949: *The Thunderstorm.* U.S. Govt. Printing Office, 287 pp.

Carey, L. D., and S. A. Rutledge, 1996: A multiparameter radar case study of the microphysical and kinematic evolution of a lightning producing storm. *Meteor. Atmos. Phys.,* **59,** 33–64.

———, and ———, 1998: Electrical and multiparameter radar observations of a severe hailstorm. *J. Geophys. Res.,* **103,** 13 979–14 000.

Carte, A. E., 1979: Some comparisons between hailstorms on the Transvaal Highveld and those elsewhere. *J. Rech. Atmos.,* **13,** 309.

Changnon, S. A., Jr., 1977: Scales of hail. *J. Appl. Meteor.,* **16,** 626–648.

Charlton, R. B., B. M. Kachman, and L. Wojtiw, 1995: Urban hailstorms: A view from Alberta. *Natural Hazards,* **12,** 29–75.

Cheng, L., and D. C. Rogers, 1988: Hailfalls and hailstorm feeder clouds—An Alberta case study. *J. Atmos. Sci.,* **45,** 3533–3545.

Chisholm, A. J., and J. H. Renick, 1972: The kinematics of multicell and supercell Alberta hailstorms. Alberta Hail Studies, 1972. Research Council of Alberta Hail Studies Rep. 72-2, 24–31.

Conway, J. W., and D. S. Zrnić, 1993: A study of embryo production and hail growth using dual-Doppler and multiparameter radars. *Mon. Wea. Rev.,* **121,** 2511–2528.

Court, A., and J. F. Griffiths, 1985: Thunderstorm climatology. *Thunderstorm Morphology and Dynamics,* E. Kessler, Ed., University of Oklahoma Press, 9–38.

Doswell, C. A. III, and D. W. Burgess, 1993: Tornadoes and tornadic storms: A review of conceptual models. *The Tornado: Its Structure, Dynamics, Prediction, and Hazards, Geophys. Monogr.,* No. 79, Amer. Geophys. Union, 161–172.

Dye, J. E., C. A. Knight, V. Toutenhoofd, and T. W. Cannon, 1974: The mechanism of precipitation formation in northeastern Colorado cumulus, III. Coordinated microphysical and radar observations and summary. *J. Atmos. Sci.,* **31,** 2152–2159.

———, B. E. Martner, and L. J. Miller, 1983: Dynamical-microphysical evolution of a convective storm in a weakly-sheared envirenment. Part I: Microphysical observations and interpretation. *J. Atmos. Sci.,* **40,** 2083–2096.

Fankhauser, J. C., 1988: Estimates of thunderstorm precipitation efficiency from field measurements in CCOPE. *Mon. Wea. Rev.,* **116,** 663–684.

Farley, R. D., 1987: Numerical modeling of hailstorms and hailstone growth. Part III: Simulation of an Alberta hailstorm—Natural and seeded cases. *J. Climate Appl. Meteor.,* **26,** 789–812.

———, and H. D. Orville, 1986: Numerical modeling of hailstorms and hailstone growth. Part I: Preliminary model verification and sensitivity tests. *J. Climate Appl. Meteor,* **25,** 2014–2035.

Federer, B., and A. Waldvogel, 1978: Time-resolved hailstone analyses and radar structure of Swiss storms. *Quart. J. Roy. Meteor. Soc.,* **104,** 69–90.

———, J. Jouzel, and A. Waldvogel, 1978: Hailstone trajectories determined from crystallography, deuterium content and radar backscattering. *Pageophys.,* **116,** 112–129.

———, B. Thalmann, and J. Jouzel, 1982: Stable isotopes in hailstones. Part II: Embryo and hailstone growth in different storms. *J. Atmos. Sci.,* **39,** 1336–1355.

Foote, G. B., 1984: A study of hail growth utilizing observed storm conditions. *J. Climate Appl. Meteor.,* **23,** 84–101.

———, 1985: Aspects of cumulonimbus classification relevant to the hail problem. *J. Rech. Atmos.,* **19,** 61–74.

———, and J. C. Fankhauser, 1973: Airflow and moisture budget beneath a northeast Colorado hailstorm. *J. Appl. Meteor.,* **12,** 1330–1353.
———, and C. A. Knight, Eds., 1977: *Hail: A Review of Hail Science and Hail Suppression. Meteor Monogr.,* No. 38, Amer. Meteor. Soc., 277 pp.
Garcia-Garcia, F., and R. List, 1992: Laboratory measurements and parameterizations of supercooled water skin temperatures and bulk properties of gyrating hailstones. *J. Atmos. Sci.,* **49,** 2058–2073.
Grenier, J.-C., P. Admirat, and S. Zair, 1983: Hailstone growth trajectories in the dynamic evolution of a moderate hailstorm. *J. Climate Appl. Meteor.,* **22,** 1008–1021.
Hallett, J., R. I. Sax, D. Lamb, and A. S. Ramachandra Murty, 1978: Aircraft measurements of ice in Florida cumuli. *Quart. J. Roy. Meteor. Soc.,* **104,** 631–651.
Heymsfield, A. J., 1983: Case study of a hailstorm in Colorado. Part IV: Graupel and hail growth mechanisms deduced through particle trajectory calculations. *J. Atmos. Sci.,* **40,** 1482–1509.
———, and M. R. Hjelmfelt, 1984: Processes of hydrometeor development in Oklahoma convective clouds. *J. Atmos. Sci.,* **41,** 2811–2835.
Höller, H., V. N. Bringi, J. Hubbert, M. Hagen, and P. F. Meischner, 1994: Life cycle and precipitation formation in a hybrid-type hailstorm revealed by polarimetric and Doppler radar measurements. *J. Atmos. Sci.,* **51,** 2500–2522.
Hubbert, J., V. N. Bringi, L. D. Carey, and S. Bolen, 1998: CSU-CHILL polarimetric radar measurements from a severe hail storm in eastern Colorado. *J. Appl. Meteor.,* **37,** 749–775.
Huschke, R. E., Ed., 1959: *Glossary of Meteorology.* Amer. Meteor. Soc., 638 pp.
Johns, R. H., and C. A. Doswell III, 1992: Severe local storms forecasting. *Wea. Forecasting,* **7,** 588–612.
———, and J. A. Hart, 1998: The occurrence and non-occurrence of large hail with strong and violent tornado episodes: Frequency distributions. Preprints, *19th Conf. on Severe Local Storms,* Minneapolis, MN, Amer. Meteor. Soc., 285–286.
Johnson, D. E., P. K. Wang, and J. M. Straka, 1994: A study of microphysical processes in the 2 August 1981 CCOPE supercell storm. *Atmos. Res.,* **33,** 93–123.
Kelly, D. L., J. T. Schaefer, and C. A. Doswell III, 1985: Climatology of nontornadic severe thunderstorm events in the United States. *Mon. Wea. Rev.,* **113,** 1997–2014.
Kennedy, P. C., and S. A. Rutledge, 1995: Dual-Doppler and multiparameter radar observations of a bow-echo hailstorm. *Mon. Wea. Rev.,* **123,** 921–943.
Knight, C. A., 1984: Radar and other observations of two vaulted storms in northeast Colorado. *J. Atmos. Sci.,* **41,** 258–271.
———, 1987: Precipitation formation in a convective storm. *J. Atmos. Sci.,* **44,** 2712–2726.
———, and N. C. Knight, 1974: Drop freezing in clouds. *J. Atmos. Sci.,* **31,** 1174–1176.
———, and N. C. Knight, 1979: Results of a randomized hail suppression experiment in northeast Colorado. Part V: Hailstone embryo types. *J. Appl. Meteor.,* **18,** 1583–1588.
———, and K. R. Knupp, 1986: Precipitation growth trajectories in a CCOPE storm. *J. Atmos. Sci.,* **43,** 1057–1073.
———, and L. J. Miller, 1993: First radar echoes from cumulus clouds. *Bull. Amer. Meteor. Soc.,* **74,** 179–188.
———, T. Ashworth, and N. C. Knight, 1978: Cylindrical ice accretions as simulations of hail growth: II. The structure of fresh and annealed accretions. *J. Atmos. Sci.,* **35,** 1997–2009.
———, G. B. Foote, and P. W. Summers, 1979: Results of a randomized hail suppression experiment in northeast Colorado. Part IX: Overall discussion and summary in the context of physical research. *J. Appl. Meteor.,* **18,** 1629–1639.
———, N. C. Knight, and K. A. Kline, 1981: Deuterium contents of storm inflow and hailstone growth layers. *J. Atmos. Sci.,* **38,** 2485–2499.
———, L. J. Miller, N. C. Knight, and D. Breed, 1982: The 22 June 1976 case study: Precipitation formation. *Hailstorms of the Central High Plains,* Vol. II, *Case Studies of the National Hail Research Experiment,* C. A. Knight and P. Squires, Eds., Colorado Associated Universities Press, 61–89.
Knight, N. C., 1981: Climatology of hailstone embryos. *J. Appl. Meteor.,* **20,** 750–755.
———, and M. English, 1980: Patterns of hailstone embryo type in Alberta hailstorms. *J. Rech. Atmos.,* **14,** 325–332.
Kubesh, R. J., D. J. Musil, R. D. Farley, and H. D. Orville, 1988: The 1 August 1981 CCOPE storm: Observations and modelling results. *J. Appl. Meteor.,* **27,** 216–243.
Krauss, T. W., and J. D. Marwitz, 1984: Precipitation processes within an Alberta supercell hailstorm. *J. Atmos. Sci.,* **41,** 1025–1034.
Lamb, D., 2001: Rain production in convective storms. *Severe Convective Storms, Meteor. Monogr.,* No. 50, Amer. Meteor. Soc., 299–321.
Lesins, G. B., and R. List, 1986: Sponginess and drop shedding of gyrating hailstones in a pressure-controlled icing wind tunnel. *J. Atmos. Sci.,* **43,** 2813–2825.
Levi, L., and A. N. Aufdermaur, 1970: Crystallographic orientation and crystal size in cylindrical accretions of ice. *J. Atmos. Sci.,* **22,** 443–452.
———, and L. Lubart, 1998: Modelled spongy growth and shedding process for spheroidal hailstones. *Atmos. Res.,* **47–48,** 59–68.
List, R., 1959: Wachstum von Eis-Wassergemischen im Hagelversuchskanal. *Helv. Phys. Act.,* **32,** 293–296.
———, 1985: Properties and growth of hailstones. *Thunderstorm Morphology and Dynamics,* E. Kessler, Ed., University of Oklahoma Press, 259–276.
Lopez, R. E., and J.-P. Aubagnac, 1997: The lightning activity of a hailstorm as a function of changes in its microphysical characteristics inferred from polarimetric radar observations. *J. Geophys. Res.,* **102,** 16 799–16 813.
Lucas, C., E. J. Zipser, and M. A. LeMone, 1994a: Convective available potential energy in the environment of oceanic and continental clouds: Correction and comments. *J. Atmos. Sci.,* **51,** 3829–3830.
———, ———, and ———, 1994b: Vertical velocity in oceanic convection off tropical Australia. *J. Atmos. Sci.,* **51,** 3183–3193.
———, ———, and ———, 1996: Reply. *J. Atmos. Sci.,* **53,** 1212–1214.
Ludlam, F. H., 1958: The hail problem. *Nubila,* **1,** 12–96.
MacGorman, D. R., and D. W. Burgess, 1994: Positive cloud-to-ground lightning in tornadic storms and hailstorms. *Mon. Wea. Rev.,* **122,** 1671–1697.
Macklin, W. C., 1977: The characteristics of natural hailstones and their interpretation. *Hail: A Review of Hail Science and Hail Suppression, Meteor. Monogr.,* No. 38, Amer. Meteor. Soc., 65–88.
Marwitz, J. D., 1972a: Precipitation efficiency of thunderstorms on the High Plains. *J. Rech. Atmos.,* **6,** 367–370.
———, 1972b: The structure and motion of severe hailstorms. Part I: Supercell storms. *J. Appl. Meteor.,* **11,** 166–179.
———, 1972c: The structure and motion of severe hailstorms. Part II: Multicell storms. *J. Appl. Meteor.,* **11,** 180–188.
———, 1972d: The structure and motion of severe hailstorms. Part III: Severely sheared storms. *J. Appl. Meteor.,* **11,** 189–201.
Meaden, G. T., 1995: The fall of ice-coated coke, clinker, coal, grit and dust from the hailstorm-cumulonimbus of 5 June 1983 over Poole, Bournemouth and neighbouring regions. *J. Meteor.,* **20,** 367–380.
Michaud, L. M., 1996: Comments on "Convective available potential energy in the environment of oceanic and continental clouds." *J. Atmos. Sci.,* **53,** 1209–1211.
Miller, L. J., and J. C. Fankhauser, 1983: Radar echo structure, air motion and hail formation in a large stationary multicellular thunderstorm. *J. Atmos. Sci.,* **40,** 2399–2418.

———, J. E. Dye, and B. E. Martner, 1983: Dynamical-microphysical evolution of a convective storm in a weakly-sheared environment. Part II: Airflow and precipitation trajectories from Doppler radar observations. *J. Atmos. Sci.,* **40,** 2097–2109.

———, J. D. Tuttle, and C. A. Knight, 1988: Airflow and hail growth in a severe Northern High Plains supercell. *J. Atmos. Sci.,* **45,** 736–762.

———, ———, and G. B. Foote, 1990: Precipitation production in a large Montana hailstorm: Airflow and particle growth trajectories. *J. Atmos. Sci.,* **47,** 1619–1646.

Moore, J. T., and J. P. Pino, 1990: An interactive method for estimating maximum hailstone size from forecast soundings. *Wea. Forecasting,* **5,** 508–525.

Morgan, G. M., Jr., 1982: Precipitation at the ground. *Hailstorms of the Central High Plains.* Vol. 1, *The National Hail Research Experiment,* C. A. Knight and P. Squires, Eds., Colorado Associated Universities Press, 59–79.

———, and P. W. Summers, 1985: Hailfall and hailstorm characteristics. *Thunderstorm Morphology and Dynamics,* E. Kessler, Ed., University of Oklahoma Press, 237–257.

Musil, D. J., and P. L. Smith, 1986: Aircraft penetrations of Swiss hailstorms—an update. *J. Weather Mod.,* **18,** 108–111.

———, and ———, 1989: Interior characteristics at mid-levels of thunderstorms in the Southeastern United States. *Atmos. Res.,* **24,** 149–167.

———, A. J. Heymsfield, and P. L. Smith, 1986: Microphysical characteristics of a well-developed weak echo region in a High Plains supercell thunderstorm. *J. Climate Appl. Meteor.,* **25,** 1037–1051.

———, S. A. Christopher, R. A. Deola, and P. L. Smith, 1991: Some interior observations of southeastern Montana hailstorms. *J. Appl. Meteor.,* **30,** 1596–1612.

Nelson, S. P., 1983: The influence of storm flow structure on hail growth. *J. Atmos. Sci.,* **40,** 1965–1983.

———, 1987: The hybrid multicellular-supercellular storm—an efficient hail producer. Part II: General characteristics and implications for hail growth. *J. Atmos. Sci.,* **44,** 2060–2073.

———, and S. K. Young, 1979: Characteristics of Oklahoma hailfalls and hailstorms. *J. Appl. Meteor.,* **18,** 339–347.

———, and N. C. Knight, 1987: The hybrid multicellular-supercellular storm—an efficient hail producer. Part I: An archetypal example. *J. Atmos. Sci.,* **44,** 2042–2059.

Paluch, I. R., 1978: Size sorting of hail in a three-dimensional updraft and implications for hail suppression. *J. Appl. Meteor.,* **17,** 763–777.

Pflaum, J. C., 1984: New clues for decoding hailstone structure. *Bull. Amer. Meteor. Soc.,* **65,** 583–593.

Prodi, F., G. Santachiara, and A. Franzini, 1986: Properties of ice accreted in two-stage growth. *Quart. J. Roy. Meteor. Soc.,* **112,** 1057–1080.

Rasmussen, E. N., and J. M. Straka, 1998: Variations of supercell morphology. Part I: Observations of the role of upper-level storm-relative flow. *Mon. Wea. Rev.,* **126,** 2406–2421.

Rasmussen, R. M., and A. J. Heymsfield, 1987: Melting and shedding of graupel and hail. Part III: Investigation of the role of shed drops as hail embryos in the 1 August CCOPE severe storm. *J. Atmos. Sci.,* **44,** 2783–2803.

Raymond, D. J., and A. M. Blyth, 1989: Precipitation development in a New Mexico thunderstorm. *Quart. J. Roy. Meteor. Soc.,* **115,** 1397–1423.

Rogers, R. R., and M. K. Yau, 1989: *A Short Course in Cloud Physics.* 3d ed. Elsevier, 290 pp.

Roos, D. V.D. S., H. Schooling, and J. C. Vogel, 1977: Deuterium in hailstones collected on 29 November 1972. *Quart. J. Roy. Meteor. Soc.,* **103,** 751–767.

Smith, P. L., D. J. Musil, A. G. Detwiler, and R. Ramachandran, 1999: Observations of mixed-phase precipitation within a CaPE thunderstorm. *J. Appl. Meteor.,* **38,** 145–155.

Sulakvelidze, G. K., B. I. Kiziriya, and V. V. Tsykunov, 1974: Progress of hail suppression work in the U.S.S.R. *Weather and Climate Modification,* W. N. Hess, Ed., Wiley, 410–431.

Takahashi, T., 1987: Hawaiian hailstorms, 30 January 1985. *Bull. Amer. Meteor. Soc.,* **68,** 1530–1534.

Vali, G., D. C. Rogers, and J. E. Dye, 1982: Aerosols, cloud nuclei and ice nuclei. *Hailstorms of the Central High Plains.* Vol. 1. *The National Hail Research Experiment,* C. A. Knight and P. Squires, Eds., Colorado Associated Universities Press, 35–58.

Vivekanandan, J., D. S. Zrnić, S. M. Ellis, R. Oye, A. V. Ryzhkov, and J. Straka, 1999: Cloud microphysics retrieval using S-band dual-polarization radar measurements. *Bull. Amer. Meteor. Soc.,* **80,** 381–406.

Wakimoto, R. M., W-C. Lee, H. B. Bluestein, C-H. Liu, and P. H. Hildebrand, 1996: ELDORA observations during VORTEX 95. *Bull. Amer. Meteor. Soc.,* **77,** 1465–1481.

Wang, C., and J. S. Chang, 1993: A three-dimensional numerical model of cloud dynamics, microphysics, and chemistry 2. A case study of the dynamics and microphysics of a severe local storm. *J. Geophys. Res.,* **98D,** 14 845–14 862.

Weisman, M. L., and J. B. Klemp, 1986: Characteristics of isolated convective storms. *Mesoscale Meteorology and Forecasting,* P. S. Ray, Ed., Amer. Meteor. Soc., 331–358.

Woodward, B., 1959: The motion in and around isolated thermals. *Quart. J. Roy. Meteor. Soc.,* **86,** 144–151.

Young, K. C., 1977: A numerical examination of some hail suppression concepts. *Hail: A Review of Hail Science and Hail Suppression, Meteor. Mongr.,* No. 38, Amer. Meteor. Soc., 195–214.

———, 1993: *Microphysical Processes in Clouds.* Oxford University Press, 427 pp.

Ziegler, C. L., P. S. Ray, and N. C. Knight, 1983: Hail growth in an Oklahoma multicell storm. *J. Atmos. Sci.,* **40,** 1768–1792.

Zrnić, D. S., V. N. Bringi, N. Balakrishnan, K. Aydin, V. Chandrasekar, and J. Hubbert, 1993: Polarimetric measurements in a severe hailstorm. *Mon. Wea. Rev.,* **121,** 2223–2238.

REVIEW PANEL RESPONSE

The panel considers the chapter to be a good exposition of the current state of scientific knowledge of hailstorm microphysics, with appropriate consideration of kinematic and dynamic factors. Applications (e.g., climatology, forecasting, detection and measurement, or hail suppression) receive less attention. The panel wishes to augment the chapter with some discussion of these and a few other topics.

6R.1. Climatology

There is some useful information available on the climatology of large hail. Kelly et al. (1985), using *Storm Data* and other supplementary sources, show the hourly, monthly, and geographical distribution of large hail within the 48 contiguous states of the United States for the period 1955–83. This was done for both "giant" (diameter equal to or greater than ¾ in., but less than 2 in.) and "enormous" (diameter greater than 2 in.) sized hail. Using *Storm Data* as a data source, Polston (1996) has also shown the distribution of 4 in. or greater diameter hail events for the 48 contiguous states for the period 1955–94. Changnon (1999) discusses the distribution of large hail in his climatology of hail risk.

6R.2. Hail at the ground

Hail has been observed at the ground since earliest times; one would expect such observations to reveal much about hailstorms and hailstone growth. The authors of this chapter point out, though, that this is largely not the case, even after several decades of dedicated research and numerous storm studies.

There are two principal aspects of observations of hail at the ground. The first, oldest, and most widespread of these is the location, frequency of occurrence, size, and severity of hailfalls that provides a climatology of hail. This information is crucial to understanding and dealing with the hazard of hailstorms as related to the broad meteorological characteristics that spawn these phenomena. This aspect is discussed in section 6R.1 above and in section 6R.7.

The second aspect relates to the detailed study of the hailstones themselves: their shapes, sizes, numbers, structure, centers, content, and isotopic and crystal make-up. The authors review and assimilate the extensive and detailed research works and present, in a concise and straightforward manner, the complex relationships that affect hailstone growth. This review provides a renewed vision of hailstone formation and growth hypotheses within the hailstorm.

One topic omitted is any discussion of the relationship of hailstone size distribution. Hailstone size distributions have been an integral part of measurements of hail in many research programs (e.g., Douglas 1963; Federer and Waldvogel 1978; Spahn and Smith 1976; Morgan 1982). Such observations provide important ground truthing for radar measurements and base data for storm intercomparisons; they also offer a potential of additional insight into hailstone growth mechanisms and hailstorm dynamics, as noted by Cheng and English (1983) and Cheng et al. (1985).

6R.3. Hail suppression

In discussing the possibilities for the deliberate suppression of hail through cloud seeding, the chapter authors limit their comments to the *beneficial competition* concept: additional hail embryos produced by glaciogenic seeding compete for the available supercooled water. The desired end result is a modified ice hydrometeor spectrum comprising greater numbers of smaller particles, which have an increased probability of melting prior to completion of their descent. There are, however, other useful concepts; their consideration helps us to understand the various ways in which hailfall might be modified and also serves to illustrate the great complexity of deep convection. Most operational hail suppression programs presently seek to apply these concepts through treatment of supercooled, but ice-free, cumulus congestus found on the flanks of mature storms, rather than the mature cells themselves (e.g., see Smith et al. 1997). Thus, the intent is to modify the cloud regions that would produce the hail embryos, not the main hail growth environment. Though the scale of the targeted clouds (cumulus congestus versus cumulonimbus) is significantly smaller, the task remains daunting.

A more complete list of hail suppression concepts, including beneficial competition, follows. There are many variations on these themes; the list is neither comprehensive nor exclusive.

1) *Beneficial competition.* Seeding must increase the embryo concentration such that resulting hailstones are small enough that they melt almost entirely during descent through the subcloud layer. The warmer the subcloud layer, the easier this task becomes. Though glaciogenic agents have historically been used for this purpose, it has been recently postulated that promotion of coalescence through hygroscopic seeding could provide a much larger population of raindrops within developing updrafts, which could in turn become frozen-drop embryos. Numerical modeling reported by Farley et al. (1996) suggests that the greatest numbers of competing embryos could be obtained through hygroscopic seeding, followed by glaciogenic seeding to ensure nucleation of the resulting drops.
2) *Early rainout.* This concept is based upon the premise that seeding will accelerate precipitation development, resulting in the "rainout" of still-small hydrometeors from convective turrets that have not yet developed updrafts strong enough to support the growth of hail. Ice-phase hydrometeor development begins sooner, causing particles to gain sufficient mass to fall through developing updrafts. This results in the formation of a precipitation shaft from what would have otherwise been rain-free cloud bases. The accelerated development of accretors would reduce cloud supercooled water content (from which hail might otherwise grow), and the subcloud precipitation shaft might alter and interfere with the subcloud storm inflow.
3) *Trajectory lowering.* In this concept, the developing precipitation embryos do not precipitate before encountering stronger updrafts, but, because of their greater mass, follow an altered (presumably lower) trajectory through the mature stages of the storm. The term "trajectory lowering" is not new (Dennis 1980); however, given the evidence supporting hail growth only within a relatively narrow temperature range (as summarized by the chapter authors), this concept might better be termed "trajectory altering." It is possible that a higher trajectory, as might be produced by generating many small ice particles, could also reduce hail production. Because those lower-mass hydrometeors are more likely to be transported far aloft and end up within the very cold cloud anvil, an altered higher trajectory may be undesirable if seeding is also

intended to increase rainfall. Trajectory lowering might be the result of unsuccessful attempts to achieve "early rainout"; the two concepts are closely related.

4) *Dynamic effects.* Treatment with glaciogenic agents of supercooled cumulus congestus within the flanking line will release more of the available convective instability in numerous, smaller turrets over a larger area, rather than within the local area of the dominant hail-producing updraft. While the resultant latent heat release would invigorate the treated congestus turrets, the ice concentration is also increased, leaving less supercooled liquid water to fuel subsequent latent heat release or build hailstones as the updraft matures (or when the treated cell is absorbed by the mature cell). Other dynamic effects might include enhanced downdrafts (as in early rainout) that might disrupt (undercut or cut off) the mature updraft, or the generation of outflow boundaries that trigger new convection.

A fifth concept, *glaciation of supercooled cloud water,* is no longer considered viable by most (World Meteorological Organization 1996). This is because of the very large amounts of seeding agent required to achieve complete glaciation, and the difficulty in dispersing the requisite agent within a large cloud volume (the dominant updraft). In addition, conversion of the updraft's cloud water to small, subprecipitation-sized ice would likely serve only to enhance the development of the anvil, potentially reducing total storm precipitation.

For any of these concepts to be effective, the right amount of seeding agent must be placed in the correct place at the correct time in the cloud's development. Failure to meet any of these three treatment requirements could compromise the effort, and knowledge of the "correct" circumstances is fragmentary. These concepts are predicated on the treatment of supercooled, ice-free cumulus congestus, therefore the "right clouds" and "right time" are broadly defined by the concept. The quantity of seeding agent required and its subsequent dispersion within the supercooled cloud volume are less certain, although recent research has been helpful in determining the latter (Reinking and Martner 1996; Stith et al. 1996).

A number of hailstorm studies (Brown and Meitín 1994; Eyerman-Torgerson and Brown 1995; Kaufman and Brown 1996) show progressions of cells, each of which, in turn, becomes dominant within its multicell storm, reinforcing the conclusion of the chapter authors that the "archetypal supercell" is rare. This is encouraging, as these hail suppression concepts are consistent with the idea that most hail embryos form relatively early in each cell's lifetime and grow into hail within the same cell as it matures.

Because of the common threads running through all of the concepts, and indeed, through the linkages of one cloud process to another, a seeding effect related to a single concept likely may not be achieved exclusive of all the others. Likewise, the case-by-case outcome of any treatment intended to reduce hailfall is not yet certain (Orville 1996). In spite of these uncertainties, a significant number of operational hail suppression efforts continue to be conducted in Africa, Asia, Europe, North America, and South America (World Meteorological Organization 1996). Historically, most of these programs have been intended to reduce crop hail damage. Recently, however, the desire to reduce property and casualty insurance claims has provided the impetus for new programs, particularly in larger urban areas.

6R.4. Numerical storm modeling

The chapter authors suggest that "the standard for numerical modeling must be high," but then, "it does not appear . . . practical to test models adequately by comparing their results with observations." Surely if models cannot be tested through comparison with observations, then it is fundamentally impossible to determine whether or not the standard of modeling is high! Perhaps what is needed is a methodology for comparing model predictions with incomplete observational data (might this be an analog of asynchronous data assimilation into NWP models?) or, better still, an objective approach to making qualitative comparisons between models and observations. It boils down to the old question about whether you would rather have a watch that was right twice a day or one that was right only once in ten thousand years. On the basis of a strict quantitative comparison, one would presumably choose the former (i.e., a watch that didn't work), whereas a more qualitative comparison would lead one to choose the latter (one that gains about a second a month). This story also illustrates the fact that it is, of course, very important to choose the right quantitative or qualitative basis for making comparisons between models and observations.

Hailstorm and hailstone modeling are progressing, although at a relatively slow pace. The papers by Farley and associates in the late 1980s (Farley and Orville 1986; Farley 1987a,b; Kubesh et al. 1988) show the development of hail in realistic storm airflow conditions. The formation of hailstone spectra throughout the hailstorm is depicted in some of the papers. The work by P. K. Wang and associates (Johnson et al. 1993, 1995) extended the hailstone and hailstorm modeling to three dimensions, using the microphysical techniques developed by Farley.

The papers just referenced treated the storm and the hailstones in a completely interactive fashion. Other studies have used prescribed motion and cloud water fields to trace the hailstone growth and trajectory (Miller et al. 1988, 1990). Realistic hailstone sizes and positions of fallout resulted.

6R.5. Radar detection and measurement

Many radar techniques or algorithms have been developed for linearly polarized radars [especially the Next Generation Weather Radar (NEXRAD) or Weather Surveillance Radar (WSR)-88D] using reflectivity and radial velocity data to infer the presence of hail in storms. A few of the more successful techniques also use other storm or environment information. The current NEXRAD Hail Detection Algorithm (Witt et al. 1998; Witt 1999) adds temperature profile information to provide forecasts of the probability of hail (of any size), the probability of severe hail (>¾ in.), and the maximum expected hail size. The algorithm assumes that high reflectivities (>40 dBZ) above the 0°C level probably indicate hail, and the algorithm gives more weight to very high reflectivities (>50 dBZ) above the −20°C level. Using the same assumption, the Layer Reflectivity Maximum (the maximum reflectivity from a specified height interval) NEXRAD products have been used to infer the existence of hail aloft. The mid- (24–33 kft) or high (33–60 kft) layer is used, depending on the height of the freezing level. Vertically integrated liquid (VIL) has long been used as a warning index for hail (and other severe weather). A relationship involving VIL and echo top (>18 dBZ) height, known as VIL density (Amburn and Wolf 1996), has shown good correlation with imminent hail falls in a few local studies.

Besides using radial velocity data to determine if a storm is a supercell (i.e., contains a persistent mesocyclone) and therefore is statistically likely to produce hail, storm-top divergence signatures have been correlated to maximum hail size potential (Witt and Nelson 1991). This technique assumes that the storm-top divergence is a measure of the updraft velocity, and that the updraft velocity is highly correlated to the maximum hailstone size. A three-body scatter spike (or hail flare echo; see Lemon 1995; Zrnić 1987; Wilson and Reum 1986, 1988 for a description) is most likely large hail; however, this phenomenon is much rarer than hail reports.

Experimental algorithms using polarimetric radar information have also been developed to infer the presence of hail in storms. Examples include the work of Höller et al. (1994) and Straka and Zrnić (1993). These multiparameter algorithms provide indications of particle types and sizes within the storms, though the observations tend to be reflectivity weighted and may thus obscure important smaller particles. The potential ability of these methods to delineate, at least in rough terms, the spatial distribution of graupel and hail in a storm offers hope of significantly increasing our understanding of important hailstorm processes.

Swiss investigators in Grossversuch IV found a strong correlation between the radar reflectivity factor and the hailstone kinetic energy flux (Waldvogel et al. 1978a,b). They make a strong case for the utility of this relationship in estimating hail impact energy, on both analytical and experimental grounds, though observations from the National Hail Research Experiment did not exhibit the same strong correlation. The difference is not fully resolved, but such a correlation should be useful in portraying major hail swaths to assist in damage surveys or the like.

6R.6. Storm microphysics

Another argument supports the observation that there are "higher percentages of frozen-drop embryos in hailstones from clouds with warmer cloud bases." If the cloud base is low, the updraft speed at the 0°C level may be substantial, capable of suspending small (or perhaps not so small) hailstones. The chapter authors argue that balanced hailstones in an updraft that increases with height are in an unstable equilibrium. This is true for hailstones that are not growing, but melting hailstones may be stably balanced because their terminal velocity will diminish as they are displaced downward (in part because of increasing air density and in part because of decreasing mass). If their terminal velocity diminishes faster than the updraft, they will be displaced back upward. Once displaced above the freezing level, rapid spongy growth and diminishing air density could (in principle) cause their terminal velocity to increase faster than the updraft, leading to a return to the freezing level. Thus it might be possible for ensembles of spongy hailstones to be trapped near the freezing level, alternately growing and partly melting to produce large drop embryos. There is no indication that anyone has looked for such a situation. Would they exhibit a bright band or some other characteristic radar signature?

This explanation could also possibly account for the higher frequency of frozen-drop embryos in large hailstones. Big hailstones mean strong updrafts. If raindrop embryos can be produced on an "assembly line" within the updraft at 0°C, surely they would outnumber graupel embryos, which must somehow conspire to be delivered into the updraft either from the side or from above (if the updraft is tilted). In any event, many of the graupel embryos would likely begin their growth as hailstones above the freezing level, leading to shorter growth times and smaller hail (or no hail at all if the hailstones with graupel embryos do not grow sufficiently to fall back through the updraft).

More generally, stability will depend on the relative magnitudes of dU/dz (the vertical gradient in updraft speed) and dv/dz (the vertical gradient of hailstone terminal velocity). For a growing hailstone, it is possible to show that the latter can be expressed by

$$\frac{dv}{dz} \approx \frac{wg}{3\rho_a C_d(U - v)} + \frac{vg}{2RT},$$

where w is liquid water content$_{\text{eff}}$, R is the specific gas constant for air, and T is air temperature. A growing hailstone that is displaced upward ($U - v > 0$) will return to its balance level if dv/dz exceeds dU/dz. This is possible since both terms on the right-hand side above are positive. If a growing hailstone is displaced downward ($U - v < 0$), then the first term on the right above is negative and is likely to dominate the second term. Hence, dv/dz will likely be negative, meaning that the terminal velocity will increase as the hailstone falls; this is an unstable situation. On the other hand, it is conceivable that the first term, though negative, will be quite small (if w is small, for example), and dv/dz could be positive due to the air density effect (the second term). If dU/dz were not too big, the growing hailstone could therefore be stable with respect to downward displacements as well. Perhaps this is quibbling, but it does suggest that the chapter authors' argument about instability of balance points below the level of maximum updraft speed need not always be true. In particular, it may not be true if the hailstone melts when displaced downward and grows when displaced upward (i.e., if the balance level is at 0°C).

6R.7. Synoptic-scale observations and hail forecasting

The chapter authors state that nearly all severe storms are associated with hail, either aloft or at the ground. Given the definition of a supercell that the authors use, it appears that the occurrence of hail at the surface varies from one supercell environment to another. Johns and Sammler (1989) found that the size of hailstones reported at the surface during tornado outbreaks (where it is assumed that mesocyclones are present) varies with the environment. Tornado outbreak environments displaying strong instability (typically found in the central United States) generally produce larger hailstones at the surface than those outbreak environments displaying weak instability (typically occurring east of the Great Plains region). Further, Johns and Doswell (1992) have noted that in some tornado outbreak situations (particularly in the southeastern United States in the cool season and those cases associated with tropical cyclones in the warm season), no large (¾ in. in diameter or greater) hail is reported at the ground. These general observations concerning the variations in hail size and occurrence with mesocyclone tornado situations could be helpful to forecasters considering the likelihood of hail occurrence with organized severe storms.

The chapter authors consider the fundamental question of why some convective storms produce hail at the ground, while others (at least externally similar) do not. The need for strong updrafts to support the growth of large hail seems well established. However, the converse, that strong updrafts necessarily lead to large hail, is less certain. Neither updraft speeds in storms nor maximum hailstone sizes are well observed, so the strength of the correlation remains putative. Nevertheless, much of the methodology for forecasting hail concerns maximum hail sizes and depends upon attempts to estimate the strength of storm updrafts.

Neither is maximum hailstone size alone a reliable indicator of potential damage, at least to crops. Other factors such as the number concentrations of the stones and the associated wind speeds can also be important. The maximum sizes may be more useful indicators of property damage.

Johns and Doswell (1992) discuss the state of hail forecasting in the early 1990s; little has changed since to improve hail forecasting. Climatology (diurnal, seasonal, and geographical) is still an important factor in the current forecasting process concerning large hail, and refinements to this factor are referenced above and in section 6R.1. Techniques based primarily on thermodynamic conditions have been developed and refined (e.g., Moore and Pino 1990) and have had reasonable success in "pulse" storm environments. However, as the chapter authors state, the complexities of hail formation in sheared environments make it difficult to forecast occurrence and size of hail with much accuracy for organized self-perpetuating storms.

REFERENCES

Amburn, S., and P. Wolf, 1996: VIL density as a hail indicator. Preprints, *18th Conf. on Severe Local Storms.* San Francisco, CA, Amer. Meteor. Soc., 581–585.

Brown, R. A., and R. J. Meitín, 1994: Evolution and morphology of two splitting thunderstorms with dominant left-moving members. *Mon. Wea. Rev.*, **122,** 2052–2067.

Changnon, S. A., 1999: Data and approaches for determining hail risk in the contiguous United States. *J. Appl. Meteor.*, **38,** 1730–1739.

Cheng, L., and M. English, 1983: A relationship between hailstone concentration and size. *J. Atmos. Sci.*, **40,** 204–213.

———, ———, and R. Wong, 1985: Hailstone size distributions and their relationship to storm thermodynamics. *J. Climate Appl. Meteor.*, **24,** 1059–1067.

Dennis, A. S., 1980: *Weather Modification by Cloud Seeding.* Academic Press, 267 pp.

Douglas, R. H., 1963: Recent hail research: A review. *Severe Local Storms, Meteor. Monogr.*, No. 27, Amer. Meteor. Soc., 157–167.

Eyerman-Torgerson, K. L., and R. A. Brown, 1995: The hail spike signature in the Carson, North Dakota hailstorm of 11–12 July 1989. Preprints, *27th Conf. on Radar Meteorology,* Vail, CO, Amer. Meteor. Soc., 80–82.

Farley, R. D., 1987a: Numerical modeling of hailstorms and hailstone growth. Part II: The role of low density riming growth in hail production. *J. Climate Appl. Meteor.,* **26,** 234–254.

———, 1987b: Numerical modeling of hailstorms and hailstone growth. Part III: Simulation of an Alberta hailstorm—Natural and seeded cases. *J. Climate Appl. Meteor.*, **26,** 789–812.

———, and H. D. Orville, 1986: Numerical modeling of hailstorms and hailstone growth. Part I: Preliminary model verification and sensitivity tests. *J. Climate Appl. Meteor.,* **25,** 2014–2035.

———, H. Chen, H. D. Orville, and M. R. Hjelmfelt, 1996: The numerical simulation of the effects of cloud seeding on hailstorms. Preprints, *13th Conf. on Planned and Inadvertent Weather Modification,* Atlanta, GA, Amer. Meteor. Soc., 23–30.

Federer, B., and A. Waldvogel, 1978: Time-resolved hailstone analyses and radar structure of Swiss storms. *Quart. J. Roy. Meteor. Soc.,* **104,** 69–90.

Höller, H., V. N. Bringi, J. Hubbert, M. Hagen, and P. F. Meischner, 1994: Life cycle and precipitation formation in a hybrid-type hailstorm revealed by polarimetric and Doppler radar measurements. *J. Atmos. Sci.,* **51,** 2500–2522.

Johns, R. H., and W. R. Sammler, 1989: A preliminary synoptic climatology of violent tornado outbreaks utilizing radiosonde standard level data. Preprints, *15th Conf. on Weather Analysis and Forecasting,* Monterey, CA, Amer. Meteor. Soc., 196–201.

———, and C. A. Doswell III, 1992: Severe local storms forecasting. *Wea. Forecasting,* **7,** 588–612.

Johnson, D. E., P. K. Wang, and J. M. Straka, 1993: Numerical simulation of the 2 August 1981 CCOPE supercell storm with and without ice microphysics. *J. Appl. Meteor.,* **32,** 745–759.

———, ———, and ———, 1995: A study of microphysical processes in the 2 August 1981 CCOPE supercell storm. *Atmos. Res.,* **33,** 93–123.

Kaufman, C. A., and R. A. Brown, 1996: Relationship between cloud-to-ground lightning and the evolution of the Elgin storm on 11–12 July 1989. Preprints, *18th Conf. on Severe Local Storms,* San Francisco, CA, Amer. Meteor. Soc., 483–487.

Kelly, D. L., J. T. Schaefer, and C. A. Doswell III, 1985: Climatology of nontornadic severe thunderstorm events in the United States. *Mon. Wea. Rev.,* **113,** 1997–2014.

Kubesh, R. J., D. J. Musil, R. D. Farley, and H. D. Orville, 1988: The 1 August 1981 CCOPE storm. Observations and modeling results. *J. Climate Appl. Meteor.,* **27,** 216–243.

Lemon, L. R., 1995: Recognition of the radar "three-body scatter spike" as a large hail signature. Preprints, *27th Conf. on Radar Meteorology,* Vail, CO, Amer. Meteor. Soc., 533–538.

Miller, L. J., J. D. Tuttle, and C. A. Knight, 1988: Airflow and hail growth in a severe northern High Plains supercell. *J. Atmos. Sci.,* **45,** 736–762.

———, ———, and G. B. Foote, 1990: Precipitation production in a large Montana hailstorm: Airflow and particle growth trajectories. *J. Atmos. Sci.,* **47,** 1619–1646.

Moore, J. T., and J. P. Pino, 1990: An interactive method for estimating maximum hailstone size from forecast soundings. *Wea. Forecasting,* **5,** 508–525.

Morgan, G. M., 1982: Precipitation at the ground. *The National Hail Research Experiment,* Vol. I, *Hailstorms of the Central High Plains,* C. A. Knight and P. Squires, Eds., Colorado Associated Universities Press, 59–79.

Orville, H. D. 1996: A review of cloud modeling in weather modifications. *Bull. Amer. Meteor. Soc.,* **77,** 1535–1555.

Polston, K., 1996: Synoptic patterns and environmental conditions associated with very large hail events. Preprints, *18th Conf. on Severe Local Storms,* San Francisco, CA, Amer. Meteor. Soc., 349–356.

Reinking, R. F., and B. E. Martner, 1996: Feeder-cell ingestion of seeding aerosol from cloud base determined by tracking radar chaff. *J. Appl. Meteor.,* **35,** 1402–1415.

Smith, P. L., L. R. Johnson, D. L. Priegnitz, B. A. Boe, and P. W. Mielke Jr., 1997: An exploratory analysis of crop hail insurance data for evidence of cloud seeding effects in North Dakota. *J. Appl. Meteor.,* **36,** 463–473.

Spahn, J. F., and P. L. Smith Jr., 1976: Some characteristics of hailstone size distributions inside hailstorms. Preprints, *17th Conf. on Radar Meteorology,* Seattle, WA, Amer. Meteor. Soc., 187–191.

Straka, J. M., and D. S. Zrnić, 1993: An algorithm to deduce hydrometeor types and contents from multiparameter radar data. Preprints, *26th Int. Conf. on Radar Meteorology,* Norman, OK, Amer. Meteor. Soc., 513–515.

Stith, J. L., J. Scala, R. F. Reinking, and B. E. Martner, 1996: Combined use of three techniques for studying transport and dispersion in cumuli. *J. Appl. Meteor.,* **35,** 1387–1401.

Waldvogel, A., W. Schmid, and B. Federer, 1978a: The kinetic energy of hailfalls. Part I: Hailstone spectra. *J. Appl. Meteor.,* **17,** 515–520.

———, and ———, 1978b: The kinetic energy of hailfalls. Part II: Radar and hailpads. *J. Appl. Meteor.,* **17,** 1680–1693.

Wilson, J. W., and D. Reum, 1986: "The hail spike": Reflectivity and velocity signature. Preprints, *23d Conf. on Radar Meteorology,* Snowmass, CO, Amer. Meteor. Soc., 62–65.

———, and ———, 1988: The flare echo: Reflectivity and velocity signature. *J. Atmos. Oceanic Technol.,* **5,** 197–205.

Witt, A., 1999: A volumetric reflectivity parameter for the identification of severe hail. Preprints, *29th Int. Conf. on Radar Meteor.,* Montreal, PQ, Canada, Amer. Meteor. Soc., 105–108.

———, and S. P. Nelson, 1991: The use of single-Doppler radar for estimating maximum hailstone size. *J. Appl. Meteor.,* **30,** 425–431.

———, M. D. Eilts, G. J. Stumpf, J. T. Johnson, E. D. Mitchell, and K. W. Thomas, 1998: An enhanced hail detection algorithm for the WSR-88D. *Wea. Forecasting,* **13,** 286–303.

World Meteorological Organization, 1996: Meeting of experts to review the present status of hail suppression. World Meteorological Organization Weather Modification Program Rep. 26, 39 pp.

Zrnić, D. S., 1987: Three-body scattering produces precipitation signature of special diagnostic value. *Radio Sci.,* **22,** 76–86.

Chapter 7

Convectively Driven High Wind Events

ROGER M. WAKIMOTO

University of California, Los Angeles, Los Angeles, California

REVIEW PANEL: Jim Wilson (Chair), N. Andrew Crook, Mark Hjelmfelt, Robert Johns, Kevin Knupp, Peggy LeMone, Ron Przybylinski, Bradley Small, and Marilyn Wolfson

---

7.1. Introduction

On a day when the potential instability is sufficient, it is possible to initiate storms from rising air parcels. One estimate of the intensity of these buoyant plumes (also an indicator of the severity of the storm) is based on parcel theory (Bluestein et al. 1988, 1989; Holton 1992). These rising parcels of air rapidly cool until saturation occurs. Further lifting results in condensation and, in a short period of time, precipitation develops. It is often at this stage that another fundamental element of a storm commonly forms: the convective downdraft.

The appearance of the downdraft completes the circuit of convective overturning by cooling and drying the boundary layer. Once the downdraft reaches the surface, it spreads out and produces a gust front at its leading edge (e.g., Charba 1974; Goff 1976; Wakimoto 1982). The maintenance of the storm is often aided by the advancement of this frontal boundary as it undercuts and lifts the warm, moist ambient air to its level of free convection (LFC) and thereby initiates new convection (e.g., Rotunno et al. 1988; Fovell and Dailey 1995). In some instances, however, the intensity of these downward motions and subsequent outflow winds within these storms can result in severe damage (Fujita 1981; Fujita and Wakimoto 1981). These high wind events are referred to as *downbursts* and *microbursts* (Fujita and Byers 1977). Microbursts can result in considerable crop, tree, and structural damage and have been identified as a causal factor in a number of aircraft accidents (Fujita and Byers 1977; Fujita and Caracena 1977; Fujita 1985, 1986). Forecasting and detection of these intense wind events is a challenging problem in operational meteorology.

It has been recognized that mesoscale convective systems can produce mesoscale downdrafts that result in widespread wind damage, often resulting in casualties (e.g., Ludlam 1970; Fujita 1985). This damage results from clusters of downbursts and can extend over several hundred kilometers (Fujita and Wakimoto 1981). Johns and Hirt (1987) and Przybylinski (1995) have referred to these events as *derechos* (Hinrichs 1888). Derechos appear to account for much of the damage owing to nontornadic winds driven by convection. Indeed, the winds can be so intense that the damage is often mistakenly attributed to tornadoes (e.g., Fujita and Wakimoto 1981).

The above and a number of other studies have illustrated the importance of the downdraft, and the outflows it produces, in severe storm research. A search of the literature reveals that much of the focus on convective storm structure has been on the updraft, with relatively fewer studies on the processes that lead to downdraft development (e.g., Knupp and Cotton 1985; Doswell 1994). Indeed, extensive reviews on severe local storms (e.g., Atlas et al. 1963; Ray 1986; Kessler 1986) have emphasized damage produced by tornadic winds, not outflows from downdrafts. However, recent work has placed a greater emphasis on understanding the downdraft. This is in large part due to increased interest in two specific phenomena: the microburst (Fujita 1985) and mesoscale downdrafts associated with the stratiform region of large convective systems (Houze 1977; Zipser 1969, 1977). This chapter presents a comprehensive summary of downdrafts associated with convective storms, with an emphasis on those events that may lead to destructive winds. A climatology of damaging wind events is presented in section 7.2. Section 7.3 discusses the fundamentals of downdrafts. Section 7.4 presents the various downdraft types and the leading edge of a thunderstorm outflow known as the *gust front.* Important forecast and detection issues regarding high-wind events are presented in section 7.5, and concluding remarks are presented in section 7.6.

7.2. Climatology of damaging wind events

The climatology of storms that produce high winds is not as well documented as that of storms associated with flash flooding, tornadoes, and lightning. A comprehensive study has been presented by Kelly et al. (1985) based on reports of 75 626 severe thunderstorms that occurred in the United States from 1955 through 1983. No attempt has been made to subdivide the wind events into microbursts and derechos.

Thunderstorm wind gusts are primarily a summertime phenomenon (Fig. 7.1), with June and July exhibiting the most activity. In an effort to stratify the relative strength of these events, three categories were created (Table 7.1). Violent thunderstorm gusts were defined as those greater than 33.5 m s^{-1}. Reported wind speeds between 25.8 and 33.5 m s^{-1} are categorized as strong thunderstorm gusts; the final grouping includes damaging wind events where no associated velocity is given. Only gusts where the velocity was measured or estimated by a trained observer have been included in this category.

The diurnal variation of these three categories is shown in Fig. 7.2 using normalized solar time (NST) (Kelly et al. 1978). The NST conversion removes biases associated with discontinuities at time zone boundaries and allows for direct comparison of data from different seasons and locations. Figure 7.2 reveals a late afternoon peak in wind gust. However, there is significant activity between midnight and noon with a secondary peak evident in all three distributions.

TABLE 7.1. Severe thunderstorm wind gusts, 1955–83. From Kelly et al. (1985).

	Gust speed	Annual number	Percent
Damaging	unknown	1114	70
Strong	25.8–33.5 m s^{-1}	375	23
Violent	>33.5 m s^{-1}	113	7
Total		1602	

The geographical distribution of severe thunderstorm wind gusts is shown in Fig. 7.3. Kelly et al. (1985) state that two major frequency axes are present. It should be noted that there is a high probability of a population bias in this figure, with the high plains and Arizona underestimating the true number of wind events. One axis curves southeastward from southern Minnesota across Iowa, Illinois, Indiana, and Ohio. The other starts in central Texas and crosses Oklahoma and Kansas before turning eastward to the Kansas City region. This second axis subsequently turns northward along the Missouri River before ending in northwestern Iowa. The latter axis is believed to be related to the well-known track of severe thunderstorms known as *tornado alley*. The former axis of strong wind gusts will be discussed in the section related to derechos. The maximum over the Nebraska panhandle was hypothesized to be related to microbursts from high-based thunderstorms. The results shown

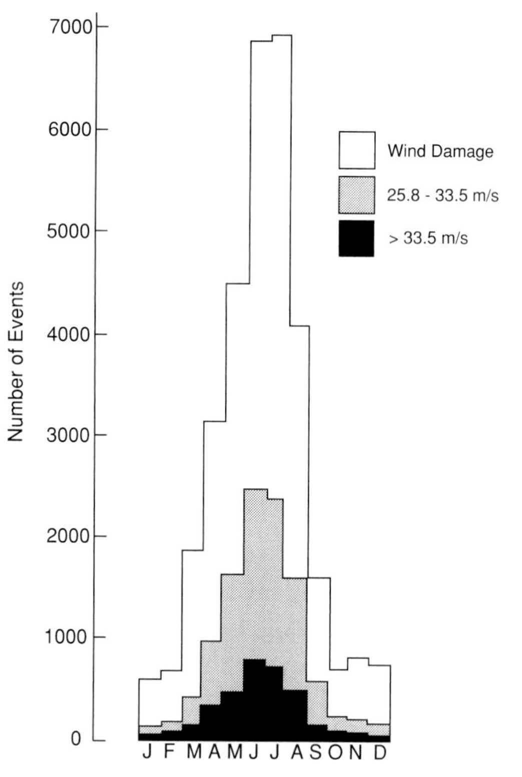

FIG. 7.1. Monthly distribution of occurrences of thunderstorm-related wind damage. Based on a figure from Kelly et al. (1985).

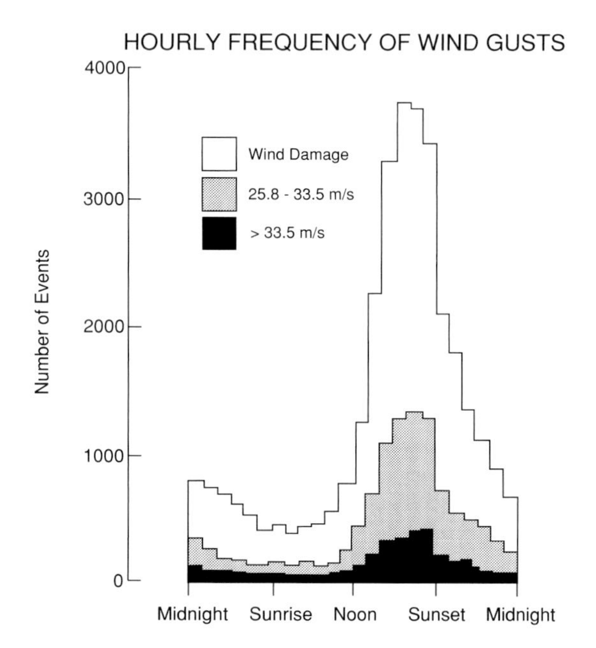

FIG. 7.2. Hourly distribution in normalized solar time (NST) of thunderstorm related wind damage. Based on a figure from Kelly et al. (1985).

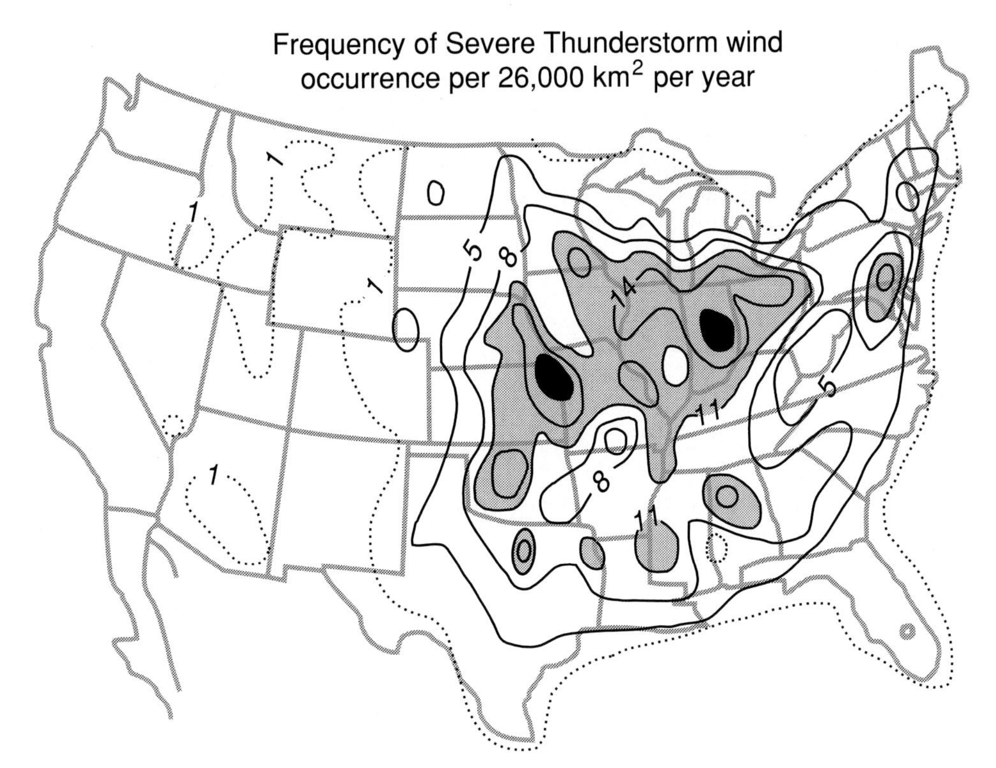

FIG. 7.3. Frequency of any severe thunderstorm wind occurrence per 26 000 km^2 per year. Dashed black lines are isopleths of one. Values greater than 11 and 17 are shaded gray and black, respectively. Based on a figure from Kelly et al. (1985).

in Fig. 7.3 are consistent with an annual distribution of high-wind events shown by Fujita (1981).

a. Downbursts/microbursts

The *downburst* has been defined as an area of strong winds produced by a downdraft over an area from <1 to 10 km in horizontal dimension. Downbursts can be further subdivided into *macrobursts* and *microbursts* with the following definitions.

Microburst: Small downburst, less than 4 km in outflow diameter at the ground, with peak winds lasting only 2–5 min. They may induce dangerous tailwind and downflow wind shears that can reduce aircraft performance.

Macroburst: Large downburst with 4 km or larger outflow diameter at the ground, with damaging winds lasting 5–20 min. Intense macrobursts cause tornado-force damage up to F3 (Fujita 1981) intensity.

The microburst definition has been modified by radar meteorologists by requiring the peak-to-peak differential Doppler velocity across the divergent center to be greater than 10 m s^{-1} (Wilson et al. 1984). While many of these events do not produce damaging winds at the surface, the small temporal and spatial scales of the microburst are particularly hazardous to aircraft operations during takeoff and landing (Fujita and Caracena 1977).

There is no climatological information on the annual occurrence of microbursts in the United States; however, comprehensive field programs have provided some insights into their characteristics. It should be noted that these programs were conducted in the spring/summer months (except for FLOWS, which operated from May 1984 to November 1985) during peak thunderstorm season (Fujita 1985; Wolfson et al. 1985). Results from the Northern Illinois Meteorological Research on Downburst (NIMROD) project (Fujita 1978, 1985), Joint Airport Weather Studies (JAWS) project (McCarthy et al. 1982; Fujita 1985), FAA/Lincoln Laboratory Operational Weather Studies (FLOWS) project (Wolfson et al. 1985), and Microburst and Severe Thunderstorm (MIST) project (Dodge et al. 1986; Atkins and Wakimoto 1991) indicate that microbursts are common phenomena (see Fig. 7.4 for project locations). Wind-shear events were detected on 60%–80% of the days during which thunderstorms occurred. A staggering total of 186 microbursts occurred during the JAWS project over an 86-day period (Wakimoto 1985).

The diurnal variations of microbursts during NIMROD, JAWS, FLOWS, and MIST reveal a pro-

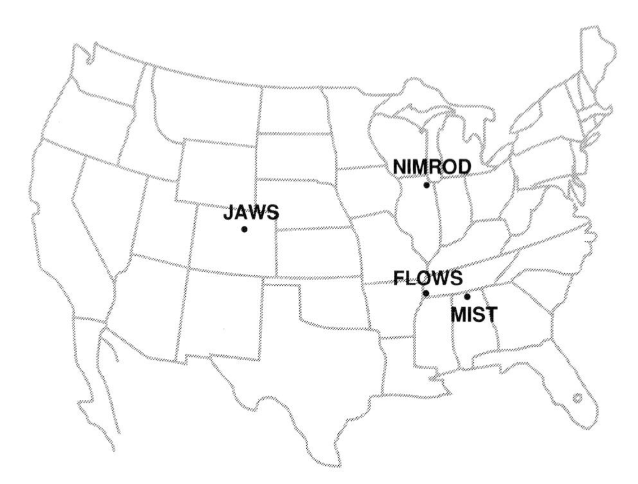

FIG. 7.4. Locations of the NIMROD, JAWS, MIST, and FLOWS field experiments.

nounced maximum in the afternoon with solar heating providing much of the forcing for convective instability. This is especially true over the high plains where there is a general lack of thunderstorm development during the night and early morning hours (Cook 1939; Beckwith 1957). However, the results from NIMROD and, to a certain extent, FLOWS suggest more nocturnal thunderstorm/microburst activity over part of the eastern United States (Fujita 1985; Wolfson et al. 1985). In general, the microbursts observed during these experiments exhibit an exponential decrease in probability of occurrence as the peak outflow wind speed increases. An example from JAWS is shown in Fig. 7.5, where the tail of the distribution occurs near 30 m s^{-1} and the mode of the distribution is between 13 and 15 m s^{-1}. This is consistent with the observational results from FLOWS, NIMROD, and MIST.

One of the most important characteristics of the microburst is the small temporal scale of the peak winds. The peak in the histogram of the number of microbursts during NIMROD and JAWS versus the duration of peak winds (>10 m s^{-1}) is approximately 2.5 min and decays exponentially for longer durations (Fujita 1985). There were only three microbursts during JAWS and one during NIMROD with peak winds lasting greater than 7 min. In contrast, the duration of the peak winds in FLOWS appeared to be uniformly distributed from 1.5 to 9 min. Wolfson et al. (1985) suggest that there was a greater tendency for FLOWS microbursts, which initially were less than 4 km in diameter, to expand to macroburst size, unlike observations from other experiments. The longer sustained winds within the macroburst would explain the FLOWS distribution.

It has been known for a number of years that outflows from thunderstorms are often responsible for a temporary respite from the hot, humid conditions of the ambient environment (e.g., Humphreys 1914). While this cooling trend in temperature was true for the microburst outflows observed during MIST (Atkins and Wakimoto 1991), a large number of microbursts have been associated with rises in temperature. Forty percent of the microbursts observed during NIMROD and JAWS were accompanied by temperature rises (Fujita 1985). The percentage drops to 11 for the microbursts observed during FLOWS (Wolfson et al. 1985). This initially unexpected trend in temperature is examined in section 7.3.

The generality of the microburst characteristics has been verified by other investigators in Japan (Ohno et al. 1996) and Australia (Potts 1991). In particular, the shape of the distribution of the diurnal variation of microburst activity and the histogram of number of microbursts versus peak wind speed presented by Ohno et al. (1996) are remarkably similar to those presented in Figs. 7.2 and 7.5, respectively.

b. Derechos

There has been a growing interest in the events that spawn multiple clusters of downbursts first defined by Fujita and Wakimoto (1981). These clusters are composed of microbursts and macrobursts and are frequently associated with widespread wind damage and large hail; they also occasionally produce tornadoes (Fujita and Wakimoto 1981; Johns and Hirt 1987; Przybylinski 1995). Johns and Hirt (1987) have referred to these events as *derechos* (Hinrichs 1888) and have defined them to be any family of downburst clusters produced by an extratropical mesoscale convective weather system. They also developed the following four criteria to identify these events.

1. There must be a concentrated area of reports consisting of convectively induced wind damage and/or convective gusts >26 m s^{-1} (50 kt). This area must have a major axis length of at least 400 km.

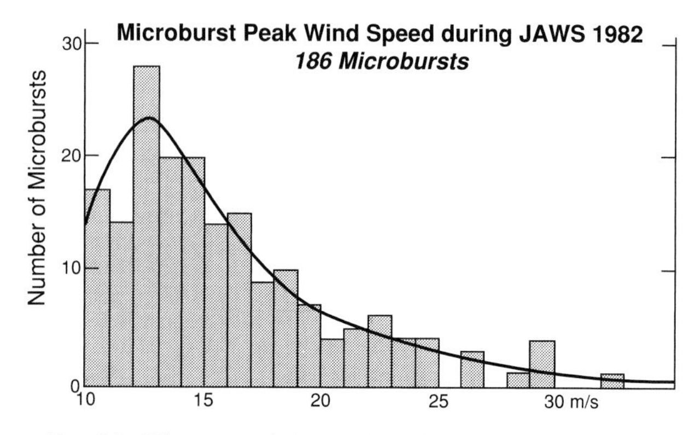

FIG. 7.5. Histogram of the number of microbursts vs the peak wind speed during JAWS 1982 microbursts. Based on a figure from Fujita (1985).

2. The reports within this area must also exhibit a nonrandom pattern of occurrence. That is, the reports must show a pattern of chronological progression, whether as a singular swath (progressive) or as a series of swaths (serial).
3. Within the area there must be at least three reports, separated by 64 km or more, of either F1 or greater damage (Fujita 1981) and/or convective gusts of 33 m s^{-1} (65 kt) or greater.
4. No more than 3 h can elapse between successive wind damage (gust events).

Johns and Hirt (1987) have plotted the total number of derechos during the months of May through August for the period 1980–83 (Fig. 7.6). There may be a population bias in this presentation, similar to the results presented in Fig. 7.3. A total of 70 derecho cases were identified during the study period. Johns (1993) and Przybylinski (1995) have stated that the derechos shown in Fig. 7.6 should be classified as warm-season events associated with a tendency for weak synoptic forcing. Note that the axis of highest frequency in Fig. 7.6 closely resembles the one documented for total severe thunderstorm wind gusts (Fig. 7.3). It is also related to the axis discussed by Johns (1982) for outbreaks of severe weather associated with northwest flow. This is not surprising, since the northwest flow events are primarily a warm-season phenomenon. Johns and Hirt (1987) have shown that 79% of all derecho cases begin during the 12-h period between 1600 and 0400 UTC in an area over the United States east of the Rocky Mountains. The duration of the derechos ranges from 2.3 to 20.0 h, with an average lifetime of 9.2 h. Johns (1993) also identified dynamic pattern events that are associated with derechos occurring in the winter season. A recent study by Bentley and Mote (1998) expands upon the results of Johns and Hirt (1987). They suggest that there is considerable spatial variability of the distribution of derechos from year to year, which may be related to annual changes in the synoptic pattern.

Total Number of Derechos during May-August (1980-1983)

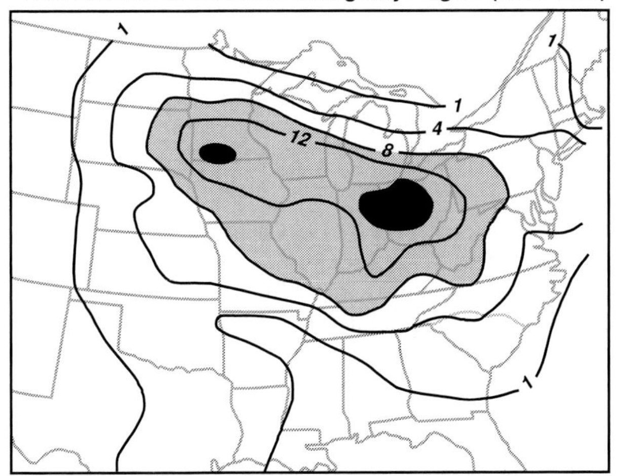

FIG. 7.6. Total number of derechos occurring during the months of May through August for the period 1980–83. Values greater than 8 and 16 are shaded gray and black, respectively. Based on a figure from Johns and Hirt (1987).

7.3. Fundamentals of the downdraft

It is instructive to understand the forcing mechanisms that create a downdraft and outflow before entering into a description of the phenomenology of destructive wind events. These forcing mechanisms are incorporated into the vertical momentum equation, which can be used to elucidate the basic structure of the downdraft and how it differs from the updraft.

a. Downdraft versus updraft

It is apparent that there are at least three characteristic differences between the updraft and the downdraft.

1. While the updraft is typically slightly supersaturated, the downdraft is often appreciably subsaturated owing to the inability for condensate cooling by evaporation, melting, or sublimation to completely offset the warming from adiabatic compression (e.g., Kamburova and Ludlam 1966; Zipser 1969; Das and Subbarao 1972; Betts and Silva Dias 1979; Brown 1979; Leary and Houze 1979; Srivastava 1985; Proctor 1989; Gilmore and Wicker 1998; Igau et al. 1999). This is a change from earlier studies (e.g., Foster 1958) suggesting that the downdraft parcel followed the moist adiabat defined by the surface wet-bulb temperature, even though Byers and Braham (1949) made several observations of humidity dips in the presence of heavy thunderstorm rainfall. The saturated wet adiabat can only be approached by the downdraft if it is weak, the mean drop size is small, or the rainfall is heavy.
2. The microphysical details of the liquid or solid condensate are more important for the downdraft than the updraft. These details are of paramount importance in quantifying the latent cooling within the downdraft (e.g., Hookings 1965, 1967; Kamburova and Ludlum 1966; Srivastava 1985, 1987; Proctor 1989). These authors suggest that small raindrops are more conducive to stronger downdrafts owing to the greater surface area exposed to the environment. Smaller drops also have greater curvature, which results in a larger equilibrium vapor pressure and, hence, lower relative humidity. This will augment the evaporative potential. Proctor (1989) has proposed that the type of frozen precipitation will have varying impacts on the downdraft speeds depending upon the character of the subcloud lapse rate.

3. For scales greater than 1 km, positive buoyancy in the updraft is much greater than the negative buoyancy in the downdraft (e.g., Klemp and Wilhelmson 1978; Schlesinger 1980). Indeed, Wei et al. (1998), in a surprising result, have shown that the average buoyancy in tropical downdrafts was positive. Accordingly, individual parcel vertical excursions tend to be less than 4 km for downdrafts and often greater than 10 km (i.e., the depth of the troposphere) for updrafts (e.g., Raymond et al. 1991).

b. Vertical momentum equation

The fundamental forcing mechanisms for the thunderstorm downdraft are contained within the inviscid vertical momentum equation

$$\frac{d\bar{w}}{dt} = \underset{1}{-\frac{1}{r}\frac{\partial \bar{p}'}{\partial z}} + g\left[\underset{2}{\frac{\theta'_v}{\theta_{v0}}} - \underset{3}{\frac{c_v p'}{c_p p_0}} - \underset{4}{(r_c + r_r + r_i)}\right], \quad (7.1)$$

where $\bar{w}$ = mean vertical velocity, p = pressure, θ_v = virtual potential temperature, c_p = specific heat at constant pressure, c_v = specific heat at constant volume, r_c = mixing ratio of cloud water, r_r = mixing ratio of rain water, and r_i = mixing ratio of ice water. The primes denote departures from a basic state (subscript 0), which varies only in height.

Term 1 is the vertical gradient of perturbation pressure, term 2 represents thermal buoyancy accounted for in parcel theory, and term 3 is the perturbation pressure buoyancy. Condensate loading of cloud, rain, and ice water are represented in term 4. In addition, entrainment of environmental air with cloudy air or precipitation has been shown to be an important factor in downdraft dynamics. Srivastava (1985) describes the formulation of this effect in the vertical momentum equation.

To facilitate the physical interpretation of the terms shown in Eq. (7.1), a simple calculation can be performed to assess their various weights (Table 7.2). Radar reflectivity factor was used as an estimate of the precipitation loading. The air density was assumed to be 1.2×10^{-3} g cm^{-3}, the basic-state temperature is

TABLE 7.2. Equivalent forcing in the vertical momentum equation for typical radar reflectivity factors.

Z (dBZ)	(g kg^{-1})	(K)	(mb km^{-1})
20	4.10×10^{-2}	−0.01	4.82×10^{-3}
30	1.45×10^{-1}	−0.04	1.71×10^{-2}
40	5.15×10^{-1}	−0.16	6.06×10^{-2}
50	1.83	−0.55	2.15×10^{-1}
60	6.47	−1.94	7.61×10^{-1}

300 K, and the following equation was applied to estimate the rainwater density based on echo intensity (Battan 1973):

$$Z = 2.4 \times 10^4 M^{1.82}, \quad (7.2)$$

where Z = radar reflectivity factor in mm^6 m^{-3} and M = rainwater density in g m^{-3}.

The importance of phase change can be shown (e.g., Srivastava 1985) by considering a case where the water mixing ratio is evaporated completely. A water content of 1 g kg^{-1} is approximately equivalent to a temperature deficit of 0.30°C in the buoyancy term in Eq. (7.1). If this water mixing ratio (assume rainwater although it is also valid for cloud water) is evaporated completely, the resulting temperature deficit is

$$\theta'_v = L\frac{r_r}{c_p} \approx 2.5 \text{ K}, \quad (7.3)$$

where L = latent heat of evaporation.

Thus, by evaporating the water, the temperature deficit has increased by a factor of 8.3, or nearly an order of magnitude. The conclusion is that the evaporation of raindrops, rather than water loading, would be more effective in accelerating downdrafts.

1) VERTICAL GRADIENT OF PERTURBATION PRESSURE

The vertical gradient of perturbation pressure is generally small for most downdrafts; however, its effect becomes significant in intense cumulonimbi and mesoscale convective systems. Strong downdrafts can be generated owing to rapid pressure falls within the low-level mesocyclone associated with supercell thunderstorms (Klemp and Rotunno 1983). Moreover, vertical pressure gradients can dictate the propagation of the supercell by promoting preferred regions of updrafts and downdrafts (Newton and Newton 1959; Schlesinger 1980; Rotunno and Klemp 1982). Newton and Newton (1959) recognized that the pressure gradient could, under certain conditions, oppose and dominate the buoyancy effects.

2) THERMAL BUOYANCY

The effect of thermal buoyancy on an air parcel is well understood. In fact, in the absence of pressure effects, and in light rain situations, it is often convenient to view the maintenance of the downdraft as the competing forces of cooling due to phase changes of condensate versus dry adiabatic warming due to compression. The importance of using virtual temperature in the vertical momentum equation has been highlighted by Droegemeier and Wilhelmson (1985), Srivastava (1985), Proctor (1989), and Igau et al. (1999). The former three studies all show that downdraft intensity increases with higher relative humidities at low levels by increasing the virtual temperature difference between the parcel and the environment. This also suggests one mechanism to explain the observed cases of warming within microbursts based on temperature alone. Srivastava (1985) states that the virtual temperature perturbation can be negative even though the temperature rises within the downdraft. Observations of substantial warming of a few degrees Celsius, however, may be a result of a heat burst, which is a strong downdraft impinging on a low-level, nocturnal stable layer (Johnson et al. 1989; Bernstein and Johnson 1994). It is not known which of these two mechanisms is more conducive to producing warm downdrafts at the surface.

More recent studies have examined the additional cooling added by the phase change as a result of melting (Leary and Houze 1979; Srivastava 1987; Knupp 1988; Proctor 1988) and sublimation (Proctor 1988; Wakimoto et al. 1994) in enhancing downdraft speeds. Indeed, Szeto et al. (1988a,b) have shown that the mesoscale circulations can be driven by melting snow alone. They also propose that these circulations propagate away from the source region and may have a dynamic effect on the environment remote from the precipitation source.

3) Perturbation pressure buoyancy

List and Lozowski (1970) stated that perturbation pressure effects, ignored in classical parcel theory, could be significant in deep convective clouds. Although their discussion was not in the context of parcel theory, Newton and Newton (1959) had already recognized the importance of these effects in influencing the propagation of intense convection. While the vertical gradient of perturbation pressure has been accounted for in recent studies of convective storms, the pressure buoyancy term has received less emphasis. Pressure buoyancy states that an air parcel will accelerate upward if it is at a lower pressure than its surroundings. This effect was examined by Schlesinger (1980) and shown to be relatively weak in comparison to thermal buoyancy and the pressure gradient effects. However, its magnitude could be appreciable where the updraft penetrates the tropopause. The pressure buoyancy at the tropopause works in concert with the pressure gradient force to counteract the very strong negative thermal buoyancy from overshooting tops (Schlesinger 1980).

4) Condensate loading

Brooks (1922) first suggested that downdrafts were composed of air initially dragged downward by the weight of precipitation particles, then cooled by evaporation. Subsequent studies have reinforced the concept that condensate loading can contribute to the initiation of the downdraft (Byers and Braham 1949; Knupp and Cotton 1985; Knupp 1988; Roberts and Wilson 1989; Kingsmill and Wakimoto 1991). However, it is now apparent in a number of cases that the microphysical details of the condensate are also important. Numerical calculations suggest that the maintenance of a downdraft by falling precipitation is a function of drop size, rain intensity, and downdraft speed (Hookings 1965; Kamburova and Ludlam 1966; Srivastava 1985; Proctor 1989).

5) Entrainment

The effect of entrainment on updrafts has been extensively studied over the years (Byers and Braham 1949). The mixing of environmental air into a rising parcel of air decreases the positive buoyancy and, therefore, the updraft speeds. Unlike rising parcels of air, however, for downdrafts there currently exist two rather diverse opinions on the effects of entrainment. Malkus (1955), Haman (1973), Heymsfield et al. (1978), Paluch (1979), Knupp and Cotton (1982), Betts (1984), Knupp (1987), Kingsmill and Wakimoto (1991), and Raymond et al. (1991) suggest that entrainment of dry air promotes downdrafts by evaporation or sublimation of cloudy air and precipitation. This mechanism for downdraft production has been well documented within severe storms, especially when the entrained region corresponds to the level of minimum equivalent or wet-bulb potential temperature (θ_e or θ_w; Normand 1946; Newton 1950; Browning and Ludlam 1962; Browning 1977; Ogura and Liou 1980; Betts 1984; Fovell and Ogura 1988; Rotunno et al. 1988).

In contrast, Srivastava (1985), using a one-dimensional cloud model, argues that mixing of environmental air reduces the negative buoyancy by decreasing the virtual potential temperature difference. Furthermore, Srivastava (1985) has shown that stronger downdrafts develop when the environmental relative humidity is high, since the ambient air is virtually warmer than the descending air parcel. This apparently contradicts the expectation that a higher relative humidity should produce weaker downdrafts owing to the lesser potential for evaporative cooling. However, without entrainment, the relative humidity of the downdraft air is essentially determined by its initial condition rather than by the environmental relative humidity. Therefore, the vertical velocities are determined purely by the virtual temperature differences between the descending parcel of air and the environment, which is greater when the environmental relative humidity is high. His results have been reproduced using three-dimensional simulations by Droegemeier and Wilhelmson (1985) and Proctor (1989).

The resolution of these two opposing theories may lie in the height where the entrainment occurs and whether a downdraft is being initiated or maintained. For the former studies, entrainment of potentially dry air may be important in initiating a downdraft; however, Srivastava (1985) illustrates that this effect may be detrimental to the downdraft at lower levels. Proctor's (1989) results support this concept by proposing that strong downdrafts require high relative humidity at low levels but dry air near the melting level. It should be noted that simply increasing the relative humidity from the surface to midlevels can also lead to the development of a more intense thunderstorm. Subsequently, this storm can produce a stronger downdraft even before the effects of entrainment are considered.

In summary, studies suggest that most downdrafts are driven by cooling from phase changes. Although condensate loading typically has a secondary effect, it can be important in initiating downward motion. Entrainment can also initiate downdrafts; however, evidence has been presented that suggests it can negate downward velocities at lower levels. Perturbation pressure effects are prominent in strongly sheared environments, while the perturbation pressure buoyancy term is relatively small.

7.4. Types of downdrafts and resultant outflows

Numerous types of downdrafts have been documented during the past 50 years. Reviews by Ludlam (1980) and Knupp and Cotton (1985) have found it useful to subdivide these events into those associated with nonprecipitating convection and cumulonimbus convection. Only the latter will be considered in this paper since the former typically do not reach the ground and are of weaker intensity. However, Malkus (1955), MacPherson and Isaac (1977), Warner (1970), Heymsfield et al. (1978), and Blyth et al. (1988) consistently found pronounced downdrafts near and beyond the cumulus cloud edges. They were often comparable in magnitude to the updraft and located in the area downshear of the cloud. Entrainment of environmental air and subsequent evaporation of cloud was hypothesized as the forcing mechanism producing the negative buoyant parcel.

Zipser (1969, 1977), Houze (1977), and Miller and Betts (1977) proposed a discrimination between convective downdrafts and mesoscale downdrafts associated with cumulonimbus convection. The spatial length scale that separates these two downdraft types is somewhat amorphous, however; the former and latter represent downdrafts associated with the convective and the stratiform regions of mesoscale convective systems, respectively. Consistent with these past studies, this section is subdivided along similar lines, for example, the microburst is nominally associated with isolated convective events while the derechos are distinctly mesoscale. When the downdraft reaches the surface, it spreads out and undercuts the warm, ambient air. This outflow is an important component in severe storm dynamics, and its leading edge is referred to as the *gust front.* This section will also summarize the current conceptual model of this phenomenon.

a. Convective downdrafts

1) REAR-FLANK AND FORWARD-FLANK DOWNDRAFT

The supercell is a relatively rare but extremely violent thunderstorm. It is often associated with damaging hail and the most intense tornadoes. Owing to these severe weather characteristics, this storm has been the subject of numerous studies in the literature. These studies have revealed the existence of two downdrafts (Fig. 7.7). The *forward-flank downdraft* occurs downwind and within the precipitation and is associated with a wet-bulb potential temperature that results from mixing of updraft (high θ_w) and midlevel environmental (low θ_w) air (Lemon and Doswell 1979). The outflow from this downdraft forms a relatively weak discontinuity at the surface on the forward and right flanks of the radar echo (Fig. 7.7). Although the downdraft speeds are not intense, modeling studies (Klemp 1987) have shown that the leading edge of the outflow plays an important role in producing baroclinically generated horizontal vorticity that is first tilted into the vertical and then stretched, leading to the development of the low-level mesocyclone.

The strongest downdraft associated with the supercell is the *rear-flank downdraft* (e.g., Charba and Sasaki 1971; Lemon 1976; Barnes 1978). This downdraft is defined as the one that supports the storm

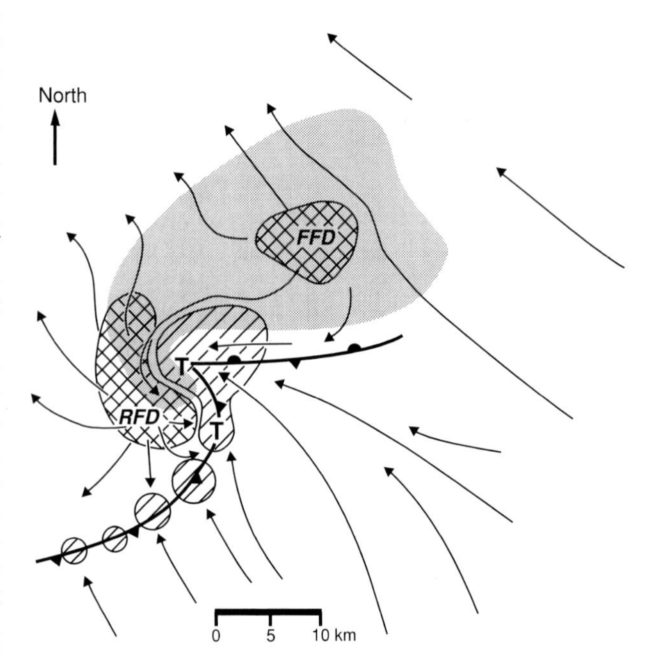

FIG. 7.7. Schematic plan view of the supercell thunderstorm at the surface adapted from Lemon and Doswell (1979) and Davies-Jones (1986). The gray shade encompasses the radar echo. The gust front structure is depicted using a sold line and frontal symbols. Surface positions of the updraft are hatched. Forward-flank downdraft (FFD) and rear-flank downdraft (RFD) are crosshatched. Streamlines relative to the ground are also indicated. Likely tornado locations are shown by the T's.

outflow behind the convergence line on the right flank (Fig. 7.7). Observations show that θ_w values within this downdraft are equivalent to those found at midlevels (3–5 km AGL) within the environment (Lemon and Doswell 1979). This level was first proposed by Browning (1964) and later confirmed by aircraft measurements (Fankhauser 1971) and analysis of Doppler radar observations and numerical simulations (Klemp et al. 1981). More recent work by Gilmore and Wicker (1998), however, has shown that, owing to entrainment, surface values of θ_w can depart significantly from their midlevel values. While the source region for this downdraft has been agreed upon, the forcing mechanism has been debated within the literature.

The rear-flank downdraft was widely thought to result from a dynamical interaction between the environmental wind at midlevels and the storm (Lemon and Doswell 1979). This interaction would produce a downward-directed pressure force that initiates the downdraft. The downdraft is subsequently enhanced and maintained through precipitation drag and evaporative cooling. However, numerical studies (Klemp and Rotunno 1983) and observational studies (Carbone 1983; Brandes 1984; Hane and Ray 1985) suggest that the downward-directed pressure gradient that initiates the downdraft is induced by vorticity intensification near the ground. As the low-level mesocyclone intensifies, it lowers the pressure locally since it is in approximate cyclostrophic balance. This dynamically induced low pressure near the surface subsequently draws the air down from above. The distinction between these two mechanisms is crucial to understanding the development of the low-level mesocyclone. In the former, mesocyclone intensification at low levels was attributed to vertical vorticity transport in the developing rear-flank downdraft. In the latter, the mesocyclone first intensifies and subsequently induces the rear-flank downdraft.

2) Occlusion downdraft

Klemp and Rotunno (1983) propose that two different types of downdrafts exist within the rear flank of a supercell, which are illustrated in the numerical simulations by Wicker and Wilhelmson (1995) shown in Fig. 7.8. Labeled in Fig. 7.8a is the main rear-flank downdraft, which appears to be driven by evaporating precipitation. There is a mesoscale area of high perturbation pressure near the downdraft owing to the weight of the cool air consistent with this proposed mechanism. Klemp and Rotunno (1983) argue that nontornadic supercell storms are often observed to persist for long periods of time with this downdraft and its subsequent gust front apparent; accordingly, this particular storm-scale downdraft is not uniquely linked to tornadogenesis within the supercell. They define the *occlusion downdraft,* which is part of the rear-flank downdraft but dynamically distinct. This downdraft is induced within the model by low pressure associated with strong low-level rotation, which is in turn associated with the developing mesocyclone. This relationship is clearly seen with the collocation of the occlusion downdraft with the mesolow in Fig. 7.8a. This distinction between the rear-flank downdraft and the occlusion downdraft is not often emphasized in review articles on supercell storm structure. These two downdrafts often merge to produce a continuous area of downward motion as shown in Fig. 7.8b.

The existence of the occlusion downdraft has been illustrated in numerical simulations. Recent high-resolution data on supercells collected by an airborne platform called ELDORA (Electra Doppler Radar) (Wakimoto et al. 1996) have provided some of the first detailed observational evidence of its structure. The synthesized wind field shown in Fig. 7.9 was for the pass immediately before tornadogenesis. The vertical cross section through the mesolow is shown in Fig. 7.9a. The agreement between the negative vertical velocities and downward-directed pressure gradient derived from the Doppler syntheses (Fig. 7.9b) provides evidence that the occlusion downdraft is dynamically driven by the strong low-level rotation. Precipitation loading was estimated from the observed radar reflectivity values ($\sim$45 dBZ) and found to be one order of magnitude smaller than the pressure gradient force (see Table 7.2). Knupp and Cotton (1985) and Knupp (1987) have suggested that only downdrafts associated with precipitation systematically reach the surface. The occlusion downdraft is one important exception to this rule. Occlusion downdrafts have been documented with cold fronts (Carbone 1983) and gust fronts (Wilson 1986; Mueller and Carbone 1987).

The occlusion downdraft can attain speeds that result in microburst damage at the surface. Indeed, intense downdrafts near tornadoes have been reported for a number of years (e.g., van Tassel 1955). Hall and Brewer (1959) noted a fan-shaped damage pattern in fallen trees near a tornado track, which they attributed to a combination of a strong downdraft (microburst) and storm translation (Orf and Anderson 1999). Alberty et al. (1980) and Fujita (1978, 1989, 1992) have documented cases of "twisting" microbursts (Fujita 1985) near tornadoes, which are probably a result of the occlusion downdraft embedded within the low-level mesocyclone. The downdraft shown in Fig. 7.9 was associated with speeds >30 m s^{-1}.

3) Microburst

A series of aerial surveys in the 1970s and 1980s confirmed the existence of divergent wind patterns in crops and forests (Fujita 1978, 1981, 1985; Fujita and Byers 1977; Forbes and Wakimoto 1983; Wakimoto 1983; Wakimoto and Bringi 1988) that could produce tornado-force damage up to F3 intensity (Fujita and Wakimoto 1981). These strong winds caused consider-

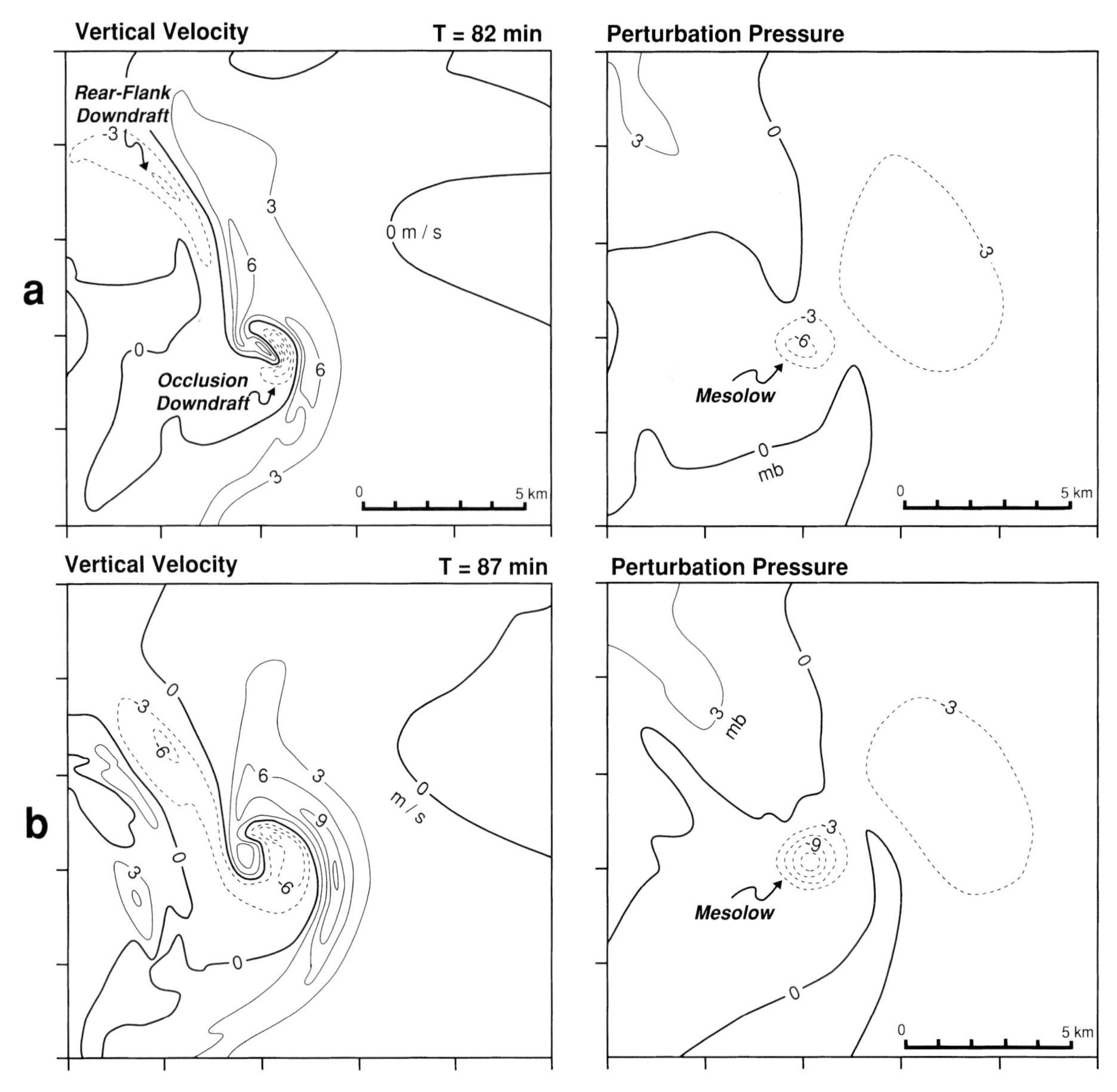

FIG. 7.8. Horizontal cross section from a fine-grid simulation at (a) 82 and (b) 87 min. Vertical velocity is at 250 m and perturbation pressure is at 100 m. Based on a figure from Wicker and Wilhelmson (1995).

able crop, tree, and structural damage, and were identified as a causal factor in a number of aircraft accidents (Fujita and Byers 1977; Fujita and Caracena 1977; Fujita 1985, 1986). Fujita (1985) analyzed a 61+ m s^{-1} (120+ kt) wind recorded at Andrews Air Force Base associated with the strongest microburst ever recorded at an airport (Fig. 7.10). According to news reports, Air Force One, with President Reagan on board, landed at 1404 EDT. Winds began increasing from a northwesterly direction at 1409:20, reaching a peak at 1410:45. Subsequently, the wind speed rapidly decreased to only 1 m s^{-1} when the eye of the microburst passed over the anemometer. A second peak in wind speed occurred at 1413:40 from a southeasterly direction. In the United States, during the period 1974 to 1985, microburst winds were a factor in at least 11 civil transport accidents and incidents involving over 400 fatalities and 145 injuries (Proctor 1988).

An example of the horizontal wind field of a microburst is shown in Fig. 7.11. The flow is characterized by strongly diverging outflow in all directions away from the microburst center. The typical scenario for a microburst-related accident is shown in Fig. 7.12. When an aircraft flies through a microburst during takeoff it first encounters a headwind component from the microburst outflow. This headwind increases lift by increasing the relative airflow over the wing. The plane may then pitch up, and the pilot may attempt to compensate by leveling off. But only a matter of seconds later the plane encounters a decreasing head-

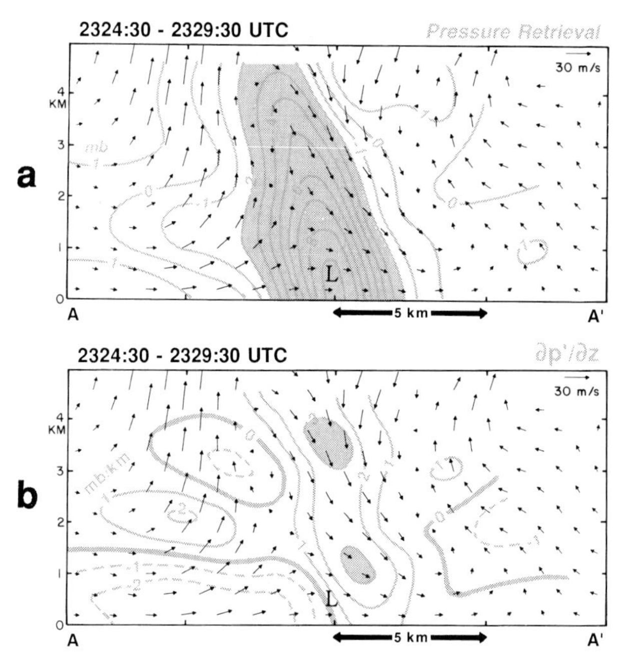

FIG. 7.9. Vertical cross section through a supercell depicting the wind field superimposed onto (a) the isobars through the mesolow, and (b) the vertical pressure gradient. Isobars are drawn as gray lines, with values <-2 mb shaded gray. Positive and negative isopleths of constant pressure gradient are drawn as gray and dashed gray lines, with values >3 mb km^{-1} shaded gray. From Wakimoto et al. (1998).

wind, downdraft (within the center of the microburst), and then a strong tailwind. Accordingly, the plane has lost lift and could find itself flying too low and with insufficient airspeed to avoid a crash.

As stated by McCarthy and Serafin (1984), efforts to improve air safety with regard to the hazard of microbursts are in three areas: education, training, and improved technology. Education encompasses efforts to broaden pilot awareness through the use of films, lectures, and required pilot proficiency testing. Training encompasses efforts to help pilots recognize microburst conditions from the cockpit and to follow appropriate flying procedures for coping with a microburst when caught in one (Elmore et al. 1986). Technology requires advanced sensing systems to detect and warn a pilot of the presence of a severe wind shear well before a possible encounter. These three objectives have been largely met and result in a modern success story for useful applications of scientific research. Section 7.5 of this review will focus on the detection and forecasting methods used to identify microbursts.

Fujita (1985) provided one of the earliest conceptual models of the descending microburst (Fig. 7.13). Besides a shaft of strong downward velocity at its center, the microburst is characterized, when it strikes the ground, by strong divergence at its center and an accelerating outburst of strong winds in an overturning rotor (Kessinger et al. 1988; Parsons and Kropfli 1990; Mahoney and Elmore 1991; Orf et al. 1996) propagating away from the center of the microburst. The highest winds are usually associated with these rotors (Proctor 1988), with the peak speeds occurring in the lower portion of the ring vortex, where outflow speeds are enhanced by the circulation of the ring. The mechanism for intensifying the outflow winds is vortex stretching as the ring expands. These intense

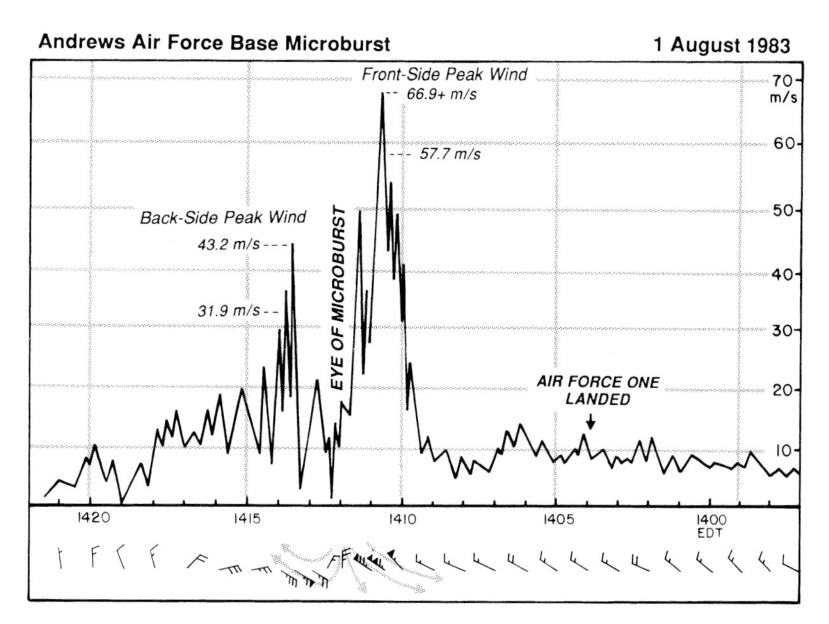

FIG. 7.10. Wind speed and direction trace recorded by an anemometer located near the runway at Andrews Air Force Base on 1 Aug 1983. Based on a figure from Fujita (1985).

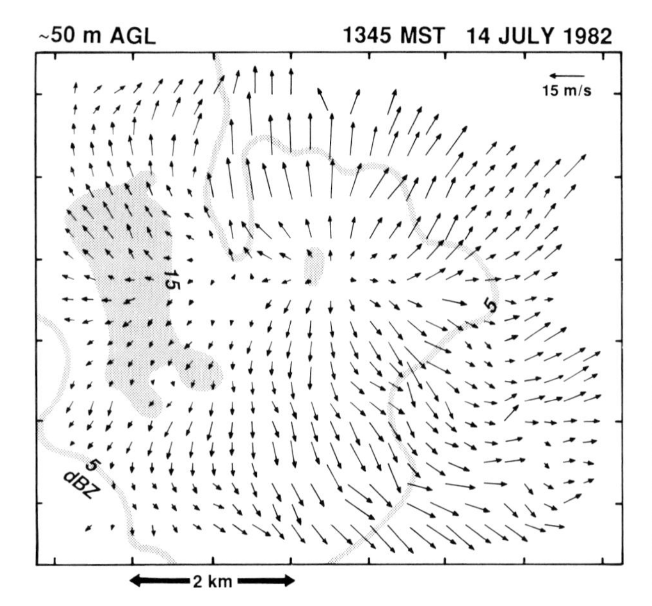

FIG. 7.11. Horizontal winds and radar reflectivity at the lowest analysis level (~50 m AGL) for a microburst at 1345 MST on 14 Jul 1982. Contours of radar reflectivity are drawn as gray lines. Based on a figure from Hjelmfelt (1988).

horizontal circulations can negatively impact aircraft performance (Fujita 1986; Mahoney 1988; Wakimoto et al. 1996). An example of this rotor can be seen in Fig. 7.14.

The microburst in three dimensions is schematically illustrated in Fig. 7.15 (Fujita 1985). A key feature in the figure is that a number of microbursts are associated with small-scale cyclonic (<4 km) circulations aloft. In fact, it is common for most microbursts to rotate, with the strongest microbursts associated with stronger rotation (Rinehart et al. 1995). The magnitude of the vorticity within these misocyclones can be comparable to the vorticity present in the parent mesocyclone associated with a tornado (Kessinger et al. 1988). Several authors have hypothesized that this rotation may actually increase downburst magnitude (Emanuel 1981; Fujita 1985; Wakimoto 1985). In contrast, Kessinger et al. (1988) and Proctor (1989) have shown that the vertical pressure force associated with the rotation tends to oppose the downward acceleration below cloud base. Indeed, sensitivity studies by Proctor (1989) have shown that the rotating microburst is weaker and lags the nonrotating case, contradicting the observations of Rinehart et al. (1995). It is the location of the minimum in perturbation pressure that differentiates the enhancement or retardation of the downward velocities in the occlusion downdraft and the rotating microburst, respectively. In the former, the pressure minimum is located at or near the surface, while in the latter, it is located substantially above the surface. Parsons and Weisman (1993) propose that rotating downbursts are most prevalent when the direction of the wind shear vector in the mid levels of the tropo-

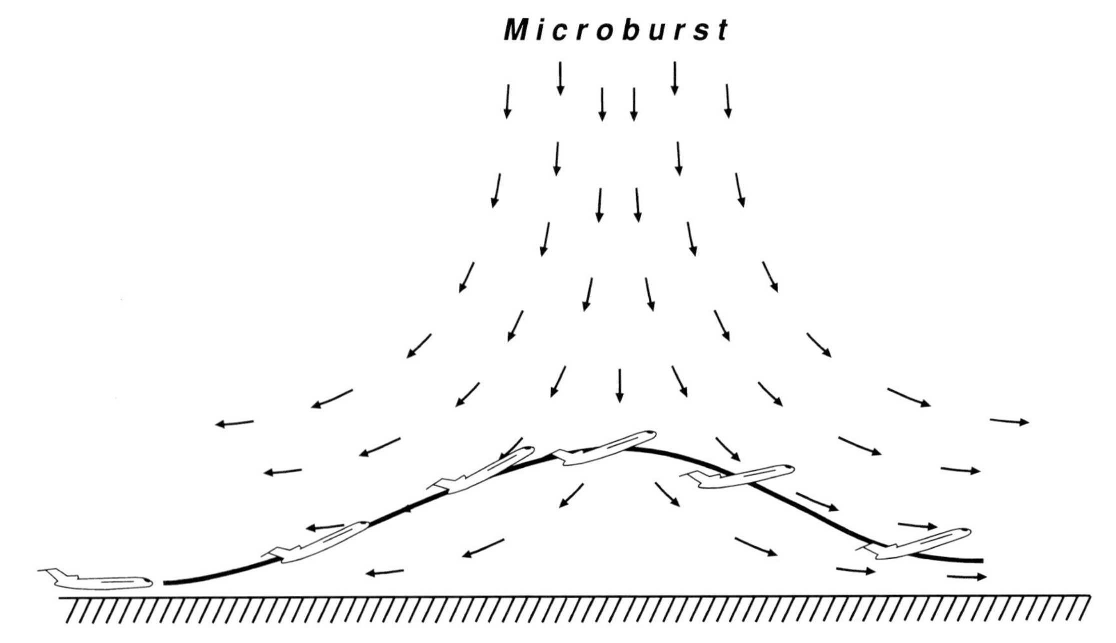

FIG. 7.12. Schematic diagram illustrating the impact of a microburst on aircraft performance during takeoff. The airplane first encounters a headwind and first experiences added lift. This is followed in short succession by a decreasing headwind component, a downdraft, and finally a strong tailwind, which may lead to an impact with the ground. Composite drawing based on numerous studies of aircraft accidents by Fujita and Caracena (1977), Fujita and Byers (1977), and Fujita (1978, 1985, 1986).

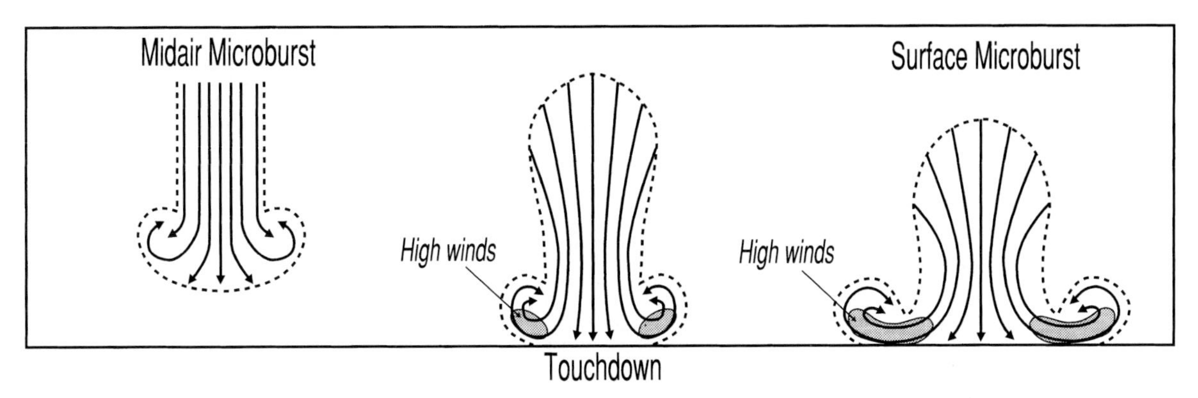

FIG. 7.13. Conceptual model of a microburst hypothesized to explain ground-damage patterns. Three stages of development are shown. A midair microburst may or may not descend to the surface. If it does, the outburst winds develop immediately after its touchdown. Based on a figure from Fujita (1985).

sphere varies with height. While rotation does not appear to enhance microburst downdraft speeds, it will be shown in section 7.5 that it can be used as a useful precursor for detecting microbursts.

Results from a variety of studies (e.g., Wilson et al. 1984; Fujita 1985; Wakimoto 1985; Mielke and Carle 1987; Hjelmfelt et al. 1989; Proctor 1989; Roberts and Wilson 1989; Wolfson et al. 1990) reveal that microburst winds are associated with a continuum of rain rates that range from heavy thunderstorm precipitation to virga shafts either from altocumuli or from clouds that have been referred to as shallow high-based cumulonimbi (Brown et al. 1982). Fujita (1985) and Wakimoto (1985) subdivided microbursts into dry/low-reflectivity and wet/high-reflectivity microbursts. The following definitions are used to describe these two phenomena (Fujita and Wakimoto 1981; Wilson et al. 1984; Fujita 1985).

Dry/low-reflectivity microburst: A microburst associated with <0.25 mm of rain or a radar echo <35 dBZ in intensity.

Wet/high-reflectivity microburst: A microburst associated with >0.25 mm of rain or a radar echo >35 dBZ in intensity.

(i) Low-reflectivity or dry microbursts

Numerical calculations (Hookings 1965; Kamburova and Ludlam 1966; Harris 1977; Srivastava 1985, 1987; Proctor 1989) have shown the sensitivity of downdraft intensity as a function of drop size, rain

FIG. 7.14. A sequence of photographs showing a curl of dust behind the leading edge of a microburst outflow on 15 Jul 1982. Photos taken by B. Waranauskas (from Fujita 1985).

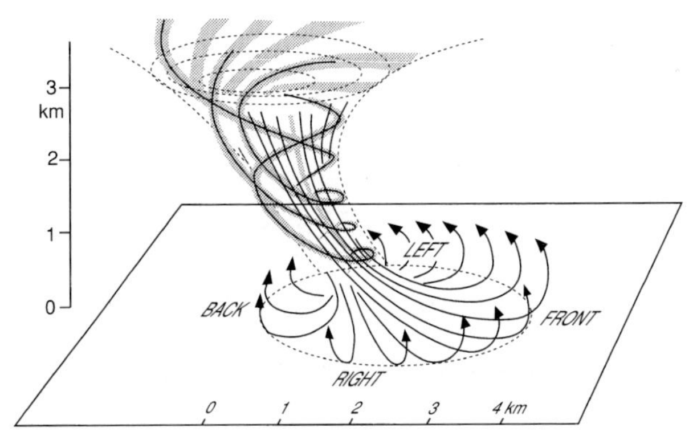

FIG. 7.15. Three-dimensional visualization of a microburst. Based on a figure from Fujita (1985).

intensity, and subcloud lapse rate. One of their conclusions is that when the environmental lapse rate is approximately equal to the dry adiabatic lapse rate, then the rates of evaporation place little restriction on the downdraft magnitude, and even in light precipitation, strong downdrafts may be generated. These results confirmed the earlier speculations by Braham (1952) and Krumm (1954). In the absence of pressure effects and in light rain situations (i.e., dry microbursts), the maintenance of the downdraft can be viewed as the competing forces of cooling due to phase changes of condensate versus dry adiabatic warming due to compression. When the subcloud lapse rate is dry adiabatic, any cooling by evaporation of the condensate results in a negative temperature perturbation that will maintain a downdraft. This effect increases with the depth of the dry adiabatic lapse rate. Compressional warming in a descending parcel can counteract this cooling when the subcloud lapse rate is less than dry adiabatic.

These results have been confirmed by observations of virga shafts from weakly precipitating cloud systems producing low-reflectivity microbursts (McCarthy and Serafin 1984; Wilson et al. 1984; Fujita 1985, Wakimoto 1985; Mahoney and Rodi 1987). An example of a parent cloud that produced a low-reflectivity microburst is shown in Fig. 7.16. A striking feature in the JAWS Project results presented by Wilson et al. (1984) and Hjelmfelt (1988) was that the strongest microburst during the project ($\sim$50 m s^{-1} Doppler velocity differential) was associated with only a 25-dBZ echo at a height of 500 m AGL. Indeed, no precipitation was detected during low-level passes by aircraft through microburst downdrafts by Mahoney and Rodi (1987) and Wakimoto et al. (1994). All low-reflectivity microbursts are particularly hazardous

FIG. 7.16. An altocumulus cloud that spawned a microburst on 14 Jul 1982. Virga was pendant from cloud base. Photograph by B. Smith (from Wakimoto 1985).

to aircraft since the parent cloud and pendant virga shafts appear innocuous (e.g., Mielke and Carle 1987; McNulty 1991; Wakimoto et al. 1994). The weak echoes and low precipitation rates result in little or no temperature changes at the surface (Fujita 1985; Srivastava 1985; Proctor 1989).

It has been widely accepted that cooling owing to phase changes is the primary forcing mechanism for the low-reflectivity microburst. Downdraft speeds are also sensitive to the type of precipitation within the parent cloud. Mahoney and Rodi (1987) and Hjelmfelt et al. (1989) reported on in situ aircraft measurements of graupel in downdrafts of low precipitation microbursts and performed calculations and model simulations that indicated that sublimation, melting, and evaporation of hail/graupel were important in cooling the downdraft parcels. More recent numerical simulations (Proctor 1989) and in situ observations (Wakimoto et al. 1994) have provided evidence that snowflakes are even more effective in producing low-reflectivity microbursts via sublimation. Sensitivity studies revealed that snowflakes produced downdraft speeds that were nearly twice as strong as those generated by hail. The sublimation process is important for three apparent reasons: 1) the numerous low-density snow particles readily sublimate, with much of the snow content depleted before melting into rain; 2) the latent heat of sublimation is greater than the latent heat of either evaporation or melting; and 3) the cooling from sublimation takes place at a relatively high altitude within the deep adiabatic layer, allowing the downdraft to accelerate through a deep column (Fig. 7.17). The prevalence of dry microbursts over the high plains is largely attributed to item 3. The deep, dry adiabatic layer is characteristic of this geographic region; therefore, the integrated negative potential energy is large, even though the temperature deviation at any particular height is small (Fig. 7.17).

Wakimoto et al. (1994) presented a series of photographs taken of virga shafts associated with a low-reflectivity echo to determine if there were identifiable features associated with microburst downdrafts (Fig. 7.18). The microbursts were shown to develop at the location where the virga shafts were, visually, the lowest and most opaque. As the downdraft intensifies, sublimation rapidly depletes the hydrometeors. As a result, the maximum negative vertical velocities are collocated with relatively low reflectivity and with transparent regions of the virga shafts. Interestingly, the radar reflectivity data, combined with cloud pictures, revealed the location of the radar bright band to be at the visible termination of the virga shafts (i.e., the melting level). This presents strong evidence that the virga shafts in the present case are ice particles rather than raindrops, confirming recent speculations by Fraser and Bohren (1992).

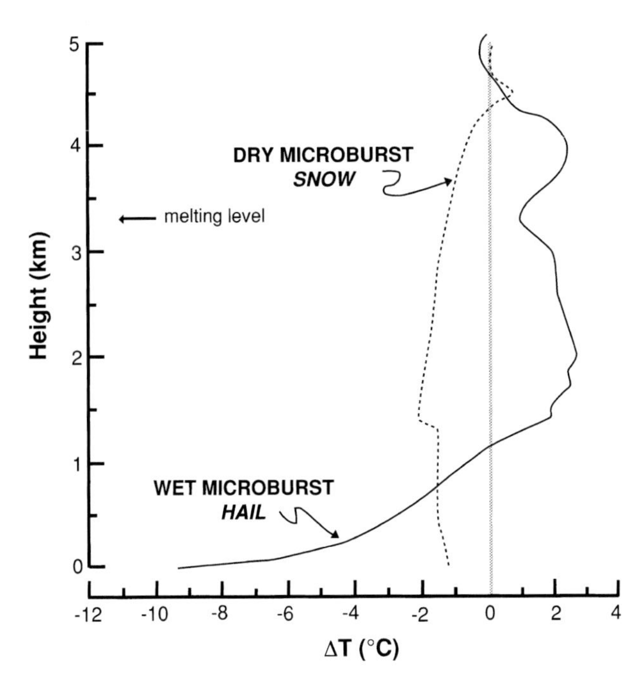

FIG. 7.17. Comparison of the axial profiles for cooling owing to phase changes of condensate for the dry microburst with snow and the wet microburst with hail. Location of the melting level based on the environmental sounding is shown. Based on a figure from Proctor (1989).

(ii) High-reflectivity or wet microbursts

In more stable lapse rates, the downdraft tends to be warmer than the environment. In such a case, the drag of the condensed water may become important in driving the downdraft, especially at high levels. Accordingly, higher rainwater mixing ratios (i.e., radar reflectivities) are required for microbursts to develop. This effect is well illustrated in Fig. 7.19. Each point on the figure represents a microburst event identified by Doppler radar. Microbursts can be seen to occur most frequently at lapse rates of temperature exceeding ~8.5 K km^{-1}. Practically no microbursts occurred for lapse rates of temperature <7.5 K km^{-1}. Furthermore, for lapse rates of temperature exceeding 8.5 K km^{-1}, microburst occurrence was practically independent of radar reflectivity. At lapse rates less than ~8.0 K km^{-1}, however, the few microbursts that occurred were associated with radar reflectivities in excess of 45 dBZ. Model simulations from Srivastava (1985) and Proctor (1989) are consistent with the Doppler radar observations shown in Fig. 7.19. The shaded area in Fig. 7.20 represents the microburst regime (downdraft speeds >20 m s^{-1}). The vertical dashed line separates dry from wet microbursts using 35 dBZ as the transition zone. Progressively higher water contents are needed to produce downburst winds as the environmental stability increases. This often manifests itself in radar reflectivity images as a prominent descending precipitation core (e.g., Roberts and Wilson

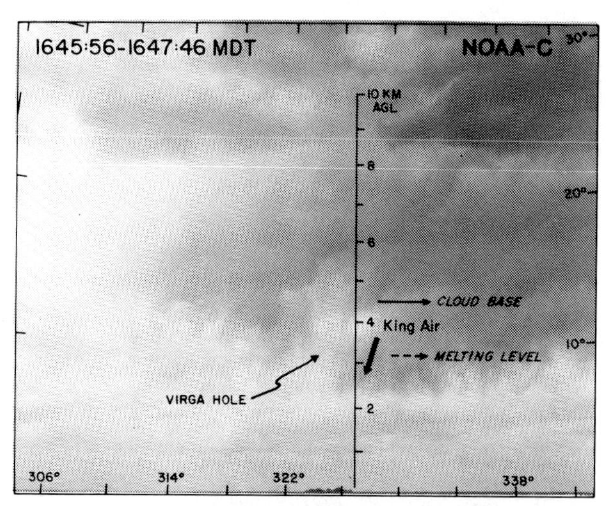

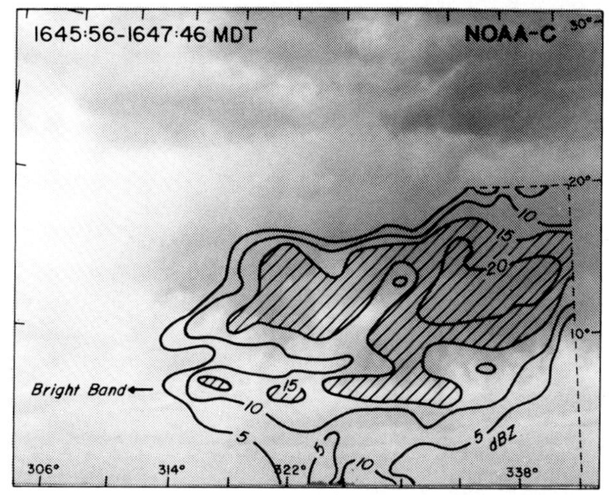

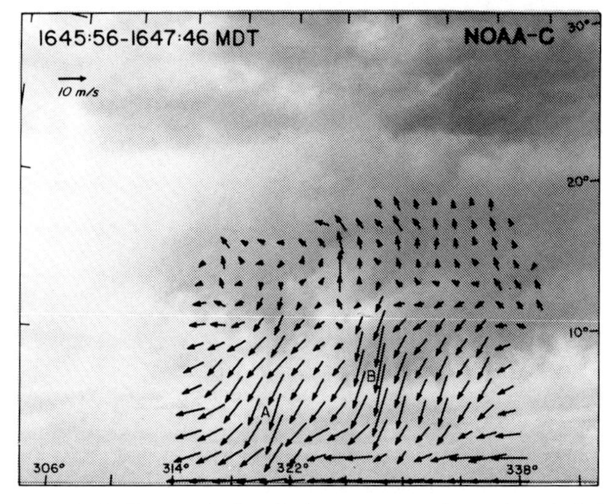

FIG. 7.18. Photograph of a microburst storm superimposed on vertical cross sections through the center of the microburst downdrafts depicting radar reflectivity and storm-relative dual-Doppler winds. The location and measured wind from the Wyoming King Air penetration is shown in the top panel. The cloud base and the environmental melting level determined from a nearby sounding are indicated in the figure. The location of the bright band is also indicated. The height grid is valid at the location of the microburst downdrafts. From Wakimoto et al. (1994).

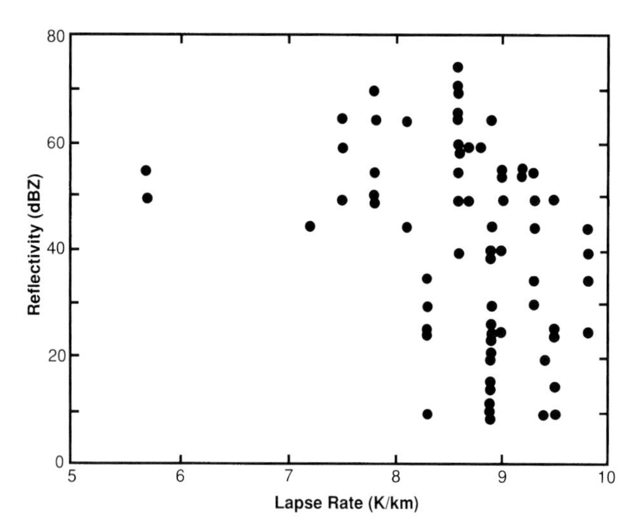

FIG. 7.19. Plot of microburst occurrence as a function of radar reflectivity and environmental lapse rate. Based on a figure from Srivastava (1985).

1989; Wakimoto and Bringi 1988; Kingsmill and Wakimoto 1991).

Similar to the results shown for low-reflectivity microbursts, frozen condensate is important for producing strong wet microbursts (Srivastava 1987; Wakimoto and Bringi 1988; Proctor 1989; Tuttle et al. 1989; Kingsmill and Wakimoto 1991). Wakimoto and Bringi (1988) presented dual-polarization radar measurements that suggest that melting played a key role in producing a wet microburst (Fig. 7.21). A narrow region of near-zero differential reflectivity Z_{dr} within the region of maximum reflectivity of the thunderstorm can be seen in the figure. This Z_{dr} hole indicates a narrow shaft of hail within the heavy shower of large raindrops (near-zero Z_{dr} is indicative of ice and positive Z_{dr} is a signature of rain). The narrow shaft of ice was coincident with the surface location of a microburst shown by the arrow.

A comparison of wet and dry microburst simulations is shown in Fig. 7.17. The dry and wet microbursts were driven by snow and hail, respectively. Note that the temperature deviations for the two cases are different. The dry microburst is associated with a smaller cooling, but it occurs through a deep column. The temperature deviation for the wet microburst case is largest at the ground but diminishes rapidly with height until it becomes warmer than the ambient air above 1 km. The warming aloft for the hail case indicates the importance of precipitation loading in the early stages of the wet microburst, which overcomes the positive temperature buoyancy. It is hypothesized that hail produces stronger microbursts in more stable environments, since the downdraft is maintained only at lower elevations (i.e., the cooling effect is delayed) where it is less likely to be depleted of negative buoyancy due to compressional heating (Proctor 1989).

A typical sequence of events associated with a microburst based on dual-Doppler analyses has been shown by Wilson et al. (1984) and Hjelmfelt (1987). A sequence of dual-Doppler winds overlaid on cloud photographs is shown in Fig. 7.22 for a wet micro-

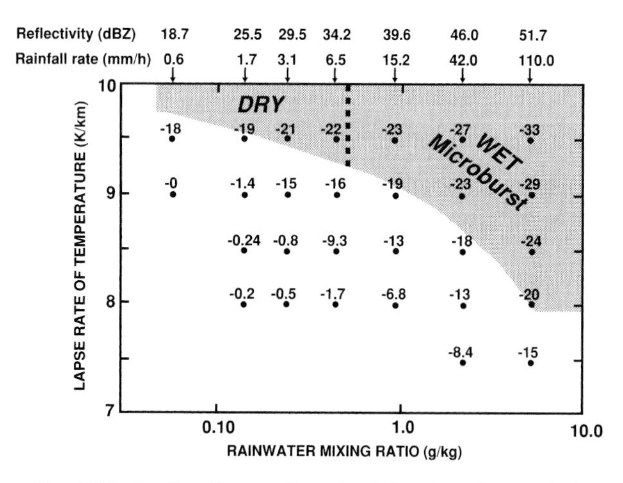

FIG. 7.20. Results of a one-dimensional time-dependent nonhydrostatic cloud model of a downdraft. Plotted numbers are vertical air velocity (m s^{-1}) at a level of 3.7 km below the top of the downdraft as a function of the lapse rate in the environment and total liquid water mixing ratio at the top of the downdraft. Numbers on top scale indicate the radar reflectivity and rain rate at the top of the downdraft. Shaded area represents microbursts (>20 m s^{-1}). Vertical dashed line separates dry (<35 dBZ) from wet (>35 dBZ) microbursts. Based on a figure from Srivastava (1985) that was modified by Houze (1993). Reprinted with the permission of Academic Press.

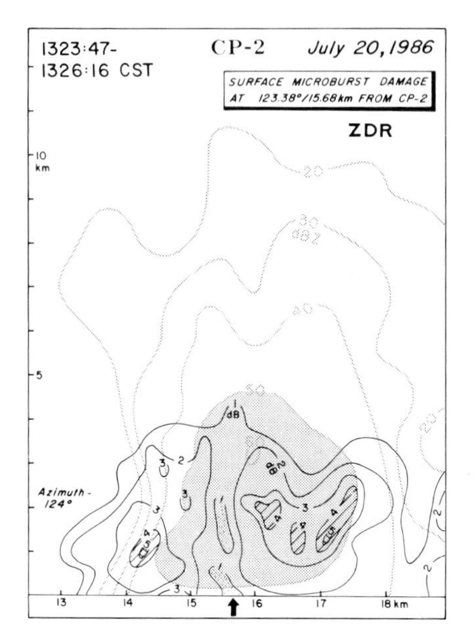

FIG. 7.21. Vertical cross section showing dual-polarization Doppler radar measurements obtained in a thunderstorm in northern Alabama. Reflectivity data are presented in dBZ. Differential reflectivity Z_{DR} are shown in dB units. Arrow indicates location of the center of a surface microburst. From Wakimoto and Bringi (1988).

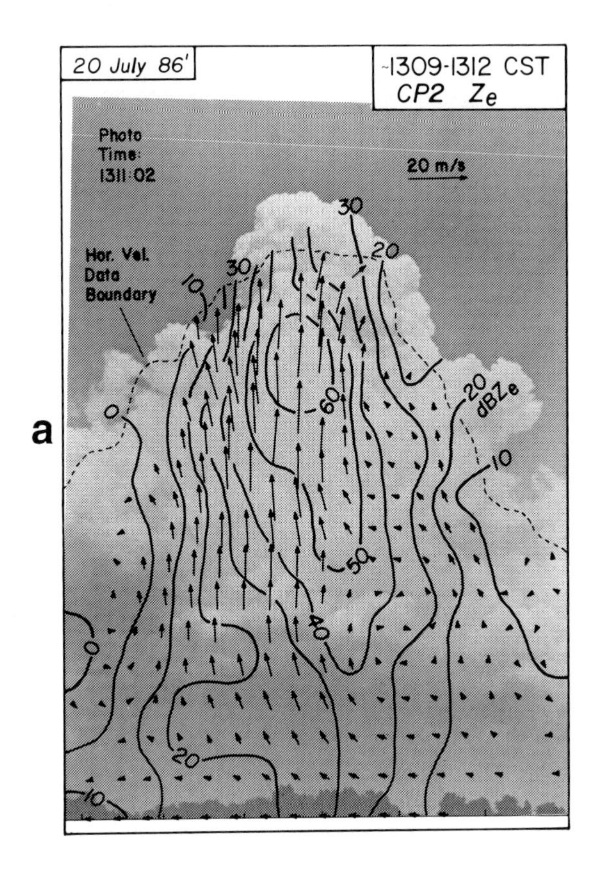

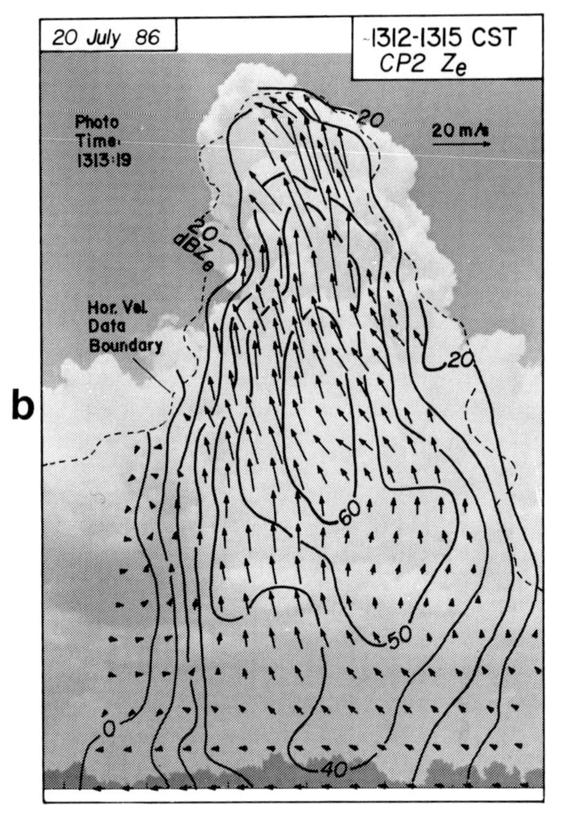

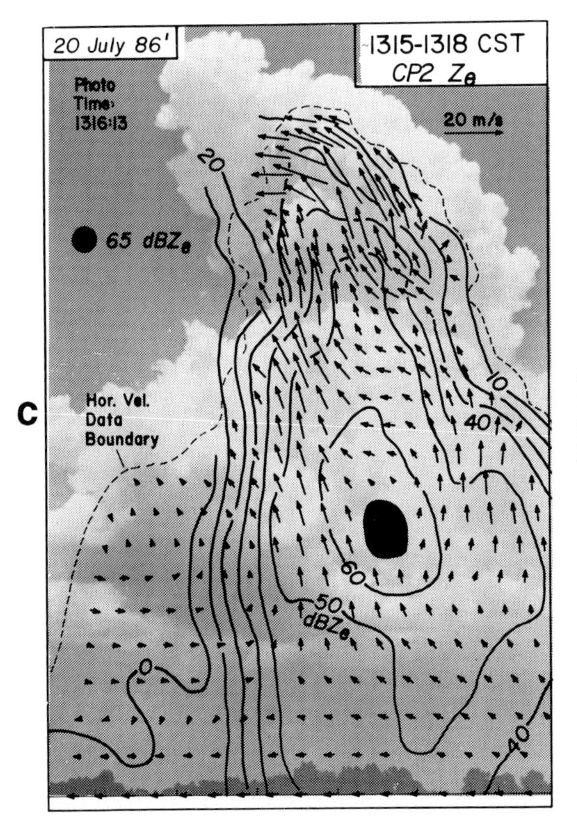

FIG. 7.22. Time sequence of photographs of a thunderstorm at 1306–1309, 1312–1315, and 1315–1318 CST on 20 Jul 1986 superimposed on vertical cross sections of radar reflectivity and storm-relative wind fields. Photographs taken by K. Knupp (from Kingsmill and Wakimoto 1991).

burst. A prominent precipitation core is seen at midlevels within the thunderstorm (Fig. 7.22a). As time progresses, the core begins a descent to lower levels (Fig. 7.22b, c); however, note that the vertical velocities are still positive. Negative vertical velocities appear soon after the core exits the cloud base and enters the near dry adiabatic subcloud layer. This scenario suggests that microbursts are initiated by precipitation loading and are subsequently maintained by cooling (Srivastava 1985, 1987; Knupp 1988; Proctor 1989). The dominance of negative thermal buoyancy at low levels, combined with the decreasing effect of precipitation loading, can result in a displacement between the location of the maximum downdraft and the maximum in radar reflectivity (Knupp and Cotton 1985; Parsons and Kropfli 1990).

An important observation was noted in the simulations shown in Fig. 7.17. Although the downdraft associated with the dry microburst was much deeper and almost twice as intense as that of the wet microburst, both produced identical outflow speeds. This was a result of the cold air for the wet microburst being primarily situated at low levels. Although this cooling at low levels is not able to translate into strong enhancement of the downdraft, it may contribute to the outflow speeds through enhanced horizontal pressure gradient forces (Krueger et al. 1986). This result emphasizes an important aspect of high winds within convective storms: the nonlinear relationship between vertical velocities and outflow speeds. Foster (1958) examined this relationship and suggested that w and u were nearly identical, with outflow speeds being slightly less than the downdrafts. Results by Wakimoto (1985) suggest that downdrafts associated with small cold pools (e.g., dry microbursts) are more likely to produce outflows of approximately the same intensity. Proctor (1989) states that the relationship between u and w is more complex and is sensitive to environmental conditions, radius of the downdraft, and precipitation type. Therefore, it may be difficult to estimate peak outflow speeds given only the maximum downdraft velocities.

4) GUST FRONT

The outflow of downdraft air from thunderstorms is an important phenomenon in convective storm dynamics. It can aid the development of new thunderstorm cells and it can cut off old cells from their supply of buoyant air. The leading edge of the outflow is often referred to as the *gust front*. An example of a gust front is shown in Fig. 7.23. The large-scale outflow depicted in the figure developed from intense thunderstorm activity located in the western section of the domain. Figure 7.24 is a schematic vertical cross section of the basic feature of a mature outflow. The bold solid line depicts the transition zone between the cool outflow air and the warmer environment, while

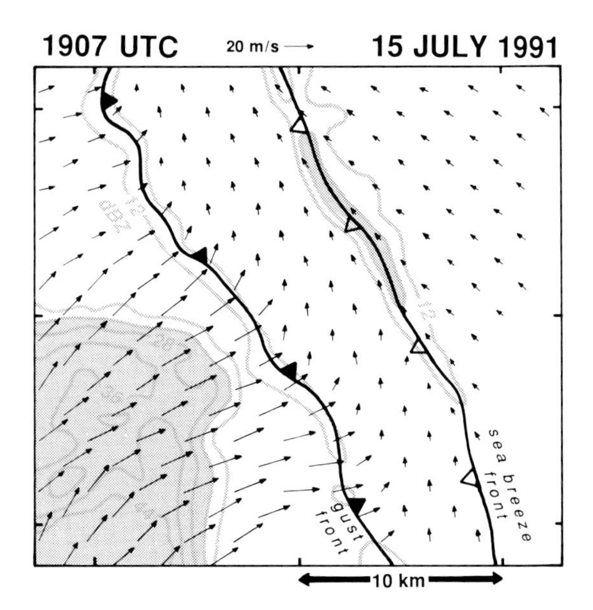

FIG. 7.23. Dual-Doppler synthesis of a gust front and sea breeze front at 0.1 km AGL. Contours of radar reflectivity are shown as gray lines, with values greater than 20 dBZ shaded gray. Based on a figure from Kingsmill (1995).

the arrows show the flow field relative to the gust front. As cold air approaches the front from within the outflow, much of it is deflected upward (counterclockwise) by a solenoidally (or baroclinically) induced circulation (Mitchell and Hovermale 1977; Droegemeier and Wilhelmson 1987). Air closer to the ground turns downward under the influence of surface friction (marked "backflow" in Fig. 7.24). Warm environmental air approaching the gust front is forced over the cold air pool, and often forms a shelf, roll, rope, or arc cloud. In arid regions, the outflow may only be apparent by the great amounts of dust and sand swept up by turbulence within the onrushing gust front head (e.g., Idso 1974).

Separating the body of the outflow from the elevated head is a turbulent wake, which is in turn followed by several waves distorting the shape of the upper boundary of the dense outflow. Droegemeier and Wilhelmson (1987) have shown that the Richardson number in this region supports the development of Kelvin–Helmholtz instability (Fig. 7.25). A Kelvin–Helmholtz billow originates in the head and propagates rearward, gradually damping and dissipating, while a new circulation forms in the head as a result of baroclinic generation. These waves have been shown to exist observationally atop thunderstorm outflows (e.g., Mueller and Carbone 1987; Mahoney 1988; Intrieri et al. 1990; Wakimoto et al. 1996). Mahoney (1988) and Intrieri et al. (1990) have suggested that the descending branch of the Kelvin–Helmholtz circulation may occasionally reach the surface, producing a surface divergence and a microburst-like signature.

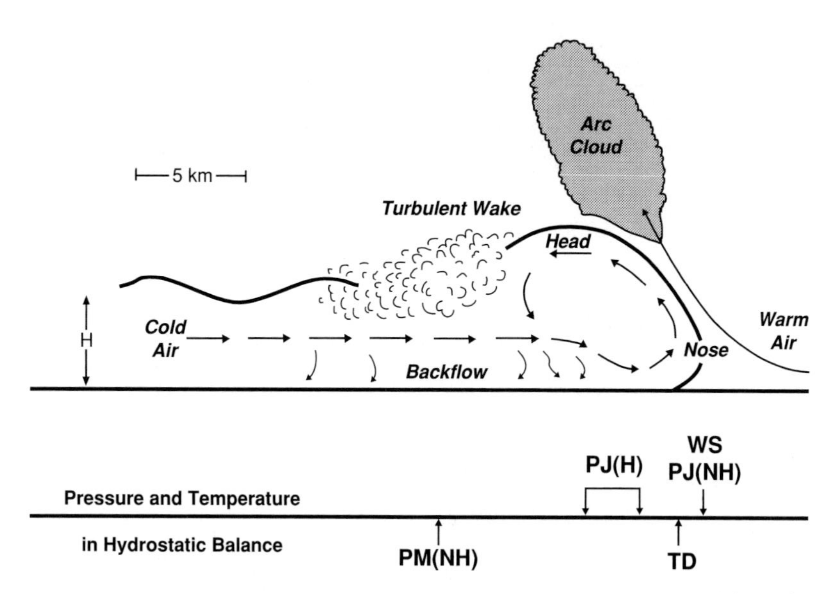

FIG. 7.24. Schematic cross section through the gust front of a thunderstorm. Prior to the arrival of the cold air, the wind begins to shift (WS) and the pressure increases or jumps (PJ-NH) due to a dynamic deceleration between the cold and warm air masses. The passage of the cold air, often called the temperature break or drop (TD), is accompanied by a hydrostatic increase in pressure (PJ-H) behind the outflow head. H is some characteristic depth of the gravity current usually taken as the height of the following flow far behind the head. The main body of the outflow is characterized by pressure and temperature that are in quasi-hydrostatic balance. From Droegemeier and Wilhelmson (1987); based on earlier studies of Charba (1974), Goff (1975), Wakimoto (1982), and Koch (1984).

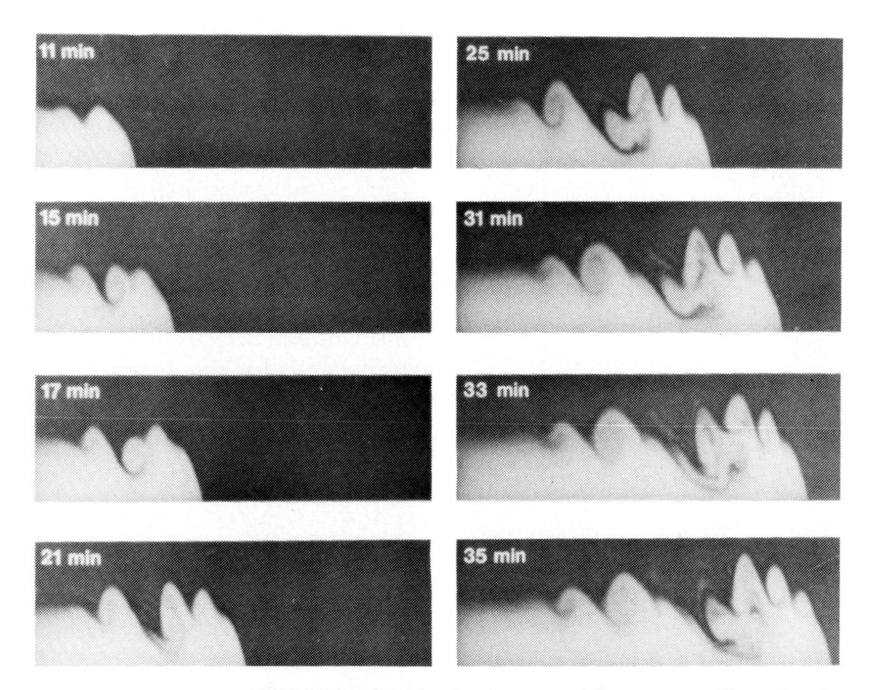

FIG. 7.25. Numerical model simulation of a thunderstorm outflow structure illustrating the sequence of wave generation on top of the outflow layer. Negative perturbation potential temperature (cold air) is shown by shading. The whiter the shading, the colder the air. The lowest values, occurring near the surface, are about −5°C. Results from Droegemeier and Wilhelmson (1987).

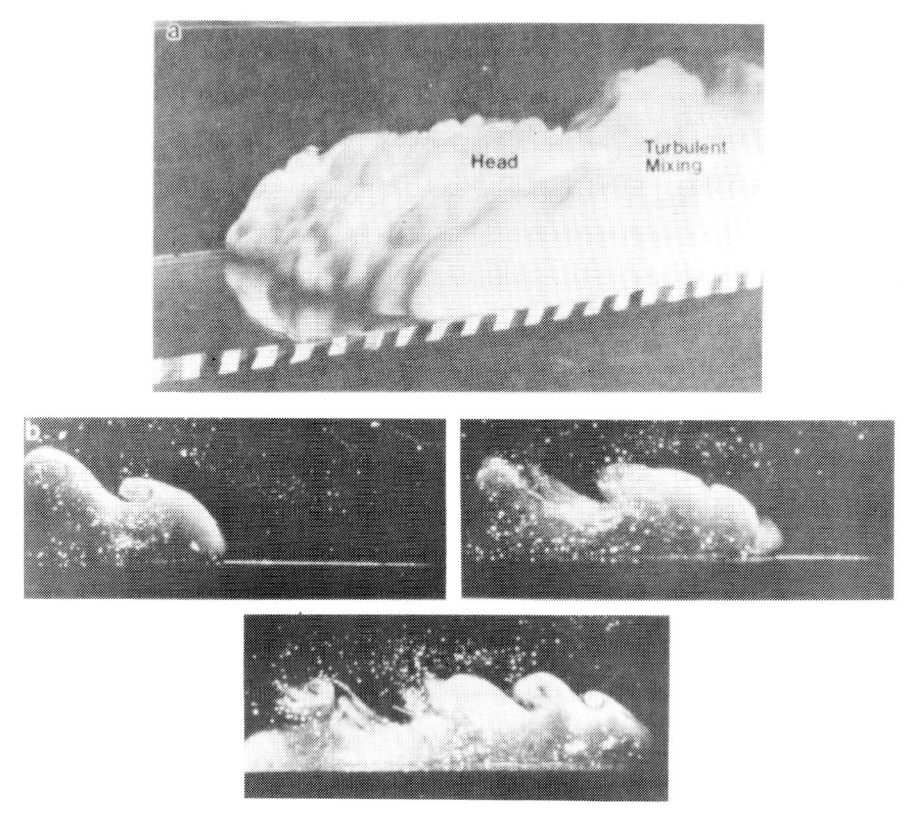

FIG. 7.26. Examples of gravity currents observed in laboratory tanks. From Simpson (1969). Reprinted with the permission of the Royal Meteorological Society.

The outflow from thunderstorms has been shown to be dynamically similar to the gravity (or density) current (Fig. 7.26; e.g., Charba 1974; Simpson 1987). A *gravity current* is defined as a mass of high-density fluid flowing along a horizontal bottom and displacing ambient fluid of lesser density. It is driven by the horizontal pressure gradient acting across the sharp lateral interface separating the two fluids. Many of the basic properties of the outflows (e.g., Kelvin–Helmhotz waves) have been recreated using laboratory tank experiments of gravity currents (Middleton 1966; Simpson 1969, 1972, 1987; Simpson and Britter 1979, 1980; Britter and Simpson 1978, 1981). The sea breeze, shown in Fig. 7.23, is another example of a gravity current.

The theoretical propagation speed of the front has been shown by Benjamin (1968) to be represented by

$$V_f = \left(2gH\frac{\Delta\rho}{\rho}\right)^{1/2}, \tag{7.4}$$

where V_f is the propagation speed of the front in a calm environment, $\Delta\rho$ is the positive density difference between the gravity current and the surrounding fluid, and H is the characteristic depth of the outflow. This equation is valid only for inviscid, incompressible, steady flows in an unstratified environment of infinite depth. It is common in meteorological situations to write the equation as

$$V_f = k\left(gH\frac{\Delta\rho}{\rho}\right)^{1/2}, \tag{7.5}$$

where k is the internal Froude number that has a value of $\sqrt{2}$ for steady, inviscid flow and appears to have a value closer to 0.75 for many gravity current phenomena seen in the atmosphere (e.g., Wakimoto 1982; Droegemeier and Wilhelmson 1987; Mahoney 1988). This suggests that turbulent mixing and surface friction are significant for atmospheric gravity currents resulting in reduced frontal speeds.

As shown in Fig. 7.24, the pressure ridge precedes the passage of the gust front (PJ-NH). The fact that the ridge occurs in advance of the temperature drop indicates that it cannot be hydrostatically induced. The location of this ridge is at the center of surface convergence near the front, implying that it is dynamically induced (Wakimoto 1982; Droegemeier and Wilhelmson 1987). The second maximum of pressure to the rear of the front is hydrostatic since it simply reflects the weight of the fluid above (PJ-H). The minimum of pressure (PM-NH) appears to be nonhy-

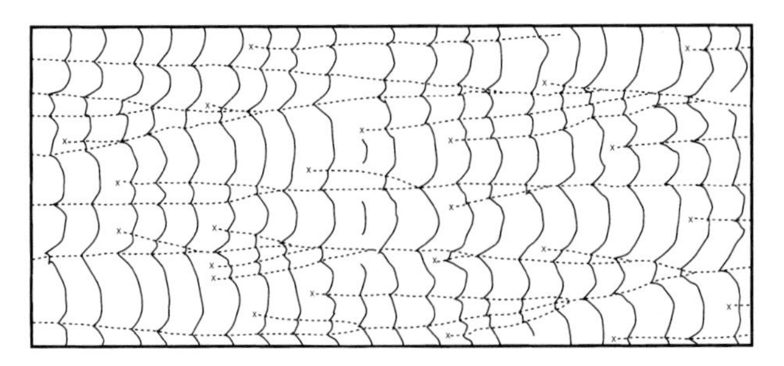

FIG. 7.27. Drawing from a shadowgraph series in a horizontal plane showing the evolution of lobes from a tank experiment of a gravity current. The dashed lines show the continuity of clefts and the x points where new clefts appear. Time interval is 0.5 s. $\Delta\rho/\rho$ is 1%. Based on a figure from Simpson (1972). Reprinted with the permission of Cambridge University Press.

drostatic and associated with dynamical effects of the turbulent wake (Fig. 7.24).

In addition to the turbulent structure associated with Kelvin–Helmholtz instability shown in Figs. 7.25 and 7.26, along-frontal variability has been documented in the laboratory experiments (see top panel of Figs. 7.26 and 7.27). These clefts and lobes at the leading edge of the gravity current are believed to be caused by gravitational instability of the less dense fluid, which is overrun by the nose of the current (Simpson 1987). Similar structures have been noted in observations of gusts fronts (McCaul et al. 1987; Mueller and Carbone 1987; Weckwerth and Wakimoto 1992); however, it is not certain that their generating mechanism is the same. McCaul et al. (1987) and Mueller and Carbone (1987) provide detailed kinematic windfields illustrating that the inflections are associated with vorticity maxima. These structures may be caused by barotropic instability along a strong horizontal shear zone. Other studies of alongfrontal variability at the leading edge of a gravity current suggest that they may be caused by the interaction of the fronts with horizontal convective rolls that develop within the convective boundary layer (Wilson et al. 1992; Wakimoto and Atkins 1994; Atkins et al. 1995; Dailey and Fovell 1999; Rao et al. 1999). Horizontal convective rolls are a common form of boundary layer convection (e.g., Kuettner 1959; LeMone 1973; Brown 1980), consisting of counterrotating helices aligned nearly parallel to the mean boundary layer shear direction.

It is common knowledge that deep outflows associated with a large cold pool can result in strong surface winds in the direction of the moving front. However, another burst of strong surface winds can result from a less appreciated mechanism. Fujita (1955), Pedgley (1962), and Schaefer et al. (1985) have identified the surface pressure field for squall lines accompanied by a trailing stratiform region. A schematic illustrating the pressure and wind analyses is shown in Fig. 7.28. The prominent features are the mesohigh owing to rain-cooled downdrafts (Fujita 1959, 1963) and the wake low. The wake low is positioned toward the rear of the area of precipitation-cooled air with surface flow converging into it. A surface difluence axis occurs to the rear of the thunderstorm mesohigh center with air accelerating through the mesohigh toward the leading convective line.

More recent work by Johnson and Hamilton (1988), Stumpf et al. (1991), and Loeher and Johnson (1995) has clarified several aspects of the features shown in Fig. 7.28. In the formative stage of the system, the pressure field to the rear of the mesohigh associated with the squall line is relatively flat with only weak stratiform precipitation present. During the developing to mature stages of the squall line, a trailing stratiform region forms, and a pronounced wake low appears at its back edge (Fig. 7.29). The wake low is attributed to subsidence warming (Williams 1963; Zipser 1977) and is a surface manifestation of a descending rear-inflow jet (also shown by Zhang and Gao 1989). The warming is maximized at the back edge of the precipitation area where there is insufficient evaporative cooling to offset strong adiabatic warming (Fig. 7.29).

Loeher and Johnson (1995) have shown that the pressure gradient between the mesohigh and wake low

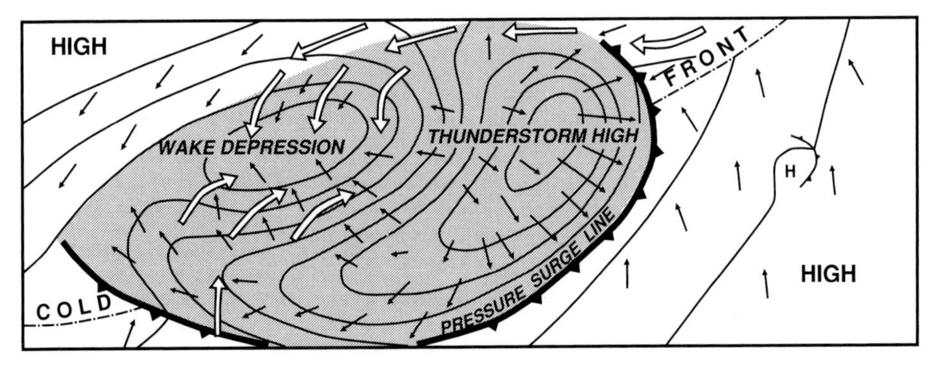

FIG. 7.28. Schematic of surface pressure field in a squall-line thunderstorm. Small arrows indicate surface wind; large arrows indicate relative flow into the wake. Gray shading indicates the extent of the precipitation-cooled air. Based on a figure from Fujita (1955).

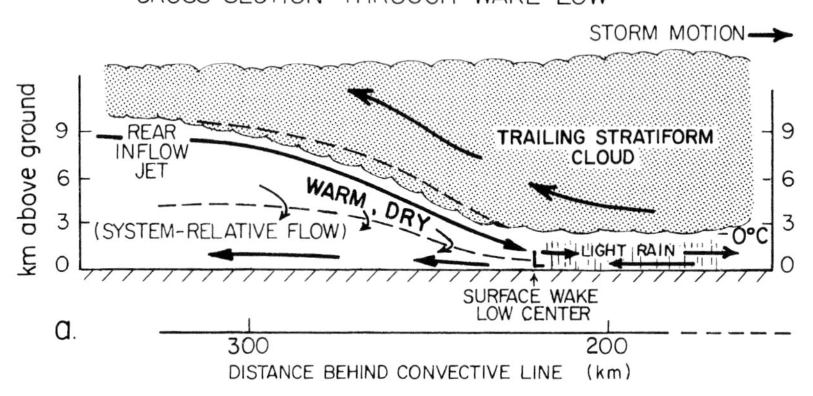

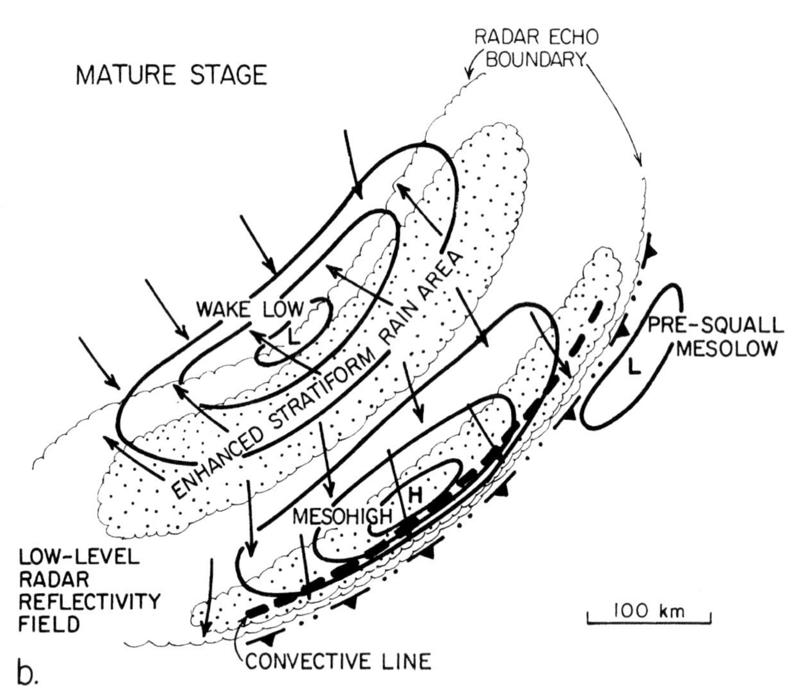

FIG. 7.29. Schematic cross section through wake low (a) and surface pressure and wind fields and precipitation distribution during squall line mature stage (b). Winds in (a) are system relative with the dashed line denoting zero relative wind. Arrows indicate streamlines with those in (b) representing actual winds. Note that horizontal scales differ in the two schematics. From Johnson and Hamilton (1988).

shown schematically in the figure is typically 2 mb/10 km but can be as large as 5 mb/10 km. This intense pressure gradient drives strong horizontal flow through the wake low (shown in the figure), which often exceeds 25 m s^{-1}. Such strong winds in a direction diametrically opposed to the flow immediately behind the gust front have not received much attention in the literature. The origin of the rear-inflow jet will be discussed in the section on mesoscale downdrafts.

5) DOWNDRAFTS ALOFT—OVERSHOOTING AND MIDLEVEL

Aircraft and Doppler radar observations have revealed the existence of several downdraft types within convective storms that exist aloft but rarely reach the surface. It is the downdraft associated with precipitation that systematically reaches the surface (the occlusion downdraft being an important exception). The most common downdrafts referred to are the *overshooting* and *midlevel* downdrafts. Both of these downdrafts can be associated with intense speeds. While these downdrafts do not result in damaging wind at the surface, they are included here for completeness.

(i) Overshooting downdrafts

The thermodynamics of overshooting downdrafts have been illustrated by Newton (1966). Updrafts within severe convective storms ascend to heights exceeding the equilibrium level defined by parcel theory. This is called *overshooting*, and the rising air quickly becomes negatively buoyant and descends. Very strong downward velocities can be expected, which are confined to the upper levels. Some of the best visual documentation of overshooting tops from satellite images and aircraft has been presented by Fujita (1974, 1992) and Fujita et al. (1976). Fujita (1974) estimated peak downdrafts speeds of 41 m s^{-1} with the collapse of an overshooting cloud top. Tripoli and Cotton (1986) associate the most intense downdrafts in a storm simulation with the collapse of overshooting tops. These downdrafts similarly overshoot their equilibrium level, leading to buoyancy oscillations that slowly decay with time. Other measurements of overshooting downdrafts >8 m s^{-1} near the storm top have been presented by Battan (1980). Comparable values were noted in numerical simulations by Schlesinger (1984).

During the early stages of research on the microburst, Fujita and Byers (1977) and Fujita and Caracena (1977) proposed that the microburst originates in the upper troposphere from the collapse of an overshooting cloud top. Fujita and Caracena (1977) offered compelling evidence (see their Fig. 30), based on a dual-Doppler analysis by Kropfli and Miller (1976), that the downdrafts extend over the entire depth of the storm. Subsequent thermodynamic and radar studies (e.g., Betts and Silva Dias 1979; Raymond et al. 1991) have argued that such a descent is highly unlikely. Downdrafts that reach the surface are not expected to originate from a level much higher than the minimum equivalent potential temperature (Betts and Silva Dias 1979). Since the analysis presented by Fujita and Caracena (1977) was based on a wind synthesis produced during the infancy of multi-Doppler radar techniques, the derived vertical velocities could have been in error.

(ii) Midlevel downdraft

Downdrafts located in the mid- and upper levels of storms have only recently been discussed in the literature (Heymsfield and Schotz 1985; Knupp and Cotton 1985; Smull and Houze 1987a; Knupp 1987; Fovell and Ogura 1988; Kingsmill and Wakimoto 1991; Raymond et al. 1991; Biggerstaff and Houze 1993; Smull and Augustine 1993; Yuter and Houze 1995a,b). An example of this downdraft is shown in Fig. 7.30. The downdraft velocities at midlevels in the figure were the strongest observed during the storm's life cycle, exceeding those associated with a microburst at low levels. The downdraft did not reach the surface and was located near the gradient of radar reflectivity (i.e., cloud edge). There is a tendency for midlevel downdrafts to be situated in areas of low reflectivity or near the cloud edge (Schlesinger 1978; Knupp and Cotton

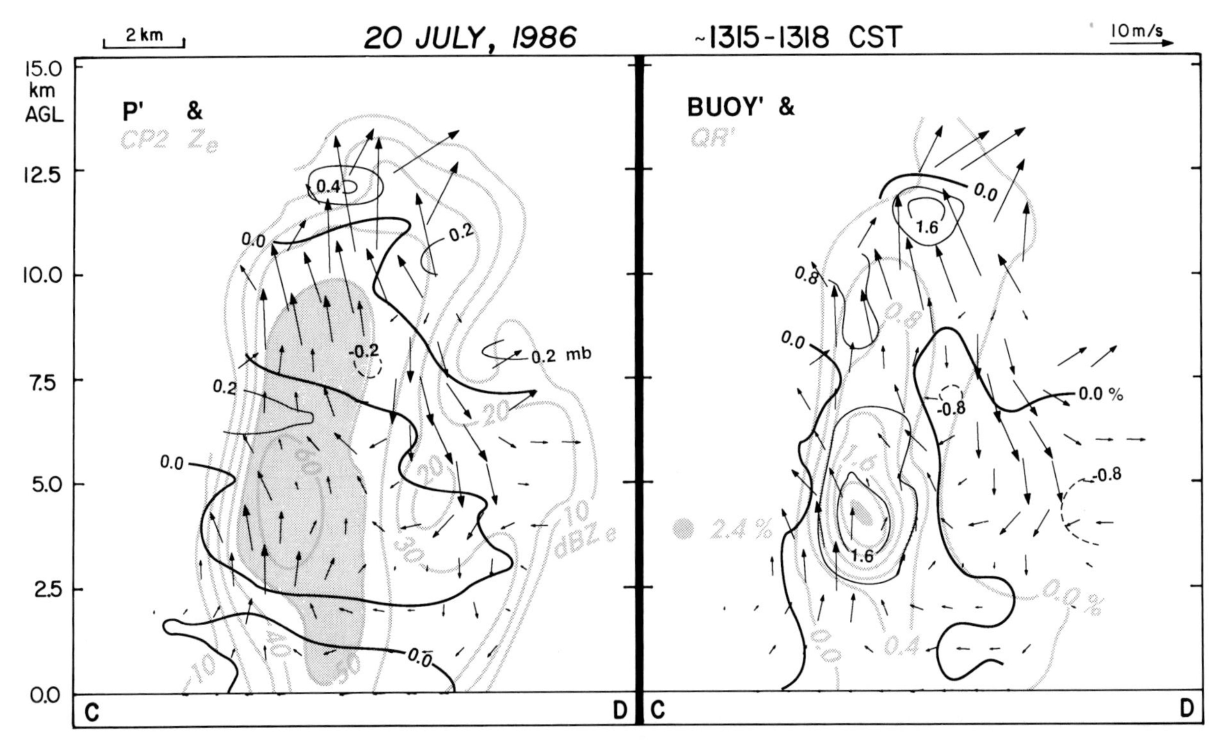

FIG. 7.30. Vertical cross section of storm-relative velocity, perturbation pressure, thermal buoyancy, radar reflectivity factor, and precipitation loading for the thunderstorm on 20 Jul 1986. Pronounced midlevel downdraft is evident in the figure. From Kingsmill and Wakimoto (1991).

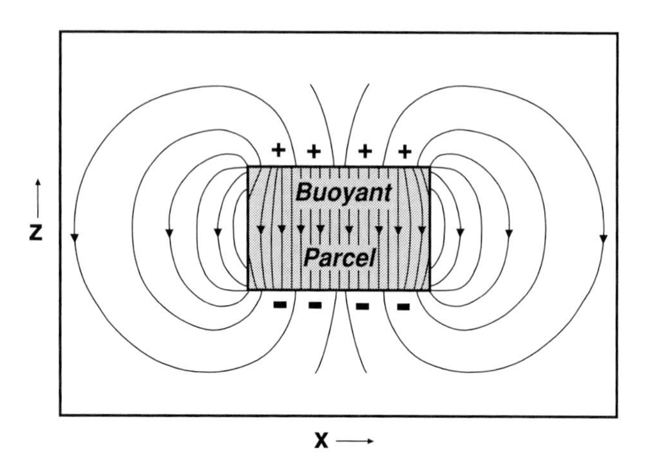

FIG. 7.31. Vector field of buoyancy pressure gradient force for a uniformly buoyant parcel of finite dimensions in the x–z plane. The plus and minus signs indicate locations of positive and negative pressure perturbations. Based on a figure from Houze (1993). Reprinted with the permission of Academic Press.

1982; Knupp 1987; Raymond et al. 1991; Yuter and Houze 1995a,b). The low-level downdraft tends to be located in zones of high reflectivity, since precipitation loading and phase changes of condensate are its main forcing mechanism (recall that the occlusion downdraft and dry microburst are two exceptions).

There appear to be two mechanisms that maintain the midlevel downdrafts. Yuter and Houze (1995b) note that the midlevel downdrafts were concentrated at levels that corresponded to the strongest upper-level updrafts and were associated with weak mean reflectivity, an indication that the downdrafts were not forced by precipitation loading. These observations suggest that the downdrafts were mechanically forced (Heymsfield and Schotz 1985; Smull and Houze 1987a; Fovell and Ogura 1988; Raymond et al. 1991; Yuter and Houze 1995b). This mechanism is a result of the pressure gradient forces required to maintain mass continuity in the presence of buoyant parcels and is illustrated in Fig. 7.31.

Buoyancy of an air parcel will produce a perturbation pressure field indicated by the plus and minus signs shown on the figure. The lines in Fig. 7.31 indicate buoyancy pressure gradient acceleration (BPGA). Within the parcel, the lines of the BPGA are downward. There is a divergence of the BPGA field at the top of the parcel, and convergence at the bottom. Outside the parcel, lines of force are up above the parcel, downward in the regions to the sides of the parcel, and upward just below the parcel. These lines indicate the directions of forces acting to produce the compensating motions in the environment that are required to satisfy mass continuity when the buoyant parcel moves upward. This effect can be especially important near the tops of growing clouds, where rising towers are actively pushing environmental air out of the way. Not surprisingly, Sun et al. (1993) have shown that these downdrafts are forced downward below their equilibrium level; accordingly, there is a tendency for these downdrafts to be warm. Indeed, the thermal buoyancy of the averaged downdrafts was nearly indistinguishable from that of the neighboring averaged updrafts (see also Igau et al. 1999).

The mechanism described above appears to explain the origins of the midlevel downdrafts under environmental conditions that are characterized by weak vertical shear of the horizontal wind. When the wind shear becomes stronger, pressure perturbations produced by dynamic forcing can become important (e.g., Rotunno and Klemp 1982) and facilitate the initiation of a midlevel downdraft. This effect has been termed *wake entrainment* by Knupp and Cotton (1985) and is schematically shown in Fig. 7.32. Airflow is diverted by high perturbation pressure around the upshear edge, while a wake forms in the downshear edge (Heymsfield et al. 1978). This wake low is dominated by cloud-scale inflow and can rapidly entrain environmental air. Ramond (1978), Tripoli and Cotton (1980), Betts (1982), Knupp and Cotton (1982), Paluch and Breed (1984), and Raymond et al. (1991) have suggested that the mixing of this environmental air with cloudy air can promote evaporative cooling and induce a downdraft. The radar echo in the area of the wake low can be associated with an indentation (or weak-echo notch) that indicates possible sublimation or evaporation of precipitation particles (Knupp and Cotton 1982). The magnitude of the pressure perturbations described in Fig. 7.32 was considered by Rotunno and Klemp (1982) in the following equation:

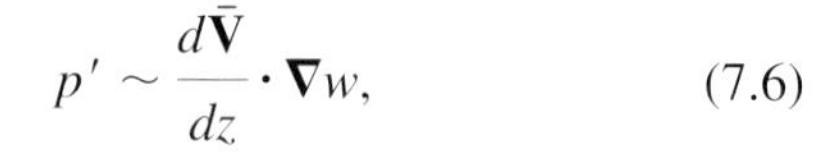

$$p' \sim \frac{d\bar{\mathbf{V}}}{dz} \cdot \nabla w, \qquad (7.6)$$

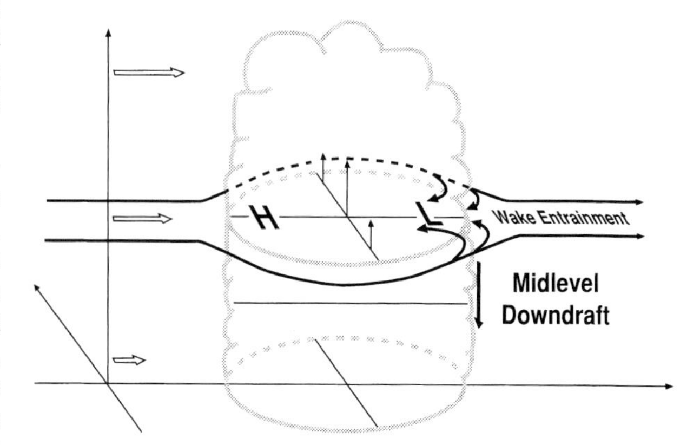

FIG. 7.32. Schematic diagram illustrating wake entrainment within the downshear flank of a convective cloud. The symbols H and L represent high and low pressure perturbations. These perturbations, along with the vertical vorticity patterns, are produced by vertical motion interacting with the environmental flow. The locations of a region of possible wake entrainment and midlevel downdraft are indicated on the figure. Based on a figure from Knupp and Cotton (1985).

where $\bar{\mathbf{V}}$ is the horizontal environmental wind vector and w is the vertical motion. The magnitude of p' is thus dependent on the magnitudes of wind shear and horizontal gradients of updraft speed (e.g., LeMone et al. 1988a,b).

In practice, it may be difficult to determine which of the two triggering mechanisms induces the midlevel downdraft air in storms that grow in moderate wind shear situations. Theory predicts that both mechanisms will occur at heights where the updrafts are vigorous; however, mechanically forced downdrafts appear to encircle an updraft and are pervasive throughout the storm (Yuter and Houze 1995b), while wake entrainment preferentially develops on the downshear sides of the updraft (Schlesinger 1978; Knupp 1987).

Normand (1946) and Newton (1950) proposed, based on observations of minimum of θ_e at midlevels (typically 500–600 mb), that cold downdraft air reaching the surface originated several kilometers above the ground. By invoking a thermodynamic analysis, Betts and Silva Dias (1979) argued that downdrafts that reach the surface are not expected to originate from a level much higher than the minimum equivalent potential temperature. The shallowness of the low-level downdraft has subsequently been confirmed by numerous investigators (Fankhauser 1976; Kropfli and Miller 1976; Battan 1980; Knupp and Cotton 1985; Knupp 1987; Kingsmill and Wakimoto 1991; Raymond et al. 1991). In light of these results, it is instructive to revisit the classical model (Fig. 7.33) of the mature stage of a nonsevere thunderstorm by Byers and Braham (1949). Although their intent was to depict the airflow at an instant of time, rather than the trajectories, they suggested that downdrafts reaching the surface are continuous up to 8–9 km. In contrast, Raymond et al. (1991) state that downdrafts descending through 9 km are almost nonexistent. Kingsmill and Wakimoto (1991), Biggerstaff and Houze (1993), and Yuter and Houze (1995a) have suggested that the midlevel and low-level downdrafts form separately and, if located one above the other, may then connect into a draft that extends through the depth of the storm. This is believed to be the situation described in Fig. 7.33, although it has never been discussed before, nor is it clear that the vertical alignment of these two downdrafts is common in this type of thunderstorm. Note that the low-level downdraft is embedded within heavy precipitation while the midlevel downdraft is near the cloud edge in the figure. This is consistent with the recent studies on these two phenomena.

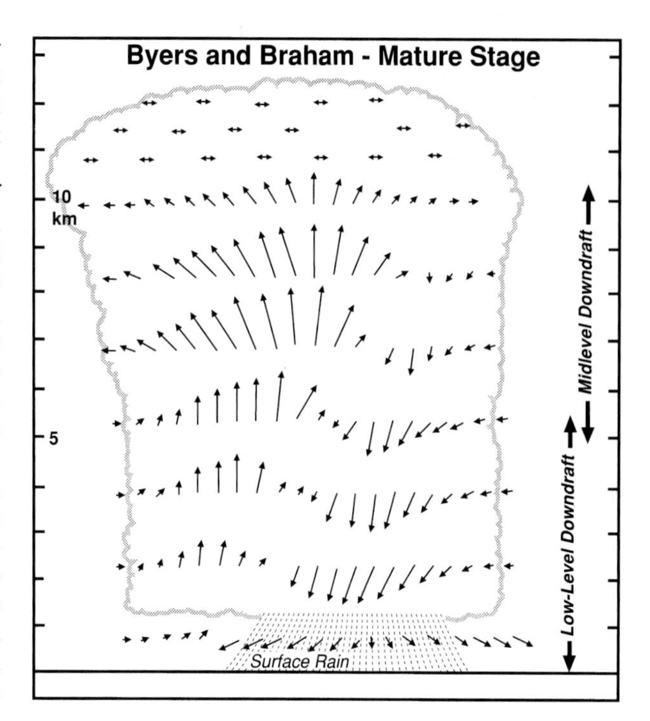

FIG. 7.33. Mature stage of the three-stage model of the updraft–downdraft circulations in a thunderstorm cell forming under weak-shear conditions. The typical heights where the precipitation-associated low-level downdraft and the midlevel downdraft occur are indicated on the figure. Based on a figure from Byers and Braham (1949).

b. Mesoscale downdraft

An important storm type that is responsible for much of the warm-season rainfall and severe weather over the central portion of the country is the mesoscale convective system (MCS). An MCS is a precipitation system that has a horizontal scale ~100 km or more and includes significant convection during some part of its lifetime. The MCSs include the mesoscale convective complex described by Maddox (1980) and Zipser (1982), as well as other squall lines and groups of convective storms. The focus in this paper is the squall lines with extensive anvils characterized by stratiform precipitation region. These types of storms have been identified both in the mid latitudes (Newton 1950; Fujita 1955) and in the Tropics (Hamilton and Archbold 1945; Zipser 1969). The conceptual model through the squall line is shown in Fig. 7.34, based on numerous studies (Newton 1950; Fujita 1955; Zipser 1969, 1977; Houze 1977; Gamache and Houze 1982; Smull and Houze 1985, 1987a,b; Rutledge et al. 1988; Johnson and Hamilton 1988). The heavy black line in the figure indicates the boundary of the storm as seen by a radar. The light scalloped line indicates the extent of the clouds. The gray shading indicates regions of enhanced radar reflectivity. LeMone et al. (1984) have shown that some tropical systems can deviate from the schematic model shown in Fig. 7.34. These tropical systems exhibit a stronger rearward tilt (30° relative to the surface).

Intense convective downdrafts are associated with the strong cells within the squall line. These are the low-level downdrafts that are driven by precipitation loading and evaporative cooling. The downdrafts

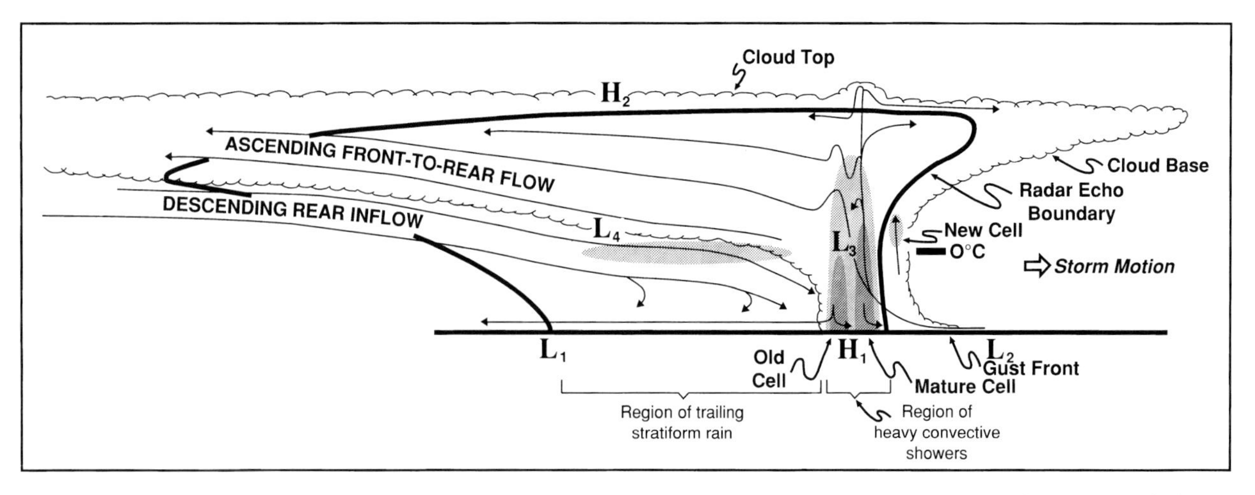

FIG. 7.34. Conceptual model of a squall line with a trailing stratiform region viewed in a vertical cross section oriented perpendicular to the convective line. Based on a figure from Houze et al. (1989).

spread out in the boundary layer behind the gust front and toward the rear of the system. There is a region at mid- to upper levels where downdrafts are prevalent immediately to the rear of the strongest updraft cells (Heymsfield and Schotz 1985; Smull and Houze 1987a). These downdrafts are mechanically forced, and their net effect can lead to a column of mean downward motion in the zone of minimum radar reflectivity (Sommeria and Testud 1984; Chong et al. 1987). Biggerstaff and Houze (1991) refer to this region as the *transition zone.*

The trailing region of stratiform precipitation is characterized by a marked radar bright band immediately below the melting level and a layer of subsiding rear-to-front flow (Newton 1950; Ragette 1973; Ogura and Liou 1980; Smull and Houze 1987b; Chalon et al. 1988; Schmidt and Cotton 1989), which enters the region just below the trailing stratiform cloud. Above the rear inflow, within the stratiform cloud, is a layer of upward-sloping, front-to-rear flow emanating from the upper portions of the convective line. This flow, which contains mesoscale ascending motion, advects ice particles detrained from the convective cells rearward. Since the fall speed of the ice particles is generally larger than the speed of the ascending air motion (Rutledge et al. 1988), the ice particles slowly fall as they are carried rearward, growing by vapor deposition in the ascending front-to-rear flow. These particles eventually reach warmer air and form aggregates. The aggregates fall through the 0°C level, below which they melt, producing the radar bright band and the region of heavier stratiform rain. Numerical simulations by Fovell and Ogura (1988) and Tao and Simpson (1989) have shown that the production of these particles in the upper levels of the convective region, and their advection rearward by the front-to-rear ascent, are crucial to the formation of the trailing stratiform region.

Generally, the mean vertical motion is zero at a height of 0–2 km above 0°C; upward above this level, corresponding to the front-to-rear ascent; and downward below, corresponding to the subsiding rear inflow air. The typical vertical motions within these mesoscale updrafts and downdrafts are <0.5 m s^{-1}. The existence of mesoscale downdrafts within the stratiform region of MCSs has been discussed by Zipser (1969, 1977), Houze (1977), Miller and Betts (1977), Gamache and Houze (1983), Srivastava et al. (1986), and Brandes and Ziegler (1993). These downdrafts are driven by evaporative cooling (Ryan and Carstens 1978; Brown 1979) and enhanced by melting (Leary and Houze 1979) and sublimation (Stensrud et al. 1991), and appear to be characteristic of squall-line systems (Houze 1989). It should be noted that rear inflow of dry air does not occur in all squall lines (Smull and Houze 1987b). Therefore, the mesoscale downdraft and the rear inflow can be considered separate entities (Biggerstaff and Houze 1991).

The rear inflow exhibits a large variety of structure and strengths. Smull and Houze (1987b) studied 18 MCSs and found that three of the cases were associated with strong rear inflow (>10 m s^{-1} of relative flow), five had moderate rear inflow (5–10 m s^{-1}), and ten, referred to as *stagnation zone cases,* had little or no rear inflow. Smull and Houze (1987b) hypothesized that rear-inflow jets were generated internally by the convective system itself, as opposed to being the manifestation of upstream variations in the preconvective winds. Recent radar analyses by Klimowski (1994) support the idea that the rear inflow systematically develops within the MCS. The forcing mechanism was suggested to be a hydrostatically induced minimum in perturbation pressure that develops under the warm convective updraft current that slopes over the cold surface outflow (LeMone 1983; LeMone et al. 1984; Roux 1988). A positively buoyant plume is characterized by high and low perturbation pressure above and below the maximum buoyancy, respectively (see Fig. 7.31). Weisman (1993) confirmed that the mesolow is largely dominated by the buoyancy pressure, with a small contribution from the dynamic pressure. There is also a contribution from the pressure minimum within a density current outflow head (Fovell and Ogura 1988; Weisman et al. 1988). The latter pressure is dynamically induced by cyclostrophically balanced flow within the overturning flow.

LaFore and Moncrieff (1989) have created a comprehensive schematic diagram illustrating how the rear inflow, buoyancy, and pressure minimum are related (Fig. 7.35). The trailing stratiform cloud layer is positively buoyant. Immediately to the rear of the convective region, the cloud layer is thicker and the buoyancy is greater. The buoyancy decreases and the cloud layer thins toward the rear of the system. Therefore, the minimum pressure perturbation is greater in the interior of the convective system, just to the rear of the convective region. The resulting horizontal gradient of perturbation pressure across the stratiform region accelerates the air from the rear to the front. As ice particles fall from the stratiform precipitation aloft into this rear-to-front flow, the air is cooled by phase changes and tends to subside while flowing horizontally toward the convective region.

Weisman (1993) has provided another physical explanation that is equivalent to the one described in Fig. 7.35. His schematic diagram (Fig. 7.36) is based on the conceptual framework proposed by Rotunno et al. (1988). Initially, a convective cell evolving in a vertically sheared environment flow leans downshear in response to the ambient vertical wind shear (Fig. 7.36a). As a cold pool develops beneath the convection, the horizontal buoyancy gradients along the edge of the cold pool generate circulation that, along the downshear edge of the cold pool, is of the opposite

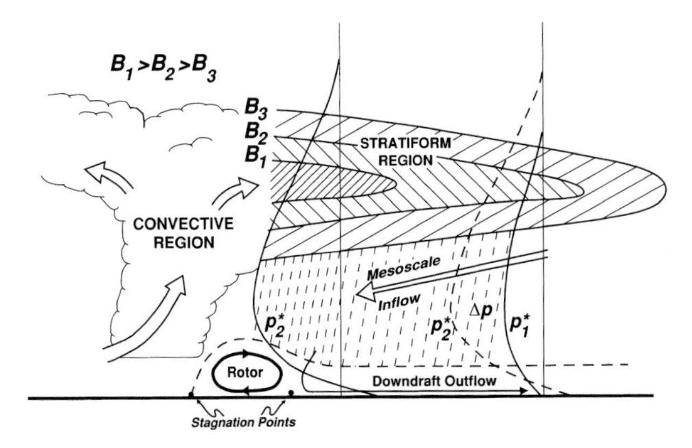

FIG. 7.35. Schematic of the relation between buoyancy (B) and the pressure perturbation fields (p^*) in the mesoscale region of the squall line. The difference between the pressure perturbation at the back (p_1^*) and leading portion (p_2^*) of the stratiform region is indicated by Δp. Based on a figure from LaFore and Moncrieff (1989).

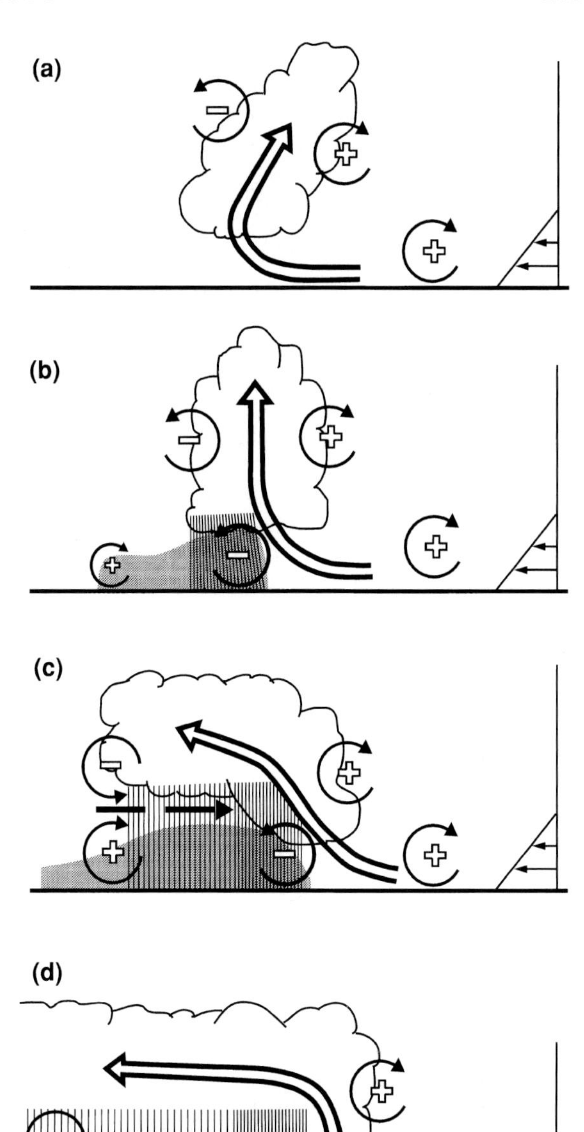

FIG. 7.36. Four stages in the evolution of an idealized bow echo. (a) An initial updraft leans downshear in response to the ambient vertical wind shear, which is shown on the right. (b) The circulation generated by the storm-induced cold pool balances the ambient shear, and the system become upright. (c) The cold pool circulation overwhelms the ambient shear, and the system tilts upshear, producing a rear-inflow jet. (d) A new steady state is achieved whereby the circulation of the cold pool is balanced by both the ambient vertical wind shear and the elevated rear-inflow jet. The updraft current is denoted by the thick double-lined flow vector, with the rear-inflow current in (c) and (d) denoted by the thick dashed vector. The shading denotes the surface cold pool. The thin, circular arrows depict the most significant sources of horizontal vorticity, which are either associated with the ambient shear or are generated within the convective system. Regions of lighter or heavier rainfall are indicated by the more sparsely or densely packed vertical lines, respectively. Based on a figure from Weisman (1993).

sense to the circulation inherent in the ambient shear. When this cold pool circulation balances the circulation in the ambient shear, deep lifting is produced at the cold pool edge, resulting in the production of strong, upright convective cells (Fig. 7.36b). As the cold pool circulation continues to strengthen, it eventually overwhelms the ambient shear, and the convective circulation begins to tilt rearward over the cold air (Fig. 7.36c). This represents the beginning phase of the trailing stratiform precipitation region.

It is during this upshear-tilted phase of the system that a significant rear-inflow jet is present. Horizontal buoyancy gradients along the rear edge of the buoyant plume aloft and cold pool near the surface generate horizontal vorticity that accelerates the flow from rear to front at midlevels, as shown in Fig. 7.36c. If the rear-inflow current descends to the surface, it may increase surface convergence along the gust front and enhance the convective system through triggering of new convective cells. The mechanism described in terms of buoyancy-produced horizontal vorticity is equivalent to that described via the diagnosed pressure analysis in that this configuration of the buoyancy field also implies a minimum in the buoyancy-derived pressure field at midlevels that accelerates the rear-inflow current.

The upshear tilt shown in Fig. 7.36c often signals the decay of the convective system, as lifting along the leading edge of the gust front becomes shallower than during the earlier phases of system evolution. However, this tendency toward decay can be superseded by the development of an elevated rear-inflow jet (Fig. 7.36d). In this scenario, the rear-inflow jet is characterized by the opposite sense of horizontal vorticity beneath the jet level as that generated by the cold pool, and may counteract some of the negative influence of the cold pool circulation and again promote deep lifting at the leading edge of the gust front. Descending jets are characterized by an updraft current that ascends gradually above a spreading surface cold pool, with light to moderate convective and stratiform rainfall extending well behind the leading edge of the cold pool. This structure is often associated with a decaying system, wherein the gust front lifting is not strong or deep enough to regenerate new convective cells and mesoscale circulation slowly weakens. The above discussion can vary significantly for tropical systems (e.g., LeMone et al. 1984) where cold pools are generally shallower and the convective inhibition (CIN) is less.

DERECHOS AND BOW ECHOES

Of particular interest to the severe storm community are those types of mesoscale downdrafts that produce intense winds that result in a family of downbursts. These events have been referred to as *derechos* and can result in significant property damage and loss of life (Fujita and Wakimoto 1981; Johns and Hirt 1987; Przybylinski 1995). The pronounced derechos are produced by long-lived convective systems that take the form of 60–100 km long bow-shaped segments of cells (Fujita 1978; Forbes and Wakimoto 1983; Schmidt and Cotton 1989; Przybylinski 1995). Nolan (1959) first identified the line echo wave pattern (LEWP) where parts of the line of echoes appeared to accelerate. Hamilton (1970) noted that the accelerated part of the line was often accompanied by damaging straight-line winds.

Fujita (1978) coined the term *bow echo* to identify a bow-shaped line of echoes that appeared to be associated with downbursts and derechos. Other investigators have subsequently confirmed this relationship (Fujita 1981; Forbes and Wakimoto 1983; Wakimoto 1983; Schmidt and Cotton 1989; Przybylinski 1995; Businger et al. 1998). Houze et al. (1990) note that symmetrically shaped mesoscale convective systems are frequent producers of damaging wind reports. The echo pattern with these systems appears to assume a bow shape. As shown in Fig. 7.37, the system usually

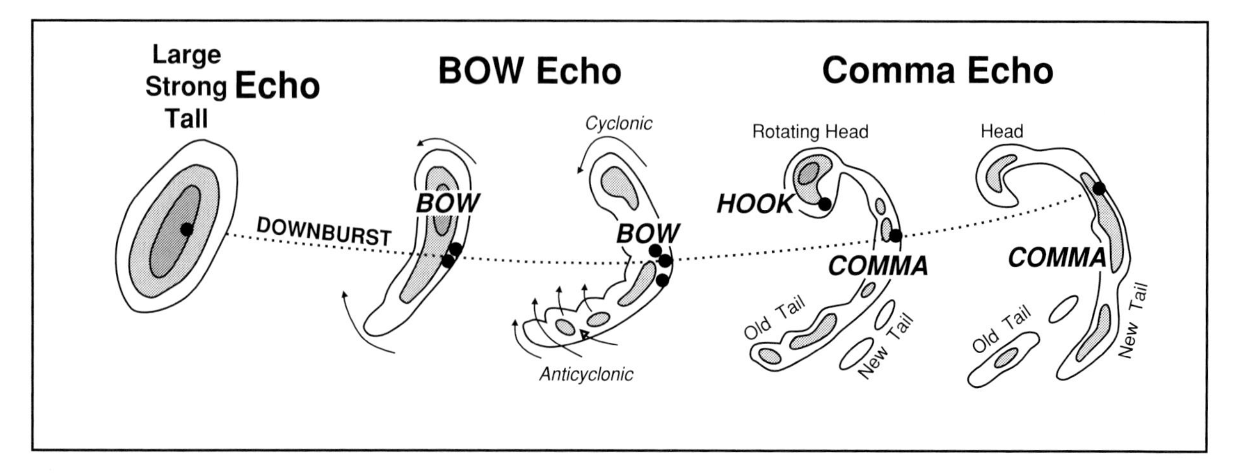

FIG. 7.37. A typical morphology of radar echoes associated with bow echoes that produce strong and extensive downbursts. Black dots indicate possible location of tornadoes. Based on a figure from Fujita (1978).

begins as a single large and strong convective echo that may be relatively isolated or may be part of a more extensive squall line. As the strong surface winds develop, the initial cell evolves into a bow-shaped segment of cells, with the strongest winds occurring at the apex of the bow. The cells at the ends of the segment may appear to move rearward relative to the center of the bow. A strong rear-inflow jet with its core at the apex of the bow is thought to be associated with the bulging and acceleration of the radar echoes. This concentrated jet was proposed by Fujita to be the source of the damaging winds. Rear-inflow notches along the trailing edge of the echo accompany the rear inflow (Smull and Houze 1987b) and indicate where the downbursts winds are the strongest (e.g., Wilson et al. 1980; Przybylinski 1995).

Cyclonic and anticyclonic circulations are noted in the figure along the flanks of the bow. During the decaying stage, the system often evolves into a comma-shaped echo with a cyclonically rotating head. The appearance of the rear-inflow jet in dictating the shape of the bow echo and the midlevel vortices has been numerically simulated by Weisman (1992, 1993). Skamarock et al. (1994) demonstrated that mesoscale vortex formation was a normal component of the evolution of any finite extent convective line, even in the presence of weak vertical wind shear. An example of a bow echo that produced significant damage at the surface is shown in Fig. 7.38. The thick black line represents the radar viewing angle through a weak-echo channel. Strong outflow winds can be seen near this channel.

A proposed mechanism for the origin of the midlevel vortices within the bow echo is shown in Fig. 7.39. An isolated updraft growing in a vertically sheared environment will deform the horizontal vortex lines by tilting them upward, producing cyclonic and anticyclonic vortices on the flanks of the cell (Fig. 7.39a). The subsequent development of a downdraft will deform this pattern considerably as the downdraft gradients provide an additional source of tilting and the downdraft accelerations stretch the vertical vorticity previously produced by the updraft. Weisman (1993) refers to the latter two counterrotating circulations as the *bookend vortices.* Lee et al. (1992) also suggested a similar scenario for the development of midlevel vortices within a bow echo. Numerical simulations have shown that, without Coriolis forcing, the bookend vortices were of equal strength. In simulations that include Coriolis forcing, the cyclonic vortex became stronger (e.g., Weisman 1993; Trier et al. 1997; Weisman and Davis 1998). These results are consistent with observations of the rotating head shown in the schematic model in Fig. 7.37. The process described in Fig. 7.39 is fundamental to the development of the supercell (e.g., Klemp 1987); however, unlike the supercell simulations, the storm does not split but remains part of a line of storms.

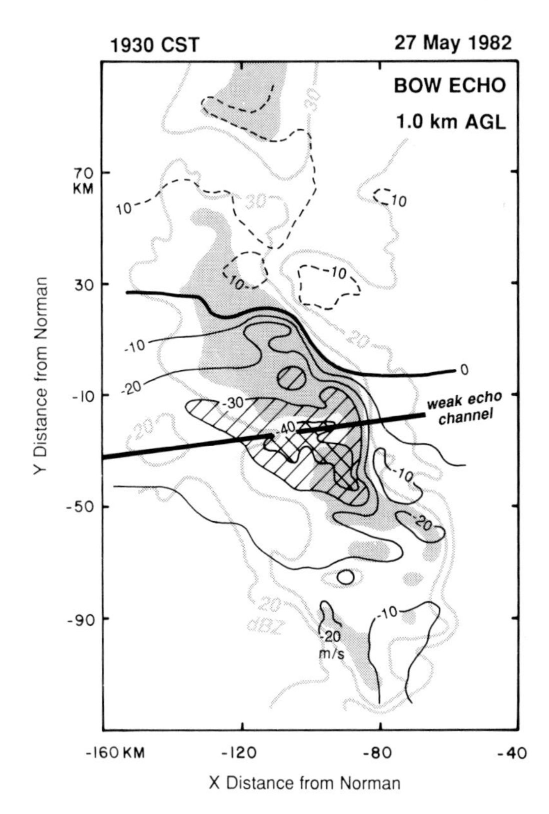

FIG. 7.38. Single-Doppler analysis of a bow echo on 27 May 1982 as viewed with the Norman, OK, radar. Velocities toward and away from the radar are shown as solid and dashed black lines, respectively. Thick black line represents the radar viewing angle. Radar reflectivity is drawn as gray lines, with values greater than 40 dBZ shaded gray. Based on a figure from Burgess and Smull (1990).

The rear-inflow jet is produced by the buoyancy gradients described in Fig. 7.35. The presence of the bookend vortices helps focus and intensify the rear inflow into the core of the developing system. Through an analogy with idealized two-dimensional vortex couplets, Weisman (1993) suggests that these bookend vortices may contribute 30%–50% of the resultant rear-inflow strength during the system's mature phase. The magnitude of this effect decreases with increasing distance between the vortices. Jorgensen and Smull (1993) have provided some detailed observations of the cyclonic bookend vortex within a bow echo.

Recent numerical simulations by Weisman and Davis (1998) reveal a more complex scenario for the generation and characteristics of mesoscale vortices within convective lines. Bookend vortices (also referred to as *line-end vortices*) are produced behind the ends of the convective line for all vertical wind shears considered in their study. Smaller pairs of vortices, however, can form at the end of bow-shaped convective elements embedded within the overall line when the strength and depth of ambient shear are increased. The line-end vortices appear to be primarily forced by

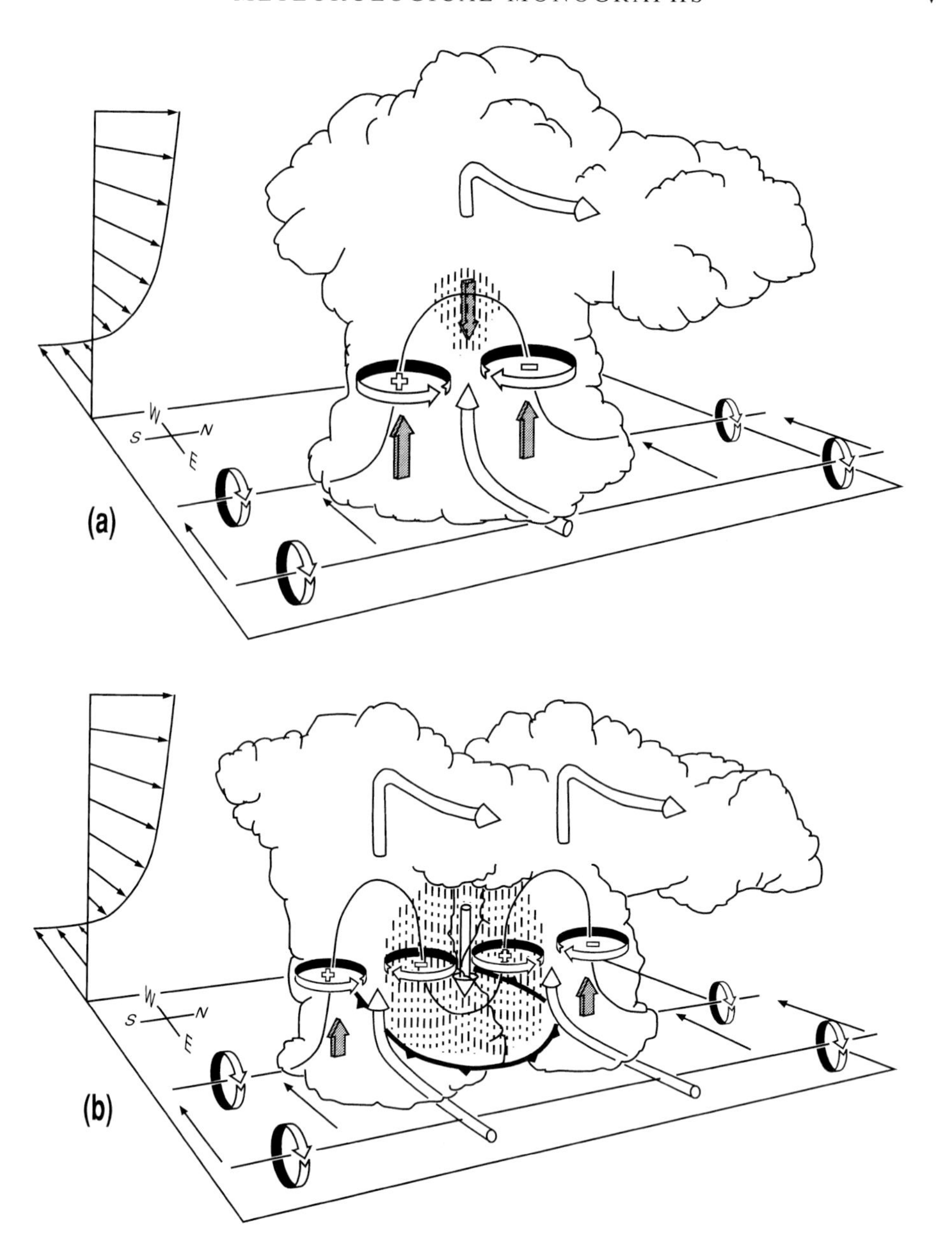

FIG. 7.39. Schematic depicting how a typical vortex tube contained within (westerly) environmental shear is deformed as it interacts with a convective cell (viewed from the southeast). Cylindrical arrows show the direction of cloud-relative airflow, and heavy solid lines represent vortex lines with the sense of rotation indicated by circular arrows. Shaded arrows represent the forcing influences that promote new updraft and downdraft growth. Vertical dashed lines denote regions of precipitation. (a) Initial stage: vortex tube loops into the vertical as it is swept into the updraft. (b) Splitting stage: downdraft forming between the splitting updraft cell tilts vortex tubes downward, producing two vortex pairs. The barbed line at the surface marks the boundary of the cold air spreading out beneath the storm. Based on a figure from Klemp (1987). Reprinted with permission from the *Annual Review of Fluid Mechanics*.

tilting of the horizontal vorticity generated by the cold pool/updraft interface. In contrast, the smaller-scale vortices seem to initiate by tilting of the environmental shear within individual updraft–downdraft couplets. The presence of these smaller-scale vortices may provide the mechanism whereby a segment of a quasi-linear squall line becomes bow-shaped while producing damaging winds (e.g. Przybylinski 1995). Important insights on this storm type are just beginning to be realized. In the future, multi-Doppler radar observations are needed to complement these numerical simulations.

The ability of the convective system to produce strong, damaging winds at the surface is a sensitive function of the convective available potential energy (CAPE) and the vertical wind shear. Bow echoes appear to develop in environments in which the CAPE is at least 2000 $m^2 s^{-2}$ and the vertical wind shear is at least 20 m s^{-1} over the lowest 5 km AGL. In addition, the development of such systems is especially favored if most of this vertical shear is confined to the lowest 2.5 km AGL. Observations of bow echoes, however, have been shown to develop in a wide range of CAPE environments (e.g., Leduc and Joe 1993). Future work could include defining strong wind regimes in terms of the Richardson number.

Numerical simulations by Weisman (1993) reveal that several types of convective systems produce strong winds, with the bow echo being the prominent echo type. This appears to be consistent with the observations of Przybylinski (1995), who proposes at least four types of radar signatures associated with derechos. These echoes all appear to be associated with reflectivity notches suggesting that rear inflow is present.

The most comprehensive information on damage caused by bow echoes has been presented by Fujita (1978, 1981), Fujita and Wakimoto (1981), and Forbes and Wakimoto (1983). This review has shown that mesoscale downdrafts can produce outflow winds exceeding 30 m s^{-1} in numerical simulations. Fujita (1978) has shown evidence of measured gusts in excess of 50 m s^{-1}. The damage surveys listed above, however, suggest that within these large swaths of destruction are embedded pockets of intense damage that occur on much smaller scales. These smaller swaths may be associated with the convective-scale downdraft and are often located in the areas subject to the strongest winds.

7.5. Detection and forecasting

An important aspect of severe storm research is the ability to transfer this knowledge into improved forecasting and detection of thunderstorm initiation and subsequent high-wind events. Most of the emphasis has been placed on forecast periods of less than 12 h (commonly called *nowcasts;* Browning 1982). Detection has been greatly improved by new remote sensing techniques, especially with the use of high-resolution satellite imagery and Doppler radar data. This section highlights three areas: thunderstorm initiation, downbursts, and derechos. Thunderstorm initiation by itself is not a high-wind event; however, making site-specific nowcasts of convection is important and necessary in order to enhance forecasts of destructive winds.

a. Thunderstorm initiation

The initiation of convective storms by organized lines of convergence within the boundary layer has been known for some time. Byers and Braham (1949), using data collected during the Thunderstorm Project, observed surface convergence 30 min prior to radar echo appearance. Subsequently, there have been a number of studies, especially over Florida, that have examined the relationship between thunderstorm activity and surface convergence caused by the sea breeze front (e.g., Byers and Rodebush 1948; Gentry and Moore 1954; Pielke 1974).

Gurka (1976), Purdom (1973, 1982), Matthews (1981), and Weaver et al. (1994) have shown the utility of using geostationary satellite data to identify cloud arc lines that were often generated by outflows and other convergent wind phenomena. Purdom (1982) performed an extensive study of over 9850 convective storms over the southeastern United States with respect to storm generation mechanism and intensity (Fig. 7.40). Generation mechanisms were classified into four categories: 1) merger, that is, development on an arc cloud as it moved into a nonprecipitating cumulus region; 2) intersection, that is, development where two arc clouds came into contact; 3) local forcing, that is, development due to some local mechanism not involving arc clouds; and 4) indeterminable, that is, development whose generation mechanism could not be clearly determined, such as new storms from beneath a cirrus deck of clouds. The results in Fig. 7.40 show that, early in the day, local forcing dominates as the convective generation mechanism. Later in the day,

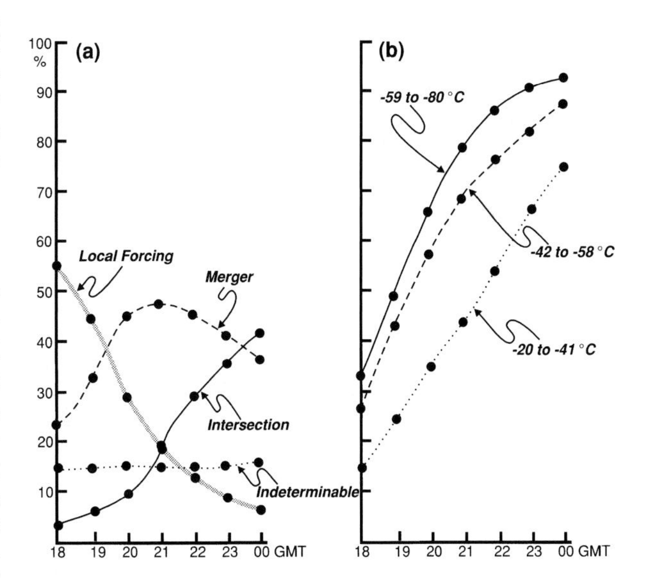

FIG. 7.40. (a) Distribution of convective generation mechanisms for all storms with tops colder than −20°C vs time for the summer of 1979 over the southeast United States. The sample contains over 9850 storms. (b) Percentage of storms in various temperature ranges due to arc cloud intersections and mergers for a given hour. For example: Of those storms with tops between −59° and −80°C at 0000 GMT, 93% are due to merger and intersection of arc clouds. Based on a figure by Purdom (1982). Reprinted with the permission of Academic Press.

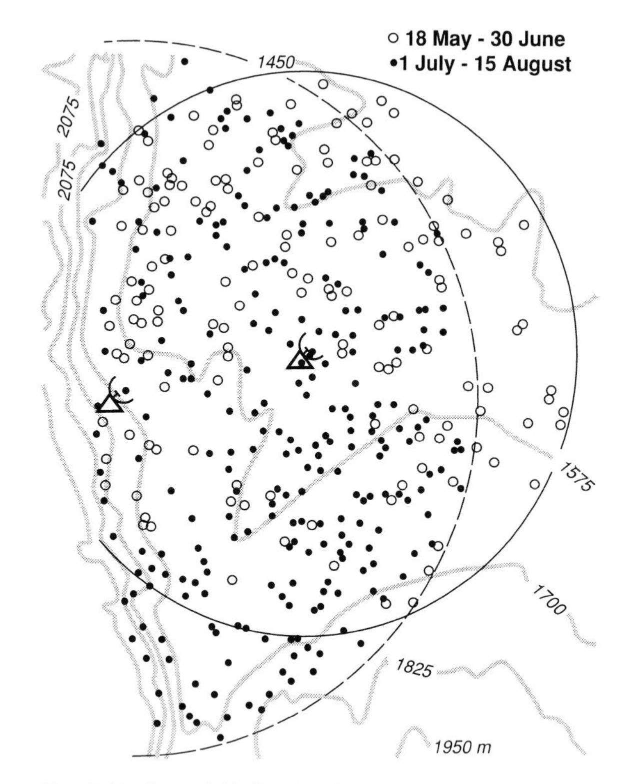

FIG. 7.41. Storm initiation locations (open and closed circles). The solid arc encloses the study area from 18 May to 30 Jun 1984 and the dashed arc 1 Jul to 15 Aug 1984. The locations of the radars for these two intervals are indicated on the figure. Contours are elevation in meters. Based on a figure from Wilson and Schreiber (1986).

when the most intense convection has developed, the dominant mechanisms are mergers and intersections of arc clouds (93% for cloud tops between −59° and −80°C).

Traditionally, the summertime thunderstorm (i.e., the Byers and Braham schematic model) has been thought to develop randomly in space in the absence of large-scale forcing. These storms were also believed to develop within homogeneous air masses. Using a sensitive Doppler radar capable of detecting winds in the optically clear planetary boundary layer, Wilson and Schreiber (1986) plotted thunderstorm initiation (precipitation echo ≥30 dBZ) near the Denver, Colorado, area. The geographical spread of convective activity is shown in Fig. 7.41. At first glance, the locations appear to support a stochastic distribution of thunderstorm occurrence; however, a more careful look at the data suggests an association with convergence lines. These convergence lines or boundaries are usually observable on a radar reflectivity display as thin (or fine) lines of enhanced reflectivity and on the velocity display as a line of strong radial or azimuthal gradient in velocity. An example of fine lines associated with a sea breeze and gust front was shown in Fig. 7.23. Three boundary types were identified in their study: moving, stationary, and colliding. A boundary was classified as moving if the reflectivity thin line or velocity convergence line showed discernible and persistent motion over a 15–30-min period prior to storm development. Figure 7.42 shows, for all of the storms, the distance to the nearest boundary at the time the storm first reached 30 dBZ. The results in the figure strongly suggest that most of the storms develop near a boundary. Hence, the storms are not random, and there is evidence to suggest that they are not developing in a uniform air mass.

The likelihood that a deep cold pool will initiate a new convective cell has been examined by Thorpe et al. (1982) and Rotunno et al. (1988). Numerical simulations point to the importance of the role of low-level wind shear in the environment ahead of the line of thunderstorms in new storm development. A schematic diagram illustrating this effect with the use of horizontal vorticity is shown in Fig. 7.43. In the no-shear case, a buoyant parcel rising over the gust front must follow a rearward sloping path. If the buoyancy is a maximum in the center of the parcel, counterrotating vortices on either side of the parcel would develop and the core of the parcel would rise vertically. However,

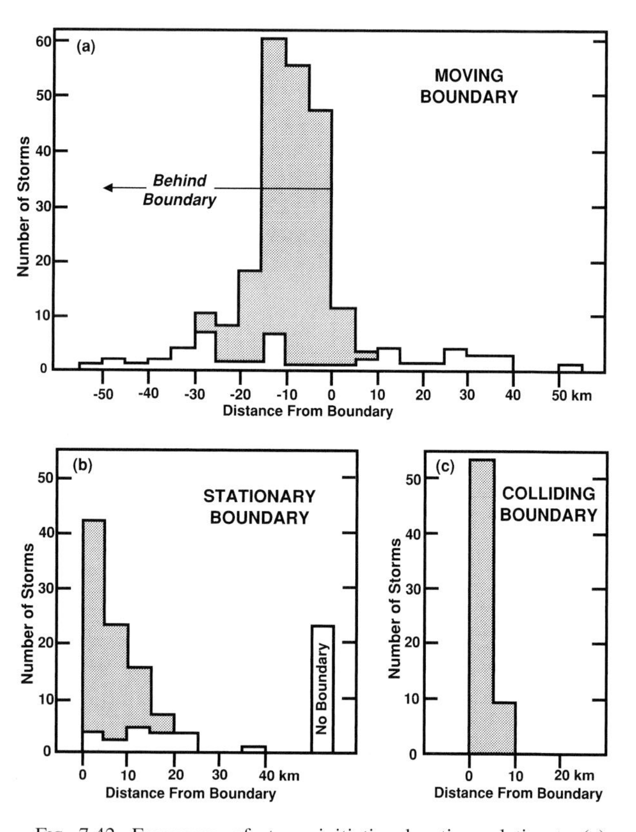

FIG. 7.42. Frequency of storm initiation location relative to (a) moving boundaries, (b) stationary boundaries, and (c) colliding boundaries. The shading represents cases that were subjectively classified as being boundary initiated. Based on a figure from Wilson and Schreiber (1986).

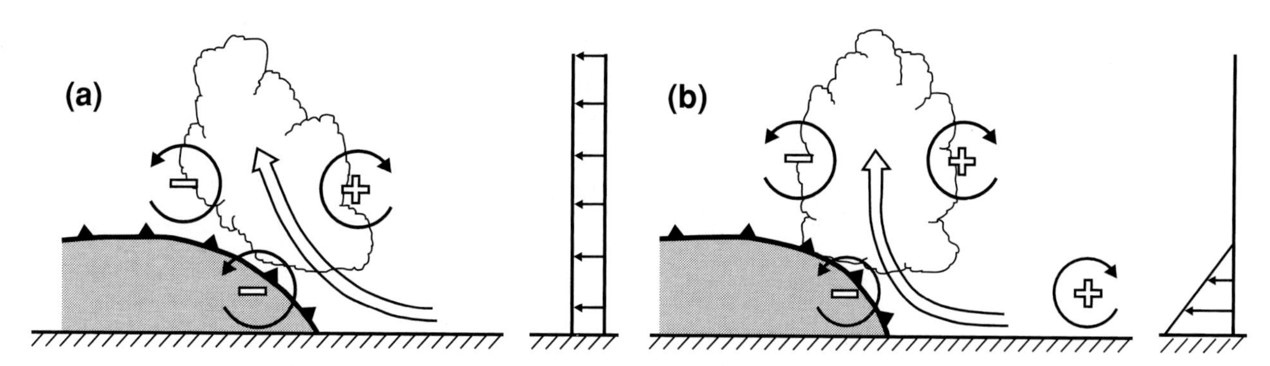

FIG. 7.43. Schematic diagram indicating horizontal vorticity in the vicinity of a long-lived squall line. Profile of the horizontal wind component normal to the line is shown on the right of each panel. Frontal symbol marks outflow boundary. (a) Case in which there is no wind shear normal to the line in the environment. (b) Case in which there is low-level wind shear ahead of the gust front. Based on a figure from Rotunno et al. (1988).

the buoyancy gradient across the gust front constitutes an additional contribution. Hence, the vertical rise associated with the parcel is superimposed on the negative vorticity associated with the gust front. This leads to a prominent slope in the flow reflecting the dominance of negative vorticity (Fig. 7.43a). In the case of low-level shear in the environmental flow (Fig. 7.43b), the initial vorticity of the inflow air is positive. It is then possible that this vorticity will balance the negative vorticity by the gust front that is baroclinically generated so that the parcel rises vertically. This latter case has been referred to as "optimal" since the parcel has no horizontal motion and, therefore, all the convective available potential energy can be converted to kinetic energy of the updraft. There have been several attempts (e.g., Carbone et al. 1990; Wilson et al. 1992; Wilson and Megenhardt 1997) to verify the importance of this effect in the initiation of new convection but to date no comprehensive dataset has been collected that can fully test this mechanism. Mueller et al. (1993) found insufficient evidence to determine whether a favorable low-level shear vector was a major factor in promoting storm initiation. Moncrieff and Liu (1999) suggest that the direction of the surface wind and the low-level shear vector in the ambient flow must be taken into account when identifying areas where convection may initiate.

The one category identified by Purdom (1982) and Wilson and Schreiber (1986) that has garnered the most interest is the colliding boundaries. Colliding boundaries produce enhanced lifting and an increased probability of convection near the point of intersection that reduces the forecast uncertainty of geographical locations most likely to be affected. In addition, colliding boundaries tend to result in major lines of thunderstorms (Fankhauser et al. 1995) and severe weather (e.g., Wilson 1986; Holle and Maier 1980; Weaver and Nelson 1982). The first signs of convection in all cases tend to appear 20–30 min after collision (Wilson and Schreiber 1986; Intrieri et al. 1990). Stensrud and Maddox (1988) have cautioned that colliding surface outflows in a conditionally unstable atmosphere do not always produce convection. Wilson et al. (1992) have shown that deep convection may develop where horizontal convective rolls intersect a convergence line. Kingsmill (1995) proposed that convection may develop preferentially at the locations where Helmholtz or shearing instability form along the boundary. These areas appear to be characterized by increased convergence and upward motion.

Droegemeier and Wilhelmson (1985) used numerical simulations to examine the sequence of events during outflow collision (Fig. 7.44). A nearly symmetric region of lifting $w > 0$ is present along the gust fronts as shown schematically in Fig. 7.44a. Consider an arbitrary area A between the two outflows. Mass continuity implies that air will flow out of A as the outflows approach each other. Both vertical and horizontal motions are present within A, but the largest velocities are horizontal, since work must be done against gravity (i.e., static stability) to lift parcels to the condensation level, but horizontal motion requires no work. As the outflows move closer together (Fig. 7.44b), cloud development along the outflow will not begin at the point where the two outflows first meet (i.e., the center of area A). Rather, the large increase in horizontal velocity out of A creates two regions of enhanced horizontal convergence where first convection should develop. Results presented by Mahoney (1988) appear to support the existence of this horizontal flow. This may not apply to situations for which the cold pool depth is significantly less than the height of the convective boundary layer, as is common over the high plains, since less work is necessary to displace parcels in the vertical direction.

Lee et al. (1991) and Mueller et al. (1993) have shown that the intensity of convection along boundaries is greatly affected by alterations in cross-line values of boundary layer moisture or convergence and by variations in the vertical wind shear profile. Sensi-

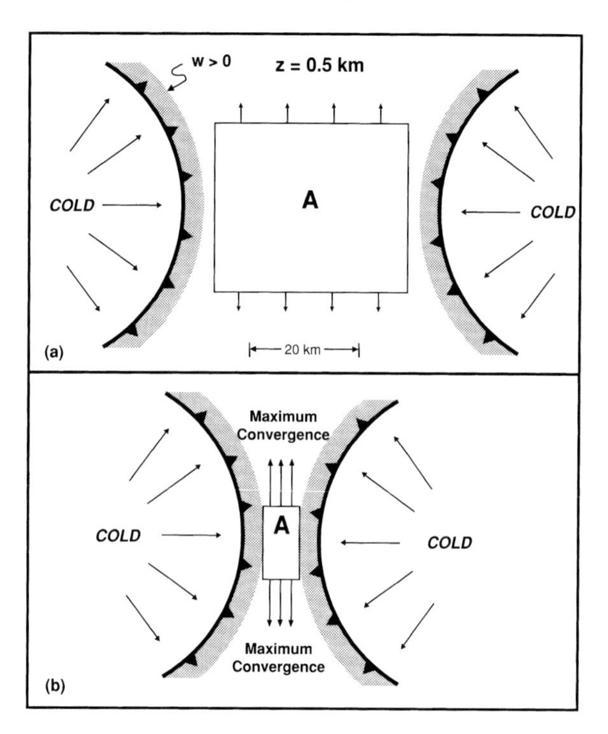

FIG. 7.44. Schematic diagram of colliding boundaries. (a) Two outflows ~40 km apart and moving toward each other. The gust fronts are indicated by the bold solid lines with barbs, and the regions of upward motion along the gust fronts are shaded. An arbitrary area A is shown by the box, and the small arrows indicate horizontal flow out of A as the outflows approach each other. (b) The outflows are now ~10 km apart. The regions of maximum horizontal convergence due to the rapid flow out of A from mass continuity are indicated. Based on a figure from Droegemeier and Wilhelmson (1985).

tivity studies (Lee et al. 1991; Crook 1996) illustrate that changes in mixing ratio and temperature of 1 g kg^{-1} and 3 K, respectively, have large effects on the developing convection. Weckwerth et al. (1996) showed that variations in water vapor mixing ratio values of 1.5–2.5 g kg^{-1} were commonly observed in the mixed boundary layer over distances of only a few kilometers. These variations were caused by horizontal convective rolls, with the largest mixing ratios confined to the regions of roll updraft.

Experimental nowcasts (30 min) by Wilson and Mueller (1993), based primarily on Doppler radar observations, showed that human forecaster results were better than persistence or extrapolation forecasts. Difficulties were encountered in the precise timing and placement of the location of storm initiation as well as in forecasting the evolution of existing storms. Better measurements of the spatial variation of moisture within the boundary layer as well as its depth are seen as important for improving nowcasts. For a comprehensive report on the status of nowcasting thunderstorms, the reader is referred to Wilson et al. (1998).

b. Microbursts

1) FORECASTS

Dry and wet microbursts typically occur under environments characterized by weak wind shear (Johns and Doswell 1992). The thermodynamic profile determined by an upper-air sounding is particularly useful for identifying when strong convectively induced winds are likely to occur. Two types of profiles associated with strong outflow are the "inverted V" profile that supports the development of dry or low-reflectivity microbursts (Beebe 1955; Wakimoto 1985) and the weakly capped wet or high-reflectivity microburst profile (Caracena and Maier 1987; Atkins and Wakimoto 1991). The dry microburst profile shown in Fig. 7.45 is characterized by a deep dry adiabatic subcloud layer from near the surface to the midlevels, a dry lower layer, and a moist midtropospheric layer (Krumm 1954; Caracena et al. 1983; Wakimoto 1985). The instabilities are often marginal, therefore the convection is usually weak. Forecasting schemes can be based on the 1200 UTC sounding and expectations for solar heating and maximum surface temperatures later in the day. Such schemes have been used successfully to identify the potential for strong wind shear events (Wilson et al. 1984; Johns and Doswell 1992).

Wet microbursts profiles, shown in Fig. 7.46, typically display high moisture values through a deep, surface-based layer, with the top of the moist layer sometimes extending beyond 4–5 km AGL. Relative humidities above the moist layer are low. The dry adiabatic subcloud layer may be only 1.5 km deep and the capping inversion is weak. Atkins and Wakimoto (1991) have shown that the vertical profile of equivalent potential temperature can be useful in identifying environmental conditions capable of supporting wet microbursts (Fig. 7.46). The difference of θ_e between the surface and midlevels equal to or greater than 20°C appears to be a characteristic profile during wet microburst events. In addition to the morning sounding, predicted moisture and thermal advection patterns must be taken into consideration. The Geostationary Operational Environmental Satellite (GOES) 6.7-mm water vapor imagery can be used to detect the influx of midlevel dry air. Ellrod (1989) has shown that deriving vertical soundings from radiance measurements from the GOES satellite can effectively update the changing thermodynamic environment.

Forecasting peak updraft speeds in deep convection based on the representative sounding data has been reasonably successful (Bluestein et al. 1988); however, there have been frequent debates concerning the usefulness of predicting downdraft speeds and subsequent outflow speeds (e.g., Doswell et al. 1982). As previously mentioned in section 7.3, this is primarily related to 1) the dependence of the downdraft to the microphysics of the precipitation particles, 2) the fact that these downdrafts are largely subsaturated, and 3)

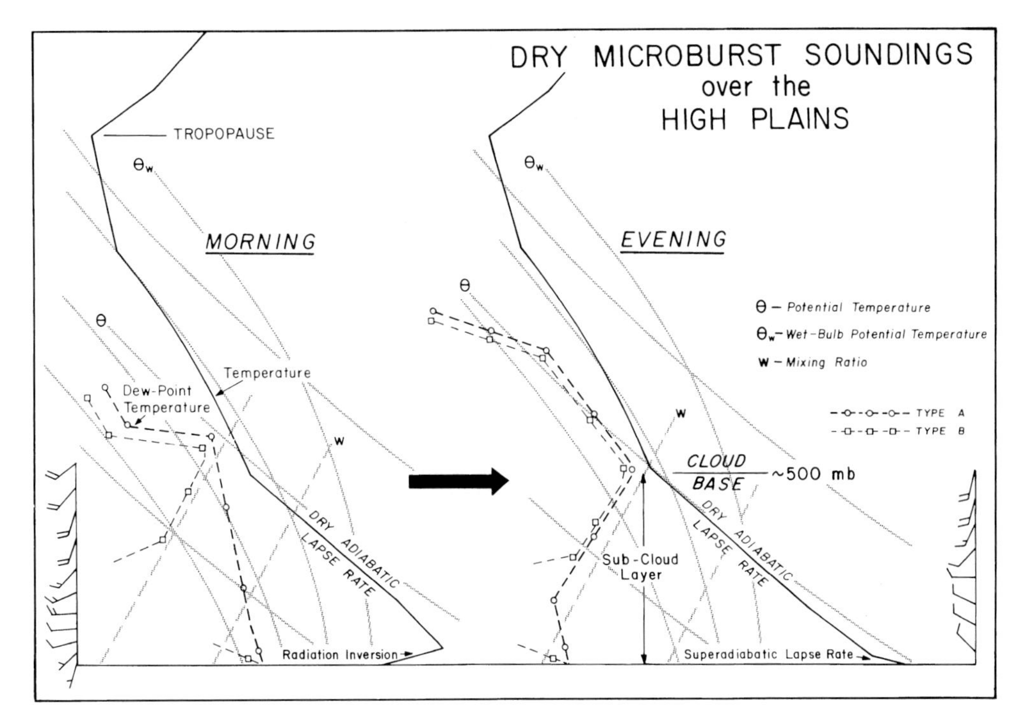

FIG. 7.45. Schematic of the characteristics of the thermodynamic profile of the morning and evening soundings favorable for dry microburst activity over the high plains. From Wakimoto (1985).

the nonlinear relationship between downdrafts and outflow speeds. Foster (1958) assumed moist adiabatic descent (i.e., saturated conditions) for his calculations of peak gusts. Predicted maximum outflows based on knowledge of the height of the melting level, the environmental mixing ratio, and the mean lapse rates have been attempted by Proctor (1989) and McCann (1994). These schemes were developed to predict outflow speeds accompanying wet microbursts using algorithms based on studies that examined the thermodynamics and dynamics of downdrafts. Accordingly, they should be more effective in forecasting high winds than schemes that are based on empirical methods (e.g., Fawbush and Miller 1954) or restrictive assumptions (e.g., Williams 1959). Forecasting peak outflow speeds remains a critical but unresolved problem.

2) DETECTION

Microburst forecasts, as discussed above, are needed to alert forecasters to the potential for intense surface winds. The forecaster carefully monitors the situation by using a combination of high-resolution satellite imagery and Doppler radar data to identify possible triggering mechanisms that may initiate thunderstorms. Once convection develops, the most useful tool to identify short-term precursors that help accurately locate microbursts has been single-Doppler radar. Roberts and Wilson (1989) have undertaken a comprehensive study to identify radar signatures that indicate the development of a downdraft capable of producing a microburst. Descending reflectivity cores (Fig. 7.22), increasing radial convergence within the cloud (in response to an accelerating downdraft), misocyclone rotation, and weak-echo reflectivity notches (in response to entrainment of low θ_e air) were found to be important microburst precursors. These events often occurred 2–6 min prior to initial surface outflow.

Descending cores and reflectivity notches are most evident with storms that are characterized by moderate to high reflectivity, while rotation and increasing radial convergence were observed for all cases. While rotation appears to be commonly associated with microburst downdrafts (e.g., Rinehart et al. 1995), the results in section 7.4 have shown that it does not contribute to downdraft acceleration. In addition, Roberts and Wilson (1989) suggest that radars equipped with dual-polarization capability (Z_{dr}) may be useful in locating strong downdrafts (Wakimoto and Bringi 1988). Near-zero values of Z_{dr} embedded within a precipitation core indicate the likely presence of hail, graupel, or snow. This precipitation type can enhance downdraft intensities (e.g., Srivastava 1987).

Important contributions toward the development of detection algorithms have been made for the Terminal Doppler Weather Radar (TDWR) system and the Integrated Terminal Weather System (ITWS). Merritt et al. (1989) describe wind shear, microburst, and gust front detection schemes used to provide timely warn-

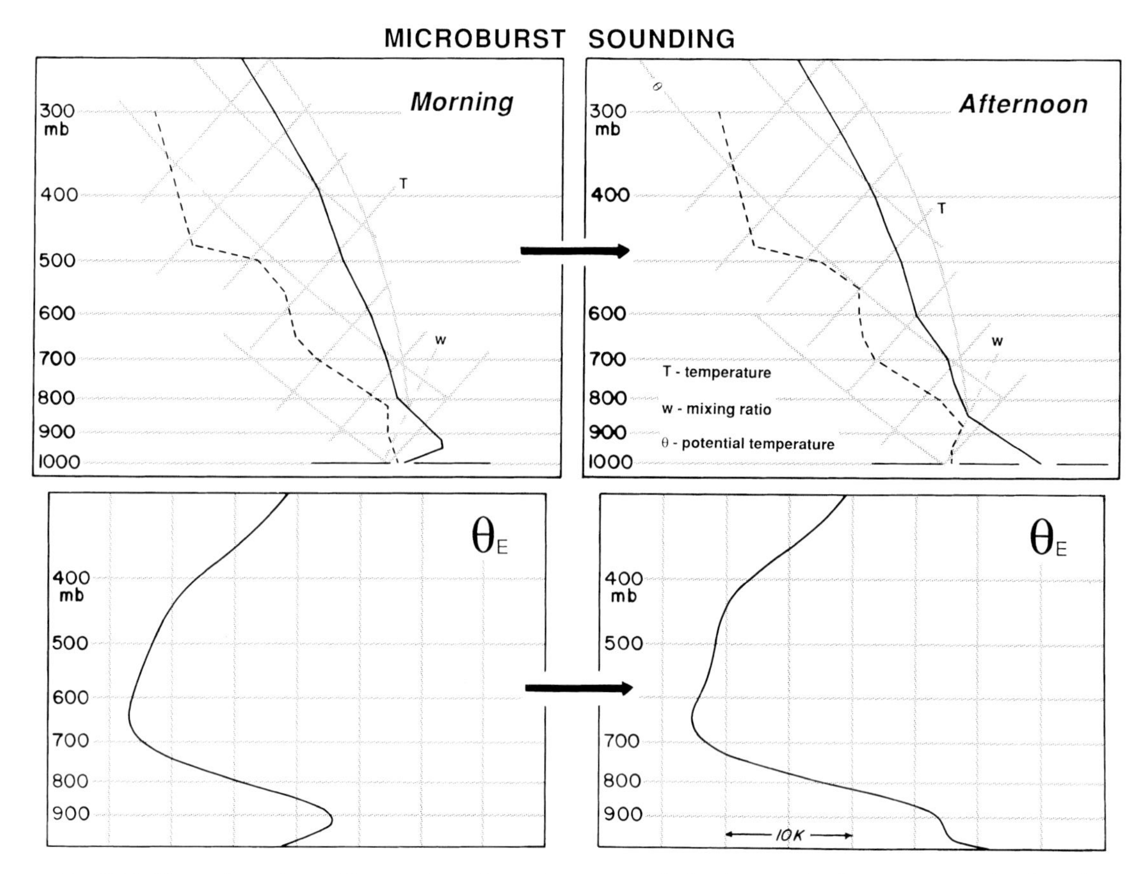

FIG. 7.46. Thermodynamic model summarizing the environment conducive for wet microburst occurrence in a humid region. From Atkins and Wakimoto (1991).

ings to air traffic controllers and pilots. Wolfson et al. (1994) discuss the formulation of a detection algorithm based on identifying features aloft within a thunderstorm to provide short-term warnings of an impending microburst.

c. *Derechos*

1) FORECASTS

Derechos typically develop in situations where the instability and vertical wind shear are moderate to strong. In addition, the synoptic pattern may also play a significant role in determining if severe outflow winds are possible. Moderate to strong vertical wind shear promotes long-lived deep convection and may result in wind damage over an extensive area, unlike the isolated, short-lived microburst event (e.g., Fujita and Wakimoto 1981; Johns and Hirt 1987; Przybylinski 1995). The primary radar signature associated with these events is the bow echo.

There are two basic synoptic patterns associated with bow echoes: 1) the warm-season pattern (Johns 1984) and 2) the dynamic pattern (Johns 1993). The warm-season pattern is commonly observed during late spring and summer. Convection develops in an area of significant warm air advection and instability, along a quasi-stationary front (Maddox and Doswell 1982; Johns 1984). Johns (1984) notes that these bow echoes develop in synoptic regimes characterized by westerly or northwesterly flow. At the 500-mb level, a shortwave trough approaches the genesis region. Warm advection maxima are present near the genesis region at both the 850- and 700-mb levels. Except near the genesis region, winds at 850 mb blow parallel to the upper-level winds and to the path of the derecho. Once formed, the bow echoes tend to move along the front. The dynamic pattern is usually associated with an extensive squall line with bow echoes forming along the line. This pattern can occur at any time of the year but appears to be least common during mid- and late summer. It has many aspects of the classic Great Plains tornado outbreak, since it involves a strong migrating low pressure system (Johns and Doswell 1992). Ambient instability values may vary widely in these situations.

The bow echoes that produce surface wind damage all appear to be characterized by strong rear inflow that results in a weak-echo notch. Weisman (1992), using a numerical cloud model, varied shear conditions and CAPE to isolate the cases that produce strong rear inflow. He concluded that a linear shear profile confined to low levels appeared to be most conducive to bow echo formation. The results of these simulations are shown in Fig. 7.47. In general, the model suggests that the rear-inflow strength increases with increasing CAPE and increasing wind shear. Figure 7.48, from Weisman (1993), presents the maximum ground-relative wind realized at the lowest level at 240 min. The strongest winds in the model generally occur for both strong shears and CAPE. If 35 m s^{-1} or greater can be used as an indicator for winds capable of producing significant damage, then it can be seen that this occurs for all cases of CAPE greater than 2000 m^2 s^{-2} and low-level shears greater than 15 m s^{-1}. Weisman (1993) notes that some of these cases do not necessarily evolve into the idealized bow echo. As previously mentioned, these criteria may be best illustrated by combining CAPE and wind shear into a Richardson number.

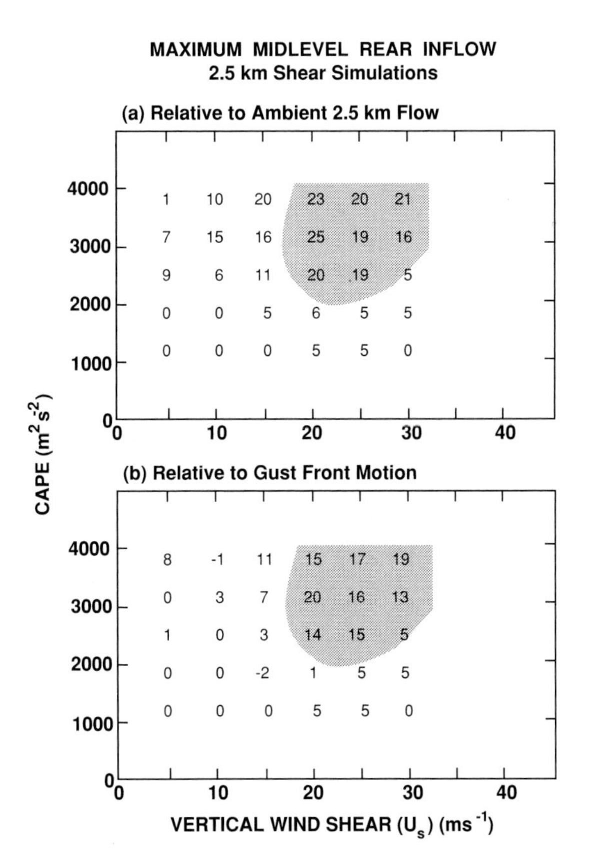

FIG. 7.47. Average maximum magnitude of the midlevel rear inflow (m s^{-1}) observed for the 2.5-km shear simulations between 200 and 240 min presented (a) relative to the ambient 2.5-km flow and (b) relative to the motion of the gust front. U_S represents the maximum magnitude of the ambient wind for the wind profile. The shaded region depicts the cases that produced elevated rear-inflow jets. Based on a figure from Weisman (1992).

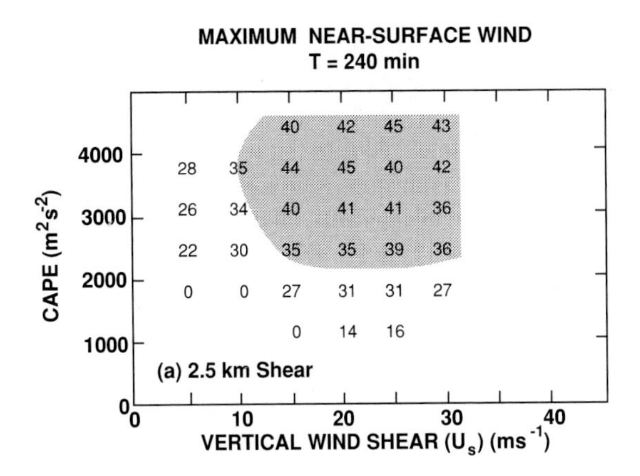

FIG. 7.48. Maximum near-surface (350 m AGL) wind (m s^{-1}) for the 2.5-km shear experiment at $t = 240$ min. U_S represents the maximum magnitude of the wind for each wind profile. The shaded region depicts the cases when the near-surface winds are greater than 35 m s^{-1}. Based on a figure from Weisman (1993).

2) DETECTION

Detection of derechos often resides in the ability to identify bow echoes in real time (e.g., Wilson et al. 1980; Fujita 1981; Schmidt and Cotton 1989). The morphology of the echo closely follows the model proposed by Fujita (1978) shown in Fig. 7.37. Przybylinski (1995) suggests that there are four different echo types, all associated with strong low-level reflectivity gradients along the leading edge of the bow and pronounced channels of weak echo. Wilson et al. (1980), Fujita (1981), and Burgess and Smull (1990) have shown that these channels are typically the location of the strongest outflow speeds. It should be noted that observations suggest that there are rear inflows associated with bow-shaped echoes that do not descend to the surface (Klimowski and Hjelmfelt 1998). The controlling factors that inhibit or augment the descent of the rear inflow to the surface are still not well understood. Accordingly, the operational criteria for derecho detection are still being determined and, with the recent deployment of the WSR-88Ds, there is currently an increased emphasis on obtaining better climatology of bow echoes around the country.

Fujita (1978) and Fujita and Wakimoto (1981) have noted detectable changes in the cloud-top temperature depicted by enhanced infrared imagery. In a number of cases, reduced areas of cold cloud-top temperature or the formation of a large warm area embedded within the thunderstorm anvil were analyzed shortly after macrobursts developed at the surface. The conclusion was that overshooting cloud-top area decreased significantly during the time surface wind damage occurred, giving the false impression that the storm system was decaying. Subsequent studies of derechos have primarily focused on detection using Doppler radar data. It is unknown whether such cloud-top signatures have been noted or whether they can be used as a precursor to warn for downburst damage.

7.6. Concluding remarks

High winds from convective phenomena have a large impact on public safety and comfort. The bulk of this review has focused on those events that result from intense downdrafts. It is well known that tornadoes, hurricanes, and winter cyclones can also produce destructive winds, some of which are described in other chapters in this monograph. The combination of observational, theoretical, and numerical research over the past two decades on strong downdrafts has resulted in new findings and improved understanding that have been effectively transferred into the operational community. Table 7.3 attempts to summarize the types of downdrafts and their dominant forcing mechanisms.

Several new techniques have been implemented that have aided forecasters/nowcasters (e.g., see reviews by Schaefer 1986; Johns and Doswell 1992; McNulty 1995; Przybylinski 1995). In concert with this research has been the availability of new datasets from remote sensing instruments, which help fill the temporal and spatial data gaps that have impeded accurate and timely warnings of severe winds. In the future, the recently deployed operational Doppler radars should improve the climatological database on these phenomena and result in modification of the detection and warning criteria.

While a number of important and pioneering studies on convectively driven downdrafts have been undertaken in recent years, several areas still require further investigation.

1. Detailed observations of the forward-flank downdraft and gust front are still lacking. These analyses would be useful complements to the high-resolution numerical simulations presented in the literature (e.g., Klemp 1987; Wicker and Wilhelmson 1995).
2. The kinematic structure of the bow echo/derecho based on Doppler radar studies is still incomplete. Moreover, these storms have been shown to be associated with tornadoes (indicated by the black dots in Fig. 7.37), which appear to be located near the comma head and along the intense horizontal shear zones near the leading edge of the outflow (e.g., Forbes and Wakimoto 1983; Wakimoto 1983; Przybylinski 1995). The evolution of this tornado is largely unknown.
3. Although mesoscale downdrafts associated with derechos can be intense, detailed damage surveys suggest that much of the destruction may be attributable to smaller-scale downdrafts. The relative contributions of mesoscale and convective-scale downdrafts in producing damaging winds are unknown.
4. The robustness of the downdraft model depicted in Fig. 7.33 is still debatable. There have been relatively few numerical and observational examinations of the Byers and Braham-type thunderstorm, in stark contrast to the plethora of studies on severe convection (e.g., squall lines, supercells).
5. The controlling factors that inhibit or augment the descent of the rear inflow to the surface within a bow echo are not well understood. Such information is necessary to improve the operational detection of derechos.
6. Forecasting of thunderstorm initiation remains a challenging problem. Emanuel et al. (1995) suggest that improved measurements of land surface properties, especially soil moisture, and better understanding of land–atmosphere interaction, are needed. In addition, improvements in the quality and quantity of atmospheric water vapor measurements are necessary, given the sensitivity of convection initiation to this parameter (e.g., Crook 1996).
7. Microphysical processes have been shown to strongly modulate the development of the convective downdraft. Recent work has improved our understanding of the influence of precipitation type on the evolution of the microburst and the mesoscale downdraft. Additional measurements are needed, however, in order to validate parameterizations of the spectrum of downdrafts in mesoscale numerical models.

TABLE 7.3. Downdraft types and primary forcing mechanisms.

Type	Forcing mechanism
Rear-flank downdraft	Evaporative cooling, condensate loading
Forward-flank downdraft	Evaporative cooling, condensate loading
Occlusion downdraft	Vertical pressure gradient force
Microburst	Evaporative cooling, melting, condensate loading
Overshooting downdraft	Buoyancy driven
Midlevel downdraft	(a) mechanically forced, (b) evaporative cooling
Mesoscale downdraft	Evaporative cooling, melting, condensate loading

Acknowledgments. I wish to acknowledge my former advisor, the late Professor T. Theodore Fujita, for making this review possible by providing the inspiration, guidance, and critical insight on many atmospheric problems throughout my graduate student career. This review was supported by the National Science Foundation under Grants ATM 9221951, 9422499, and 9801720. Excellent comments from a review panel [Andrew Crook, Mark Hjelmfelt, Robert Johns, Kevin Knupp, Peggy LeMone, Ron Przybylinski, Brad Smull, Marilyn Wolfson, and Jim Wilson (Chair)] significantly improved the manuscript.

REFERENCES

Alberty, R. L., D. W. Burgess, and T. T. Fujita, 1980: Severe weather events of 10 April 1979. *Bull. Amer. Meteor. Soc.,* **61,** 1033–1034.

Atkins, N. T., and R. M. Wakimoto, 1991: Wet microburst activity over the southeastern United States. *Wea. Forecasting,* **6,** 470–482.

———, ———, and T. M. Weckwerth, 1995: Observations of the sea-breeze front during CaPE. Part II: Dual-Doppler and aircraft analysis. *Mon. Wea. Rev.,* **123,** 944–969.

Atlas, D., and Coauthors, 1963: *Severe Local Storms. Meteor. Monogr.,* No. 27, Amer. Meteor. Soc., 247 pp.

Barnes, S. L., 1978: Oklahoma thunderstorms on 29–30 April 1970. Part I: Morphology of a tornadic storm. *Mon. Wea. Rev.,* **106,** 673–684.

Battan, L. J., 1973: *Radar Observations of the Atmosphere.* University of Chicago Press, Chicago, 324 pp.

———, 1980: Observations of two Colorado thunderstorms by means of a zenith-pointing Doppler radar. *J. Appl. Meteor.,* **19,** 580–592.

Beckwith, W. B., 1957: Characteristics of Denver hailstorms. *Bull. Amer. Meteor. Soc.,* **38,** 20–30.

Beebe, R. G., 1955: Types of airmasses in which tornadoes occur. *Bull. Amer. Meteor. Soc.,* **36,** 349–350.

Benjamin, T. B., 1968: Gravity currents and related phenomena. *J. Fluid Mech.,* **31,** 209–248.

Bentley, M. L., and T. L. Mote, 1998: A climatology of derecho-producing mesoscale convective systems in the central and eastern United States, 1986–95. Part I: Temporal and spatial distribution. *Bull. Amer. Meteor. Soc.,* **79,** 2527–2540.

Bernstein, B. C., and R. H. Johnson, 1994: A dual-Doppler radar study of an OK PRE-STORM heat burst event. *Mon. Wea. Rev.,* **122,** 259–273.

Betts, A. K., 1982: Saturation point analysis of moist convective overturning. *J. Atmos. Sci.,* **39,** 1484–1505.

———, 1984: Boundary layer thermodynamics of a high plains severe storm. *Mon. Wea. Rev.,* **112,** 2199–2211.

———, and M. F. Silva Dias, 1979: Unsaturated downdraft thermodynamics in cumulonimbus. *J. Atmos. Sci.,* **36,** 1061–1071.

Biggerstaff, M. I., and R. A. Houze Jr., 1991: Kinematic and precipitation structure of the 10–11 June 1985 squall line. *Mon. Wea. Rev.,* **119,** 3035–3065.

———, and ———, 1993: Kinematics and microphysics of the transition zone of a midlatitude squall-line system. *J. Atmos. Sci.,* **50,** 3091–3110.

Bluestein, H. B., E. W. McCaul, G. P. Byrd, and G. R. Woodall, 1988: Mobile sounding observations of a tornadic storm near the dryline: The Canadian, Texas, storm of 7 May 1986. *Mon. Wea. Rev.,* **116,** 1790–1804.

———, ———, ———, ———, G. Martin, S. Keighton, and L. C. Showell, 1989: Mobile sounding observations of a thunderstorm near the dryline: The Gruver, Texas storm complex of 25 May 1987. *Mon. Wea. Rev.,* **117,** 244–250.

Blyth, A. M., W. A. Cooper, and J. B. Jensen, 1988: A study of the source of entrained air in Montana cumuli. *J. Atmos. Sci.,* **45,** 3944–3964.

Braham, R. R., Jr., 1952: The water and energy budgets of the thunderstorm and their relation to thunderstorm development. *J. Meteor.,* **9,** 227–242.

Brandes, E. A., 1984: Vertical vorticity generation and mesocyclone sustenance in tornadic thunderstorms: The observational evidence. *Mon. Wea. Rev.,* **112,** 2253–2269.

———, and C. L. Ziegler, 1993: Mesoscale downdraft influences on vertical vorticity in a mature mesoscale convective system. *Mon. Wea. Rev.,* **121,** 1337–1353.

Britter, R. E., and J. E. Simpson, 1978: Experiments on the dynamics of a gravity current head. *J. Fluid Mech.,* **88,** 223–240.

———, and ———, 1981: A note on the structure of the head of an intrusive gravity current. *J. Fluid Mech.,* **112,** 459–466.

Brooks, C. F., 1922: The local, or heat thunderstorm. *Mon. Wea. Rev.,* **50,** 281–287.

Brown, J. M., 1979: Mesoscale unsaturated downdrafts driven by rainfall evaporation: A numerical study. *J. Atmos. Sci.,* **36,** 313–338.

———, K. R. Knupp, and F. Caracena, 1982: Destructive winds from shallow, high-based cumulonimbi. Preprints, *12th Conf. on Severe Local Storms,* San Antonio, TX, Amer. Meteor. Soc., 272–275.

Brown, R. A., 1980: Longitudinal instabilities and secondary flows in the planetary boundary layer: A review. *Rev. Geophys. Space Phys.,* **18,** 683–697.

Browning, K. A., 1964: Airflow and precipitation trajectories within severe local storms which travel to the right of the mean wind. *J. Atmos. Sci.,* **21,** 634–639.

———, 1977: The structure and mechanisms of hailstorms. *Hail: A Review of Hail Science and Hail Suppression, Meteor. Monogr.,* No. 38, Amer. Meteor. Soc., 1–43.

———, 1982: *Nowcasting.* Academic Press, 256 pp.

———, and F. H. Ludlam, 1962: Airflow in convective storms. *Quart. J. Roy. Meteor. Soc.,* **88,** 117–135.

Burgess, D. W., and B. F. Smull, 1990: Doppler radar observations of a bow echo with a long-track severe windstorm. Preprints, *16th Conf. on Severe Local Storms,* Kananaskis Park, AB, Canada, Amer. Meteor. Soc., 203–208.

Businger, S., T. Birchard, K. Kodama, P. A. Jendrowski, and J.-J. Wang, 1998: A bow echo and severe weather associated with a Kona low in Hawaii. *Wea. Forecasting,* **13,** 576–591.

Byers, H. R., and H. R. Rodebush, 1948: Causes of thunderstorms of the Florida peninsula. *J. Meteor.,* **5,** 275–280.

———, and R. R. Braham Jr., 1949: *The Thunderstorm.* U.S. Government Printing Office, 287 pp.

Caracena, F., and M. W. Maier, 1987: Analysis of a microburst in the FACE meteorological mesonetwork in southern Florida. *Mon. Wea. Rev.,* **115,** 969–985.

———, J. McCarthy, and J. A. Flueck, 1983: Forecasting the likelihood of microbursts along the front range of Colorado. Preprints, *13th Conf. on Severe Local Storms,* Tulsa, OK, Amer. Meteor. Soc., 261–264.

Carbone, R., 1983: A severe frontal rainband. Part II: Tornado parent vortex circulation. *J. Atmos. Sci.,* **40,** 2639–2654.

———, J. W. Conway, N. A. Crook, and M. W. Moncrieff, 1990: The generation and propagation of a nocturnal squall line. Part I: Observations and implications for mesoscale predictability. *Mon. Wea. Rev.,* **118,** 26–49.

Chalon, J.-P., F. Roux, G. Jaubert, and J.-P. Lafore, 1988: The west African squall line observed on 23 June during COPT81. *J. Atmos. Sci.,* **45,** 2744–2763.

Charba, J., 1974: Application of a gravity current model to analysis of squall line gust front. *Mon. Wea. Rev.,* **102,** 140–156.

———, and Y. Sasaki, 1971: Structure and movement of the severe thunderstorms of 3 April 1964 as revealed from radar and surface mesonetwork data analysis. *J. Meteor. Soc. Japan,* **49,** 191–214.

Chong, M., P. Amayenc, G. Scialom, and J. Testud, 1987: A tropical squall line observed during the COPT 81 experiment in west Africa. Part I: Kinematic structure inferred from dual-Doppler radar data. *Mon. Wea. Rev.,* **115,** 670–694.

Cook, A. W., 1939: The diurnal variation of summer rainfall in Denver. *Mon. Wea. Rev.,* **67,** 95–98.

Crook, N. A., 1996: Sensitivity of moist convection forced by boundary layer processes to low-level thermodynamic fields. *Mon. Wea. Rev.,* **124,** 1767–1785.

Dailey, P. S., and R. G. Fovell, 1999: Numerical simulation of the interaction between the sea-breeze front and horizontal convective rolls. Part I: Offshore ambient flow. *Mon. Wea. Rev.,* **127,** 520–534.

Das, P., and M. C. Subbarao, 1972: The unsaturated downdraught. *Ind. J. Meteor. Geophys.,* **23,** 135–144.

Davies-Jones, R. P., 1986: Tornado dynamics. *Thunderstorms: A Social and Technological Documentary,* Vol. II, 2d ed., E. Kessler, Ed., University of Oklahoma Press, 197–236.

Dodge, J., J. Arnold, G. Wilson, J. Evans, and T. T. Fujita, 1986: The Cooperative Huntsville Meteorological Experiment (COHMEX). *Bull. Amer. Meteor. Soc.,* **67,** 417–419.

Doswell, C. A., 1994: Extreme convective windstorms: Current understanding and research. Report of the Proceedings of the U.S.-Spain Workshop on Natural Hazards, 44–55. [Available from the Iowa Institute of Hydraulic Research, University of Iowa, Iowa City, IA 52242.]

———, J. T. Schaefer, D. W. McCann, T. W. Schlatter, and H. B. Wobus, 1982: Thermodynamic analysis procedures at the National Severe Storms Forecast Center. Preprints, *Ninth Conf. Weather Forecasting and Analysis,* Seattle, WA, Amer. Meteor. Soc., 304–309.

Droegemeier, K. K., and R. B. Wilhelmson, 1985: Three-dimensional numerical modeling of convection produced by interacting thunderstorm outflows. Part I: Control simulation and low-level moisture variations. *J. Atmos. Sci.,* **42,** 2381–2403.

———, and ———, 1987: Numerical simulation of thunderstorm outflow dynamics. Part I: Outflow sensitivity experiments and turbulence dynamics. *J. Atmos. Sci.,* **44,** 1180–1210.

Ellrod, G., 1989: Environmental conditions associated with the Dallas microburst storm determined from satellite soundings. *Wea. Forecasting,* **4,** 469–484.

Elmore, K. L., J. McCarthy, W. Frost, and H. P. Chang, 1986: A high resolution spatial and temporal multiple Doppler analysis of a microburst and its application to aircraft flight simulation. *J. Climate Appl. Meteor.,* **25,** 1398–1452.

Emanuel, K. A., 1981: A similarity theory for unsaturated downdrafts within clouds. *J. Atmos. Sci.,* **38,** 1541–1580.

———, and Coauthors, 1995: Report of the first prospectus development team of the U.S. Weather Research Program to NOAA and the NSF. *Bull. Amer. Meteor. Soc.,* **76,** 1194–1208.

Fankhauser, J. C., 1971: Thunderstorm-environment interactions determined from aircraft and radar observations. *Mon. Wea. Rev.,* **99,** 171–192.

———, 1976: Structure of an evolving hailstorm. II. Thermodynamic structure and airflow in the near environment. *Mon. Wea. Rev.,* **104,** 576–587.

———, N. A. Crook, J. Tuttle, L. J. Miller, and C. G. Wade, 1995: Initiation of deep convection along boundary layer convergence lines in a semitropical environment. *Mon. Wea. Rev.,* **123,** 291–313.

Fawbush, E. J., and R. C. Miller, 1954: A basis for forecasting peak wind gusts in non-frontal thunderstorms. *Bull. Amer. Meteor. Soc.,* **35,** 14–19.

Forbes, G. S., and R. M. Wakimoto, 1983: A concentrated outbreak of tornadoes, downbursts, and microbursts, and implications regarding vortex classification. *Mon. Wea. Rev.,* **111,** 220–235.

Foster, D. S., 1958: Thunderstorm gusts compared with computed downdraft speeds. *Mon. Wea. Rev.,* **76,** 91–94.

Fovell, R. G., and Y. Ogura, 1988: Numerical simulation of a midlatitude squall line in two dimensions. *J. Atmos. Sci.,* **45,** 3846–3879.

———, and P. S. Dailey, 1995: The temporal behavior of numerically simulated multicell-type storms. Part I: Mode of behavior. *J. Atmos. Sci.,* **52,** 2073–2095.

Fraser, A. B., and C. F. Bohren, 1992: Is virga rain that evaporates before reaching the ground? *Mon. Wea. Rev.,* **120,** 1565–1571.

Fujita, T. T., 1955: Results of detailed synoptic studies of squall lines. *Tellus,* **7,** 405–436.

———, 1959: Precipitation and cold air production in mesoscale thunderstorm systems. *J. Meteor.,* **16,** 454–466.

———, 1963: Analytical mesometeorology. A review. *Meteor. Monogr.,* No. 5, Amer. Meteor. Soc., 77–125.

———, 1974: Overshooting thunderheads observed from ATS and Learjet. SMRP Research Paper 117, University of Chicago, Department of the Geophysical Sciences, 51 pp.

———, 1978: Manual of downburst identification for project NIMROD. SMRP Research Paper 156, University of Chicago, 104 pp. [NTIS PB-2860481.]

———, 1981: Tornadoes and downbursts in the context of generalized planetary scales. *J. Atmos. Sci.,* **38,** 1511–1534.

———, 1985: The downburst. SMRP Research Paper 210, University of Chicago, 122 pp. [NTIS PB-148880.]

———, 1986: DFW microburst on August 2, 1985. SMRP Research Paper 217, University of Chicago, 154 pp. [NTIS PB 86-131638.]

———, 1989: The Teton-Yellowstone tornado of 21 July 1987. *Mon. Wea. Rev.,* **117,** 1913–1940.

———, 1992: Mystery of severe storms. WRL Research Paper 239, University of Chicago, 298 pp. [NTIS PB 92-182021.]

———, and H. R. Byers, 1977: Spearhead echo and downbursts in the crash of an airliner. *Mon. Wea. Rev.,* **105,** 129–146.

———, and F. Caracena, 1977: An analysis of three weather-related aircraft accidents. *Bull. Amer. Meteor. Soc.,* **58,** 1164–1181.

———, and R. M. Wakimoto, 1981: Five scales of airflow associated with a series of downbursts of 16 July 1980. *Mon. Wea. Rev.,* **109,** 1438–1456.

———, G. S. Forbes, and T. A. Umenhofer, 1976: Close-up view of 20 March 1976 tornadoes: Sinking cloud tops to suction vortices. *Weatherwise,* **29,** 116–145.

Gamache, J. F., and R. A. Houze Jr., 1982: Mesoscale air motions associated with a tropical squall line. *Mon. Wea. Rev.,* **110,** 118–135.

———, and ———, 1983: Water budget of a mesoscale convective system in the tropics. *J. Atmos. Sci.,* **40,** 1835–1850.

Gentry, R. C., and P. L. Moore, 1954: Relation of local and general wind interaction near the sea coast to time and location of air-mass showers. *J. Meteor.,* **11,** 507–511.

Gilmore, M. S., and L. J. Wicker, 1998: The influence of midtropospheric dryness on supercell morphology. *Mon. Wea. Rev.,* **126,** 943–958.

Goff, R. C., 1976: Vertical structure of thunderstorm outflows. *Mon. Wea. Rev.,* **104,** 1429–1440.

Gurka, J. J., 1976: Satellite and surface observations of strong wind zones accompanying thunderstorms. *Mon. Wea. Rev.,* **104,** 1484–1493.

Hall, F., and R. D. Brewer, 1959: A sequence of tornado damage patterns. *Mon. Wea. Rev.,* **87,** 207–216.

Haman, K., 1973: On the updraft-downdraft interaction in convective clouds. *Acta. Geophys. Polonica,* **31,** 216–233.

Hamilton, R. E., 1970: Use of detailed intensity radar data in mesoscale surface analysis of the 4 July 1969 storm in Ohio. Preprints, *14th Conf. on Radar Meteor.,* Tucson, AZ, Amer. Meteor. Soc., 339–342.

Hamilton, R. A., and J. W. Archbold, 1945: Meteorology of Nigeria and adjacent territory. *Quart. J. Roy. Meteor. Soc.,* **71,** 231–265.

Hane, C. E., and P. S. Ray, 1985: Pressure and buoyancy fields derived from Doppler radar data in a tornadic thunderstorm. *J. Atmos. Sci.,* **42,** 18–35.

Harris, F. I., 1977: The effects of evaporation at the base of ice precipitation layers: Theory and radar observations. *J. Atmos. Sci.,* **34,** 651–672.

Heymsfield, A. J., P. N. Johnson, and J. E. Dye, 1978: Observations of moist adiabatic ascent in northeast Colorado cumulus congestus clouds. *J. Atmos. Sci.,* **35,** 1689–1703.

Heymsfield, G. M., and S. Schotz, 1985: Structure and evolution of a severe squall line over Oklahoma. *Mon. Wea. Rev.,* **113,** 1563–1589.

Hinrichs, G., 1888: Tornadoes and derechos. *Amer. Meteor. J.,* **5,** 306–317, 341–349.

Hjelmfelt, M. R., 1987: The microburst of 22 June 1982 in JAWS. *J. Atmos. Sci.,* **44,** 1646–1665.

———, 1988: Structure and life cycle of microburst outflows observed in Colorado. *J. Appl. Meteor.,* **27,** 900–927.

———, H. D. Orville, R. D. Roberts, J. P. Chen, and F. J. Kopp, 1989: Observational and numerical study of a microburst line-producing storm. *J. Atmos. Sci.,* **46,** 2731–2743.

Holle, R. L., and M. W. Maier, 1980: Tornado formation from downdraft interactions in the FACE mesonet-work. *Mon. Wea. Rev.,* **108,** 1010–1028.

Holton, J. R., 1992: *An Introduction to Dynamic Meteorology,* Academic Press, 511 pp.

Hookings, G. A., 1965: Precipitation-maintained downdrafts. *J. Appl. Meteor.,* **4,** 190–195.

———, 1967: Hail-maintained downdrafts. *J. Appl. Meteor.,* **6,** 589–591.

Houze, R. A., Jr., 1977: Structure and dynamics of a tropical squall-line system. *Mon. Wea. Rev.,* **105,** 1540–1567.

———, 1989: Observed structure of mesoscale convective systems and implications for large-scale heating. *Quart. J. Roy. Meteor. Soc.,* **115,** 425–461.

———, 1993: *Cloud Dynamics.* Academic Press, 570 pp.

———, S. A. Rutledge, M. I. Biggerstaff, and B. F. Smull, 1989: Interpretation of Doppler weather radar displays of midlatitude mesoscale convective systems. *Bull. Amer. Meteor. Soc.,* **70,** 608–619.

———, B. F. Smull, and P. Dodge, 1990: Mesoscale organization of springtime rainstorms in Oklahoma. *Mon. Wea. Rev.,* **118,** 613–654.

Humphreys, W. J., 1914: The thunderstorm and its phenomena. *Mon. Wea. Rev.,* **42,** 348–380.

Idso, S. B., 1974: Thunderstorm outflows: Different perspectives over arid and mesic terrain. *Mon. Wea. Rev.,* **102,** 603–604.

Igau, R. C., M. A. LeMone, and D. Wei, 1999: Updraft and downdraft cores in TOGA COARE: Why so many buoyant downdraft cores? *J. Atmos. Sci.,* **56,** 2232–2245.

Intrieri, J. M., A. J. Bedard, and R. M. Hardesty, 1990: Details of colliding thunderstorm outflows as observed by Doppler lidar. *J. Atmos. Sci.,* **47,** 1081–1098.

Johns, R. H., 1982: A synoptic climatology of northwest flow severe weather outbreaks. Part I: Nature and significance. *Mon. Wea. Rev.,* **110,** 1653–1663.

———, 1984: A synoptic climatology of northwest flow severe weather outbreaks. Part II: Meteorological parameters and synoptic patterns. *Mon. Wea. Rev.,* **112,** 449–464.

———, 1993: Meteorological conditions associated with bow echo development in convective storms. *Wea. Forecasting,* **8,** 294–299.

———, and W. D. Hirt, 1987: Derechos: Widespread convectively induced windstorms. *Wea. Forecasting,* **2,** 32–49.

———, and C. A. Doswell III, 1992: Severe local storms forecasting. *Wea. Forecasting,* **7,** 588–612.

Johnson, R. H., and P. J. Hamilton, 1988: The relationship of surface pressure features to the precipitation and air flow structure of an intense midlatitude squall line. *Mon. Wea. Rev.,* **116,** 1444–1472.

———, S. Chen, and J. J. Toth, 1989: Circulations associated with a mature-to-decaying midlatitude mesoscale convective system. Part I: Surface features—Heat bursts and mesolow development. *Mon. Wea. Rev.,* **117,** 942–959.

Jorgensen, D. P., and B. F. Smull, 1993: Mesovortex circulations seen by airborne Doppler radar within a bow-echo mesoscale convective system. *Bull. Amer. Meteor. Soc.,* **74,** 2146–2157.

Kamburova, P. L., and F. H. Ludlam, 1966: Rainfall evaporation in thunderstorm downdrafts. *Quart. J. Roy. Meteor. Soc.,* **92,** 510–518.

Kelly, D. L., J. T. Schaefer, R. P. McNulty, C. A. Doswell III, and R. F. Abbey, 1978: An augmented tornado climatology. *Mon. Wea. Rev.,* **106,** 1172–1183.

———, ———, and C. A. Doswell III, 1985: Climatology of nontornadic severe thunderstorm events in the United States. *Mon. Wea. Rev.,* **113,** 1997–2014.

Kessinger, C. J., D. B. Parsons, and J. W. Wilson, 1988: Observations of a storm containing misocyclones, downbursts, and horizontal vortex circulations. *Mon. Wea. Rev.,* **116,** 1959–1982.

Kessler, E., 1986: *Thunderstorms: A Social and Technological Documentary,* Vol. II., 2d ed., University of Oklahoma Press, 603 pp.

Kingsmill, D. E., 1995: Convection initiation associated with a sea-breeze front, a gust front, and their collision. *Mon. Wea. Rev.,* **123,** 2913–2933.

———, and R. M. Wakimoto, 1991: Kinematic, dynamic, and thermodynamic analysis of a weakly sheared severe thunderstorm over northern Alabama. *Mon. Wea. Rev.,* **119,** 262–297.

Klemp, J. B., 1987: Dynamics of tornadic thunderstorms. *Ann. Rev. Fluid Mech.,* **19,** 369–402.

———, and R. B. Wilhelmson, 1978: Simulation of right- and left-moving storms produced through storm splitting. *J. Atmos. Sci.,* **35,** 1097–1110.

———, and R. Rotunno, 1983: A study of the tornadic region within a supercell thunderstorm. *J. Atmos. Sci.,* **40,** 359–377.

———, R. B. Wilhelmson, and P. S. Ray, 1981: Observed and numerically simulated structure of a mature supercell thunderstorm. *J. Atmos. Sci.,* **38,** 1558–1580.

Klimowski, B. A., 1994: Initiation and development of rear inflow within the 28–29 June 1989 North Dakota mesoconvective system. *Mon. Wea. Rev.,* **122,** 765–779.

———, and M. R. Hjelmfelt, 1998: Climatology and structure of high wind-producing mesoscale convective systems over the northern High Plains. Preprints, *19th Conf. on Severe Local Storms,* Minneapolis, MN, Amer. Meteor. Soc., 444–447.

Knupp, K. R., 1987: Downdrafts within high plains cumulonimbi. Part I: General kinematic structure. *J. Atmos. Sci.,* **44,** 987–1008.

———, 1988: Downdrafts within high plains cumulonimbi. Part II: Dynamics and thermodynamics. *J. Atmos. Sci.,* **45,** 3965–3982.

———, and W. R. Cotton, 1982: An intense, quasi-steady thunderstorm over mountainous terrain. Part II: Doppler radar observations of the storm morphological structure. *J. Atmos. Sci.,* **39,** 343–358.

———, and ———, 1985: Convective cloud downdraft structure: An interpretive survey. *Rev. Geophys.,* **23,** 183–215.

Koch, S. E., 1984: The role of an apparent mesoscale frontogenetical circulation in squall line initiation. *Mon. Wea. Rev.,* **112,** 2090–2111.

Kropfli, R. A., and L. J. Miller, 1976: Kinematic structure and flux quantities in a convective storm from dual-Doppler radar observations. *J. Atmos. Sci.,* **33,** 520–529.

Krueger, S. K., R. M. Wakimoto, and S. J. Lord, 1986: Role of ice-phase microphysics in dry microburst simulations. Preprints, *23d Conf. on Radar Meteor.,* Snowmass, CO, Amer. Meteor. Soc., R73–R76.

Krumm, W. R., 1954: On the cause of downdrafts from dry thunderstorms over the plateau area of the United States. *Bull. Amer. Meteor. Soc.,* **35,** 122–125.

Kuettner, J. P., 1959: The band structure of the atmosphere. *Tellus,* **2,** 267–294.

Lafore, J.-P., and M. W. Moncrieff, 1989: A numerical investigation of the organization and interaction of the convective and stratiform regions of tropical squall lines. *J. Atmos. Sci.,* **46,** 521–544.

Leary, C. A., and R. A. Houze Jr., 1979: Melting and evaporation of hydrometeors in precipitation from the anvil clouds of deep tropical convection. *J. Atmos. Sci.,* **36,** 669–679.

Leduc, M., and P. Joe, 1993: Bow echoes storms near Toronto, Canada associated with very low buoyant energies. Preprints, *17th Conf. on Severe Local Storms,* St. Louis, MO, Amer. Meteor. Soc., 573–576.

Lee, B. D., R. D. Farley, and M. R. Hjelmfelt, 1991: A numerical case study of convection initiation along colliding convergence boundaries in northeast Colorado. *J. Atmos. Sci.,* **48,** 2350–2366.

Lee, W.-C., R. M. Wakimoto, and R. E. Carbone, 1992: The evolution and structure of a "bow-echo-microburst" event. Part II: The bow echo. *Mon. Wea. Rev.,* **120,** 2211–2225.

Lemon, L. R., 1976: The flanking line, a severe thunderstorm intensification source. *J. Atmos. Sci.,* **33,** 686–694.

———, and C. A. Doswell, 1979: Severe thunderstorm evolution and mesocyclone structure as related to tornadogenesis. *Mon. Wea. Rev.,* **107,** 1184–1197.

LeMone, M. A., 1973: The structure and dynamics of horizontal roll vortices in the planetary boundary layer. *J. Atmos. Sci.,* **30,** 1077–1091.

———, 1983: Momentum transport by a line of cumulonimbus. *J. Atmos. Sci.,* **40,** 1815–1834.

———, G. M. Barnes, and E. J. Zipser, 1984: Momentum flux by lines of cumulonimbus over the tropical oceans. *J. Atmos. Sci.,* **41,** 1914–1932.

———, G. M. Barnes, J. C. Frankhauser, and L. F. Tarleton, 1988a: Perturbation pressure fields measured by aircraft around the cloud-base updraft of deep convective clouds. *Mon. Wea. Rev.,* **116,** 313–327.

———, L. F. Tarleton, and G. M. Barnes, 1988b: Perturbation pressure at the base of cumulus clouds in low shear. *Mon. Wea. Rev.,* **116,** 2062–2068.

List, R., and E. P. Lozowski, 1970: Pressure perturbations and buoyancy in convective clouds. *J. Atmos. Sci.,* **27,** 168–170.

Loeher, S. M., and R. H. Johnson, 1995: Surface pressure and precipitation life cycle characteristics of PRE-STORM mesoscale convective systems. *Mon. Wea. Rev.,* **123,** 600–621.

Ludlam, D. M., 1970: *Early American Tornadoes.* Amer. Meteor. Soc., 97 pp.

Ludlam, F. H., 1980: *Cloud and Storms: The Behavior and Effect of Water in the Atmosphere.* The Pennsylvania State University Press, 405 pp.

MacPherson, J. I., and G. A. Isaac, 1977: Turbulent characteristics of some Canadian cumulus clouds. *J. Atmos. Sci.,* **16,** 81–90.

Maddox, R. A., 1980: Mesoscale convective complexes. *Bull. Amer. Meteor. Soc.,* **61,** 1374–1387.

———, and C. A. Doswell III, 1982: An examination of jetstream configurations, 500 mb vorticity advection and low-level thermal advection patterns during extended periods of intense convection. *Mon. Wea. Rev.,* **110,** 184–197.

Mahoney, W. P., 1988: Gust front characteristics and the kinematics associated with interacting thunderstorm outflows. *Mon. Wea. Rev.,* **116,** 1474–1491.

———, and A. R. Rodi, 1987: Aircraft measurements on microburst development from hydrometeor evaporation. *J. Atmos. Sci.,* **44,** 3037–3051.

———, and K. L. Elmore, 1991: The evolution and fine-scale structure of a microburst-producing cell. *Mon. Wea. Rev.,* **119,** 176–192.

Malkus, J. S., 1955: On the formation and structure of downdrafts in cumulus clouds. *J. Meteor.,* **12,** 350–354.

Matthews, D. A., 1981: Observations of a cloud arc triggered by thunderstorm outflow. *Mon. Wea. Rev.,* **109,** 2140–2157.

McCann, D. W., 1994: WINDEX—A new index for forecasting microburst potential. *Wea. Forecasting,* **9,** 532–541.

McCaul, E. W., H. B. Bluestein, and R. J. Doviak, 1987: Airborne Doppler lidar observations of convective phenomena in Oklahoma. *J. Atmos. Oceanic Technol.,* **4,** 479–497.

McCarthy, J., J. W. Wilson, and T. T. Fujita, 1982: The joint airport weather studies project. *Bull. Amer. Meteor. Soc.,* **63,** 15–22.

———, and R. Serafin, 1984: The microburst: Hazard to aircraft. *Weatherwise,* **37,** 121–127.

McNulty, R. P., 1991: Downbursts from innocuous clouds: An example. *Wea. Forecasting,* **6,** 148–154.

———, 1995: Severe and convective weather: A central region forecasting challenge. *Wea. Forecasting,* **10,** 187–202.

Merritt, M. W., D. Klingle-Wilson, and S. D. Campbell, 1989: Wind shear detection with pencil-beam radars. *Lincoln Lab. J.,* **2,** 483–510.

Middleton, G. V., 1966: Experiments on density and turbidity currents. *Can. J. Earth. Sci.,* **3,** 523–546.

Mielke, K. B., and E. R. Carle, 1987: An early morning dry microburst in the great basin. *Wea. Forecasting,* **2,** 169–174.

Miller, M. J., and A. K. Betts, 1977: Traveling convective storms over Venezuela. *Mon. Wea. Rev.,* **105,** 833–848.

Mitchell, K. E., and J. B. Hovermale, 1977: A numerical investigation of a severe thunderstorm gust front. *Mon. Wea. Rev.,* **105,** 657–675.

Moncrieff, M. W., and C. Liu, 1999: Convection initiation by density currents: Role of convergence, shear, and dynamical organization. *Mon. Wea. Rev.,* **127,** 2455–2464.

Mueller, C. K., and R. E. Carbone, 1987: Dynamics of a thunderstorm outflow. *J. Atmos. Sci.,* **44,** 1879–1898.

———, J. W. Wilson, and N. A. Crook, 1993: The utility of sounding and mesonet data to nowcast thunderstorm initiation. *Wea. Forecasting,* **8,** 132–146.

Newton, C. W., 1950: Structure and mechanism of the prefrontal squall line. *J. Meteor.,* **7,** 210–222.

———, 1966: Circulations in large sheared cumulonimbus. *Tellus,* **18,** 699–713.

———, and H. R. Newton, 1959: Dynamical interaction between large convective clouds and environment with vertical shear. *J. Meteor.,* **16,** 483–496.

Nolan, R. H., 1959: A radar pattern associated with tornadoes. *Bull. Amer. Meteor. Soc.,* **40,** 277–279.

Normand, C. W. B., 1946: Energy in the atmosphere. *Quart. J. Roy. Meteor. Soc.,* **72,** 145–167.

Ogura, Y., and M.-T. Liou, 1980: The structure of a midlatitude squall line: A case study. *J. Atmos. Sci.,* **37,** 553–567.

Ohno, H., O. Suzuki, and K. Kusunoki, 1996: Climatology of downburst occurrence in Japan. Preprints, *18th Conf. on Severe Local Storms,* San Francisco, CA, Amer. Meteor. Soc., 87–90.

Orf, L. G., and J. R. Anderson, 1999: A numerical study of traveling microbursts. *Mon. Wea. Rev.,* **127,** 1244–1258.

———, ———, and J. M. Straka, 1996: A three-dimensional numerical analysis of colliding microburst outflow dynamics. *J. Atmos. Sci.,* **53,** 2490–2511.

Paluch, I. R., 1979: The entrainment mechanism in Colorado cumuli. *J. Atmos. Sci.,* **36,** 2467–2478.

———, and D. W. Breed, 1984: A continental storm with a steady, adiabatic updraft and high concentrations of small ice particles: 6 July 1976 case study. *J. Atmos. Sci.,* **41,** 1008–1024.

Parsons, D. B., and R. A. Kropfli, 1990: Dynamics and fine structure of a microburst. *J. Atmos. Sci.,* **47,** 1674–1692.

———, and M. L. Weisman, 1993: A numerical study of a rotating downburst. *J. Atmos. Sci.,* **50,** 2369–2385.

Pedgley, D. E., 1962: A meso-synoptic analysis of the thunderstorms on 28 August 1958. *Geophys. Mem.,* **106,** 74 pp.

Pielke, R. A., 1974: A three-dimensional numerical model of the sea breezes over south Florida. *Mon. Wea. Rev.,* **102,** 115–139.

Potts, R. J., 1991: Microburst observations in tropical Australia. Preprints, *25th Int. Conf. on Radar Meteor.,* Paris, France, Amer. Meteor. Soc., J67–J72.

Proctor, F. H., 1988: Numerical simulations of an isolated microburst. Part I: Dynamics and structure. *J. Atmos. Sci.,* **45,** 3137–3160.

———, 1989: Numerical simulations of an isolated microburst. Part II: Sensitivity experiments. *J. Atmos. Sci.,* **46,** 2143–2165.

Przybylinski, R. W., 1995: The bow echo: Observations, numerical simulations, and severe weather detection methods. *Wea. Forecasting,* **10,** 203–218.

Purdom, J. F. W., 1973: Meso-highs and satellite imagery. *Mon. Wea. Rev.,* **101,** 180–181.

———, 1982: Subjective interpretations of geostationary satellite data for nowcasting. *Nowcasting,* K. Browning, Ed., Academic Press, 149–166.

Ragette, G., 1973: Mesoscale circulations associated with Alberta hailstorms. *Mon. Wea. Rev.,* **101,** 150–159.

Ramond, D., 1978: Pressure perturbations in deep convection. *J. Atmos. Sci.,* **35,** 1704–1711.

Rao, P. A., H. E. Fuelberg, and K. K. Droegemeier, 1999: High resolution modeling of the Cape Canaveral area land-water circulations and associated features. *Mon. Wea. Rev.,* **127,** 1808–1821.

Ray, P. S., 1986: *Mesoscale Meteorology and Forecasting.* Amer. Meteor. Soc., 793 pp.

Raymond, D. J., R. Solomon, and A. M. Blyth, 1991: Mass flux in New Mexico mountain thunderstorms from radar and aircraft measurements. *Quart. J. Roy. Meteor. Soc.,* **117,** 587–621.

Rinehart, R. E., A. Borho, and C. Curtiss, 1995: Microburst rotation: Simulations and observations. *J. Appl. Meteor.,* **34,** 1267–1285.

Roberts, R. D., and J. W. Wilson, 1989: A proposed microburst nowcasting procedure using single-Doppler radar. *J. Appl. Meteor.,* **28,** 285–303.

Rotunno, R., and J. B. Klemp, 1982: The influence of the shear-induced pressure gradient on thunderstorm motion. *Mon. Wea. Rev.,* **110,** 136–151.

———, ———, and M. L. Weisman, 1988: A theory for strong, long-lived squall lines. *J. Atmos. Sci.,* **45,** 463–485.

Roux, F., 1988: The west African squall line observed on 23 June 1981 during COPT 81: Kinematics and thermodynamics of the convective region. *J. Atmos. Sci.,* **45,** 406–426.

Rutledge, S. A., R. A. Houze Jr., M. I. Biggerstaff, and T. Matejka, 1988: The Oklahoma-Kansas mesoscale convective system of 10–11 June 1985: Precipitation structure and single-Doppler radar analysis. *Mon. Wea. Rev.,* **116,** 1409–1430.

Ryan, B. F., and J. C. Carstens, 1978: A comparison between a steady-state downdraft model and observations behind squall lines. *J. Appl. Meteor.,* **17,** 395–400.

Schaefer, J. T., 1986: Severe thunderstorm forecasting: A historical perspective. *Wea. Forecasting,* **1,** 164–189.

———, L. R. Hoxit, and C. F. Chappell, 1985: Thunderstorms and their mesoscale environment. *Thunderstorm Morphology and Dynamics,* E. Kessler, Ed., U.S. Government Printing Office, 113–130.

Schlesinger, R. E., 1978: A three-dimensional numerical model of an isolated thunderstorm. Part I: Comparative experiments for variable ambient wind shear. *J. Atmos. Sci.,* **35,** 690–713.

———, 1980: A three-dimensional numerical model of an isolated thunderstorm. Part II: Dynamics of updraft splitting and meso-vortex couplet evolution. *J. Atmos. Sci.,* **37,** 395–420.

———, 1984: Mature thunderstorm cloud-top structure and dynamics: A three-dimensional numerical simulation study. *J. Atmos. Sci.,* **41,** 1551–1570.

Schmidt, J. M., and W. R. Cotton, 1989: A high plains squall line associated with severe surface winds. *J. Atmos. Sci.,* **46,** 281–302.

Simpson, J. E., 1969: A comparison between laboratory and atmosphere density currents. *Quart. J. Roy. Meteor. Soc.,* **95,** 758–765.

———, 1972: Effects of the lower boundary on the head of a gravity current. *J. Fluid Mech.,* **53,** 759–768.

———, 1987: *Gravity Currents: In the Environment and the Laboratory.* Ellis Horwood, 244 pp.

———, and R. E. Britter, 1979: The dynamics of the head of a current advancing over a horizontal surface. *J. Fluid Mech.,* **94,** 478–497.

———, and ———, 1980: A laboratory model of an atmospheric mesofront. *Quart. J. Roy. Meteor. Soc.,* **106,** 485–500.

Skamarock, W. C., M. L. Weisman, and J. B. Klemp, 1994: Three-dimensional evolution of simulated long-lived squall lines. *J. Atmos. Sci.,* **51,** 2563–2584.

Smull, B. F., and R. A. Houze Jr., 1985: A midlatitude squall line with a trailing region of stratiform rain: Radar and satellite observations. *Mon. Wea. Rev.,* **113,** 117–133.

———, and ———, 1987a: Dual-Doppler radar analysis of a midlatitude squall line with a trailing region of stratiform rain. *J. Atmos. Sci.,* **44,** 2128–2148.

———, and ———, 1987b: Rear-inflow in squall lines with trailing stratiform precipitation. *Mon. Wea. Rev.,* **115,** 2869–2889.

———, and J. A. Augustine, 1993: Multiscale analysis of a mature mesoscale convective complex. *Mon. Wea. Rev.,* **121,** 103–132.

Sommeria, G., and J. Testud, 1984: COPT81: A field experiment designed for the study of dynamics and electrical activity of deep convection in continental tropical regions. *Bull. Amer. Meteor. Soc.,* **65,** 4–10.

Srivastava, R. C., 1985: A simple model of evaporatively driven downdraft: Application to microburst downdraft. *J. Atmos. Sci.,* **42,** 1004–1023.

———, 1987: A model of intense downdrafts driven by the melting and evaporation of precipitation. *J. Atmos. Sci.,* **44,** 1752–1773.

———, T. J. Matejka, and T. J. Lorello, 1986: Doppler radar study of the trailing anvil region associated with a squall line. *J. Atmos. Sci.,* **43,** 356–377.

Stensrud, D. J., and R. A. Maddox, 1988: Opposing mesoscale circulations: A case study. *Wea. Forecasting,* **3,** 189–204.

———, ———, and C. L. Ziegler, 1991: A sublimation-initiated mesoscale downdraft and its relation to the wind field below a precipitating anvil cloud. *Mon. Wea. Rev.,* **119,** 2124–2139.

Stumpf, G. J., R. H. Johnson, and B. F. Smull, 1991: The wake low in a midlatitude mesoscale convective system having complex convective organization. *Mon. Wea. Rev.,* **119,** 134–158.

Sun, J., S. Braun, M. I. Biggerstaff, R. G. Fovell, and R. A. Houze Jr., 1993: Warm upper-level downdrafts associated with a squall line. *Mon. Wea. Rev.,* **121,** 2919–2927.

Szeto, K. K., C. A. Lin, and R. E. Stewart, 1988a: Mesoscale circulations forced by melting snow. Part I: Basic simulations and dynamics. *J. Atmos. Sci.,* **45,** 1629–1641.

———, R. E. Stewart, and C. A. Lin, 1988b: Mesoscale circulations forced by melting snow. Part II: Application to meteorological features. *J. Atmos. Sci.,* **45,** 1642–1650.

Tao, W.-K., and J. Simpson, 1989: Modeling study of a tropical squall-line convective line. *J. Atmos. Sci.,* **46,** 177–202.

Thorpe, A. J., M. J. Miller, and M. W. Moncrieff, 1982: Two-dimensional convection in nonconstant shear: A model of midlatitude squall lines. *Quart. J. Roy. Meteor. Soc.,* **108,** 739–762.

Trier, S. B., W. C. Skamarock, and M. A. LeMone, 1997: Structure and evolution of the 22 February 1993 TOGA COARE squall line: Organization mechanisms inferred from numerical simulation. *J. Atmos. Sci.,* **54,** 386–407.

Tripoli, G. J., and W. R. Cotton, 1980: A numerical investigation of several factors contributing to the observed variable intensity of deep convection over South Florida. *J. Appl. Meteor.,* **19,** 1037–1063.

———, and ———, 1986: An intense, quasi-steady thunderstorm over mountainous terrain. Part IV: Three-dimensional numerical simulation. *J. Atmos. Sci.,* **43,** 894–912.

Tuttle, J. D., V. N. Bringi, H. D. Orville, and F. J. Kopp, 1989: Multiparameter radar study of a microburst: Comparison with model results. *J. Atmos. Sci.,* **46,** 601–620.

van Tassel, E. L., 1955: The North Platte Valley tornado outbreak of June 27, 1955. *Mon. Wea. Rev.,* **83,** 255–264.

Wakimoto, R. M., 1982: The life cycle of thunderstorm gust fronts as viewed with Doppler radar and rawinsonde data. *Mon. Wea. Rev.,* **110,** 1060–1082.

———, 1983: The West Bend, Wisconsin storm of 4 April 1981: A problem in operational meteorology. *J. Climate Appl. Meteor.,* **22,** 181–189.

———, 1985: Forecasting dry microburst activity over the high plains. *Mon. Wea. Rev.,* **113,** 1131–1143.

———, and V. N. Bringi, 1988: Dual-polarization observations of microbursts associated with intense convection: The 20 July storm during the MIST Project. *Mon. Wea. Rev.,* **116,** 1521–1539.

———, and N. T. Atkins, 1994: Observations of the sea-breeze front during CaPE. Part I: Single-Doppler, satellite, and cloud photogrammetry analysis. *Mon. Wea. Rev.,* **122,** 1092–1114.

———, C. J. Kessinger, and D. E. Kingsmill, 1994: Kinematic, thermodynamic, and visual structure of low-reflectivity microbursts. *Mon. Wea. Rev.,* **122,** 72–92.

———, W.-C. Lee, H. B. Bluestein, C.-H. Liu, and P. H. Hildebrand, 1996: ELDORA observations during VORTEX 95. *Bull. Amer. Meteor. Soc.,* **77,** 1465–1481.

———, C.-H. Liu, and H. Cai, 1998: The Garden City, Kansas, storm during VORTEX 95. Part I: Overview of the storm's life cycle and mesocyclogenesis. *Mon. Wea. Rev.,* **126,** 372–392.

Warner, J., 1970: The microstructure of cumulus cloud. Part III. The nature of the updraft. *J. Atmos. Sci.,* **27,** 682–688.

Weaver, J. F., and S. P. Nelson, 1982: Multiscale aspects of thunderstorm gust fronts and their effects on subsequent storm development. *Mon. Wea. Rev.,* **110,** 707–718.

Weaver, J., J. F. W. Purdom, and S. B. Smith, 1994: Comments on nowcasts of thunderstorm initiation and evolution. *Wea. Forecasting,* **9,** 658–662.

Weckwerth, T. M., and R. M. Wakimoto, 1992: The initiation and organization of convective cells atop a cold-air outflow boundary. *Mon. Wea. Rev.,* **120,** 2169–2187.

———, J. W. Wilson, and R. M. Wakimoto, 1996: Thermodynamic variability within the convective boundary layer due to horizontal convective rolls. *Mon. Wea. Rev.,* **124,** 769–784.

Wei, D., A. M. Blyth, and D. J. Raymond, 1998: Buoyancy of convective clouds in TOGA COARE. *J. Atmos. Sci.,* **55,** 3381–3391.

Weisman, M. L., 1992: The role of convectively generated rear-inflow jets in the evolution of long-lived mesoconvective systems. *J. Atmos. Sci.,* **49,** 1826–1847.

———, 1993: The genesis of severe, long-lived bow echoes. *J. Atmos. Sci.,* **50,** 645–670.

———, and C. A. Davis, 1998: Mechanisms for the generation of mesoscale vortices within quasi-linear convective systems. *J. Atmos. Sci.,* **55,** 2603–2622.

———, J. B. Klemp, and R. Rotunno, 1988: Structure and evolution of numerically simulated squall lines. *J. Atmos. Sci.,* **45,** 1990–2013.

Wicker, L. J., and R. B. Wilhelmson, 1995: Simulation and analysis of tornado development and decay within a three-dimensional supercell thunderstorm. *J. Atmos. Sci.,* **52,** 2675–2703.

Williams, D. T., 1959: A theoretical estimate of draft velocities in a severe thunderstorm. *Mon. Wea. Rev.,* **87,** 65–68.

———, 1963: The thunderstorm wake of May 4, 1961. National Severe Storms Project Rep. 18, U.S. Dept. of Commerce, Washington, D.C., 23 pp. [NTIS PB 168223].

Wilson, J. W., 1986: Tornadogenesis by nonprecipitation induced wind shear lines. *Mon. Wea. Rev.,* **114,** 270–284.

———, and W. E. Schreiber, 1986: Initiation of convective storms at radar-observed boundary-layer convergence lines. *Mon. Wea. Rev.,* **114,** 2516–2536.

———, and C. K. Mueller, 1993: Nowcasts of thunderstorm initiation and evolution. *Wea. Forecasting,* **8,** 113–131.

———, and D. L. Megenhardt, 1997: Thunderstorm initiation, organization, and lifetime associated with Florida boundary layer convergence lines. *Mon. Wea. Rev.,* **125,** 1507–1525.

———, R. Carbone, H. Baynton, and R. Serafin, 1980: Operational application of meteorological Doppler radar. *Bull. Amer. Meteor. Soc.,* **61,** 1154–1168.

———, R. D. Roberts, C. Kessinger, and J. McCarthy, 1984: Microburst wind structure and evaluation of Doppler radar for airport wind shear detection. *J. Climate Appl. Meteor.,* **23,** 898–915.

———, G. B. Foote, N. A. Crook, J. C. Fankhauser, C. G. Wade, J. D. Tuttle, C. K. Mueller, and S. K. Krueger, 1992: The role of boundary-layer convergence zones and horizontal rolls in the initiation of thunderstorms: A case study. *Mon. Wea. Rev.,* **120,** 1785–1815.

———, N. A. Crook, C. K. Mueller, J. Sun, and M. Dixon, 1998: Nowcasting thunderstorms: A status report. *Bull. Amer. Meteor. Soc.,* **79,** 2079–2099.

Wolfson, M. M., J. T. Distefano, and T. T. Fujita, 1985: Low-altitude wind shear in the Memphis, TN area based on mesonet and LLWAS data. Preprints, *14th Conf. on Severe Local Storms,* Indianapolis, IN, Amer. Meteor. Soc., 322–327.

———, and Coauthors, 1990: Characteristics of thunderstorm-generated low altitude wind shear. *29th Conf. on Decision and Control,* Honolulu, HI, IEEE, 682–688.

———, R. L. Delanoy, B. E. Forman, R. G. Hallowell, M. L. Pawlak, and P. D. Smith, 1994: Automated microburst wind-shear prediction. *Lincoln Lab. J.,* **7,** 399–426.

Yuter, S. E., and R. A. Houze Jr., 1995a: Three-dimensional kinematic and microphysical evolution of Florida cumulonimbus. Part I: Spatial distribution of updrafts, downdrafts, and precipitation. *Mon. Wea. Rev.,* **123,** 1921–1940.

———, and ———, 1995b: Three-dimensional kinematic and microphysical evolution of Florida cumulonimbus. Part II: Frequency distributions of vertical velocity, reflectivity, and differential reflectivity. *Mon. Wea. Rev.,* **123,** 1941–1963.

Zhang, D.-L., and K. Gao, 1989: Numerical simulation of intense squall line during 10–11 June 1985 PRE-STORM. Part II: Rear inflow, surface pressure perturbations and stratiform precipitation. *Mon. Wea. Rev.,* **117,** 2067–2094.

Zipser, E. J., 1969: The role of organized unsaturated convective downdrafts in the structure and rapid decay of an equatorial disturbance. *J. Appl. Meteor.,* **8,** 799–814.

———, 1977: Mesoscale and convective-scale downdrafts as distinct components of squall-line circulation. *Mon. Wea. Rev.,* **105,** 1568–1589.

———, 1982: Use of a conceptual model of the life-cycle of mesoscale convective systems to improve very-short-range forecasts. *Nowcasting,* K. Browning, Ed., Academic Press, 191–204.

Chapter 8

Rain Production in Convective Storms

DENNIS LAMB

The Pennsylvania State University, University Park, Pennsylvania

REVIEW PANEL: Hans Verlinde (Chair), Bill Cotton, Tom McKee, William Woodley, and Roelof Bruintjes

---

8.1. Introduction

Convective systems can, in addition to generating strong winds, hail, and lightning, produce large amounts of rainfall. In some cases, it is the heavy rain itself that distinguishes the storm, even when the rain is accompanied by one or more of the other defining features of severe storms (Maddox et al. 1979; Changnon 1999). Whereas rain is usually a favorable product of moist convection in the atmosphere, excessive amounts or rates of rainfall can lead to surface erosion, property or crop damage, and devastating floods (e.g., Pontrelli et al. 1999). Rain production carries with it atmospheric implications, as well, so it is important that we understand the physical processes that give rise to precipitation-size hydrometeors in convective systems (Ludlam 1963).

The production of rain requires the coherent action of many processes acting over broad ranges of space scales and timescales. As Hobbs (1981) aptly reminds us, atmospheric phenomena typically operate within a hierarchy of spatial scales, in some cases spanning nearly 15 orders of magnitude. The precipitation-generating parts of a cloud, for instance, are often substantially smaller than the storm system itself and exhibit distinctive structural patterns (see, e.g., McAnelly and Cotton 1986). The atmospheric motions set up by the macro- and mesoscale storm structure provide the dynamical and thermodynamical setting in which condensate forms and ultimately yields precipitation on the ground. At the same time, the collective action of the many microphysical processes throughout the storm forces adjustments in the larger-scale mass, energy, and momentum fields in ways that often lead to positive feedback and amplification of the system as a whole (Maddox 1983). Such nonlinear interactions of processes operating over diverse scales is a characteristic feature of mesoscale convective systems in the atmosphere (Doswell 1987).

The inherently complex nature of convective systems demands that we find simplifying ways to visualize their organizing patterns on the various scales and to relate the physical processes acting on one scale to those operating on another. The conceptual framework set up by Doswell (1987) offers some aid for heavy rainstorms, as it does for severe weather in general, by distinguishing the separate roles played by large-scale flows (establishing a favorable thermodynamic environment) and mesoscale processes (providing a mechanism for convective initiation). Fortunately, it is not difficult to realize that, regardless of its particular structure, any given convective system involves vertical motions that lead to cooling/warming of local air parcels by virtue of their expansion/compression associated with the rapid changes in pressure. Both the production of condensate (the material source for rain) and the release of latent heat (an energy source), for instance, depend crucially on the upward motion of moist air (Forbes 1990). We will therefore find it convenient to view the vertical motion field as the main link between the storm-scale structure and the microscale processes responsible for precipitation formation.

In the sections that follow, we will look at how storms "utilize" the moisture brought in by the storm-scale motions and form the large hydrometeors that eventually comprise rainfall. After an overview of some macroscopic features of rainstorms, the in-cloud processes actually responsible for the generation of cloud water, ice, and rain are considered. Finally, some attention will be paid to the efficiency with which the overall processing of water vapor into rain takes place in rainstorms.

8.2. General features of rainstorms and their environments

Convective systems that yield significant rainfall require somewhat specific climatic conditions. Typically, convective storms are spawned within humid regions because it is the moisture, initially in the form of water vapor, that provides both the source material for precipitation and an important fraction of the energy that drives the convection (Newton 1981).

Moisture abundance in the atmosphere increases strongly with the temperature of any surfaces capable of evaporating water, such as the tropical oceans and foliage-rich land surfaces, so it is not surprising that most heavy rain events occur generally at low latitudes or in the warm season of the midlatitudes (Maddox et al. 1979; Ludlam 1980, p. 226).

Heavy rain is commonly associated with localized deep convection and severe weather, but that need not always be the case. A listing of some noteworthy rainstorms, as given in Table 8.1, exhibits the diversity that exists. The storms mentioned here are considered notable in the sense that they have been worthy of scrutiny by either the press or the research community. This compilation is not meant to be comprehensive, but illustrative of the variety of heavy rain events found in nature. The entries in Table 8.1 represent, for the most part, point-rainfall amounts. The specified depth of rain is therefore that which fell into an individual gauge over the duration of the indicated storm. It is important to recognize that the dominant portion of heavy rain typically falls during a small fraction of the storm duration (Angel and Huff 1999). The selected storms show that significant rain events occur throughout the world and over a broad range of time periods.

The amount of rain that falls from a particular storm reflects in some sense the longevity of the storm. Some storms may yield rain at a high rate, but they tend to be short-lived. On average, the longer a storm persists in a given vicinity, the larger the total rainfall tends to be (Doswell et al. 1996). The relationship between rain depth and storm duration has been investigated several times in the past (e.g., Jennings 1950; Fletcher 1950; Paulhus 1965; Ludlam 1980, p. 300). Because any storm releases a specific amount of rain at a given spot on the earth over a well-defined period of time, we can, in principle, plot each measured rainfall amount against the recorded duration. In the tradition of Paulhus (1965) and Ludlam (1980, p. 300), the depth–duration relationships of some representative heavy rain events (those listed in Table 8.1) have been plotted in Fig. 8.1. The axes have been scaled logarithmically to allow the wide ranges in both depth and duration to be presented uniformly. Note that specific rain rates show up as parallel straight lines on such a plot and range between about 0.1 and 10^3 mm h^{-1}. Light rains of long duration yield points scattered across the lower right-hand part of the graph, whereas more intense, short-lived events tend to show up as points farther toward the left-hand side of such plots. Data from regions of the world that are characterized by persistent or monosoonal flows of moist air occupy the upper right-hand part of the graph. It has been suggested (Fletcher 1950) that all rain data are enveloped by a power-law relationship, expressible as

$$P_{\max} = c\Delta t^n, \tag{8.1}$$

where $P_{\max}$ is the maximum precipitation depth that can occur over duration Δt, and where c and n are empirical parameters. It appears that a value for n = 0.5, indicative of a square root relationship between depth and duration (shown by the dashed straight line in Fig. 8.1), fits the world-record data remarkably well (Fletcher 1950; Henderson 1993), in which case c = 363 mm $h^{-0.5}$ (or 14.3 in. $h^{-0.5}$) when depth is expressed in millimeters (alternatively, inches) and duration is in hours. The physical basis for this limiting relationship between rain totals and storm duration has yet to be established, although it seems reasonable to expect that part of the answer lies in the natural connection that exists between the size and longevity of most atmospheric phenomena (Forbes and Greenfield 1992).

TABLE 8.1. Some notable storms arranged in order of increasing duration. After Fletcher (1950) and Ludlam (1980, p. 300).

Location	Date	Depth (mm)	Duration (h)	Intensity (mm h^{-1})	Reference
Unionville, MD	4 July 1956	31.2	0.017	1835	Engelbrecht and Brancato (1959)
Holt, MO	22 June 1947	305	0.7	436	Lott (1954)
Hampstead, United Kingdom	14 August 1975	170	2.5	68	Keers and Wescott (1976)
Madrid, Spain	25 June 1995	79	3	26	Wheeler (1995)
Big Thompson Canyon, CO	31 July 1976	305	4	76	Maddox et al. (1977)
Rapid City, SD	9 June 1972	380	4	95	Wagner (1972), Maddox et al. (1978)
Johnstown, PA	19–20 July 1977	128	12	11	Hoxit et al. (1978)
Belouve, La Reunion	27–28 February 1964	1690	18.5	91	Paulhus (1965)
Chicago, IL	17–18 July 1996	430	24	18	Changnon and Kunkel (1999)
London, United Kingdom	16–17 August 1977	113	24	4.7	Haggett (1980)
South Island, New Zealand	10–12 March 1982	1810	72	25	Henderson (1993)
Upper Mississippi, USA	Summer 1993	875	3600	0.24	Williams (1994)
Cherrapunji, India	1860–1861	41 000	18 000	2.3	Jennings (1950)

Note: Duration represents the time interval over which the indicated amount of rain fell, not necessarily the full length of the storm. The intensity is simply the average rate of rainfall, derived from the ratio of the previous two columns.

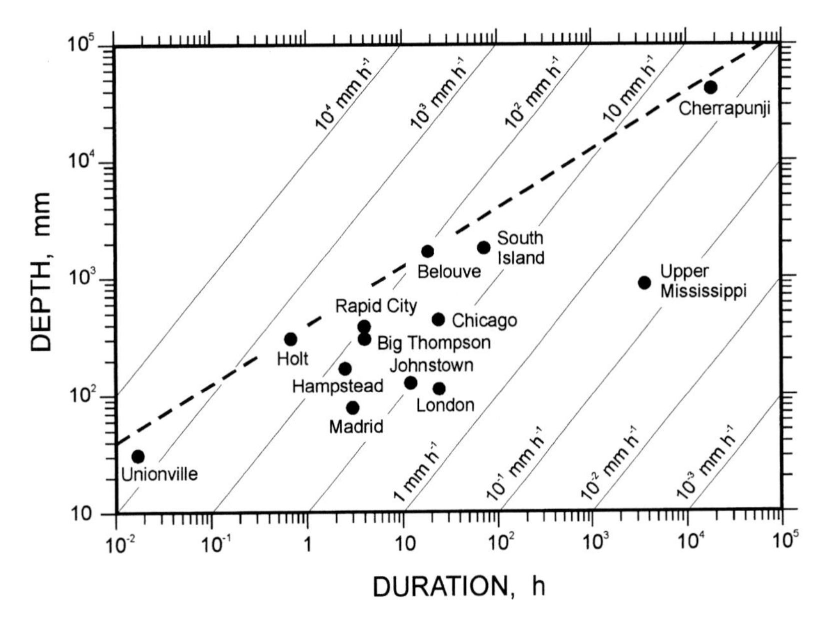

FIG. 8.1. Point-rainfall amounts (DEPTH) for specified time intervals (DURATION) from selected rainstorms. The thin solid lines are isopleths of the indicated rain rate, whereas the dashed line represents the upper bound of the depth–duration envelope for world-record events. Adapted from Ludlam (1980, p. 300).

Storms have been classified in the past either as *small-scale systems,* when they are characterized by strong upward motions that are comparable in magnitude to the horizontal winds prevailing at the time, or as *large-scale systems,* when they exhibit relatively weak upward motions and horizontal winds that are subject to the earth's rotation (Ludlam 1963). While this scale-dependent classification may be useful for some purposes, we must note that individual storms of all scales often involve both strong and weak upward motion. It is now recognized that convective elements (with strong vertical motions) are commonly embedded in large-scale flows and that stratiform regions (with weak vertical motions) are often associated with deep convection (Houze 1981).

A study of specific storms yields clues as to the meteorological processes responsible for rain formation. The rain event that occurred within a small region around Holt, Missouri, on 22 June 1947, for instance, represents an extreme, in fact a world record, case of a huge rain rate for a short period of time (Jennings 1950). In the brief interval of 42 min, 305 mm (12 in.) of rain fell from this "explosive" storm and contributed to major local flooding, as well as to flooding in St. Louis several days later (Lott 1954). Whereas Lott's meteorological analysis concluded that deep convection was associated with a squall line ahead of a traditional surface cold front, the reanalysis by Locatelli and Hobbs (1995) has suggested that the "cold front" was actually a "dry trough" and that frontogenesis aloft (at about 700 hPa) was the mechanism responsible for releasing the latent instability of the air. Locatelli and Hobbs do not address how the storm energy became so focused in the vicinity of Holt, but presumably differential advection, accentuated by the large pressure gradient that existed toward the northern end of the squall-line trough, contributed significantly to the strength of the updraft and intensity of the rainfall (Lott 1954). This storm serves as a good example of how Doswell's (1987) triad of ingredients (low-level moisture, conditional instability, and lifting mechanism) needed for severe convection in general applies to heavy rain events in particular.

Differential advection, wherein warm, moist air underruns potentially warmer, but drier, air, is common in regions where arid plateaus are adjacent to moist lowlands (Carlson and Ludlam 1968; Carlson et al. 1983). The overrunning warm plume from the elevated terrain creates a lid, or capping inversion, that temporarily suppresses convection, allowing latent instability to build up, only to be released violently near the lateral edge of the elevated mixed layer (Carlson 1991, pp. 457, 463). Carlson and Ludlam (1968) have shown that geographical situations suitable for the development of severe storms and heavy rain arise in western Europe and tropical western Africa, as well as in the middle parts of the United States. Whereas in the United States the Mexican plateau generates the dry mixed layer that stays aloft and the Gulf of Mexico provides the low-level moisture, in western Europe the analogous features are the tablelands of Spain or the Sahara of northern Africa and the warm waters of the Mediterranean Sea or even soil moisture in France (Carlson and Ludlam 1968).

The very localized and intense rainstorm at Hampstead (United Kingdom) on 14 August 1975 may be an example of such differential advection. It had been suggested that local topography contributed to the precise location of this storm, which led to serious flooding in the greater London area (Keers and Wescott 1976). However, an alternative view was considered by Miller (1978): Results from a three-dimensional (3D) model with parameterized bulk microphysics showed the pattern of airflow in the system as a whole to persist substantially longer than the times required for air parcels or even individual cells to move through it. Such a situation can arise when the wind shears directionally with height in an appropriate manner. As shown schematically in Fig. 8.2, the veering of the environmental wind probably caused the low-level inflow air (lightly shaded arrow) to turn toward the right as it ascended, while at the same time the midlevel air (darker arrow) that contributed to the downdraft was forced to turn toward the left before fanning out behind the gust front. A 3D, interlocking pattern was thereby established that allowed the downdraft air to counter the inflowing air just enough to make the system roughly stationary.

A conceptual picture of the importance of directional shear emerged following the use of alternative wind profiles in the modeling effort of Miller (1978). As shown in Fig. 8.3, two extreme situations are conceivable for a given upper-level wind direction, which serves as the main determinant of the elongation and orientation of the rain region and its associated mesohigh area. As shown in situation (a), the absence of directional shear results in a very small zone of convergence, one that is insufficient to buffer the thrust of inflowing air and maintain the storm in its location. In situation (b), however, when moist surface air flows into the system at an angle that is more or less perpendicular to the midlevel air, a broad convergence zone results that creates an effective barrier to the flow of low-level air, thereby inhibiting the movement of the storm. Miller's study, while aimed mainly at explaining a specific event, nevertheless aligns closely with other studies of storm structure (e.g., Browning 1964) and illustrates the importance of the 3D organization of the airflow to the localization and steady-state character of such storms and their ability to focus large amounts of rain on unfortunate communities.

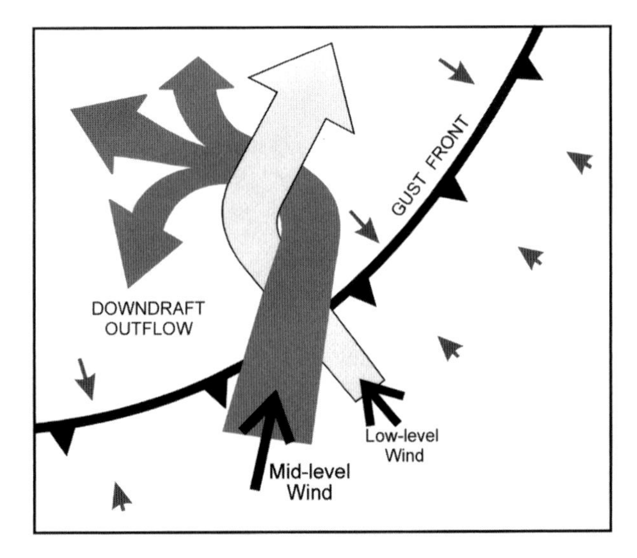

FIG. 8.2. The airflow pattern in the Hampstead storm of 14 August 1975. The lighter, outlined arrow indicates the ascending flow of moist, initially low-level air; the darker arrow shows the flow of midlevel air that contributes to the downdraft and the formation of the gust front. Adapted from Miller (1978).

Whereas Miller (1978) downplayed orographic influences on the Hampstead storm, the effects of topography can nevertheless play important, sometimes crucial, roles in the development of convective storms and in the damaging consequences of heavy rainfall on the ground (Pontrelli et al. 1999). Under suitable conditions, as dictated by the large-scale pressure, moisture, and temperature fields, the Front Range of the Rocky Mountains of Colorado can experience torrential rains and flash floods. An excellent example is the rainstorm that formed over the watershed of the Big Thompson River on 31 July 1976. The strong low-level easterly winds that followed a frontal passage through central Colorado late in the day pushed a moist, conditionally unstable air mass against the Front Range (Caracena et al. 1979). As the air was forced to rise by the orography, the convective instability of the moist air was released, causing large updraft speeds and explosive storm development. An estimated 254 mm (10 in.) of rain fell at the Glen Comfort station in little more than 3 h, causing devastating flash floods that left 139 dead (Caracena et al. 1979). Neither large hail nor strong winds accompanied the storm, so this convective system was a rainstorm in the strictest sense. Photographs (e.g., Fig. 8.4) and radar cross sections (see Fig. 8.5) showed that the updraft tilted slightly to the northwest despite negative vertical wind shear in the environment, a direct result of the orography (Maddox et al. 1977). Apparently, when the terrain is the mechanism for lifting and the winds aloft are weak, strong easterly momentum of the low-level air tends to be preserved as the air is transported vertically, producing an updraft structure that slopes along the initial flow direction. Although it exaggerates the vertical dimension greatly, Fig. 8.5 schematically depicts the internal flows of air early in this convective system. The tilted flow pattern facilitated the release of the condensate produced in the updraft, permitting the cloud to grow extremely high and to sustain steady-state conditions with little overall storm movement (Caracena et al. 1979). In situations such as this, where moist and conditionally unstable air is forced against significant

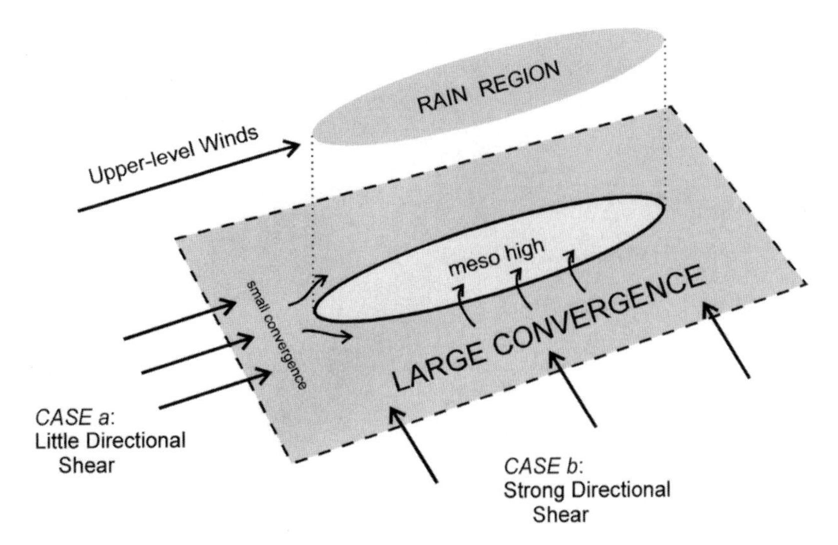

FIG. 8.3. Alternative orientations of the low-level flow relative to a mid/upper-level wind aligned parallel to the major axis of an elongated storm. Case (a) occurs when little or no directional shear is present, resulting in only small convergence. Case (b) prevails when the winds veer strongly with height, causing large convergence and uplift of moist air. Adapted from Miller (1978).

FIG. 8.4. Photograph of the large cloud associated with the Big Thompson Canyon storm on 31 July 1976. The photo, reproduced here with permission as it appeared in the report by Maddox et al. (1977), was taken by John Asztalos.

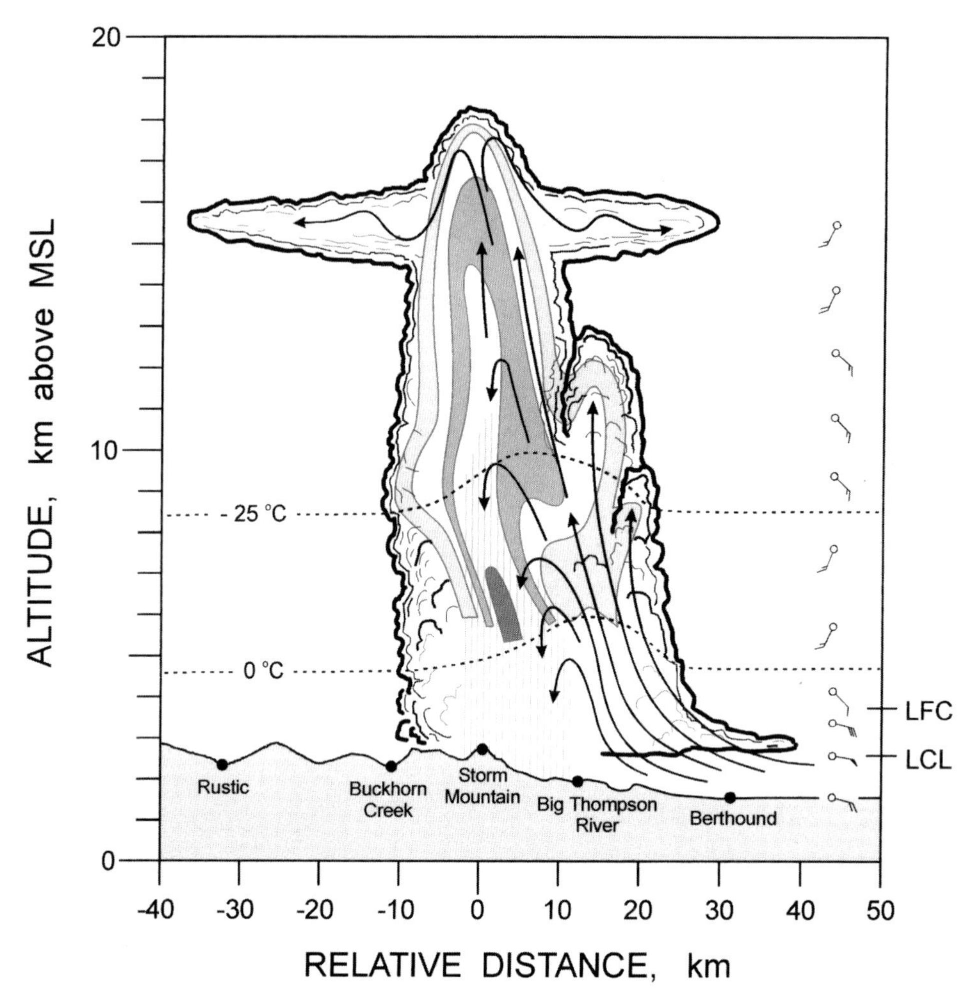

FIG. 8.5. Schematic depiction of one of the initial cells that composed the Big Thompson storm. The shaded isopleth regions represent 10-dBZ reflectivity ranges from the Grover radar, the strongest values (greater than 55 dBZ) being found relatively low in the center of the storm. The altitudes of the lifting condensation level (LCL) and the level of free convection (LFC) are shown along the right-hand border. Wind data (each full barb representing 10 knots) indicate strong flow from the east in the lowest levels and weak, variable flow aloft. The arrows suggest that the likely motion of air through the storm sloped upward toward the northwest. Diagram reproduced with permission from Maddox et al. (1977).

terrain barriers, orography serves both to provide the necessary lifting mechanism and to anchor the storm to a fixed geographical location.

Other locations in the world are of course subject to the consequences of orography. New Zealand is an island nation that is repeatedly exposed to strong flows of moist air. By contrast with many other places in the world, rainfall in the Southern Alps of South Island tends to have an exceptionally large orographic, rather than convective, component (Henderson 1993). The three-day storm in March 1982 (see Table 8.1) yielded near-record amounts of rainfall at one site some 10 km upwind of the mountain crest: 1810 mm (71 in., or nearly 6 ft). Such extremely strong orographic influences on rainfall are probably not unique to New Zealand, but this example suffices to illustrate that a significant uplift of moist air, regardless of the cause, can be very effective in setting the stage for the transformation of water vapor into rain.

Large rain amounts can also occur with weather systems of extended duration or repetitive action. In such cases, large-scale atmospheric motions are especially important, and the rates of rainfall tend to be modest. Notable in this regard are the record amounts of convective rain that have been recorded several times in Cherrapunji, India (Jennings 1950). Whether one considers daily or weekly amounts of rain, or even accumulations over the two-year period from 1860 to 1861, data from this region frequently fall close to the depth–duration "envelope," the dashed curve in Fig. 8.1. Typically, such extreme events are related to monsoonal flows that are strongly influenced by topography. Updraft speeds are typically modest compared with those in severe local storms driven by large values of convective potential energy, but the process of rain formation is repeated over and over, and very effectively.

The "Great Flood of 1993" in North America represents a case in which anomalous large-scale patterns of air pressure and winds persisted and permitted the repetitive formation of convective storms (Williams 1994). During the summer of 1993 large sections of the northern United States and southern Canada experienced the most prolonged rainfall of the century (Rannie and Blair 1995). Rainfall amounts between two and four times normal led to severe flooding of the Missouri, upper Mississippi, and Red (Manitoba) Rivers and many of their tributaries. Statistical analyses by Kunkel et al. (1994) show that monthly, as well as the 2-, 3-, 4- and 12-month rainfall totals in the greater upper Mississippi River Basin exceeded all previous amounts on record. These areally averaged depths are plotted on the graph given in Fig. 8.1 and show that devastating consequences can arise even when the rainfall is well below world-record proportions. Still, in this region of the world, events such as these are not likely to recur more than once in every 200 years (Kunkel et al. 1994).

The conditions that spawned the large amounts of rain over such a broad area of North America in 1993 were established on the large scale by an anomalous circulation pattern in the Northern Hemisphere (Williams 1994; Bell and Janowiak 1995). Following a period during late winter and early spring in which an unusually strong ridge was established over western North America, possibly associated with the positive phase of the North Pacific teleconnection pattern (Barnston and Livezey 1987), the midlatitude flow became more zonal during May and June, providing a duct for cyclone propagation into the midwest from the North Pacific (Bell and Janowiak 1995). By mid-July the wave pattern amplified, leading to the formation of a deep trough over the western United States and the establishment of a baroclinic zone lying farther south than normal for that time of year (Rannie and Blair 1995). A quasi-stationary frontal boundary lay across the Midwest on more than 40 days during June and July, along which short-wave disturbances could propagate and stimulate convective activity (Kunkel et al. 1994; Bell and Janowiak 1995). During this same period the anticyclonic circulations around the Bermuda high were particularly strong and funneled unusually large fluxes of moisture northward from the Gulf of Mexico at low levels (Williams 1994; Bell and Janowiak 1995). This overall circulation pattern was similar to those of previous heavy rain periods in North America (Maddox et al. 1979), but 1993 was characterized by the persistence of the pattern (Kunkel et al. 1994). Rannie and Blair (1995) have shown particular parallels that existed between the 1993 and 1958 rainy periods. In addition, based on similar flooding patterns recorded in the diaries of westward-moving pioneers in 1849, these authors suggest that pressure and flow patterns conducive to flooding in central North America may not be an uncommon state of the atmosphere. Whether or not these types of circulation patterns are associated with recurring El Niño events, volcanic eruptions, or greenhouse warming cannot yet be established (Williams 1994). Displacements of the circulation patterns from their normal configurations may simply reflect the chaotic nature of the atmospheric system (Bell and Janowiak 1995).

Convective systems often develop in response to suitable large-scale forcing conditions, regardless of the cause of the particular pattern. The local weather throughout the flood-plagued region in 1993, for instance, was not especially severe in the usual sense, but the thunderstorms, taken collectively, were able to produce copious amounts of rain. In general, organizational structures can develop on intermediate scales that foster a synergism of activity and lead to complicated interactions with local surface conditions (Giorgi et al. 1996). Individual thunderstorm cells may merge with one another or interact in ways that stimulate the overall growth of the convective system (Maddox 1983). In July 1993, in particular, the association of the quasi-stationary frontal boundary with a strong, diverging jet stream aloft provided a focus for moisture convergence, the release of substantial potential instability, and the outbreak of organized mesoscale convective systems (Bell and Janowiak 1995). If sufficiently large, mesoscale convective complexes (MCCs) form that develop their own cyclonic circulations and significantly modify the large-scale environment (Maddox et al. 1981; Maddox 1983; cf. Cotton et al. 1983).

As a class, MCCs exhibit a number of characteristics in common, some of which are important for rain production (Cotton et al. 1989; McAnelly and Cotton 1989). Because they are relatively large systems of long duration (Maddox 1980, 1983; McAnelly and Cotton 1989), MCCs represent nearly ideal meso-α structures with embedded meso-β convective elements that can process huge amounts of moisture into rain over large geographical areas (Maddox et al. 1979; McAnelly and Cotton 1986; Tollerud and Collander 1993). In the central United States, it appears that MCCs account for an appreciable fraction of the normal warm-season rainfall, perhaps as much as 70% or more in some areas (Fritsch et al. 1986; Kane et al. 1987). Analyses of satellite data have shown that MCCs are about as frequent in the midlatitudes of both North and South America, with those in South America tending to be even larger in size than those in North America (Velasco and Fritsch 1987). Velasco and Fritsch (1987) have also shown that MCCs tend to form downwind of major mountain barriers where low-level jets of moisture frequently develop. MCCs exhibit a strong diurnal cycle, with the maximum strength occurring at night (Maddox et al. 1979; Maddox 1980, 1983; Velasco and Fritsch 1987; Tollerud and Collander 1993), probably a consequence of the acceleration of the low-level jet that occurs when the planetary boundary layer becomes decoupled from the surface after sunset (Blackadar and Buajitti 1957; Wetzel et al. 1983; Helfand and Schubert 1995). Although the large sizes of MCCs tend to limit the average intensity of rainfall to modest proportions, rain rates in excess of 13 mm h^{-1} for extended periods are characteristic (Tollerud and Collander 1993). The intensities of convection and rainfall are probably linked to the typical alignment of the upper-level air flows with the surface frontal zone that stimulates the release of potential instability, a large-scale configuration that favors the formation of slow-moving precipitation outflow boundaries perpendicular to the low-level inflow of moisture (see Fig. 8.3 and Miller 1978; Cotton and Anthes 1989, p. 669).

Some MCCs are long-lived and produce significant weather for several days as they propagate across the United States (Wetzel et al. 1983). A particularly long-lived MCC started in South Dakota and eventually developed into a tropical storm over the Atlantic Ocean, but not before causing devastating floods in Johnstown, Pennsylvania, during the night of 19 July 1977 (Hoxit et al. 1978; Bosart and Sanders 1981). Long-lived systems such as this undergo the normal diurnal cycle of intensifying at night and weakening during the daytime, a pattern that leads to intermittency in the precipitation tracks and possible difficulties in forecasting (Bosart and Sanders 1981; McAnelly and Cotton 1986; Kane et al. 1987). This particular system became temporarily dispersed as it passed near the Great Lakes and was exposed to less favorable moisture and instability conditions, but then it became rejuvenated as it moved eastward and encountered a rich supply of low-level moisture (Bosart and Sanders 1981). Dynamical processes on the meso-α scale seemed to interact with the convective-scale processes in a way that was favorable for development of a nearly stationary outflow boundary aligned perpendicular to the low-level moisture flux from the southwest (Hoxit et al. 1978; Zhang and Fritsch 1986). Again, we see a possible parallel between the flow organization in the Johnstown storm and that of the Hampstead storm, the stationarity of which was linked to directional shear of the environmental wind in association with warm-air advection (see Fig. 8.3 and related discussion). Nevertheless, the terrain of southwestern Pennsylvania is complex, so one must be careful not to rule out topographical influences in this region, which is so noted for floods (Ludlam 1989).

We thus see that abundant rain can be produced whenever moist air ascends at appreciable rates in the atmosphere. The uplift of the air typically occurs within convective elements when conditional instability is released by either dynamical or orographic means. Particularly important rain events take place when the uplift occurs for extended periods or repetitively over one spot. Again, suitable dynamical organization or topography can serve to hamper the movement of a storm and lead to nearly steady-state conditions over extended periods, with possibly devastating consequences. In the sections that follow we look at the processes that take place within individual convective elements to convert water vapor into the large water drops that ultimately make up rain.

8.3. Condensate production

The condensate in a cloud, whether in the form of liquid or solid particles, must be produced initially from water vapor as air cools. In all cases of importance to rain production, this cooling is achieved when air rises toward lower pressures in the atmosphere and loses some of its internal energy by doing work on the environment. In turn, the lowered temperature in a parcel of air decreases the equilibrium vapor pressures of the liquid and solid forms of water, allowing the partial pressure of water vapor to exceed, at least temporarily, the local equilibrium or saturation value. It is this "excess" vapor, the amount over and above the equilibrium value, that can be condensed out to form the liquid and solid particles that make up a cloud. The supersaturation at any given instant is simply the ratio of the excess vapor pressure to the equilibrium value at the local temperature. In cloud physics it is helpful to think of the cooling that takes place during the uplift of moist air as making excess water available, alternatively, as generating supersaturation, while at the same time the actual formation of condensate depletes vapor and destroys supersaturation (e.g., Young 1993, p. 251). Under many conditions in the atmosphere, the tendency toward supersaturation generation is largely offset by the rapid condensation of excess vapor onto the many cloud particles that have formed (Kessler 1969). Whereas such a tendency to deviate little from equilibrium can be exploited to simplify some problems in studies of cloud dynamics, exceptions occur, especially when precipitation is involved (Cotton and Anthes 1989, p. 136). The condensate in a cloud, forming mainly on the many small cloud droplets, is perhaps best viewed as the source material upon which the much larger precipitation particles feed. This conversion of condensate to precipitation may well be an integral part of the interaction of scales observed in some mesoscale convective phenomena (Ludlam 1980, p. 270; Knupp and Cotton 1987; Krueger 1988; Feingold et al. 1991).

The amount of condensate that can form in a parcel of air is related to several factors, but especially to the vapor density in the air at the start of condensation and the temperature change that occurs during ascent above the initial condensation level. If we consider an idealized reference process, namely, adiabatic ascent without loss of condensate, the maximum liquid water mixing ratio at any given temperature level is readily calculated as the difference between the saturation vapor mixing ratios at the initial condensation temperature (i.e., at cloud base) and at the given temperature (see Cotton and Anthes 1989, p. 17, for details). For practical reasons it is often more convenient to use liquid water concentrations, in which case the atmospheric density must be considered as well as temperature. Kessler (1969) has treated such situations extensively and showed that a maximum in the liquid water concentration or "content" must exist at some level. Figure 8.6 shows how the liquid water mass content m_c varies with altitude for a number of cloud-base temperatures. Note that the vertical distribution of adiabatic liquid water content depends only on the thermodynamic properties of the air, not on the updraft profile (Kessler 1969). At low altitudes, m_c increases with height because the saturation mixing ratio decreases strongly as the temperature decreases, allowing for the relatively large production of condensate. At high altitudes, on the other hand, condensate production is weak compared with the falloff of air density with increasing height. The maximum value of m_c occurs at the so-called *compensation level,* where the generation of condensate is just offset by the decreasing density of the air with altitude. Kessler (1975) noted that the maximum liquid water content in the atmosphere should be about 8 g m^{-3} and occur at altitudes between about 5 and 10 km. It proves significant that typical liquid water contents in nonprecipitating clouds are often substantially less than the adiabatic values (Warner 1955; Warner and Squires 1958), whereas in large convective storms, liquid water contents in excess of 20 g m^{-3} can be found (Kyle and Sand 1973).

The development of condensate in real clouds is more complicated than any idealized thermodynamic

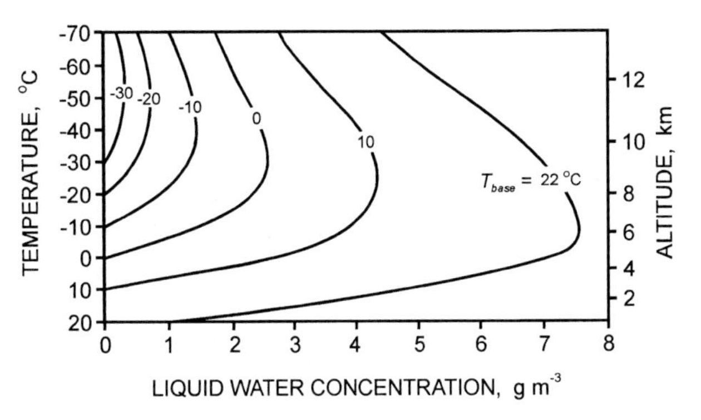

FIG. 8.6. Vertical profiles of condensate concentration in adiabatic clouds of various cloud-base temperatures (T_{base}). Adapted from Kessler (1969).

treatment would suggest. Whereas the rate of condensate production in an adiabatic parcel does depend upon the rate of uplift and can even be related to rainfall rates (Fulks 1935), equilibrium conditions are still assumed, so such rate relationships amount to little more than a kinematic transformation between altitude and time (Kessler 1969). The development of real clouds does of course depend on the vertical motion field for physical reasons, in part because the relaxation to equilibrium is not instantaneous. Finite supersaturations must exist in all developing clouds because of this inherent time lag and because nucleation processes are involved that impose thresholds on the supersaturation for changes in phase.

The supersaturation in a cloud, in effect, connects the condensate production to the evolving thermodynamic conditions in the vapor field. Squires (1952) was the first to derive a quantitative relationship between the rate of supersaturation development and the environmental parameters. If we let w be the local updraft speed, then the rate with which the supersaturation s changes with time in a Lagrangian parcel is given by an equation of the form

$$\frac{ds}{dt} = Q_1 w - Q_2 \frac{dm_c}{dt}, \tag{8.2}$$

where Q_1 and Q_2 are slowly varying functions of the environmental variables. The first term on the right-hand side represents the generation of supersaturation and exhibits a clear linear dependence on the updraft speed. Supersaturation would continue to build up in a rising parcel of air at the rate $Q_1 w$ were it not for the depletion of vapor because of condensate formation, as specified by the second term. A qualitative "solution" to Eq. (8.2) can be found by noting how the two terms vary in relationship to each other as the parcel ascends. However, the actual evolution of supersaturation is complicated, even under relatively simplified scenarios. Figure 8.7 shows one possible situation from Young (1974) in which w = 3 m s^{-1} throughout the model run. Initially, before any condensate has formed, the second term is zero, and s increases linearly with time as the condensation level is first crossed (at t = 0). Condensate begins to form on an increasing number of droplets in response to the increasing supersaturation (Twomey 1959), causing the magnitude of the second term to increase and counteract the positive effect of the first term. The rate of increase of s with time falls off and eventually becomes zero, yielding a maximum in magnitude, s_{max}, very early in the process (t < 1 min in Fig. 8.7). It is at this point, when $s = s_{max}$, that the number concentration in the cloud is established (depicted as n in Fig. 8.7). Eventually, the rates of condensate production and vapor depletion become large enough to cause s to decrease with time. As long as the updraft remains positive, however, the supersaturation must

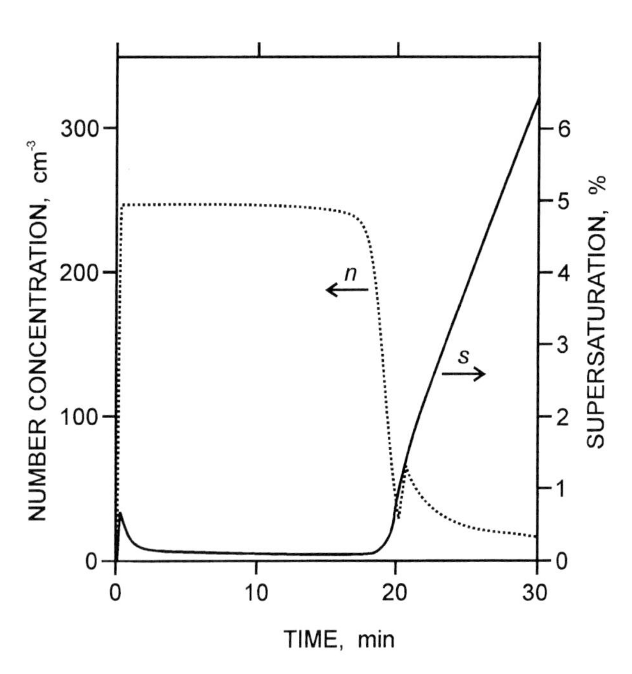

FIG. 8.7. Cloud droplet number concentration (n, dotted curve) and supersaturation (s, solid curve) in a Lagrangian parcel of air ascending at 3 m s^{-1} from cloud base at 15°C. Adapted from Young (1974).

also remain positive, for it is the supersaturation that drives the formation of condensate in the first place. Throughout much of the condensational growth phase of a convective element, a quasi-steady-state condition becomes established in which the continual generation of excess vapor is just balanced by the depletion of vapor by condensation. Condensate then forms at a rate that is closely linked to the updraft speed:

$$\frac{dm_c}{dt} = \frac{Q_1}{Q_2} w. \tag{8.3}$$

The corresponding supersaturation during this quasi-stationary period of growth can likewise be shown to be related to the updraft speed (Paluch and Knight 1984; Politovich and Cooper 1988; Young 1993, p. 253). At some point, as the drops become sufficiently large, other processes may occur that also affect the supersaturation. As shown on the right-hand half of Fig. 8.7, if the concentration of drops were to decrease, as it would during collision-coalescence and rain formation, for instance, the collective surface area of the cloud drops available for condensation would decrease, diminishing the rate of vapor depletion and allowing the supersaturation to increase dramatically. We see even in simple cases that a strong interaction exists between the form and abundance of condensate and the vapor field from which it is formed.

Condensation in the atmosphere occurs readily only because of the presence of an aerosol. Without the particles of the atmospheric aerosol the supersaturation in a rising parcel would grow to hundreds of percent before the liquid phase would form spontaneously (see Pruppacher and Klett 1997, chapter 7). Many of the nonaqueous particles that compose the aerosol are hygroscopic and offer particularly good sites on which condensation can take place. The fundamental role that these particles play in cloud formation arises from the strong affinity that soluble salts and strong acids have for water at the molecular level. Even small soluble particles less than 0.1 μm in diameter are able to compensate effectively for the otherwise very high equilibrium vapor pressures characteristic of highly curved liquid surfaces. The combination of reduced vapor pressure resulting from the solute in an aerosol particle and increased vapor pressure arising from drop curvature yields a nonlinear dependence of the equilibrium vapor pressure on the radius of the solution droplet. The equilibrium saturation ratio S_K over a solution droplet of radius r, obtainable from Köhler theory (Pruppacher and Klett 1997, p. 172), may be expressed as

$$S_K \equiv \frac{e_c}{e_s} = a_w \exp\left(\frac{A}{r}\right), \tag{8.4}$$

where e_c and e_s are, respectively, the partial pressures of water in equilibrium with the embryonic cloud drop and with a pure, flat surface of liquid water, a_w is the activity of water in the solution, and A is a parameter that is a function of the liquid–vapor interfacial free energy (Pruppacher and Klett 1997, p. 172; Chen 1994). The water activity, arising from the solute–water interaction, and the exponential factor, accounting for the effect of drop curvature, both depend on drop radius in opposing senses, so S_K exhibits a maximum at a certain critical radius. The magnitude of this maximum or critical supersaturation (often designated s_c) depends inversely on the solute content of the aerosol particle, so large aerosol particles are able to "activate" and form cloud drops most readily as the supersaturation in a parcel rises. Only a subset of the total aerosol population thus serves as the nuclei of cloud condensation (CCN), and the number of drops formed initially in a cloud depends on the evolving supersaturation in the updraft (Twomey and Squires 1959; Hudson 1993). The size distribution of the hygroscopic CCN in an updraft is important, as it can sometimes influence the microphysical evolution of the cloud in substantial ways (Cooper et al. 1997).

Following the activation of the cloud condensation nuclei, the resulting cloud droplets grow by the deposition of water vapor from the interstitial gas phase. Initially, while they are still only a few micrometers in radius, the droplets grow rapidly, limited mainly by the rates with which vapor can diffuse to their surfaces and the enthalpy of condensation can be dissipated by thermal conduction back to the air (Howell 1949;

Mason 1971, p. 122). As the radius of a droplet increases, the radial gradients of vapor and temperature at the surface decrease and further limit the linear growth rate, making it exhibit an inverse relationship with size:

$$\frac{dr}{dt} = G \times (s - s_K) \times \frac{1}{r}. \tag{8.5}$$

Here, G is a growth parameter that acts like a diffusivity and varies mainly with the temperature and pressure (Rogers and Yau 1989, p. 104). Note that a droplet grows only to the extent that the interstitial supersaturation (s) exceeds the equilibrium value (s_K) that applies to the droplet at its instantaneous size and solute content. Calculations based on Eq. (8.5) show that individual droplets experiencing a steady supersaturation of 1% would require several hundred seconds to grow to a radius of 20 μm from typical CCN (Pruppacher and Klett 1997, p. 511). Larger updraft speeds would yield larger supersaturations and larger growth rates, perhaps, but then the time available for growth would be proportionately shortened (Srivastava 1989). Droplets large enough to fall against the updraft would still not be expected to form within reasonable time periods. While necessary for cloud and rain formation, condensational growth is a relatively slow process and so not sufficient for producing rain.

All real clouds involve large populations of particles all competing in effect for the same, time-varying interstitial vapor. Initially, prior to condensation, a broad range of aerosol sizes is available, but only the most active particles (meaning those with the largest solute contents) serve as the sites for continued condensation. Such droplets grow at the rate given by Eq. (8.5), deplete some of the vapor, and thereby affect the supersaturation that is common to all particles in the local population. Figure 8.8 presents results of the numerical calculations performed by Mordy (1959) when the vapor budget [Eq. (8.2)] and a representative population of NaCl aerosol particles were taken into account simultaneously. The smallest particles in the population vary little in size during ascent because the actual supersaturation (dashed curve) never gets above the critical supersaturations of such particles. On the other hand, the radii of the larger particles (those with solute contents equal to or larger than about 10^{-18} mol) can be seen to increase markedly as they ascend through the condensation level. Particularly noteworthy in this presentation is the very sudden growth of the smallest of the activated particles, a result of the inverse relationship that exists between growth rate and particle size, as given by Eq. (8.5). The tendency for the growth rate to decrease as the droplets grow therefore causes the spectrum of any population of droplets to become narrower with time. Thus, the process of condensation is not only relatively slow, it

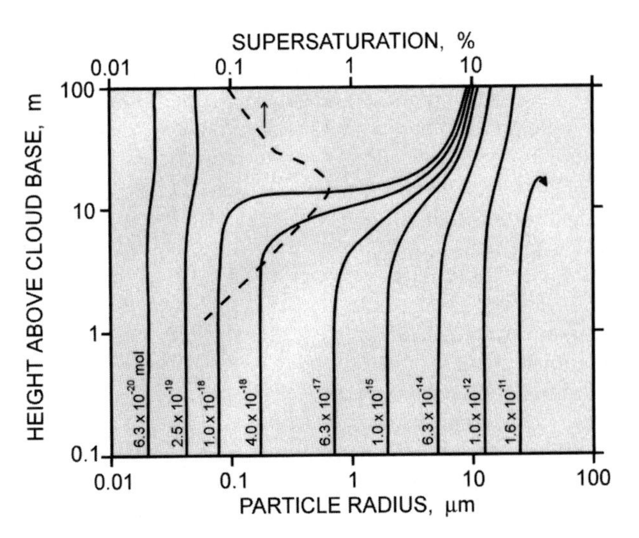

FIG. 8.8. Growth of a population of droplets from a given distribution of NaCl aerosol particles in a parcel ascending adiabatically at 15 cm s^{-1}, as computed by Mordy (1959) and presented by Byers (1965). The solute contents of the particles are indicated near the bases of the solid curves, each of which shows the altitude dependence of the radius of a particle having the given solute content. The arrow on the rightmost curve identifies particles that develop fall speeds in excess of the imposed updraft speed during growth. The dashed curve gives the calculated supersaturation in the parcel at each altitude.

also tends to produce a population of cloud droplets of relatively uniform size.

Comparisons of results from adiabatic condensation models with observations in real clouds show some important differences. Measurements of either the liquid water content, that is, the composite mass of condensate per unit volume, or the number concentration of droplets in nonprecipitating clouds, indicate that adiabatic values are seldom attained (Warner 1955; Squires 1958b). The drop spectra observed in cumulus clouds tend to be rather broad, often bimodal distributions that are not predictable from adiabatic condensation theory. In fact, typical convective clouds sometimes exhibit huge variabilities of their microphysical properties on remarkably small space scales (Squires 1956; Baker 1992). Although further refinements of the microscopic processes taking place within a given parcel should be pursued (Srivastava 1989; Fukuta 1992, 1993), we almost certainly need to go beyond parcel concepts for explanations. Convective clouds are inherently turbulent systems, so it is natural to expect that many of the microphysical characteristics of clouds can be understood only by considering the interactions of convective elements with their immediate surroundings and with the storm system as a whole (Jonas 1996; Shaw et al. 1998).

The entrainment of environmental air into the body of a cloud has received much attention in attempts to reconcile the observations of cloud microstructure with theory (see reviews by Reuter 1986; Blyth 1993;

Jonas 1996; Telford 1996). Included among the numerous influences turbulent mixing can have on cloud properties are a reduction of the amount of condensate available for precipitation growth, temporal variations in the supersaturation experienced by a droplet population, enhanced condensational growth of selected drops because of diminished competition for the available vapor, activation of CCN well above cloud base, a general broadening of the drop size distribution, and stimulation of ice formation (e.g., Telford 1975; Baker and Latham 1979; Paluch and Knight 1984; Austin et al. 1985; Telford et al. 1987; Politovich and Cooper 1988; Cooper 1989). Although most studies of entrainment have been limited to small, nonprecipitating clouds, indications are that the entrainment process involves mixing over very broad ranges of both space scales and timescales (Baker 1992). It is thus reasonable to expect that the findings of the available studies are general enough to permit their application to strong convective systems (Raymond and Blyth 1992). It will therefore be accepted that turbulent mixing, either between a convective element and its clear-air environment or with neighboring cloudy regions, has an important effect on the microphysical characteristics of large convective systems and their ability to produce rain.

8.4. Rain formation

Rain forms in a cloud whenever liquid hydrometeors become large enough to fall with appreciable speeds toward the ground (Houze 1993, p. 94). The formal definition of rain may vary with the purpose, but conventionally we take the minimum radius of the water drops to be 250 μm (Huschke 1959, p. 464). The terminal fall speed of such hydrometeors is about 2 m s^{-1} (Beard 1976), so collisional interactions with the much smaller cloud droplets are likely to be important within the cloud and cause the raindrops to grow and lag ever farther behind the rising air in a convective element (Bowen 1950). The rapid accretion of condensate by the large particles is an important growth mechanism and contributes effectively to the amount of rainfall reaching the ground (Kessler 1969; Beard and Ochs 1984).

The microphysical processes by which the sizes and phases of the hydrometeors change throughout the life cycle of a cloud must be considered if we are to understand rain formation. The rapid collection of condensate by falling raindrops, for instance, can occur only after some of the hydrometeors have attained minimal sizes. How these "precipitation embryos" form in any given storm system is nontrivial, however, and is indeed the subject of ongoing research. The effectiveness with which the moisture brought into a system by the convective air currents is converted into rain depends crucially on the rates with which such embryos can form, so we will need to understand the processes by which the size distribution of condensate evolves.

The onset of active precipitation requires the microstructure of the cloud to change appreciably from that which characterized the incipient cumulus. Houghton (1950) and Squires (1958a), among others, recognized early that clouds represent colloidal suspensions that tend to be stable initially, especially when droplet concentrations are high (as in most continental clouds) or when drop spectra are narrow (as in adiabatic parcels). At least a few relatively large particles are needed to unlock this colloidal stability (Johnson 1993). The two mechanisms generally thought to be responsible for forming the required embryos that initiate precipitation, that is, drop coalescence and ice-crystal growth, depend on the environmental conditions during cloud formation (Houghton 1950; Young 1993, p. 7). Each of these processes will be treated separately below, although we must bear in mind that both mechanisms are often operating, sometimes synergistically, in any given storm system (Houghton 1950; Braham 1964, 1968, 1986).

a. Warm-rain process

In "warm" clouds, liquid drops interact among themselves and with their environment to produce precipitation-size drops. The ice phase is not involved, even though the cloud may extend above the thermodynamic freezing level (Braham 1964). The warm-rain process has been reviewed extensively by Beard and Ochs (1993) and Rasmussen (1995), so only essential points will be brought out here. In this subsection, the nature of drop–drop interactions will be reviewed, and the need for drops of minimal size will be emphasized. Then some possible ways will be explored by which such drops might be produced in rainstorms.

Drop–drop interactions that lead to rapid hydrometeor growth require that the drops first collide with each other and then coalesce (Braham 1968). This collection process is inherently discrete and stochastic in nature (Telford 1955). In general, the hydrodynamic collection of one drop by another requires that their sizes and fall speeds differ so that the larger drop may overtake the other, smaller drop in its path. When an appropriate accounting is taken of all of the pair-wise interactions in a population of drops, the so-called stochastic collection equation (SCE) results (Twomey 1966; Berry 1967; Pruppacher and Klett 1997, p. 460):

$$\frac{\partial n(m)}{\partial t} = \frac{1}{2}\int_0^m K(m_x, m - m_x) \times n(m_x) \times n(m - m_x) \times dm_x - n(m)\int_0^\infty K(m, m_x) \times n(m_x)dm_x. \quad (8.6)$$

Here, $n(m)$ is the continuous mass distribution function, and the gravitational collection kernel is

$$K(r_L, r_S) = E(r_L, r_S)\pi \times (r_L + r_S)^2 \times (v_L - v_S), \quad (8.7)$$

when the radius of the large drop is r_L (corresponding to the collector drop of mass m) and the radius of the small drop is r_S. The corresponding fall speeds are given by v_L and v_S, and the collection efficiency is $E(r_L, r_S)$, which is actually a product of the collision and coalescence probabilities (Beard and Ochs 1993). The collection efficiency is a complicated and still uncertain function of the drop radii (Klett and Davis 1973; Böhm 1992; Beard and Ochs 1993) and possibly of the turbulence structure of the cloud (Pinsky et al. 1999).

Solutions to the SCE have been demonstrated numerous times in the past and show how the drop spectrum evolves with time in a cloud parcel. An example of how the mass distribution function changes from a relatively narrow, monomodal function initially to a bimodal distribution some minutes later resulted from the computations of Berry and Reinhardt (1974), as shown here in Fig. 8.9. A relatively few drops in the large-drop tail of the initial distribution fortuitously collect some smaller drops and grow, thereby increasing their collection kernels and the likelihood for subsequent collection events. This growth process is by nature nonlinear and leads to the eventual development of a well-defined second mode in the distribution. As time passes, the large-drop mode grows at the expense of the original small-drop mode and contributes importantly to precipitation development and other properties of a cloud (Ochs 1978; Johnson 1982).

Once the drop spectrum has evolved for relatively long times by the collision-coalescence process, a clear separation between the two modes of the drop distribution persists, and the large drops tend to retain their identity as such. The individual large drops then collect the much smaller cloud droplets by a more nearly continuous accretion mechanism (Beard and Ochs 1993). The individual raindrops then grow in mass M at a rate approximated by

$$\frac{dM}{dt} = K \times m_c, \quad (8.8)$$

where m_c is the mass concentration of cloud water (condensate), and the collection kernel can now be approximated by

$$K = \epsilon \pi r_L^2 v_L, \quad (8.9)$$

where ϵ is the coalescence efficiency, the residual contributor to the collection efficiency once near-unity (geometrical) collision efficiency has been achieved (Beard and Ochs 1984). In the form given in Eq. (8.9), K can be seen to depend explicitly on the cross-sectional area of the collector drop. Moreover, the fall speed of the drop (v_L) also increases more or less proportionally to the drop radius in the intermediate size range, so drops growing by coalescence increase their mass almost exponentially (Pruppacher and Klett 1997, pp. 417, 617). Such large growth rates, limited mainly by the eventual breakup of the drops (Young 1975; Low and List 1982), are needed to account for heavy rain rates in convective systems.

The fundamental problem associated with the warm-rain mechanism as described here is the slowness with which it gets under way unless large particles are already present (Beard and Ochs 1993). Although no rigid thresholds exist in the sizes of either the collector or collected drops below which the collision efficiency is truly zero, its magnitude is nevertheless small ($E <$ 1%) when the collector-drop radius $r_L < 15$ μm or the collected-drop radius $r_S < 8$ μm (Klett and Davis 1973; Beard and Ochs 1993). In addition, the cross sections and fall speeds of the drops are initially small, meaning that the magnitudes of the collection kernel (K) are then very small also. The coalescence process needs the presence of some drops in the population in excess of about 25 μm radius for it to get started effectively (Beard and Ochs 1993), although this barrier may be smaller when both condensation and coalescence are proceeding at the same time (Kogan 1991) or when the air is sufficiently turbulent (Pinsky et al. 1999). Truly rapid rain development, however, may need embryos with radii of 35 μm or greater (Johnson 1993; Rasmussen 1995).

A number of mechanisms have been proposed over the years for generating coalescence embryos. The most straightforward way of producing large drops early in the rain formation process is an influx of "giant" nuclei (aerosol particles with radii between 1 and 10 μm) or "ultragiant" particles (radii > 10 μm) into the cloud with the updraft (Johnson 1976, 1982; Beard and Ochs 1993). Clearly, if and when such large particles are available during cloud formation, then the problem of warm-rain initiation becomes moot. Diffi-

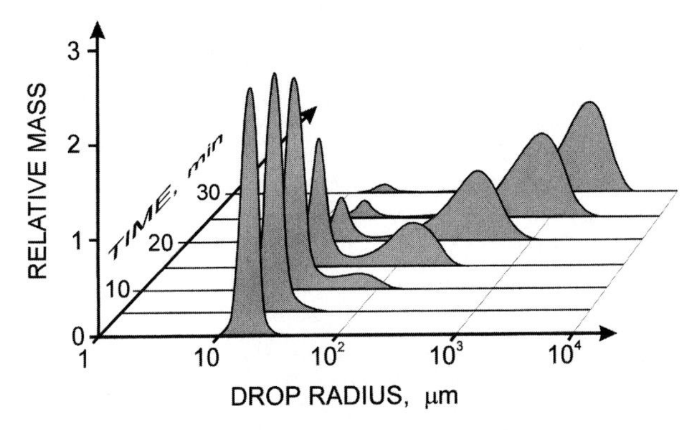

FIG. 8.9. The evolution of drop size distribution due to stochastic collection, as computed by Berry and Reinhardt (1974).

culties with this mechanism exist, however. Moreover, the hypothesis cannot be tested readily, in part because of the extreme difficulty of sampling air containing very low number concentrations (Beard and Ochs 1993), leading some to resort to indirect techniques (Woodcock and Blanchard 1955). Still, some aircraft measurements over the central United States have detected particles as large as 25 μm in radius (Hobbs et al. 1985). Such large particles have relatively large fall speeds and so tend to be of local origins: insoluble soil particles over continental regions, sea salt particles over the oceans (Hobbs et al. 1985; Blanchard 1985). Soluble, hygroscopic particles are naturally favorable because of their potential to grow as they enter the cloud, but the largest particles grow by vapor deposition only slowly (Ludlam 1951), and their large masses may not allow them to rise far above the surface (Beard and Ochs 1993). Nevertheless, the direct influence that giant and ultragiant particles potentially have on the initiation of the coalescence process, coupled with other uncertainties in our understanding of cloud processes, means that we should not entirely rule out the importance of externally derived coalescence embryos to the warm-rain process (Beard and Ochs 1993; Levin et al. 1996).

The intermixing of air parcels of different histories, invoked earlier as a mechanism for explaining the broadening of the cloud drop spectrum in general, probably also plays important roles in producing the relatively large coalescence embryos needed to generate "warm" rain. Although the details may be argued (e.g., Paluch and Knight 1984), considerable evidence exists that the entrainment of dry air into a cloud alters the drop spectrum in ways that are consistent with the need to form a few large drops. Through inhomogeneous, or localized mixing scenarios, for instance, some fraction of the cloud droplet population evaporates, leaving the remaining droplets to enjoy enhanced condensational growth (Baker et al. 1984). We may need to think beyond the entrainment of dry environmental air only into the cloud, however, and consider mixing and turbulence more generally (Pinsky and Khain 1997; Shaw et al. 1998). Some entrained air may indeed be cloud air that was previously detrained from the same or a neighboring cloud, the effect being the blending of new and old parcels of air (Mason and Jonas 1974; Roesner et al. 1990; Kogan 1991). Internal mixing, or the recycling of air within the cloud, is also important (Telford and Chai 1980), as are pulsations in the updraft structure that allow successive thermals to intermix (Mason and Jonas 1974). It may even be that statistical fluctuations of drop number concentration and supersaturation along the various pathways followed by parcels within a convective element suffice, independent of traditional mixing scenarios, to account for the enhanced condensational growth of cloud droplets (Politovich and Cooper 1988; Cooper 1989; Politovich 1993). Inertial effects of small-scale turbulence on the relative velocities of approaching drops may also prove to enhance the collection process (Pinsky and Khain 1997), but as yet no consensus can be reached on the general origin of the coalescence embryos in warm clouds.

b. Cold-rain process

"Cold" clouds are characterized by microphysical processes that involve the solid, or ice, phase of water (Young 1993, p. 7). Whereas ice is a necessary component of the cold-rain process, the liquid phase still plays important roles in the evolution of the ice phase in many supercooled clouds and in the development of precipitation (Koenig 1963; Braham 1964, 1968; Hallett et al. 1978; Willis and Hallett 1991; Rangno and Hobbs 1994). It was once thought that most of the rain from midlatitude weather systems must be initiated by ice-phase processes (Bergeron 1935; Houghton 1950), but we now realize that such a generalization is unwarranted (Braham 1964, 1986). Nevertheless, stratiform clouds and convective systems with cool cloud bases do seem to involve the ice phase to a large, if not dominant, extent (Knight et al. 1974; Dye et al. 1974; Rutledge and Hobbs 1984; Waldvogel et al. 1987; Zhang 1989; Khain and Sednev 1995). Here, we develop the concepts of rain formation in cold clouds, but we must continue to rely on the principles of warm-cloud processes rather strongly.

The formation of a new phase (ice) within a supercooled cloud creates a mixed-phase zone in which the "warm" and "cold" microphysical processes become complicated and intertwined. It is therefore convenient to utilize a diagram, such as that shown in Fig. 8.10, for depicting the diverse interactions that can occur between the various types of cloud particles, both liquid and solid. (Also see Braham 1968; Lin et al. 1983; Cotton and Anthes 1989, p. 117, for more complete diagrams.) Note that the interaction diagram shown here is of the type sometimes used for parameterizing the microphysics in "bulk" terms, that is, as distinct categories of water substance, such as "small" (nonsedimenting) or "large" (sedimenting) liquid particles and as "small" or "large" ice particles. Large ice particles are often further subdivided into snow (grown by vapor deposition or aggregation) and graupel (grown from snow by riming or formed directly by the freezing of supercooled raindrops). Water vapor is separately categorized to represent the parent source of all forms of condensed water. The "rain" category suggests that liquid precipitation appearing at the ground can arise from the direct fallout of large drops (typical for convective regions of severe storms) as well as from the melting of graupel (cool-base convection) or snow (stratiform anvil regions). Other categories can be devised as needed for specific applications, and size dependencies can be imposed on some or all of the categories as well (Houze 1993, p. 97). Even so,

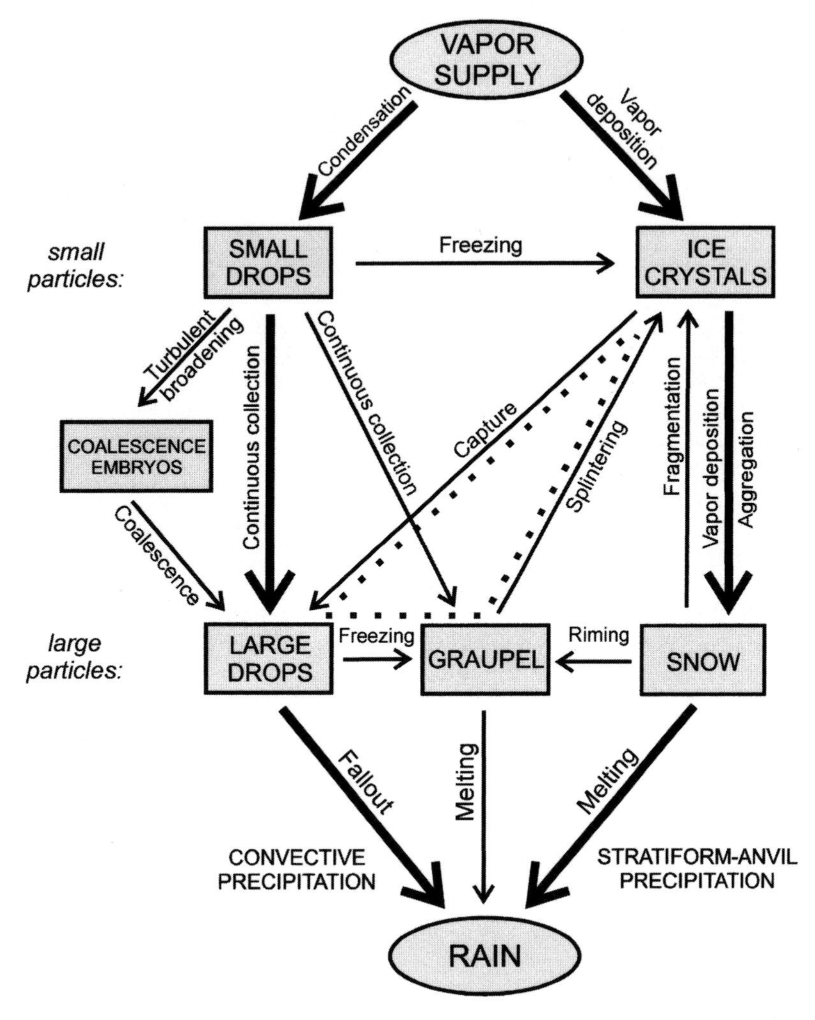

FIG. 8.10. A simplified schematic diagram of possible interactions between various categories of cloud particles. The bold arrows show the dominant pathways by which water vapor is transformed into rain via the "warm-rain" process (left-hand side) and via the "cold-rain" process (right-hand side) in heavy rain situations. The dotted set of arrows near the center of the diagram identifies a likely cyclical process for generating secondary ice particles via the rime-splintering mechanism of Hallett and Mossop (1974). The form of the diagram was adapted from Rutledge and Hobbs (1984).

one should recognize that resorting to clearly defined categories of water substance and classes of interactions necessarily leads to some oversimplification of the processes that actually take place in atmospheric clouds.

The interactions that occur between the different types of cloud particles represent the various physical processes of cloud and precipitation formation. Key interactions have been accentuated by bold arrows in Fig. 8.10. The main pathway taken by water during warm-rain formation is depicted by the set of bold arrows toward the left-hand side of the diagram. As discussed in the previous section, water vapor first condenses onto soluble aerosol particles (CCN), thereby forming the many small drops that compose cloud water. This condensate is then collected by the large drops, derived initially from coalescence embryos, to form the rain that eventually reaches the ground. This collision-coalescence mechanism is probably the main contributor to rain from convective storms, especially those low-echo-centroid systems that exhibit maximum radar reflectivities near or even below the freezing level (as in the Big Thompson Canyon storm; Maddox et al. 1977; Caracena et al. 1979). Collision-coalescence also forms the supercooled drops at the larger end of the size distribution that enhance cloud glaciation and contribute to the production of rain in "cold" clouds.

The main pathway involved in the cold-rain process is shown by the bold arrows along the right-hand side of Fig. 8.10. Vapor depositing onto the small ice crystals causes them to grow large enough to sediment and begin colliding with other crystals. Aggregation of the crystals leads to snowflakes that may enter the warmer parts of the cloud, where they melt and fall out as rain (which tends to occur in the stratiform regions of larger, mesoscale convective systems). Alternatively, the snow may first collect supercooled cloud water and form graupel particles, as occurs in convective regions (McAnelly and Cotton 1986; Johnson and Hamilton 1988; Steiner and Smith 1998).

The growth of graupel particles by the collection of supercooled cloud water is key to an effective cold-rain process in some convective storms, such as those forming along cold fronts (Rutledge and Hobbs 1984), as well as contributing to the glaciation and electrification of clouds in general. Given a population of graupel particles several hundred micrometers in radius and an abundant supply of supercooled cloud water, the riming process becomes an efficient way to convert condensate into precipitation-size particles. Mathematically, one can represent the growth of a graupel particle in much the same way that one does for the growth of raindrops. Equation (8.8) becomes equally valid for expressing the growth rates of the liquid and solid precipitation particles, once one has suitably adjusted the magnitude of the collection kernel (K). Graupel particles are naturally rigid, compared with liquid drops, so the coalescence efficiency tends to be larger, probably close to unity. Moreover, this rigidity prevents ice particles from deforming hydrodynamically, so their fall speeds increase steadily with increasing particle size (Auer 1972), unlike raindrops, whose fall speeds tend to reach a plateau of about 10 m s^{-1} because of shape distortion (Beard 1976). When caught in strong updrafts rich in condensate, graupel and small hail can be expected to grow rapidly and nonlinearly until the particles enter a downdraft and experience a diminished supply of condensate. As these ice particles fall into the lower parts of the cloud, they melt and become raindrops. If their melted sizes exceed the thresholds for mechanical stability, the newly formed drops may break up and generate a new population of precipitation embryos and contribute to a possible warm-rain chain reaction (Langmuir 1948; Low and List 1982).

The origins of the graupel—indeed, of any of the initial ice particles in a supercooled cloud—have been the subject of much ongoing research (for overviews, see Beard 1992; Rasmussen 1995). Whereas ice can form spontaneously from the freezing of cloud droplets once the temperature has dropped to about $-35°C$ or lower, its formation lower in the mixed-phase cloud (i.e., at higher temperatures), where abundant concentrations of liquid water and vapor are available, is the main interest from the standpoint of rain formation. Of the two main ways by which large ice particles (the graupel category) can arise, namely, ice crystal riming and drop freezing (Pruppacher and Klett 1997, p. 39), the first is by far the slowest. This traditional mechanism of graupel formation starts with small ice particles that probably formed initially via primary nucleation events, such as the freezing of supercooled cloud droplets that contacted certain insoluble aerosol particles (Cotton 1972; Young 1974; Beard 1992; Rangno and Hobbs 1994). The small ice particles must then grow, usually by vapor deposition (Bergeron 1935) and possibly aggregation, until they achieve a size large enough to permit them to fall significantly against the background cloud and begin collecting supercooled cloud droplets. Given suitable cloud water concentrations, subsequent riming of the snow eventually leads to graupel particles. This pathway for generating the graupel particles that can contribute effectively to rainfall is relatively slow, but in convective clouds lacking an active warm-rain process, the crystal riming mechanism offers the best alternative (e.g., Dye et al. 1974; Cannon et al. 1974).

The frozen drop pathway for forming graupel particles can be very effective when the depth of the warm cloud is large and the collision-coalescence process is operating. Observations in summertime convective systems have typically shown that graupel particles form readily when supercooled raindrops exist near the $-5°$ to $-15°C$ levels (Brown and Braham 1963; Koenig 1963; Hallett et al. 1978; Willis and Hallett 1991). Some thought has been given to the generation of primary ice by mechanical means at these relatively high temperatures during the collision of large supercooled drops (Koenig 1965; Czys 1989), but it is not clear that such a mechanism is really needed to account for the first ice in the cloud. Nevertheless, the freezing of supercooled raindrops leads to enhanced accretional sweepout of cloud water and possibly to invigoration of the updraft through augmented latent-heat effects (Rosenfeld and Woodley 1997).

When suitable capabilities are available for sampling very small ice particles (maximum dimensions <100 μm) from aircraft, observations have shown that large concentrations of columnar ice crystals coexist with the graupel particles (Hallett et al. 1978). Such data strongly suggest that the large ice particles (graupel) arose from the freezing of supercooled raindrops by the hydrodynamic capture of the small ice crystals. In such cases, primary ice nuclei need to serve only as a trigger for cloud glaciation (Hallett et al. 1978; Lamb et al. 1981; Mossop 1985). As suggested by the dotted paths in Fig. 8.10, a regenerative feedback loop can form when large supercooled raindrops are present in a given volume of cloud between the $-3°$ and $-8°C$ levels (Hallett et al. 1980; Lamb et al. 1981). Any small cloud-ice particles, regardless of their origins, can be "captured" by the supercooled raindrops (Cotton 1970; Lew and Pruppacher 1983), which then freeze to form graupel particles "instantly." Of key importance to the feedback mechanism is the fact that the riming process not only contributes to the growth of the graupel particles, but it also produces secondary ice particles, splinters that rapidly grow by vapor deposition and add to the population of small ice particles (Hallett and Mossop 1974; Mossop 1985). These new ice crystals can lead to the freezing of additional raindrops and the formation of new riming centers that produce still more splinters. Under appropriate cloud conditions (e.g., modest updraft speeds), the number concentrations of small ice crystals and graupel particles tend to increase exponentially for a time in localized regions of the cloud, stimulating both the glaciation of the cloud and its buoyancy, with time constants well under a minute (Lamb et al. 1981). Not all clouds permit such prolific ice-multiplication mechanisms to operate, of course. Indeed, some observations suggest that high concentrations of ice are produced even when the conditions for rime-splintering are marginal or absent, so we are left with considerable uncertainty regarding the origins of ice (Rangno and Hobbs 1991, 1994). Whereas active glaciation in a cloud may not always contribute to the rain appearing on the ground (Reisin et al. 1996), the additional latent heat released as a result of the phase change can stimulate overall cloud development (Caracena et al. 1979; Lucas et al. 1995).

8.5. Precipitation efficiency

The efficiency with which clouds produce rain varies greatly and depends on numerous environmental and microphysical factors. When defined as the ratio of the rate of rain reaching the ground to the flow of water vapor entering the cloud through its base (Marwitz 1972a), the precipitation efficiency can range from zero in nonprecipitating clouds to greater than unity for short times in intense, time-dependent systems (Cotton et al. 1989). Precipitation efficiencies based on storm totals avoid such nonphysical values, but the literature provides a mixture of definitions. In this section, some of the physical factors contributing to the efficiency with which clouds remove water from the atmosphere are discussed.

Convective systems that yield significant rain rates tend to be active both dynamically and microphysically. However, not all storms that produce intense rainfalls are necessarily efficient. In fact, some of the earliest studies showed that ordinary thunderstorms transform less than about 20% of the inflowing vapor into rain on the ground (Braham 1952). Small systems tend to be eroded relatively easily by entrainment of dry environmental air, meaning that much of the inflowing vapor is simply reevaporated at upper levels in the troposphere, especially if the updraft speeds are large (Ludlam 1980, p. 271). Evaporation of rain in the downdraft also causes a significant loss of rainwater, although this conversion of liquid to vapor can represent an important energy source for the storm dynamics (Braham 1952; Ludlam 1980, p. 270).

Efficient rainstorms utilize the water vapor flowing in through cloud base in effective ways. Generally, we expect larger systems to be more efficient if for no other reason than that the entrainment losses tend to be relatively less. More of the condensate then remains available for feeding the growth of the raindrops or large ice particles. Such a situation may have contributed to the large rainfall in the Big Thompson storm (Caracena et al. 1979). At the same time, however, one must not lose sight of the fact that entrainment and turbulent mixing are likely to play important roles in the generation of coalescence embryos, as discussed earlier. Such embryos must then be able to grow rapidly to raindrop sizes (typically 1 mm or larger in radius) and sweep up the condensed water within a reasonable depth of cloud. We can gain some insight into rain production in efficient systems by estimating the depth of cloud needed for a given population of raindrops to sweep a given cloud volume clean of all condensed water (mainly cloud droplets). If a given raindrop has radius r_L, its effective cross-sectional area is $A_R = E_c \pi r_L^2$, where E_c is the collection efficiency. We may establish a rough criterion for complete sweepout by requiring that the integrated cross-sectional area from all the raindrops throughout a depth H_c be just equal to a unit geometrical area (in a horizontal plane). In integral form, this criterion becomes

$$\int_0^{H_c} \int_{r_0}^{\infty} A_R n_R(r_L) dr_L dz = 1, \qquad (8.10)$$

where r_0 is the radius of the smallest sedimenting particle (typically 100 μm) and n_R is the number concentration of raindrops within the size interval r_L and $r_L + dr_L$. At this point, we could apply a commonly used exponential size-distribution function (e.g., that from Marshall and Palmer 1948) and then solve for H_c. However, it suffices here to assume the cloud to be composed of a monodisperse population of raindrops of average radius $\bar{r}_L$, in which case the criterion reduces to $A_R N_R H_c = 1$, enabling us to estimate

$$H_c = \frac{1}{\pi \bar{r}_L^2 E_c N_R}, \qquad (8.11)$$

where $N_R = \int_{r_0}^{\infty} n_R(r_L) dr_L$ is the total number concentration of raindrops. With typical values ($\bar{r}_L = 1$ mm, $E_c = 1$, $N_R = 10^3$ m^{-3}), we find $H_c = 320$ m, a relatively thin slice of cloud. Even allowing for collection efficiencies well under unity (Beard and Ochs 1984), we see that the rain in a mature cloud can sweep up the available condensate within a depth of something like a kilometer or less. The key to understanding efficient rain production is therefore not so much the microphysics of condensate removal once raindrops have formed but rather the generation of raindrops in the first place, within appropriate, condensate-rich parts of the cloud.

The kinematic structure of any rainstorm serves both to supply the water vapor to the cloud from the large scale and to provide the setting in which the precipitation microphysics can operate effectively. As we have seen in section 8.2, the vertical wind shear is an important contributor to storm organization. Past studies (e.g., Marwitz 1972a) have suggested that the magnitude of the wind shear significantly affects the efficiency of precipitation formation in midlatitude storms. When the estimated precipitation efficiency is plotted against the observed, cloud-averaged vertical wind shear, as shown in Fig. 8.11, we find a robust inverse relationship between these two parameters, at least for the selected set of storms in the United States. In cases of large wind shear, the precipitation efficiency can be as low as 5%, perhaps because of rapid detrainment and evaporation of cloud water into the dry air aloft (Marwitz 1972a, b). When the wind shear is weak, on the other hand, the updrafts tend to be erect, possibly allowing the precipitation elements to fall through the condensate-rich parts of the cloud (Ferrier et al. 1996). Whereas Caracena et al. (1979) have claimed that the Big Thompson storm, estimated to have a precipitation efficiency of 85%, fits this inverse relationship with wind shear, they also noted that the wind shear was confined mainly to the lowest levels and that the environmental wind structure differed substantially from that associated with typical high plains thunderstorms. The high precipitation efficiency in this case may actually have been aided by the slight westward tilt in the updraft, which allowed the mature raindrops to be released from the updraft before they adversely affected the kinematic structure of the storm (Maddox et al. 1977; Caracena et al. 1979). It is unlikely that wind shear alone controls the precipitation efficiency, as other environmental factors (e.g., ambient humidity) and mesoscale organization are probably important in many situations (Fankhauser 1988; Ferrier et al. 1996).

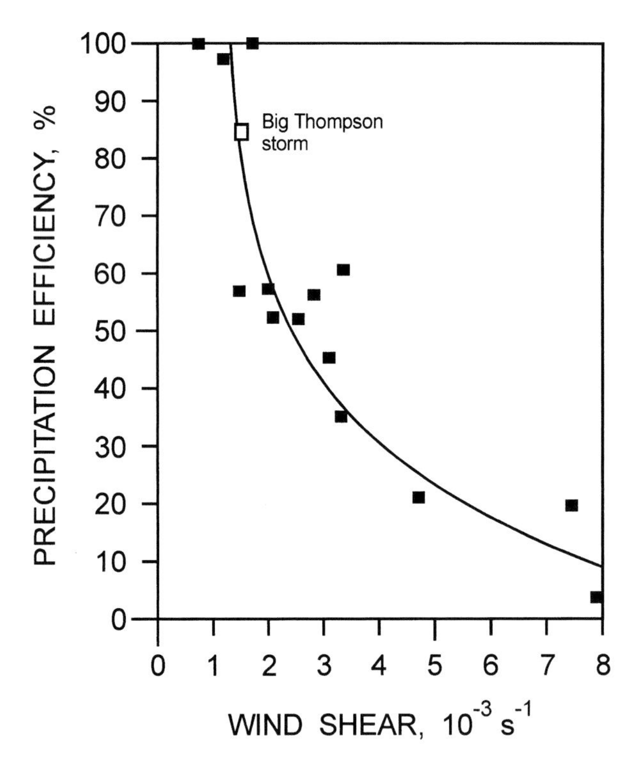

FIG. 8.11. Scatter diagram of precipitation efficiency against wind shear for thunderstorms on the High Plains (filled symbols) and for the Big Thompson storm (open symbol). Wind shear is based on the difference in wind speeds at cloud base and at the level of buoyancy equilibrium. Adapted from Marwitz (1972b).

Extreme rain events pose interesting problems for considerations of precipitation efficiency. It is natural to assume that the efficiency of rain production is high in storms that yield huge, record-breaking rainfalls, but such may not always be the case. Unfortunately, rare events are seldom studied at the level of detail needed to yield accurate estimates of their water budgets. We must also distinguish between the areal and temporal averaging that takes place in the calculation of precipitation efficiency and the measurement of rainfall at specified points on the earth's surface. Point-rainfall data, such as those discussed in relation to Table 8.1 and Fig. 8.1, reflect not only the efficiency with which a storm converts water vapor into rain but also the duration/movement of the storm and the location of the measurement site relative to the main rain shaft (Ludlam 1980, p. 272). During transient conditions, even ordinary storms can yield large 1-min rainfalls and rain rates in excess of 1000 mm h^{-1} (Engelbrecht and Brancato 1959; Elford 1956; Ludlam 1980, p. 273). Storms yielding large rainfalls over protracted times (such as an hour or more) very likely have stable flow patterns that favor persistent rain production. It often becomes very difficult to separate true efficiency of rain formation from the numerous other factors that contribute to heavy point-rainfall totals.

8.6. Summary and conclusions

Rain production in convective systems has been looked at from diverse points of view in this chapter. The large-scale features of the atmosphere are crucial for providing the general environment in which rainstorms form. The abundance of atmospheric moisture, both the material source of the rain itself and the energy basis for vigorous convection, depends on surface fluxes that vary with climatic zones and possible changes in the global hydrological cycle (Frei et al. 1998). Moisture is often advected great distances by the synoptic-scale motions into a region, where it may be forced to rise with vertical wind currents that are driven by mesoscale pressure gradients and low-level convergence of the air. The vertical motions become large in specific locations as potential instability is released by the phase changes occurring in the cloud. The thermodynamic properties of the atmosphere that determine the degree of stability of the atmosphere also set limits on the amount of condensate, confining heavy rain events generally to warm seasons or regions of the globe. Inside a cloud, the rising water vapor is transformed into rain via a succession of physical processes. The spectral evolution of the particles, from small droplets or ice crystals that simply follow the air motions to large hydrometeors capable of falling against strong updrafts, involves nucleation, vapor deposition, and collisional interactions that operate interactively on the microscale. The ranges of space scales and timescales involved in the formation of convective rain are enormous.

In order to gain a perspective of rain development, we may need to come back to the concern, expressed in the introduction, over the interactions of scales. That is, we may need to look for phenomena that couple the microscale processes with the storm structure as a whole (Braham 1968; Ludlam 1980, p. 270). If one considers an active system like that shown in Figs. 8.4 and 8.5, for instance, one is immediately impressed with the hierarchy of scales involved. Superimposed on the large, mean motions are a myriad of smaller-scale motions, each in itself transient, that nevertheless contribute to a relatively stationary flow pattern in a statistical sense. As one turret begins to decay, another forms in its wake or slightly toward its upshear side. In line with the view put forth by Blyth et al. (1988), one can imagine a scenario in which each small thermal rises and mixes in environmental air near its top. This air, with an already modified microstructure, then enters the interior of another convective element. As these mixed parcels blend with successively larger convective units, size-sorting mechanisms and stochastic drop trajectories interact in complex ways that depend on some general features of the larger-scale flows, much as Kogan (1993) has tried to show. Drop coalescence events increase in frequency to the point where modest raindrops form readily, possibly freeze via one or more of the conceivable cold-cloud microphysical mechanisms, and help the upper reaches of the cloud to glaciate and grow in vigor. The subsequent accretional sweepout of condensate leads to the rapid growth of the precipitation particles, imposing a substantial load on the rising air. The downdrafts and shafts of rain that form modify the flow pattern in the storm and help stimulate the influx of new moisture that replenishes the condensate upon which the rain feeds. A cycle of interaction, involving all scales, creates a system of awesome proportions and potentially devastating effects. The details will always be complex, but the hierarchical scales of motions and physical phenomena ensure that heavy rain can fall from many strong convective systems when conditions are suitable.

Future research must focus on reducing the uncertainties in our understanding about rainstorms from several points of view. We need to learn how to identify key signatures in the synoptic and mesoscale weather patterns in order to provide adequate warnings of potential flooding events (e.g., Massacand et al. 1998). Increased understandings of the return frequency of heavy rain events will almost certainly lead to improved building codes and flood-mitigation strategies (Angel and Huff 1999; Changnon 1999). On the cloud scale, we need to understand the origins of precipitation embryos, those particles that serve to get the collection process started and maintained throughout the life cycle of a storm. Are they really giant particles from the environment, or are they cloud droplets that have enjoyed some favorable set of conditions within an evolving cloud? What is the real role of turbulent mixing in convective systems, and how does it affect the microphysics, the cloud-scale dynamics, and the mass and energy budgets of the system? The origin of ice in supercooled clouds is likewise fraught with uncertainties. What are the mechanisms of initial ice formation, and how does the ice phase propagate throughout cold clouds and influence rainfall? Many of the scientific questions confronting us today may well be a result of the limited observational tools we have available for investigating complicated and time-dependent systems that involve an enormous range of spatial scales. Stronger coordination between observational programs and modeling efforts may well lead to substantial scientific progress in the future and improvements in our ability to predict heavy rain from convective systems.

Acknowledgments. The author would like to express his gratitude to Lee Grenci and Hans Verlinde for stimulating discussions and helpful suggestions during the preparation of the manuscript. The comments and changes suggested by the panel reviewers were also of great help.

REFERENCES

Angel, J. R., and F. A. Huff, 1999: Record flood-producing rainstorms of 17–18 July 1996 in the Chicago metropolitan area. Part II: Hydrometeorological characteristics of the rainstorms. *J. Appl. Meteor.,* **38,** 266–272.

Auer, A. H., 1972: Distribution of graupel and hail with size. *Mon. Wea. Rev.,* **100,** 325–328.

Austin, P. H., M. B. Baker, A. M. Blyth, and J. B. Jensen, 1985: Small-scale variability in warm continental cumulus clouds. *J. Atmos. Sci.,* **42,** 1123–1138.

Baker, B. A., 1992: Turbulence entrainment and mixing in clouds: A new observational approach. *J. Atmos. Sci.,* **49,** 387–404.

Baker, M. A., and J. Latham, 1979: The evolution of droplet spectra and rates of production of embryonic raindrops in small cumulus. *J. Atmos. Sci.,* **36,** 1612–1615.

Baker, M. B., R. J. Breidenthal, T. W. Charlarton, and J. Latham, 1984: The effects of turbulent mixing in clouds. *J. Atmos. Sci.,* **41,** 299–304.
Barnston, A. G., and R. E. Livezey, 1987: Classification, seasonality, and persistence of low-frequency atmospheric circulation patterns. *Mon. Wea. Rev.,* **115,** 1083–1126.
Beard, K. V., 1976: Terminal velocity and shape of cloud and precipitation drops aloft. *J. Atmos. Sci.,* **33,** 851–864.
———, 1992: Ice initiation in warm-base convective clouds: An assessment of microphysical mechanisms. *Atmos. Res.,* **28,** 125–152.
———, and H. T. Ochs, 1993: Warm-rain initiation: An overview of microphysical mechanisms. *J. Appl. Meteor.,* **32,** 608–625.
———, and ———, 1984: Collection and coalescence efficiencies for accretion. *J. Geophys. Res.,* **89,** 7165–7169.
Bell, G. D., and J. E. Janowiak, 1995: Atmospheric circulation associated with the midwest floods of 1993. *Bull. Amer. Meteor. Soc.,* **76,** 681–695.
Bergeron, T., 1935: On the physics of clouds and precipitation. *Proes-Verg. Assoc. Met. UGGI,* **Part 2**, 156–178.
Berry, E. X., 1967: Cloud droplet growth by collection. *J. Atmos. Sci.,* **24,** 688–701.
———, and R. L. Reinhardt, 1974: An analysis of cloud drop growth by collection: Part II. Single initial distributions. *J. Atmos. Sci.,* **31,** 1825–1831.
Blackadar, A. K., and K. Buajitti, 1957: Theoretical studies of diurnal wind variations in the planetary boundary layer. *Quart. J. Roy. Meteor. Soc.,* **83,** 486–500.
Blanchard, D. C., 1985: The oceanic production of atmospheric sea salt. *J. Geophys. Res.,* **90,** 961–963.
Blyth, A. M., 1993: Entrainment in cumulus clouds. *J. Appl. Meteor.,* **32,** 626–641.
———, W. A. Cooper, and J. B. Jensen, 1988: A study of the source of entrained air in Montana cumuli. *J. Atmos. Sci.,* **45,** 3944–3964.
Böhm, J. P., 1992: A general hydrodynamic theory for mixed-phase microphysics. Part II: Collision kernels for coalescence. *Atmos. Res.,* **27,** 275–290.
Bosart, L. F., and F. Sanders, 1981: The Johnstown flood of July 1977: A long-lived convective system. *J. Atmos. Sci.,* **38,** 1616–1642.
Bowen, E. G., 1950: The formation of rain by coalescence. *Austr. J. Sci. Res.,* **A3,** 193–213.
Braham, R. R., 1952: The water and energy budgets of the thunderstorm and their relation to thunderstorm development. *J. Meteor.,* **9,** 227–242.
———, 1964: What is the role of ice in summer rain-showers? *J. Atmos. Sci.,* **21,** 640–645.
———, 1968: Meteorological bases for precipitation development. *Bull. Amer. Meteor. Soc.,* **49,** 343–353.
———, 1986: Coalescence-freezing precipitation mechanism. Preprints, *10th Conf. on Planned and Inadvertent Weather Modification,* Arlington, VA, Amer. Meteor. Soc., 142–145.
Brown, E. N., and R. R. Braham, 1963: Precipitation particle measurements in cumulus congestus. *J. Atmos. Sci.,* **20,** 23–28.
Browning, K. A., 1964: Airflow and precipitation trajectories within severe local storms which travel to the right of the winds. *J. Atmos. Sci.,* **21,** 634–639.
Byers, H. R., 1965: *Elements of Cloud Physics.* University of Chicago Press, 191 pp.
Cannon, T. W., J. E. Dye, and V. Toutenhoofd, 1974: The mechanism of precipitation formation in northeast Colorado cumulus. II. Sailplane measurements. *J. Atmos. Sci.,* **31,** 2148–2151.
Caracena, F., R. A. Maddox, L. R. Hoxit, and C. F. Chappell, 1979: Mesoanalysis of the Big Thompson storm. *Mon. Wea. Rev.,* **107,** 1–17.
Carlson, T. N., 1991: *Mid-latitude Weather Systems.* Harper Collins Academic, 507 pp.
———, and F. H. Ludlam, 1968: Conditions for the occurrence of severe local storms. *Tellus,* **20,** 203–226.
———, S. G. Benjamin, G. S. Forbes, and Y-F. Li, 1983: Elevated mixed layers in the regional severe storm environment: Conceptual model and case studies. *Mon. Wea. Rev.,* **111,** 1453–1473.
Changnon, S. A., 1999: Record flood-producing rainstorms of 17–18 July 1996 in the Chicago metropolitan area. Part III: Impacts and responses to the flash flooding. *J. Appl. Meteor.,* **38,** 273–280.
———, and K. E. Kunkel, 1999: Record flood-producing rainstorms of 17–18 July 1996 in the Chicago metropolitan area. Part I: Synoptic and mesoscale features. *J. Appl. Meteor.,* **38,** 257–265.
Chen, J-P., 1994: Theory of deliquescence and modified Köhler curves. *J. Atmos. Sci.,* **51,** 3505–3516.
Cooper, W. A., 1989: Effects of variable droplet growth histories on droplet size distributions. Part I: Theory. *J. Atmos. Sci.,* **46,** 1301–1311.
———, R. T. Bruintjes, and G. K. Mather, 1997: Calculations pertaining to hygroscopic seeding with flares. *J. Appl. Meteor.,* **36,** 1449–1469.
Cotton, W. R., 1970: A numerical simulation of precipitation development in supercooled cumuli. Ph.D. dissertation, Pennsylvania State University, 178 pp.
———, 1972: Numerical simulation of precipitation development in supercooled cumuli. Part II. *Mon. Wea. Rev.,* **100,** 764–784.
———, and R. A. Anthes, 1989: *Storm and Cloud Dynamics.* Academic Press, 880 pp.
———, R. L. George, P. J. Wetzel, and R. L. McAnelly, 1983: A long-lived mesoscale convective complex. Part I: The mountain-generated component. *Mon. Wea. Rev.,* **111,** 1893–1918.
———, M. S. Lin, R. L. McAnelly, and C. J. Tremback, 1989: A composite model of mesoscale convective complexes. *Mon. Wea. Rev.,* **117,** 765–783.
Czys, R. R., 1989: Ice initiation by collision-freezing in warm-based cumuli. *J. Appl. Meteor.,* **28,** 1098–1104.
Doswell, C. A., 1987: The distinction between large-scale and mesoscale contributions to severe convection: A case study example. *Wea. Forecasting,* **2,** 3–16.
———, H. E. Brooks, and R. A. Maddox, 1996: Flash flood forecasting: An ingredients-based methodology. *Wea. Forecasting,* **11,** 560–581.
Dye, J. E., C. A. Knight, V. Toutenhoofd, and T. W. Cannon, 1974: The mechanism of precipitation formation in northeastern Colorado cumulus. III. Coordinated microphysical and radar observations and summary. *J. Atmos. Sci.,* **31,** 2152–2159.
Elford, C. R., 1956: A new one-minute rainfall record. *Mon. Wea. Rev.,* **84,** 51–52.
Engelbrecht, H. H., and G. N. Brancato, 1959: World record one-minute rainfall at Unionville, Maryland. *Mon. Wea. Rev.,* **87,** 303–306.
Fankhauser, J. C., 1988: Estimates of thunderstorm precipitation efficiency from field measurements in CCOPE. *Mon. Wea. Rev.,* **116,** 663–684.
Feingold, G., Z. Levin, and S. Tzivion, 1991: The evolution of raindrop spectra. Part III: Downdraft generation in an axisymmetrical rainshaft model. *J. Atmos. Sci.,* **48,** 315–330.
Ferrier, B. S., J. Simpson, and W.-K. Tao, 1996: Factors responsible for precipitation efficiencies in midlatitude and tropical squall simulations. *Mon. Wea. Rev.,* **124,** 2100–2125.
Fletcher, R. D., 1950: A relation between maximum observed point and areal rainfall values. *Trans. Amer. Geophys. Union,* **31,** 344–348.
Forbes, G. S., 1990: Precipitation in Pennsylvania. *Water Resources in Pennsylvania: Availability, Quality, and Management,* S. K. Majumdar, E. W. Miller, and R. R. Parizek, Eds., Pennsylvania Academy of Science, 41–59.
———, and R. Greenfield, 1992: Modelling, remote sensing, and prediction of natural disasters: An Overview. *Natural and Technological Disasters: Causes, Effects and Preventative Measures,* S. K. Majumdar, G. S. Forbes, E. W. Miller, and

R. F. Schmaltz, Eds., Pennsylvania Academy of Science, 35–48.

Frei, C., C. Schär, D. Lüthi, and H. C. Davies, 1998: Heavy precipitation processes in a warmer climate. *Geophys. Res. Lett.,* **25,** 1431–1434.

Fritsch, J. M., R. J. Kane, and C. R. Chelius, 1986: The contribution of mesoscale convective weather systems to the warm-season precipitation in the United States. *J. Climate Appl. Meteor.,* **25,** 1333–1345.

Fukuta, N., 1992: Theories of competitive cloud droplets growth and their applications to cloud physics studies. *J. Atmos. Sci.,* **49,** 1107–1114.

———, 1993: Water supersaturation in convective clouds. *Atmos. Res.,* **30,** 105–126.

Fulks, J. R., 1935: Rate of precipitation from adiabatically ascending air. *Mon. Wea. Rev.,* **63,** 291–294.

Giorgi, F., L. O. Mearns, C. Shields, and L. Mayer, 1996: A regional model study of the importance of local versus remote controls of the 1988 drought and the 1993 flood over the central United States. *J. Climate,* **9,** 1150–1162.

Haggett, C. M., 1980: Severe storm in the London area—16–17 August 1977. *Weather,* **35,** 2–11.

Hallett, J., and S. C. Mossop, 1974: Production of secondary ice particles during the riming process. *Nature,* **249,** 26–28.

———, R. I. Sax, D. Lamb, and A. S. Ramachandra Murty, 1978: Aircraft measurements of ice in Florida cumuli. *Quart. J. Roy. Meteor. Soc.,* **104,** 631–651.

———, D. Lamb, and R. I. Sax, 1980: Geographical variability of ice phase evolution in supercooled clouds. *J. Rech. Atmos.,* **14,** 317–324.

Helfand, H. M., and S. D. Schubert, 1995: Climatology of the simulated Great Plains low-level jet and its contribution to the continental moisture budget of the United States. *J. Climate,* **8,** 784–806.

Henderson, R. D., 1993: Extreme storm rainfalls in the southern Alps, New Zealand. *Extreme Hydrological Events: Precipitation, Floods and Droughts,* IAHS, 113–120.

Hobbs, P. V., 1981: The Seattle workshop on extratropical cyclones: A call for a national cyclone project. *Bull. Amer. Meteor. Soc.,* **62,** 244–254.

———, D. A. Bowdle, and L. F. Radke, 1985: Particles in the lower troposphere over the high plains of the United States. Part I: Size distributions, elemental compositions and morphologies. *J. Climate Appl. Meteor.,* **24,** 1344–1356.

Houghton, H. G., 1950: A preliminary quantitative analysis of precipitation mechanisms. *J. Meteor.,* **7,** 363–369.

Houze, R. A., 1981: Structure of atmospheric precipitation systems: A global survey. *Radio Sci.,* **16,** 671–689.

———, 1993: *Cloud Dynamics.* Academic Press, 570 pp.

Howell, W. E., 1949: The growth of cloud drops in uniformly cooled air. *J. Meteor.,* **6,** 134–149.

Hoxit, L. R., and Coauthors, 1978: Meteorological analysis of the Johnstown, Pennsylvania, flash flood, 19–20 July 1977. NOAA Tech. Rep. ERL 401-A PCL 43, 71 pp.

Hudson, J. G., 1993: Cloud condensation nuclei. *J. Appl. Meteor.,* **32,** 596–607.

Huschke, R. E., 1959: *Glossary of Meteorology.* Amer. Meteor. Soc., 638 pp.

Jennings, A. H., 1950: World's greatest observed point rainfalls. *Mon. Wea. Rev.,* **78,** 4–5.

Johnson, D. B., 1976: Ultragiant urban aerosol particles. *Science,* **194,** 941–942.

———, 1982: The role of giant and ultragiant aerosol particles in warm rain initiation. *J. Atmos. Sci.,* **39,** 448–460.

———, 1993: The onset of effective coalescence growth in convective clouds. *Quart. J. Roy. Meteor. Soc.,* **119,** 925–933.

Johnson, R. H., and P. J. Hamilton, 1988: The relationship of surface pressure features to the precipitation and air flow structure of an intense midlatitude squall line. *Mon. Wea. Rev.,* **116,** 1444–1472.

Jonas, P. R., 1996: Turbulence and cloud microphysics. *Atmos. Res.,* **40,** 283–306.

Kane, R. J., C. R. Chelius, and J. M. Fritsch, 1987: Precipitation characteristics of mesoscale convective weather systems. *J. Climate Appl. Meteor.,* **26,** 1345–1357.

Keers, J. F., and P. Wescott, 1976: The Hampstead storm—14 August 1975. *Weather,* **31,** 2–10.

Kessler, E., 1969: *On the Distribution and Continuity of Water Substance in Atmospheric Circulations. Meteor. Monogr.,* No. 31, Amer. Meteor. Soc., 84 pp.

———, 1975: Condensate content in relation to sloping updraft parameters. *J. Atmos. Sci.,* **32,** 443–444.

Khain, A. P., and I. L. Sednev, 1995: Simulation of hydrometeor size spectra evolution by water-water, ice-water and ice-ice interactions. *Atmos. Res.,* **36,** 107–138.

Klett, J. D., and M. H. Davis, 1973: Theoretical collision efficiencies of cloud droplets at small Reynolds numbers. *J. Atmos. Sci.,* **30,** 107–117.

Knight, C. A., N. C. Knight, J. E. Dye, and V. Toutenhoofd, 1974: The mechanism of precipitation formation in northeastern Colorado cumulus. I. Observations of the precipitation itself. *J. Atmos. Sci.,* **31,** 2142–2147.

Knupp, K. R., and W. R. Cotton, 1987: Internal structure of a small mesoscale system. *Mon. Wea. Rev.,* **115,** 629–645.

Koenig, L. R., 1963: The glaciating behavior of small cumulonimbus clouds. *J. Atmos. Sci.,* **20,** 29–47.

———, 1965: Drop freezing through drop breakup. *J. Atmos. Sci.,* **22,** 448–451.

Kogan, Y. L., 1991: The simulation of a convective cloud in a 3-D model with explicit microphysics. Part I: Model description and sensitivity experiments. *J. Atmos. Sci.,* **48,** 1160–1189.

———, 1993: Drop size separation in numerically simulated convective clouds and its effect on warm rain formation. *J. Atmos. Sci.,* **50,** 1238–1253.

Krueger, S. K., 1988: The role of entrainment by falling raindrops in microbursts. *Proc. 15th Conf. on Severe Local Storms,* Baltimore, MD, Amer. Meteor. Soc., J103–J106.

Kunkel, K. E., S. A. Changnon, and J. R. Angel, 1994: Climatic aspects of the 1993 Upper Mississippi River Basin Flood. *Bull. Amer. Meteor. Soc.,* **75,** 811–822.

Kyle, T. G., and W. Sand, 1973: Water content in convective storms. *Science,* **180,** 1274–1276.

Lamb, D., J. Hallett, and R. I. Sax, 1981: Mechanistic limitations of the release of latent heat during the natural and artificial glaciation of deep convective clouds. *Quart. J. Roy. Meteor. Soc.,* **107,** 935–954.

Langmuir, I., 1948: The production of rain by a chain reaction in cumulus clouds at temperatures above freezing. *J. Meteor.,* **5,** 175–192.

Levin, Z., E. Ganor, and V. Gladstein, 1996: The effects of desert particles coated with sulfate on rain formation in the eastern Mediterranean. *J. Appl. Meteor.,* **35,** 1511–1523.

Lew, J. K., and H. R. Pruppacher, 1983: A theoretical determination of the capture efficiency of small columnar ice crystals by large cloud drops. *J. Atmos. Sci.,* **40,** 139–145.

Lin, Y-L., R. D. Farley, and H. D. Orville, 1983: Bulk parameterization of the snow field in a cloud model. *J. Climate Appl. Meteor.,* **22,** 1065–1092.

Locatelli, J. D., and P. V. Hobbs, 1995: A world record rainfall at Holt, Missouri: Was it due to cold frontogenesis aloft? *Wea. Forecasting,* **10,** 779–785.

Lott, G. A., 1954: The world-record 42-minute Holt, Missouri, rainstorm. *Mon. Wea. Rev.,* **82,** 50–59.

Low, T. B., and R. List, 1982: Collision, coalescence and breakup of raindrops. Part I: Experimentally established coalecence efficiencies and fragment size distributions in breakup. *J. Atmos. Sci.,* **39,** 1591–1606.

Lucas, C., E. J. Zipser, and B. S. Ferrier, 1995: Warm-pool cumulonimbus and the ice phase. Preprints, *Conf. on Cloud Physics,* Dallas, TX, Amer. Meteor. Soc., 318–320.

Ludlam, D. M., 1989: The Johnstown flood: Our most infamous natural disaster. *Weatherwise,* **42,** 88–92.
Ludlam, F. H., 1951: The production of showers by the coalescence of cloud droplets. *Quart. J. Roy. Meteor. Soc.,* **77,** 402–417.
———, 1963: Severe local storms: A review. *Severe Local Storms, Meteor. Monogr.,* No. 27, Amer. Meteor. Soc., 1–30.
———, 1980: *Clouds and Storms.* The Pennsylvania State University Press, 405 pp.
Maddox, R. A., 1980: Mesoscale convective complexes. *Bull. Amer. Meteor. Soc.,* **61,** 1374–1387.
———, 1983: Large-scale meteorological conditions associated with midlatitude, mesoscale convective complexes. *Mon. Wea. Rev.,* **111,** 1475–1493.
———, F. Caracena, L. R. Hoxit, and C. F. Chappell, 1977: Meteorological aspects of the Big Thompson flash flood of 31 July 1976. NOAA Tech. Rep. ERL 388-APCL 41, 83 pp.
———, L. R. Hoxit, C. F. Chappell, and F. Caracena, 1978: Comparison of meteorological aspects of the Big Thompson and Rapid City flash floods. *Mon. Wea. Rev.,* **106,** 375–389.
———, C. F. Chappell, and L. R. Hoxit, 1979: Synoptic and meso-α scale aspects of flash flood events. *Bull. Amer. Meteor. Soc.,* **60,** 115–123.
———, D. J. Perkey, and J. M. Fritsch, 1981: Evolution of upper tropospheric features during the development of a mesoscale convective complex. *J. Atmos. Sci.,* **38,** 1664–1674.
Marshall, J. S., and W. M. Palmer, 1948: The distribution of raindrops with size. *J. Meteor.,* **5,** 165–166.
Marwitz, J. D., 1972a: Precipitation efficiency of thunderstorms on the high plains. *J. Rech. Atmos.,* **6,** 367–370.
———, 1972b: The structure and motion of severe hailstorms. Part III: Severely sheared storms. *J. Appl. Meteor.,* **11,** 189–201.
Mason, B. J., 1971: *The Physics of Clouds.* Clarendon Press, 671 pp.
———, and P. R. Jonas, 1974: The evolution of droplet spectra and large droplets by condensation in cumulus clouds. *Quart. J. Roy. Meteor. Soc.,* **100,** 23–38.
Massacand, A. C., H. Wernli, and H. C. Davies, 1998: Heavy precipitation on the alpine southside: An upper-level precursor. *Geophys. Res. Lett.,* **25,** 1435–1438.
McAnelly, R. L., and W. R. Cotton, 1986: Meso-beta-scale characteristics of an episode of meso-alpha-scale convective complexes. *Mon. Wea. Rev.,* **114,** 1740–1770.
———, and ———, 1989: The precipitation of mesoscale convective complexes over the central United States. *Mon. Wea. Rev.,* **117,** 784–808.
Miller, M. J., 1978: The Hampstead storm: A numerical simulation of a quasi-stationary cumulonimbus system. *Quart. J. Roy. Meteor. Soc.,* **104,** 413–427.
Mordy, W., 1959: Computations of the growth by condensation of a population of cloud droplets. *Tellus,* **11,** 16–44.
Mossop, S. C., 1985: The origin and concentration of ice crystals in clouds. *Bull. Amer. Meteor. Soc.,* **66,** 264–273.
Newton, C. W., 1981: Pseudo-cold-fronts in the USA. *PAGEOPH,* **119,** 594–611.
Ochs, H. T., 1978: Moment-conserving techniques for warm cloud microphysical computations. Part II. Model testing and results. *J. Atmos. Sci.,* **35,** 1959–1973.
Paluch, I. R., and C. A. Knight, 1984: Mixing and the evolution of cloud droplet size spectra in a vigorous continental cumulus. *J. Atmos. Sci.,* **41,** 1801–1815.
Paulhus, J. L. H., 1965: Indian Ocean and Taiwan rainfalls set new records. *Mon. Wea. Rev.,* **93,** 331–335.
Pinsky, M. B., and A. P. Khain, 1997: Turbulence effects on droplet growth and size distributions in clouds—A review. *J. Aerosol Sci.,* **28,** 1177–1214.
———, ———, and M. Shapiro, 1999: Collisions of small drops in a turbulent flow: Problem formulation and preliminary results. Part I: Collision efficiency. *J. Atmos. Sci.,* **56,** 2585–1600.
Politovich, M. K., 1993: A study of the broadening of droplet size distributions in cumuli. *J. Atmos. Sci.,* **50,** 2230–2244.
———, and W. A. Cooper, 1988: Variability of the supersaturation in cumulus clouds. *J. Atmos. Sci.,* **45,** 1651–1664.
Pontrelli, M. D., G. Bryan, and J. M. Fritsch, 1999: The Madison County, Virginia, flash flood of 27 June 1995. *Wea. Forecasting,* **14,** 384–404.
Pruppacher, H. R., and J. D. Klett, 1997: *Microphysics of Clouds and Precipitation.* 2d ed. Kluwer Academic Publishers, 954 pp.
Rangno, A. L., and P. V. Hobbs, 1991: Ice particle concentrations and precipitation development in small polar maritime cumuliform clouds. *Quart. J. Roy. Meteor. Soc.,* **117,** 207–241.
———, and ———, 1994: Ice particle concentrations and precipitation development in small cumuliform clouds. *Quart. J. Roy. Meteor. Soc.,* **120,** 573–601.
Rannie, W. F., and D. Blair, 1995: Historic and recent analogues for the extreme 1993 summer precipitation in the North American mid-continent. *Weather,* **50,** 193–200.
Rasmussen, R. M., 1995: A review of theoretical and observational studies in cloud and precipitation physics: 1991–1994. *Rev. Geophys. (Suppl.),* 795–809.
Raymond, D. J., and A. M. Blyth, 1992: Extension of the stochastic mixing model to cumulonimbus clouds. *J. Atmos. Sci.,* **49,** 1968–1983.
Reisin, T., Z. Levin, and S. Tzivion, 1996: Rain production in convective clouds as simulated in an axisymmetric model with detailed microphysics. Part II: Effects of varying drops and ice initiation. *J. Atmos. Sci.,* **53,** 1815–1837.
Reuter, G. W., 1986: A historical review of cumulus entrainment studies. *Bull. Amer. Meteor. Soc.,* **67,** 151–154.
Roesner, S., A. I. Flossmann, and H. R. Pruppacher, 1990: The effect on the evolution of the drop spectrum in clouds of the preconditioning of air by successive convective elements. *Quart. J. Roy. Meteor. Soc.,* **116,** 1389–1403.
Rogers, R. R., and M. K. Yau, 1989: *A Short Course in Cloud Physics.* 3d ed. Pergamon Press, 290 pp.
Rosenfeld, D., and W. L. Woodley, 1997: Cloud microphysical observations of relevance to the Texas cold-cloud conceptual seeding model. *J. Wea. Mod.,* **29,** 56–69.
Rutledge, S. A., and P. V. Hobbs, 1984: The mesoscale and microscale structure and organization of clouds and precipitation in midlatitude cyclones. *J. Atmos. Sci.,* **41,** 2949–2972.
Shaw, R. A., W. C. Reade, L. R. Collins, and J. Verlinde, 1998: Preferential concentration of cloud droplets by turbulence: Effects on the early evolution of cumulus cloud droplet spectra. *J. Atmos. Sci.,* **55,** 1965–1976.
Squires, P., 1952: The growth of cloud drops by condensation. I: General characteristics. *Austr. J. Sci. Res.,* **A5,** 59–86.
———, 1956: The micro-structure of cumuli in maritime and continental air. *Tellus,* **8,** 443–444.
———, 1958a: The microstructure and colloidal stability of warm clouds. *Tellus,* **10,** 256–271.
———, 1958b: The spatial variation of liquid water and droplet concentration in cumuli. *Tellus,* **10,** 372–380.
Srivastava, R. C., 1989: Growth of cloud drops by condensation: A criticism of currently accepted theory and a new approach. *J. Atmos. Sci.,* **46,** 869–887.
Steiner, M., and J. A. Smith, 1998: Convective versus stratiform rainfall: An ice-microphysical and kinematic conceptual model. *Atmos. Res.,* **47–48,** 317–326.
Telford, J. W., 1955: A new aspect of coalescence theory. *J. Meteor.,* **12,** 436–444.
———, 1975: Turbulence, entrainment and mixing in cloud dynamics. *Pure Appl. Geophys.,* **113,** 1067–1084.
———, 1996: Clouds with turbulence; the role of entrainment. *Atmos. Res.,* **40,** 261–282.
———, and S. K. Chai, 1980: A new aspect of condensation theory. *Pure Appl. Geophys.,* **118,** 720–742.
———, ———, and S. Ionescu-Niscov, 1987: Comments on "Ice particle concentrations in clouds." *J. Atmos. Sci.,* **44,** 903–910.
Tollerud, E. I., and R. S. Collander, 1993: Mesoscale convective systems and extreme rainfall in the central United States.

Extreme Hydrological Events: Precipitation, Floods and Droughts, IAHS, 11–19.

Twomey, S., 1959: The nuclei of natural cloud formation. Part II: The supersaturation in natural clouds and the variation of cloud droplet concentration. *Geofis. Pur. Appl.,* **43,** 243–249.

———, 1966: Computations of rain formation by coalescence. *J. Atmos. Sci.,* **23,** 405–411.

———, and P. Squires, 1959: The influence of cloud nucleus population on the microstructure and stability of convective clouds. *Tellus,* **11,** 408–411.

Velasco, I., and J. M. Fritsch, 1987: Mesoscale convective complexes in the Americas. *J. Geophys. Res.,* **92,** 9591–9613.

Wagner, A. J., 1972: Weather and circulation of June 1972: A month with two major flood disasters. *Mon. Wea. Rev.,* **100,** 692–699.

Waldvogel, L. Klein, D. J. Musil, and P. L. Smith, 1987: Characteristics of radar-identified big drop zones in Swiss hailstorms. *J. Climate Appl. Meteor.,* **26,** 861–877.

Warner, J., 1955: The water content of cumuliform cloud. *Tellus,* **7,** 449–457.

———, and P. Squires, 1958: Liquid water content and the adiabatic model of cumulus development. *Tellus,* **10,** 390–394.

Wetzel, P. J., W. R. Cotton, and R. L. McAnelly, 1983: A long-lived mesoscale convective complex. Part II: Evolution and structure of the mature complex. *Mon. Wea. Rev.,* **111,** 1919–1937.

Wheeler, D., 1995: Madrid flooded, but the drought continues. *J. Meteor.,* **20,** 329–333.

Williams, J., 1994: The great flood. *Weatherwise,* Feb.–March 1994, 18–20.

Willis, P. J., and J. Hallett, 1991: Microphysical measurements from an aircraft ascending with a growing isolated maritime cumulus tower. *J. Atmos. Sci.,* **48,** 283–300.

Woodcock, A. H., and D. C. Blanchard, 1955: Test of the salt-nuclei hypothesis of rain formation. *Tellus,* **7,** 437–448.

Young, K. C., 1974: The evolution of drop spectra through condensation, coalescence and breakup. Preprints, *Conf. on Cloud Physics,* Tucson, AZ, Amer. Meteor. Soc., 95–98.

———, 1975: The evolution of drop spectra due to condensation, coalescence and breakup. *J. Atmos. Sci.,* **32,** 965–973.

———, 1993: *Microphysical Processes in Clouds.* Oxford University Press, 427 pp.

Zhang, D.-L., 1989: The effect of parameterized ice microphysics on the simulation of vortex circulation with a mesoscale hydrostatic model. *Tellus,* **41A,** 132–147.

———, and J. M. Fritsch, 1986: Numerical simulation of the meso-β scale structure and evolution of the 1977 Johnstown flood. Part I: Model description and verification. *J. Atmos. Sci.,* **43,** 1913–1943.

Chapter 9

Mesoscale Convective Systems

J. M. Fritsch and G. S. Forbes

The Pennsylvania State University, University Park, Pennsylvania

Review Panel: Ed Tollerud (Chair), Bob Maddox, Brad Small, Morris Weisman, Jerry Schmidt, and John Brown

---

9.1. Introduction

According to Zipser (1982), a mesoscale convective system (MCS) is a weather feature that exhibits moist convective overturning contiguous with or embedded within a mesoscale circulation that is at least partially driven by the convective processes. Such a broad definition includes a wide variety of mesoscale phenomena ranging from short-lived indiscriminate aggregates of a few thunderstorms to well-organized squall lines to long-lived tropical storms and even hurricanes.

Because of the broad temporal and spacial spectra of MCSs, it is neither possible nor desirable to try to discuss here all systems that fit under the MCS-definition umbrella. Rather, we choose to concentrate on the largest portion of the spectrum of MCSs, excluding, however, tropical storms. This would include systems that are dynamically mesoscale, that is, systems with a Rossby number of order 1, and systems with a radius comparable to the Rossby radius of deformation λ_R (Schubert et al. 1980; Cotton et al. 1989), where

$$\lambda_R = \frac{C_N}{(\zeta + f)^{1/2}(2VR^{-1} + f)^{1/2}}, \qquad (9.1)$$

ζ is the vertical component of relative vorticity, f is the coriolis parameter, C_N is the phase speed of an inertia-gravity wave, and V is the tangential component of the wind at the radius of curvature R. For such systems, both ageostrophic advection and rotational influences are important (Lilly and Gal-Chen 1983). Cotton et al. (1989) found that, for midlatitude MCS environments, λ_R is about 300 km. For midlatitude systems of this scale, it typically takes about 3 to 6 h for Coriolis influences to become significant. Moreover, Cotton et al. also demonstrated that mesoscale convective complexes (MCCs), a subset of mesoscale convective systems that exhibit a large, circular (as observed by satellite), long-lived cold cloud shield, typically fulfill these dynamical criteria. Likewise, as shown by Skamarock et al. (1994) and Davis and Weisman (1994), some large, long-lived squall lines also satisfy these dynamical criteria and, not surprisingly, exhibit dynamical properties similar to those of MCCs. Consequently, the material selected for this chapter will concentrate on the properties of MCCs but will also occasionally include some discussion of large MCSs and squall lines.

Following Maddox (1980), the definition of an MCC is based on satellite imagery and requires that the convective system's cold cloud shield exhibit certain characteristics (Table 9.1). Principally, the system's cold cloud shield must surpass a size threshold, persist above the threshold size for at least 6 h, and exhibit a nearly circular shape at the time of maximum extent. Life cycle phases for MCCs are similar to those defined by Maddox (1980) and Zipser (1982) and are defined in Table 9.2. Systems are categorized into two types, depending upon the dynamical mechanism whereby the convection and the large cold cloud shields, characteristic of MCCs, are produced. Type 1 events result when a mesoscale ribbon of low-level potentially unstable air is forced to ascend in a frontal zone or other baroclinic region (e.g., a warm or stationary frontal overrunning situation). Type 2 events occur in more barotropic environments and depend on the moist-downdraft production of a surface-based cold pool and its interaction with the ambient vertical wind shear to produce mesoscale ascent and the associated large stratiform cloud shield characteristic of MCCs. The distinction between these two types of events is that type-1 systems form and develop largely as a consequence of externally imposed forcing (e.g., by mesoscale components of synoptic-scale baroclinic systems such as fronts), whereas type-2 systems depend more strongly on features and processes imposed by the convection itself (e.g., the moist-downdraft-generated mesoscale cold pool). These two types of events are discussed more fully in later sections.

TABLE 9.1. Definition of mesoscale convective complex.

Size:	A—Cloud shield with continuously low infrared (IR) temperature $\leq -33°C$; must have an area $\geq 10^5$ km^2 B—Interior cold cloud region with temperature $\leq -52°$; must have an area $\geq 0.5 \times 10^5$ km^2
Initiate:	Size definitions A and B are first satisfied
Duration:	Size definitions A and B must be met for a period ≥ 6 h
Maximum extent:	Contiguous cold cloud shield (IR temperature $\leq -33°$) reaches maximum size
Shape:	Eccentricity (minor axis/major axis) ≥ 0.7 at time of maximum extent
Terminate:	Size definitions A and B no longer satisfied

The definition of a squall line is far broader than that of an MCC and includes many events that would not satisfy the condition that their length scale be comparable to λ_R. For example, Bluestein and Jain (1985) used radar-based criteria to identify squall lines in a study of springtime events in Oklahoma. They defined a line as a group of related or similar convective precipitation echoes that form a pattern exhibiting a length-to-width ratio of at least 5 to 1. Lines had to be at least 50 km long and less than 50 km in width, and had to persist for at least 15 min. Similarly, Hane (1986), in a review and summary of the structure and dynamics of extratropical squall lines and rainbands, pointed out that usage of the term *squall line* has been generally applied to any line of thunderstorms, with or without squalls (Bluestein and Jain noted that some squall lines may not even contain buoyant elements!). It is not surprising then that, based on the results of Bluestein and Jain (1985) and Bluestein et al. (1987), lines are three to five times as common as MCCs and exhibit a great breadth of scales and physical processes. These broad-based squall line definitions can result in overlap with the MCC definition since many, but not all, MCCs contain lines of thunderstorms. Visible satellite images of several squall lines and an MCC are shown in Fig. 9.1.

9.2. Climatological characteristics

Mesoscale convective complexes tend to cluster in favored locations around the world. This tendency is most easily seen by examining their global distribution (Fig. 9.2). From this distribution, it is evident that

1. most MCCs occur over land,
2. development occurs in zonally elongated belts of easterlies or westerlies, and
3. most populations are downstream (typically within 1500 km) of north–south mountain ranges.

It is particularly interesting that MCCs rarely occur in several regions where deep moist convection is extremely common, for example, in the southeastern United States and the Amazon River basin. However, these regions do commonly experience squall lines (Kousky 1980; Greco et al. 1992; Garstang et al. 1994).

One of the more intriguing aspects of MCCs is that all populations are predominantly nocturnal (Maddox 1980; Velasco and Fritsch 1987; Miller and Fritsch 1991; Laing and Fritsch 1993a,b). Thunderstorms usually first develop in the mid- to late afternoon, then increase in areal coverage as sunset approaches. The attendant cold-cloud shield typically attains MCC initiation size in the early nighttime hours, reaches maximum extent after midnight, and dissipates a few hours after sunrise. Figure 9.3 shows the life cycle distribution for the global population (Laing and Fritsch 1997).

Although some MCCs produce cold cloud shields that extend over a million square kilometers, most systems are about one-third of this size (Fig. 9.4). Also, as expected, the size and duration of MCSs and MCCs are positively correlated, that is, larger systems tend to last longer (Tollerud et al. 1992; Laurent 1996; Laing and Fritsch 1997, Fig. 9.5). The typical MCC cold-cloud shield equals or exceeds the 10^5 km^2 threshold for around 11 h, although some systems have continued for up to 36 h. The large contiguous cloud shields sometimes form from mergers of several initially distinct meso-γ- and meso-β-scale elements, while in other instances they simply grow from a single convective system (Houze et al. 1990).

A distinguishing factor in the longevity and size distributions of MCSs appears to be the strength of the early growth. The critical importance of the early growth was quantified by Tollerud et al. (1992) in their analysis of the relationship between early growth rates and cold cloud shield size. They found that the growth rate of the cold cloud shield in the first three hours of the life cycle of an MCS was a strong predictor of the eventual size and duration of the system (Fig. 9.5). Similarly, Cotton and McAnelly (1992) identified a meso-β-scale convective cycle that occurs early in the growth phase and argue that it distinguishes between the early stage evolution of an MCC and other shorter-lived MCSs.

The early growth period also appears to be the most crucial time for the development of severe weather

TABLE 9.2. Life cycle phases of MCCs.

Phase	Definition
Initiation	Time of first storms until the contiguous cold cloud shield reaches initiation size criteria
Development	Initiation time until the cold cloud shield reaches maximum extent
Mature	Time of maximum extent until termination (size definitions no longer satisfied)
Dissipation	Period following termination lasting until new convection ceases to form

(a)

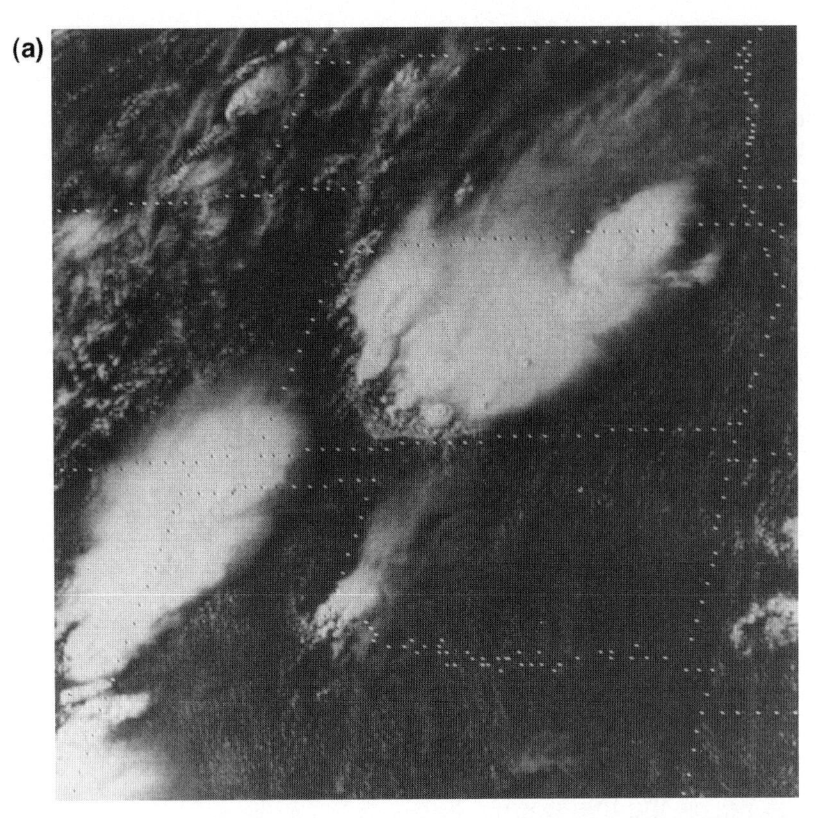

(b)

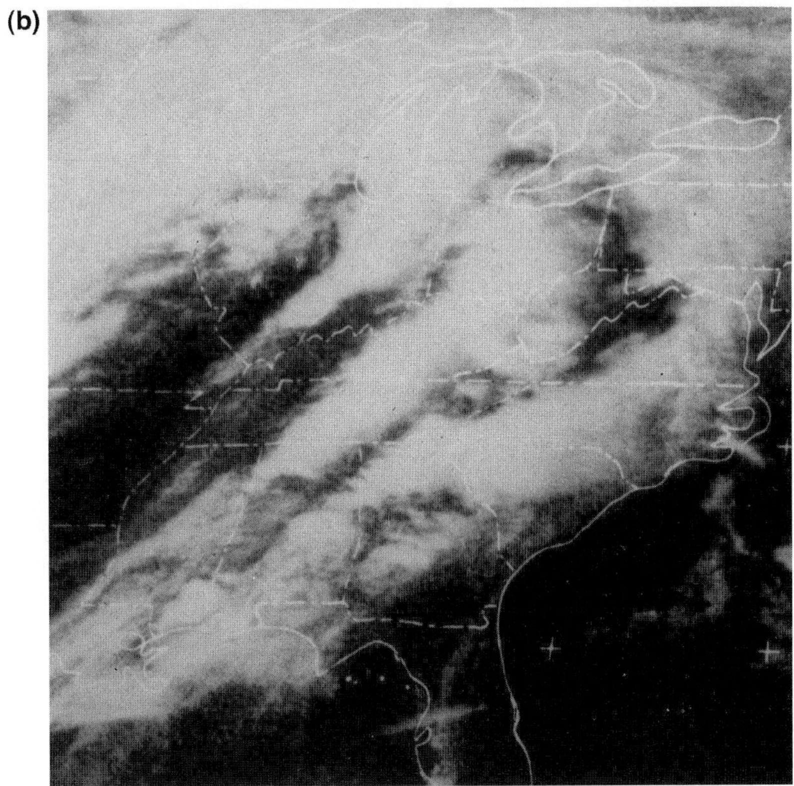

FIG. 9.1. Visible satellite images for (a) a developing MCC over Kansas and a squall line over the Texas panhandle at 2200 UTC 25 May 1978; (b) three squall lines at 2100 UTC 3 April 1974. These lines were responsible for the greatest tornado outbreak in U.S. history.

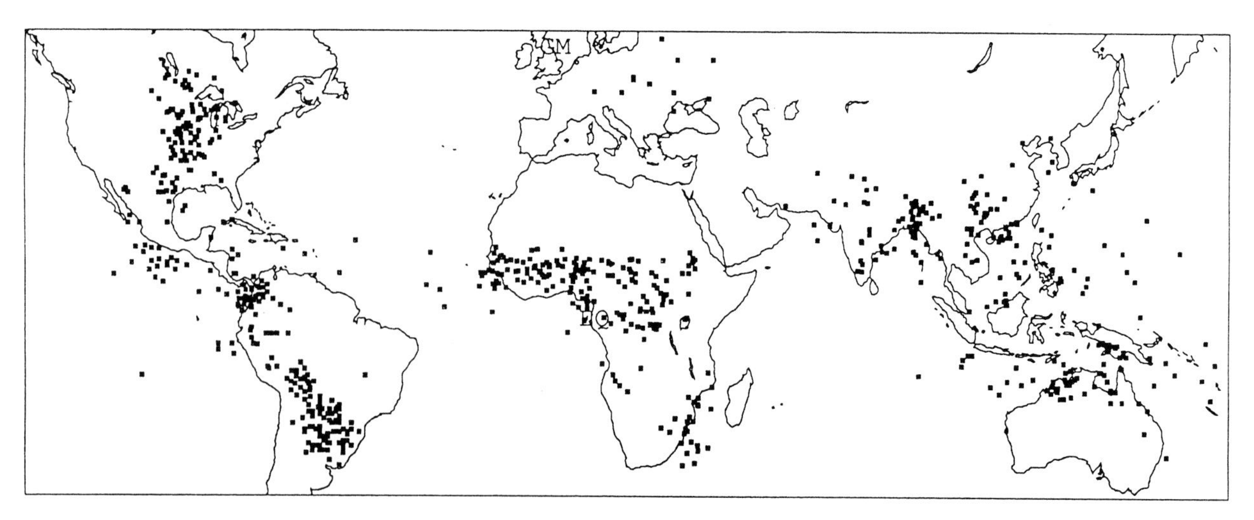

FIG. 9.2. Locations of MCCs based upon 1–3-yr regional samples of satellite imagery (from Laing and Fritsch 1997). Locations are shown for the time of maximum extent of the cold-cloud shield. Additional data concerning the samples are given in the appendix.

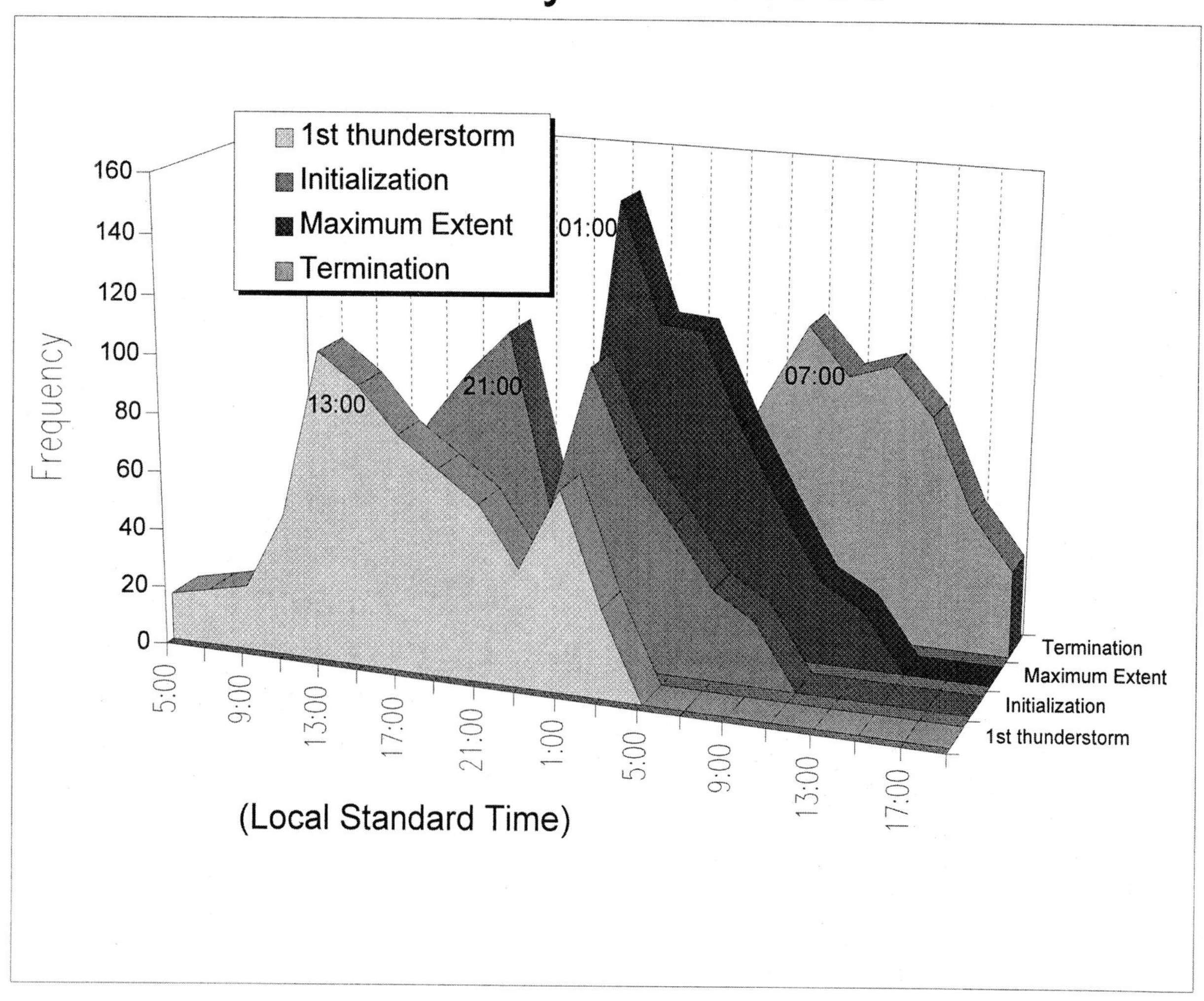

FIG. 9.3. Distribution of the times of occurrence of stages in MCC life cycle. From the global distribution of Laing and Fritsch (1997).

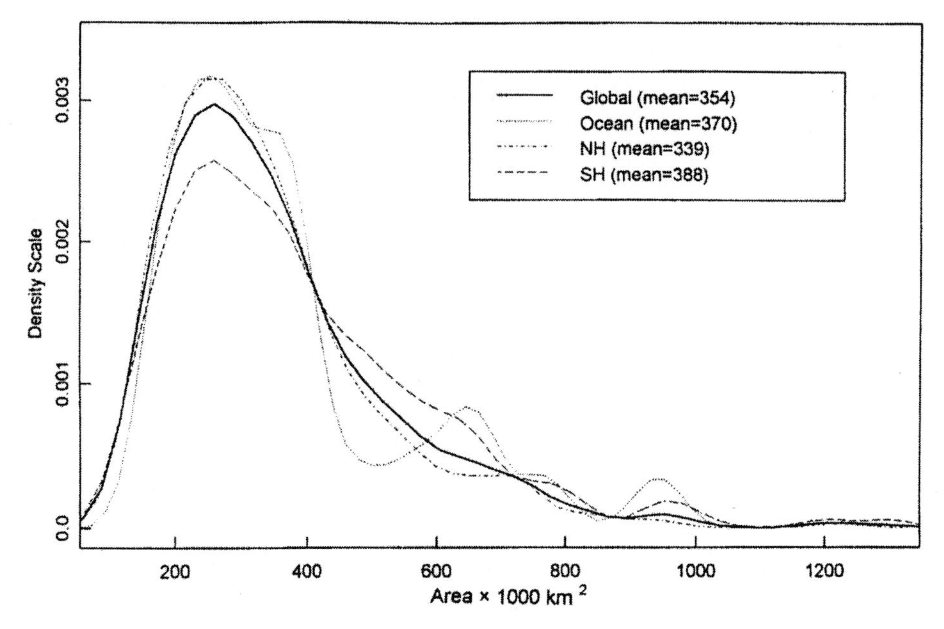

FIG. 9.4. Frequency distribution of the areal extent of MCC cold cloud shields. From Laing and Fritsch (1997).

(high wind, tornadoes, and hail). For example, in an analysis of 12 nocturnal MCCs, Maddox et al. (1986) found that over 80% of the severe weather occurred during the initiation period. Likewise, in a comprehensive analysis of Oklahoma mesoscale convective precipitation systems, Houze et al. (1990) found that tornado and hail reports were biased toward the early stages of the systems' development. Similarly, Tollerud and Collander (1993a) investigated 350 MCCs that occurred in the United States during the 10-yr

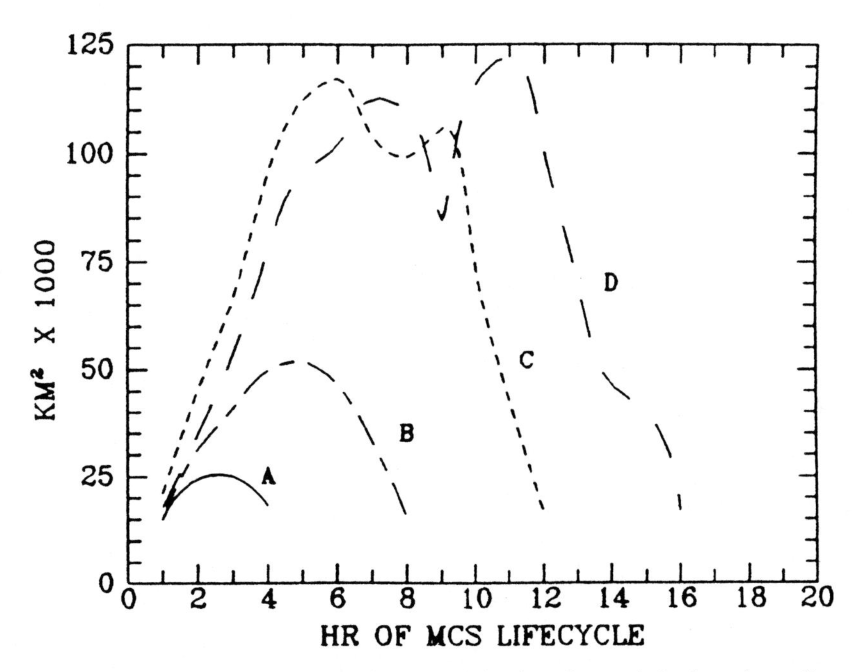

FIG. 9.5. Relationship between MCS size and duration. Curves A–D show the median −52°C cloud top areas of MCSs that last 4, 8, 12, and 16h, respectively. Curves C and D show early growth of areal extent at a much more rapid rate for long-lasting MCSs than for shorter-duration MCSs A and B. From Tollerud et al. (1992).

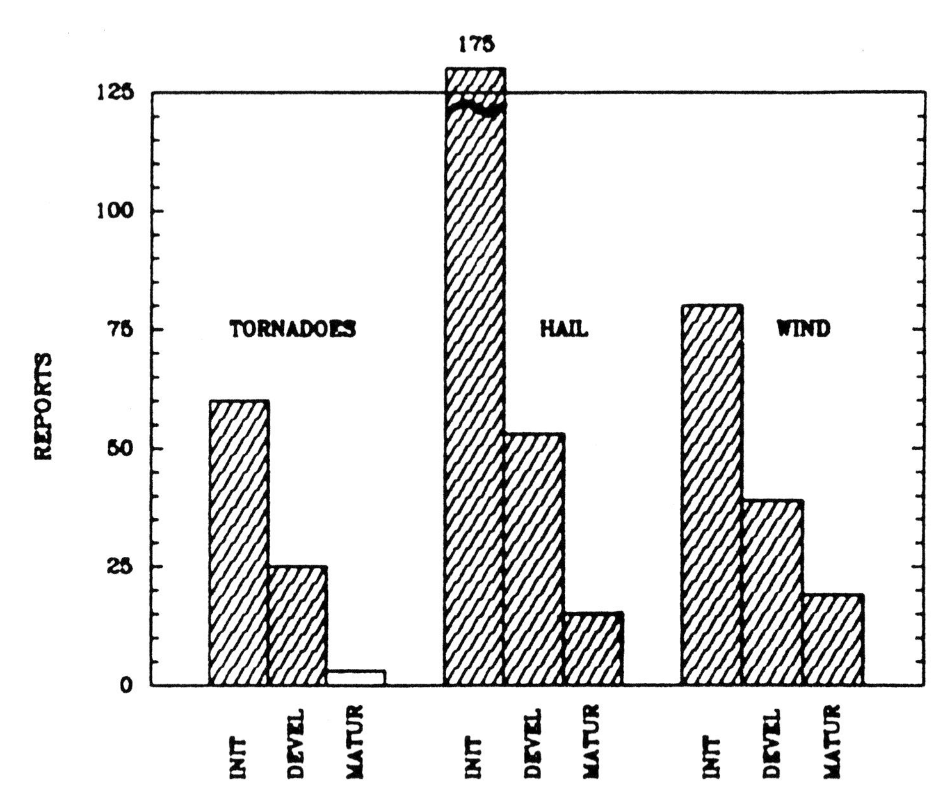

FIG. 9.6. Number of reports of severe weather associated with stages of MCC life cycle in the United States, 1978–87. From Tollerud and Collander (1993a).

period 1978–87 and found that an overwhelming fraction of the severe weather occurs in the initiation phase (Fig. 9.6). On the other hand, nearly half of the severe wind reports occur during the developing and mature stages. Although it is not known what fraction of the global population of MCSs produces severe weather, about half of the systems that occur in the central United States exhibit some type of severe activity (Houze et al. 1990). Of this number, about twice as many severe events occur in linearly organized systems compared to the more circular MCCs (Tollerud and Bartels 1988). Nevertheless, as shown in Table 9.3, MCCs in the United States are prolific producers of severe events. Moreover, Maddox (1980) points out that one in every five MCCs produces injuries or death.

Although severe weather tends to occur in the initiation phase of the life cycle of MCSs, the heaviest rainfall generally occurs in the development phase

TABLE 9.3. Percent frequency of 103 MCCs reporting severe weather. Weather reports from *Storm Data;* from Merritt (1985).

None reported	Tornado	Hail ≥ 19 mm	Wind ≥ 25 m s^{-1}	Heavy rain or flash flooding
15	39	62	62	49

(the period between the time that the cold cloud shield reaches Maddox's criteria for MCC initiation and the time of maximum extent; see Table 9.1 and Fig. 9.3). McAnelly and Cotton (1989), in a study of 122 MCCs, found that the peak average precipitation rate usually happens after initiation but several hours before the rain area and volumetric rain rate reach their maximum values (Fig. 9.7). Likewise, Collander's (1993) investigation of the hourly precipitation associated with 350 MCCs showed that the maximum frequency of observations of heavy rainfall (amounts ≥25 mm h^{-1}) most commonly occurs during the period 2–6 h after initiation (Fig. 9.8). As shown in several studies (Kane et al. 1987; McAnelly and Cotton 1989; Collander 1993), these large rain rates and heavy rainfall amounts most frequently occur on the equatorward flank of the cold cloud shield. For example, Fig. 9.9 shows the mean precipitation for 74 MCCs in the United States. Clearly, the heaviest amounts favor the southwestern quadrant. And, as expected, the distribution of hourly heavy rainfall observations also shows a maximum in the southwestern flank (Fig. 9.10).

In an analysis of the hourly precipitation data for the 10-yr period 1978–87 for the plains region of the United States, MCCs accounted for 20%–40% of the record hourly rain rates during May through August,

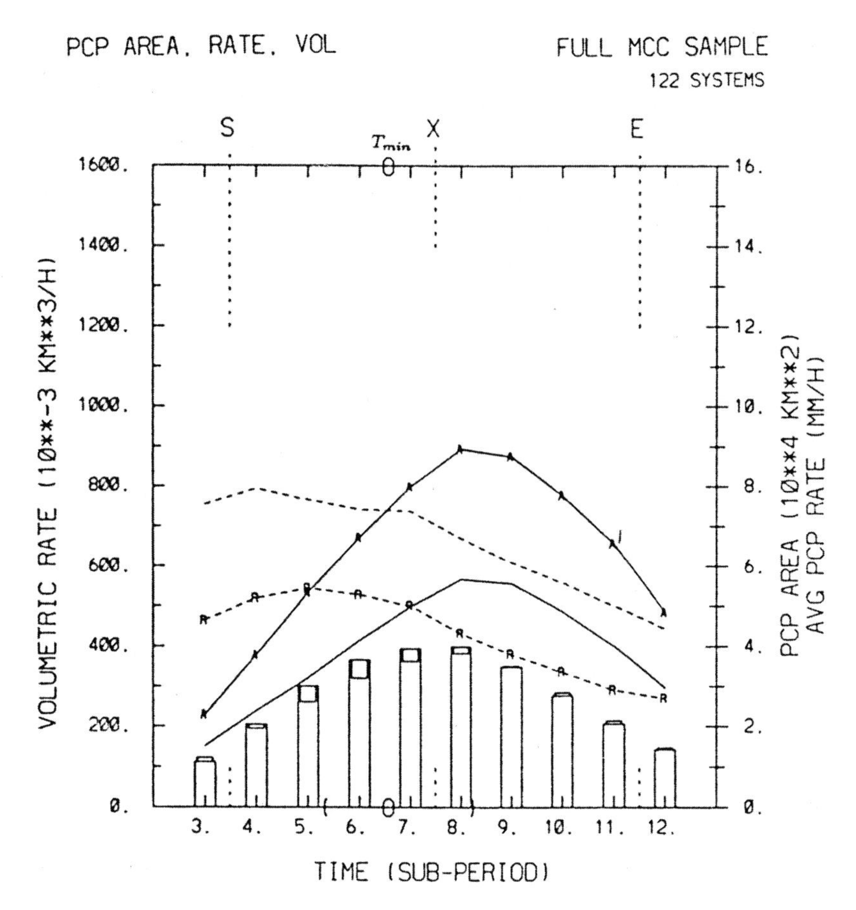

FIG. 9.7. Precipitation characteristics of 122 MCCs. Solid curves show that precipitating area peaks much later than average rain intensity (dashed curves). Curves labelled A and R are from rain gauges with 0.25-mm resolution, whereas unlabelled curves are from coarser-resolution 2.5-mm gauges. Volumetric rain rate for 0.25 (2.5) mm resolution is shown by wide (thin) bars. Labels S, X, and E give reference times of start, maximum, and end of MCC life cycle. T_{min}(O) marks the time of the mesoscale thermal minimum (coldest MCC cloud top). From McAnelly and Cotton (1989).

even though they only accounted for less than 7% of the total number of observations of measurable rain (Tollerud and Collander 1993b). Thus, MCCs produce extreme rainfall at a rate far greater than that produced by other types of weather systems. This agrees with previous studies (eg., Simpson and Woodley 1971; Simpson et al. 1980, 1993) that have shown that precipitation efficiency and rain rates increase significantly as individual deep convective clouds merge into mesoscale convective systems. Finally, it is important to note that the total rainfall from two or three MCCs, or even a single large MCC, can rival and even exceed that from landfalling hurricanes and tropical storms (Fritsch et al. 1986).

9.3. Large-scale environment

Analyses of many MCC events (e.g., Maddox 1983; Merritt 1985; Kane et al. 1987; Cotton et al. 1989; Augustine and Howard 1991; Laing 1996) show that certain thermodynamic patterns and dynamical features are usually present in their large-scale environment. For example, inspection of the mean synoptic environment for 12 MCCs in the United States (Fig. 9.11) reveals that these large convective systems form in the vicinity of the terminus of a low-level jet that is supplying high-θ_e air to the MCC genesis region (Augustine and Caracena 1994). The jet is oriented nearly perpendicular to a pronounced horizontal thermal gradient and slopes upward as it overruns a surface-based layer of cooler air (Trier and Parsons 1993; Augustine and Caracena 1994; Rochette and Moore 1996). A weak midlevel short wave is approaching the genesis area and enhances the low-level convergence associated with the low-level jet (Maddox 1983; Cotton et al. 1989). Warm advection dominates the lower troposphere while diffluence is the rule at high levels. As shown by Laing and Fritsch (2000),

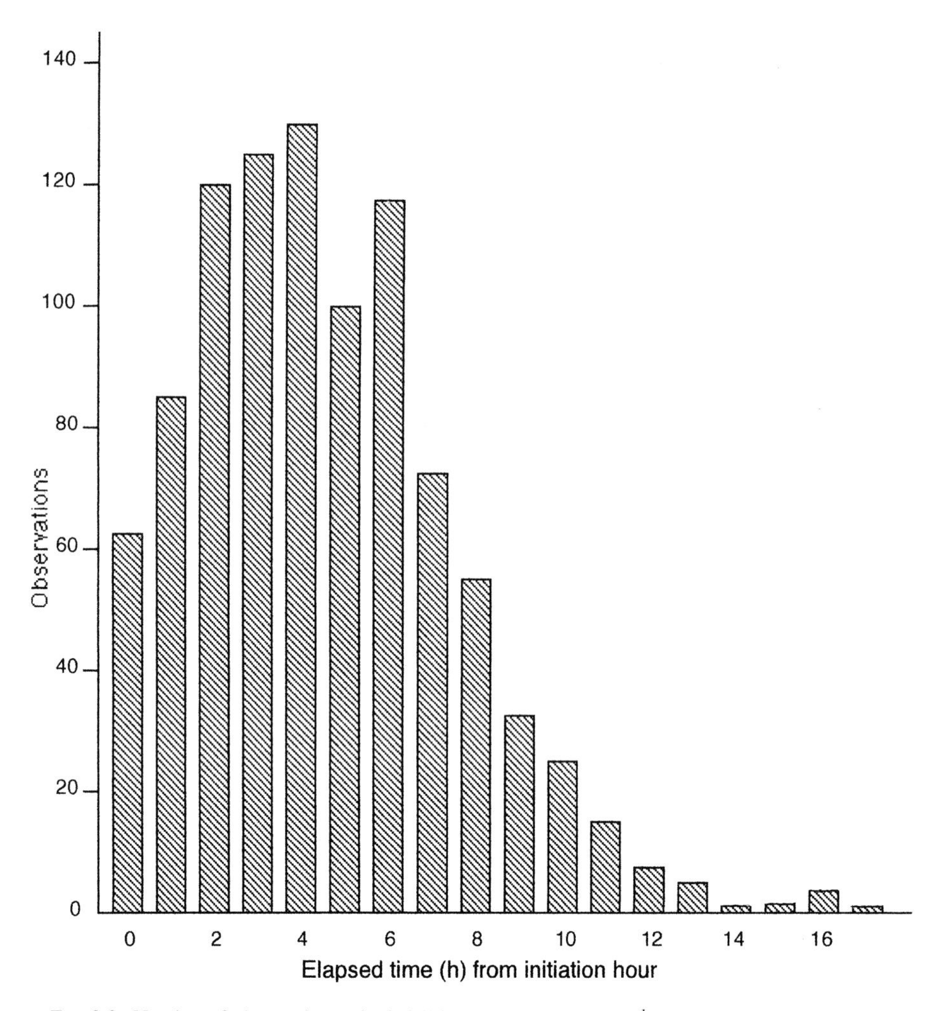

FIG. 9.8. Number of observations of rainfall in excess of 25 mm h^{-1} in 350 MCCs in the United States for each hour following MCC initiation. From Collander (1993).

the large-scale environments of other MCC populations around the world are generally very similar to the U.S. environment.

The pattern shown in Fig. 9.11 is not a particularly unusual one and could be observed over most midlatitude regions of the globe. Therefore, it is of special interest that this pattern results in MCCs only in certain regions and that these exceptionally large convective systems tend to occur at night. It has been argued that the nocturnal life cycle of convective systems may be related to the differential radiative heating between the convective cloud area and its environment, that is, that cloud tops cool while the subcloud layer warms relative to the surrounding clear air. This would tend to enhance environmental subsidence on the periphery of the cloud shield and thereby increase low-level convergence into the system (Gray and Jacobson 1977). In support of this argument, Webster and Stephens (1980) and Chen and Cotton (1988) found that radiative cooling at cloud top and warming at cloud base reduced the static stability of the stratiform anvil of mature MCSs. However, free-troposphere longwave cooling is typically small ($<0.1°C\ h^{-1}$; Dopplick 1972; Machado and Rossow 1993) compared to the convective heating rates observed in mesoscale convective systems ($\approx 5°C\ h^{-1}$; Ogura and Chen 1977) and, considering that MCCs exhibit intense and massive convective overturning for many hours, it is highly unlikely that local free troposphere radiatively driven circulations can create convective available potential energy (CAPE) fast enough to offset the rates at which it is consumed in MCC overturning. Moreover, since cloudy/clear radiative forcing occurs everywhere, but, as noted above, some regions of the globe exhibit plentiful deep convection but few nocturnal MCSs, the nocturnal life cycle of MCCs must not be directly tied to the diurnal radiation cycle, but rather must be indirectly connected through dynamical processes. Thus, there must be certain synoptic or mesoscale processes and circulations, strongly

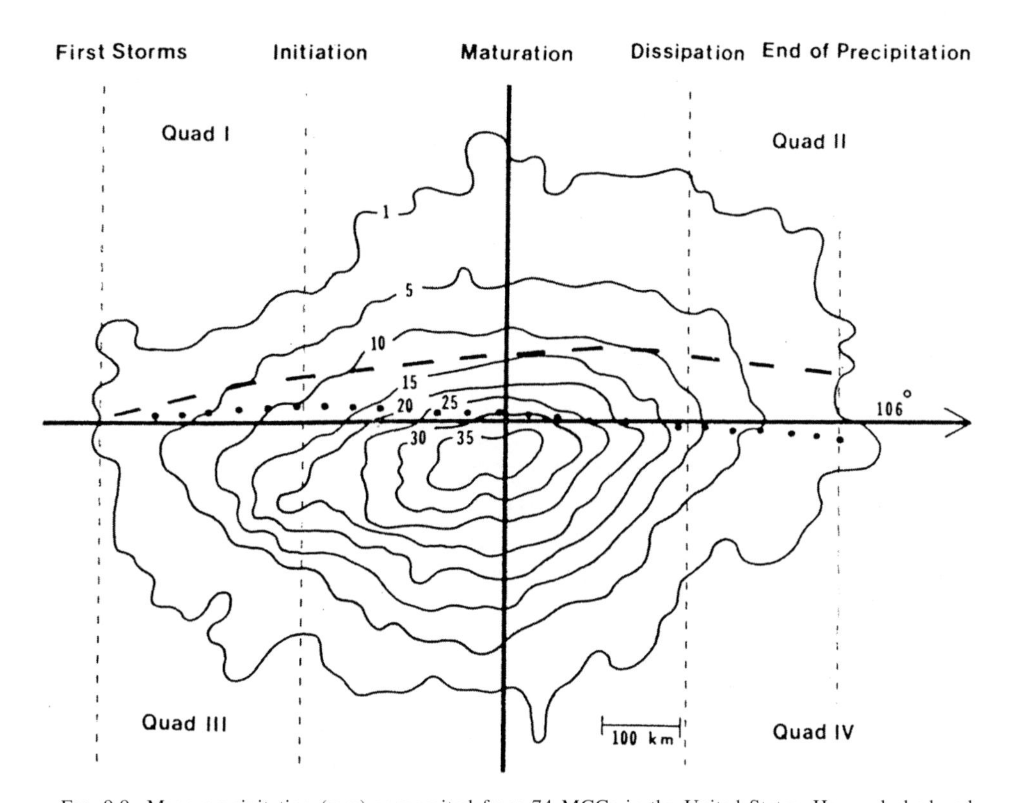

FIG. 9.9. Mean precipitation (mm) composited from 74 MCCs in the United States. Heavy dashed and dotted lines show tracks of the centroids of the −32°C and −52°C cloud shields, respectively. Thin dashed lines indicate approximate locations of the center of the MCC at various times and stages of its life cycle. In computing the mean precipitation pattern, each individual pattern was aligned with the axis of propagation of the convective system. Following Merritt (1985), the axis of propagation for each particular pattern was defined to run parallel to the 1000–500-mb thermal wind vector and to intersect the centroid of the precipitation pattern. Each precipitation pattern was rotated before compositing so that the axis of propagation coincided with the x axis (pointing in a heading of 90°). This results in all systems being aligned as if they propagated in the same direction. The origin of the coordinate system (bold solid lines) represents the centroid of the normalized and rotated precipitation patterns. From Kane et al. (1987).

linked to the radiative cycle, that either locally and rapidly create huge amounts of CAPE via horizontal advection and vertical motion, or act to overcome convective inhibition (Colby 1984; Keenan and Carbone 1992) that has prohibited release of CAPE generated during daytime heating.

Since MCCs tend to be clustered in certain favored regions of the world (Fig. 9.2), the diurnal processes and circulations that support their nocturnal life cycle may be related to the physiography of the favored regions. Such a regional favoritism is not unusual for weather phenomena. For example, it is well known that tropical cyclones only occur in areas where the sea surface temperature exceeds 26°C (Gray 1979). Even extratropical cyclones have favored locations due to local reductions in stability as a result of surface fluxes (Sanders and Gyakum 1980). It is also well known that certain mesoscale weather features have pronounced diurnal cycles. For example, mountain–valley and sea breeze circulations favor daytime convective events (Burpee 1979; Cooper et al. 1982). Therefore it is not surprising that there is a strong correlation between the nocturnal MCC population centers and regions that experience a high frequency of nocturnal low-level jets of high-θ_e air (e.g., Maddox 1983; Velasco and Fritsch 1987; Miller and Fritsch 1991; Augustine and Caracena 1994).

According to Bleeker and Andre (1951), Blackadar (1957), Holton (1967), Bonner (1968), Paegle (1978), McNider and Pielke (1981), and others, nocturnal low-level jets develop as a result of adjustments that take place as 1) the mixed layer decouples from the surface as the surface cools and 2) horizontal temperature differences develop as a result of sloping terrain (e.g., the Great Plains) and an east–west gradient in the Bowen ratio (ratio of surface sensible to latent heat flux). Inasmuch as these processes are independent of the dynamics of traveling disturbances, the low-level wind accelerations that they produce provide a significant enhancement to the low-level warm advection and convergence that would normally be present as a result of an approaching short wave and/or a synoptic-

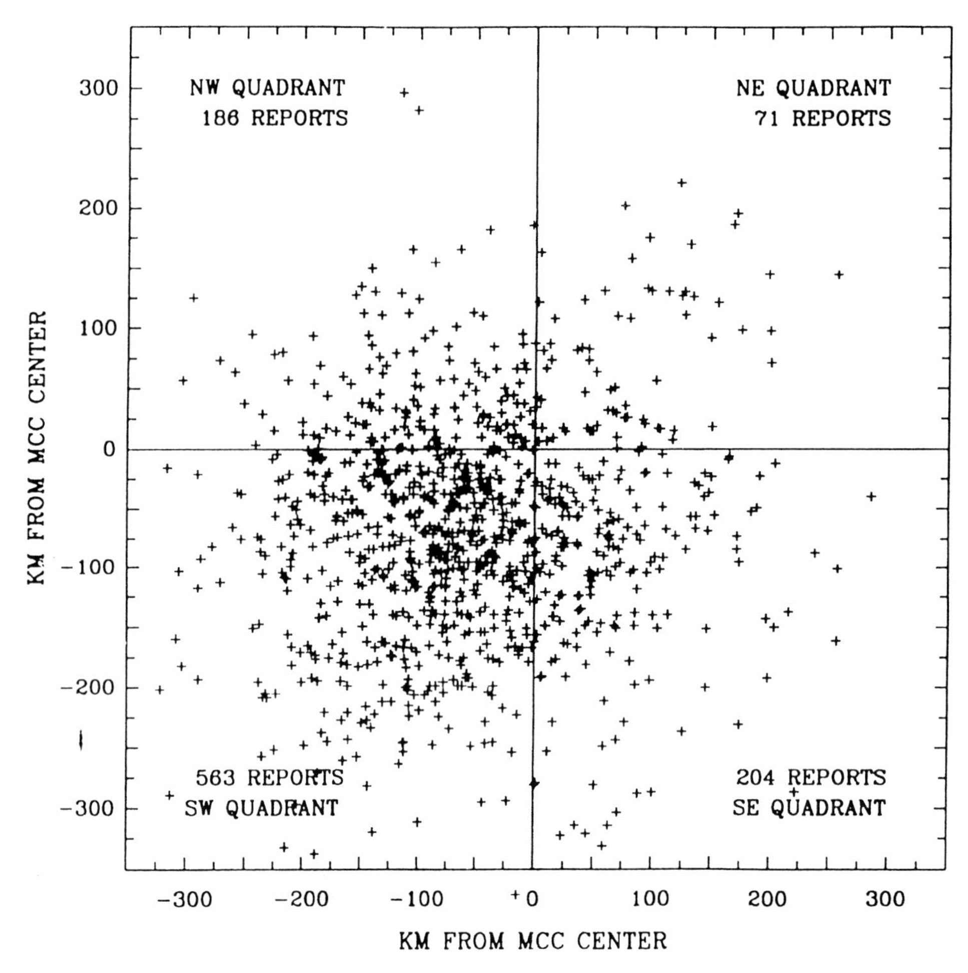

FIG. 9.10. Locations, relative to MCC center, of occurrence of hourly precipitation in excess of 25 mm for MCCs in the United States, 1978–87. From Collander (1993).

scale circulation (e.g., the return flow from the Bermuda high). Because the jets typically are confined to the lowest 100–150 mb above the ground, the high-θ_e air that they transport can more readily underrun colder air aloft compared to what would occur with a deep tropospheric disturbance alone. This buildup of convective instability is further enabled by the necessity of veering wind direction with height in warm advection flow that is approximately geostrophic in direction. The westerly component of flow above the core of the low-level jet necessitated by this warm advection often advects a capping drier residual layer of low static stability over the low-level high-θ_e air, enabling this underrunning. Further, in the typical traveling disturbance, the warm advection extends to higher levels than that associated with physiographically produced low-level jets, thus making it more difficult to destabilize the environment via differential advection or vertical motion. The "advantage" of nocturnal low-level jets to destabilize the atmosphere by underrunning meridional temperature gradients is obviously enhanced in regions where midlevel baroclinic environments are present. It is noteworthy that MCCs tend to be concentrated in latitudinal belts of easterlies or westerlies where baroclinic conditions prevail. Moreover, considering that the processes leading to the formation of low-level jets are inherently ageostrophic, the development of MCCs may be limited geographically generally to those regions in which such ageostrophic features can frequently be produced.

In addition to the underrunning of cold air aloft, Augustine and Howard (1988), Trier and Parsons (1993), and Augustine and Caracena (1994) have shown that pronounced low-level convergence occurs in the terminus region of developing low-level jets and that the vertical motion from this convergence contributes significantly to destabilizing the local environment prior to the onset of deep convection. Moreover, Tripoli and Cotton (1989) have shown how mountain/plains solenoidal circulations provide additional forcing in the region downwind of mountain barriers. Therefore, when the processes that form low-level jets and the mountain/plains solenoidal circulations phase with favorable synoptic-scale circulations (e.g., over-

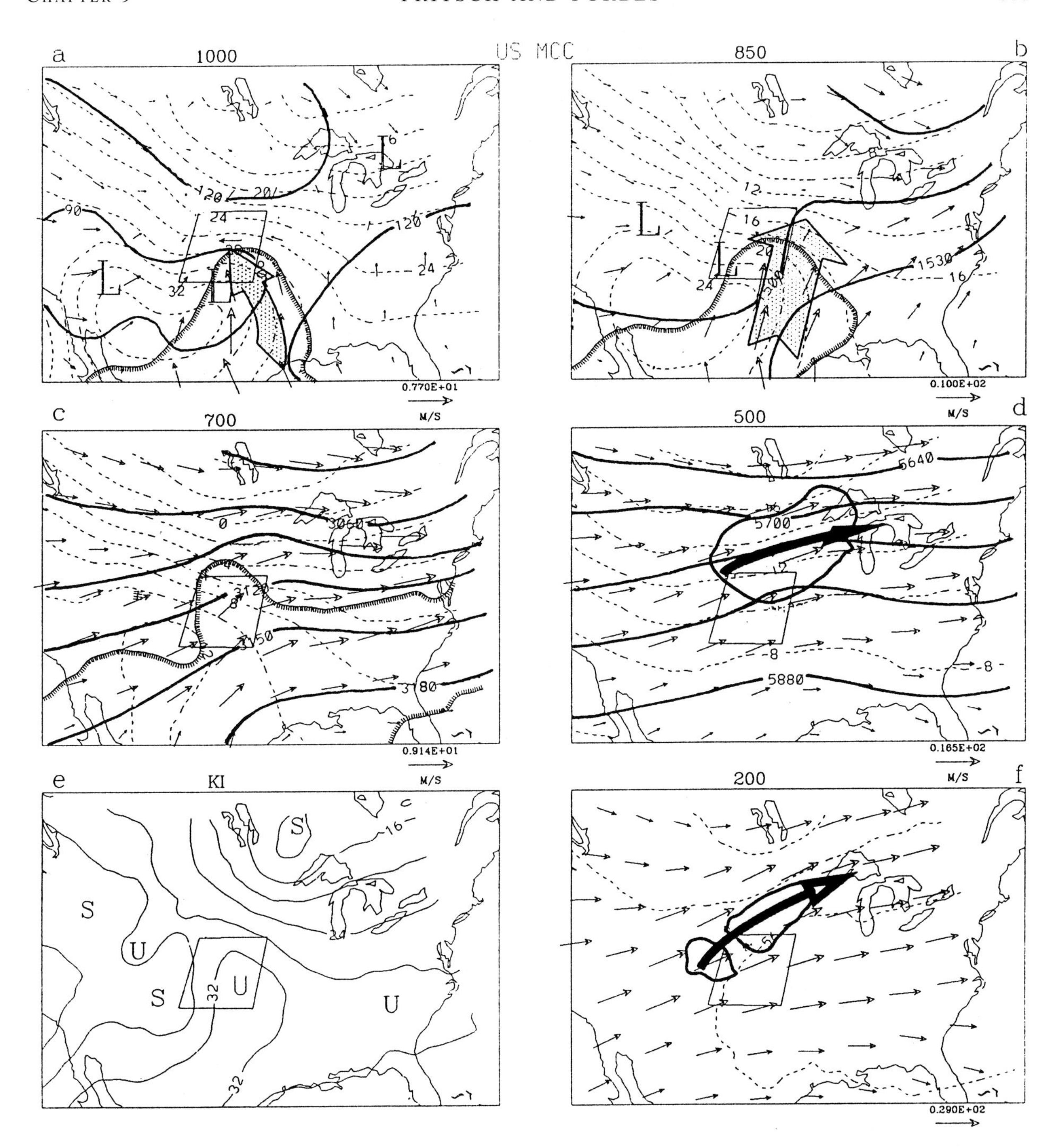

FIG. 9.11. Analyses of the mean genesis environment of U.S. MCCs: (a) 1000 mb, (b) 850 mb, (c) 700 mb, (d) 500 mb, (e) *K* index, and (f) 200 mb. Velocity vectors (m s^{-1}) are plotted at every other grid point and the quadrilateral in the center indicates the approximate genesis region. Heights (m) and *K*-index values are solid contours, isotherms (°C) are dashed, wind speed maxima are heavy solid with arrows, and hachured areas are (a) $\theta_e > 350$ K, (b) $\theta_e > 338$ K, (c) mixing ratio > 5 g kg^{-1}. U and S identify the relatively unstable and stable areas, respectively, in the *K*-index map. Broad stippled arrows in (a) and (b) highlight the low-level inflow of high-θ_e air and the 850-mb warm advection, respectively. Large bold arrows in (d) and (f) depict jet-streak axes. From Laing and Fritsch (2000).

running of a quasi-stationary front) and with the forcing from an approaching short wave, exceptionally large destabilization rates can occur for several hours. As demonstrated by Augustine and Caracena (1994), the lifting and destabilization are especially strong when a low-level jet intersects the thermally direct circulation associated with frontogenetic forcing (Sawyer 1956). The magnitude and rate of production of CAPE in such situations can be far greater than that which occurs in other areas of the world where the regional physiographic features do not support the development of nocturnal low-level jets and destabilization must occur from direct radiative effects and/or from the quasigeostrophic forcing of traveling disturbances and synoptic features alone (Nicolini et al. 1993).

Trier and Parsons (1993) explored these issues when they investigated one of several MCCs that occurred during the Oklahoma–Kansas PRE-STORM field program in 1985. (The PRE-STORM program was specifically designed to observe the structure and dynamics of mesoscale convective weather systems; see Cunning 1986.) The case that they investigated formed in an environment that was very similar to the mean MCC conditions shown in Fig. 9.11. They utilized the PRE-STORM high-density sounding network to construct a north–south cross section (Fig. 9.12) through an east–west quasi-stationary front. The cross section was constructed using soundings taken approximately 2 h before a developing MCC moved into the region. Their analysis clearly shows an upward-sloping low-

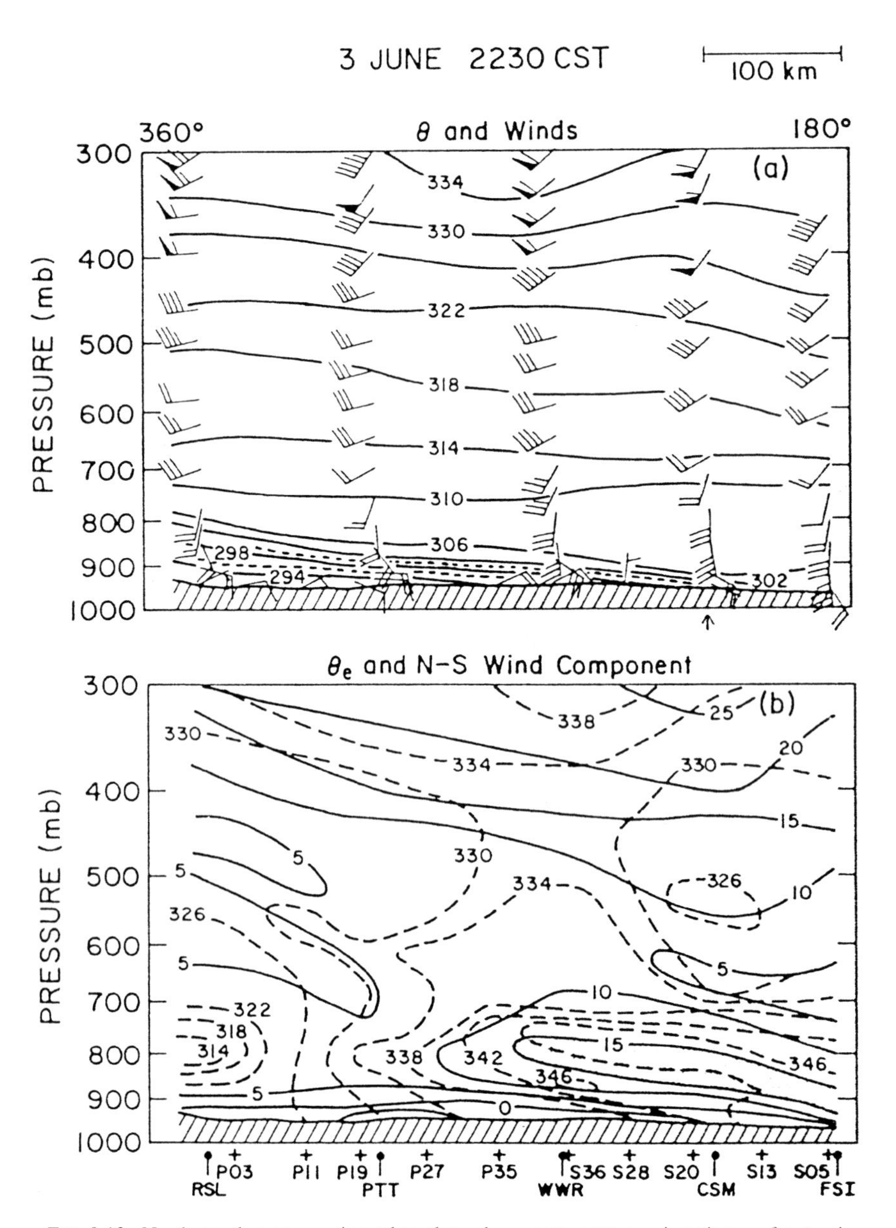

FIG. 9.12. North–south cross section taken through an east–west quasi-stationary front using PRE-STORM high-density soundings, about 2 h prior to MCC formation. (a) Horizontal winds and potential temperatures; (b) v-component (meridional) wind speeds (solid lines) and equivalent potential temperatures (dashed). Shaded regions in (b) indicate southerly wind components in excess of 15 m s^{-1}. Arrow at the base of (a) indicates the location of the surface front. From Trier and Parsons (1993).

level jet overrunning the quasi-stationary front and underrunning a north–south midtropospheric potential temperature gradient. As the system moved through the region, the strongest convective overturning occurred near Pratt, Kansas (station PTT in Fig. 9.12). The effect of the low-level jet on the stability at Pratt, just prior to the arrival of the system, is shown in Fig. 9.13. Note that the low-level θ_e values increased over 20° in 90 min, while midlevel values increased only marginally.

As noted above, an important feature in the typical MCC genesis environment is a surface-based frontal or baroclinic zone (Fig. 9.11). This feature facilitates slantwise ascent of the lowest few hundred millibars of warm sector air. Portions of the overrunning layer eventually saturate, and scattered deep convective overturning ensues over a broad zone, typically several hundred km in the along-front direction (e.g. Figs. 9.14a,b and 9.15a,b). In some instances, the saturated slantwise ascent contributes to the production of the large cold cloud shields characteristic of MCCs (e.g., Fig. 9.15 and Smull and Augustine 1993). These frontal overrunning events will hereafter be referred to as *type-1 events.*

Although the formation of most MCCs occurs in conjunction with frontal lifting, a significant number of events occur in warm sector environments without the benefit of the synoptic-scale frontal forcing (Kane et al. 1987; Johnson et al. 1989; Geldmeier and Barnes 1997). These events rely on the formation of surface-based moist-downdraft-generated cold pools originating from deep convective storms typically rooted in a well-mixed boundary layer to provide the layer lifting and slantwise ascent characteristic of MCCs. These events will hereafter be termed *type-2 events.* As shown by Rotunno et al. (1988), Weisman et al. (1988), and Weisman (1992), convective overturning in type-2 events tends to slope downshear, upshear, or remain upright, depending upon the relative strengths of the low-level vertical wind shear (Δu) and the downdraft-generated cold pool (Fig. 9.16). Typically, these three states of slope represent stages in the life cycle of the convective system, as the cold pool tends to strengthen over time. Significantly, it is not until after a mature cold pool develops that the slantwise front-to-rear ascent, characteristic of the flow that produces much of the MCC cold cloud shield, begins to appear (Fig. 9.17a, b).

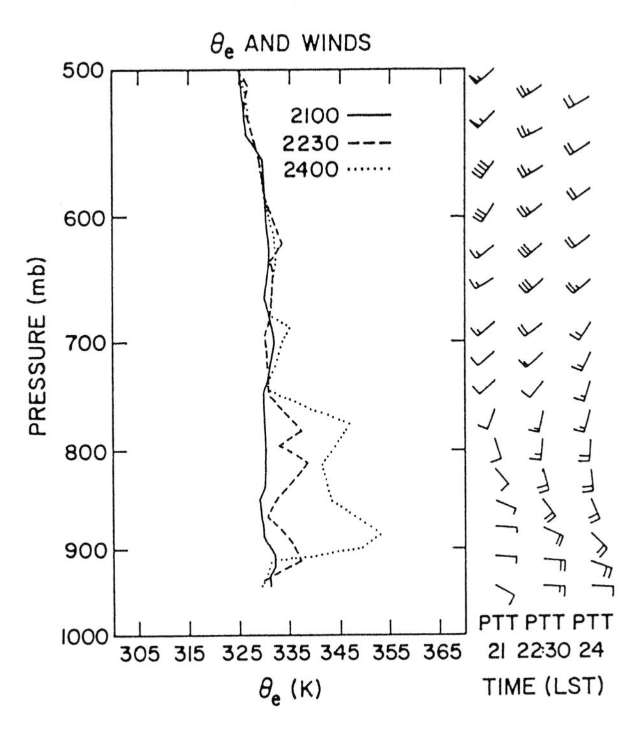

FIG. 9.13. Vertical profiles of equivalent potential temperature θ_e and horizontal winds from soundings at Pratt, KS, at 2100 CST 3 June, 2230 CST 3 June, and 0000 CST 4 June 1985. Winds are plotted with standard meteorological convention; full barb = 5 m s^{-1}. From Trier and Parsons (1993).

Following Weisman (1992), the strength of the cold pool (C) is obtained by integrating the buoyancy (B) over the depth (H) of the cold pool. Thus, the cold pool strength can be expressed as

$$C^2 = 2\int_0^H (-B)dz, \tag{9.2}$$

where

$$B = g\left[\frac{\theta'}{\overline{\theta}} + 0.61(q_v - \overline{q_v}) - q_c - q_r\right], \tag{9.3}$$

θ is the potential temperature, and q_v, q_c, and q_r are the mixing ratios of water vapor, cloud water, and rainwater, respectively. According to Rotunno et al. (1988), the optimal situation for producing long-lived squall lines occurs when the ratio (σ) of the cold pool strength to the shear is equal to one, that is,

$$\sigma = \frac{C}{\Delta u} = 1, \tag{9.4}$$

where the shear is for the layer corresponding to the depth of the cold pool. (Note that this formulation ignores the nonnegligible v component of the shear.) The optimal state is the one that produces the deepest lifting at the leading edge of the cold pool, and thus is the state where the cold pool is most readily able to continually retrigger deep convective cells. In those instances where $\sigma > 1$, the system tilts upshear and forms a slantwise front-to-rear flow resembling the frontal overrunning within typical MCC environments (Figs. 9.16 and 9.17). This slantwise circulation produces a large cold cloud shield reminiscent of that produced by MCCs. Similar structures and circulations were obtained by Fovell and Ogura (1989) in their

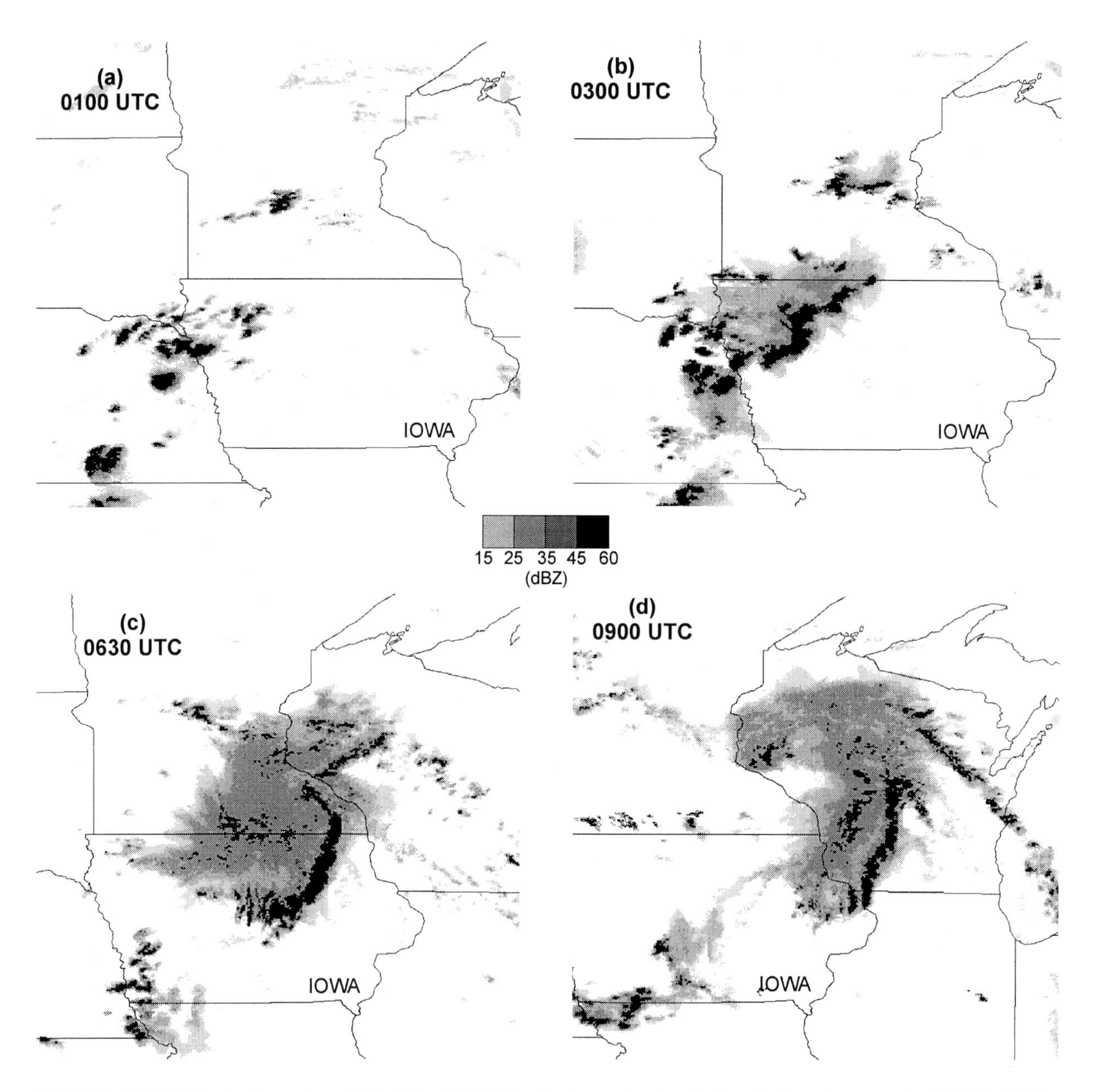

FIG. 9.14. WSR-88D radar observations (reflectivity, dBZ) of the development and evolution of a frontal overrunning MCC. (a) 0100, (b) 0300, (c) 0630, and (d) 0900 UTC 24 June 1998.

two-dimensional simulations of multicell convective storms. Interestingly, the same structures and circulations were also produced by Pandya and Durran (1996) by introducing a steady thermal forcing into a dry simulation. The thermal forcing was scaled and structured to duplicate the latent heat released in a simulated moist convective line, that is, the latent heating and cooling structure that they prescribed already represented the mature upshear-tilted feature that results from the cold-pool circulation overwhelming the ambient shear as evident in Figs. 9.16 and 9.17. Significantly, they did not obtain a realistic flow field when they used a vertically stacked heating and cooling pattern. Nevertheless, their results suggest that the "far field" circulation of squall line/MCC type systems is the result of gravity waves forced primarily by the low-frequency components of the latent heating and cooling in the leading line. Schmidt and Cotton (1990) and Mapes (1993) also present evidence that gravity waves play a key role in shaping the structure and circulation of mesoscale convective systems, even in events where there is no surface-based well-mixed layer. In particular, using numerical simulations, Schmidt and Cotton (1989, 1990) explored a derecho-

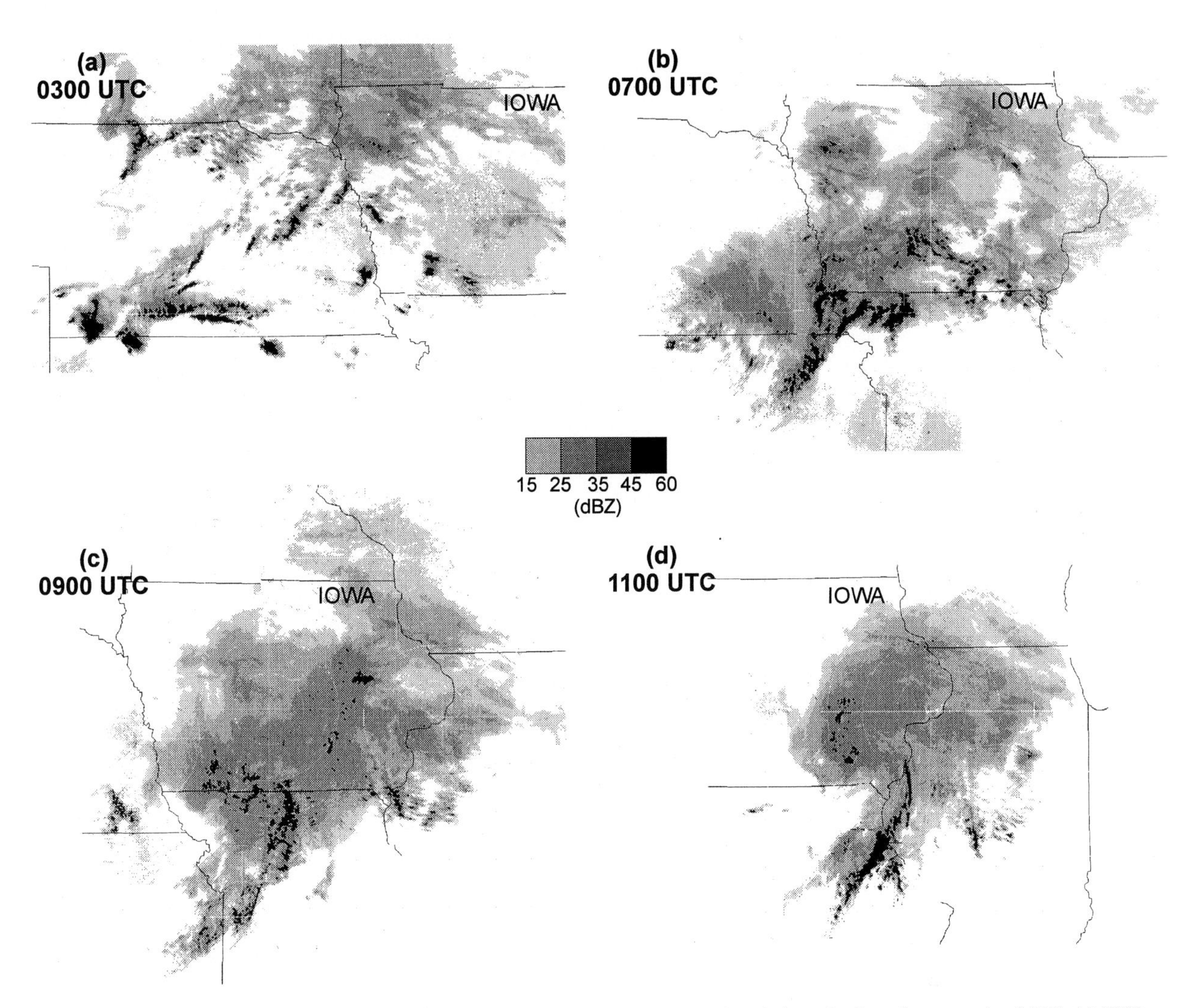

FIG. 9.15. WSR-88D radar observations (reflectivity, dBZ) of the development and evolution of a frontal overrunning MCC. (a) 0300, (b) 0700, (c) 0900, and (d) 1100 UTC 23 May 1996.

like event wherein the environment was characterized by an absolutely stable surface-based layer (generated by previous convection) overrun by a moist well-mixed layer. Their results show how the propagation of the convective system was the result of gravity wave propagation within the surface-based stable layer and that the wave propagation was very sensitive to vertical wind shear.

It is interesting to compare the MCC and squall line environments within the framework of the Rotunno et al. (1988) conceptual model for long-lived convective systems. Recall that, within this framework, the vorticity of the ambient shear opposes and nearly balances the vorticity of a convectively generated surface-based cold pool in order for long-lived squall lines to occur (Fig. 9.18a). Typically, the cold pool is generated locally by two main branches of moist downdrafts: 1) the up–down branch, which initially ascends from the prestorm low levels and then descends within the precipitation region, and 2) the midlevel branch, which originates from above the boundary layer and descends within the precipitation region (see Fig. 9.17; see also Knupp and Cotton 1985; Knupp 1987). The environment of the cold-pool-forced squall line events differs from the type-1 MCC environment in a number of important ways. First, in the type-1 MCC event, the convection originates from an elevated layer of high-θ_e air overrunning a quasi-stationary surface-based frontal zone (Fig. 9.18b). Therefore, a moist-downdraft-generated cold pool is not necessary to lift a layer of conditionally unstable air to saturation. Moreover, analyses of frontal events reveal that the moist downdrafts from the deep convective elements that develop from the frontal overrunning sometimes do not penetrate through the large-scale low-level layer of cold air to the surface (Fortune et al. 1992; Smull and Augustine 1993; Trier and Parsons 1993). Although mesohighs sometimes still form, the cold

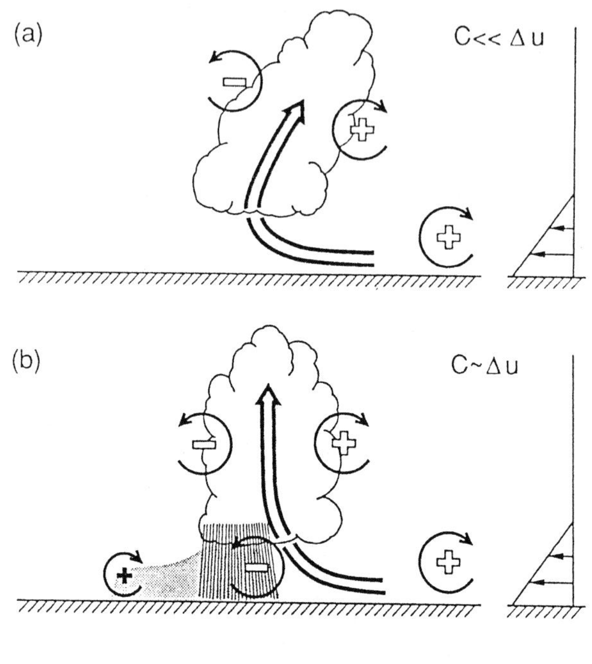

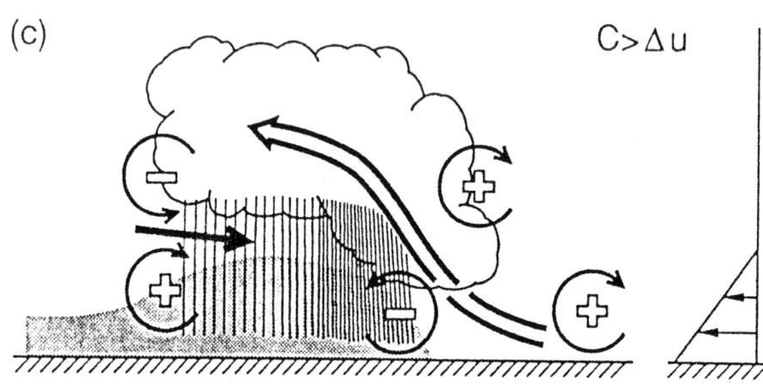

FIG. 9.16. Evolution of storm updraft tilt in an environment with low-level wind shear, shown at right in the diagrams. (a) Updraft (wide double arrow) initially leans downshear in response to ambient vertical wind shear. (b) Circulation generated by storm-induced cold pool (shaded) balances the ambient shear, and updraft becomes upright. (c) Cold-pool circulation (C) overwhelms the ambient shear (Δu) and updraft tilts upshear. Thick solid arrow in (c) indicates rear-inflow jet. Cloud edge is scalloped; rainfall intensity is depicted by density of vertical lines. Thin circling arrows with + or − symbols indicate circulations affiliated with ambient wind shear and cold pool. From Weisman (1992).

pool strength (as indicated by the temperature difference between the moist downdraft air and the ambient environment) is much weaker than in type-2 events. This is especially true for the portion of the cold pool associated with the mesoscale rain areas (see Loehrer and Johnson 1995; Kain and Fritsch 1998). Therefore, the concept of ambient vorticity (associated with vertical shear) opposing buoyancy-generated cold pool vorticity is not as readily applicable as in some warm sector (type 2) environments.

A second important difference between the typical MCC environment and the squall line conceptual model is the configuration of vertical wind shear relative to the orientation and direction of propagation of the deep convection. In the squall line model, the shear-induced vorticity opposes the cold-pool-induced vorticity (Fig. 9.18a). However, in the MCC environment, as a consequence of the low-level jet, the ambient shear in the cold-pool layer has the same sign as that which would be expected from any convectively generated cold pool that would develop (Fig. 9.18c). According to Rotunno et al. (1988) and Weisman (1992), this arrangement of shear and cold-pool-induced vorticity would tend to produce slantwise upshear ascent and a large stratiform cloud layer. An additional but as yet unexplored factor is the effect of the stronger veering in the more baroclinic type 1 events compared to the weaker directional shear in the type-2 events.

The presence of the strong warm advection characteristic of MCC environments in general suggests that, in addition to the interaction of the convectively generated cold pool with the ambient shear, the large-scale environment itself can contribute significantly to the slantwise ascent that comprises the stratiform cloud region. Specifically, the warm advection implies, and observations confirm (e.g., Maddox 1983; Augustine and Howard 1991; Smull and Augustine 1993; Trier and Parsons 1993; Augustine and Caracena 1994), that there is sloping isentropic lift as an inherent part of the MCC environment. Thus, considering the sign of the environmental shear and the presence of the sloping isentropic lift, the type-1 MCC environment is conducive to producing convection wherein large stratiform anvils are favored. This, of course, is characteristic of MCCs and helps to distinguish them from narrow convective lines and from mesoscale convective clusters composed of clearly discernible individual elements.

In addition to the differences between the MCC and squall line environments, there is also an important difference in how their mesoscale structures evolve. In squall lines, a line of convection typically forms first, and then the slantwise ascent and associated stratiform region develop as the cold pool strengthens. As the system matures, its northern flank tends to evolve into a larger-scale balanced mesovortex from which the convective line emanates. The line then appears as a component of an overall larger-scale system (Skamarock et al. 1994; Davis and Weisman 1994). On the other hand, in some MCCs, convection initially appears as a disorganized agglomeration of meso-γ-scale and meso-β-scale elements (see Figs. 9.14a,b and 9.15a,b). These elements are the first convective features to appear within a broad area of larger-scale forcing, frequently associated with frontal overrunning. The disorganized convection is sometimes embedded within a broad but irregularly shaped stratiform region (Fig. 9.15). With continued forcing by the larger scale, the intervening cloud-free regions fill in and the entire system gradually amalgamates into a more coherent and organized contiguous structure with cyclonic rotation (Figs. 9.14c, 9.15c). As the system organizes and the cyclonic circulation emerges, the convection organizes into a highly coherent line, typically arcing southward from the center of rotation (Figs. 9.14c,d and 9.15d). The contiguous system with its midlevel mesovortex and convective line trailing southward then resembles the commonly observed comma-shaped cloud. Note in Fig. 9.15 how the appearance of a well-defined convective line *follows* the development of the mesovortex. This is opposite to the scenario simulated by Davis and Weisman (1994), wherein the formation of the vortex occurs after the appearance of the line.

It was noted in section 9.2 that linearly organized systems (squall lines) produce about twice the number of severe weather events as MCCs (Tollerud and Bartels 1988). In view of the fact that the intense convective overturning with type-1 MCCs often has its roots in an elevated layer overrunning colder, more stable air, and that the moist downdrafts have difficulty penetrating this cold layer to reach the surface, it is clear that a sizable fraction of severe events related to moist downdrafts may be eliminated due to the protective cold layer. This is especially true of supercell thunderstorms wherein the cold downdraft outflow plays an important role in the formation of the tornado-bearing mesocyclone (Rotunno and Klemp 1985; Klemp 1987). It is also true for damaging microburst winds, since the downdraft and outflow speeds are proportional to the negative buoyancy in the subcloud layer (Foster 1958; Rose 1996; Wakimoto 1999). This buoyancy is much reduced or even eliminated by the presence of the low-level cold air. Note, however, that this argument does not apply to certain derecho environments wherein convectively generated pressure gradients associated with gravity waves provide an alternative mechanism for accelerating low-level winds over meso-β-scale regions (Johns and Hirt 1987; Johns 1993; Schmidt and Cotton 1989).

A similar "protective" mechanism is operative during the latter stages of type-2 MCCs. Recall that type-2 MCCs form in environments characterized by surface-based well-mixed boundary layers and that the development of the large cold cloud shield characteristic of MCCs is intimately tied to the lifting forced by the moist-downdraft-generated cold pool. Once a well-developed cold pool is established, subsequent convective overturning occurs as the well-mixed layer overruns the cold pool. Thus, as the system matures, it becomes progressively more two-dimensional, and it is therefore more difficult to produce the individual well-organized three-dimensional cloud-scale circulations such as those associated with supercells. It also becomes more difficult for new moist downdrafts to accelerate all the way to the surface since they must penetrate the existing cold pool. Thus, the propensity for mesocylone/tornado formation and for strong winds from cloud-scale penetrative moist downdrafts is reduced. This agrees with the Maddox et al. (1986), Houze et al. (1990), and Tollerud and Collander (1993a) empirical results showing that severe weather is strongly biased toward the initiation phase. Of course, other mitigating factors are at work during the latter stages of the life cycle, principally the reduction in CAPE as the boundary layer cools. Also, it is important to note that, in some instances, even though the overall magnitudes of the moist downdrafts may be weaker, such downdrafts may still contribute to and enhance the surface outflow as long as midlevel low θ_e air is reaching the surface and as long as that air is substantially colder than the ambient air ahead of the outflow. This is because these downdrafts contribute to the strength of the mesoscale outflow (as opposed to that from individual thunderstorms) associated with bow-echo or derecho type events (see chapter 3).

9.4. Structure

Houze et al. (1990) used radar data to examine the internal structure of 55 springtime MCSs (some of which were MCCs) that were responsible for major rain events in Oklahoma. They found that, during their mature phase, virtually all MCSs are composed of deep convective elements and an associated region of stratiform rain.[1] They stressed that the convective and stratiform rain areas comprise a continuous spectrum of mesoscale structures. They also noted that, within most systems, there is a tendency for the strongest convective elements to be organized into a line, and that the line is followed by a region of stratiform rain. Figure 9.19 shows what Houze et al. termed *symmetric* and *asymmetric* forms of these line-type systems. Based on the work of Skamarock et al. (1994), it is now recognized that the symmetric and, asymmetric forms represent sequential stages in the life cycle of line-type MCSs. Moreover, Skamarock et al. also demonstrated that northern, cyclonic and, sometimes, southern, anticyclonic midlevel vortices develop at the line ends as a consequence of the finite size of a line-type convective system. The vortices tend to be of relatively equal strength during the early symmetric phase of the system and then evolve to a dominant cyclonic vortex during the later asymmetric phase. The evolution to an asymmetric structure is the result of

[1] According to Yuter and Houze (1995), stratiform conditions are present when

$$|w| \ll |V_{ice}|,$$

where w is the vertical air velocity, and V_{ice} is the terminal fall speed of precipitation-sized ice crystals and snow particles (~ 1–3 m s^{-1}). Convective conditions are present when

$$|w| \gg |V_{ice}|.$$

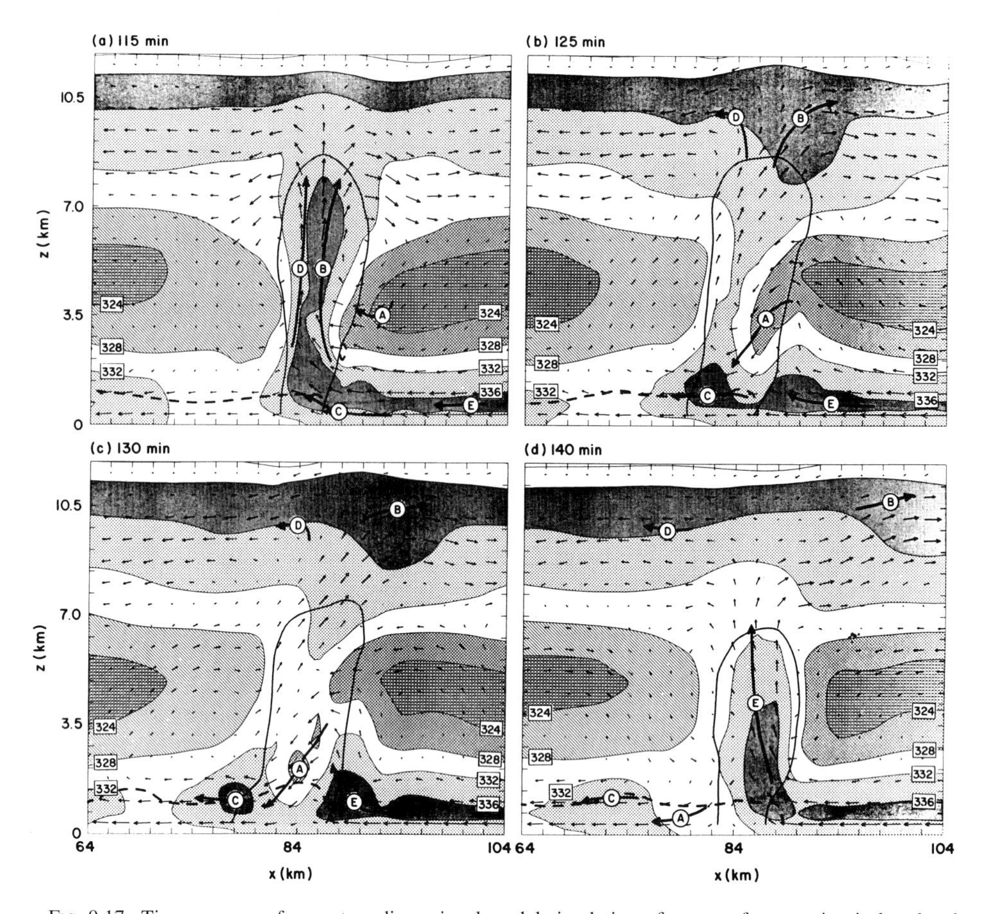

FIG. 9.17. Time sequence from a two-dimensional model simulation of stages of convection in low-level shear. Contours and shading denote equivalent potential temperature θ_e. Vectors are scaled such that one grid interval equals 16 m s^{-1}. Thick dashed line represents -1 K perturbation potential temperature. Solid line is the 2 g kg^{-1} rainwater contour. Letters indicate parcel trajectories, marked at the center of forward and backward calculations where the arrow shaft extends upwind and downwind of the letter. (a)–(d): oscillating cell development during transition between stages (a) and (b) of Fig. 9.16, with updraft becoming upright. Forward and backward trajectory calculations are each five minutes in duration. (a) At 115 min, updraft is fully developed, high-θ_e air is transported upward. Updraft and rainwater fields lean downshear. Parcels D and B originated on the front side earlier. (b) At 125 min, downshear-leaning rainwater field evaporates into mid- and low-level air on the front side; parcel C flows through the rain; parcel A descends from the front side. (c) At 130 min, front-side air from mid- and low levels contributes to the cold pool. (d) At 140 min, a new cell is triggered (parcel E). (e)–(h): transition from stage (b) to (c) of Fig. 9.16, as cold pool surges downshear and updraft tilts upshear; forward and backward trajectory calculations each 15 minutes in duration. Parcels are not the same ones as in (a)–(d). (e) At 240 min, cold pool begins to dominate the flow (parcel C originated from low levels downshear). (f) At 270 min, rain evaporates into rear-side midlevel air and induces its descent (parcels A and F). (g) At 300 min, rear-side midlevel air pours to the ground. (h) At 330 min, cold pool continues to surge downshear as the system weakens. Note that the cold pool contains both rear- and front-side parcels. From Rotunno et al. (1988).

Coriolis influences, which favor the development of the northern cyclonic vortex and, as a result of system-scale midlevel convergence, weaken the southern anticyclonic vortex. The relative low pressure associated with the balanced midlevel cyclonic circulation results primarily as a hydrostatic consequence of a positive temperature anomaly above it. The warm temperature anomaly is composed largely of high θ_e air that ascended through the convective portion of the system.

Also shown in Fig. 9.19 are the typical mesoscale surface pressure features observed with MCSs that exhibit line-type organization (Zhang and Gao 1989; Loehrer and Johnson 1995). Radar-derived cross sections through such convective systems generally show the leading convective line followed by slantwise front-to-rear ascent and rear-to-front descent (Fig. 9.20). A mesohigh usually appears in the rain-cooled downdraft air, and a wake low, formed by adiabatic

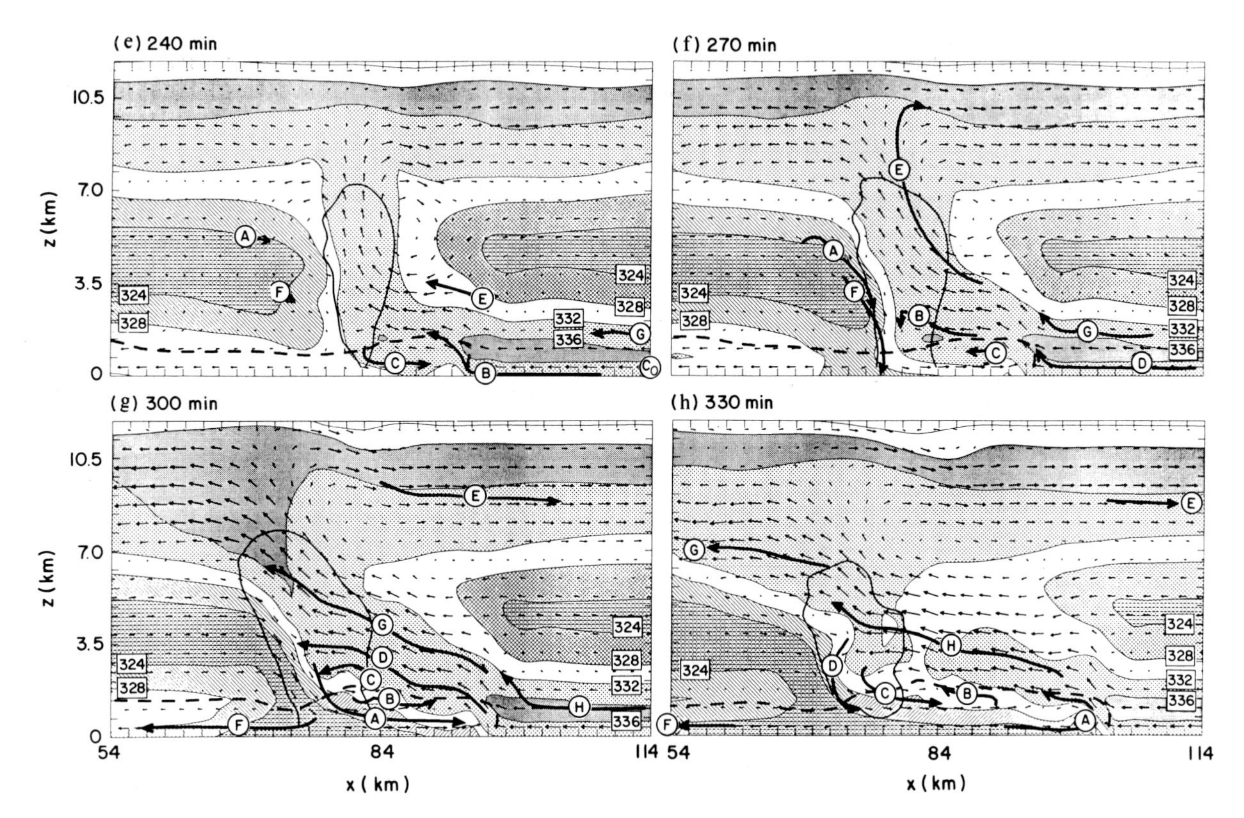

FIG. 9.17. (*continued*)

compressional warming, typically develops at the rear edge of the stratiform precipitation (see Johnson and Hamilton 1988). This conceptual model of an MCS resembles the squall line depictions constructed by Newton (1950), Brunk (1953), Fujita (1955), Pedgley (1962), and Zipser (1977). Similar structures were observed in African and Australian MCSs (Chong et al. 1987; Mapes and Houze 1992; Keenan and Carbone 1992) and in simulations by Pandya and Durran (1996), wherein gravity waves as a result of the low-frequency components of prescribed steady thermal forcing in the leading line generated the observed structures.

Although the radar-observed internal structure of some MCCs resembles that of line-type structures, Maddox (1980) pointed out that the evolution of surface weather events during MCC passage often differs significantly from that typically observed with squall line passage. In particular, he noted that the temperature, dewpoint, and wind changes tend to be more precipitous at the onset of squall line passage compared to that of an MCC. He also emphasized that the period of steady rain and rain showers associated with MCCs persisted much longer than that commonly observed with squall lines. These observations are consistent with the large cold cloud shields of MCCs and the fact that they often develop on the cool side of a surface front (see section 9.3; see also Trier and Parsons 1993). Without strong moist downdrafts penetrating to the surface, the mechanisms whereby lines form as a result of the strong lifting along sharply defined outflow boundaries (Purdom 1976; Weisman et al. 1988) are considerably weakened or eliminated. Correspondingly, Houze et al. (1990) found that the radar echoes with MCCs did not favor any particular type of organization. Rather, there was a tendency for the convective elements to exhibit a mixture of various sizes, shapes, and orientations (e.g., Fig. 9.21). Similarly, other investigators have documented numerous examples of more or less random distributions of the convective elements within MCCs, especially for type-1 events (e.g., McAnelly and Cotton 1986, 1989; Fortune et al. 1992; Trier and Parsons 1993; Rochette and Moore 1996). It is important to note, however, that early in the life cycle of some type-1 MCCs the convective activity appears disorganized but then evolves into a well-defined wavelike structure (Figs. 9.14 and 9.15). Fortune et al. (1992) and Smull and Augustine (1993) documented a type-1 system that produced two convective lines and a large stratiform region (Fig. 9.22a). The north–south line in Fig. 9.22a was perpendicular to a quasi-stationary front that stretched across northern Oklahoma. The extreme southern portion of the north–south line extended into the warm sector and exhibited the strongest convection

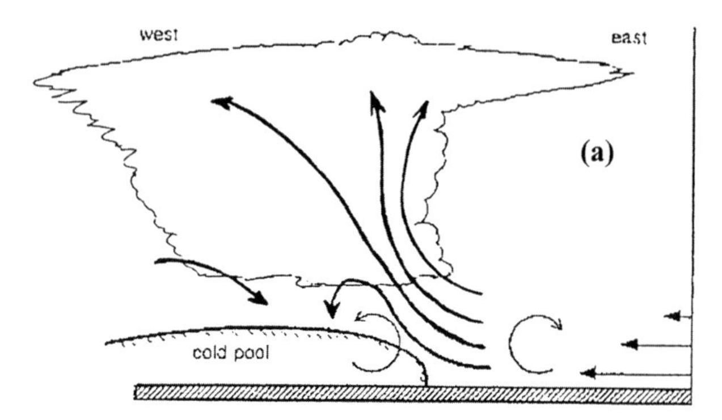

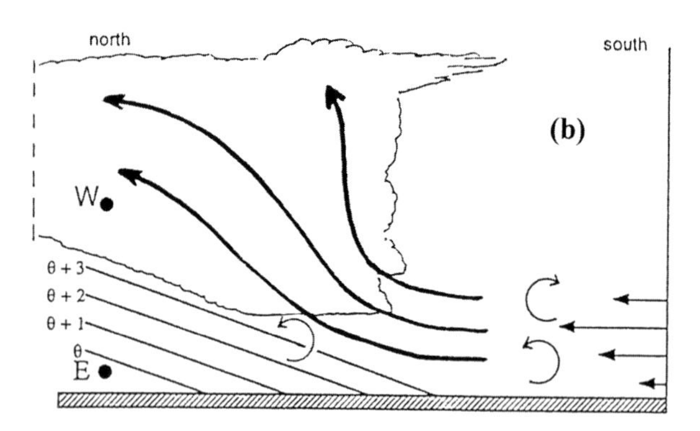

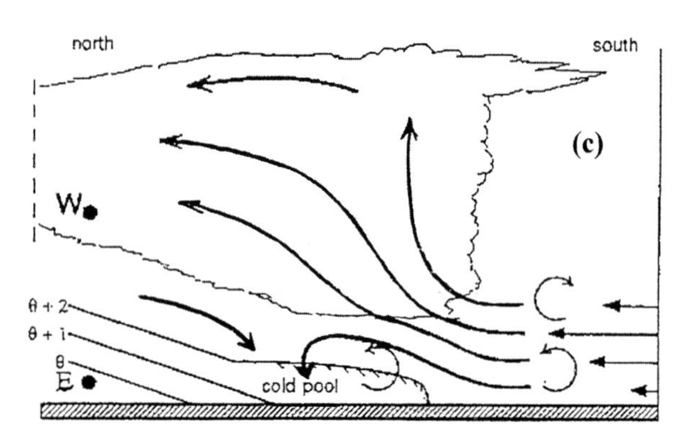

FIG. 9.18. Convective-mesoscale feedback processes in contrasting environments. (a) Low-level wind shear as in Figs. 9.16 and 9.17 supports vertical circulation. (b) Low-level jet overruns a frontal zone. The low-level shear is opposite that in (a). Moist downdrafts do not penetrate the surface-based synoptic-scale cold air so a mesoscale cold pool does not develop. (c) Cold pool forms on cool side of frontal boundary where synoptic-scale cold air is shallow; low-level shear vorticity is the same sign as buoyancy-induced vorticity and favors rapid development of stratiform region. Bold lines with arrowheads depict the flow in the plane of the figure. The letters W and E indicate the presence of westerly and easterly wind components, respectively.

with the system. Nevertheless, intense convective overturning also developed in the overrunning layer above the cool sector. Cross sections perpendicular to the two lines (Fig. 9.22b,c) show that both lines produced updraft velocities in excess of 15 m s^{-1}; however, there was no low-level gust front or convective downdraft exceeding 2 m s^{-1} anywhere along the east–west line, despite the presence of pronounced precipitation. On the other hand, downdrafts were stronger in the north–south line and a gust front materialized along the southernmost portion of the line, which had extended into the warm sector. Thus, it is clear that both cloud and mesoscale processes play important roles in organizing MCCs, but how these processes depend upon the large-scale environment is not well understood.

Since numerical experiments have shown that MCS-induced long-lived balanced dynamics (Bartels et al. 1997) do not prevail until after the appearance of an extensive stratiform cloud (e.g., Zhang and Fritsch 1988; Weisman 1992; Skamarock et al. 1994; Davis and Weisman 1994), and since Houze et al. (1990) found that no particular configuration of radar echoes was favored within MCSs and MCCs, it appears that the spatial arrangement of convective elements (e.g., whether or not there is a line) may not be a crucial factor in establishing the formation of a long-lived balanced system. Rather, the processes that determine the shape, extent, and duration of the stratiform region may be more important. In particular, since the Rossby radius is directly a function of the static stability (Schubert et al. 1980), the transformation of a dry statically stable environment into a near-neutral moist static stability condition within the stratiform region decreases the Rossby radius for the convective system and thereby enhances the likelihood that a balanced long-lived mesoscale system will emerge (Chen and Frank 1993).

Although convective elements in MCCs often do not exhibit a preferred organization, case studies of the stratiform region frequently display similar reflectivity and circulation patterns. In particular, as the MCC cold cloud shield matures, cyclonically spiraling bands often appear in the stratiform region (Fig. 9.23; see Smull and Houze 1985; Leary and Rappaport 1987; Brandes 1990; Bartels and Maddox 1991; Fritsch et al. 1994; Bartels et al. 1997). Doppler wind analyses of the stratiform rain region (e.g., Brandes 1990) indicate that development of these bands is a reflection of the formation of a midlevel vortex, one of the characteristic signatures of how MCCs modify their mesoscale environment. Similar circulations have developed in the stratiform rain areas of numerical model simulations of MCCs (e.g., Zhang and Fritsch 1987, 1988a). The tendency of MCSs to strongly modify their environment is discussed in section 9.6.

9.5. Movement

The movement of convective systems is composed of the sum of an advective component, given by the mean motion of the convective cells, and a propaga-

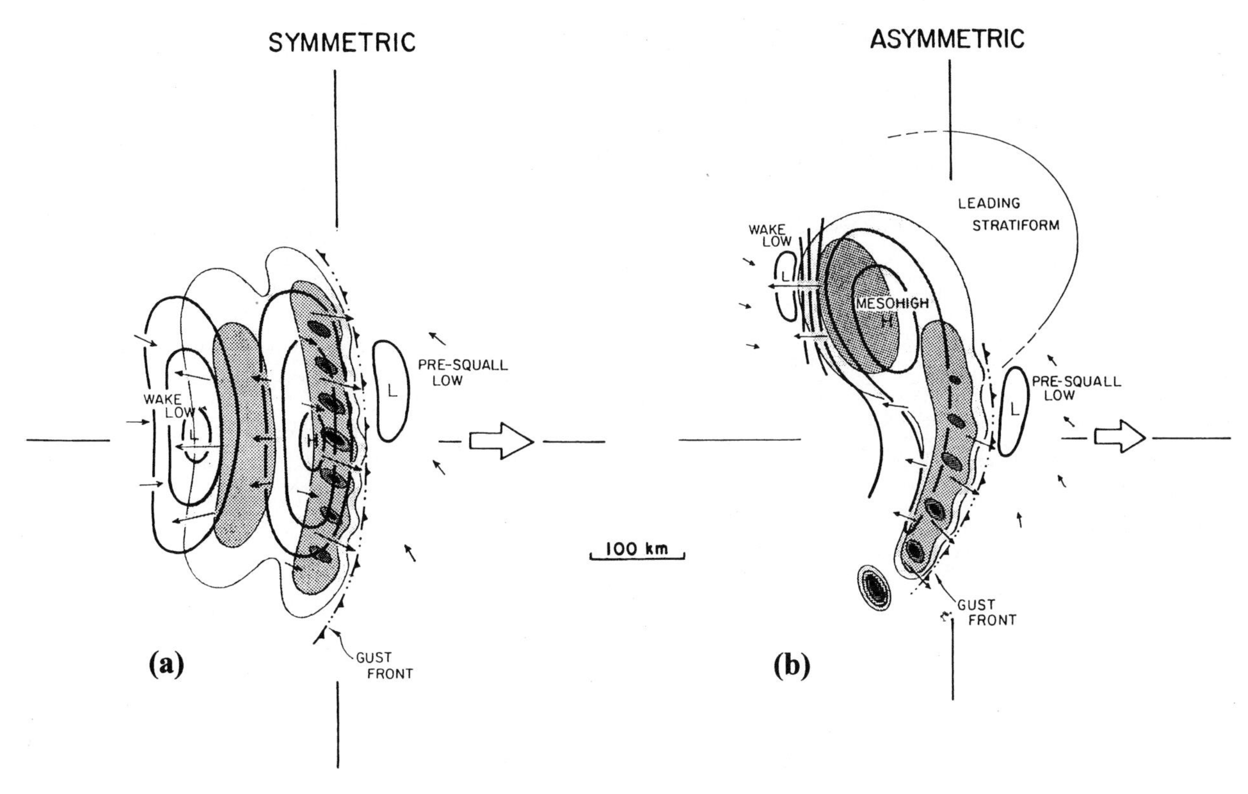

FIG. 9.19. Conceptual model of the surface pressure, flow, and precipitation fields associated with (a) symmetric and (b) asymmetric stages of the MCS life cycle. Levels of shading denote increasing radar reflectivity, with darkest shading corresponding to convective cell cores (adapted from Houze et al. 1990). Pressure is in 1-mb increments. Small arrows represent the surface flow. Length of the arrow is proportional to the wind speed found at its center. The large arrows represent the storm motion. From Loehrer and Johnson (1995).

tion component, defined by the rate and location of new convective cell formation relative to existing cells (Newton and Katz 1958; Newton and Newton 1959; Bluestein and Jain 1985). Merritt and Fritsch (1984) and Corfidi et al. (1996) applied this concept to the movement of the heavy rain producing meso-β-scale elements (MBEs) of mesoscale convective complexes. They examined 103 MCSs (of which 99 satisfied the criteria for an MCC) and found, as did the early radar studies of Brooks (1946) and Byers and Braham (1949), that individual cells within MCSs tended to be swept downwind with the mean flow in the cloud layer

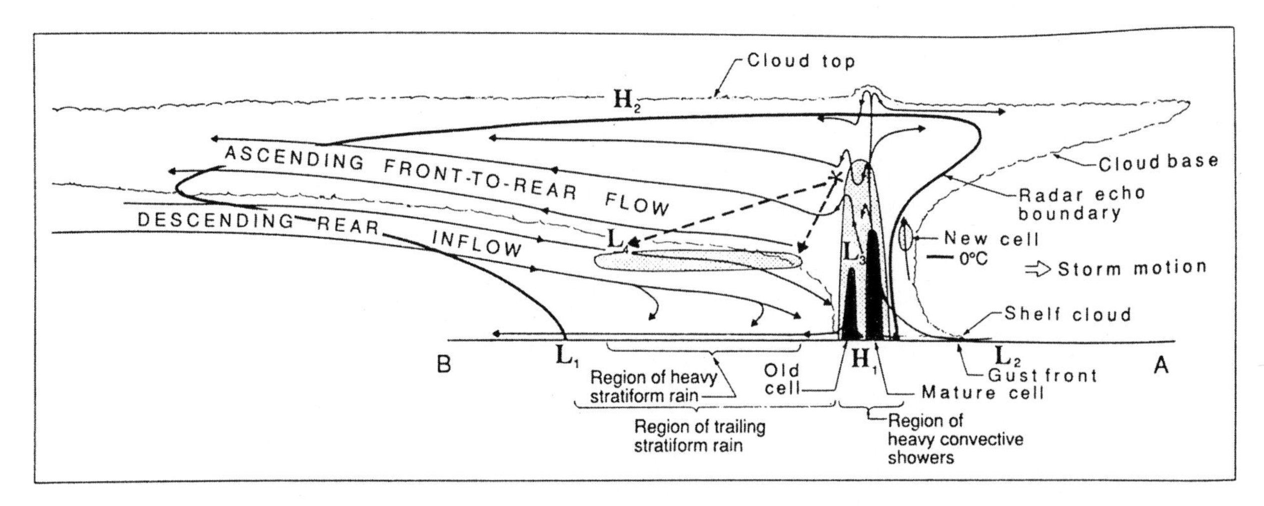

FIG. 9.20. Conceptual model of a squall line with a trailing stratiform region, viewed in a vertical cross section oriented perpendicular to the convective line and parallel to the motion of the line. From Houze et al. (1989).

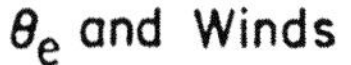

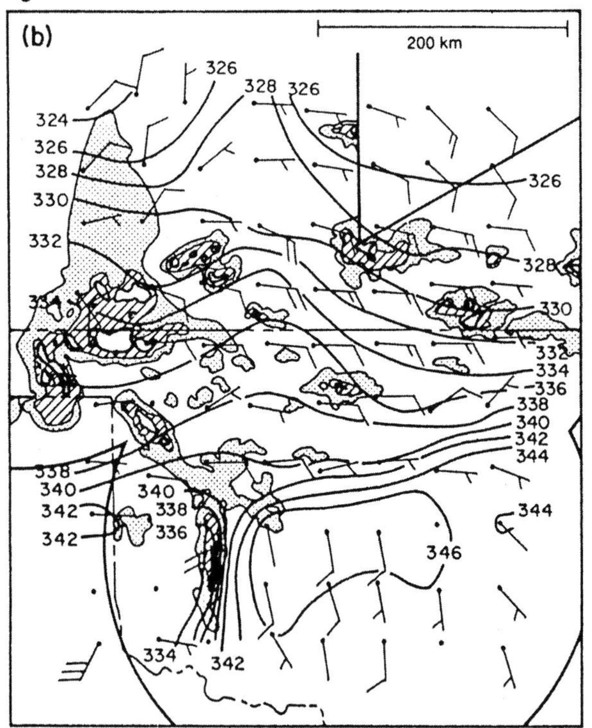

FIG. 9.21. Mesoanalysis of surface equivalent potential temperature θ_e at 0100 CST 4 June 1985. Shading indicates radar reflectivity at 15-, 30-, and 40-dBZ thresholds, from a RADAP-II composite (0.5° elevation) of the Oklahoma City, OK (OKC); Wichita, KS (ICT); and Garden City, KS (GCK) National Weather Service (NWS) WSR-57 radars. Bold solid lines indicate borders of radar coverage. Surface winds are plotted with standard meteorological convention. From Trier and Parsons (1993).

($\mathbf{V}_{CL}$). However, the MBEs tended to move at roughly half the speed of the supporting meso-α-scale system (McAnnelly and Cotton 1986, 1989) and often departed significantly from the direction of the mean flow. Corfidi et al. (1996) defined the departure of the observed movement of the MBEs ($\mathbf{V}_{MBE}$) from the mean cloud-layer flow as the propagation component ($\mathbf{V}_{PROP}$) of the MBEs (Fig. 9.24). Further examination of the 103 MCSs revealed that $\mathbf{V}_{PROP}$ tended to be directed opposite to the low-level flow supplying the system with high-θ_e air. Usually this inflow was dominated by a low-level jet ($\mathbf{V}_{LLJ}$). Thus, using these relationships, they constructed a simple model for predicting the movement of the heavy rain areas of MCSs. This model is presented in Fig. 9.24 and is mathematically expressed as

$$\mathbf{V}_{MBE} = \mathbf{V}_{CL} - \mathbf{V}_{LLJ}. \tag{9.5}$$

Following Fankhauser (1964), they calculated the mean flow in the cloud layer from

$$\mathbf{V}_{CL} = \frac{\mathbf{V}_{850} + \mathbf{V}_{700} + \mathbf{V}_{500} + \mathbf{V}_{300}}{4}, \tag{9.6}$$

where the direction and magnitude of the wind at each level on the right-hand side is taken to be representative of the 900–800-mb, 800–600-mb, 600–400-mb, and 400–200-mb layers, respectively. Estimates of the speed and direction of the low-level jet are obtained following the criteria of Bonner (1968). Only wind maxima at or below 1.5 km above ground level are considered low-level jets, since the thunderstorm cells of developing systems are most likely to ingest air from near the surface.

Once estimates for $\mathbf{V}_{CL}$ and $\mathbf{V}_{LLJ}$ are in hand, the magnitude of $\mathbf{V}_{MBE}$ is determined using the following simple geometric relationship:

$$|\mathbf{V}_{MBE}| = \{|\mathbf{V}_{CL}|^2 + |-\mathbf{V}_{LLJ}|^2 - 2(|\mathbf{V}_{CL}| \times |-\mathbf{V}_{LLJ}|) \cos \phi\}^{1/2}, \tag{9.7}$$

where ϕ is the angle between the mean cloud-layer wind and the low-level jet (see Fig. 9.24). Similarly, MBE direction can be determined using an elementary relationship for the angle ψ between the direction of MBE movement and the mean cloud-layer wind,

$$\psi = \arcsin\left(\frac{|-\mathbf{V}_{LLJ}| \sin \phi}{|\mathbf{V}_{MBE}|}\right), \tag{9.8}$$

where MBE speed is given by Eq. (9.7). Corfidi et al. (1996) used soundings taken during the initiation and development phases of the systems to test the conceptual model. Their results, shown in Fig. 9.25, indicate considerable skill in forecasting MBE movement. For example, they calculated that, for the typical MCC in the United States, the average error in the location of the center of the MBE activity at the time the system dissipates would be about 100–150 km. Moreover, for most of the system's life cycle, the typical location error would be less than 100 km, which is well within the approximately 300-km width of the heavy rain (> 13 mm) area of most MCCs (see Fig. 9.9).

Chappell (1986) used the same conceptual model proposed by Merritt and Fritsch (1984) to argue that certain configurations of the environmental flow favor the occurrence of very heavy rains and flash floods. In particular, he recognized from their model that when the low-level jet is similar in magnitude and direction to the mean flow in the cloud layer, the propagation component of the MBEs will be directed opposite to the movement of individual cells. Thus, systems will tend to remain nearly stationary or even propagate upstream (backbuild), especially if the low-level jet is stronger than the mean flow in the cloud

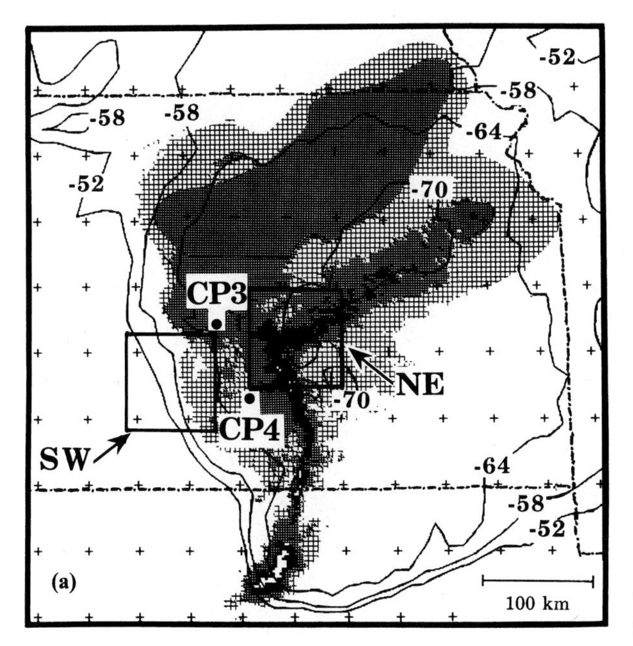

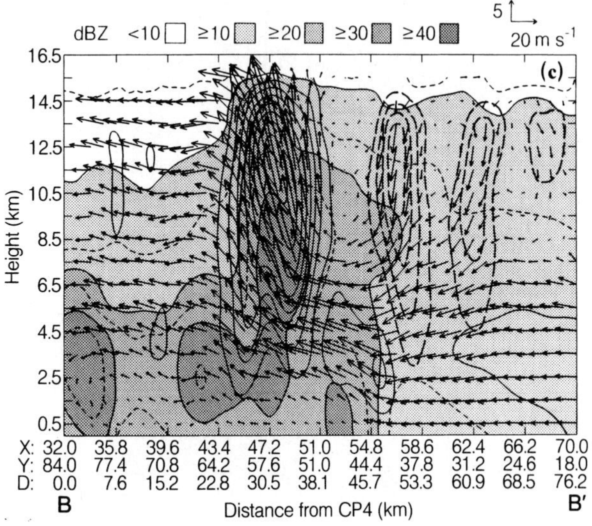

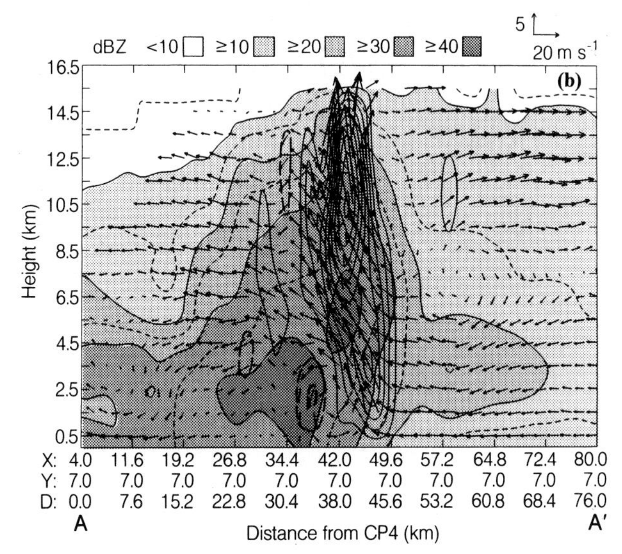

FIG. 9.22. Radar depiction of a frontal overrunning MCC at 0107 UTC 4 June 1985 over Kansas. (a) Near-surface radar reflectivity is shown in shading from radars at locations labelled CP-2 and CP-3 and supplemented by data from adjacent NWS radars. Cloud-top temperatures are contoured, derived from 0100 UTC satellite data. (b) East–west vertical cross section across the north–south convective line in the NE domain of (a). Shading denotes reflectivity at 10-dBZ intervals; vectors represent system-relative flow (m s^{-1}) in plane of section. Heavy contours denote vertical airspeed w (at 2 m s^{-1} intervals beginning at +/− 2 m s^{-1}; dashed negative). The X and Y values indicate distances east–west and north–south from CP-4 radar; D indicates distance from left edge of diagram. (c) As in (b), except across the northeast–southwest convective band in region NE of (a). From Smull and Augustine (1993).

layer (Fig. 9.24). Shi and Scofield (1987) further refined Merritt and Fritsch's conceptual model for backbuilding systems by introducing the configuration of the thermodynamic fields (Fig. 9.26). Specifically, they examined nine backward-propagating MCSs and found that eight of them propagated along and north of the axis of the maximum mean value of θ_e in the cloud layer (850–300 mb). Since the deep-tropospheric axis of maximum θ_e is likely dominated by the highest θ_e values at low levels, their results simply imply that the MCSs tend to propagate toward the highest θ_e air feeding the convective updrafts. It is well known that situations with slow moving, quasi-stationary, or backbuilding systems where individual cells form and "train" over the same region are a favored mechanism for the production of locally heavy rainfall (Simpson and Woodley 1971; Maddox et al. 1979; Chappell 1986; Shi and Scofield 1987).

9.6. Modification of the regional environment

As noted in section 9.1, MCCs are sufficiently large and long-lived that the earth's rotation becomes a factor in determining their dynamical structure and circulation. As such, their structure and circulation should tend to follow the balanced dynamical constraints derivable from potential vorticity concepts. According to Haynes and McIntyre (1987), the mass-integrated potential vorticity in a layer between two isentropic surfaces remains constant even if mass is transported into or out of the layer. Therefore, removal (addition) of mass from (to) a layer will necessitate an

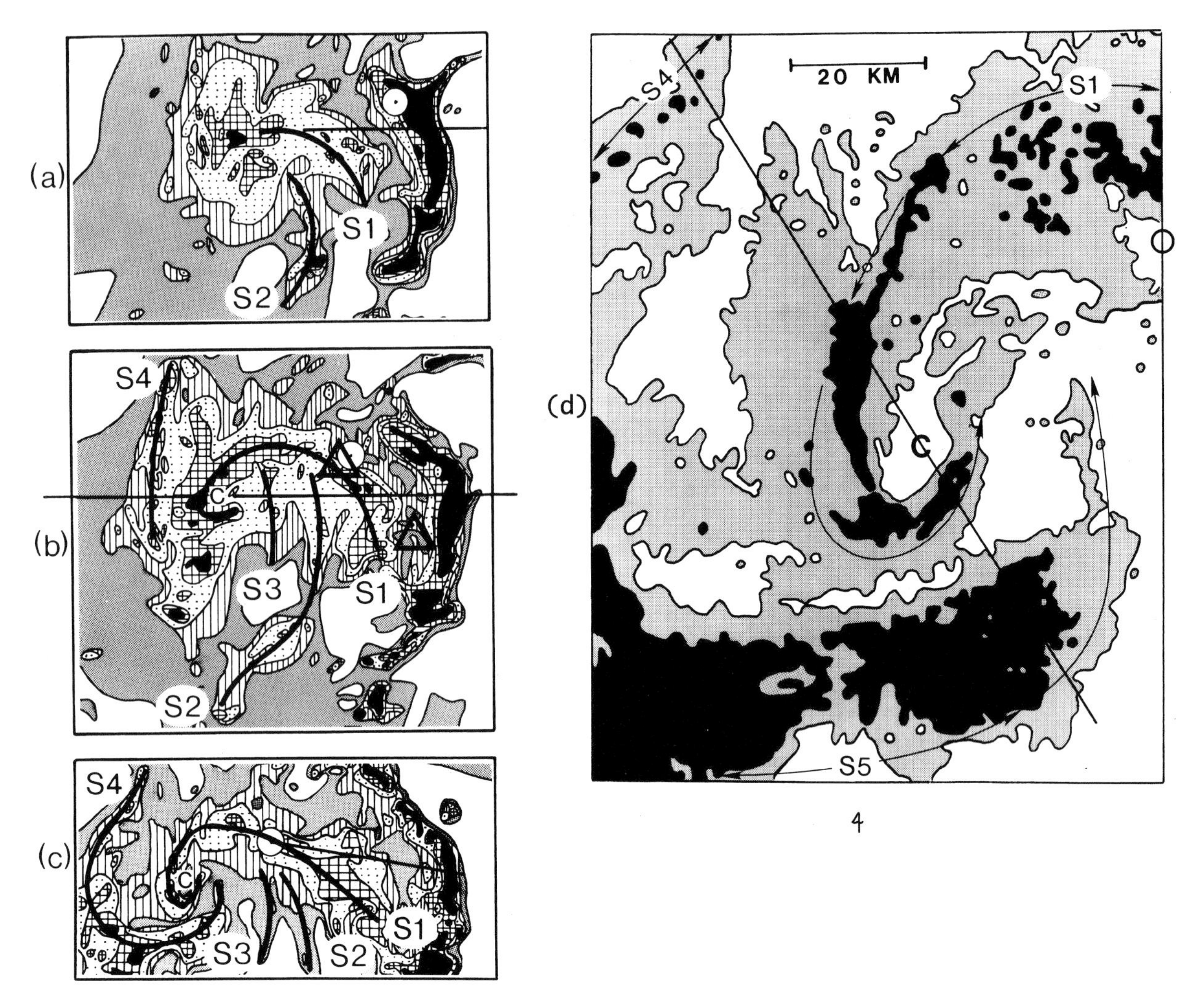

FIG. 9.23. Spiraling rainbands in the stratiform region of an MCC. (a) 1500, (b) 1600, (c) 1700, (d) 1710 CDT 8 June 1980 in Texas. Reflectivity thresholds are at 11, 20, 24, 28, and 32 dBZ in (a)–(c), and at 19 and 26 dBZ in (d). Panels extend 275 km from west to east in (a)–(c). Lines and labels denote various rainbands. From Leary and Rappaport (1987).

increase (decrease) in the potential vorticity. Raymond and Jiang (1990) recognized this relationship and noted that moist convection transports mass across isentropic surfaces from low levels to high levels. Thus, they reasoned that the massive moist convective transports by MCSs should have a profound effect on their large-scale environment and result in a low-level positive potential vorticity anomaly with a negative potential vorticity anomaly above. This notion can be refined somewhat by recognizing that much of the air that ascends moist adiabatically to high levels in MCCs originates not just from the low-level surface-based mixed layer but from a thick midlevel layer typically extending from the surface-based mixed layer to approximately 6 km above ground level (see, e.g., Ogura and Liou 1980; Smull and Houze 1985, 1987; Rotunno et al. 1988; Smull and Augustine 1993). Moreover, there is also a substantial moist adiabatic transport of mass from midlevels to low levels by the cloud and mesoscale moist downdrafts (e.g., Smull and Houze 1987; Fovell and Ogura 1988; Johnson and Hamilton 1988; Johnson et al. 1989; Zhang et al. 1989; Brandes and Ziegler 1993). Thus, the origination layer for most air parcels that participate in moist overturning in MCSs can be most accurately characterized as coming more from midlevels (considered here to be the layer between 2 and 6 km) than from low levels. Correspondingly, high levels where moist updrafts detrain would be predominantly mass deposition layers. Low levels act as both a source (for updrafts) and a sink (for downdrafts). Therefore, according to Raymond and Jiang's reasoning, a negative potential vorticity anomaly should develop at high levels and a positive anomaly, underlaid by relatively cold air, should form at midlevels.

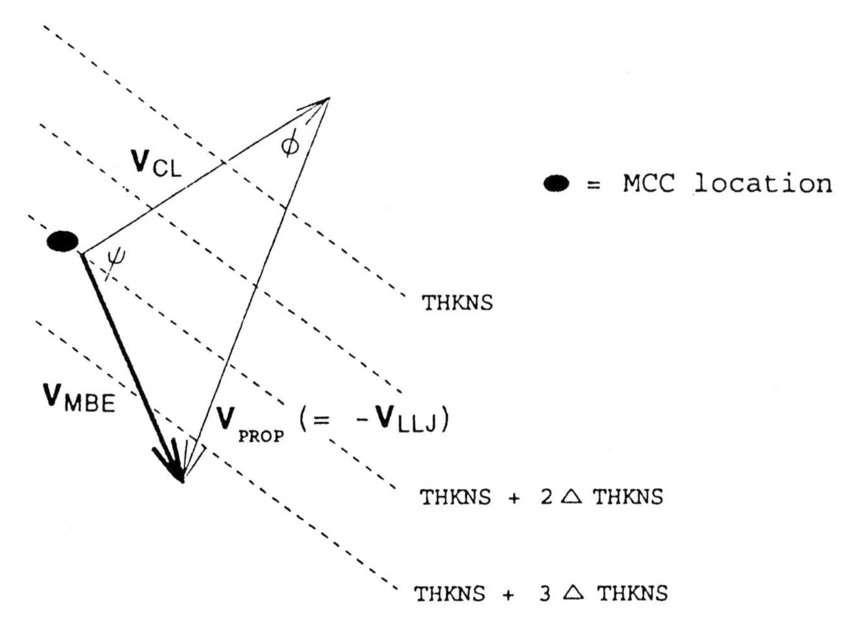

FIG. 9.24. Conceptual model of the movement ($\mathbf{V}_{MBE}$) of a meso-β-scale element (MBE) within an MCC as the vector sum of the mean flow in the cloud layer ($\mathbf{V}_{CL}$) and the propagation component ($\mathbf{V}_{PROP}$). The magnitude and direction of $\mathbf{V}_{PROP}$ are assumed to be equal and opposite to those of the low-level jet ($\mathbf{V}_{LLJ}$). Dashed lines indicate a typical relationship of the 850–300-mb thickness pattern to the environmental flow and MBE movement during MCC events. From Corfidi et al. (1996).

In agreement with these potential vorticity theory arguments, many studies (e.g., Houze 1989; Cotton et al. 1989) have shown that the characteristic vertical heating profile with MCSs tends to cool the atmosphere near tropopause levels, warm the upper troposphere, and cool the lower troposphere. This form of heating profile favors vertical changes in stability that support development of a positive potential vorticity anomaly at midlevels and a negative anomaly at high levels. Still further, calculations of the vertical distribution of divergence in MCSs show pronounced convergence at midlevels and strong divergence near the tropopause and in the lower troposphere (Maddox 1983; Cotton et al. 1989; Johnson and Bartels 1992; Smull and Augustine 1993). From the stretching term of the baroclinic vorticity equation, this distribution of divergence implies vorticity spinup at midlevels and spindown at high levels. Therefore, both stability and vorticity tendencies work in favor of generating a positive potential vorticity anomaly at midlevels and a negative anomaly at high levels.

An overwhelming body of observational evidence indicates that mesoscale convective systems, especially MCCs, generate environmental changes that strongly agree with the potential vorticity concepts outlined above. Cross sections through MCCs (Fig. 9.27; Wetzel et al. 1983) clearly show a warm upper troposphere with cold layers above and below. Correspondingly, it is well known that MCSs frequently exhibit a low-level cold pool and mesohigh (e.g., Bosart and Sanders 1981; Menard and Fritsch 1989; Brandes 1990; Johnson and Bartels 1992; Fritsch et al. 1994; Loehrer and Johnson 1995). Similarly, numerous investigators have documented the development of a pronounced anticyclone with an attendant cold dome near tropopause levels (e.g., Fig. 9.28; see Leary and Thompson 1976; Houze 1977; Leary 1979; Gamache and Houze 1982; Fritsch and Maddox 1981; Maddox 1983; Wetzel et al. 1983; Velasco and Fritsch 1987; Zhang and Fritsch 1988b; Cotton et al. 1989). And, as expected, potential vorticity analyses of MCS events exhibit a midlevel maximum and a high-level minimum (e.g., Fig. 9.29; Fritsch et al. 1994; Bartels et al. 1997).

Of most interest has been the formation of long-lived midlevel mesovortices, a reflection of MCS-created midlevel positive potential vorticity anomalies. These warm-core mesoscale cyclonic circulations were first noticed with tropical MCSs (Leary and Thompson 1976; Houze 1977; Leary 1979; Gamache and Houze 1982, 1985; Houze and Rappaport 1984) and are often seen in the cloud debris of MCCs (e.g., Fig. 9.30; Menard and Fritsch 1989; Bartels and Maddox 1991; Johnson and Bartels 1992; Bartels et al. 1997). The development of such vortices is thought to be an inherent process characteristic of MCCs (Velasco and Fritsch 1987; Menard and Fritsch 1989; King 1996). Their significance to severe weather stems from their ability to induce subsequent cycles of deep moist convective overturning (Raymond and Jiang 1990;

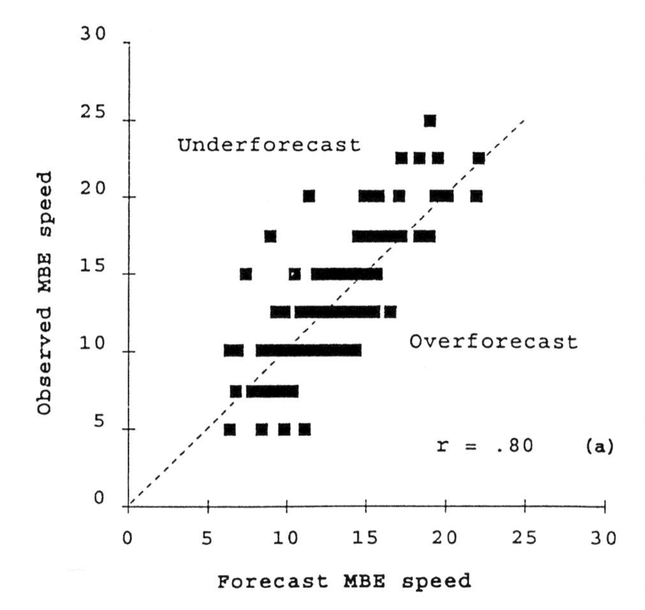

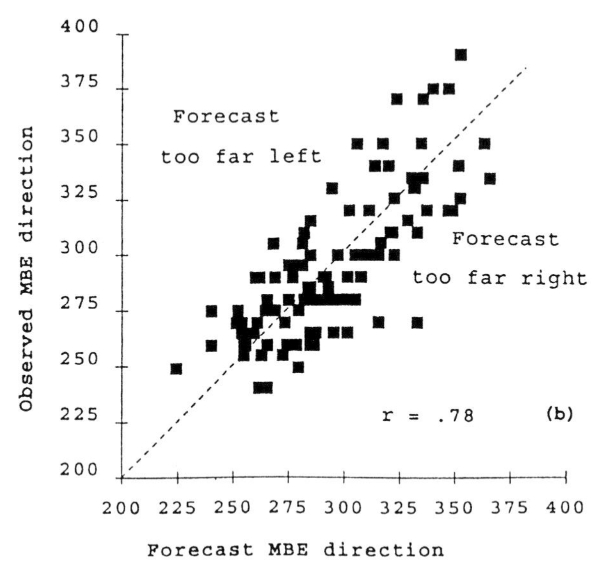

FIG. 9.25. Scatterplots of (a) forecast MBE speed vs observed and (b) forecast MBE direction vs observed for 103 events using the technique of Fig. 9.24. Speeds are rounded to the nearest 2.5 m s^{-1}; directions are rounded to the nearest 5° azimuth. Dashed lines indicate perfect forecasts. From Corfidi et al. (1996).

Fritsch et al. 1994; Davis and Weisman 1994) that sometimes produces extreme rainfall (e.g., > 300 mm: Bosart and Sanders 1981; > 250 mm: Fritsch et al. 1994; > 700 mm: Caracena and Fritsch 1983). It is also of interest that their dynamical structure and circulation resemble those of tropical disturbances, and that maritime MCCs have developed into tropical storms and hurricanes (Velasco and Fritsch 1987; Miller and Fritsch 1991).

Raymond and Jiang (1990) and Davis and Weisman (1994) demonstrated that the upward motion thought necessary for convective redevelopment within a midlevel mesovortex can only occur in the presence of ambient vertical shear. As theorized by Raymond and Jiang (1990) and as observed by Fritsch et al. (1994), the vertical motion arises primarily from the differential movement of air (relative to the vortex) along the sloping isentropic surfaces associated with the vortex (Fig. 9.29). Davis and Weisman (1994) and Skamarock et al. (1994), who provide an excellent discussion of the balanced dynamics of convectively generated mesovortices and squall lines, simulated the interaction of an idealized mesovortex with various configurations of environmental vertical shear. They found that weak shear confined to low levels provides an optimal environment for maintaining the vortex while still providing an opportunity for convective redevelopment. This is precisely the type of environment within which the long-lived convectively generated mesovortex documented by Fritsch et al. (1994) formed, underwent five convective redevelopments, tripled its detectable size, and survived for three days as it traveled from Colorado to Quebec.

In addition to their tendency to develop balanced midlevel mesovortices, MCCs also produce many other effects on their near environment. For example, Webster and Stephens (1980) presented evidence that the large and deep cloud layer with organized convective systems acts to absorb and reradiate outgoing longwave radiation, thereby resulting in local nocturnal warming of the lower troposphere. Conversely, Maddox and Heckman (1982) used 10 MCC events to document how MCCs severely reduce incoming shortwave radiation and alter the sensible and latent heat fluxes at the surface. In particular, they used the errors in Model Output Statistics (MOS) daytime maximum

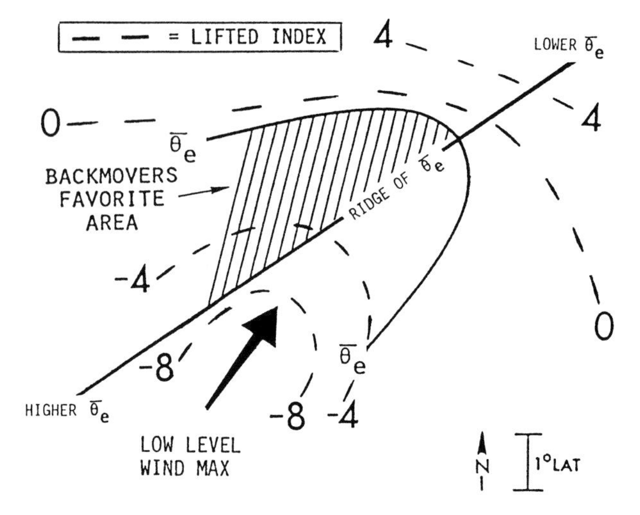

FIG. 9.26. Favored area (stippled) for the occurrence of backward-propagating MCSs. The area is typically located along and north of the axis of maximum mean θ_e values in the 850–300-mb layer. From Shi and Scofield (1987).

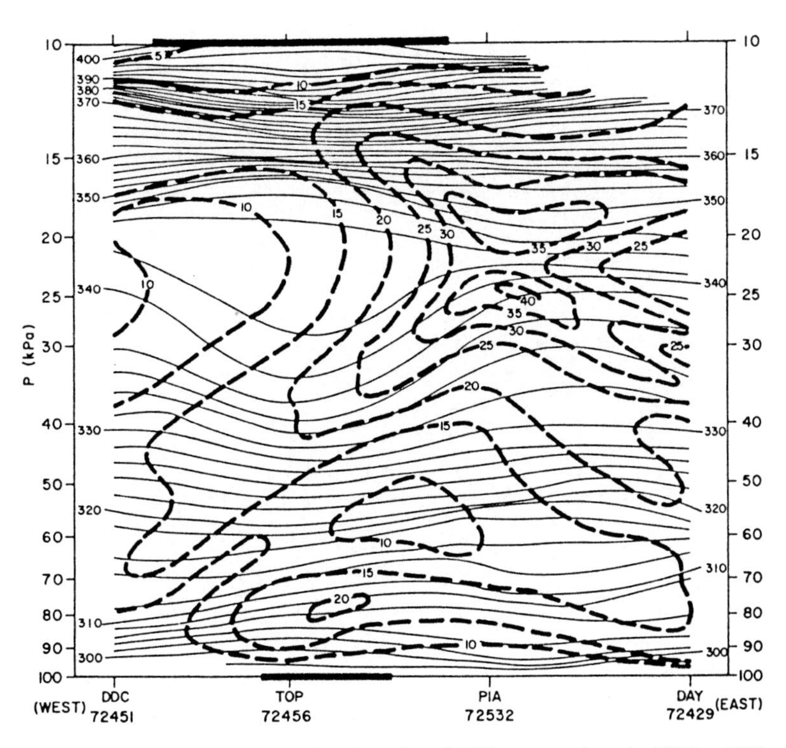

FIG. 9.27. East–west cross section through an MCC at approximately 40°N at 1200 UTC 5 August 1977. Potential temperatures (solid, K) and wind speeds (dashed, m s^{-1}) are based upon radiosondes from Dodge City, KS (DDC); Topeka, KS (TOP); Peoria, IL (PIA); and Dayton, OH (DAY). Thick horizontal line at top of the figure indicates the extent of the cross section covered by the MCC cloud shield (IR temperatures colder than −32°C). Thick line at the bottom shows the extent of the radar-detected meso-β-scale echo. From Wetzel et al. (1983).

temperature forecasts (Klein and Hammons 1975; Carter et al. 1979) as an objective measure of how the afternoon high temperatures were reduced by the cloudiness, wet ground, and moist downdrafts from the 10 MCCs. Figure 9.31 presents an example of the MOS temperature errors (observed temperatures minus the MOS forecast values) for a typical event. Diabatically forced low-level cooling over this great an area and of this magnitude can shift the location of lower-tropospheric baroclinic zones and can thereby alter the phase–height relationship of traveling baroclinic disturbances. Zhang and Harvey (1995) showed that such a shift occurred with the PRE-STORM 10–11 June convective system and upper-level short wave and was instrumental in significantly deepening and changing the location of the surface cyclone associated with the upper-level traveling disturbance.

Along with the modifications to their near environment, MCSs also strongly alter their large-scale environment. For example, Stensrud (1996) estimated the effect of multiple MCSs on the large scale by computing the difference between numerical model simulations of a four-day period that included 15 MCSs over the central United States and a simulation that was identical with the exception that the diabatic effects of the convective systems were omitted. His results show that the cumulative effects of the MCSs have a profound effect on the synoptic scale. A truly massive anticyclonic perturbation is imposed near tropopause levels (Fig. 9.32a) and a correspondingly large cyclonic perturbation is introduced at low levels (Fig. 9.32b). These changes strongly resemble the more local changes caused by individual large convective systems. Stensrud also showed that the convective systems produce a positive feedback such that the large-scale environment is more favorable for additional convection, a result that would tend to extend the lifetime of the convective region and the potential for severe weather.

9.7. Summary

Mesoscale convective systems are globally ubiquitous and play a key role in the earth's hydrologic cycle. In the United States and many other areas of the world, they produce copious rainfall and a significant fraction of the reports of heavy rain, flash floods, and severe weather.

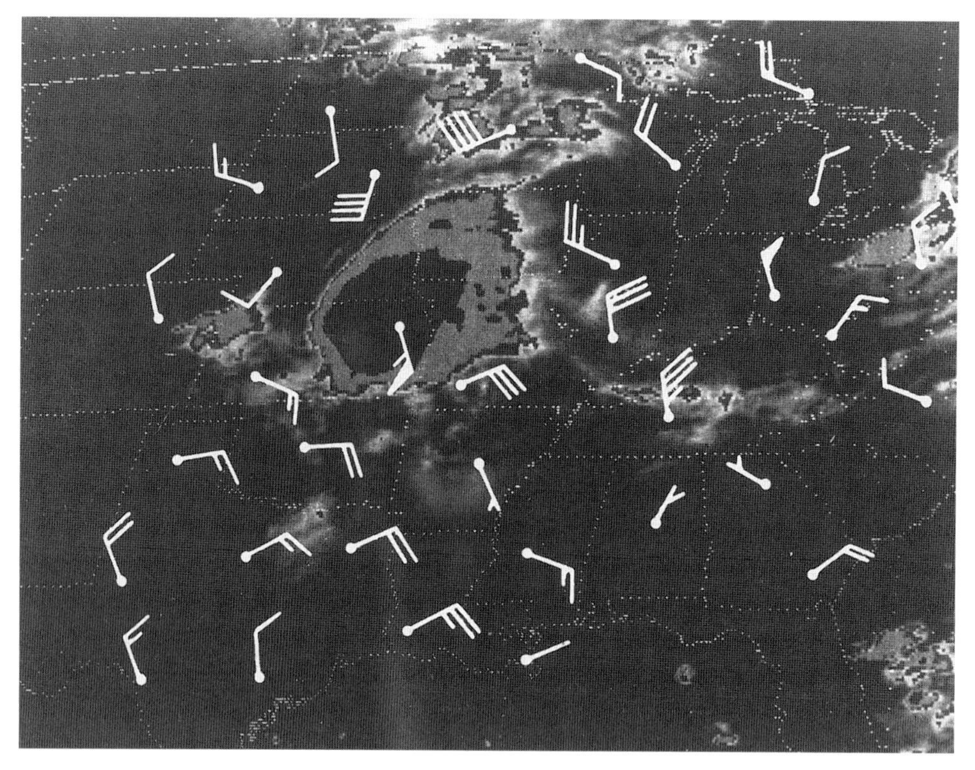

FIG. 9.28. Infrared satellite imagery of an MCC at 1200 UTC 28 June 1979. Wind barbs depict upper-level MCC-induced circulation, as deduced from the vector errors of the 12-h LFM predicted 200-mb wind field. Full wind barb = 5 m s^{-1}; flag = 25 m s^{-1}. From Fritsch and Brown (1982).

Typically, these systems undergo a life cycle that commences with the appearance of individual thunderstorms during the late afternoon or evening and then evolves into a complex structure of stratiform clouds, precipitation, and embedded areas of moist convection. A typical system produces a cold cloud shield that covers hundreds of thousands of square kilometers and persists for six or more hours. The largest systems, mesoscale convective complexes, are predominantly nocturnal, producing their peak cold cloud shields shortly after midnight local time. Severe weather occurs most commonly during the early portion of their life cycle, in conjunction with intense individual thunderstorms. Heavy rains and flash flooding tend to occur later as mesoscale circulations develop and the environment moistens.

The large-scale environment and forcing mechanisms for producing MCSs vary greatly. However, for the very large long-lived systems (MCCs), there appear to be two kinds of environments that characterize most events. In what we term type-1 events, the convective mass flux emanates from an elevated layer overrunning the cool side of a surface-based front or baroclinic zone. Type-2 events form in more barotropic environments wherein the convective mass flux emanates from a surface-based well-mixed layer. In type-2 events, the convection itself, through the action of moist downdrafts, creates a mechanism for subsequent lifting and organization of the convective overturning. Significantly, a key feature common to both types of events is that the MCS structure and characteristics do not develop unless a broad and deep layer of conditionally unstable air is brought to saturation. In type-1 events, slantwise overrunning forced by large-scale baroclinic features facilitates the layer lifting. In type-2 events, the layer-lifting mechanism is sensitive to the relative strengths of vertical wind shear and the moist-downdraft-generated cold pool. Other environmental features common to most significant events are low-level jets, weak midlevel short waves, and pronounced lower-tropospheric warm advection. One of the factors that distinguishes most MCC population centers from other regions of convective activity is that these favored areas have a mechanism (the terrain/radiation forced low-level jet) that can transport low-level warm moist air much faster and farther than what would occur due to quasigeostrophic forcing from traveling disturbances and synoptic features alone. This results in greater CAPE, thereby enhancing the size and strength of the thunderstorms that comprise the convective systems.

Virtually all MCSs are composed of deep convective elements and an associated region of stratiform clouds and precipitation. The convective elements have a tendency to be organized into lines, although

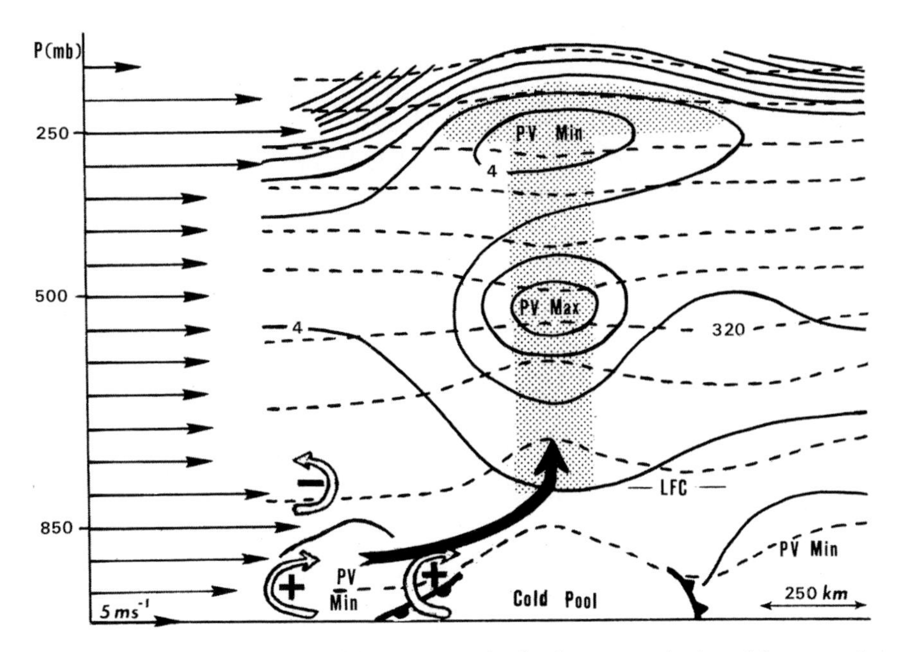

FIG. 9.29. Conceptual diagram of the structure and redevelopment mechanism of the mesoscale warm core vortex. Thin arrows along the ordinate indicate the vertical profile of the environmental wind. Open arrows with plus or minus signs indicate the sense of the vorticity component perpendicular to the plane of the cross section produced by the cold pool and by the environmental vertical wind shear. Thick solid arrow indicates updraft axis created by the vorticity distribution. Frontal symbols indicate outflow boundaries. Dashed lines are potential temperature (5 K intervals) and solid lines are potential vorticity (2×10^{-7} m^2 s^{-1} K kg^{-1} intervals). The system is propagating left to right at about 5–8 m s^{-1} and is being overtaken by high-θ_e air in the low-level jet. Air overtaking the vortex ascends isentropic surfaces, reaches its level of free convection (LFC), and thereby initiates deep convection. Adapted from Rotunno et al. (1988), Raymond and Jiang (1990), and Fritsch et al. (1994).

FIG. 9.30. Visible satellite image of an MCC-generated mesovortex on 1830 UTC 8 July 1982. From Menard and Fritsch (1989).

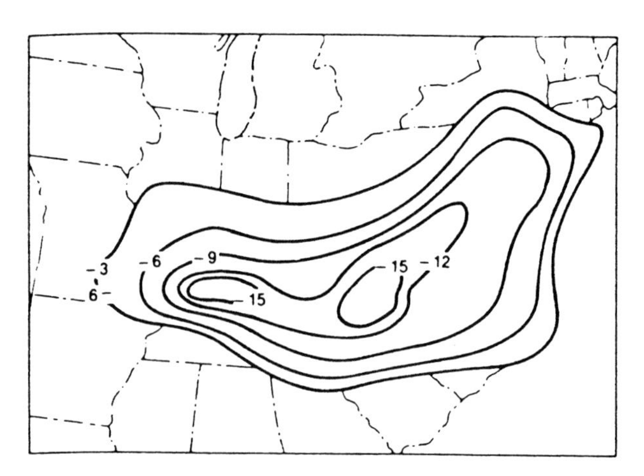

FIG. 9.31. Errors (°F) in MOS maximum temperature forecasts for 3 July 1980. From Maddox and Heckman (1982).

many systems, especially in large type-1 events, display more or less random distributions of the convective structures. As expected, the convective portions of the systems produce the bulk of the severe weather and flash flooding. Stratiform regions typically display lighter rainfall, with mesoscale ascent in the upper troposphere and descent below. The ascending branch of the circulation is anomalously warm relative to the larger-scale environment, whereas the descending branch, chilled by melting and evaporation, is characteristically cool. At midlevels, the stratiform region frequently exhibits pronounced convergence and cyclonically spiraling bands.

As with individual thunderstorms, the movement of most MCSs is the result of the combined contributions of

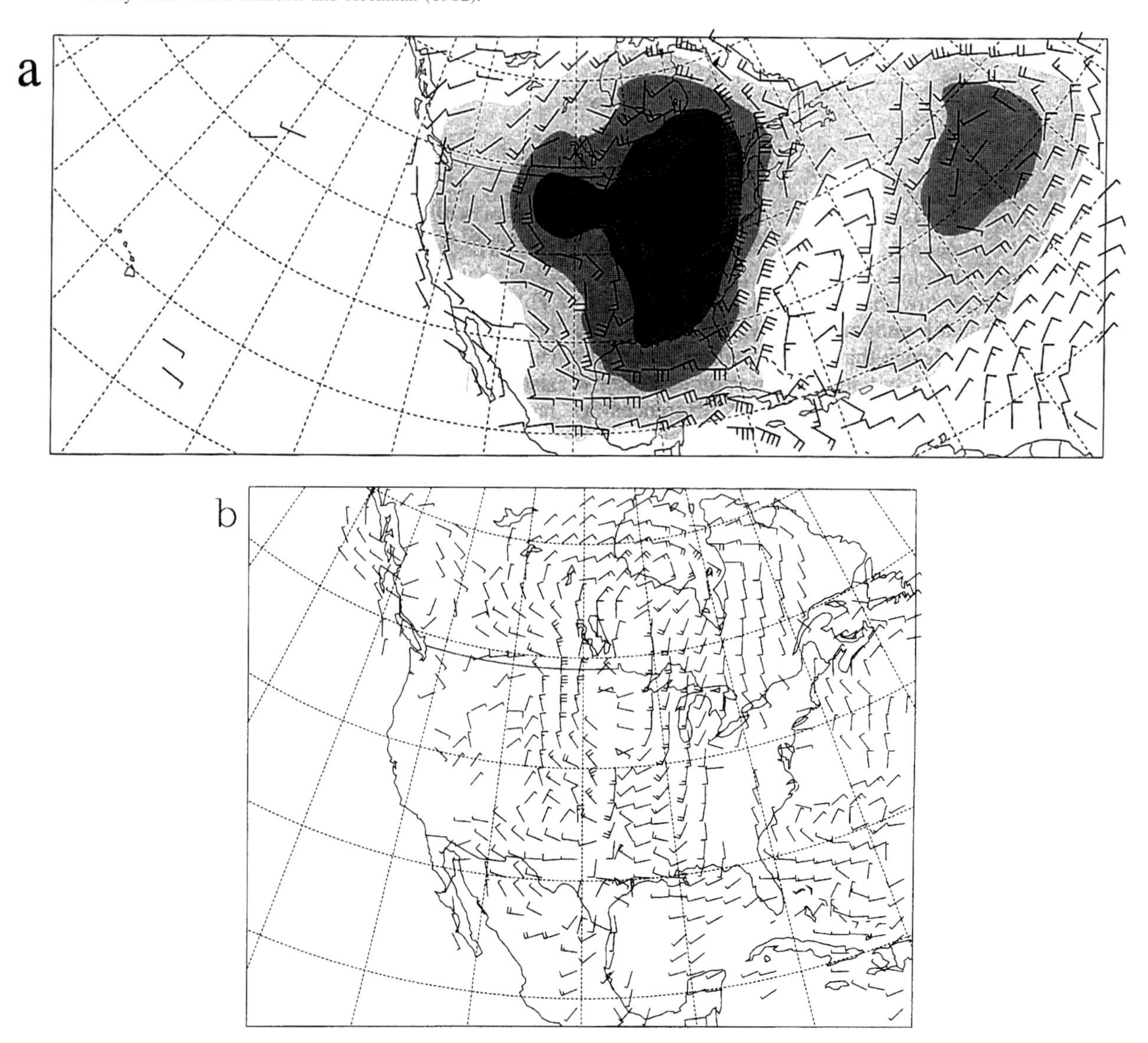

FIG. 9.32. Modifications to the large-scale circulation as a result of multiple MCSs over the central United States during a 4-day period ending 0000 UTC 15 May 1982. (a) Differences in 200-mb height (shaded at 40-m intervals, beginning at 20 m) and winds (full barb is 5 m s^{-1}) at 200 mb. (b) Wind differences at 850 mb. From Stensrud (1996).

advection and propagation. This is especially true of the intense convective rain areas responsible for the bulk of the severe weather and flooding. The advective component correlates strongly with the mean flow in the cloud layer. The departure from the mean cloud-layer flow is strongly influenced by the speed and direction of the low-level inflow of high-θ_e air feeding the deep convective overturning. New growth favors this inflow area and the systems therefore tend to propagate toward the source of the high-θ_e air, that is, in a direction opposite the low-level inflow. It has been shown that a reasonable estimate of the speed and direction of MCSs can be obtained from the vector difference between the mean flow in the cloud layer and the low-level jet.

More recent studies have shown that the tendency for MCSs to propagate toward the source of the high-θ_e air feeding the convection (usually southward in the United States) is augmented or possibly even dominated by the interaction of the MCS-generated cold pool with Coriolis accelerations. Specifically, Skamarock et al. (1994) found that, relative to numerical simulations of squall lines without Coriolis effects, the with-Coriolis simulations showed greater anticyclonic turning of the gust front winds. This leads to enhanced convergence and deeper lifting along the right (left) flank of Northern (Southern) Hemisphere convective lines. Additional convection as a result of the enhanced forcing reenforced the cold outflow, thereby promoting a further rightward (leftward) advancement of a Northern (Southern) Hemisphere line. Conceivably, the conceptual model presented in section 9.5 could be modified or enhanced by introducing a measure of cold pool strength, especially for forecasting the movement of warm sector squall lines.

Finally, there is a wealth of evidence that shows that MCSs substantially alter their large-scale environment. The most commonly observed changes show that the systems generate a positive potential vorticity anomaly at midlevels and a negative anomaly near the tropopause. These features are balanced structures and have been observed to persist for several days. They are typically manifested as a midlevel mesovortex, a high-level anticyclone, and a surface-based cold pool and mesohigh. Under certain conditions, the positive potential vorticity anomaly can be instrumental in initiating new MCSs, sometimes with devastating consequences in the form of very heavy rainfall and flash floods.

Acknowledgments. The authors gratefully acknowledge the contributions of Bill Frank, Peter Bannon, John Clark, Jack Kain, Rob Rogers, and Craig Bishop. We are also grateful to George Bryan for providing several of the Doppler radar figures and to Ed Tollerud, the chapter editor, for his diligence, patience, and encouragement as he guided the complex review process. This work was supported by NSF Grant ATM-9806309.

APPENDIX
MCC Population

TABLE 9.A1. Mesoscale convective complex population included in the global dataset. The first number not in parentheses in the third column indicates the average number of MCCs per season. The number in parentheses is the difference between the maximum annual occurrence and the minimum annual occurrence.

Region, period of study	Geographic domain	Avg. number per season (max-min annual occurrence)	Reference
South, Central America, May 1981–Apr 1983	40°S–30°N, 120°–28°W	96 (47)	Velasco and Fritsch (1987)
Western Pacific region, 1983, 1985	50°S–50°N, 90°E–170°W	82 (14)	Miller (1990)
United States, 1986–87	25°–50°N, 130°–70°W	51 (14)	Augustine and Howard (1991)
Africa, 1986–87	40°S–35°N, 45°W–45°E	97 (7)	Laing (1992)
Indian subcontinent, Apr–Dec 1988	0–50°N, 45–110°E	49	Laing (1992)
Europe, 1986–87	35–55°N, 20°W–45°E	6 (1)	Laing and Fritsch (1997)

REFERENCES

Augustine, J. A., and K. W. Howard, 1988: Mesoscale convective complexes over the United States during 1985. *Mon. Wea. Rev.*, **116,** 685–701.

———, and ———, 1991: Mesoscale convective complexes over the United States during 1986 and 1987. *Mon. Wea. Rev.*, **119,** 1575–1589.

———, and F. Caracena, 1994: Lower-tropospheric precursors to nocturnal MCS development over the Central United States. *Wea. Forecasting.*, **9,** 116–135.

Bartels, D. L., and R. A. Maddox, 1991: Midlevel cyclonic vortices generated by mesoscale convective systems. *Mon. Wea. Rev.*, **119,** 104–118.

———, J. M. Brown, and E. I. Tollerud, 1997: Structure of a midtropospheric vortex induced by a mesoscale convective system. *Mon. Wea. Rev.*, **125,** 193–211.

Blackadar, A. K., 1957: Boundary layer wind maxima and their significance for the growth of nocturnal inversions. *Bull. Amer. Meteor. Soc.*, **38,** 283–290.

Bleeker, S., and M. J. Andre, 1951: On the diurnal variation of precipitation, particularly over central U.S.A. and its relation to large-scale orographic circulation systems. *Quart. J. Roy. Meteor. Soc.*, **77,** 260–271.

Bluestein, H. B., and M. H. Jain, 1985: Formation of mesoscale lines of precipitation: Severe squall lines in Oklahoma during the spring. *J. Atmos. Sci.*, **42,** 1711–1732.

———, G. T. Marx, and M. H. Jain, 1987: Formation of mesoscale lines of precipitation: Nonsevere squall lines in Oklahoma during the spring. *Mon. Wea. Rev.*, **115,** 2719–2727.

Bonner, W. D., 1968: Climatology of the low level jet. *Mon. Wea. Rev.*, **96,** 833–850.

Bosart, L. F., and F. Sanders, 1981: The Johnstown flood of July 1977: A long-lived convective storm. *J. Atmos. Sci.*, **38,** 1616–1642.

Brandes, E. A., 1990: Evolution and structure of the 6–7 May 1985 mesoscale convective system and associated vortex. *Mon. Wea. Rev.,* **118,** 109–127.

———, and C. L. Ziegler, 1993: Mesoscale downdraft influences on vertical vorticity in a mature mesoscale convective system. *Mon. Wea. Rev.,* **121,** 1337–1353.

Brooks, H. B., 1946: A summary of some radar thunderstorm observations. *Bull. Amer. Meteor. Soc.,* **27,** 557–563.

Brunk, I. W., 1953: Squall lines. *Bull. Amer. Meteor. Soc.,* **34,** 1–9.

Burpee, R. W., 1979: Peninsular-scale convergence in the south Florida seabreeze. *Mon. Wea. Rev.,* **107,** 852–860.

Byers, H. R., and R. R. Braham Jr., 1949: *The Thunderstorm.* U.S. Government Printing Office, 287 pp.

Caracena, F., and J. M. Fritsch, 1983: Focusing mechanisms in the Texas Hill Country flash floods of 1978. *Mon. Wea. Rev.,* **111,** 2319–2332.

Carter, G. M., J. P. Dallavalle, A. L. Forst, and W. H. Klein, 1979: Improved automated surface temperature guidance. *Mon. Wea. Rev.,* **107,** 1263–1274.

Chappell, C. F., 1986: Quasi-stationary convective events. *Mesoscale Meteorology and Forecasting,* P. Ray, Ed., Amer. Meteor. Soc., 289–310.

Chen, S., and W. R. Cotton, 1988: The sensitivity of a simulated extratropical mesoscale convective system to longwave radiation and ice-phase microphysics. *J. Atmos. Sci.,* **45,** 3897–3910.

Chen, S. S., and W. M. Frank, 1993: A numerical study of the genesis of extratropical convective mesovortices. Part I: Evolution and dynamics. *J. Atmos. Sci.,* **50,** 2401–2426.

Chong, M., P. Amayenc, G. Scialom, and J. Testud, 1987: A tropical squall line observed during the COPT 81 Experiment in West Africa. Part I: Kinematic structure inferred from dual-Doppler radar data. *Mon. Wea. Rev.,* **115,** 670–694.

Colby, F. P., Jr., 1984: Convective inhibition as a predictor of convection during AVE-SESAME II. *Mon. Wea. Rev.,* **112,** 2239–2252.

Collander, R. S., 1993: A ten-year summary of severe weather in mesoscale convective complexes, Part 2: Heavy rainfall. Preprints, *17th Conf. on Severe Local Storms,* St. Louis, MO, Amer. Meteor. Soc., 638–641.

Cooper, H. J., M. Garstang, and J. Simpson, 1982: The diurnal interaction between convection and peninsular-scale forcing over south Florida. *Mon. Wea. Rev.,* **110,** 486–503.

Corfidi, S. F., J. H. Merritt, and J. M. Fritsch, 1996: Predicting the movement of mesoscale convective complexes. *Wea. Forecasting,* **11,** 41–46.

Cotton, W. R., and R. L. McAnelly, 1992: Early growth of mesoscale convective complexes: A meso-β scale cycle of convective precipitation? *Mon. Wea. Rev.,* **120,** 1851–1877.

———, M.-S. Lin, R. L. McAnelly, and C. J. Tremback, 1989: A composite model of mesoscale convective complexes. *Mon. Wea. Rev.,* **117,** 765–783.

Cunning, J., 1986: The Oklahoma-Kansas preliminary regional experiment for STORM-Central. *Bull. Amer. Meteor. Soc.,* **67,** 1478–1486.

Davis, C. A., and M. L. Weisman, 1994: Balanced dynamics of mesoscale vortices produced in simulated convective systems. *J. Atmos. Sci.,* **51,** 2005–2030.

Devlin, K. I., 1995: Application of the 85Ghz ice scattering signature to a global study of mesoscale convective systems. M.S. thesis, Texas A&M University, 99 pp.

Dopplick, T. G., 1972: Radiative heating of the global atmosphere. *J. Atmos. Sci.,* **29,** 1278–1294.

Fankhauser, J. C., 1964: On the motion and predictability of convective systems. National Severe Storms Prediction Center Rep. 21, 34 pp.

Fortune, M. A., W. R. Cotton, and R. L. McAnelly, 1992: Frontal-wave-like evolution in some mesoscale convective complexes. *Mon. Wea. Rev.,* **120,** 1279–1300.

Foster, D. S., 1958: Thunderstorm gusts compared with computed downdraft speeds. *Mon. Wea. Rev.,* **86,** 91–94.

Fovell, R. G., and Y. Ogura, 1988: Numerical simulation of a squall line in two dimensions. *J. Atmos. Sci.,* **45,** 3846–3879.

———, and ———, 1989: Effect of vertical wind shear on numerically simulated multicell storm structure. *J. Atmos. Sci.,* **46,** 3144–3176.

Fritsch, J. M., and R. A. Maddox, 1981: Convectively-driven mesoscale pressure systems aloft. Part I: Observations. *J. Climate Appl. Meteor.,* **20,** 9–19.

———, and J. M. Brown, 1982: On the generation of convectively-driven mesohighs aloft. *Mon. Wea. Rev.,* **110,** 1554–1563.

———, R. A. Maddox, and A. G. Barnston, 1981: The character of mesoscale convective complex precipitation and its contribution to the warm season rainfall in the U.S. Preprints, *Fourth Conf. on Hydrometeorology,* Reno, NV, Amer. Meteor. Soc., 94–99.

———, R. J. Kane, and C. H. Chelius, 1986: The contribution of mesoscale convective weather systems to the warm season precipitation in the United States., *J. Climate Appl. Meteor.,* **25,** 1333–1345.

———, J. D. Murphy, and J. S. Kain, 1994: Warm core vortex amplification over land. *J. Atmos. Sci.,* **51,** 1781–1806.

Fujita, T., 1955: Results of detailed synoptic studies of squall lines. *Tellus,* **4,** 405–436.

Gamache, J. F., and R. A. Houze, 1982: Mesoscale air motions associated with a tropical squall line. *Mon. Wea. Rev.,* **110,** 118–135.

———, and ———, 1985: Further analysis of the composite wind and thermodynamic structure of the 12 September GATE squall line. *Mon. Wea. Rev.,* **113,** 1241–1259.

Garstang, M., H. L. Massie Jr., J. Halverson, S. Greco, and J. Scala, 1994: Amazon coastal squall lines. Part I: Structure and kinematics. *Mon. Wea. Rev.,* **122,** 608–622.

Geldmeier, M. F., and G. M. Barnes, 1997: The "footprint" under a decaying tropical mesoscale convective system. *Mon. Wea. Rev.,* **125,** 2879–2895.

Gray, W. M., 1979: Hurricanes: Their formation, structure and likely role in the tropical circulation. Supplement to *Meteorology over the Tropical Oceans,* D. B. Show, Ed., James Glaisher House, 155–218.

———, and R. W. Jacobson, 1977: Diurnal variation of deep convective system in the Tropics. *Mon. Wea. Rev.,* **105,** 1171–1188.

Greco, S., and Coauthors, 1992: Rainfall and surface kinematic conditions over Central Amazonia during ABLE 2B. *J. Geophys. Res.,* **95,** 17 001–17 014.

Hane, C. E., 1986: Extratropical squall lines and rainbands. *Mesoscale Meteorology and Forecasting,* P. Ray, Ed., American Meteorological Society, 359–389.

Haynes, P. H., and M. E. McIntyre, 1987: On the evolution of vorticity and potential vorticity in the presence of diabatic heating and frictional or other forces. *J. Atmos. Sci.,* **44,** 828–841.

Holton, J. R., 1967: The diurnal boundary layer wind oscillation above sloping terrain. *Tellus,* **19,** 199–205.

Houze, R. A., 1977: Structure and dynamics of a tropical squall-line system. *Mon. Wea. Rev.,* **105,** 1540–1567.

———, 1989: Observed structure of mesoscale convective systems and implications for large scale heating. *Quart. J. Roy. Meteor. Soc.,* **115,** 425–461.

———, and E. N. Rappaport, 1984: Air motions and precipitation structure of an early summer squall line over the eastern tropical Atlantic. *J. Atmos. Sci.,* **41,** 553–574.

———, S. A. Rutledge, M. I. Biggerstaff, and B. F. Smull, 1989: Interpretation of Doppler weather radar displays of midlatitude mesoscale convective systems. *Bull. Amer. Meteor. Soc.,* **70,** 608–619.

———, B. F. Smull, and P. Dodge, 1990: Mesoscale organization of springtime rainstorms in Oklahoma. *Mon. Wea. Rev.,* **118,** 613–654.

Johns, R. H., 1993: Meteorological conditions associated with bow echo development in convective storms. *Wea. Forecasting,* **8,** 294–299.

———, and W. D. Hirt, 1987: Derechos: Widespread convectively induced windstorms. *Wea. Forecasting,* **2,** 32–49.

Johnson, R. H., and P. J. Hamilton, 1988: The relationship of surface pressure features to the precipitation and air flow structure of an intense midlatitude squall line. *Mon. Wea. Rev.,* **116,** 1444–1472.

———, and D. L. Bartels, 1992: Circulations associated with a mature-to-decaying mid-latitude mesoscale convective system. Part II: Upper-level features. *Mon. Wea. Rev.,* **120,** 1301–1320.

———, S. Chen, and J. J. Toth, 1989: Circulations associated with a mature-to-decaying midlatitude mesoscale convective system. Part I: Surface features—Heat bursts and mesolow development. *Mon. Wea. Rev.,* **117,** 942–959.

Kain, J. S., and J. M. Fritsch, 1998: Multiscale convective overturning in mesoscale convective systems: Reconciling observations, simulations and theory. *Mon. Wea. Rev.,* **126,** 2254–2273.

Kane, R. J., C. R. Chelius, and J. M. Fritsch, 1987: The precipitation characteristics of mesoscale convective weather systems. *J. Climate Appl. Meteor.,* **26,** 1323–1335.

Keenan, T. D., and R. E. Carbone, 1992: A preliminary morphology of precipitation systems in tropical northern Australia. *Quart. J. Roy. Meteor. Soc.,* **118,** 283–326.

King, R. G., 1996: Mid-level vorticity in mesoscale convective systems. Atmospheric Science Paper 606, Department of Atmospheric Science, Colorado State University, 92 pp.

Klein, W. H., and G. A. Hammons, 1975: Maximum/minimum temperature forecasts based on model output statistics. *Mon. Wea. Rev.,* **103,** 796–806.

Klemp, J. B., 1987: Dynamics of tornadic thunderstorms. *Ann. Rev. Fluid Mech.,* **19,** 369–402.

Knupp, K. R., 1987: Downdrafts within High Plains cumulonimbi. Part I: General kinematic structure. *J. Atmos. Science,* **44,** 987–1008.

———, and W. R. Cotton, 1985: Precipitating convective cloud downdraft structure: An interpretive survey. *Rev. Geophys.,* **23,** 183–215.

Kousky, V. E., 1980: Diurnal rainfall variation in northeastern Brazil. *Mon. Wea. Rev.,* **108,** 488–498.

Laing, A. G., 1992: Mesoscale convective complexes over Africa and the Indian subcontinent. M.S. thesis, The Pennsylvania State University. [Available from Dept. of Meteorology, 503 Walker Building, The Pennsylvania State University, University Park, PA 16802.]

———, 1996: A global climatology of mesoscale convective complexes. Doctoral dissertation, The Pennsylvania State University, 158 pp.

———, and J. M. Fritsch, 1993a: Mesoscale convective complexes in Africa. *Mon. Wea. Rev.,* **121,** 2254–2263.

———, and J. M. Fritsch, 1993b: Mesoscale convective complexes over the Indian monsoon region. *J. Climate,* **6,** 911–919.

———, and ———, 1997: The global population of mesoscale convective complexes. *Quart. J. Roy. Meteor. Soc.,* **123,** 389–405.

———, and ———, 2000: The large scale environments of the global populations of mesoscale convective complexes. *Mon. Wea. Rev.,* **128,** 2756–2776.

Laurent, H., 1996: Tracking of convective cloud clusters from Meteosat data. *Tenth Meteosat Scientific Users' Conference,* Cascais, Portugal, European Organisation for the Exploitation of Meteorological Satellites, 187–191.

Leary, C. A., 1979: Behavior of the wind field in the vicinity of a cloud cluster in the intertropical convergence zone. *J. Atmos. Sci.,* **36,** 631–639.

———, and R. O. Thompson, 1976: A warm-core disturbance in the western Atlantic during BOMEX. *Mon. Wea. Rev.,* **104,** 443–452.

———, and E. N. Rappaport, 1987: The life cycle and internal structure of a mesoscale convective complex. *Mon. Wea. Rev.,* **115,** 1503–1527.

Lilly, D. K., and T. Gal-Chen, 1983: *Mesoscale Meteorology: Theories, Observations and Models.* NATO Advanced Science Institute Series, D. Reidel, 781 pp.

Loehrer, S. M., and R. H. Johnson, 1995: Surface pressure and precipitation life cycle characteristics of PRE-STORM mesoscale convective systems. *Mon. Wea. Rev.,* **123,** 600–621.

Machado, L. A. T., and W. B. Rossow, 1993: Structural characteristics and radiative properties of tropical cloud clusters. *Mon. Wea. Rev.,* **121,** 3234–3260.

Maddox, R. A., 1980: Mesoscale convective complexes. *Bull. Amer. Meteor. Soc.,* **61,** 1374–1387.

———, 1983: Large-scale meteorological conditions associated with mid-latitude, mesoscale convective complexes. *Mon. Wea. Rev.,* **111,** 1475–1493.

———, and B. E. Heckman, 1982: The impact of mesoscale convective weather systems upon MOS temperature guidance. Preprints, *Ninth Conf. on Weather Forecasting and Analysis,* Seattle, WA, Amer. Meteor. Soc., 214–218.

———, C. F. Chappell, and L. R. Hoxit, 1979: Synoptic and meso-α-scale aspects of flash flood events. *Bull. Amer. Meteor. Soc.,* **60,** 115–123.

———, K. W. Howard, D. L. Bartels, and D. M. Rodgers, 1986: Mesoscale convective complexes in the middle latitudes. *Mesoscale Meteorology and Forecasting,* P. Ray, Ed., Amer. Meteor. Soc., 390–413.

Mapes, B., 1993: Gregarious tropical convection. *J. Atmos. Sci.,* **50,** 2026–2037.

———, and R. A. Houze, 1992: An integrated view of the 1987 Australian monsoon and its mesoscale convective systems. I: Horizontal structure. *Quart. J. Roy. Meteor. Soc.,* **118,** 927–963.

McAnelly, R. L., and W. R. Cotton, 1986: Meso-β-scale characteristics of an episode of meso-α-scale convective complexes. *Mon. Wea. Rev.,* **114,** 1740–1770.

———, and ———, 1989: The precipitation life cycle of mesoscale convective complexes over the central United States. *Mon. Wea. Rev.,* **117,** 784–808.

McNider, R. T., and R. A. Pielke, 1981: Diurnal boundary-layer development over sloping terrain. *J. Atmos. Sci.,* **38,** 2198–2212.

Menard, R. D., and J. M. Fritsch, 1989: An MCC-generated inertially stable warm core vortex. *Mon. Wea. Rev.,* **117,** 1237–1260.

Merritt, J. H., 1985: The synoptic environment and movement of mesoscale convective complexes over the United States. M.S. thesis, The Pennsylvania State University, 129 pp. [Available from Dept. of Meteorology, 503 Walker Building, The Pennsylvania State University, University Park, PA 16802.]

———, and J. M. Fritsch, 1984: On the movement of the heavy precipitation areas of mid-latitude mesoscale convective complexes. Preprints, *10th Conf. on Weather Forecasting and Analysis,* Tampa, FL, Amer. Meteor. Soc., 529–536.

Miller, D., 1990: Mesoscale convective complexes in the western Pacific region. M.S. thesis, The Pennsylvania State University. [Available from Dept. of Meteorology, 503 Walker Building, The Pennsylvania State University, University Park, PA 16802.]

———, and J. M. Fritsch, 1991: Mesoscale convective complexes in the western Pacific region. *Mon. Wea. Rev.,* **119,** 2978–2992.

Newton, C. W., 1950: Structure and mechanism of the prefrontal squall line. *J. Meteor.,* **7,** 210–222.
———, and S. Katz, 1958: Movement of large convective rainstorms in relation to winds aloft. *Bull. Amer. Meteor. Soc.,* **39,** 129–136.
———, and H. R. Newton, 1959: Dynamical interactions between large convective clouds and environment with vertical shear. *J. Meteor.,* **16,** 483–496.
Nicolini, M., K. M. Waldron, and J. Paegle, 1993: Diurnal oscillation of low-level jets, vertical motion, and precipitation: A model case study. *Mon. Wea. Rev.,* **121,** 2588–2610.
Ogura, Y., and Y.-L. Chen, 1977: A life history of an intense mesoscale convective storm in Oklahoma. *J. Atmos. Sci.,* **34,** 1458–1476.
———, and M.-T. Liou, 1980: The structure of a midlatitude squall line: A case study. *J. Atmos. Sci.,* **37,** 553–567.
Paegle, J., 1978: The transient mass-flow adjustment to heated atmospheric circulations. *J. Atmos. Sci.,* **35,** 1678–1688.
Pandya, R. E., and D. R. Durran, 1996: The influence of convectively generated thermal forcing on the mesoscale circulation around squall lines. *J. Atmos. Sci.,* **53,** 2924–2951.
Pedgley, D. E., 1962: A meso-synoptic analysis of the thunderstorms on 28 August 1958. Geophys. Mem. 106, Met Office 711, 74 pp.
Purdom, J. F. W., 1976: Some uses of high-resolution GOES imagery in the mesoscale forecasting of convection and its behavior. *Mon. Wea. Rev.,* **104,** 1474–1483.
Raymond, D. J., and H. Jiang, 1990: A theory for long-lived mesoscale convective systems. *J. Atmos. Sci.,* **47,** 3067–3077.
Rochette, S. M., and J. T. Moore, 1996: Initiation of an elevated mesoscale convective system associated with heavy rainfall. *Wea. Forecasting,* **11,** 443–457.
Rose, M. A., 1996: Downbursts. *Natl. Wea. Dig.,* **21,** 11–17.
Rotunno, R., and J. B. Klemp, 1985: On the rotation and propagation of simulated supercell thunderstorms. *J. Atmos. Sci.,* **42,** 271–292.
———, ———, and M. L. Weisman, 1988: A theory for strong, long-lived squall lines. *J. Atmos. Sci.,* **45,** 463–485.
Sanders, F., and J. R. Gyakum, 1980: Synoptic-dynamic climatology of the "bomb." *Mon. Wea. Rev.,* **108,** 1589–1606.
Sawyer, J. S., 1956: The vertical circulation at meteorological fronts and its relation to frontogenesis. *Proc. Roy. Soc. London,* **A234,** 346–362.
Schmidt, J. M., and W. R. Cotton, 1989: A high plains squall line associated with severe surface winds. *J. Atmos. Sci.,* **46,** 281–302.
———, and ———, 1990: Interactions between upper and lower tropospheric gravity waves on squall line structure and maintenance. *J. Atmos. Sci.,* **47,** 1205–1222.
Schubert, W. H., J. J. Hack, P. L. Silva Dias, and S. R. Fulton, 1980: Geostrophic adjustment in an axisymmetric vortex. *J. Atmos. Sci.,* **37,** 1464–1484.
Shi, J., and R. A. Scofield, 1987: Satellite-observed mesoscale convective system (MCS) propagation characteristics and a 3–12 hour heavy precipitation forecast index. NOAA Tech. Memo NESDIS 20, U.S. Dept. of Commerce, Washington, DC, 43 pp.
Simpson, J., and W. L. Woodley, 1971: Seeding cumulus in Florida: New 1970 results. *Science,* **172,** 117–126.
———, N. E. Wescott, R. J. Clerman, and R. A. Pielke, 1980: On cumulus mergers. *Arch. Meteor. Geophys. Bioklimatol.,* **A29,** 1–40.
———, Th. D. Keenan, B. Ferrier, R. H. Simpson, and G. J. Holland, 1993: Cumulus mergers in the Maritime Continent Region. *Meteor. Atmos. Phys.,* **51,** 73–99.
Skamarock, W. C., M. L. Weisman, and J. B. Klemp, 1994: Three-dimensional evolution of simulated long-lived squall lines. *J. Atmos. Sci.,* **51,** 2563–2584.
Smull, B. F., and R. A. Houze, 1985: A midlatitude squall line with a trailing region of stratiform rain: Radar and satellite observations. *Mon. Wea. Rev.,* **113,** 117–133.
———, and ———, 1987: Dual-Doppler radar analysis of a midlatitude squall line with a trailing region of stratiform rain. *J. Atmos. Sci.,* **44,** 2128–2148.
———, and J. A. Augustine, 1993: Multiscale analysis of a mature mesoscale convective complex. *Mon. Wea. Rev.,* **121,** 103–132.
Stensrud, D. J., 1996: Effects of persistent, midlatitude mesoscale regions of convection on the large-scale environment during the warm season. *J. Atmos. Sci.,* **53,** 3503–3527.
Tollerud, E. I., and D. Bartels, 1988: A comparative study of the environment of severe-weather-producing mesoscale convective systems: MCCs, meso-beta systems, and large convective lines. Preprints, *15th Conf. on Severe Local Storms,* Baltimore, MD, Amer. Meteor. Soc., 544–547.
———, and R. S. Collander, 1993a: A ten-year summary of severe weather in mesoscale convective complexes. Part I: High wind, tornadoes, and hail. Preprints, *17th Conf. on Severe Local Storms,* St. Louis, MO, Amer. Meteor. Soc., 533–537.
———, and R. S. Collander, 1993b: A ten-year synopsis of record station rainfall produced in mesoscale convective complexes. Preprints, *13th Conf. on Weather Analysis and Forecasting,* Vienna, VA, Amer. Meteor. Soc., 430–433.
———, J. A. Augustine, and B. D. Jamison, 1992: Cloud top characteristics of mesoscale convective systems in 1986. Preprints, *Sixth Conf. on Satellite Meteorology and Oceanography,* Atlanta, GA, Amer. Meteor. Soc., J3–J7.
Trier, S. B., and E. B. Parsons, 1993: Evolution of environmental conditions preceding the development of a nocturnal mesoscale convective complex. *Mon. Wea. Rev.,* **121,** 1078–1098.
Tripoli, G. J., and W. R. Cotton, 1989: Numerical study of an observed orogenic mesoscale convective system. Part I: Simulated genesis and comparison with observations. *Mon. Wea. Rev.,* **117,** 273–304.
Velasco, I., and J. M. Fritsch, 1987: Mesoscale convective complexes in the Americas. *J. Geophys. Res.,* **92,** 9591–9613.
Wakimoto, R., 2001: Convectively drive, high wind events. *Severe Convective Storms, Meteor. Monogr.,* No. 50, Amer. Meteor. Soc. 255–298.
Webster, P. J., and G. L. Stephens, 1980: Tropical upper-tropospheric extended clouds: Inferences from winter MONEX. *J. Atmos. Sci.,* **37,** 1521–1541.
Weisman, M. L., 1992: The role of convectively generated rear-inflow jets in the evolution of long-lived mesoconvective systems. *J. Atmos. Sci.,* **49,** 1826–1847.
———, J. B. Klemp, and R. Rotunno, 1988: Structure and evolution of numerically simulated squall lines. *J. Atmos. Sci.,* **45,** 1990–2013.
Wetzel, P. J., W. R. Cotton, and R. L. McAnnelly, 1983: A long-lived mesoscale convective complex. Part II: Evolution and structure of the mature complex. *Mon. Wea. Rev.,* **111,** 1919–1937.
Yuter, S. E., and R. A. Houze, 1995: Three-dimensional kinematic and microphysical evolution of Florida cumulonimbus. Part I: Spatial distribution of updrafts, downdrafts, and precipitation. *Mon. Wea. Rev.,* **123,** 1921–1940.
Zhang, D.-L., and J. M. Fritsch, 1987: Numerical simulation of the meso-β scale structure and evolution of the 1977 Johnstown Flood. Part II: Inertially-stable warm-core vortex and the mesoscale convective complex. *J. Atmos. Sci.,* **44,** 2593–2612.
———, and ———, 1988a: A numerical investigation of a convectively-generated, inertially-stable, extra-tropical, warm-core mesovortex over land. Part I: Structure and evolution. *Mon. Wea. Rev.,* **116,** 2660–2687.

———, and ———, 1988b: Numerical sensitivity experiments on the structure, evolution and dynamics of two mesoscale convective systems. *J. Atmos. Sci.,* **45,** 261–293.

———, and K. Gao, 1989: Numerical simulation of an intense squall line during the 10–11 June 1985 PRE-STORM. Part II: Rear inflow, surface pressure perturbations and stratiform precipitation. *Mon. Wea. Rev.,* **117,** 2067–2094.

———, and R. Harvey, 1995: Enhancement of extratropical cyclogenesis by a mesoscale convective system. *J. Atmos. Sci.,* **52,** 1107–1127.

———, K. Gao, and D. B. Parsons, 1989: Numerical simulation of an intense squall line during the 10–11 June 1985 PRE-STORM. Part I: Structure and evolution. *Mon. Wea. Rev.,* **117,** 960–994.

Zipser, E. J., 1977: Mesoscale and convective-scale downdrafts as distinct components of squall-line structure. *Mon. Wea. Rev.,* **105,** 1568–1589.

———, 1982: Use of a conceptual model of the life cycle of mesoscale convective systems to improve very-short-range forecasts. *Nowcasting,* K. Browning, Ed., Academic Press, 191–204.

Chapter 10

Severe Local Storms in the Tropics

GARY BARNES

Department of Meteorology, University of Hawaii, Honolulu, Hawaii

REVIEW PANEL: E. S. Zipser (Chair), B. Hanstrum, E. Williams, M. Douglas, and H. Thompson

10.1. Introduction

The Tropics evoke romantic images of spectacular thunderstorms that tower above an aquamarine sea (Fig. 10.1). These impressive-looking cumulonimbi might lead one to believe that they also have vigorous circulations and thus have the potential to spawn severe weather. Surface station climatologies, satellite imagery, and lightning detection networks all demonstrate that the tropical landmasses have the highest frequency of deep convection in the world. Despite this high count, I will argue that all types of severe weather [tornadoes, hail with a diameter $\geq$ 20 mm (0.75 in.), and straight-line winds $\geq$ 26 m s^{-1} (50 knots or 93 km h^{-1})], save waterspouts, are rare over the tropical oceans and sporadic over tropical lands. Expansive regions of the Tropics are dominated by marginal conditional instability, dry midlevel air, and subsidence (e.g., the trades), while in the near-equatorial trough, the shear of the horizontal wind is weak, and the conditional instability, though it can reach high values, is spread through a deep troposphere, resulting in only modest updraft velocities. Severe local storms (SLS) thrive on the rapid juxtaposition of air with differing characteristics in the horizontal and especially the vertical. For much of the Tropics, trajectories over homogeneous warm seas erase these differences and soften the gradients conducive for SLS. Over the tropical continents, however, the conditions that favor SLS occur with greater frequency.

I believe that the body of evidence found in the literature or in other sections of this monograph will support the following tenets.

1) Tornadoes of F3 or greater intensity require a substantial updraft in the presence of strong low-level shear.
2) Strong straight-line winds require a dry layer in the lower to midtroposphere to promote cooling from evaporation and melting.
3) Large hail needs a strong updraft and a region of supercooled water.
4) More modest vortices such as waterspouts require horizontal shears in the boundary layer that can be created in a cloud line.

These tenets are clearly an oversimplification, but they do highlight the premise that each type of SLS has different minimum needs. Of course, a cumulonimbus (Cb) forming in strong shear and substantial conditional instability may spawn all three types of SLS. In the Tropics, instability is often distributed evenly through a deep troposphere (16–18 km). Though the magnitude of the instability may be high, the updrafts rarely achieve the velocity or the acceleration to form SLS. Shear in the upper troposphere may promote the genesis of deep convection, but not necessarily SLS.

Tropical cyclones (TCs) have earned the reputation for being the primary threat to life and resources in the Tropics. I will not discuss the vigorous sustained winds that are found in the TC on the meso-β scale, but I will discuss tornadoes that form in the environment created by the TC. Besides tropical cyclones, there are a number of more benign planetary- and synoptic-scale features that shift the odds toward SLS, which I will briefly discuss.

Issues that I will explore include the following.

1) What are the planetary- and synoptic-scale phenomena that create an environment favorable for the genesis and maintenance of deep convection and SLS? Conversely, are there regions where SLS occur that are not embedded in larger-scale phenomena?
2) What regions witness hail, severe straight-line winds, and intense vortices, what is their frequency of occurrence, and do these events have a preferred season?
3) Is the frequency of SLS correlated with the frequency of thunderstorms or cumulonimbi? If not, then what are the mechanisms that limit the frequency of SLS in the Tropics?

FIG. 10.1. The large cumulonimbus over Bathurst and Melville Islands, north of Darwin, Northern Territory, Australia, affectionately known as Hector.

4) How should one attempt to forecast SLS in the Tropics? What remote sensing techniques are the most useful for SLS forecasting?

A limiting aspect of this review is the lack of any meteorological observational network in the Tropics that can rival what is found in the United States or Europe. This inadequate record will haunt these findings. My intention is that the review will serve as the first step toward understanding SLS in the Tropics.

A universally accepted meteorological definition of the Tropics remains elusive. I shall define the Tropics as the latitudinal belt from the Tropic of Cancer to the Tropic of Capricorn, but will include territory farther north in India to the Himalayan foothills. Wet and dry seasons may exist, but there is no time when the region has persistent winter conditions at sea level. I will also discuss summertime Florida peninsula conditions. Florida is blessed with a surfeit of observations and thus serves as a proxy for other data-sparse tropical regions.

10.2. Planetary-scale features of the Tropics

There is an array of tropical planetary-scale phenomena (Hastenrath 1991) that alter thermodynamic and kinematic conditions, which contradicts the popularly held notion that the Tropics are homogeneous and unchanging. These phenomena offer a range of environments for SLS that are a function of both their own life cycle and the annual cycle. Figures 10.2a, 10.2b, 10.2c, and 10.2d are an attempt at quickly summarizing the "where" and "when" aspects of planetary- and synoptic-scale phenomena in the Tropics. Planetary surface features include, but are not limited to, the trades, the near-equatorial trough and convergence zone, the South Pacific convergence zone, and the monsoon. Note that global averages suffer from blending two disparate regions, one monsoonal, the other oceanic. I shall first discuss the primary lower-troposphere features, then move to the upper-level phenomena that may affect convection and severe weather.

a. The trades

The trade wind region (see Fig. 10.2) covers the largest portion of the Tropics and the mean conditions found there do not readily support deep convection (Riehl 1954, 1979; Malkus 1956; Malkus and Riehl 1964; Augstein et al. 1973; Albrecht 1984). The trade regions are dominated by a lower-tropospheric inversion, low-level divergence, and midtropospheric subsidence; soundings from this region are conditionally unstable for only a shallow layer below the trade wind inversion, and are dry neutral to absolutely stable through much of the mid- and upper troposphere (e.g., Malkus 1956; Riehl 1979; Kloesel and Albrecht 1989). The east and central portions of the Atlantic and Pacific Ocean basins are where the trades are at their driest and most stable, with the inversion being low

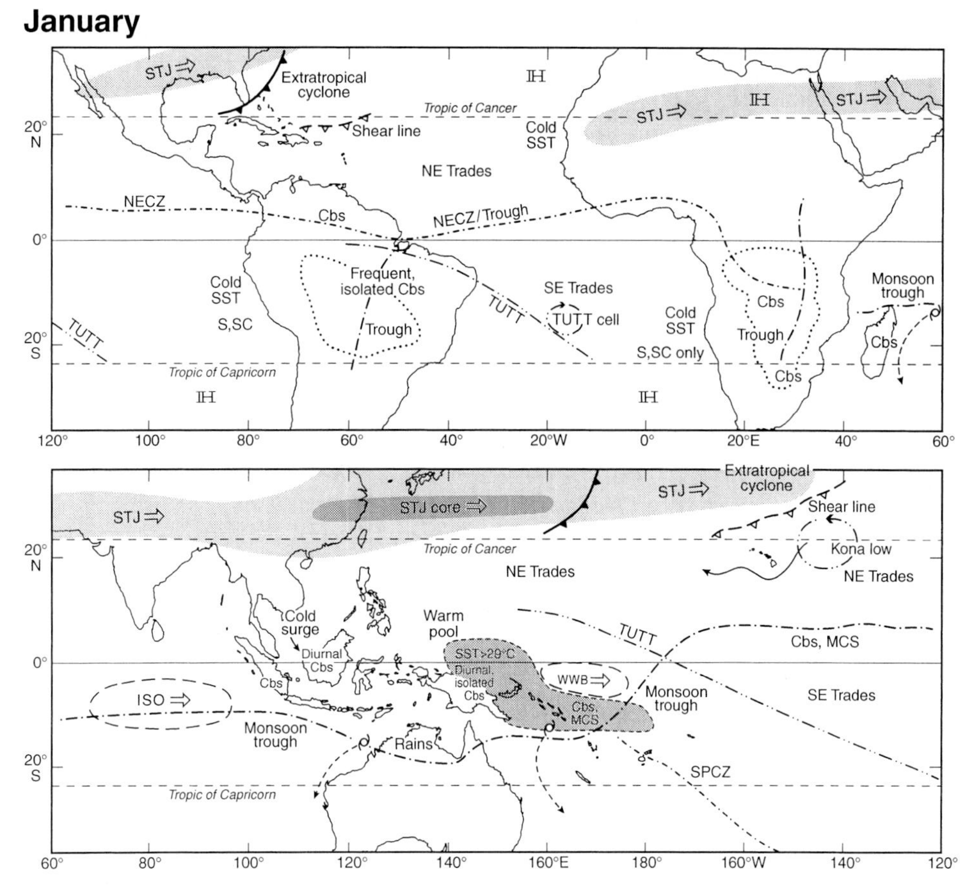

FIG. 10.2. Schematics of the primary planetary- and synoptic-scale features found at the surface in the Tropics in January for (a) the Atlantic, (b) the Pacific and, in July, (c) for the Atlantic and (d) the Pacific.

and strong with a large drop of humidity occurring across the inversion (Fig. 10.3). The air below the trade wind inversion moistens, warms, and deepens as one travels westward and equatorward and away from the western edge of the continents. Along a typical parcel trajectory to the equator the temperature difference across the inversion layer steadily erodes from about 5°C to less than 2°C and specific humidity below the trade wind inversion increases substantially. However, even at latitudes within 10° of the equator, half the soundings from the central Pacific analyzed by Kloesel and Albrecht (1989) show either a strong trade wind inversion with a base at 850 mb and top at 800 mb or simply subcloud layer air that is too dry to support deep convection. Soundings from Lihue, Hawaii, in the trades, and from Majuro, in the equatorial trough, demonstrate the large change in conditional instability and relative humidity in the midlevels as one moves southwest toward the equator (Fig. 10.4).

Sample soundings from the Atlantic Tradewind Experiment (ATEX 1969), conducted in the central Atlantic, show only a shallow layer with little convective available potential energy (CAPE). Observations in the eastern Pacific from Riehl (1979) and from ATEX (Augstein et al. 1973) for five days show strong lower-troposphere divergence and mean downward motion (Fig. 10.5). A convex wind profile with the maximum speed in the middle of the trade wind layer is often found, but this results in weak shear ($\sim 1. \times 10^{-3}\ s^{-1}$) over a layer less than 500 m (Riehl et al. 1951; Augstein et al. 1973). Based on the studies of LeMone et al. (1984a), we do not expect a dynamic pressure perturbation of the magnitude that would aid in the maintenance or intensification of deep convection. On the western sides of the ocean basins the influence of the subtropical highs is diminished and more substantial convective clouds increase in frequency.

b. Near-equatorial convergence zone or equatorial trough

The term *near-equatorial convergence zone* (NECZ, see Fig. 10.2 schematic) is an attempt at a more encompassing, and grammatically correct, label for the Intertropical Convergence Zone (ITCZ). The NECZ is the floor for the upward branch of the Hadley cell. Here, in the doldrums, large-scale gentle ascent is not

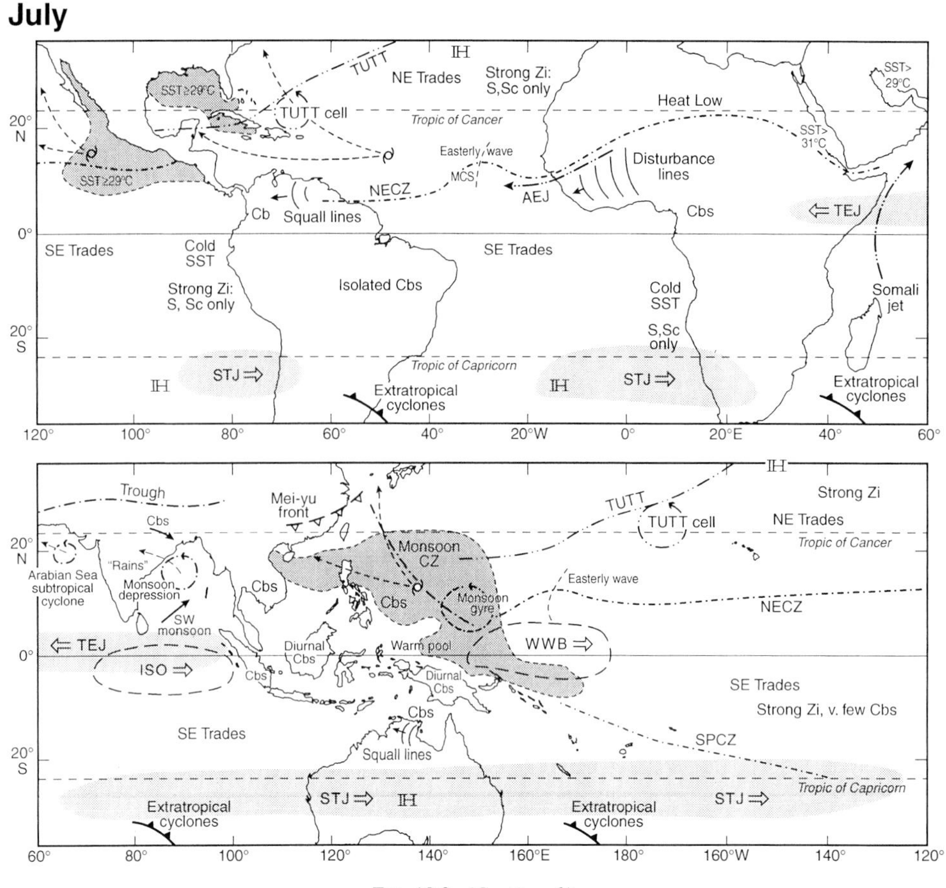

FIG. 10.2. (*Continued.*)

the dominant mechanism for vertical exchange; instead "hot towers" or cumulonimbi serve as the channels for the upward transport of heat and moisture (Riehl and Malkus 1958). The convergence is between the northeast and southeast trades when the zone is near the equator and between trades at higher latitudes and the westerly or monsoon flow when the zone is farther away from the equator (McBride 1987). If one prefers pressure to convergence, then *equatorial trough* and *monsoon trough* would be the appropriate terminology. The distinguishing characteristic for the monsoon trough is westerly flow on the equator side for the summer hemisphere. The term may seem like an unnecessary complication, but the monsoon trough will have much larger vorticity associated with it than the convergence caused by the confluence of the northeast and southeast trades.

The schematic for the ocean basins (Figs. 10.2b, 10.2d) shows that the NECZ migrates with—albeit with a lag of about two months—the annual march of the sun's zenithal latitude. The center of the confluence is biased toward the Northern Hemisphere because of the greater amount of land mass found there (Riehl 1954, 1979). Note that the monsoon trough, which stretches from 30°W to 170°E by Ramage's (1971) definition, can be oriented more northwest to southeast. The upper-level cloudiness, and the boundary layer convergence and pressure trough at the surface, are not collocated (Sadler 1975b; Sadler et al. 1987). The trough is a broad feature (1000 km in the north–south direction) while the convergence zone and accompanying deep clouds are usually focused over a much smaller latitudinal belt (100–200 km).

At times a large monsoon gyre (Lander 1994) can form in the western Pacific, which appears to be a result of interaction between the monsoon southwesterlies and the midlatitudes. This gyre may survive for two to three weeks, is about 2500 km in diameter, and has a band of deep convective clouds on its south and east sides. Small tropical cyclones are spawned in this band and the gyre itself can evolve into a large tropical cyclone (Lander 1994).

Soundings reveal that the typical shear and instability for the NECZ is lower than what is believed to be needed for F3 or better tornadoes, and for large hail (e.g., Barnes and Sieckman 1984; Firestone and Albrecht 1986; Williams and Renno 1993; Lin and Johnson 1996). The deep Tropics over the central Pacific rarely possess severe sounding conditions (CAPE > 2200 m^2 s^{-2}, low-level shears $> 4.0 \times$

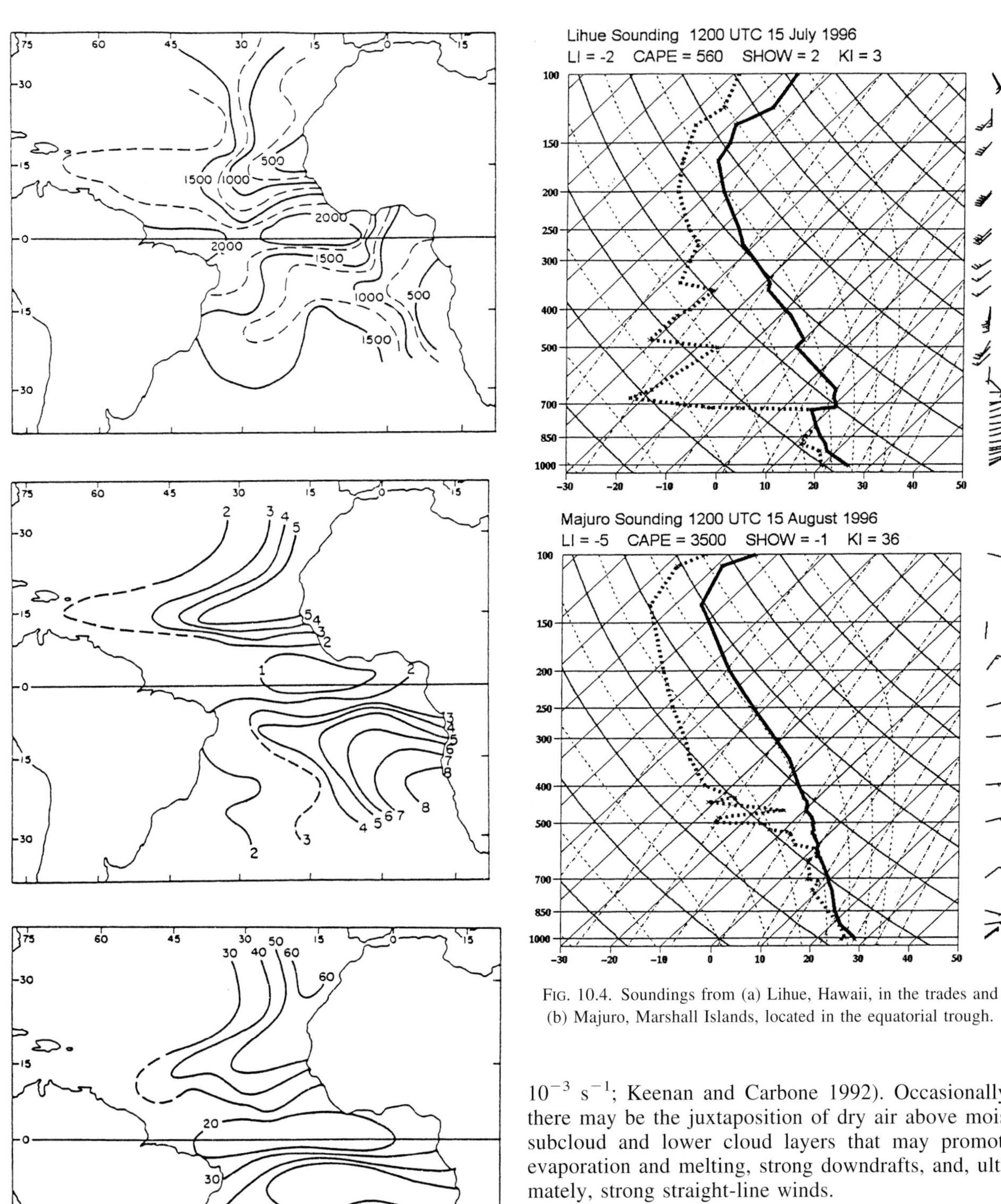

FIG. 10.3. Trends in (a) inversion height (m), (b) temperature jump across the inversion (°C), and (c) relative humidity jump across the inversion over the Atlantic Ocean. From Riehl (1979).

FIG. 10.4. Soundings from (a) Lihue, Hawaii, in the trades and (b) Majuro, Marshall Islands, located in the equatorial trough.

10^{-3} s^{-1}; Keenan and Carbone 1992). Occasionally, there may be the juxtaposition of dry air above moist subcloud and lower cloud layers that may promote evaporation and melting, strong downdrafts, and, ultimately, strong straight-line winds.

c. South Pacific convergence zone

The South Pacific convergence zone (SPCZ, see Figs. 10.2b, 10.2d) was first detected by Bergeron using surface analyses; based on satellite pictures it appears to be one of the earth's most expansive and persistent cloud bands (Vincent 1994). The SPCZ stretches east-southeastward from New Guinea to about 30°S, 120°W. Near the equator it merges with

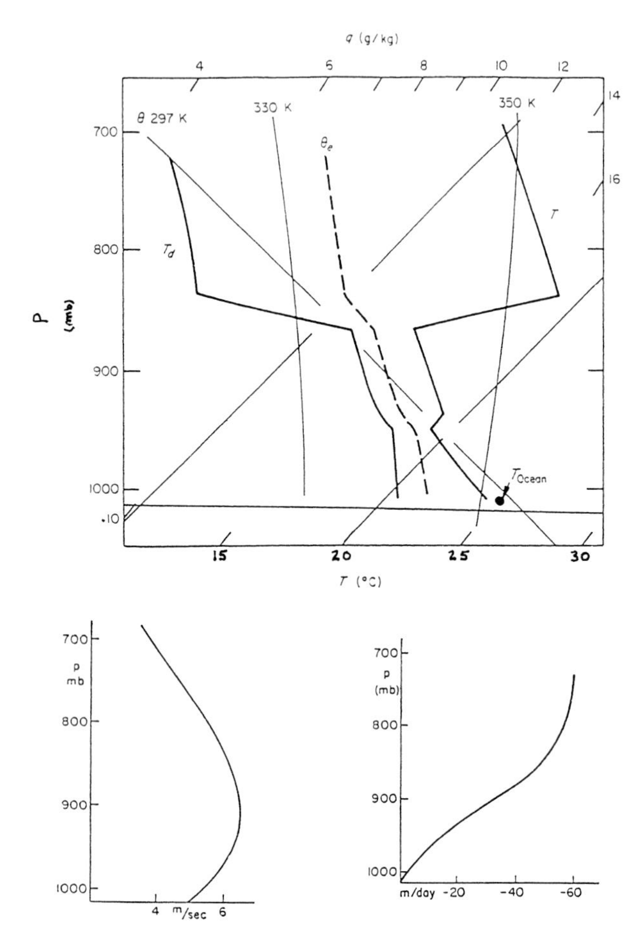

FIG. 10.5. (a) Tephigram of the mean ATEX sounding 7–12 February 1969, (b) vertical cross section of the wind speed (m s^{-1}), and (c) vertical motion (m day^{-1}). From Riehl (1979) and Augstein et al. (1973).

the NECZ. During the austral summer it is essentially a monsoon trough and reaches its most intense stage; during the austral winter it is a convergence zone between the South Pacific high and the anticyclone that dominates the Australian continent. The cloudiness maximum associated with the SPCZ can also be due to upper-level troughs that become stationary (Ramage 1995). During the Southern Hemisphere summer, tropical depressions can develop within the SPCZ and may be stationary for a number of days before moving east or southeast (Vincent 1982, 1985).

Cross sections through the SPCZ reveal that it is convergent below 750 mb, divergent above 400 mb, and warmer, moister, and more sheared than the surrounding environment (Fig. 10.6). The shear is primarily above 500 mb where it has less effect on the dynamics of SLS. There has not been a study of the SPCZ with the horizontal resolution to determine the detailed structure of embedded convectively active regions.

d. Monsoons

The monsoon, using the stringent definition of Ramage (1971) that is based on flow, not precipitation, covers a large portion of the globe (Fig. 10.7) and is often decomposed into smaller facets based on a particular region (e.g., Indian, Australian, northwest Pacific). It is a result of the reversal of the temperature gradients caused by the heating and cooling of the Asian continent and especially the Tibetan Plateau. Cold oceanic upwelling along the north and west coasts of South America and along the west coast of southern Africa prevents monsoon development in those locations (Ramage 1971). During the boreal summer an upper-level anticyclone, centered over the Tibetan Plateau, overlies the southwesterly flow that is about 5 km thick. Only a relatively shallow layer of the southwesterlies nearest the sea is moist, with the remaining southwesterlies consisting of dry, warm air originating from Arabia and Africa. During the austral summer low-level flow from the west-northwest (WNW) is overlaid by an anticyclone centered near northern Australia.

Contrary to popular belief, the monsoon is not a period of thunderstorms. Despande (1964) and Ramage (1971; Fig. 10.8a) show that as precipitation increases, thunderstorms decrease. Chatterjee and Prakash (1990), using eight years of 3.2-cm radar data from near Delhi, show that the percentage of convective clouds with tops above 8 km decreases as the monsoon is established. The virtual potential temperature difference (Fig. 10.8b) from 850 to 300 mb decreases by 10°–12°C as the temperature profile moves closer to a moist adiabatic lapse rate. Shear does increase through this same layer, but the increase is primarily due to the development of the upper-level tropical easterly jet (TEJ). Supporting these results, albeit for Australia, are Williams et al. (1992), who show that lightning is about 10 times more frequent in continental or break monsoon conditions than it is for the active stages of the monsoon. Wet-bulb potential temperature near the surface is reduced by 1°–2°C during the active monsoon; this leads to a reduction of convective instability, less ice, and far less lightning. It is during the premonsoon season that convective instability can grow to impressive magnitudes and some of the more vigorous convective events take place. Zipser (1994) also shows that peak thunderstorm activity over West Africa does not occur during the maximum rain.

Once the monsoon is established, the shear in the lower troposphere is modest. Figure 10.9 shows two meridional cross sections, one through India (73°E) during the boreal summer and the other through Australia (140°E) during the austral summer (Davidson et al. 1983). Note that the shear of the u component is less than 3.0×10^{-3} s^{-1} below 500 mb for both regions. Again, the TEJ does produce stronger shears,

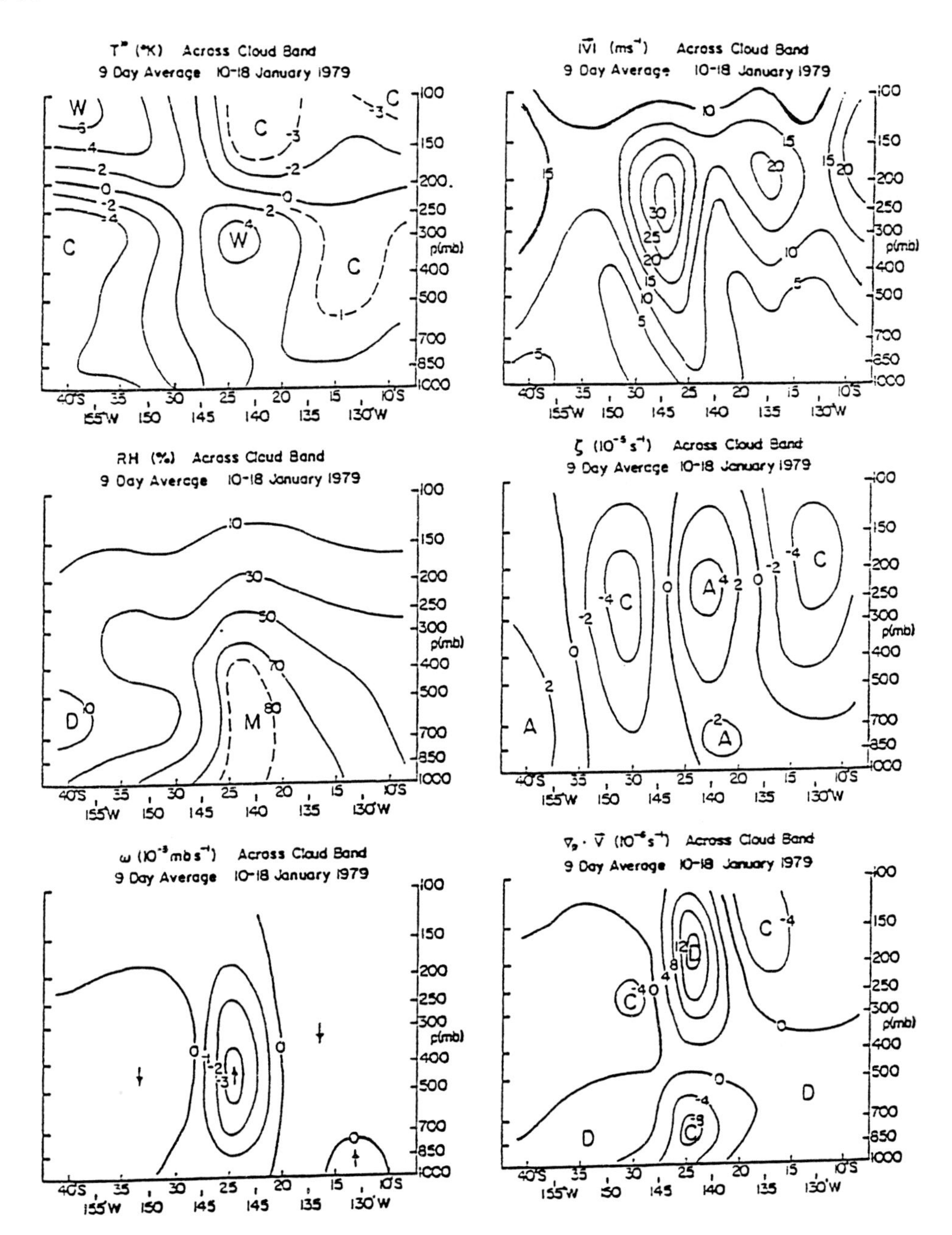

FIG. 10.6. Vertical cross sections of (a) temperature deviations (K), (b) wind speed (m s^{-1}), (c) relative humidity (%), (d) relative vorticity ($\times 10^{-5}$ s^{-1}), (e) vertical p velocity (mb s^{-1}), and (f) divergence ($\times 10^{-6}$ s^{-1}) normal to the SPCZ cloud band. From Vincent (1982).

but only in the upper troposphere. Gunn et al. (1989) support the findings for the Australian region based on Australian Monsoon Experiment data.

Views of oceanic conditions during the monsoon are available from the Monsoon Experiment (MONEX-79). Ray and Bedi (1985) use the rawinsondes from three arrays of ships in the Arabian Sea and the Bay of Bengal to depict the mean zonal and meridional winds over the ocean surrounding India during the monsoon season; the shear of either component is less than 1.0×10^{-3} s^{-1} from the surface to 700 mb. The lapse rate of moist static energy is about 15 kJ kg^{-1} from the surface to 600 mb; this is a modest change that would not support the formation of vigorous downdrafts via evaporative cooling.

Most of the severe weather in northern India occurs in the period between the winter and summer monsoons and is related to westerly waves superimposed on the warm and moist air originating over the Bay of Bengal (e.g., Ramage 1971, 1995; Lal 1990).

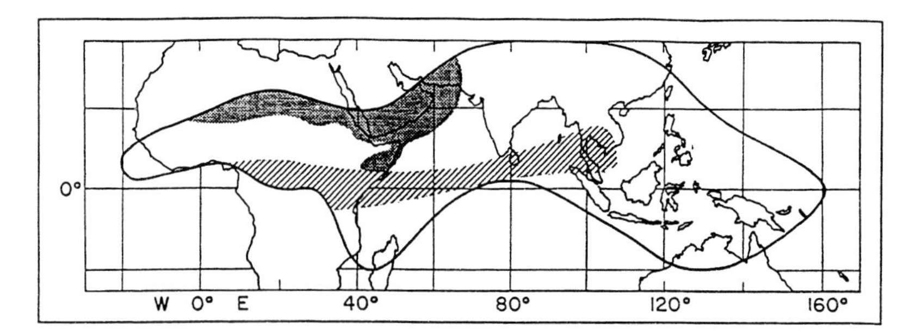

FIG. 10.7. The region of the world regarded as monsoonal, based on the definition of Ramage (1972). Within the enclosed area, deserts (average rainfall less than 250 mm) are shaded. Area with a bimodal annual rainfall variation is hatched. From Ramage (1995).

e. 40–50-day or intraseasonal oscillation (ISO)

The 40–50-day tropical intraseasonal oscillation (ISO; Madden and Julian 1972, 1994) migrates eastward around the globe approximately along the equatorial plane. A region of deep convection that leads low surface pressure and 850-hPa westerly winds are a few of its traits when it is located in the Indian Ocean and western Pacific. The area of enhanced convection is approximately 3000–4000 km across, within 15° of the equator, and contains numerous mesoscale convective systems (MCS; Chen et al. 1996) (Fig. 10.10). The oscillation does not generally support deep clouds east of the date line as the environment becomes less supportive of convection (e.g., SSTs cool, convergence weakens, CAPE decreases). The cloud activity associated with the 40–50-day oscillation is strongest in the summer hemisphere. Mean brightness temperature for the intensive flux array (IFA) that was centered at about 3°S, 155°E over the west Pacific warm pool throughout COARE (Lin and Johnson 1996) reveals that the coolest temperatures aloft do occur in the center of the cloud field, but this is interpreted to be the cold anvils that continue to slowly rise, not necessarily vigorous convection directly below the lowest temperatures. Infrared temperatures as much as 50°C higher are found before and after the ISO (sensors measure the cloud temperatures in the lower troposphere), suggesting that the subsidence and drying associated with the circulation strongly modulate deep convection activity.

Lin and Johnson (1996) use TOGA COARE rawinsonde data to depict the kinematic structure of the 40–50-day oscillation over the IFA. Two events can be seen in the filtered u-component winds (Fig. 10.11a) around mid-December and late January. When the low-level westerlies occur (called a *westerly wind burst*), the shear increases substantially (Fig. 10.11b). In the core of the ISO, shear in the u component can exceed $6.5 \times 10^{-3}\ s^{-1}$ from the surface to 850 mb. This shear is double that found in the eastern Atlantic during GATE (e.g., Frank 1978; Barnes and Sieckman 1984). Shear in the v component for the IFA region does not vary much (Lin and Johnson 1996). The strong low-level shear supports the formation and maintenance of squall lines. However, convective activity appears to lead the core of the westerly winds by 10–15 days, suggesting that most of the deep convection does not exist in the more strongly sheared conditions.

Recently Anyamba et al. (1997), in contrast to most other work on the ISO (e.g., Salby and Hendon 1994), detect a modulation in the deep clouds and lightning activity over the continents that appear to be in phase with the ISO. One can imagine that a combination of the large-scale forcing of the ISO, in conjunction with the sensible heat inputs from the land, results in deeper convection, and perhaps an increase in SLS.

f. Heat lows

Surface heat lows form above large tropical desert regions in the summer (the Sahara and Arabian deserts, northern Australia, and Namibia; see Fig. 10.2c). The excessive insolation heats the surface, and high sensible heat flux from the surface to the atmosphere results in a trough. The trough is shallow and usually under the influence of the subtropical highs that are the descending branch of the Hadley cell. The subsidence yields a strong inversion and very dry conditions aloft, making deep convective clouds extremely rare, despite the low-level convergence induced by the heat trough. Heat lows are often several thousand kilometers in dimension, but their lifetime of months, and their anchoring to continents, make them candidates for a planetary-scale label.

g. Tropical upper-tropospheric trough

The tropical upper-tropospheric trough (TUTT), diagnosed by Sadler (1975a) in global climatological analyses, develops during the summer over the Atlantic and Pacific Oceans. It is most easily detected in the resultant wind fields as a region with cyclonic flow in the 200–300-hPa layer (Fig. 10.12), and is found south of the subtropical ridge and north of the subequatorial

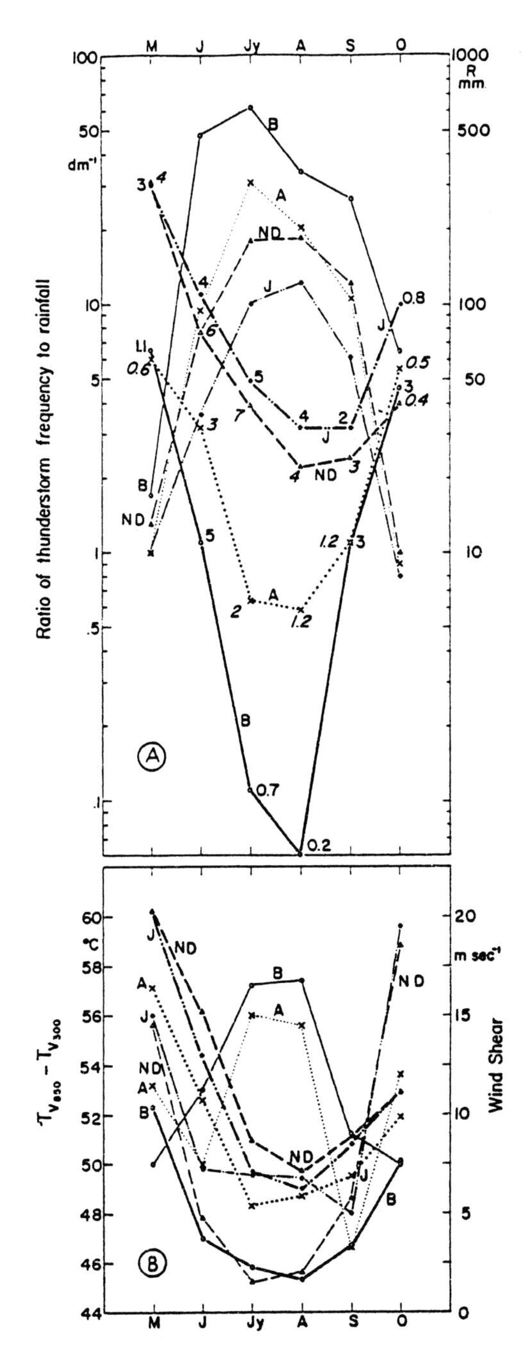

FIG. 10.8. (a) Ratio of thunderstorm frequency to rainfall in decimeters shown with heavy lines; numbers denote thunderstorm frequencies and light lines denote rainfall. (b) Difference in virtual temperature between 850 and 300 mb (heavy lines) and wind shear (light lines) through the same layer. From Ramage (1995).

ridge (20°–30° latitude). The feature appears in response to heating of the continents and is oriented WSW–ENE in the Northern Hemisphere. The TUTT can result in a reduction of upper subsidence, which then results in the formation of deeper clouds in the trade wind belt. The TUTT is the locus of a series of cold core lows, often called *TUTT cells,* which give rise to deep convection, and which I shall discuss in the synoptic-scale section.

h. Jet streams: subtropical, tropical easterly, and low level

Upper-level jet streams have been implicated in the genesis and maintenance of deep convective activity and severe storm occurrence (e.g., Ramaswamy 1956; Uccellini and Johnson 1979; Reiter 1961). The subtropical jet stream (STJ), created by partial conservation of angular momentum, occurs at the upper boundary between tropical and midlatitude air (e.g., Krishnamurti 1961). The magnitude of the STJ is closely correlated with the Hadley cell circulation, which is strongest in winter. During the winter the polar and the subtropical jets can combine when a strong trough dips into the low latitudes. About 80% of the time the STJ is located between 25° and 30° north of the equator, but Sadler (1975a) shows the mean annual location of the STJ dipping into the Tropics (south of the Tropic of Cancer) near Hawaii, West Africa, and northeast India (see Fig. 10.12). The strong shear associated with the STJ is concentrated in the upper 100–250-mb layer of the troposphere. Shear induced by the STJ in the lower troposphere is weak, hence the STJ probably plays only a limited role in creating dynamic pressure perturbations (e.g., Rotunno and Klemp 1985; LeMone et al. 1988b) that are crucial for supercell formation and maintenance.

The upper-level outflow from deep convection originating within the NECZ can interact with an amplifying trough to the north to produce a synoptic-scale band of upper-level cloud called a *tropical plume* (McGuirk and Ulsh 1990). This narrow plume may stretch thousands of kilometers downstream. The plume collapses when the tropical moisture source and the STJ become separated. The plumes are generally triggered to the east of a midlatitude trough and simply serve as a tracer for the subtropical jet (Schroeder 1983).

The tropical easterly jet (TEJ; see strong flow over southern India, Fig. 10.12) is generated by sensible heating and the barrier effect of the Tibetan Massif, and prodigious amounts of latent heat released by the convection embedded within the monsoon. The TEJ is best developed in the Northern Hemisphere south of Asia; there appears to be a reinforcement of the TEJ over equatorial West Africa that may be due to the heating of the Sahara. Easterly flow first appears in May and becomes intense through the summer. By October the easterly flow has weakened and jet characteristics disappear. The TEJ has winds exceeding 50 m s^{-1} in the 100–200-mb layer and, like the STJ, is most strongly sheared in the upper troposphere.

Low-level jets (Stull 1988) often develop where there are persistent large temperature gradients and subsidence above the boundary layer (Ramage 1995).

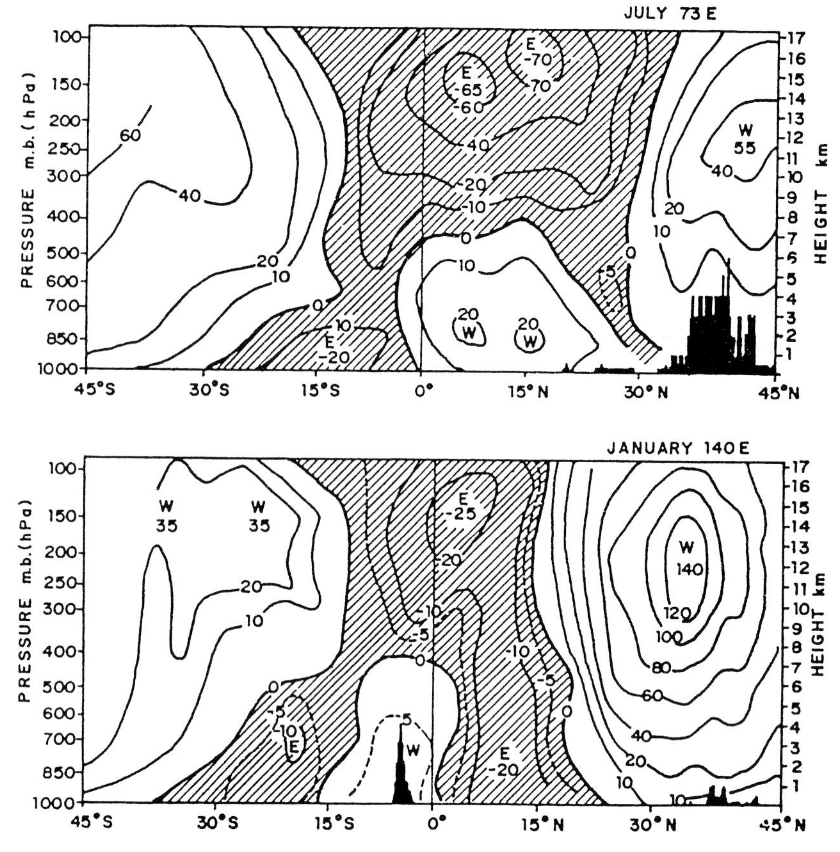

FIG. 10.9. Climatological cross sections of the zonal wind component for 73°E (through India, Himalayas depicted by dark blocky shading) in July and for 140°E in January (through Australia, with high terrain of New Guinea prominent). Wind speeds in knots. From Davidson et al. (1983).

Locations include desert littorals, especially where cold waters are upwelling (e.g., the west coast of South America, east coast of Africa). Mountains along the coast, as well as the subsidence inversion, contribute to the maintenance and intensity of low-level jets. The Somali jet (Fig. 10.2a global schematic) is one of the more intense found, with wind speeds reaching 30 m s^{-1} at 1500-m altitude; this yields shears that can exceed 10.0×10^{-3} s^{-1} from the jet core to the surface. The Somali jet is most often found in stable conditions that do not allow for the formation of deep convection.

The reversed horizontal temperature gradient between the hotter Sahara and the cooler, moister Sahel in the summer produces the African easterly jet (AEJ) in the 600–700-mb layer. The jet and its attendant shear partially modify the MCSs that form over West Africa and the ocean region immediately downstream. When I discuss disturbance lines (Hamilton and Archbold 1945) over Africa, this feature will receive more discussion.

Low-level jets occur south of the Mei-yu front in the spring and early summer over south China. These are most often associated with a lee trough developing behind the Tibetan Plateau (Chen and Chen 1995). Since they are chiefly associated with midlatitude waves, these jets are discussed in chapter 2 of this volume.

10.3. Synoptic-scale features of the Tropics

Palmer (1951b) describes the perturbation approach to tropical meteorology, developed by the University of Chicago, whereby one focuses on the "weather making" phenomena rather than on climatological means. The easterly wave was the feature of interest then; now we see that the Tropics have a variety of synoptic and mesoscale perturbations that may alter the shear, conditional instability, and the divergence profile, and thus affect the occurrence of SLS.

a. Easterly waves

Easterly waves, which modulate the pattern and intensity of convection in the NECZ, have been detected in the tropical north Atlantic (e.g., Dunn 1940; Riehl 1954; Burpee 1972) and in the western Pacific (e.g., Reed and Recker 1971; Yanai 1963; Wallace

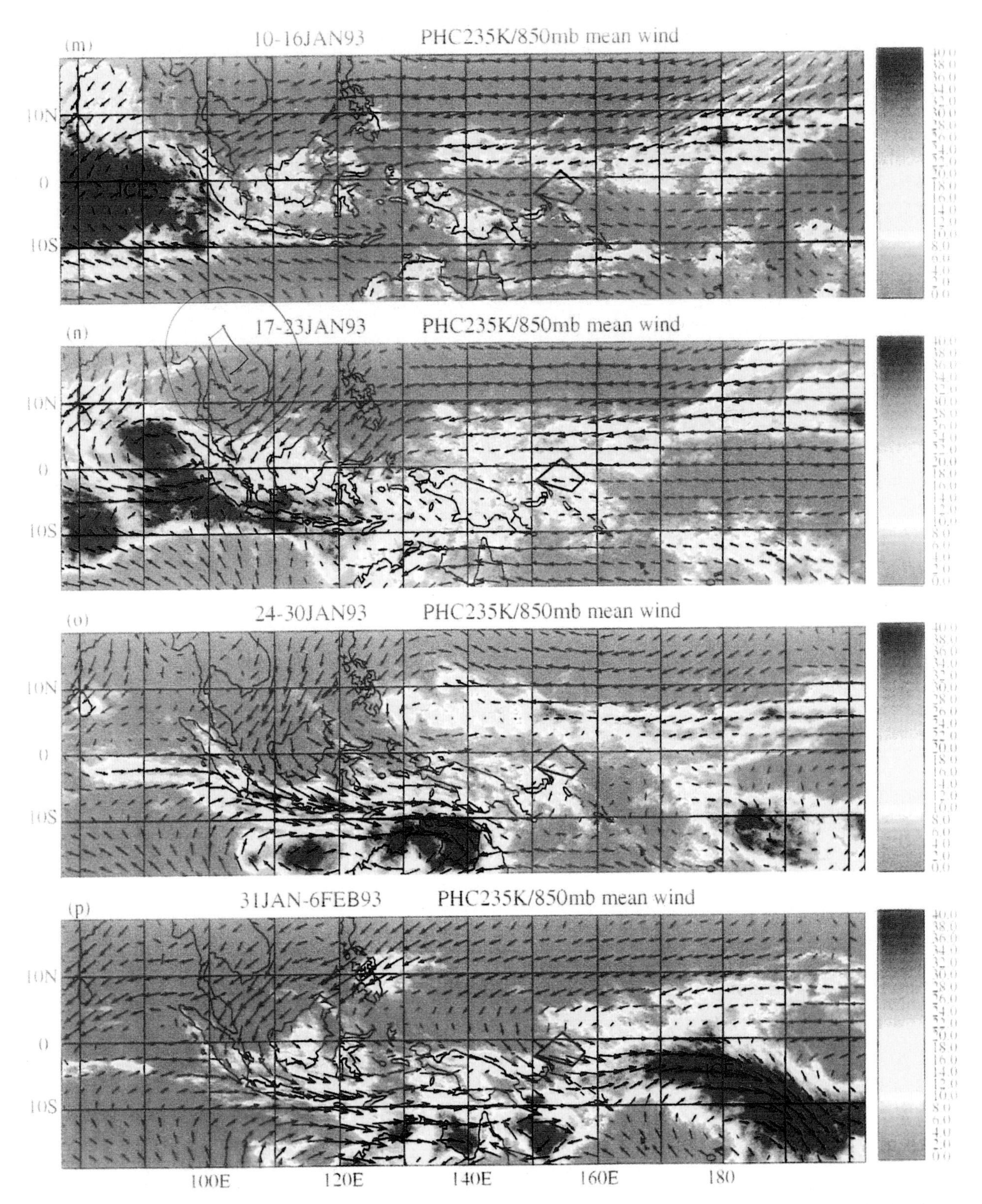

FIG. 10.10. Weekly percent of high cloudiness calculated as the portion of the week that each 10 km × 10 km pixel in the GMS IR image had a value of 235 K or less, from 10 January to 6 February 1993. Weekly mean winds from ECMWF global wind analysis with 2.5° resolution. Vectors are scaled such that 20 m s^{-1} is equal to 2.5°. From Chen et al. (1996).

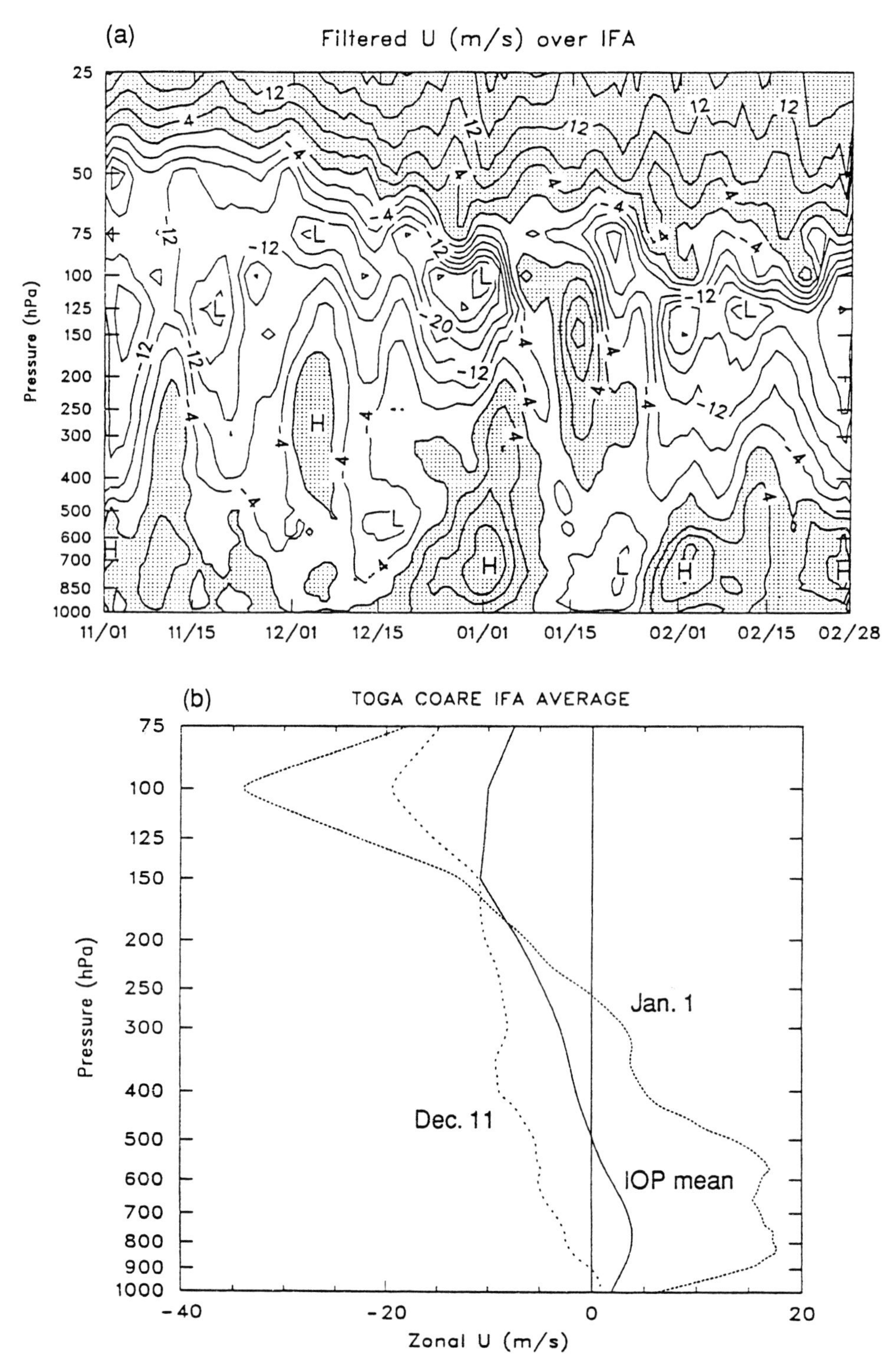

FIG. 10.11. (a) Five-day running mean time series of zonal wind (m s^{-1}) at 0000 UTC over the IFA from TOGA COARE. Contour intervals of 4 m s^{-1} with westerlies shaded. (b) The intensive operations period mean vertical profile of the zonal wind and daily average u profiles for the IFA on 11 December and 1 January, with positive values depicting westerlies. From Lin and Johnson (1996).

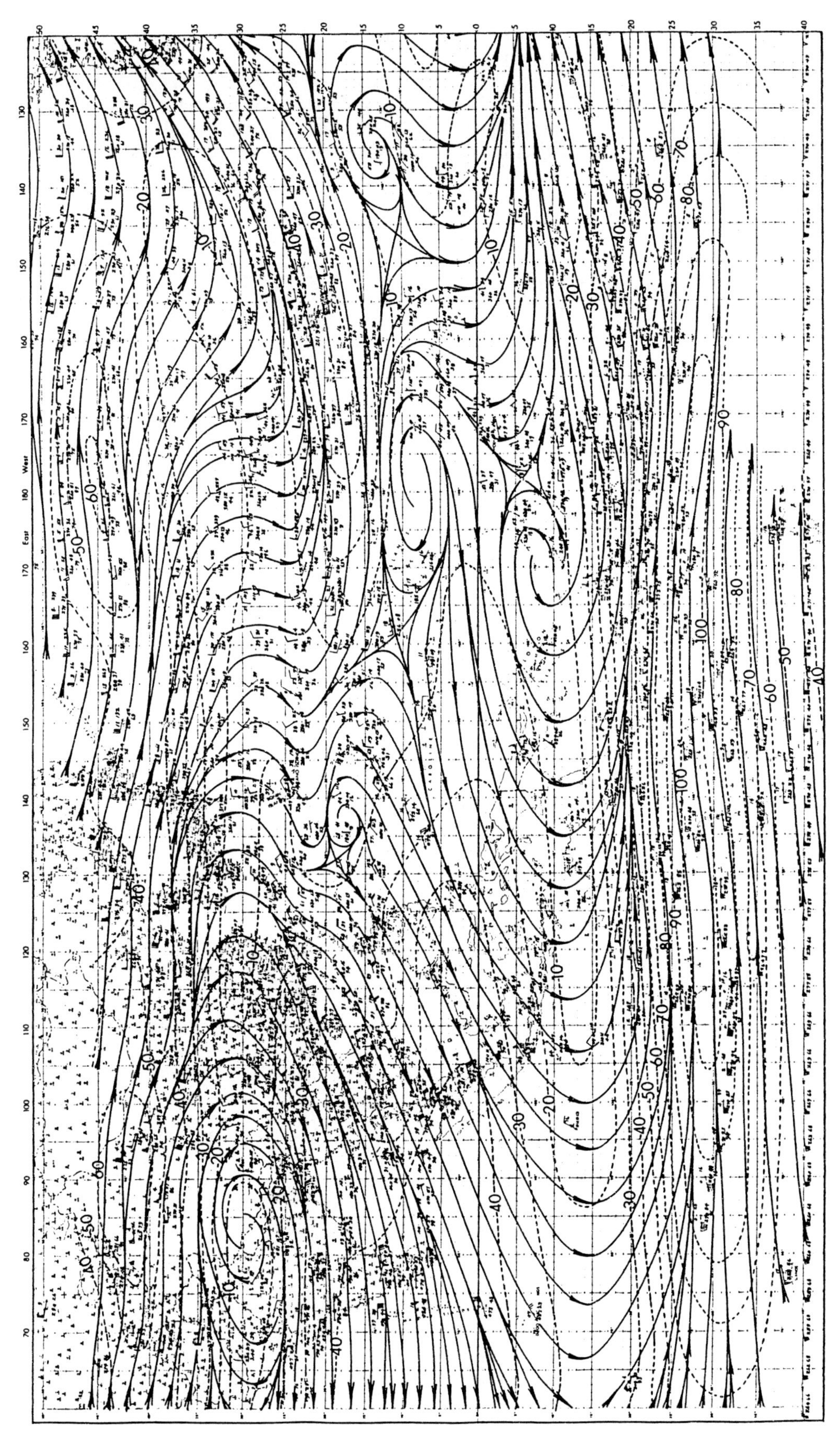

FIG. 10.12. July 200-mb flow from 60°E to 140°W. Isotachs (m s^{-1}) are dashed. From Sadler et al. (1975).

1971); the most convincing descriptions are based on composite upper-air analyses based on GATE data (e.g., Reed et al. 1977; Norquist et al. 1977; Thompson et al. 1979). The Atlantic waves are difficult to track once they interact with the strong diurnal and topographic forcing found over Central America. In the eastern Pacific, the stable conditions at very low latitudes suppress the convective clouds that are used to identify and track the wave axis. The lack of data has also hindered identification of the waves in the eastern Pacific, but Nitta and Takayabu (1985) found disturbances that propagate toward the WNW. Mixed Rossby–gravity waves have been detected

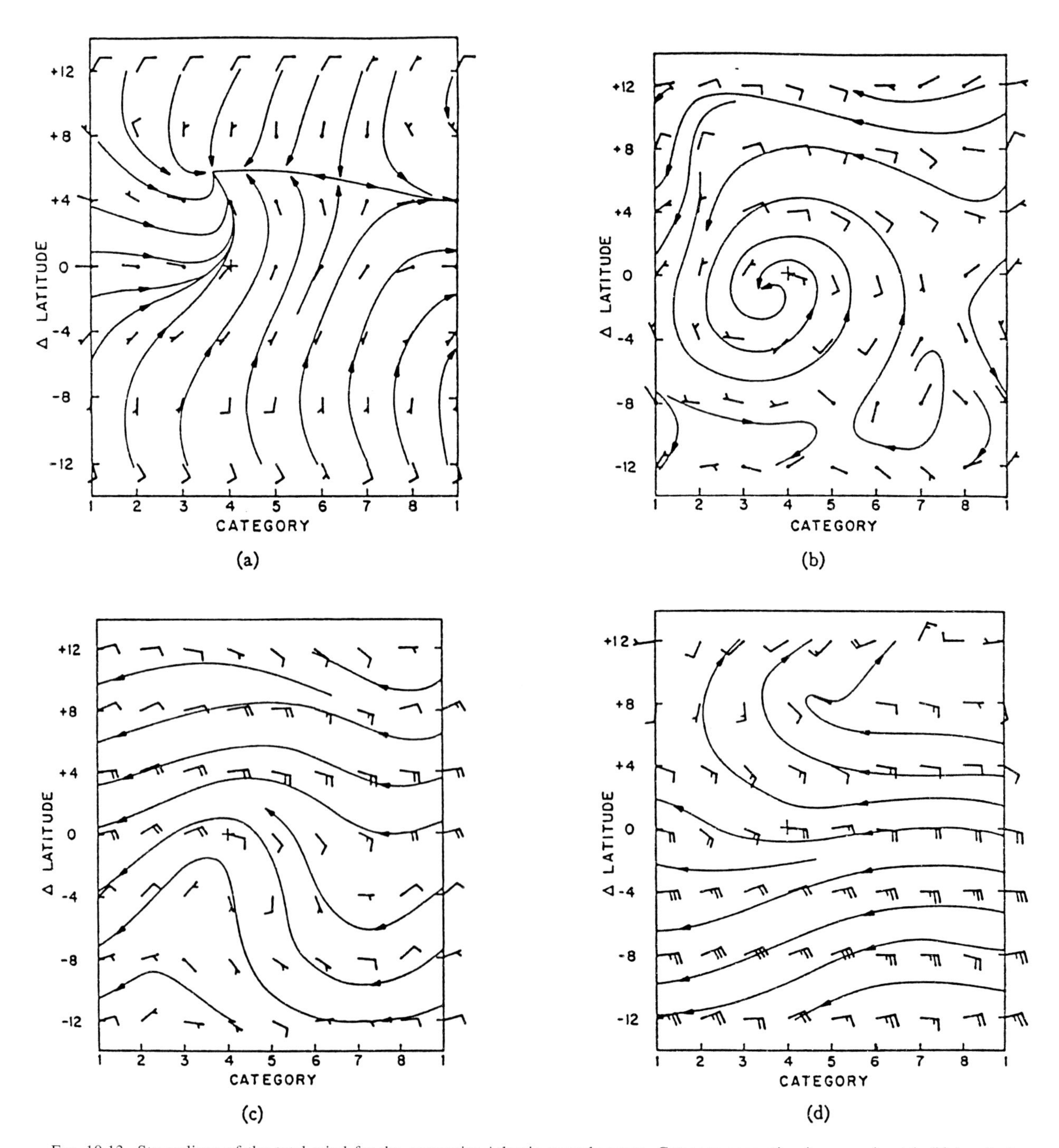

FIG. 10.13. Streamlines of the total wind for the composite Atlantic easterly wave. Category separation is approximately 3° longitude, cross marks center of disturbance at 700 mb. One full barb corresponds to 5 m s^{-1}. (a) Surface, (b) 850 mb, (c) 700 mb, and (d) 200 mb levels are shown. Categories 2, 4, 6, 8, correspond to the regions of maximum northerlies, the trough, maximum southerlies, and the ridge, respectively. From Reed et al. (1977).

by Liebmann and Hendon (1990) within 10° of the equator. These waves move west and have periods of less than 6 days.

Atlantic waves propagate westward at 5–8 m s^{-1}, have a mean wavelength of 2500 km, and have a period of 3.5 days (Reed et al. 1977). The total wind field (Fig. 10.13) composited from several waves sampled during GATE for various levels as a function of wave phase and latitude shows a cyclonic pattern at 850 mb centered near the trough center, maximum wind speeds and a well-developed wave at 700 mb, and large meridional variations in the surface flow. Additional analyses of the composite fields reveal upward vertical velocities from the northerlies to the trough (Fig. 10.14a), and maximum shear in the u component north of the center of the trough (Fig. 10.14b). The wind shear from the surface to 650 mb, ahead of the trough, can reach 4.0×10^{-3} s^{-1} (Barnes

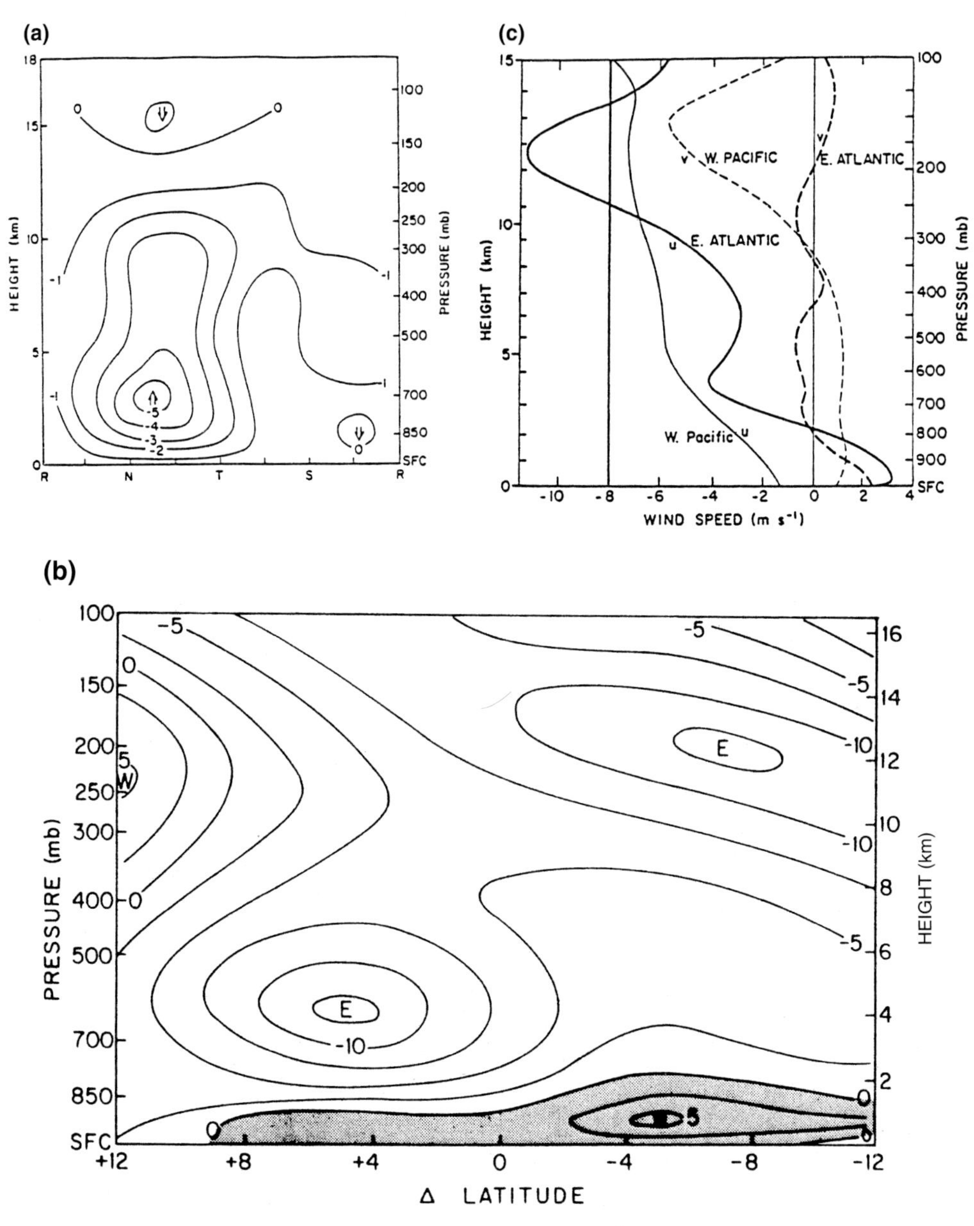

FIG. 10.14. (a) Vertical cross section of the vertical velocity (mb h^{-1}) along the center, or reference latitude of the wave, with R, N, T, and S denoting ridge, northerlies, trough, and southerlies. (b) Mean zonal wind field from 23 August to 19 September 1974. "Zero" latitude corresponds to the average latitude of the easterly waves. Westerlies are shaded. From Reed et al. (1977). (c) Mean zonal (heavy solid) and meridional (heavy dashed) wind components at the center of the GATE B-scale network during Phase III and corresponding profiles for the Kwajalein–Enewetak–Pohnpei triangle in the western Pacific. From Thompson et al. (1979).

and Sieckman 1984). Behind the trough shears decrease to about 1.0×10^{-3} s^{-1}. Comparison of the mean conditions for the eastern Atlantic (GATE) and the central Pacific (near Kwajalein) by Thompson et al. (1979) show that the low-level shear is weak, especially for the Pacific (Fig. 10.14c); these magnitudes are well below that needed for SLS (Keenan and Carbone 1992).

Ahead of the trough CAPE is about 1200 m^2 s^{-2}; it decreases to 700 m^2 s^{-2} in and immediately behind the trough (Thompson et al. 1979). Some of this modulation is due to convective overturning as well as wave-scale influence. These values do not favor strong updrafts and accompanying vortex stretching. The dry, dust-laden air just east of the trough (Carlson 1969) is a potential source for vigorous downdrafts and strong outflows, which do appear ahead of the trough in squall lines.

b. Subtropical cyclones

Simpson (1952) first defined subtropical cyclones as separate phenomena that originate from 1) an occluded extratropical cyclone that is trapped in low latitudes by a warm high pressure cell, or 2) a midtropospheric disturbance that gradually builds a low at the surface. They appear during the winter season over the North Pacific (called Kona lows) and North Atlantic; an aspect that distinguishes them from an extratropical cyclone is their isolation from a source of low-level cold air.

Subtropical cyclones also form in the summer in the Arabian Sea. These may be the decay phase of a monsoon depression or the result of a midlevel vorticity maximum developing through a deeper layer. Miller and Keshavamurthy (1968) offer one of the few detailed views of a subtropical cyclone that was located over the west coast of India. Examination of the

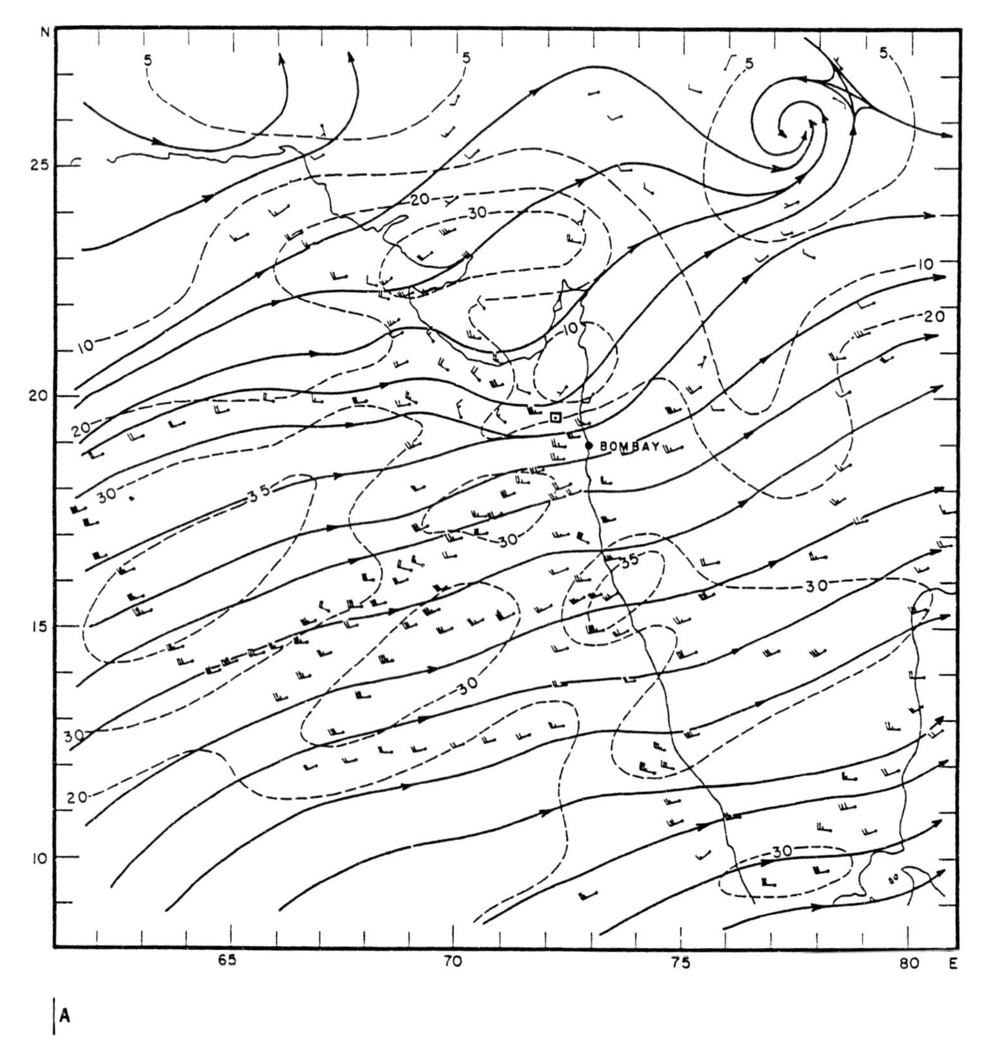

FIG. 10.15. Composite kinematic analyses (2–9 July) for (a) 900-m flow and (b) 700 mb. Streamlines are solid thin lines and isotachs are dashed, in knots. From Miller and Keshavamurthy (1968).

kinematic fields (Figs. 10.15a, 10.15b with ~700-m and 700-mb flows) reveals an area of very strong shear to the north of the cyclone center (defined by the 600-mb level). In this region there are moist southwesterlies overlain by drier northeast flow by 700 mb; shears exceed 7.0×10^{-3} s^{-1} over the lowest 3 km. This lower-troposphere shear is of the magnitude that has been implicated in the formation of supercells. However, most deep convection is located near the center and well to the south of the center and in a region of much weaker shear.

Kona lows in the eastern North Pacific can deliver a great deal of rain (Fig. 10.16c) in a meridional band of convective clouds several hundred kilometers east and southeast of the center. Often the center of these Pacific Ocean cyclones has layered clouds and light winds, and light rain only. Large-scale uplifting associated with an intense subtropical cyclone will remove the subsidence inversion typical of the trade wind region and allow for the conditional instability to be enhanced. CAPE can shift from the background trade value of 50–200 m^2 s^{-2} to greater than 1000 m^2 s^{-2} over the Pacific.

In some instances a Kona low is capable of triggering SLS. In February of 1997 a Kona low produced thunderstorms that yielded severe straight-line winds and hail reaching 12.5 mm in diameter (Morrison 1999). The upper-level jet stream was implicated in the Kona low development. Shear from the surface to 700 mb was about 4.5×10^{-3} s^{-1}, and the lifted index, using a surface parcel, reached values cooler than -8°C on the convective east side of the low during the mature stage.

Hybrid storms that form in the Gulf of Mexico (Hagemeyer 1997) have similarities to subtropical cyclones in that they form along an old frontal boundary that has lost its thermal gradients over the sea, and can produce heavy rain to the north and east away from the cyclone center. The hybrids appear to have their wind maxima at a lower altitude than the subtropical

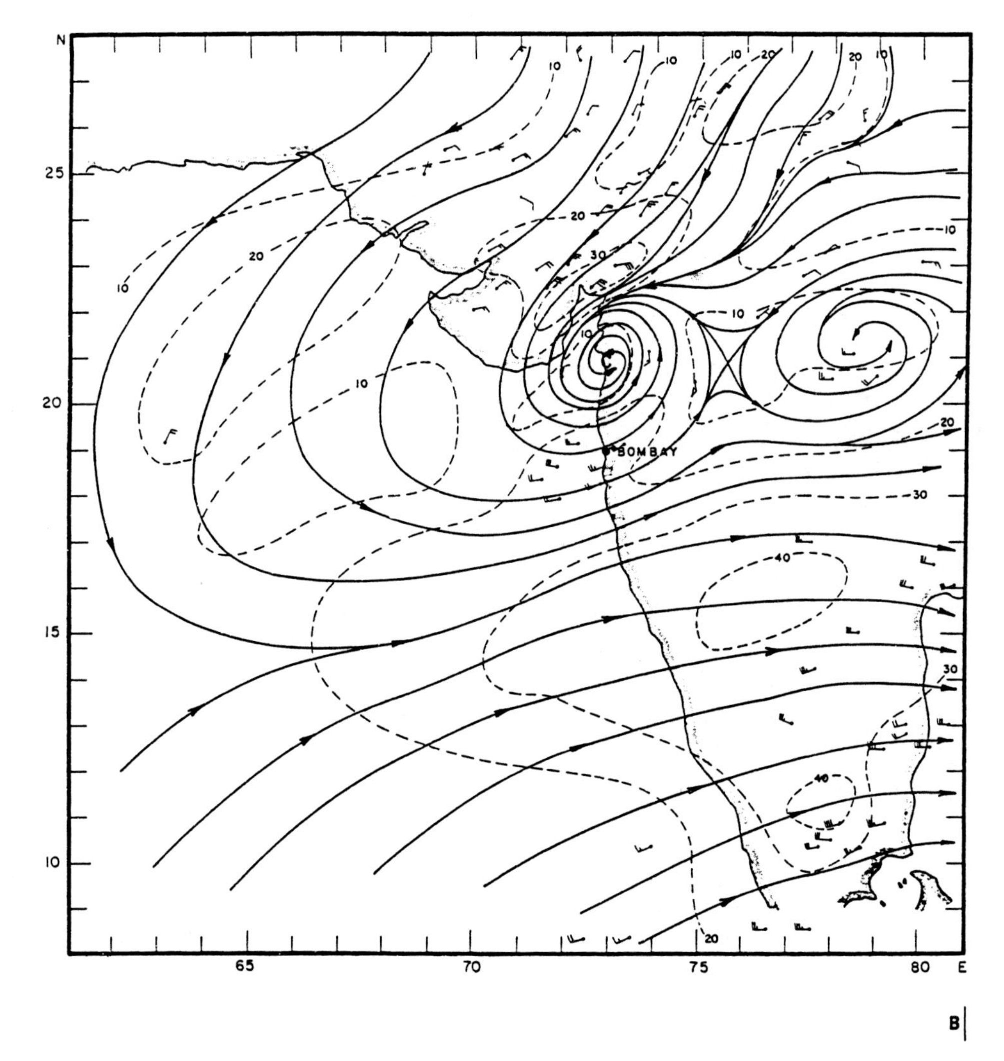

FIG. 10.15. (*Continued.*)

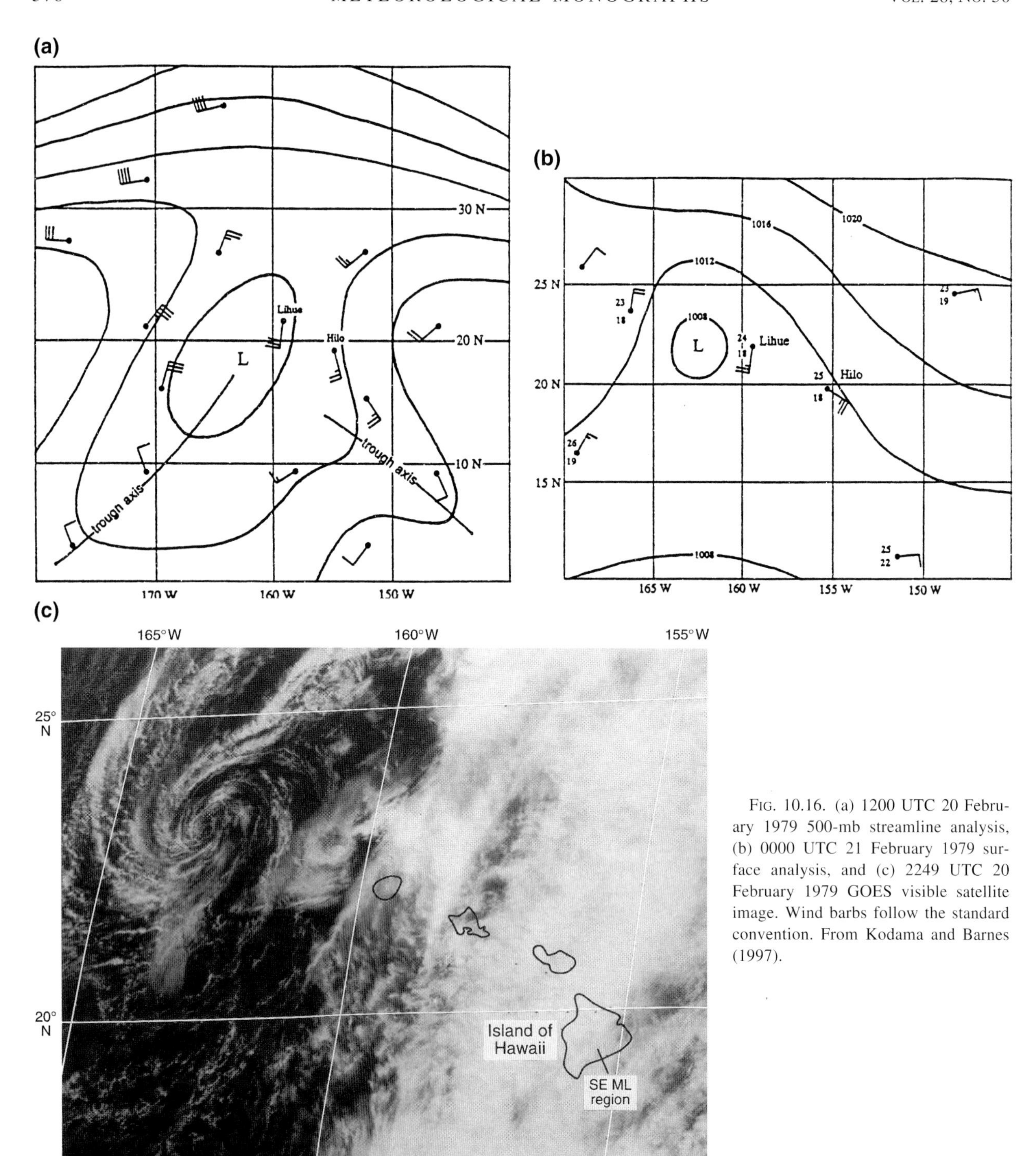

FIG. 10.16. (a) 1200 UTC 20 February 1979 500-mb streamline analysis, (b) 0000 UTC 21 February 1979 surface analysis, and (c) 2249 UTC 20 February 1979 GOES visible satellite image. Wind barbs follow the standard convention. From Kodama and Barnes (1997).

cyclone (e.g., Simpson 1952; Miller and Keshavamurthy 1968).

c. Monsoon depression

Approximately two monsoon depressions form per month over and near the northern Bay of Bengal during the active stage (low-level southwesterlies) of the monsoon (Koteswaram and George 1958). The vortices are most intense over the Bay of Bengal and weaken slowly as they move over land (Douglas 1992a,b). Despite this weakening, Rao (1976) and Douglas (1992a,b) demonstrate that these depressions

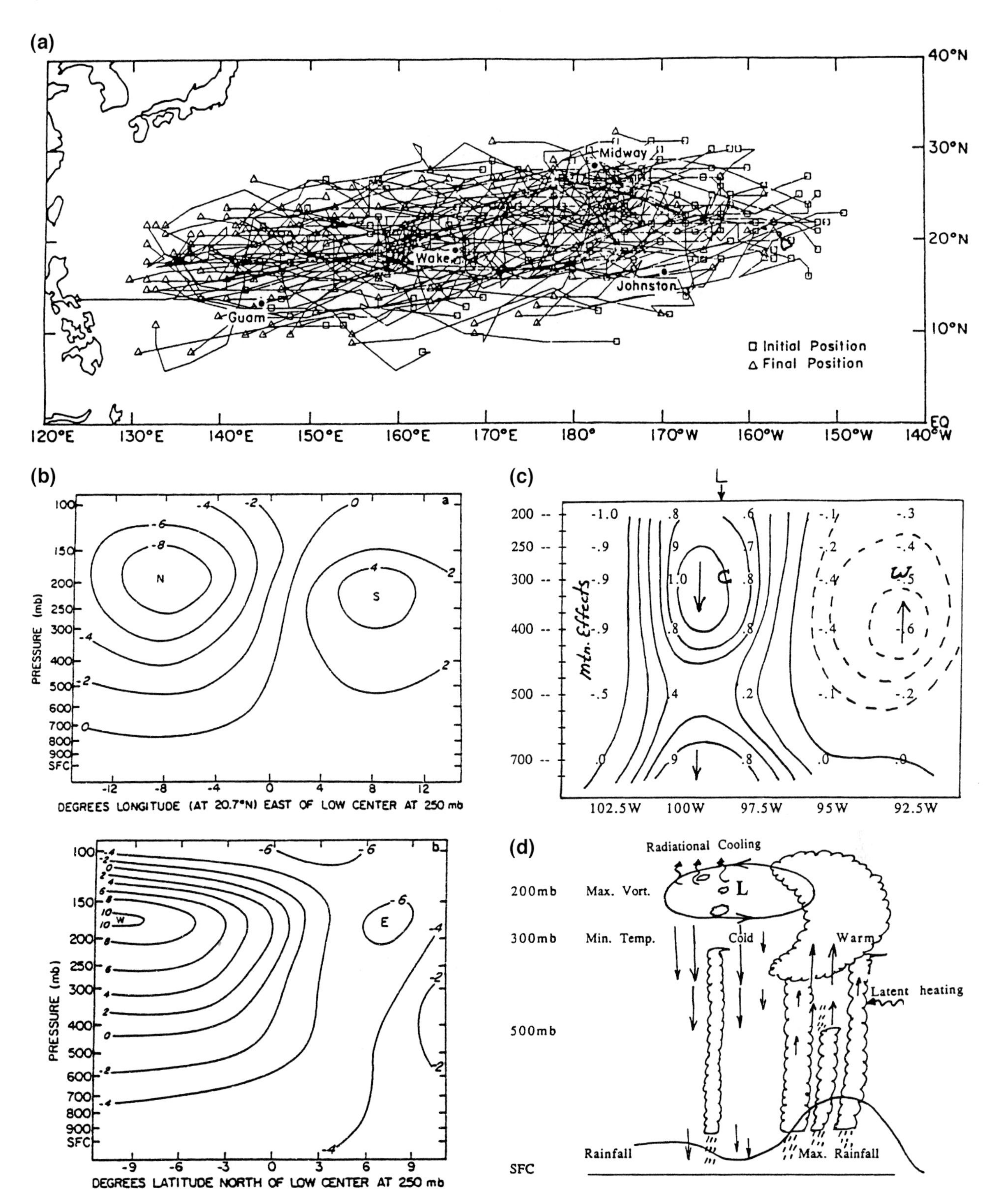

FIG. 10.17. (a) Tracks of TUTT cells in the Pacific from four summers. (b) The meridional wind for an east–west cross section and the zonal wind for a north–south cross section through the composite cold core in m s^{-1}. From Kelley and Mock (1982). A TUTT cell. (c) Pressure–longitude section of anomaly vertical velocity along 30°N for 2 August 1988. Units are in microbars per second, arrows indicate direction of motion, center of TUTT cell marked by "L." From Whitfield and Lyons (1992). (d) Schematic of the vertical circulation, cloudiness, and precipitation.

are responsible for much of the summer rains over India. These lows look like an inverted extratropical cyclone because the warm air is now to the north and the winds aloft are strong from the east. Central pressures are 2–10 mb below background (Ramage 1962; Douglas 1992a). There is little evidence for airmass discontinuities within the depression (Koteswaram and George 1958; Krishnamurti and Hawkins 1970; Sanders 1984). Development occurs when there is a superposition of a wave in the easterlies aloft with a surface trough.

Douglas (1992a) shows that the maximum winds at 850 mb are over 5° from the center of the circulation. The shear from 850 to 500 mb is $\leq 1.5 \times 10^{-3}$ s^{-1}; cross sections of virtual temperature do not support the concept that conditions are becoming more convectively unstable. The cooling to the southwest, where most rain is found, is due to reduced insolation, cooling from the evaporation of rain, and larger-scale ascent (Godbole 1977). The depressions yield much rain but little SLS.

Over the Arabian Sea, a vortex often forms during the early stage of the southwest monsoon. This onset vortex is similar to the monsoon depressions (e.g., Douglas 1992a) in wind and thermodynamic structure. Winds at 850 mb can exceed 25 m s^{-1} 600–700 km from the center, but apparently the shear is concentrated in the surface layer where it has little effect on cumulonimbi. Overall the virtual temperature variations around the vortex do not shift in favor of more unstable conditions (Douglas 1992a), supporting the earlier results that the monsoon yields copious amounts of rain, but very rarely vigorous convection.

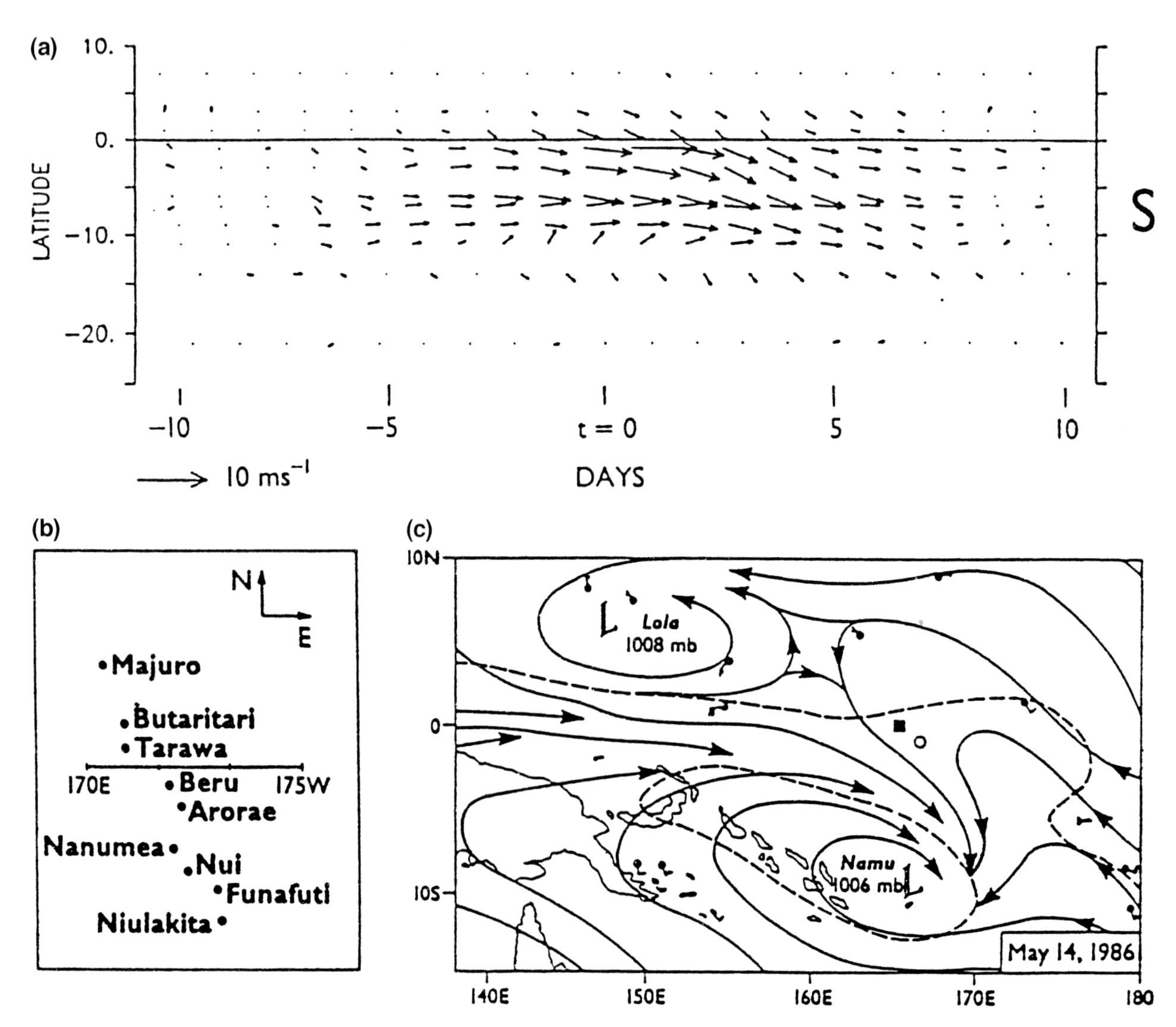

FIG. 10.18. (a) Composite zonal wind anomalies for westerly wind bursts centered south of the equator (S events). (b) Mid-Pacific stations used to diagnose WWBs. From Harrison and Giese (1991). (c) Surface streamline analysis from the Australian Royal Meteorological Service showing two cyclones and a westerly wind burst. The dashed line is the 7.5 m s^{-1} isotach. From McPhaden et al. (1988).

d. Upper-level cold core lows—TUTT cells

Analyses by Palmer (1951a), Dean (1956), and Ramage (1962) revealed the existence of cyclones best developed in the 200–300-mb layer. These cold core lows weaken downward and often have no effect below 700 mb (Frank 1970). A broad region located over the ocean basins during the summer, the tropical upper tropospheric trough (TUTT, Sadler 1976), spawns these upper-tropospheric cyclonic cells. Sadler (1975a) assembled a global climatology of the upper troposphere that demonstrated the intensity and seasonality (summer months) of both the trough and the TUTT cells.

A composite study of the cells found over the Pacific by Kelley and Mock (1982) reveals that the cells 1) track westward at 3–5 m s^{-1} and slowly move into lower latitudes (Fig. 10.17a); 2) survive for 7–14 days; 3) are 2000–3000 km in diameter; 4) have their largest kinematic signals in the 100–400-mb layer (Fig. 10.17b); and 5) produce clouds on their south and east sides and clearing on their north and west sides (Figs. 10.17c, 10.17d). The TUTT enhances the shear only in the upper troposphere. Cyclonic flow near the surface is weak, and rarely is there a closed isobar more than 4 mb less than the surroundings.

TUTT cells do appear as if they have little effect below 700 mb, but in fact they interrupt the subsidence associated with the subtropical high. This creates a favorable thermodynamic environment for the warm and moist trade flow, especially when the TUTT cell moves over progressively lower latitudes. In the core of the TUTT it is possible to generate an isolated Cb, and a line of Cbs can form in a crescent pattern centered on the east side of the center (Kodama and Barnes 1997). Whitfield and Lyons (1992) and Kodama and Barnes (1997) demonstrate the significant effect a TUTT cell may have on rainfall.

It is unlikely that the TUTT cell provides an environment for supercell development given that the shear in the lower troposphere remains quite weak. Near Hawaii, the TUTTs produce short-lived ordinary cells, based on reflectivity patterns the author has examined from the WSR-88D radars.

e. Westerly wind episodes or bursts

There are, in the western and central Pacific, episodes of westerly winds, also more sensationally labelled *westerly wind bursts* (WWBs), that are believed to play an important role in the Madden–Julian or 40–50-day intraseasonal oscillation (Lau et al. 1989). Harrison and Giese (1991) have examined western and central Pacific islands surface wind records and have classified episodes by the latitude of the maximum westerly wind if they meet the following criteria: 1) there is a westerly wind that is greater than 5 m s^{-1}; 2) this anomaly is present at two or more nearby islands; and 3) the anomaly survives at least 3 days. These episodes have a narrow meridional extent of 3°–5°, but a much larger zonal scale of 15°–30°. A typical WWB will last from 10 to 20 days. Harrison and Giese catalogued 169 episodes from 1956–80 (6–7 per annum) and have offered composite views of their structure (Figs. 10.18a, 10.18b). Besides the 40–50-day intraseasonal oscillation, these events may be caused by cold surges from the Asian continent, and may be correlated with the formation of a typhoon–tropical cyclone pair to the north and south of the westerlies (Fig. 10.18c). The episodes appear yearround but have a broad frequency maximum from September to February in the central Pacific. The WWB will increase the vorticity on either flank of the wind maximum, alter the low-level vertical wind shear, and advect moister air eastward. SSTs become better mixed during an episode resulting in 0.3°–0.4°C cooling of the surface temperatures (McPhaden et al. 1988).

f. Tropical cyclones

Tropical cyclones, which often achieve synoptic scales, increase the low-level wind shear, which favors the formation of Cbs with at least some supercell characteristics (Black et al. 1986), and attendant tornadoes (Pearson and Sadowski 1965; Hill et al. 1966; Orton 1970; Novlan and Gray 1974; Gentry 1983; McCaul 1987, 1991; Spratt et al. 1997). Novlan and Gray (1974) argue that for TC landfalls for the United States, about 10% of the deaths are caused by tornadoes. The data for these studies were collected when a hurricane made landfall either along the U.S. Gulf coast or along the Atlantic coast, hence the author must beg that the reader stretch the boundaries of the Tropics to discuss these findings, which appear in the intense vortices section. Bogner et al. (2000) provide evidence that the shear over the ocean in a hurricane is only about half that found over land, suggesting that

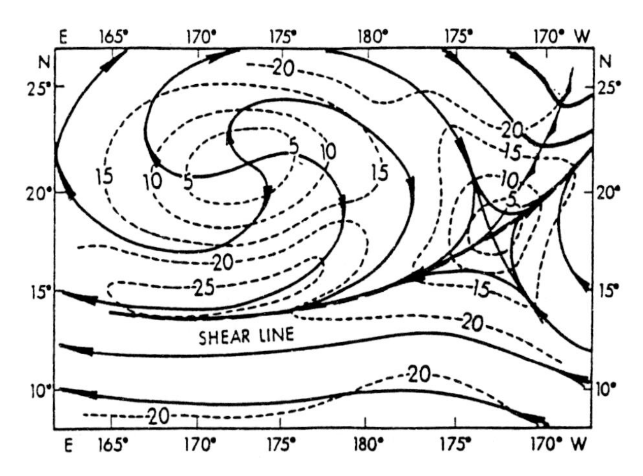

FIG. 10.19. Model of the surface kinematic pattern associated with a cold front extending into a tropical ocean shear line. Isotachs (dashed) in knots. From Ramage (1995), in turn adapted from Palmer et al. (1955).

(a)

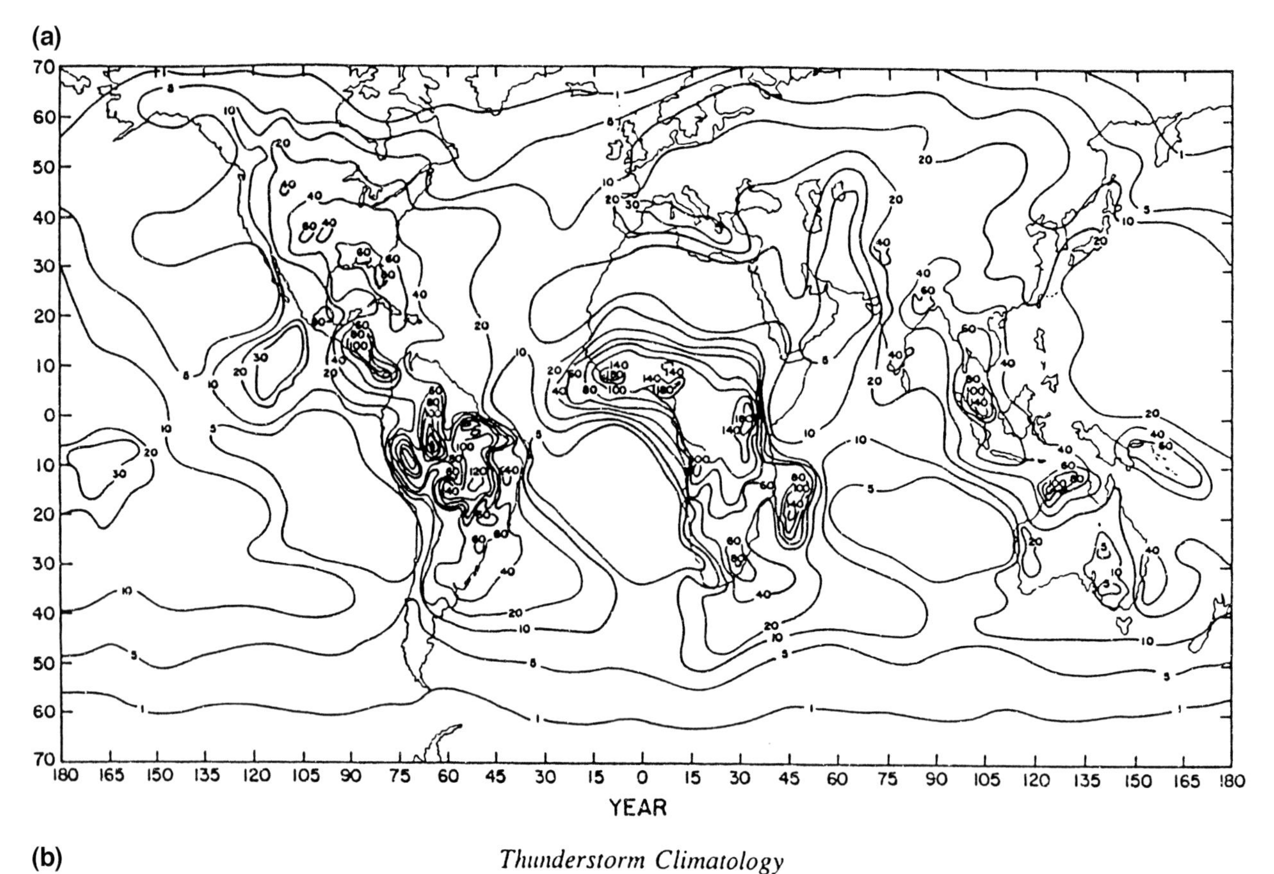

(b)

Thunderstorm Climatology

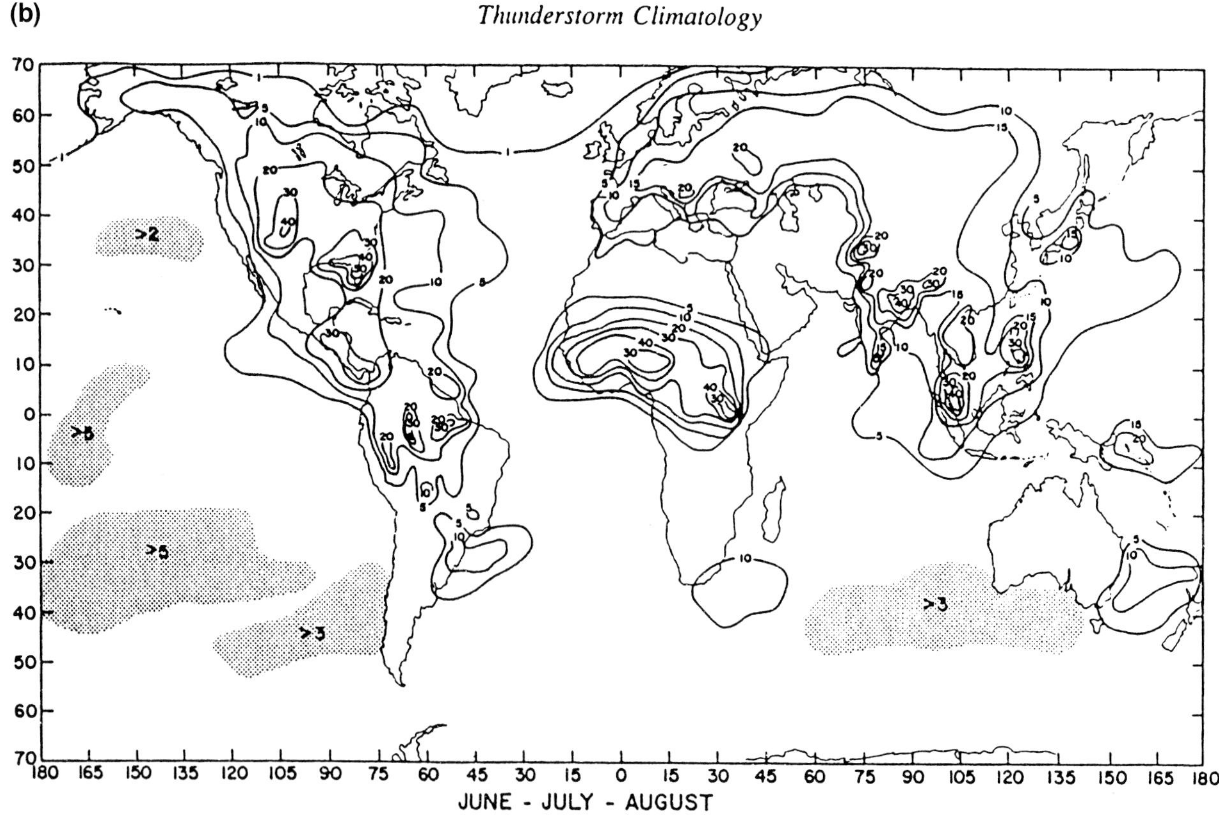

FIG. 10.20. The number of thunderstorm days, (a) annual, (b) June–August, (c) December–February. From WMO (1953).

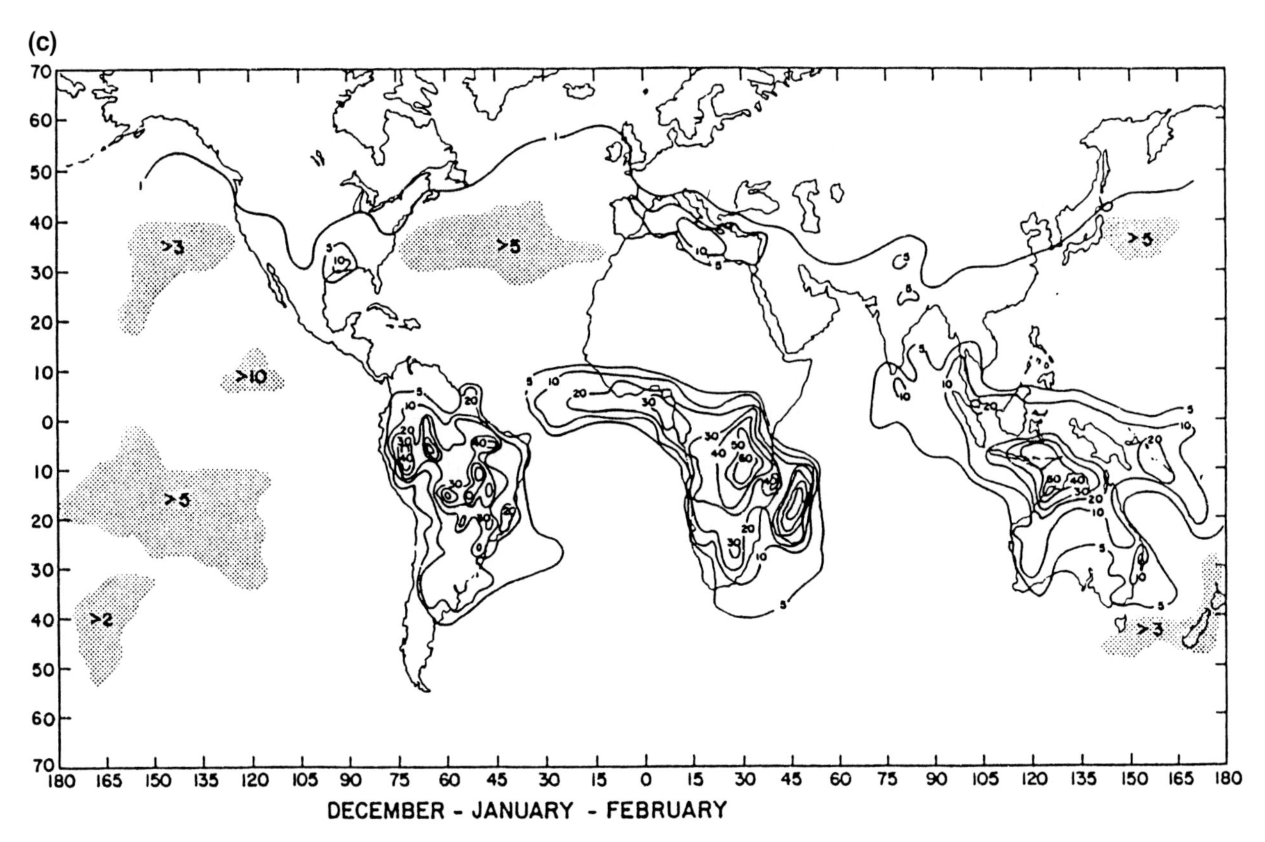

FIG. 10.20. (*Continued.*)

major changes in this parameter occur at landfall, presumably due to surface friction (Powell et al. 1998). In the tornado section, the details of tornadoes induced by tropical cyclones will be presented.

g. Extratropical troughs

Invasions of extratropical troughs are responsible for more SLS than all other previously described synoptic tropical phenomena combined (Fujita 1973; Gupta and Ghosh 1982; Mandal and Saha 1983; Alfonso and Naranjo 1996; Hagemeyer 1997). Regions that are influenced more by frontal and trough passages include northeast India and the Caribbean. Farther from the continents, sensible and latent fluxes from the warm oceans destroy the strong frontal gradients. Details of the role that extratropical cyclones play in SLS may be found in chapter 2 of this monograph or in Barnes and Newton (1985) and Doswell (1980). Examples of severe weather embedded within an extratropical trough will be presented in later sections that cover occurrence of SLS.

Skirting the northern edge of the Tropics along southern China and Taiwan is a convergence zone caused by the interaction of the westerlies with the developing southwest monsoon. The result is the Mei-yu front, which can deliver copious amounts of rain to southern China in May and along the Yangtze River basin (~30°N) in later summer. Some of these fronts, especially those in late June, appear to consist of a series of mesoscale cyclones that do not exhibit a strong thermal gradient (Ninomiya and Murakami 1987) and appear to have characteristics more like the near-equatorial convergence zone (Chen and Chang 1980). However, often what is labeled the Mei-yu front is actually a typical midlatitude feature with baroclinic characteristics (Chen 1993).

h. Shear lines

A late stage in the life of a cold front is the formation of a shear line. There is no temperature discontinuity at the surface, but there remains a zone of low-level convergence where the now much-modified air behind the front interacts with the trade winds. Ramage (1995) presents a schematic from Palmer et al. (1955) of a shear line (Fig. 10.19), which usually coincides with a line of convective clouds. The low-level convergence associated with a shear line in conjunction with mountainous terrain can produce continuous rain with impressive totals, but many shear lines do not harbor any deeper clouds than cumulus congestus. A few shear lines do yield deep convection, but evidence of supercells along a shear line is presently lacking.

i. Surges

Surges are accelerations in the winds that may be caused by an anticyclone trailing a winter storm, or by a strengthening of the Hadley circulation (Ramage 1995). At times the acceleration of air occurs ahead of the cold front, masking the location of the front. Accelerations can occur in monsoon or trade flow without any evidence of frontal influence.

During the winter, cold surges, which are related to Rossby wave passage, affect East Asia (e.g., Chang et al. 1979; Lau 1982), the Central American region (Schultz et al. 1998; Schultz et al. 1997), Australia (e.g., Smith et al. 1995), and South America (e.g., Fortune and Kousky 1983). A large-amplitude trough in the westerlies, mountain ranges that channel the flow equatorward, and convergence in the upper troposphere that builds the anticyclone are all contributing factors to the strong winds, cold air advection, and decreasing SST observed in the lower latitudes (Ramage 1995; Schultz et al. 1998).

Evidence for SLS along the leading edge of the surge, where strong convergence is present, has not been documented. Even the cold surge associated with the so-called Superstorm of 1993 (Schultz et al. 1997) triggered mostly extensive orographic rains along the eastern slopes of the Sierra Madre, though some remarkably cold tops ($T < -90°C$) were noted over Colombia in South America. Convection over the Maritime Continent is enhanced by cold surges moving across the South China Sea (Lau and Chang 1987); Ramage (1995) notes that the surges may produce extensive areas of stratiform rain with intensified deep convection equatorward.

Another type of surge that can occur throughout the year is produced when there is rapid acceleration in trade or monsoon flow. This also produces convergence along the leading edge and divergence on the trailing side of the pulse; a cyclonic–anticyclonic couplet also forms around the flanks of the surge (Ramage 1995). Surges in the monsoon flow would be synonymous with WWBs. There is enhanced rain ahead of a surge on the side that exhibits cyclonic shear, but reports of severe weather within a surge are not known.

10.4. Thunderstorm frequency

a. Thunderstorm days and lightning frequency

Mean annual number of thunderstorm days (WMO 1953; Fig. 10.20a) confirms that the Tropics are a hotbed of deep convection. All stations that report more than 200 thunderstorm days annually are in the Tropics (Portig 1963; Court and Griffiths 1985). Six of these stations are in Africa within 6° of the equator; the other two are found in Indonesia and Brazil. Comparison of the counts for the boreal winter and the boreal summer (Figs. 10.20b, 10.20c) show that even the deep Tropics undergo significant annual variations, with Mexico, the Sahel, Kenya, Tanzania, south and southeast Asia, and northern Australia all demonstrating annual fluctuations. Some remarkably strong gradients in thunderstorm days are associated with variations in topography or land–sea contrasts (Orville and Henderson 1986). The combination of terrain adjacent to the sea or a large lake produces some dramatic results (e.g., Madagascar, Indonesia, Central America, Lake Victoria). Contrasting the high counts over the tropical landmasses are the observations over the tropical oceans.

For the tropical ocean only the NECZ supports a thunderstorm frequency of more than 20 thunderstorm days per year. Sanders and Freeman (1986) use two different ship report schemes, one compiled for the Naval Research Laboratory (NRL) and the other for the WMO, to estimate the percentage frequency of thunderstorms for 10° squares. Both schemes (Figs. 10.21a, 10.21b) show that the tropical oceans have many more thunderstorms than do the high latitudes, but the frequency is still quite low when compared to land reports. Coastal waters (e.g., near Central America, the Gulf of Guinea, Indonesia) reach values of 4%–5% or 15–20 days a year for the NRL dataset; the WMO (1953) dataset yields values 3–5 times lower because it excludes sightings of distant lightning. The "blue water" portions of the oceans have very few thunderstorms; in fact, it is difficult to resolve major features such as the equatorial trough with thunder reports. The dramatic difference between tropical land and sea lightning frequency (Orville and Spencer 1979) supports the contention that SLS is more frequent over land. Here we note that higher CAPE results in more vigorous updrafts, which is correlated with lightning frequency (e.g., Williams et al. 1992).

"Thunderstorm days" is a notoriously inaccurate statistic for a variety of reasons (Court and Griffiths 1985). However, the objective data from the Defense Meteorological Satellite Program (DMSP; e.g., Orville and Henderson 1986) confirm that the tropical landmasses have a high frequency of thunderstorms. Figures 10.22a, 10.22d show the local midnight lightning flashes for 365 consecutive nights. Maxima appear over South America, equatorial Africa, the Maritime Continent, and northern Australia. The objective sampling from the satellite sensor confirms the low count of thunderstorms over the tropical oceans gathered from ship reports.

Neither lightning flash counts nor surface station climatologies can be distilled to number of Cbs, but one can identify the following temporal patterns.

1) Equatorial Africa: maximum in AMJ, minimum in NDJ.
2) South Africa: maximum in DJ, minimum in MJJA.
3) South America: Amazon has a minimum in JJA, and a weaker annual cycle for the northern parts of South America.

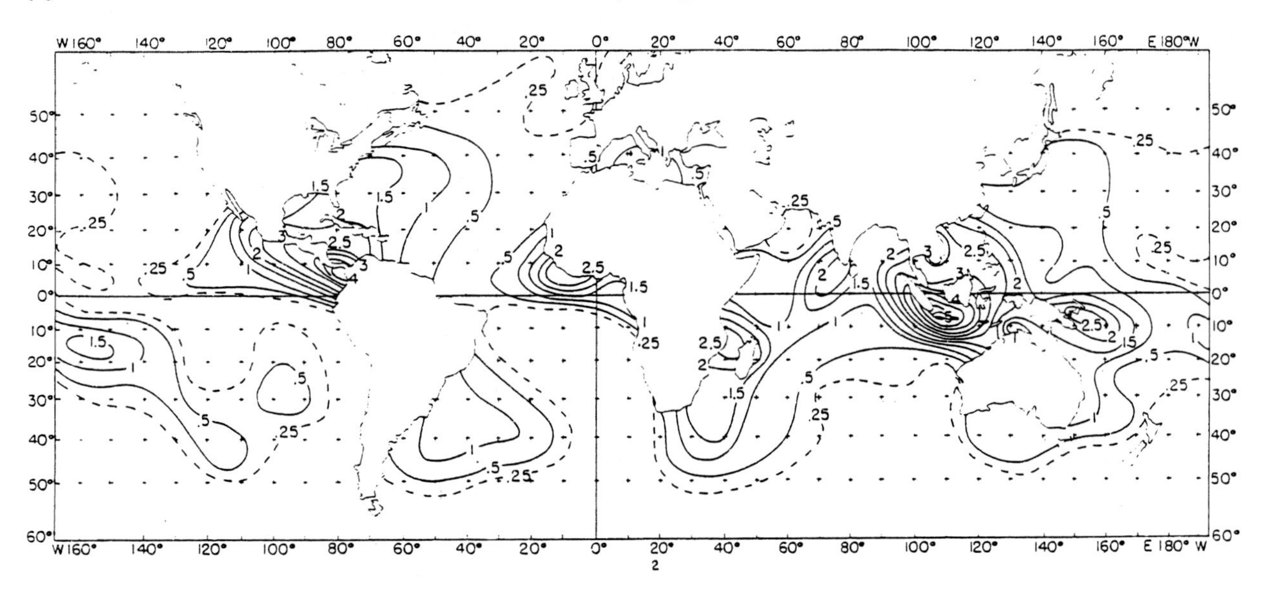

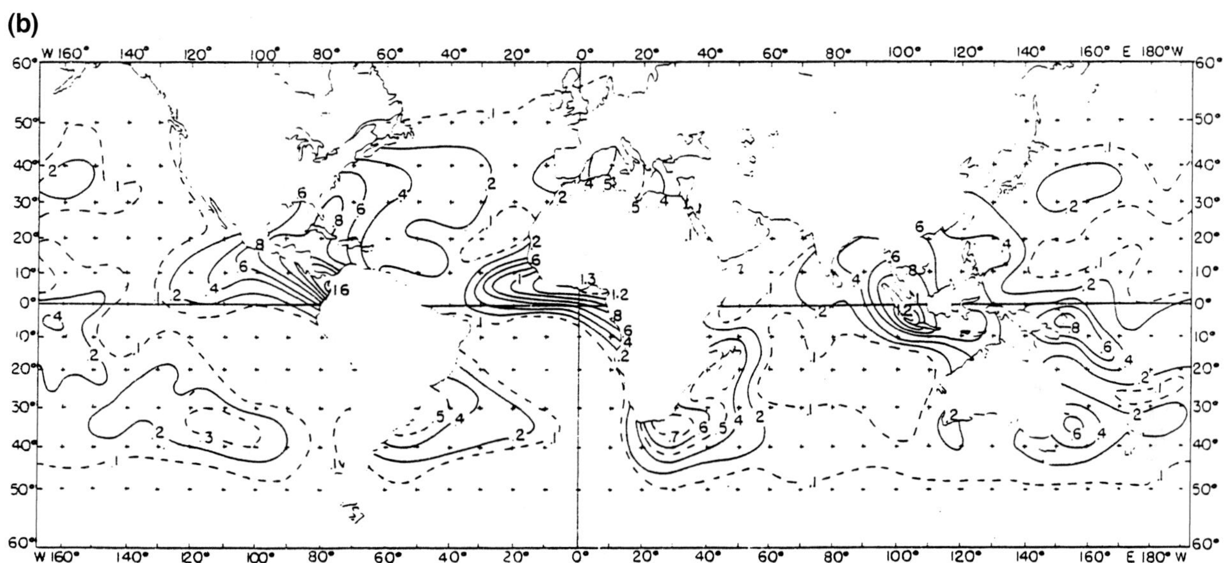

FIG. 10.21. Percentage frequencies of thunderstorms and/or lightning in surface observations from (a) the NRL sample with solid isopleths at intervals of 0.5% and (b) the WMO sample with solid isopleths at intervals of 0.2%. From Sanders and Freeman (1986).

4) Central America: boreal winter minimum, summer maximum.
5) Indonesia: flatter annual cycle signal than Africa.
6) India: maximum in AM, minimum in DJF.
7) North Australia: maximum in austral summer, minimum in winter.
8) Oceania: low counts, weak annual cycle. The land-to-ocean flash ratio, normalized to unit area, is 7.7. This count is only for the local midnight distribution of flashes.

Does lightning serve as proxy for severe weather? Reap and MacGorman (1989) show that a positive flash density for a given area in Oklahoma (flash rate $\geq$ 30 per 2300 km^{-2}) is correlated with an increase in reported severe weather in that area; MacGorman and Burgess (1994) make a case that most storms with high positive flash rates also yield either hail or tornadoes. Hodanish et al. (1998) show a correlation between total flashes and tornadic activity over Florida. Williams et al. (1992) find approximately 10 times more lightning per unit mass of rain from continental tropical Cbs compared to convection in monsoonal systems. They argue that subtle increases in wet-bulb equivalent potential temperature at the

(a)

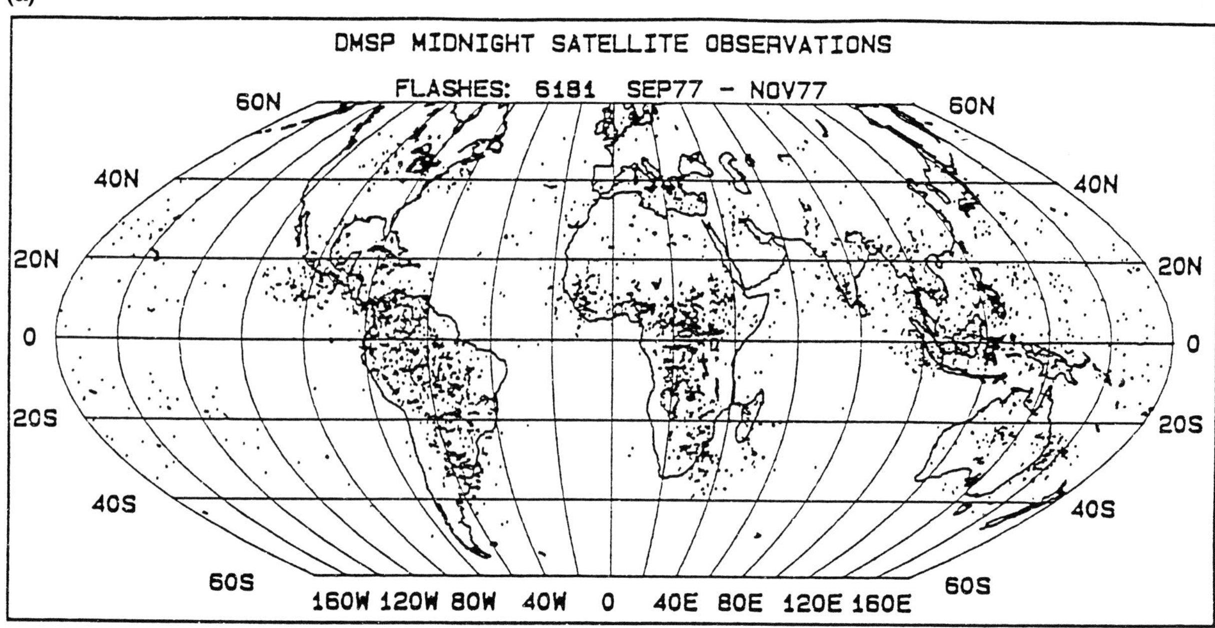

(b)

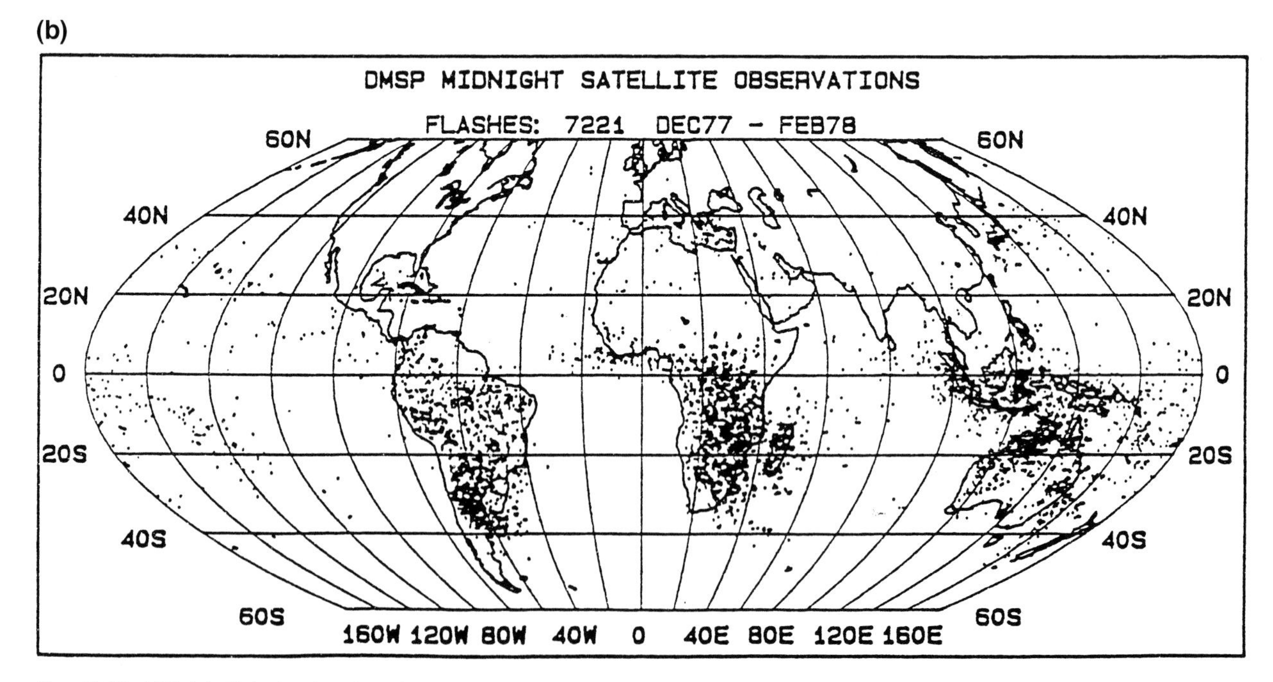

FIG. 10.22. Midnight lightning locations for (a) fall, (b) winter, (c) spring, and (d) summer of 1977. From Orville and Henderson (1986).

surface lead to both enhanced CAPE and more ice phase precipitation, which is crucial for lightning activity. Here the increase in lightning is a clue as to when a system has more vigorous updrafts and presumably is more capable of spawning severe weather.

However, there are other studies that support the argument that lightning may not help identify which Cbs yield SLS. Supercells accompanied by heavy precipitation also occur with negative flashes dominating. Positive flashes are detected in decaying storms, in stratiform regions of MCSs, and in nonsevere thunderstorms (e.g., Williams et al. 1989; Rutledge et al. 1990; Engholm et al. 1990). Kumar (1992) found that the majority of thunderstorms at Lucknow, India, located in the foothills of the Himalayas, occur during the monsoon when SLS is rare. Perez et al. (1997) examine cloud-to-ground (CG) flashes in tornadic storms from 30 min before tornado touchdown to 30 min after tornado dissipation and arrive at the conclu-

(c)

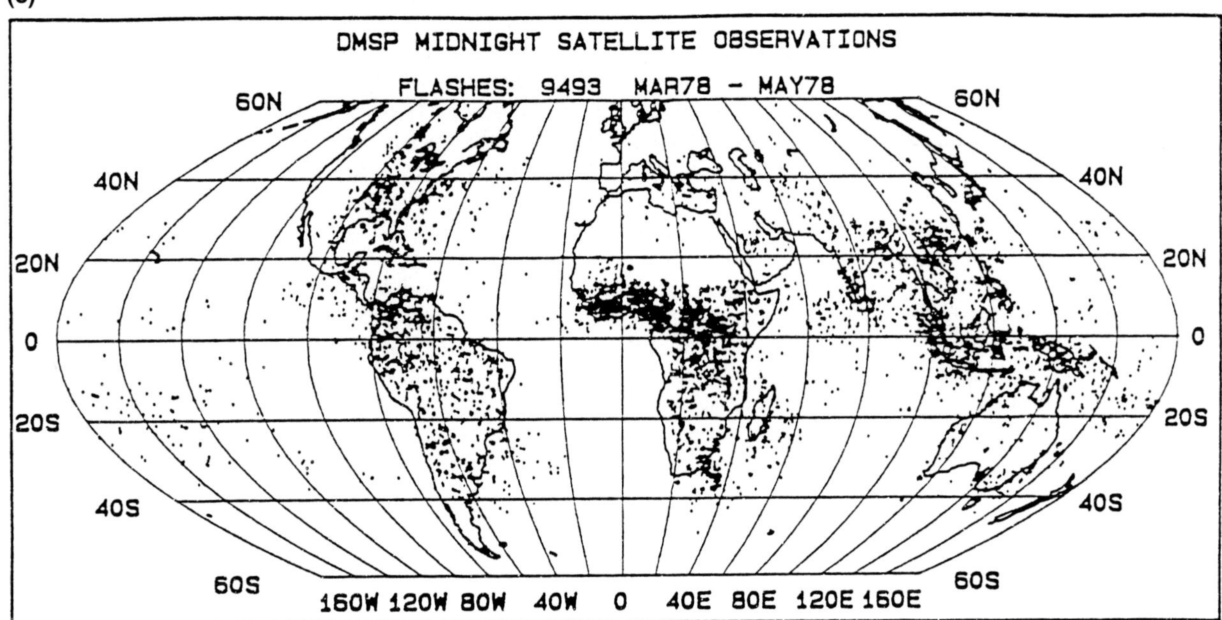

(d)

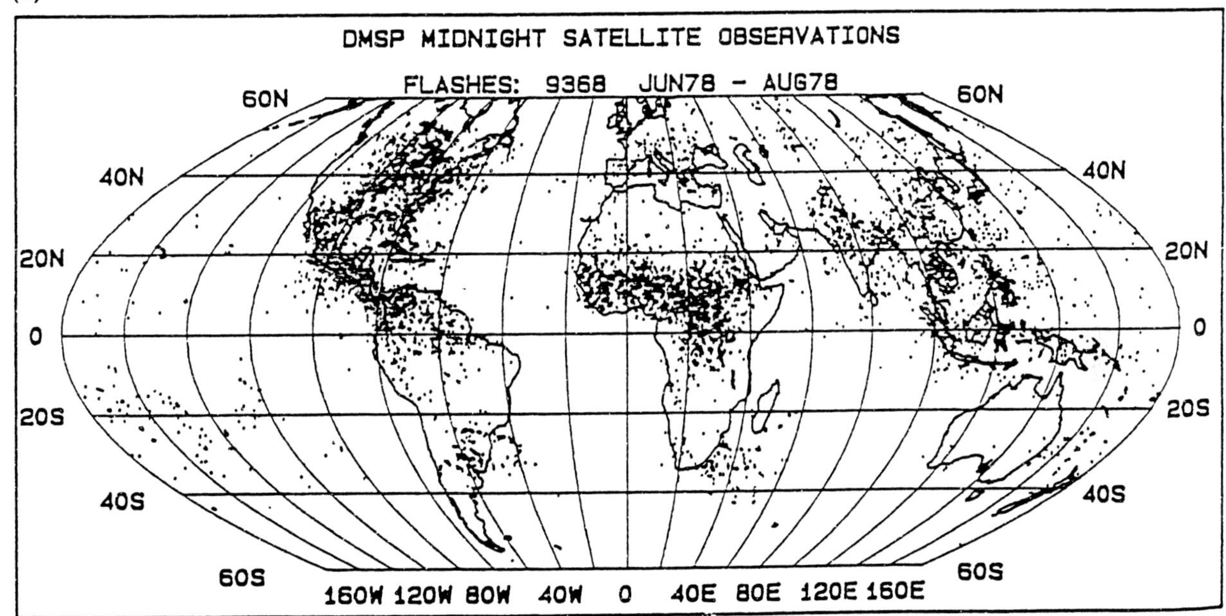

FIG. 10.22. (*Continued.*)

sion that there is no practical way to identify tornado formation. Branick and Doswell (1992) argue that the positive ground flashes may be due to the location of a given storm with respect to other, prior thunderstorms that generated negative cloud-to-ground flashes.

Lightning detection is an inexpensive and objective way to determine if deep convection is occurring. In certain regions, where lightning is uncommon, it serves as an early warning that a situation deserves vigilance. At this time the application of lightning for detection of SLS awaits further research.

b. Other proxies for Cbs and SLS

Cloud-top temperature, derived from satellites sensing outgoing longwave radiation (OLR), continues to serve as a locator for deep convection. In the Tropics it is common to find large areas with infrared (IR) temperatures less than 208 K, equivalent to 15–16 km height (e.g., Mapes and Houze 1993). Complications arise as an MCS ages toward and beyond the mature stage. Typically the anvil clouds from an MCS will achieve their coldest IR temperatures well behind the active leading edge of convection (e.g., Maddox 1980; Geldmeier and Barnes 1997), supporting the hypothesis that the continued slow mesoscale ascent, not vigorous convective-scale drafts, is producing these large areas. Simply identifying the coldest tops will lead to a terrible false alarm rate for SLS.

There are other schemes to estimate severe weather remotely. Heymsfield and Blackmer (1988) discuss the existence of the V-shaped region of lower equivalent blackbody temperatures with the point of the "V" at or just upwind of the cloud top. Regions of higher temperatures are found downstream. The presence of a V is correlated with severe weather about 70% of the time, and the appearance of the V precedes the severe weather by about 30 min (McCann 1983). Not all severe weather events have a V, but most of the V cases are correlated with severe weather (Adler et al. 1985). All the V cases possessed strong updrafts and strong vertical shear of the horizontal winds. These studies are for midlatitude wind environments, not the Tropics.

Highly reflective cloud (HRC) is an attempt to circumvent the problems that OLR has in differentiating active Cbs from thick cirrus. The brightness and shape of clouds observed with the visible portion of the spectrum are used to identify deep convection (e.g., Oswaldo 1985; Kilonsky and Ramage 1976). A combination of HRC and OLR eliminates most of the cold tops that are due to continued mesoscale ascent of anvil clouds. There have been no attempts at applying HRC in the Tropics for SLS detection.

OLR has been the basis for the identification and study of a type of MCS known as a *mesoscale convective complex* (MCC; Maddox 1980, 1983). MCCs are known to be linked with severe weather, at least over the United States, though their greatest threat is that of heavy rains (Maddox 1980; Fritsch et al. 1986). MCCs may provide clues as to where severe weather may exist in the Tropics, especially in the absence of a synoptic entity.

Table 10.1 from Laing and Fritsch (1993a) is a summary of MCC frequency, time of formation, maximum extent and decay, and duration and size for several tropical regions, as well as for the better-known U.S. situation. MCCs over Africa, the Americas, the western Pacific, and India exhibit similar traits with a maximum size usually about 2–3 × 10^5 km^2, a modal duration of 9 h, and a mean duration of 10–12 h. Most MCCs form over land and propagate toward the high θ_e air in the subcloud layer. MCCs are predominantly nocturnal, reaching definition size

TABLE 10.1. Mean characteristics of MCCs in Africa, the Americas, the western Pacific and the Indian subcontinent from Laing and Fritsch (1993a).

		Time (LST)					Area (×10^3 km^2)	
							Temperature (BB)	
	Average systems per season	First storm	Initiation	Maximum observed extent	Termination	Duration (h)	−33°C	−54°C
Midlatitude South Africa	11	1230	2030	0200	0700	10.5	388	234
Low-latitude Africa	86	1430	2100	0200	0830	11.5	307	183
Total: Africa (1986, 1987)	97	1400	2100	0200	0830	11.5	316	189
Indian subcontinent								
Midlatitude	33	1430	2200	0600	1000	12.0	403	237
Low-latitude	16	1300	2200	0500	0900	11.0	353	209
Total: Indian subcontinent (April–December 1988)	49	1400	2200	0600	1000	12.0	388	229
Australia	20	1400	0000	0600	1000	10.0	338	156
New Guinea	7	1630	2300	0430	0930	10.5	255	118
China/South China Sea	21	1400	2230	0600	0930	11.0	323	149
Bangladesh/northeast India								
Bay of Bengal	10	1400	0000	0800	1030	10.5	399	183
Miscellaneous	24	1400	2300	0730	1030	11.5	343	158
Total: western Pacific region (1983, 1985)	82	1400	2300	0630	1000	11.0	336	155*
United States (1978, 1981, 1982)	34	1500	2100	0130	0630	9.5	299	
Midlatitude South America	39	1900	2130	0300	0900	11.5	485	
Low-latitude Americas	57	2300	0200	0530	1030	8.5	320	
over land	28	2230	0100	0530	0930	8.5	323	
over sea	29	0000	0230	0630	1130	9.0	316	
Total: Americas	130	2000	2300	0400	0900	10.0	364	

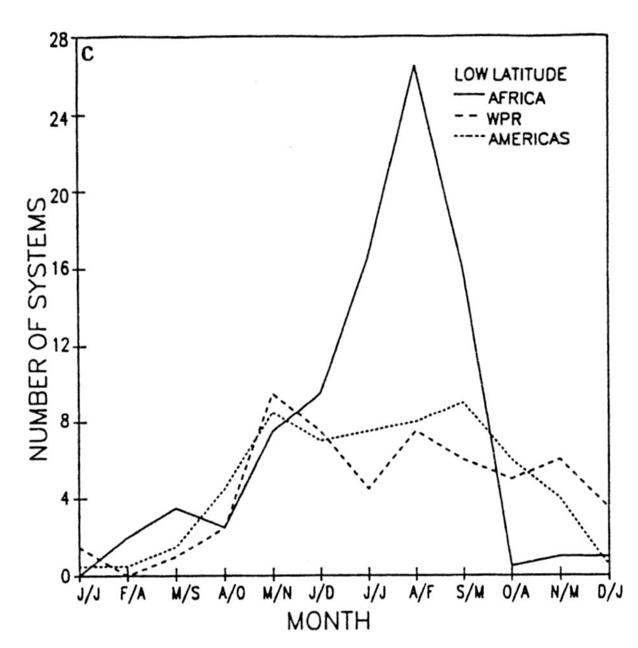

FIG. 10.23. Frequency of MCCs in the low latitudes. The x axis is adjusted to match the summers together. From Laing and Fritsch (1993a).

around 2300 LST, maximum size just prior to sunrise, and dissipating a few hours thereafter.

Figure 10.23 shows the period of formation for Africa, the western Pacific, and the Americas. Seasonal dependency is strong for Africa, less so for the Americas and the western Pacific. The low-latitude MCCs are distributed uniformly throughout the warm season while late spring and early summer are the favored times for midlatitude MCCs. The deep Tropics clearly have an important annual cycle of some of the criteria necessary for MCC formation.

Where are MCCs found? The great majority (80%) of the western Pacific MCCs form over land or within 250 km of the coast; Thailand, Burma, Bangladesh, and northeastern India are active regions (Miller and Fritsch 1991; Laing and Fritsch 1993b). Preferred genesis regions in Africa include the latitudinal band from 5°S to 18°N covering central Africa, and along the southeast coast of Mozambique (Laing and Fritsch 1993a). The MCCs over the low-latitude Americas exist in three regions: 1) Colombia and the Gulf of Fonseca, 2) Bolivia and northern Argentina, and 3) over the ocean south of Acapulco (Fig. 10.24; Velasco

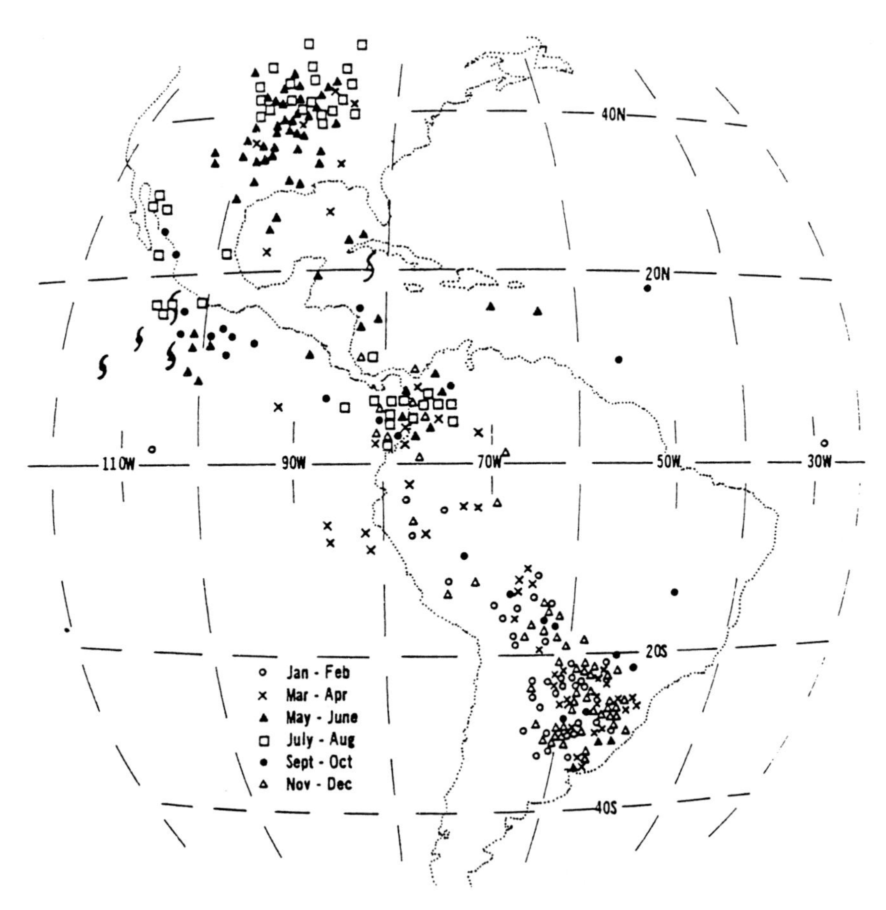

FIG. 10.24. Geographic and monthly distribution of MCCs in and around the Americas. Locations are for the MCC cold cloud shields at the time of maximum extent. Hurricane symbols indicate an MCC that developed into a tropical storm. From Velasco and Fritsch (1987).

and Fritsch 1987). The first two regions are subject to strong orographic influences. Surprisingly, the Amazon is nearly devoid of MCCs. Here is a region known for numerous Cbs but virtually no MCCs, and only very rare severe weather.

There is a serious deficiency of surface observations for the Amazon, which limits our ability to detect SLS. Ecologists studying the rain forest have mentioned the existence of tree blow-down areas, but a census of these events cannot be found. One could argue that the rain forest, given its high growth rates and limited exposure to high winds, would be susceptible to damage at thresholds below severe. Conversely, there may be a crucial factor missing that affects both MCC and SLS formation. Critical factors that rarely occur over the Amazon include 1) a low-level wind maximum, 2) dry midlevel air, and 3) vorticity, either in the vertical or horizontal. High CAPE and roughly barotropic conditions are the norm for the Amazon (Williams and Renno 1993). In the middle of the wet season there exists a deeper layer of moisture that is not conducive to cooling via evaporation and subsequent strong downdrafts and outflows. Other locations with few MCCs and maybe rare SLS occurrence include the equatorial rain forest in Africa (Laing and Fritsch 1993a) and the west Pacific equatorial trough.

Given that MCC formation favors the warmer season in the Tropics, it would seem plausible that there are recognizable planetary-scale features that play a role in their formation. For equatorial Africa and India it is the monsoon that provides the needed environment for MCCs.

Recently, a Special Sensor Microwave/Imager (SSM/I) on the Defense Meteorological Polar Orbiter Satellite has been employed to assess the intensity of MCSs (Mohr and Zipser 1996). The 85-GHz channel reacts to the presence of large ice particles when found through a deep layer and thus reveals the larger convective regions (13 $\times$ 15 km resolution). A variable known as *polarization corrected temperature* (PCT) is used to assess the coldness or intensity of the MCS. Figures 10.25a, 10.25b (Mohr and Zipser 1996) show that some of the more intense MCSs, as determined by the SSM/I, are correlated with active areas of SLS. These areas include northern Australia and southern Brazil in January, and northeastern India and the Sahel in July. The 85-GHz channel may well lead to a better sorting of deep convection with and without severe weather. This work with the 85-GHz channel also verifies much of the previously cited work on MCCs by Fritsch and colleagues.

10.5. SLS examples and patterns

In this section I will discuss examples of the three types of severe weather: strong straight-line winds, hail, and tornadoes and waterspouts. These examples are a poor substitute for a real climatology of SLS in the Tropics, but the best that one can do given the paucity of tropical data and the time constraints of this work.

a. Straight-line winds exceeding 25 m s^{-1}

Wind records, though a more objective statistic than thunderstorm days, must be interpreted cautiously. Anemometer design, maintenance, exposure, height above ground, and recording frequency strongly impact wind speed estimates. There is no standard for sustained or peak/gust winds worldwide. The landscape around most long-running metropolitan observation sites has undergone considerable evolution; often sites just a few kilometers apart differ in their sustained and peak estimates.

1) EXTRATROPICAL WAVES

The robust gradients of extratropical systems have the best chance for survival when they can rapidly move into the Tropics from continental regions. The Caribbean and northeastern India–Bangladesh are two regions that are affected.

Alfonso (1988) has examined the frequency of severe weather in Cuba and notes that such events occur yearround. In the spring (March–May), extratropical systems (shortwave troughs and cold fronts) are responsible for the majority of the severe weather. In the summer and early fall (July–September), tropical cyclones, upper-level troughs, and mesoscale forcing become important. In June both sets of phenomena can occur; this results in the maximum of severe local storms (SLS) in Cuba. The most damaging events do occur in the cool season and are associated with prefrontal squall lines.

A prefrontal squall on 13 March 1993 was responsible for 10 deaths and over one billion dollars worth of damage to Cuba (Alfonso and Naranjo 1996) when it produced straight-line winds exceeding 55 m s^{-1}, but no tornadoes. The parent cyclone, labeled a superstorm (e.g., Dickinson et al. 1997), was the deepest extratropical cyclone observed over the Gulf of Mexico over the preceding 40 years. A well-recognized pattern is seen at the surface (Fig. 10.26a) and at 300 hPa (Fig. 10.26b). The maximum gusts observed for the event are shown in Fig. 10.26c. Lifted indices were -3 to -5 K, there was high q through a deep layer, and there was a shortwave trough recognizable on a larger deep long-wave trough (Fig. 10.26b) to initiate the squall line. Radar showed that the squall line could be divided into a series of bow echoes and cloud tops reached over 16 km; reflectivity maxima were as high as 70 dBZ. Cells were moving from 257° at 28 m s^{-1}. A swath of F2 (Fujita scale) damage 10–20 km wide and about 80 km long was produced over 1.5 h, but reports actually cover over 400 km of the island (Fig. 10.26c). This squall line had what the authors believe to be significant differences from the usual severe weather events for Cuba; these differences include no occlusion at the surface and little relationship with the STJ.

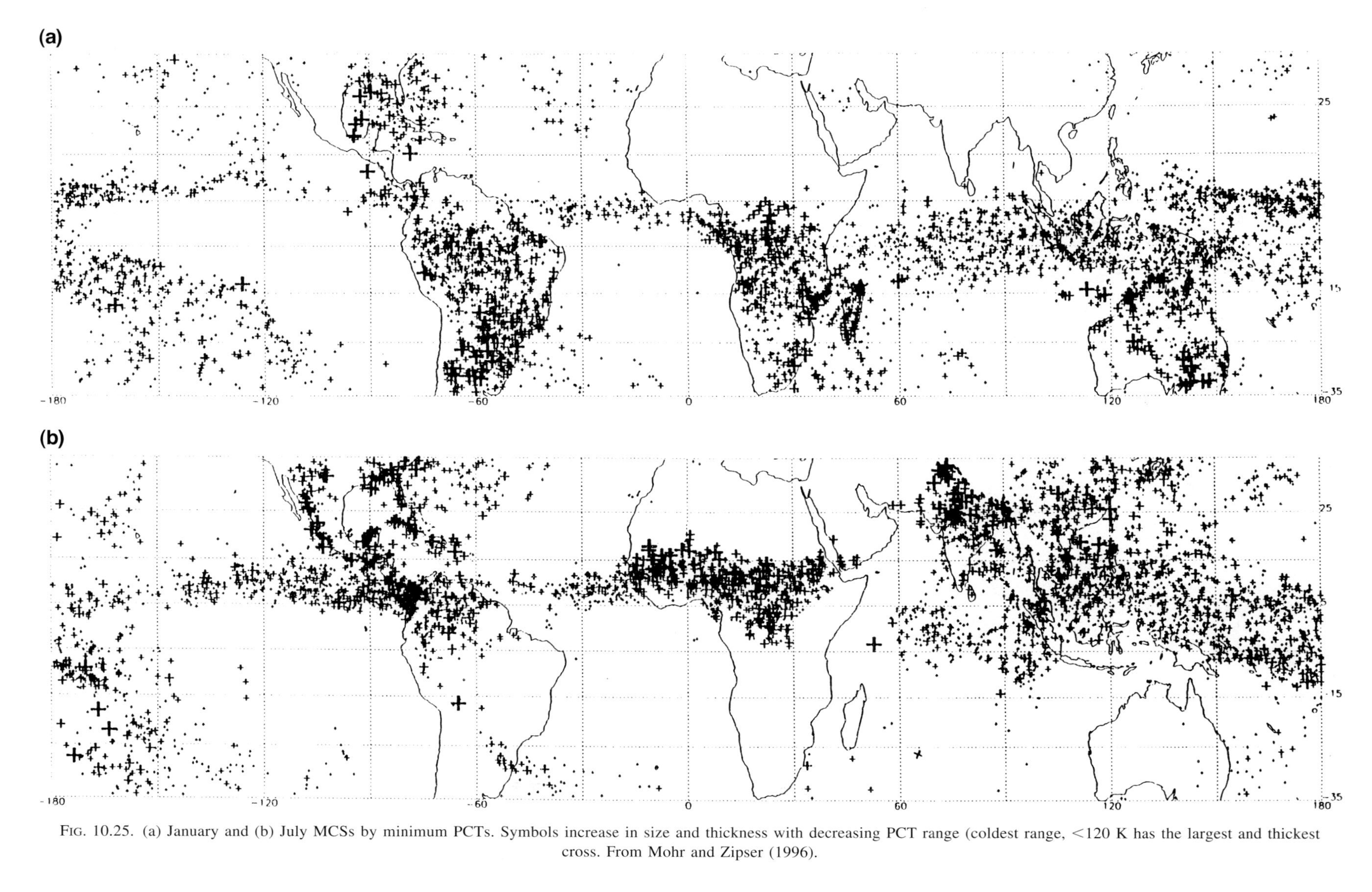

FIG. 10.25. (a) January and (b) July MCSs by minimum PCTs. Symbols increase in size and thickness with decreasing PCT range (coldest range, <120 K has the largest and thickest cross. From Mohr and Zipser (1996).

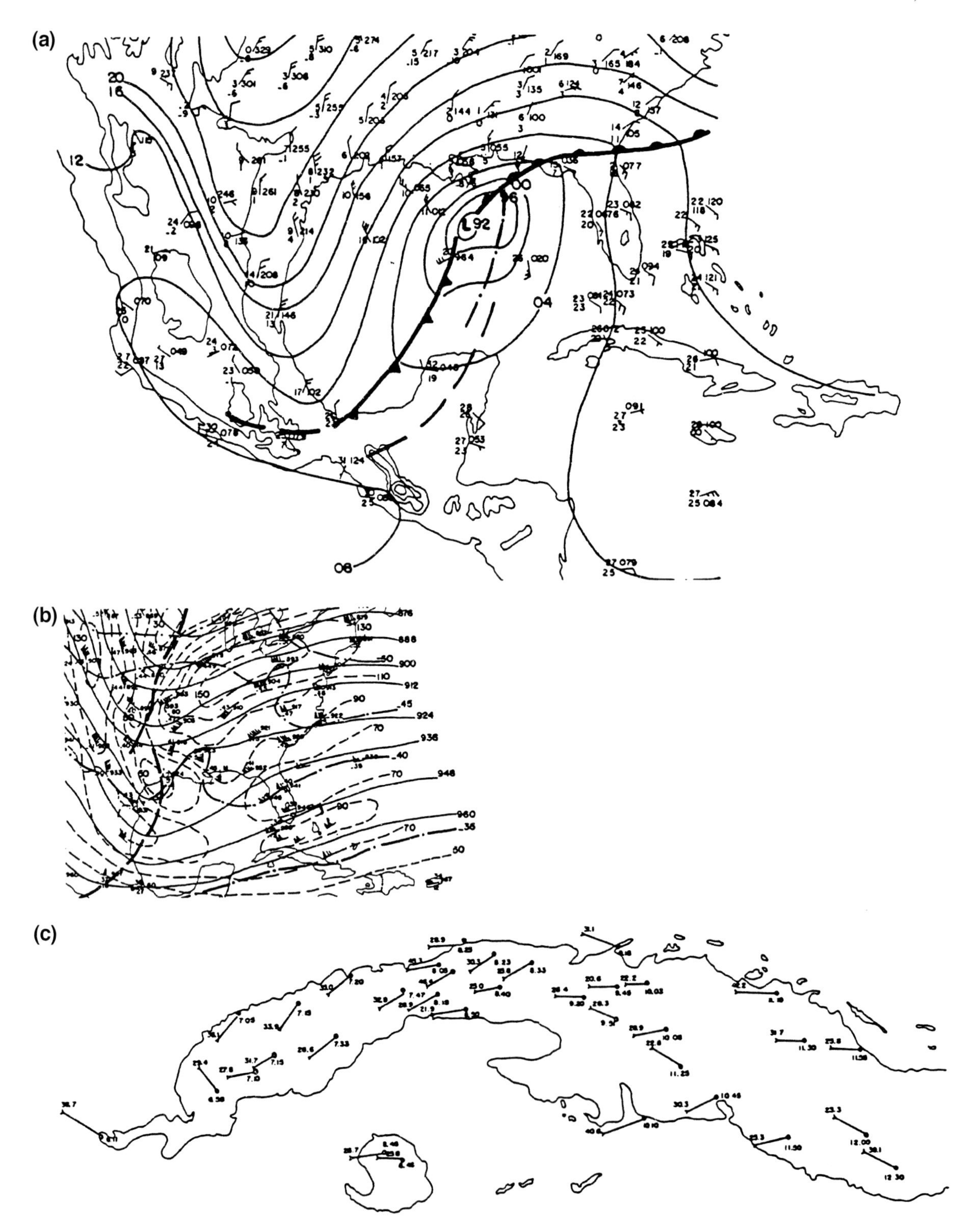

FIG. 10.26. (a) Sea level pressure fields for 0000 UTC March 1993. Isobars every 4 hPa, temperature (top) and dewpoint (bottom) in Celsius. A full wind barb is equal to 10 knots. (b) Isotachs (dashed) and geopotential heights (solid) at 300 hPa. (c) Highest wind gusts (m s^{-1}) and the time they were recorded across Cuba for 13 March 1993. Taken from Alphonso and Naranjo (1996).

SLS embedded in extratropical cyclones yield all forms of severe weather in northwestern India and Bangladesh. Hoddinott (1986) discusses "nor'westers," which develop often in the afternoons of April and May and move to the southeast through Bengal (Bangladesh and eastern India) before decaying at night. Shallow humid flow from the Bay of Bengal, which Hoddinott labels the *Chota* or *little monsoon*, undercuts the cold and dry winter northwesterlies aloft. Surface outflows reaching 39 m s^{-1} have been recorded; Hoddinott reports aircraft being flipped over. His drawings made in 1945 of the phenomena match the popular schematics of strong oceanic squall lines that we have been graced with over the last 25 years. Some of the pressure changes are surprisingly large (13 mb; Hoddinott 1986). The conditions mimic those for the U.S. Midwest—moist and warm southerlies undercutting cooler, dry westerlies with strong shear.

2) SOUTH AMERICA—NO CLEAR SYNOPTIC TRIGGER (AND LITTLE DOCUMENTED SLS)

The northern part of South America (equator to 10°N) is often devoid of significant synoptic-scale forcing. The remaining larger-scale agents for severe straight-line winds are the squall lines embedded in the easterly flow. Examination of the meridional wind component suggests that the passage of the squall is not associated with an easterly wave (Betts et al. 1976). In 1972 the second Venezuelan International Meteorological and Hydrological Experiment (VIMHEX) collected data on squall lines that moved westward at speeds of 11–16 m s^{-1}. These lines exist with modest CAPE of 800–1000 m^2 s^{-2} and shear in the u component of 2.5×10^{-3} s^{-1} (Betts et al. 1976). Such magnitudes would support only ordinary cell formation, based on bulk Richardson number calculations. The θ_e difference is over 20 K from the surface to the 650-mb level, suggesting that substantial cooling of midlevel air via evaporation is possible; nonetheless, evidence for straight-line winds of severe levels is not forthcoming (Miller and Betts 1977). Miller and Betts (1977) show that these squall lines have propagation speeds faster by a few meters per second than the maximum gust recorded at the surface.

Garstang et al. (1994), Greco et al. (1990), Molion (1987), and Kousky (1980) discuss squall lines that form in the coastal region of the Amazon Basin. These systems can survive up to 48 h, attain a length scale that reaches 1000 km (Greco et al. 1990), and may propagate all the way to the Andes Mountains (Molion 1987). The lines move at speeds of 12 to nearly 17 m s^{-1} (Molion 1987; Greco et al. 1990). However, even with these speeds, the peak gusts remain modest, usually 10–15 m s^{-1} (Garstang et al. 1994).

There is at least one reference that supports the existence of strong straight-line winds in northern South America. Court and Griffiths (1985) remark on the thunderstorms in Suriname referred to as the *sibiboesie* (literally "broom and forest"—sweeping clean), which generate winds in excess of 30 m s^{-1} and cooling of 6°–8°C. These occur with a trough to the north and strong surface heating. The trough may supply drier air in the midlevels that favors evaporatively driven outflows. Further documentation for these events has not been discovered.

3) EASTERLY WAVES—WEST AFRICA AND THE EASTERN ATLANTIC

West Africa is famous for its disturbance (squall) lines (Hamilton and Archbold 1945; Eldridge 1957; Leroux 1983). During the summer, hot and moist monsoonal flow from the southwest undercuts the air of the heat low over the Sahara. The African Easterly Jet (AEJ) in the midtroposphere, formed by the reversal of the temperature gradients when the Sahara warms due to insolation, creates the shear and serves as a source of high momentum for downdrafts when water is evaporated into this dry air. The disturbance lines have a favored genesis region over the highlands of Mali and generally beyond the northern limit of low-level clouds associated with the monsoon (Aspliden et al. 1976). Peak formation time is late afternoon, and they move west and southwest at speeds of 11–20 m s^{-1}. Such high speeds would suggest that wind along the gust front could reach severe levels, but the sparse observations do not provide evidence for such magnitudes very often. The genesis of squall lines and numerous Cbs is linked to the easterly waves, with the maximum development at and ahead of the trough defined from the 700-mb flow (Fig. 10.27).

Fortune (1980) uses satellite visible imagery to identify arc lines over West Africa moving at 15–25 m s^{-1}.

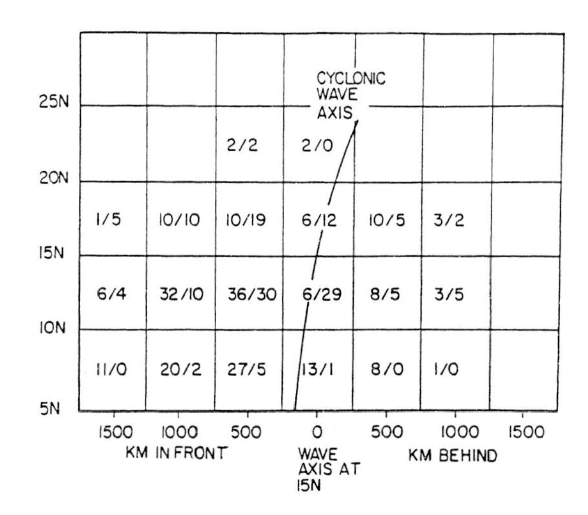

FIG. 10.27. Number of Cb images on days with frequent appearances of Cbs (first value) and organized phenomena or disturbance lines (second value) in each 5° square relative to the cyclonic axis at 700 mb of the easterly wave for all three phases of GATE. From Aspliden et al. (1976).

Correlations between surface winds and the speed of cloud arcs (Gurka 1976) show that for well-defined arcs near Cbs, the maximum gust is equal to or slightly greater in magnitude than the arc speed. It is likely that the fastest disturbance lines do produce surface outflows that achieve severe status, but the sorry state of observations does not allow one to compile any statistics.

The Convection Profonde Tropicale Experiment in 1981 (COPT), conducted in the northern portion of the Ivory Coast, supplies a superior view of these disturbance lines through the use of dual-Doppler radar (e.g., Testud and Chong 1983; Roux 1988). Roux and Ju (1990) categorize the mean environmental conditions as 1780 $m^2\ s^{-2}$ for CAPE; 64 for the bulk Richardson number (multicell); LCL and LFC of 700 m and 1450 m, respectively; and shear from 2 to 6 km of 2–3. $\times 10^{-3}\ s^{-1}$. These values, though high, are somewhat unfavorable for supercell or tornado genesis. COPT squalls can occur in dry or moist environments; the latter inhibit evaporative cooling and thus strong downdrafts and outflows at the surface. The COPT region is biased somewhat toward the moister conditions aloft. Roux and Ju (1990) use COPT 81 Doppler radar data to show a squall that moves at 13 $m\ s^{-1}$ and generates a maximum near-surface wind of the same magnitude.

Chong et al. (1987) analyze a COPT 81 squall (Figs. 10.28a,b) that does not yield severe straight-line winds, though many of its characteristics would lead one to believe that it should. This system had rain rates of 60 $mm\ h^{-1}$ along its leading edge, a pressure rise of 2 mb, a cooling of 4°C, and maximum wind gusts of 15.0 $m\ s^{-1}$ (Fig. 10.28c). Maximum updrafts reached 13 $m\ s^{-1}$ at 2.5-km altitude. Maximum downdrafts are only $-4\ m\ s^{-1}$. There is strong and deep (3 km) rear-to-front relative flow, again a trait that is usually implicated in strong straight-line wind production. Note that the squall formed in a region with a strong low-level jet and shear greater than $3.5 \times 10^{-3}\ s^{-1}$ in the lower troposphere (Fig. 10.28d). Lafore et al. (1988) have conducted three-dimensional simulations of the disturbance lines; generally the winds behind the gust front rarely exceed the gust front speed itself by more than 4 $m\ s^{-1}$. The Doppler analyses of these systems show that there is strong rear-to-front flow descending from midlevels to the leading edge during the mature stage.

COPT did occur south of the majority of disturbance lines studied by Aspliden et al. (1976). North of 10° latitude, beyond the COPT observations, there would be a deeper subcloud layer that could be cooled by evaporation, yielding stronger downdrafts, greater divergence, and stronger surface outflows. While deep Cbs form throughout West Africa during the southwest monsoon flow, there is a tendency to see the 700-mb jet better developed farther north.

Observations over the eastern Atlantic show that severe straight-line winds are rare. Houze (1977), Zipser (1977), Frank (1978), Zipser et al. (1981), Houze and Betts (1981), Gamache and Houze (1982), Barnes and Sieckman (1984), and Houze and Rappaport (1984) are but a few of the efforts describing GATE MCSs or the environment in which they grow and mature. Even the strongest squall lines sampled do not spawn winds of severe magnitudes. The 4–5 September squall yields wind speeds less than 12 $m\ s^{-1}$ (Houze 1977); the 12 September squall, which has the reputation as being the strongest line to pass through the ship array during GATE, yields speeds slightly in excess of 15 $m\ s^{-1}$ (Zipser 1977; Gamache and Houze 1982). Zipser et al. (1981) find maximum speeds of 10–11 $m\ s^{-1}$ with aircraft flying at 150-m altitude through a slow-moving MCS. Much of this wind was from the south and parallel to the leading edge of the eastward moving MCS.

The author reexamined the wind record for the USCGS *Dallas* for all three phases of GATE and found that there were 66 instances where convection and attendant outflows reached the ship; there were no cases where the three-minute mean speeds exceeded 12 $m\ s^{-1}$. Easterly waves enhance convective activity in and ahead of the trough, but this convection consists of ordinary cells that have difficulty making conditions dangerous at the surface.

4) Monsoonal systems

The Monsoon Experiments (MONEX, winter and summer phases), Equatorial Mesoscale Experiment 1987 (EMEX), Australian Monsoon Experiment 1987 (AMEX), Taiwan Area Mesoscale Experiment 1987 (TAMEX), and studies around Darwin, Australia, such as the Down Under Doppler and Electricity Experiment 1987–88 (DUNDEE) offer views of systems during active and break stages of the monsoon.

In the summer Taiwan is under the influence of the monsoon. Jorgensen et al. (1991) examined an MCS, just east of southern Taiwan, that exhibited no rear inflow and no strong convective-scale downdrafts; both are associated with strong straight-line winds. A gust front was not detected by the aircraft. The WP-3D Doppler showed velocities slightly in excess of 8 $m\ s^{-1}$.

Tao et al. (1991) examined a squall line near Taiwan and found an environment that is only slightly more unstable than the GATE squall line environment. Evidence for only ordinary cells along the leading edge was found.

The many reports from India are organized as a function of increasing latitude. Gupta (1987) uses 10 years of data from flat, humid Car Nicobar Island (9.1°N) in the Andaman Sea to show that thunderstorm frequency is highest in the premonsoon and

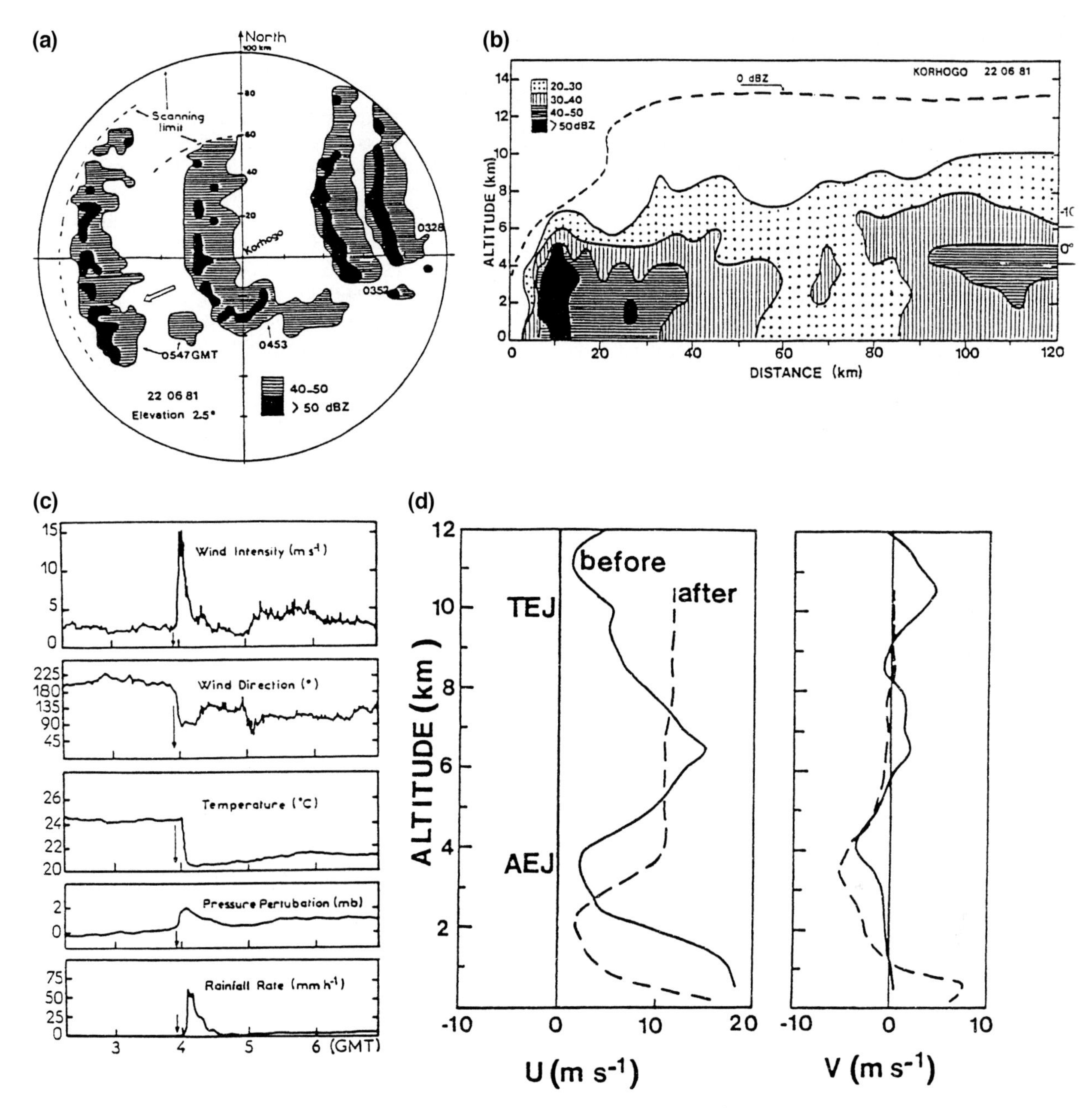

FIG. 10.28. (a) Horizontal and (b) vertical views of the reflectivity field for a disturbance line; (c) a series of surface variables associated with a passage of a disturbance line, and (d) the vertical wind profiles (u, v, m s^{-1}) that support the formation and maintenance of a West African disturbance line. From Chong et al. (1987).

early monsoon; the island has never witnessed severe winds or hail.

In southern India severe straight-line winds are extremely rare. Reddy (1985) has studied squalls (here apparently defined as gusty winds that surpass the mean flow by about 8 m s^{-1}) at Madras airport (13.5°N). He finds, based on 19 years of data, that 90% of the "squalls" are under 21 m s^{-1} and only two events were greater than 27.8 m s^{-1} (100 km h^{-1}). Nearly 75% of the squalls were embedded in the southwest monsoon. The squalls were associated with Cbs and have a temperature decrease and a pressure rise, suggesting that they are an outflow from convection.

Sivaramamakrishnan (1989) notes that no severe straight-line winds were produced from the 44 thunderstorms that occurred at the India space port, Sriharikota—which is about 13.7°N and on the east coast of India—from April through September.

Banerjee et al. (1983) conducted a survey of squalls for 21 stations distributed over India for a five-year record. They classify squalls as *light* ($\leq$14 m s^{-1}), *moderate* (>14–<22), and *severe* ($\geq$22 m s^{-1}). Some of their findings regarding severe squalls are

1) a rarity of strong winds during the postmonsoon and winter months;
2) a maximum in occurrence during the premonsoon and early monsoon months (AMJ);
3) the three stations with 20 or more severe squalls over five years are Calcutta (22.3°N) and Jamshedpur (22.5°N), both in the northeast, and Nagpur (21.1°N), located almost in the center of the subcontinent; and
4) the severe winds most often come from the northwest.

Mukherjee et al. (1983) examine the conditions that favor strong straight-line winds with radar and soundings in the Calcutta area. They note that cloud top does not serve as a predictor for the severity of the wind gust. Shear of the zonal wind in the 850–500- and 500–300-mb layers also proved to be ineffective as a forecast tool. They did not, however, combine shear of the total wind, instability, and potential for evaporation in the mid- to lower troposphere to find a predictor, which one might expect to have led to an improvement. Significant moisture advection from the south at 850 mb did occur in 8 of 12 cases. Other works that discuss squalls include Dekate and Bajaj (1966), Alvi and Punjabi (1966), and Ramakrishnan and Rao (1954).

Kamble and Kanoujia (1990) review the phenomena known as *multiple squalls,* which are repeated gusts 8 m s^{-1} above ambient conditions. For central India the events are concentrated (~85%) in the premonsoon (hot weather season from March to May) and early monsoon, and are associated with thunderstorms. From 1969 to 1978 Nagpur, Bhopal, and Jagdalpur (19°–23°N) recorded 11 events with speeds in excess of 22 m s^{-1}; only four of these events achieve severe status by the U.S. definition. Just about half of the multiple squalls occur in the late afternoon, and most last for 5 min.

On 31 May 1986, Karachi, Pakistan (24.5°N), was hit by a dust storm that contained winds over 31 m s^{-1} (Middleton and Chaudhary 1988). These winds lasted for only 5 min before heavy rain, but the winds and near-zero visibility within the dust wall caused 12 deaths, extensive damage to the modest homes, and disruption of utility service for half of the city, the country's largest. A monsoon depression created the favorable environment for this *andhi*—a low-precipitation, high-wind thunderstorm of northwest India. Joseph et al. (1980) offer more information on andhis.

Extreme winds (winds of 20 m s^{-1} or more) outside of those associated with a TC have been shown to be associated with thunderstorms for 13 stations in Thailand (Atkinson 1966). Severe straight-line winds are generally found in the mountainous regions during the premonsoon period.

Northern Australia's weather is strongly modulated by the monsoon. During the austral summer, heating of the Australian continent produces low pressure in the lower troposphere, which in turn results in low-level westerlies; above these there are easterlies associated with the upper-level anticyclone. This wind pattern enhances the wind shear from the surface to 700 mb to magnitudes greater than 5.0×10^{-3} s^{-1}. Additionally, the low-level air traveling over the warm Timor and Arafura Seas is a source of moisture; CAPE often exceeds 2000 m^2 s^{-2} and can reach as high as 4000 m^2 s^{-2}. Dry adiabatic subcloud layers can often be deep (2500–3500 m), which is beneficial for evaporation, strong downdrafts, and vigorous straight-line winds. Squall lines, and in some instances splitting storms, have been noted by forecast offices in the Northern Territory and Queensland (Gill and Kersemakers 1992). Some of the squalls are akin to the nor'westers of India and Bangladesh, except that they move from the southeast.

Northern Australia cloud lines (NACLs) have been classified by Drosdowsky and Holland (1987). These lines typically move about 5 m s^{-1}, consist of cumulus congestus, and are not associated with major synoptic weather systems. The most vigorous NACLs that do contain deep cumulonimbi have been tracked using the satellite criteria of Aspliden et al. (1976); they move to the west or northwest at noticeably higher speeds (9–13 m s^{-1}) than the shallower lines. Duration of the strong systems is usually 9–21 h. These vigorous lines can deliver straight-line winds that reach severe magnitudes; Drosdowsky (1984) and Drosdowsky and Holland (1987) note that severe squall lines occur during transition and monsoon break phases, which have more shear, more instability, and drier air in the midlevels. Maximum frequency for these events is in January and February. The sparse population of tropical Australia likely leads to many SLS events going unreported.

The *morning glory,* a photogenic low-roll cloud of northern Australia (Smith et al. 1982), rarely produces severe weather conditions. This shallow cloud has the characteristics of an undular bore that is propagating on the nocturnal low-level stable layer. The morning glory can move at speeds reaching 15 m s^{-1}, which would lead one to believe that wind speeds of similar magnitude are possible. This solitary wave can pose a hazard to low-flying aircraft (Christie and Muirhead 1983). Smith et al. (1986) discuss southerly nocturnal wind surges over northeastern Australia that, besides triggering an eastward-moving morning glory, can trigger deep convection, given an unstable environment.

The weather around Indonesia is also under the influence of the monsoon. Analyses by Ishihara and Yanagisawa (1982) of a squall line moving at 7 m s^{-1}, with cloud tops to 14 km, just northwest of the head of New Guinea, reveal no winds greater than about 12 m s^{-1}. The shear for this squall is about 3.0×10^{-3} s^{-1}.

5) Microbursts in the Tropics

A *microburst* is a strong downdraft that produces intense divergence over a few km. Aircraft are vulnerable at landing and takeoff to the rapid change from headwind to tailwind; crashes can occur even when the winds fall well short of severe magnitudes. Byers and Braham (1949) were perhaps the first to discuss this threat to aircraft safety.

Fujita (1985) discusses two crashes in the Tropics that are due to thunderstorm outflows that were below severe wind speeds. The sparse data do not support the belief that either the Pago-Pago, American Samoa (14°S, 171°W), crash on 30 January 1974 or the Kano, Nigeria (12°N, 9°E), crash on 24 June 1956 encountered wind speeds that reached 26 m s^{-1}.

Palam, New Delhi's major airport, was affected by a microburst on 27 June 1985 that yielded a 40–13 m s^{-1} couplet (Kishan and Lal 1989). The thunderstorm that produced the microburst moved at 23 m s^{-1} and had a top reaching 17 km. Williams (chapter 13, this volume) discusses microbursts around Darwin and finds that the great majority are nonsevere.

6) Convection over the Pacific

Convection over the west Pacific warm pool is modulated by the 30–50-day oscillation. The strong shear that can occur when the westerly winds appear offers an increased likelihood that severe winds gusts may occur.

Perhaps one of the strongest squall lines sampled during TOGA COARE was on 22 February 1993. This squall line was bow-shaped, with strong rear inflow, a low-level vortex, and a well-developed cold dome, and moved eastward at 12 m s^{-1} (Jorgensen et al. 1997). The environment was characterized by moderate instability (CAPE = 1440 m^2 s^{-2}) and wind shear in the lowest 200 mb of 6.7×10^{-3} s^{-1}. Maximum wind speeds reached near 20 m s^{-1} both near the leading edge and about 100 km behind. The system appeared to be a weak cousin of a bow echo or derecho (e.g., Johns and Hirt 1983, 1987; Przybylinski 1995; Weisman 1993).

Zipser (1969) used Line Islands Experiment 1967 (LIE) aircraft observations to diagnose the mesoscale downdraft associated with the stratiform rain portion of a squall line that was moving west at 15 m s^{-1}; wind speeds reaching 24 m s^{-1} were observed. Again, this system existed in strong low-level wind shear.

Both examples lend credence to the idea that only those oceanic squall lines in strong shear ($>5.0 \times 10^{-3}$ s^{-1}) tend to yield winds that approach the severe threshold.

7) Other relevant studies

Florida is not in the Tropics, but has been the subject of numerous studies of deep convection [e.g., Thunderstorm Project, Florida Area Cumulus Experiment (FACE), Convection and Precipitation/Electrification Experiment (CaPE)]. The summer circulations over Florida are akin to lower-latitude situations when no upper-level westerly trough is present.

The results from several studies support the notion that severe straight-line winds are rare. A part of the Thunderstorm Project was conducted near Orlando in 1946. Byers and Braham (1949) do not report any winds from the many ordinary cells they sampled that reached severe magnitudes. A Cape Kennedy survey (Carter and Schuknecht 1969) shows that the average gust (8–10 m s^{-1}) is well below the threshold for a severe rating. The absolute peak gust has exceeded the severe threshold only in the spring (26 m s^{-1}), under the influence of an extratropical system.

Caracena and Maier (1987) document a microburst in FACE where winds of 20–30 m s^{-1} were estimated, based on damage to sugarcane. This microburst was considered "wet," that is, embedded in heavy rain; however, their analyses show that evaporation of rain, not water loading, is responsible for the event. A dry layer aloft was viewed as an important ingredient.

b. Hail

Hail is labeled severe when it is greater than 20 mm or three-quarters of an inch in diameter (Huschke 1959). Unfortunately, the majority of hail records from the Tropics record little more than the occurrence of hail on a given day. We shall proceed to discuss these records with the questionable assumption that knowledge of where hail of any size falls also provides clues as to where the stones that reach severe status fall.

Usually hail reports come from two sources: weather station reports that cover a small fraction of a given region, and insurance claims, mostly from crop damage. Of course there is no report for events after harvest, nor would there be if there is no insurance, which is the situation for most underdeveloped countries in the Tropics. There are essentially no data from a dense network of hail pads or cubes from tropical locations. Most hail swaths cover very small areas ($\sim$1 square km), so counts are certainly too low.

Morgan (1982) shows that only a small percentage of the hail days account for the majority of the damage. This we can assume to be from hail with a diameter greater than 20 mm (diameter $>$ 8 mm is used to differentiate hail from graupel). Table 10.2 from Morgan (1982) lists a number of regions that are prone to hail and the percent of the hail that reaches small, medium, and large sizes. The two regions that fall within or close to the Tropics have disparate distributions. Northern India appears to witness severe

TABLE 10.2. Percentage frequency of occurrence of maximum hail size within specified ranges per observed point hail fall in various regions ranked in order of decreasing frequency of large hail.

Regions	Maximum hail diameter (cm) <1	1–3	>3
Northern India	37	38	25
Southwestern France	40	46	14
Northeast Colorado	33	60	7
Oklahoma	49	43	7
Central Alberta	49	45	6
Denver area, Colorado	44	52	4
New England	60	36	4
Transvaal, South Africa	54	43	3
Central Illinois	70	27	3
Kericho, Kenya	74	26	0.3
NHRE Network	31	69	<0.1

hail more frequently than any other location, whether it be midlatitude or tropical, while Kericho, Kenya, rarely has hail greater than 3 cm in diameter. The smaller hail is still a serious problem, as it often falls on the delicate tea plants grown on the mountainsides of Kenya and Tanzania.

Hull (1958), Frisby (1966), and Frisby and Sansom (1967) supply a great deal of information about hail days in the Tropics. Most of this information is difficult to condense in that different countries have differing record lengths, reporting procedures, and station densities. Some countries only offer a total for the entire country, not for each station. Sometimes the station records for a given country will vary widely. For example, data from Mozambique generate suspicion in that all the stations with a long record have a low frequency per annum while almost all the short-record stations have a much higher frequency. Obviously changing observation techniques and continued development of a region lead to higher counts. The author will take the risk of trying to summarize the many tables from Frisby (1966) and Frisby and Sansom (1967) in an attempt to identify hail frequency as a function of elevation and season.

1) ELEVATION AND HAIL FREQUENCY

Frisby (1966) analyzed stations by elevation throughout the Tropics and found that, on average, stations below 1200 m experience fewer than two hail days per year; stations above have three to eight per year. The quality of the correlation between elevation and hail occurrence demonstrates that other factors are also at work. Frisby and Sansom (1967) analyzed 41 stations with identified elevations from Brazil (Fig. 10.29a), mostly from 18° to 22°S latitude, and show a clear trend of increasing frequency beyond a threshold of 600–800 m elevation. Below this threshold there is no correlation. The Brazilian data span 17 years. I have estimated the frequency per annum for each station. Stations from Ghana, Nigeria, and Cameroon (Fig. 10.29b), from 3° to 10°N latitude, also have a clear increase of hail with elevation, and again the possibility of a threshold near 600–700 m where the positive correlation becomes obvious. Four stations from Angola (9°–15°S latitude) also exhibit a positive correlation between station elevation and hail fre-

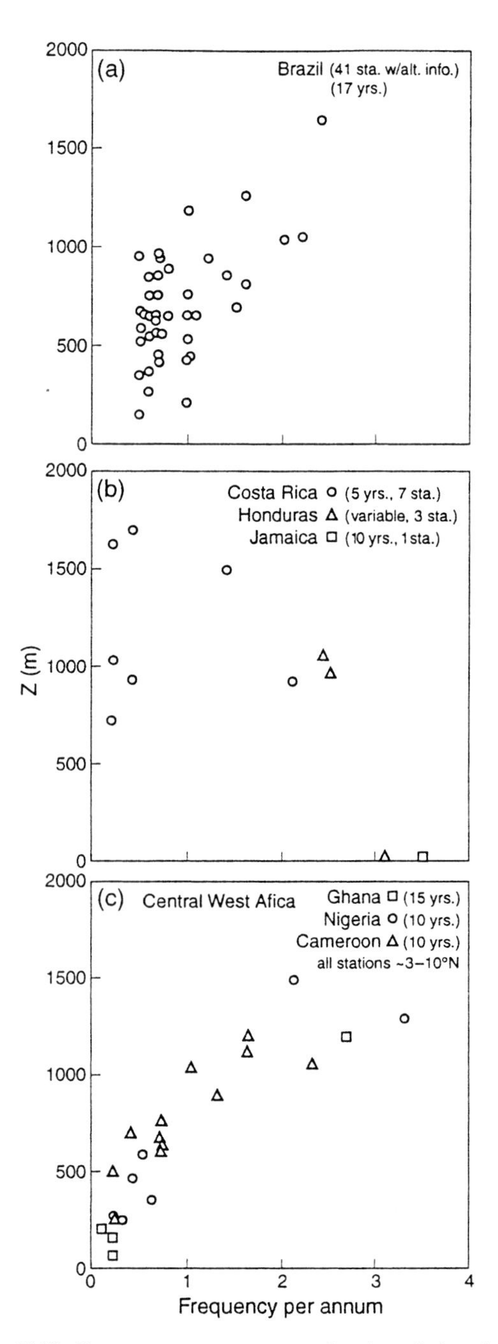

FIG. 10.29. Frequency per annum as a function of elevation for (a) 41 stations in Brazil, (b) Central America from 9° to 17°N, (c) central West Africa stations from 3° to 10°N.

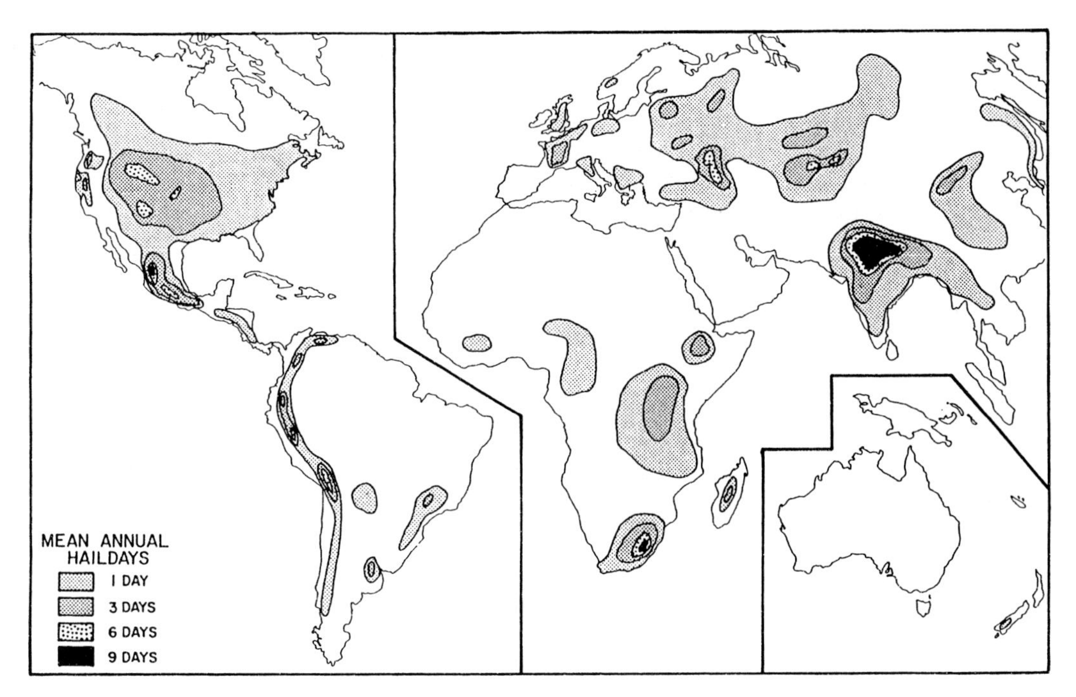

FIG. 10.30. Mean annual hail days at a point. From Frisby and Sansom (1967).

quency, but Angola does not have enough stations to identify a threshold.

Data from other countries do not exhibit a positive correlation with height above sea level. Data from Costa Rica, Honduras, and Jamaica (Fig. 10.29c), all located in the latitudinal band from 9° to 17°N, show that stations near sea level have the highest frequency. The consistency of reporting from the meteorological offices at the major cities near the coast versus those more isolated stations may partially account for this unexpected result. Stations from Ecuador and Peru ranging from 2600 to 3900 m have a higher rate of occurrence (~10 per annum) than that found for other lower stations, but no clear trend with height.

Figure 10.30 from Frisby and Sansom (1967) shows that the frequency of hail in the Tropics is closely tied to high terrain (e.g., Kenya, Ethiopia, along the Andes, and Madagascar). Over high terrain, the depth of air warmer than freezing is decreased, allowing for a higher survival rate of the small stones. Larger hail, with its much greater fall velocity and mass, is little affected by melting below the freezing level. If there is more large hail at higher elevations, then it is due to other factors, such as mountain valley wind patterns, that result in larger and stronger updrafts.

Some areas that do not fit the elevation–hail frequency pattern include Bangladesh, which is about 400 km from the Himalaya foothills. Bangladesh and northeastern India are influenced by extratropical troughs.

Flora (1956), in his book *Hailstorms of the United States,* included a chapter covering hail in other countries. Here are listed some of his findings, but classified as *near or at sea level, intermediate,* and *high terrain.*

1. Near sea level (0–200-m elevation): For West Africa only one storm was reported along the coast from 1921 to 1934. Fiji reported three events from 1880 to 1918. Panama had three hailstorms in twelve years. Hail in Mexico at low elevation has not been reported (Veracruz, Tampico). The *Jamaican* was hit by a hailstorm in the Gulf of Mexico on 17–18 March 1906. Stones up to 6.4 cm in diameter were measured.

   Reports of hail in the Philippines are rare, with only three storms over 41 years. Hailstones are small and therefore of little consequence. Indang, in Cavite Province, did suffer some slight damage during a storm.

   Puerto Rico has suffered damage (Lemons 1942), especially to the delicate tobacco crop. Events occurred in April, when we might expect extratropical influence to be important. The Bahamas witnessed a storm in February where the stones reached a depth of 6 in. Hail has been recorded in Panama within 9° of the equator.

   One concludes that events do occur at sea level to about 10° away from the equator, but they are very rare, and are usually associated with a vigorous extratropical circulation.
2. Intermediate elevation (200–500 m): Flora notes that India has particularly good records because hail damage results in a remittance of the land tax. Detailed reports are filed by revenue collectors. Almost all storms occur during the northeast or dry

monsoon. The number of storms from 1883 to 1897 are December—57, January—91, February—112, March—241, April—203, and May—72. This works out to 52 storms per year for these 6 months.

India and Bangladesh suffer more deaths from hail than other locales due to the combination of large stones and a dense agrarian population that lacks adequate shelter. Newspaper articles report on catastrophic storms that kill hundreds and injure thousands of people. On 30 April 1888, 246 were killed in the Moradabad and Beheri districts. The hail swath reached 150 miles. On 17–18 March 1939 in Hyderabad, a storm damaged a few thousand homes, killed 1500 livestock, and destroyed several thousand acres of crops. Reports of gigantic stones the size of elephants (Buist 1885) are amusing and probably due to stones melting together after they fell. More careful studies (e.g., Hariharan 1950) did document a 3.4-kg stone.

Extratropical systems are responsible for almost all the hailstorms in northeastern India and Bangladesh. Once the southwest monsoon engages, both low-level shear and instability are reduced.

3. Mountains and high plateaus (>500 m): In West Africa 321 storms were recorded in the high terrain of French West Africa from 1921 to 1934. This is 23 storms per year, which is competitive with some of the high-frequency locations found at higher latitudes.

On the high plateau of central Mexico (~20°N), hail falls about six times a year in Mexico City, seven times a year in Puebla, four times a year in Morelia, and three times a year in Guanajuato.

Several high-terrain locations in the Tropics do suffer frequent hailstorms, but this should not be used to infer that these storms yield stones of the size that would be characterized as severe.

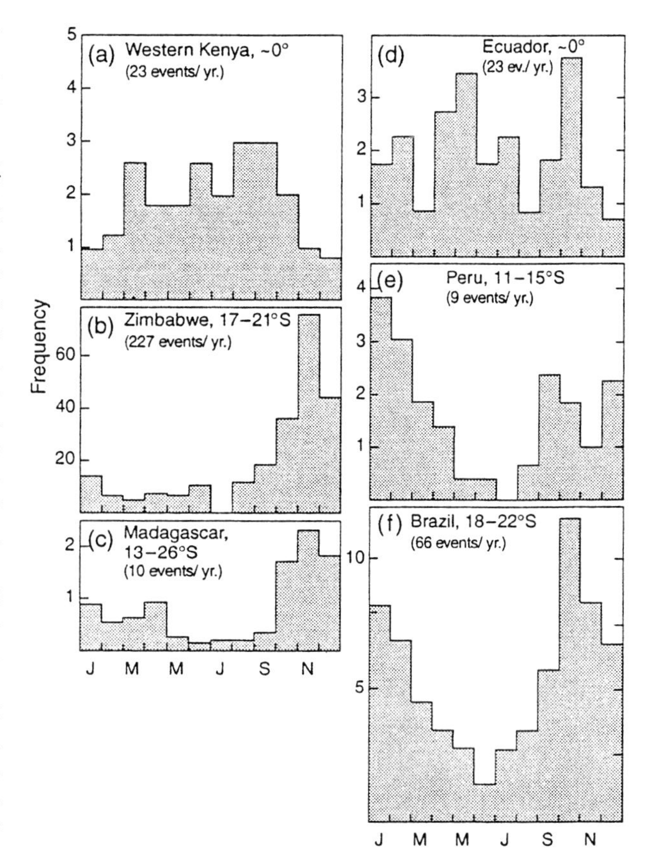

FIG. 10.31. The annual cycle of hail for stations in the Southern Hemisphere: (a) Kenya, (b) Zimbabwe, (c) Madagascar, (d) Ecuador, (e) Peru, and (f) Brazil.

2) Hail frequency and season

When hail occurs provides clues as to which synoptic or mesoscale phenomena are playing the primary role in the development of favorable conditions. Frisby and Sansom (1967) again provide much of the raw data; here it is reduced to the frequency of hail per month per station or frequency per month per country. This latter statistic is an act of desperation since several countries did not list the number of stations that recorded hail. Hail frequency per annum for the entire country will exhibit much higher values than those for a particular station.

Figures 10.31a–f show the results for stations in the Southern Hemisphere. In Kenya, centered on the equator, events are witnessed throughout the year, while Zimbabwe (17°–21°S) and Madagascar (13°–26°S) have a minimum in the austral winter and a maximum in austral spring and early summer. For South America (Figs. 10.31d–f), Ecuador (along the equator) shows the influence of the migration of the NECZ, while Peru (11°–15°S) and Brazil (18°–22°S) have a peak of activity in the austral spring and summer and a minimum in the winter. Annual patterns for both continents support the conjecture that spring extratropical cyclones, associated with a deeper 500-mb trough and warm and moist low-level air, produce the needed environment.

Australia has few hailstorms, though a few have cut impressive swaths of destruction. Visher (1922) examined records from 1908 to 1917 and found 10 hailstorms from 13° to 16°S, all at sea level. Townsville (~19°S) has reported damaging hail. Yates and Hanstrum (1996) report on a left-moving member of a splitting storm that yielded hail the size of golf balls. This occurred 100 km southeast of Broome, located along the northwest coast of western Australia.

For stations in the Northern Hemisphere one sees peak hail occurrence in the boreal spring and summer. The frequency per station per month is generally low in Central America and the Caribbean (Figs. 10.32a–c). In Asia, Burma (17°–24°N) and Thailand (13°–20°N) also have a boreal spring maximum with no hail during the summer.

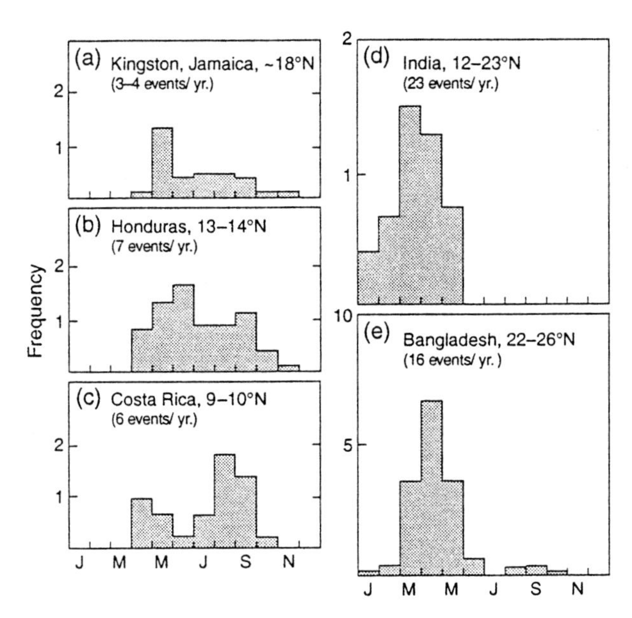

FIG. 10.32. The annual cycle of hail for stations in the Northern Hemisphere: (a) Jamaica, (b) Honduras, (c) Costa Rica, (d) India, and (e) Bangladesh.

On 30 January 1985 hail with an average diameter of 18 mm fell on the island of Hawaii during the passage of a cold front (Takahashi 1987). In February 1997 Cbs embedded in a Kona low dropped 12.5-mm diameter hail. Hail falls on Hawaii 5–10 times per year but it is well below severe size and covers only a small area, most typically at high elevations found on the island of Hawaii.

Ramage (1995) notes that south Florida has had only three hail events with hail reaching the severe size threshold over the last 50 years. Neumann (1965) analyzed one of these storms, which delivered hail up to 75 mm in diameter in Miami; this happened on 29 March 1963 when the axis of the upper cold low passed by.

India and Bangladesh are different from other Northern Hemisphere tropical stations in that hail is observed (Figs. 10.32d, 10.32e) in the winter and premonsoon seasons, with virtually no events in the summer after the onset of the southwest monsoon. Chaudhury and Mazumdar (1983) show that the expectancy percentage for hail, defined as the number of days with hail divided by the number of days with thunder, decreases from 5% to less than 2% from March to May for northeastern India and Bangladesh. Through this period, which is considered to be the premonsoon, CAPE increases at Calcutta from 2000 to 4550 $m^2 s^{-2}$, but shear decreases in the layer from 6 to 12 km from $2.8 \times 10^{-3} s^{-1}$ to $0.4 \times 10^{-3} s^{-1}$. The height of the freezing level also increases from 4330 to 5140 m through the premonsoon period.

Statistics from Prasad and Pawar (1985) for the areas east and west of the Ghats, sounding analyses from Calcutta by Kanjilal et al. (1989), and seasonal variations studied by Biswas and Gupta (1989) are evidence for the decreased instability during the monsoon (after a May peak) for the southern and central portions of India. In these regions there are maxima in thunderstorm frequency at the start and sometimes at the end of the monsoon, but not during the heart of the event. To the north, in New Delhi (Gupta and Sharma 1983) and Lucknow (Kumar 1992), thunderstorms are at a maximum during the monsoon. Recall that the inclusion of northeastern India and northern Bangladesh is stretching the boundaries of the Tropics somewhat.

Nizamuddin (1993) reviews the annual reports of the Indian Meteorological Department from 1982 to 1989 and finds 228 hail days (about 29 per year) of moderate to severe intensity; hail size is often compared to mangoes, lemons, and tennis balls. Nizamuddin (1993) shows a concentration of hail occurrences in the central highlands and the Himalayan region. Ramanamurthy (1983; Fig. 10.33) shows the frequency of hail for a 100-year period; there is one event per year in northeast India, but 5–10 per year in northern India and in the Himalayas. Eliot (1899) looked at the sizes of stones for 597 hail falls in India and found that 153 of these yielded stones of 3.0 cm diameter or greater. These events killed 250 people and caused extensive damage to the winter wheat crop. Primary season is the premonsoon but north and central India have winter events. The premonsoon is when midlatitude troughs tap into the moist and very hot (>40°C) boundary layer air originating over the Bay of Bengal.

Chowdhury and Banerjee (1983) examine eight years' worth of data for northeastern India, which is a

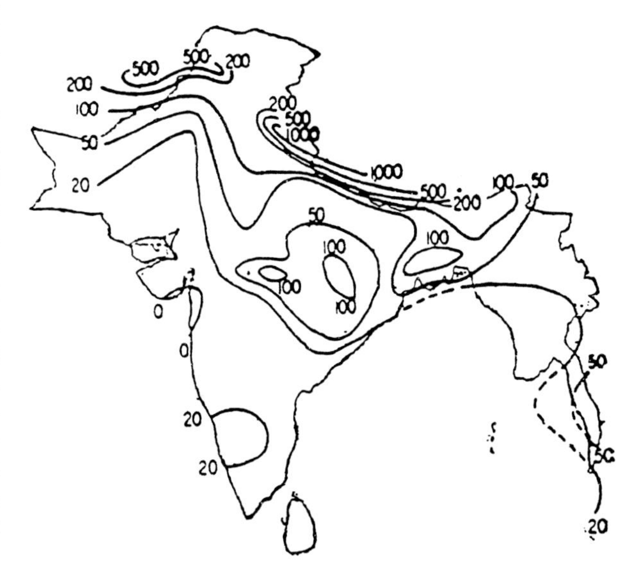

FIG. 10.33. Hail occurrences for a 100-yr period. From Ramanamurthy (1983).

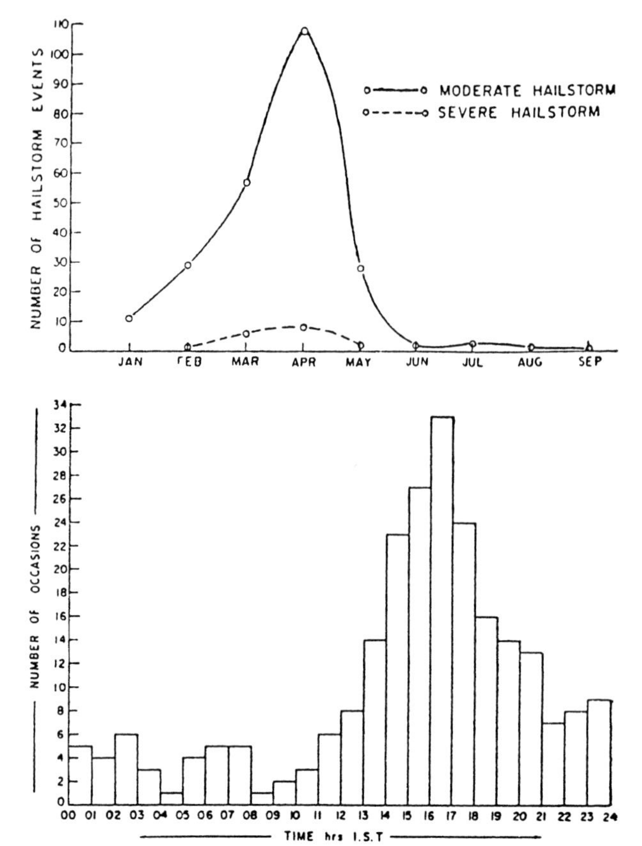

FIG. 10.34. (a) Monthly distribution of moderate and severe hail for India, and (b) diurnal variation of hailstorms. From Chowdhury and Banerjee (1983).

vital agricultural region. They find that 30 hailstorms occur per annum. The yearly pattern demonstrates the influence of westerly troughs in the premonsoon season, with a maximum in April (Fig. 10.34a). Chowdhury and Banerjee (1983) find a jet-stream maximum, upper-level divergence accompanying the trough. The storms do tend to be initiated by insolation, with the maximum in the late afternoon (Fig. 10.34b), and they develop over hilly terrain (200–500-m elevation). Shear in the 850–300-mb layer is more than four times greater than average for the premonsoon season (22 vs 5 m s^{-1}) and the freezing level descends from 585 to 625 mb. For the premonsoon above 850 mb, temperatures are lower than normal, but dewpoint values are essentially the same for premonsoon hailstorms as for the climatological mean.

Misra and Prasad (1980) show that hailstorm frequency is very low for the west coast and central inland India from 15° to 20°N. There has been at least one event, near Bombay (Chakrabarty and Bhowmik 1993) in April 1988, associated with an extratropical system that pushed freezing levels below 610 mb.

Pandharinath and Bhavanarayana (1990) describe hailstorms from 11 to 13 March 1981 that killed 18 people and 13 000 livestock, and damaged 85 000 houses and 33 000 acres of crops in the Andhra Pradesh (~16°N). Hail mass ranged from 100 to 1200 g. The storm environment included marked upper-level divergence and a strong jet stream, a trough in the westerlies, attendant low surface pressure, strong shear reaching 5.5×10^{-3} s^{-1} (Fig. 10.35), and low-level convergence of moist air from the Bay of Bengal. Wind speed from 850 to 200 mb changes over 40 m s^{-1}.

Almost all of South Africa lies south of the Tropic of Capricorn but deserves some brief discussion. Simpson (1976) notes that an area of 7000 km^2 centered on Nelspruit (25°S, 750-m elevation, called the low veld) has about 50 hail days per annum. Garstang et al. (1987) show that hail is a frequent problem in the high veld (elevations > 1500 m) in South Africa. Often an eastward-moving trough at 850 mb yields westerly or northerly flow that carries the high veld mixed layer over the escarpment and the low veld. The juxtaposition of this now elevated mixed layer above the low veld mixed layer creates an inversion over the low veld, inhibiting convection there. The easterly flow near the surface of the low veld transports moist air toward the higher terrain. Finally, diurnal heating of the high veld, and the formation of upslope flow, generates hailstorms.

Hail in South Africa needs the interaction of three scales: synoptic, which produces the eastward moving trough; the mesoscale, which forms the capping inversion over the escarpment; and local-scale heating, which helps generate the easterly flow under the synoptic/mesoscale westerlies.

3) SUMMARY OF HAIL FINDINGS

Tropical hail remains an infrequent occurrence at low elevations, which seems to lend credence to the belief that a higher freezing level (4.5–5.0 km) results in melting of the hailstones before they reach the ground. However, large hail does not melt readily, so the long fall (0.0°C to the surface distance) does not explain the rarity of large, damaging hail at low elevations. It is likely that the mesoscale flows, organized by differential heating that develops around steep gradients of terrain, lead to more vigorous updrafts and the increased likelihood that hail of a reasonable size can be created. These larger stones can survive their descent through the warmer atmosphere below.

Frisby's (1966) partitioning of the Tropics into three belts—23.5°–10°N, 10°N–10°S, and 10°S–23.5°S—apparently works well. The higher-latitude belts had hail chiefly during the spring months, when winter storms interact with warmer, moister, low-level air. Equatorial trough stations recorded hail more randomly through the year.

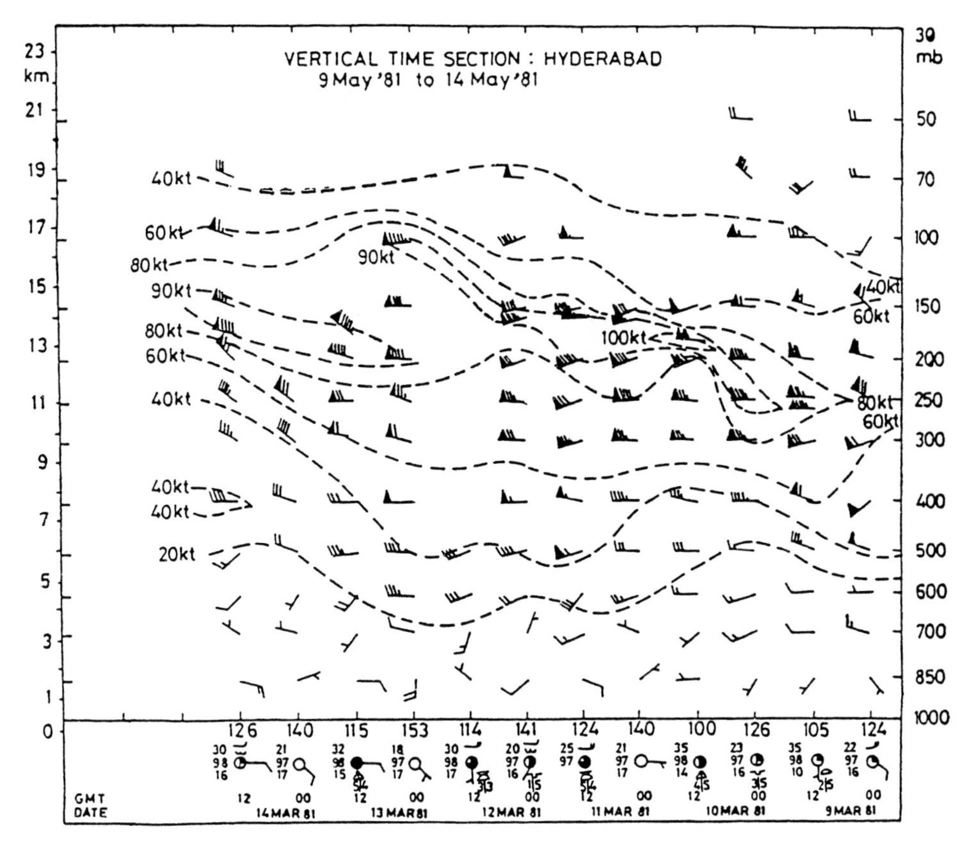

FIG. 10.35. Time-height cross section of the wind field over Hyderabad with surface reports in standard format found below. Time runs from left to right. From Pandharinath and Bhavanarayana (1990).

c. Tornadoes and waterspouts

Tornadoes and waterspouts have been classified according to their intensity (e.g., Fujita 1981). I will bifurcate the population of rotating columns of air into those associated with a supercell and those associated with weaker, shorter-lived convection. Waterspouts (Golden 1971, 1974a,b) and landspouts (Bluestein 1985; Wilson 1986; Wakimoto and Wilson 1989) rarely achieve F3 magnitudes. Shearing instability along convergence boundaries is often the source of vorticity for the weaker so-called landspouts; stretching of the column by an accelerating parcel within cumulus congestus or developing Cb intensifies the vortex to tornado strength. The majority of waterspouts usually have a line of cumulus congestus clouds creating an environment favorable for genesis (e.g., Golden 1974a,b) and rarely achieve intensities beyond F1; their more benign nature invites a separation from the higher F-scale tornadoes that are intimately tied to a supercell circulation.

Waterspouts have been defined in the *Glossary of Meteorology* (Huschke 1959) as simply a tornado over water; therefore a vortex can instantly change its label by translating over the opposite surface.

1) TORNADOES—ASSOCIATED WITH EXTRATROPICAL INFLUENCES

Invading extratropical waves are the most effective way to create an environment favorable for tornadogenesis of high F-scale vortices (e.g., Fawbush and Miller 1954; Barnes and Newton 1985; Fig. 10.36). The intense tornado hot spots that appear in the Tropics occur where extratropical waves and terrain work together (e.g., northeastern India, southernmost Brazil, and Australia). The juxtaposition of dry, vigorous westerlies above hot and moist low-level air flowing toward the poles creates the buoyancy and shear for supercell formation. Terrain seems to enhance the effect of the waves by aiding in the juxtaposition of air with contrasting histories. Northeastern India and Bangladesh, just downstream of the Tibetan Massif, witness numerous intense tornadic storms.

The invasions of these waves into the Tropics are met with considerable resistance. Sensible and latent heat fluxes from the warm seas rapidly weaken the temperature and moisture gradients in the boundary layer. The subtropical deserts, with their high Bowen ratios, also tend to destroy discontinuities found within an extratropical wave (e.g., the Sahara serves as a buffer for the Sahel and equatorial Africa). Radiation

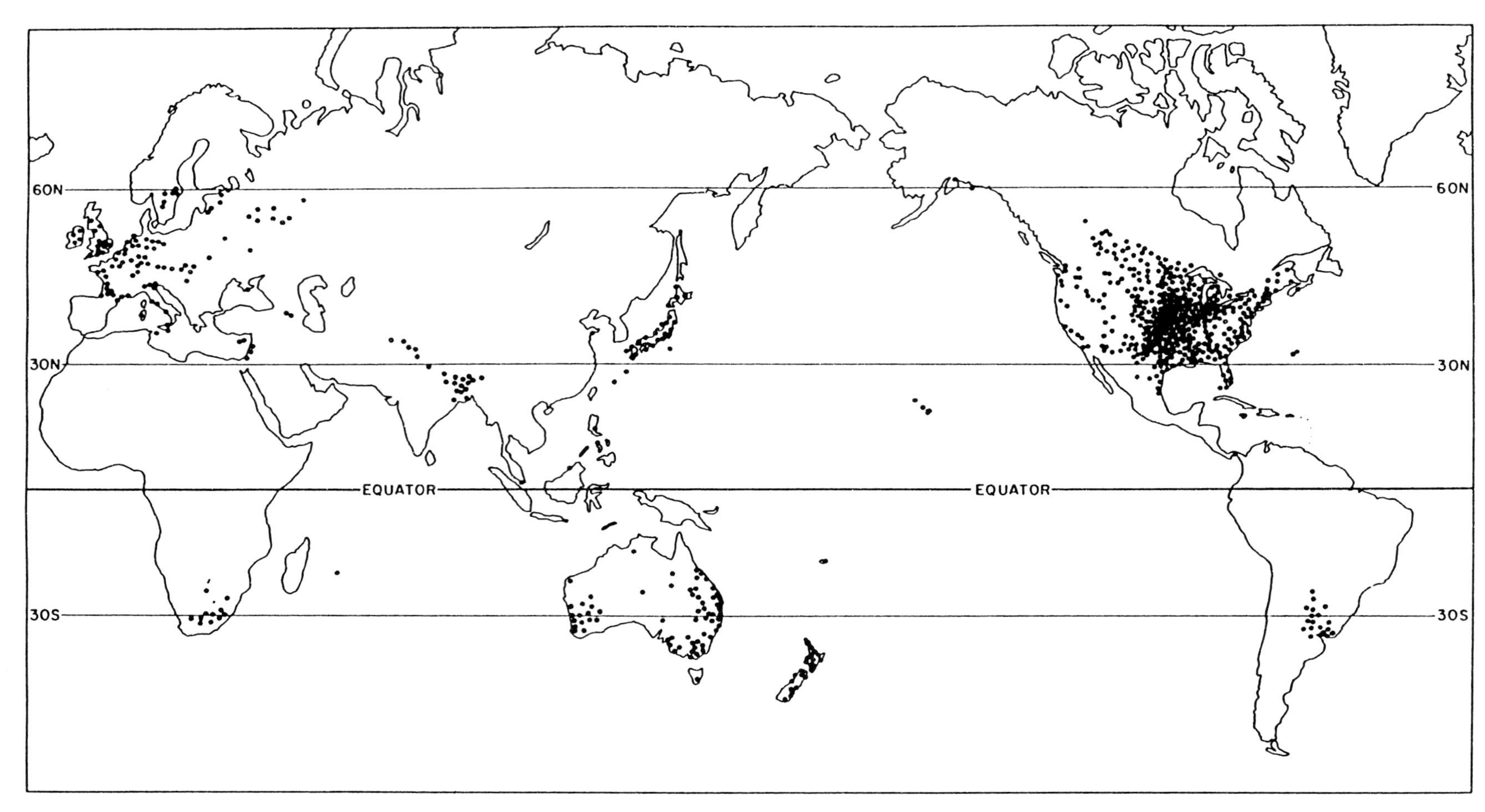

FIG. 10.36. Density plot of tornadoes over a 4-yr period (1963–66). In the United States the number exceeds 2600. From Fujita (1973).

inputs are large in the low latitudes and rapidly modify cold air masses.

The Southern Hemisphere landmasses see a lower frequency of SLS, perhaps due to the smaller amplitude of the troughs found there. Southern Brazil, at least theoretically, should be a region plagued by tornadoes, as it is somewhat of a mirror image of the United States with mountains to the west and a moist air source from the Amazon in the return flow of the high. However, Brazil does not have a warm sea directly equatorward, such as the Gulf of Mexico. During the austral winter an anticyclonic circulation present over the Amazon, in part maintained by the convection found there, inhibits the equatorward progress of the westerly waves (Fig. 10.37, from Dean 1971) and reduces the tornado frequency greatly (see Fig. 10.36).

The majority of the tornadoes in India, known as *Hatishnura* or *elephant's trunk* (Asnani 1993), occur during the premonsoon season (Nandi and Mukherjee 1966; Mukherjee and Bhattacharjee 1972; Ghosh 1982; Gupta and Ghosh 1980; Singh 1981). Singh (1981) reports that 42 tornadoes occurred in India between 1951 and 1980; 33 of these occurred between March and May in the premonsoon period, and 27 were confined to west Bengal and Bangladesh. Most were spawned by thunderstorms forming ahead of a trough, and strong wind shear in the lower troposphere was present. Preconditions match those of the Gulf Coast region of the United States; see Chapter 2 of this volume.

Gupta and Ghosh (1982) have documented 35 tornadoes from 1876 to 1978, about one every three years (Fig. 10.38), with the majority occurring in the premonsoon months. Only five tornadoes were noted

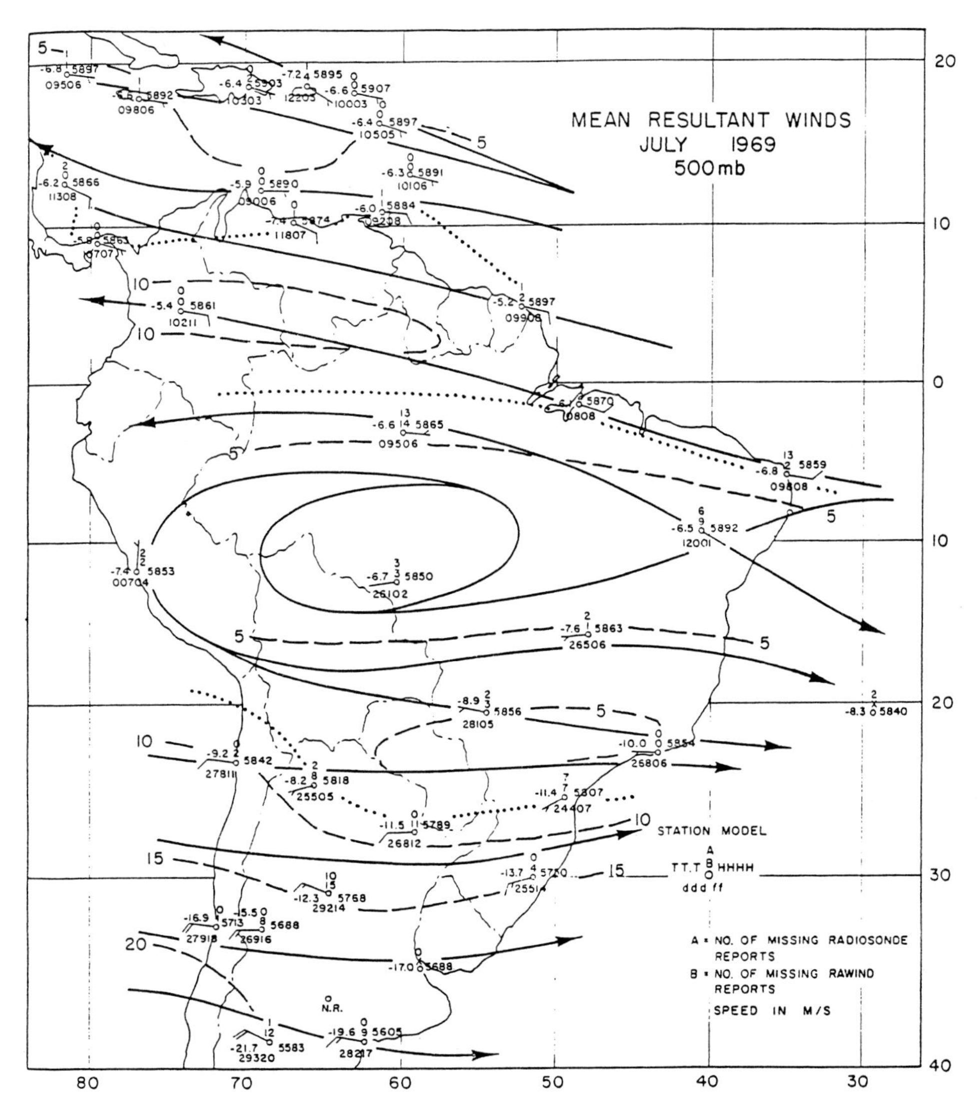

FIG. 10.37. Mean resultant winds at 500 mb for July 1969. From Dean (1971).

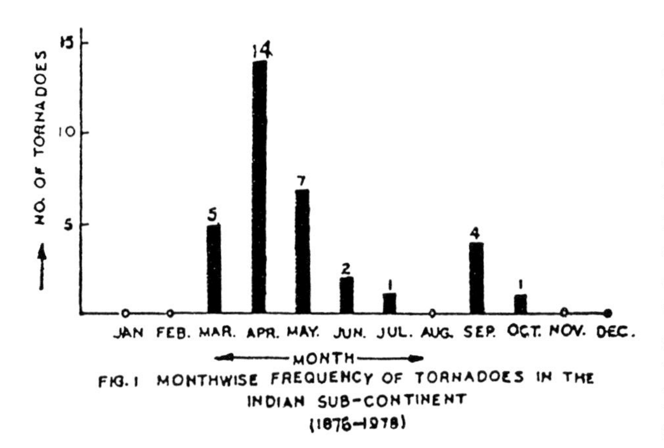

FIG. 10.38. Monthly frequency of tornadoes for India, from 1876 to 1978. From Gupta and Ghosh (1982).

during the monsoon months. Northeastern India and Bangladesh account for 25 of the tornadoes. Approximately 86% occurred during the afternoon and evening hours. Some of the more disastrous events include the 4 April 1978 tornado that killed 150 in Orissa, the 3 March 1978 North Delhi tornado that killed 28, and the 4 April 1895 tornado near Dacca that took 24 lives. More recently, in May 1996, a tornado killed 50 people and injured more than 2500 in Bangladesh. Estimated wind speeds were only 60 m s^{-1}, but the mud and straw houses simply did not offer any protection. A tornado, with winds estimated to be 56 m s^{-1}, killed more than 440 people and injured 33 000 in many villages near Rampur, in northern Bangladesh. The terrible losses were due in part to the lack of substantial shelter.

Prasad (1990) presents evidence for a supercell in the western part of Bihar (northeastern India) in October that killed 20, injured 517, destroyed nearly 2800 homes, and damaged over 17 000 others. The supercell apparently survived for two hours and covered 200 km on a north-northeast track. The supercell formed within a weak depression that did contain surprisingly strong shear of 8.6×10^{-3} s^{-1} (25 m s^{-1}) from the surface to 700 mb.

Flora (1954) mentions a tornado that occurred on April 30 1888. The path length reached 150 miles in the Delhi and Peshawar regions of northern India. Despite this impressive path length, most of the 230 reported deaths were caused by very large hail.

Mandal and Saha (1983) review eight tornadoes that occurred in India from 1963 to 1981. Most of these tornadoes achieved F3 (71–92 m s^{-1}) magnitudes. These tornadoes produced a mean track length of 17 km and moved along the ground at a mean speed of 16–17 m s^{-1}; diameters of the tornadoes ranged from 30 to 1000 m.

The Tibetan Massif does complicate the conditions over northeastern India and Bangladesh. Mandal and Basandra (1978) and Mandal and Saha (1983) show that the tornadoes over northwestern India match the classic model (Barnes and Newton 1985), which includes an upper-level trough to the west, mid- and upper-level jet streams, a synoptic surface cyclonic disturbance to the west, and movement of the tornado from southwest to northeast. In contrast, the tornadoes in northeastern India move from the northwest to the southeast. Most of these cases do not have a clearly defined upper-level trough, which is probably due to trough deformation caused by the Tibetan Plateau. The surface charts reveal a weak cyclonic circulation, not a well-developed extratropical low invading the lower latitudes. All but one tornado are associated with a subtropical westerly jet stream, often with a magnitude exceeding 40 m s^{-1}, and a midlevel wind maximum. The midlevel flow is from the west at 25–30 m s^{-1}, resulting in shears from near the surface to 500 mb in excess of 5×10^{-3} s^{-1}. The midlevel wind maxima provide a favorable tilting term for vorticity and usually transport a layer of air with nearly a dry adiabatic lapse rate over northeastern India, promoting instability and strong downdraft formation.

Viswanadham and Sud (1991) noted an anticyclonic tornado in southeastern Madhya Pradesh, 95 km south-southeast of Raipur (~20°N). The tornado moved to the south-southwest and covered 10 km. Over 17 000 trees were damaged; the tornado was assigned an F2 ranking. Synoptic conditions did not include the standard upper-level trough, but there was a depression at the surface 500 km to the east-southeast.

Reliable counts of tornadoes in South America suffer from a lack of sophisticated observing equipment. Brazil, for instance, did not have a 3-cm radar available for weather observation until 1974 (Chu 1975).

All but a very small section of Argentina lies south of the Tropic of Capricorn. Schwarzkopf (1982) surveyed newspaper reports from 1930 to 1979 and found 646 instances of damage due to winds; 12% were certain tornadoes. This works out to be about one to two tornadoes per year. All of the events occurred in a region from 25° to 40°S (Fig. 10.39), and east of the 65° meridian, which coincides with a 2500-m mountain ridge. Austral summer is the preferred time for these tornadoes. Many of the tornadoes have a track from southwest to northeast, which is similar to events in northeastern India. Less than 6% of the tornadoes were ≥F3 intensity. The synoptic analyses reveal that extratropical cyclones help organize conditions, and afternoon heating is important in the development of the parent Cb. The Argentina experience shows a decrease toward the Tropics, supporting the conjecture that extratropical wave influence decreases substantially as one moves equatorward in the Southern Hemisphere.

Southern Brazil is occasionally affected by midlatitude circulations; for instance, Chu (1975) presents evidence for a dryline near São Paulo, Brazil. The

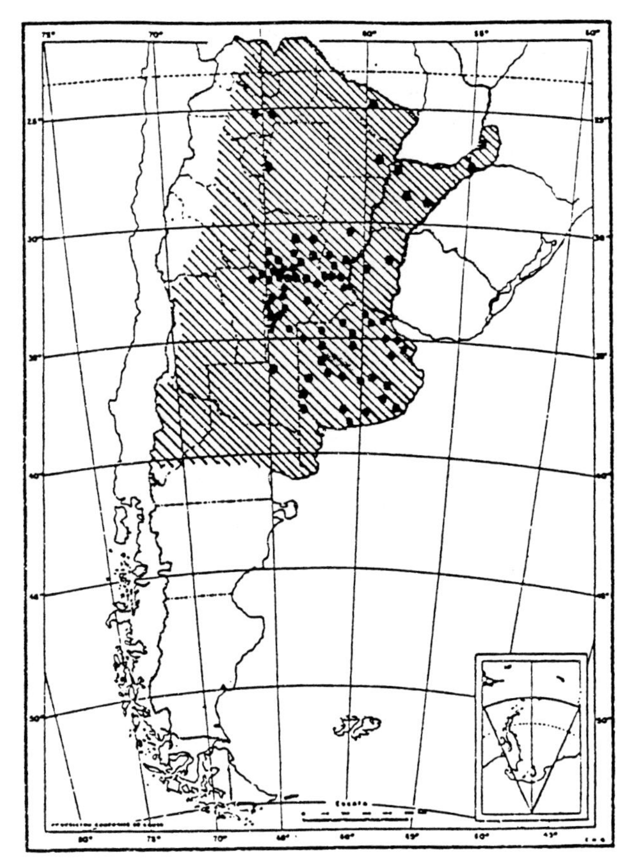

FIG. 10.39. Distribution of tornadoes in Argentina with the shaded region where damaging winds from all phenomena have been recorded. From Schwarzkopf (1982).

thunderstorms did not produce any severe weather. The dryline was well ahead of a cold front; aloft there was a strong subtropical jet stream.

About a third of Australia lies within the Tropics. Clarke (1962) attempts to correct for population sparseness, short records, and confusing nomenclature used to describe tornadoes in Australia (e.g., "cock-eyed bob"), and argues that tornadoes are as frequent as, but less intense than, those found in North America. The crucial assumption for this conclusion is his correction for the low population density. MacDonald et al. (1975) and Allen (1980) show that population densities of at least 100 per square mile are needed to obtain accurate counts. Supporting this hypothesis are Hanstrum et al. (1998), who note that more than 700 tornadoes have been reported in Australia since the arrival of Europeans. From 1987 to 1996 162 events have been recorded; this increase in tornado frequency is due to the development of a spotter network.

Minor and Peterson (1979) identified regions in Australia that are susceptible to tornadoes; these include regions such as the Queensland Highlands, which are within the Tropics (Fig. 10.40). They list 36 tornadoes that received wider attention, but only two of these occurred equatorward of the Tropic of Capricorn. Allen (1980) notes that reports of tornadoes equatorward of 20°S latitude are rare. Hanstrum et al. (1998) show that virtually all of the tornadoes that occur during the austral winter are south of the Tropic of Capricorn; northern, tropical Australia is under the influence of the Asian southwest monsoon.

Minor and Peterson (1979) discuss one case at Port Hedland (~20°S), western Australia, on 17 December 1975, which would be just prior to the onset of the Australian monsoon. This is when instability is reaching its greatest levels and substantial shear is present due to the overlying westerlies.

The southern section of Taiwan is within the Tropic of Cancer. Wang (1979) documents mechanisms that may trigger tornadoes and the approximate number associated with each larger-scale system. From 1951 to 1978 squall lines with a supporting subtropical jet generated 9 tornadoes, the warm sector of a extratropical cyclones spawned 23, and typhoons yielded 9. Near Kaohsiung (~22.6°N), located on the coast of the southwest plains, there is a "tornado nest" where 26 of the 41 tornadoes have occurred (Fig. 10.41). Changes in surface friction and attendant changes in the low-level shear might be a cause for this concentration.

Strong frontal systems entering the Caribbean can spawn tornadoes. On 26 December 1940 a prefrontal squall line that developed a tornado killed 20 and injured 200 in Bejucal, near Havana, Cuba (Ortiz and Ortiz 1940).

Florida is entirely north of the Tropic of Cancer but the central and southern parts of the peninsula have dry and wet seasons like much of the Tropics. More importantly, there are both quality observations for and analyses of the Florida situation that will provide insight into tornado formation within the Tropics.

Hagemeyer (1997) concentrates on tornado outbreaks in Florida, where 1448 tornadoes have been recorded from 1950 to 1994 (58 per year) south of 30° latitude. Outbreaks are defined as at least four tornadoes in four hours or less from the same triggering phenomena. These outbreaks account for less than 4% of the tornado days but are responsible for 62% of the tornado injuries. Outbreaks are associated with three types of synoptic phenomena (Fig. 10.42); their variation with the annual cycle matches expectation.

An example of an outbreak is the tornadoes of 22–23 February 1998 that occurred in east central Florida (Sharp et al. 1998), causing 42 deaths and injuring 260 others. This outbreak was associated with a strong upper-level trough and a surface low to the northwest that left Florida in the warm sector of the circulation, which included a low-level southerly jet reaching 30 m s^{-1}. The F3 tornadoes occurred at night, demonstrating that the Cbs did not need any more instability that might be gained by surface heating of the peninsula.

Figure 10.42 from Hagemeyer (1997) may lull one into thinking that the summer is quiet, but the total

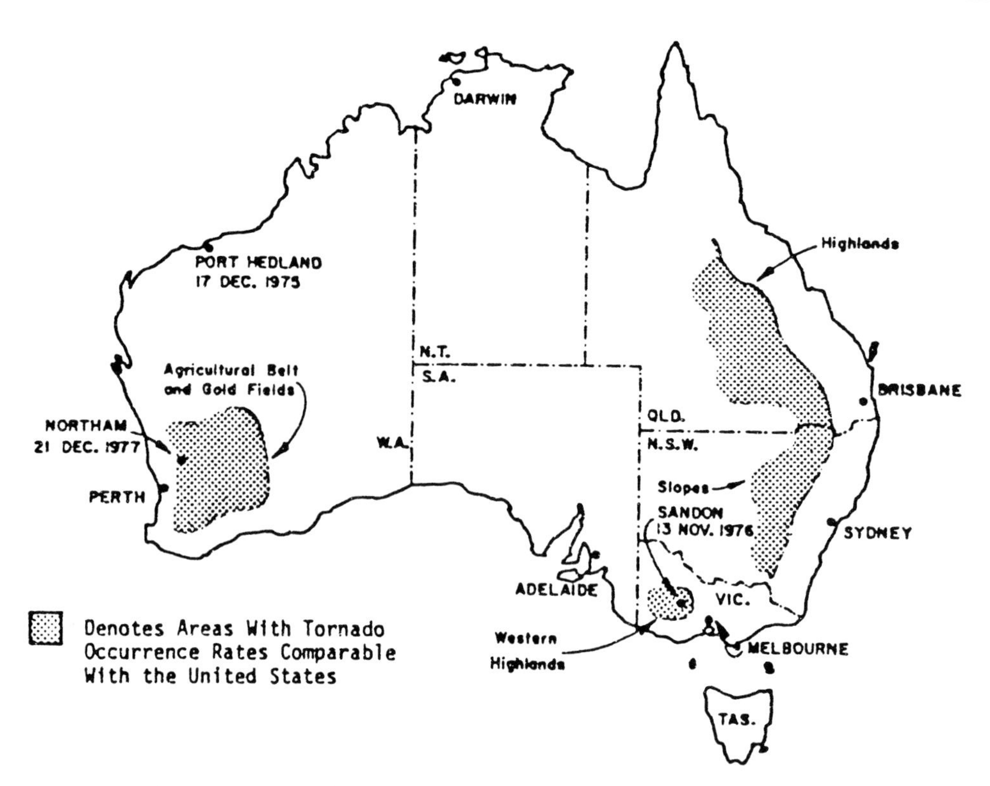

FIG. 10.40. Locations of documented tornado events. From Minor and Peterson (1979).

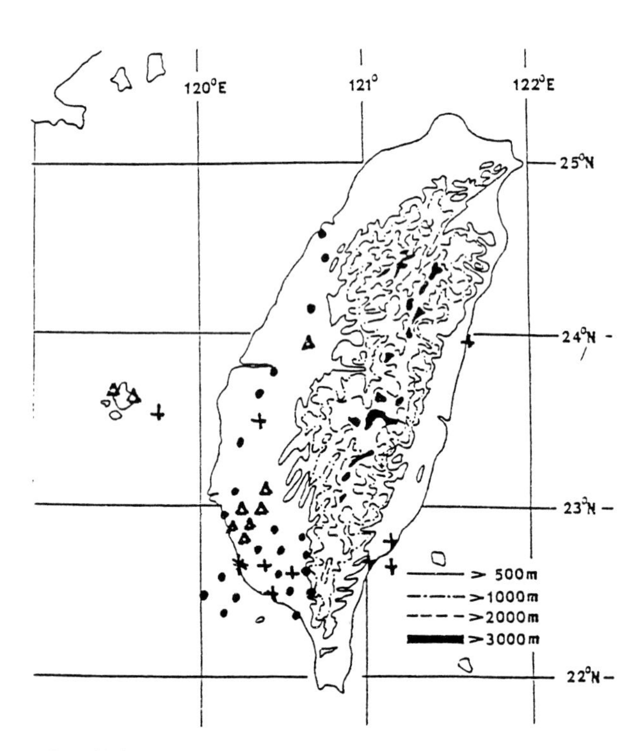

FIG. 10.41. Topography of Taiwan with known tornadoes; triangles are those associated with squall lines, crosses are those associated with typhoons, and dots are those associated with fronts. From Wang (1979).

counts of tornadoes (Fig. 10.43) demonstrate that Cbs without obvious synoptic-scale support can yield a tornado. The distribution of tornadoes with the annual cycle shows that the outbreaks occur during the peak of the annual distribution.

The hourly distribution of tornadoes for tropical versus extratropical outbreaks (Fig. 10.44) supports the contention that the tropical systems, with their afternoon and early evening maximum, need the extra instability from insolation. Tornadoes embedded in extratropical cyclones also peak during the day, but more than half of the outbreaks occur before noon.

Hagemeyer (1997) notes that severe straight-line winds almost always accompany outbreaks (28 of 35 cases); every tornado outbreak since 1979 has had severe straight-line winds. Large hail occurred in only 6 of the 35 events. During the outbreaks, severe weather is usually seen first on the west coast and steadily moves to the east or southeast. The appearance of SLS near the coast supports the hypothesis that changes in friction, shear, and convergence are shifting rapidly to favor tornado development. Outbreaks associated with tropical systems tended to have their tornadoes move northward.

The proximity environment for the extratropical outbreaks, as depicted by the few upper-air soundings that met the criteria set by Hagemeyer (1997), have the following values: LI $= -3°C$, K $= 31°C$, CAPE $= 995\ m^2\ s^{-2}$, BRN shear $= 95\ m\ s^{-2}$, BRN =

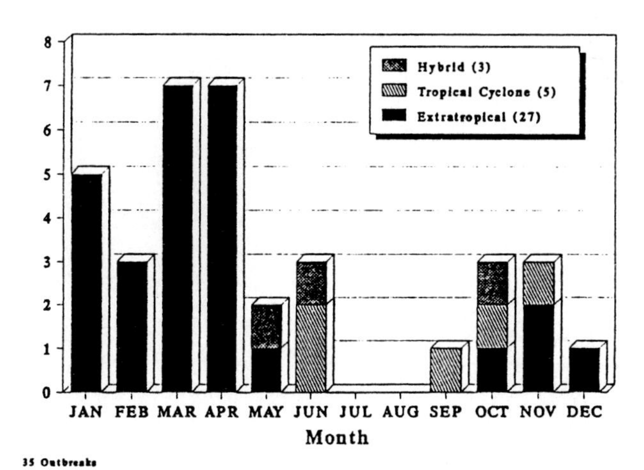

FIG. 10.42. Monthly distribution of peninsular Florida tornado outbreaks, 1950–94. From Hagemeyer (1997).

14, and helicity from 0 to 3 km = 405 m s^{-2}. The instability is surprisingly low, though the BRN is indicative of supercells. The composite soundings for both the extratropical and the tropical-hybrid groups show strong shear in the lower troposphere. Winds decrease above 750 mb in the tropical-hybrid group, indicative of the warm core of the tropical cyclone. Cb motion is to the right of the winds found below 6 km. Hagemeyer (1997) presents schematics for each case, depicting the key wind and thermodynamic features associated with a Florida tornado outbreak (Figs. 10.45, 10.46, and 10.47). In all three cases the tornadoes formed in moist flow from the south and ahead or to the east of the surface low pressure. Note that in the hybrid (Fig. 10.47), there is no warm front.

Three other studies illustrate the Florida cases. Collin (1983) shows an example of a leading cause of severe weather in the fringes of the Tropics: the invasion of a well-developed extratropical wave. The deep trough with its attendant positive vorticity advection, a strong subtropical jet aloft with diffluence, and strong convergence at 850 mb were all present. This environment helped spawn 18 tornadoes, damaging straight-line winds, and hail in northern and central Florida. Hiser (1968) demonstrates the importance of a trough and lower-tropospheric shear. Holle and Maier (1980) show how a tornado may form from the collision of two outflows. This situation is common in Florida (e.g., Fig. 10.43), and difficult to forecast without close radar coverage that reveals the surface flows.

2) THE HURRICANE ENVIRONMENT

Novlan and Gray (1974), in their study of both tornadic and nontornadic hurricanes, identified aspects that affect tornadogenesis. Factors leading to more tornadoes include 1) vertical shear of the horizontal wind greater than 20 m s^{-1} from the surface to 850 mb (Fig. 10.48), and 2) an intense and/or intensifying inner core (estimated by minimum sea level pressure or maximum winds in the eyewall) prior to landfall (Fig. 10.49). CAPE was much less for hurricanes than values typically found for tornadoes over the Great Plains. The location of the tornadoes favors the right side of the hurricane (Fig. 10.50). Most hurricanes that do spawn tornadoes produce fewer than 10, but Hurricane Beulah (1967) yielded more than 115 when it made landfall on the Texas coast (Orton 1970).

McCaul (1991) analyzed nearly 1300 rawinsondes launched in hurricanes that produced tornadoes. Most of the tornadoes had an intensity of F0–F2 and were located in the right front quadrant (Fig. 10.51). These soundings were categorized by radial distance to the hurricane circulation center, quadrant, and distance to the tornado. Observations were used that are within 185 km and 3 h of the tornado. CAPE (Fig. 10.52)

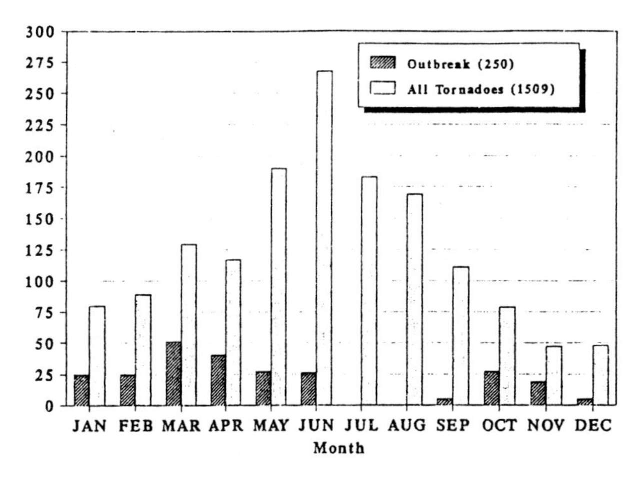

FIG. 10.43. Monthly distribution of all peninsular Florida tornadoes and outbreak tornadoes, 1950–94. From Hagemeyer (1997).

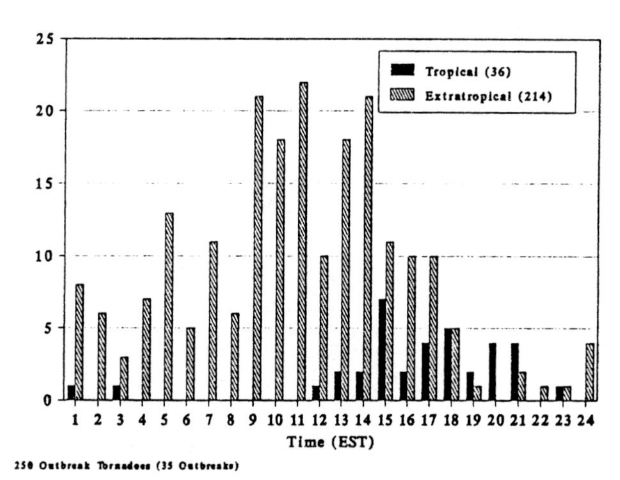

FIG. 10.44. Hourly distribution of all peninsular Florida tornadoes and outbreak tornadoes, 1950–94. From Hagemeyer (1997).

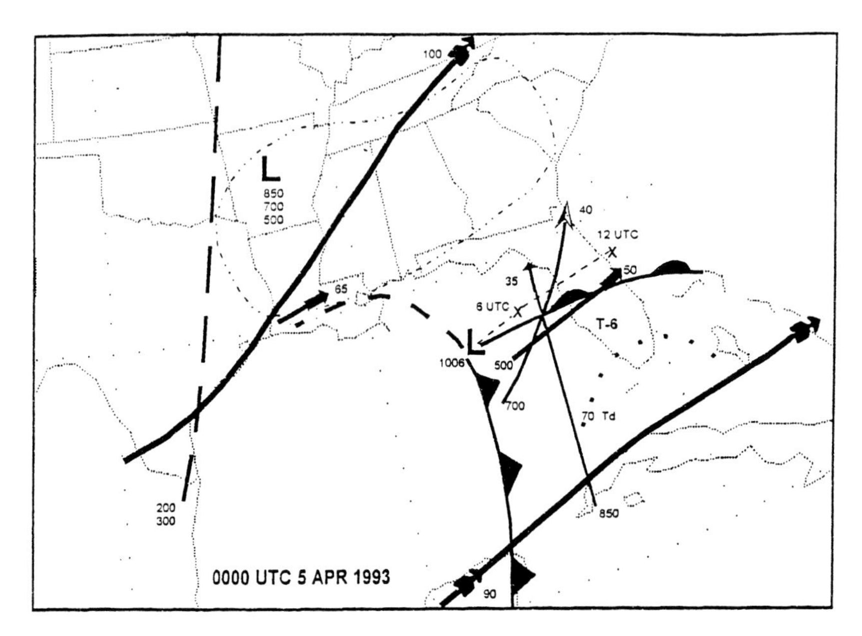

FIG. 10.45. Composite synoptic analysis for 0000 UTC 5 April 1993. Surface fronts, areas of high and low pressure, and upper-level troughs are shown by convention notation with levels in mb indicated. Axes of maximum winds at 850, 700, 500, and 250 mb are indicated by the symbols →, ➣, ➔, and ➧, respectively. Maximum winds are in knots. The track of the surface low is noted with the light dashed line. The leading edge of the highest dewpoint is marked by the heavy dotted line over south Florida. The approximate location of the first tornado is indicated by the "T", 6 h after map time. From Hagemeyer (1997).

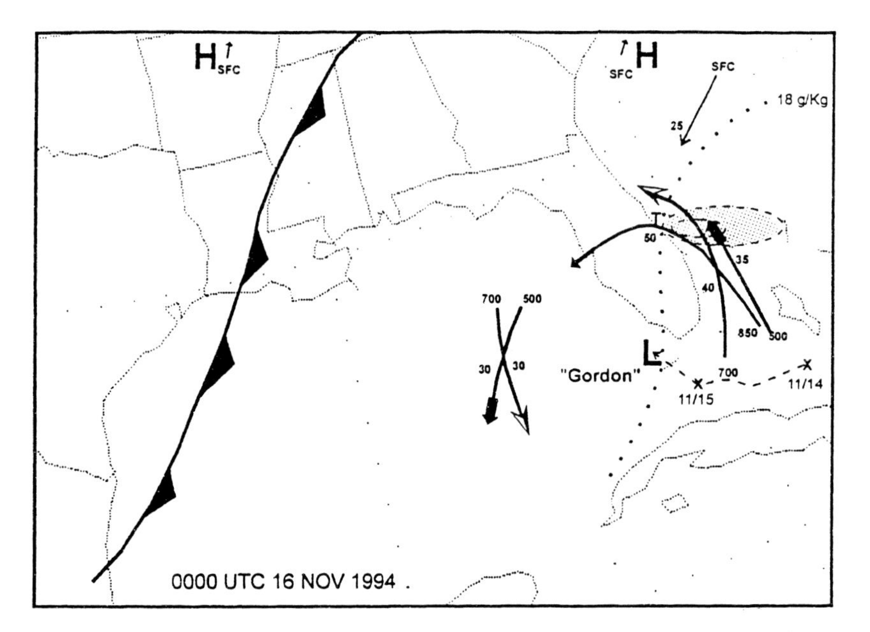

FIG. 10.46. Composite analysis for 0000 UTC 16 November 1994 for Tropical Storm Gordon. Surface fronts and areas of high and low pressure are shown by the conventional scheme. Axes of maximum winds at 850, 700, and 500 mb are indicated by the symbols →, ➣, and ➔, respectively. Maximum winds are in knots. The track of the surface low is noted with the light dashed line. The leading edge of the highest mixing ratio is marked by the heavy dotted line over south Florida. The approximate location of the first tornado is indicated by the "T", at about map time, and the shaded zone depicts upward motion just east of Florida. From Hagemeyer (1997).

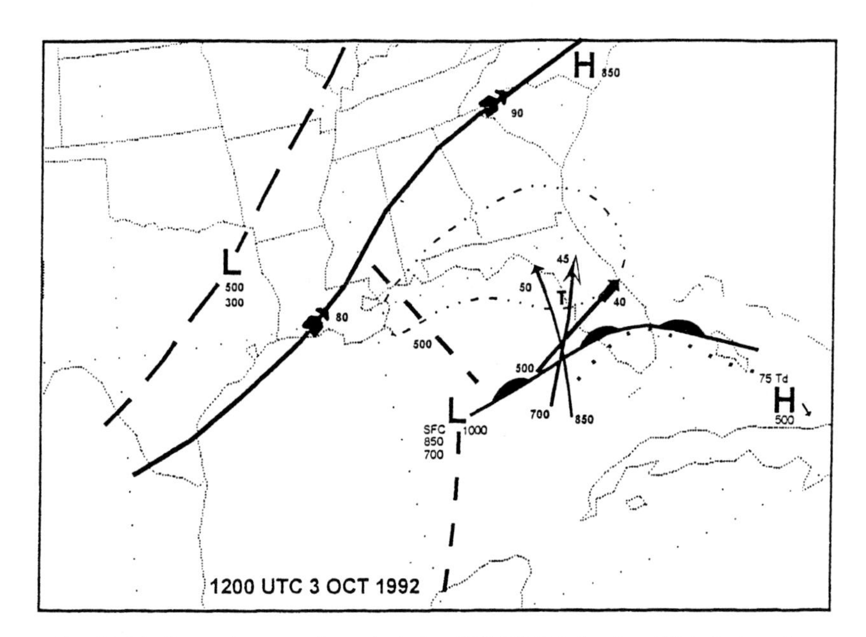

FIG. 10.47. Composite synoptic analysis for 1200 UTC 3 October 1992. Surface fronts, areas of high and low pressure, and upper-level troughs are shown by conventional notation with levels in mb indicated. Axes of maximum winds at 850, 700, 500, and 250 mb are indicated by the symbols →, ➢, ➔, and ➧, respectively. Maximum winds are in knots. The track of the surface low is noted with the light dashed line. The leading edge of the highest dew point is marked by the heavy dotted line over south Florida. The approximate location of the first tornado is indicated by the "T", 2 h after map time. The dashed-dot line encloses the rain-cooled air associated with the cyclone. From Hagemeyer (1997).

decreases as one approaches the center and is highly asymmetrical, with the highest values to the right of the track. The shear (Fig. 10.53) is also strongest to the right of the track and reaches high values from 200 to 500-km radial distance. The two fields combine in the bulk Richardson number (Fig. 10.54) and predict that the region to the right of the track from 200 to 400 km is most favorable for supercell formation. Note that the CAPE for the TC modified environment can be an order of magnitude less than that estimated for Great Plains tornadoes. The strong asymmetry is partially due to the trajectory of the inflow on the left side of the track, which is under the influence of the continent; the right side more often has a long flow over warm water. There is also an interaction with the larger-scale flow in which the TC is embedded, which affects the shear.

All the prior work has relied on soundings made over land after the tropical cyclone has made landfall. Bogner et al. (2000) use 136 Omega dropwindsondes (ODWs) to extend an analysis similar to McCaul's for six TC in a purely oceanic environment. The kinematic and thermodynamic fields are much more symmetrical over the ocean. CAPE (Fig. 10.55) and BRN shear (Fig. 10.56) combine to identify a ring from 50- to 250-km radial distance where supercells would have the best chance to form. This includes the eyewall, as well as inner and outer rainbands that are convectively active (e.g., Barnes et al. 1983). Comparison with the work of Novlan and Gray (1974) and McCaul (1991) demonstrates that the shear over the ocean is only half that found over land. Friction and turbulence apparently have a major effect on the wind profile. CAPE values are more similar over land and sea, though there is more instability over the ocean.

Spratt et al. (1997) have used the WSR-88D radars in Florida to detect mesocyclones in outer rainbands within Tropical Storm Gordon (1994) and Hurricane Allison (1995). These mesocyclones are only 3–4 km deep and 1–2 km in diameter, shallower and smaller than those observed over the Great Plains. Mesocyclones with shears $\geq 10 \times 10^{-3}\ \mathrm{s}^{-1}$ in the lowest 2 km were diagnosed as those most likely to generate a tornado. Remember that tropical cyclones have no upper-level jet, so arguments that depend on such a feature are certainly not applicable here.

Wang (1979) shows that typhoons (see Fig. 10.41), especially the right front and rear quadrants, can generate tornadoes over Taiwan. Flores and Bagalot (1969) note that the Philippines has the greatest frequency of TC in the world; one suspects that there are many instances where tornadoes were responsible for deaths and damage but the evidence has been masked by the TC itself.

Tropical Australia witnesses tornadoes within the tropical cyclones as they come ashore along the northern portions of Australia (Yates and Hanstrum 1996).

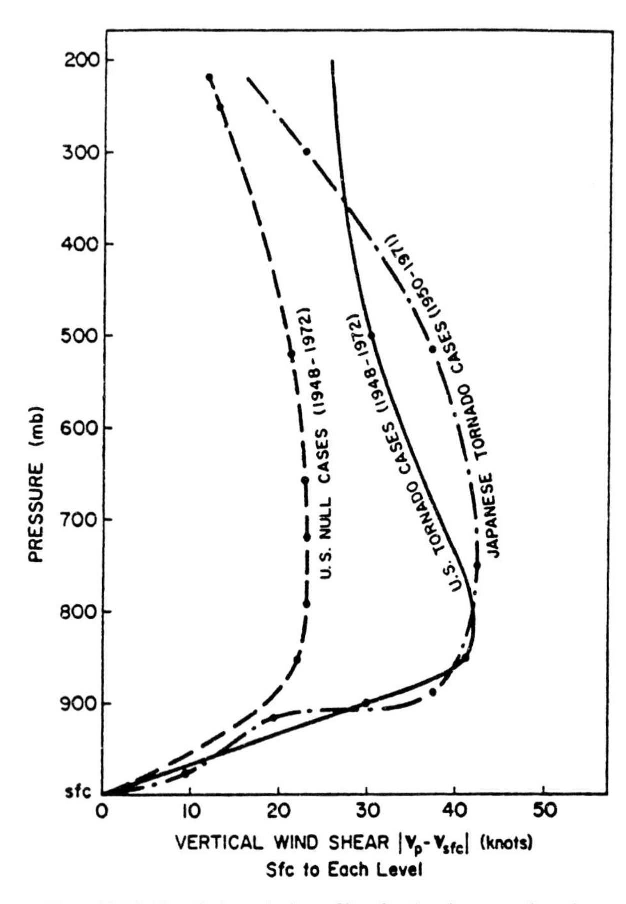

FIG. 10.48. Proximity wind profiles for hurricane and typhoon tornadoes and right front quadrant wind profiles for hurricanes without tornadoes. From Novlan and Gray (1974).

The Queensland forecast office reports that tornadoes are quite common in decaying TCs.

Hagemeyer (1997) offers a schematic (Fig. 10.46) of the flow for Hurricane Gordon (1994) that shows tornado formation well away from the circulation center and in the right front flank of the storm.

3) Tornadoes not linked to a synoptic feature

The vast majority of tornadoes in Florida occur singly; 763 tornado days of the total 998 had only one reported tornado (Hagemeyer 1997) and are not tied to a synoptic feature. Many of these tornadoes are associated with a sea breeze-induced Cb line that has interacting gust fronts. The issue is that regions in the Tropics with vigorous sea breeze circulations in the presence of conditional instability may generate weak tornadoes, but there are simply no observational networks to detect the event. Likely regions with this type of tornado activity include the Maritime Continent with its strong diurnal forcing, a function of the land–sea and mountain circulations that work in concert.

4) Waterspouts

Accurate waterspout frequency estimates are compromised by several factors that include their short lives, no damage swath to assess their occurrence or strength, unspectacular parent clouds, and few observers at sea. Sightings at night are nearly impossible. Population growth for a region has a major impact on tornado and especially waterspout counts. Wolford (1960) noted that Florida averaged six intense vortices per year from 1916 to 1958, but Senn et al. (1969) counted 50 tornadoes and 132 waterspouts for the first 11 months of 1968.

Perhaps the most encompassing attempt to understand waterspout frequency is by Gordon (1951), who examined British ship observations from 1900 to 1947. He presents a map of waterspouts per 10 000 ship observations (Fig. 10.57). Tropical maxima appear along the Atlantic NECZ, the Gulf of Mexico, the Bay of Bengal, and the southeast Brazilian coast. Tropical minima are collocated with the eastern sides of the subtropical highs and, surprisingly, to the west and north of Australia. The region southwest of Mexico also has few waterspouts despite the warm water, strong convergence, and higher background vorticity present.

The majority of waterspout studies are mesoscale or case study in nature. For a small portion of the trade wind region, Hawaii, Price and Sasaki (1963) and Schroeder (1976) show that waterspouts and weak tornadoes occur chiefly during the winter months, when the influence of a deep trough in the upper-tropospheric westerlies limits subsidence and destroys the trade wind inversion (Fig. 10.58). The tendency to see the majority of the reports near airports and military installations (Fig. 10.59a, 10.59b) suggests that many waterspouts go unseen and thus unrecorded. Waterspouts are rare around Hawaii when there is a strong inversion at lower than average heights ($\sim$2.2 km). Inconsequential, short-lived waterspouts have been seen in areas where topography has induced vortex shedding.

In the deeper Tropics, Smith (1947) documented eight waterspouts that formed as part of a squall line near Singapore (2°N), and Wales-Smith (1969) had observations to identify environmental conditions for a waterspout that came ashore in Singapore to include low shear in the lower troposphere, SST $>$ 29°C, low-level convergence, large conditional instability, and apparently sharp horizontal gradients in temperature and moisture near the surface. The high counts for Singapore suggest that if Indonesia were to establish an observational network, it would record numerous events.

In the lower Persian Gulf (24°N), several waterspouts have been recorded recently (Davey 1987; Jackson 1987), in contrast to the historic record that would make one conclude that waterspouts are not a

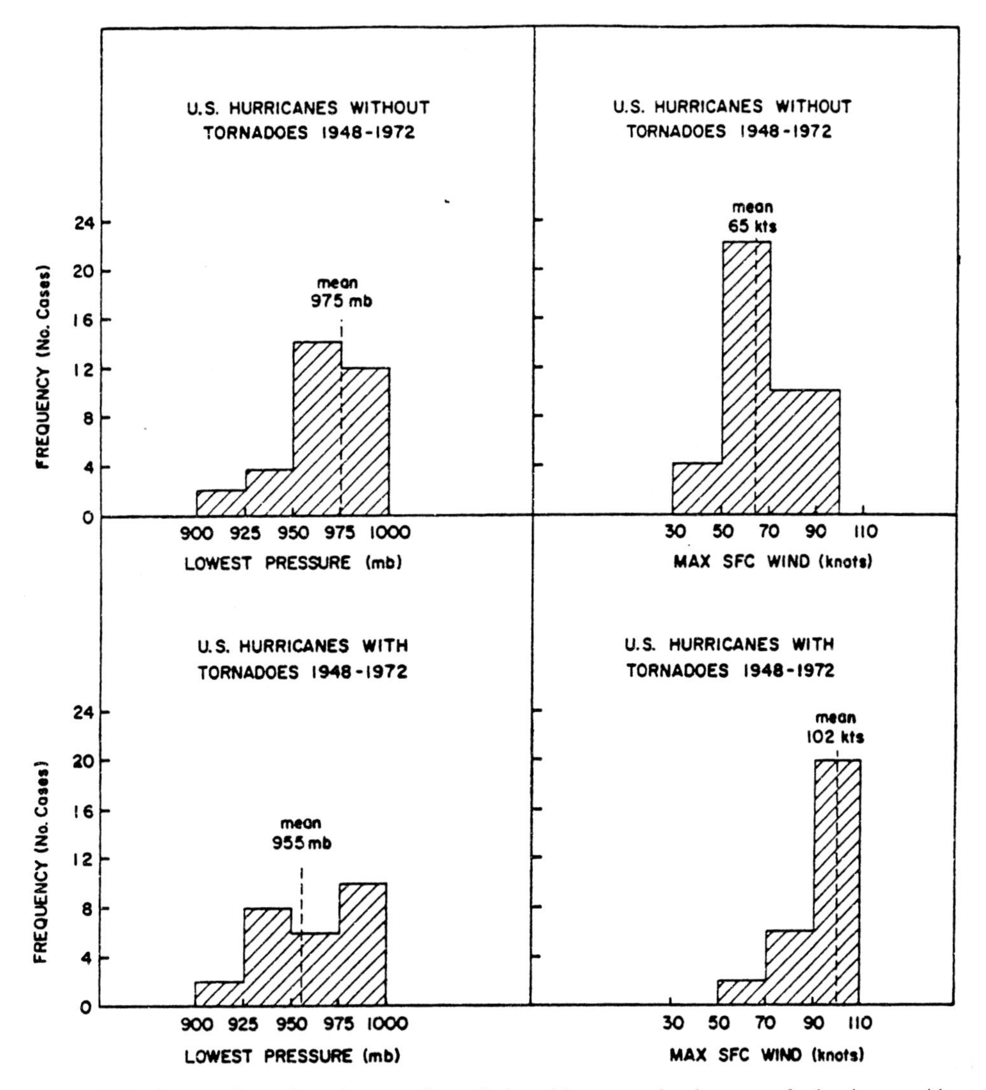

FIG. 10.49. A comparison of maximum surface winds and lowest sea level pressure for hurricanes with and without tornadoes at landfall. From Novlan and Gray (1974).

possibility. Some of the waterspouts can be intense; Davey (1987) discusses an F1 event near Dab, United Arab Emirates, in March 1986. The sightings in the Persian Gulf are in the winter during the passage of a westerly trough. Visibility around troughs can be less than 1000 m due to sand that is carried aloft by the freshening winds with the trough (Davey 1987), which lowers the chances of seeing waterspouts. The summers are dominated by strong subsidence and are virtually cloudless.

Kingwell and Butterworth (1986) document a waterspout that lasted for 10 min near Elcho Island (12°S), west of Gove, Australia. Here the parent Cb was embedded in a mesoscale vortex that was part of the monsoon trough and thus had an adequate supply of vorticity.

The Lower Florida Keys Waterspout Project, conducted from May to September 1969, recorded 390 waterspouts and funnels within 50 nautical miles of Key West (24°N). Golden (1974b) noted that a synoptic-scale disturbance such as a weak trough or an easterly wave favored waterspout development, as did weak wind shear in the lower troposphere and an unstable sounding to at least 850 mb. The frequency of funnels and waterspout sightings increased from 1–2 per month in the winter to greater than 60 per month during the summer. Most sightings were over shallow, and therefore warm, water. Golden (1974a) used aircraft to show that occasionally these waterspouts reach dangerous intensities with tangential speeds exceeding 70 m s^{-1}. Typical tangential velocities are <30 m s^{-1} (Leverson et al. 1977).

How do waterspouts marshal such high levels of vorticity given that the parent cloud is often a cumulus congestus that survives for less than half an hour? Golden (1974b) notes that 95% of the waterspouts in

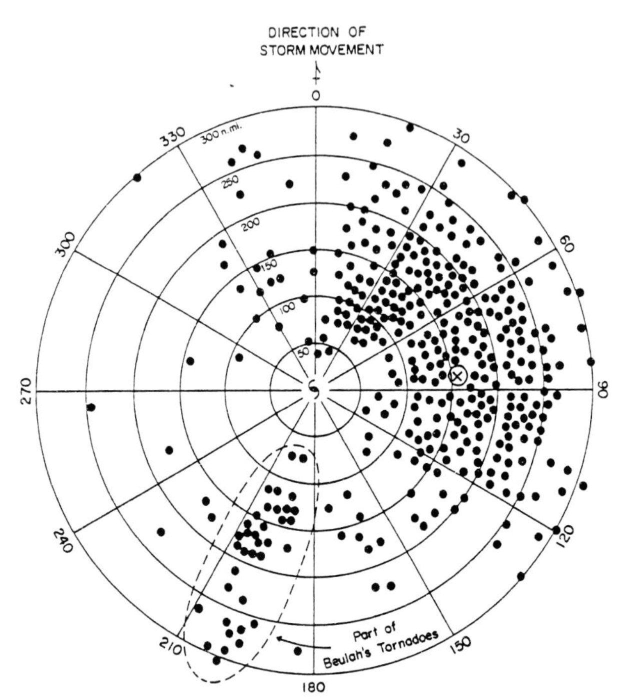

FIG. 10.50. Plan view of 373 U.S. tornadoes associated with hurricanes from 1948 to 1972 with respect to center and direction of motion. The "X" is the centroid of all tornadoes. From Novlan and Gray (1974).

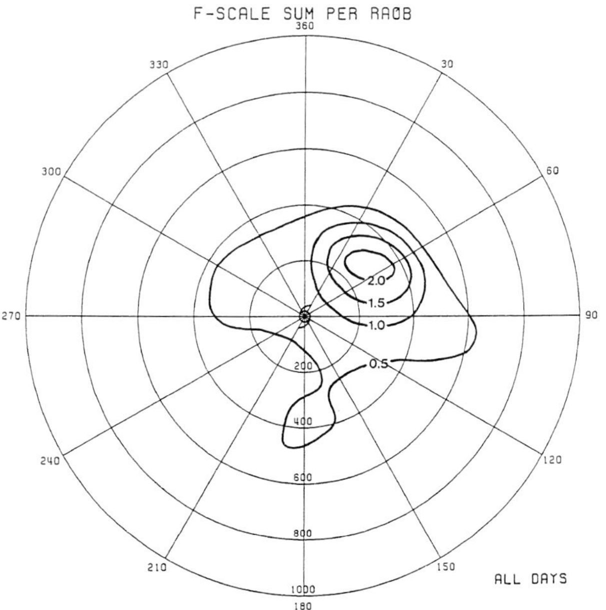

FIG. 10.51. Objectively analyzed values of F-scale sum per raob; analysis grid mesh is spaced at 100-km intervals. Range rings are every 200 km with storm moving toward 360°. From McCaul (1991).

the Keys were seen under cloud lines, where outflows from different cells could interact. Simpson et al. (1986) studied two days in GATE; one day waterspouts formed where outflows collided and produced new convection (Fig. 10.60), which confirms the observation by Golden (1974a,b) that cloud lines are a vital part of the waterspout formation process. The soundings that support waterspout formation contrast those that favor strong tornadoes in that they have less shear through the first 3 km (Fig. 10.61); the thermodynamics shows instability in the lower troposphere (Fig. 10.62). The lower troposphere can become more unstable due to gust-front lifting of a layer (Fig. 10.62). Model runs using the conditions sampled from both days demonstrate that the atmosphere needs to be preconditioned, that is, further destabilized and sheared, and that this occurs along the leading edge of a cloud line. Intense vortex stretching is an important mechanism; on day 186 the interaction of two outflows led to convergence reaching -1.0×10^{-2} s^{-1}. The cloud needs to have the updraft and downdraft in juxtaposition at low levels with a vortex stretched along the interface, and/or in the updraft core.

On day 261 of GATE Simpson and McCumber (1982) estimate the shear to be 2.8×10^{-3} s^{-1} in the layer from 0.8 to 4.0 km. This is larger than most waterspout situations. The observed funnel is only 50 m across with peripheral velocities of 4–5 m s^{-1}. Numerical simulations demonstrate that the vertical velocity needs to be an order of magnitude larger than what the observations support. Tilting, then stretching/convergence are believed to be the primary mechanisms to spin up the air, but the magnitude of the

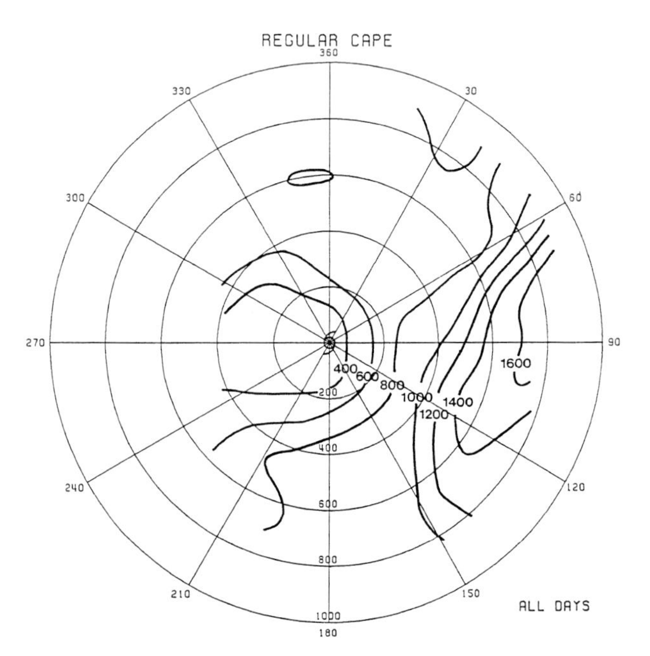

FIG. 10.52. Spatial distribution of CAPE for the 1296 raobs. Contour units are joules per kilogram. Other conventions follow Fig. 10.51. From McCaul (1991).

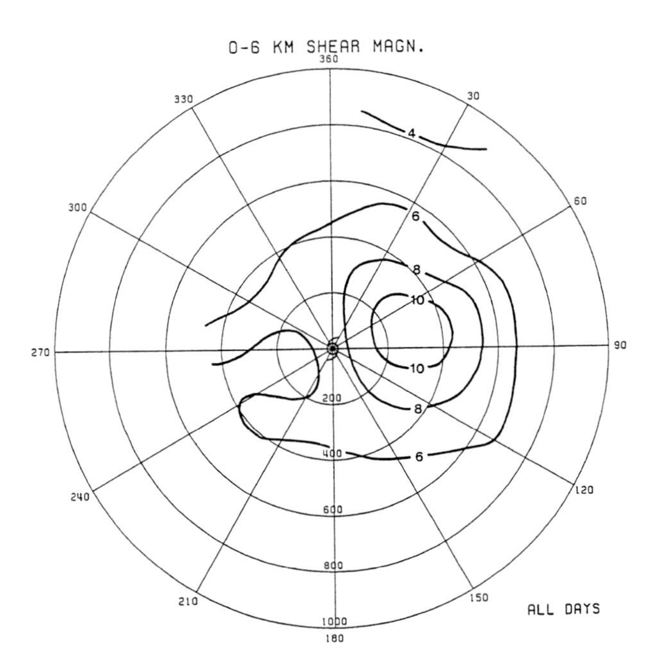

FIG. 10.53. Spatial distribution of 0–6-km BRN shear magnitude; conventions follow Fig. 10.51. From McCaul (1991).

initial vorticity field continues to be too low to explain the waterspout's characteristics.

Ways of producing more instability and shear in the lower troposphere include 1) more sensible heat flux due to unusually warm water (SST > 30°C), 2) lifting the layer, and 3) collision of gust fronts that considerably enhances the horizontal shear.

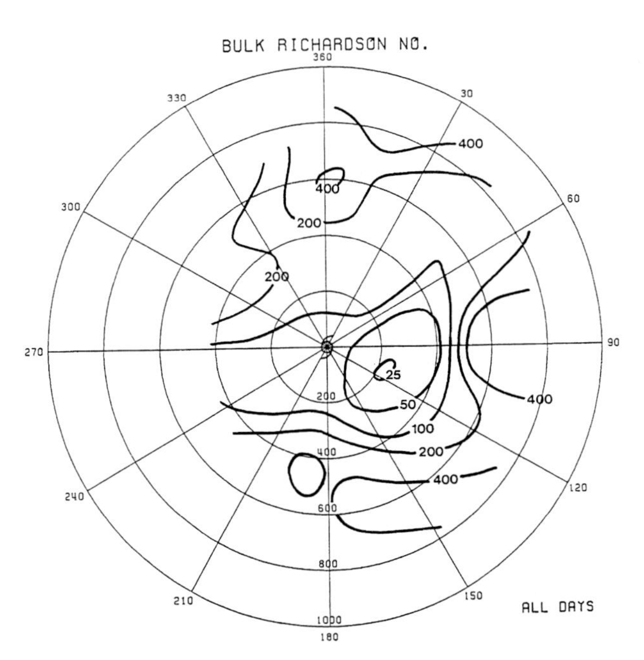

FIG. 10.54. Spatial distribution of the BRN; conventions follow Fig. 10.51. From McCaul (1991).

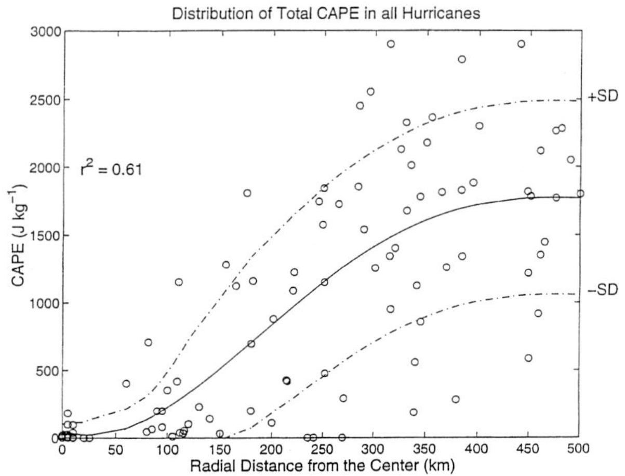

FIG. 10.55. CAPE as a function of radial distance from the TC center. The dashed lines show range dependent one standard deviation. From Bogner et al. (2000).

10.6. Why is SLS so rare in the Tropics?

a. Land and sea biases

Nearly three-quarters of the Tropics are covered by oceans. Above the sea the tropical surface layer adjusts quickly to the underlying conditions, resulting in small air–sea temperature gradients. During fair weather, the fluxes of latent heat (80–120 W m^{-2}) are modest and the fluxes of sensible heat (10 W m^{-2}) are inconsequential, save for the windiest conditions. The small sensible heating does not promote the development of vigorous eddies in the mixed layer. Diminutive diurnal variations of the fluxes favor a more uniform rate of release of convective

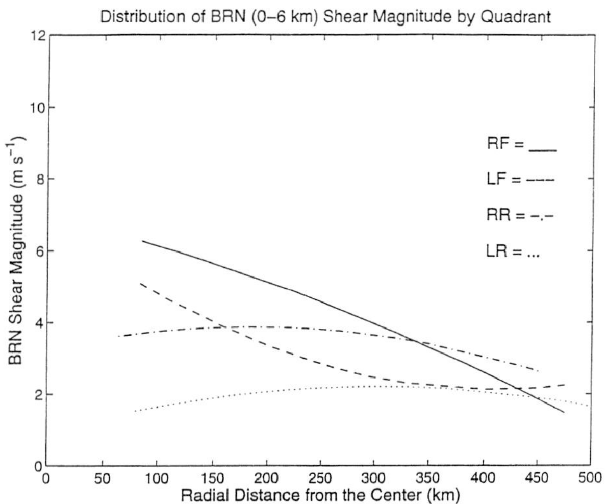

FIG. 10.56. BRN shear as a function of radial distance and quadrant from the TC center, with RF, right front (solid); LF, left front (dashed); RR, right rear (dashed-dot); and LR, left rear (dotted). From Bogner et al. (2000).

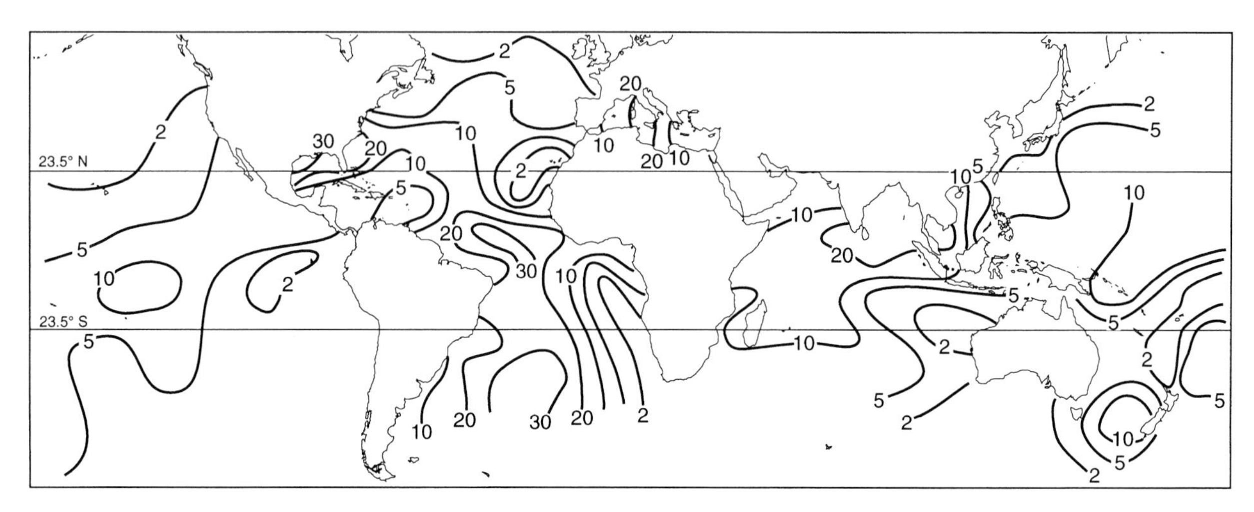

FIG. 10.57. Spatial distribution of waterspout reports per 10 000 ship observations. From Gordon (1951).

instability. Sea-surface temperature gradients on the horizontal are essentially synoptic to planetary in scale, limiting the creation of strong mesoscale horizontal gradients of temperature and moisture that contribute to the genesis or maintenance of strong convection.

Over much of the tropical oceans, planetary variations of the winds (e.g., NECZ, SPCZ, ISO) are the chief agents for the initialization of deep convection. The heavy reliance of deep convection on this scale tends to limit the frequency of the exciting situations characterized by differential advection, which produces sharp gradients in the horizontal and vertical that favor SLS. Instead, we often have deep layers with similar characteristics converging over the oceans.

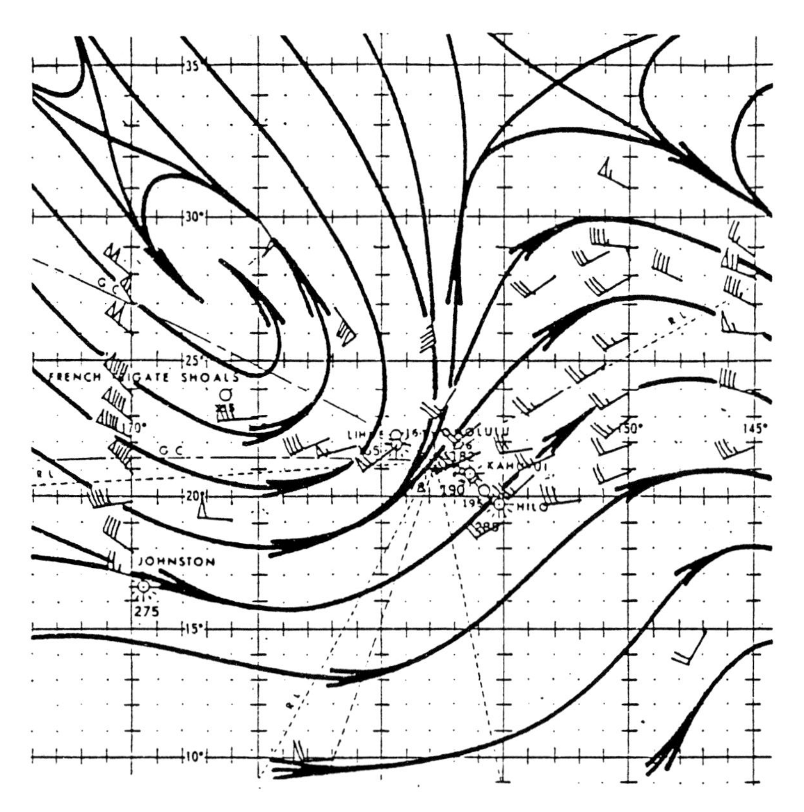

FIG. 10.58. 250-mb flow over Hawaii for 0000 UTC 26 November 1975. Winds are in knots. From Schroeder (1976).

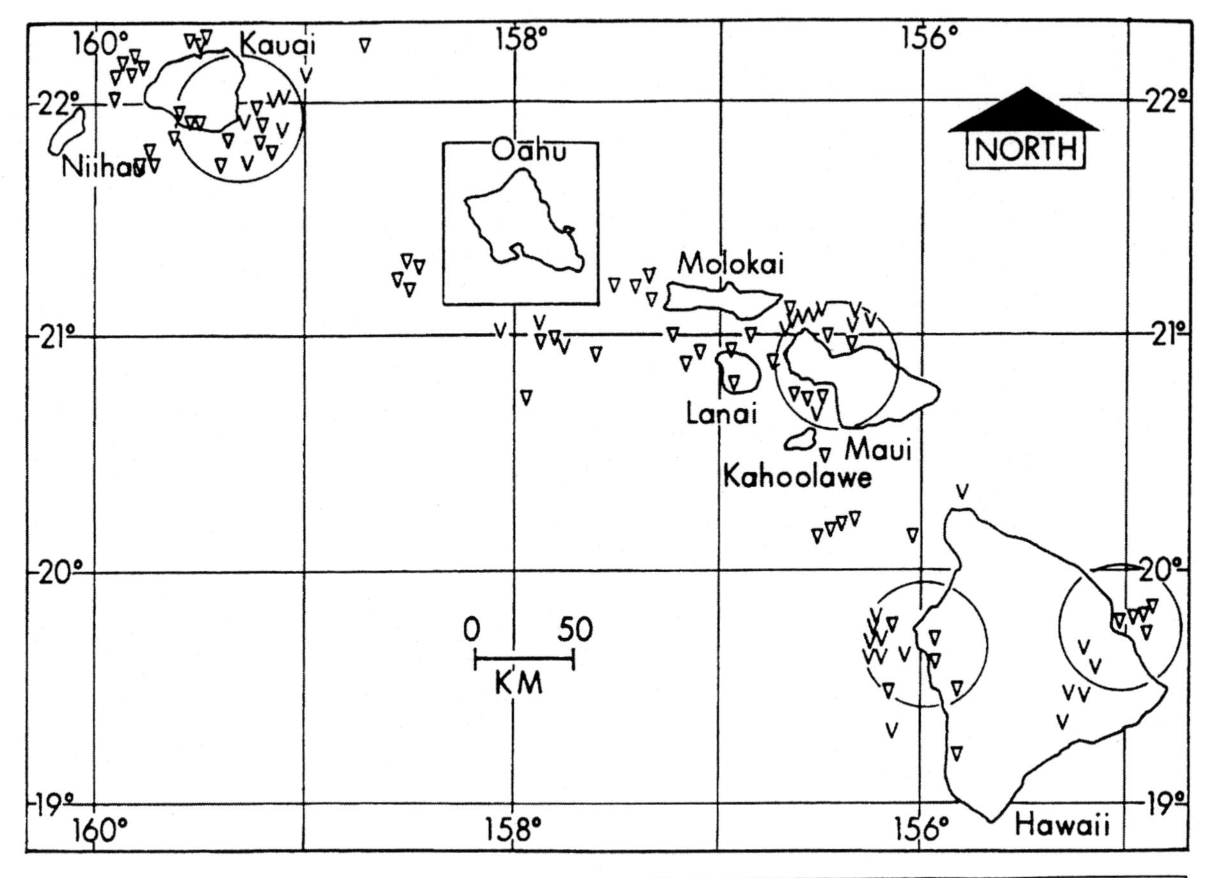

FIG. 10.59. Locations of reported funnels aloft (V) and waterspouts/tornadoes (▽) for the Hawaiian islands, 1961–74. 30-km radius circles are drawn around the major airports. The island of Oahu is enlarged at right. From Schroeder (1976).

SLS thrive on strong vertical and horizontal gradients of moisture, temperature, and wind, which have a greater chance of forming over tropical landmasses. Sensible heat fluxes from the land are often 10–30 times greater than those found for environmental conditions over the sea; this flux also undergoes a strong diurnal variation that allows the boundary layer to store, then release energy into the entire troposphere at a nonsteady rate (e.g., Carlson and Ludlam 1968).

Mountain ranges preferentially block flow from some directions and enhance it from others, which can lead to a juxtaposition of warm, moist air flowing from one direction under cooler, drier air from another direction. Subtle variations in trajectories over differing land surfaces can result in significant gradients.

HYPOTHESIS: VORTEX INTENSIFICATION

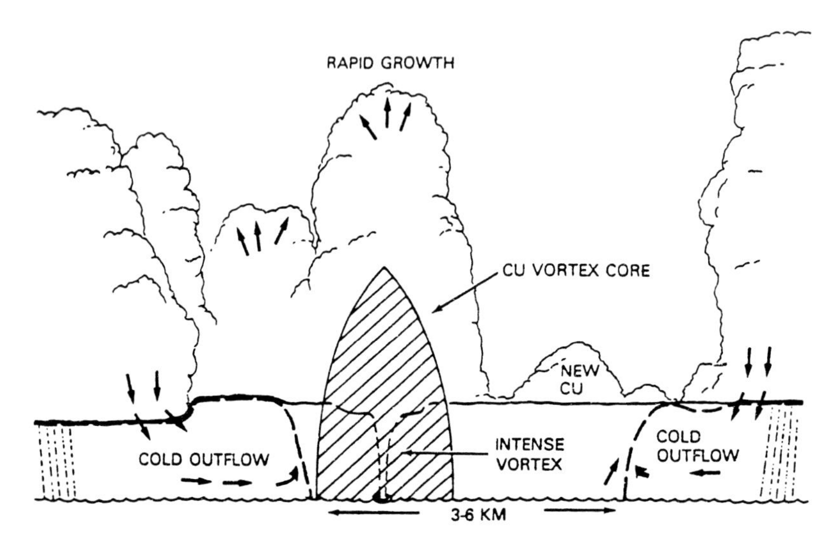

FIG. 10.60. Schematic of cumulus–outflow interactions that lead to an invigorated cumulus that produces a parent vortex. The shaded region outlines the vortex core, the circle at the surface within the core is the sea surface dark spot, and the dashed lines show where the condensation funnel will appear as the pressure drops. From Simpson et al. (1986).

Mesoscale circulations such as mountain–valley and land–sea breezes often serve as a trigger to convection, though primarily over land and coastal waters. Such mesoscale forcing creates some remarkable thunderstorm-day gradients. Kampala, Uganda, just north of Lake Victoria, has 242 thunderstorm days a year, while Mbarara, on the west side of the lake, has only 7 (Lumb 1970)!

Land–sea contrasts help create a more favorable environment for SLS during the premonsoon for Asia and Australia, and during the monsoon for Australia as the moist westerlies slide beneath drier easterlies. Farther away from the coast instability decreases as the moisture is mixed out through a deeper layer.

In a more speculative vein, there may be differences in the horizontal scale of the updraft between land and sea. Deeper boundary layers over land may favor a greater diameter for the initial updraft (e.g., LeMone 1989) that is more apt to survive entrainment. Greater

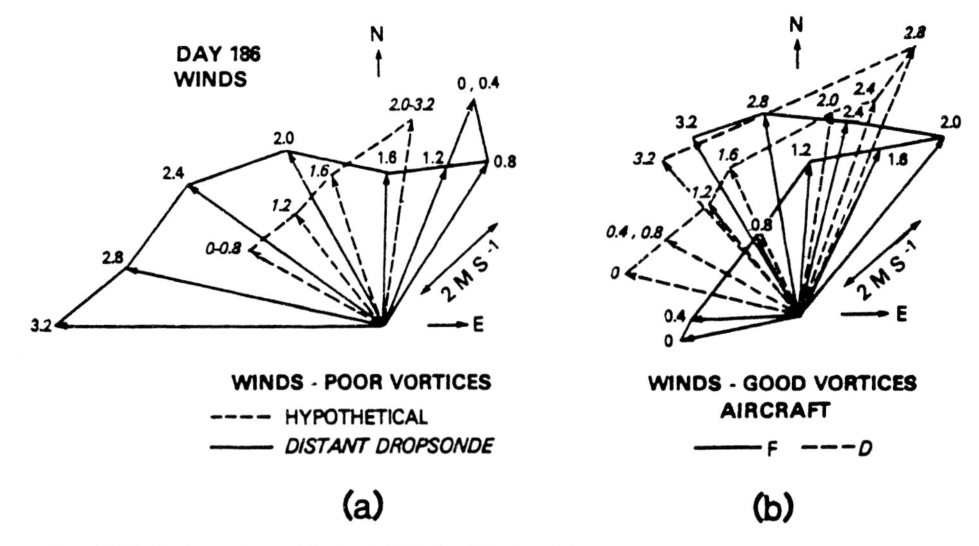

FIG. 10.61. Hodographs used in the GATE day 186 simulations up to 3.2-km altitude. Hodographs in (a) failed to produce a parent vortex, while those in (b) did. From Simpson et al. (1986).

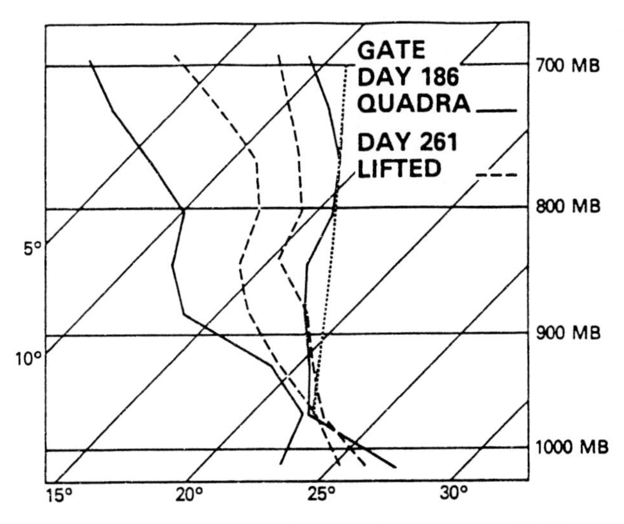

FIG. 10.62. Tephigram showing soundings used for GATE day 186 (solid) and modified day 261 (dashed) by hypothesized gust front effects. The dotted curve is a moist adiabat through the level of cloud base on day 186. The sounding on this day is from the *Quadra* at 1200 UTC. From Simpson et al. (1986).

friction over land enhances the shear in the lower troposphere. When a sizable updraft interacts with this shear the development of a dynamic pressure perturbation in the lower cloud favors supercells.

b. The nature of tropical convection

Climatologies derived from surface data, thunderstorm reports, and satellite sensors demonstrate that land regions such as equatorial Africa and the Amazon, and ocean belts such as the NECZ, have some of the highest frequencies of deep convection in the world. None of these regions appear to have a corresponding peak in SLS. An examination of observations in tropical cumulonimbi provides clues as to why this is so.

Aircraft have probed a variety of mesoscale convective systems and tropical cyclones over all the major tropical oceans in numerous experiments (e.g., LIE, BOMEX, GATE, MONEX, EMEX, TAMEX, COARE and the continuing NOAA/AOML/HRD hurricane studies). Statistics describing the updrafts from these flights, especially when compared to midlatitude observations, reveal that clouds originating over the tropical oceans have noticeably weaker maximum and mean vertical velocities, and smaller updraft diameters. LeMone and Zipser (1980) define the most prominent vertical velocity features sampled by the aircraft as cores that consist of at least 5 consecutive seconds of flight ($\geq$500 m) with $|w| \geq 1.0$ m s^{-1}. Even the top 10% of the cores sampled during GATE (Fig. 10.63a) have a peak 1-s magnitude less than 10 m s^{-1} for all the layers sampled and a mean magnitude barely reaching 5 m s^{-1}; the size is equally unimpressive, with only the 10% largest cores having a diameter greater than 2–3 km (Fig. 10.63b). Comparison of these strongest cores with samples in hurricanes and with the Thunderstorm Project (Fig. 10.64) shows that the updrafts and downdrafts in the tropical oceanic environment lack the muscle found in midlatitude ordinary and multicell convection. Thunderstorm Project updrafts are 3–4 times stronger than their tropical counterparts at the same level. Hurricane cores and drafts are also weak, even in intense TC (Jorgensen et al. 1985). Why is the typical tropical oceanic Cb so weak? There are several candidate mechanisms that may work alone or together that hinder the development of strong updrafts, which in turn limits SLS development.

c. Factors that hinder the development of vigorous updrafts

Water loading can seriously hinder updraft speeds. The typical mixing ratio for a parcel originating in the mixed layer in the equatorial trough is 16–20 g kg^{-1}. For every 3 g kg^{-1} of liquid water mixing ratio, buoyancy is reduced by an equivalent of ~1°C. Most tropical soundings show that the virtual temperature excess in the lower and midtroposphere is 3°–5°C; by 600 mb the liquid water load would be approaching 9 g kg^{-1}, reducing buoyancy by 60%–100%. Betts (1982) and Xu and Emanuel (1989) argue that tropical convection is essentially moist-neutral due to water loading.

Several observations lead one to doubt the efficacy of water loading. First, no one has measured such high liquid water contents. Second, tropical clouds often develop a reflectivity core between 3 and 5 km, which is evidence that collision and coalescence is the primary precipitation formation mechanism. Heavy precipitation rates from even modest cumuli (e.g., Hawaiian trade showers) lend credence to the argument that few clouds carry their maximum liquid burden into the upper troposphere. Finally, if calculations of CAPE were to include the latent heat of fusion, then the extra buoyancy due to freezing counters any reduction in buoyancy caused by water loading (Williams and Renno 1993).

Perhaps a more basic issue is the magnitude of the virtual temperature excess (ΔT_v) in tropical clouds. The few experiments (e.g., EMEX, TAMEX, COARE) graced with radiometers, which are largely uncompromised by sensor wetting, show only modest ΔT_v. Jorgensen and LeMone (1989) show that, for 92 updraft cores sampled during TAMEX (Fig. 10.65), it was rare to see ΔT_v greater than 4°C. When these modest values are further reduced using observations of the liquid water content, as was done for the updraft cores sampled in an MCS in TAMEX, then one can see that updrafts rarely achieve the theoretical estimates of buoyancy derived from parcel theory (Fig.

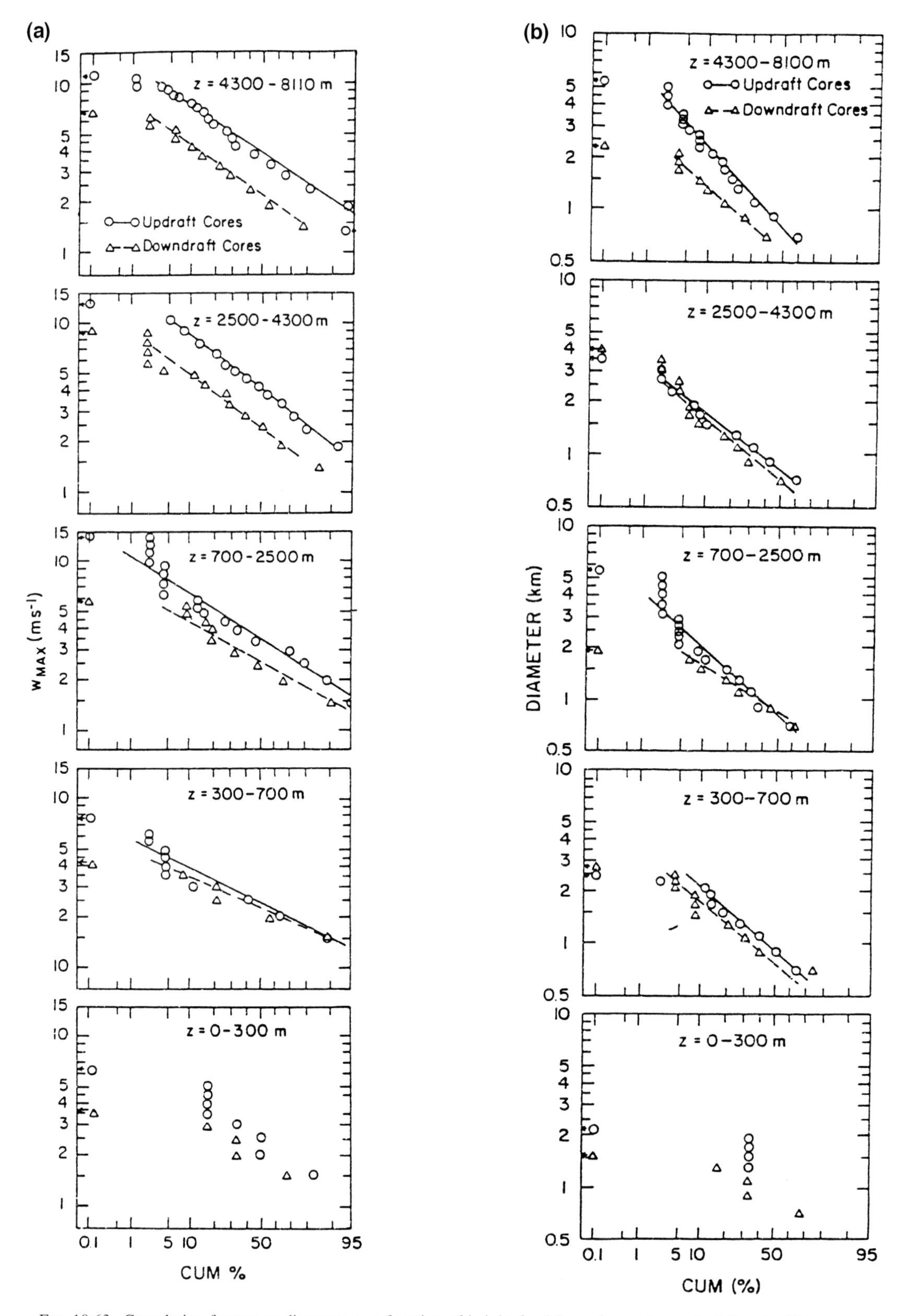

FIG. 10.63. Cumulative frequency diagrams as a function of height for (a) maximum 1-s updraft found within a core and (b) diameter of a core. Data are from GATE aircraft. From Zipser and LeMone (1980).

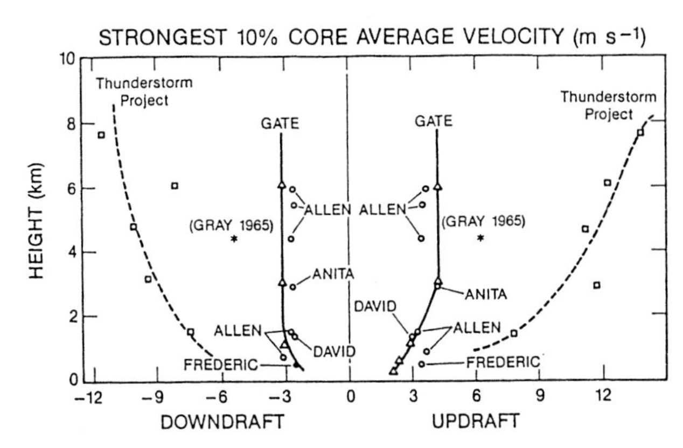

FIG. 10.64. Strongest 10% average vertical velocity in updraft and downdraft cores as a function of height. GATE values are denoted by triangles, Thunderstorm Project values by squares, and hurricane values by circles and storm name. Observations in hurricanes by Gray also indicated. From Jorgensen et al. (1985).

10.66). This updraft core vertical profile is derived from estimated average ΔT_v, and cloud droplet and rain droplet content for each level, and does not include the effects of entrainment, which are, of course, always present and apparently significant given the difficulty of finding an undiluted updraft. These findings are reinforced by Lucas et al. (1994b), who estimate a mean ΔT_v of +0.6°C for updraft cores sampled during EMEX.

Is the amount of instability less for tropical than for midlatitude environments? Many severe weather outbreaks in the midlatitudes have CAPE values from 2000 to 4000 $m^2\ s^{-2}$ (see chapters 2, 11 in this

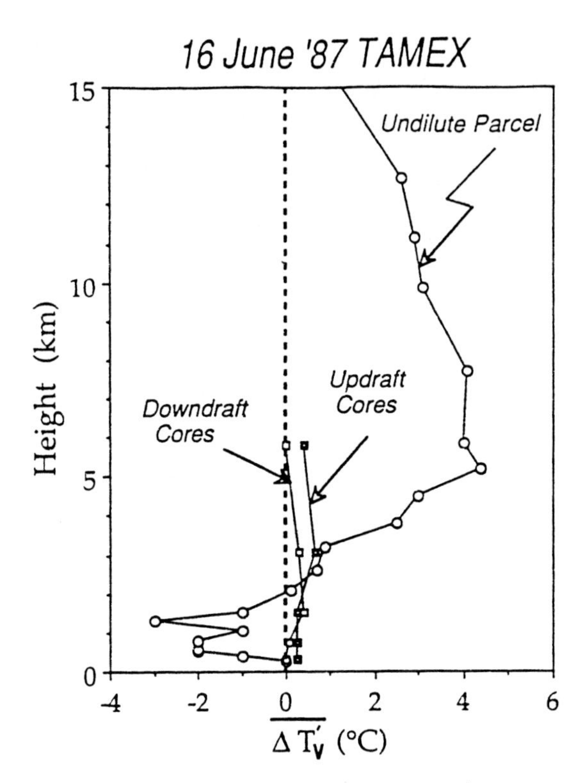

FIG. 10.66. Core average virtual temperature excess ($\Delta T_v = T_v - T_{v1}$, where T_{v1} is the leg mean) in updraft and downdraft cores as a function of altitude from in situ measurements using a CO_2 radiometer, and the value that would be expected from adiabatic assumptions based on the proximity sounding (open circles). For the computation of the sounding T_v excess, the parcel is assumed to start at 982 mb with a T = 26.8 C and a mixing ratio = 20.4 g kg^{-1}. From Jorgensen and LeMone (1989).

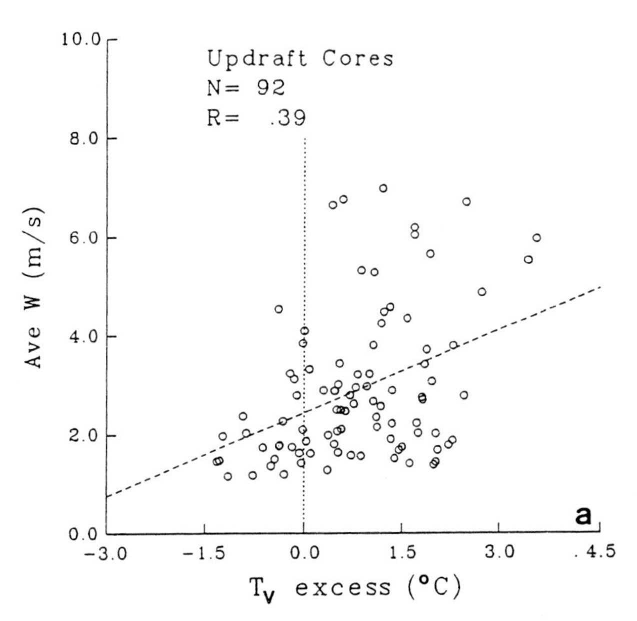

FIG. 10.65. Average vertical velocity as a function of virtual temperature excess for 92 cores sampled during TAMEX. From Jorgensen and LeMone (1989).

volume). Some continental tropical regions witness very impressive instabilities. Keenan and Carbone (1992) and Williams et al. (1992) estimated CAPE exceeding 3000 $m^2\ s^{-2}$ in the premonsoon flow near Darwin, Australia, which supports the 20-km cloud tops often seen there. Golding (1993) simulates the convection over Bathurst and Melville Islands with this instability and predicts vertical velocities in excess of 30 m s^{-1}. Such strong updrafts are often associated with SLS. A caution is that some investigators calculate CAPE based on a surface parcel (e.g., Williams and Renno 1993) while others use mean conditions in the lowest 500 m or the subcloud layer (e.g., Barnes and Sieckman 1984). Often the subcloud layer is not perfectly mixed, and the first few values from a rawinsonde can be suspect unless the launch crew allows the sonde to equilibrate with environment conditions instead of with the balloon shelter.

Typical CAPE magnitudes in the Tropics are less than 2000 $m^2\ s^{-2}$, though many of these reports are for oceanic environments (e.g., Zipser and LeMone 1980; Barnes and Sieckman 1984; Alexander and Young 1992; Williams and Renno 1993, see their Table 2). Studies that cover both continental and ocean tropical environments show a difference of only a few hundred meters squared per second squared (Williams and Renno 1993). CAPE values for the Thunderstorm Project are about 1950 $m^2 s^{-2}$ (Lucas et al. 1994a); again, this magnitude is only slightly higher than oceanic conditions. Soundings over Venezuela (Miller and Betts 1977) and over West Africa (Roux et al. 1984) also have CAPEs comparable to or only slightly larger than oceanic conditions.

Lucas et al. (1994b) hypothesize that the shape of the CAPE has a critical effect on updraft intensity. Most tropical soundings have their CAPE spread through a deep layer (12–14 km), but the virtual temperature difference between parcel and environment at any level is 3°–5°C. For midlatitude situations (Fig. 10.67), the virtual temperature difference can be much larger in the lower troposphere, allowing for greater acceleration of the parcel. Compounding this is the water loading contrast for each situation. For tropical clouds there is potentially greater water loading, given that the initial parcel specific humidity is usually 2–6 g kg^{-1} more than in the midlatitudes; in some situations, such as the high plains, it may be a factor of 2 greater. An ascending air parcel may accelerate due to the latent heat of fusion and shedding of water in the upper troposphere, but this probably has little effect on the magnitude of the updraft in the lower troposphere, and thus on the production of SLS, given the importance of the updraft in the lower to middle troposphere (e.g., Weisman and Klemp 1982, 1984).

Entrainment reduces buoyancy, but could it be responsible for having a greater effect on the magnitude of tropical versus midlatitude updrafts? Initially one could argue that the moister midlevel conditions found

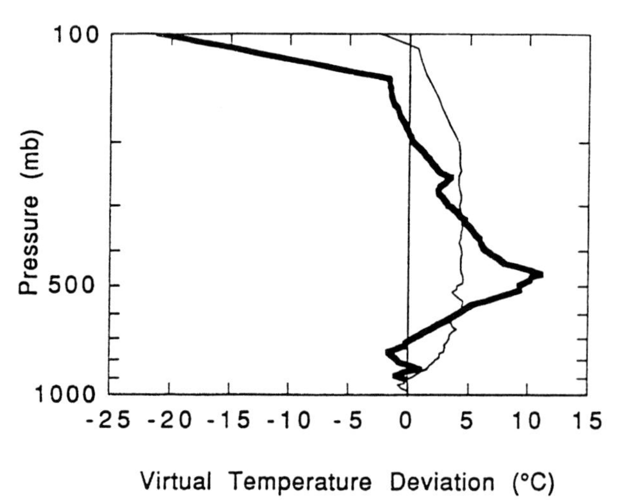

FIG. 10.67. Virtual temperature of air parcel minus environment predicted for moist adiabatic ascent of the lowest 50 mb for an oceanic (thin) and continental (thick) sounding. Oceanic sounding from EMEX, continental sounding from Oklahoma before passage of a squall line. From Lucas et al. (1994b).

in the low latitudes would lessen the impact of entrainment. Boatman and Auer (1983) offer a scheme to demonstrate the effect of dry versus moist air entraining into the cloud; Kodama and Barnes (1997) apply this technique to Hawaiian heavy rain situations to show that drier air strongly cools as it becomes saturated, which significantly limits the buoyancy. Given this result, tropical clouds should be more vigorous than their midlatitude cousins if they grow in a moister midlevel environment.

If, however, the rate of entrainment is inversely proportional to the radius of the updraft (e.g., Levine 1959; Malkus 1960; Simpson and Wiggert 1971), and tropical clouds have smaller radii, then we may have a cause for the weak vertical velocities in tropical clouds. Once a Cb establishes a channel of air flowing into the cloud, it is difficult for entrainment at the top of the cloud to account for the dilution far down into the plume. Barnes et al. (1996) use a mass flux technique to confirm that mixing is occurring laterally, and the rate of mixing is high for clouds sampled in Florida during the summer. Certainly there is mixing at cloud top as well (e.g., Blyth et al. 1988; Paluch 1979), but if lateral entrainment and detrainment is ongoing then the smaller-radii tropical clouds would suffer a greater reduction of buoyancy than their midlatitude counterparts. Data from the Thunderstorm Project do show that midlatitude clouds have radii that are two to three times larger than tropical oceanic clouds (Lucas et al. 1994b).

Why would tropical Cbs have smaller updraft radii? LeMone (1989), using observations of convective clouds over Montana, hypothesizes that the updraft diameter is related to the shear through cloud base, especially prior to the development of a gust front. Greater shear may also lead to stronger, more long-lasting gravity waves generated in the cloud and boundary layer (see Clark et al. 1986; Balaji and Clark 1988). Less shear in the Tropics limits the positive feedback of the gravity waves and results in smaller diameters that undergo more rapid entrainment than their fatter midlatitude counterparts. This intriguing hypothesis remains to be investigated in the Tropics.

Initially clouds may have a horizontal scale that is related to the size of the eddies in the subcloud layer. These eddies are often scaled to the depth of the subcloud layer (Stull 1988). For the tropical oceanic environment a typical subcloud layer depth is 500–600 m, while for the midlatitudes depths can be 1000–3000 m.

The interaction of the updraft with the ambient shear of the horizontal wind produces a pressure perturbation that intensifies the inflow to a Cb. The numerical simulations of a cumulonimbus in either strong unidirectional shear (e.g., Klemp and Wilhelmson 1978; Schlesinger 1978) or in strong shear with varying direction (Schlesinger 1980; Rotunno and Klemp 1982) show the importance of the pressure

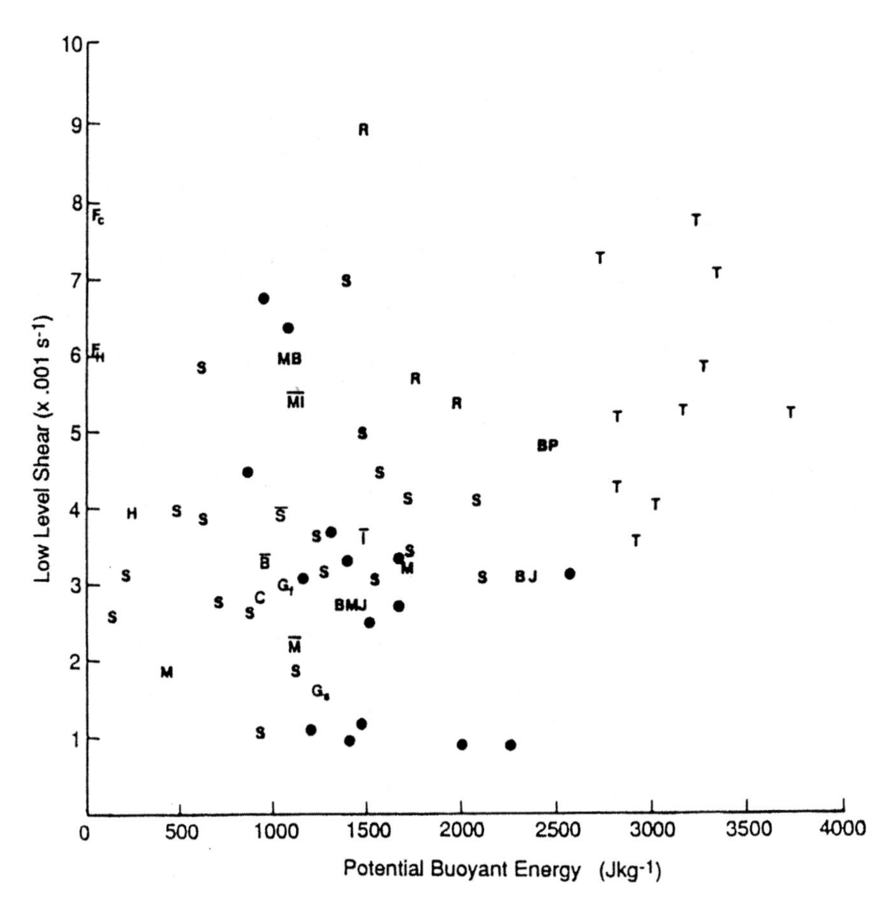

FIG. 10.68. Variation of buoyant energy and low-level shear during the 1987–88 wet season in Darwin, Australia, and from many other locations. For Darwin there is monsoon (M), mean monsoon (MI), microburst storm (MB), continental convection (C), squall lines (S), and break monsoon (B) conditions. For midlatitude situations there are North American storms (●), North American tornadic (T), mesocyclone—no tornado (R), cold fronts (F), isolated supercell storms for Oklahoma (BP), severe squall lines in Oklahoma (BJ), and Oklahoma nonsevere lines (BMJ). From other parts of the tropics there are GATE fast and slow lines (G_F, G_S), and hurricane rainbands (H). From Keenan and Carbone (1992).

perturbation. Observations at cloud base verify the existence of a strong pressure perturbation, 90° out of phase with the updraft, when there is strong shear (e.g., LeMone et al. 1988b; Barnes 1995) and no perturbation when there is no shear (LeMone et al. 1988a). Observations in the Tropics (Miller and Betts 1977; Zipser 1977; Houze 1977; Frank 1978; Barnes and Sieckman 1984; Chalon et al. 1988; Jorgensen et al. 1991; Alexander and Young 1992; Keenan and Carbone 1992; Lin and Johnson 1996) demonstrate that the shear is much less than that found for supercell midlatitude situations (Newton 1966; Marwitz 1972; LeMone et al. 1988b; Miller et al. 1990) or in the aforementioned numerical simulations of such storms. With only a weak pressure perturbation (e.g., LeMone et al. 1984b) the tropical oceanic MCSs have one less force contributing to strong updrafts.

Keenan and Carbone (1992) compare the low-level shear and CAPE of convective systems around Darwin, Australia to severe and nonsevere events from a myriad of tropical and midlatitude systems (Fig. 10.68). Darwin itself is a useful site as it comes under the influence of premonsoon, monsoon, break-monsoon, and tropical cyclone environments. In Fig. 10.68 the low-level shear magnitude is taken to be representative of the depth of the naturally occurring shear layer, which is the lowest 20%–35% of the depth of the convection. Tropical cases generally have a combination of low CAPE and weak shear, while the severe events from whatever region have higher values of both of these parameters. Note that a bulk Richardson number could mislead unless a threshold value for either CAPE or shear is first considered. CAPE, for most of the undisturbed deep Tropics, is 1000–2000 m^2 s^{-2}, and though respectable, is not as much as the extreme environments for SLS over the U.S. Midwest. Bluestein et al. (1987) note that the squall lines in the central United States that do not produce severe weather have less CAPE (1300–1400 $m^2\ s^{-2}$), wind shear from the surface to 6 km of 3.3–3.6 × $10^{-3}\ s^{-1}$, and bulk Richardson numbers from 50 to 75. Bluestein and Jain (1985) find much larger values for CAPE and shear for the squall lines that generate SLS; the majority of tornado proximity soundings manifest stronger shear (Maddox 1976) than is typically found in the Tropics. In the Tropics virtually all of the trades fall below these values, and most conditions over the ocean and in the equatorial trough are also in this range. During the premonsoon, shear and CAPE can reach the magnitudes to support SLS.

Much of the deep Tropics offers little resistance to vertical exchange and therefore much less buildup of conditional instability to the magnitudes that alarm forecasters (e.g., CAPE > 2500 $m^2\ s^{-2}$). Fawbush and Miller (1952), in their examination of tornado proximity soundings, note that a strong low-level inversion allows for the buildup of instability to dangerous levels. Today convective inhibition energy (CIN) is a parameter that is regularly calculated for a sounding. A full study of CIN has not been done for the Tropics, but I suspect that those regions with low CIN have regular overturning of the atmosphere (e.g., NECZ, the Amazon) and will not develop the extreme CAPE values that support the formation of SLS. Too much CIN will, of course, prevent convection. Some CIN, coupled with the appropriate trigger, is likely the right combination that favors vigorous convection and SLS production.

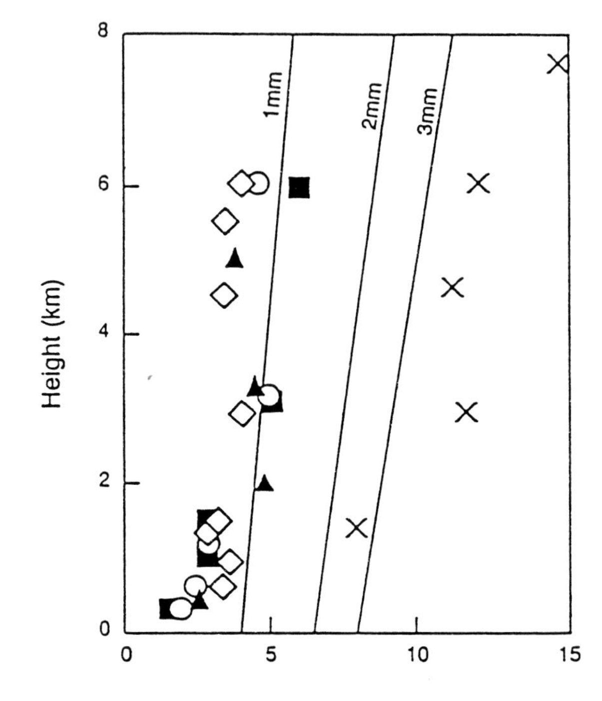

FIG. 10.69. Average vertical velocity in the strongest 10% updraft cores over the tropical oceans with triangles for EMEX, circles for GATE, diamonds for hurricanes, and squares for TAMEX. The X marks Thunderstorm Project data. Sloped lines are the terminal velocities for 1-, 2-, and 3-mm diameter drops as a function of height. From Zipser and Lutz (1994).

d. Why large hail is rare

To exceed the severe size threshold a hailstone must either be recycled through an updraft (Browning and Foote 1976) or stay in a strong updraft that contains plenty of liquid water (Miller et al. 1988). Supercells are the chief culprits that deliver damaging hail (Nelson and Young 1979). The largest growth rates for a hail embryo occur in an environment with supercooled water (Knight et al. 1982). The growth to large size appears to be more correlated with the particle trajectory that results in a long residence time in the storm circulation rather than the intensity of the updraft. Do the Tropics meet these requirements? Except in the rarest of cases, the typical tropical Cb fails to supply the magnitude of the updraft, the appropriate arrangement of convective cells such as a flanking line, and the amount of supercooled water required to yield damaging hail.

The prior sections have demonstrated that tropical clouds typically contain weak updrafts. Vertical velocity profiles for the typical midlatitude Cbs can support raindrops of 3-mm diameter, but the tropical oceanic clouds fail to support even 1-mm diameter drops (Fig. 10.69; Zipser and Lutz 1994); thus hail embryos will fall out of the updraft and away from the region of high liquid water content before they grow substantially.

Is there a surfeit of supercooled water available in tropical Cbs? Vertical profiles of radar reflectivity are in contrast for tropical oceanic, tropical continental, and midlatitude convection. For tropical oceanic profiles there is a rapid decrease in reflectivity above the freezing level (Fig. 10.70; Zipser and Lutz 1994). This is interpreted as a lack of large graupel, which in turn is due to a lack of supercooled water, which in turn is due to the weak vertical velocities in these clouds. Hurricane microphysics studies also demonstrate the rarity of copious amounts of supercooled water (Willoughby et al. 1985). Bounded weak echo regions and hook echoes, both evidence for strong updrafts and a circulation favorable for large hail growth, are very rare commodities in the low latitudes.

Tropical cumulonimbi are often tilted well away from the vertical. Malkus (1952) and LeMone et al. (1984a) show that the slope of the updraft can be quite shallow, in the range of 60°–70° from the vertical. Such large tilts will tend to have most potential hailstones falling far from the core of the updraft where the most liquid water would be found.

The organization of most tropical MCSs does not favor the recycling of stones. Leary and Houze (1979)

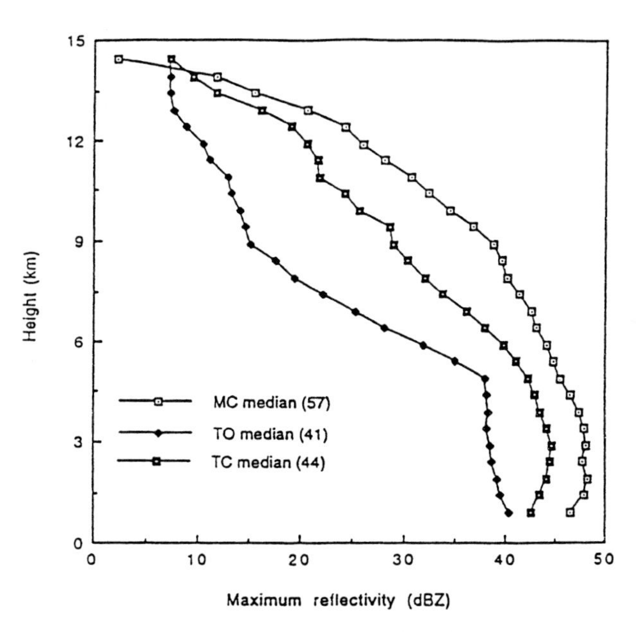

FIG. 10.70. Median reflectivity as a function of height through 41 tropical oceanic (TO), 44 tropical continental (TC), and 57 midlatitude continental MCS clouds (MC). From Zipser and Lutz (1994).

show that MCSs usually have their new cells forming ahead of the mature cells. If the mature cells tilt backward, away from the leading edge (LeMone et al. 1984b), then any stones that fall would have essentially no chance of being lifted in a new updraft. The rear-inflow jet that exists in the stronger squall lines (see COPT studies) descends as it approaches the leading edge and eventually feeds the downdraft; it is unlikely that any hail embryos could be recycled back to the updraft portion of the system.

Supercells, which do have a large updraft of sufficient magnitude to produce large hail, and are responsible for the majority of the damaging events, may be distinguished by their strong pressure perturbation that enhances the vertical velocity, especially in the lower troposphere. LeMone et al. (1988a,b) estimate the pressure perturbation for Cbs and congestus in the midlatitudes for strong and weak shear conditions; the clouds in low shear have no detectable pressure perturbation at cloud base and thus do not have a key forcing mechanism that contributes to the formation of sustained, strong updrafts that yield large hail.

e. Rarity of severe straight-line winds

Strong outflows, especially those emanating from microbursts, are often driven by cooling due to melting and evaporation of hydrometers (Brown et al. 1982; Fujita and Wakimoto 1981; Wakimoto 1985; Knupp 1987; Srivastava 1985; Parsons and Kropfli 1990). In the high plains of the United States, deep, dry adiabatic, low relative humidity subcloud layers serve as an environment for microbursts (Krumm 1954; Wakimoto 1985). There the subcloud layer can be 2500–3500 m thick.

The tropical oceans provide a startling contrast to these high plains conditions with a subcloud layer that is usually 500–600 m thick and has a relative humidity of 80%–85%. For much of the annual cycle, the tropical land areas also have low cloud bases and moist subcloud layers (e.g., the Amazon, equatorial Africa). Mechanisms other than evaporation below cloud base must play a greater role if we are to witness strong straight-line winds in tropical environments.

Microbursts, sampled during the Microburst and Severe Thunderstorm (MIST) Project conducted in northern Alabama in the summer, offer more clues as to why wind speeds exceeding the severe threshold are rare in the Tropics. The environmental conditions for the Cb on 20 July 1986 that produced a microburst had conditions more like the high plains than the Tropics with a dry adiabatic subcloud layer that was 2 km deep, an LI = −7, and dry air above 570 mb (Wakimoto and Bringi 1988). The storm was characterized by high reflectivities (>60 dB*Z*). The maximum single-Doppler velocity differential across the center of the burst was 30 m s^{-1}; this is impressive, but failed to reach the severe threshold. Dual-polarization radar observations indicate the presence of hail; the melting of this hail is the key ingredient of the formation of the *wet microburst,* one that is accompanied by heavy rain (Wolfsen 1983).

Kingsmill and Wakimoto (1991) analyzed another Cb from MIST that formed in very weak shear (0.5×10^{-3} s^{-1}) and high CAPE (2300 J kg^{-1}). The storm produced a microburst that appeared to be generated initially by water loading and later maintained by thermal effects, but the outflow winds did not exceed 12 m s^{-1}. Such winds would pose a problem for aircraft during landing and takeoff but are far from the magnitude needed to reach severe status. These results confirm the numerical simulations by Srivastava (1985) that identify the importance of a dry adiabatic subcloud layer and a surfeit of small drops to produce a microburst.

Caracena and Maier (1987) catalog a wet microburst in the FACE array. Within a rain gush of 80 mm h^{-1} there were gusts of 20 m s^{-1}, cooling of 8°–9°C, and a pressure spike of 2.4 mb. Conditions that favored the development of the microburst include a shortwave trough aloft; this last feature is, of course, not usually found in the deep Tropics. The sounding showed an elevated dry layer centered at 425 mb, which is susceptible to cooling from evaporation. Calculations show that the water loading effect is an order of magnitude less than the evaporation. The authors mention the interesting aside to their study that the Cb responsible for the Big Thompson flood in Colorado contained rain rates of 250 mm h^{-1}, which should supply an abundance of water loading, but the storm yielded no appreciable outflow.

It seems that high rain rates are not a useful predictor for strong outflows; what is crucial is a source of dry air that can be cooled and where water and ice can surrender their latent heats.

Those tropical regions that might support higher based Cbs with a dry adiabatic and lower relative humidity subcloud layer will have microbursts and perhaps accompanying outflows that reach severe magnitudes; candidate regions are the African Sahel, northeastern India, northern Australia, and the higher terrain of Bolivia, Mexico, and Central America. These conditions are best developed in the premonsoon season, and during the monsoon if the lower-level flow does not reach into the midtroposphere.

f. Factors that contribute to the rarity of tornadoes

Absolute vorticity can be increased through four terms: divergence or stretching, tipping or tilting, solenoidal effects, and friction. The majority of midlatitude numerical simulation (Klemp and Wilhelmson 1978; Schlesinger 1978) and observational (Newton 1966; Marwitz 1972; Fankhauser 1988) studies implicate the tilting term for the formation of supercells; here a strong vertical velocity gradient in the horizontal coupled with strong shear of the horizontal wind in the vertical in the lower troposphere are the critical ingredients.

The weakness of the tropical oceanic updrafts leads to less stretching of vorticity. Profiles of w as a function of height (Fig. 10.64) show that a near-constant vertical velocity is the typical situation, which limits the magnitude of the tilting term. CAPE in the Tropics is not often due to large differences in temperature through a thin layer; instead values become large simply because the instability is found throughout a deep troposphere. This instability is rarely realized because of entrainment.

The profiles of du/dz and dv/dz presented earlier show shears in many of the unstable regions to be less than 3.0×10^{-3} s^{-1} for a wide range of features including monsoons, the NECZ, the Amazon Basin, easterly waves, upper- and midlevel cyclones, and monsoon depressions. Invading extratropical cyclones and tropical cyclones are the primary larger-scale phenomena that do create high shear environments. In the case of a TC the shear is found through a shallow layer of 1.5–2.5 km due to the warm core nature of the TC.

The solenoidal term is usually negligible for all large-scale tropical phenomena, but, on the gust-front scale, this term contributes significantly to the vorticity. Numerical studies (e.g., Klemp and Rotunno 1983) support the argument that vorticity along the forward flank outflow boundary is an important factor contributing to tornadogenesis. This horizontal vorticity generated along the outflow is tilted into the vertical by the updraft and contributes to tornado intensity. Outflows over the tropical oceans are, however, shallow, with only modest temperature decreases (e.g., Houze 1977; Zipser 1977; Barnes and Garstang 1982). The result is an outflow that does not boost the vorticity to the magnitude required to form an intense, low-level circulation. Outflows over drier continental tropical areas would have a stronger contribution from the solenoidal term.

The friction term is small, and there is no reason to assume that it would play a greater or lesser role in the Tropics than it is currently believed to play in the midlatitudes.

10.7. Conclusions

a. Caveats

Dense quality controlled observation networks capable of detecting SLS are rare over tropical land regions and nonexistent over the oceans. The conclusions drawn here must be tempered by this limitation, and this chapter can serve only as a rudimentary foundation. It is expected that, with the advent of increasingly sophisticated satellite sensors, the placement of 10-cm radars, and the establishment of weather station networks in the developing countries, our understanding of SLS will improve, and the frequency of severe local storms will increase.

b. Summary

Describing the tropical atmosphere as an unchanging homogeneous body of warm, moist air leads to a terrible deception. There is a wide range of tropical environments due to the annual variability of planetary features and the more focusing effects of synoptic-scale phenomena. Identification of the larger-scale forcing is the first step in determining the odds of deep convection and severe local storms, save for waterspouts that are regularly spawned by lines of cumulus congestus.

There are portions of the Tropics that are prone to SLS.

1. Northeastern India and Bangladesh suffer all types of SLS in the premonsoon season due to the juxtaposition of midlatitude westerlies above hot humid southerlies. Here the invasion of extratropical waves is the important trigger during the premonsoon season.
2. The high terrain of East Africa is prone to hail, which is related to the annual cycle of the equatorial trough and local lake–mountain induced gradients. This hail is usually below 2 cm diameter but nonetheless is a serious problem given the susceptibility of the delicate crops growth there.
3. Tropical cyclones create strong low-level shear that, when coupled with modest CAPE, can produce tornadoes of F0–F2 intensity, most often in the right quadrants of the cyclone. Hurricanes produce an environment that contrasts the standard high shear, large CAPE of the U.S. Mid-

west, but are nonetheless effective at spawning tornadoes.
4. Immediately downstream of the continents extratropical troughs can still maintain the horizontal and vertical gradients that support vigorous convection (e.g., the Caribbean, South China Sea). SLS events would tend to occur in the winter and spring when troughs have large meridional amplitude and strong temperature gradients.
5. Over western Africa and especially in the Sahel, disturbances or squall lines can produce strong straight-line winds. Here the moist southwest monsoon and the African easterly jet, modulated by easterly waves, are the contributing larger-scale phenomena.
6. Tropical Australia, throughout the year, but especially during the premonsoon, can have the combination of a deep dry adiabatic subcloud layer and wind shear that can produce squall lines and even some splitting cells. The most frequent form of SLS from these systems is strong straight-line winds, but tornadoes also occur. For tropical Australia extratropical systems are not a major catalyst for the development of a favorable SLS environment.
7. Waterspouts appear to be common, but rarely account for a disaster at sea. Their reported frequency is grossly underestimated. They occur throughout the Tropics, but especially in the Bay of Bengal, the Java Sea, the Atlantic equatorial trough, and the Caribbean.

Much of the Tropics has an extremely low risk for SLS. Subsidence in the trade wind regions and cold waters along the eastern portions of the ocean basins prevent even modest cumulonimbi from occurring. There is rarely moisture to fuel clouds over the Sahara and the Arabian peninsula. Large hail is extremely rare over the Tropics at sea level.

Other expansive regions of the Tropics have a surfeit of deep convection, but do not apparently yield a corresponding high frequency of SLS. Equatorial Africa, the Amazon, and the equatorial trough are active according to outgoing longwave radiation surveys but these environments often lack a critical ingredient to support any type of severe weather.

What are the missing ingredients? These are discussed in the physics section of this chapter and plenty of insights can be found elsewhere in this volume. Most tropical Cbs have weak updrafts, small variation of the updraft with height in the lower to middle troposphere, and modest updraft diameters. Most tropical Cbs develop in an environment characterized by weak shear in the lower troposphere. The absence of strong shear limits the strength of the dynamic pressure perturbation in the Cb, which in turn hinders vertical velocities. The weak updrafts hinder 1) hail growth to large size, and 2) vortex stretching that favors tornadogenesis. Weak shear also implies that the initial horizontal momentum for the downdraft is modest and thus does not favor vigorous straight-line winds. The Tropics, more usually than not, have a positive correlation between moist midlevels and instability (contrast the NECZ with the trades), which is favorable for deep clouds, but not for Cbs that produce strong outflows via evaporative cooling.

Instability in the Tropics, especially regions like the Amazon, equatorial Africa, and the near-equatorial trough, is often respectable, but rarely is it extreme. Here the vertical profile of the instability (shape of the CAPE) is such that much less acceleration of the updraft occurs in the lower troposphere. Seasons with heavy rains (e.g., the monsoon) yield floods but rarely SLS. There is a shift to a moist-adiabatic lapse rate through a deeper layer of the troposphere that diminishes buoyancy.

The tropical atmosphere contains strong upper-level jet streams, but the resulting upper-level shear's role in SLS is suspect. Upper-level jet streams play an important role in the divergence profile and Cb formation, but SLS does not seem to occur without accompanying shear in the lower troposphere.

c. Speculation

Johns and Doswell (1992) review tried and new techniques for the forecasting of SLS that could be applied more thoroughly to the Tropics, but observations of deep convection, especially over tropical landmasses, must be improved. Certain continental regions lack meteorological infrastructure (e.g., Sahel Africa), are sparsely populated (e.g., northern Australia), or fit both descriptions (e.g., the Amazon). The first two regions have premonsoon conditions that favor SLS and thus are likely to harbor much more SLS than is currently reported.

Some regions have remarkable horizontal gradients, the fruit of variations in terrain, the juxtaposition of water and land, or a combination of these, that regularly produce deep convection. The author has found few reports of SLS from Indonesia and only some documentation about frequent though small hail in East Africa, but argues for caution, given the state of the observational network in these regions. Analyses during the summer for Florida demonstrate that Cb outflows can interact to spin up weak tornadoes and waterspouts. One suspects that the Maritime Continent may harbor at least as much SLS as Florida, given its considerable instability and mesoscale forcing.

d. Scientific challenges

Have we been able to accurately define the environment for deep convection, especially when it is part of an MCS? Do clouds that grow within an MCS have less susceptibility to the negative effects of entrainment? A more careful examination of clouds that grow in isolation versus those in an MCS may reveal how clouds modify their environment in ways that enhance convective intensity. The atmosphere also may be preconditioned by fair-weather mesoscale circulations that may alter the odds in favor of SLS occurrence (e.g., the Denver convergence vorticity zone; Szoke and Brady 1989).

Can the community develop techniques to forecast SLS occurrence that are based solely on remote platforms such as satellite sensors? For the near future, the Tropics will depend on this avenue more than any other.

Acknowledgments. This work was partially supported by NSF Grant ATM-9619398 and JIMAR, University of Hawaii. The author benefited from the panel reviews and from the devils' advocate role played by Garpee Barleszie.

REFERENCES

Adler, R. F., M. Markus, and D. D. Fenn, 1985: Detection of severe midwest thunderstorms using geosynchronous satellite data. *Mon. Wea. Rev.,* **113,** 769–781.

Albrecht, B. A., 1984: A model study of the downstream variations of the thermodynamic structure of the trade winds. *Tellus,* **36A,** 187–202.

Alexander, G. D., and G. S. Young, 1992: The relationship between EMEX mesoscale precipitation feature properties and their environmental characteristics. *Mon. Wea. Rev.,* **120,** 554–560.

Alfonso, A. P., 1988: Severe local storms in Cuba: Climatology and fundamentals for forecasting. Ph.D. dissertation, Academia de Ciencias de Cuba, 112 pp.

———, and L. R. Naranjo, 1996: The 13 March 1993 severe squall line over western Cuba. *Wea. Forecasting,* **11,** 89–102.

Allen, S. C., 1980: A preliminary Australian tornado climatology. Tech Rep. 39, Australian Bureau of Meteorology, 14 pp.

Alvi, S. M. A., and K. G. Punjabi, 1966: Diurnal and seasonal variations of squalls in India. *Indian J. Meteor. Geophys.,* **17,** 206–216.

Anyamba, E., E. Williams, J. Susskind, A. Fraser-Smith, and M. Fullekrug, 1997: The manifestation of the Madden-Julian oscillation in global deep convection and in the Schumann Resonance intensity. Preprints, *22d Conf. On Hurricanes and Tropical Meteorology,* Ft. Collins, CO, Amer. Meteor. Soc., 227–231.

Asnani, G. C., 1993: *Tropical Meteorology.* Indian Institute of Tropical Meteorology, Pune, India, 1202 pp.

Aspliden, C. I., Y. Tourre, and J. B. Sabine, 1976: Some climatological aspects of West African disturbance lines during GATE. *Mon. Wea. Rev.,* **104,** 1029–1035.

Atkinson, G. D., 1966: A preliminary estimate of extreme wind speeds in Thailand. 1st Wea. Wing Tech. Study 3, 24 pp.

Augstein, E., H. Riehl, F. Ostapoff, and V. Wagner, 1973: Mass and energy transports in an undisturbed Atlantic trade wind flow. *Mon. Wea. Rev.,* **101,** 101–111.

Balaji, V., and T. L. Clark, 1988: Scale selection in locally forced convective fields and the initiation of deep cumulus. *J. Atmos. Sci.,* **45,** 3188–3211.

Banerjee, A. K., A. Chowdhury, and H. R. Ganesan, 1983: Some characteristic features of squall over India. *Vayu Mandal,* **13,** 110–119.

Barnes, G. M., 1995: Updraft evolution: A perspective from cloud base. *Mon. Wea. Rev.,* **123,** 2693–2715.

———, and M. Garstang, 1982: Subcloud layer energetics of precipitating convection. *Mon. Wea. Rev.,* **110,** 102–117.

———, and K. Sieckman, 1984: The environment of fast- and slow-moving tropical mesoscale convective cloud lines. *Mon. Wea. Rev.,* **112,** 1782–1794.

———, E. J. Zipser, D. P. Jorgensen, and F. D. Marks, 1983: Mesoscale and convective structure of a hurricane rainband. *J. Atmos Sci.,* **40,** 2125–2137.

———, J. C. Fankhauser, and W. D. Browning, 1996: Evolution of the vertical mass flux and diagnosed net lateral mixing in isolated convective clouds. *Mon. Wea. Rev.,* **124,** 2764–2784.

Barnes, S. L., and C. W. Newton, 1985: Thunderstorms in the synoptic setting. *Thunderstorm Morphology and Dynamics,* E. Kessler, Ed., University of Oklahoma Press, 75–112.

Betts, A. K., 1982: Saturation point analysis of moist convective overturning. *J. Atmos. Sci.,* **39,** 1484–1505.

———, R. W. Grover, and M. Moncrieff, 1976: Structure and motion of tropical squall lines over Venezuela. *Quart. J. Roy. Meteor. Soc.,* **102,** 395–404.

Biswas, B., and K. Gupta, 1989: Heights of Cb clouds around Calcutta airport—Diurnal and seasonal variations. *Mausam,* **40,** 169–174.

Black, P. G., F. D. Marks Jr., and R. A. Black, 1986: Supercell structure in tropical cyclones. *Proc. 23d Conf. on Radar Meteorology,* Snowmass, CO, Amer. Meteor. Soc., JP255–JP259.

Bluestein, H. B., 1985: The formation of a "landspout" in a "broken line" in Oklahoma. Preprints, *14th Conf. on Severe Local Storms,* Indianapolis, IN, Amer. Meteor. Soc., 267–270.

———, and M. H. Jain, 1985: Formation of mesoscale lines of precipitation: Severe squall lines in Oklahoma during the spring. *J. Atmos. Sci.,* **42,** 1711–1732.

———, G. T. Marx, and M. H. Jain, 1987: Formation of mesoscale lines of precipitation: Nonsevere squall lines in Oklahoma during the spring. *Mon. Wea. Rev.,* **115,** 2719–2727.

Blyth, A. M., W. A. Cooper, and J. B. Jensen, 1988: A study of the source of entrained air in Montana cumuli. *J. Atmos. Sci.,* **45,** 3944–3964.

Boatman, J. F., and A. H. Auer, 1983: The role of cloud top entrainment in cumulus clouds. *J. Atmos. Sci.,* **40,** 1517–1534.

Bogner, P. B., G. M. Barnes, and J. L. Franklin, 2000: Conditional instability and shear for six hurricanes over the Atlantic Ocean. *Wea. Forecasting,* **15,** 192–207.

Branick, M. L., and C. A. Doswell III, 1992: An observation of the relationship between supercell structure and lightning ground strike polarity. *Wea. Forecasting,* **7,** 143–149.

Brown, J. M., K. R. Knupp, and F. Caracena, 1982: Destructive winds from shallow, high-based cumulonimbi. Preprints, *12th Conf. on Severe Local Storms,* San Antonio, TX, Amer. Meteor. Soc., 272–275.

Browning, K. A., and G. B. Foote, 1976: Airflow and hailgrowth in supercell storms and some implications for hail suppression. *Quart. J. Roy. Meteor. Soc.,* **102,** 499–533.

Buist, 1885: Remarkable hailstones in India, March 1841 and May 1885. British Assn. Rep. 34.

Burpee, R. W., 1972: The origin and structure of easterly waves in the lower troposphere of North Africa. *J. Atmos. Sci.,* **29,** 77–90.

Byers, H. R., and R. R. Braham, 1949: *The Thunderstorm.* U.S. Government Printing Office, Washington, DC, 287 pp.

Caracena, F., and M. Maier, 1987: Analysis of a microburst in the FACE meteorological mesonetwork in southern Florida. *Mon. Wea. Rev.,* **115,** 969–985.

Carlson, T. N., 1969: Some remarks on African disturbances and their progress over the tropical Atlantic. *Mon. Wea. Rev.,* **97,** 716–726.

———, and F. H. Ludlam, 1968: Conditions for the formation of severe local storms. *Tellus,* **20,** 203–226.

Carter, E. A., and L. A. Schuknecht, 1969: Peak wind statistics associated with thunderstorms at Cape Kennedy, Florida. NASA Contr. Rep. 61304, 32 pp.

Chakrabarty, K. K., and S. K. Bhowmik, 1993: Study of unusual hailstorm over Bombay. *Mausam,* **44,** 292–295.
Chalon, J. P., G. Jaubert, F. Roux, and J. P. Lafore, 1988: The West African squall line observed on 23 June 1981 during COPT81: Mesoscale structure and transports. *J. Atmos. Sci.,* **45,** 2744–2763.
Chang, C.-P., J. E. Erickson, and K. M. Lau, 1979: Northeasterly cold surges and near-equatorial disturbances over the winter MONEX area during December, 1974: Synoptic aspects. *Mon. Wea. Rev.,* **107,** 812–829.
Chatterjee, R. N., and P. Prakash, 1990: Radar study of thunderstorms around Delhi during the monsoon season. *Mausam,* **41,** 161–165.
Chaudhury, A. K., and A. B. Mazumdar, 1983: Nor'westers and the synoptic climatology of hot weather season in northeast India. *Vayu Mandal,* **13,** 48–55.
Chen, G. T.-J., and C.-P. Chang, 1980: The structure and vorticity budget of an early summer monsoon trough (Mei-Yu) over southeastern China and Japan. *Mon. Wea. Rev.,* **108,** 942–953.
Chen, S. S., R. A. Houze Jr., and B. E. Mapes, 1996: Multiscale variability of deep convection in relation to large-scale circulation in TOGA COARE. *J. Atmos. Sci.,* **53,** 1380–1409.
Chen, X. A., and Y.-L. Chen, 1995: Development of low-level jets during TAMEX. *Mon. Wea. Rev.,* **123,** 1695–1719.
Chen, Y.-L., 1993: Some synoptic-scale aspects of the surface fronts over southern China during TAMEX. *Mon. Wea. Rev.,* **121,** 50–64.
Chong, M., P. Amayenc, G. Scialom, and J. Testud, 1987: A tropical squall line observed during the COPT81 experiment in West Africa. Part I: Kinematic structure inferred from dual-Doppler radar data. *Mon. Wea. Rev.,* **115,** 670–694.
Chowdhury, A., and A. K. Banerjee, 1983: A study of hailstorms over northeast India. *Vayu Mandal,* **13,** 91–95.
Christie, D. R., and K. J. Muirhead, 1983: Solitary waves: A hazard to aircraft operating at low altitudes. *Austr. Meteor. Mag.,* **31,** 97–109.
Chu, P.-S., 1975: A meso-synoptic analysis of convective thunderstorms and an associated dryline in south-central Brazil. Preprints, *Ninth Conf. on Severe Local Storms,* Norman, OK, Amer. Meteor. Soc., 181–186.
Clark, T. L., T. Hauf, and J. P. Kuettner, 1986: Convectively forced internal gravity waves: Results from two-dimensional numerical experiments. *Quart. J. Roy. Meteor. Soc.,* **112,** 899–925.
Clarke, R. H., 1962: Severe local wind storms in Australia. CSIRO, Div. of Meteor. Phys., Tech Paper 13, 56 pp.
Collin, W. P., 1983: Severe thunderstorm and tornado outbreak over Florida—February 1–2, 1983. Preprints, *13th Conf. on Severe Local Storms,* Tulsa, OK, Amer. Meteor. Soc., 17–20.
Court, A., and J. F. Griffiths, 1985: Thunderstorm climatology. *Thunderstorm Morphology and Dynamics,* E. Kessler, Ed., University of Oklahoma Press, 9–39.
Davey, B. J., 1987: Tornadic waterspout at the Jebel Ali Sailing Club. *Meteor. Mag.,* **116,** 129–137.
Davidson, N. E., J. L. McBride, and B. J. McAvaney, 1983: The onset of the Australian monsoon during winter MONEX. *Mon. Wea. Rev.,* **112,** 1697–1708.
Dean, G., 1956: The 1955 mean monthly wind circulation over the tropical central Pacific area. University of California Institute of Geophysics Rep., 16 pp.
———, 1971: The three-dimensional wind structure over South America and associated rainfall over Brazil. Rep. LAFE-164, Dept. of Meteor., Florida State University, 173 pp.
Dekate, M. V., and K. K. Bajaj, 1966: Study of thundersqualls over Santacruz airport. *Indian J. Meteor. Geophys.,* **17,** 217.
Despande, D. V., 1964: Heights of Cb clouds over India during the southwest monsoon season. *Indian J. Meteor. Geophys.,* **15,** 47–54.
Dickinson, M. J., L. F. Bosart, W. E. Bracken, G. J. Hakim, D. E. Schultz, M. A. Bedrick, and K. R. Tyle, 1997: The March 1993 superstorm cyclogenesis: Incipient phase synoptic- and convective-scale flow interaction and model performance. *Mon. Wea. Rev.,* **125,** 3041–3072.
Doswell, C. A., 1980: Synoptic-scale environments associated with High Plains severe thunderstorms. *Bull. Amer. Meteor. Soc.,* **61,** 1388–1400.
Douglas, M. W., 1992a: Structure and dynamics of two monsoon depressions. Part I: Observed structure. *Mon. Wea. Rev.,* **120,** 1524–1547.
———, 1992b: Structure and dynamics of two monsoon depressions. Part II: Vorticity and heat budgets. *Mon. Wea. Rev.,* **120,** 1548–1564.
Drosdowsky, W., 1984: Structure of a northern Australian squall line system. *Austr. Meteor. Mag.,* **32,** 177–183.
———, and G. J. Holland, 1987: North Australian cloud lines. *Mon. Wea. Rev.,* **115,** 2645–2659.
Dunn, G. E., 1940: Cyclogenesis in the Tropical Atlantic. *Bull. Amer. Meteor. Soc.,* **21,** 215–229.
Eldridge, R. H., 1957: A synoptic study of West African disturbance lines. *Quart. J. Roy. Meteor. Soc.,* **83,** 303–314.
Eliot, J., 1899: Hailstorms in India during the period 1883–1897 with a discussion of their distribution. *India Meteor. Memoirs* **6** (4), 237–315.
Engholm, C. D., E. R. Williams, and R. M. Dole, 1990: Meteorological conditions associated with positive cloud-to-ground lightning. *Mon. Wea. Rev.,* **118,** 470–487.
Fankhauser, J. C., 1988: Estimates of precipitation efficiency from field measurements in CCOPE. *Mon. Wea. Rev.,* **116,** 663–684.
Fawbush, E. J., and R. C. Miller, 1952: A mean sounding representative of the tornadic airmass environment. *Bull. Amer. Meteor. Soc.,* **33,** 303–307.
———, and ———, 1954: The types of airmasses in which North American tornadoes form. *Bull. Amer. Meteor. Soc.,* **35,** 154–165.
Firestone, J. K., and B. A. Albrecht, 1986: The structure of the atmospheric boundary layer in the central equatorial Pacific during January and February of FGGE. *Mon. Wea. Rev.,* **114,** 2219–2231.
Flora, S. D., 1954: *Tornadoes of the United States.* University of Okla. Press, 221 pp.
———, 1956: *Hailstorms of the United States.* University of Okla. Press, 201 pp.
Flores, J. F., and V. F. Bagalot, 1969: Climate of the Philippines. *Climates of Northern and Eastern Asia,* World Survey of Climatology, Vol. 8, Elsevier, 159–213.
Fortune, M., 1980: Properties of African squall lines inferred from time-lapse satellite photography. *Mon. Wea. Rev.,* **108,** 153–168.
———, and V. E. Kousky, 1983: Two severe freezes in Brazil: Precursors and synoptic evolution. *Mon. Wea. Rev.,* **111,** 181–196.
Frank, N. L., 1970: On the energetics of cold lows. *Proc. Symp. on Tropical Meteorology,,* Honolulu, HI, Amer. Meteor. Soc., EIV-1–EIV-6.
Frank, W. M., 1978: The life cycles of GATE convective systems. *J. Atmos. Sci.,* **35,** 1256–1264.
Frisby, E. M., 1966: Hail incidence in the tropics. U.S. Army Electr. Comm. Tech Rep. 2768.
———, and H. W. Sansom, 1967: Hail incidence in the Tropics. *J. Appl. Meteor.,* **6,** 339–354.
Fritsch, J. M., R. J. Kane, and C. R. Chelius, 1986: The contribution of mesoscale convective weather systems to the warm season precipitation in the United States. *J. Climate Appl. Meteor.,* **25,** 1333–1345.
Fujita, T. T., 1973: Tornadoes around the world. *Weatherwise,* **26,** 56–62, 78–83.
———, 1981: Tornadoes and downbursts in the context of generalized planetary scales. *J. Atmos. Sci.,* **38,** 1511–1534.
———, 1985: *The Downburst.* University of Chicago Press, 122 pp.

———, and R. M. Wakimoto, 1981: Five scales of airflow associated with a series of downbursts on 16 July 1980. *Mon. Wea. Rev.,* **109,** 1438–1456.

Gamache, J. F., and R. A. Houze Jr., 1982: Mesoscale air motions associated with a tropical squall line. *Mon. Wea. Rev.,* **110,** 118–135.

Garstang, M., B. E. Kelbe, G. D. Emmitt, and W. E. London, 1987: Generation of convective storms over the escarpment of northeastern South Africa. *Mon. Wea. Rev.,* **115,** 429–443.

———, H. L. Massie Jr., J. Halverson, S. Greco, and J. Scala, 1994: Amazon coastal squall lines. Part I: Structure and kinematics. *Mon. Wea. Rev.,* **122,** 608–622.

Geldmeier, M. F., and G. M. Barnes, 1997: The "footprint" under a decaying tropical mesoscale convective system. *Mon. Wea. Rev.,* **125,** 2879–2895.

Gentry, R. C., 1983: Genesis of tornadoes associated with hurricanes. *Mon. Wea. Rev.,* **111,** 1793–1805.

Ghosh, A. K., 1982: A tornado over Orissa in April 1978. *Mausam,* **33,** 235–240.

Gill, J., and M. Kersemakers, 1992: A severe tropical squall line observed over northern Australia. Preprints, *Third Australian Severe Thunderstorm Workshop,* Gympie, Queensland, Bureau of Meteorology.

Godbole, R. V., 1977: The composite structure of the monsoon depression. *Tellus,* **29,** 25–40.

Golden, J. H., 1971: Waterspouts and tornadoes over South Florida. *Mon. Wea. Rev.,* **99,** 146–154.

———, 1974a: Life cycle of Florida Keys waterspouts. *J. Appl. Meteor.,* **13,** 676–692.

———, 1974b: Scale interaction implications for the waterspout life cycle. *J. Appl. Meteor.,* **13,** 693–709.

Golding, B. W., 1993: A numerical investigation of tropical island thunderstorms. *Mon. Wea. Rev.,* **121,** 1417–1433.

Gordon, A. H., 1951: Waterspouts. *Weather,* **6,** 364.

Greco, S., R. Swap, M. Garstang, S. Ulanski, M. Shipman, R. C. Harriss, and R. Talbot, 1990: Rainfall and kinematic conditions over central Amazonia during ABLE 2B. *J. Geophys. Res.,* **95,** 17 001–17 014.

Gunn, B. W., J. L. McBride, G. J. Holland, T. D. Keenan, and N. E. Davidson, 1989: The Australian summer monsoon circulation during AMEX II. *Mon. Wea. Rev.,* **117,** 2554–2574.

Gupta, H. N., and S. K. Ghosh, 1980: North Delhi tornado of 17 March 1978. *Mausam,* **31,** 93–100.

———, and S. K. Ghosh, 1982: Reported cases of tornadoes in Indian subcontinent. *Vayu Mandal,* **12,** 57–60.

———, and R. C. Sharma, 1983: A study of thunderstorms and hailstorms over New Delhi. *Mausam,* **34,** 390–392.

Gupta, R. K., 1987: On the rain/thunderstorm activity over Car Nicobar. *Mausam,* **38,** 229.

Gurka, J. J., 1976: Satellite and surface observations of strong wind zones accompanying thunderstorms. *Mon. Wea. Rev.,* **104,** 1484–1493.

Hagemeyer, B. C., 1997: Peninsular Florida tornado outbreaks. *Wea. Forecasting,* **12,** 399–427.

Hamilton, R. A., and J. W. Archbold, 1945: Meteorology of Nigeria and adjacent territory. *Quart. J. Roy. Meteor. Soc.,* **71,** 231–264.

Hanstrum, B. N., G. A. Mills, and A. Watson, 1998: Cool-season tornadoes in Australia. Part I: Synoptic climatology. Preprints, *19th Conf. on Severe Local Storms,* Minneapolis, MN, Amer. Meteor. Soc., 97–100.

Hariharan, P. S., 1950: Sizes of hailstones. *Indian J. Meteor. Geophys.,* **1** (1), 73.

Harrison, D. E., and B. S. Giese, 1991: Episodes of surface westerly winds as observed from islands in the western tropical Pacific. *J. Geophys. Res.,* **96,** 3221–3237.

Hastenrath, S., 1991: *Climate Dynamics of the Tropics.* Atmospheric Sciences Library, Kluwer Academic, 488 pp.

Heymsfield, G. M., and R. H. Blackmer Jr., 1988: Satellite-observed characteristics of midwest severe thunderstorms. *Mon. Wea. Rev.,* **116,** 2200–2224.

Hill, E. L., W. Malkin, and W. A. Schulz Jr., 1966: Tornadoes associated with cyclones of tropical origin—Practical features. *J. Appl. Meteor.,* **5,** 745–763.

Hiser, H. W., 1968: Radar and synoptic analysis of the Miami tornado of June 17, 1959. *J. Appl. Meteor.,* **7,** 892–900.

Hodanish, S., J. Sharp, E. Williams, B. Boldi, A. Matlin, M. Weber, S. Goodman, and R. Raghavan, 1998: Comparisons between total lightning data, mesoscyclone strength, and storm damage associated with the Florida tornado outbreak of February 23, 1998. Preprints, *19th Conf. on Severe Local Storms,* Minneapolis, MN, Amer. Meteor. Soc., 681–684.

Hoddinott, M. H. O., 1986: Thunderstorm observations in west Bengal. *Weather,* **41,** 1–5.

Holle, R. L., and M. W. Maier, 1980: Tornado formation from downdraft interaction in the FACE network. *Mon. Wea. Rev.,* **108,** 1010–1027.

Houze, R. A., Jr., 1977: Structure and dynamics of a tropical squall-line system. *Mon. Wea. Rev.,* **105,** 1540–1567.

———, and A. K. Betts, 1981: Convection in GATE. *Rev. Geophys. Space Phys.,* **19,** 541–576.

———, and E. N. Rappaport, 1984: Air motions and precipitation structure of an early summer squall line over the eastern tropical Atlantic. *J. Atmos. Sci.,* **41,** 553–574.

Hull, B. B., 1958: Hail size and distribution. Environmental Protection Research Division, U.S. Army, Tech. Rep. EP-83, 89 pp.

Huschke, R. E., Ed., 1959: *Glossary of Meteorology.* Amer. Meteor. Soc., 638 pp.

Ishihara, M., and Z. Yanagisawa, 1982: Structure of a tropical squall line observed in the western tropical Pacific during MONEX. *Meteor. Geophys.,* **33,** 117–135.

Jackson, C. C. E., 1987: Vortex phenomena in the United Arab Emirates. *Weather,* **42,** 302–308.

Johns, R. H., and W. D. Hirt, 1983: The derecho . . . a severe weather producing system. Preprints, *13th Conf. on Severe Local Storms,* Tulsa, OK, Amer. Meteor. Soc., 178–181.

———, and ———, 1987: Derechos: Widespread convectively induced windstorms. *Wea. Forecasting,* **2,** 32–49.

———, and C. A. Doswell, 1992: Severe local storms forecasting. *Wea. Forecasting,* **7,** 588–612.

Jorgensen, D. P., and M. A. LeMone, 1989: Vertical velocity characteristics of oceanic convection. *J. Atmos. Sci.,* **46,** 621–640.

———, E. J. Zipser, and M. A. LeMone, 1985: Vertical motions in intense hurricanes. *J. Atmos. Sci.,* **42,** 839–856.

———, M. A. LeMone, and B. J.-D. Jou, 1991: Precipitation and kinematic structure of an oceanic mesoscale convective system. Part I: Convective line structure. *Mon. Wea. Rev.,* **119,** 2608–2637.

———, ———, and S. B. Trier, 1997: Structure and evolution of the 22 February 1993 TOGA COARE squall line: Aircraft observations of precipitation, circulation, and surface energy fluxes. *J. Atmos. Sci.,* **54,** 1961–1985.

Joseph, P. V., D. K. Raipal, and S. N. Deka, 1980: "Andhi", the convective dust storm of northwest India. *Mausam,* **31,** 431–442.

Kamble, V. P., and M. L. Kanoujia, 1990: Multiple squalls over central India. *Mausam,* **41,** 451–454.

Kanjilal, T., B. Basu, A. Roy, and M. C. Sinha, 1989: Potential convective instability analysis on squall days at Calcutta. *Mausam,* **40,** 409–412.

Keenan, T. D., and R. E. Carbone, 1992: A preliminary morphology of precipitating systems in tropical northern Australia. *Quart. J. Roy. Meteor. Soc.,* **118,** 283–326.

Kelley, W. E., and D. R. Mock, 1982: A diagnostic study of upper tropospheric cold lows over the western North Pacific. *Mon. Wea. Rev.,* **110,** 471–480.

Kilonsky, B. J., and C. S. Ramage, 1976: A technique for estimating tropical open-ocean rainfall from satellite observations. *J. Appl. Meteor.,* **15,** 972–975.

Kingsmill, D. E., and R. M. Wakimoto, 1991: Kinematic, dynamic and thermodynamic analysis of a weakly sheared thunderstorm over Northern Alabama. *Mon. Wea. Rev.,* **119,** 262–297.

Kingwell, J., and I. J. Butterworth, 1986: A waterspout near Elcho Island, northern Australia. *Weather,* **41,** 154–156.

Kishan, D., and B. Lal, 1989: A case study of possible occurrence of a microburst cell on 27 June 1985 over Palam airport. *Mausam,* **40,** 221–224.

Klemp, J. B., and R. B. Wilhelmson, 1978: The simulation of three-dimensional convective storm dynamics. *J. Atmos. Sci.,* **35,** 1070–1096.

———, and R. Rotunno, 1983: A study of the tornadic region within a supercell thunderstorm. *J. Atmos. Sci.,* **40,** 359–377.

Kloesel, K. A., and B. A. Albrecht, 1989: Low-level inversions over the tropical Pacific—Thermodynamic structure of the boundary layer and the above-inversion moisture structure. *Mon. Wea. Rev.,* **117,** 87–101.

Knight, C. A., W. A. Cooper, D. W. Breed, I. R. Paluch, P. L. Smith, and G. Vali, 1982: Microphysics. *Hailstorms of the Central High Plains,* C. A. Knight and P. Squires, Eds., Colorado Associated University Press, 151–193.

Knupp, K. R., 1987: Downdrafts within High Plains cumulonimbi. Part I: General kinematic structure. *J. Atmos. Sci.,* **44,** 987–1008.

Kodama, K., and G. M. Barnes, 1997: Heavy rain events over the south facing slopes of Hawaii: Attendant conditions. *Wea. Forecasting,* **12,** 347–367.

Koteswaram, P., and C. A. George, 1958: On the formation of monsoon depressions in the Bay of Bengal. *Indian J. Meteor. Geophys.,* **9,** 9–22.

Kousky, V. E., 1980: Diurnal variation of rainfall in northeast Brazil. *Mon. Wea. Rev.,* **108,** 488–498.

Krishnamurti, T. N., 1961: The subtropical jet stream of winter. *J. Meteor.,* **18,** 172–191.

———, and R. S. Hawkins, 1970: Mid-tropospheric cyclones of the southwest monsoon. *J. Appl. Meteor.,* **9,** 442–458.

Krumm, W. R., 1954: On the cause of downdrafts from dry thunderstorms over the plateau area of the United States. *Bull. Amer. Meteor. Soc.,* **35,** 122–125.

Kumar, A., 1992: A climatological study of thunderstorms at Lucknow airport. *Mausam,* **43,** 441–444.

Lafore, J.-P., J.-L. Redelsperger, and G. Jaubert, 1988: Comparison between a three-dimensional simulation and Doppler-radar data of a tropical squall line: Transports of mass, momentum, heat and moisture. *J. Atmos. Sci.,* **45,** 3483–3500.

Laing, A. G., and J. M. Fritsch, 1993a: Mesocale convective complexes in Africa. *Mon. Wea. Rev.,* **121,** 2254–2263.

———, and ———, 1993b: Mesocale convective complexes over the Indian monsoon region. *J. Climate,* **6,** 911–919.

Lal, R., 1990: Forecasting of severe convective activity over Lucknow in pre-monsoon months. *Mausam,* **41,** 455–458.

Lander, M., 1994: Description of a monsoon gyre and its effects on the tropical cyclones in the western North Pacific during August 1991. *Wea. Forecasting,* **9,** 640–654.

Lau, K.-M., 1982: Equatorial response to northeasterly cold surges as inferred from satellite imagery. *Mon. Wea. Rev.,* **110,** 1306–1313.

———, and C. P. Chang, 1987: Planetary scale aspects of the winter monsoon and atmospheric teleconnections. *Monsoon Meteorology,* C.-P. Chang and T. N. Krishnamurti, Eds., Oxford University Press, 161–202.

———, Li Peng, C. H. Sui, and T. Nakazawa, 1989: Dynamics of super cloud clusters, westerly wind bursts, 30–60 day oscillations and ENSO: A unified view. *J. Meteor. Soc. Japan,* **67,** 205–219.

Leary, C. A., and R. A. Houze Jr., 1979: The structure and evolution of convection in a tropical cloud cluster. *J. Atmos. Sci.,* **36,** 437–457.

LeMone, M. A., 1989: The influence of vertical wind shear on the diameter of cumulus clouds in CCOPE. *Mon. Wea. Rev.,* **117,** 1480–1491.

———, and E. J. Zipser, 1980: Cumulonimbus vertical velocity events in GATE. Part I: Diameter, intensity, and mass flux. *J. Atmos. Sci.,* **37,** 2444–2457.

———, G. M. Barnes, and E. J. Zipser, 1984a: Momentum flux by lines of cumulonimbus over the tropical oceans. *J. Atmos. Sci.,* **41,** 1914–1932.

———, E. J. Szoke, and E. J. Zipser, 1984b: The tilt of the leading edge of mesoscale tropical cloud lines. *Mon. Wea. Rev.,* **112,** 510–519.

———, L. F. Tarleton, and G. M. Barnes, 1988a: Perturbation pressure at the base of cumulus clouds in low shear. *Mon. Wea. Rev.,* **116,** 2062–2068.

———, G. M. Barnes, J. C. Fankhauser, and L. F. Tarleton, 1988b: Perturbation pressure fields measured by aircraft around the cloud base updraft of deep convective clouds. *Mon. Wea. Rev.,* **116,** 313–327.

Lemons, H., 1942: Hail in high and low latitudes. *Bull. Amer. Meteor. Soc.,* **23,** 61–68.

Leroux, M., 1983: *Le Climat de L'Afrique Tropical.* Editions Champion, Paris, 633 pp.

Leverson, V. H., P. C. Sinclair, and J. H. Golden, 1977: Waterspout wind, temperature, and pressure structure deduced from aircraft measurements. *Mon. Wea. Rev.,* **105,** 725–733.

Levine, J., 1959: Spherical vortex theory of bubble-like motion in cumulus clouds. *J. Meteor.,* **16,** 653–662.

Liebmann, B., and H. H. Hendon, 1990: Synoptic-scale disturbances near the equator. *J. Atmos. Sci.,* **47,** 1463–1479.

Lin, X., and R. H. Johnson, 1996: Kinematic and thermodynamic characteristics of the flow over the West Pacific Warm Pool during TOGA-COARE. *J. Atmos. Sci.,* **53,** 695–715.

Lucas, C., E. J. Zipser, and M. A. LeMone, 1994a: Convective available potential energy in the environment of oceanic and continental clouds: Correction and comments. *J. Atmos. Sci.,* **51,** 3829–3830.

———, ———, and ———, 1994b: Vertical velocity in oceanic convection off tropical Australia. *J. Atmos. Sci.,* **51,** 3183–3193.

Lumb, F. E., 1970: Topographic influences on thunderstorm activity near Lake Victoria. *Weather,* **25,** 404–409.

MacDonald, J. R., K. C. Mehta, and J. E. Minor, 1975: Development of windspeed risk model for the Argonne National Laboratory site. Institute for Disaster Research, Texas Tech University, 68 pp.

MacGorman, D. R., and D. W. Burgess, 1994: Positive cloud-to-ground lightning in tornadic storms and hailstorms. *Mon. Wea. Rev.,* **122,** 1671–1697.

Madden, R. A., and P. R. Julian, 1972: Description of global-scale circulation cells in the tropics with a 40–50 day period. *J. Atmos. Sci.,* **29,** 1109–1123.

———, and ———, 1994: Observations of the 40–50 day tropical oscillation—A review. *Mon. Wea. Rev.,* **122,** 814–837.

Maddox, R. A., 1976: An evaluation of tornado proximity wind and stability data. *Mon. Wea. Rev.,* **104,** 133–142.

———, 1980: Mesoscale convective complexes. *Bull. Amer. Meteor. Soc.,* **61,** 1374–1387.

———, 1983: Large scale meteorological conditions associated with mid-latitude mesoscale convective complexes. *Mon. Wea. Rev.,* **111,** 1475–1493.

Malkus, J. S., 1952: The slopes of cumulus clouds in relation to external wind shear. *Quart. J. Roy. Meteor. Soc.,* **78,** 530–542.

———, 1956: On the maintenance of the trade winds. *Tellus,* **8,** 335–350.

———, 1960: Penetrative convection and an application to hurricane cumulonimbus towers. *Cumulus Dynamics,* C. E. Anderson, Ed., Pergamon Press, 65–84.

———, and H. Riehl, 1964: *Cloud Structure and Distributions over The Tropical Pacific Oceans.* University of California Press, 229 pp.

Mandal, G. S., and M. L. Basandra, 1978: Tornado over Punjab. *Indian J. Meteor. Hydrol. Geophys.,* **29,** 547–554.

———, and S. K. Saha, 1983: Characteristics of some recent north Indian tornadoes. *Vayu Mandal,* **13,** 74–80.

Mapes, B. E., and R. A. Houze Jr., 1993: Cloud clusters and superclusters over the oceanic warm pool. *Mon. Wea. Rev.,* **121,** 1398–1415.

Marwitz, J. D., 1972: The structure and motion of severe hailstorms. Part I: Supercell storms. *J. Appl. Meteor.,* **11,** 166–179.

McBride, J. L., 1987: The Australian summer monsoon. *Monsoon Meteorology,* C. P. Chang and T. N. Krishnamurti, Eds., Oxford University Press, 203–231.

McCann, D. W., 1983: The enhanced-V: A satellite observable severe storm signature. *Mon. Wea. Rev.,* **111,** 887–894.

McCaul, E. W., Jr., 1987: Observations of the Hurricane "Danny" tornado outbreak of 16 August 1985. *Mon. Wea. Rev.,* **115,** 1206–1223.

———, 1991: Buoyancy and shear characteristics of hurricane-tornado environments. *Mon. Wea. Rev.,* **119,** 1954–1978.

McGuirk, J. P., and D. J. Ulsh, 1990: Evolution of tropical plumes in VAS water vapor imagery. *Mon. Wea. Rev.,* **118,** 1758–1766.

McPhaden, M. J., H. P. Freitag, S. P. Hayes, B. A. Taft, Z. Chen, and K. Wyrtki, 1988: The response of the equatorial Pacific to a westerly wind burst. *J. Geophys. Res.,* **C9** (93), 10 589–10 603.

Middleton, N. J., and Q. Z. Chaudhary, 1988: Severe dust storm at Karachi, 31 May 1986. *Meteor. Mag.,* **116,** 298–301.

Miller, D., and J. M. Fritsch, 1991: Mesoscale convective complexes in the western Pacific region. *Mon. Wea. Rev.,* **119,** 2978–2992.

Miller, F. R., and R. N. Keshavamurthy, 1968: Structure of an Arabian Sea summer monsoon system. *Int. Indian Ocean Expedition, Meteor. Monogr.,* East-West Center Press, 94 pp.

Miller, L. J., J. D. Tuttle, and C. A. Knight, 1988: Airflow and hail growth in a severe northern High Plains supercell. *J. Atmos. Sci.,* **45,** 736–762.

———, ———, and G. B. Foote, 1990: Precipitation production in a large Montana hailstorm: Airflow and particle growth trajectories. *J. Atmos. Sci.,* **47,** 1619–1646.

Miller, M. J., and A. K. Betts, 1977: Traveling convective storms over Venezuela. *Mon. Wea. Rev.,* **105,** 833–848.

Minor, J. E., and R. E. Peterson, 1979: Characteristics of Australian tornadoes. Preprints, *11th Conf. on Severe Local Storms,* Kansas City, MO, Amer. Meteor. Soc., 208–215.

Misra, P. K., and S. K. Prasad, 1980: Forecasting hailstorms over India. *Mausam,* **31,** 385–396.

Mohr, K. I., and E. J. Zipser, 1996: Defining mesoscale convective systems by their 85-GHz ice-scattering signatures. *Bull. Amer. Meteor. Soc.,* **77,** 1179–1189.

Molion, L. C. B., 1987: On the dynamic climatology of the Amazon Basin and associated rain-producing mechanisms. *The Geophysiology of Amazonia: Vegetation and Climate Interactions,* R. Dickerson, Ed., John Wiley, 391–407.

Morgan, G. M., 1982: Precipitation at the ground. *Hailstorms of the High Plains,* Vol. I, *The National Hail Research Experiment,* Colorado Associated University Press, 57–79.

Morrison, I., 1999: The structure and evolution of a kona low. M.S. thesis, Dept. of Meteor., University of Hawaii, 42 pp.

Mukherjee, A. K., and P. B. Bhattacharjee, 1972: The Diamond Harbor tornado. *Indian J. Meteor. Geophys.,* **23,** 227–230.

Mukherjee, B. K., S. K. Sharma, R. D. Reddy, and Bh. V. Ramaanamurty, 1983: A study of thundersqualls in Calcutta region. *Vayu Mandal,* **13,** 3, 4, 24.

Nandi, J., and A. K. Mukherjee, 1966: A tornado over northwest Assamand adjoining west Bengal on 19 April 1963. *Indian J. Meteor. Geophys.,* **17,** 421–426.

Nelson, S. P., and S. K. Young, 1979: Characteristics of Oklahoma hailfalls and hailstorms. *J. Appl. Meteor.,* **18,** 339–347.

Neumann, C. J., 1965: Mesoanalysis of a severe south Florida hailstorm. *J. Appl. Meteor.,* **4,** 161–171.

Newton, C. W., 1966: Circulations in a large sheared cumulonimbus. *Tellus,* **18,** 699–713.

Ninomiya, K., and T. Murakami, 1987: The early summer rainy season (Baiu) over Japan. *Monsoon Meteorology,* C. P. Chang and T. N. Krishnamurti, Eds., Oxford University Press, 93–121.

Nitta, T., and Y. Takayabu, 1985: Global analysis of the lower troposhere disturbances in the tropics during the northern summer of the FGGE year. Part II: Regional characteristics of the disturbances. *Pure Appl. Geophys.,* **123,** 272–292.

Nizamuddin, S., 1993: Hail occurrences in India. *Weather,* **48,** 90–92.

Norquist, D. C., E. E. Recker, and R. J. Reed, 1977: The energetics of African wave disturbances as observed during Phase III of GATE. *Mon. Wea. Rev.,* **105,** 334–342.

Novlan, D. J., and W. M. Gray, 1974: Hurricane spawned tornadoes. *Mon. Wea. Rev.,* **102,** 476–488.

Orton, R., 1970: Tornadoes associated with Hurricane Beulah on September 19–23, 1967. *Mon. Wea. Rev.,* **98,** 541–547.

Orville, R. E., and D. W. Spencer, 1979: Global lightning flash frequency. *Mon. Wea. Rev.,* **107,** 934–943.

———, and R. W. Henderson, 1986: Global distribution of midnight lightning: September 1977 to August 1978. *Mon. Wea. Rev.,* **114,** 2640–2653.

Ortiz, R., and R. Ortiz Jr., 1940: The Bejucal disaster. *Diaro de la Marina,* Deciembre 31, p. 1.

Oswaldo, G., 1985: *Atlas of Highly Reflective Clouds for the Global Tropics: 1971–1983.* U.S. Dept. of Commerce, 365 pp. [Available from U.S. Govt. Printing Office, Washington, DC 20402.]

Palmer, C. E., 1951a: On high-level cyclones originating in the tropics. *Trans. Amer. Geophys. Union,* **33,** 683–696.

———, 1951b: Tropical Meteorology. *Compendium of Meteorology,* Amer. Meteor. Soc., 859–880.

———, C. W. Wise, L. J. Stempson, and G. H. Duncan, 1955: *The Practical Aspect of Tropical Meteorology.* AWS Manual 105-48 Vol. 1, Air Wea. Service, Scott AFB, IL.

Paluch, I. R., 1979: The entrainment mechanisms in Colorado cumuli. *J. Atmos. Sci.,* **36,** 2467–2478.

Pandharinath, N., and V. Bhavanarayana, 1990: Hailstorm over Telangana. *Mausam,* **41,** 433–438.

Parsons, D. B., and R. A. Kropfli, 1990: Dynamics and fine structure of a microburst. *J. Atmos. Sci.,* **47,** 1674–1692.

Pearson, A. D., and A. F. Sadowski, 1965: Hurricane-induced tornadoes and their distribution. *Mon. Wea. Rev.,* **93,** 461–464.

Perez, A. H., L. J. Wicker, and R. E. Orville, 1997: Characteristics of cloud-to-ground lightning associated with violent tornadoes. *Wea. Forecasting,* **12,** 428–437.

Portig, W. H., 1963: Thunderstorm frequency and amount of precipitation in the tropics, especially in the African and Indian monsoon regions. *Arch. Meteor. Geophys. Bioklimatol.,* **13,** (Series B), 21–35.

Powell, M. D., S. H. Houston, P. Dodge, C. Samsury, and P. G. Black, 1998: Surface wind fields of 1995 Hurricanes Erin, Opal, Luis, Marilyn, and Roxanne at landfall. *Mon. Wea. Rev.,* **126,** 1259–1273.

Prasad, S. K., 1990: Tornado over Chapra and neighborhood in Bihar on 19 October 1987. *Mausam,* **41,** 496–499.

———, and B. C. Pawar, 1985: Climatological studies of thunderstorms to the west and east of the western Ghats in the state of Maharashtra and Goa: Part I. *Mausam,* **36,** 107–110.

Price, S., and R. I. Sasaki, 1963: Some tornadoes, waterspouts and other funnel clouds of Hawaii. *Mon. Wea. Rev.,* **91,** 175–190.

Przybylinski, R. W., 1995: The bow echo: Observations numerical simulations and severe weather detection methods. *Wea. Forecasting,* **10,** 203–218.
Ramage, C. S., 1962: The subtropical cyclone. *J. Geophys. Res.,* **67,** 1401–1411.
———, 1971: *Monsoon Meteorology.* Academic Press, 296 pp.
———, 1995: *Forecasters Guide to Tropical Meteorology.* Air Wea. Service, Scott Air Force Base, IL, 444 pp.
Ramanamurthy, Bh. V., 1983: Some cloud physical aspects of local severe storms. *Vayu Mandal,* **13,** 3–11.
Ramarkrishnan, K. P., and G. Rao, 1954: Squalls in India. *Indian J. Meteor. Geophys.,* **5,** 337.
Ramaswamy, C., 1956: On the subtropical jet stream and its role in the development of large scale convection. *Tellus,* **8,** 26–60.
Rao, Y. P., 1976: *Southwest Monsoon. Meteor. Mongr.,* No. 1, India Meteor. Dept.
Ray, T. K., and H. S. Bedi, 1985: The thermodynamic and kinematic structure of the troposphere over the Arabian Sea and the Bay of Bengal during the 1979 monsoon season. *Mausam,* **36,** 417–422.
Reap, R. M., and D. R. MacGorman, 1989: Cloud to ground lightning: Climatological characteristics and relationships to model fields, radar observations, and severe local storms. *Mon. Wea. Rev.,* **117,** 518–535.
Reddy, E. V. S., 1985: A study of squalls at Madras Airport. *Mausam,* **36,** 91–96.
Reed, R. J., and E. E. Recker, 1971: Structure and properties of synoptic-scale wave disturbances in the equatorial western Pacific. *J. Atmos. Sci.,* **28,** 1117–1133.
———, D. C. Norquist, and E. E. Recker, 1977: The structure and properties of African wave disturbances as observed during Phase III of GATE. *Mon. Wea. Rev.,* **105,** 317–333.
Reiter, E. R., 1961: *Jet-Stream Meteorology.* University of Chicago Press, 515 pp.
Riehl, H., 1954: *Tropical Meteorology.* McGraw-Hill, 392 pp.
———, 1979: *Climate and Weather in the Tropics.* Academic Press, 611 pp.
———, and J. S. Malkus, 1958: On the heat balance of the equatorial trough zone. *Geophysica,* **6,** 503–537.
———, J. S. Malkus, T.-C. Yeh, and N. E. LaSeur, 1951: The northeast trade of the Pacific Ocean. *Quart. J. Roy. Meteor. Soc.,* **77,** 598–626.
Rotunno, R., and J. B. Klemp, 1982: The influence of the shear-induced pressure gradient on thunderstorm motion. *Mon. Wea. Rev.,* **110,** 136–151.
———, and ———, 1985: On the rotation and propagation of simulated supercell thunderstorms. *J. Atmos. Sci.,* **42,** 271–292.
Roux, F., 1988: The West African squall line observed on 23 June 1981 during COPT 81: Kinematics and thermodynamics of the convective region. *J. Atmos. Sci.,* **45,** 406–426.
———, and S. Ju, 1990: Single-Doppler observations of a West African squall line on 27–28 May 1981 during COPT 81: Kinematics, thermodynamics, and water budget. *Mon. Wea. Rev.,* **118,** 1826–1854.
———, J. Testud, M. Payen, and B. Pinty, 1984: West African squall-line thermodynamic structure retrieved from dual-Doppler radar observations. *J. Atmos. Sci.,* **41,** 3104–3121.
Rutledge, S. A., C. Lu, and D. R. MacGorman, 1990: Positive cloud-to-ground lightning in meoscale convective systems. *J. Atmos. Sci.,* **47,** 2085–2100.
Sadler, J. C., 1975a: The upper tropospheric circulation over the global tropics. University of Hawaii Rep. UHMET-75-05, 35 pp.
———, 1975b: The monsoon circulation and cloudiness over the GATE area. *Mon. Wea. Rev.,* **103,** 369–387.
———, 1976: A role of the tropical upper tropospheric trough in the early season typhoon development. *Mon. Wea. Rev.,* **104,** 1266–1278.
———, M. Lander, A. M. Hori, and L. K. Oda, 1987: *Tropical Marine Climatic Atlas.* Dept. of Meteor., University of Hawaii.
Salby, M. L., and H. H. Hendon, 1994: Intraseasonal behavior of clouds, temperature, and motion in the tropics. *J. Atmos. Sci.,* **51,** 2207–2224.
Sanders, F., 1984: Quasi-geostrophic diagnosis of the monsoon depression of 5–8 July 1979. *J. Atmos. Sci.,* **41,** 538–552.
———, and J. C. Freeman, 1986: Thunderstorms at sea. *Thunderstorm Morphology and Dynamics,* E. Kessler, Ed., University of Oklahoma Press, 41–58.
Schlesinger, R. E., 1978: A three-dimensional numerical model of an isolated thunderstorm. Part I: Comparative experiments for variable ambient wind shear. *J. Atmos. Sci.,* **35,** 690–713.
———, 1980: A three-dimensional numerical model of an isolated thunderstorm. Part II: Dynamics of updraft splitting and mesovortex couplet evolution. *J. Atmos. Sci.,* **37,** 395–420.
Schroeder, T. A., 1976: Hawaiian waterspout-tornado of 26 November 1975. *Weatherwise,* **29,** 172–177.
———, 1983: The subtropical jet stream and severe local storms—a view from the tropics. Preprints, *13th Conf. on Severe Local Storms,* Tulsa, OK, Amer. Meteor. Soc., 161–162.
Schultz, D. M., W. E. Bracken, L. F. Bosart, G. J. Hakim, M. A. Bedrick, M. J. Dickinson, and K. R. Tyle, 1997: The 1993 superstorm cold surge: Frontal structure, gap flow, and tropical impact. *Mon. Wea. Rev.,* **125,** 5–39; Corrigendum, **125,** 662.
———, ———, and ———, 1998: Planetary and synoptic-scale signatures associated with Central American cold surges. *Mon. Wea. Rev.,* **126,** 5–27.
Schwarzkopf, M. L., 1982: Severe storms and tornadoes in Argentina. Preprints, *12th Conf. on Severe Local Storms,* San Antonio, TX, Amer. Meteor. Soc., 59–62.
Senn, H. V., C. L. Courtright, and H. W. Hiser, 1969: Preliminary characteristics of South Florida tornadoes in the summer of 1968. Preprints, *Sixth Conf. on Severe Local Storms,* Chicago, IL, Amer. Meteor. Soc., 192–196.
Sharp, D. W., A. J. Cristaldi, S. M. Spratt, and B. C. Hagemeyer, 1998: Multifaceted general overview of the east central Florida tornado outbreak of 22–23 February, 1998. Preprints, *19th Conf. on Severe Local Storms,* Minneapolis, MN, Amer. Meteor. Soc., 140–143.
Simpson, J. S., 1976: Report on the hail suppression program at Nelspruit, Transvaal, republic of South Africa. NCAR Tech Rep. 7100-76/5, 85 pp.
———, and V. Wiggert, 1971: 1968 Florida cumulus seeding experiment: Numerical models results. *Mon. Wea. Rev.,* **99,** 87–118.
———, and M. McCumber, 1982: Three-dimensional simulations of cumulus congestus clouds on GATE day 261. *J. Atmos. Sci.,* **39,** 126–145.
———, B. R. Morton, M. C. McCumber, and R. S. Penc, 1986: Observations and mechanisms of GATE waterspouts. *J. Atmos. Sci.,* **43,** 753–782.
Simpson, R. H., 1952: Evolution of the Kona Storm: A subtropical cyclone. *J. Meteor.,* **9,** 24–35.
Singh, R., 1981: On the occurrence of tornadoes and their distribution in India. *Mausam,* **32,** 307–314.
Sivaramakrishnan, T. R., 1989: An analytical study of thunderstorms over Sriharikota. *Mausam,* **40,** 489–491.
Smith, R. K., N. Crook, and G. Roff, 1982: The morning glory: An extraordinary atmospheric undular bore. *Quart. J. Roy. Meteor. Soc.,* **108,** 937–956.
———, M. J. Coughlan, and J.-L. Lopez, 1986: Southerly nocturnal wind surges and bores in northeastern Australia. *Mon. Wea. Rev.,* **114,** 1501–1518.
———, N. J. Tapper, and D. R. Christie, 1995: Central Australian cold fronts. *Mon. Wea. Rev.,* **123,** 16–38.
Smith, W. H., 1947: A series of waterspouts in the Straits of Singapore. *Weather,* **2,** 235.
Spratt, S. M., D. W. Sharp, P. Welsh, A. Sandrik, F. Alsheimer, and C. Paxton, 1997: A WSR-88D assessment of tropical cyclone outer rainband tornadoes. *Wea. Forecasting,* **13,** 479–501.

Srivastava, R. C., 1985: A simple model of evaporatively driven downdraft: Application to microburst downdraft. *J. Atmos. Sci.,* **42,** 1004–1023.

Stull, R. B., 1988: *An Introduction to Boundary Layer Meteorology.* Kluwer Academic, 666 pp.

Szoke, E. J., and R. H. Brady, 1989: Forecasting implications of the 26 July 1985 northeastern Colorado tornadic thunderstorm case. *Mon. Wea. Rev.,* **117,** 1834–1860.

Takahashi, T., 1987: Hawaiian hailstones—30 January 1985. *Bull. Amer. Meteor. Soc.,* **68,** 1530–1534.

Tao, W.-K., J. Simpson, and S.-T. Soong, 1991: Numerical simulation of a subtropical squall line over the Taiwan Strait. *Mon. Wea. Rev.,* **119,** 2699–2723.

Testud, J., and M. Chong, 1983: Three-dimensional wind field analysis from dual-Doppler radar data. Part I: Filtering, interpolating and differentiating the raw data. *J. Climate Appl. Meteor.,* **22,** 1204–1215.

Thompson, R. M., Jr., S. W. Payne, E. E. Recker, and R. J. Reed, 1979: Structure and properties of synoptic-scale wave disturbances in the intertropical convergence zone of the eastern Atlantic. *J. Atmos. Sci.,* **36,** 53–72.

Uccellini, L. W., and D. R. Johnson, 1979: The coupling of upper and lower tropospheric jet streaks and implications for the development of severe convective storms. *Mon. Wea. Rev.,* **107,** 682–703.

Velasco, I., and J. M. Fritsch, 1987: Mesoscale convective complexes in the Americas. *J. Geophys. Res.,* **92,** 9591–9613.

Vincent, D. G., 1982: Circulation features over the South Pacific during 10–18 January 1979. *Mon. Wea. Rev.,* **110,** 981–993.

———, 1985: Cyclone development in the South Pacific Convergence Zone during FGGE, 10–17 January 1979. *Quart. J. Roy. Meteor. Soc.,* **111,** 155–172.

———, 1994: The South Pacific Convergence Zone (SPCZ): A review. *Mon. Wea. Rev.,* **122,** 1949–1970.

Visher, S. S., 1922: Hail in the tropics. *Bull. Amer. Meteor. Soc.,* **3,** 117–118.

Viswanadham, S., and A. M. Sud, 1991: Anticyclonic tornado near Raipur in southeast Madhya Pradesh. *Mausam,* **42,** 430–431.

Wakimoto, R. M., 1985: Forecasting dry microburst activity over the High Plains. *Mon. Wea. Rev.,* **113,** 1131–1143.

———, and V. N. Bringi, 1988: Dual-polarization observations of microbursts associated with intense convection: The 20 July storm during the MIST project. *Mon. Wea. Rev.,* **116,** 1521–1539.

———, and J. W. Wilson, 1989: Non-supercell tornadoes. *Mon. Wea. Rev.,* **117,** 1113–1140.

Wales-Smith, B. G., 1969: A waterspout at Singapore. *Weather,* **24,** 348–354.

Wallace, J. M., 1971: Spectral studies of tropospheric wave disturbances in the tropical western Pacific. *Rev. Geophys. Space Phys.,* **9,** 557–612.

Wang, G. C. Y., 1979: Tornadoes in Taiwan. Preprints, *11th Conf. on Severe Local Storms,* Kansas City, MO, Amer. Meteor. Soc., 216–221.

Weisman, M. L., 1993: The genesis of severe, long-lived bow echoes. *J. Atmos. Sci.,* **50,** 645–670.

———, and J. B. Klemp, 1982: The dependence of numerically simulated convective storms on vertical wind shear and buoyancy. *Mon. Wea. Rev.,* **110,** 504–520.

———, and ———, 1984: The structure and classification of numerically simulated convective storms in directionally varying shears. *Mon. Wea. Rev.,* **112,** 2479–2498.

Whitfield, M. B., and S. W. Lyons, 1992: An upper tropospheric low over Texas during summer. *Wea. Forecasting,* **7,** 89–106.

Williams, E., and N. Renno, 1993: An analysis of the conditional instability of the tropical atmosphere. *Mon. Wea. Rev.,* **121,** 21–36.

Williams, E. R., M. E. Weber, and R. E. Orville, 1989: The relationship between the lightning type and state of convective development of thunderclouds. *J. Geophys. Res.,* **94,** 13 213–13 220.

———, S. A. Rutledge, S. G. Geotis, N. Renno, E. Rasmussen, and T. Rickenbach, 1992: A radar and electrical study of tropical "hot towers." *J. Atmos. Sci.,* **49,** 1386–1395.

Willoughby, H. E., D. P. Jorgensen, R. A. Black, and S. L. Rosenthal, 1985: Project STORMFURY: A scientific chronicle 1962–1983. *Bull. Amer. Meteor. Soc.,* **66,** 505–514.

Wilson, J. W., 1986: Tornadogenesis by nonprecipitation induced wind shear lines. *Mon. Wea. Rev.,* **114,** 270–284.

WMO, 1953: *World Distribution of Thunderstorm Days.* WMO No. 21, TP. 6 and supplement.

Wolfsen, M., 1983: Doppler radar observations of an Oklahoma downburst. Preprints, *21st Conf. on Radar Meteorology,* Edmonton, AB, Canada, Amer. Meteor. Soc., 590–595.

Wolford, L. V., 1960: Tornado occurrences in the United States. USWB. Tech. Paper 20, Washington, DC, 71 pp. [Available from National Technical Information Service, 5285 Port Royal Road, Springfield, VA 22161.]

Xu, K.-M., and K. A. Emanuel, 1989: Is the tropical atmosphere conditionally unstable? *Mon. Wea. Rev.,* **117,** 1471–1479.

Yanai, M., 1963: A preliminary survey of large-scale disturbances over the tropical Pacific region. *Geofys. Int.,* **3,** 73–84.

Yates, A. L., and B. N. Hanstrum, 1996: Organized severe convection in the outer convergence bands of tropical cyclones in northwestern Australia. Preprints, *Fifth Australian Severe Thunderstorm Conf.,* Avoca Beach, NSW, Australia, Bureau of Meteorology.

Zipser, E. J., 1969: The role of organized unsaturated convective downdrafts in the structure and rapid decay of an equatorial disturbance. *J. Appl. Meteor.,* **8,** 799–814.

———, 1977: Mesoscale and convective-scale downdrafts as distinct components of squall-line structure. *Mon. Wea. Rev.,* **105,** 1568–1589.

———, 1994: Deep cumulonimbus cloud systems in the Tropics with and without lightning. *Mon. Wea. Rev.,* **122,** 1837–1851.

———, and M. A. LeMone, 1980: Cumulonimbus vertical velocity events in GATE. Part II: Synthesis and model core structure. *J. Atmos. Sci.,* **37,** 2458–2469.

———, and K. Lutz, 1994: The vertical profile of radar reflectivity of convective cells: A strong indicator of storm intensity and lightning probability? *Mon. Wea. Rev.,* **122,** 1751–1759.

———, R. J. Meitin, and M. A. LeMone, 1981: Mesoscale motion fields associated with a slowly moving GATE convective band. *J. Atmos. Sci.,* **38,** 1725–1750.

Chapter 11

Severe Local Storms Forecasting

ALAN R. MOLLER

National Weather Service, Fort Worth, Texas

REVIEW PANEL: Robert H. Johns (Chair), John Monteverdi, Kevin Pence, Ron Przybylinski, Ed Szoke, and Steve Weiss

11.1. Introduction

Atmospheric hazards are the deadliest of all natural disasters in the United States (White and Haas 1975; Petak and Atkisson 1982). Local convective storm–induced severe weather and floods result in approximately 400 fatalities annually, with excessive heat and cold accounting for about 600 deaths and hurricanes for more than 30 fatalities each year (Riebsame et al. 1986). Convective storm mortality rates have stabilized, or even decreased, in the United States and other industrialized countries, but they have generally increased in developing nations.

Economic losses from convective storms have risen in most countries, including the United States (Kates 1980). Riebsame et al. (1986) estimate that American severe storms and floods result in between two and five billion dollars of damage annually. Furthermore, over 80% of the presidential disaster declarations in the last several decades resulted from severe storms and flooding. The potential for loss of life and property from convective storms is enormous (Table 11.1); hence the National Weather Service (NWS) has no greater responsibility than the preparation and dissemination of information about impending severe weather.

The NWS has defined local convective storms to be *severe* when they contain one or more of the following phenomena:

(a) tornadoes;
(b) damaging winds, or gusts greater than or equal to 26 m s^{-1} (50 kt); and
(c) hail size greater than or equal to 1.9 cm (¾ in.) in diameter.

This paper focuses on these threats to life and property. Flash flood forecasting is handled elsewhere in this volume.

Johns and Doswell (1992, hereafter JD92) describe severe local storm forecasting as consisting of two parts: anticipation of severe weather potential in the storm environment and recognition of severe convective storms once they form. Doswell et al. (1993, hereafter DWJ93) similarly assess tornado forecasting, explaining that anticipation is a forecast issue whereas recognition of severe storms is a warning concern. The JD92 and DWJ93 papers review severe local storm and tornado forecasting from the national perspective of the Storm Prediction Center (SPC).[1] These papers describe the evolution of understanding that is shifting severe storm forecasting (i.e., the process that results in the issuance of SPC outlooks, watches, and mesoscale updates) from the empirical into the scientific realm.

Local NWS weather forecast offices (WFOs) disseminate warnings, statements, and short-term forecasts for imminent severe thunderstorm and tornado threats. Ideally, warnings are issued prior to severe weather occurrence, thus minimizing the loss of life and property. However, the NWS issues many warnings in response to reports of existing severe weather. Research efforts have broadened our knowledge of convective storm radar morphology and evolution since the 1963 severe local storm monograph (Burgess and Lemon 1990; Burgess et al. 1993). Furthermore, the completion of the WSR-88D radar network (Crum and Alberty 1993; Klazura and Imy 1993) provides the NWS with the most powerful and accurate severe storm detection tool in operational use. The improved science and technology give the NWS an unprecedented opportunity to increase warning proficiency.

This paper reviews current severe local storm forecast and detection (i.e., events that trigger severe local storm warnings) procedures from the perspective of an NWS WFO. Moreover, the discussion includes severe weather information dissemination and public preparedness, because even the best forecasts and warnings are ineffective without well-written severe weather products and appropriate public response (Mi-

[1] The Storm Prediction Center (SPC) was previously known as the National Severe Storms Forecast Center (NSSFC).

TABLE 11.1. Selected, notable worldwide severe local storm events (tornadoes, hail, and windstorms) 1970–99. Damage given in terms of 1998 U.S. dollars.

Date	Location	Event	Deaths	Damage
11 Sep 1970	Gulf of Venice, Italy	Tornado	47	
21 Feb 1971	Mississippi	Tornadoes	119	$82 million
10 Jan 1973	San Justo, Argentina	Tornado	50	
30 Jan 1974	Pago Pago, Samoa	Microburst[a]	96	
3 Apr 1974	Southern, eastern United States; southern Canada	Tornadoes	315	$2.2 billion
24 Jun 1975	New York City	Microburst[a]	112	
1 Apr 1977	Madaripur, Bangladesh	Tornadoes	500	
4 Apr 1977	New Hope, Georgia	Hail[b]	68	
2 Jun 1978	Jablah Coast, Syria	Hail[c]		
10 Apr 1979	Wichita Falls area	Tornadoes	60	$1.2 billion
9 Jul 1982	New Orleans	Microburst[a]	152	
28 Mar 1984	Georgia, South Carolina, North Carolina	Tornadoes	57	$80 million
9 Jun 1984	Belyanitsky, Ivanova, and Balino, USSR	Tornadoes	400	
31 May 1985	Ohio, Pennsylvania, Ontario	Tornadoes	88	$275 million
2 Aug 1985	Dallas-Fort Worth	Microburst[a]	137	
13 Apr 1986	Brahmabaria, Bangladesh	Hail	92	
31 Jul 1987	Edmonton, Alberta, Canada	Tornado	27	$548 million
26 Apr 1989	North-central Bangladesh	Tornado	1300	
11 Jul 1990	Denver	Hail		$800 million
8 May 1992	Sichuan Province, China	Hail	25	
19 Apr 1995	Guangzhou Province, China	Hail/tornado	37	
5 May 1995	Dallas-Fort Worth		20[e]	$2 billion
15 Jul 1995	New York, Vermont, New Hampshire, Massachusetts, Maine	Derecho[d]	6	$275 million
13 May 1995	Tangail, Bangladesh	Tornado	700	
23 Feb 1998	Florida	Tornadoes	42	$106 million
24 Mar 1998	Midnapore, India	Tornado	200	
15 Dec 1998	Umtatay, South Africa	Tornadoes	22	
18 Jan 1999	Mount Ayliff, South Africa	Tornadoes	21	
14 Apr 1999	Sydney, Australia	Hail		$1 billion
3 May 1999	Oklahoma/Kansas	Tornadoes	48	$1.2 billion

[a] Aircraft accidents resulting from microbursts.
[b] Aircraft accident resulting from hail ingestion in engines.
[c] Hail destroyed crops from over 40 villages; thousands faced possible starvation.
[d] Over 1 billion board feet of timber destroyed in more than 1 million affected acres of timber.
[e] One lightning and 19 flash flood deaths; 175 hail injuries.

leti and Sorenson 1990). The elements of forecasting, detection, dissemination, and public response comprise an integrated warning system (IWS; Leik et al. 1981). The success (or failure) of the IWS depends on the functional synergy of these four components.

In section 11.2, we discuss forecast philosophy and provide an overview of severe weather forecast techniques. The meteorological processes that produce deep, moist convection are defined, and the environmental factors that determine (or influence) the occurrence of severe convective storm events are discussed. The second IWS element, severe local storm detection, composes section 11.3. Important connections between successful forecasts and subsequent detection of severe local storms are stressed. Further, the changing, though ever complementary, warning roles of the WSR-88D radar network and storm spotters are addressed. Section 11.4 is a discussion of multiple agency preparedness planning that elicits the most favorable public response. We review the multitiered severe weather dissemination system (i.e., outlooks, watches, warnings, statements, short-range forecasts), as applied by the NWS, news media, and local emergency preparedness officials, in section 11.5. The various severe storm information needs of the three main IWS user groups are described in this section. A summary discussion of the IWS is presented in section 11.6.

11.2. Severe local storm forecasting

a. Forecast philosophy and methodology

Weather forecasting consists of combining a diagnosis with a prognosis (Doswell 1986a,b). Forecasters initially diagnose the current atmospheric state by analyzing various weather processes and integrating their understanding of these processes into a coherent picture of the weather. The diagnosis is then combined with numerical model output to create a trend (prognosis), and forecasters use their knowledge of atmospheric processes to develop a feedback between the ongoing diagnosis and forecast steps (Doswell 1986a). Numerical prediction models also start with a diagnosis and develop a trend. The major distinction between numerical models and human forecasters is the lack of feedback with the former. Each time a model is run, it is restricted to the same linear methodology of variable input/output, equally weighing the four-dimensional data.

The human forecast process is more complicated. Humans use pattern recognition to assimilate the complexity of the forecast (Allen 1981). This process combines all of the available data (regardless of their relevance to the numerical model governing equations), experience, theory, and concepts into a four-dimensional image of the atmosphere (Doswell 1986b). Doswell further elucidates that when armed with basic physical understanding, the human forecaster can make correct assessments and predictions even with limited data—something no strictly objective approach can accomplish. Of course, if the human forecaster does not possess the scientific understanding, the intuitive human process will produce terribly incorrect predictions! Neural networks give promise that machines can make some "intuitive" decisions, but results so far suggest that artificial intelligence will augment, rather than replace, human decision making in the forecast process (e.g., Stumpf et al. 1998).

Diagnosis becomes increasingly more important than numerical model prognosis for diminishing lead times (i.e., day 1 severe weather; McGinley 1986; DWJ93). Numerical forecast models are the primary input for longer-term forecasting (day 2 severe weather). The diagnostic routine includes surface and sounding data analyses, used in conjunction with remotely sensed data from satellite, radar, profilers, and lightning detection systems. Forecasters must also integrate local climatology and the effects of terrain (mountains, valleys, oceans, and lakes) into the diagnosis.

Many forecasters begin their daily diagnosis with careful hand analysis of upper-air and surface charts. Those who engage in hand analysis agree with Doswell (1986b) that the process enhances their comprehension of the data. Moreover, when forecasters blend extensive diagnostic work with a pursuit of physical understanding, they are more likely to develop valid conceptual models (four-dimensional atmospheric images). Daily diagnosis eventually provides the operational testing and revision process required to improve these models (Doswell and Maddox 1986). A working diagnostic routine produces additional benefits. Once forecasters develop diagnostic skills, they can more correctly assess the accuracy of a numerical model's initial conditions (Hales 1979), then adjust the numerical model's output and determine subsequent effects on the local sensible weather. While it may not be necessary for all forecasters to hand-analyze maps in order to make a valid diagnosis, forecasters must devise a diagnostic routine (see Doswell 1982 for a thorough discussion of the diagnosis of preconvective environments).

The era of rapid technological advance greatly benefits severe weather forecasters. The surface analysis routine is streamlined by use of objectively analyzed parameters such as pressure, moisture, temperature, equivalent potential temperature, virtual potential temperature gradients, and moisture convergence. Derived numerical model parameters (e.g., inflow-level moisture convergence and ageostrophic vertical circulations forced by upper-level jet streaks and frontogenesis) can enhance the forecasting of convective storm initiation and severe weather potential. The accuracy of model-forecasted vertical motions can often be evaluated by quasigeostrophic theory, through the use of Q-vectors (Hoskins et al. 1978; Barnes and Colman 1994). Mesoscale models (horizontal grid spacing on the order of 25 km), used in conjunction with nonhydrostatic numerical cloud models (horizontal grid spacing of about 1 km), can in principle be run every few hours with new data (Brooks et al. 1993). Forecasters who are familiar with mesoscale numerical model characteristics (e.g., convective parameterization schemes; Cortinas and Stensrud 1995) can evaluate model performance by making comparisons against observations. Ensemble mesoscale and cloud-scale model runs, made with a number of varying initial conditions, may give forecasters an idea of the spectrum of possible solutions on a severe storm day (Brooks et al. 1992).

The new technologies and numerical models can be misapplied. Forecasters occasionally become so transfixed by the visual esthetics of the numerical model data, or so impressed by model performance, that they rush into the prognosis phase before making a thorough diagnosis (e.g., Doswell and Maddox 1986; Snellman 1991). The author's most regrettable severe storm forecast mistakes have arisen from ignoring data that were relevant to the daily diagnosis and/or failing to complete the diagnosis on what initially appeared to be a benign weather day.

The great forecast challenge is to assimilate great quantities of weather data, conceptualize a model of the environment, and extrapolate forward in time (McGinley 1986). Indeed, the copious amount of data passing through a modernized and associated restructured (MAR) NWS office (or private sector weather forecast office) already produces data overload. Forecasters must know (through education, training, and experience) which information to extract from the data stream on a given day. The forecasting agency, through research and development, must provide forecasters with the tools that enhance their ability to apply scientific principles to the task (Doswell and Maddox 1986). Successful severe weather forecasters must sustain a passion for learning and a penchant for hard work throughout their careers (Miller 1972).

b. The severe local storm forecast challenge

There are four concerns that a severe weather forecaster must address (McNulty 1995). The first is that of thunderstorm potential within the spatial and temporal domain of the forecast. If there is even a small chance that thunderstorms will occur, forecasters should next use their knowledge of environmental and convective storm interactions to determine a range of potential thunderstorm types. This information allows the forecaster to ascertain overall severe thunderstorm potential and, finally, which of the severe storm phenomena (and/or flash flood–producing rainfall) are possible.

The forecast process used by SPC involves parameter evaluation, pattern recognition,[2] and climatology (DWJ93). Parameter evaluation and chart pattern recognition change as more is learned about storm-scale processes and the interactions between the storm-scale and the large-scale environment (JD92). Perhaps chart pattern recognition, which has evolved as "particular collections of elements are associated with particular groups of events" (DWJ93), is most commonly used to describe soundings and synoptic map patterns that have been associated with major severe weather episodes (e.g., Miller 1972). When chart pattern recognition (or any of the other methodologies) is used in isolation, serious false alarm and probability of detection errors result. For instance, rare events, such as upper Midwest tornado outbreaks in January, will be poorly forecast if climatology is used without evaluation of the relevant parameters. Chart pattern recognition that is backed by scientifically based parameter evaluation (e.g., assessment of vertical wind shear profiles and instabilities when analysis suggests a possible supercell tornado outbreak) maximizes the scientific forecast process. As our knowledge increases, forecasters will eventually advance into what has been called *ingredients-based forecasting* (Doswell et al. 1996).

The remaining forecast discussion blends an SPC approach, particularly the use of scientific parameter evaluation, with McNulty's (1995) methodology. The reader is referred to Doswell (1982), DWJ93, JD92, and McNulty (1995) for further discussion of severe local storm forecast techniques, and to Schaefer (1986) for a historical perspective on severe weather forecasting.

c. Convective storm initiation and movement

There are three necessary ingredients to achieve deep moist convection (Doswell 1987):

1. a moist layer of sufficient depth in the low or midtroposphere,
2. a sufficiently steep lapse rate to allow for a substantial "positive area," and
3. sufficient lifting of a parcel from the moist layer to allow it to reach its level of free convection (LFC).

[2] The SPC expression refers to associative relationships between map (and sounding) patterns and events such as tornadoes. Thus, we will use the term *chart pattern recognition,* to avoid confusion with the use of *pattern recognition* in section 11.2a.

1) CONVECTIVE STORM INSTABILITY

Forecasters use soundings of temperature lapse rates and moisture to determine the stability of the thermodynamic stratification. A full sounding analysis should be completed, rather than a single-level, index-based analysis (e.g., the lifted index; Galway 1956), since indices may be unrepresentative of the overall static stability (JD92). Techniques that use more of the information in a sounding have been developed, such as the skew T–log$p$ Hodograph Analysis and Research Program (called SHARP, Fig. 11.1) (Hart and Korotky 1991). These procedures incorporate the old single-level indices and convective available potential energy (CAPE; Moncrieff and Miller 1976), a value of potential buoyancy that is integrated through the positive area from the LFC to the equilibrium level (EL). After forecasters diagnose the current state of instability, they must predict subsequent changes resulting from differential temperature and moisture advection and vertical motions (JD92). Interactive programs such as SHARP enable forecasters to modify soundings and attempt to account for expected spatial and temporal variations in the storm environment. Since radiosondes are released only twice a day, accurately deducing the stability structure between sounding times (0000 and 1200 UTC) is a formidable task.

When calculating CAPE, an important consideration concerns which parcel to lift—the surface parcel, the parcel with the greatest CAPE, or a parcel that is averaged through some well-mixed layer (Williams and Renno 1993). Doswell and Rasmussen (1994) suggest using the most unstable parcel in the lowest 300 mb of a sounding as the standard for CAPE. This allows for consistency when comparing soundings and is applicable when surface-based parcels are clearly inappropriate (e.g., with elevated convection—updrafts not rooted in the surface layer—which usually occurs at night and/or on the cool side of a low-level thermal boundary). However, using a "most unstable" point parcel, as opposed to a mixed-layer parcel, can be unrepresentative at times (B. Johns 1999, personal communication). If consistency is not always possible, meteorologists should at least explain how they calculate a particular value of CAPE.

CAPE can be an overly simplistic measure of instability (Zipser and LeMone 1980). For instance, Lucas et al. (1994a,b) show that oceanic convection often has weaker vertical velocities than continental convection for environments with equivalent CAPEs. They note that the positive area on oceanic soundings is generally "thin" with small instability, but is maintained through a large fraction of the atmosphere. The positive area on land soundings can often be described as "short and wide" with large instability, but is spread over a shallower depth of the atmosphere. Blanchard (1998) notes that CAPE is a vertically integrated measure of parcel buoyant energy (with units of

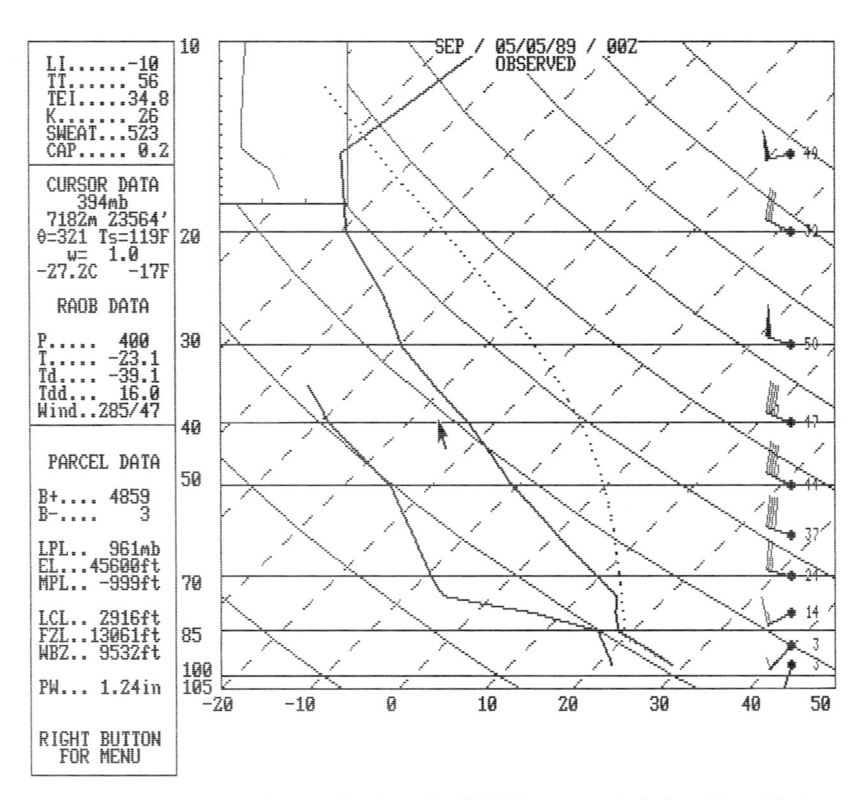

FIG 11.1. Thermodynamic sounding from the SHARP program, including (dotted line) an ascent curve for a surface parcel at 0000 UTC 5 May 1989 from Stephenville, TX (from Moller et al. 1994).

J kg^{-1} rather than degrees), and that comparisons of CAPE with standard instability indices such as the lifted index show only moderate correlations. Further, as suggested by the work of Lucas, and that of Zipser and LeMone, CAPE is sensitive to both the magnitude of buoyancy and the depth of integration. Blanchard suggests that a normalized CAPE [CAPE divided by the depth of the free convective layer (FCL)] can account for the variations in CAPE attributable to differences in the depth of the FCL. When evaluating an atmosphere where CAPE is characterized by a low EL and low-topped convection, normalized CAPE (NCAPE) appears to be more important than total CAPE (Davies 1993b; Wicker and Cantrell 1996, hereafter WC96).

Vertical motions produced by midlevel baroclinic disturbances act to steepen midlevel lapse rates (Doswell et al. 1985). This destabilization process is augmented by surface heating on the elevated Mexican Plateau (Carlson et al. 1983) and the east slopes of the Rocky Mountains (Doswell et al. 1985), often producing lapse rates exceeding 7.5°–8.0°C km^{-1} in the 700- to 500-mb layer. The unstable and capped "loaded gun" sounding often associated with severe weather is formed when winds advect this air mass and/or a traveling midlevel disturbance carries the destabilization process over an area of abundant low-level moisture. This usually occurs over the plains, but on occasion the steep lapse rate, elevated mixed-layer (EML; Lanicci and Warner 1991) air can travel as far from the source region as New England (Johns and Dorr 1996). Forecasters can anticipate sudden changes in CAPE by analyzing midlevel lapse rates, the quality and depth of low-level moisture (elevation-sensitive variables such as equivalent potential temperature should be used), and the proximity of these parameters to each other from day to day.

2) CONVECTIVE STORM INHIBITION

(i) Convective inhibition energy

Convective inhibition (CIN), a value of static stability, is integrated through the negative area of the sounding. In a convective storm situation, CIN is usually relegated to the capping inversion [i.e., the negative area through which a parcel must ascend to reach the level of free convection (LFC); see Colby 1984; Bluestein 1993].

Capping inversions are among the most difficult of atmospheric phenomena to forecast. Subtle vertical motions over a period of hours (not routinely observed and poorly handled by numerical models) cause substantial changes in the configuration of the cap, often making the difference between no storms and significant severe weather. When the capping inversion is weak or nonexistent, widespread deep moist convection (DMC) can form before large potential buoyant energy builds, reducing the probability of severe storm development. Weak CIN occurring with large initial CAPE can result in early formation of severe storms, which might then suppress additional storm formation. Thus, CIN cannot be handled as an independent parameter.

When the cap is too strong, DMC is suppressed (Carlson and Ludlam 1968). When expected convection does not occur, an unforeseen cap is frequently involved. For instance, Stensrud and Maddox (1989) document a case where well-defined, merging outflow boundaries from two mesoscale convective systems (MCS) (Zipser 1982) failed to produce additional convection in a very unstable atmosphere. The apparent cause was interacting MCS vertical circulations, which resulted in unfavorable midlevel convergence (i.e., the descending branch of a coupled vertical circulation) above the intersecting boundaries and subsequent strengthening of the capping inversion.

Careful analysis of surface variables such as visibility, temperature, and dewpoint tendencies can infer the presence of capping inversions (McGinley 1986). Rapid temperature rises imply shallow or capped layers, and a decrease in visibilities may indicate an increasing cap and/or increasing moisture. Once the cap is broken, deep mixing results in slower temperature increases and falling dewpoints. Strong west to east increases in temperature change, not related to advection, may indicate a breaking capping inversion and often delineate where DMC will develop (McGinley et al. 1987).

(ii) Other factors suppressing convection

Factors other than a capping inversion can result in a "nonevent." On a high-risk day during the verification of rotation in tornadoes experiment (VORTEX) (Rasmussen et al. 1994), a vertical circulation along a dryline was too shallow to support DMC in a strongly sheared environment. Rising, uncapped parcels detrained into dry westerly flow before reaching either the lifting condensation level (LCL) or the LFC (Ziegler and Rasmussen 1998). Events such as this are poorly understood and possibly not even detectable much of the time. Detailed studies of convective initiation are needed to improve our understanding of such events.

3) CONVECTIVE STORM LIFTING MECHANISMS

The lift needed to initiate deep, moist convection is provided by mesoscale and storm-scale processes. Synoptic-scale baroclinic disturbances do not act as DMC "triggers"; rather, they act to destabilize the thermodynamic structure and weaken the capping inversion through slow, relatively weak lift (Doswell 1987). Convergence along low-level boundaries (e.g., convective outflow, frontal, sea or lake breeze, and dryline boundaries) results in the mesoscale lift that deepens the moist layer, erases any remaining cap, and allows thermals to reach the LFC (McGinley 1986; Doswell 1987).

Purdom (1982) used satellite imagery to demonstrate the association of boundary-layer convergence lines and DMC initiation. Wilson and Schreiber (1986) show that more than 80% of Denver-area thunderstorms form in close proximity to radar-detected boundaries (Fig. 11.2). The WSR-88D low-level reflectivity and spectrum width products can reveal boundaries within a range of about 100 km. Radar should be used in conjunction with surface data and satellite imagery to locate and track boundaries.

Because of their uniquely high temporal and spatial resolution, surface data are important in forecasting DMC initialization. However, forecasters must be aware of surface data limitations when the surface parcel is not representative of the actual lifted parcel, or when other processes above the surface affect convective potential (e.g., vertical mixing that results in depletion of rich but shallow moisture). Forecasters monitor DMC potential through evaluation of CAPE, CIN, surface moisture convergence, pressure (altimeter) changes, temperature advection, and potential temperature advection in programs that use objective analysis of hourly surface data [e.g., the ADAP program, (Bothwell 1988) and the more recently deployed Local Analysis and Prediction System (LAPS; McGinley et al. 1991), which is used in the NWS AWIPS system]. Moisture flux convergence is especially important in short-term forecasting since it usually precedes DMC formation by a few hours (Hudson 1971). DMC becomes more likely when peak values within a moisture convergence center increase rapidly over a period of several hours.

Forecasters can use mesoscale and synoptic-scale numerical models to estimate DMC initiation. The gridded datasets from these models include forecasts of moisture convergence (at different levels), CAPE, and CIN.[3] Moisture convergence forecasts above the surface are particularly important in forecasting elevated thunderstorms (Doswell 1982).

4) CONVECTIVE STORM MOTION

Density-weighted mean wind profiles (generally estimated through the depth of a storm) give reasonable

[3] The use of operational mesoscale and synoptic-scale numerical models to forecast thermodynamic fields such as CAPE and CIN should be treated with considerable caution, as these fields are subject to large errors and uncertainty (e.g., Weiss 1996). Frequent model failure in forecasting these thermodynamic fields (and, thus, convective initiation) is often related to poor handling of low-level moisture (Weiss et al. 1998).

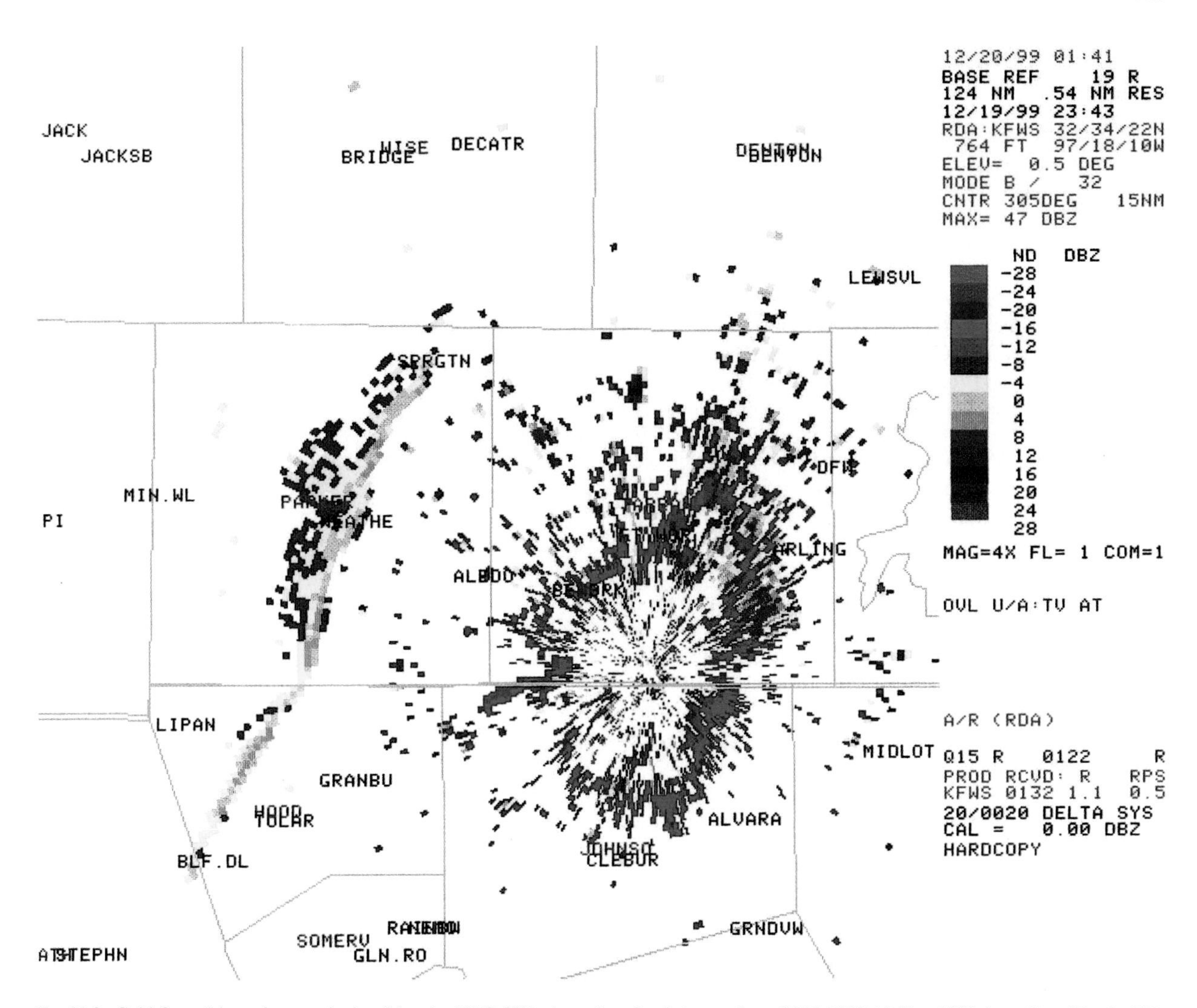

FIG 11.2. Cold frontal boundary as depicted by the FWS-88D clear air reflectivity mode at 2343 UTC 19 Dec 1999 from Fort Worth, TX.

estimates of storm motion. Nevertheless, total storm motion is the sum of two vectors, the mean velocity of the cells constituting the storm system and the propagation velocity due to the formation of new cells on the periphery of the storm (Chappell 1986). Propagational effects on a storm's motion are minimal when the environmental steering winds are very strong. In weaker wind environments, updraft propagation often results in more significant storm motion accelerations. When preferred development is on the front flank, the storms move faster than expected. Rear-flank updraft regeneration causes the system to slow, remain nearly stationary, or even back up (McNulty 1995). Preferred new development on the side flanks causes the storm complex to move either to the right or left of the mean flow.

Storm motion can be influenced by the orientation of low-level boundaries (Weaver 1979; Wade and Foote 1982). Updraft regeneration is favored on (or near) the boundary, resulting in storm motion more closely related to boundary orientation than to the mean wind field. Corfidi (1998) shows that MCSs generally move in the direction of greatest system-relative convergence, which typically is on the down-shear (forward) side of the MCS. However, some MCSs become quasi-stationary or even back up through propagation. Forward-traveling MCSs usually have a source of lower tropospheric unsaturated air available for evaporational cooling and mesohigh formation, which may assist in the production of damaging winds (Corfidi 1998). MCSs that back up or remain nearly stationary generally occur in nearly saturated environments, where outflow production is weaker and the threat of heavy rainfall is greater.

DMC may move into the local area through propagation, even when mean winds do not indicate the likelihood. Propagation tends to occur in the direction of the maximum storm-relative low-level moist inflow. By monitoring the total motion of approaching storms, static stability, and factors that would either support or suppress convection, forecasters will have a reasonable idea about the future of approaching storms.

d. Convective storm type and environmental conditions

Real storms are difficult to classify, because convection is represented better as a continuous spectrum than as a collection of distinct entities (Vasiloff et al. 1986). Most storms are hybrids, occupying the indistinct areas between more prototypical storms. The difficulties in classifying hybrid storms make it critical that forecasters understand, and are able to modify, conceptual models of four-dimensional storm structure. The arrangement of storm types within the storm spectrum is based broadly on two ingredients, the buoyancy and the character of the vertical wind shear. Thunderstorms can be considered as being *unorganized convection* (convective storms generally associated with weak shear, weak wind environments), or *organized convection* (storms generally in a moderate to strong shear, strong wind environment; Moller et al. 1994). Further, given a fixed amount of instability, observations and numerical models show that convective storm persistence increases with the vertical wind shear, at least to some ill-defined limit, beyond which intense shears become destructive to convection.

Weak shear, unorganized storms are short-lived, with new convective cells not developing in any consistent location relative to the cells preceding them. They may consist of one major cumulus tower (single-cell storm), or a series of towers in close juxtaposition (unorganized multicell storm). The new cells form in seemingly random fashion as a result of outflow interactions. The weak shear associated with unorganized storms is typified by the hodograph in Fig. 11.3a. Severe storms within these environments have been called *pulse storms* (Wilk et al. 1979). Severe events occur in a quick burst, usually when the rainy downdraft collapses through, or undercuts, the updraft.

Enhanced organization of convective storms occurs with increasing shear and instability, with successive cells developing in preferred locations relative to their predecessors. Organized storms tend to occur in clusters (multicell storms) and line segments (squall lines). The best documented multicell storm exhibits new cell growth on the right flank (relative to storm motion in the Northern Hemisphere), with dissipating cells being shed on the left flank (Chisholm and Renick 1972; Marwitz 1972). The vertical wind shear for this type of storm is considerably stronger than for unorganized storms (Fig. 11.3b). With the storm moving slower than the mean wind in the storm-bearing layer, there can be considerable storm-relative (SR) wind near the storm top, resulting in precipitation being blown downwind. Other multicell storm types (often undocumented in the literature) have been observed within the WSR-88D radar network. Most of these storms are characterized by new cell formation on the front, rear, or left flanks. Squall lines favor cell formation on the downshear (usually east) storm flank, mainly attributable to the long, nearly continuous gust front. Considerable vertical wind shear also typifies this storm type (Fig. 11.4). Severe weather with organized multicell storms occurs as episodic events associated with updraft cycles.

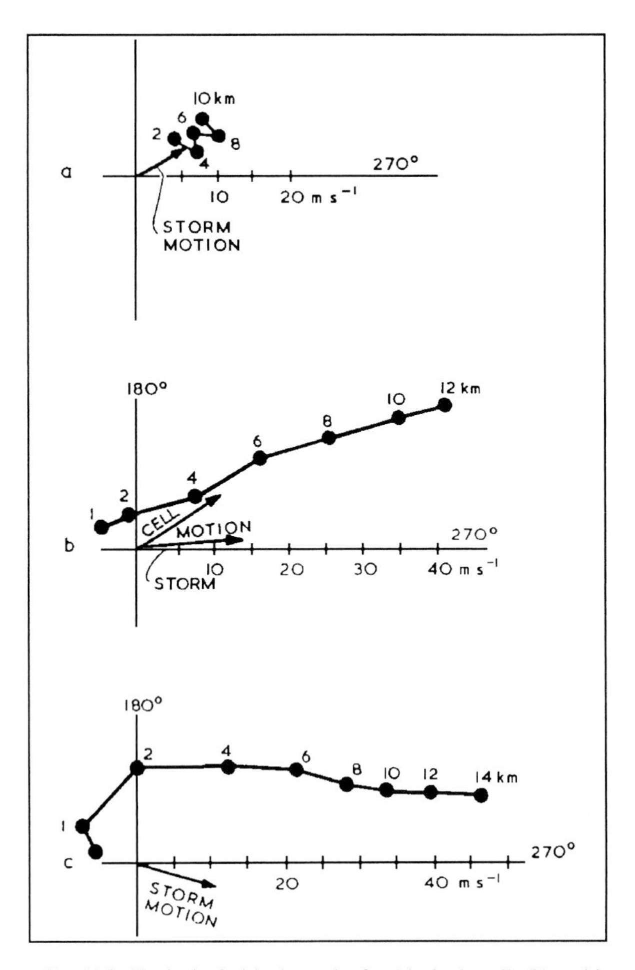

FIG 11.3. Typical wind hodographs for (a) single cell, (b) multicell, and (c) supercell storms, observed during the Alberta Hail Studies Project (from Chisholm and Renick 1972).

Certain combinations of vertical shear and instability (which I shall describe later) can result in the strongest and possibly best organized storm, the *supercell.* Supercell storms are capable of producing the most violent of hail, wind, and tornado events (Moller et al. 1994); thus they are the most important storm type to forecast and detect. The quantitative evaluation of physical processes (notably the perturbation pressure distribution), made possible by numerical cloud models, has led to a revised supercell definition,[4] based on the presence of a strong correlation between vertical vorticity and vertical motion (i.e., a deep,

[4] The first definition of a supercell involved the three-dimensional reflectivity radar structure.

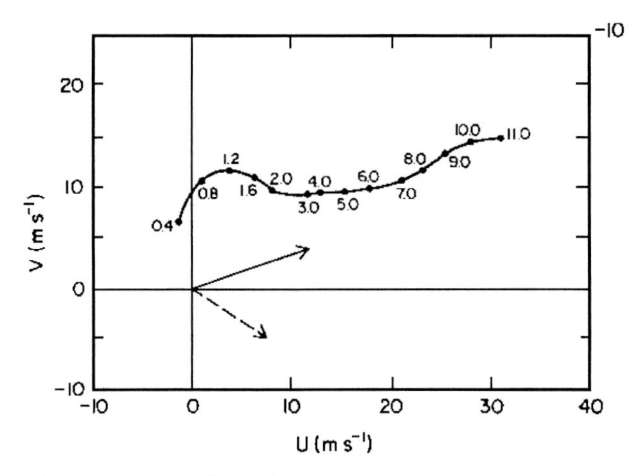

FIG 11.4. Composite hodograph for 40 severe squall lines in Oklahoma. Cell motion is plotted as a solid vector; line motion is plotted as a dashed vector (from Bluestein and Jain 1985).

persistent rotating updraft called a mesocyclone; Weisman and Klemp 1984).

Observations (Chisholm and Renick 1972; Fankhauser and Mohr 1977; Burgess and Curran 1985) and cloud models (Weisman and Klemp 1982, 1984, 1986; Klemp 1987) show that vertical wind shear is more crucial than buoyancy for mesocyclone formation. The pattern of vertical wind shear in Fig. 11.3c (a hodograph that turns clockwise with height) allows a right-moving (RM) storm (i.e., a storm that moves to the right of the mean wind) to tilt existing horizontal vorticity, arising from the vertical wind shear, into the vertical. The resulting helical updraft is seen as a mesocyclone. Numerical cloud simulations demonstrate that a straight-line hodograph (Fig. 11.3b) also can produce supercells when vertical wind shears are strong (i.e., 20 m s^{-1}, or greater, over the lowest 4–6 km; Weisman 1996). The initial storm often splits into two cells that are nearly mirror images of one another; a left-moving (LM) storm, often exhibiting anticyclonic updraft rotation, and an RM storm with a mesocyclone (Schlesinger 1980; Weisman and Klemp 1982, 1986). The LM storm subsequently moves off the hodograph to the left, and the RM storm to the right, of the mean flow.

Perhaps the only scientifically supported storm classification scheme is to categorize storms as either supercells or nonsupercells (Browning 1977). The presence of the mesocyclone makes supercells dynamically different from other forms of convection (Weisman and Klemp 1984). A storm spectrum based on differentiating between multicell and supercell storm types is problematic, since supercells frequently display multicell characteristics (Nelson 1987; Moller et al. 1990; see also chapter 6 of this volume).

Observations of the spatial relationship between the rainy downdraft area and the updraft support a supercell spectrum (Figs. 11.5–11.7). It consists of classic supercells, where precipitation is thrown miles downwind of the updraft (except for the hook echo, in which a small amount of precipitation wraps around the rear side of the mesocyclone); high- (or heavy-) precipitation (HP) supercells, where precipitation falls close to the mesocyclone and significant amounts of rain and hail are ingested into the rear side of the mesocyclone circulation (Moller et al. 1990); and low-precipitation (LP) supercells, where the precipitation area is downwind of the updraft and contains meager amounts of liquid precipitation (Bluestein and Parks 1983). Because of the lack of precipitation, strong hydrometeor-induced downdrafts and surface cold pools (storm outflow) are not observed with LP supercells (Bluestein and Woodall 1990; Rasmussen and Straka 1998).

Miniature supercells (called *MS storms*) (WCS96) represent a supercell size variation. Davies (1993b) suggests that MS storms have tops less than 30 000 feet; Guerrero and Read (1993) describe southeast Texas MS storms that were roughly one-half the horizontal size of a typical supercell (the MS storms were about 6.5 km wide and 13 km long). Supercells can have low tops, but rather typical horizontal dimensions; it has been suggested that these storms should be called *low-topped supercells* (J. Monteverdi 1997, personal communication). The MS and low-topped storms occur with relatively low equilibrium levels, hence the lower tops. In a numerical simulation, WC96 demonstrate that the shallow nature of MS storms results in limited downwind transport of precipitation, which can account for the small horizontal size. The MS storms are most common in the cool season, particularly in the eastern United States (e.g., Vescio et al. 1993; Murphy and Woods 1992; McCaul 1987). However, they have also been documented in Texas (Foster et al. 1995a), the Los Angeles basin (Hales 1993), the Central Valley of California (low-top supercells, Monteverdi and Quadros 1994), Wisconsin (Wakimoto 1983), and Illinois (Kennedy et al. 1990).

The recent number of MS case studies suggests that they are not uncommon. Moreover, several years of WSR-88D observations indicate that mesocyclones (and other convective storm circulations) are not as rare as previously thought. Although supercells are most common in the central United States, they occur east of the Appalachians and west of the Continental Divide (Moller et al. 1994). They have also been documented in England (Browning 1964), France (Dessens and Snow 1989), Australia (Hanstrum et al. 1998), Japan (Niino et al. 1993; Suzuki et al. 1996), China (Zixiu et al. 1993), and Switzerland (Schmid et al. 1996).

Davies and Johns (1993) used observational data to quantify the deviant motion of supercells. They concluded that cyclonic supercells move about 20° to the right and at 85% of the mean wind speed (MWS) for MWS greater than 15 m s^{-1}; otherwise motion is about 30° to the right and at 75% of the MWS. The

FIG 11.5. Schematics for a classic supercell showing (a) a plan view from above, and (b) an idealized view of the storm by a surface observer to the storm's east (from Moller et al. 1994).

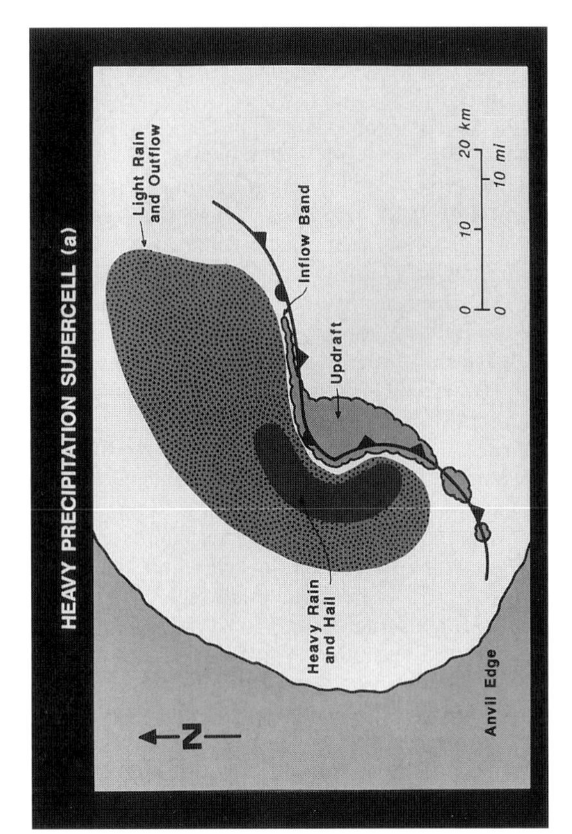

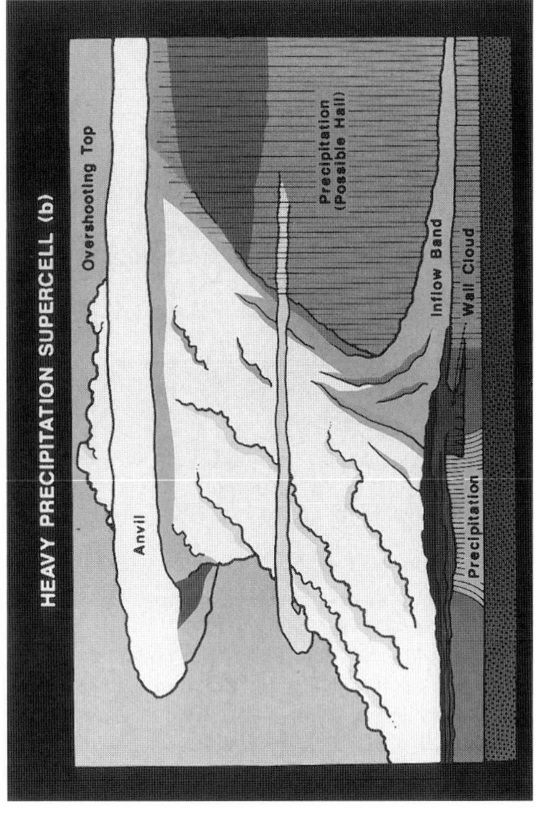

FIG 11.6. Same as in Figs. 11.5a,b except for a heavy- (or high-) precipitation supercell (from Moller et al. 1994).

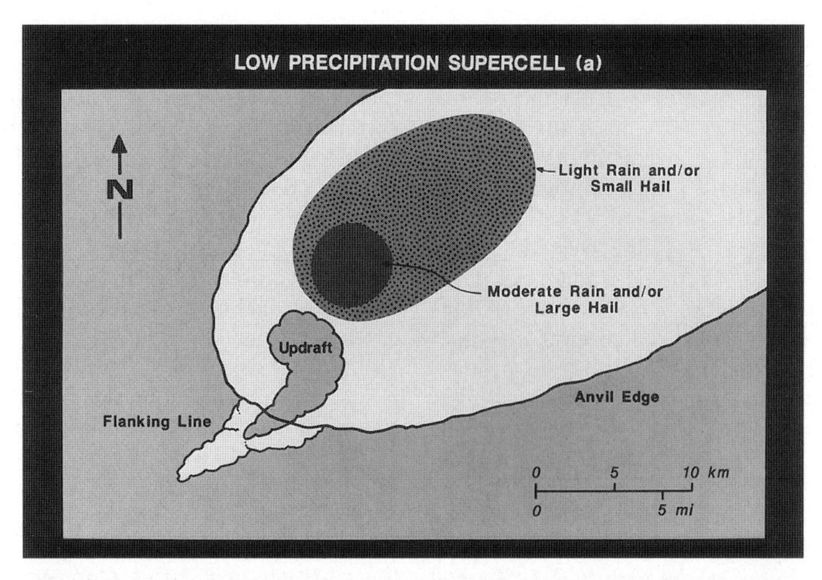

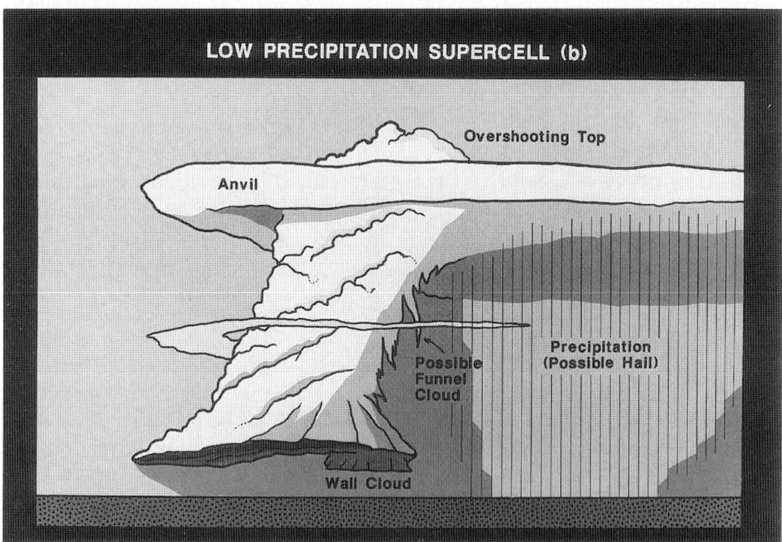

FIG 11.7. Same as in Figs. 11.5a,b except for a low-precipitation supercell (from Moller et al. 1994).

method works well with many events but is not robust with northwest flow aloft and many HP supercells (Bunkers et al. 1998). Therefore, Bunkers et al. propose a "dynamic method" of supercell motion prediction, based on numerical model results (Rotunno et al. 1988; Weisman 1993). The dynamic method is defined as 1) 8 m s^{-1} either to the left (LM supercell) or to the right (RM supercell) of the 0–6-km mean wind, and 2) constrained along a line that both is perpendicular to the 0–6-km mean vertical wind shear vector and passes through the 0–6-km mean wind. This method improves motion estimates of many HP supercells. However, it assumes that supercell motion is largely governed by internal processes; external features such as outflow boundaries also affect the motion of supercells (Weaver 1979; Moller et al. 1990).

e. Severe convective wind storm forecasting

Fujita and Byers (1977) designated strong downdrafts as *downbursts,* but the term has evolved to mean any damaging (or potentially damaging) winds produced by downdrafts. Fujita (1978) noted the different scales of downbursts, with macrobursts > 4 km and microbursts < 4 km in diameter. Macrobursts typically last for ten minutes or longer, whereas microbursts generally evolve in less than ten minutes (Wolfson 1990).

1) SEVERE WIND STORMS IN WEAK SHEAR ENVIRONMENTS

Downbursts are the most common severe event associated with weak shear storms. Many of these events are microbursts, possibly because of the small size of pulse storms. Pulse storm microbursts can occur in both moist and relatively dry atmospheres, respectively producing wet and dry microbursts (Fujita 1985). At midlatitudes these events typically occur when the midlevel westerlies have retreated north during the warm season. Dry microbursts predominate in arid and semiarid areas (e.g., the western United States) and wet downbursts are most common in moist areas, such as the southeastern United States (Atkins and Wakimoto 1991) and Japan (Ohno et al. 1996).

The vertical thermodynamic profile is the primary signal used in forecasting pulse storm downbursts (JD92). The "inverted V" or "Type A" profile (Beebe 1955) is characterized by a dry-adiabatic layer from near the surface to the midlevels, a very dry lower layer, and a moist midtropospheric layer (Fig. 11.8). The resulting dry microbursts (also called *low-reflectivity microbursts*) are associated with high LCLs and LFCs and weak instabilities (Krumm 1954; Caracena et al. 1983; Wakimoto 1985). Forecast procedures, based on morning soundings and expected diurnal heating, have been developed for dry microbursts (Caracena et al. 1983; Wakimoto 1985), and used successfully in operational forecasts (Sohl 1987). Low-reflectivity microburst storms in general are particularly hazardous to aircraft since the parent cloud and virga shaft often appear innocuous (see chapter 7 of this volume).

On occasion, momentum can push dry downdraft air past its thermal equilibrium point so that it is warmer than the ambient environment when it reaches the surface (Williams 1963). Such a heatburst is typical of a dissipating storm (Johnson 1983). When heatbursts produce damaging winds, they often cause public confusion, since they generate considerable warming and can occur some distance from discernible storms.

Wet microbursts occur with moderately unstable conditions and rich low-level moisture that typically extends beyond 4–5 km AGL (Fig. 11.9; Read 1987). The low relative humidities above the moist layer play a role in microburst production through dry air entrainment and subsequent evaporative cooling. As a result of the relatively dry air aloft, Atkins and Wakimoto (1991) found that on wet microburst days the afternoon equivalent potential temperature difference between the surface and the midlevels is usually equal to or greater than 20°C. On inactive days the difference is typically equal to or less than 13°C.

2) Severe wind storms in moderate to strong shear environments

The self-perpetuating nature of these storm systems supports long-lived, or repetitive, wind events that produce a majority of severe wind reports east of the Rocky Mountains (JD92). The vertical thermodynamic and wind profiles and the synoptic and mesoscale

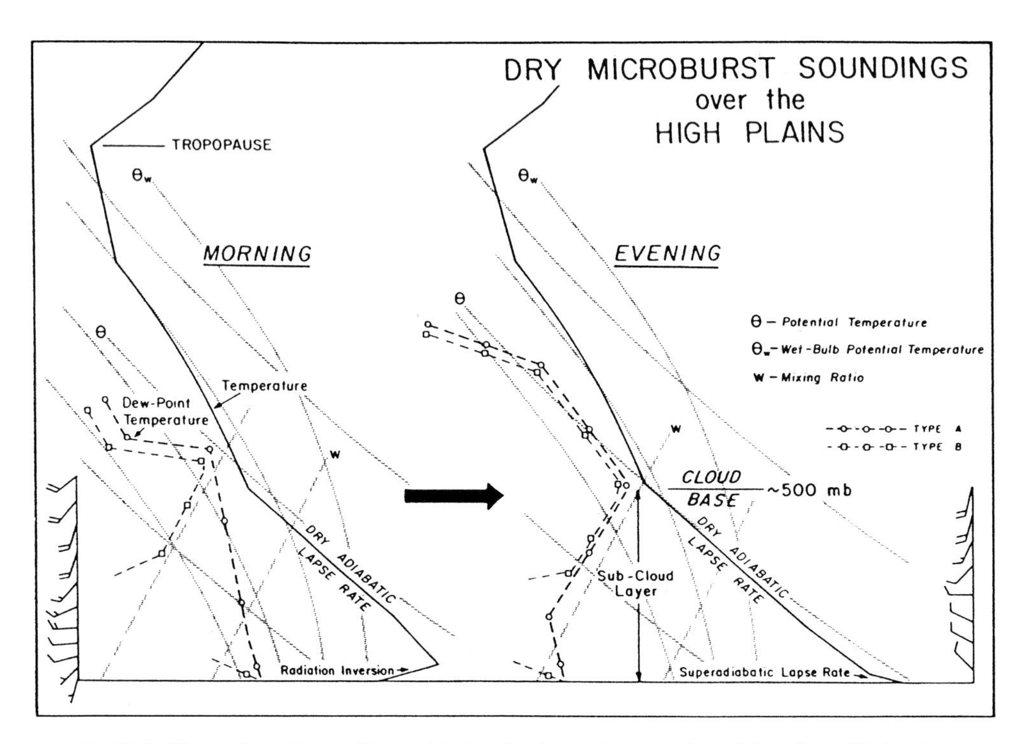

FIG 11.8. Thermodynamic sounding model showing the environmental conditions favorable for dry microbursts over the U.S. High Plains (from Wakimoto 1985).

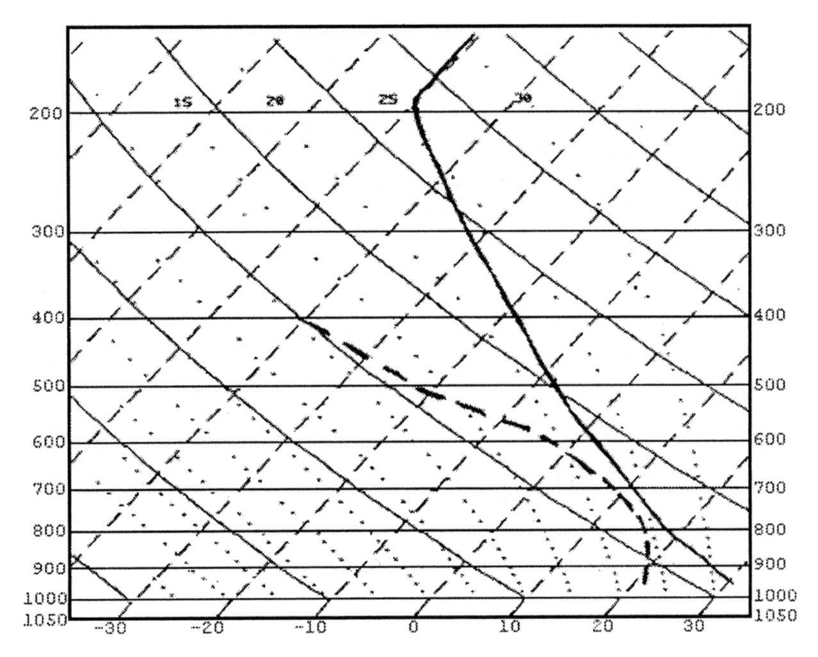

FIG 11.9. Thermodynamic sounding model for north Texas wet microbursts (from Read 1987).

surface features play crucial roles in the development of damaging winds.

There is considerable range in the size and structure of the various storm complexes that spawn downbursts in strong shear environments (Fig. 11.10). Isolated windstorms can be produced by supercells (Fig. 11.10a). Damaging supercell winds usually occur in the rear-flank downdraft (RFD), which is the mesocyclone's outflow region. Extreme downbursts (outflow winds exceeding 50 m s^{-1}) can occur with HP supercells (Moller et al. 1990; Brooks and Doswell 1993). These storms usually form in very unstable environments with high boundary-layer moisture content and weak SR midlevel winds (Brooks et al. 1994a, hereafter BDW94). Multicell storms can produce downbursts, mainly when midlevel conditions are characterized by low humidities and strong winds.

Bow echoes (Fujita 1978; Figs. 11.10b,c,d) account for a majority of the casualties and damage resulting from nontornadic U.S. windstorms (Johns and Hirt 1987, hereafter known as JH87). Bow echo storms are capable of producing extremely damaging winds of greater than 50 m s^{-1}. Life cycles of supercell storms can include bow echo shapes; however, bow echoes more often form along convective lines. Large bow echoes may contain smaller bows that are more transient in nature. Further, large bow echo complexes may contain supercells, most frequently along the convective line's south or southwest flanks (Przybylinski and DeCaire 1985; Schmidt and Cotton 1989),

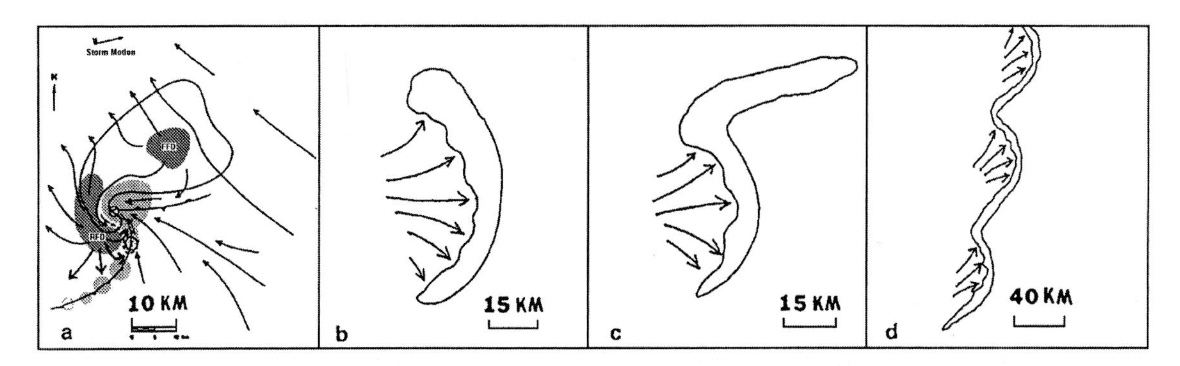

FIG 11.10. (a) Schematic of flow associated with a supercell storm (from Lemon and Doswell 1979) showing forward-flank downdraft (FFD) and the more intense rear-flank downdraft (RFD). (b) Schematic of downdraft flow associated with a relatively large bow echo. (c) Schematic of downdraft flow associated with a line-echo wave pattern. (d) As in (b) except for extensive squall line with embedded bow echoes and line echo wave patterns (from Johns and Doswell 1992).

which can produce tornadoes (Smith and Partacz 1985; Moller et al. 1990). Other bow echo tornadoes arise from nonsupercell circulations (Forbes and Wakimoto 1983). Long-lived bow echoes (or a series of bow echoes) may produce an extended wind event called a *family of downburst clusters* (Fujita and Wakimoto 1981), or a *derecho,* when the track length exceeds 400 km (JH87).

An elevated rear-inflow jet (RIJ; Augustine and Zipser 1987; Smull and Houze 1987) forms as thunderstorms develop into a bowing line segment. In advanced stages of the complex, subsidence warming beneath the descending RIJ may result in a surface wake low (Fujita 1955). Strong pressure gradients between the low and the convectively induced mesohigh can produce damaging gradient winds that are not caused by convective storm cells (e.g., Stumpf et al. 1991).

Bow echoes include warm-season and dynamic pattern types. Warm-season events (Fig. 11.11a) occur as isolated storms or short lines with multiple bows (JH87). They may be most common from the northern plains into the Ohio River valley (JH87) and in the southern plains (Bentley and Mote 1998); nevertheless, they also occur elsewhere. Moreover, they usually travel along quasi-stationary fronts, in areas of low-level warm advection and considerable moisture pooling. The environments are very unstable, with average maximum CAPEs (using a lower 100-mb mixed parcel) of 4500 J kg^{-1} (Johns et al. 1990). Numerical models show that the high instability plays a role in bow echo maintenance in weak to moderate dynamic forcing (Weisman 1993). Winds aloft are often northwesterly (Johns 1984) and oriented nearly parallel to the surface boundary. Midlevel wind speeds are moderate, about 20 m s^{-1} at 500 mb. Vertical shear in modeled bow echoes is strong, especially in the lowest 2.5 km AGL, averaging about 20 m s^{-1} (Rotunno et al. 1988; Weisman 1993). However, an observational study shows that slightly more than 50% of 115 long-lived bow echoes had 0–2-km shears of less than 10 m s^{-1} (Evans 1998). The vertical wind profile for both types of bow echo is typically more unidirectional than with tornado outbreaks (Johns 1993).

Dynamic pattern bow echoes involve strong baroclinic disturbances and winds aloft (average 500 mg winds of 38 m s^{-1}), but weaker instabilities than their warm-season counterparts. Despite marginal instabilities and limited hodograph curvature, significant tornadoes can occur (Przybylinski 1988; Johns 1993). These complexes usually form near cold fronts and are associated with more extensive squall lines than their warm-season counterparts (Fig. 11.11b). They may be most common in the southeastern United States and along the Atlantic seaboard, and generally have shorter tracks than warm-season events (Bentley and Mote 1998). Dynamic patterns are least common in mid- and late summer. Readers should consult Przybylinski (1995) for a detailed review of bow echoes.

3) Automated downburst forecast algorithms

Early algorithms, incorporating observed sounding data to predict severe wind gusts at the surface, focused on the evaporative cooling and momentum transfer terms (Fawbush and Miller 1954; Prosser and Foster 1966). However, the success of these algorithms was limited (Doswell et al. 1982). Anthes (1977) accounted for midlevel dry air entrainment into the downdraft. Pino and Moore (1989) combine these elements into an algorithm that updates with time, using the latest hourly surface data and advected

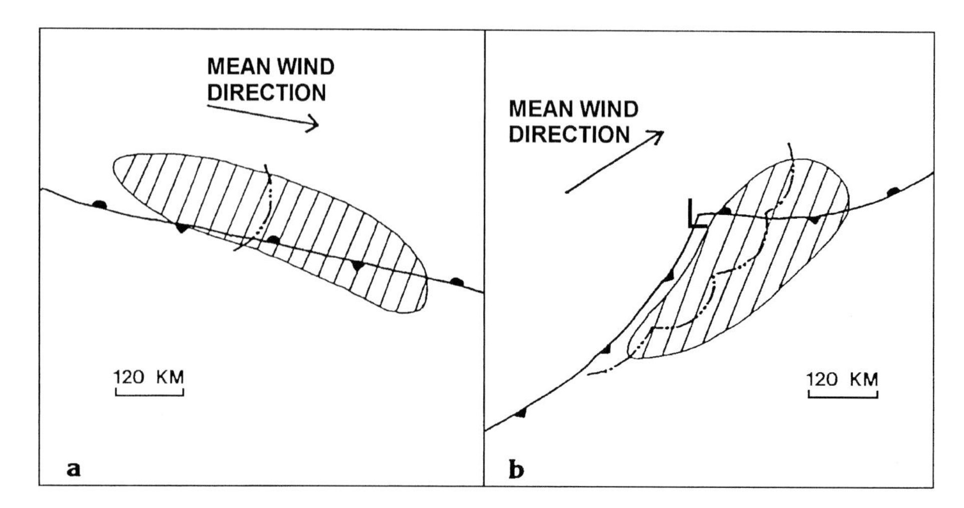

FIG 11.11. (a) Schematic of features associated with a warm season bow echo event (sometimes called a *progressive* derecho). (b) As in (a), except for a dynamic bow echo event (sometimes called a *serial* derecho; from Johns and Hirt 1987).

thermodynamic parameters aloft. Their algorithm has demonstrated some increase in skill over the others.

f. Severe convective hail storm forecasting

Strong thunderstorm updrafts are necessary for the production of large hail (Morgan and Summers 1985); thus early hail size forecast techniques concentrated on buoyancy derived from radiosonde data (Fawbush and Miller 1953; Foster and Bates 1956). These routines have provided only limited success (Doswell et al. 1982). A more recent algorithm (Moore and Pino 1990) has shown some improvement by accounting for the negative effects of water loading and entrainment on updraft strength. Part of the problem in using thermodynamic profiles is the sparsity of radiosonde data. Furthermore, factors that cannot be analyzed from soundings are important contributors to the production of large hail, such as the transit time of hail embryos through hail-growth zones and storm precipitation efficiency (see chapter 6 of this volume). These effects favor the production of the largest hail in the supercell and severe multicell storms that occur in strongly sheared wind environments. An external factor that influences hail size is the melting of hailstones after they fall below the freezing level [which is approximated by the environmental wet-bulb zero (WBZ) level]. Generally, small stones fall more slowly and have proportionally more melting area, thus melting more than do large stones.

Numerical simulations strongly suggest that the interaction of supercell updrafts with environmental winds creates perturbation pressure gradients and resultant vertical accelerations that contribute substantially to updraft speeds (Weisman and Klemp 1984). Estimates are that about 50% of the total updraft velocity may arise from this effect. Therefore, forecasters must recognize that supercell hail size can be much larger than is predicted by operational algorithms (JD92). The most destructive hailstorms are supercells (e.g., Nelson 1987; Moller et al. 1990, 1994). Some HP supercells, occurring with large values of CAPE (often greater than 4000 J kg^{-1} in a mixed boundary layer), produce unusually extensive areal coverage of large hail, resulting in catastrophic damage (Moller et al. 1990). These storms often occur with southerly surface winds and west or northwest flow aloft, resulting in large directional shear. They frequently move along stationary thermal boundaries that are oriented nearly parallel to the midlevel flow.

There is a perception that eastern U. S. supercells generally do not produce large hail. Many exceptions have been noted (e.g., Lenning et al. 1996), usually when instabilites are large. Nonetheless, some tornadic supercells do not produce large hail, particularly those that occur in the eastern United States during the autumn (Johns and Hart 1998). It is speculated that many of these events occur in weak CAPE environments. Other "nonlarge hail" tornadic supercell storms are summer events spawned by the remnants of tropical storms in the southeastern United States. These storms clearly occur with relatively weak CAPEs. Conversely, a significant number of weak instability supercells do produce large hail, including so-called postfrontal supercells that have been documented in California (Monteverdi and Quadros 1994). It is likely that cold environmental temperatures and vertical pressure perturbation effects contribute to increased hail size in these cases.

g. Tornado forecasting

Most tornadoes, especially those that are strong and violent (i.e., F2 or greater on the Fujita damage scale), are associated with supercell storms. However, an important fraction of tornadoes occurs with nonsupercell storms (e.g., Brady and Szoke 1989; Wakimoto and Wilson 1989). Nonsupercell tornadic storms often exhibit Doppler radar circulations that are weaker than the mesocyclone strength threshold (vertical vorticity larger than or equal to $10^{-2} s^{-1}$) and usually smaller; in some cases the only circulation present is that of the tornado itself. These storms also do not contain reflectivity structures normally associated with supercells, such as bounded weak echo regions. The environments that produce supercells are becoming better understood, largely because of observations and cloud modeling experiments. It has been considerably more difficult to find atmospheric signals that reliably discriminate between tornadic and nontornadic supercell events.

1) SUPERCELL TORNADO FORECASTING

Large values of instability and vertical shear accompany most midlevel disturbances that are associated with supercell tornadoes. However, many other systems with similar values of instability and shear fail to produce tornadoes. Further, some tornadic supercells occur at the margins, with a weak shear–large CAPE combination (e.g., Jarrell, Texas, 27 May 1997), or with strong shear and weak CAPE (the 1988 Raleigh, North Carolina, tornado; Korotky 1990). Other notable strong shear–weak CAPE events include postfrontal supercells occurring with robust midlevel disturbances in California (Braun and Monteverdi 1991), supercells associated with tropical storms approaching landfall (Weiss 1987; McCaul 1991), and decaying tropical storms that have moved inland (McCaul 1991). Also, MS storms can occur with apparently weak values of CAPE, mainly because of the shallow nature of the FCL. Thus, Davies (1993b), WC96, and Blanchard (1998) suggest using NCAPE in forecasting MS storms. WC96's numerical model shows that surface to 500 mb CAPEs of 600 J kg^{-1} can support tornadic MS storms when the vertical wind shear is strong (at least 30 m s^{-1} through a depth of 6 km).

Early estimates were that about 50% of mesocyclones produce tornadoes (Burgess and Lemon 1990). Observations from VORTEX and the WSR-88D network reduce this figure to 20% or less. These considerations suggest that supercell forecasting is not equivalent to supercell tornado forecasting (JD92). Further, numerical cloud models (Weisman and Klemp 1982, 1984) and theoretical studies (Davies-Jones 1984; Rotunno and Klemp 1985) show that low-level shear in unstable environments is the source of midlevel (3–10 km) mesocyclones (MLMs). Tornadoes appear to be more closely connected to the formation of low-level (below 1 km AGL) mesocyclones (LLMs) (Brooks et al. 1994a), which may originate from the baroclinic generation of vorticity within evaporatively cooled downdrafts (Rotunno and Klemp 1985; Davies-Jones and Brooks 1993).

The vertical wind shear of greatest importance for MLMs is in the storm's inflow layer. Bluestein et al. (1989) suggest that in unstable, "loaded gun" type environments, entrainment into the updraft core is minimized above the LFC. Therefore, the inflow layer depth roughly coincides with LFC height (Davies and Johns 1993). Bluestein and Parks (1983) found that the average spring LFC height in the southern plains is slightly above 2 km. Thus, studies associating inflow layer wind shear with supercells have generally used 0–2- or 0–3-km shear values.

Weisman and Klemp (1982) find that the bulk Richardson number [BRN, which combines deep layer shear (0–6 km) and buoyancy into one term; Moncrieff and Green 1972] is useful in forecasting the storm type of numerically modeled storms. BRN values between 15 and 45 favor the occurrence of modeled supercells in a homogeneous environment (Weisman and Klemp 1984). A frequently used measure of shear is storm-relative environmental helicity (SREH; Davies-Jones et al. 1990), which is graphically equivalent to minus twice the area swept out by the inflow layer SR wind vector on a hodograph. Increasing positive values of SREH denote increasing streamwise vorticity in the storm inflow layer, resulting in updraft rotation (Davies-Jones 1984). Using a dataset of Oklahoma tornadoes, Davies-Jones et al. (1990) suggest approximate mean 0–3-km SREH values for weak, strong, and violent tornadoes of 280, 330, and 530 $m^2 s^{-2}$, respectively. However, a substantial number of significant tornadoes nationally occur with SREH below 200 $m^2 s^{-2}$ (Kerr and Darkow 1996; Rasmussen and Blanchard 1998). Operational experience suggests that supercells storms are possible when SREH values exceed a value around 100–150 $m^2 s^{-2}$ (Moller et al. 1994).

SREH is subject to rapid temporal and spatial change (Burgess 1988; Moller et al. 1994; Markowski et al. 1998c), such that a shear structure favorable (unfavorable) for mesocyclones may change quickly to a nonsupercell (supercell) environment. Some SREH fluctuations occur because of diurnal low-level wind oscillations (e.g., diurnal low-level jet accelerations can result in large values of SREH during the late night and early morning hours;[5] Maddox 1993). VORTEX data reveals substantial SREH variability during each of four events, particularly near thermal boundaries (Markowski et al. 1998a,c). Davies-Jones (1993) found that SREH varies as much as an order of magnitude within 100 km and 3 h. Nevertheless, Droegemeier et al. (1993), using results from a cloud model, show that SREH is superior to BRN in predicting net updraft rotation. They suggest that BRN can be used to forecast storm type, since it is independent of storm motion, and SREH can be used to describe the rotational potential of storms once their motion is known.

Efforts have been made to combine SREH and CAPE (Rasmussen and Wilhelmson 1983; Johns et al. 1993), including the energy-helicity index (EHI) (Hart and Korotky 1991). The EHI was derived from defining the curves within the parameter space that would enclose the greatest number of tornado events (see Fig. 1 of Davies 1993a). However, according to BDW94, the primary importance of low-level shear/CAPE scatter diagrams may be to differentiate between nonsupercell and supercell events, rather than between tornadic and nontornadic episodes [Fig. 11.12 (from BDW94) and Fig. 11.13 (from Turcotte and Vigneax 1987)]. Another study shows that the likelihood of significant tornadoes increases with EHI, although the false alarm rate is high (Rasmussen and Blanchard 1998). Differences in methodologies of these studies may be the primary reason for the disparate results.

The premise of LLM formation and sustenance is that a balance exists between baroclinic vorticity generation (first seen in the RFD; Davies-Jones and Brooks 1993) and RFD outflow effects. Further, MLMs may help produce the correct conditions for LLMs (BDW94). BDW94's numerical model shows a horizontal redistribution of rain, modulated by relative strengths of the MLM circulation and SR midlevel winds. When SR midlevel winds are weak, there is sufficient low-level baroclinity (originating from rain falling near the updraft) for LLM formation, but strong outflow undercuts the LLM, inhibiting tornadoes. When the SR winds are strong, rain is blown downstream and there is insufficient low-level baroclinity for LLMs. The appropriate balance of SR winds aloft and MLM strength results in LLM formation. Brooks et al. (1996b, hereafter BDC94) test this hypothesis with observational data. They derive a parameter space

[5] In the absence of dynamic forcing that is sufficient to maintain the strength of the low-level jet during the day, this may partially explain the secondary maximum of early morning tornadoes near the Gulf Coast (Kelley et al. 1978; Maddox 1993; Knupp and Garinger 1993).

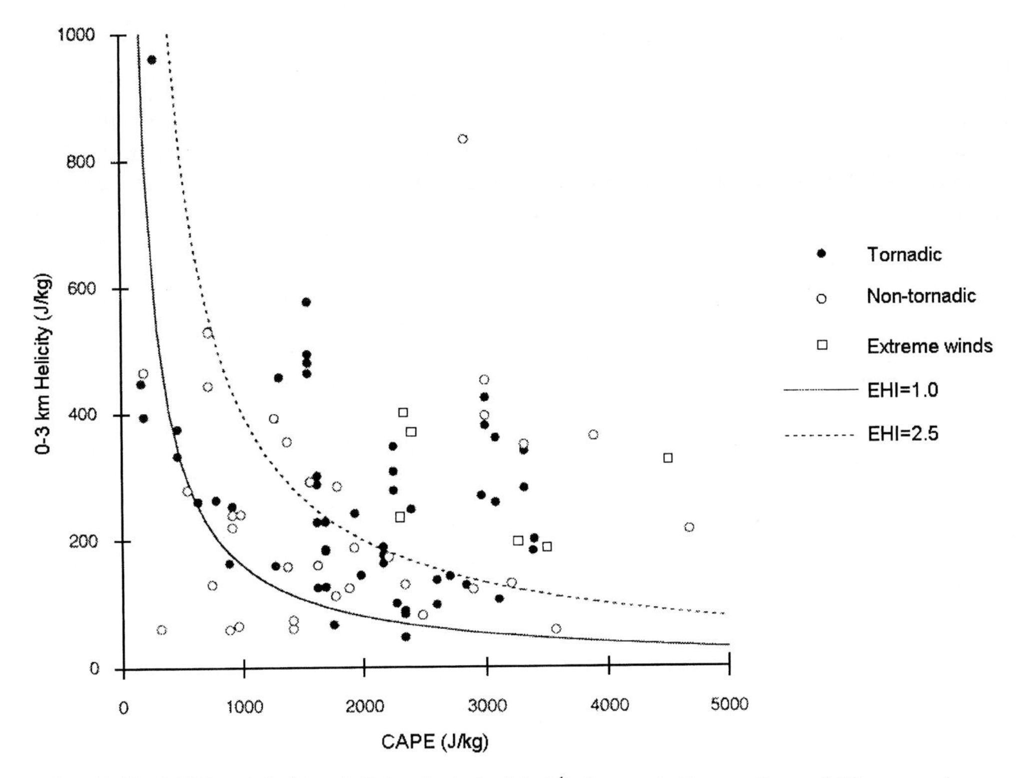

FIG 11.12. CAPE and 0–3-km helicity (both in J kg^{-1}) for proximity soundings of 92 mesocyclones. Energy-helicity index (EHI) isopleths of 1.0 and 2.5 are indicated by solid and dashed lines, respectively (from Brooks et al. 1994b).

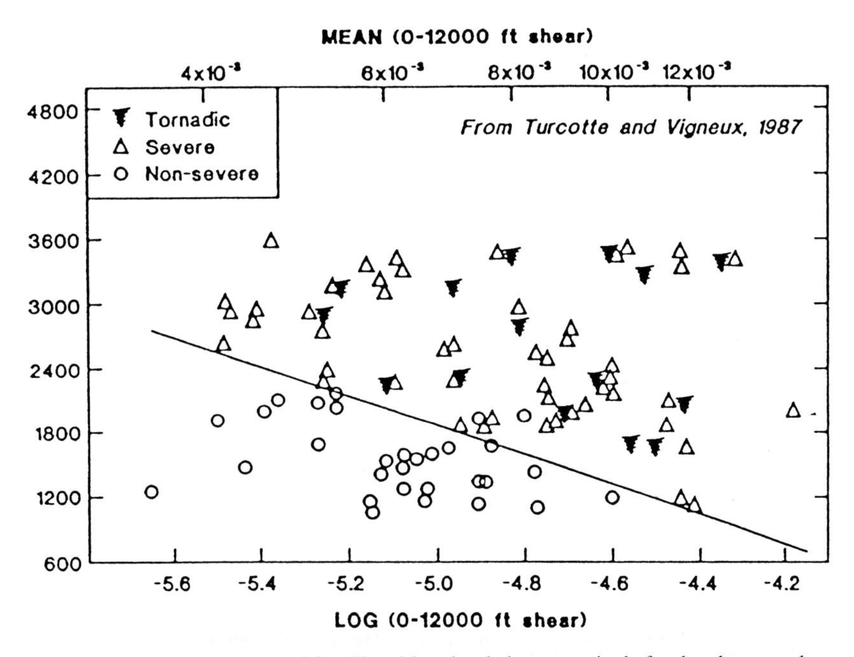

FIG 11.13. Scatter diagram of CAPE and low-level shear magnitude for thunderstorm days and associated weather events in Quebec Weather Centre's forecast area for 1985–87. Sloping line shows approximate division between severe and nonsevere storms (from Turcotte and Vigneux 1987).

diagram (Fig. 11.14), with the results confirming the three regions suggested by BDW94. The area represented by a combination of maximum water vapor content (q-max) and high SREH/weak SR winds is dominated by extreme wind, nontornadic events (mainly HP supercells); the lower q-max area includes the environmental conditions for LP supercells (typically nontornadic; Bluestein and Parks 1983). Strong LLMs and tornadoes form in the middle for a given value of q-max, and this area slopes toward lower values of SREH/SR winds as q-max increases. The authors stress that the results should be considered guidelines, rather than thresholds.

Thompson (1998) applies the concepts of SR midlevel wind and low-level inflow strength (Droegemeier et al. 1993) to LLM/tornado forecasting. Results from Eta model forecasts indicate an increased likelihood of classic supercells and tornadic LLMs with 850-mb SR inflow of greater than or equal to 10 m s^{-1} and SR midlevel flow of 10 to 25 m s^{-1}. Using a different dataset, Rasmussen and Straka (1998) find that SR anvil-level winds, rather than midlevel winds, offer the best short-term forecast signal to differentiate between the precipitation distributions near HP and classic supercell updrafts. They hypothesize that the weak anvil-level SR winds associated with HP storms allow more hydrometeors to be reingested into the updraft, leading to much greater precipitation rates in the updraft region. They also conclude that seeding of a supercell updraft by upshear anvils and storm towers can convert a classic or LP storm into an HP storm.

Stensrud et al. (1997) use a mesoscale model to test the physically based parameters of CAPE, SREH, and bulk Richardson number shear (BRNSHR), the denominator of BRN. They find CAPE to be a reliable indicator of thunderstorm potential; CAPE, SREH, and model-produced convection to help delineate supercell from nonsupercell environments; and BRNSHR values (a proxy for SR midlevel winds) between 40 and 100 $m^2 s^{-2}$ to be associated with storms that produce LLM tornadoes. Markowski et al. (1998b) find that 0–2-km or deeper SREH is a strong indicator for supercell formation and that supercell tornadoes are more likely when a majority of the SREH is concentrated in the shallow 0–1-km layer.

Following the numerical results of Klemp and Rotunno (1983), Wicker (1996) postulates that the front-flank gust front can lead to LLM formation. His numerical model results show that equivalent directional alignments of the low-level environmental horizontal vorticity vector and the baroclinic horizontal vorticity vector along the supercell's front flank gust front can result in a strong LLM. LLM formation is delayed, or even prevented, when the two vectors are perpendicular to each other (Fig. 11.15). Thus, LLMs may depend on detailed (and difficult to detect) wind structures in the lowest several hundred meters. Markowski et al. (1998a) hypothesize that the forward flank baroclinic zone generally has insufficient horizontal vorticity for updraft tilting and stretching into a tornadic LLM. Citing that 70% of the 1994–95 VORTEX-studied tornadoes occurred near an external

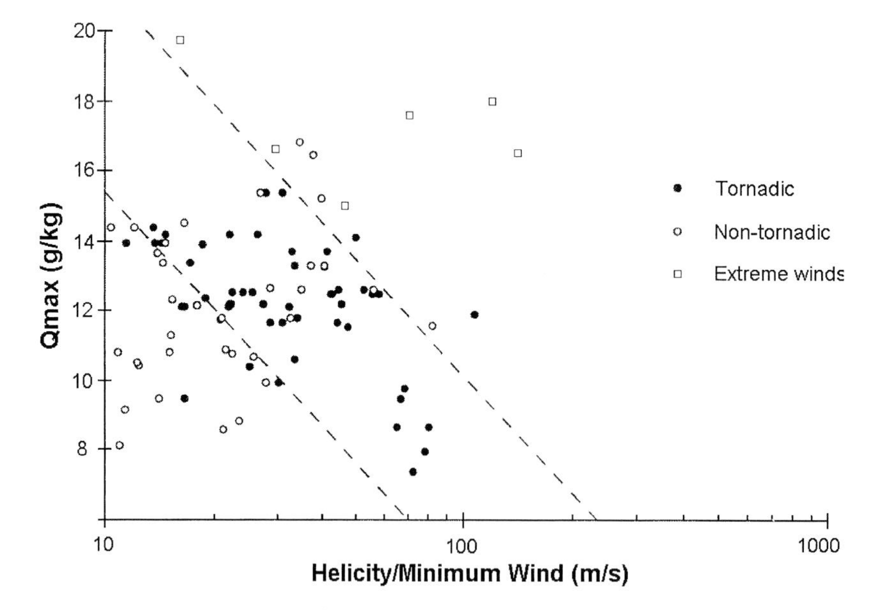

FIG 11.14. The H/vmin (m s^{-1}) and maximum water vapor content in the boundary layer [qmax (g kg^{-1})] for proximity soundings of 92 mesocyclones. Vmin is the minimum value of the storm-relative wind between 2- and 9-km altitude, averaged over a depth of 1 km. Note that horizontal axis is logarithmic. Dashed lines delineate three regions of environmental conditions as discussed in text (from Brooks et al. 1994b).

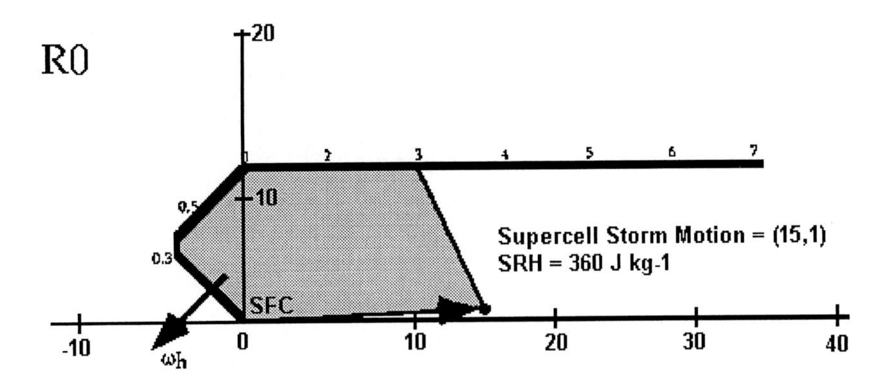

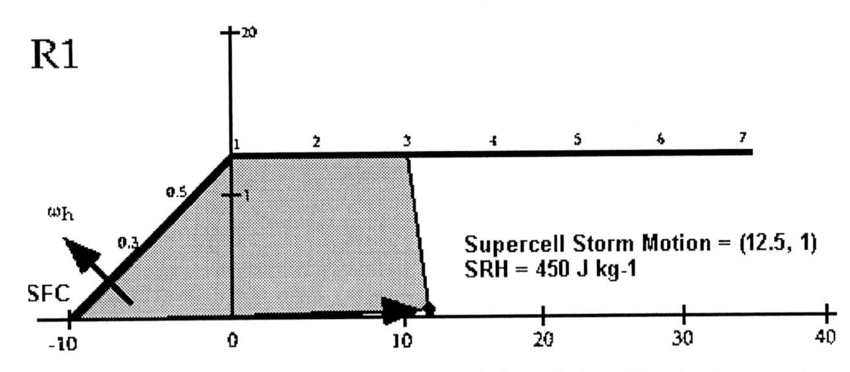

FIG 11.15. Two hodographs used for the numerical simulation of low-level mesocyclone formation. Case R0: the environmental vorticity vector (solid arrow) is parallel to the baroclinic generation of vorticity vector (not shown), and low-level mesocyclogenesis results. Case R1: the environmental vorticity vector is perpendicular to the baroclinic generation of vorticity vector and low-level mesocyclogenesis does not occur (from Wicker 1996).

boundary (usually an outflow boundary from previous convection), they show that LLMs can form via baroclinic enhancement provided by the extremely rich SREH environment immediately to the cool side of the boundary (Fig. 11.16). They conclude that subsequent factors (probably arising from RFD formation) then lead to tornadogenesis. A problem for forecasters is in determining which boundaries have sufficient buoyancy and vertical shear within the cool-side sector to support supercell tornado formation and sustenance.

Factors may inhibit tornadoes in otherwise favorable environments. Lifting condensation levels (LCLs) generally above 1200 m may result in a decreasing likelihood of tornadoes, because of low boundary-layer relative humidities that result in evaporative cooling and premature undercutting of mesocyclones by strong outflow (Rasmussen and Blanchard 1998; Davies 1998). Moreover, the relative orientations of thunderstorm-initiating boundaries and anvil-level winds may suppress tornadic storms when the winds are parallel to the boundary (LaDue 1998). Anvil-born precipitation seeds adjacent updrafts, weakening them prematurely. This alignment of a cold front and anvil-level winds can also result in significant front-to-rear anvil orientation when the front moves rapidly, resulting in additional rain falling behind the boundary. According to LaDue, this accelerates the cold outflow and can result in rapid transition from isolated supercells to a nontornadic squall line. Finally, an atmosphere typified by vertical shear favorable for tornadoes may contain an instability profile with the most unstable parcels located above the strong shear layer, a situation that occurs frequently at night. Elevated parcels that are lifted into the updraft will not possess the strong vertical vorticity that lower-based parcels would have, thereby greatly diminishing the threat of tornadoes.

2) NONSUPERCELL TORNADO FORECASTING

Nonsupercell tornadoes (NST) can occur in a wide variety of environments that support convective storms. Therefore, they can be particularly difficult to forecast. Many occur within weak shear environments, and their formation appears to be heavily influenced by the effects of topography on the organization of mesoscale boundaries. Other NSTs form in strong shear environments, often along convective lines (Forbes and Wakimoto 1983; Knupp et al. 1996). Short-term forecasting for both weak shear and moderate to strong shear NSTs is critically dependent on

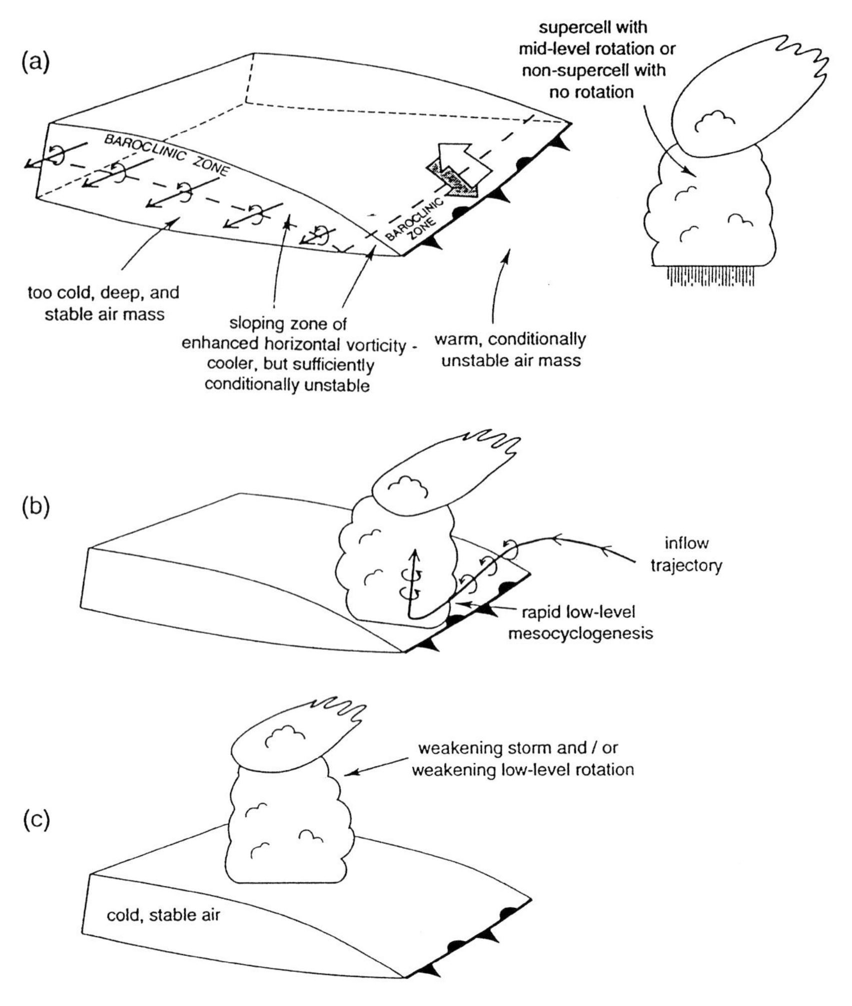

FIG 11.16. A conceptual model of how updraft–boundary interaction may lead to low-level mesocyclogenesis: (a) a supercell on the warm side of a boundary; (b) a tornadic supercell on the immediate cool side of the boundary, where vorticity augmentation via baroclinic generation is maximized; (c) a supercell storm above an air mass that is too cold and deep to support tornadoes (from Markowski et al. 1998a).

mesoscale analysis, climatology, and an understanding of what is known about NST formation.

Golden and Purcell (1978) first noted that NS waterspouts occur in weak shear environments. Bluestein (1985) makes an analogy between waterspouts and a variety of weak shear NST, which he calls a *landspout*. Wilson (1986) and Wakimoto and Wilson (1989) describe small NSTs that form along convergence lines in northeastern Colorado. Brady and Szoke (1989) document a landspout that formed along intersecting boundaries, when preexisting vertical vorticity along the intersection was stretched by a rapidly growing towering cumulus cloud (Fig. 11.17). Colorado NSTs usually form in the presence of a zone of converging winds that often develops into a circulation called the *Denver convergence-vorticity zone*, or DCVZ (Szoke and Augustine 1990). The DCVZ arises from the interaction of southeast surface flow and a terrain escarpment southeast of Denver. Landspouts tend to form where the trough intersects outflow boundaries, or where nearby outflow boundaries intersect each other. Wakimoto and Wilson (1989) show that nearby Doppler radar can pick up the small-scale, clear-air circulations that lead to NSTs up to 20 minutes ahead of tornado formation. The circulations, often called *misocyclones* (Fujita 1981), are smaller, weaker, and typically more shallow than mesocyclones. Forecasters have little to focus on other than atmospheres usually typified by some amount of CAPE, the presence of terrain-induced boundaries (e.g., the DCVZ), and/or colliding mesoscale boundaries. These conditions indicate only a small probability of NST occurrence, since similar conditions frequently occur without NST formation.

In a numerical model of NSTs along an advancing thunderstorm cold pool, Lee and Wilhelmson (1997a)

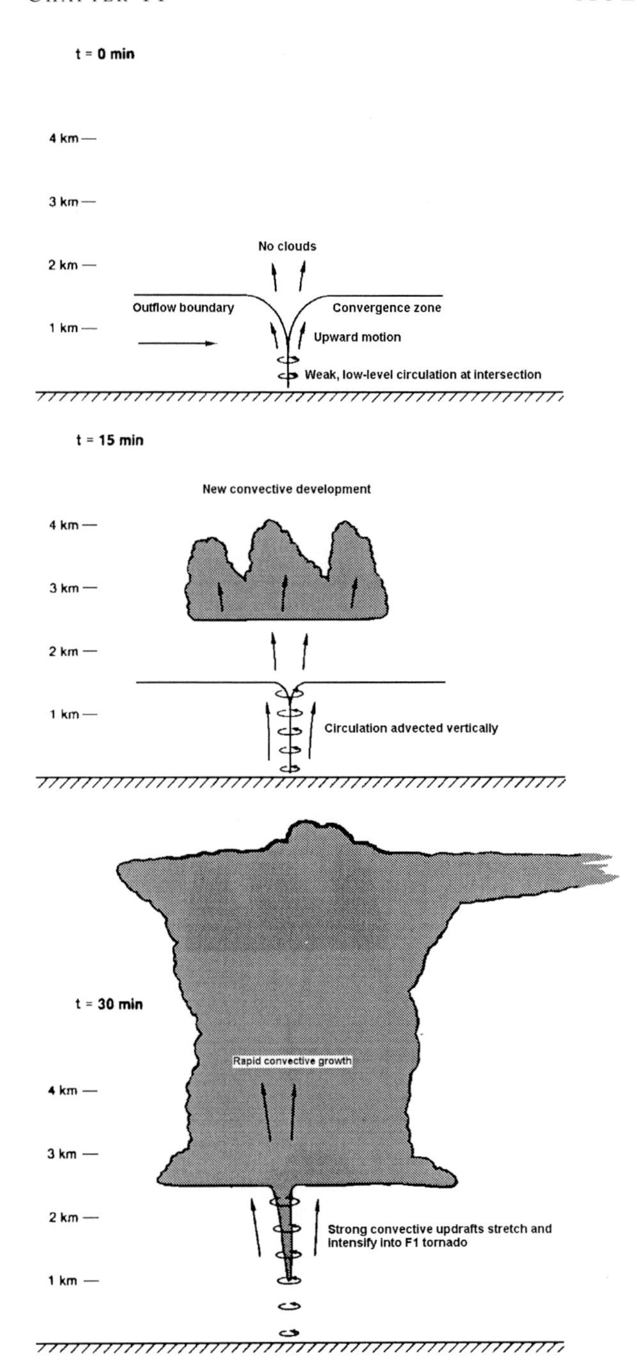

FIG 11.17. Conceptual model of misocyclone and landspout tornado evolution (from Brady and Szoke 1989).

found that shearing instabilities along the boundary can produce a number of small, weak circulations. These misocyclones (usually a diameter of 1 km or less) undergo a merger phase, creating larger misocyclones that eventually become tornadic when stretched by rapidly forming thunderstorm updrafts (Lee and Wilhelmson 1997b). The existence of misocyclone mergers has been verified by Doppler radar observations of Colorado landspouts (Wakimoto and Wilson 1989). A mixture of NSTs and supercell tornadoes can occur in high shear events, increasing the level of forecast difficulty. Knupp et al. (1996) found that northern Alabama tornadoes can occur within (or ahead of) line segments. While a few of the storms exhibited robust supercell characteristics, most were either nonsupercells or hybrid storms. Similar events have been noted elsewhere [e.g., Indiana (Przybylinski 1988) and California (Carbone 1983)].

A poorly understood NST is the *cold core tornado* (Cooley 1978). These NSTs form under cold pools associated with midlevel low pressure circulations, generally to the rear of surface cold fronts. Residual low-level moisture and moderately strong lapse rates are associated with these rare tornadoes. Vertical stretching of existing vertical vorticity may be partially responsible for cold core tornadoes (Doswell and Burgess 1993). However, many purported cold core tornadoes occur with large vertical wind shears, and likely are supercell tornadoes.

Perhaps the most transient NST is the *gustnado.* Gustnadoes typically form along developing lobes or cusps of outflow boundaries (Idso 1975; Doswell 1985). They are often smaller and more shallow than landspouts, and they seem not to depend on the superposition of updrafts above (Doswell and Burgess 1993). Storm chase observations suggest that gustnadoes are frequently too weak to produce damage. More documentation is needed; nevertheless, gustnadoes appear to have less strength potential than landspouts because of the lack of vorticity stretching by appreciable updrafts. Forecasters have little ability to either forecast or warn for gustnadoes.

3) CHART PATTERN RECOGNITION

Chart pattern recognition has been particularly important in identifying the "classic" tornado outbreaks [Miller's Type B pattern (Miller 1972) or what DWJ93 calls a *synoptically evident tornado day* (see Fig. 10a of DWJ93)]. These events are characterized by deep surface lows that are associated with progressive and unusually strong midlevel baroclinic disturbances. Tornado outbreaks may be most common during the strengthening phase of these disturbances, when an approaching, intense jet streak forces the midlevel height trough into a negative tilt.[6]

Mesoscale areas of rapid pressure falls and subsequent backing and strengthening of surface winds[7]

[6] The important ingredient is the wind acceleration associated with the mid- and upper-level jet maximum, rather than the negative tilt of the upper trough. The wind acceleration enhances the vertical wind shear and may promote vertical motions through transverse ageostrophic circulations.

[7] These wind accelerations are isallobaric responses, caused by a number of possible events, including jet streak–induced ageostrophic transverse circulations. Forecasters can use radiosonde data or numerical model gridded data to analyze Q vectors and validate the

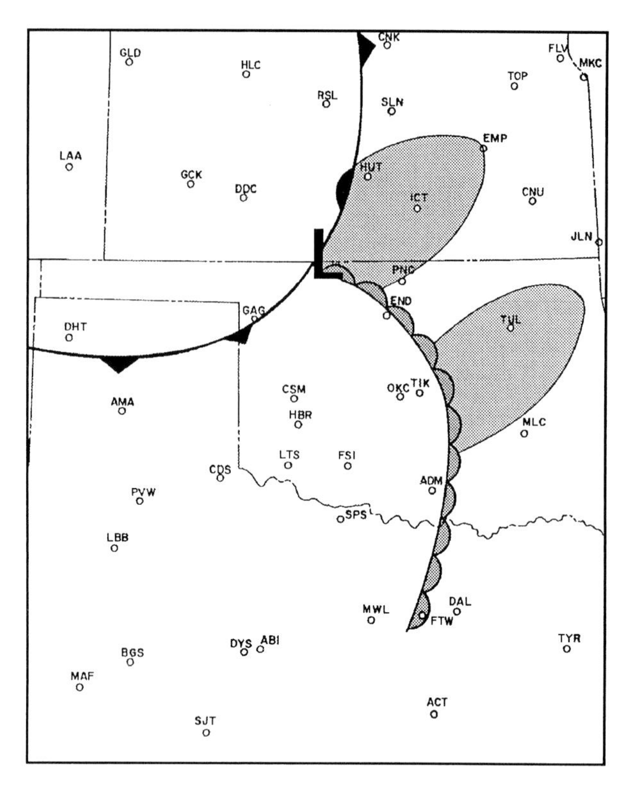

FIG 11.18. Schematic surface pressure and frontal chart with favored areas of supercell tornado formation in the southern plains (from Moller 1979).

frequently occur several hours prior to tornadoes (Miller 1972; Tegtmeier 1974; Moller 1979). These pretornado surface tendencies in the southern plains (and elsewhere) often occur on the moist side of deepening surface lows [often a subsynoptic-scale low (SSL)] and eastward-bulging dry lines (Fig. 11.18; Tegtmeier 1974; Moller 1979). Surface thermal ridges are often located to the immediate west of the threat area during these tornado events (Fig. 11.19). Progressive southern plains tornado outbreaks tend to occur early in the spring, with synoptically evident systems. In cases involving damaging, long-track tornadoes or tornado families, very small, translating areas of rapid pressure falls often occur in the inflow area ahead of the parent storm (Moller 1979). Localized, less transient tornado outbreaks most often occur late in the spring, when moderate winds aloft encounter strong instabilities, often near favorable terrain-gradient features such as the northwest Texas Caprock escarpment (Moller 1979). Although the accompanying midlevel winds are only "moderate" (generally 10–20 m s^{-1}), SR winds can be strong because of slow storm motion.

McGinley and Ray (1985) use a three-dimensional numerical model over sloping terrain to study SSLs associated with the late spring plains events. The modeled SSL grows and intensifies through a period of induced heating; when heating is eliminated, the SSL does not intensify to the same degree. Moreover, SSL intensification increases convergence and low-level vertical wind shear, priming the atmosphere for supercell formation. These modeling results are similar to many observed tornadic events in the southern plains over the past 30 years [the thermal ridge typically extends into the western portion of the SSL, with the eastern thermal gradient (Fig. 11.19) coincident with the threat area east of the SSL (Fig. 11.18)]. Localized differential heating commonly occurs with these SSL and dryline bulge events, in part because of the juxtaposition of different vegetation types, wet and dry soils, and irrigated crops in the region (Hane et al. 1997; Ziegler et al. 1997).

Wilczak and Glendening (1988) use a mixed-layer 2D model of the northeastern Colorado DCVZ to show that vertical vorticity may arise from the intersection of a surface baroclinic zone with the strong terrain gradient. These results may apply to other high plains regions, including the sloping terrain along the Texas Caprock escarpment. Moller (1979) and others have noted a propensity for SSL tornado events in this area. Many of the Caprock events have occurred near the Canadian River valley and an adjacent stretch of

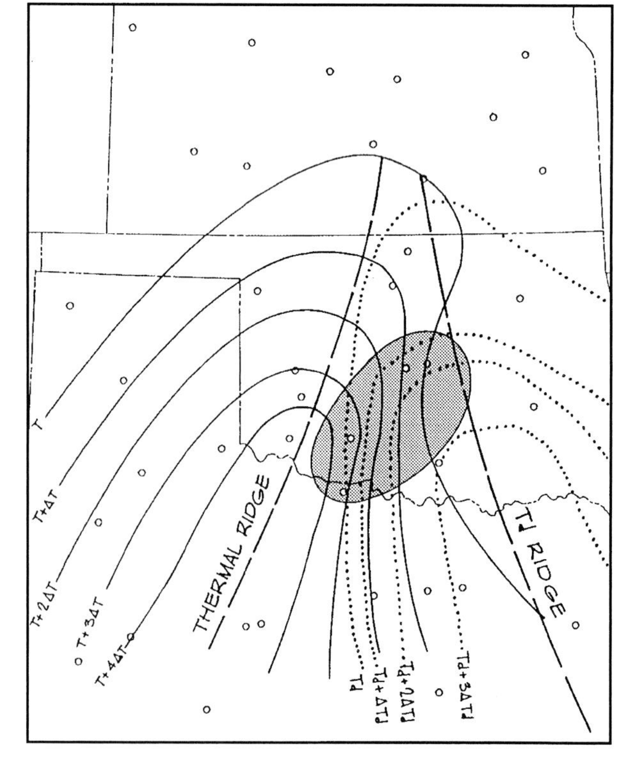

FIG 11.19. Schematic surface thermodynamic chart with favored areas of supercell tornado formation in the southern plains (from Moller 1979).

presence of ageostrophic circulations associated with jet maxima, rather than relying on the quadrapole conceptual model that assumes the presence of these circulations (Barnes and Colman 1994).

higher terrain to the river's immediate south, and near other east- to west-oriented canyons and valleys along various high plains escarpments. Perhaps the small low pressure centers result in very large values of 0–1-km SREH that may be required for tornadic supercells (Markowski et al. 1998b).

Chart pattern recognition is important in other local tornado outbreaks [e.g., the role of the DCVZ in northeast Colorado NST events (Brady and Szoke 1989; Szoke and Augustine 1990)]. Supercell tornado events in the high plains (Colorado and Wyoming) are most likely to occur in postfrontal environments typified by easterly upslope flow and strong vertical wind shear profiles (Doswell 1980). California supercell tornadoes are usually postfrontal events, occurring either with robust and progressive upper-level troughs (Braun and Monteverdi 1991; Monteverdi and Quadros 1994) or closed upper lows (Hales 1985). These California tornadic storms frequently form along postfrontal surface troughs characterized by converging northwesterly flow and topographically channeled southerly flow. A preliminary study of peninsular Florida tornado outbreaks identifies synoptic patterns that include surface lows in the central Gulf that are either extratropical, tropical, or hybrid (surface lows with tropical and extratropical characteristics) in nature (Hagemeyer 1997). A common element in all three types of outbreaks is the apparent rapid spinup of tornadoes in the coastal zone, followed by a diminishing threat of tornadoes further inland.

Stationary thermal boundaries can assist in thunderstorm intensification (Purdom 1976) and tornado formation (Maddox et al. 1980). Recent studies (e.g., Korotky 1990; Businger et al. 1991; Vescio et al. 1993) show that localized tornado events in North Carolina often occur near subtle thermal and moisture boundaries. Markowski et al. (1998a) conclude that when no other evidence of a boundary exists (because of lack of data or the subtle nature of the feature), forecasters can occasionally infer its presence by noting where the WSR-88D accumulated precipitation algorithm indicates that earlier storms laid down a swath of soil-wetting precipitation.

Pattern recognition must be augmented by parameter evaluation to minimize false alarms. Furthermore, daily diagnosis can increase the probability of detection by revealing tornado events that do not match these prototypical map patterns. New data sources [e.g., a statewide surface data mesonet in Oklahoma (Crawford et al. 1992)], profilers,[8] and WSR-88D VAD winds are very useful in short-term forecasting of severe weather parameters and map patterns.

In summary, MLM (supercell) formation is followed by LLMs and tornadoes only about 20% of the time. It may be that soundings that appear favorable for supercell tornadoes frequently are not representative of the inflow to nontornadic events, or that large-scale environments almost never contain sufficient conditions for tornadic supercells (Rasmussen and Blanchard 1998). When supercell tornadoes occur, it appears that the baroclinic generation of horizontal vorticity, arising from external thermal boundaries and/or the storm's forward-flank gust front, leads to LLM formation. Subsequent RFD development increases the likelihood of tornadoes.

Short-term clues that supercell tornadoes may be imminent include observations of supercell-external outflow boundary interactions, backing and strengthening supercell inflow winds associated with rapid pressure falls (often occurring near surface low pressure centers or dryline bulges), and Doppler radar observations of near-range LLMs. Forecasters should monitor these environmental conditions and use CAPE, NCAPE, SREH, 0–6-km shear, and EHI to forecast supercells. Measures of anvil-, mid-, and inflow-level SR winds; EHI; BRNSHR; 0–1-km SREH; and the height of the LCL can provide useful supercell tornado forecast information. The relative orientations of anvil-level winds and thunderstorm-initiating boundaries can also be influential to the formation and sustenance of tornadic storms. *It is crucial that forecasters use parameter measures in a probabilistic, rather than a deterministic, manner.*

Nonsupercell tornado forecasting is often possible in areas that experience these events with some regularity (e.g., northeastern Colorado and Florida), but is generally more difficult elsewhere. Nevertheless, forecasters must understand the processes that can result in NSTs, such that they are not completely surprised by such events.

11.3. Severe local storm warnings

Forecasters who correctly diagnose the day's severe weather potential are more likely to make correct warning decisions through prudent use of information from the WSR-88D radar and severe storm spotters (Quoetone and Huckabee 1995). For instance, on a high-risk day with large values of shear and instability, forecasters may issue tornado warnings when storms first become supercellular, whereas in most cases, they will wait for additional evidence (i.e., strengthening and deepening of the mesocyclone

[8] Profilers were critical to the short-term forecasting of the 3 May 1999 tornado outbreak in central Oklahoma. The 2300 UTC radiosonde data at Oklahoma City did not reveal a sudden mid- and upper-level wind acceleration that was revealed by a profiler only 35 miles to the south. The wind acceleration, poorly forecast by the numerical models, was noted by SPC forecasters earlier in the day at profiler sites in southern and eastern New Mexico. Its presence was critical to decision making that upgraded the severe storm risk from slight to moderate, and, finally, high risk (NOAA Staff 1999).

through several volume scans and/or spotter reports of strong low-level rotation and inflow). Furthermore, knowledge of the environment and its effect on severe weather potential allows forecasters to determine appropriate WSR-88D strategies on a given day (Andra and Foster 1992). Without this knowledge, forecasters can be left with nearly useless Advanced Weather Interactive processing System (AWIPS) procedures, such as datasets that are designed for supercell tornado scanning strategies when pulse storms are the main threat.

These prewarning considerations are elements of *situation awareness* (SA; Bunting 1998). It is defined by Endsley (1988) as "perception of the environmental elements in a volume of space and time, comprehension of their meaning, and projection of their status in the near future." Efforts to increase SA for aviation in-flight emergencies have focused on anticipating a range of possible outcomes, initiating planning early in the event, and increasing communications among all members of the team (Orasanu 1994). NWS severe storm warning SA applies to meteorological and sociological factors. For example, forecasters must be aware of the potential effects of severe storms on outdoor gatherings such as boating and stadium events, and have the means to alert emergency managers and/or event officials about the threatening weather.

WFOs can implement SA through planning, training, and awareness of potential decision-making pitfalls [e.g., worker fatigue, too many storms to monitor, and nonmeteorological distractions such as equipment failure (Bunting 1998)]. Bunting suggests that WFOs engaged in a warning event should be proactive about internal staffing issues. This includes 1) having sufficient personnel in place, each with well-defined roles (e.g., radar operator, communications officer, warning composer, and mesoscale analyst) to maximize performance and avoid distraction; 2) encouraging open

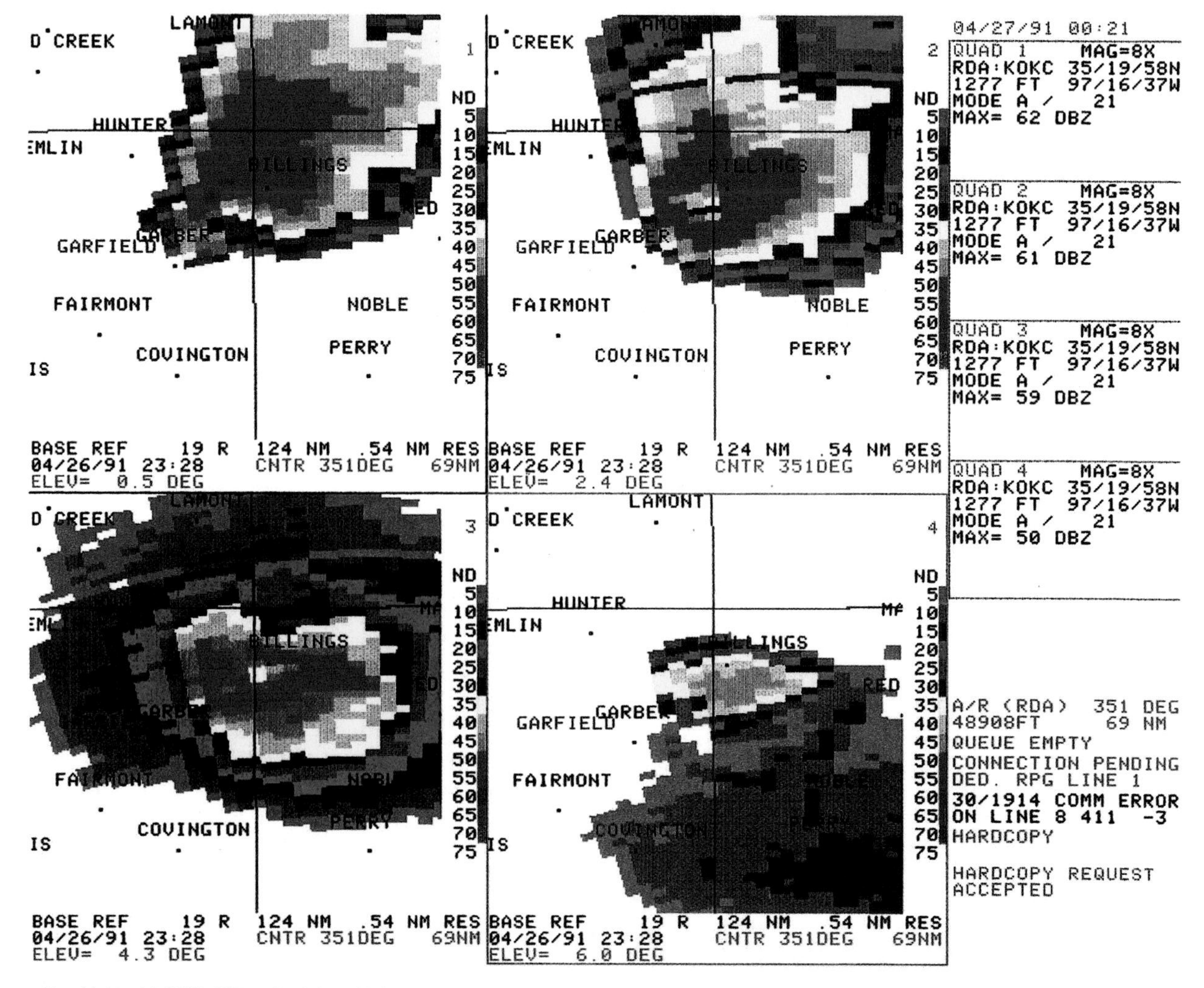

FIG 11.20. (a) WSR-88D reflectivity of 26 Apr 1991 Red Rock, OK, tornadic storm showing low-level (0.5° elevation) hook echo in upper left panel, midlevel bounded weak echo region (BWER) and echo overhang in upper right panel (2.4°) and lower left panel (4.3°), and storm top displaced above the BWER and overhang in lower right panel (6.0°).

communication among the warning operations team members; and 3) assigning a warning coordinator who monitors overall office actions and assigns tasks as needed. Having warning team members change duties midway through an event, or relieving members during an extended warning episode, combats fatigue. Experience at the Forth Worth (FTW) WFO validates SA as being critical for warning success.

The WSR-88D is more sensitive than past radars and far superior as a severe storm detection tool. To maximize effectiveness, a meteorologist *must* be dedicated to WSR-88D surveillance during severe weather. Radar operators use reflectivity and velocity data from each volume scan to determine the evolution of three-dimensional storm structure (Fig. 11.20) and combine this information with radar algorithm output (Lemon et al. 1992; Andra and Foster 1992). These radar data are combined with spotter reports and other information to make appropriate warning decisions.

WSR-88D radar algorithms were developed in the late 1970s and early 1980s. Advances in computer technology, imaging process techniques, and knowledge of severe storm radar signatures have made it possible to improve these algorithms, which has been an ongoing project at the National Severe Storm Laboratory (NSSL), as part of the Warning Decision Support System (WDSS). The system directs meteorologists toward the most critical storms and displays the information about these storms in a readily available format (Eilts et al. 1996a). WDSS includes the Radar Algorithm Display Software (RADS), which gives specific information pertaining to each identified storm cell, including extremely important four-dimensional cell trends (Fig. 11.21) and improved cell tracking. WDSS has been tested at several NWS offices (including Forth Worth) and forecasters have found it indispensable when used with manual radar interrogation.

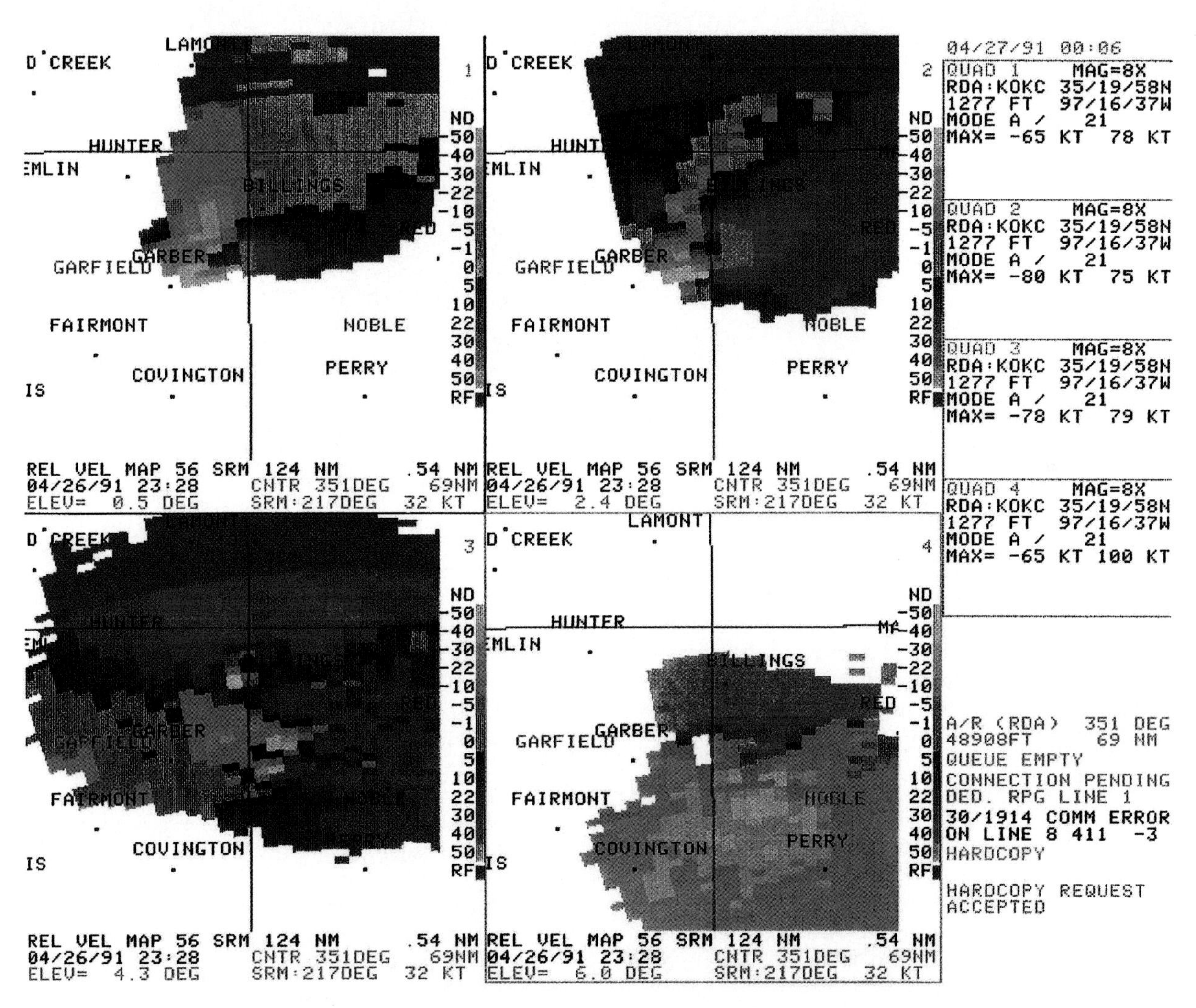

FIG. 11.20. (*Continued.*) (b) Same as in (a) except for storm-relative velocity. Convergent rotation is seen in upper-left panel, pure rotation in the upper-right panel, divergent rotation in lower-left panel, and storm-top divergence in lower-right panel.

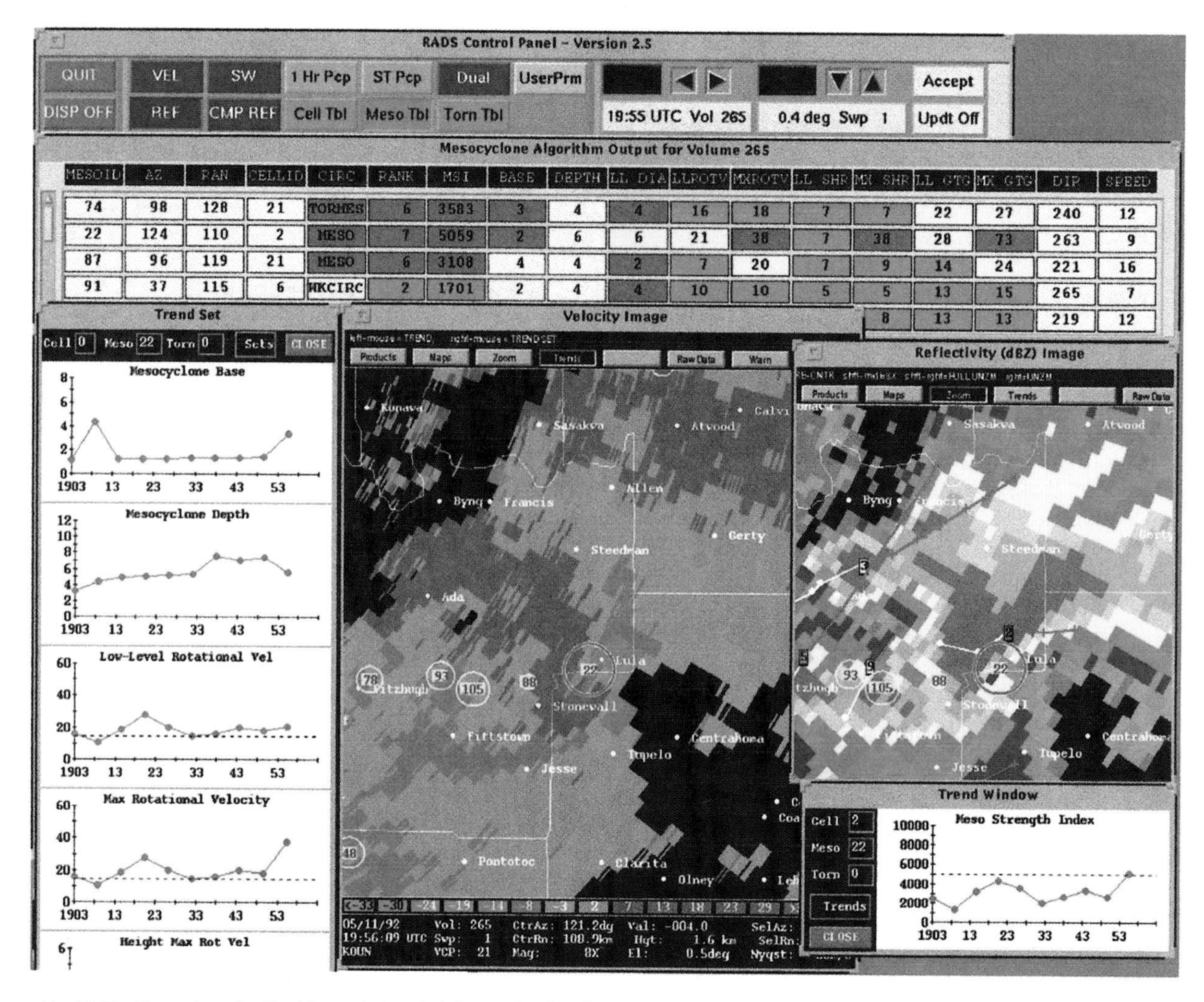

FIG 11.21. Examples of cell tables and time–height trends of various reflectivity and storm-relative velocity attributes from the Radar Algorithm Display Software (RADS) of the Warning Decision Support System (WDSS).

Storm spotters will remain a critical component of severe storm detection for many years. Spotters augment radar data by supplying information that the radar cannot provide. WSR-88D and spotter strategies are described in this section. For more detailed discussions of radar severe storm and tornado detection, the reader should consult Burgess and Lemon (1990, hereafter BL90) and Burgess et al. (1993, hereafter BDD93).

a. Radar limitations

Radar limitations include those affecting all radar systems (radar horizon, aspect ratio, and attenuation) and some indigenous to the Doppler mode (the "Doppler dilemma"). The radar horizon limitation occurs because radar beams do not bend at the same rate that the earth curves, but typically increase in height as they travel away from the radar (BDD93). Therefore, radar beams overshoot distant, low-topped storms, and some MLMs will not be detected beyond 200 km. Beyond about 80 km, radar will overshoot cloud base, meaning that LLMs will not be sampled. Radar data beyond these ranges must be supplemented by spotter information (e.g., reports of cloud-base rotation and tornadoes; BDD93). Aspect ratio problems occur when the radar resolution volume size becomes significantly larger than storm features such as mesocyclone radius (about a factor of 3 for a Rankine combined vortex; BL90). Attenuation (resulting in reduced reflectivities) is substantially worse with C-band radars than with S-band radars such as the WSR-88D (Allen et al. 1981).

Prohibitive costs limit the network to single WSR-88Ds, rather than overlapping, multiple-Doppler radars; thus velocity data retrieval is limited to the radial component. Fortunately, the flows within severe storms, particularly strong vertical drafts and areas of rotation, can be approximated by simple models (e.g., the Rankine vortex for mesocyclones), which have easily definable single-Doppler signatures (BL90). The Doppler dilemma is a wavelength-dependent sampling conflict that exists between the maximum unambiguous range of the radar and the maximum unambiguous velocity measured by the radar. When the radar pulse repetition frequency (PRF) is high, return from two or more trips (different ranges) may be overlaid, a situation commonly referred to as *range folding.* When the PRF is low, the range is relatively long and range folding is absent, but radial velocities frequently exceed the bounds of the fundamental interval over which they can be uniquely measured, a situation referred to as *velocity aliasing* (BL90). Forecasters have some control over the Doppler dilemma through manipulation of PRF values.

b. Radar detection of severe storms in weak shear environments

Weak shear pulse storms often differ from their nonsevere counterparts in that first cell height is higher (6–9 km; Chisholm and Renick 1972; BL90). In the mature stage, the area of 50+ dBZ reflectivity is higher and more persistent than in nonsevere storms, while maintaining continuity upon descending to the ground (Fig. 11.22). A study of North Texas downburst-producing pulse storms indicates that virtually all occurred with reflectivities > 50 dBZ extending above 10 km (Read and Elmore 1989). The duration of severe weather (most commonly downbursts, but occasionally hail ranging up to 4.5 cm) is brief, usually occurring immediately prior to complete cell collapse. Dry microburst storms occur in weak CAPE and weak shear environments; hence they contain very weak updrafts and seldom produce large hail.

Doppler velocities occasionally indicate strong midlevel convergence prior to a rapidly descending reflectivity core and downburst (Roberts and Wilson 1989). When observed, this midlevel signature allows less than ten minutes' warning lead time for both wet and dry microburst storms. The NSSL downburst algorithm keys on midlevel convergence, midlevel rotation within the area of descent, and, to a lesser extent, storm-top divergence (which is at a maximum prior to storm collapse; Eilts et al. 1996b). Once a downburst reaches the surface, Doppler base velocities often show a strongly divergent low-level signature, but at maximum ranges only of 40–60 km. When hail occurs with pulse storms, it reaches ground coincident with the highest radar reflectivities. Warning lead times for pulse storms are shorter than for more organized storms; nevertheless, downburst algorithms can add minutes to local aviation and severe thunderstorm warnings.

c. Radar detection of severe storms in moderate to high shear environments

1) THREE-DIMENSIONAL STORM STRUCTURE

The 3D radar reflectivity structure of severe storms has long been documented in various global locations (e.g., Browning 1964; Chisholm and Renick 1972).

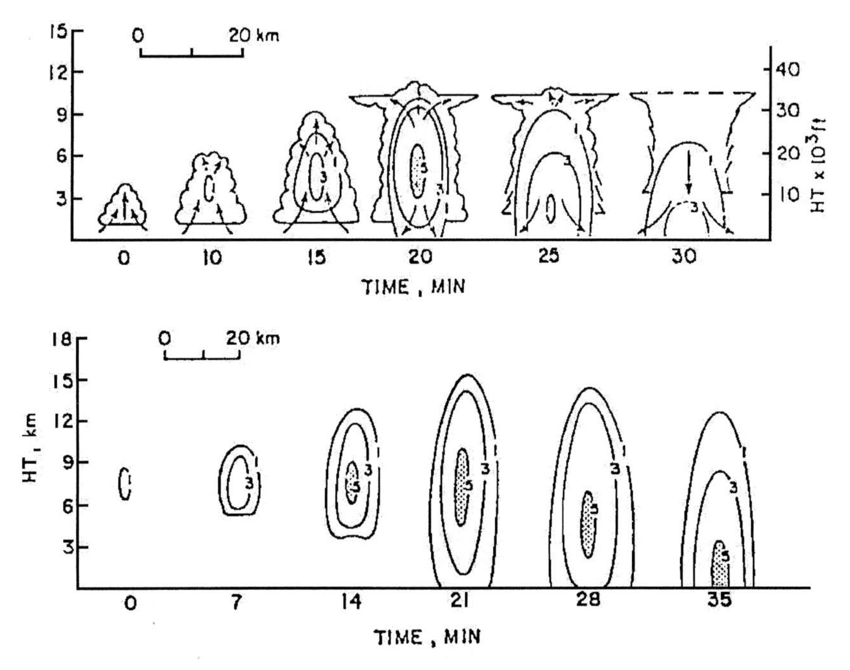

FIG 11.22. (a) Ordinary, weak shear thunderstorm evolution seen in vertical section. Counters are log Z, greater than 50 dBZ dark hatching. (b) Pulse severe thunderstorm evolution seen in vertical section. Contours and hatching are as in (a). Note the symmetry and vertical updraft stance of this storm type (from Chisholm and Renick 1972).

Nevertheless, radar-based warnings were dominated by low-level signatures until Lemon (1980) devised the volumetric radar scan method, which is the basis for the WSR-88D automated scan strategy. Given that the strongest updrafts are most likely to produce severe weather, the significance of volume scans is that changes in the shape and intensity of the 3D reflectivity structure reflect changes in the updraft's strength. Strong updrafts result in pronounced reflectivity cores exhibiting large values of reflectivity aloft and a core of descending precipitation downwind from the updraft. Forecasters must understand that it is the 3D updraft orientation that is directly affected by vertical wind shear, not the shape of the 3D radar echo. Weak updrafts in sheared environments have considerable downshear slope, whereas stronger updrafts are more erect (Fig. 11.23) and persistent. When upshear tilt of the strong reflectivity core aloft is present, it is indicative of sloping echo overhang, partially suspended above an intense updraft and partially descending around the periphery of the updraft while encountering backing winds with decreasing height (Figs. 11.23b,c).

The significant severe storm radar features (Figs. 11.23b and 11.23c) include 1) a strong horizontal reflectivity gradient at low levels (often accompanied by a reflectivity concavity that suggests the presence of strong low-level inflow winds intruding into the updraft–FFD interface region); 2) a weak echo region (WER, partially coincident with the strong updraft) or, with the most intense storms, a bounded weak echo region (BWER, Figs. 11.20 and 11.23c); 3) a high-reflectivity, sloping echo overhang above the WER; and 4) top displacement above the right-rear storm flank. The evolution in Fig. 11.23 is typical of isolated multicell and supercell storms occurring with considerable vertical shear and southwest flow aloft. Perhaps variations of this structure are just as common. In northwest flow aloft, the right-rear flank updraft (with midlevel overhang and top displacement) often resides on the west or northwest storm flank. With southerly flow aloft, the updraft may be located on the south or southeast flank. Severe storms occurring in environments with weak SR, midlevel winds (e.g., HP supercells) have front flank updrafts; hence radar WERs, overhang, and top displacement are front flank features. Intense front flank updrafts are topped by strong reflectivity overhangs that tilt in a downshear direction from the descending reflectivity core. It is this 3D radar configuration that has been mistaken by forecasters as being caused by precipitation being blown downwind by strong SR winds. Such radar structures do occur, but in the form of

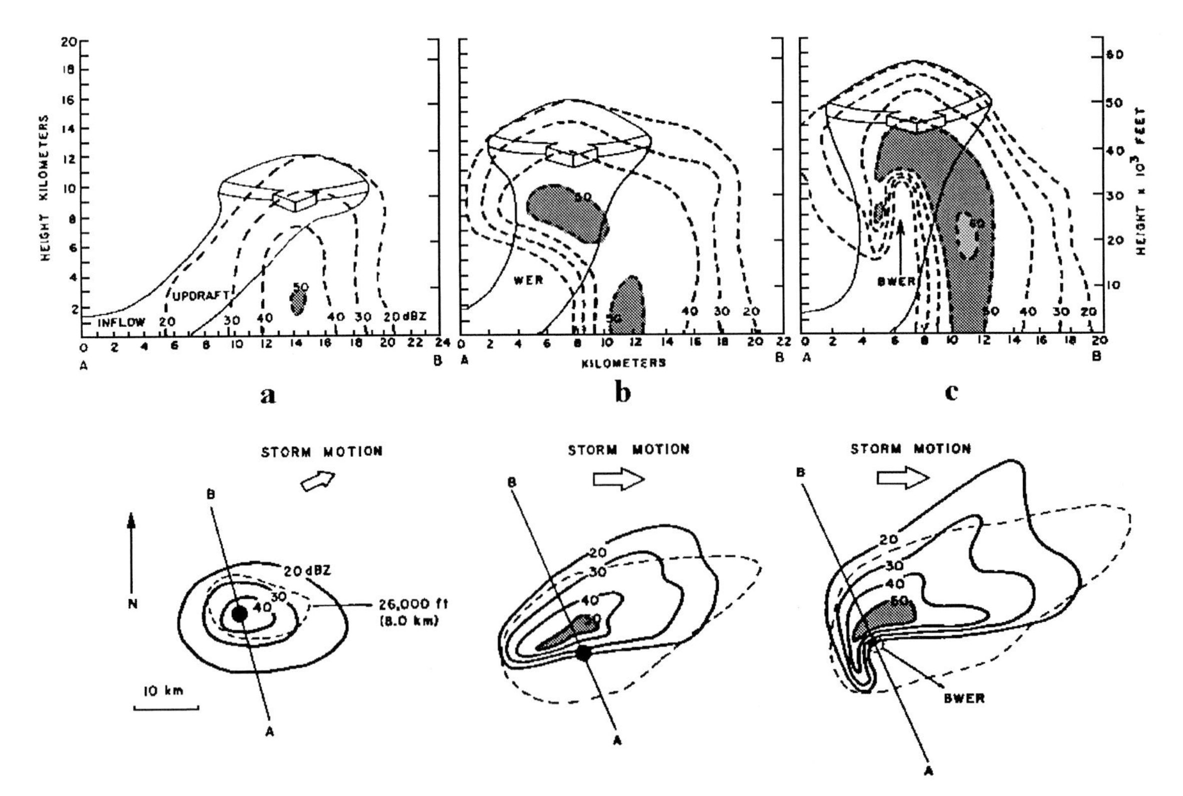

FIG 11.23. Multicell to supercell evolution. (a) Schematic view of a vertical cross section (upper, looking west) and a plan view (lower) of a nonsevere multicell storm. (b) As in (a), but for a severe multicell storm with a strong updraft. (c) As in (a) but for a supercell storm (from Lemon 1980).

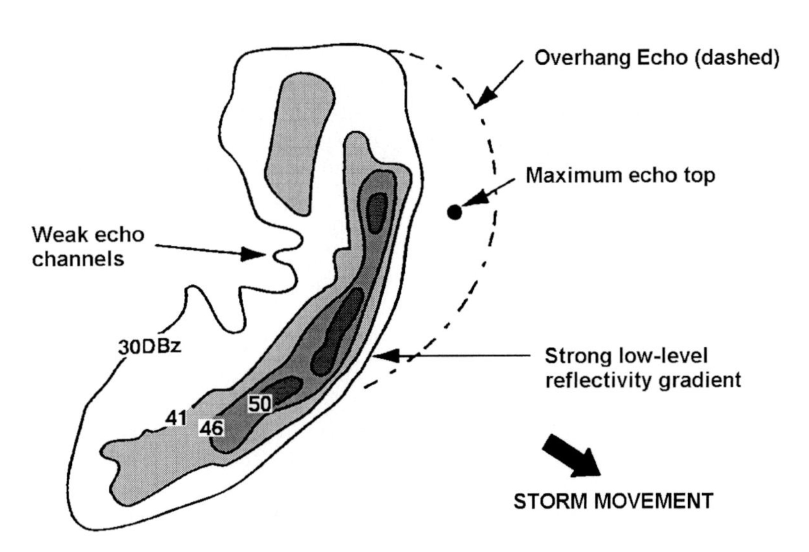

FIG 11.24. Schematic diagram of a plan view of a "distinctive" bow echo, exhibiting front flank overhang and top displacement above the updraft (from Przybylinski and Gery 1983).

weak reflectivity anvil echoes, not strong reflectivity updraft overhangs.

These front flank updraft and reflectivity characteristics occur with severe storms embedded within squall lines. Przybylinski and Gery (1983, hereafter PG83) show that echo overhang and top displacement are present on the leading edge of "distinctive bow echoes" (Fig. 11.24). Weak bow echo updrafts (and line segment updrafts in general) often tilt upshear with height, because of an imbalance between inflow and stronger outflow. Large hail and downbursts are possible with the stronger, more erect bow echo updrafts. Damaging winds are the primary threat with upshear-tilted updrafts. Numerically modeled storms show that severe bow echoes exhibit the upshear tilt prior to the most active (erect updraft) phase, and again in the weakening stage (Weisman 1992).

Warning forecasters must properly interpret the various 3D radar echo morphologies to estimate updraft strength and determine the severe weather potential. Moreover, they must be able to modify these conceptual models quickly, since they will often encounter variations (mainly hybrid storms) when using high-resolution WSR-88D data.

2) HAIL DETECTION

Perhaps hail is the severe phenomenon most easily "detected"[9] when using the Lemon technique (BL90). Given favorable atmospheric conditions, large hail is likely with storms that exhibit large reflectivities aloft (50 dB*Z* above 8 km AGL; BL90), especially in the presence of midlevel overhang and echo top displacement. The most damaging hail falls are associated with the well-developed BWERs of supercells. Low-level reflectivities can be indicative of ongoing hail events. Spring southern plains storms almost always produce hail when low elevation reflectivities are 60 dB*Z* or greater (Witt 1996). Similar observations were made by Geotis (1963) in New England and by Waldvogel and Federer (1976) in Switzerland. Another signature that is observed with severe hail storms is the *flare echo* (Wilson and Reum 1986) or *three-body scatter signature* (Zrnic 1987). Flare echoes are more common on horizontally polarized radars (such as the WSR-88D). Examples can be found in Lemon (1998) and Nielson-Gammon and Read (1995). These signatures are observed in the midlevels, often resulting in up to 30 minutes of warning lead time (Lemon 1998).

Digital processing techniques used in severe storm detection were generated as a result of the development of the Video Integrator and Processor (VIP), associated digital data, and early research of Wilson (1970), Saffle (1976), and Elvander (1977). Greene and Clark (1972) developed the quantity known as *vertically integrated liquid* (VIL), which is actually integrated precipitation, much of it in the ice phase (BL90). Devore (1983) and Winston and Ruthi (1986) improved VIL, using the results effectively in Oklahoma warnings. NSSL algorithms have changed VIL from a vertically integrated *grid point* product to a *cell-based* product that accounts for high-reflectivity storm overhang.

NSSL research has improved the WSR-88D reflectivity-based hail detection algorithm (HDA; Witt et al. 1998). HDA output includes probabilities of hail

[9] Detection is implied by the presence of radar structures strongly associated with the occurrence of hail, such that positive warning lead times can be achieved.

(POH) and severe hail (POSH), and maximum estimated hail size (MEHS). POSH uses cell-based, rather than grid-based, reflectivities, similar to the improved VIL product. Environmental data are incorporated into POSH, including the height of the 45-dBZ echo above the freezing level, a temperature-weighted vertical integration (since growth for severe hail usually occurs near −20°C), and low-level environmental thermodynamics for hail melting considerations. Although storm-top divergence is a reliable indicator of maximum hail size, a combination of velocity aliasing errors, range folding, and coarse vertical sampling degrade the storm-top divergence product. Thus, as of 2000, it was not being utilized in calculating MEHS.

Initial NSSL tests show a 50% reduction in false alarm rates for POSH. The algorithm performed well, especially for MEHS, at WFO Forth Worth tests in the 1990s. When MEHS failed, environmental shears were generally weak, and 3D radar structures, although with large reflectivity and VIL values, exhibited little overhang and top displacement. Therefore, a human–machine mix provided confidence in using the hail algorithm products effectively. These hail algorithms were developed in the southern plains, and modifications likely will be needed to use them elsewhere (the main problem in some areas appears to be a relatively high number of false alarms). Perhaps algorithms are most useful during major severe weather outbreaks, when the time factor limits human radar interrogation of a great number of storms.

3) Downburst detection

Severe convective winds are more difficult to detect than hail. Nevertheless, Doppler velocity data, reflectivity patterns, and algorithms provide useful information. The NSSL downburst algorithm [based mainly on midlevel convergence (Eilts et al. 1996b)] can provide lead time, but it is limited to maximum velocity range (230 km) from the radar site. Low-level base velocity detection of downbursts (offering no lead time) is limited to ranges of less than about 50 km by the radar horizon. Radar cell shapes (bow echoes, etc.) can imply a downburst threat, particularly when used with 3D reflectivity structure (e.g., distinctive bow echoes, PG83). PG83 identify rear inflow notches (alternately called *RINs* or *weak echo channels*; Fig. 11.24) that are related to the presence of damaging winds, probably as a result of descending rear-inflow jets. RINs and the occurrence of damaging winds often precede the maximum distortion of bow-shaped reflectivity structures; thus identification of RINs may be necessary for achieving warning lead time.

Supercells (mainly HP storms) can produce extreme wind events. Lemon and Burgess (1993) note the presence of a deep convergence zone (DCZ) at the interface of downdraft and updraft in an Oklahoma HP supercell that produced extensive severe weather, including downbursts. Lemon and Parker (1996, hereafter LP96) detected a DCZ in their study of an extreme (55 m s^{-1} winds and 11 × 17 cm maximum hail size) HP supercell in northern Oklahoma. The DCZ extended to a depth of 10 km and a length of 50 km, and was characterized by strong, deep, and long-lived Doppler velocity convergence. LP96 theorize that deep-layer, mass convergence–related flow accelerations in the DCZ could have combined with negative buoyancy to produce the intense winds. Detection of DCZs may help discriminate between "ordinary" and extreme intensity HP supercells.

Przybylinski et al. (1995) identify a bow echo velocity feature they call *midaltitude radial convergence* (MARC), with convergence values reaching 25 s^{-1}. The MARC appears to be similar to the DCZ, although it was observed from 3 to 7 km (i.e., not as low as the DCZ's surface roots). Significant to warning considerations, the MARC was detected 24 minutes prior to the first occurrence of damaging winds. A follow-up study of MARCs (Schmocker et al. 1996) showed that the lead time for six bow echo/embedded supercell cases varied from 10 to 30 minutes.

Three-dimensional radar reflectivity characteristics and Doppler velocities, particularly midlevel convergence signatures, can help identify developing downbursts. Moreover, forecasters will have more confidence in their warning decisions by combining the severe weather output of WSR-88D algorithms with manual 3D radar analysis and environmental severe weather diagnosis. Even though wind and hail are treated as separate topics, many storms produce both damaging wind and hail, sometimes with catastrophic effects. The most dangerous of these storms is the supercell. A critical consideration is determining when tornado warnings should be issued for supercells.

d. Tornado detection

The concept for the WSR-88D emerged from Doppler radar tornado detection research in the 1960s and 1970s (e.g., Donaldson et al. 1969; Brown et al. 1978). It was discovered that high-resolution Doppler radar can detect two supercell circulations: mesocyclones and smaller, more intense tornadic vortex signatures (TVS) (Brown et al. 1978). Mesocyclones came to be associated with the rotating updraft, whereas TVSs became closely identified with the tornado (Burgess et al. 1975; Brown et al. 1978). Further, Donaldson (1970) proposed that size, shear strength, height, and temporal criteria be used to define a mesocyclone. Similar criteria categories were used for TVSs.

1) Mesocyclone signatures

For operational considerations, a Doppler radar circulation is called a mesocyclone when it has the following attributes: 1) a core diameter of 10 km or

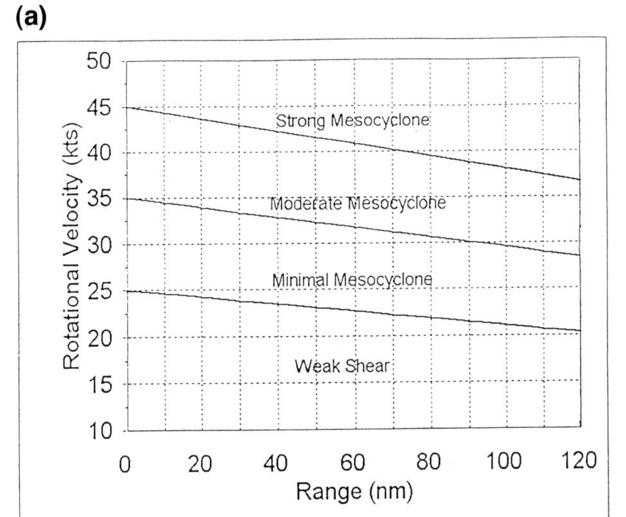

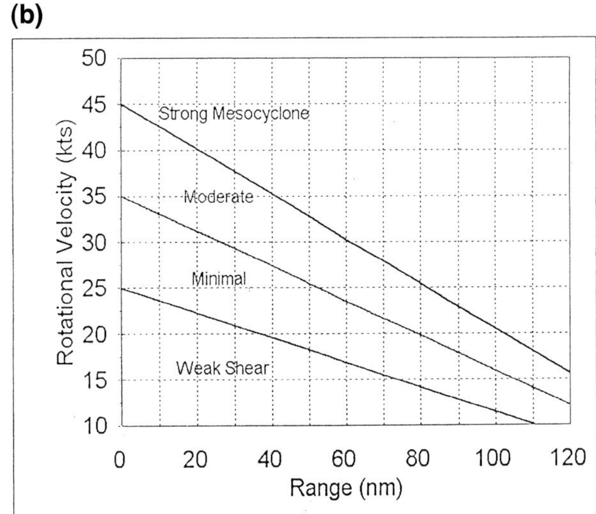

FIG 11.25. WSR-88D mesocyclone strength guidelines: (a) for mesocyclones with an approximate diameter of 6.3 km; (b) same as (a) but for a diameter of 1.9 km.

less, 2) range-dependent rotational velocity that equals or exceeds the minimum strength shown in Fig. 11.25, 3) vertical extent to 4 km, and 4) persistence for at least two volume scans. These criteria are arbitrary, which is somewhat of a problem; however, they have been engineered to include most of the storms that nearly everyone would call a supercell (C. Doswell 1996, personal communication). Because of numerous sampling problems, these criteria should be used in a probabilistic sense.

WSR-88D SR algorithms can detect "false" mesocyclones, often in areas of strong reflectivity and/or velocity gradients. Forecasters should use 3D analysis to verify the presence of supercell reflectivity structures (e.g., BWERs and extensive overhangs) whenever SR velocities suggest mesocyclones. Moreover, beyond the maximum velocity range of 230 km, the radar's superior reflectivity resolution often allows supercell detection. At intermediate to maximum reflectivity range, supercells frequently exhibit elliptical shapes, noticeably larger and with higher reflectivity values than neighboring nonsupercell storms (note the 4–10-km level echoes in Fig. 11.26). At intermediate ranges, the WSR-88D may detect BWERs, overhang, and top displacements in particularly tall supercells. Forecasters must rely on reflectivity data, spotters, and even satellite imagery (e.g., presence of the so-called enhanced-V signature, usually associated with supercell storms; McCann 1983) when outright mesocyclone detection is impossible (Jones et al. 1985).

A typical mesocyclone core[10] life cycle is composed of the following three stages (Burgess et al. 1982).

[10] The core is that part of the mesocyclone swirl with solid body rotation (i.e., the tangential velocity is proportional to the radius).

Organizing stage. The mesocyclone's growth stage begins in the midlevels near 5 km AGL (at least with mesocyclones following the cascade paradigm), with convergence beneath the mesocyclone base as it builds up and down with time.

Mature stage. This stage begins as the low-level mesocyclone forms and reaches maximum strength. Mesocyclone velocities have their highest values, and the vertical extent of the mesocyclone is perhaps through about-two thirds of the storm height. During the mature stage, the mesocyclone, previously embedded within the updraft, assumes a divided structure; that is, the mesocyclone center becomes located in the area separating the occluding rear-flank downdraft from the updraft (Lemon and Doswell 1979). Tornadoes are most likely at this time.

Dissipating stage. Weakening velocities and rapid decrease in height mark the demise of the mesocyclone. The circulation diminishes to a shallow depth and is associated with divergence.

Burgess et al. (1995) compare the life cycles of 18 tornado-producing MS storms (mostly from the northeastern United States), and Oklahoma mesocyclones from the 1982 study. They find that MS mesocyclones are smaller and weaker in the first two stages. In the organizing stage, Oklahoma mesocyclone core velocities and diameters are 20 m s^{-1} and 5.4 km, respectively, whereas MS mesocyclones had corresponding values of 13 m s^{-1} and 3.9 km. The mature stage values are for deeper mesocyclones (extending through the mid- and low level), with Oklahoma values of 25 m s^{-1} and 6.0 km (midlevels) and 23 m s^{-1} and 5.4 km (low levels). The MS values are 17 m s^{-1} and 3.7 km (midlevels) and 15 m s^{-1} and 3.5 km (low levels). The Oklahoma and MS mesocyclones have similar strength values in the dissipating stage.

(a)

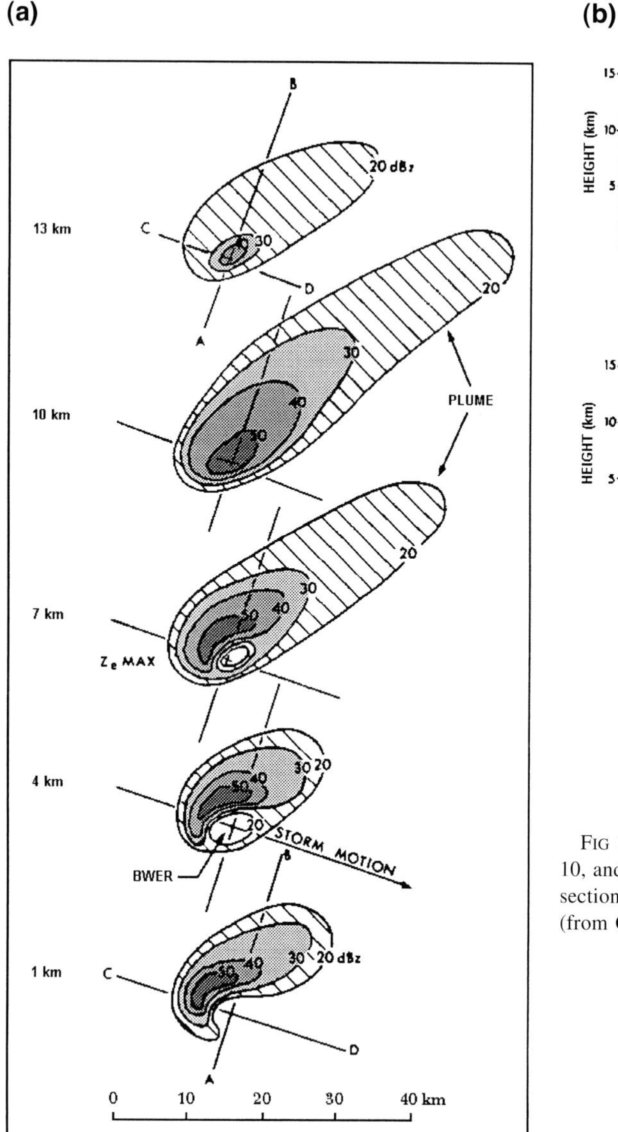

(b)

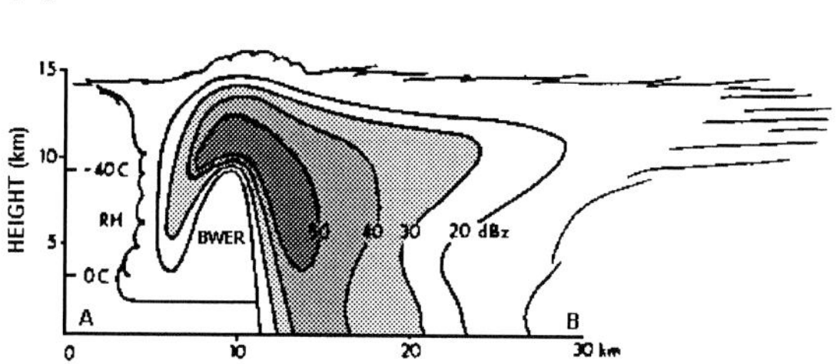

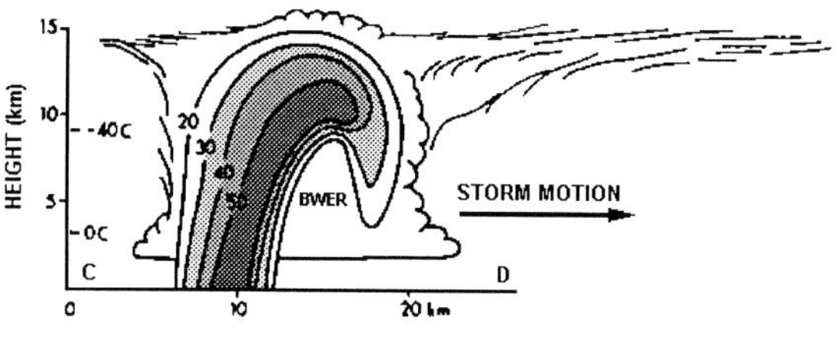

FIG 11.26. (a) Radar plan view of a supercell storm at 1, 4, 7, 10, and 13 km AGL, and (b) vertical sections of the storm along section A–B (top) and C–D (bottom). Reflectivities are in dBZ (from Chisholm and Renick 1972).

Most supercells have only one mesocyclone core. However, cyclic supercells produce a series of mesocyclone cores and sometimes a corresponding family of tornadoes (Figs. 11.27 and 11.28). The first mesocyclone in Fig. 11.27 follows the cascade evolution, with relatively long organizing and mature stages. Mesocyclone cores 2 and 3 form quickly through a large depth. This represents a different paradigm than the cascading of mesocyclone core 1. Burgess et al. (1982) show that the latter mesocyclones form along the gust front, several miles east of the preceding mesocyclone (Fig. 11.28). Low-level baroclinic vorticity generation (followed by tilting and intense updraft stretching) could be the reason that subsequent mesocyclones form with little lead time. *Once a cyclic, tornado-producing storm has been identified, warning forecasters must treat it as a tornadic storm until the reflectivity echo diminishes substantially in size and intensity. Thus, tornado warnings probably are required, regardless of short-term fluctuations in mesocyclone rotational strength. Forecasters should surmise that a cyclic storm continues to be supercellular if its large and intense reflectivity core remains intact. The rapid spinup of secondary mesocyclones and the fact that cyclic storms seem to be most common in outbreaks of particularly dangerous (i.e., strong and violent) tornadoes are ample reason to err on the side of caution. Warning forecasters should be especially sensitive to this axiom when a cyclic tornadic supercell is approaching a community of any size.*

Tornadic supercells may follow either of the mesocyclone paradigms. Research reveals that tornadoes

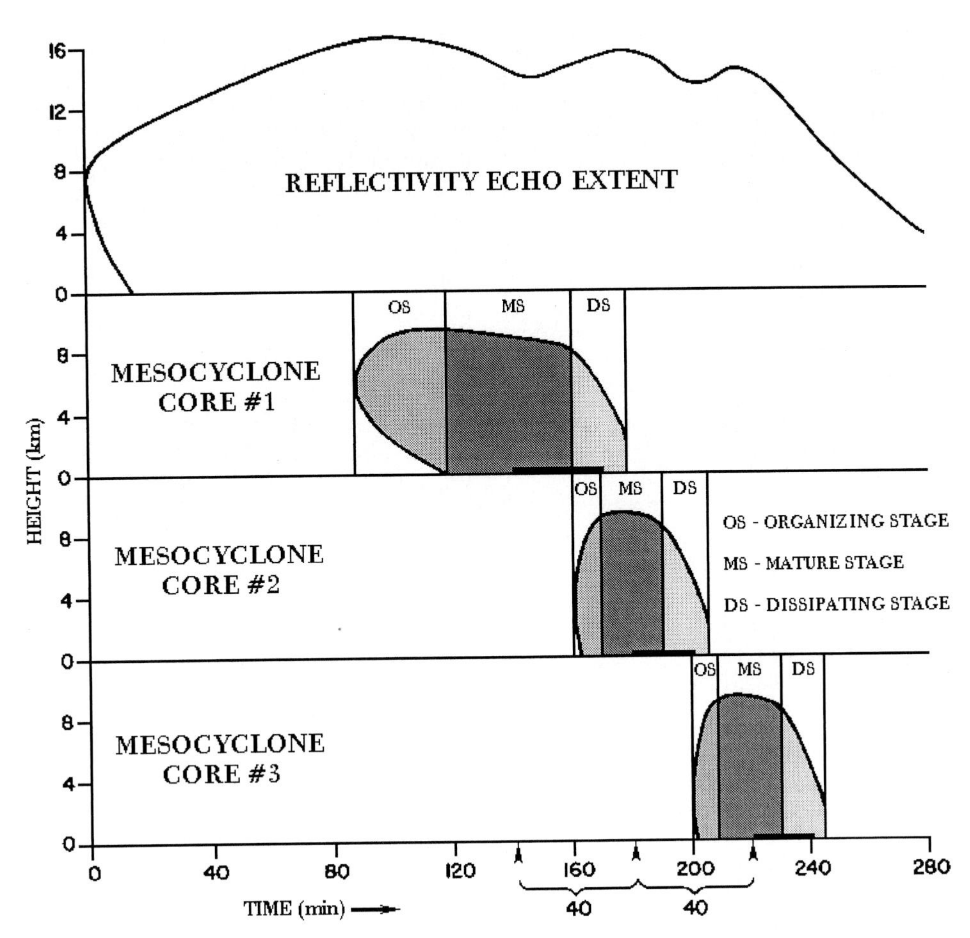

FIG 11.27. Time–height evolution of radar echo (top) and (bottom) mesocyclone cores. Solid horizontal bars are tornado occurrence; tornado formation interval is indicated along bottom (from Burgess et al. 1982).

that form quickly throughout their depth can occur when storms interact with external factors, such as orthogonally intersecting thermal boundaries (e.g., Fig. 11.29) that supply the baroclinic vorticity necessary for LLM formation. Atkins and Wakimoto (1995) document a large F3 tornado (sampled by a VORTEX Doppler radar equipped aircraft) that formed quickly through all levels within a rapidly forming cumulonimbus that was immediately west of a mature supercell. The tornado formed when the supercell's outflow interacted with the developing storm. There was limited evidence of a brief midlevel mesocyclone, which weakened prior to tornado formation.

Naturally, warning lead time with these tornadoes is considerably less than with the cascade paradigm events. However, both types of tornadic mesocyclones typically exhibit higher rotational velocities (and stronger shears) than nontornadic mesocyclones (Fig. 11.30; Burgess et al. 1982), which helps when identifying the most likely mesocyclone candidates for tornadogenesis. Mesocyclone detection algorithms have been hampered by high false alarm rates and only moderately successful probability of detection scores (Burgess et al. 1993). Ongoing NSSL attempts to improve the algorithm have shown a gradual increase in performance (Stumpf et al. 1998). Further, the improved algorithms incorporate new circulation classifications that help identify low-topped (MS) supercells and low-level circulations (not exceeding a height of 3 km) that may be tornadic. Nevertheless, although the probability of detection has increased somewhat, false alarms remain rather high.

2) TORNADIC VORTEX SIGNATURES

A TVS may be thought of as resulting from two velocity profiles, one small (tornado) and one large (mesocyclone), overlaid on each other (BDD93). For operational considerations, a TVS is defined as a rare, gate-to-gate[11] azimuthal shear (Fig. 11.31), associated with tornadic rotation, that is larger than or equal to established criteria for shear, vertical extent, and persistence. Shear is the most important criterion, and it is

[11] A circulation with TVS intensity may show maximum shear across more than two gates when in close proximity to the radar.

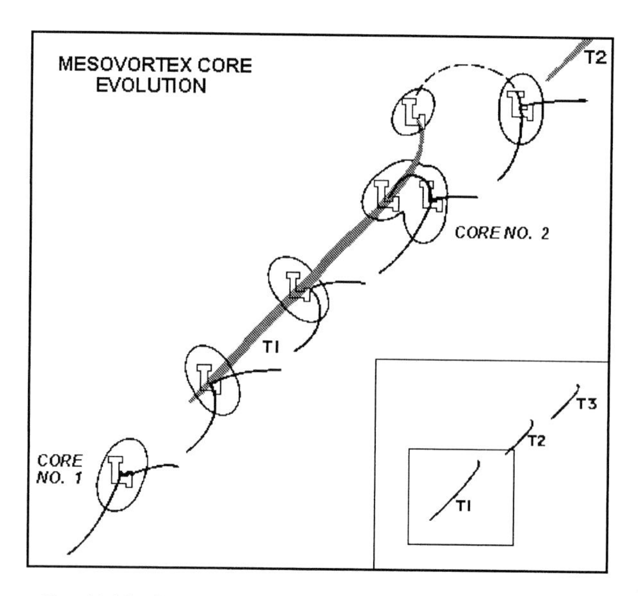

FIG 11.28. Conceptual model of mesocyclone core evolution. Thick lines are low-level wind discontinuities, and tornado tracks (exaggerated for size indication) are shaded. Insert shows tornado family tracks and the small square is the region expanded in the figure (from Burgess et al. 1982).

figured by adding the maximum outbound and inbound velocities of the small couplet and dividing by the distance between the two (the distance is essentially across one gate, and for simplicity's sake defaults to a value of 1). TVS rotational velocities are greater than or equal to 22.5 m s^{-1} for distances less than 60 km, and greater than or equal to 17.5 m s^{-1} for distances from 60 to 100 km. Less important are the vertical criterion (600 m or so) and temporal criterion (two volume scans), since some tornadoes that form in the low levels are very shallow. Tornadic winds are unlikely to be sampled beyond 100 km because of the aspect ratio problem and overshooting of the near-ground portion of the TVS.

The cascade paradigm was noted with the first observed TVSs (Brown et al. 1978). However, not all TVSs follow the cascade life cycle. Trapp et al. (1999) examine TVSs from 52 tornadic storms at 15 WSR-88D sites. They found that slightly more than 50% followed the cascade life cycle[12] (which the authors call a *descending TVS*), whereas the others formed almost uniformly over a depth of several km, or only in the lowest levels (nondescending TVS). Detection of descending TVSs will provide tornado warning lead time, while nondescending TVSs will not. Finally, line segment (squall line and bow echo) tornadoes are almost always of the nondescending variety, whereas supercell tornadoes can be of either type (Trapp et al. 1999).

Ongoing NSSL attempts to improve existing WSR-88D TVS algorithms (called the *tornado detection algorithm*, or TDA; Mitchell et al. 1998) have shown promise in that both supercell and nonsupercell tornadoes can be detected. The TDA is capable of 1) tracking persistent tornadic vortices, 2) forecasting their future positions, and 3) producing trends of various vortex measures. However, TDA false alarm rates can be large, and numerous radar sampling problems exist. Therefore, the TDA is not a self-sufficient means upon which to issue a tornado warning, but should be considered a complementary guidance tool in the tornado warning decision process (Mitchell et al. 1998). Knowledge of the environmental conditions and dedicated manual interrogation of WSR-88D data are also required.

A study of near WSR-88D tornadoes reveals that accelerated low-level inflow and strengthening convergence often occur in the SR velocity field below cloud base, immediately before tornadogenesis (Burgess and Magsig 1998). Strong mesocyclone rotation is often observed above these low-level features. A near-radar, 3D reflectivity phenomenon may foreshadow TVS detection at times. Foster et al. (1995a) describe a Texas MS storm where high-reflectivity echo aloft wrapped cyclonically around a WER and formed a BWER, preceding TVS (and tornado) formation by a few minutes. Similar radar features were observed

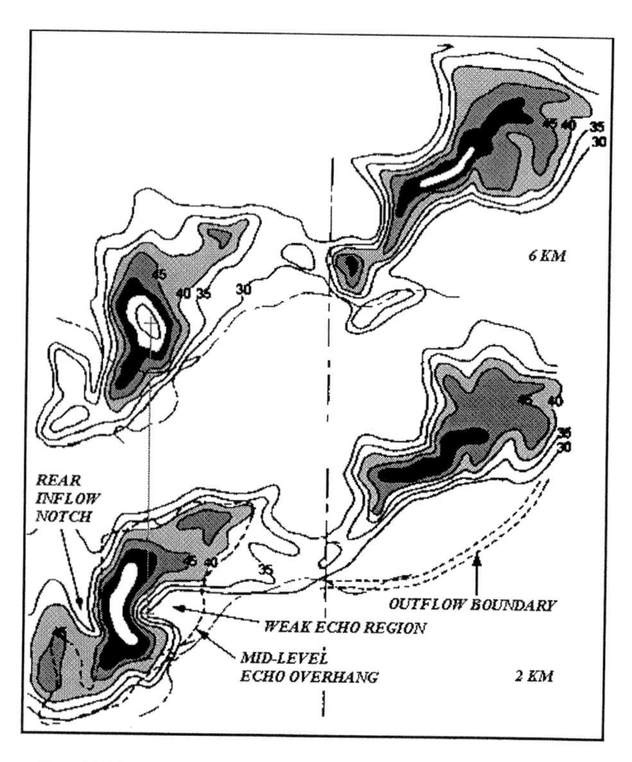

FIG 11.29. Example of a trailing supercell that produced a tornado while interacting with the outflow boundary of another thunderstorm (from Przybylinski et al. 1993).

[12] The authors conclude that, at times, descending TVSs may be an artifice of radar sampling.

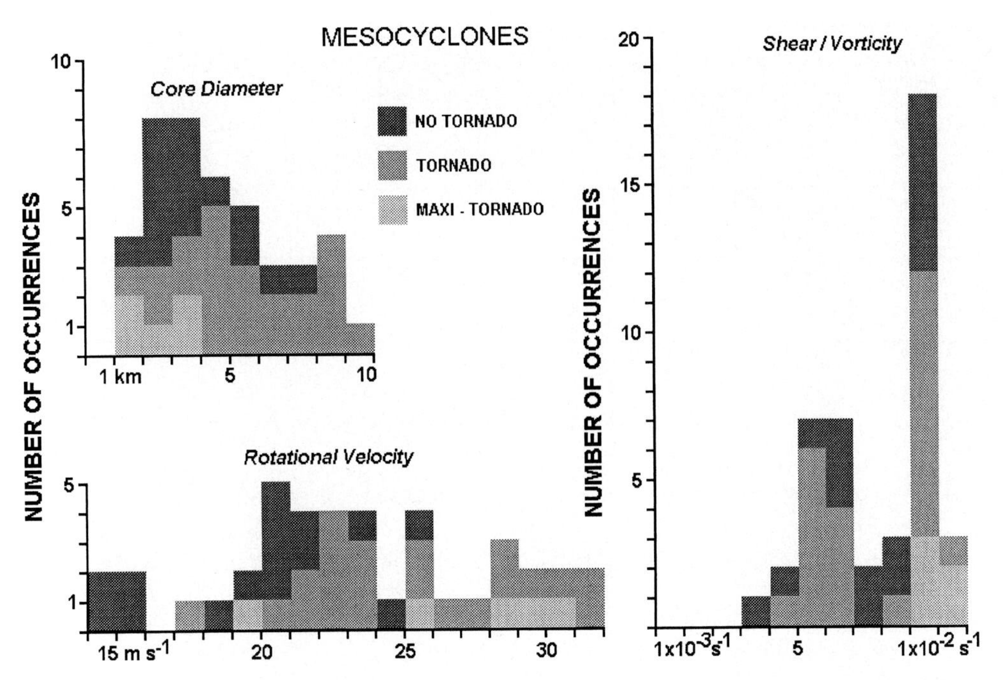

FIG 11.30. Characteristic parameters of mature supercell mesocyclones for tornadic and nontornadic storms. Tornado implies F0–F3 and maxi-tornado implies F4–F5 tornadoes (from Burgess et al. 1982).

with violent tornadoes that struck Tulsa, Oklahoma, in 1993 and Lancaster, Texas, in 1994, with each event occurring within 50 km of WSR-88D radars (M. Foster 1997, personal communication).

3) NONSUPERCELL (MISOCYCLONE) SIGNATURES

Misocyclone tornadoes are generally weak and short-lived; nevertheless, they are capable of causing fatalities and significant damage. Warning lead time for NSTs is limited by the small size of misocyclones (generally a diameter of 1 km or less) and by the fact that NSTs often form early in the parent storm's life cycle, sometimes before a radar echo is observed (BDD93). Further, the small, shallow nature of most misocyclones precludes their detection beyond 50 km (BDD93). Wakimoto and Wilson (1989) suggest that intersecting boundaries and boundaries characterized by strong horizontal shear and/or wavelike inflections should be recognized as areas of potential tornadogenesis. Their study of Colorado landspouts indicates that average warning lead times of 10–15 minutes can be attained for misocyclones that form close to the radar site. A few Oklahoma NSTs that have been documented offered virtually no lead time for warnings, possibly because of greater distances from the radar than with the Colorado NSTs (Burgess and Donaldson 1979). NSSL continues to test new automated vortex recognition procedures (e.g., the TDA) in an attempt to improve NST detection. Generally, spotter reports will trigger most NST warnings.

e. Severe storm spotters

Although a few spotter groups were active in the 1930s and 1940s (mainly as a response to the 1925 tristate tornado and a threat to military installations during World War II; Doswell et al. 1999), the NWS began organizing spotter groups in the 1950s. Some thought that the development of the WSR-57 radar network would eliminate the need for spotters. However, it has become obvious that radar and spotters are complementary. Radar detects storms and signatures of possible severe weather, whereas spotters are needed to detect actual severe events. The importance of spotters increased as communities took a more active role in disaster planning. Metropolitan spotter networks became more sophisticated as loosely coordinated groups of citizens, relying heavily on telephone communications, evolved into well-organized amateur radio networks. Storm spotters in rural areas include some amateur radio operators; however, the majority of rural spotters are law enforcement officials and volunteer firemen. Most storm spotters report from their homes, whereas others become mobile during severe storm threats. Mobile spotters attempt to maintain visual contact with a particular storm's updraft area, watching for evolving mesocyclone signatures (i.e., wall clouds, tornadoes, funnel clouds, and inflow bands) and visual indications of possible microbursts (rain and dust curls adjacent to rain shafts, etc.).

The increased reliance on sophisticated spotter networks resulted in a need for improvements in spotter

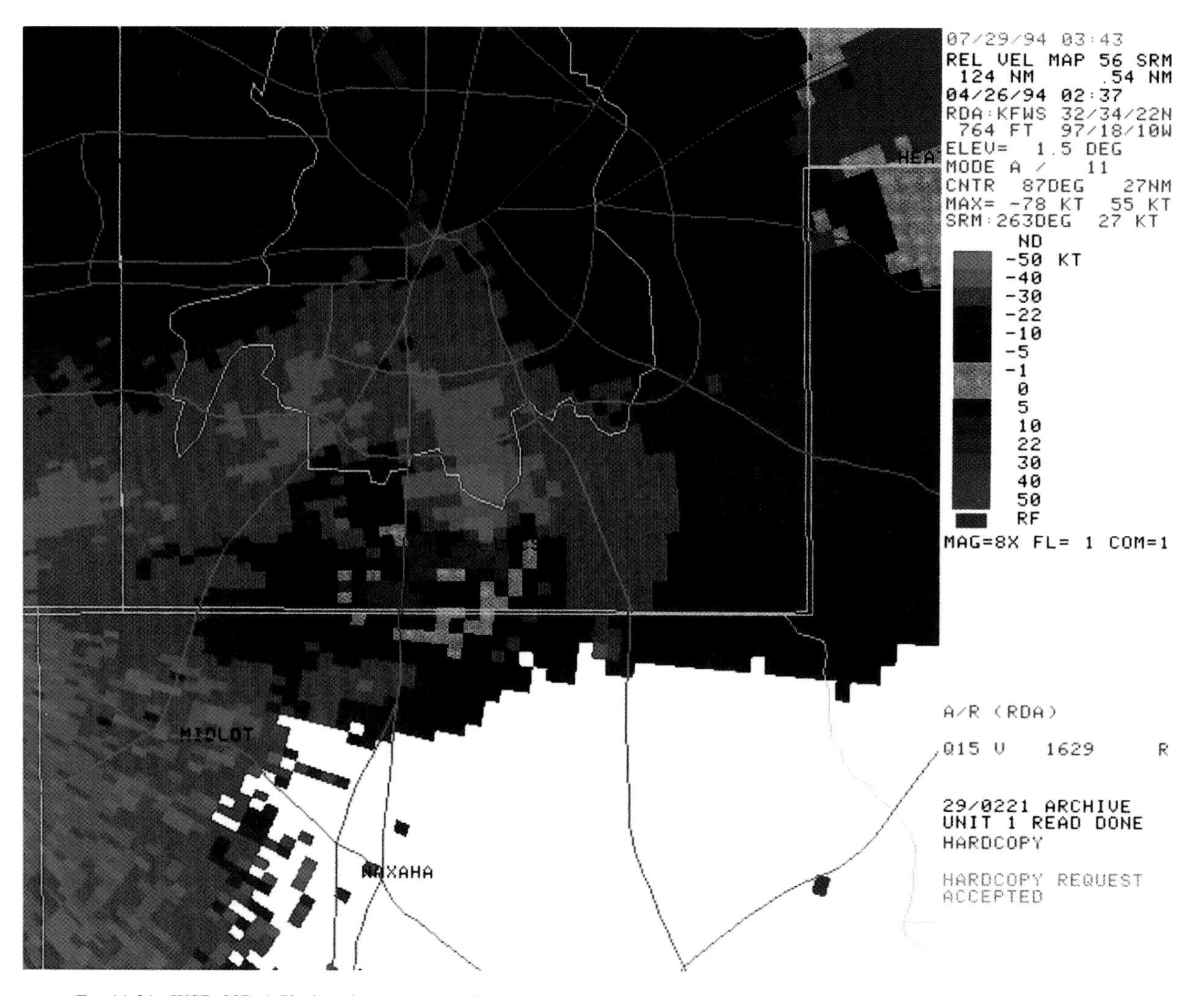

FIG 11.31. WSR-88D 1.5° elevation, storm-relative velocity display from Fort Worth, TX, of a TVS as an F4 tornado moved through Lancaster, TX, on 25 Apr 1994.

training programs. Results from the NSSL/Oklahoma University Tornado Intercept Project (Golden and Morgan 1972) were incorporated into an effort to modernize spotter training in the late 1970s. A basic storm spotter training movie and extensive slide set resulted (Moller 1978; Lemon et al. 1979). In time, storm spotters asked for more comprehensive information on storm structure and evolution. They were having problems distinguishing true indications of threatening weather from cloud features that only coincidentally resembled things such as wall clouds. The 1970s training may have concentrated on wall clouds to the extent that spotters were looking for tornadoes whenever any cloud lowering was sighted. Many spotters were left confused and NWS forecasters hesitant to issue subsequent warnings after dealing with nontornadic wall clouds and wall cloud lookalikes. Hence, an advanced spotter training slide set was developed in 1988 (Moller and Doswell 1988), followed by an accompanying video in 1995. An updated basic slide set (replacing the 1979 set) was produced in 1996 by G. Woodall.

Advanced training has emphasized more scientific topics, such as prototypical thunderstorm taxonomy, the atmospheric ingredients that result in different storm types, and the severe weather potential of these storms. Other topics include the evolution of mesocyclone flow patterns, including visual RFD identification and recognition of its life cycle during tornadic events. This information is intended to help spotters pass on timely IWS reports of tornado potential (e.g., that a mesocyclone has been undercut by cold outflow, or that an RFD is occluding, possibly leading to tornadogenesis). Advanced spotters learn that HP supercells account for most of the tornadoes that are enshrouded in rain, and that these tornadoes can be difficult to identify and hazardous to the spotter. Subjects such as Doppler radar, downbursts, and nonsupercell tornadoes are also presented in the advanced training. Storm spotter response to advanced training has been very positive.

The role of severe local storm spotters evolves as scientific understanding and technology advance. WSR-88D radars and enhanced storm detection methods suggest that more warnings will be issued before severe events are reported by storm spotters. Spotters will continue to give first indications of severe storm events at great ranges from radar sites (where radar limitations are maximized) and with events such as NSTs. However, spotters will be more involved in providing verification of severe events, rather than passing along reports that trigger warnings. Ground truth gives forecasters confidence that they made a correct warning decision and that storms following a similar evolution are also likely to become severe. Moreover, dissemination of severe reports through the warning system lets IWS users know that severe weather is occurring; thus appropriate life-saving measures can be put into effect (e.g., activating all-important local siren systems).

This was accomplished 35 minutes prior to the 1994 Lancaster, Texas, F4 tornado (Foster et al. 1995b). With a tornado warning in effect, two strongly worded severe weather statements were disseminated by WFO FTW, detailing amateur radio spotter reports of initial tornadoes that were approaching Lancaster, an indication that the probabilistic threat was increasing (a short time later a developing radar TVS heralded formation of a larger and more intense tornado; Fig. 11.31). Lancaster emergency officials sounded the outdoor warning sirens, and media meteorologists and announcers passed the information to an attentive public. Local officials and Lancaster citizens said that the wording in the severe weather statements was crucial to favorable public response during the emergency.

Many WFOs, emergency preparedness agencies, and electronic media stations use storm spotters for multiple purposes, including winter weather and flood events. Moreover, communication routes have increased between spotters and WFOs via e-mail, World Wide Web sites, NWS SKYWARN newsletters, and the Emergency Manager Weather Information (EMWIN) service. These factors have made it easier to recruit and train spotters in areas where severe convective storms are relatively rare, such as the northeastern United States. Severe storm spotter recruitment and training is but one component of the critical multiagency severe storm preparedness program.

11.4. Severe convective storm preparedness

There are three primary user groups in the IWS program (Friday 1988). These are

1. news media and private sector meteorologists,
2. emergency management officials (EMO) and storm spotters, and
3. the general public.

Sociological studies of natural disasters have shown that the NWS *must* form a partnership with these users and present them with an incisive meteorological risk analysis in order to prepare them for weather hazards (Riebsame et al. 1986). Preparedness efforts at many NWS offices have shown that achieving this goal is a vital but time-consuming and never-ending process (Stagno 1985; Moller et al. 1993).

The three IWS user groups consist of people with various knowledge levels and different needs for severe weather information (these factors weigh heavily in the composition of different severe storm dissemination products; see section 11.5). IWS groups must first be convinced of the necessity of severe weather preparedness (Stagno 1985). Of course, this is easier to accomplish in high-risk areas. After community officials have accepted the idea of severe storm preparedness, the work begins. This includes the implementation of training and education programs (e.g., severe storm safety education in schools, safety training of the general public via electronic and print media, and business and hospital safety programs). Technical severe storm training programs are established for storm spotters, EMOs, and the media. Coordination with the media is established through media symposiums, NWS visits, and interviews.

NWS officials typically provide the technical training for storm spotters and serve as experts in severe weather-related concerns. EMOs, media meteorologists, and NWS personnel all expend considerable effort in informing the public about the threat of severe weather and safety procedures. *Training is not the only objective of these meetings; the face-to-face contact is critical in developing the familiarity and trust that is needed in a functioning IWS.* The keys to developing a sound preparedness program are hard work and innovative thought and planning. If a certain technique does not work, preparedness trainers must try a different approach until they succeed (Stagno 1985). The users must be well informed; IWS officials cannot let them develop faulty assumptions, such as "Killer tornadoes don't happen here" or "Warnings are issued all the time and nothing has ever actually happened to us." Furthermore, funding for emergency preparedness and meteorological support is constantly being threatened by tightening budgets. The storm warning and preparedness community must work as a team to successfully articulate the urgency of the severe weather threat so that vital links are not eliminated from the IWS equation.

The risk of severe weather varies considerably across the nation. Many IWS members perceive that most of the truly dangerous convective events are relegated to the area of the plains states popularly known as "Tornado Alley." However, a substantial number of severe storm disasters occur outside of the plains. For example, tornado death tolls are higher in many states east of the Mississippi River than in the

Great Plains. The focus in many NWS offices away from Tornado Alley is, rightfully, on events other than severe convective storms. Nevertheless, in addition to whatever the most common threat may be, there remains a finite danger from severe convective storms, and flash flood events are possible anywhere. Development of sound preparedness planning is vital, no matter where the NWS office is located. Fortunately, the position of a full-time warning and coordination meteorologist at each NWS WFO makes this a realistic goal.

11.5. Severe convective storm dissemination products

As a WFO develops a preparedness program, it must determine how to disseminate relevant information to the three principal IWS user groups. Dissemination routes and products are ever more critical in an era of diminishing staffing, since it will become more important to reach as many people as possible with a single written or broadcast (NOAA Weather Radio, Weather Wire, etc.) release. Information exchange between emergency managers, storm spotters, media meteorologists, and the NWS also occurs over telephones, amateur radio, and pager systems. The most important concern is providing appropriately worded warnings to the public, to ensure that a maximum number of lives are saved and injuries prevented.

Mileti and Sorenson (1990) review about 200 studies of natural disaster warning systems from a social science perspective. They conclude that successful warning messages contain consistent, accurate, and clear information; guidance of what to do; at-risk locations; and confidence or certainty in tone. It is highly desirable that a unified warning message come from multiple channels (e.g., the electronic media, emergency broadcast stations, sirens, tone-alert radios, personal contact, and possibly, telephone ring-down systems) to have maximum impact. It is undesirable to give conflicting warnings to the public; the IWS must be a unified "team" effort. According to Mileti and Sorenson, in order to minimize rumors and misinformation, it is critical to send (read) the warning message as often as possible. The authors conclude that people must hear, understand, and believe the warning message, and the message must be personalized for them. This accomplished, citizens will decide what to do and carry out their response decisions. The authors strongly advise that public response to warnings must be monitored, with warning adjustments made as needed.

The NWS three-tiered system of severe local storm products includes outlooks, watches, and warnings. Severe weather statements and nowcasts (more often called *short-term forecasts*) provide important update information between issuances of the three major releases. Severe thunderstorm and tornado watches contain information concerning where and when the watch is valid, the definition of a watch, and meteorological information relevant to the particular watch. A very small number of watches are issued with statements that the weather situation is "particularly dangerous" (say, with a chance of very damaging derecho or tornado events). The watch program was being redesigned at the time of this writing, as part of the process of transferring some of the watch issuance responsibility from the SPC to WFOs. Watches in the new program likely will not be as restrictive as in the past (e.g., watches will probably be issued with as many as six outline points, rather than the current limit of four). The remainder of this discussion centers on severe local storm outlooks, warnings, and statements.

The Norman, Oklahoma, WFO pioneered severe weather outlooks in the 1980s, calling their product the Oklahoma Thunderstorm Outlook. IWS users (mainly EMOs, news media, and spotters) gave resounding support to the outlook, hence other NWS offices started their own severe weather outlooks (SWO). WFO FTW (and other WFOs) tailors its outlook with the media, EMOs, and storm spotters in mind. These users have many tasks to perform (e.g., most spotters have jobs and EMOs are charged with nonweather-related community duties) and cannot watch the weather continuously through the day. Therefore, the primary intent of the SWO is to give these users an early projection of when, where, and what types of severe weather and/or flash flooding are expected, so that these people can plan their day accordingly.

The FTW SWO is based on the SPC "day 1" severe weather outlook, using the SPC terminology of *slight, moderate,* or *high risk.* The first SWO of the day is issued at 0600 LST, with updates at noon and 1500 LST (if needed). FTW's SWOs contain some technical information, limited to initial positions and expected movements of upper-level disturbances and pertinent low-level boundaries, and basic reasoning for severe weather type expectations (e.g., a high vertical wind shear and instability day, resulting in a major tornado threat). The above descriptive wording is used for these phenomena, rather than more technical expressions such as storm-relative helicity, which would take far longer to explain in spotter training sessions.

There are many reasons behind the inclusion of relevant technical information (Moller et al. 1993). Several decades of spotter training have disclosed that serious storm spotters want to know the essential science of severe storms (thus, the advanced spotter training includes information such as the role of vertical wind shear and instability in supercell and tornado formation). Second, this background forecast information validates the "what, where, and when expected" portion of the outlook. Knowledgeable EMOs, media meteorologists, and spotters are more apt to understand NWS expectations and uncertainties about severe weather potential when they listen to a series of outlooks that describe the evolution of severe storm potential. Electronic media meteorologists and weather announcers, who have the critical task of sensitizing the general public in a particularly dangerous situation, will understand when to do so. Within the NWS, these outlooks have begun to prepare WFO meteorologists for severe weather forecasting in the modernized era, when the forecast emphasis is placed on short-range, mesoscale events (Friday 1988). Finally, SWOs have provided motivation to pursue scientific publications and other self-training exercises, thereby enhancing the quality of the product.

An example of an SWO follows.

SEVERE WEATHER OUTLOOK FOR NORTH TEXAS

NATIONAL WEATHER SERVICE FORECAST OFFICE, FORT WORTH, TEXAS ISSUED: 12 NOON CDT FRIDAY MAY 15 2000

A MODERATE RISK OF SEVERE THUNDERSTORMS CONTINUES THROUGH THIS EVENING FOR THAT PORTION OF NORTH TEXAS WEST OF A SHERMAN-FORT WORTH-WACO LINE. A SLIGHT RISK OF THUNDERSTORMS . . . MAINLY TONIGHT . . . EXISTS EAST OF THIS LINE.

AN UPPER-LEVEL DISTURBANCE WAS APPROACHING NORTH TEXAS FROM THE WEST AT NOON. AN ASSOCIATED DRYLINE . . . MOVING EAST AT 25 MPH . . . EXTENDED FROM A SURFACE LOW NEAR CHILDRESS TEXAS . . . TO MIDLAND AND MARFA. TO THE EAST . . . A STATIONARY THUNDERSTORM OUTFLOW BOUNDARY EXTENDED FROM NEAR THE SURFACE LOW TO FORT WORTH AND ATHENS.

THUNDERSTORMS WILL FORM ALONG THE DRYLINE AFTER 3 PM . . . WITH THE THUNDERSTORMS AND DRYLINE MOVING EAST TO NEAR A GAINESVILLE-FORT WORTH-HAMILTON LINE BY 6 PM. SOME OF THESE STORMS WILL CONTAIN LARGE HAIL . . . DAMAGING WINDS . . . AND POSSIBLY A FEW TORNADOES. THE GREATEST THREAT OF TORNADOES WILL BE NEAR THE INTERSECTION OF THE DRYLINE AND THE OUTFLOW BOUNDARY . . . WHERE THE VERTICAL WIND SHEAR WILL BE MAXIMIZED.

THE THREAT OF TORNADOES WILL DIMINISH BY MIDEVENING AS THE UPPER DISTURBANCE MOVES NORTHEAST AND VERTICAL WIND SHEARS DECREASE ACROSS NORTH TEXAS. BY THIS TIME THE DRYLINE WILL BECOME STATIONARY NEAR A PARIS-WACO LINE. HIGH INSTABILITIES WILL RESULT IN A CONTINUED THREAT OF LARGE HAIL AND DAMAGING WINDS TONIGHT ACROSS EAST TEXAS. HEAVY RAINS MAY ALSO OCCUR IN EAST TEXAS TONIGHT . . . AND A FLASH FLOOD WATCH MAY BE ISSUED LATER TODAY FOR PORTIONS OF EAST TEXAS.

STORM SPOTTERS AND EMERGENCY MANAGEMENT OFFICIALS WILL BE ACTIVATED THIS AFTERNOON IN WEST AND CENTRAL SECTIONS OF NORTH TEXAS AS CONDITIONS WARRANT. STORM SPOTTERS EAST OF THE DALLAS-FORT WORTH AREA MAY BE ACTIVATED WHEN THUNDERSTORMS APPROACH THE AREA THIS EVENING.

Virtually no technical information is included in statements, watches, and warnings, since the public has little need for such information, and there is no time for lengthy meteorological discussion once severe weather begins. However, for the benefit of the public and special users such as spotters and EMOs, attempts are made to include concise information about short-range expectations of severe weather. Such a nowcast may be worded as follows.

SHORT-TERM FORECAST

NATIONAL WEATHER SERVICE OFFICE, FORT WORTH, TEXAS

ISSUED: 535 PM CDT FRIDAY MAY 15 2000

A LARGE, SEVERE THUNDERSTORM WAS APPROACHING WEATHERFORD AT 535 PM THIS STORM WILL PRODUCE BASEBALL-SIZE HAIL AND 60 MPH WINDS IN THE WEATHERFORD TO ALEDO AREA THROUGH 615 PM . . . AS IT MOVES EAST AT 30 MPH. DAMAGING HAIL AND WIND ARE LIKELY TO ENTER TARRANT COUNTY . . . INCLUDING FORT WORTH . . . AFTER 615 PM

A SEVERE THUNDERSTORM WARNING REMAINS IN EFFECT TIL 630 PM FOR PARKER COUNTY.

A more dangerous event, such as a damaging tornado, may require even more brevity and dramatic wording. In such rare events, forecasters, EMOs, and media meteorologists can save scores of lives with well-written warnings and critical local warning systems (e.g., sirens). It is best to avoid "generic" warnings or statements in such situations and to treat the situation as an extraordinary event, conveying a sense of urgency to the public.

TORNADO WARNING

NATIONAL WEATHER SERVICE OFFICE, FORT WORTH, TEXAS

ISSUED: 630 PM CDT FRIDAY MAY 15 2000

THE NATIONAL WEATHER SERVICE HAS ISSUED A TORNADO WARNING FOR TARRANT COUNTY UNTIL 730 PM.

AT 630 PM DOPPLER RADAR DETECTED A DEVELOPING TORNADO 5 MILES WEST OF FORT WORTH. AMATEUR RADIO STORM SPOTTERS REPORT STRONG CLOUD BASE ROTATION AT THE INTERSECTION OF INTERSTATE HIGHWAYS 20 AND 30.

A TORNADO IS IMMINENT . . . RESIDENTS OF WEST FORT WORTH SHOULD TAKE SHELTER IN REINFORCED STRUCTURES NOW!

Frequent reports from a spotter network will allow for vital updates during a serious weather event.

SEVERE WEATHER STATEMENT
NATIONAL WEATHER SERVICE OFFICE, FORT WORTH, TEXAS
642 PM CDT FRIDAY MAY 15 2000
. . . DAMAGING TORNADO MOVING THROUGH WEST FORT WORTH . . . NEAR THE INTERSTATE 35 CORRIDOR . . .
. . . TAKE INTERIOR SHELTER NOW! . . .
A TORNADO WARNING REMAINS IN EFFECT FOR TARRANT COUNTY TIL 730 PM.

There are three types of "goodness" of weather forecasts (Murphy 1993). They are 1) Consistency—How well does a forecast correspond to the forecaster's best judgement of the weather? 2) Value—What are the benefits (or losses) to users? 3) Quality—How well do the forecasts and observations correspond to each other? The IWS for severe convective storms (and flash floods) can result in the highest possible forecasting value: the prevention of fatalities and injuries. It has been estimated that on the order of 10 000 tornado deaths have been prevented since the inception of the NWS watch/warning program in the early 1950s (Doswell et al. 1999). Moreover, both human injury and monetary damage can be reduced when mitigation procedures are included in preparedness planning. For instance, new building designs, relatively inexpensive retrofit for existing structures, and inclusion of fortified interior shelters in schools and homes can save lives and money in tornado- and windstorm-prone areas (McDonald 1993; Marshall 1993; Harris et al. 1993).

Interactive feedback among NWS and news media meteorologists, and other IWS users, combined with scientific education and training, can help IWS meteorologists improve the consistency of their life-saving severe storm forecasts and warnings. Finally, proper application of the ever-improving technology and severe storm sciences allow the meteorological community to improve the quality of their severe storm forecasts and warnings.

11.6. Discussion

Severe local storms pose a formidable worldwide threat to people and property. An integrated warning system, consisting of severe storm forecasting, severe storm detection, warning dissemination, and public response, can minimize the devastating effects of tornadoes, large hail, damaging winds, and flash flooding. Integrated warning system failure results from a breakdown of one or more of these four basic components.

National Weather Service severe storm forecasting in the United States is shared by the national Storm Prediction Center and local weather forecast offices. Severe storm forecasting is a critical first step in the integrated warning system, since accurate severe storm warning dissemination is impossible without prior knowledge of environmental conditions. Furthermore, accurate forecasts allow media meteorologists, emergency preparedness officials, and storm spotters to plan their workday around the expected timing of a severe storm threat.

Meteorological forecast routines must start with a thorough diagnosis of pertinent atmospheric processes. Extensive diagnostic work that is accompanied by a quest for physical understanding allows forecasters to develop, and revise, four-dimensional atmospheric models. Forecasters who attain this working level will be more successful in combining their diagnosis with objective synoptic and mesoscale numerical model output to develop a trend. It is imperative that forecasters use sound analysis and diagnosis of observations along with numerical model output to formulate forecasts. The rapidly changing nature of meteorological parameters in severe storm outbreaks invariably necessitates diagnostic scrutiny of observational data. Successful forecasting of the 3 May 1999 tornado outbreak in central Oklahoma was dependent on wind profiler observations of an upper-level jet stream that was deeper, stronger, and lower in the atmosphere than was depicted by any numerical model forecasts.

While values of certain scientific parameters may best describe the causal ingredients for severe storms, a lack of complete physical understanding and profound spatial and temporal sparsity of meteorological data necessitate the use of conceptual modeling, meteorological chart pattern recognition, climatology, persistence, and geographical considerations in the forecast process. Forecasters first evaluate the ingredients of moisture, instability, and lift to determine the likelihood of storm formation.[13] Convective storms are dependent on the mesoscale lift associated with low-level boundaries and significant terrain features. Although synoptic-scale lift is insufficient for storm initiation, it provides thermodynamic destabilization through hours of relatively weak lift and subsequent steepening of lapse rates prior to the onset of deep convection. Forecasters diagnose the values of vertical wind shear and instability to determine the threat level of various types of convective storms. Knowledge of a range of possible storm types and evaluation of certain parameters allow forecasters to ascertain the probability of damaging winds, large hail, and tornadoes.

[13] The precise timing and location of convective storm initiation are poorly understood. Therefore, a critically important research project called the Thunderstorm Initiation Mobile Experiment (TIMEX) is in the formative stage at the publication time of this monograph.

In weak shear environments, short-lived pulse storms can result in damaging downbursts, marginally severe hail, and infrequent nonsupercell tornadoes. Larger values of vertical wind shear often result in longer-lived, more organized storm structures, such as multicell, squall line, and supercell storms. The most dangerous thunderstorms are supercells, bow echo complexes, microburst-producing storms (which are particularly hazardous to aircraft), and flash flood-producing storms. Supercells are characterized by the presence of a persistent, strongly rotating updraft, called a *mesocyclone.* They produce most strong to violent tornadoes and significant falls of large hail. Moreover, supercells can produce intense rainfall rates over a short period of time and extreme downburst windstorms. Bow echoes are responsible for a significant fraction of extreme intensity windstorms. Some bow echoes contain supercells; these complexes pose an even greater threat of devastating weather. The details of how these storms produce their severe weather events are still somewhat speculative; further research is needed. Because of the lack of definitive causal knowledge and the paucity of observations, severe local storm forecasting must be viewed in a probabilistic, rather than deterministic, sense.

Historically, severe storm detection has been achieved through combined use of weather radar and severe storm spotters. The WSR-88D radar system has dramatically increased NWS WFO capabilities to detect the most dangerous severe storms. Nevertheless, information gained through use of the more precise radar data has resulted in a need for drastic revision of existing conceptual models. For instance, as late as 1990 it was believed that about 50% of all supercell storms produce tornadoes. The WSR-88D network has revealed that more storms are supercellular than was previously thought, and it is likely that less than 20% of all supercells produce tornadoes. Additional research is needed to help warning meteorologists and storm spotters better discriminate between tornadic and nontornadic supercells, thereby decreasing the tornado warning false alarm rate.

Ever-improving versions of WSR-88D automated algorithms help warning meteorologists to identify the storms with the greatest severe weather potential. Nonetheless, these algorithms are not perfect, and WFO radar operators must be dedicated to four-dimensional reflectivity and velocity surveillance of storm morphology and trends. Furthermore, forecasters must have an acute sense of situation awareness to be effective during severe storm warning episodes. Knowledge of environmental conditions is necessary in making appropriate warning decisions on any given day. For instance, meteorological conditions on a particular day may favor supercells with midlevel mesocyclones, but without organized low-level rotation. Forecasters who correctly diagnose the situation may consider tornado warnings inappropriate. However, should a supercell encounter a low-level thermal boundary (often subtle in nature and hard to detect without scrupulous diagnostic work), low-level mesocyclone formation and tornadogenesis may occur rapidly.

Situation awareness also includes sociological concerns, such as forecaster cognizance of outdoor events that can result in considerable vulnerability to severe storms. Local weather forecast offices must also be attentive to staffing concerns, such as maximizing staff performance and minimizing distraction through adherence to well-defined work assignment. Worker fatigue during prolonged warning events can be minimized by changing work assignments midway through a severe weather episode.

Because of their ability to provide ground-truth information, storm spotters will remain a critical element of the severe storm warning program for years to come. Spotters and radar are complementary in that the limitations of radar data (spotters) are often canceled by the strengths of spotters (radar data). Radar can detect internal storm reflectivity and velocity structures that are invisible to storm spotters. Spotters can observe specific severe events that are not directly detectable by radar.

Finally, the integrated warning system fails when severe local storm warning information is disregarded by the public. Therefore, warning dissemination products must contain consistent, accurate, and clear information, and offer guidance on what to do, location of risk areas, and confidence of tone. Sociological studies indicate that it is highly desirable that a unified warning message come from multiple channels, including the electronic media, emergency broadcast stations, sirens and other community warning systems, tone-alert radios, and personal contact. Citizens react most favorably to warning messages when they hear consistent information coming from different sources. *Thus, conflicting warnings are undesirable. In order for the IWS to succeed and elicit favorable public response, a spirit of cooperation and teamwork must be established among members of the news media, private sector meteorologists, local emergency management officials, storm spotters, and the National Weather Service.*

Acknowledgments. I am grateful to reviewers Ed Szoke, Ron Przybylinski, Steve Weiss, John Monteverdi, Kevin Pence, and review committee chairman Bob Johns. Their input was extremely comprehensive, resulting in many improvements to the final paper. I am indebted to Jason Jordan for his extensive help in preparing the figures. Skip Ely, Mike Foster, and Chuck Doswell provided many insightful discussions. Tim Vasquez, John Finch, Harald Richter, Herbert Fiala, and Bill McCaul contributed important information relating to the contents of Table 11.1.

REFERENCES

Allen, G., 1981: Aiding the weather forecasting: Comments and suggestions from a decision analytic perspective. *Austr. Meteor. Mag.,* **29,** 25–29.

Allen, R. H., D. W. Burgess, and R. J. Donaldson, 1981: Attenuation problems associated with a 5-cm radar. *Bull. Amer. Meteor. Soc.,* **62,** 807–810.

Andra, D., and M. P. Foster, 1992: Operational strategies for the WSR-88D radar. Preprints, *Symp. on Weather Forecasting,* Atlanta, GA, Amer. Meteor. Soc., 122–124.

Anthes, R. A., 1977: A cumulus parameterization scheme utilizing a one-dimensional cloud model. *Mon. Wea. Rev.,* **105,** 270–286.

Atkins, N. T., and R. M. Wakimoto, 1991: Wet microburst activity over the southeastern United States: Implications for forecasting. *Wea. Forecasting,* **6,** 470–482.

———, and ———, 1995: Observations on the origins of rotation: The Newcastle tornado during VORTEX'94. Preprints, *27th Conf. on Radar Meteorology,* Vail, CO, Amer. Meteor. Soc., 1–3.

Augustine, J. A., and E. J. Zipser, 1987: The use of wind profilers in a mesoscale experiment. *Bull. Amer. Meteor. Soc.,* **68,** 4–17.

Barnes, S. L., and B. R. Coleman, 1994: Diagnosing an operational numerical model using q-vector and potential vorticity concepts. *Wea. Forecasting,* **9,** 85–102.

Beebe, R. G., 1955: Types of air masses in which tornadoes occur. *Bull. Amer. Meteor. Soc.,* **36,** 349–350.

Bentley, M. L., and T. L. Mote, 1998: A climatology of derecho-producing mesoscale convective systems in the central and eastern United States, 1986–1995. Part I: Temporal and spatial distribution. *Bull. Amer. Meteor. Soc.,* **79,** 2527–2540.

Blanchard, D. O., 1998: Assessing the vertical distribution of convective available potential energy. *Wea. Forecasting,* **13,** 870–877.

Bluestein, H. B., 1985: The formation of a "landspout" in a "broken-line" squall line in Oklahoma. Preprints, *14th Conf. Severe Local Storms,* Indianapolis, IN, Amer. Meteor. Soc., 267–270.

———, 1993: *Synoptic-Dynamic Meteorology in Midlatitude.* Vol. II, *Observations and Theory of Weather Systems,* Oxford University Press, 594 pp.

———, and C. R. Parks, 1983: A synoptic and photographic climatology of low-precipitation supercells in the Southern Plains. *Mon. Wea. Rev.,* **111,** 2034–2046.

———, and M. H. Jain, 1985: Formation of mesoscale lines of precipitation: Severe squall lines in Oklahoma during the spring. *J. Atmos. Sci.,* **42,** 1711–1732.

———, and G. R. Woodall, 1990: Doppler-radar analysis of a low-precipitation severe storm. *Mon. Wea. Rev.,* **118,** 1640–1664.

———, E. W. McCaul Jr., G. P. Byrd, and G. R. Woodall, 1989: Mobile sounding observations of a tornadic thunderstorm near the dryline: The Gruver, Texas storm complex of 25 May 1987. *Mon. Wea. Rev.,* **117,** 244–250.

Bothwell, P. D., 1988: Forecasting convection with the AFOS Data Analysis Program (ADAP). NOAA Tech. Memo. SR-122, NWS Southern Region, Fort Worth, TX, 92 pp.

Brady, R. H., and E. J. Szoke, 1989: A case study of non-mesocyclone tornado development in northeast Colorado: Similarities to waterspout formation. *Mon. Wea. Rev.,* **117,** 843–856.

Braun, S. A., and J. Monteverdi, 1991: An analysis of a mesocyclone-induced tornado occurrence in northern California. *Wea. Forecasting,* **6,** 13–31.

Brooks, H. E., and C. A. Doswell III, 1993: Extreme winds in high-precipitation supercells. Preprints, *17th Conf. on Severe Local Storms,* St. Louis, MO, Amer. Meteor. Soc., 56–60.

———, ———, and R. A. Maddox, 1992: On the use of mesoscale and cloud-scale models in operational forecasting. *Wea. Forecasting,* **7,** 120–132.

———, D. J. Stensrud, and J. V. Cortinas, 1993: The use of mesoscale models for convective storm forecasting. Preprints, *13th Conf. on Weather Forecasting and Analysis,* Vienna, VA, Amer. Meteor. Soc., 301–304.

———, C. A. Doswell III, and R. B. Wilhelmson, 1994a: The role of midtropospheric winds in the evolution and maintenance of low-level mesocyclones. *Mon. Wea. Rev.,* **122,** 126–136.

———, ———, and J. Cooper, 1994b: On the environments of tornadic and nontornadic mesocyclones. *Wea. Forecasting,* **9,** 606–618.

Brown, R. A., L. R. Lemon, and D. W. Burgess, 1978: Tornado detection by pulsed Doppler radar. *Mon. Wea. Rev.,* **106,** 29–38.

Browning, K. A., 1964: Airflow and precipitation trajectories within severe local storms which travel to the right of the winds. *J. Atmos. Sci.,* **21,** 634–639.

———, 1977: The structure and mechanisms of hailstorms. *Hail: A Review of Hail Science and Hail Suppression, Meteor. Monogr.,* No. 38, Amer. Meteor. Soc., 1–43.

Bunkers, M. J., B. A. Klimowski, J. W. Zeitler, R. L. Thompson, and M. L. Weisman, 1998: Predicting supercell motion using hodograph techniques. Preprints, *19th Conf. on Severe Local Storms,* Minneapolis, MN, Amer. Meteor. Soc., 611–614.

Bunting, W. F., 1998: Maintaining situation awareness in a modernized National Weather Service warning environment. Preprints, *19th Conf. on Severe Local Storms,* Minneapolis, MN, Amer. Meteor. Soc., 592–594.

Burgess, D. W., 1988: The environment of the Edmond, Oklahoma tornadic storm. Preprints, *15th Conf. on Severe Local Storms,* Baltimore, MD, Amer. Meteor. Soc., 292–295.

———, and R. J. Donaldson Jr., 1979: Contrasting tornadic storm types. Preprints, *11th Conf. on Severe Local Storms,* Kansas City, MO, Amer. Meteor. Soc., 189–192.

———, and E. B. Curran, 1985: The relationship of storm type to environment in Oklahoma on 26 April 1984. Preprints, *14th Conf. on Severe Local Storms,* Indianapolis, IN, Amer. Meteor. Soc., 208–211.

———, and L. R. Lemon, 1990: Severe thunderstorm detection by radar. *Radar in Meteorology,* D. Atlas, Ed., Amer. Meteor. Soc., 619–627.

———, and M. A. Magsig, 1998: Recent observations of tornado development at near range to WSR-88D radars. Preprints, *19th Conf. on Severe Local Storms,* Minneapolis, MN, Amer. Meteor. Soc., 756–759.

———, L. R. Lemon, and R. A. Brown, 1975: Evolution of a tornado signature and parent circulation as revealed by single Doppler radar. Preprints, *16th Conf. on Radar Meteorology,* Amer. Meteor. Soc., Houston, TX, 40–43.

———, V. T. Wood, and R. A. Brown, 1982: Mesocyclone evolution statistics. Preprints, *12th Conf. on Severe Local Storms,* San Antonio, TX, Amer. Meteor. Soc., 422–424.

———, R. J. Donaldson Jr., and P. R. Desrochers, 1993: Tornado detection and warning by radar. *The Tornado: Its Structure, Dynamics, Prediction, and Hazards, Geophys. Monogr.,* No. 79, Amer. Geophys. Union, 203–221.

———, R. R. Lee, S. S. Parker, D. L. Floyd, and D. L. Andra, 1995: A study of mini supercells observed by WSR-88D radars. Preprints, *27th Conf. on Radar Meteorology,* Vail, CO, Amer. Meteor. Soc., 4–6.

Businger, S., W. H. Bauman III, and G. F. Watson, 1991: The development of the Piedmont front and associated severe weather on March 13 1986. *Mon. Wea. Rev.,* **119,** 2224–2251.

Caracena, F., J. McCarthy, and J. A. Flueck, 1983: Forecasting the likelihood of microbursts along the front range of Colorado. Preprints, *13th Conf. on Severe Local Storms,* Tulsa, OK, Amer. Meteor. Soc., 261–264.

Carbone, R. E., 1983: A severe frontal rainband. Part II: Tornado parent vortex circulation. *J. Atmos. Sci.,* **40,** 2639–2654.

Carlson, T. N., and F. H. Ludlam, 1968: Conditions for the occurrence of severe local storms. *Tellus,* **20,** 204–225.

———, S. G. Benjamin, G. S. Forbes, and Y. F. Li, 1983: Elevated mixed layer in the regional severe storm environment: Conceptual model and case studies. *Mon. Wea. Rev.,* **111,** 1453–1473.

Chappell, C. F., 1986: Quasi-stationary convective events. *Mesoscale Meteorology and Forecasting,* P. Ray, Ed., Amer. Meteor. Soc., 289–310.

Chisholm, A. J., and J. H. Renick, 1972: The kinematics of multicell and supercell Alberta hailstorms. Alberta Hail Studies, 1972, Research Council of Alberta Hail Studies Rep. 72–2, 24–31.

Colby, F. P., Jr., 1984: Convective inhibition as a predictor of convection during AVE-SESAME II. *Mon. Wea. Rev.,* **112,** 2239–2252.

Cooley, J. R., 1978: Cold air funnel clouds. *Mon. Wea. Rev.,* **106,** 1368–1372.

Corfidi, S. F., 1998: Forecasting MCS mode and motion. Preprints, *19th Conf. on Severe Local Storms,* Minneapolis, MN, Amer. Meteor. Soc., 626–629.

Cortinas, J. V., Jr., and D. J. Stensrud, 1995: The importance of understanding mesoscale model parameterization schemes for weather forecasting. *Wea. Forecasting,* **10,** 716–740.

Crawford, K. C., F. V. Brock, R. L. Elliot, G. W. Cuperus, S. J. Stadler, H. L. Johnson, and C. A. Doswell III, 1992: The Oklahoma mesonetwork—A 21st century project. Preprints, *Eighth Int. Conf. on Interactive Information and Processing Systems for Meteorology, Oceanography, and Hydrology,* Atlanta, GA, Amer. Meteor. Soc., 27–33.

Crum, T. D., and R. L. Alberty, 1993: The WSR-88D and the WSR-88D operational support facility. *Bull. Amer. Meteor. Soc.,* **74,** 1669–1686.

Davies, J. M., 1993a: Hourly helicity, instability, and EHI in forecasting supercell tornadoes. Preprints, *17th Conf. on Severe Local Storms,* St. Louis, MO, Amer. Meteor. Soc., 107–111.

———, 1993b: Small tornadic supercells in the Central Plains. Preprints, *17th Conf. on Severe Local Storms,* St. Louis, MO, Amer. Meteor. Soc., 305–309.

———, 1998: On BRN shear and CAPE associated with tornadic environments. Preprints, *19th Conf. on Severe Local Storms,* Minneapolis, MN, Amer. Meteor. Soc., 599–602.

———, and R. H. Johns, 1993: Some wind and instability parameters associated with strong and violent tornadoes. 1: Wind shear and helicity. *The Tornado: Its Structure, Dynamics, Prediction, and Hazards, Geophys. Monogr.,* No. 79, Amer. Geophys. Union, 573–582.

Davies-Jones, R. P., 1984: Streamwise vorticity: The origin of updraft rotation in supercell storms. *J. Atmos. Sci.,* **41,** 2991–3006.

———, 1993: Helicity trends in tornado outbreaks. Preprints, *17th Conf. on Severe Local Storms,* St. Louis, MO, Amer. Meteor. Soc., 56–60.

———, and H. E. Brooks, 1993: Mesocyclogenesis from a theoretical perspective. *The Tornado: Its Structure, Dynamics, Prediction, and Hazards, Geophys. Monogr.,* No. 79, Amer. Geophys. Union, 105–114.

———, D. W. Burgess, and M. Foster, 1990: Test of helicity as a tornado forecast parameter. Preprints, *16th Conf. on Severe Local Storms,* Kananaskis Park, AB, Canada, Amer. Meteor. Soc., 588–592.

Dessens, J., and J. T. Snow, 1989: Tornadoes in France. *Wea. Forecasting,* **4,** 110–132.

Devore, D. R., 1983: The operational use of digital radar. Preprints, *13th Conf. on Severe Local Storms,* Tulsa, OK, Amer. Meteor. Soc., 21–24.

Donaldson, R. J., Jr., 1970: Vortex signature recognition by a Doppler radar. *J. Appl. Meteor.,* **9,** 661–670.

———, G. M. Armstrong, A. C. Chmela, and M. J. Kraus, 1969: Doppler radar investigation of air flow and shear within severe thunderstorms. Preprints, *Sixth Conf. on Severe Local Storms,* Chicago, IL, Amer. Meteor. Soc., 146–154.

Doswell, C. A., III, 1980: Synoptic-scale environments associated with High Plains severe thunderstorms. *Bull. Amer. Meteor. Soc.,* **61,** 1388–1400.

———, 1982: The operational meteorology of convective weather. Vol. I: Operational mesoanalysis. NOAA Tech. Memo. NWS NSSFC-5, 158 pp.

———, 1985: The operational meteorology of convective weather. Vol. II: Storm-scale analysis. NOAA Tech. Memo. ERL ESG-15.

———, 1986a: The human element in weather forecasting. *Natl. Wea. Dig.,* **11,** 6–18.

———, 1986b: Short range forecasting. *Mesoscale Meteorology and Forecasting,* P. Ray, Ed., Amer. Meteor. Soc., 689–719.

———, 1987: The distinction between large-scale and mesoscale contribution to severe convection: A case study example. *Wea. Forecasting,* **2,** 3–16.

———, and R. A. Maddox, 1986: The role of diagnosis in weather forecasting. Preprints, *11th Conf. on Weather Forecasting and Analysis,* Kansas City, MO, Amer. Meteor. Soc., 177–182.

———, and D. W. Burgess, 1993: Tornadoes and tornadic storms: A review of conceptual models. *The Tornado: Its Structure, Dynamics, Prediction, and Hazards, Geophys. Monogr.,* No. 79, Amer. Geophys. Union, 161–172.

———, and E. N. Rasmussen, 1994: The effect of neglecting the virtual temperature correction on CAPE calculations. *Wea. Forecasting,* **9,** 625–629.

———, J. T. Schaefer, D. W. McCann, T. W. Schlatter, and H. B. Wobus, 1982: Thermodynamic analysis procedures at the National Severe Storms Forecast Center. Preprints, *Ninth Conf. on Weather Forecasting and Analysis,* Seattle, WA, Amer. Meteor. Soc., 304–309.

———, F. Caracena, and M. Magnano, 1985: Temporal evolution of 700–500 mb lapse rate as a forecasting tool—A case study. Preprints, *14th Conf. on Severe Local Storms,* Indianapolis, IN, Amer. Meteor. Soc., 398–401.

———, S. J. Weiss, and R. H. Johns, 1993: Tornado forecasting: A review. *The Tornado: Its Structure, Dynamics, Prediction, and Hazards, Geophys. Monogr.,* No. 79, Amer. Geophys. Union, 557–571.

———, H. E. Brooks, and R. A. Maddox, 1996: Flash flood forecasting: An ingredients-based methodology. *Wea. Forecasting,* **11,** 560–581.

———, A. R. Moller, and H. E. Brooks, 1999: Storm spotting and public awareness since the first tornado forecasts of 1948. *Wea. Forecasting,* **14,** 544–557.

Droegemeier, K. K., S. M. Lazarus, and R. Davies-Jones, 1993: The influence of helicity on numerically simulated convective storms. *Mon. Wea. Rev.,* **121,** 2005–2029.

Eilts, M. D., and Coauthors, 1996a: Severe weather warning decision support system. Preprints, *18th Conf. on Severe Local Storms,* San Francisco, CA, Amer. Meteor. Soc., 536–540.

———, J. T. Johnson, E. D. Mitchell, R. J. Lynn, P. Spencer, S. Cobb, and T. M. Smith, 1996b: Damaging downburst prediction algorithm for the WSR-88D. Preprints, *18th Conf. on Severe Local Storms,* San Francisco, CA, Amer. Meteor. Soc., 541–545.

Elvander, R. C., 1977: Relationships between radar parameters observed with objectively defined echoes and reported severe weather occurrences. Preprints, *Tenth Conf. on Severe Local Storms,* Omaha, NE, Amer. Meteor. Soc., 73–76.

Endsley, M. R., 1988: Design and evaluation for situation awareness enhancement. *Proc. 32d Annual Meeting of the Human Factors Society,* Santa Monica, CA, Human Factors Soc., 97–101.

Evans, J. S., 1998: An examination of observed shear profiles associated with long-lived bow echoes. Preprints, *19th Conf. on Severe Local Storms,* Minneapolis, MN, Amer. Meteor. Soc., 30–33.

Fankhauser, J. C., and C. G. Moore, 1977: Some correlations between various sounding parameters and hailstorm characteristics in northeast Colorado. Preprints, *Tenth Conf. on Severe Local Storms,* Omaha, NE, Amer. Meteor. Soc., 218–225.

Fawbush, E. J., and R. C. Miller, 1953: A method for forecasting hail size near the earth's surface. *Bull. Amer. Meteor. Soc.,* **34,** 235–244.

———, and ———, 1954: A basis for forecasting peak wind gusts in nonfrontal thunderstorms. *Bull. Amer. Meteor. Soc.,* **35,** 14–19.

Forbes, G. S., and R. M. Wakimoto, 1983: A concentrated outbreak of tornadoes, downbursts, and microbursts, and implications regarding vortex classification. *Mon. Wea. Rev.,* **111,** 220–235.

Foster, D. S., and F. Bates, 1956: A hail size forecasting technique. *Bull. Amer. Meteor. Soc.,* **37,** 135–140.

Foster, M. P., A. R. Moller, L. J. Wicker, and L. Cantrell, 1995a: The rapid evolution of a tornadic small supercell: Observations and simulation. Preprints, *14th Conf. on Weather Analysis and Forecasting,* Dallas, TX, Amer. Meteor. Soc., 323–328.

———, ———, R. P. Kleyla, G. F. Ely, and J. P. Stefkovich, 1995b: The April 25, 1994 Dallas County tornadoes: Interfacing preparedness, spotters, the media, and WSR-88D observations. Preprints, *14th Conf. on Weather Analysis and Forecasting,* Dallas, TX, Amer. Meteor. Soc., (J2)19–(J2)24.

Friday, E. W., 1988: The National Weather Service severe storm program: Year 2000. Preprints, *15th Conf. on Severe Local Storms,* Baltimore, MD, Amer. Meteor. Soc., J1–J8.

Fujita, T. T., 1955: Results of detailed synoptic studies of squall line. *Tellus,* **4,** 405–436.

———, 1978: Manual of downburst identification for project NIMROD. Satellite and Mesometeorology Research Paper 156, University of Chicago, Dept. of Geophysical Sciences, 104 pp.

———, 1981: Tornadoes and downbursts in the context of generalized planetary scales. *J. Atmos. Sci.,* **38,** 1511–1534.

———, 1985: The downburst: Microburst and macroburst. Satellite and Mesometeorology Project, University of Chicago, Dept. of Geophysical Sciences, 122 pp.

———, and H. R. Byers, 1977: Spearhead echo and downburst in the crash of an airliner. *Mon. Wea. Rev.,* **105,** 129–146.

———, and R. M. Wakimoto, 1981: Five scales of airflow associated with a series of downbursts on 16 July 1980. *Mon. Wea. Rev.,* **109,** 1438–1456.

Galway, J. G., 1956: The lifted index as a predictor of latent instability. *Bull. Amer. Meteor. Soc.,* **37,** 528–529.

Geotis, S. G., 1963: Some radar measurements of hailstorms. *J. Appl. Meteor.,* **2,** 270–275.

Golden, J. H., and D. Purcell, 1978: Life cycle of the Union City, Oklahoma tornado and comparison with waterspouts. *Mon. Wea. Rev.,* **106,** 3–11.

———, and B. Morgan, 1972: The NSSL-Notre Dame tornado intercept program. *Bull. Amer. Meteor. Soc.,* **53,** 1178–1180.

Greene, D. R., and R. A. Clark, 1972: Vertically integrated liquid water: A new analysis tool. *Mon. Wea. Rev.,* **100,** 548–552.

Guerrero, H., and W. Read, 1993: Operational use of the WSR-88D during the November 21, 1992 Southeast Texas tornado outbreak. Preprints, *17th Conf. on Severe Local Storms,* St. Louis, MO, Amer. Meteor. Soc., 399–402.

Hagemeyer, B. C., 1997: Peninsular Florida tornado outbreaks. *Wea. Forecasting,* **12,** 399–427.

Hales, J. E., Jr., 1979: A subjective assessment of model initial conditions using satellite imagery. *Bull. Amer. Meteor. Soc.,* **60,** 206–211.

———, 1993: Topographically induced enhancement and its role in the Los Angeles tornado maximum. Preprints, *17th Conf. on Severe Local Storms,* St. Louis, MO, Amer. Meteor. Soc., 98–101.

Hane, C. E., H. B. Bluestein, T. M. Crawford, M. E. Baldwin, and R. A. Rabin, 1997: Severe thunderstorm development in relation to along-dryline variability: A case study. *Mon. Wea. Rev.,* **125,** 231–251.

Hanstrum, B. N., G. A. Mills, and A. Watson, 1998: Cool-season tornadoes in Australia. Part I: Synoptic climatology. Preprints, *19th Conf. on Severe Local Storms,* Minneapolis, MN, Amer. Meteor. Soc., 97–100.

Harris, H. W., K. C. Mehta, and J. R. McDonald, 1993: Design for occupant protection on schools. *The Tornado: Its Structure, Dynamics, Prediction, and Hazards, Geophys. Monogr.,* No. 79, Amer. Geophys. Union, 545–553.

Hart, J. A., and W. D. Korotky, 1991: The SHARP Workstation-v1.50. A Skew T/Hodograph Analysis and Research Program for the IBM and Compatible PC. User's Manual. NOAA/NWS, 62 pp.

Hoskins, B. J., I. Draghici, and H. C. Davies, 1978: A new look at the omega-equation. *Quart. J. Roy. Meteor. Soc.,* **104,** 31–38.

Hudson, H. R., 1971: On the relationship between horizontal moisture convergence and convective cloud formation. *J. Meteor.,* **10,** 755–762.

Idso, S. B., 1975: Whirlwinds, density currents, and topographic disturbances: A meteorological melange of intriguing interactions. *Weatherwise,* **28,** 61–65.

Johns, R. H., 1984: A synoptic climatology of northwest flow severe weather outbreaks. Part II: Meteorological parameters and synoptic patterns. *Mon. Wea. Rev.,* **112,** 449–464.

———, 1993: Meteorological conditions associated with bow echo development in convective storms. *Wea. Forecasting,* **8,** 294–299.

———, and W. D. Hirt, 1987: Derechos: Widespread convectively induced windstorms. *Wea. Forecasting,* **2,** 32–49.

———, and C. A. Doswell III, 1992: Severe local storm forecasting. *Wea. Forecasting,* **7,** 588–612.

———, and R. A. Dorr Jr., 1996: Some meteorological aspects of strong and violent tornado episodes in New England and eastern New York. *Natl. Wea. Dig.,* **20,** 2–12.

———, and J. A. Hart, 1998: The occurrence and non-occurrence of large hail with strong and violent tornado episodes: Frequency distributions. Preprints, *19th Conf. on Severe Local Storms,* Minneapolis, MN, Amer. Meteor. Soc., 283–286.

———, K. W. Howard, and R. A. Maddox, 1990: Conditions associated with long-lived derechos—An examination of the large scale environment. Preprints, *16th Conf. on Severe Local Storms,* Kananaskis Park, AB, Canada, Amer. Meteor. Soc., 408–412.

———, J. M. Davies, and P. W. Leftwich, 1993: Some wind and instability parameters associated with strong and violent tornadoes. Part II: Variations in the combinations of wind and instability parameters. *The Tornado: Its Structure, Dynamics, Prediction, and Hazards, Geophys. Monogr.,* No. 79, Amer. Geophys. Union, 583–590.

Johnson, B. C., 1983: The heat burst of 29 May 1976. *Mon. Wea. Rev.,* **111,** 1776–1797.

Jones, W. A., K. K. Keeter, and L. L. Lee, 1985: Some considerations for long-range radar evaluation of potentially tornadic thunderstorms. Preprints, *14th Conf. on Severe Local Storms,* Indianapolis, IN, Amer. Meteor. Soc., 139–142.

Kates, R. W., 1980: Climate and society: Lessons from recent events. *Weather,* **35,** 1–25.

Kelley, D. L., J. T. Schaefer, R. P. McNulty, C. A. Doswell III, and R. F. Abbey Jr., 1978: An augmented tornado climatology. *Mon. Wea. Rev.,* **106,** 1172–1183.

Kennedy, P. C., N. E. Westscott, and R. W. Scott, 1990: Single Doppler observations of a mini-tornado. Preprints, *16th Conf. on Severe Local Storms,* Kananaskis Park, AB, Canada, Amer. Meteor. Soc., 209–212.

Kerr, B. W., and G. L. Darkow, 1996: Storm-relative winds and helicity in the tornadic thunderstorm environment. *Wea. Forecasting,* **11,** 489–505.

Klazura, G. E., and D. A. Imy, 1993: A description of the initial set of analysis products available from the NEXRAD WSR-88D system. *Bull. Amer. Meteor. Soc.,* **74,** 1293–1311.

Klemp, J. B., 1987: Dynamics of tornadic thunderstorms. *Ann. Rev. Fluid. Mech.,* **19,** 369–402.

———, and R. J. Rotunno, 1983: A study of the tornadic region within a supercell thunderstorm. *J. Atmos. Sci.,* **40,** 359–377.
Knupp, K. R., and L. P. Garinger, 1993: The gulf coast region morning tornado phenomenon. Preprints, *17th Conf. on Severe Local Storms,* St. Louis, MO, Amer. Meteor. Soc., 20–24.
———, R. L. Clymer, and B. Geerts, 1996: Preliminary classification and observational characteristics of tornadic storms over northern Alabama. Preprints, *18th Conf. on Severe Local Storms,* San Francisco, CA, Amer. Meteor. Soc., 447–450.
Korotky, W. D., 1990: The Raleigh tornado of November 28, 1988: The evolution of the tornadic environment. Preprints, *16th Conf. on Severe Local Storms,* Kananaskis Park, AB, Canada, Amer. Meteor. Soc., 532–537.
Krumm, W. R., 1954: On the cause of downdrafts from dry thunderstorms over the plateau area of the United States. *Bull. Amer. Meteor. Soc.,* **35,** 122–126.
LaDue, J. G., 1998: The influence of two cold fronts on storm morphology. Preprints, *19th Conf. on Severe Local Storms,* Minneapolis, MN, Amer. Meteor. Soc., 324–327.
Lanicci, J. M., and T. T. Warner, 1991: A synoptic climatology of the elevated mixed-layer inversion over the southern Great Plains in spring. Part I: Structure, dynamics, and seasonal evolution. *Wea. Forecasting,* **6,** 181–197.
Lee, B. D., and R. B. Wilhelmson, 1997a: The numerical simulation of non-supercell tornadogenesis. Part I: Initiation and evolution of pre-tornadic misocyclone circulations along a dry outflow boundary. *J. Atmos. Sci.,* **54,** 32–60.
———, and ———, 1997b: The numerical simulation of non-supercell tornadogenesis. Part II: Evolution of a family of tornadoes along a weak outflow boundary. *J. Atmos. Sci.,* **54,** 2387–2415.
Leik, R. K., T. M. Carter, and J. P. Clark, 1981: Community response to natural hazard warning. U.S. Dept. of Commerce. 77 pp. [NTIS PB82–111287.]
Lemon, L. R., 1980: Severe thunderstorm radar identification techniques and warning criteria: A preliminary report. NOAA Tech. Memo. NWS NSSFC-1, 60 pp. [NTIS PB273049.]
———, 1998: The radar "three-body scatter spike": An operational large-hail signature. *Wea. Forecasting,* **13,** 327–340.
———, and C. A. Doswell III, 1979: Severe thunderstorm evolution and mesocyclone structure as related to tornadogenesis. *Mon. Wea. Rev.,* **107,** 1184–1197.
———, and D. W. Burgess, 1993: Supercell associated deep convergence zone revealed by a WSR-88D. Preprints, *26th Int. Conf. on Radar Meteorology,* Norman, OK, Amer. Meteor. Soc., 206–208.
———, and S. Parker, 1996: The Lahoma storm deep convergence zone: Its characteristics and role in storm dynamics and severity. Preprints, *18th Conf. on Severe Local Storms,* San Francisco, CA, Amer. Meteor. Soc., 70–75.
———, A. R. Moller, and C. A. Doswell III, 1979: A slide series supplement to Tornado: A spotter's guide. NOAA-NWS training slide series.
———, L. R. Ruthi, and L. Quoetone, 1992: WSR-88D Effective operational applications of a high data rate. Preprints, *Symp. on Weather Forecasting,* Atlanta, GA, Amer. Meteor. Soc., 173–180.
Lenning, E., J. Korotky, and H. E. Fuelberg, 1996: The Thomasville, Georgia supercell hailstorm of 28 January 1995. Preprints, *18th Conf. on Severe Local Storms,* San Francisco, CA, Amer. Meteor. Soc., 91–93.
Lucas, C., E. J. Zipser, and M. A. LeMone, 1994a: Vertical velocity in oceanic convection off tropical Australia. *J. Atmos. Sci.,* **51,** 3182–3193.
———, ———, and ———, 1994b: Convective available potential energy in the environment of oceanic and continental clouds: Corrections and comments. *J. Atmos. Sci.,* **51,** 3829–3930.
Maddox, R. A., 1993: Diurnal low-level jet wind oscillation and storm-relative helicity. *The Tornado: Its Structure, Dynamics, Prediction, and Hazards, Geophys. Monogr.,* No. 79, Amer. Geophys. Union, 591–598.
———, L. R. Hoxit, and C. F. Chappel, 1980: A study of tornadic thunderstorm interactions with thermal boundaries. *Mon. Wea. Rev.,* **108,** 322–336.
Markowski, P. M., E. N. Rasmussen, and J. M. Straka, 1998a: The occurrence of tornadoes in supercells interacting with boundaries during VORTEX-95. *Wea. Forecasting,* **13,** 852–859.
———, J. M. Straka, and E. N. Rasmussen, 1998b: A preliminary investigation of the importance of helicity location in the hodograph. Preprints, *19th Conf. on Severe Local Storms,* Minneapolis, MN, Amer. Meteor. Soc., 230–233.
———, ———, ———, and D. O. Blanchard, 1998c: Variability of storm-relative helicity during VORTEX. *Mon. Wea. Rev.,* **126,** 2959–2971.
Marshall, T. P., 1993: Lessons learned from analyzing tornado damage. *The Tornado: Its Structure, Dynamics, Prediction, and Hazards, Geophys. Monogr.,* No. 79, Amer. Geophys. Union, 495–499.
Marwitz, J. D., 1972: The structure and motion of severe hailstorms. Part II: Multi-cell storms. *J. Appl. Meteor.,* **11,** 180–188.
McCann, D. W., 1983: The enhanced-V, a satellite observable severe storm signature. *Mon. Wea. Rev.,* **111,** 887–894.
McCaul, E. W., Jr., 1987: Observations of the Hurricane "Danny" tornado outbreak of 16 August 1985. *Mon. Wea. Rev.,* **115,** 1206–1223.
———, 1991: Buoyancy and shear characteristics of hurricane-tornado environments. *Mon. Wea. Rev.,* **119,** 1954–1978.
McDonald, J. R., 1993: Damage mitigation and occupant safety. *The Tornado: Its Structure, Dynamics, Prediction, and Hazards, Geophys. Monogr.,* No. 79, Amer. Geophys. Union, 523–528.
McGinley, J., 1986: Nowcasting mesoscale phenomena. *Mesoscale Meteorology and Forecasting,* P. Ray, Ed., Amer. Meteor. Soc., 657–688.
———, and P. S. Ray, 1985: Numerical modeling of a thermal frontal disturbance. Preprints, *14th Conf. on Severe Local Storms,* Indianapolis, IN, Amer. Meteor. Soc., 194–197.
———, K. Runk, and J. Alleca, 1987: Use of time changes of surface data parameters in short range forecasting. *Proc., Symp. Mesoscale Analysis and Forecasting,* Vancouver, BC, Canada, 313–319.
———, S. Albers, and P. Stamos, 1991: Validation of a composite convective index as defined by a real time local analysis system. *Wea. Forecasting,* **6,** 337–356.
McNulty, R. P., 1995: Severe convective weather: A Central Region forecasting challenge. *Wea. Forecasting,* **10,** 187–202.
Meaden, G. T., 1981: Whirlwind formation at a sea breeze front. *Weather,* **36,** 47–48.
Mileti, D. S., and J. H. Sorenson, 1990: Communication of emergency public warnings: A social science perspective and state-of-the-art assessment. U.S. Dept. of Energy ORNL-6609.
Miller, R. C., 1972: Notes on analysis and severe-storm forecasting procedures of the Air Force Global Forecast Center. Air Weather Service Tech. Rep. 200 (rev.), Air Weather Service Headquarters, Scott Air Force Base, IL, 190 pp.
Mitchell, E. D., S. V. Vasiloff, G. J. Stumpf, A. Witt, M. D. Eilts, J. T. Johnson, and K. W. Thomas, 1998: The National Severe Storms Laboratory tornado detection algorithm. *Wea. Forecasting,* **13,** 352–366.
Moller, A. R., 1978: The improved NWS storm spotter's training program at Ft. Worth, TX. *Bull. Amer. Meteor. Soc.,* **59,** 1574–1582.
———, 1979: The climatology and synoptic meteorology of Southern Plains' tornado outbreaks. Master's thesis, University of Oklahoma, Norman, 70 pp.
———, and C. A. Doswell III, 1988: A proposed advanced storm spotters' training program. Preprints, *15th Conf. on Severe Local Storms,* Baltimore, MD, Amer. Meteor. Soc., 173–177.
———, C. A. Doswell, and R. Przybylinski, 1990: High-precipitation supercells: A conceptual model and documentation. Pre-

prints, *16th Conf. on Severe Local Storms,* Kananaskis Park, AB, Canada, Amer. Meteor. Soc., 52–57.

———, M. P. Foster, and C. A. Doswell III, 1993: Some considerations of severe local storm product dissemination in the modernized and restructured National Weather Service. Preprints, *17th Conf. on Severe Local Storms,* St. Louis, MO, Amer. Meteor. Soc., 375–379.

———, C. A. Doswell III, M. P. Foster, and G. R. Woodall, 1994: The operational recognition of supercell thunderstorms environments and storm structures. *Wea. Forecasting,* **9,** 327–347.

Moncrieff, M. W., and J. S. A. Green, 1972: The propagation and transfer properties of steady convective overturning in shear. *Quart. J. Roy. Meteor. Soc.,* **98,** 336–352.

———, and M. J. Miller, 1976: The dynamics and simulation of tropical cumulonimbus and squall lines. *Quart. J. Roy. Meteor. Soc.,* **102,** 373–394.

Monteverdi, J., and J. Quadros, 1994: Convective and rotational parameters associated with three tornado episodes in northern and central California. *Wea. Forecasting,* **9,** 285–300.

Moore, J. T., and J. P. Pino, 1990: An interactive method for estimating maximum hailstone size from forecast soundings. *Wea. Forecasting,* **5,** 508–525.

Morgan, G. M., Jr., and P. W. Summers, 1985: Hailfall and hailstorm characteristics. *Thunderstorms: A Social, Scientific, and Technological Documentary.* Vol. 2. *Thunderstorm Morphology and Dynamics,* 2d ed., E. Kessler, Ed., University of Oklahoma Press, 237–275.

Murphy, A. H., 1993: What good is a forecast? An essay on the nature of goodness in weather forecasts. *Wea. Forecasting,* **8,** 281–293.

Murphy, T. W., and V. S. Woods, 1992: A damaging tornado from low-topped convection. Preprints, *Symp. on Weather Forecasting,* Atlanta, GA, Amer. Meteor. Soc., 195–201.

Nelson, S. P., 1987: The hybrid multicellular-supercellular storm—An efficient hail producer. Part II: General characteristics and implications for hail growth. *J. Atmos. Sci.,* **44,** 2060–2073.

Nielson-Gammon, J. W., and W. L. Read, 1995: Detection and interpretation of left-moving severe thunderstorms using the WSR-88D: A case study. *Wea. Forecasting,* **10,** 127–140.

Niino, H., O. Suzuki, T. Fujitani, H. Nirasawa, H. Ohno, I. Takayabu, and N. Kinoshita, 1993: An observational study of the Mobara tornado. *The Tornado: Its Structure, Dynamics, Prediction, and Hazards, Geophys. Monogr.,* No. 79, Amer. Geophys. Union, 511–519.

NOAA Staff, 1999: Oklahoma/southern Kansas tornado outbreak of May 3, 1999. Service Assessment. U.S. Department of Commerce, National Oceanic and Atmospheric Administration, National Weather Service, 51 pp.

Ohno, H., O. Suzuki, and K. Kusunoki, 1996: Climatology of downburst occurrence in Japan. Preprints, *18th Conf. on Severe Local Storms,* San Francisco, CA, Amer. Meteor. Soc., 87–90.

Orasanu, J., 1994: Shared problems models and flight crew performance. *Aviation Psychology in Practice,* N. Johnson, N. McDonald, and R. Fuller, Eds., Aldershot, 255–285.

Petak, W. J., and A. A. Atkisson, 1982: *Natural Hazard Risk Assessment and Public Policy.* Springer-Verlag, 489 pp.

Pino, J. P., and J. T. Moore, 1989: An interactive method for estimating hailstone size and convectively-driven wind gusts from forecast soundings. Preprints, *12th Conf. on Weather Analysis and Forecasting,* Monterey, CA, Amer. Meteor. Soc., 103–106.

Prosser, N. E., and D. S. Foster, 1966: Upper air sounding analysis by use of electronic computer. *J. Appl. Meteor.,* **5,** 296–300.

Przybylinski, R. W., 1988: Radar signatures associated with the 10 March 1986 tornado outbreak over Indiana. Preprints, *15th Conf. on Severe Local Storms,* Baltimore, MD, Amer. Meteor. Soc., 253–256.

———, 1995: The bow echo: Observations, numerical simulations, and severe weather detection methods. *Wea. Forecasting,* **10,** 203–218.

———, and W. J. Gery, 1983: The reliability of the bow echo as an important severe weather signature. Preprints, *13th Conf. on Severe Local Storms,* Tulsa, OK, Amer. Meteor. Soc., 270–273.

———, and D. M. DeCaire, 1985: Radar signatures associated with the derecho. One type of mesoscale convective system. Preprints, *14th Conf. on Severe Local Storms,* Indianapolis, IN, Amer. Meteor. Soc., 228–231.

———, T. J. Shea, D. L. Ferry, E. H. Goetsch, R. R. Czys, and N. E. Wescott, 1993: Doppler radar observations of high-precipitation supercells over the mid-Mississippi valley region. Preprints, *17th Conf. on Severe Local Storms,* St. Louis, MO, Amer. Meteor. Soc., 158–163.

———, Y. J. Lin, G. K. Schmocker, and T. J. Shea, 1995: The use of real-time WSR-88D, profiler, and conventional data sets in forecasting a northeastward moving derecho over eastern Missouri and central Illinois. Preprints, *14th Conf. on Weather Analysis and Forecasting,* Dallas, TX, Amer. Meteor. Soc., 335–342.

Purdom, J. F. W., 1976: Some uses of high-resolution GOES satellite imagery in the mesoscale forecasting of convection and its behavior. *Mon. Wea. Rev.,* **104,** 1474–1483.

———, 1982: Subjective interpretations of geostationary satellite data for nowcasting. *Nowcasting,* K. Browning, Ed., Academic Press, 149–166.

Quoetone, E. M., and K. L. Huckabee, 1995: Anatomy of an effective warning: Event anticipation, data integration, feature recognition. Preprints, *14th Conf. on Weather Analysis and Forecasting,* Dallas, TX, Amer. Meteor. Soc., 420–425.

Rasmussen, E. N., and R. B. Wilhelmson, 1983: Relationships between storm characteristics and 1200 GMT hodographs, low-level shear, and stability. Preprints, *13th Conf. on Severe Local Storms,* Tulsa, OK, Amer. Meteor. Soc., J5–J8.

———, and D. O. Blanchard, 1998: A baseline climatology of sounding-derived supercell and tornado forecast parameters. *Wea. Forecasting,* **13,** 1148–1164.

———, and J. M. Straka, 1998: Variations in supercell morphology. Part I: Observations of the role of upper-level storm-relative flow. *Mon. Wea. Rev.,* **126,** 2406–2421.

———, ———, R. Davies-Jones, C. A. Doswell III, F. H. Carr, M. D. Eilts, and D. R. McGorman, 1994: Verification of the origins of rotation in tornadoes experiment: VORTEX. *Bull. Amer. Meteor. Soc.,* **75,** 995–1006.

Read, W. L., 1987: Observed microbursts in the NWS Southern Region during 1986—Four case studies. NOAA Tech. Memo. NWS SR-121, 10–34.

———, and J. T. Elmore, 1989: Summer season severe downbursts in North Texas: Forecast and warning techniques using current NWS technology. Preprints, *12th Conf. on Weather Analysis and Forecasting,* Monterey, CA, Amer. Meteor. Soc., 142–147.

Riebsame, W. E., H. F. Diaz, and M. Price, 1986: The social burden of weather and climate hazards. *Bull. Amer. Meteor. Soc.,* **67,** 1378–1388.

Roberts, R. D., and J. W. Wilson, 1989: A proposed microburst nowcasting procedure using single-Doppler radar. *J. Appl. Meteor.,* **28,** 285–303.

Rotunno, R. J., and J. B. Klemp, 1985: On the rotation and propagation of simulated supercell thunderstorms. *J. Atmos. Sci.,* **42,** 271–292.

———, ———, and M. L. Weisman, 1988: A theory for strong, long-lived squall lines. *J. Atmos. Sci.,* **45,** 463–485.

Saffle, R. E., 1976: DIRIDER products and field operation. Preprints, *17th Conf. on Radar Meteorology,* Seattle, WA, Amer. Meteor. Soc., 555–559.

———, 1986: Severe thunderstorm forecasting: A historical perspective. *Wea. Forecasting,* **1,** 164–189.

Schlesinger, R. E., 1980: A three-dimensional numerical model of an isolated thunderstorm. Part II: Dynamics of updraft splitting and mesovortex couplet evolution. *J. Atmos. Sci.,* **37,** 396–420.

Schmid, W., S. Humbeli, B. Messner, and W. Linder, 1996: On the formation of supercell storms. Preprints, *18th Conf. on Severe Local Storms,* San Francisco, CA, Amer. Meteor. Soc., 451–454.

Schmidt, J. M., and W. R. Cotton, 1989: A High Plains squall-line associated with severe surface winds. *J. Atmos. Sci.,* **46,** 281–302.

Schmocker, G. K., R. W. Przybylinski, and Y. J. Lin, 1996: Forecasting the initial onset of damaging downburst winds associated with a mesoscale convective system (MCS) using the mid-latitude radial convergence (MARC) signature. Preprints, *15th Conf. on Weather Analysis and Forecasting,* Norfolk, VA, Amer. Meteor. Soc., 306–311.

Smith, B. E., and J. W. Partacz, 1985: Bow-echo induced tornado at Minneapolis on 26 April 1984. Preprints, *14th Conf. on Severe Local Storms,* Indianapolis, IN, Amer. Meteor. Soc., 81–84.

Smull, B. F., and R. A. Houze Jr., 1987: Rear inflow in squall lines with trailing stratiform precipitation. *Mon. Wea. Rev.,* **115,** 2869–2889.

Snellman, L. W., 1991: An old forecaster looks at modernization—Pros and cons. *Natl. Wea. Dig.,* **16,** 2–5.

Sohl, C. J., 1987: Observed microbursts in the NWS Southern Region during 1986—Four case studies. NOAA Tech. Memo. NWS SR-121, 1–9.

Stagno, R. P., 1985: New directions in warning and preparedness activities: A modern approach using public relations to reach a weather hungry public. Preprints, *14th Conf. on Severe Local Storms,* Indianapolis, IN, Amer. Meteor. Soc., J7–J9.

Stensrud, D. J., and R. A. Maddox, 1989: Opposing mesoscale circulations: A case study. *Wea. Forecasting,* **3,** 189–204.

———, J. V. Cortinas Jr., and H. E. Brooks, 1997: Discriminating between tornadic and nontornadic thunderstorms using mesoscale model output. *Wea. Forecasting,* **12,** 613–632.

Stumpf, G. J., R. H. Johnson, and B. F. Smull, 1991: The wake low in the mid-latitude mesoscale convective system having complex organization. *Mon. Wea. Rev.,* **119,** 134–158.

———, A. Witt, E. D. Mitchell, P. L. Spencer, J. T. Johnson, M. D. Eilts, K. W. Thomas, and D. W. Burgess, 1998: The National Severe Storms Laboratory mesocyclone detection algorithm for the WSR-88D. *Wea. Forecasting,* **13,** 304–326.

Suzuki, O., H. Niino, H. Ohno, and H. Nirasawa, 1996: Tornado producing bonsai supercells associated with typhoon 9019. Preprints, *18th Conf. on Severe Local Storms,* San Francisco, CA, Amer. Meteor. Soc., 344–348.

Szoke, E. J., and J. A. Augustine, 1990: A decade of tornado occurrence with a surface mesoscale flow feature—The Denver cyclone. Preprints, *16th Conf. on Severe Local Storms,* Kananaskis Park, AB, Canada, Amer. Meteor. Soc., 554–559.

Tegtmeier, S., 1974: The role of the subsynoptic low pressure system in severe weather forecasting. M.S. thesis, University of Oklahoma, 66 pp.

Thompson, R. L., 1998: ETA storm-relative winds associated with tornadic and non-tornadic supercells. *Wea. Forecasting,* **13,** 125–137.

Trapp, R. J., E. D. Mitchell, G. A. Tipton, D. W. Effertz, A. I. Watson, D. L. Andra Jr., and M. A. Magsig, 1999: Descending and nondescending tornadic vortex signatures detected by WSR-88Ds. *Wea. Forecasting,* **14,** 625–639.

Turcotte, V., and D. Vigneux, 1987: Severe thunderstorm and hail forecasting using derived parameters from standard RAOBS data. Preprints, *Second Workshop on Operational Meteorology,* Halifax, NS, Canada, Atmos. Environ. Service/Canadian Meteor. and Oceanogr. Society, 142–153.

Vasiloff, S. V., E. A. Brandis, R. P. Davies-Jones, and P. S. Ray, 1986: An investigation of the transition from multicell to supercell storms. *J. Climate Appl. Meteor.,* **25,** 1022–1036.

Vescio, M. D., K. K. Keeter, P. Badgett, and A. J. Riordan, 1993: A low-top weak-reflectivity severe weather episode along a thermal/moisture boundary in eastern North Carolina. Preprints, *17th Conf. on Severe Local Storms,* St., MO, Amer. Meteor. Soc., 628–632.

Wade, C. G., and G. B. Foote, 1982: The 22 July 1976 case study: Low-level outflow and mesoscale influences. *Hailstorms of the Central High Plains,* Vol. 2, C. Knight and A. Squires, Eds., Colorado Associated University Press, 115–130.

Wakimoto, R. M., 1983: The West Bend, Wisconsin storm of 4 April 1981: A problem in operational meteorology. *J. Appl. Meteor.,* **22,** 181–189.

———, 1985: Forecasting dry microburst activity over the High Plains. *Mon. Wea. Rev.,* **113,** 1131–1143.

———, and J. W. Wilson, 1989: Non-supercell tornadoes. *Mon. Wea. Rev.,* **117,** 1113–1140.

Waldvogel, A., and B. Federer, 1976: Large raindrops and the boundary between rain and hail. Preprints, *17th Conf. on Radar Meteorology,* Seattle, WA, Amer. Meteor. Soc., 167–172.

Weaver, J. F., 1979: Storm motion related to boundary layer convergence. *Mon. Wea. Rev.,* **107,** 612–619.

Weisman, M. L., 1992: The role of convectively generated rear-inflow jets in the evolution of long-lived meso-convective systems. *J. Atmos. Sci.,* **49,** 1826–1847.

———, 1993: The genesis of severe, long-lived bow echoes. *J. Atmos. Sci.,* **50,** 645–670.

———, 1996: On the use of vertical wind shear versus helicity in interpreting supercell dynamics. Preprints, *18th Conf. on Severe Local Storms,* San Francisco, CA, Amer. Meteor. Soc., 200–204.

———, and J. B. Klemp, 1982: The dependence of numerically simulated convective storms in directionally varying wind shears. *Mon. Wea. Rev.,* **112,** 504–520.

———, and ———, 1984: The structure and classification of numerically simulated convective storms in directionally varying wind shears. *Mon. Wea. Rev.,* **112,** 2479–2498.

———, and ———, 1986: Characteristics of isolated convection. *Mesoscale Meteorology and Forecasting,* P. S. Ray, Ed., Amer. Meteor. Soc., 331–358.

Weiss, S. J., 1987: Some climatological aspects of forecasting tornadoes associated with tropical cyclones. Preprints, *17th Conf. on Hurricanes and Tropical Meteorology,* Miami, FL, Amer. Meteor. Soc., 160–163.

———, 1996: Operational evaluation of the MesoETA model for the prediction of severe local storms. Preprints, *18th Conf. on Severe Local Storms,* San Francisco, CA, Amer. Meteor. Soc., 367–371.

———, G. S. Mankin, and K. E. Mitchell, 1998: ETA model forecasts of low-level moisture return from the Gulf of Mexico during the cool season. Preprints, *19th Conf. on Severe Local Storms,* Minneapolis, MN, Amer. Meteor. Soc., 673–676.

White, G. F., and J. E. Haas, 1975: *Assessment of Research on Natural Hazards.* Oxford University Press, 487 pp.

Wicker, L. J., 1996: The role of near surface wind shear on low-level mesocyclone generation and tornadoes. Preprints, *18th Conf. on Severe Local Storms,* San Francisco, CA, Amer. Meteor. Soc., 225–229.

———, and L. Cantrell, 1996: The role of vertical buoyancy distributions in miniature supercells. Preprints, *18th Conf. on Severe Local Storms,* San Francisco, CA, Amer. Meteor. Soc., 115–119.

Wilczak, J. M., and J. W. Glendening, 1988: Observations and mixed-layer modeling of a terrain-induced mesoscale gyre: The Denver cyclone. *Mon. Wea. Rev.,* **116,** 2688–2711.

Wilk, K. E., L. R. Lemon, and D. W. Burgess, 1979: Interpretation of radar echoes from severe thunderstorms: A series of illustrations with extended captions. Prepared for training of FAA ARTCC Coordinators, National Severe Storms Laboratory, Norman, Oklahoma, 55 pp.

Williams, D. T., 1963: The thunderstorm wake of May 4 1961. Tech. Memo. ERL NSSL-18, National Severe Storms Laboratory, Norman, OK, 23 pp.

Williams, E., and N. Renno, 1993: An analysis of the conditional instability of the tropical atmosphere. *Mon. Wea. Rev.,* **121,** 21–36.

Wilson, J. W., 1970: Operational measurement of rainfall with the WSR-57: Review and recommendations. Preprints, *14th Conf. on Radar Meteorology,* Tucson, AZ, Amer. Meteor. Soc., 257–263.

———, 1986: Tornadogenesis by nonprecipitation induced wind shear lines. *Mon. Wea. Rev.,* **114,** 270–284.

———, and D. Reum, 1986: "The hail spike": a reflectivity and velocity signature. Preprints, *23d Conf. on Radar Meteorology,* Snowmass, CO, Amer. Meteor. Soc., R62–R65.

———, and W. E. Schreiber, 1986: Initiation of convective storms by radar-observed boundary layer convergent lines. *Mon. Wea. Rev.,* **114,** 2516–2536.

Winston, H. A., and L. J. Ruthi, 1986: Evaluation of RADAP II severe-storm detection algorithms. *Bull. Amer. Meteor. Soc.,* **67,** 145–150.

Witt, A., 1996: The relationship between low-elevation WSR-88D reflectivity and hail at the ground using precipitation observations from the VORTEX project. Preprints, *18th Conf. on Severe Local Storms,* San Francisco, CA, Amer. Meteor. Soc., 183–185.

———, M. D. Eilts, G. J. Stumpf, J. T. Johnson, E. D. Mitchell, and K. W. Thomas, 1998: An enhanced hail detection algorithm for the WSR-88D. *Wea. Forecasting,* **13,** 286–303.

Wolfson, M. M., 1990: Understanding and predicting microbursts. Preprints, *16th Conf. on Severe Local Storms,* Kananaskis Park, AB, Canada, Amer. Meteor. Soc., 340–351.

Ziegler, C. L., and E. N. Rasmussen, 1998: The initiation of moist convection on the dryline: Forecasting issues from a case study perspective. *Wea. Forecasting,* **13,** 1106–1131.

———, T. J. Lee, and R. A. Pielke Sr., 1997: Convective initiation at the dryline: A modeling study. *Mon. Wea. Rev.,* **125,** 1001–1026.

Zipser, E. J., 1982: Use of a conceptual model of the life-cycle of mesoscale convective systems to improve very short-range forecasts. *Nowcasting,* K. Browning, Ed., Academic Press, 191–204.

———, and M. A. LeMone, 1980: Cumulonimbus vertical velocity events in GATE. Part II: Synthesis and model core structure. *J. Atmos. Sci.,* **37,** 2458–2469.

Zixiu, X., W. Pengyun, and L. Xuefang, 1993: Tornadoes of China. *The Tornado: Its Structure, Dynamics, Prediction, and Hazards, Geophys. Monogr.,* No. 79, Amer. Geophys. Union, 435–444.

Zrnić, D. S., 1987: Three-body scattering produces precipitation signature of special diagnostic signature. *Radio Sci.,* **22,** 76–86.

Chapter 12

Flash Flood Forecast and Detection Methods

ROBERT S. DAVIS

National Weather Service Forecast Office, Pittsburgh, Pennsylvania

REVIEW PANEL: Josh Korotky (Chair), Fred Glass, Norman Junker, Mathew Kelsch, James Moore, and Steve Zubrick

12.1. Introduction

Flash floods occur within minutes or hours of excessive rainfall, a dam or levee failure, or a sudden discharge of water previously held upstream by an ice jam. Flash floods can carry large debris, rip out trees, destroy buildings and bridges, trigger catastrophic mud slides, and scrape out new channels. Rapidly rising floodwaters can attain heights exceeding 10 m (~30 ft). Most flood deaths are associated with flash floods. Most flash floods are caused by slow-moving thunderstorms, thunderstorms repeatedly moving over the same location, or excessive rainfall from hurricanes and other tropical systems. An examination of severe weather fatalities by Wood (1994) indicates that, since 1960, flash floods have killed more people in the United States (around 150 fatalities per year) than any other severe weather phenomena, including tornadoes and lightning (Graziano 1998). The problem is not limited to the United States; excessive rainfall and flash floods can occur anywhere in the world, as long as favorable meteorological and hydrological factors coexist.

Given the scope and magnitude of the problem, it is not difficult to understand the need for accurate and reliable flash flood forecasts. It may be less apparent, however, that the number of people at risk from flash floods is increasing due to population growth and urban expansion. Urbanization creates favorable conditions for flash floods because the impervious character of metropolitan environments promotes runoff. Population growth is placing more people at risk for a number of reasons. More people are moving to flood-prone regions, prompting residential and commercial development where flash floods are likely to occur. Also, an increasing number of people are vacationing along streams and in the mountains, where flash floods are more likely. Even when flash floods are forecast well and detected with adequate lead time, public response to the flash flood threat is still a problem. This may be because people fail to realize the potential danger associated with heavy rain and flash floods, especially the force applied to an object by moving water. The fact that many people have died while in their cars during flash floods implies that they may not have realized the impending danger. By contrast, most people recognize that tornadoes bring only death and destruction, so they respond with much greater urgency to tornado warnings.

Flash floods represent formidable forecast and detection challenges. The forecast challenge is augmented because flash floods are not caused uniquely by meteorological phenomena. Flash floods result when favorable meteorological and hydrological circumstances coexist. Although heavy rainfall is necessary, a given amount and duration of rainfall may or may not result in a flash flood, depending on the hydrologic characteristics of the watershed where rainfall accumulates. The amount of runoff and magnitude of stream flow depend on the distribution of rainfall, rainfall rate, and duration of high rainfall rates in the drainage basin, which is a function of storm location and movement within the basin. Physical characteristics of the basin (e.g., soil saturation and permeability, geography/slope, urbanization, vegetation) further modulate the runoff potential. Detecting flash floods can be a considerable challenge because most flash floods are not associated with clearly distinguishable storm types. Rather, flash floods tend to be associated with ordinary convection that produces excessive rainfall amounts over a particular area. To further hamper detection, flash floods are often identified with storms that also produce other types of important severe weather. If tornadoes and destructive winds are prominent, flash flood producing rainfall may not be detected, or viewed as an immediate threat. Moreover, the public's response may be more focused on the severe weather threat.

This chapter examines flash flood forecast and detection methods from the perspective of a Weather Service Forecast Office (WFO). A primary goal of the chapter is to present flash floods as the product of both meteorological and hydrological circumstances. Section 12.2 concentrates on flash flood characteristics, including causative factors of flash flooding. Section 12.3 focuses on the role of watersheds, while section 12.4 describes the basin factors that determine the hydrological response of a watershed to rainfall. Section 12.5 describes the meteorological factors that determine the scale, duration, and intensity of rainfall events. Section 12.6 examines the organization of convective systems that produce flash floods. Section 12.7 discusses the synoptic and mesoscale aspects of flash floods, and presents flash flood forecast methods. Finally, section 12.8 addresses remote sensing systems and the role of flash flood detection applications such as the Areal Mean Basin Estimated Rainfall (AMBER) program (Davis 1993).

12.2. Defining the flash flood problem

This section will review the characteristics that distinguish flash floods and will summarize the National Weather Service (NWS) flash flood watch and warning program.

a. Characteristics of flash floods

The simplest definition of a flash flood is "too much water, too little time" (Gruntfest and Huber 1991). While the preceding is certainly descriptive, it also underscores the complexity of arriving at a definition that will serve all interests. For example, what exactly do we mean when we say "too much water" or "too little time"? The answer can vary by geographic location and a number of hydrological, geographic, and meteorological factors.

In general, flash flood definitions incorporate high velocity flows that occur in a short period of time (Gruntfest and Huber 1991). Flash floods are distinguished from slow-rise stream floods because flash floods tend to evolve on the same time- and spatial scales as the intense precipitation, leading to short warning and response times. Slow-rise floods cause significant damage annually in the United States, but flash floods cause most of the flood-related fatalities.

Flash floods can be grouped into four general categories: 1) those resulting from excessive rainfall in natural watersheds, 2) those resulting from intense rainfall in catchments altered by humans (including urban drainage areas), 3) those resulting from dam or levee failure, and 4) those resulting from ice jams and flash floods enhanced by snowmelt. The flash flood threat due to urbanization is increasingly recognized as a serious and growing problem because vegetation has been removed from urban environments, bridges and culverts constrict flow, and building/paving have greatly expanded impermeable surfaces (Federal Emergency Management Agency 1987).

Winkler (1988) discussed the factors that mark the difference between catastrophic flash floods and fewer significant events. She found that there was no standard definition of an extreme precipitation event. Rather, the potential for a storm to produce flash flooding depends on previous rainfall, topography, rainfall rate, duration of rainfall, ground cover, and season. In a survey of 50 professionals and experts from a wide variety of disciplines, Gruntfest and Huber (1991) found the following conclusions: 1) rainfall intensity was the primary influence on the occurrence and severity of flash floods (especially in the western United States); 2) rainfall amount, duration, and previous rainfall were considered much more important in the East and Midwest while rainfall rate was seen as the primary catalyst for flash floods in the West; and 3) urbanization was expected to have an increasing impact on the occurrence and severity of flash floods. The respondents of the survey also proposed that any definition of a flash flood should emphasize the human impact due to the short warning and response time, and the resulting high potential for fatalities.

Considering the issues, a flash flood may be described as the intense hydrological response of a watershed or urban area to excessive rainfall, a dam or levee failure, a combination of rainfall and snowmelt, or a sudden release of impounded water by an ice jam. Characteristics of this response include 1) a high velocity flow of water that follows shortly (within minutes to a few hours) after the causative event, and 2) a very short warning and response time, resulting in a high potential for fatalities.

1) FLASH FLOODS VERSUS SLOW-RISE STREAM FLOODS

Flash floods are distinguished from slow-rise stream floods by the rapid rate of rise in the water level of impacted streams or urban drainage systems. Consequently, flash flood warnings typically provide short warning lead times, resulting in very little time for citizens and agencies to respond to the threat. Slow-rise floods cause significant damage annually in the United States, but flash floods are responsible for most of the flood-related fatalities.

The rate of rise in water level is directly related to the rainfall rate (intensity). Rainfall rate is considered light (<2.8 mm h^{-1}), moderate (2.8–7.5 mm h^{-1}), and heavy (>7.5 mm h^{-1}) by Bluestein (1993). These rates are typical with synoptic-scale cyclones that occur in the cool season, but they are not sufficient to produce flash floods. Rainfall rates that produce flash floods could be characterized as light (<20 mm h^{-1}), moderate (20–50 mm h^{-1}), and heavy (>50 mm h^{-1}). Only deep moist convection (DMC) can produce rainfall rates sufficiently high to produce flash flooding.

Synoptic-scale cyclones persist for days and can maintain rainfall rates of 5 mm h^{-1} or more for the lifetime of the storm. A rainfall rate of 5 mm h^{-1} for a 24-h period yields 120 mm of rainfall and may produce considerable slow-rise stream flooding, but not flash flooding (unless accompanied by rapid snowmelt). Slow-rise stream flooding usually provides adequate time for citizens and agencies to respond to the flood threat. The extreme rainfall rates associated with DMC can accumulate the same 120 mm of rainfall in less than an hour. Large accumulations of rainfall over such a short period can lead to disastrous flash flooding, with very little time for citizens to respond to warnings. Different combinations of rainfall intensity and duration result in a spectrum of stream flooding, with slow-rise flooding (and longer response time) at one end of the spectrum and flash flooding (with shorter response time) at the other end.

2) Flash floods caused by ice jams or aided by snowmelt

Synoptic-scale storms produce a deep snowpack across the Rocky Mountains and northern United States. An extensive and deep snowpack increases the runoff produced by melting snow in the winter and spring across the Rocky Mountains and the northern United States. Heavy spring rains falling on a melting snowpack can produce disastrous flash flooding. Since the snow cover may extend over very large geographic areas, the potential exists for widespread flash flooding in addition to river flooding.

Early in January of 1996, much of Pennsylvania and the mountains of West Virginia and Maryland received 60–120 cm of snow, with a water equivalent near 150 mm. Warming temperatures on 16–17 January ripened the snowpack. A combination of warm thermal advection, high surface dewpoints, and significant winds melted all of the snow in about 12 h, while 50–75 mm of warm rain fell into the melting snowpack. Since the ground was completely saturated by the melting snow, a tremendous amount of runoff was produced on 18–19 January 1996, resulting in what was described as a flash flood on the Ohio River at Pittsburgh (National Oceanic and Atmospheric Administration 1998). Severe flash flooding also occurred in the eastern portion of Pennsylvania (Leathers et al. 1998).

The stratiform rainfall rates typical of a cool season synoptic-scale storm are usually well below the 25 mm h^{-1} typically needed to produce flash flooding. Although stratiform rainfall rates of 2–10 mm h^{-1} acting over 24–48 h can produce serious slow-rise stream and river flooding, flash flooding is not likely. If, however, embedded convection briefly increases the rainfall rate to 25 mm h^{-1} or higher over a rapidly melting snowpack, serious flash flooding can result. A slow stream rise to near bankfull may suddenly become a rapid stream rise of 1–2 m over bankfull. Flash floods enhanced by snowmelt occur primarily in the late winter and spring, when the snowpack is at its greatest depth.

A melting snowpack may also contribute to flash floods produced by ice jams on creeks and rivers. Thick layers of ice often form in the winter on streams and rivers. Melting snow and/or warm rain running into the streams may lift and break this ice, allowing large chunks of ice to jam against bridges or other structures. This causes rapid water rises behind the ice jam and serious flash flooding downstream, if the water is suddenly released. The rising water can push immense pieces of ice onto the shore and through houses and buildings, causing considerable structural damage.

3) Flash floods from dam and levee failures

The failure of earthen dams and levees usually results from the structure being overtopped by rising water. Failure may also occur when poor construction permits water to leak through the structure. Impounded water is released suddenly when a dam or levee fails, inundating homes and businesses as a raging torrent moves downstream. Considerable damage resulted from levee failure during the great flood of 1993 (National Oceanic and Atmospheric Administration 1994).

Dam failures have extracted a grim toll of human life in the history of the United States. The greatest loss of life occurred on 31 May 1889, when the South Fork Dam 14 miles upstream from Johnstown, Pennsylvania, on the Little Conemaugh River burst (Degan and Degan 1984). Heavy rainfall on the night of 30 May 1889 filled the dam to capacity, and water started flowing over the crest of the 22-m-high earthen dam about 1530 UTC 31 May 1889. The dam gave way at 1915 UTC, releasing 20 million tons of water in less than 45 min. More than 2000 people lost their lives in the Little Conemaugh River valley from the dam site to the city of Johnstown, Pennsylvania. The flood wave hit the city of Johnstown at 2010 UTC.

Johnstown, Pennsylvania, was again struck with severe flash flooding on 19–20 July 1977, when heavy rainfall inundated the Conemaugh River watershed (Fig. 12.1). A total of 85 people lost their lives, with the majority of the fatalities in the three small watersheds shown in Fig. 12.2. Almost half of the fatalities resulted from the failure of the Laurel Run Dam. This earthen dam was located about 10 km north of Johnstown, several kilometers upstream along Laurel Run from the city of Tanneryville (Armbruster 1978). The dam broke about 0635 UTC 20 July 1977, causing great destruction in Tanneryville. Just four months after the Johnstown flood, an earthen dam near Toccoa Falls, Georgia, failed at 0530 UTC on 06 November 1977 (Land 1978), killing 39 people. More recently, two fatalities occurred when the Timber Lake Dam in western Virginia failed on 23 June 1995 (Kane et al. 1996).

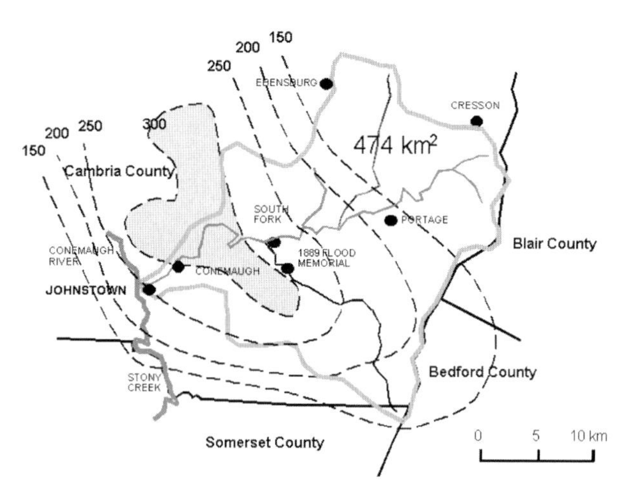

FIG. 12.1. Bucket survey of rainfall (mm) for the 19–20 Jul 1977 Johnstown, PA, flash flood (adapted from the National Oceanic and Atmospheric Administration 1977). The watershed of the Little Conemaugh River is shown. The Flood Memorial site is the location of the South Fork dam that failed on 31 May 1889, killing over 2000 people in the Conemaugh River valley.

b. NWS flash flood products

The NWS flash flood program (National Weather Service 1990) provides state/local emergency management agencies and the public with three primary products, including the flash flood watch, flash flood warning, and flash flood statements. Flash flood watches and warnings are issued for individual counties. A flash flood watch (warning) alerts agencies and the public to counties where flash flooding is possible (occurring or imminent). Flash flood statements are issued to provide additional information about ongoing watches and warnings, or to describe areas of potential flooding before watches or warnings are issued.

The success of a flash flood watch or warning is measured by *lead time*: the time from the issuance of the watch or warning to the first observed flooding. If the watch or warning is issued after flooding is observed, the watch or warning is said to have negative lead time.

A flash flood watch is usually issued within 12 h of potential flooding, and usually covers a period of 6 to 24 h. The flash flood watch may be issued for many counties or for a single county, depending on the areal extent of the flash flood threat. Flash flood watches are intended to increase agency and public readiness for possible flash flood warnings. Due to the short-fused nature of flash floods, short warning lead times require a rapid response from agencies and citizens.

A flash flood warning is issued for a particular county when flash flooding is impending or ongoing. While a severe thunderstorm or tornado warning is generally issued for a period of 1 h or less, a flash flood warning is typically issued for at least 3 h. Although flash floods are short duration events with a relatively high peak discharge (Cahail 1987), the entire process of a flash flood lasts longer than the initial surge of water. For example, streams take a significant amount of time to rise from a bankfull condition to peak flooding. The recession period from peak flooding back to a bankfull condition takes longer than the time of rise. This process from rise to peak and then recession seldom takes less than 3 h. The purpose of the flash flood warning is to produce a potential life-saving response by those who receive the warning.

Flash flood statements may be issued while a flash flood watch or warning is in effect, to provide critical new information not contained in the initial watch or warning. Consider the following situation. A warning is issued initially because a stream goes out of bank and inundates part of a highway, closing the road. Later, an additional period of heavy rainfall occurs in the same watershed, causing a rapid 4 m rise on the same creek, this time threatening homes along the highway. A flash flood statement under these circumstances will indicate the immediacy of the threat and prompt residents to act decisively to save their lives.

It is important to note that flash flood warnings and statements will only be effective if residents first receive the warning, then understand the correct life-

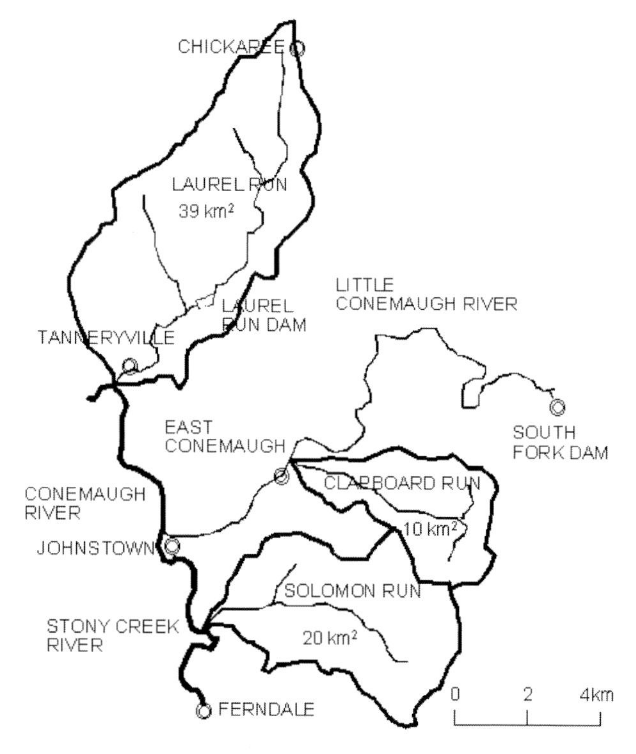

FIG. 12.2. Most of the 85 fatalities during the 19–20 Jul 1977 Johnstown, PA, flash flood occurred in the three small watersheds shown: Laurel Run, Clapboard Run, and Solomon Run. Unlike the river flooding caused by the collapse of the South Fork Dam in 1889, the 1977 flash flood was the result of intense rainfall on small watersheds. The area of the watersheds is shown in km^2.

FIG. 12.3. Delineation of a watershed boundary (McCuen 1998).

saving response to a particular threat, and have a plan of action that requires very little forethought. Recall, flash flood warnings generally offer short lead times and require rapid response.

12.3. The role of watersheds

A watershed can be thought of as a system that converts rainfall to runoff (Singh 1992). For a given rainfall event, only the rain that falls within the watershed boundary determines the volume, rate, and concentration of runoff, and thus the extent and intensity of flash flooding. Consequently, the key to determining the flash flood threat of any particular basin is quantifying the distribution of rain in the watershed. The hydrological characteristics of the basin receiving rainfall then determine the stream response.

A *watershed* is defined as an area of land that drains to a single outlet and is separated from other watersheds by a divide (Bedient and Huber 1992). Given that water flows downhill, the divide (boundary) of the watershed describes the land area that contributes water to the outlet during rainfall (McCuen 1998). Figure 12.3 shows a watershed that drains to the outlet at point A. Note that water flows to the outlet from a number of smaller areas, which define smaller watersheds. Thus, larger watersheds can be divided into smaller watersheds, and the smaller watersheds each have their own outlet and divide. In Fig. 12.3 two stream tributaries drain to the outlet at point B. The area bounded by the dashed contour and the shaded area defines the watershed for point B.

a. Subdividing watersheds

Larger watersheds are subdivided into smaller constituent basins in order to resolve the scale where flash floods occur. For example, the Pine Creek watershed (Fig. 12.4) covers 175 km^2 and drains to a basin outflow point just south of the city of Etna, Pennsylvania, where the creek flows into the Allegheny River. Notice how the larger Pine Creek watershed can be subdivided into eight smaller watersheds, each covering significantly less area, and each draining to its individual outflow point. During flash flooding in Pittsburgh, Pennsylvania, on 30 May 1986, nine fatalities occurred within the 16 km^2 area encompassing the Little Pine Creek basin (#101). A very small portion of the Pine Creek watershed was impacted by the heavy rainfall that fell in less than 2 h (Fig. 12.4). Although flash floods also occur in larger watersheds, they occur with much greater frequency in smaller watersheds. This emphasizes the importance of subdividing watersheds down to a scale sufficiently small to resolve flash floods. The geographic area of the Pine Creek watershed is much too large to resolve the drainage characteristics and flash flood potential of streams in the Little Pine Creek basin, where flooding actually occurred. Consequently, watersheds are subdivided down to a minimum basin area (MBA) for rainfall analysis. This will be discussed further in the next section.

b. The national stream database

Given the importance of subdividing watersheds down to a scale where flash floods occur, a national initiative is under way at the National Severe Storms Laboratory (NSSL) to provide 121 WFOs with a database of stream watersheds for use in flash flood applications. The National Basin Delineation Project (NBD) will define approximately one million watersheds across the United States, using an MBA of 5 km^2 (www.nssl.noaa.gov/teams/wester/basins). The stream basin delineation process is automated by using a

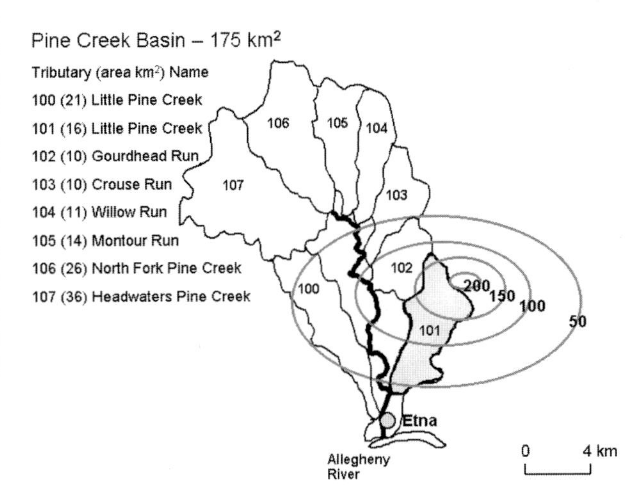

FIG. 12.4. The bucket survey of the rainfall that produced the Etna, PA, flash flood of 30 May 1986 is shown on the Pine Creek watershed (from the National Climatic Data Center 1986). A maximum of 200 mm of rainfall was observed in about 1.5 h. All nine fatalities occurred in the Little Pine Creek Watershed (101).

Geographic Information System (GIS; Jendrowski and Davis 1998). The NBD stream database is a foundation for the Flash Flood Monitoring and Prediction (FFMP) program, part of the System for Convective Analysis and Nowcasting (SCAN) (Smith et al. 1998) in the Advanced Weather Interactive Processing System (AWIPS) of the NWS. The FFMP is based on research and a proof of concept originating from the AMBER program (Davis and Jendrowski 1996). AMBER is a flash flood application that accumulates areal average values of precipitation derived from the Weather Surveillance Radar 1988 Doppler (WSR-88D) Digital Hybrid Scan Reflectivity (DHR) product. Areal averages from AMBER are computed and updated on the scale of small flash flood streams for each radar volume scan. After the implementation of FFMP, the rainfall estimates will be mapped directly on the NBD watersheds, providing a direct connection between the meteorological and hydrological factors of flash flooding.

The NBD program will identify each defined watershed segment with a unique Pfafstetter number (Verdin and Verdin 1999). The Pfafstetter number not only provides a unique identification number, it also provides information on the hydrological connectivity of all defined streams. For example, all defined watersheds in the Mississippi River basin will begin with the number 8. All basins within the Ohio River basin (contained within the Mississippi River basin) will begin the number 86. An additional digit is added to the Pfafstetter number for each additional subdivision of basins until watersheds to the scale of the minimum basin area (MBA) are reached. The subdivisions are accomplished by dividing each basin into nine parts (numbers 1 to 9) in upstream order with basin 1 closest to the outflow point of the watershed and 9 assigned to the headwater area of the basin. The Pfafstetter coding defines a complete upstream–downstream order for all watersheds.

To complete the stream basin delineation, larger watersheds must be subdivided into smaller constituent basins, defined to the scale of the MBA. The MBA determines the smallest watersheds to be included in the database and defines how larger streams are subdivided into smaller hydrological components. The area of the MBA is designed to be small enough to detect the scale of historical flash floods. Table 12.1 lists some historical flash floods that have produced fatalities across the United States and shows the area of the watersheds producing the flash flooding. It is noteworthy that many killer flash floods have occurred on watersheds with areas less than 40 km^2, and severe flash flooding is possible on watersheds as small as 5 km^2.

The importance of detecting rainfall on the scale of an MBA cannot be overstated. Of the 85 fatalities that occurred during the Johnstown, Pennsylvania, flood of 1977, 59 people lost their lives in two small watersheds (Table 12.1), Laurel Run and Solomon Run (Fig. 12.2). During this event, the city of Tanneryville was virtually wiped out by the failure of the Laurel Run Dam. The basin area above the dam is only 25 km^2. In 1986, a flash flood near Etna, Pennsylvania, claimed nine lives in a watershed encompassing only 16 km^2. If the MBA is too large, important flash floods on small watersheds may not be detected.

TABLE 12.1. Watershed area of selected basins that have produced fatal flash floods.

Date	Area (km^2)	Stream name	City, state	Deaths
12 Sep 1977	75	Brush Creek	Kansas City, MO	25
20 Jul 1977	39	Laurel Run	Tanneryville, PA	39
20 Jul 1977	20	Solomon Run	Johnstown, PA	20
30 May 1986	16	Little Pine Creek	Etna, PA	9
14 Jun 1990	32	Pipe Creek	Shadyside, OH	13
14 Jun 1990	30	Wegee Creek	Shadyside, OH	11
14 Jun 1990	5	Cumberland Run	Shadyside, OH	2
5 May 1995	30	Turtle Creek	Dallas, TX	15
29 Jul 1997	29	Spring Creek	Fort Collins, CO	5

After receiving the hydrologic database from the NBD, individual WFOs can further subdivide their watersheds by examining the scale of flash flood events that have occurred within their county warning areas (CWAs). Examination of historical flash floods in a WFO warning area can indicate the basin areas that have produced flash flooding. For example, a plot of the locations where people were swept away during the Shadyside, Ohio, flash flood of 1990 is shown in the disaster survey report (National Weather Service 1991). This plot clearly indicates that flash flooding occurred on three very small watersheds, Pipe Creek, Wegee Creek, and Cumberland Run, and shows the need for an MBA of about 5 km^2.

12.4. Factors that determine the hydrologic response of watersheds

A given amount of rainfall may or may not result in a flash flood, depending on the hydrologic condition of the watershed (drainage basin) where the rain falls. For example, the flash flood threat may be relatively high when heavy rainfall concentrates in a small watershed, especially if the basin has received rainfall recently. The same amount of rainfall may pose little or no threat if it falls across several basins, or if the affected watershed has a large capacity to absorb water. Doswell et al. (1996) present a case where significant rainfall accumulations (greater than 180 mm) on 7–8 September 1989 produced only minor flooding across parts of Iowa, because all of the affected watersheds were dry from a two-year drought. If the same event had occurred on saturated watersheds, the consequences could have been much worse.

Water from rainfall either travels overland to the stream channel, infiltrates into the soil layer, or evaporates back to the atmosphere (Bedient and Huber 1992). Overland flow is referred to as *surface runoff;*

water from surface runoff contributes most of the flow during flash flooding. Water that infiltrates into the soil mantle is referred to as *interflow*. Interflow travels laterally within the top layers of soil into the stream channel, but this process is much slower than the surface runoff and does not contribute directly to flash floods (Ponce 1989). Surface runoff occurs when the rainfall rate exceeds both the depression storage capacity and infiltration capacity of the soil. *Infiltration capacity* refers to the soil's ability to absorb water, and *depression storage* indicates the holding capacity of ground cavities and other depressions (Viessman et al. 1977). Greater surface runoff is possible when the infiltration capacity is low. Although precipitation deposits water directly into lakes and streams within individual watersheds, this contribution is relatively small compared to the total volume of water falling in the watershed.

The amount of runoff and magnitude of stream flow depend on the temporal and spatial distribution of high rainfall rates in the watershed. Given a particular rainfall, the flash flood threat is determined by physical characteristics of the basin: 1) size, shape, and topography of the basin; 2) size, shape, and condition of the stream channel; 3) infiltration capacity and saturation of the soil; 4) vegetation and land use (especially urbanization); and 5) season.

Flash floods represent the hydrologic response of a watershed to rainfall. This response may range from no response, to minor basement flooding, inundation of campers in tents along a stream, enough water on a highway to float an automobile or truck, or a flood wave severe enough to destroy homes and businesses. Although a thorough treatment of hydrology is beyond the scope of this chapter, a basic understanding of relevant hydrologic processes is fundamental to estimating the flash flood threat. This section concentrates on the rainfall characteristics and hydrologic processes that promote flash flooding.

a. Rainfall intensity and duration

Since surface runoff occurs when the rainfall intensity becomes greater than the infiltration capacity, the amount of surface runoff depends largely on the rainfall intensity. For rainfall events of equal magnitude, a short period of intense rainfall generally produces greater runoff than a longer period of less intense rainfall (Sweeney 1992). Thus, convective rainfall produces greater runoff than stratiform rainfall, because convective rainfall rates are much greater than stratiform rainfall rates. For example, a convective storm may produce more than 125 mm of rain in an hour; a stratiform event may take many hours to produce the same storm total. Rainfall of 125 mm in one hour will produce much greater runoff than 125 mm of rain spread over 8–10 h. If high rainfall rates persist over a particular area, the infiltration capacity will be exceeded quickly, resulting in flash flood producing runoff.

b. Average basin rainfall

The average rainfall depth on a watershed is known as the *average basin rainfall* (ABR). When ABR increases, the surface runoff and flash flood response of a watershed also increase. The product of ABR and watershed area determines the volume of water generated by rainfall. Some of this volume is lost to infiltration, evaporation, and interception, but the remaining volume moves into streams as runoff. A given volume of water may or may not cause flash flooding, depending on the time distribution of the flash flood runoff (McCuen 1998). Flash flooding is likely when a large volume of rain accumulates in a watershed in a short time, but flooding will be less severe if that same volume of rain accumulates over a longer period of time. Physical basin characteristics, combined with the distribution of runoff in the watershed, determine the speed of the basin's hydrologic response.

c. Physical characteristics of watersheds

Physical characteristics of a watershed have a large impact on the hydrological response of a basin to rainfall. These characteristics include basin area and length, slope of the watershed and stream channel, and geometric shape of the watershed and stream channel. The watershed area is important because it is related to the potential for rainfall to produce a volume of water that is then available for flood runoff. The hydrological length (measured along the flow path, from the basin divide to the outlet) is used in time-of-concentration (TC) calculations, which indicate how quickly water flows along the main flow path. The slopes of the watershed and stream channel contribute to the flood magnitude, which reflects the momentum of runoff. The geometric shape of the watershed and stream channel indicates how water converges at the outlet point.

Watershed slope (i.e., channel slope, measured as the change in elevation along the main stream channel) and the steepness of the stream valley walls largely determine the speed of the flash flood response of a stream channel. Steep slopes produce higher velocity surface runoff, which retards infiltration. High velocity surface runoff resulting from steep terrain will reach the stream channel more quickly than lower velocity flows associated with more level terrain. Thus, watersheds with steep slopes and steep valley walls produce the most rapid stream rises and the sharpest flash flood crests. On the other hand, watersheds with shallow slopes and shallow valley walls may take days to rise, rather than hours or minutes. Consequently, greater (lesser) rainfall amounts are needed to produce flash floods in watersheds with shallow (steep) slopes.

Table 12.2 indicates the watershed slope of several flash flood basins depicted in Table 12.1. Roaring Run was added as an example of a basin with an extreme

TABLE 12.2. Watershed slope of selected flash flood producing streams.

Stream name	Length (m)	Drop (m)	Drop/length (m/m × 1000)	Time of concentration (min)
Roaring River	10 622	1173	110.4	57
Cumberland Run	4184	116	27.7	47
Little Pine Creek	8562	165	19.3	91
Pipe Creek	12 682	225	17.7	130
Wegee Creek	12 070	200	16.6	128
Spring Creek	13 056	164	12.6	151
Brush Creek	15 417	101	6.6	221

slope. The Roaring Run watershed contains the Lawn Lake Dam that failed on 15 July 1982 (National Climatic Data Center 1982b), flooding the town of Estes Park, Colorado. The destructive power of water moving rapidly down this extreme slope is demonstrated by the size of the boulders deposited below the dam in Rocky Mountain National Park. Some of the boulders in the alluvial fan along Roaring Run measure more than 6 m (~20 ft) in diameter.

Watersheds with a slope of 10.0 or higher may be regarded as basins with high flash flood potential. Basins with slopes of 5.0–10.0 might be considered marginal. The Brush Creek watershed in Kansas City, Missouri, is included as an example of a small watershed (75 km^2) with relatively flat topography. Despite the shallow slope of the watershed, two significant flash floods have occurred in this watershed, one in 1977 and a second in 1998. Both of these flash floods will be examined in more detail later in the chapter.

Streams in watersheds with slopes of 5.0 or less respond slowly and typically produce flooding or ponding of water over larger geographic areas. The primary flooding problem in shallow sloping basins is from inundation of low-lying areas. Although the slow movement of water in these basins usually precludes life-threatening flash flooding, widespread damage to homes, businesses, and agriculture is still possible. Large areas of the midwestern and southeastern United States contain gently sloping watersheds.

d. Urban effects

As watersheds become increasingly urbanized, more people are at risk from flash flooding. Large geographic areas have been transformed from field, meadow, or wooded land to residential and commercial developments, roads, driveways, and parking lots. Vegetation has been removed from urban environments, bridges and culverts constrict flow, and building/paving have greatly expanded impervious surfaces (Federal Emergency Management Agency 1987). Impervious surfaces prevent infiltration, causing much greater surface runoff and a greatly increased volume of water reaching stream channels. The surface runoff collects in storm drains, which are quickly overwhelmed by intense rainfall. Otherwise, rapid runoff feeds high-velocity flow directly into storm drains, surface storm channels, and modified stream channels.

Rapid stream rises in highly urbanized areas can produce deadly flash floods. For example, 125 mm of rainfall in less than 1 h caused 16 fatalities and over $1 billion in property damage in Dallas, Texas, on 5 May 1995 (Baeck and Smith 1998). The Fort Collins, Colorado, flash flood of 28 July 1997 resulted in five fatalities and $200 million in damage (Petersen et al. 1999) when over 250 mm of rainfall occurred in less than 4 h. On 4–5 October 1998 the Kansas City, Missouri, business district suffered 15 fatalities and extreme flash flood damage after 125 mm of rain fell in 75 min. The Kansas City flash flood struck in the same Brush Creek watershed that flooded in 1977 (Larson and Vochatzer 1978). The 1977 flash flood was associated with 406 mm of rainfall, resulting in 25 fatalities and $90 million in property damage. Details of the rainfall characteristics and ABR associated with these three flash floods will be examined in more detail in section 12.8.

e. Soil, soil moisture, and ground cover

Low porosity soils such as clay have a lower infiltration capacity than higher porosity soils such as sand. Rocky areas and shallow soils are largely resistant to infiltration and allow greater runoff than nonrocky or deep soils. Due to the large fraction of paved areas, urban environments allow very little infiltration, resulting in large surface runoff and a heightened flash flood hazard.

Soil moisture has a large influence on surface runoff. If the soil is already holding water from previous rainfall, the infiltration capacity may be used up quickly during subsequent rainfall, resulting in considerable runoff. For a given rainfall, flash flooding may or may not occur, depending on whether or not the soil is saturated from previous rainfall. Antecedent rainfall is less of a factor in the desert areas of the West, however, because desert soil quickly recovers its infiltration capacity.

Vegetation is important because it intercepts rainfall, inhibits surface runoff, and allows more time for infiltration. Dense, low-lying vegetation is particularly effective at intercepting rainfall. On the other hand, wildfires remove vegetation and retard infiltration. A severe wildfire on 19 May 1996 in the Buffalo Creek watershed reduced the infiltration capacity in the burn area (shaded area of Fig. 12.5). Intense rainfall concentrated in the burn area (see ABR in Fig. 12.5), resulting in a deadly flash flood in the city of Buffalo Creek, Colorado, on 13 July 1996 (Fulton 1999).

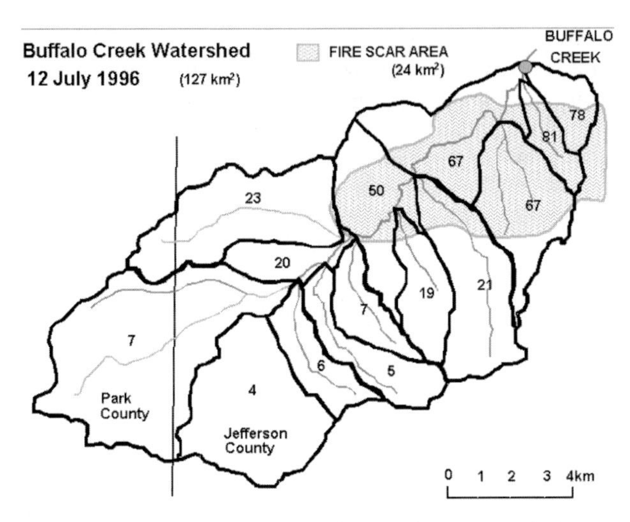

FIG. 12.5. Tributaries for the Buffalo Creek watershed near Buffalo Creek, CO. The ABR (mm) from 0200 to 0400 UTC on 13 Jul 1996 is plotted in each watershed segment. The majority of the rainfall fell in about 1 h. The circle at the mouth of the watershed is the town of Buffalo Creek. The gray shaded area is the burn scar that resulted from a forest fire several months before the flash flood. The solid black lines are watershed segment boundaries. Note that the largest ABR occurred in the watersheds containing the burn area.

f. Seasonal effects

Seasonal differences have a large impact on the surface runoff potential. Given a particular rainfall rate, duration, and ABR, more surface runoff will result in the winter than in the summer. Lower temperatures in the winter result in higher water viscosity and lower infiltration. Frozen ground further reduces the infiltration capacity, causing greater runoff. Vegetation decreases in the winter, resulting in less evapotranspiration, less interception, and greater water retention, resulting in greater runoff potential.

12.5. Factors that determine the character and scale of rainfall

Simply stated, the heaviest precipitation occurs where the rainfall rate is highest for the longest period of time (Chappell 1993). Key ingredients for flash floods include 1) high rainfall rates, 2) long duration of high rainfall rates over a particular area, and 3) basin hydrological circumstances that promote rapid runoff, resulting in flash flooding for the given rainfall. High rainfall rates are produced when air with considerable water vapor content ascends in convective updrafts, especially if the environment promotes efficient production of rainfall. High rainfall rates will persist over a particular area if convective systems (made up of a number of convective cells) move slowly, if they become quasi-stationary, if they become anchored in elevated terrain, or if they redevelop persistently upstream and traverse downstream locations in succession. In the latter case, convective systems may move across the same general region for several hours or days. If favorable large-scale conditions persist, convective systems may redevelop with the diurnal cycle and traverse common locations for an extended period, as was the case during the Midwest river and flash floods of 1993. High rainfall rates and flash floods occur most frequently with DMC, so this section will concentrate on meteorological processes and storm types that contribute to flash flood producing convective rainfall. Several topics in this section are covered in great detail elsewhere in this monograph, so they will only be summarized here for completeness.

a. Terrain effects

Mountains and complex terrain (including small hills, ridges, and escarpments) impose a significant influence on the focusing, forcing, and anchoring of convective precipitation (Banta 1990). Primary terrain effects include orographic lifting, thermally driven upslope flows, and obstacle effects (e.g., blocking, flow deflection, and gravity waves). Terrain effects occur with a range of large-scale and subsynoptic-scale flows. For example, flow tends to slow down when it approaches elevated terrain. The flow-retarding effects of the terrain may produce strong mass convergence and precipitation upstream of the barrier. Flow may also be diverted around terrain features, resulting in mass convergence, greater instability, and precipitation on the lee side of the barrier. Flow over terrain features may generate gravity waves, which induce areas of enhanced lift and may influence the downstream distribution of clouds and precipitation (Bruintjes et al. 1994). Terrain also functions as an elevated heat and moisture source (Banta 1990; Orville 1965, 1968). Increased heating near the slope surface induces relative low pressure, which induces upslope flow. Evapotranspiration from vegetation and soil (Oke 1987) contributes to the buoyancy of the low-level air (Orville 1968). In addition to focusing and forcing precipitation, terrain can play an important role in anchoring convection, resulting in locally high rainfall accumulations. The relationship between terrain and quasi-steady precipitation systems will be discussed later.

Orographically forced rainfall can be produced by both convective and nonconvective processes. Nonconvective rainfall results when persistent synoptic-scale flow forces weakly stable air to rise in hilly or mountainous terrain. The vertical motion associated with this upslope flow may be sufficient to produce locally heavy rainfall amounts despite the relative stability of the moist air (Doswell et al. 1998). Rainfall can persist for several days if the synoptic-scale flow continues to feed moist air into the terrain.

Orographically forced convective rainfall is likely when strong persistent flow forces moist unstable air to the level of free convection (LFC), initiating buoyant ascent. Destabilization associated with forced ascent can produce convection if the moisture content is sufficiently high, even in a weakly stable atmosphere. A number of disastrous flash floods have been associated with orographically forced convection, especially when vigorous lifting of easterly moist low-level flow is induced by the eastern slopes of the Rocky Mountains or the Appalachians. If the mid- and upper-level flow is weak, intense convective rainfall can last for several hours over any particular location, as individual storms drift very slowly along the sloped topography.

Heavy orographic precipitation in the winter is produced mainly by extratropical cyclones (Cotton and Anthes 1989). Although these storms do not typically produce intense convective rainfall rates, they can produce considerable rainfall and flash floods when they interact with elevated terrain for a long period of time. For example, substantial rainfall often follows when sustained strong winds associated with synoptic-scale storms drive moisture up the mountains of California and the Pacific Northwest. On 3–5 January 1982, upslope winds produced widespread excessive rainfall near Santa Cruz, California (National Climatic Data Center 1982a). Rainfall amounts of 610 mm accumulated over a period of 28–30 h in the higher elevations near Santa Cruz, resulting in flash flooding. Orographically enhanced rainfall due to strong synoptic-scale storms occurs primarily in the late fall, winter, and spring, and is a significant flash flood producing mechanism in the coastal mountain ranges of Washington, Oregon, and California.

Flash flood producing rainfall also occurs when the strong east winds of a nor'easter force moist air up the mountainous terrain of the northeastern United States. For example, an intense cyclone fed by tropical moisture from the southern Atlantic produced record high storm total rainfall, over 300 mm, during 19–23 October 1996 across much of New England (Keim 1998). Hourly rainfall rates at Portland, Maine, on 20–21 October 1996 were generally 12 mm h^{-1} or less, typical of the rates produced by a synoptic-scale storm. However, rainfall rates exceeded 25 mm h^{-1} locally for several hours due to embedded convection fed by the tropical moisture, resulting in widespread flooding.

b. The role of large-scale and mesoscale processes

Doswell (1987) provides an insightful foundation for understanding the relationship between large-scale and mesoscale processes as they relate to DMC. Large-scale processes are associated with synoptic-scale systems such as short-wave troughs and satisfy quasigeostrophic assumptions. Modest but persistent vertical ascent and horizontal advections ahead of shortwave troughs establish a favorable thermodynamic environment by increasing the low-level moisture, reducing (or eliminating) the convective inhibition, and decreasing the static stability over relatively large geographic areas. Large-scale processes also contribute to the vertical wind shear profile, which influences the physical and dynamical attributes of convection (i.e., convective organization and storm type). Mesoscale processes are linked to both larger and smaller scales, but quasigeostrophic balance assumptions are not valid on the mesoscale. Subsynoptic-scale processes provide a lifting mechanism for convective initiation and further increase the convective available potential energy (CAPE) and vertical wind shear locally. Lifting mechanisms are usually associated with low-level boundaries (e.g., strong convergence associated with fronts, convective outflows, etc.) and/or terrain. The low-level jet (LLJ) plays a prominent role in supplying moisture and may contribute to the convective initiation of elevated convection (Maddox et al. 1979; Junker et al. 1999; Glass et al. 1995).

c. Processes that determine rainfall rate

1) DEEP MOIST CONVECTION

High rainfall rates are possible when the environment supports DMC. For DMC to occur, environmental processes must provide a renewable supply of moisture with relatively high water vapor content, an unstable lapse rate, and subsynoptic-scale lifting mechanisms or topographic forcing sufficiently strong to force parcels to their level of free convection (LFC; Doswell 1985). Although DMC is possible during the cool season, it occurs with much greater frequency during the warm season, when air with high moisture content and buoyant instability more naturally coincides with low-level forcing mechanisms that are strong enough to produce convective updrafts.

Given an environment favorable for DMC, high rainfall rates are promoted by low-level air with high moisture content and convective updrafts sufficiently large and vigorous to process a large quantity of moisture. In general, the moisture processed in a convective updraft is proportional to the condensation rate and, thus, the rainfall intensity (Doswell et al. 1996). Flash floods are most often associated with DMC because convective updrafts are capable of processing the greatest amount of water vapor mass.

Since convective updrafts can quickly use up locally available moisture (Doswell et al. 1996), a mechanism is needed to replenish moisture so ongoing convection can continue long enough to produce locally high rainfall accumulations. An LLJ is often observed with convective systems that produce locally high rainfall accumulations (Maddox et al. 1979; Junker et al. 1999; Glass et al. 1995) and appears to be an important mechanism for generating extreme rainfall rates. The LLJ continually replenishes the supply of warm moist air processed by convective systems, allowing regenerative convection to produce high rainfall rates for an extended period of time.

2) Precipitation efficiency

Rainfall rates are modulated by the precipitation efficiency of a convective system, which indicates the extent to which the water vapor flux passing through a cloud base is converted to rainfall reaching the ground (Marwitz 1972a,b). Precipitation efficiency is determined by numerous environmental and microphysical processes; it can range from zero in nonprecipitating clouds to greater than unity for a short period (Cotton and Anthes 1989). It is noteworthy that some storms produce intense flash flood producing rainfall even though they do so inefficiently, for example, the heavy precipitation (HP) supercell. Braham (1952) found that isolated thunderstorms transform less than 20% of their inflow water vapor to rainfall.

Nearly all the water vapor entering a convective system condenses when it rises in convective updrafts, but only a fraction reaches the ground as rain (Doswell et al. 1996). The remainder evaporates when it is carried away by winds at the anvil level of the system (Ludlam 1980), or evaporates in downdrafts before reaching the ground (Braham 1952; Ludlam 1980). Consequently, less rain will reach the ground when updrafts are excessively strong, or when excessive evaporation occurs in storm downdrafts.

Entrainment of dry air reduces precipitation efficiency because it introduces unsaturated air into the convection, which promotes evaporation (Cotton and Anthes 1989). In general, isolated cells are susceptible to greater entrainment and reduced precipitation efficiency. For small convective systems, much of the inflow water vapor evaporates at higher tropospheric levels, especially if the updraft speeds are large (Ludlam 1980, p. 271). The environments of larger convective systems are much closer to saturation, resulting in less entrainment for embedded cells and more efficient conversion of inflowing water vapor flux to rainfall. Thus, precipitation efficiency increases (decreases) in humid (dry) environments, where evaporation is inhibited (enhanced). This underscores the importance of relative humidity as a modulator of precipitation efficiency. If the low-level storm environment remains nearly saturated, evaporation is diminished, resulting in greater precipitation efficiency and higher rainfall rates.

In general, precipitation efficiency is not an important issue when the large-scale environment supports regenerative or supercell convection, excluding the high-based convection that develops over the interior Rocky Mountains (Doswell et al. 1996). Convective rainfall rates are sufficiently high to cause flash floods even when the precipitation efficiency is low, especially if the convection becomes quasi-stationary, if many cells or storm systems move in succession over a particular area, or if an HP supercell slowly traverses a small watershed. Heavy precipitation supercells will be discussed in greater detail later.

3) Rainfall production

Cloud microphysics exert a strong influence on the production of rainfall and, thus, on the potential for high rainfall rates. In general, the potential for high rainfall rates is greater for cloud systems with deep warm cloud layers (i.e., the layer from the LCL to the parcel freezing level), but modulated by the residence time of parcels in this layer. Since the residence time for parcels in a layer is related to the updraft speed in the layer, cloud systems with moderate updrafts generally produce rainfall more efficiently than cloud systems with very strong updrafts. Large warm cloud layers increase the potential for heavy rain and flash flooding because cloud droplets have more time to interact in a warm rain process (i.e., collision-coalescence). This process converts a significant portion of the inflow water vapor flux to rain relatively low in the cloud, resulting in a higher precipitation efficiency (Braham 1964; Beard and Ochs 1993; Rasmussen 1995).

The Dallas, Texas, flash flood of 5 May 1995 demonstrates that extreme rainfall rates are possible when warm rain processes predominate. Rainfall estimates from a mesonet in the Dallas metropolitan area indicated peak 15-min rainfall accumulations greater than 60 mm (Baeck and Smith 1998). A combination of extreme rainfall rates and a dramatic lowering of the vertical center of mass in the storm (storm centroid) indicate that warm rain processes prevailed during the period of flash flood producing rainfall.

Cloud systems with low freezing levels and corresponding shallow warm cloud layers are more likely to produce most of their precipitation through ice processes, which typically are not associated with efficient rainfall production.

4) Vertical wind shear

Large vertical shear of the horizontal wind promotes entrainment and evaporation, which reduce precipitation efficiency. In fact, the precipitation efficiency can be as low as 5%, due to the rapid detrainment and evaporation of cloud water aloft (Marwitz 1972a). We will see later, however, that vertically sheared environments are associated with well organized and supercell convection. Organized convection can evolve into larger convective systems; the low-level environments of larger systems often contain abundant moisture, which retards evaporation and entrainment. Although supercell storm environments promote evaporation,

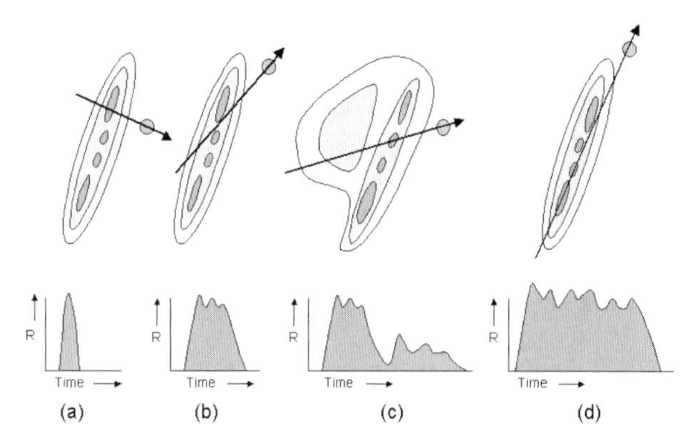

FIG. 12.6. Schematic showing how different types of convective systems with different motions affect the rainfall rate (R) at a point (indicated by a shaded circle) as a function of time; contours and shading indicate radar reflectivity. (a) A convective line is passing the point with a motion nearly normal to the line; (b) the line is moving past the point with a large component tangent to the line itself; (c) the line has a trailing region of moderate precipitation but is otherwise similar to (b); (d) the motion of the line has only a small component normal to the line but is otherwise similar to (c). Total rainfall experienced at the point is the shaded area under the R vs time graphs. From Doswell et al. (1996).

supercells still produce flash floods, due to the tremendous amount of water vapor mass that they process (Moller et al. 1990).

d. Processes that determine rainfall duration

Many convective systems generate high rainfall rates without producing flash floods. Systems that produce flash floods generate high rainfall rates over a particular area long enough to cause streams to leave their banks. In general, rainfall duration is related to 1) the areal extent of convective rainfall, 2) the movement of this area, and 3) the rate of new cell growth upstream (Chappell 1986). The potential for flash flood producing rainfall is enhanced when convective systems move slowly or become quasi-stationary, allowing many cells/storms to reach maturity/train over the same location. This is most likely when 1) large areas of intense rainfall are aligned with low-level boundaries oriented roughly along the axis of the storm motion (i.e., parallel to the mean cloud layer shear; Fig. 12.6), and 2) unstable air and low-level forcing mechanisms continue to generate new convective cell growth upstream of maturing convective cells. If the mean flow is oriented slightly toward the colder air, maturing cells will tend to move away from the boundary while leaving a quasi-stationary outflow on the rear flank of the convective system. This promotes moist boundary-relative flow along the rear flank, enhances the low-level moisture flux convergence, promotes repetitive cell growth along the rear (upstream) flank, and reinforces the low-level boundary. Upstream regeneration of new cells is favored when the strongest instability, moisture transport, and moisture flux convergence remain upstream (Chappell 1986), providing a continuous supply of moisture and low-level forcing for new cells on the system flank.

While many flash floods are associated with slow-moving or quasi-stationary convective systems, flash floods are also associated with significant large-scale weather systems that produce rapid storm movements. Regardless of the storm motion, most flash floods require upstream regeneration of new cells and low-level boundaries or fronts aligned with the mean flow. In situations where fast storm movements are prompted by strong winds aloft, a steady progression of rapidly moving storms or storm systems can pass over the same location for an extended period.

1) THE ROLE OF STORM MOTION

To understand how high rainfall rates can be sustained over a particular location, it is important to recognize that convective storms are composed of several (perhaps many) convective cells, each undergoing a similar evolution (Fig. 12.7). When the convection becomes quasi-stationary, convective cells will tend to develop just upstream of previous cells and then mature just downstream over a common area. In Fig. 12.7, each cell in the sequence develops in strong convergence associated with the outflow from previous cells and matures downstream over a common area.

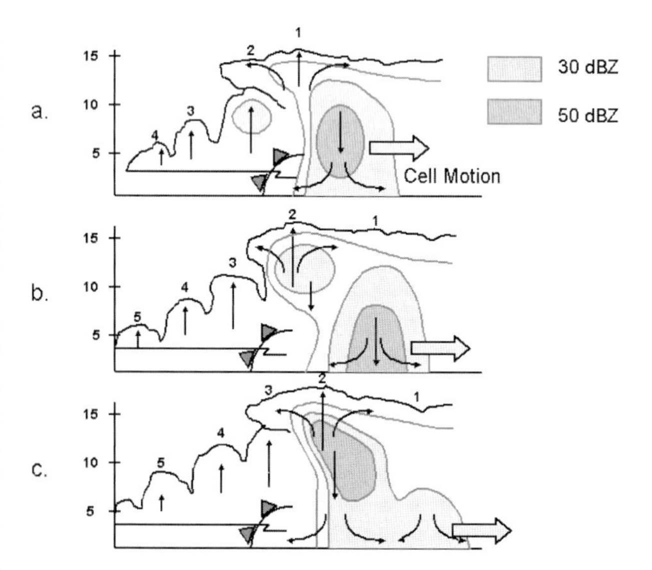

FIG. 12.7. Schematic showing the three stages in the evolution of a multicell thunderstorm system. Cells are labeled with numbers 1, 2, 3, etc.; thin arrows indicate the updrafts, downdrafts, and divergence at the storm top and surface associated with each cell; the frontal symbol indicates the low-level boundary; the cell motion is indicated by the heavy arrow; hatched lines show radar reflectivity and are labeled in dBZ. Note that the cells are moving left to right while the outflow boundary remains fixed in place. From Doswell et al. (1996).

Note that the cells are moving from left to right while the outflow boundary remains stationary.

In general, the movement of convective systems is determined by two processes: 1) individual cells move downstream roughly with the mean wind of the storm layer, generally in the direction of the cloud layer shear/thickness (Merrit and Fritsch 1984); and 2) propagation effects, whereby new cells develop in a preferred location relative to previous cells. Overall movement of convective systems is determined by the vector sum of these two contributions (Chappell 1986). Thus, the movement of a storm system is determined by the rate at which new cells develop along a particular flank of the system, and how quickly they move with the mean flow.

(i) Internal and external forcing mechanisms

New cell growth is determined by a number of processes, both internal and external to the convection (Doswell et al. 1996, 1998). Internal processes are generated by the convection; these include outflow boundaries generated by existing and previous cells. External processes are preexisting features, which include old outflow boundaries, fronts, etc. Internally produced outflow boundaries provide a storm-scale source of intense lift when the boundary-relative moist inflow is large (Doswell et al. 1996). This is most likely when the ambient flow is strong but the outflow boundary is moving slowly or is quasi-stationary. For example, locally prolonged high rainfall rates are likely when convective cells move along a slow-moving outflow boundary, leaving a quasi-stationary segment of the boundary on the rear flank of the convection (Doswell et al. 1996). Moist boundary-relative flow along the rear flank enhances the low-level moisture flux convergence and promotes repetitive cell growth. Outflows from subsequent cells reinforce the boundary, allowing the system to persist for many hours. The Johnstown, Pennsylvania, flash flood of 1977 (Hoxit et al. 1978) was associated with this process.

(ii) Modes of cell propagation

In general, the motion of a convective system will be greater than individual cell motions if the net propagation is aligned in the same general direction as the cell motions, a condition known as *forward propagation.* Conversely, the motion of a convective system will be less than the individual cell motion if the net propagation is aligned in the opposite direction of the cell motion, a condition known as *backward propagation.*

Propagation effects may operate with a direction and speed nearly opposite the cell movement (Chappell 1986), producing an overall system that is quasi-stationary or slow-moving (Fig. 12.8). When this happens, new cells will tend to regenerate just upstream, then mature downstream over a common location (see Fig. 12.7), producing a long duration of high rainfall rates. This was the case during the Pitcairn, Pennsylvania, flash flood of 1 July 1997 (Davis 1998), when a number of cells matured over a very small watershed. When propagation acts in the same direction as the mean steering flow, convective cells or storms will tend to move downstream, sometimes rapidly. Fast-moving storms can produce widespread flash flooding under the right circumstances because they produce a larger ground footprint of rain than slower-moving storms (Davis and Jendrowski 1996).

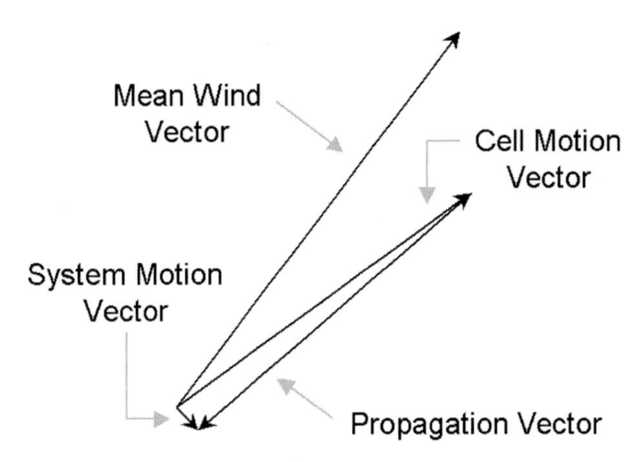

FIG. 12.8. Vector representation of the mean tropospheric wind vector, cell motion vector, cell propagation vector, and system motion vector. The diagram shows the near cancellation between cell motion and propagation, resulting in slow system movement. Note that slow system movement is still possible when the mean wind is significant. From Chappell (1986).

A steady progression of storms may pass over the same general area as long as 1) the ingredients for DMC remain in place along- and especially upstream, and 2) the axis of convection remains nearly stationary. The axis of convection will remain nearly stationary if it is associated with a low-level boundary aligned roughly parallel to the mean steering flow. This was the case during widespread flash flooding in the Ohio Valley during 26–28 June 1998 (details in section 12.8), when numerous convective storms traversed the same region over several days. It should be noted that there is a spectrum of propagation effects acting on a variety of spatial scales between the two prototypes discussed here. For example, larger MCSs sometimes regenerate upstream and then traverse downstream locations, a condition known as *supertraining* (Doswell et al. 1996).

2) MESOSCALE BETA ELEMENTS

When forecasting flash flood producing rainfall, it is more important to track the movement of the storm flank with active convection, rather than the storm system centroid (Chappell 1986). When convective systems become large, the active storm flank is defined by meso-β-scale elements (MBEs). MBEs are associated with the most vigorous updrafts and coincide with the heaviest rainfall (McAnelly and Cotton 1985). Corfidi et al. (1996) present a conceptual model termed the *vector approach* for predicting MBE movement, based on the vector sum of the mean flow in the cloud-bearing layer, and a vector representing the propagation component (Fig. 12.9). In the Corfidi approach, the vector representing the propagation component was defined as a vector opposite but equal in magnitude to the LLJ. Corfidi (1998) defined the caveats and improved the technique for situations where there is significant dry air in the midlevels of the troposphere or the subcloud layer. These environments are conducive to forward-propagating MCSs such as bow echoes or squall lines. Other factors, such as the CAPE/instability distribution, convective inhibition, low-level equivalent potential temperature configuration, presence of low-level boundaries, and terrain gradients, can modulate storm propagation and subsequently affect storm motion (Juying and Scofield 1989; Moore et al. 1993).

3) Terrain as an anchoring mechanism

Terrain can play an important role in anchoring convection and promoting high precipitation rates. This happens when persistent and moist low-level flow becomes focused and lifted by the topography, resulting in a quasi-stationary area of convective development. New cell growth will continue to regenerate over a common area as long as the axis of moist low-level flow remains unchanged. If the steering flow is weak, maturing cells will move slowly away from where new cells are forming, resulting in quasi-stationary convection (Maddox et al. 1980).

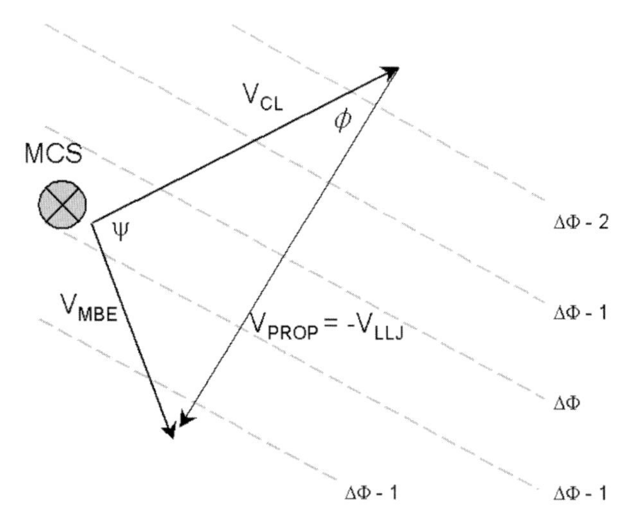

FIG. 12.9. Conceptual model of MBE movement ($\mathbf{V}_{MBE}$) as the vector sum of the mean flow in the cloud layer ($\mathbf{V}_{CL}$) and the propagation component ($\mathbf{V}_{PROP}$). The magnitude and direction of $\mathbf{V}_{PROP}$ are assumed to be equal but opposite to those of the low-level jet ($\mathbf{V}_{LLJ}$). The angles ϕ and φ are used to calculate $\mathbf{V}_{MBE}$ given observed values of $\mathbf{V}_{CL}$ and $\mathbf{V}_{LLJ}$. Dashed lines indicate the 850–300-mb thickness. From Corfidi et al. (1996).

When the environment favors DMC, the threat of flash flooding can be especially high if convective outflows or other low-level mesoscale boundaries become anchored in terrain. Quasi-stationary convection associated with these convective-scale interactions can cause extremely high rainfall rates to persist over an area, resulting in disastrous flash flooding. For example, the thunderstorms that produced 200–300 mm of rainfall near Johnstown, Pennsylvania, on 20 July 1977 were anchored in terrain by a nearly stationary outflow boundary for almost 8 h (Hoxit et al. 1978), resulting in severe flash flooding and 85 fatalities. On 28 July 1997 a convective cluster became quasi-stationary when it interacted with an outflow boundary and local topography near Fort Collins, Colorado. The anchored convection generated more than 250 mm of rainfall in the Spring Creek watershed that flows through Fort Collins, Colorado (Kelsch 1998), resulting in a flash flood that caused five fatalities and over $100 million in property damage to Colorado State University (Knievel and Johnson 1998). Other examples of flash flood producing storms anchored by terrain include the Big Thompson, Colorado (Maddox et al. 1978); Rapid City, South Dakota (Maddox et al. 1978; Nair et al. 1996); and Madison County, Virginia flash floods (Pontrelli et al. 1999). The Madison County, Virginia, event produced a 500-yr record flood in addition to storm totals of nearly 600 mm in 18 h (Pontrelli et al. 1999).

12.6. Flash flood producing convective systems

High rainfall rates are possible with a variety of storm evolutions. It is important to note that storms exist as a spectrum of possible evolutions, from short-lived storms composed of a few cells, to long-lived organized storm clusters composed of numerous cells, to supercell storms. Given this complexity, however, convective storms can be distinguished by their physical and dynamical organization (Weisman and Klemp 1982, 1984, 1986; Klemp 1987). *Physical organization* refers to the tendency of new updrafts (cells) to form in a consistent location relative to previous updrafts, a characteristic of ordinary convection. *Dynamical organization* is related to the development of vertical pressure gradients and storm-scale rotation, processes associated with supercell storms (Klemp 1987). Thus, from physical and dynamical arguments, convective storms can be classified as either supercell or "ordinary" (i.e., nonsupercell). Physical and dynamical differences between storm types are determined largely by characteristics of the vertical wind shear and the nature of new cell development (Doswell 1985; Weisman and Klemp 1986). Understanding the distinctions between storm types provides a foundation for determining the nature of flash flood potential on any given day, because the scale, intensity, and duration of rainfall events are often related to particular convective evolutions. Thorough discussions of storm type can be found elsewhere in this volume; the following discussion will summarize the importance of storm type as it relates to the flash flood threat.

a. Ordinary convection

The term "ordinary" is not related to the importance or potential severity of the convection; it is a statement about the processes that produce and sustain the convection. Most convective storms are ordinary, and most flash floods are produced by ordinary convection. Ordinary convection relies primarily on buoyancy potential and the availability of lifting mechanisms in the environment (Doswell 1985; Weisman and Klemp 1986). Large-scale advections determine the buoyancy potential. Lifting mechanisms are usually associated with low-level boundaries (e.g., convergence associated with fronts, convective outflows, etc.) and/or terrain.

Ordinary convection can be more or less organized, depending on the vertical structure of the wind field (Weisman and Klemp 1986), microphysical properties of the downdraft, and/or the orientation and strength of preexisting low-level boundaries. Vertical wind shear determines the storm-relative flow, the location of the precipitation cascade relative to existing updrafts, convergence associated with the low-level outflow, and, thus, formation and proximity of new updrafts relative to existing updrafts (Doswell et al. 1996). Microphysical properties of downdrafts (especially evaporation of rain) influence the strength of subsequent outflows.

Vertical wind shear and microphysical characteristics of downdrafts provide a spectrum of possibilities for convective organization. If the vertical wind shear is weak, new updrafts will tend to develop near existing updrafts. Conversely, new updrafts will be displaced some distance from existing updrafts if the vertical wind shear is stronger. Displacement of new cells from existing convection promotes convective organization, but too much displacement means that new cells will be disconnected from, and thus unaided by, previous convection (Doswell et al. 1996). Weak outflows tend to move slowly, allowing updrafts to regenerate adjacent to existing convection. Strong outflows may rush well ahead of existing convection, thereby removing the source of low-level moisture and instability for regenerative growth. Outflow strength is modulated by the low-level relative humidity, which determines the evaporation rate in convective downdrafts as they descend through the subcloud layer. Low-level environments with higher (lower) relative humidity promote lower (higher) evaporation potential, resulting in weaker (stronger) outflows and more (less) convective-scale organization.

Organized convection is likely when 1) the vertical wind shear promotes new cell growth in proximity to previous cells, 2) outflows remain optimally displaced from ongoing convection, and/or 3) ongoing convection interacts with external low-level boundaries (e.g., old outflow boundaries, fronts) and/or terrain. When convection interacts with external boundaries or terrain, organized storms can result even if the environment exhibits very little shear. Organized storms normally contain several (perhaps many) cells in various stages of evolution. Each new updraft develops in a consistent location relative to previous updrafts (Fig. 12.7), an evolution known as *discrete propagation* (Chisholm and Renick 1972; Marwitz 1972b).

Organized convection tends to form into small clusters and line segments, but may also evolve into larger mesoscale systems. When the environment supports organized DMC, storms can last for many hours, while processing a substantial amount of water vapor mass. If the low-level moisture content is high, organized convection can generate high rainfall rates very efficiently. Depending on the consistency of cell propagation and the distribution of low-level forcing mechanisms, organized storms may become quasi-stationary, or a "train" of storm clusters may move in succession over the same location. Both of these situations can result in flash flood producing rainfall if the hydrological circumstances are favorable.

b. Supercell convection

If the vertical wind shear becomes sufficiently strong in a statically unstable environment, convective updrafts may become dynamically enhanced by vertical pressure gradients, which act independently of buoyancy to produce vigorous vertical accelerations and strong storm-scale rotation (i.e., a mesocyclone) (Rotunno 1993; Klemp 1987). The resulting storm evolution is referred to as a *supercell.* Unlike the discrete formation of new updrafts associated with ordinary convection, supercell updrafts propagate continuously, giving the appearance of a single quasi-steady updraft. The resulting storm motion is usually to the right or left of the mean wind, depending on the strength and directional change of the vertical wind shear (Weisman and Klemp 1986). Supercells that propagate continuously to the right tend to move more slowly than other storms in the same environment, thus increasing the potential for sustained high rainfall rates over a given area.

Environments that favor supercells typically exhibit substantial dry air aloft, implying high evaporation potential. Given the greatly enhanced tendency for evaporation in supercell environments, precipitation efficiency is relatively low for individual supercell storms. There is a growing awareness, however, that some supercells can generate abundant rainfall, despite growing in an environment that promotes lower precipitation efficiency. Moller et al. (1990) describe a spectrum of supercell evolutions. Within this spectrum, the greatest flash flood potential is associated with HP supercells (Moller et al. 1990; Doswell and Burgess 1993; Moore et al. 1995), which tend to develop in moisture-rich environments. Substantial low-level moisture allows HP supercells to process prodigious water vapor mass; this results in extraordinary rainfall rates, but with relatively low precipitation efficiency. Catastrophic flash floods are possible with HP supercells because they often combine high rainfall rates with slower system movement.

On 1 July 1993 a series of HP supercells produced flash flooding in northeastern Missouri (Moore et al. 1995). A combination of slow system movement and echo training produced localized storm rainfall totals of 225 mm. Rainfall estimates from the WSR-88D indicated that 125 mm of the storm total rainfall occurred over 3 h. In addition to flash flooding, these storms also produced hail to 10 cm in diameter, and spawned an F2 tornado and several F0 tornadoes.

On 26 May 1978 an HP supercell generated over 250 mm of rain in 90 min just west of Canyon, Texas (Belville et al. 1979). The heavy rain was deposited on the Palo Duro Creek and the Tierra Blanca Creek watersheds, which flow into the city of Canyon, Texas. The resulting flash flood killed four people and resulted in $20 million in property damage. The storm was also accompanied by large hail and wind gusts of 40–45 m s^{-1} in the western part of the city.

The Dallas flash flood of 5 May 1995 was produced by the merger of a bow echo with a HP supercell. The bow echo produced strong winds and grapefruit-sized hail in Fort Worth, Texas, before the merger. The merging cells inundated the Dallas metropolitan area with over 100 mm of rain in 1 h. The resulting flash flood caused 17 fatalities (Baeck and Smith 1998).

c. Mesoscale convective systems

Mesoscale convective systems (MCSs; Zipser 1982; Kane et al. 1987) are covered in great detail elsewhere in this monograph, so this section will only summarize important aspects of MCSs as they relate to flash flooding. MCSs include a wide variety of organized mesoscale convective structures and may contain elements of both ordinary and supercell convection. The spectrum of MCSs ranges from relatively small short-lived storm clusters and line segments to full latitude squall lines and the mesoscale convective complex (MCC) (Maddox 1980). MCCs are essentially the largest and most persistent member of the MCS spectrum. They are predominantly nocturnal storm systems, identified in infrared (IR) satellite imagery by their large, long-lasting (>6 h), nearly circular, and cold cloud shields (Maddox 1980). It should be noted that MCCs are not qualitatively different from other MCSs (Doswell et al. 1996). The arbitrary criteria for defining MCCs were developed to distinguish them from systems with less circular cloud shields (squall lines) or from smaller-scale MCSs. Recall, however, that even small-scale MCSs (small convective storm clusters) are capable of producing significant flash floods when they become stationary or move in succession over a particular area.

MCCs and larger MCSs are important because 1) they produce 30%–70% of the warm season rainfall across a large part of the central United States (Merritt and Fritsch 1984), 2) they produce tremendous rainfall over large areas (especially at night); and 3) they are associated with widespread flash flooding and river flooding. MCCs and larger MCSs occur most frequently across the Midwest; they occur only rarely east of the Ohio and Tennessee Valleys (Tollerud et al. 1987).

Given the right set of circumstances, MCSs may become quasi-stationary, or a "train" of MCSs may move in succession over the same location, resulting in flash flood–producing rainfall. Due to the intense rainfall, watersheds often lose their capacity to absorb additional rainfall after the passage of an MCS. If another MCS passes by the same area, the intense rainfall will likely convert rapidly into surface runoff, causing streams to leave their banks.

d. Tropical storms and hurricanes

High precipitation rates in tropical storms are associated with deep tropical moisture, organized spiral band convection, and the decaying center of tropical circulations. Radar reflectivity in the convective cells of a hurricane (Burpee 1986) tend to range from 45 to 50 dBZ, with brief maximum reflectivity of 51–53 dBZ. A radar reflectivity–rainfall rate (*Z–R*) relationship defines a specific rainfall rate (*R*) in millimeters per hour for each radar reflectivity (*Z*) measured in dBZ. Using the tropical *Z–R* relationship of the WSR-88D (Ulbrich et al. 1996), rainfall rates in hurricanes would typically range from 56 to 147 mm h^{-1}. Although the time duration of rainfall in tropical storms can be measured in days, the rain is not continuous, but occurs in bursts of heavy rain associated with convection in the spiral bands and the eyewall (Marks 1990).

To the date of this publication, there have been no studies detailing the characteristics of flash floods produced when tropical storms and hurricanes move inland over the United States. Regional studies of heavy rainfall associated with land-falling tropical storms have been produced (Ward 1981), and disaster survey reports have documented heavy rainfall from specific storms, such as Hurricane Agnes (National Oceanic and Atmospheric Administration 1972). The

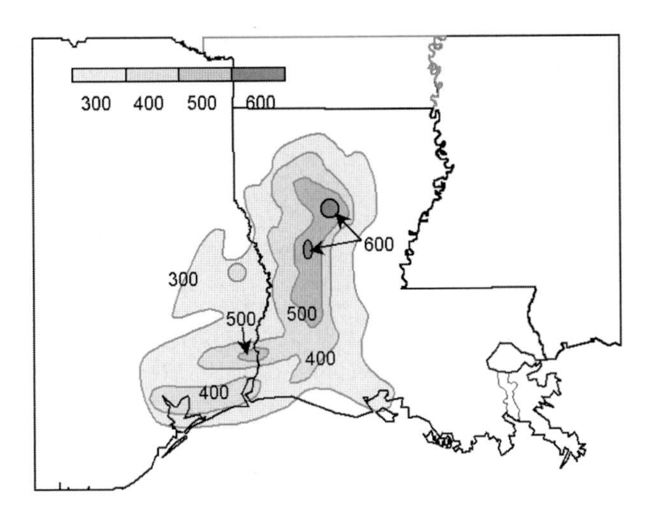

FIG. 12.10. Tropical Storm Allison observed rainfall (mm) from 26 Jun to 7 Jul 1989 in Louisiana and Texas. Adapted from National Climatic Data Center (1989).

National Hurricane Center (NHC) Web site (http://www.nhc.noaa.gov) lists the deadliest hurricanes in the United States from 1900 to 1996, but no breakdown is given for loss of life due to hurricane storm surge, tornadoes, damaging winds, or flooding. The deadliest hurricane on record was the Galveston hurricane of 1900, which caused 8000 fatalities, primarily due to storm surge. The Lake Okeechobee hurricane of 1928 caused the second greatest number of fatalities, when 1836 people died during flooding. The flooding was caused when strong hurricane winds produced 15-ft waves on the lake, inundating the homes of agricultural workers on the south side of the lake.

Weaker hurricanes (category 1) and tropical storms (storms not reaching hurricane strength) do not produce significant storm surge. Many of the fatalities and most of the damage occurring with these storms can most likely be attributed to flooding and flash flooding associated with heavy rainfall. Using the same logic, the three storms in this category producing the greatest number of fatalities are Hurricane Diane in 1955 (184 deaths), Hurricane Agnes in 1972 (122 deaths), and Tropical Storm Alberto in 1994 (30 deaths). The NHC also provides a list of the 30 most costly storms. The storms producing the greatest damage from flooding include Hurricane Agnes in 1972 ($2.1 billion), Hurricane Juan in 1985 ($1.5 billion), Hurricane Diane in 1955 ($831 million), Tropical Storm Allison in 1989 ($500 million), Tropical Storm Alberto in 1994 ($500 million), Tropical Storm Claudette in 1979 ($400 million), and Tropical Storm Gordon in 1994 ($400 million).

Many of the stronger hurricanes (category 2 and higher) produce flash flooding after moving inland and weakening. For example, the James River at Richmond, Virginia, flooded [flood crest of 8.7 m (28.6 ft)] in 1969 from the heavy rainfall produced by the remnants of the category 5 Hurricane Camille that initially came inland across Mississippi and Louisiana. The James River at Richmond, Virginia, crested at 11.1 m (36.5 ft) in 1972 from Hurricane Agnes and again at 9.4 m (30.76 ft) in 1985 from Hurricane Juan, both category 1 hurricanes. A decaying hurricane can move slowly across a large section of the eastern United States, spreading flash flooding and widespread river flooding. The slower-moving tropical storms may produce heavy rainfall for more than a week. Tropical Storm Allison produced heavy rainfall from Texas and Louisiana (Fig. 12.10) to Delaware from 26 June 1989 to 7 July 1989. The remnants of Agnes in June of 1972 (Bosart and Dean 1991) produced record flash flooding from Virginia to New York over a period of many days.

Rainfall amounts produced by slower-moving tropical storms can be extreme. In addition to the locally high rainfall amounts, the spatial extent of the rainfall can be quite large. Of all the meteorological systems that cause flash floods, tropical storms produce the largest geographic footprint of rain (Clark et al. 1980). Frequently this widespread heavy rainfall will produce record river flooding in addition to flash flooding. Alvin, Texas, received 1143 mm of rain from Tropical Storm Claudette in 1979. From 11–14 October 1981, Hurricane Norma produced a swath of more than 250 mm of rainfall from Abilene, Texas, to McAlester, Oklahoma, a distance of almost 500 km. Five different maxima of over 500 mm of rainfall were reported in this band of intense rainfall. Flood-producing rainfall from Tropical Storm Alberto (Fig. 12.11) occurred over several days in June 1993.

12.7. Forecasting flash floods

a. Synoptic and mesoscale aspects of flash floods

Maddox et al. (1979) examined 151 intense convective rainfall events and developed conceptual models

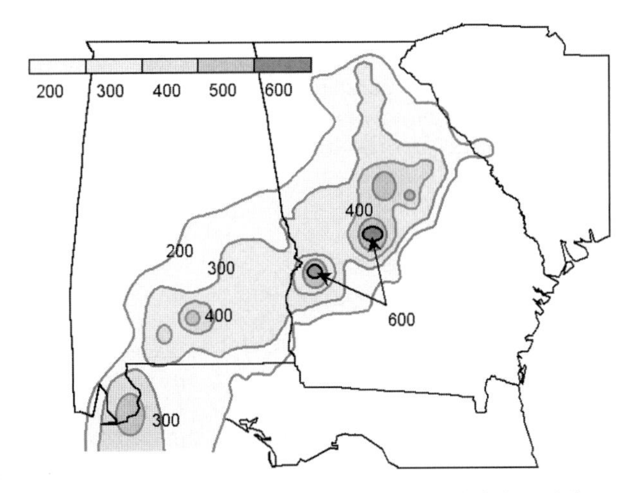

FIG. 12.11. Tropical Storm Alberto observed rainfall (mm) from 1 to 7 Jul 1994 in the southeast United States. Adapted from National Climatic Data Center (1994).

to illustrate the primary meteorological features and climatological characteristics associated with flash flooding over different regions of the United States. The study identified three conceptual models for flash flood producing rainfall over the eastern and central United States, including 1) synoptic events, 2) frontal events, and 3) mesohigh events. Although western events were treated separately in this study, Maddox et al. (1980) showed that they share many attributes with eastern and central U. S. events.

Synoptic events are typically associated with significant weather systems, usually an intense large-scale cyclone with a southwest–northeast (or west–east) oriented quasi-stationary surface front. Due to the baroclinic nature of these systems, relatively strong wind fields promote significant vertical wind shear and relatively fast storm movements, resulting in both severe and flash flood producing storms. Regardless of the orientation of the surface front, winds aloft coincide with the front, allowing a train of convective storms to pass rapidly over the same region. Synoptic events can affect several states over two to three days. Several individual flash floods may be accompanied by river flooding, depending on the movement of the large-scale frontal system, the moisture content of the warm-sector air, and the extent and degree of convective instability.

Frontal events are typically associated with west-to-east oriented boundaries (e.g., warm fronts) and weak large-scale forcing. Heavy convective rainfall typically occurs over the cool side of the boundary, aided by isentropic (i.e., warm advection) processes. This situation is distinguished from synoptic events, where storms develop and remain within the warm air mass. A weak to moderate LLJ provides a continuous supply of warm moist air for new cell growth and winds aloft parallel the front, causing convective cells and storms to mature or move in succession over the same region. The greatest flash flood threat occurs when low-level southerly flow is strong while winds veer and become weaker with height. Veering winds promote a storm movement roughly parallel to the forcing boundary, allowing a continuous inflow of moist unstable air on the right flank of ongoing convection. The greatest flash flood potential exists along and north of the boundary; the extent of heavy rainfall and flash flooding can cover several states, depending on the available moisture and instability.

Mesohigh events occur in the vicinity of quasi-stationary outflow boundaries generated by ongoing or previous convection. The heaviest rain generally falls on the cool side of the boundary, typically to the south or southwest of the mesohigh produced by the convection. Mesohigh events sometimes occur with large-scale frontal systems; at other times they occur in benign weather situations. In general, mesohigh events are much smaller in scale than frontal events. Flash floods are typically associated with cells that form upstream and mature downstream over the same area. The most notable mesohigh event was probably the Johnstown, Pennsylvania, flash flood of 19–20 July 1977. Figure 12.12 shows a regional mesoanalysis for 0300 UTC 20 July 1977 (Hoxit et al. 1978). Radar indicated that 12 individual cells matured over the Johnstown area between 0000 and 0900 UTC, producing 300 mm of rain north of Johnstown. This excessive rainfall generated a catastrophic flash flood, resulting in 85 fatalities and $200 million in property damage. Convection focused north of a preexisting outflow boundary; the outflows from each successive cell reinforced the boundary and provided forcing for continuous cell growth.

Western events often develop in weak large-scale patterns with nondescript surface features. Western events are usually associated with considerable instability, climatologically high values of precipitable water (PW), and weak wind fields. Downstream advections associated with a weak upstream shortwave trough often contribute to local destabilization while increasing the depth and magnitude of available moisture. Flash floods are associated with DMC and typically involve a complex interaction of outflow boundaries, surface fronts, and terrain (Maddox et al. 1980). The most significant difference between eastern and western events involves timing; eastern and central U. S. events are generally nocturnal, while western events occur most frequently before sunset in association with diurnal heating.

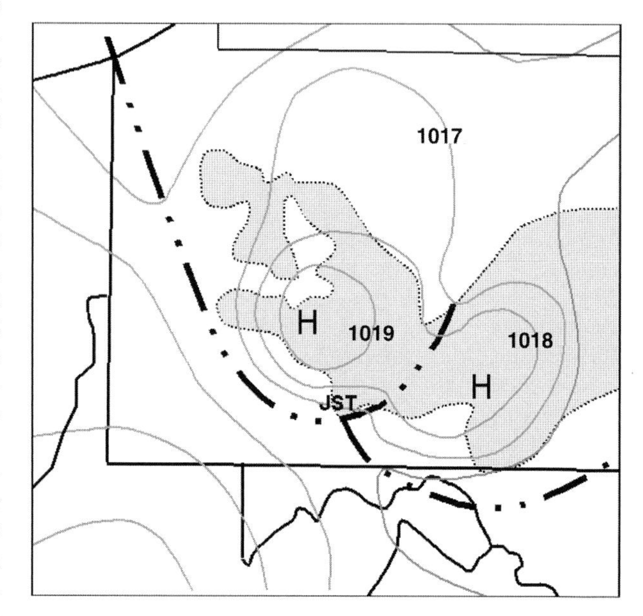

FIG. 12.12. Subjective regional mesoscale analysis of surface pressure (mb) for 0300 UTC 20 Jul 1977. Figure depicts the positions of outflow boundaries and radar echoes. Shading represents VIP 1 or greater radar echoes. Adapted from National Oceanic and Atmospheric Administration (1977).

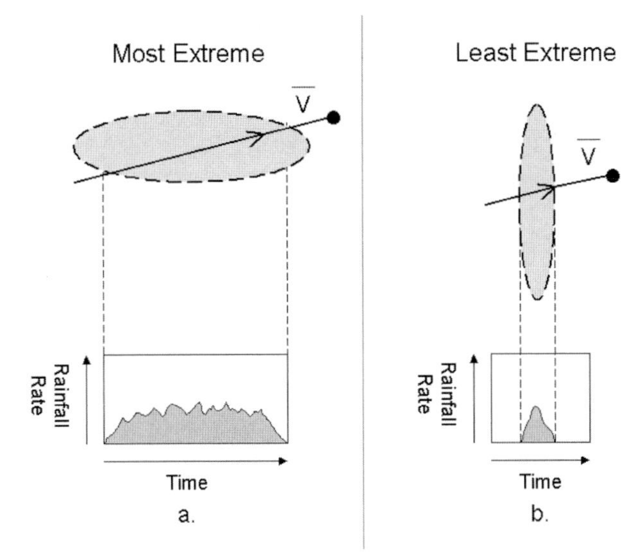

FIG. 12.13. The extent of the moisture flux convergence along the axis of the 800–300-mb mean wind and upstream of the heaviest rainfall for (a) the most extreme rainfall events and (b) the least extreme rainfall events. The figure depicts the center of the heaviest rainfall (dot), moisture flux convergence (shaded area), the axis of the moisture flux convergence (thin line), and direction of 850–300-mb mean wind vectors (arrows). In (a), a larger scale of forcing and a longer duration of intense rainfall are possible when a long axis of moisture flux convergence coincides with the mean wind. From Junker et al. (1999).

Flash flood producing storms require a persistent source of low-level moisture. The Gulf of Mexico supplies low-level moisture to a large portion of the United States between the Rocky Mountains and the Appalachians. The Atlantic Ocean provides a source of moisture for areas east of the Appalachians. Moisture for western events emanates from the southwestern (or Mexican) monsoon.

Junker et al. (1999) constructed a synoptic–dynamic climatology from 85 rainfall events that produced 24-h 2-in. (50.8 mm) or greater rainfall amounts across nine states during the Great Midwest Floods of 1993. The purpose of the research was to identify factors that determine the scale and intensity of rainfall events, and to distinguish the most extreme events from the least extreme ones. Results indicate that characteristics of the moisture field play a critical role in modulating the scale and intensity of convective rainfall events. For example, the most widespread and extreme rainfall events occur when an elongated pattern of low-level moisture flux convergence coincides with a low-level quasi-stationary boundary oriented roughly along the axis of the mean wind (Fig. 12.13). Processes associated with this pattern promoted upstream cell regeneration. Cells develop repeatedly upstream where forcing and instability remain strong, then track downstream with the mean flow. The low-level boundary focuses new cell development and the westerly flow routes storms over the same general areas. Additional insights from the study indicate that the environments of extreme rainfall events contain high values of both PW and mean relative humidity, a situation conducive to efficient rainfall production and high rainfall rates.

Composite fields by Junker et al. (1999) show the relationship between important features associated with the most extreme rainfall events. The composite maximum rainfall (Fig. 12.14) is located just downwind of an axis of strong 850-mb moisture flux (transport), in the gradient area of 850-mb moisture flux convergence. The surface-based moisture flux convergence is generally located beneath this area, indicating that convection is rooted initially in the boundary layer. Convective storms likely formed west of the rainfall maximum and then moved eastward with the mean flow. These conclusions are supported by Junker and Schneider (1997), who determined that surface moisture flux convergence developed upstream from initial storm development prior to a period of back-building. Composites also substantiate that MCSs often develop near or slightly downwind of the 850-mb equivalent potential temperature (θ_e) ridge axis and just south of the maximum θ_e advection (not shown).

Augustine and Caracena (1993) created composite fields to identify the relationship between lower-tropospheric precursors and nocturnal MCS development over the central United States. Composites and case

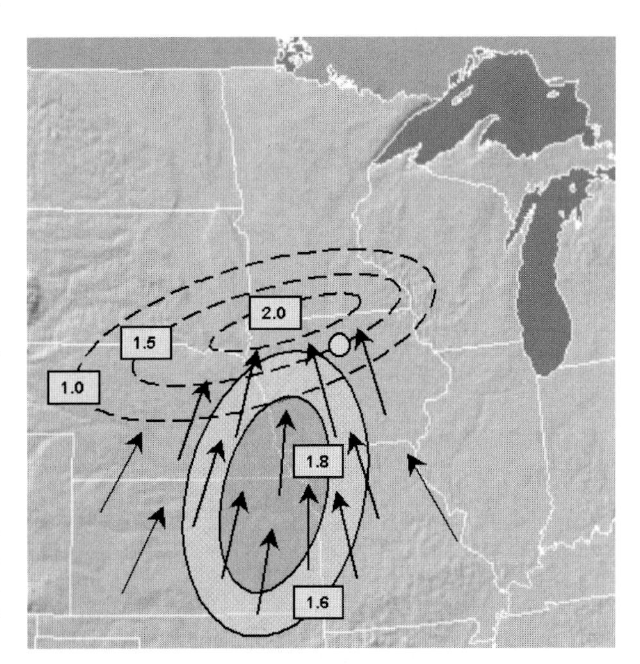

FIG. 12.14. Composite of 850-mb moisture flux convergence (dashed contours, g kg^{-1} s^{-1}), moisture flux vector directions (arrows), and moisture flux vector magnitudes (solid contours, g m kg^{-1} s^{-1}). The center of heaviest rainfall is represented by a lightly shaded circle. From Junker et al. (1999).

studies reveal that the largest MCSs occur where a nocturnal LLJ (above the location of the afternoon surface geostrophic wind maximum) interacts with a relatively deep low-level boundary maintained by significant frontogenesis (Fig. 12.15). In this region, focused vertical motion associated with frontogenesis is enhanced by warm advection and convergence, resulting in significant MCS development. Less significant MCS development is associated with more diffuse boundaries and weaker temperature gradients. Weaker boundaries are not supported by significant frontogenesis, resulting in less focused vertical motion.

In a study of heavy convective rainfall events across the mid-Mississippi Valley, Glass et al. (1995) proposed conceptual models for heavy rainfall events in the vicinity of an east–west boundary (Fig. 12.16). The greatest threat of heavy convective rainfall (indicated by the shaded region) occurs north of the boundary, where strong moisture flux convergence and warm thermal advection coincide with the LLJ and positive θ_e advection. In this region cells repeatedly develop and organize into MBEs of intense convection, then move downstream with the 850–300-mb cloud layer shear. Forward-propagating and regenerative MCSs are more likely when the LLJ–850-mb θ_e advection couplet veers significantly with time (Fig. 12.17). Backward-propagating or regenerative MCSs are more likely when the LLJ–850-mb θ_e advection couplet remains stationary or veers minimally with time (Fig. 12.18). Both of these patterns are capable of producing a train of convective storms. The heaviest rainfall typically extends downshear from the LLJ–850-mb θ_e advection couplet, parallel to the 850–300 mb thickness. The downshear extent of the heaviest rain is determined by the width and magnitude of the LLJ, the strength of the cloud layer shear, and the distribution of moisture and instability along the low-level boundary. For example, a broad LLJ creates a long axis of

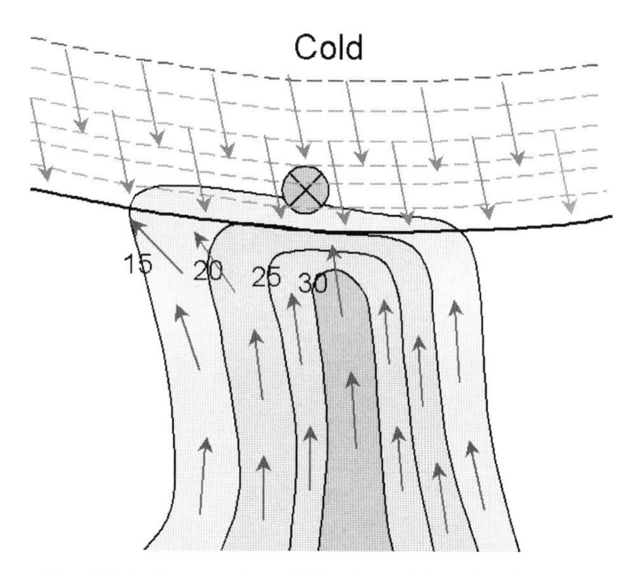

FIG. 12.15. Conceptual model for determining where large MCS development is favored. Composite fields combine late afternoon surface analysis with the 0000 UTC 850-mb analysis. Heavy dashed lines and downward-facing arrows are 850-mb isentropes and F vectors for 0000 UTC, respectively; wind magnitudes (shaded contours) and upward-facing arrows represent the late afternoon surface geostrophic winds and isotachs (kt). The stationary front marks the position of the surface boundary in late afternoon, and the shaded circle with an x indicates the probable location of a large MCS at maturity. The figure indicates that the largest MCSs occur where a nocturnal LLJ (above the location of the afternoon surface geostrophic wind maximum) interacts with a relatively deep low-level boundary maintained by significant frontogenesis. From Augustine and Caracena (1993).

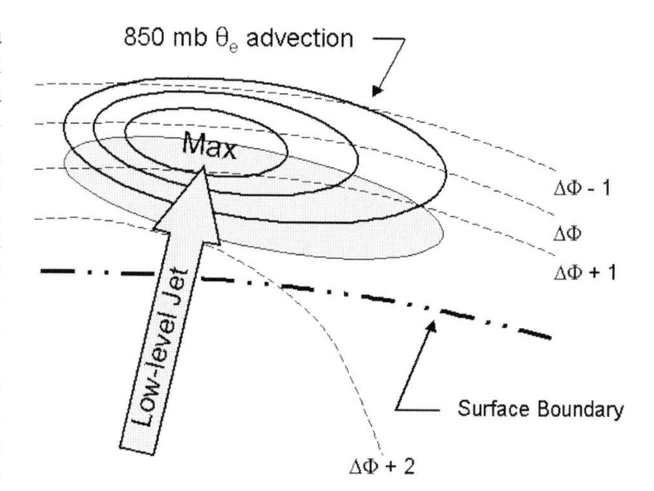

FIG. 12.16. Conceptual model illustrating the likely area for heavy convective rainfall (shaded oval) in the vicinity of a quasi-stationary low-level boundary (thick dash–dotted contour). Dashed contours represent the 850–300-mb thickness. From Glass et al. (1995).

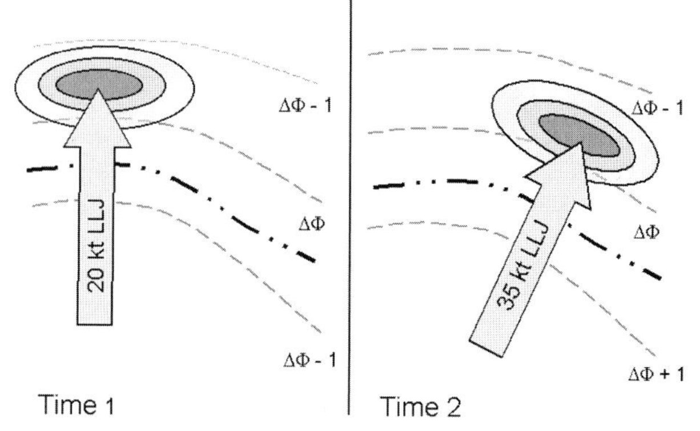

FIG. 12.17. Conceptual model illustrating the processes that promote a forward-propagating MCS. Forward-propagating MCSs are more likely when the LLJ increases and the LLJ–850-mb θ_e couplet veers with time. Thick dash–dotted contour represents quasi-stationary low-level boundary.

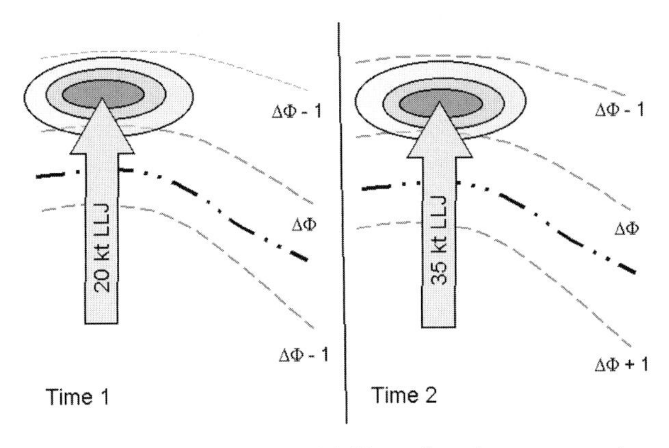

FIG. 12.18. Conceptual model illustrating the processes that promote a backward-propagating MCS. Stationary or backward-propagating MCSs are more likely when the LLJ increases but the LLJ–850-mb θ_e couplet remains stationary or veers minimally with time (Glass et al. 1995).

strong moisture flux convergence, resulting in an elongated pattern of strong low-level forcing on the right flank of ongoing convection. The LLJ also replenishes the moisture processed by convection and contributes to further destabilization of the environment. These results also agree with Junker et al. (1999), who found that the most extreme rainfall events occur along an extended axis of strong low-level moisture flux convergence in an area where the LLJ encounters a low-level boundary or front.

These studies (and others) identify a number of meteorological features often observed with flash flood events: 1) DMC is prominent (especially at night); 2) high moisture content extends through a deep layer; 3) weak to moderate midlevel vertical wind shear prevails; 4) large-scale forcing mechanisms are negligible or weak; 5) winds veer considerably with height due to the orientation of an LLJ relative to the flow aloft; 6) an LLJ supplies moisture, acts to focus/force convection, and determines the scale of rainfall; 7) winds aloft flow parallel to low-level boundaries (e.g., warm fronts and larger-scale outflows); 8) cells and/or storms mature or move in sequence over a particular area; and 9) an upstream meso-α-scale wave is generally moving into the long wave ridge. Given these observations, it is important to understand the physical properties that they demonstrate.

Synoptic event patterns are distinguished from other flash flood prototypes because they are typically associated with baroclinic, strongly forced systems that develop in moderate to strong vertical wind shear. Regardless of the apparent differences, significant weather systems share fundamental traits with other flash flood–producing situations. For example, high moisture content and DMC are critically important, and winds aloft tend to coincide with the axis of a quasi-stationary boundary (in this case a front). Moreover, flash floods are associated with intense long duration rainfall, in this case resulting from a train of convective storms. Unlike weakly forced storm situations where cells or storms move slowly, numerous convective storms develop with baroclinic systems, and many move rapidly over the same region. Due to the fast storm movements, synoptic systems generally produce a larger ground footprint of rainfall than weakly forced systems (Davis and Jendrowski 1996). Although less rain falls over any particular area from individual storms, the combined rainfall from many storms can cause widespread river flooding in addition to flash flooding.

b. The forecast process

Flash flood situations represent a spectrum of synoptic and mesoscale possibilities. Some events are strongly forced and others are weakly forced. Some are associated with strong wind fields while others exhibit very weak wind fields; subsequent storm movements may be rapid or negligible. Some situations are associated with patterns that produce severe weather and others are not. Topography is important in some events but not in others. Most flash floods are produced by ordinary convection but some are associated with supercell convection; still others occur when ordinary and/or supercell convection organizes into large MCSs. Some flash floods are associated with weak outflows while others are forced by significant frontal processes; still others are associated with a variety of boundaries between these two extremes. Some situations produce widespread flash flooding and river flooding while others produce only localized flooding.

Although flash flood prototypes represent a spectrum of possibilities, they are linked by the processes (ingredients) that make the events possible (Doswell et al. 1996). For example, flash floods are promoted by DMC, a deep moist layer, an LLJ feeding moisture into (or along) a quasi-stationary boundary or front, and a situation that encourages cells or storms to mature or move in a sequence over the same general region. Consequently, the flash flood potential is enhanced when synoptic and mesoscale processes assemble these ingredients, regardless of the actual map patterns associated with the ingredients (Doswell et al. 1996). Conceptual models grounded in physical reasoning can help forecasters choose a relevant forecast approach from a large set of diagnostic possibilities. For example, Maddox et al. (1979, 1980), Junker et al. (1999), Glass et al. (1995), and Augustine and Caracena (1993) provide conceptual models and diagnostics that reveal important underlying physical principles. Recognition of a particular situation can lead to more productive diagnostic methods and a greater understanding of the flash flood potential.

A thorough treatment of diagnostic methods is beyond the scope of this chapter. Instead, this section will present physical reasoning that should guide the forecast process. See Doswell et al. (1998) for a more thorough discussion of diagnostic methods. In general terms, forecasting flash flood potential consists of forecasting the likelihood of an event, the scale, intensity, and areal coverage of rainfall, and the expected hydrologic contribution. Therefore, the forecast process should reveal whether or not high rainfall rates of long duration are possible. This is accomplished by diagnosing the current state and forecasting a future state of atmospheric processes that promote flash flood-producing rainfall. Since most flash floods are associated with DMC, the diagnosis and forecast should reveal the current and future potential for DMC. This amounts to diagnosing and forecasting the strength and distribution of moisture, instability, vertical wind shear, and lifting mechanisms (both large and small scale).

Large-scale processes accumulate moisture and produce unstable lapse rates, which contribute to CAPE. Relatively weak but persistent large-scale lift also reduces the convective inhibition (CIN). Mesoscale lifting mechanisms are generally required to force low-level parcels to their LFC, which converts the CAPE to kinetic energy through parcel accelerations. Although upstream mesoscale waves do not trigger convection, they can reduce local stability downstream without promoting strong wind fields, resulting in slow system movement once convection develops. Upstream synoptic-scale waves are associated with downstream quasigeostrophic advections, which act to destabilize relatively large areas downstream while increasing the low-level moisture.

After establishing the present and estimating the future state of DMC, the scale, intensity, and duration of rainfall need to be assessed. The diagnostics should reveal the availability and magnitude of moisture for high rainfall rates, the role of vertical wind shear in organizing the convection, the scale and strength of meteorological and/or topographic forcing, and the relative orientation and strength of boundaries, fronts, and LLJs.

High rainfall rates are produced when considerable water vapor is processed by convective updrafts. Thus, high rainfall rates are more likely when substantial moisture extends through a deep layer. A deep moist layer (e.g., high PW) increases the water vapor mass available for DMC, encourages warm rain processes, and reduces the effects of evaporation in convective downdrafts. This contributes to high rainfall rates (through greater precipitation efficiency), promotes weaker outflows, and encourages new cell growth in proximity to previous cells. Recall, however, that locally available moisture can be depleted quickly by DMC (Doswell et al. 1996). Therefore, a mechanism is needed to restore the depleted moisture. The LLJ replenishes the moisture processed by convection, contributes to further destabilization of the environment, and provides an important source of low-level forcing for new cells (through low-level convergence).

Vertical wind shear determines the mode and organization of convective storms. Weak to moderate vertical wind shear promotes greater convective-scale organization and slower system movement. If the low-level environment is sufficiently moist, weak vertical wind shear promotes weaker outflows, and encourages new cell growth near previous cells. This process extends the lifetime of convective storms, thereby increasing the potential for long-duration rainfall. A veering wind profile (e.g., in the vicinity of a quasi-stationary front or east–west oriented boundary) creates a lengthy pattern of precipitation fallout along the axis of the upper-level flow (Fig. 12.19). Convective outflow associated with this precipitation cascade produces an extended region of low-level convergence, yielding intense and persistent rainfall along the axis of the upper-level flow (Cotton 1990).

The scale and strength of forcing are related to the orientation, width, and magnitude of the LLJ; the strength and orientation of the cloud layer wind; and the distribution of moisture and instability along the axis of a low-level forcing boundary or front. For situations involving an east–west boundary, a broad LLJ oriented into the boundary creates a long axis of strong moisture flux convergence along the boundary, resulting in an elongated pattern of strong low-level forcing (Glass et al. 1995). Recall that Junker et al. (1999) found that the most extreme rainfall events occur along an extended axis of strong low-level moisture flux convergence, in an area where the LLJ encounters a low-level boundary or front. This situation encourages convective cells and/or storms to

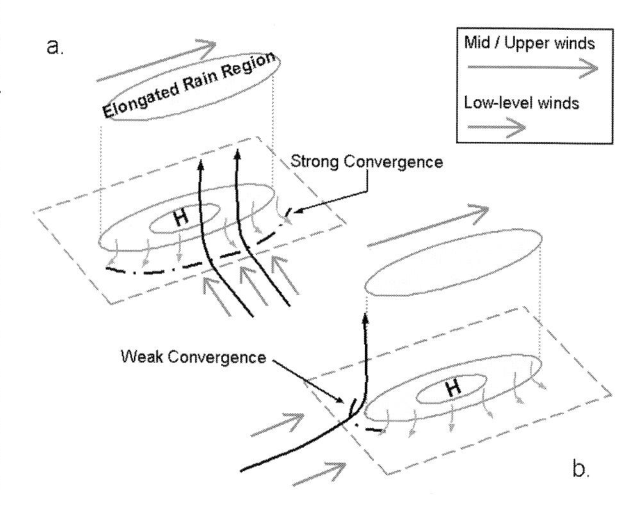

FIG. 12.19. Illustration of a multicellular storm growing in an environment where (a) the low-level winds are perpendicular to the mid- and upper-level winds and, (b) the low-level winds are parallel to the mid- and upper-level winds. From Cotton (1990).

mature or "train" over a particular region, resulting in a higher flash flood threat.

Although strongly forced synoptic systems are distinguished from other flash flood prototypes, they still share fundamental characteristics with other flash flood-producing situations. High moisture content and DMC are critically important, winds aloft coincide with the axis of a quasi-stationary frontal system, and flash floods are associated with intense long duration rainfall, in this case resulting from a train of convective storms.

It is important to recall that flash floods occur when a meteorological event coincides with favorable hydrologic circumstances. Consequently, it is important to anticipate the interaction of meteorological and hydrological contributions. Critical environmental factors include topography, previous rainfall, and expected storm motion relative to terrain features and low-level convergence zones. Given the forecast of meteorological and hydrological contributions, it is important to forecast a range of possibilities. This will enhance quick recognition of important storm-scale evolutions and prevent denial.

c. *Sounding analysis*

Rainfall is produced from two distinct microphysical processes: a warm rain process of coalescence and a cold rain (ice) process (Cotton and Anthes 1989). Figure 12.20 illustrates how a sample sounding might be divided into layers related to the occurrence of these microphysical processes. Both warm and cold

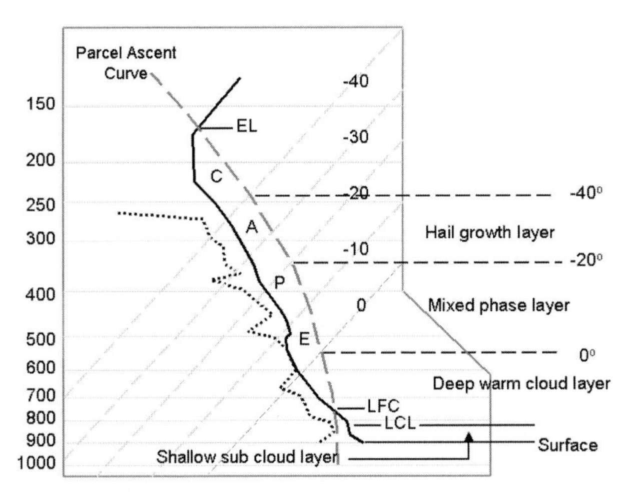

FIG. 12.20. A representative flash flood sounding is broken down into layers based on the microphysical processes responsible for rainfall production. The depth of the subcloud layer, warm cloud layer, the mixed phase layer, and the hail growth layer can be important parameters for rainfall production with warm or cold rain processes. The mean precipitable water, relative humidity, and CAPE in each layer also greatly impact the rainfall production processes.

rain processes frequently operate within the same storm. Coalescence occurs at parcel temperatures warmer than 0°C and therefore occurs in the lower levels of a convective storm. Mixed ice processes and snow production occur in the layer from 0° to −20°C. Ice processes, especially those related to the production of large hail, occur at temperatures between −20° and −40°C (Young 1993). Since warm rain processes generate the highest rainfall rates, forecasters should pay particular attention to environments that promote warm rain processes. The factors primarily responsible for warm rain processes are summarized in Pontrelli et al. (1999).

Warm rain processes promote a rapid conversion of inflow water vapor to rain relatively low in the cloud, resulting in a higher precipitation efficiency and higher rainfall rates. Consequently, deep warm cloud layers increase the potential for heavy rain and flash flooding because warm rain processes can operate over a greater cloud depth. Warm rain processes operate within a layer that extends from the lifted condensation level (LCL) to the freezing level. The depth of this warm cloud layer is particularly important in determining the potential for high rainfall rates. For example, flash flood situations are typically associated with warm cloud layers that are 3–4 km thick (Chappell 1993). Since only a limited amount of ice forms in a convective updraft until temperatures fall below −10°C, a thick warm cloud layer allows substantial time for warm rain processes to generate intense precipitation.

Moderate CAPE (i.e., 1500–2000 J kg^{-1}) is preferable to extreme CAPE because warm rain processes require time to convert condensate to raindrops (Chappel 1993). High CAPE in the low levels of the storm rapidly accelerates water vapor through the warm cloud layer, which reduces the time available for rainfall production (Young 1993). This forces much of the water vapor into high levels of the storm and promotes the production of ice and large hail. A combination of weak vertical wind shear and high precipitable water promotes higher rainfall rates because entrainment and evaporation are inhibited.

The amount of water vapor available for rainfall production is determined by the availability and magnitude of low-level moisture. For efficient rainfall production, the subcloud layer should be shallow with high relative humidity (RH) and high surface dewpoints (T_d). A deep subcloud layer with low RH will evaporate rainfall and result in strong cool downdrafts. The cold pool produced by these downdrafts can cut off the inflow of warm moist air into the storm and stop the rainfall production.

The most intense rainfall cores of DMC are not very large, usually only 1–3 km in diameter. Small intense rainfall cores can be maintained only if the effects of entrainment can be minimized. Entrainment is inhibited when high RH extends through a substantial depth of the sounding. Although high values of PW generally indicate substantial RH through a deep layer, it is important to examine the moisture profile of the entire sounding. For example, a high PW is possible with substantial midlevel dry air, indicating a primary threat from severe weather rather than flash flooding. It should be recalled, however, that supercells develop in environments that exhibit substantial dry air aloft if the vertical wind shear is sufficiently large. If supercells are possible with such a sounding, they may move into a moisture-rich environment, resulting in a possible HP supercell evolution and a correspondingly high flash flood threat. The PWs seldom reach 50 mm, but rainfall rates greater than 125 mm h^{-1} occur with some regularity in DMC. This substantiates that large quantities of water vapor must be ingested into the bottom of a storm to support observed rainfall rates of 100–200 mm h^{-1}. Thus, the most efficient rainfall-producing storms have high RH and a regenerative supply of high moisture content through a deep layer. A continuous supply of moisture for new updrafts is typically delivered by the LLJ.

An ingredients-based approach can be applied readily to atmospheric soundings. For example, key ingredients for high rainfall rates include 1) sufficient moisture and CAPE to promote DMC, 2) production of raindrops from warm rain processes, 3) moderate rather than high CAPE, 4) relatively weak vertical wind shear, and 5) a deep layer of substantial moisture. Given this understanding of the physical processes that promote high rainfall rates, an optimum sounding for establishing convective storms with a high potential for high rainfall rates should include the following (Chappell 1993):

1. A deep moist layer with atypically high precipitable water (PW);
2. Moderate CAPE (1500–2000 J kg^{-1});
3. An elongated distribution of CAPE, indicating a low, warm cloud base and a high equilibrium level; and
4. Relatively weak vertical wind shear.

It is important to realize that soundings with tropical characteristics are not limited to tropical regions. For example, a flash flood struck Rapid City, South Dakota, on 9 June 1972, killing more than 230 people (Schwarz et al. 1975). Nearly 380 mm of rain fell across the steep slopes of the Black Hills within about 4 h. Figure 12.21 illustrates the 0000 UTC 10 June 1972 sounding for Rapid City, South Dakota. Notable tropical sounding features favoring heavy convective rainfall include (Maddox et al. 1977) 1) a high freezing level (5–6 km) with low cloud bases, 2) weak vertical wind shear and light midlevel storm-relative winds, and 3) high moisture content through a deep layer. The thick warm layer allows substantial time for warm rain processes to generate intense precipitation. A combination of weak vertical wind shear and substantial deep moisture contributes to higher precipitation efficiency (less evaporation) and weaker outflow, resulting in effective cell generation and slow system movement. A flash flood event with similar sounding characteristics caused more than 140 fatalities in the Big Thompson Canyon of Colorado four years earlier (Maddox et al. 1978). These cases demonstrate the importance of recognizing situations where warm rain convective processes are likely. Should this occur, the *Z–R* relationship used by the WSR-88D may need to be changed to a "tropical" *Z–R* in order to account for the higher rainfall rates.

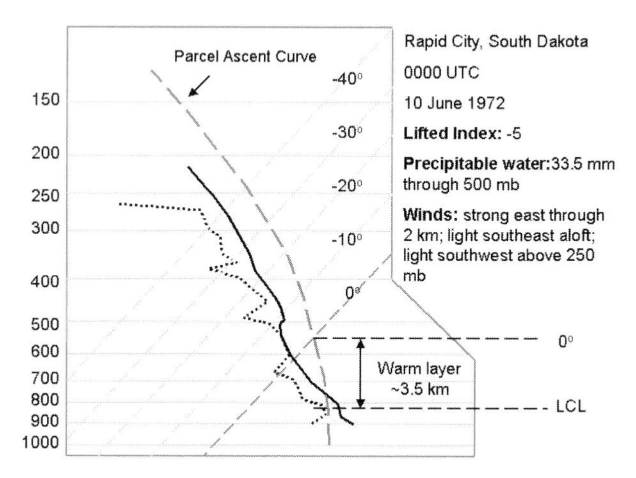

FIG. 12.21. The skew *T*–log*p* plot of 0000 UTC 10 Jun 1972, Rapid City, SD, upper-air sounding. From Maddox et al. (1978).

12.8. Detecting flash floods

The next section concentrates on using radar, satellite, and lightning observing systems to detect the heavy rainfall signatures associated with flash floods. The quantitative detection of flash floods will be detailed in the following section, which focuses on measuring rainfall to support ABR computations. The characteristics of rainfall estimates based on satellite, radar, and rain gauges will be examined in the context of computing ABR. The hydrologic response of watersheds based on observed ABR and ABR rate will then be examined for a number of flash flood cases.

Several investigators (Merritt and Fritsch 1984; Glass et al. 1995; Corfidi et al. 1996) have demonstrated that the heavy rainfall cores (MBEs) of larger MCSs are responsible for most of the flash flood producing rainfall. The heavy rainfall results when a number of MBEs traverse common ground (Merritt and Fritsch 1984). Since the propagation and speed of MBEs have such a strong influence on determining the production of flash flood producing rainfall, the detection of heavy rainfall requires remote sensing systems capable of detecting the small temporal and spatial scales of MBEs.

a. Detecting heavy rainfall signatures

Satellite and radar are the primary observing tools for detecting characteristics of the convective systems responsible for most flash floods. Satellite imagery can indicate the presence of synoptic and mesoscale systems associated with heavy rainfall. Although lightning detection systems can indicate the presence of DMC based on the occurrence of cloud-to-ground lightning strikes, heavy rainfall can also occur without cloud-to-ground lightning. The WSR-88D can show the location of the intense rainfall cores associated with DMC and specify with great accuracy the duration of rainfall. Radar does a good job of imaging the storm-scale detail of the heavy rainfall-producing cells, but satellite imagery more clearly indicates the organization of larger mesoscale features associated with MCSs. Integrating lightning data with satellite and radar imagery allows greater understanding of the development and duration of convective systems.

Much of our current knowledge of the structure and rainfall characteristics (McAnelly and Cotton 1985; Kane et al. 1987; McAnelly and Cotton 1989) of flash flood producing MCSs and MCCs (Maddox 1980) comes from studies of Geostationary Orbiting Earth Satellite (GOES) IR and visible satellite pictures. Many smaller-scale MCSs responsible for the production of heavy rainfall have been catalogued (Scofield 1985), and more than five types of convective subtle heavy rainfall signatures (SHARS) have been described by Spayd and Scofield (1983). Clark et al. (1980) showed that heavy rainfall is directly related to cold IR cloud tops. Operationally, satellite imagery often reveals areas of convective initiation before they are visible on radar. Satellite imagery also shows outflow boundaries, which tend to be favored areas for both convective initiation and intensification of existing storms systems. The satellite animation capability of AWIPS facilitates the detection of important convective systems and allows forecasters to monitor the movement and evolution of these systems.

Radar provides a more detailed view of the storm-scale structure of convective storms and systems. The vertical and horizontal extent of the heavy rainfall cores responsible for flash flood producing rainfall can be seen directly. The evolution of convective systems can be tracked in time, including the intensification of existing cells and the development of new cells aloft. Animation of radar provides specific information on the movement and propagation characteristics of convective systems. Although a graphical radar reflectivity display provides a storm-scale view of convective systems and helps in the assessment of the flash flood threat, it does not provide a quantitative measure of rainfall.

Researchers have attempted to link observed cloud-to-ground lightning strike characteristics with observations of heavy rainfall. The frequency of lightning strikes can be a good indicator of heavy rainfall when cold rain processes dominate the production of rainfall (Goodman and Buechler 1990), but provides little helpful information when rainfall is produced by warm rainfall processes. Kane (1993) studied the relationship between the areal coverage of lightning strikes and the areal distribution of rainfall, and found that 60% of all cloud-to-ground strikes were concentrated in 16% of the area that received rainfall. This small 16% area of rainfall received over half of the total volumetric water produced by the convective system. There was also an excellent temporal correlation, with the heaviest rain starting within 5 min of the peak concentration of lightning strikes. Most of the lightning strikes tended to be collocated with the high reflectivity cores of the thunderstorms. The source of all real-time lightning information is the National Lightning Detection Network (Orville 1991).

b. Measuring rainfall to support ABR computations

A quantitative approach to flash flood detection provides several advantages over the qualitative methods described in the last section. Linking defined stream watersheds to observed rainfall allows the computation of ABR and ABR rate. If the ABR in a defined watershed can be quantified with sufficient accuracy, the flash flood potential in the basin can be determined with greater accuracy and confidence. Moreover, the severity of the flash flooding can be estimated. Names of the streams, cities, towns, and highways impacted by the flooding can then be included in warnings and statements. Thus, determining accurate rainfall estimates on the scale of flash flood watersheds will enhance the flash flood warning process.

The key to detecting flash flood producing rainfall is determining the location, areal extent, and time duration of the MBEs associated with MCSs. Before ABR can be calculated, rainfall must be measured or estimated, and the rainfall estimate must be mapped into defined watersheds. Rainfall estimates are provided in some form by rain gauges, radar, and satellite. To support ABR calculations, the rainfall estimates must be available in near real time, in a digital format (as opposed to a graphic display), with a time frequency sufficient to see short-term trends of rainfall intensity, and in a grid size smaller than the MBA. The grid size of the rainfall data is a limiting factor in the computation of ABR. The smallest basin area for which ABR can be calculated is equal to the grid size of the rainfall data. Grid size is not a factor for rain gauge measurements, but it is an important consideration when dealing with satellite and radar rainfall estimates. The time frequency of the rainfall data is critically important to flash flood applications. Significant flash flooding can result from a heavy rainfall burst that lasts only 15–30 min.

TABLE 12.3. Satellite precipitation estimates prepared 30 May 1986 by SAB. Quantitative values reflect maximum or significant estimates.

County in PA	Total rainfall (mm)	Time of rainfall (UTC)	Time sent (UTC)	Time of IR data (UTC)
Washington	64	1700–2000	2030	2000
E Washington	76	1800–2100	2130	2100
W Washington	61	1800–2100	2130	2100
N Allegheny	36	1900–2100	2130	2100
E Washington	84	1800–2200	2230	2200
W Fayette	74	1900–2200	2230	2200
E Allegheny	46	1900–2200	2230	2200

The spatial and temporal characteristics of the heavy rainfall cores that produce flash flooding determine the utility of the rainfall detection capability of each of the observing tools. Flash floods frequently occur on watersheds with areas ranging from 5 to 20 km^2 over time intervals ranging from 30 to 90 min. The determination of ABR in these small watersheds requires that the spatial scale of the rainfall estimation (grid size of the rainfall estimate) must be smaller than the MBA of 5 km^2. Furthermore, the time samples must be in small enough time steps to see a trend in the rainfall over a 30–60-min time period. Time steps on the order of 5 min would be necessary to see trends in heavy rainfall in a 30-min time period.

1) SATELLITE RAINFALL ESTIMATES

For many years, manual estimation of rainfall using IR-based cloud-top temperatures has been produced by the Satellite Analysis Branch (SAB) forecasters at the National Environmental Satellite Data and Information Service (NESDIS) in support of WFO flash flood operations. The interactive flash flood analyzer (IFFA) used by SAB is a manually intensive procedure (Scofield 1987) that can be used to produce rainfall estimates for small portions of the United States. The SAB forecaster must choose the area of interest and then produce rainfall estimates for transmission to a WFO. The output to the WFO is a text product giving the name of the county, time duration of the rainfall, and an estimate of the maximum rainfall.

Table 12.3 shows the IFFA estimates that were transmitted to the Pittsburgh WFO during the three hours of the Etna, Pennsylvania, flash flood of 30 May 1986. The disaster survey report indicated that up to 200 mm of rainfall fell between 1930 and 2130 UTC (Fig. 12.4 shows the maximum rainfall). The flood crest struck Little Pine Creek at about 2130 UTC. Heavy rainfall was indicated in the correct hour, and in the area of flash flooding, but the quantitative estimate of the rainfall was about 25% of the observed maximum. The heavy rainfall associated with the flood covered a very small area (Fig. 12.4). The large grid size of the satellite rainfall estimates (12 km) and the large time steps between observations (30 min) make the resolution of these small-scale events difficult.

Satellite rainfall estimates are more accurate for widespread rainfall produced from larger MCSs, including MCCs. Figure 12.22 shows the SAB estimate for rainfall associated with intense flash flooding that occurred in the southern suburbs of Chicago, Illinois, on 17–18 July 1996. The 24-hour graphic summary of the hourly SAB estimates shows that a very large area received 200 mm rainfall, with a maximum estimate of 310 mm in Will County, Illinois. The actual maximum rainfall reported in *Storm Data* (National Climatic Data Center, 1996) was 430 mm at Aurora, Illinois. Eleven counties in northeastern Illinois were declared disaster areas.

This same synoptic system moved into western Pennsylvania on 19 July 1996 and caused record flash flooding across several counties. Brookville, in Jefferson County, received 289 mm of rainfall in 6 h. The SAB satellite estimate for this system (Fig. 12.23) was considerably lower than the estimates from the previous day near Chicago, Illinois. The WSR-88D at Pittsburgh, Pennsylvania, considerably underestimated the rainfall, indicating that the warm rain processes may have been more dominant than the previous day in Chicago. When warm rain processes dominate in DMC, cloud tops are warmer than when ice processes dominate.

The automated version of the IFFA, known as the *auto-estimator,* runs in near real time for the entire United States (Vincente et al. 1998). One drawback of the IFFA is the time necessary to create and transmit the product (about 30 min), which limits its utility for detecting rapidly developing flash floods. The IFFA is also limited to a few geographic areas, while the

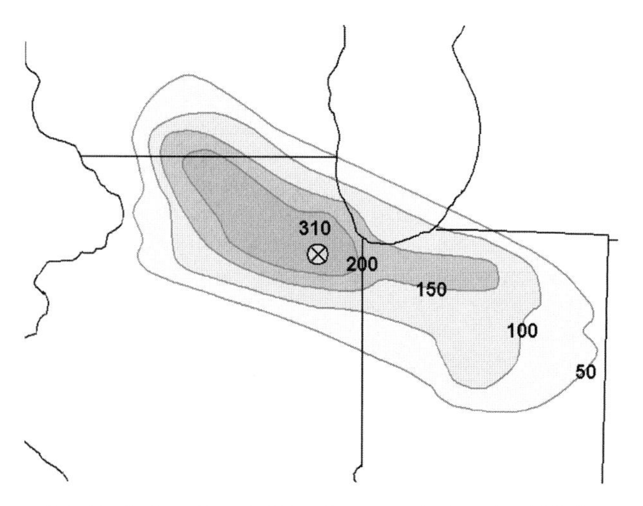

FIG. 12.22. Satellite rainfall estimate (mm) from the SAB from 1815 UTC 17 Jul 1996 to 1215 UTC 18 Jul 1996. The maximum estimated rainfall is 310 mm. The heavy rainfall resulted in record flash flooding south of Chicago, IL.

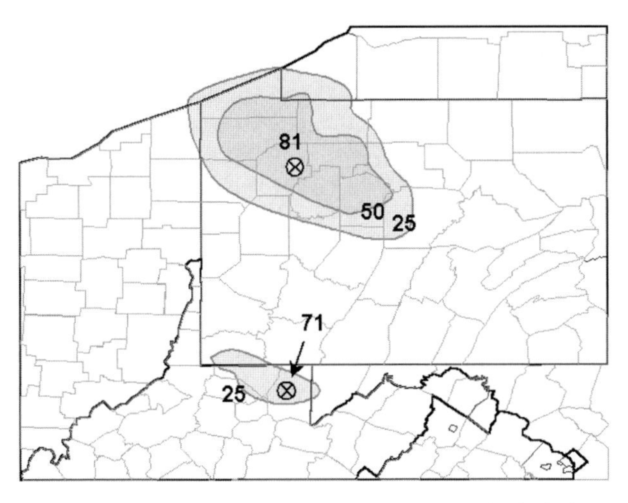

FIG. 12.23. Satellite rainfall estimate (mm) from the SAB from 1215 UTC 18 Jul to 1215 UTC 19 Jul 1996. The maximum observed rainfall is 81 mm in Venango County, PA, and 71 mm in Preston County, WV. Widespread flash flooding occurred in all the counties contained within the 50-mm isohyet. Severe flash flooding and urban flooding were reported in Monongalia County and Preston County, WV, in the area enclosed by the 25-mm isohyet to the south.

auto-estimator provides estimates for the entire United States. Rainfall estimates produced by the auto-estimator could be used to compute ABR if the output is made available in a digital format, but the large grid size (144 km^2) of the rainfall estimates will limit their flash flood utility.

2) RAIN GAUGE NETWORKS

Rain gauges provide the most accurate method of measuring rainfall at a single geographic point. The rain gauge report must be available in real time if it is to have any operational value for flash flood applications. Automatic reporting rain gauge networks such as the Integrated Flood Observing and Warning (IFLOWS) system (National Weather Service 1993) and the Automated Local Evaluation in Real Time (ALERT) are of great value to the flash flood program (Larson et al. 1995). The IFLOWS and ALERT gauges are radio reporting gauges, available in real time and programmed to report rainfall in 15-min time increments. These automated tipping bucket gauges are capable of measuring rainfall down to 1-min time intervals, and are available in real time, 24 h a day.

The U.S. Army Corps of Engineers installs and maintains a large network of automated rain gauges that report data automatically through the GOES satellite network. Some of these gauges report hourly, while others report once every 4 h in hourly time increments. This dataset would be of more utility to flash flood applications if all of these gauges reported each hour.

Each WFO maintains a volunteer network of observers called SKYWARN spotters (Hitchens and Belville 1988). Many of these observers are supplied with rain gauges and are asked to report any rainfall amounts of 25 mm or more, or any observed flooding. These observers can provide critical rainfall measurements when the information is received in a timely manner. The SKYWARN program is the most cost-effective rain gauge network available to WFOs.

After a major flash flood event, the NWS often sends a disaster survey team to conduct a "bucket survey" of the area struck by the heavy rainfall (Brooks and Stensrud 2000). The survey team searches for rain gauges, or any container that may have caught the rain, such as an unfilled child's swimming pool, garbage cans, etc. The rainfall pattern of the Little Pine Creek (Fig. 12.4) was constructed from such a bucket survey. The bucket survey provides valuable historic information about the intensity and amount of rainfall needed to produce flash floods, but this detail of rainfall information would have to be available in real time to provide help in the warning process.

Real-time rain gauge networks are most useful for flash flood detection when WSR-88D rainfall estimates at the location of each rain gauge within the network are compared to the rain gauge value to determine the relative accuracy of the WSR-88D estimate. Rain gauges can also be used to determine when to use the WSR-88D tropical *Z–R* relationship. For example, a rain-gauge reading of 50 mm in 15 min (as occurred at several rain gauges in the Dallas, Texas, mesonet on 5 May 1995) indicates an hourly rainfall rate of 200 mm h^{-1}, indicating that a warm rain process is producing high rainfall rates very efficiently. Conversely, typical maximum rainfall rates observed with a cold rain process are generally less than 125 mm h^{-1}. A tropical *Z–R* rate would be more appropriate with 200 mm h^{-1}; the standard *Z–R* relationship would be sufficient for situations dominated by a cold rain process.

Some local flood warning systems have been installed using ALERT, including the Harris County Brays Bayou watershed (Schwertz 1995) in Houston, Texas. ALERT is widely used in the western United States. An ALERT system concentrated in a specific watershed can produce excellent results (Harned 1987), but economic considerations prevent a national deployment of this type of warning system.

Detection of heavy flash flood producing rainfall with rain gauges is difficult due to the small spatial extent of MBEs. Brooks and Stensrud (2000) show that the hourly precipitation dataset of the National Climatic Data Center (NCDC), with an average gauge spacing of 50 km, fails to detect most of the heavy flash flood producing rainfall events. Examination of the 120 years of rainfall records at the Pittsburgh, Pennsylvania, observation site shows the rarity of 100 mm of rainfall in a 24-h period. In the period of record

1880–2000, 100 mm of rainfall has never been observed in a 24-h calendar day. In fact, more than 75 mm of rainfall in 24 h has been reported in only nine days out of the 43 830 days of record. The two highest 24-h rainfall amounts occurred with dying tropical storms. Many flash floods have been observed within a 100-mile radius of Pittsburgh, but the odds of catching the small intense rain cores of an MBE with a rain gauge network are very slight.

3) Radar rainfall estimates

The WSR-88D precipitation algorithm (Fulton et al. 1998) uses the lowest four reflectivity scans (hybrid scan reflectivity) to create digital rainfall estimates in real time. The reflectivity is converted to a rainfall rate using a user-selectable *Z–R* relationship. The rainfall rates from two consecutive volume scans (5–6 min for each scan) are used to produce an average rainfall rate. The rate is then multiplied by the time interval between volume scans to produce a 5–6-min rainfall estimate. These estimates are then used to generate WSR-88D rainfall products. Users of graphic WSR-88D rainfall products need to be aware of the reflectivity sampling process. Radar rainfall estimates are most accurate when the reflectivity is sampled low in the precipitation cascade (close to the ground). The terrain-based hybrid scan (Fulton et al. 1998) provides the lowest possible sampling in the precipitation cascade without ground clutter contamination of the data.

The WSR-88D precipitation algorithm provides WFOs with a graphic display of rainfall on political county map backgrounds, and two digital rainfall products that may be processed on a computer external to the WSR-88D (Fulton et al. 1998). The graphic display products provide the WFO with a visual display of the spatial distribution of rainfall within a county. The display products are available for three different time durations of rainfall, including the 1-h rainfall product (OHP), the 3-h precipitation product (THP), and the storm total precipitation product (STP), with a variable time duration based on the actual storm duration. Although the graphic display of WSR-88D rainfall provides a quantitative estimate of rainfall, ABR in defined watersheds cannot be determined with confidence on the graphic display.

The Digital Precipitation Array (DPA) and the Digital Hybrid Scan Reflectivity (DHR) are digital rainfall products generated by the WSR-88D. The Hydrologic Rainfall Analysis Project (HRAP) grid (Schaake 1989) of the DPA product has a grid size that varies as a function of latitude, 3.8 km at 30° to 4.5 km at 50°. Each grid box has an area of approximately 16 km^2. The rainfall estimates for each grid box are created by averaging the rainfall data from the 1° by 1 km polar grid of the hybrid scan reflectivity. The rainfall data are created in 1-h increments. The large grid size and large time increment of the DPA data greatly decrease its utility for ABR calculations.

Davis and Drzal (1991) showed that averaging the polar radar grid into the HRAP grid for the Etna, Pennsylvania, flash flood of 1986 reduced the maximum observed rainfall in the Little Pine Creek watershed from 142 mm to about 91 mm. The HRAP grid is a subset of the Manually Digitized Radar (MDR) grid that was used from the late 1970s until the introduction of the WSR-88D in the early 1990s to produce manually coded radar reports. To show the scales involved, the Pine Creek watershed is shown on the HRAP grid in Fig. 12.24. The MDR grid is the large grid, with a single MDR grid covering most of Allegheny County. The HRAP grid was created by dividing each MDR grid into 100 parts. This figure indicates that the computation of ABR in small watersheds such as Pine Creek is not possible using the MDR grid. Even the relatively large grid size of the HRAP grid would make ABR computations in a watershed the size of Little Pine Creek (shaded watershed in Fig. 12.24) difficult.

The DHR product of the WSR-88D provides radar reflectivity in the WSR-88D polar grid of 1° by 1 km. The DHR grid size is a function of radar range and varies from 0.43 km^2 at 25 km to 2.61 km^2 at 150 km. The bin area is an order of magnitude below the grid size of DPA and allows computation of rainfall in small flash flood watersheds down to 3 km^2. The DHR product is available at each volume scan of the radar, allowing the production of rainfall estimates in 5-min time durations.

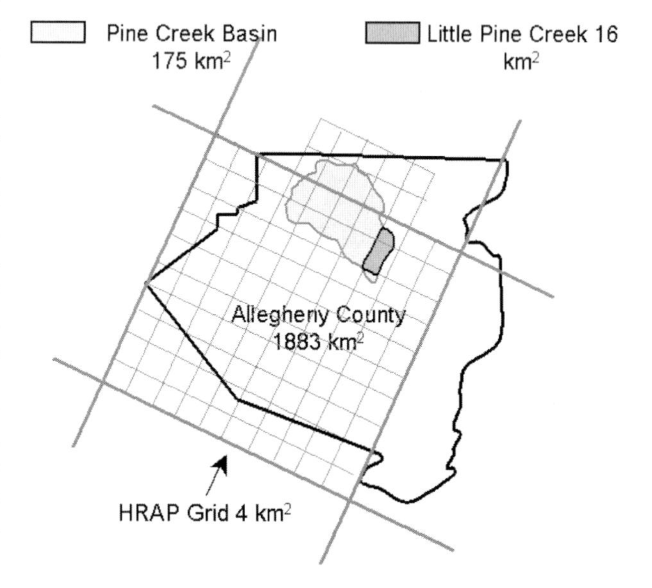

FIG. 12.24. A single MDR grid box over Allegheny County, PA. The Pine Creek watershed (light gray shading) is shown in northern part of the county. The MDR grid is divided into 10 × 10 rows to produce the HRAP grid. Little Pine Creek watershed is shown as the dark area within the Pine Creek basin.

The WSR-88D precipitation algorithm must be run on a computer external to the radar to convert the DHR reflectivity into 5-min rainfall estimates. The AMBER program (Davis and Jendrowski 1996) performs this function, producing 5-min rainfall estimates for each 1-km radar range bin of the WSR-88D polar grid. The 5-min rainfall estimates can then be mapped directly into defined stream watersheds, allowing the computation of ABR for each watershed. The NBD project will provide this mapping of 1-km range bins into the defined watersheds for each WFO.

The only data input required by AMBER is the DHR product, which can be produced from Archive II tapes of the WSR-88D that store the base data reflectivity for each radar. Consequently, AMBER can be run in a playback mode for any flash flood case for which Archive II data are available. All of the examples showing ABR and ABR rate computations in the following sections are from the Pittsburgh WFO real-time AMBER database, or playback cases created by the author using Archive II data from other WSR-88Ds.

Radar rainfall estimates are subject to both the detection limitations of the radar and several forms of contamination, including ground clutter, anomalous propagation, melting level contamination, and hail contamination (Hunter 1996). The three major sources of contamination of the rainfall estimates (ground clutter, hail contamination, and melting level contamination) result in overestimates of the ABR. A more serious limitation occurs when the radar fails to detect heavy rainfall. This can occur due to 1) range limitations of the radar, 2) using a standard *Z–R* relationship when a tropical *Z–R* relationship is more appropriate, 3) poor radar calibration, 4) ground clutter suppression close to the radar, and 5) beam blockage due to high terrain. In general, reliable rainfall estimates can be expected within 150 km of the radar (Serafin and Wilson 2000). Continued refinements, especially dual polarization (Zrnić and Ryzhkov 1999), will improve radar rainfall estimates and *Z–R* selection techniques, and reduce the impact of various types of radar contamination, especially hail and melting level contamination.

c. The watershed's response to rainfall

1) AVERAGE BASIN RAINFALL

Recall that the hydrological response of a watershed to rainfall is determined by the observed ABR in the watershed. The ABR multiplied by the watershed area defines the volume of rainfall that has fallen on the watershed in a specific period of time. It is important to note, however, that the rate of ABR is as important as the amount of the ABR. A stream's response to 100 mm of ABR in 4 h will be much slower and less extreme than a stream's response to 100 mm of ABR in 1 h. Rainfall intensity, as measured by the ABR rate, actually determines the stream's hydrological response, rather than the amount of ABR.

Intense rainfall rates associated with MBEs produce the great majority of all flash floods. These flash floods typically follow from the cumulative rainfall at the ground produced by a series of MBEs moving over the same area. The spatial and temporal scales of flash floods are determined by the size of the MBEs and the length of time they dump rain over a particular area. The meso-β scale defined by Orlanski in 1975 (Fujita 1986) is for a length of 25–250 km, while the meso-γ scale is for lengths of 2–25 km. Figure 12.25 shows the size of a typical reflectivity core of a supercell with the reflectivity converted to rainfall rate in millimeters per hour using the standard WSR-88D *Z–R* relationship for convective storms. The core of 40 dB*Z* returns is clearly a meso-γ scale element (MGE) rather than a meso-β scale phenomenon. The area enclosed by the 75 mm h^{-1} rates is very close to the low end of the meso-γ scale. These small MGE rainfall cores produce the heavy rainfall that causes flash flooding. The small area enclosing the 75 mm h^{-1} rates explains why so many flash floods occur on watersheds less than 20 km^2. A single convective cell can easily inundate a small watershed, while training of many MGEs is necessary to flood larger watersheds.

If the WSR-88D tropical *Z–R* rate is used, a much larger area of heavy rainfall results (Fig. 12.26). The area enclosed by the 75 mm h^{-1} rate doubles from 20 to 44 km^2, and the area of 125 mm h^{-1} rate increases from 5 to 32 km^2, a sixfold increase. Thus, flash flooding is more easily produced on larger watersheds when the warm rain coalescence processes dominate (tropical *Z–R* used).

To determine ABR, the observed rainfall from these heavy rainfall cores must be directly mapped into the

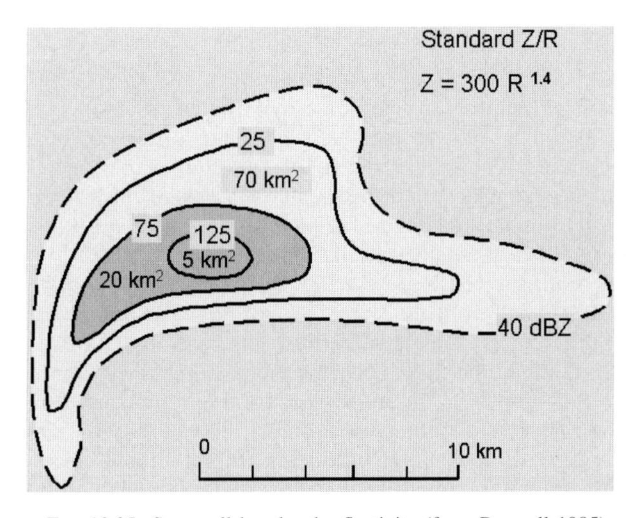

FIG. 12.25. Supercell low-level reflectivity (from Doswell 1985), converted to hourly rainfall rate (mm h^{-1}) using the WSR-88D standard convective *Z–R* relationship.

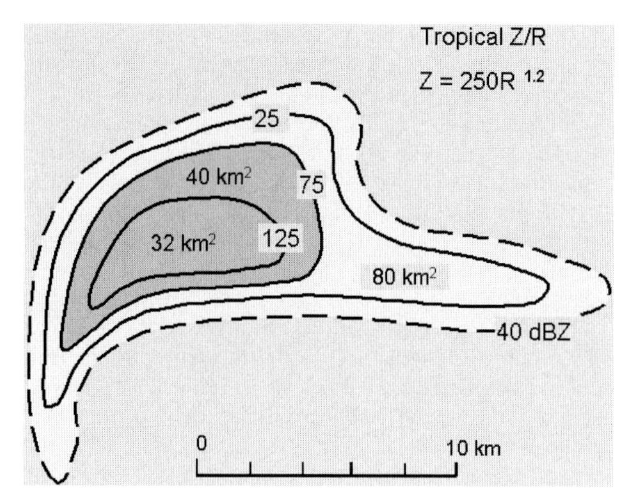

FIG. 12.26. Supercell low-level reflectivity (from Doswell 1985), converted to hourly rainfall rate (mm h^{-1}) using the WSR-88D tropical Z–R relationship.

watershed and then summed over short time steps (5 min) to show the rate of ABR accumulation (ABR rate). Figure 12.27 shows how ABR is computed from the 1-km reflectivity of the WSR-88D polar grid. Three radar bins are shown within a watershed. The reflectivity of each radar bin is converted to a 5-min rainfall amount and then multiplied by the radar bin area to create a rainfall volume. The rainfall volumes for all radar bins with a center point falling within the watershed are summed and then divided by the total area of all radar bins in the watershed to determine the ABR. The ABR rate is created by multiplying the calculated 5 min ABR by 12. It should be emphasized that the ABR estimate will only be as good as the WSR-88D rainfall estimates, and is subject to all the limitations and weaknesses of the WSR-88D detection capability.

2) ESTIMATING RUNOFF

The amount of surface runoff in a watershed is directly related to the severity of the observed flash flooding. The greater the runoff, the higher the associated stream rise, and the greater the risk of flash flood fatalities and damage. The determination of runoff is critical to the detection of flash flooding.

With the ABR provided as the primary input to the determination of the stream's hydrologic response, the next step is to determine how much of the ABR is converted directly to surface runoff. A small portion of ABR may be lost due to evaporation, transpiration, and depression storage, but compared to the ABR rate related to flash flood producing rainfall, these losses are minimal. Initially a large amount of ABR can be lost due to infiltration of rain into the soil, depending on the type and depth of soil. However, as the soil becomes saturated with water due to continued rainfall, the loss due to infiltration can decrease rapidly with time. The maximum runoff occurs when the infiltration capacity is used up.

In areas of the United States with very rocky soil or very little soil on bedrock, such as much of the desert and mountainous areas of the west or the volcanic soil of Hawaii, almost no infiltration of rainfall occurs, and almost all of the observed ABR is converted directly to runoff (Runk and Kosier 1998). In the central and eastern portions of the United States, where soil is deeper and more porous, a significant amount of ABR may initially be infiltrated into the ground. However, even in these areas, once heavy rainfall saturates the soil, nearly all additional ABR is converted directly into runoff.

Since the infiltration capacity in many areas is modulated by the degree of soil saturation, it is important to account for the effects of soil moisture before determining how much of the forecast or observed ABR will be converted to flood runoff. River forecast centers (RFCs) produce flash flood guidance (FFG) to account for the amount of ABR that will initially be absorbed into the ground during low flow stream conditions (Stallings and Wenzel 1995). FFG is defined for ABR durations of 1- and 3-h time periods. Guidance for slow-rise stream floods is also provided in some areas of the country for 6-, 12-, and 24-h time

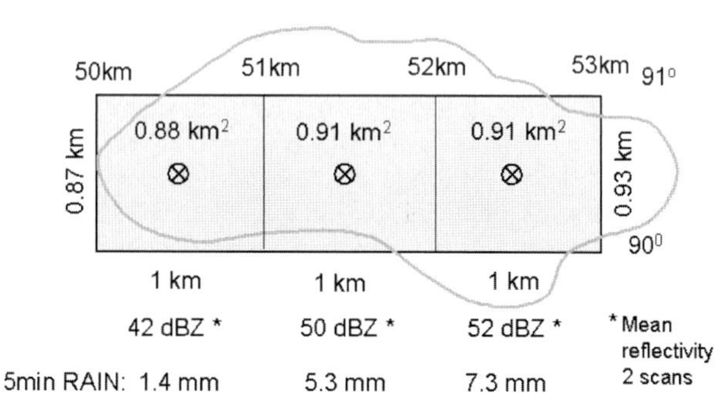

FIG. 12.27. The AMBER ABR and ABR rate computation. Three of the WSR-88D polar radar grids are shown along the 90° radar radial (beam center at 90.5°), from 50 to 53 km in range. The bin centered at 50.5 km is 1 km in length, 0.87 km in width at 50 km, and 0.89 km in width at 51 km, with a bin area of 0.88 km^2. Each bin whose centerpoint falls within a defined watershed (solid black line) is assigned to that basin for the ABR computation. The 5-min rainfall in each bin is multiplied by the bin area to produce a volume of rainfall. The ABR is produced by summing the bin volumes for each watershed, and then dividing by the total bin area for all bins in the watershed. ABR rate is computed by multiplying the ABR by 3600 and then dividing by the number of seconds between the current and previous WSR-88D volume scans. For exactly 5 min between volume scans ABR rate = ABR × 12.

periods. A single value of FFG is issued for each county. FFG is currently updated in most parts of the country once a day, but the RFCs do have plans to eventually update FFG four times a day (Fread et al. 1995).

Additional limitations to FFG are related to scale issues. For example, mean areal precipitation (MAP) areas, used by the RFC to update soil moisture content, are typically large (250–2500 km^2). Convective storms, however, produce intense rainfall over much smaller areas, resulting in a wide variation of rainfall accumulation across the larger MAP areas. Because soil moisture accounting is determined on the scale of MAP areas, FFG may not accurately represent the convective-scale rainfall distributions on small watersheds contained within these MAP areas. Consequently, FFG may be unrepresentative when convective storms deposit significant rainfall in small watersheds, or when multiple high-intensity rainfall events traverse small basins. For these reasons, FFG is most reliable during an extended period of dry weather and least effective when multiple rainfall events have occurred during the previous 24–48 h. The observation of ABR is most important to the flash flood forecaster. If the soil is close to saturation due to previous rainfall, most ABR will be converted directly to runoff.

The saturation of relatively dry soil can occur in a very short time. On 30 May 1986, a devastating flash flood struck the northern suburbs of Pittsburgh on the Little Pine Creek watershed near Etna, Pennsylvania (Fig. 12.4). Almost no rainfall had fallen in the watershed over the previous two months and FFG was quite high (81 mm for 3 h). From 1930 to 2000 UTC, 53 mm of rain fell in Little Pine Creek, completely saturating the soil. From 2000 to 2100 UTC, an additional 127 mm of rain fell on the same watershed. Almost all of this rainfall became surface runoff, producing a raging flood that caused nine people to drown in their automobiles. This flash flood is of some historical importance, as the disaster survey team recommended the creation of 1 h FFG due to the short time duration (90 min) and high-intensity rainfall rate (100 mm h^{-1}) of the observed rainfall. Prior to this flood, the 3-h FFG was the shortest time duration FFG issued by the RFC. A computation of 1-h FFG after the event showed that 49 mm was needed in 1 h to initiate flood and saturate the soil; 53 mm fell in the first 30 min of the storm.

Other factors can influence the infiltration capacity of the soil. For example, the severe wildfire that occurred on the Buffalo Creek, Colorado, watershed greatly lowered the infiltration capacity in the burn area. Intense convective rainfall converted quickly to large surface runoff in the burn area, resulting in a deadly flash flood in the city of Buffalo Creek. Figure 12.5 illustrates that the highest ABR accumulated within the burn area. Figure 12.28 shows the ABR and

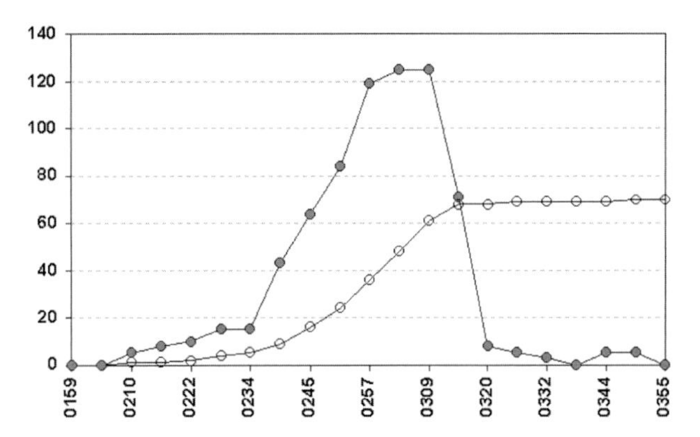

FIG. 12.28. Accumulated ABR (mm) and ABR rate (mm h^{-1}) for 13 Jul 1996 plotted for the rain gauge site at Buffalo Creek, CO. Time in UTC. For all rain gauge measurements AMBER uses the single 1-km bin that contains the rain gauge location. The Buffalo Creek rain gauge is located at 234° at 76 km from the Denver, CO, WSR-88D.

ABR rate plot for the rain gauge located in the city of Buffalo Creek. The rain gauge measured 68 mm and the WSR-88D estimated 70 mm, indicating that the radar estimates were quite good. Notice that the ABR rate reached 120 mm h^{-1}, and most of the rainfall fell in 45 min.

As previously noted, soil moisture exerts a strong influence on surface runoff. If the soil is already holding water from previous rainfall, the infiltration capacity may be close to zero, resulting in runoff nearly equal to the observed ABR. Accounting for previous rainfall is an important part of estimating the current flash flood potential of streams. During the Kansas City, Missouri, flash flood of 1977 (Larson and Vochatzer 1978), two distinct periods of intense rainfall were separated by about 12 h, resulting in a flash flood that claimed 25 lives. The first period of rainfall produced a crest around 2 m for the stream gauge located near the Plaza shopping area, but no flooding. The second period of rain fell on ground already saturated by the previous rainfall. This produced a much greater stream response (Fig. 12.29), resulting in a crest around 5 m and disastrous flash flooding.

The saturation of the soil may result from several consecutive days of heavy rainfall, as was the case during the eastern Ohio flash floods of 26–29 June 1998. The most deadly flash flood of this event occurred on the Salt Run watershed (Fig. 12.30) near Caldwell, causing four fatalities. From 0100 UTC 27 June to 0300 UTC 28 June 1998, Salt Run received 81 mm of rainfall. With the creek already near bankfull, an additional 28 mm of ABR fell from 0300 to 0400 UTC for a total of 109 mm (Fig. 12.31). Significant flash flooding occurred from this rainfall, but an additional 58 mm fell from 0515 to 0645 UTC on 28 June 1998 (Fig. 12.31). The resulting flood wave swept

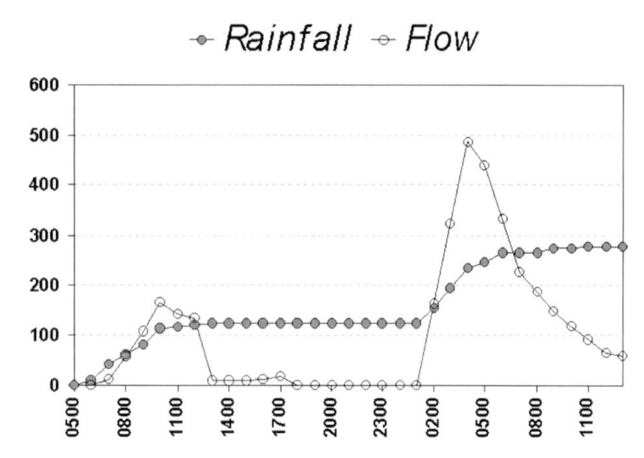

FIG. 12.29. Accumulated rainfall (mm) from a rain gauge located 4 km south of Shawnee, KS, just west of the Brush Creek watershed. The stream flow is from the stream gauge on Brush Creek at the Plaza in Kansas City, MO. Data are for the time period from 0500 UTC 12 Sep to 1200 UTC 13 Sep 1977. (Adapted from Larson and Vochatzer 1978.)

away an elderly couple and their house, and also floated two cars into the creek off Ohio State Route 47, drowning both drivers, during the night of 28 June 1998.

3) TOPOGRAPHY AND RAINFALL INTENSITY

ABR intensity strongly influences surface runoff and the watershed's response to rainfall. A short duration but intense rain event generally produces greater runoff than a rain event of equal magnitude occurring over a longer period (Sweeney 1992). The rainfall intensity combined with the topography of the watershed determines the speed of the hydrological response of the stream. If the topography is steeply sloped, surface runoff reaches the stream more quickly than if the valley walls have a shallow slope. Flash floods in steeply sloped watersheds produce not only a more rapid response, but generally a higher flood crest and more rapidly flowing water (greater destruction).

The variation of rainfall intensity or ABR rate produces a spectrum of flash flooding possibilities. Intense ABR rates of 100 mm h^{-1} or higher will produce the most extreme hydrological response, as the runoff tends to be concentrated in short time durations (15–60 min) with these extreme rainfall rates. Less intense ABR rates of 25–50 mm h^{-1} will produce a less extreme hydrological response, but may persist over longer time durations (60–180 min) and cause more widespread flooding than the other end of the intensity spectrum. The hydrological response is also controlled by the basin topography and the hydrological condition of the basin, but rainfall intensity is perhaps the most important determination of potential flash flood severity.

The Brush Creek watershed in Kansas City, Missouri/Kansas, is an example of basin topography with shallow slope compared to watersheds in the Rocky Mountains or even the Appalachians (Table 12.2). Despite the lack of steep topography, this watershed has produced two devastating flash floods, in 1977 and 1998, due to extremely intense rainfall. An AMBER playback of the 5 October 1998 flash flood produced

FIG. 12.30. Tributaries of the West Fork of Duck Creek near the city of Caldwell, OH (hatched area). The area of each watershed is given in km^2. Radar range (km) from the Pittsburgh, PA, WSR-88D to each watershed shown, 136–140 km for Salt Run. Shaded areas in Wolf Run and Dog Run are reservoirs. Dashed lines are highways.

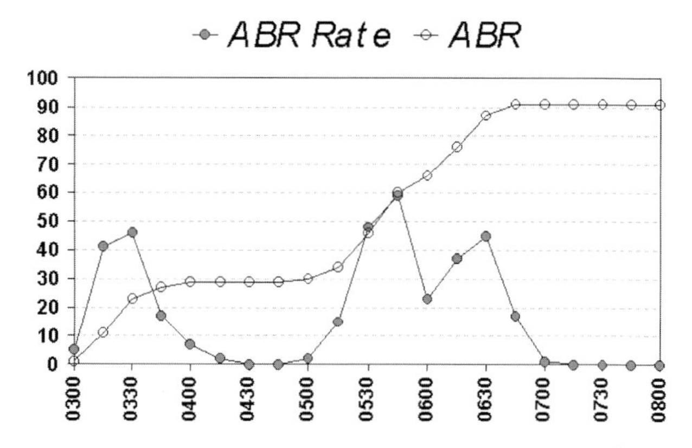

FIG. 12.31. Accumulated ABR (mm) and ABR rate (mm hr^{-1}) for the Salt Run watershed, near Caldwell, OH, from 0500 to 1100 UTC on 28 Jun 1998. Rainfall data interval 15 min.

the following distribution of ABR in the Brush Creek and Turkey Creek watersheds (Fig. 12.32). The heavy rain was preceded by several days of soaking rains. The rainfall that produced the 1977 flash flood fell over a period of about 4 h (Fig. 12.29). The ABR that caused the 1998 flood fell in about 90 min. The ABR and ABR rate plot for a portion of Brush Creek (Fig. 12.33) were typical of the plots for the other stream segments. ABR rates varied from 50 to 120 mm h^{-1} for about 80 min. The news media reported that up to 178 mm fell. In a watershed with gently sloping topography, intense rainfall rates sustained over 60–90 min can produce flash flooding.

Notice that the size of the watersheds (Fig. 12.32) is very close to the area enclosed by a single supercell (Fig. 12.26). The tropical *Z–R* of the WSR-88D was used for the playback of this flash flood. An NWS employee measured 127 mm of rainfall in 75 min just west of the Turkey Creek watershed. The AMBER radar rainfall estimate for the site of the employee's rain gauge indicated 125 mm using the tropical *Z–R* playback (Fig. 12.34) in the exact time period reported. Use of the standard convective *Z–R* would have shown only about 50% of the observed rainfall.

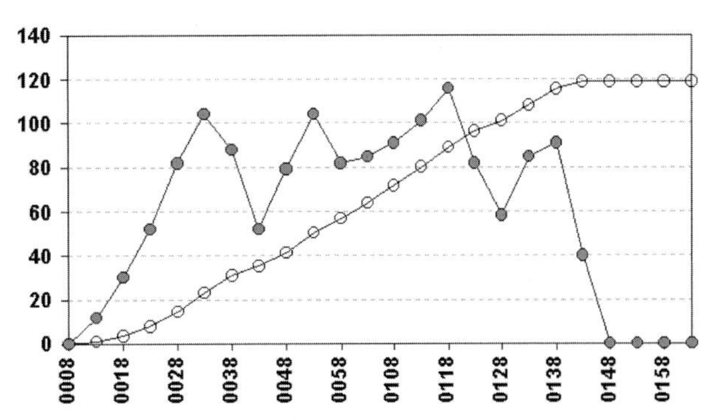

FIG. 12.33. Accumulated ABR (mm) and ABR rate (mm h^{-1}) are plotted for the Brush Creek watershed (1001) for 0008 to 0158 UTC on 5 Oct 1998. The watershed is 35–39 km from the Pleasant Hill, MO, WSR-88D.

4) THE IMPACT OF WATERSHED AREA

Several observational facts have emerged from the analysis of four flash flood seasons of ABR data at the Pittsburgh WFO, in addition to the lessons learned from AMBER playback cases of the major flash floods of the past five years across the United States. Because ABR is an areally weighted average of rainfall in a defined watershed, the area of the watershed plays a big role in the interpretation of the ABR data. The small scale of the MGE elements that produce flash floods can easily inundate an area of 5–20 km^2, while many heavy rainfall cores must pass across a larger stream watershed of 100–200 km^2 to cause flash flooding. When slow cell movement of the MGE elements occurs, small watersheds (<40 km^2) are at great risk for flash flooding with very heavy rainfall. If watersheds are not defined small enough (to the area of the MBA), there is a danger of failure to detect the flash flooding.

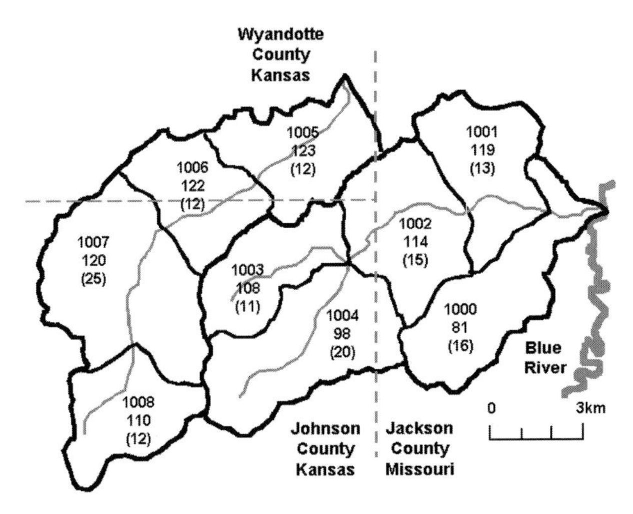

FIG. 12.32. Watershed segments for the Brush Creek (1000–1004) and Turkey Creek (1005–1008) watersheds near Kansas City, MO/KS. In each watershed segment is the watershed number (top), ABR for 0013 to 0138 UTC on 5 Oct 1998 (middle number), and the watershed area (km^2, bottom number). The thin line at the intersection of watersheds 1002–1001 is the Troost Avenue bridge across Brush Creek.

Some examples of flash flood produced by slow-moving storms on small watersheds include Bradys Bend, Pennsylvania, on 14 August 1980 (Scofield 1981); Austin, Texas, on 24 May 1981 (Maddox and Grice 1986); Etna, Pennsylvania, on 30 May 1986 (Saffle 1988); Shadyside, Ohio, on 15 June 1990

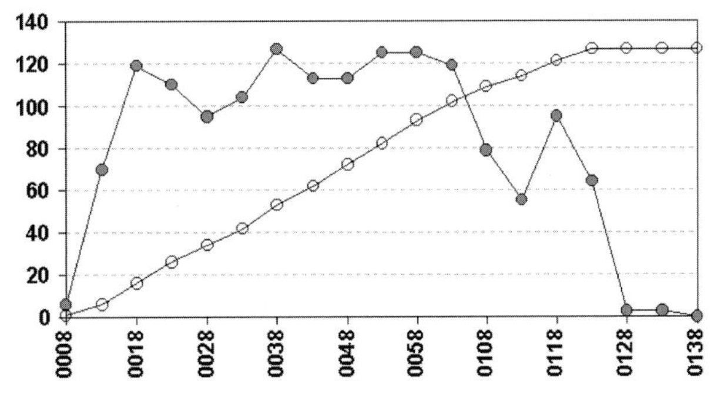

FIG. 12.34. Accumulated ABR (mm) and ABR rate (mm h^{-1}) plot for the location of an NWS rain gauge observer in the western suburbs of Kansas City, KS, for 0008 to 00138 UTC 5 Oct 1998. The observer reported 125 mm of rainfall in 75 min. The rain gauge is located at 292° and 46 km from the Pleasant Hill, MO, WSR-88D.

(National Weather Service 1991); Pitcairn, Pennsylvania, on 1 July 1997 (Davis 1998); and Forest Hills, Pennsylvania, on 18 May 1999 (Davis 2000). All of these events occurred in watersheds smaller than 40 km^2 with high rainfall rates and rainfall time durations of less than 2 h.

The flash flood that struck Pitcairn, Pennsylvania, on 1 July 1997 caused one fatality and $10 million in damage. While most of the damage occurred in the Dirty Camp Run watershed (#100, 9 km^2) shown in Fig. 12.35, severe road flooding occurred in the Abers Creek (#101, 27 km^2) basin. The ABR observed in Abers Creek (38 mm) was approximately half of the ABR for Dirty Camp Run (67 mm). But if Abers Creek is broken down into its smaller tributaries (Fig. 12.36), a different picture emerges. Over 80 mm of ABR (Fig. 12.37) occurred in the Thompson Run watershed (6 km^2 area), resulting in the severe road flooding. If Abers Creek had not been subdivided, this road flooding would have gone undetected.

Flash floods occur most often on small watersheds, but the environment occasionally supports widespread flash flooding on watersheds larger than 1000 km^2. Although flash floods occur only rarely on such large watersheds, such widespread flash flooding can lead to record river flooding. The key to predicting when flash floods are possible on large watersheds is understanding the relationship between the cell speed of intense thunderstorm cores and the area of heavy rainfall observed at the ground (Davis and Jendrowski 1996). If small intense rainfall cores remain nearly stationary or move slowly, a small surface area (40 km^2 or less) may be inundated with 50–100 mm of rain in less than 1 h. If these same storm cores move rapidly past a ground location at 25 m s^{-1}, a much smaller amount of rain (generally less than 25 mm) will fall at that location, but the rain will be spread over a much larger area. If a large number of convective cells move rapidly past a given location, widespread flash flooding can result.

The flash flooding previously mentioned near Chicago, Illinois (Fig. 12.22), and the Red Bank Creek

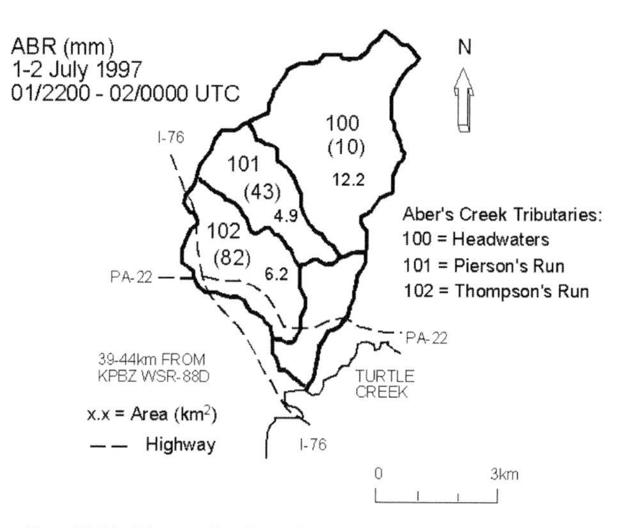

FIG. 12.36. The small tributaries (100–102) of Abers Creek are shown with ABR (mm) plotted for each tributary from 2200 UTC 1 Jul to 0000 UTC on 2 Jul 1997. Watershed area is shown in km^2. Thompson Run (watershed area 6.2 km^2) had 82 mm of ABR. Mud slides were reported at the intersection of Interstate 76 and Pennsylvania Route 22 in the western portion of the basin.

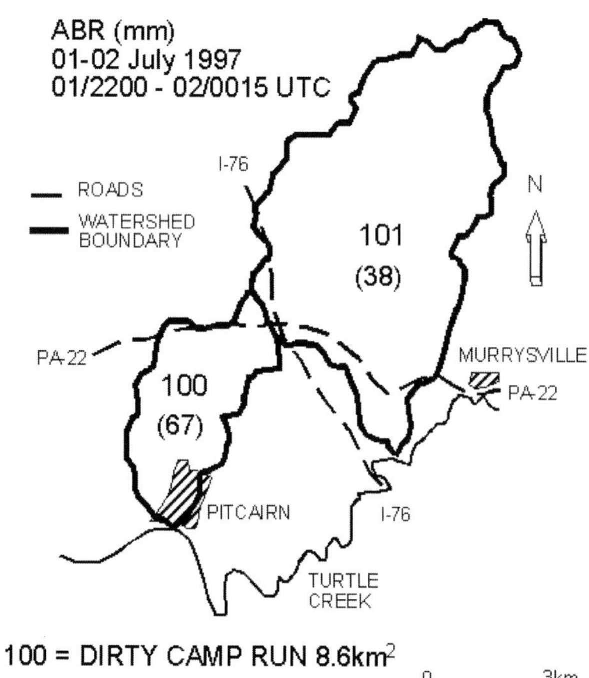

FIG. 12.35. Accumulated ABR from 2200 UTC 1 Jul to 0015 UTC on 2 Jul 1997 is shown for the Abers Creek (101, 26.9 km^2) and the Dirty Camp Run (100, 8.6 km^2) watersheds. The watersheds are at 38–44 km from the Pittsburgh, PA, WSR-88D. Highways are shown as dashed lines.

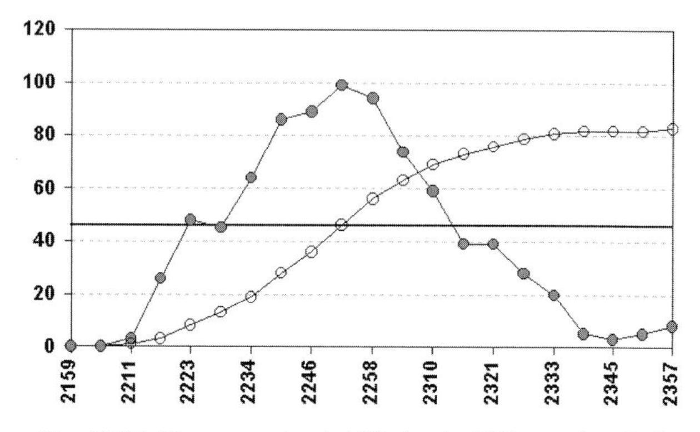

FIG. 12.37. The accumulated ABR (mm), ABR rate (mm h^{-1}), and 1-h FFG are plotted for the Thompson Run watershed for 2159 to 2357 UTC 1 Jul 1997.

flash flood in Pennsylvania (Fig. 12.23) are examples of flash floods produced by widespread convection with high cell training speeds (20–25 m s^{-1}). Flash flooding became record river flooding at St. Charles, Pennsylvania, in the Red Bank Creek watershed (1200 km^2 in area) during the flash flooding of 19 July 1996. A similar flash flood that led to record river flooding occurred during the Ohio flash floods of 26–29 June 1998.

On the midnight operational shift of 27 June 1998, forecasters at the Pittsburgh WFO were monitoring developing thunderstorms in eastern Ohio. A quasi-stationary warm front was oriented northwest to southeast across eastern Ohio and a low-level jet was transporting warm moist air northward and across the front. The mean tropospheric wind exceeded 25 m s^{-1} from the northwest, and the moist unstable air remained upstream, where low-level forcing was strongest. Given such large-scale support, numerous intense thunderstorms developed near the upstream forcing and traveled southeast along the warm front at speeds near 25 m s^{-1}. Consequently, many intense thunderstorms moved in succession over Muskingum County, Ohio. The Wakatomika Creek (585 km^2 basin area) flows into the town of Frazeysburg, Ohio (Fig. 12.38). Output from AMBER indicated that an ABR of about 25 mm accumulated in the Wakatomika Creek watershed between 0000 and 0100 UTC on 27 June 1998. Although the rain stopped at 0100 UTC, another period of heavy rain began at about 0430 UTC. The ABR rate on the Wakatomika basin (Fig. 12.39) increased to over 25 mm h^{-1} at 0600 UTC and remained

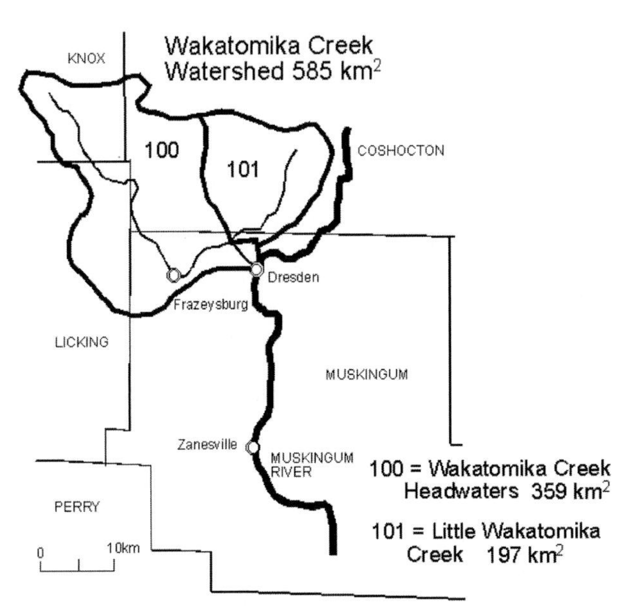

FIG. 12.38. A map of Muskingum County, OH, showing the Wakatomika Creek basin and the two watershed segments, the Little Wakatomika Creek (101), and the headwaters of the Wakatomika Creek (100) above Frazeysburg, OH.

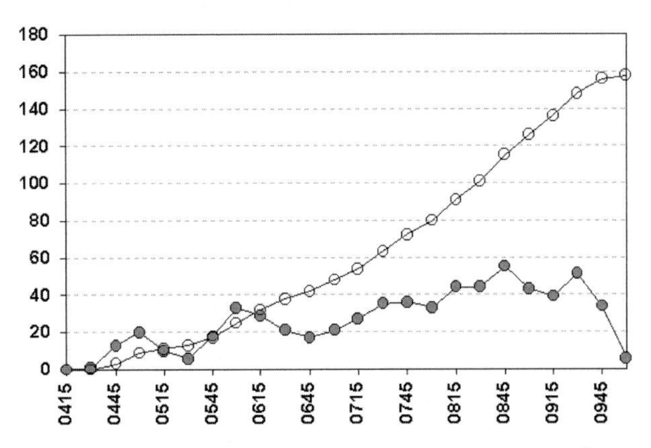

FIG. 12.39. Accumulated ABR (mm) and ABR rate (mm h^{-1}) are plotted for the Wakatomika Creek watershed near Dresden, OH, from 0415 to 0945 UTC 27 Jun 1998. The basin is 147–181 km from the Pittsburgh, PA, WSR-88D.

at this rate or higher until 0945 UTC. The ABR rate indicated 50 mm h^{-1} at 0845 and again at 0930 UTC. A flash flood warning was issued at 0700 UTC and the stream gauge at Frazeysburg rose from 0.2 m at 0600 UTC to 3.7 m at 1200 UTC, flooding much of the town. Almost 160 mm of ABR accumulated in the Wakatomika watershed.

Periods of intense convective rainfall continued during the nights of 27–28 June 1998 across portions of eastern Ohio. As a result, record river flooding occurred at Cambridge, Ohio, by the morning of 29 June 1998. Over 250 mm of rainfall fell across watersheds above Cambridge. Many counties in eastern Ohio were declared disaster areas as a result of the widespread flooding. Flash flooding over large areas can often result in river flooding. This type of widespread flash flooding is frequently observed with slow-moving land-falling tropical storms, but flash flooding with high cell training speeds is very unusual.

This case illustrates the inherent difficulty of detecting flash flood potential when cell speeds are high. For example, high cell speeds are usually associated with highly sheared environments, which also support severe and possibly tornadic thunderstorms. If forecasters are preoccupied with severe thunderstorms (Schwein and Dummer 1998), they may not respond adequately to a growing flash flood threat. AMBER makes output of ABR and ABR rate available in real time, allowing the computer to search for flash flood potential while the forecaster working the radar can concentrate on damaging winds and tornadoes. It should be noted that severe thunderstorms with damaging wind gusts did occur in association with the storms that traversed the Wakatomika watershed. To further complicate matters, some forecasters may believe that flash floods only occur when convective cells and storms move slowly. Consequently, flash flooding may not be considered a threat when cell speeds are high. The previous example shows that significant and widespread flash flooding is possible with high cell speeds, when numerous cells move along a similar path.

5) Using hydrologic connectivity to determine which watersheds have the greatest flash flood risk

Just as the path of potential destruction is included in a tornado warning or severe thunderstorm warning, the "threat area" for flash flooding should be included in a flash flood warning. Unlike tornadoes and severe thunderstorms, which can strike any geographic location, the threat area of a flash flood generally occurs along a stream or dry arroyo. The stream basins of the NBD project allow forecasters to include threat areas for specific watersheds in warnings and statements. The NBD stream database also indicates the hydrologic connectivity of stream segments, allowing forecasters to determine where the water will go once the flood wave is in the stream channel.

The importance of stream connectivity is demonstrated by the flash flood that occurred in Zion National Park on 27 July 1998. The Salt Lake City WFO issued a flash flood warning that saved at least 40 lives with several hours of lead time before the flooding. A popular hiking trail winds through the "Narrows" of Zion National Park in southern Utah. In the Narrows hikers must walk in the creek for more than one hour, with steep rock walls on both sides of the creek bed. Once in the Narrows there is virtually no escape if the stream level rises rapidly. On 27 July 1998, the Salt Lake City WFO notified the ranger station in the park that an intense thunderstorm was depositing heavy rainfall upstream of the Narrows. Due to this early notification, the ranger was able to prevent more than 40 hikers from going into the Narrows. The Salt Lake City WFO was able to determine that the storm was in the North Fork of the Virgin River by overlaying the WSR-88D rainfall estimates on the AWIPS topographic map of southern Utah.

The NBD stream database shows how stream segments are connected, allowing forecasters to determine the stream segments at greatest risk of flash flooding. Figure 12.40 shows the AMBER playback watershed segments defined for the tributaries flowing into the Narrows. Figure 12.41 shows a block diagram of the stream connectivity for each stream segment of Fig. 12.40. Although four major streams feed into the Narrows, flash flood–producing rainfall was restricted to the North Fork of the Virgin River. The effective area of rainfall is defined as the portion of the watershed that produces significant runoff. Although the entire watershed area above the Narrows is 740 km^2, only about 120 km^2 contributed runoff to the stream rise. Figure 12.42 shows the watershed segments for

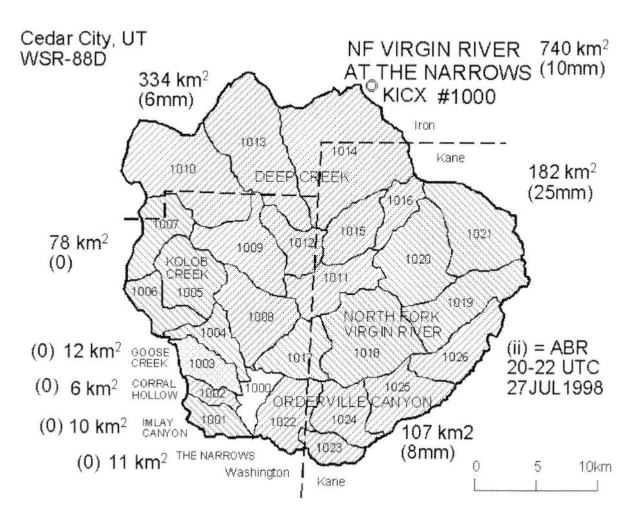

FIG. 12.40. ABR (mm) from 2000 to 2200 UTC 27 Jul 1998 is shown for the seven tributaries that flow into the Narrows, watershed segment 1000, near Cedar City, UT. The area of each tributary is shown in km^2. Notice that no rainfall fell in the Narrows or in segments 1001 through 1007. The North Fork of the Virgin River (1017–1021) had an ABR of 25 mm over its 182 km^2 area. The location of the Cedar City, UT, WSR-88D (KICX) is shown near headwaters of stream segment 1014.

the North Fork of the Virgin River and indicates the ABR for each segment. The ABR rate plot for basin 1018 (Fig. 12.43) illustrates that most of the rainfall occurred in one hour, showing a rainfall rate of 64 mm h^{-1}. FFG is not needed in the mountains of Utah because almost no infiltration of rain occurs into the rocky soil. Typically, 25 mm of rainfall in an hour causes significant flash flooding in this semi-arid mountainous terrain. Notice (Fig. 12.42) that all of the rain fell in Kane County, and no rain fell in the Narrows, which is located in Washington County.

This case highlights the importance of understanding how stream segments are connected. A forecaster considering only graphic WSR-88D rainfall products on a political county background might have warned for Kane County (where the rain accumulated) but not Washington County (where no rainfall was indicated). Both counties were correctly warned because the Salt Lake City forecaster knew that the North Fork of the Virgin River flowed into the Narrows in Washington County. The NBD database provides this critical information to all WFOs for all defined stream segments.

The timing of rainfall associated with this hydrologic connectivity can also be an important flash flood severity factor. When rainfall occurs in a headwaters area, and the rainfall-producing storm moves in a downstream direction, flooding can be enhanced in the downstream segment of the stream. On 1 July 1997 flash flooding along Dirty Camp Run (8.6 km^2) caused $10 million of damage in Pitcairn, Pennsylvania. The ABR and ABR rate plot (Fig. 12.44) shows that ABR

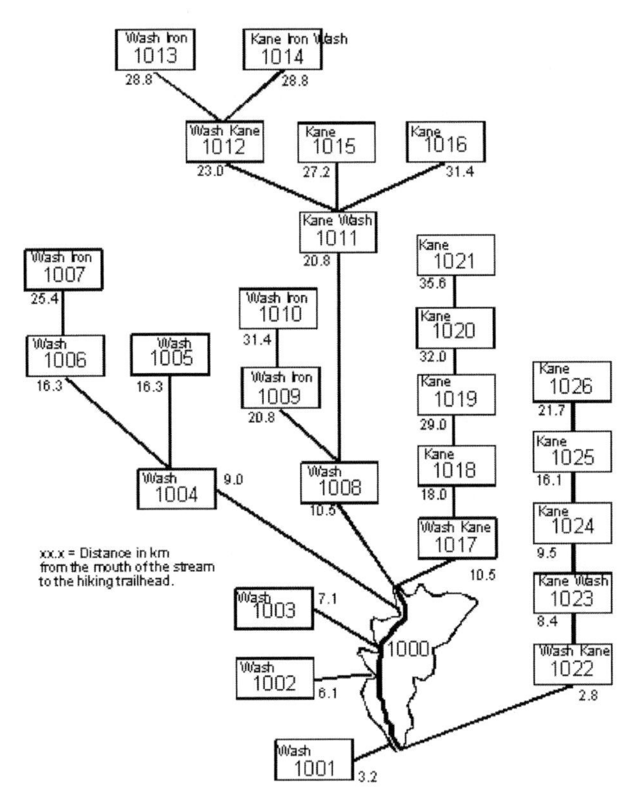

FIG. 12.41. The hydrologic connectivity is shown for each defined stream segment of Fig. 12.40. The distance in km along the stream from the outflow point (the mouth) of each stream segment to the hiking trail head (2.3 km downstream from the mouth of the Narrows) is shown. The counties containing each stream segment are listed in the top of each box [Iron, Kane, or Washington (Wash) counties in UT].

is 18 mm over FFG, not even reaching the 25 mm over FFG category typical of significant flash flooding in western Pennsylvania. The watershed is divided into two parts, the headwaters of 4.7 km^2 (Fig. 12.45) and the downstream segment containing the city of Pitcairn (3.9 km^2). The headwaters area received 30 mm over FFG with the peak of heavy rainfall occurring around 2300 UTC (Fig. 12.46), while the downstream segment just reached FFG with the peak of heavy rainfall 30 min later at 2330 UTC (Fig. 12.47). During the disaster survey, residents of Pitcairn reported that the stream through town was running near bankfull before the rain started. The water from the headwaters filled the channel, and the 50 mm of rainfall that fell on the highly urbanized downstream segment caused severe flash flooding that resulted in $10 million in damage. FFG was no longer valid for the downstream segment, because the stream was no longer at a low flow condition. The highly urbanized downstream segment produced significant runoff into a stream channel that was already at a bankfull condition.

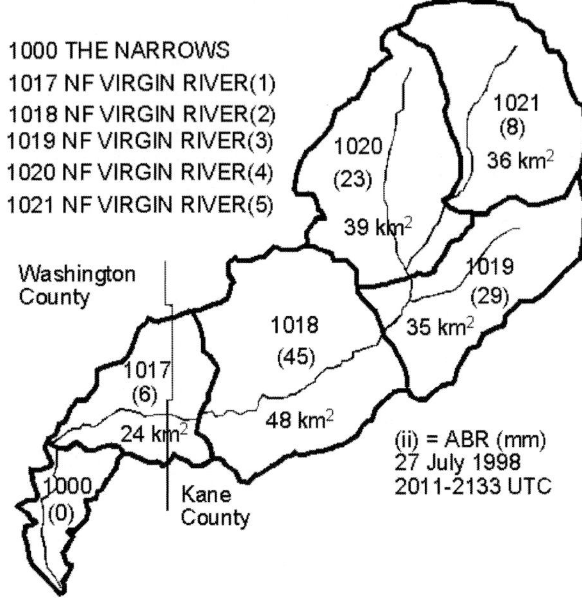

FIG. 12.42. Accumulated ABR (mm) for 2011–2133 UTC 27 Jul 1998 is plotted for the stream segments (1017 to 1021) of the North Fork of the Virgin River that feeds into the Narrows. The area of each watershed segment is shown in km^2.

6) URBAN WATERSHEDS

Although urban areas may not contain naturally flowing streams, intense rainfall can quickly overwhelm storm drainage systems because urban watersheds are dominated by buildings, streets, pavements, and parking lots. Thus, urban environments have limited land area capable of absorbing rainfall and correspondingly greater potential for converting rainfall to runoff. Urban watersheds are particularly vulnerable to flash floods, especially where drainage systems are

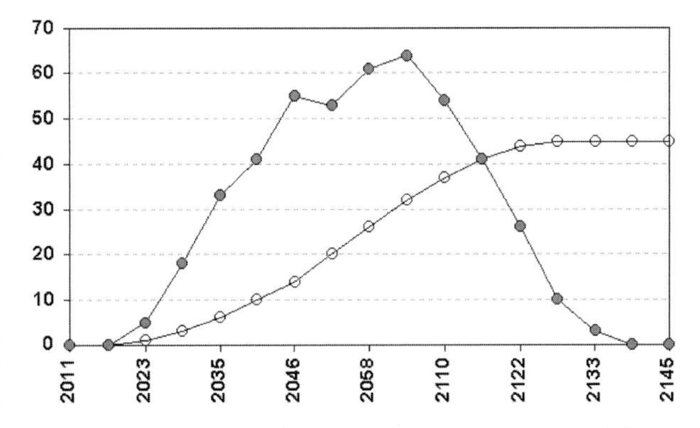

FIG. 12.43. Accumulated ABR (mm) and ABR rate (mm h^{-1}) are plotted for stream segment 1018 on the North Fork of the Virgin River for 2011–2145 UTC on 27 Jul 1998. The stream segment is 18–26 km from the Cedar City, UT, WSR-88D.

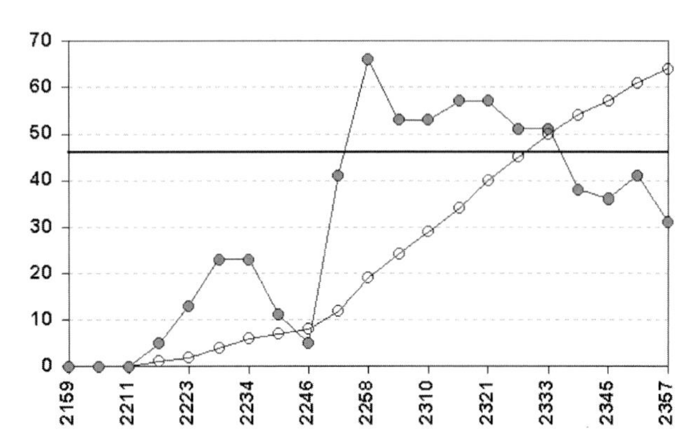

FIG. 12.44. Accumulated ABR (mm) and ABR rate (mm h^{-1}) are plotted for 2159–2357 UTC 1 Jul 1997 for the Dirty Camp Run watershed, near Pitcairn, PA. The watershed is 38–40 km from the Pittsburgh, PA, WSR-88D.

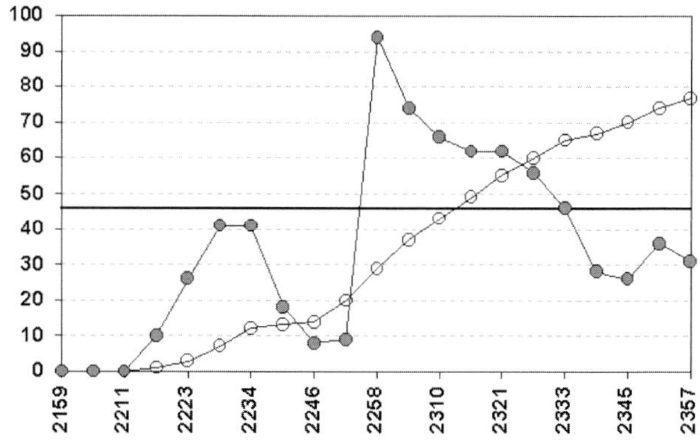

FIG. 12.46. Accumulated ABR (mm), ABR rate (mm h^{-1}), and FFG (mm h^{-1}) are plotted for Dirty Camp Run (2) from 2159 to 2357 UTC 1 Jul 1997.

inadequate. As urban development continues, the number of people at risk from urban flash floods will increase.

Some urban areas are literally built over existing watersheds. Forest Hills Run, a highly urbanized watershed in the eastern suburbs of Pittsburgh, Pennsylvania, is about 6 km in length; over 4 km of that length is buried underground in storm drains. A severe flash flood struck Forest Hills Run on 18 May 1999 (Davis 2000). The Forest Hills Run watershed is a rapidly responding small watershed (2.6 km^2) due to both its small size and its steep terrain. The plot of ABR rate (Fig. 12.48) provides valuable information about the intense rainfall rates before a significant accumulation of rain occurs in the watershed. Notice that the ABR rate exceeded 50 mm h^{-1} at 2156 UTC and the ABR equaled FFG at 2211 UTC (when minor flooding should begin), and ABR was 25 mm h^{-1} over FFG at 2236 UTC (when significant flash flooding is under way). The ABR rate provides an early alert for watersheds at risk of flooding, before flooding begins. The local Emergency Management Office in Forest Hills received the first report of water in a basement at 2215 UTC. Significant flash flooding was reported by 2300 UTC, with many cars

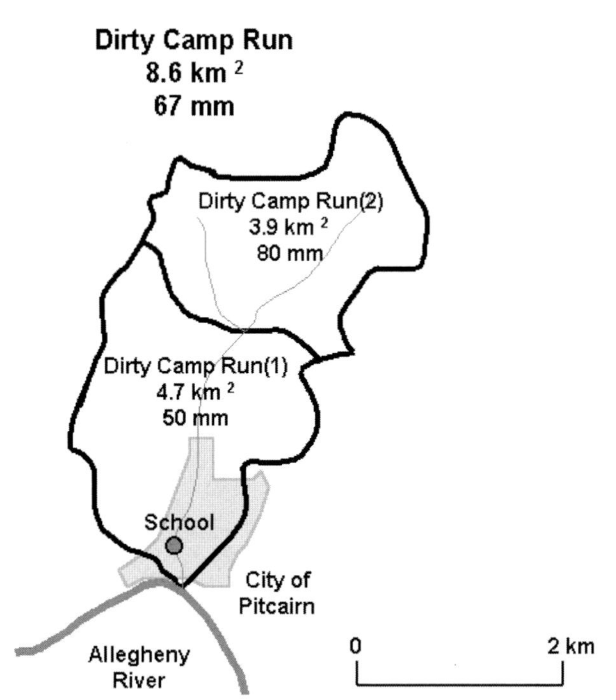

FIG. 12.45. A map showing the stream segments of the Dirty Camp Run watershed. The ABR (mm) is plotted for 2200 UTC 1 Jul to 0015 UTC 2 Jul 1997. The gray shaded area is the city of Pitcairn, PA. The Pitcairn elementary school, shown as a circle in the gray area, was severely flooded, with water flowing through the school's first-floor windows. The watershed areas are shown in km^2.

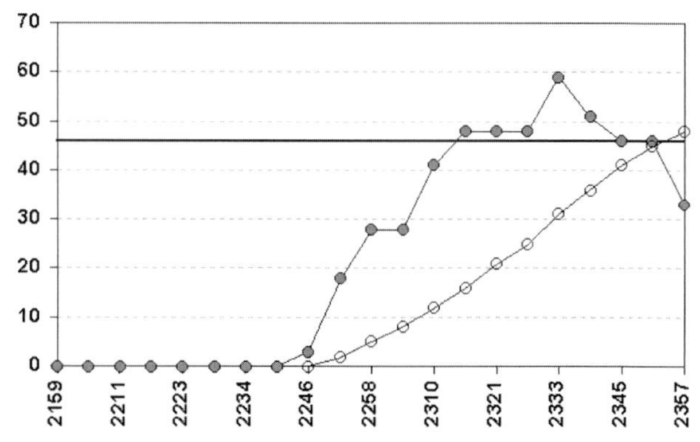

FIG. 12.47. Accumulated ABR (mm), ABR rate (mm h^{-1}), and FFG (mm h^{-1}) are plotted for Dirty Camp Run (1) from 2159 to 2357 UTC 1 Jul 1997.

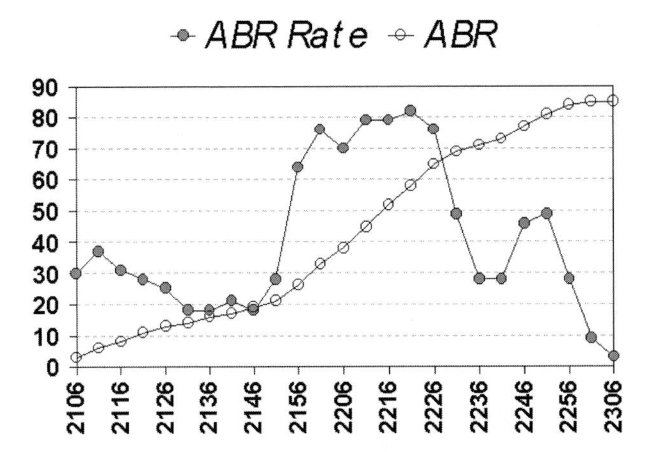

FIG. 12.48. Accumulated ABR (mm) and ABR rate (mm h^{-1}) are plotted for Forest Hills Run (6.2 km^2 in area), in Forest Hills, PA, from 2106 to 2306 UTC 18 May 1999. The watershed is 31–35 km from the Pittsburgh, PA, WSR-88D.

stranded in deep water on highways. At 0010 UTC 20 people were evacuated from 10 homes threatened by high water. There is some delay between the time the ABR greater than 25 mm over FFG is observed on radar (2236 UTC) and the onset of the significant flash flooding. There is a finite time delay for the runoff to reach the creek and produce the stream rise (about 25 min in this case).

Many urban areas contain no naturally flowing streams. AMBER can produce ABR computations for these urban areas that are defined locally by the WFO (not as part of the NBD project). The defined urban areas should be small in size, seldom more than 5 km^2. A large urban area should be divided into small segments to allow the detection of the heavy rainfall maxima common with slow-moving MGE. An example of WFO-defined urban areas in the eastern suburbs of Pittsburgh, Pennsylvania, is shown in Fig. 12.49. Notice that many of the urban areas fall along a major river and are intermingled with watershed areas. Widespread flooding occurred in these urban areas on 18–19 May 1999, during the same time period as the flooding in Forest Hills Run. All nine of these small urban areas (combined population 86 000) reported damage and stranded automobiles from flash flooding. The ABR plot for Rankin/Braddock is shown in 15-min time increments (Fig. 12.50) and in 5-min time increments (Fig. 12.51) for the period of the heaviest rainfall. Notice how the time averaging of the ABR from 15 to 5 min reduces the ABR rate.

d. Using tropical Z–R versus standard convective Z–R for DMC

The operational use of the tropical *Z–R* rate results in almost twice the rainfall computed using the normal operating mode of the WSR-88D with the convective *Z–R* for DMC. Since the WSR-88D allows only one

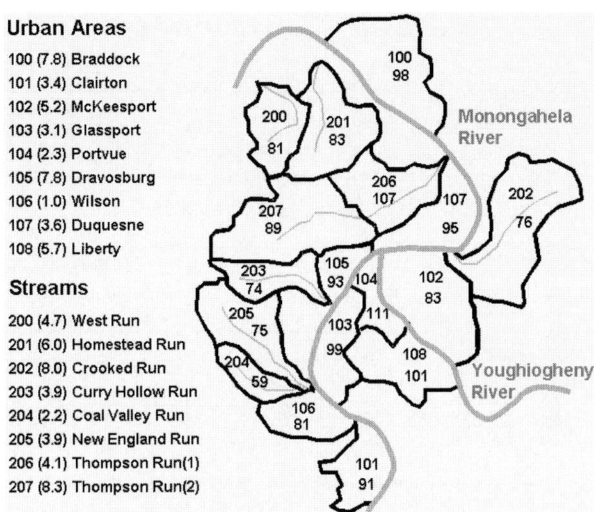

FIG. 12.49. A map showing the AMBER defined urban areas (100 to 108) and defined watersheds (200 to 207) in the eastern suburbs of Pittsburgh, PA. ABR (mm) is plotted below the segment number for 2000 UTC 18 May to 0200 UTC 19 May 1999. The urban areas are 30–39 km from the Pittsburgh, PA, WSR-88D. The area (km^2) of each urban area is shown in the key.

Z–R to be used at a time, WFOs will usually run the standard convective *Z–R* unless rainfall from a landfalling tropical storm is expected. AMBER can produce two distinct rainfall databases for both tropical and standard *Z–R*. A forecaster can choose the appropriate ABR database, based on radar–rain gauge comparisons or environmental indications such as warm satellite tops associated with a heavy rain-producing storm with few lightning strikes. The tropical *Z–R* should not be applied unless deep tropical moisture is available to the storm and depth of the coalescence

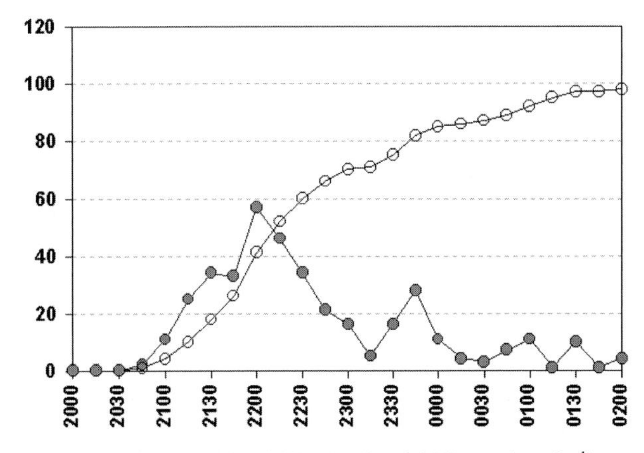

FIG. 12.50. Accumulated ABR (mm) and ABR rate (mm h^{-1}) are plotted for 2000 to 2306 UTC 18 May 1999 for the Rankin/Braddock, PA, urban area. Time interval of the ABR data is 5 min.

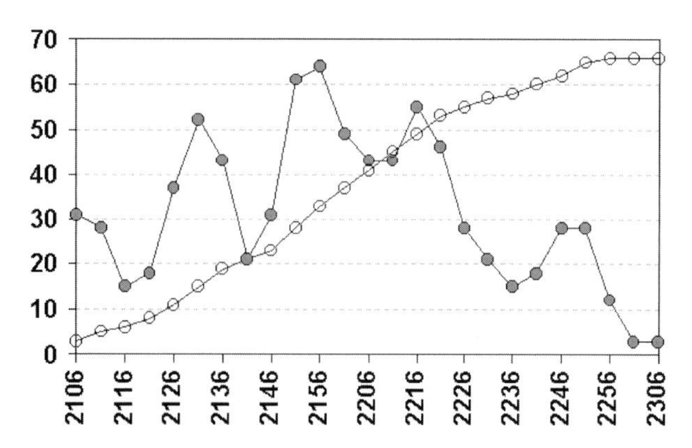

FIG. 12.51. Accumulated ABR (mm) and ABR rate (mm h^{-1}) are plotted for 2106 UTC 18 May to 0200 UTC 19 May 1999 for the Rankin/Braddock, PA, urban area. Time interval of the ABR data is 15 min.

layer (Fig. 12.20) is at least 2 km in depth, with many flash flood cases occurring with a depth of 3–4 km (Chappell 1993). Deep tropical moisture and a coalescence layer of 3–4 km do not assure that tropical rates will occur. Large values of CAPE in the coalescence layer may speed the air parcels too quickly through the layer to allow much growth by coalescence (Chappell 1993).

Radar rainfall estimates can be computed at the location of any rain gauge entered into the AMBER database. The rainfall estimate is computed at the location of the single 1-km radar range bin that contains the location of the rain gauge. The *Z–R* rainfall estimate may then be compared to the actual rain gauge measurement to determine the accuracy of the radar estimate (Davis 1997). Using the parallel ABR databases of both standard *Z–R* and tropical *Z–R* rainfall estimates, the *Z–R* gauge comparison can be used to determine which datasets are providing the best rainfall estimates. Recall that the rain gauge measurement of an NWS observer was verified by the AMBER computation at the gauge site, validating the tropical *Z–R* relationship for the Archive II playback of the Kansas City, Missouri/Kansas, flash flood.

Determining whether or not the environment favors warm rain processes is critical to using ABR effectively. For example, if the standard convective WSR-88D *Z–R* relationship is used to compute ABR in an environment favoring a warm rain event, the radar estimates will be about 50% of the actual rainfall. The flash flood on the Spring Creek watershed in Fort Collins, Colorado, on 28–29 July 1997 is a classic case of extreme rainfall rates produced by warm rain processes. Very little lightning was observed with the Fort Collins storm, and IR satellite data showed very warm tops (Kelsch 1998). When using the tropical WSR-88D tropical *Z–R* rate, the choice of maximum allowable hourly rainfall rate can be critical. For the purpose of the AMBER playback, Spring Creek was divided into three segments (100–102) as shown (Fig. 12.52). Using the playback mode of AMBER, radar rainfall estimates were computed at each WSR-88D 1-km range bin for comparison with the observed rainfall (Fig. 12.53) using a 100 mm h^{-1} hourly rainfall cap. Using a 175 mm h^{-1} hourly rainfall cap for the Fort Collins case results in a gross overestimation of the observed rainfall (Fig. 12.54). The ABR and ABR rate computed for Spring Creek (2) using the 100 mm h^{-1} cap shows that several surges of heavy rainfall (Fig. 12.55) produced the flash flooding.

Fulton (1999) noted that during both the Fort Collins and the Buffalo Creek, Colorado, flash floods, WSR-88D rainfall estimates were reliable in some parts of the radar domain, but not reliable in other parts of the same radar display. These observational differences may result from warm rain or cold rain processes dominating in different parts of the region. This case highlights the necessity for a dual ABR database for both tropical and standard *Z–R*.

The time lapse display of real-time lightning data on AWIPS may provide help in determining the application of tropical *Z–R* versus standard *Z–R* rates. Since lightning production is tied to ice processes in the thunderstorm, an additional benefit of real-time lightning data may be the identification of storms with warm rain processes predominating. Heavy rainfall was indicated by the Denver WSR-88D during the Fort Collins flash flood, while very little cloud-to-ground lightning was associated with the storm producing the heavy rainfall.

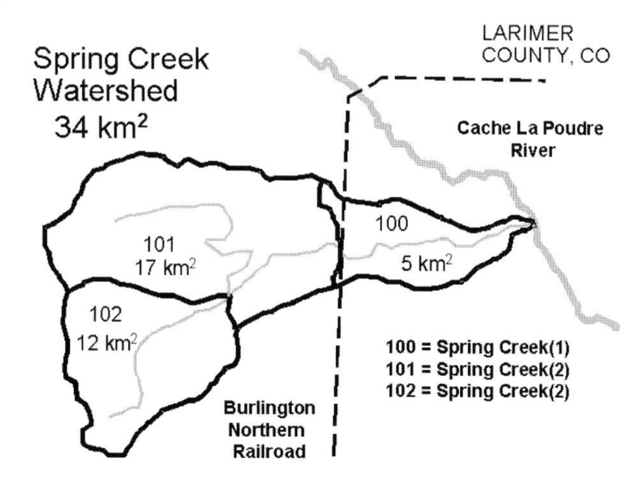

FIG. 12.52. Watershed segments (100–102) for the Spring Creek watershed in Fort Collins, CO. The area of each stream segment is shown in km^2. The dashed line shows the Burlington Northern railroad tracks located on an elevated water retention dike across Spring Creek. Four of the five fatalities during the flash flood occurred in a trailer park just downstream of the railroad tracks when the flood waters topped the dike.

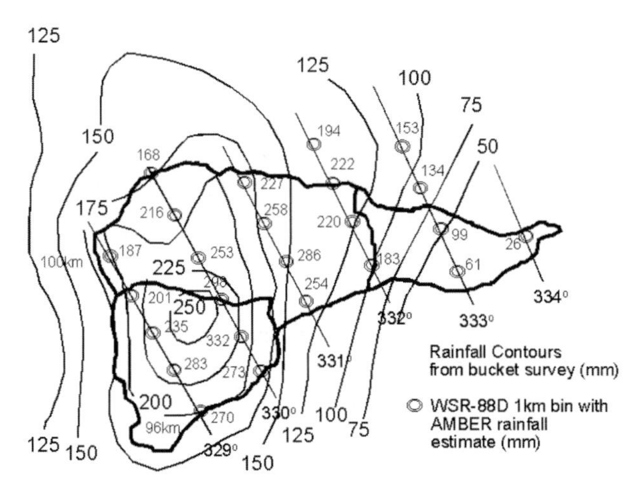

FIG. 12.53. Bucket survey rainfall (solid isohyets, mm) for the Fort Collins, CO, flash flood of 28–29 Jul 1997 (from Petersen et al. 1999). Circles show the location of the Denver, CO, WSR-88D 1-km radar bins with the radar rainfall estimate (mm) from 2300 UTC 28 Jul to 0500 UTC 29 Jul 1996 for each bin. The computations were created using the tropical *Z–R* and a maximum rainfall cap of 178 mm h^{-1}. The straight lines plotted through the radar bins are the radar radials labeled in degrees of radar azimuth. For example, the 329° radar radial at 97 km has an estimated rainfall of 283 mm, while 331° at 97 km has a rainfall estimate of 254 mm.

Rainfall rates observed by rain gauges during warm rain process events can reach 200 mm h^{-1} or more for short periods of time (15 min), but these extreme rainfall rates are difficult to maintain for long periods (30–60 min or more) over the same geographic area. During the AMBER playback of the Dallas, Texas, flash flood of 05 May 1995, allowing a maximum rate of 200 mm h^{-1} caused a significant overestimation of the observed rainfall. If the hourly rainfall cap was reduced to 125 mm h^{-1}, radar rainfall estimates are within 5% of the observed mesonet gauges. Several rain gauges in the Dallas mesonet did report 15-min rainfall in excess of 50 mm, indicating that rainfall rates in excess of 200 mm h^{-1} rates were occurring.

12.9. Summary

Forecasting and detecting flash floods represents the most difficult challenge facing NWS forecasters. Numerous environmental and microphysical processes contribute to the production of heavy rainfall, and the hydrological contribution to flash floods adds additional complexity. The production of a national dataset of flash flood watersheds with the NBD project is a key to incorporating hydrology into forecasting and detecting flash floods. The detailed WSR-88D rainfall estimates mapped into defined watersheds allow the computation of ABR and ABR rate. The application of ABR and ABR rate will dramatically improve WFO flash flood detection capabilities and will allow greater detail to be included in flash flood warnings.

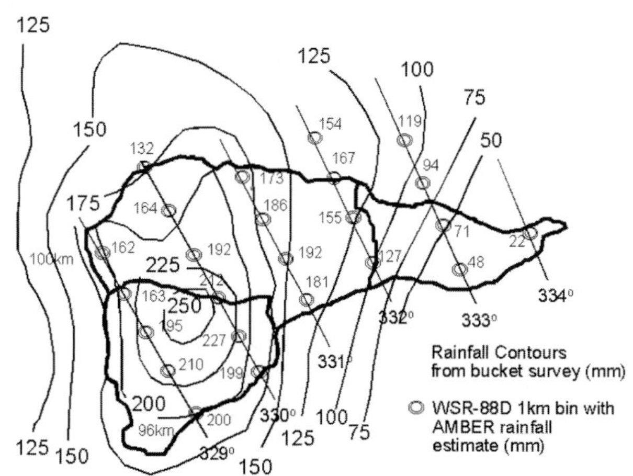

FIG. 12.54. Bucket survey rainfall (solid isohyets, mm) for the Fort Collins, CO, flash flood of 28–29 Jul 1997 (from Petersen et al. 1999). Circles show the location of the Denver, CO, WSR-88D 1-km radar bins with the radar rainfall estimate (mm) from 2300 UTC 28 Jul to 0500 UTC 29 Jul 1996 for each bin. The computations were created using the tropical *Z–R* rate and a maximum rainfall cap of 102 mm h^{-1}. The straight lines plotted through the radar bins are the radar radials labeled in degrees of radar azimuth. For example, the 320° radar radial at 97 km has an estimated rainfall of 210 mm, while 331° at 97 km has a rainfall estimate of 181 mm. These radial rainfall estimates are then averaged to produce the ABR shown in Fig. 12.55.

Continued development of radar and satellite rainfall estimates will lead to improved computation of ABR. The availability of NBD watersheds will also encourage advances in hydrologic modeling, which will improve our understanding of the processes that cause the rapid stream rises associated with many flash floods. AWIPS will allow the use of multisensor databases for the estimation of rainfall, including satellite, radar, rain gauge, and lightning networks. Locally run mesoscale models and neural net-

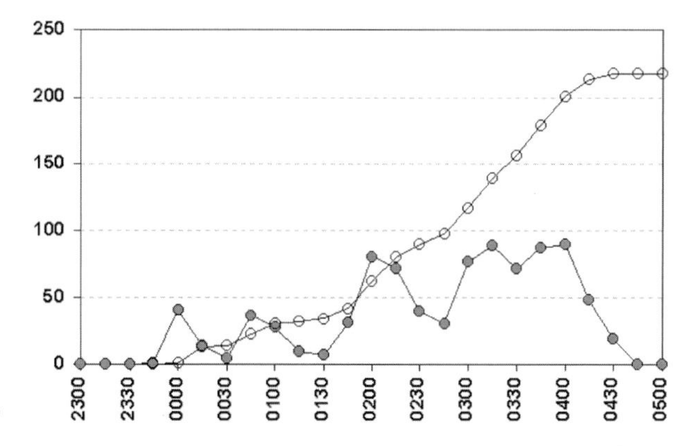

FIG. 12.55. Accumulated ABR (mm) and ABR rate (mm h^{-1}) are plotted for 2300 UTC 28 Jul to 0500 UTC 29 Jul 1996 for Spring Creek (2) watershed segment 102.

works will improve the forecasts of location, intensity, and movement of flash flood producing convective systems. These developments and advances will contribute to more accurate, detailed, and timely flash flood warnings.

Acknowledgments. The author wishes to acknowledge the panel chairman, Josh Korotky, for his extensive guidance in determining the content, organization, and readability of this chapter. This chapter could not have been completed without comprehensive reviews and insightful guidance from Mr. Fred Glass, Mr. Norman W. Junker, Mr. Matthew Kelsch, Dr. James Moore, and Mr. Steven Zubrick. Their efforts were integral to making this chapter scientifically rigorous and operationally relevant.

REFERENCES

Armbruster, J. T., 1978: Model of the flooding caused by the failure of the Laurel Run Reservoir dam, July 19–20, 1977, near Johnstown, Pennsylvania. Preprints, *Conf. on Flash Floods: Hydrometeorological Aspects,* Los Angeles, CA, Amer. Meteor. Soc., 190–193.

Augustine, J. A., and F. Caracena, 1994: Lower-tropospheric precursors to nocturnal MCS development over the central United States. *Wea. Forecasting,* **9,** 116–135.

Baeck, M. L., and J. A. Smith, 1998: Rainfall estimation by the WSR-88D for heavy rainfall events. *Wea. Forecasting,* **13,** 416–436.

Banta, R. M., 1990: The role of mountain flows in making clouds. *Atmospheric Processes over Complex Terrain, Meteor. Monogr.,* No 45, Amer. Meteor. Soc., 173–228.

Beard, K. V., and H. T. Ochs, 1993: Warm rain initiation: An overview of microphysical mechanisms. *J. Appl. Meteor.,* **32,** 608–625.

Bedient, P. B., and W. C. Huber, 1992: *Hydrology and Floodplain Analysis.* 2d ed. Addison-Wesley, 692 pp.

Belville, J. D., G. A. Johnson, A. R. Moller, and J. D. Ward, 1979: The Palo Alto Canyon storm: A combination severe weather–flash flood event. Preprints, *11th Conf. on Severe Local Storms,* Kansas City, MO, Amer. Meteor. Soc., 72–79.

Bluestein, H. B., 1993: *Synoptic-Dynamic Meteorology in Midlatitudes.* Vol. II, *Observations and Theory of Weather Systems,* Oxford University Press, 594 pp.

Bosart, L. F., and D. B. Dean, 1991: The Agnes rainstorm of June 1972: Surface features evolution culminating in inland storm redevelopment. *Wea. Forecasting,* **6,** 515–537.

Braham, R. R., 1952: The water and energy budgets of the thunderstorm and their relation to thunderstorm development. *J. Meteor.,* **9,** 227–242.

———, 1964: What is the role of ice in summer rain showers? *J. Atmos. Sci.,* **21,** 640–645.

Brooks, H. E., and D. J. Stensrud, 2000: Climatology of heavy rainfall events in the United States from hourly precipitation observations. *Mon. Wea. Rev.,* **128,** 1194–1201.

Bruntjes, R. T., T. L. Clark, and W. D. Hall, 1994: Interactions between topographic airflow and cloud/precipitation development during the passage of a winter storm in Arizona. *J. Atmos. Sci.,* **51,** 48–67.

Burpee, R. W., 1986: Mesoscale structure of hurricanes. *Mesoscale Meteorology and Forecasting,* P. Ray, Ed., Amer. Meteor. Soc., 311–330.

Cahail, S., 1987: An examination of the development and utilization of automated flood warning systems: Amherst, MA. M.S. thesis, Dept. of Geology and Geography, University of Massachusetts, 119 pp.

Chappell, C. F., 1986: Quasi-stationary convective events. *Mesoscale Meteorology and Forecasting,* P. Ray, Ed., Amer. Meteor. Soc., 289–310.

———, 1993: Dissecting the flash flood forecasting problem. Post-Print Volume, *Third National Heavy Precipitation Workshop,* NOAA Tech. Memo. NWS ER-87, 293–297.

Chisholm, A. J., and J. H. Renick, 1972: The kinematics of multicell and supercell Alberta hailstorms. Alberta Hail Studies, Research Council of Alberta Hail Studies Rep. 72-2, 24–31.

Clark, J. D., A. J. Lindner, R. Borneman, and R. E. Bell, 1980: Satellite observed cloud patterns associated with excessive precipitation outbreaks. Preprints, *Eighth Conf. on Weather Forecasting and Analysis,* Denver, CO, Amer. Meteor. Soc., 463–473.

Corfidi, S. F., 1998: Forecasting MCS mode and motion. Preprints, *19th Conf. on Severe Local Storms,* Minneapolis, MN, Amer. Meteor. Soc., 626–629.

———, J. H. Merritt, and J. M. Fritsch, 1996: Predicting the movement of mesoscale convective complexes. *Wea. Forecasting,* **11,** 41–46.

Cotton, W. R., 1990: *Storms.* Geophysical Science Series, Vol. 1, Aster Press, 158 pp.

———, and R. A. Anthes, 1989: *Storm and Cloud Dynamics.* Academic Press, 883 pp.

Davis, R. S., 1993: AMBER, a prototype flash flood warning system. Preprints, *13th Conf. on Weather Analysis and Forecasting,* Vienna, VA, Amer. Meteor. Soc., 379–383.

———, 1997: The integration of the Areal Mean Basin Estimated Rainfall (AMBER) flash flood warning system with automated rain gauge data. Preprints, *First Symp. on Integrated Observing Systems,* Long Beach, CA, Amer. Meteor. Soc., 189–196.

———, 1998: Detecting time duration of rainfall: A controlling factor of flash flood intensity. Preprints, *Special Symp. on Hydrology,* Phoenix, AZ, Amer. Meteor. Soc., 258–263.

———, 2000: Detecting flash floods in small urban watersheds. Preprints, *15th Conf. on Hydrology,* Long Beach, CA, Amer. Meteor. Soc., 233–236.

———, and W. Drzal, 1991: The potential use of WSR-88D digital rainfall data for flash flood applications on small streams. *Natl. Wea. Dig.,* **16,** 2–18.

———, and P. Jendrowski, 1996: The operational areal mean basin estimated rainfall (AMBER) module. Preprints, *15th Conf. on Weather Analysis and Forecasting,* Norfolk, VA, Amer. Meteor. Soc., 332–335.

Degan, P., and C. Degan, 1984: *The Johnstown Flood of 1889.* Eastern Acorn Press, 64 pp.

Doswell, C. A., III, 1985: *The Operational Meteorology of Convective Weather.* Vol. II, *Storm Scale Analysis,* NOAA, 240 pp.

———, 1987: The distinction between large-scale and mesoscale contribution to severe convection: A case study example. *Wea. Forecasting,* **2,** 3–16.

———, and D. Burgess, 1993: Tornadoes and tornadic storms: A review of conceptual models. *The Tornado: Its Structure, Dynamics, Prediction, and Hazards, Geophys. Monogr.,* No. 79, Amer. Geophys. Union, 161–172.

———, H. E. Brooks, and R. A. Maddox, 1996: Flash flood forecasting: An ingredients-based methodology. *Wea. Forecasting,* **11,** 560–581.

———, C. Ramis, R. Romero, and S. Alonso, 1998: A diagnostic study of three heavy precipitation episodes in the western Mediterranean region. *Wea. Forecasting,* **13,** 102–124.

Federal Emergency Management Agency, 1987: *Reducing Losses in High Risk Flood Hazard Areas.* U.S. Government Printing Office, 224 pp.

Fread, D. L., and Coauthors, 1995: Modernization in the National Weather Service river and flood program. *Wea. Forecasting,* **10,** 477–484.

Fujita, T. T., 1986: Mesoscale classifications: Their history and their application to forecasting. *Mesoscale Meteorology and Forecasting,* P. Ray, Ed., Amer. Meteor. Soc., 18–35.

Fulton, R. A., 1999: Sensitivity of WSR-88D rainfall estimates to the rain-rate threshold and rain gauge adjustment: A flash flood case study. *Wea. Forecasting,* **14,** 604–624.

———, J. P. Breidenbach, D. J. Seo, and D. A. Miller, 1998: The WSR-88D rainfall algorithm. *Wea. Forecasting,* **13,** 377–395.

Glass, F. G., D. L. Ferry, J. T. Moore, and S. M. Nolan, 1995: Characteristics of heavy convective rainfall events across the mid-Mississippi Valley during the warm season: Meteorological conditions and a conceptual model. Preprints, *14th Conf. on Weather Analysis and Forecasting,* Dallas, TX, Amer. Meteor. Soc., 34–41.

Goodman, S. J., and D. E. Buechler, 1990: Lightning-rainfall relationships. Preprints, *Conf. on Operational Precipitation Estimation and Prediction,* Anaheim, CA, Amer. Meteor. Soc., 112–117.

Graziano, T. M., 1998: The NWS end-to-end forecast process for quantitative precipitation information. Preprints, *Special Symp. on Hydrology,* Phoenix, AZ, Amer. Meteor. Soc., J35–J40.

Gruntfest, E., and C. J. Huber, 1991: Toward a comprehensive national assessment of flash flooding in the United States. *Episodes,* **14,** 26–34.

Harned, S. W., 1987: Use of alert data to prepare improved flash flood warnings at WSO Houston. Preprints, *Seventh Conf. on Hydrometeorology,* Edmonton, AB, Canada, Amer. Meteor. Soc., 127–129.

Hitchens, R. D., and J. D. Belville, 1988: The National Weather Service skywarn program. Preprints, *15th Conf. on Severe Local Storms,* Baltimore, MD, Amer. Meteor. Soc., 169–172.

Hoxit, L. R., and Coauthors, 1978: Meteorological analysis of the Johnstown, PA, flash flood, 19–20 July, 1977. NOAA Tech. Rep., ERL 401-APCL 43, 77 pp.

Hunter, S. M., 1996: WSR-88D Radar rainfall estimation: Capabilities, limitations and potential improvements. *Natl. Wea. Dig.,* **20,** No. 4, 28–36.

Jendrowski, P., and R. S. Davis, 1998: Use of geographic information systems with the areal mean basin estimated rainfall algorithm. Preprints, *Special Symp. on Hydrology,* Phoenix, AZ, Amer. Meteor. Soc., 129–133.

Junker, N. W., and R. S. Schneider, 1997: Two case studies of quasi-stationary convection during the 1993 Great Midwest Flood. *Natl. Wea. Dig.,* **21,** 5–17.

———, ———, and S. L. Fauver, 1999: A study of heavy rainfall events during the Great Midwestern Flood of 1993. *Wea. Forecasting,* **14,** 701–712.

Juying, X., and R. A. Scofield, 1989: Satellite derived rainfall estimates and propagation characteristics associated with mesoscale convective systems (MCSs). NOAA Tech. Memo. NESDIS 25, 49 pp.

Kane, R. J., 1993: Lightning-rainfall relationships in an isolated mid-Atlantic thunderstorm. Postprints, *Third National Heavy Precipitation Workshop,* Pittsburgh, PA, National Weather Service, 249–260.

———, C. R. Chelius, and J. M. Fritsch, 1987: Precipitation characteristics of mesoscale convective weather systems. *J. Climate Appl. Meteor.,* **26,** 1345–1357.

———, M. Gillen, K. Kostura, G. V. Loganathan, D. F. Kibler, and J. Warner, 1996: Hydrometeorological analysis of the Timber Lake flash flood and dam failure. Preprints, *15th Conf. on Weather Analysis and Forecasting,* Norfolk, VA, Amer. Meteor. Soc., 272–275.

Keim, B. D., 1998: Record precipitation totals from the coastal New England rainstorm of 20–21 October 1996. *Bull. Amer. Meteor. Soc.,* **79,** 1061–1067.

Kelsch, M., 1998: The Fort Collins flash flood: Exceptional rainfall and urban runoff. Preprints, *19th Conf. on Severe Local Storms,* Minneapolis, MN, Amer. Meteor. Soc., 404–407.

Klemp, J. B., 1987: Dynamics of tornadic thunderstorms. *Ann. Rev. Fluid Mech.,* **19,** 369–402.

Knievel, J. C., and R. H. Johnson, 1998: The 28 July 1997 Fort Collins flood: Synoptic and mesoscale analysis. Preprints, *19th Conf. on Severe Local Storms,* Minneapolis, MN, Amer. Meteor. Soc., 396–399.

Land, L. F., 1978: Analysis of flood resulting from the Toccoa Falls, GA, dam break. Preprints, *Conf. on Flash Floods: Hydrometeorological Aspects,* Los Angeles, CA, Amer. Meteor. Soc., 127–130.

Larson, L. W., and J. M. Vochatzer, 1978: A case study: Kansas City flood September 12–13, 1977. Preprints, *Conf. on Flash Floods: Hydrometeorological Aspects,* Los Angeles, CA, Amer. Meteor. Soc., 163–167.

———, and Coauthors, 1995: Operational responsibilities of the National Weather Service river and flood program. *Wea. Forecasting,* **10,** 465–476.

Leathers, D. J., D. R. Kluck, and S. Kroczynski, 1998: The severe flooding event of January 1996 across north-central Pennsylvania. *Bull. Amer. Meteor. Soc.,* **79,** 785–797.

Ludlam, F. H., 1980: *Clouds and Storms.* The Pennsylvania State University Press, 405 pp.

Maddox, R. A., 1980: Mesoscale convective complexes. *Bull. Amer. Meteor. Soc.,* **61,** 1374–1387.

———, and G. K. Grice, 1986: The Austin, TX, flash flood: An examination from two perspectives—Forecasting and research. *Wea. Forecasting,* **1,** 66–76.

———, F. Caracena, L. R. Hoxit, and C. F. Chappell, 1977: Meteorological aspects of the Big Thompson flash flood of 31 July 1976. NOAA Tech. Rep. ERL 388-APCL 41, 83 pp.

———, L. R. Hoxit, C. F. Chappell, and F. Caracena, 1978: Comparison of the meteorological aspects of the Big Thompson and Rapid City flash floods. *Mon. Wea. Rev.,* **106,** 375–389.

———, C. F. Chappell, and L. R. Hoxit, 1979: Synoptic and meso-α scale aspects of flash flood events. *Bull. Amer. Meteor. Soc.,* **60,** 115–123.

———, F. Canova, and L. R. Hoxit, 1980: Meteorological characteristics of flash flood events over the western United States. *Mon. Wea. Rev.,* **108,** 1866–1877.

Marks, D. F., Jr., 1990: Radar observations of tropical weather systems. *Radar in Meteorology: Battan Memorial and 40th Anniversary Radar Meteorology Conference,* D. Atlas, Ed., Amer. Meteor. Soc., 401–425.

Marwitz, J. D., 1972a: Precipitation efficiency of thunderstorms on the high plains. *J. Rech. Atmos.,* **6,** 367–370.

———, 1972b: The structure and motion of severe hailstorms. Part II: Multi-cell storms. *J. Appl. Meteor.,* **11,** 180–188.

McAnelly, R. L., and W. R. Cotton, 1985: The precipitation life cycle of mesoscale convective complexes. Preprints, *Sixth Conf. on Hydrometeorology,* Indianapolis, IN, Amer. Meteor. Soc., 197–204.

———, and ———, 1989: The precipitation life cycle of mesoscale convective complexes over the central United States. *Mon. Wea. Rev.,* **117,** 784–808.

McCuen, R. H., 1998: *Hydrologic Analysis and Design.* 2d ed. Prentice Hall, 814 pp.

Merritt, J. H., and J. M. Fritsch, 1984: On the movement of heavy precipitation areas of mid-latitude mesoscale convective complexes. Preprints, *10th Conf. on Weather Forecasting and Analysis,* Clearwater Beach, FL, Amer. Meteor. Soc., 529–536.

Moller, A. R., C. A. Doswell III, and R. Przybylinski, 1990: High precipitation supercells: A conceptual model and documentation. Preprints, *16th Conf. on Severe Local Storms,* Kananaskis Park, AB, Canada, Amer. Meteor. Soc., 52–57.

Moore, J. T., C. H. Pappas, and F. H. Glass, 1993: Propagation characteristics of mesoscale convective systems. Preprints, *17th Conf. on Severe Local Storms,* St. Louis, MO, Amer. Meteor. Soc., 538–542.

———, S. M. Nolan, F. H. Glass, D. L. Ferry, and S. M. Rochette, 1995: Flash flood-producing high-precipitation supercells in Missouri. Preprints, *Conf. on Hydrology,* Dallas, TX, Amer. Meteor. Soc., (J4) 7–(J4)12.

Nair, U. S., M. R. Hjelmfelt, and R. A. Pielke Sr., 1996: Numerical simulation of the 9–10 Jun 1972 Black Hills storm using CSU RAMS. *Mon. Wea. Rev.,* **125,** 1753–1766.

National Climatic Data Center, 1982a: *Storm Data,* **24,** No. 1, 8–9.

———, 1982b: *Storm Data,* **24,** No. 7, 14.

———, 1986: *Storm Data,* **28,** No. 5, 19–54.

———, 1989: *Storm Data,* **31,** No. 6, 6.

———, 1994: *Storm Data,* **36,** No. 7, 3.

National Oceanic and Atmospheric Administration, 1972: Preliminary reports on hurricanes and tropical storms: Hurricane Agnes. National Weather Service, 3–13.

———, 1977: Johnstown, Pennsylvania, flash flood of July 19–20, 1977. Natural Disaster Survey Rep. 77-1, United States Department of Commerce, 60 pp.

———, 1994: Natural disaster survey report: The Great Flood of 1993. National Weather Service, 128 pp.

———, 1998: Northeast floods of January 1996. Office of Hydrology, National Weather Service, 74 pp.

National Weather Service, 1990: Flood/flash flood watch and warning program. *National Weather Service Operations Manual,* National Weather Service, 28 pp.

———, 1991: Shadyside, Ohio flash floods of June 14, 1990. Natural Disaster Survey Rep. NOAA, 62 pp.

———, 1993: Integrated flood observing and warning system management guide. IFLOWS Program Office, Office of Hydrology, National Weather Service, 42 pp.

Oke, T. R., 1987: *Boundary Layer Climates.* 2d ed. University Press, 435 pp.

Orville, H. D., 1965: A numerical study of the initiation of cumulus clouds over mountainous terrain. *J. Atmos. Sci.,* **22,** 684–699.

———, 1968: Ambient wind effects on the initiation and development of cumulus clouds over mountains. *J. Atmos. Sci.,* **25,** 385–403.

Orville, R. E., 1991: Annual summary, lightning ground flash density in the contiguous United States—1989. *Mon. Wea. Rev.,* **119,** 573–577.

Petersen, W. A., and Coauthors, 1999: Mesoscale and radar observations of the Fort Collins flash flood of 28 July 1997. *Bull. Amer. Meteor. Soc.,* **80,** 191–216.

Ponce, V. M., 1989: *Engineering Hydrology.* Prentice Hall, 640 pp.

Pontrelli, M. D., G. Bryan, and J. M. Fritsch, 1999: The Madison County, VA, flash flood of 27 June 1995. *Wea. Forecasting,* **14,** 384–404.

Rasmussen, R. M., 1995: A review of theoretical and observational studies in cloud and precipitation physics: 1991–1994. *Rev. Geophys.,* Suppl., 795–809.

Rotunno, R., 1993: Supercell thunderstorm modeling and theory. *The Tornado: Its Structure, Dynamics, Prediction, and Hazards, Geophys. Monogr.,* No. 79, Amer. Geophys. Union, 57–73.

Runk, K. J., and D. P. Kosier, 1998: Post-analysis of the 10 August 1997 southern Nevada flash flood event. *Natl. Wea. Dig.,* **22,** 10–24.

Saffle, R. E., 1988: Radar rainfall estimates for the Pittsburgh Little Pine Creek flash flood of May 1986. Preprints, *15th Conf. on Severe Local Storms,* Baltimore, MD, Amer. Meteor. Soc., 198–200.

Schaake, J. C., Jr., 1989: Importance of the HRAP grid for operational hydrology. PRC/US Flood Forecasting and Hydrologic Information Seminar, NWS, 9 pp. + figures. [Available from the Office of Hydrology, National Weather Service, Silver Spring, MD 20910.]

Schwarz, F. K., L. A. Hughes, E. M. Hansen, M. S. Petersen, and D. B. Kelly, 1975: The Black Hills–Rapid City Flood of June 9–10, 1972: A description of the storm and flood. Geological Survey Professional Paper 877, U.S. Government Printing Office, 47 pp.

Schwein, N. O., and S. D. Dummer, 1998: A study of the National Weather Service evolving hydrometeorological forecast process and its impact on hydrologic forecasts and warnings. Preprints, *Special Symp. on Hydrology,* Phoenix, AZ, Amer. Meteor. Soc., 275–280.

Schwertz, D. C., 1995: Development and calibration of a flash flood model for the Brays Bayou watershed. Preprints, *Conf. on Hydrology,* Dallas, TX, Amer. Meteor. Soc., (J4)22–(J4)24.

Scofield, R. A., 1981: Satellite-derived rainfall estimates for the Bradys Bend, PA, flash flood. Preprints, *Fourth Conf. on Hydrometeorology,* Reno, NV, Amer. Meteor. Soc., 188–193.

———, 1985: Satellite convective categories associated with heavy precipitation. Preprints, *Sixth Conf. on Hydrometeorology,* Indianapolis, IN, Amer. Meteor. Soc., 42–51.

———, 1987: The NESDIS operational convective precipitation estimation technique. *Mon. Wea. Rev.,* **115,** 1773–1792.

Serafin, R. J., and J. W. Wilson, 2000: Operational weather radar in the United States: Progress and opportunity. *Bull. Amer. Meteor. Soc.,* **81,** 501–518.

Singh, V. P., 1992: *Elementary Hydrology.* Prentice Hall, 973 pp.

Smith, S. B., and Coauthors, 1998: The System for Convective Analysis and Nowcasting (SCAN). Preprints, *14th Int. Conf. on Interactive Information and Processing Systems for Meteorology, Oceanography and Hydrology,* Phoenix, AZ, Amer. Meteor. Soc., J22–J31.

Spayd, L. E., and R. A. Scofield, 1983: Operationally detecting flash flood producing thunderstorms which have subtle heavy rainfall signatures in GOES imagery. Preprints, *Fifth Conf. on Hydrometeorology,* Tulsa, OK, Amer. Meteor. Soc., 190–197.

Stallings, E. A., and L. A. Wenzel, 1995: Organization of the river and flood program in the National Weather Service. *Wea. Forecasting,* **10,** 457–464.

Sweeney, T. L., 1992: Modernized areal flash flood guidance. NOAA Tech. Memo., NWS HYDRO 44, Office of Hydrology, National Weather Service, 32 pp.

Tollerud, E., D. Rodgers, and K. Howard, 1987: Seasonal, diurnal, and geographic variations in the characteristics of heavy-rain-producing mesoscale convective complexes: A synthesis of eight years of MCC summaries. Preprints, *Seventh Conf. on Hydrometeorology,* Edmonton, AB, Canada, Amer. Meteor. Soc., 143–146.

Ulbrich, C. W., J. M. Pelissier, and L. G. Lee, 1996: Effects of variations in *Z-R* law parameters and the radar constant on rainfall rates measured by the WSR-88D. Preprints, *15th Conf. on Weather Analysis and Forecasting,* Norfolk, VA, Amer. Meteor. Soc., 316–319.

Verdin, K. L., and J. P. Verdin, 1999: A topological system for delineation and codification of the Earth's river basins. *J. Hydrol.,* **218,** 1–12.

Viessman, W., J. W. Knapp, G. L. Lewis, and T. E. Harbaugh, 1977: *Introduction to Hydrology.* Harper and Row, 704 pp.

Vincente, G. A., R. A. Scofield, and W. P. Menzel, 1998: The operational GOES infrared rainfall estimation technique. *Bull. Amer. Meteor. Soc.,* **79,** 1883–1898.

Ward, J. D., 1981: Spatial and temporal heavy rainfall patterns overland associated with weakening tropical cyclones. Preprints, *Fourth Conf. on Hydrometeorology,* Reno, NV, Amer. Meteor. Soc., 174–180.

Weisman, M. L., and J. B. Klemp, 1982: The dependence of numerically simulated convective storms in directionally varying wind shears. *Mon. Wea. Rev.,* **112,** 504–520.

———, and ———, 1984: The structure and classification of numerically simulated convective storms in directionally varying wind shears. *Mon. Wea. Rev.,* **112,** 2479–2498.

———, and ———, 1986: Characteristics of isolated convective storms. *Mesoscale Meteorology and Forecasting,* P. S. Ray, Ed., Amer. Meteor. Soc., 331–358.

Winkler, J., 1988: Climatological characteristics of summertime extreme rainstorms in Minnesota. *Ann. Assoc. Amer. Geograph.,* **78,** 57–73.

Wood, R. A., 1994: *Storm Data 1993 with Annual Summaries.* National Climatic Data Center, 80–81.

Young, K. C., 1993: *Microphysical Processes in Clouds.* Oxford University Press, 427 pp.

Zipser, E. J., 1982: Use of a conceptual model of the life cycle of mesoscale convective systems to improve very-short-range forecasts. *Nowcasting,* K. Browning, Ed., Academic Press, 191–204.

Zrnić, D. S., and A. V. Ryzhkov, 1999: Polarimetry for weather surveillance radars. *Bull. Amer. Meteor. Soc.,* **80,** 389–405.

Chapter 13

The Electrification of Severe Storms

EARLE R. WILLIAMS

Parsons Laboratory, Massachusetts Institute of Technology, Cambridge, Massachusetts

REVIEW PANEL: Steven A. Rutledge (Chair), Lawrence Carey, Walter Peterson, and Terry Schuur

---

13.1. Introduction

This review is concerned with electrification and lightning in severe weather. Based on substantial evidence that the electrical processes active in ordinary (nonsevere) thunderstorms are also present in severe storms, the initial discussion here is focused on ordinary thunderstorms. This material forms a physical basis for understanding the often marked departures in electrical behavior in severe storms, which are defined by exceedence criteria for surface wind speed [50 knots (26 m s^{-1})], hailstone diameter [¾″ (19 mm)], and/or the occurrence of a tornado.

The well-recognized variability between lightning and other manifestations of severe weather, most notably hail and tornadoes, is then explored by emphasizing basic physical relationships applicable to all three phenomena.

Vonnegut (1963) provided a lucid and well-balanced discussion of thunderstorm electrification in the previous monograph on severe storms. At that time, numerous studies of the electrical behavior of ordinary thunderstorms were available, while only a handful dealt with severe storms. In the intervening three decades, greatly expanded observations with radar, electromagnetic, and balloon methods have markedly improved this situation. This new information forms an important component of the main body of the review.

13.2. Nonsevere thunderstorms

The meteorological aspects of ordinary (i.e., nonsevere) thunderstorms have recently been reviewed (Williams 1995a). Emphasis will be given here to those physical aspects that provide connections between ordinary thunderstorms and severe thunderstorms.

Thunderstorms are the deepest convective clouds in the earth's atmosphere. They are caused by buoyancy forces set up initially when sunlight heats the earth's surface and the air in the adjacent planetary boundary layer. The life cycle of individual (nonsevere) thunderstorms was extensively investigated in the Thunderstorm Project (U.S. Weather Bureau 1947) and this remains a viable description today. The lifetime of an ordinary thunderstorm is typically less than 30 minutes.

The thermodynamic basis for thunderstorm formation is conditional instability, and the energy on which this process feeds is convective available potential energy (CAPE). The finite vertical displacement that initiates conditional instability can occur by a wide variety of mechanisms—boundary layer thermals, gust-front boundaries, orographic lifting, and frontal surfaces. Whatever the initiator, the subsequent ascent of buoyant air parcels leads to condensation of water vapor by adiabatic expansion. When this process extends to subfreezing temperature, all three phases of water substance are possible. The "mixed phase" provides conditions for the formation of ice crystals and graupel particles, which appear to be fundamental ingredients for strong electrification and lightning (Latham 1981; Illingworth 1985; Krehbiel 1986; Williams 1989b; Saunders 1995).

Perhaps the most important single distinction between ordinary thunderstorms and severe thunderstorms lies in the environment in which they grow. Ordinary thunderstorms frequently develop in barotropic environments in which lateral variations in air temperature and vertical shear of the horizontal wind are both modest. Such conditions are prevalent in the Tropics, where the majority of the world's ordinary thunderstorms are found. The updraft is nearly vertical and the accumulated condensate frequently collapses in place to create the low-level downdraft. Severe storms are often born from baroclinic environments that are often highly sheared. The updraft is tilted and is laterally displaced from the downdraft, thereby enabling a longer-lived migrating storm. The baroclinic zones in the extratropics of both hemispheres are the preferred locations for severe storms (Ludlam 1963).

Research in recent years has disclosed a strong interplay among CAPE, vertical air motion, the microphysical formation and vertical extent of ice particles, and the resulting electrical activity. Both the magnitude and vertical distribution of CAPE are important in influencing the size and vertical distribution of hydrometeors that contribute to charge separation (Williams 1995a). An important gauge of ice particle size and terminal fall speed is obtained by equating updraft speed and fall velocity at thunderstorm altitudes characterized by a "balance level" (Atlas 1966; Lhermitte and Williams 1985). Figure 13.1 shows this relationship assuming an ice-particle density of 0.9 gm cm^{-3} (more appropriate for hail than graupel, which is the more prevalent ice particle in both nonsevere and severe storms). The particle diameter that defines severe hail is also indicated.

For typical updraft speeds in nonsevere thunderclouds, 10–20 m s^{-1}, the particle diameters for levitated ice particles are 3–10 mm. Calculations of ice particle melting in particle descent from the 0°C isotherm, summarized in Fig. 13.2, suggest that such ice particles will not survive the fall to the surface without melting (Srivastava 1987).

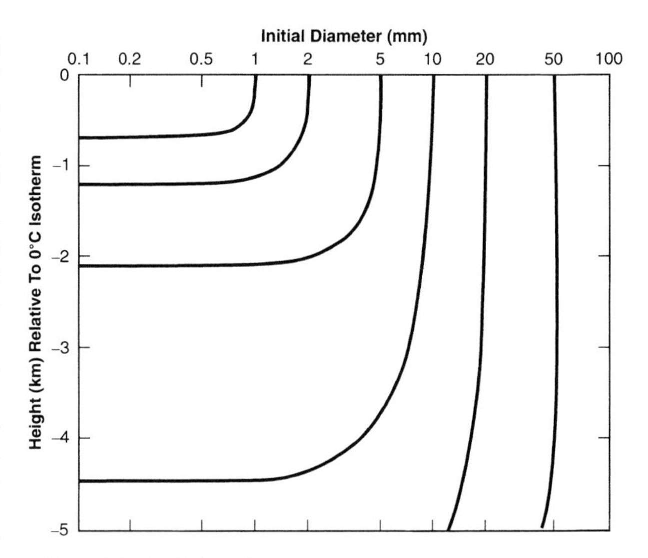

FIG. 13.2. Evolution of hailstone melting with distance below the 0°C isotherm, for particles with a range of initial diameter. The initial bulk density for all ice particles is 0.9 g cm^{-3}. The threshold diameter for severe hail (¾ in. diameter) is marked. [Based on Rasmussen and Heymsfield (1987) and Srivastava (1987)].

The evidence that ice particles are essential for strong electrification and lightning has been reviewed elsewhere (Williams 1988, 1989b). The physical basis for charge separation at the cloud scale (as distinct from the particle scale) is reasonably well understood and is illustrated schematically in Fig. 13.3. Collisions between graupel particles and ice crystals (or small graupel particles or snow) result in the selective transfer of negative charge to the larger particles. The subsequent differential separation under gravity is believed responsible for the creation of cloud-scale dipole moments. This process occurs in an updraft whose speed in nonsevere thunderclouds is of the order of the terminal fall speed of the larger ice particles (Fig. 13.1). Such a process will operate effectively even in the presence of turbulent air motions because the larger particles will always fall faster (relative to the turbulent air) than the smaller particles (Williams 1988; Vonnegut and Moore 1989; Williams 1989b). The mechanism illustrated in Fig. 13.3 also provides for the creation of dipole moment lengths of kilometers when the actual vertical displacements of charged particles in the intervals between lightning flashes are only a few tens to hundreds of meters, so long as a deep charging zone is present.

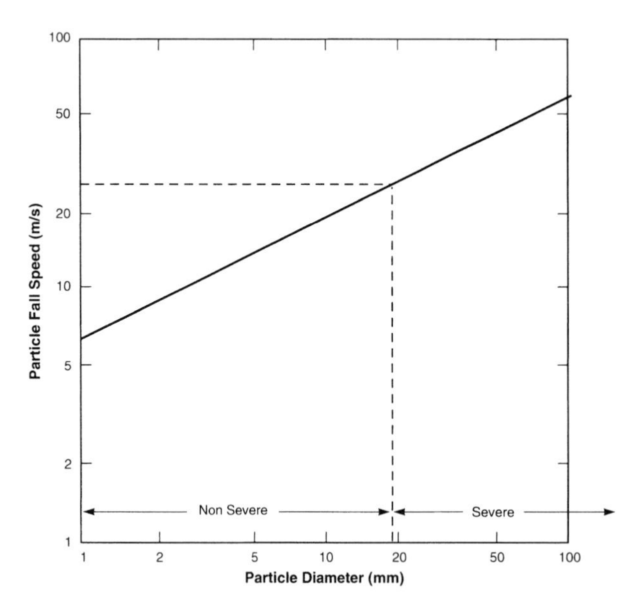

FIG. 13.1. Fall speed of ice spheres vs diameter at 400 mb and $T = -20$°C. The assumed ice bulk density is 0.9 g cm^{-3}. The threshold diameter for severe hail (¾ in. diameter) is marked.

In addition to the positive dipole that is the dominant large-scale electrical characteristic of ordinary thunderclouds, a smaller accumulation of positive charge is often present beneath the main negative charge (Simpson and Scrase 1937; Williams 1989b; Marsh and Marshall 1993; Murphy et al. 1996). This lower positive charge is plausibly created by the same macroscopic mechanism but with appropriate charge transfer at the particle scale. In addition to the charge regions formed by ice particle collisions, a negative screening layer has frequently been observed at the upper cloud boundary. The role of this rich accumulation of space charge is still not known, but it plays a vital role in the convective theory for thunderstorm electrification (Vonnegut 1963). The role of convective downdrafts in transporting negative screening layer charge to the main negative charge region has not been established.

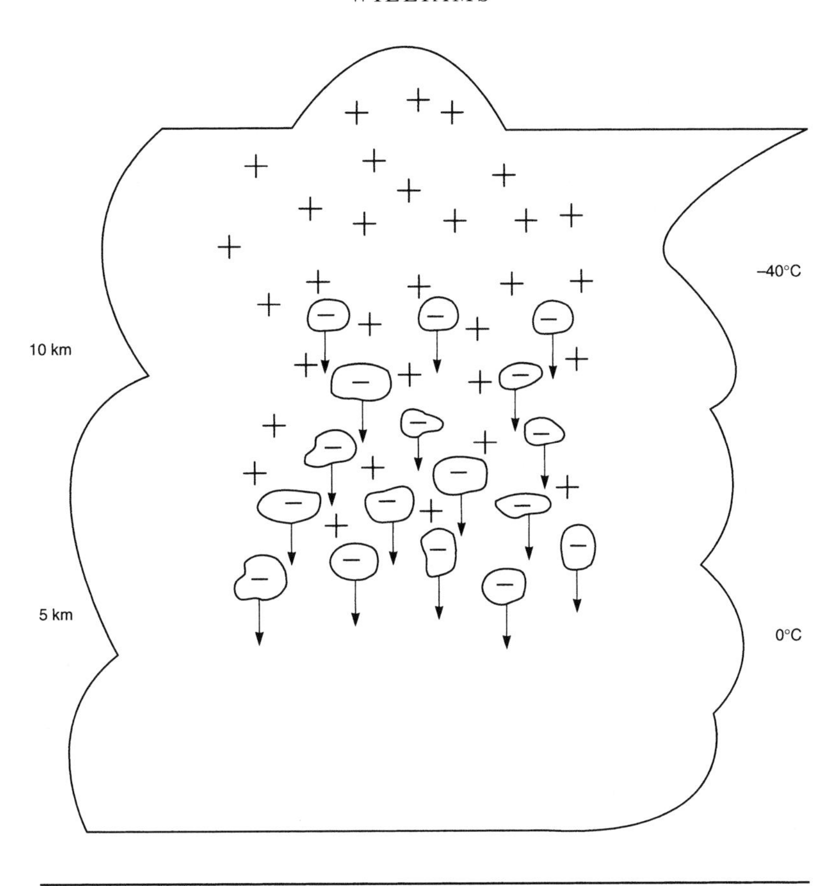

FIG. 13.3. Illustration of differential charge separation by gravity with selective transfer of negative charge to the larger ice particles (graupel) to form a large-scale positive dipole.

Remarkably, the very systematic and reproducible behavior seen at the large scale in ordinary thunderclouds is very poorly understood at the particle scale. Laboratory-scale experiments that seek to simulate microphysical conditions in thunderclouds (Takahashi 1978; Saunders et al. 1991) do show results consistent with the large-scale observations, such as the presence of large charge transfers in particle collisions, the negative charge transfer to larger riming particles over a wide range of cloud temperature (see Fig. 13.3), and the existence of a charge reversal temperature to account for the lower positive change. The polarity of the charge transfer to the larger rimed particle in the parameter space of cloud water content and temperature in two sets of laboratory experiments is shown in Fig. 13.4. Unfortunately, the laboratory results have not yet isolated the microphysical conditions responsible for selective charge transfer in particle collisions, in part because significant differences in the results from different laboratories have still not been resolved (Takahashi 1978; Saunders et al. 1991; Williams et al. 1991; Williams and Zhang 1993; Saunders 1994; Williams 1995b; Williams and Zhang 1996).

Despite these shortcomings, the systematic large-scale behavior illustrated in Fig. 13.3 is also consistent with the well-documented behavior of lightning in ordinary thunderclouds. The two predominant lightning types are *intracloud* and *cloud-to-ground* lightning, illustrated in Fig. 13.5. Intracloud lightning, the most prevalent lightning type, is recognized to link the main negative charge and the upper positive charge. The more energetic cloud-to-ground lightning transfers negative charge from the main negative charge region to earth and may often affect the lower positive charge. Ground flashes of positive polarity have been observed occasionally from the upper anvil regions of nonsevere thunderstorms (Fuquay 1982; Brook et al. 1982). A less common lightning type—the air discharge—is another intracloud lightning type linking main negative charge with lower positive charge.

A summary of maximum total lightning flash rate for nonsevere thunderstorms is shown in the top portion of Table 13.1. Values range from 1 flash per minute to about 60 flashes per minute, with 3 flashes per minute being a typical value. The more active storms are often selected for case study, and so the most common thunderstorms are not well represented by the results in Table 13.1. Recent observations of thunderstorms from space with the Optical Transient Detector and the Lightning Imaging Sensor indicate

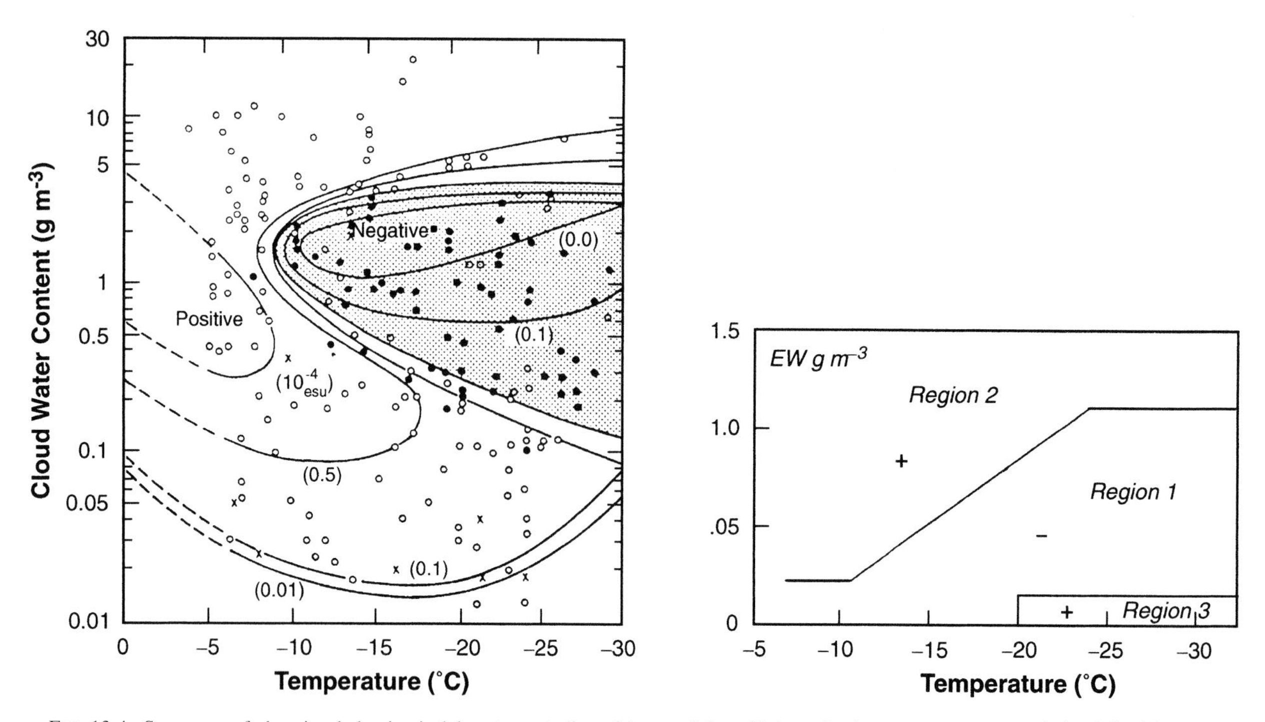

FIG. 13.4. Summary of charging behavior in laboratory studies of ice particle collisions, in the parameter space of cloud liquid water content and temperature. [From Takahashi (1978; left) and Saunders et al. (1991; right).]

mean flash rates of 1–3 flashes per minute (Williams et al. 2000).

Lightning flash rate increases dramatically with thunderstorm size, as measured with the radar cloud top. Vonnegut (1963) estimated the dependence of electrical power on cloud size based on simple considerations of electrostatics. A large number of measurements in different geographical locations show considerable variability in the relationship between instantaneous total lightning flash rate and radar cloud-top height, but on average, Vonnegut's predicted fifth-power dependence is strongly upheld for midlatitude convection. Figure 13.6 summarizes these average results. More recent results may be found in Ushio et al. (2001) and Boccipio (2001). For active clouds exhibiting lightning at rates of several flashes per minute, the maximum cloud top is closely in phase with the peak flash rate as first noted in the Thunderstorm Project report (Byers and Braham 1949). Figure 13.7 from Williams (1985) shows the evolution of these two quantities for a deep (17–18 km) Florida thunderstorm whose cloud top was continuously monitored with a rapidly scanning radar in vertical cross section. Here it is seen that the fifth

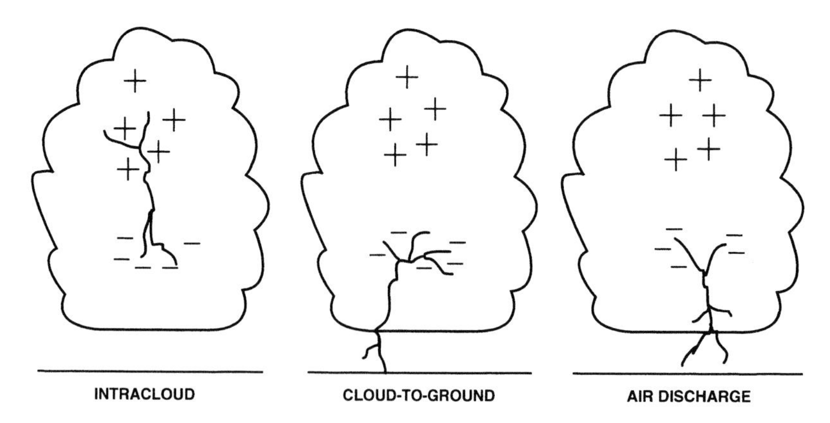

FIG. 13.5. Illustration of common lightning types in ordinary thunderstorms: (a) intracloud flash, (b) negative cloud-to-ground flash, (c) air discharge (which may involve a lower positive charge center).

TABLE 13.1. Comparisons of total flash rate: ordinary thunderstorms and severe storms.

Investigators: Location	Method	Maximum flash rate
Ordinary nonsevere thunderstorms		
Lhermitte and Krehbiel (1979): Florida	LDAR system	60 min^{-1}
Lhermitte and Williams (1985): Florida	Field change meter	4 min^{-1}
Dye et al. (1986): Montana	Aircraft	1 min^{-1}
Krehbiel (1986): Florida	Field mill	<1 min^{-1}
Nisbet (1990): Florida	LDAR; field mills	16 min^{-1}
Goodman et al. (1988): Alabama	Field change meter	22 min^{-1}
Williams et al. (1989b): Alabama	Corona points	6 min^{-1}
Buechler and Goodman (1990): Alabama	Field change meter	15 min^{-1}
Williams (1991a): Australia	Field change meter	40–60 min^{-1}
Malherbe et al. (1992): Florida	Interferometer (mean of several storms)	14 min^{-1}
Carey and Rutledge (1996): Colorado	Field change meter	10 min^{-1}
Rison et al. (1996): Florida	LDAR system	3 min^{-1}
Stanley et al. (1996): Florida	LDAR system	4 min^{-1}
Severe storms		
Blackwell-Udall tornado storm, Gunn (1956)	Field mill	12 min^{-1}
Worcester, Massachusetts, tornadic storm, Vonnegut and Moore (1958)	Visual	600 min^{-1}
Oklahoma storms, Silberg (1965)	10 KHz–30 MHz sferics	>600 strokes min^{-1}
Colorado hailstorms, Blevins and Marwitz (1968)	Visual	70 min^{-1}
Oklahoma tornadic storms, Taylor (1973a)	1 MHz sferics	>20 min^{-1}
Severe hailstorm, Pakiam and Maybank (1975)	Field mill	42 min^{-1}
Binger tornado storm, MacGorman & Nielsen (1991)	Lightning radar echoes	14 min^{-1}
Huntsville, Alabama, hailstorm, Williams (1990, unpublished)	Visual	50 min^{-1}
Florida giant, Williams (1985) and M. Brook 1980, personal communication	Field change meter	30 min^{-1}
Orlando, Florida, hailstorm, March, 1992 (M. Isaminger 1996, personal communication)	Lightning Position and Tracking System	35 ground strokes min^{-1}
Huntsville, Alabama, tornadic storm, Goodman and Knupp (1993)	Video camera	>40 min^{-1}
Oklahoma tornadic storm, Buechler et al. (2000)	Optical Transient Detector (satellite)	40–60 min^{-1}
Arkansas hailstorm, Westcott et al. (1996)	Video camera	230 strokes min^{-1}
Colorado hailstorm, Carey and Rutledge (1998)	Flat plate antenna	60 min^{-1}

power of the measured radar cloud top is a good predictor of the total flash rate throughout the storm lifetime. As in Fig. 13.7, the tendency for the lightning rate to maximize at the time of peak cloud height has also been found in the Tropics (Frost 1954; Williams 1991a).

The strong association between thunderstorm depth and electrical vigor in specific geographical locations and meteorological regimes is attributable to the gravitational energy of large ice particles in the upper regions of the storm. Such ice particle accumulations are also positively correlated with CAPE and the updraft speed (Williams et al. 1992; Rutledge et al. 1992; Zipser 1994; Petersen et al. 1996).

Shackford's (1960) early studies of the relationship between lightning activity and radar cloud-top height showed great variability. Although some of this variability is attributable to sampling problems with these early observations (Williams 1996), Shackford recognized a more fundamental problem of nonuniqueness: Two storms with equal tops but different vertical profiles of radar reflectivity exhibit different maximum flash rates. This nonuniqueness problem was remedied to some extent by Larsen and Stansbury (1974) and Marshall and Radhakant (1978), who suggested a different radar measurement of cloud size to relate with electrical activity: the area within the 30-dBZ reflectivity contour at 7-km altitude. This measure also gives emphasis to the region in altitude believed to be important to thunderstorm charge separation by ice-particle collisions. Observations of sferics rate versus *Larsen area* (the name given to this measure by Marshall and Radhakant 1978) show less scatter than Shackford's earlier cloud top–lightning plots. More recently, Carey and Rutledge (1996) have measured echo volume greater than 30 dBZ in the mixed-phase region and show excellent correlations with the total flash rate.

The nonunique relationship between radar cloud-top height and lightning flash rate has been demonstrated more recently in studies of tropical convection (Rutledge et al. 1992; Williams et al. 1992; Zipser 1994; Watson et al. 1995; Petersen et al. 1996) where differences in meteorological regime are very influential in the distribution of buoyancy and radar reflectivity. Thunderstorms embedded in widespread mon-

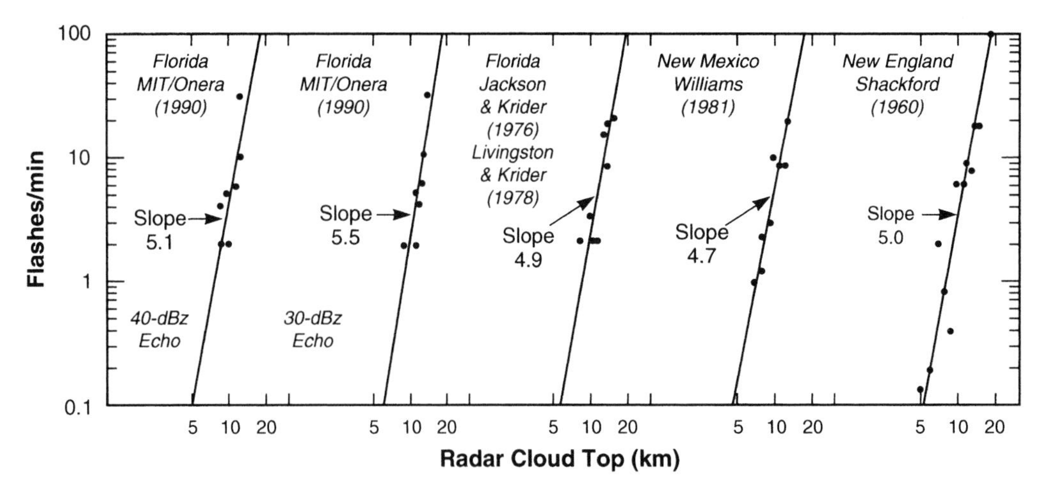

FIG. 13.6. Observations of total flash rate vs radar cloud-top height [From Williams (1985) and Malherbe et al. (1992).]

soonal precipitation were found to exhibit peak flash rates an order of magnitude less than more vigorous storms more removed from the ITCZ in the more unstable "break period" regime. These comparisons underscore the need for caution when single-parameter relationships are used to predict lightning activity.

The relative prevalence of intracloud (IC) and ground flashes (CG), the IC/CG ratio, also has a strong tendency to increase with thunderstorm size and electrical activity. Although minute-to-minute values show considerable variability, 15-minute averages of IC/CG ratio are seen to increase systematically with total flash rate in Fig. 13.8. These results are replotted from Cheze and Sauvageot (1997). The upward trend is consistent with earlier results in the Tropics reported by Rutledge et al. (1992), who interpreted their findings as a higher probability for intracloud flashes as the altitude of the charge separation process and region of high electric field rises to regions of weaker dielectric strength in the atmosphere. A possible additional explanation for the infrequent occurrence of ground flashes in the Tropics is the narrow nature of the clouds relative to their heights. Laterally extensive charge regions are more likely to provide the electrostatic energy necessary to bridge the long gap to ground. The relationship between lightning type and

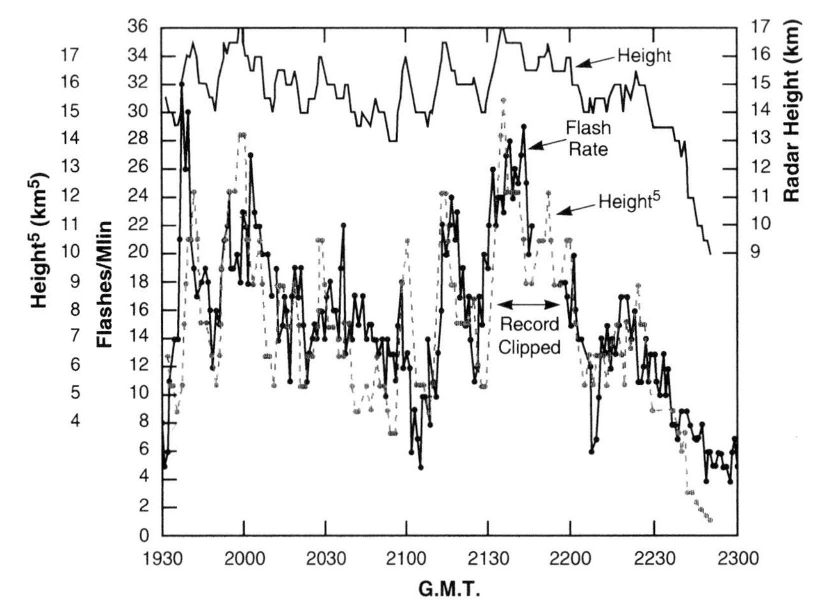

FIG. 13.7. High time resolution (1-minute update) observations of total flash rate and radar cloud-top height for a deep Florida thunderstorm [From Williams (1985), M. Brook 1980, personal communication.]

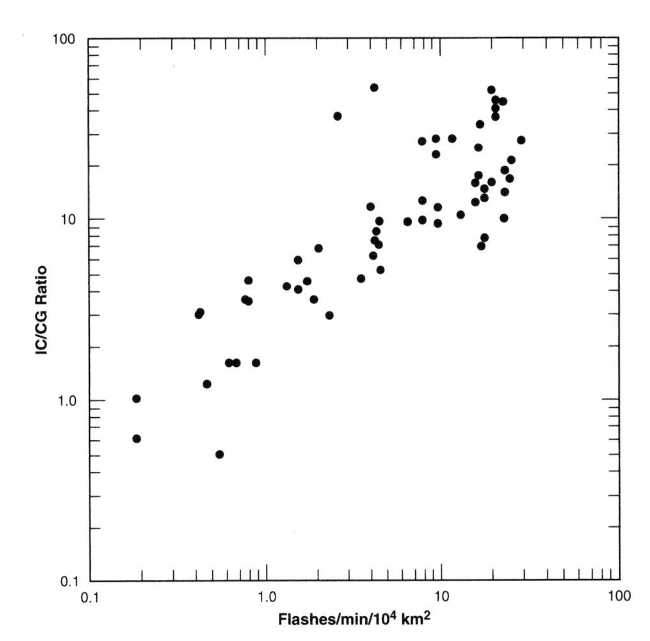

Fig. 13.8. The behavior of the intracloud/ground flash ratio vs total lightning flash rate density. (Replotted data from Cheze and Sauvageot 1997.)

the convective state of thunderstorms is better understood when both updrafts and downdrafts are considered.

Microbursts are a common feature of nonsevere thunderstorms, and occur when the strong updraft in the developing phase transitions to strong downdraft at the end of the mature phase, as discussed in the description of the Thunderstorm Project (Byers and Braham 1949). These events have presented a grave concern to commercial aviation (Fujita 1985). Episodes of high lightning rate 5–10 minutes prior to maximum microburst outflow velocity have been identified (Goodman et al. 1988; Williams et al. 1989b; Malherbe et al. 1992; Williams et al. 1995; Carey and Rutledge 1996; Hoffert and Pearce 1996). The precursory peak in lightning is identified with the updraft and the accumulation of large ice particles in the mixed phase region, which invigorate the process of charge separation illustrated earlier in Fig. 13.3. The subsequent descent and melting of this precipitation contributes to the downdraft that makes the microburst. Ground flash activity typically lags behind the initial peak in intracloud activity and has been associated with the descent of graupel and small hail through the melting level (Carey and Rutledge 1996).

Thunderstorm microbursts have now been studied with Doppler radar in several geographical areas by MIT Lincoln Laboratory. Figure 13.9 summarizes the results for maximum outflow velocity in Huntsville, Alabama; Denver, Colorado; Kansas City, Missouri; and Orlando, Florida (M. Isaminger 1996, personal communication). Given the severe storm threshold for wind speed of 50 knots (26 m s^{-1}), it is seen that only a small fraction of thunderstorm microbursts are severe.

13.3. Severe storms

a. Physical relationships useful in interpreting severe storm behavior

The relationships between lightning and other storm parameters such as cloud size, updraft speed, thermodynamic soundings, and environmental wind are complex for ordinary thunderstorms. These relationships are increasingly complex for severe storms, whose physical parameters lie in the tails of the respective distributions and which are very poorly sampled in the observations in comparison with ordinary thunderstorms. A severe storm is also born from a special combination of initial conditions and no one parameter is likely to be (nor has been shown to be) a reliable predictor of storm severity. Nevertheless, an improved understanding of relationships between lightning and other manifestations of severe storms, notably hail and tornadoes, can be achieved by considering simple physical relationships and the predominant dependence on one (or a small number) of parameters. Despite the substantial variance in all of the relationships, their scrutiny can aid in the identification of observations still to be made in improving operational forecasts.

1) Lightning flash rate and cloud size

Lightning is the central topic in this article and so it is appropriate to begin with this variable. Research has shown that the most important single parameter in controlling the total flash rate of a thunderstorm is the cloud size (Williams 1985). As noted in section 13.2, cloud size may be measured with radar as the radar cloud-top height (Shackford 1960; Williams 1985; Petersen et al. 1996), or as an area enclosed within an arbitrary reflectivity contour at a specific altitude (Larsen and Stansbury 1974; Marshall and Radhakant 1978), or as the thresholded echo volume in the mixed phase region (Carey and Rutledge 1996). Cloud size may also be estimated thermodynamically as the parcel equilibrium level, with important input from the wet-bulb potential temperature of boundary layer air. Unfortunately, little attention has been given to relating the radar and thermodynamic measures of cloud size. Thermodynamic issues will be discussed later in this section.

The strong average dependence of flash rate on measured radar cloud-top height for ordinary thunderstorms was summarized in Fig. 13.6. Though some important exceptions occur for hailstorms, as will be discussed, severe thunderstorms are generally larger

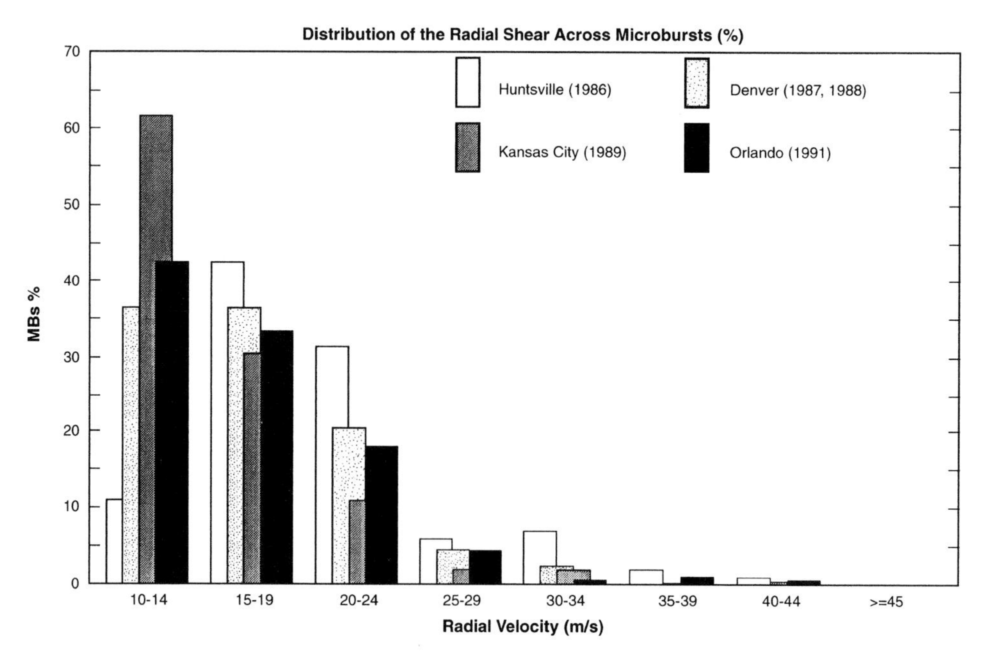

FIG. 13.9. Distribution of maximum differential radial velocity associated with surface microbursts in Denver, CO; Huntsville, AL; Kansas City, MO; and Orlando, FL (MIT Lincoln Laboratory data). For reference on the abscissa, the severe storm wind threshold is 29 m s^{-1}.

and deeper than ordinary nonsevere thunderstorms (Darrah 1978). We have no a priori reason to doubt that the strong fifth-power height dependence manifested in Fig. 13.6 will cease to be valid for the deep severe storms that have crossed some arbitrary threshold [e.g., ice particle sizes > ¾ in. (1.9 cm) diameter]. Nor is there reason to believe that the strong dependence of lightning rate on Larsen area (Marshall and Radhakant 1978) will cease to be valid in the realm of severe storms. The evidence that the total flash rates for severe storms are significantly larger, on average, than for ordinary thunderstorms (Table 13.1) supports this latter assertion.

The results for total flash rate in Table 13.1 are not exhaustive, but probably represent the majority of documented cases in the literature. Additional recent results on severe weather cases in Florida using the Lightning Detection and Ranging (LDAR) system at Kennedy Space Center (Williams et al. 1999) are included in Table 13.2. The latter results are likely the most reliable single set on severe weather because the same proven instrument was used in every case. These peak flash rates contrast sharply with those in nonsevere thunderstorms (Table 13.1). Several reasons account for the sparsity of reliable observations of total flash rate in severe weather in general. The diurnal climatology of severe weather (Doswell 1985) shows a maximum during daytime. Reliable visual observations of total flash rate are impossible, and eyewitness accounts of a frequent absence of lightning in tornadic storms (H. Bluestein 1996, personal communication; J. Wurman 1996, personal communication) are probably often attributable to the preponderance of intracloud lightning, which is completely masked to the naked eye by background sunlight. This general assertion is supported by recent observations of daytime supercells with an optical lightning detector (C. Doswell 1997, personal communication) indicating frequent intracloud activity.

A second reason for the shortage of total lightning observations is the necessity for storm proximity with instruments conventionally used to determine flash rates. When storms are strongly migratory in baroclinic environments, the deployment of instruments to follow the evolution of storm electrical activity is problematic. When the flash count is dominated by intracloud lightning, an expected behavior for severe

TABLE 13.2. Severe Florida storms: Recorded with the LDAR system at the NASA Kennedy Space Center (Williams et al. 1999).

Date	Maximum total flash rate* (flashes min^{-1})	Severe manifestation
23 April 1997	195	Tornado; hail
22 May 1997	290	1 in. hail
13 June 1997	410	Nickel hail
7 July 1997	425	Golf ball hail
11 July 1997	170	Tornado
23 August 1997	300	Wind

* The LDAR flash rate is determined by grouping LDAR sources (verified by two independent arrays of radio receivers) into flashes with the following simple rules: a new flash is declared if either 1) 300 ms elapses since the previous LDAR source or 2) the new LDAR source is greater than 5000 m from the previous source.

storms based on the results shown in Fig. 13.8, greater instrument sensitivity is required. Previous studies indicate that the electromagnetic signals from the intracloud flashes prevalent during high flash rate periods are weak in comparison with the usual cloud-to-ground flash (Taylor et al. 1984). Figure 13.10 shows the evolution of a hailstorm near Huntsville, Alabama, as observed by the MIT C-band radar and visually by the author. Times of intracloud and cloud-to-ground flashes are shown. The peak flash rate was largely dominated by small discharges in the upper anvil of the storm, which were detected visually (the storm occurred just after dusk) but only marginally so with ground-based corona point measurements (Williams et al. 1989b). The tendency for the total lightning activity to follow the height of the cloud is also apparent in Fig. 13.10.

Cloud-to-ground flash activity is now measured effectively on a routine basis with ground-based networks (Orville et al. 1985; Watson et al. 1995; Cummings et al. 1998). Holle and Maier (1982) have related ground flash rate to radar cloud-top height for ordinary thunderstorms. On average, the ground flash rate increases rapidly with storm size. However, the tendency for ground flash activity to decline at very high flash rates (Lang et al. 2000) makes extrapolation of these results to deep severe storms suspect. The physical reasons for this apparent change in lightning behavior with cloud depth may be related to electrostatic structure of thunderstorms discussed in greater detail below.

The prediction for cloud size (the vertical dimension) based on parcel theory is illustrated in the thermodynamic diagram in Fig. 13.11. The level at which the parcel buoyancy vanishes is the *level of neutral buoyancy* (LNB), though in principle the parcel may rise further to level 5 in Fig. 13.11. The wet-bulb potential temperature, which labels the wet bulb adiabat on the basis of the boundary layer properties of temperature and moisture, is a fundamental determinant of both the level of neutral buoyancy (LNB) and the overall profile of parcel buoyancy. In ordinary nonsevere thunderstorms with narrow radar updrafts, entrainment is a dominant process, and parcel overshoot of the LNB is unusual. The actual cloud tops on any given day show considerable scatter around the LNB (Cruz 1973), but with the majority of tops below this level.

Given the importance of cloud depth in lightning production, it is useful to consider the distribution of thunderstorms with size. Figure 13.12 shows the distribution of radar echo tops observed in the Thunderstorm Project in Ohio. The general behavior appears to be exponential, with the possible exception of the last data point representing storms exceeding tropopause height and showing a downturn relative to exponential behavior. The great majority of thunderstorms are small and nonsevere. The majority, but by no means all, of severe storms are in the tail of the size distribution and account for a tiny fraction of all thunderstorms.

Given the importance of wet-bulb potential temperature (θ_w) in influencing cloud size and parcel buoyancy, it is also useful to consider the distribution of observed θ_w values prior to thunderstorms (nonsevere and severe) based on U.S. Weather Bureau (1947)

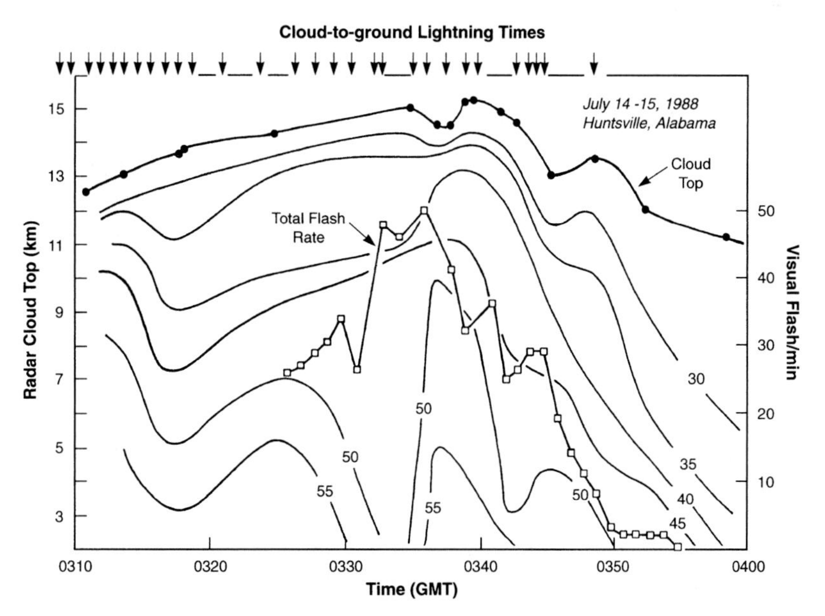

FIG. 13.10. Observations of lightning flash rate (intracloud and cloud to ground) and the vertical distribution of radar reflectivity for an electrically active hailstorm near Huntsville, AL.

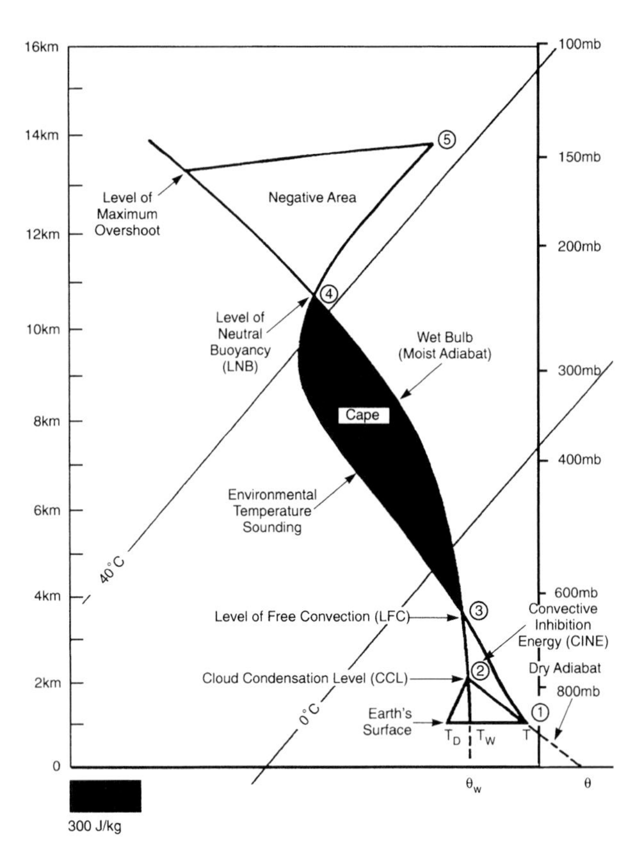

FIG. 13.11. Illustration of basic aspects of conditional instability on a thermodynamic diagram.

thunderstorm rainfall data for July 1942. As shown in Fig. 13.13, the most probable value for this distribution is near 23°C. It is important to note that enhancements of θ_w amounting to just a few degrees are

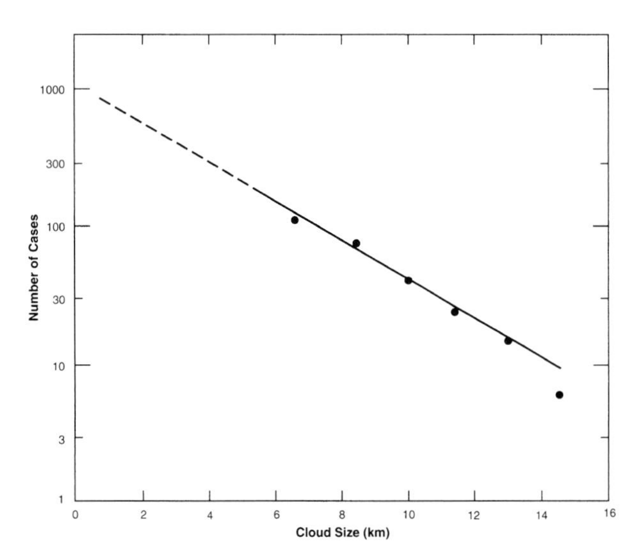

FIG. 13.12. Distribution of thunderstorm cloud top heights from the Thunderstorm Project report (Byers and Braham 1949).

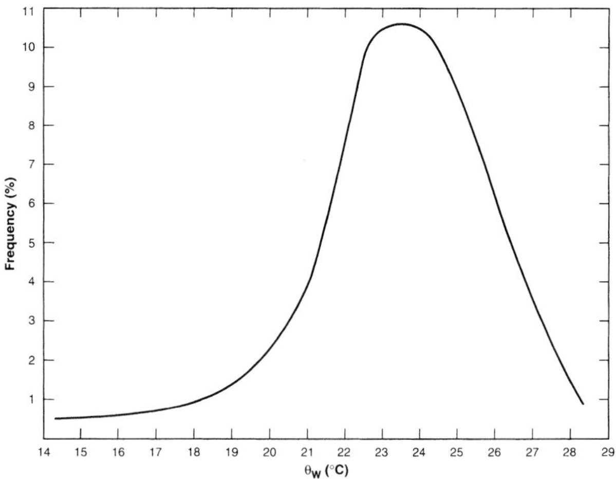

FIG. 13.13. Distribution of screen level wet bulb potential temperature values prior to U.S. thunderstorms shown on the 1630 EST weather maps in July 1942 (from U.S. Weather Bureau 1947).

associated with dramatic increases in lightning activity and storm severity. The composite sounding for supercell thunderstorms in Bluestein and Jain (1985) is about 24°C. The θ_w value for boundary layer air ingested by the Plainfield F-5 tornado storm (Seimon 1993a,b; Doswell and Brooks 1993) exceeded 27°C. The wet bulb adiabats associated with 4 in. (10 cm) diameter hail in Fawbush and Miller (1953) have θ_w values in the range of 27°–28°C. Understandably, the latter values lie in the tail of the distribution shown in Fig. 13.13 and generally occur as the result of pronounced capping inversions (Carlson and Ludlam 1968) that strongly isolate boundary layer air from air aloft.

When storms are severe and cloud buoyancy is the dominant updraft driver, observations (Bluestein et al. 1989; Zipser 2001) suggest that the parcel theory predictions for velocity are more accurate than in the nonsevere case. The tilted updraft, the broader updraft, and the rotating updraft may all play roles in this result. In such cases, parcels overshoot the level of neutral buoyancy and often the tropopause as well (Vonnegut and Moore 1958; Roach 1967), and the parent storm can therefore exhibit substantially higher maximum tops. Bonner and Kemper (1971) and Darrah (1978) have studied the relationships between severe weather and echo top height (but, unfortunately, not lightning) in the United States. In both studies, the probability for severe weather is found to increase exponentially with cloud-top height and with tropopause overshoot. Figure 13.14 shows replotted results from Darrah (1978) to illustrate the pronounced exponential behavior. The *e*-folding scale height for severe storm probability is approximately 2.7 km. Unfortunately, the dependence of actual severity for the severe weather (i.e., hailstone size, tornadic intensity) on cloud-top height was not studied. The impor-

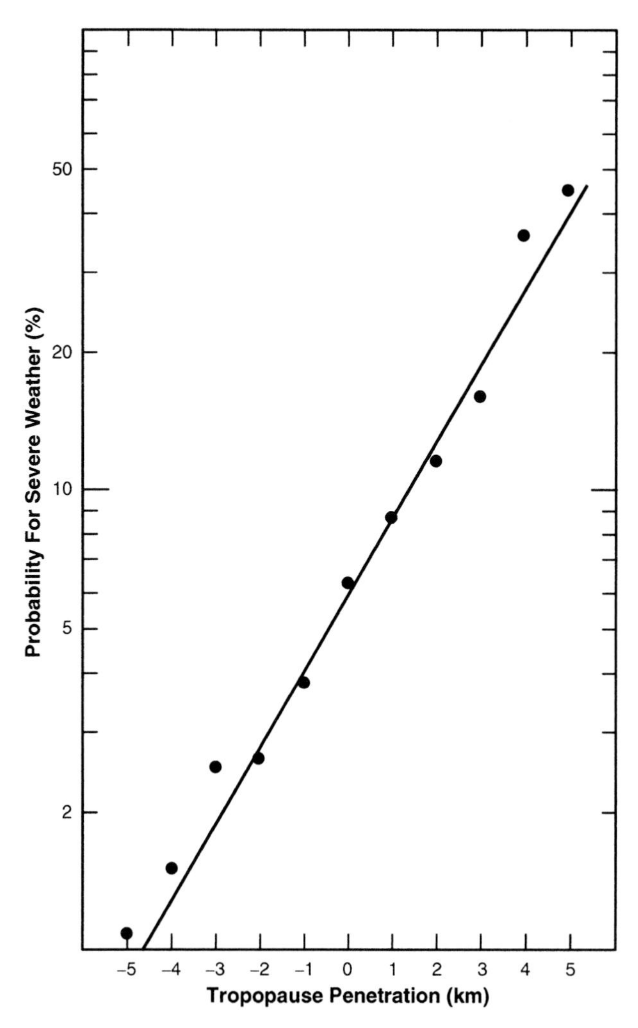

FIG. 13.14. Percentage of U.S. storm cases in the severe category versus the estimated tropopause overshoot in kilometers. (From Darrah 1978.)

tant point in the lightning context is the following tentative conclusion: If both total lightning and the probability for severe weather increase strongly with cloud size, then there should be some prospect for using total lightning as an additional useful indicator of severe weather. Recent evidence in support of this expectation is found in Williams et al. (1999).

It is important here to contrast again the behavior of weather at midlatitude and in the Tropics. The author's impression, based on radar observations of thunderstorms in tropical Australia, was that cloud-top heights were generally greater than at midlatitude, but tropopause overshoots were much less common. This behavior is more likely attributable to the upright updrafts in the barotropic tropical atmosphere than to insufficient CAPE, although the "shape" of the CAPE is also different (Williams 1995). The higher tropical tropopause may also be a factor. Severe weather is notably less prevalent in the Tropics (Ludlam 1963; see also chapter 10 of this monograph).

Observations support the idea that the gross dipole structure of clouds is also related, on average, to cloud depth. The trends with cloud size are illustrated in Fig. 13.15. Williams (1989b) noted systematic differences in the altitude (and in situ temperature) of main negative charge in small New Mexico thunderclouds and the larger Florida storms. Marshall et al. (1995) and more recently Stolzenburg (1996) have shown, on the basis of balloon soundings of electric field, that the main negative charge region in supercells in the Great Plains is higher still. A similar effect has been found in a model study of a tornadic thunderstorm (Ziegler and MacGorman 1994). These results may be fundamental in explaining the behavior of intracloud and cloud-to-ground lightning for ordinary thunderstorms shown in Fig. 13.8, as well as the extrapolation of this trend of relatively greater numbers of intracloud flashes in severe storms, as noted by MacGorman et al. (1989), Williams et al. (1999) and also evident in Ray et al. (1987). A gross storm dipole at higher altitude, simultaneously farther from the ground and with greater vertical extension into the dielectrically weaker portion of the upper troposphere, is consistent with a systematic increase in intracloud lightning.

Though more speculative, the trend in Fig. 13.15 may also provide for a more intense region of lower positive charge (Williams 1989b) in the extraordinarily deep severe storms as compared to the situation in ordinary thunderstorms simply because of the deeper zone of ice particles beneath the main negative charge. This point relates to one possible hypothesis for positive ground flash production in severe storms discussed in section 13.3c.

An apparent inconsistency with the trend shown in Fig. 13.15 is the tendency for the height of main negative charge within ordinary thunderclouds to remain at nearly constant altitude throughout their electrical lifetimes (Krehbiel 1986; Krider 1989; Murphy et al. 1996), when the cloud size is changing. The latter result is based primarily on analyses of the field changes due to lightning, but the reasons for the inconsistency are still not understood.

2) HAIL AND UPDRAFT SPEED

Probably the most important determinant of maximum hail size in severe storms is updraft speed, and despite the crucial importance of the latter parameter for both moist convection and the redistribution of water substance in the general circulation of the atmosphere, it is one of the most difficult to observe directly. The particle balance level was emphasized as a systematic aspect of graupel particle growth in ordinary thunderstorms (Atlas 1966; Lhermitte and Williams 1985) in an earlier section, and this carries over directly to the larger hailstones of severe storms

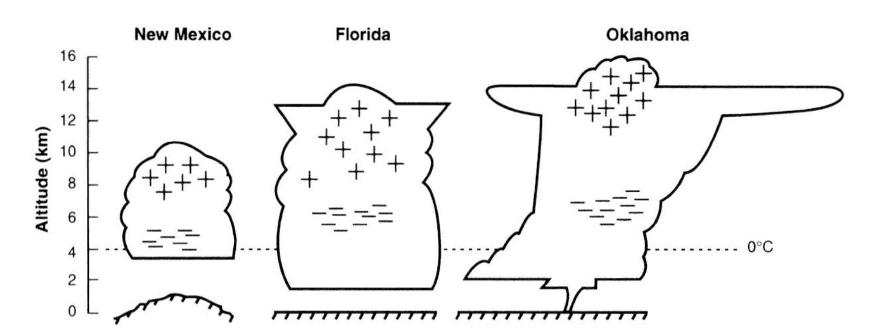

FIG. 13.15. Coarse thunderstorm charge distributions vs cloud size based on balloon soundings of electric field and measurements of the field changes at the ground accompanying lightning discharges. (a) New Mexico, (b) Florida, (c) Oklahoma.

as illustrated by the equilibrium condition in Fig. 13.16. Perhaps the best evidence for a systematic connection between hail size and updraft speed is found in the observations of Witt and Nelson (1991), replotted in log–log coordinates in Fig. 13.17. These authors had previously fit a linear relationship between maximum hailstone size and cloud-top divergence. Figure 13.17, however, suggests a power law slope closer to 2 than to 1, and hence a quadratic relationship. A doubling of updraft speed will quadruple the maximum hailstone size. The latter connection is expected for growth in a balance condition between updraft speed and terminal fall speed, as shown in Fig. 13.16. (The linearity between cloud-top divergence and updraft speed is assured kinematically so long as the cloud geometry is insensitive to convective strength.)

Additional factors are no doubt at work in influencing hailstone size, most notably updraft persistence and width (Nelson 1983), both of which are favored in baroclinic environments, and the updraft profile in relation to the mixed phase region and supercooled water supply that is the basis for hailstone growth. The systematic nature of the results in Fig. 13.17, however, points to one dominant control parameter.

It is important to emphasize the possible regional aspect of Witt and Nelson's (1991) result as these observations were obtained exclusively in the Great Plains of the United States. It would be valuable to organize similar results for regions characterized by more barotropic conditions (i.e., Florida in the United States), where the incidence of large hail on the ground is notably less frequent.

Following the idea that the sizes of precipitation ice particles are strongly influenced by draft speed, it stands to reason that storms producing large hail also produce relatively larger amounts of smaller graupel. Storms with very strong drafts are also likely to contain zones of more moderate updraft (away from

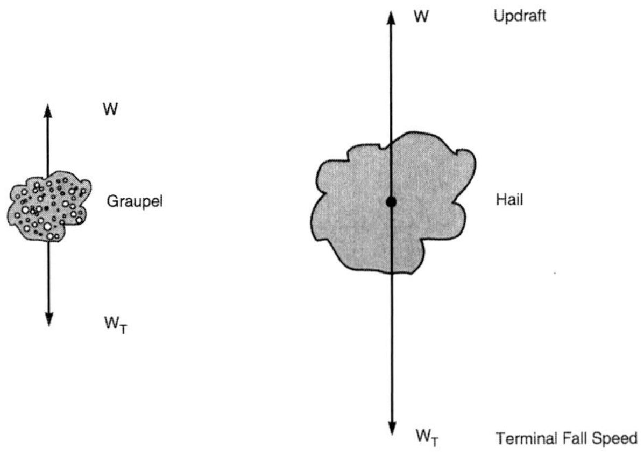

FIG. 13.16. Illustration of balance level condition for (a) graupel growth in ordinary thunderstorms and (b) hailstone growth in severe hailstorms.

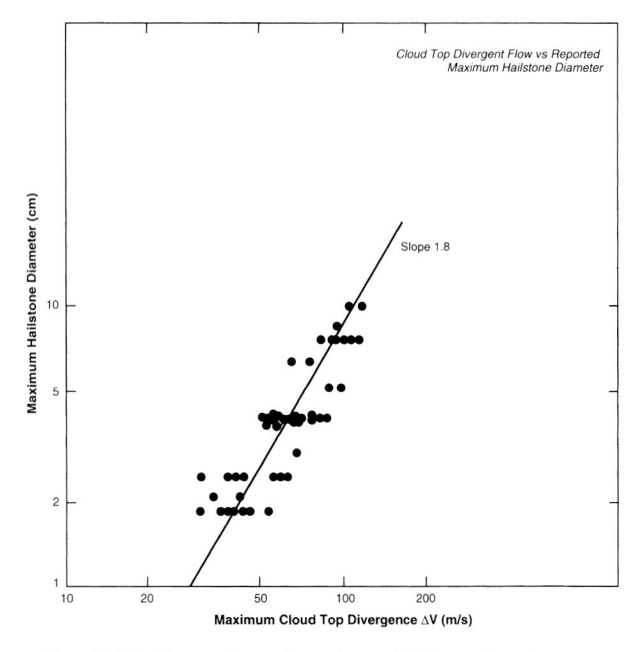

FIG. 13.17. Observations of maximum hailstone diameter versus cloud top differential radial velocity. (From Witt and Nelson 1991.) The line of best fit has a power-law slope of 1.8.

the updraft core) in which these smaller ice particles can grow. Nearly all parameterizations of particle size distributions, derived from observations, show that enhancements in the large particle "tail" are accompanied by greater numbers of smaller particles. In the case of large hail, the larger particles separate quickly from the smaller particle population, and in the presence of horizontal air currents prevalent in severe storms, substantial separations in precipitation type at ground level are common (Ludlam 1963; Doswell et al. 1990; Doswell and Burgess 1993).

An important consideration for charge separation by ice particle collisions is the surface area of the particles, which limits their electrical charge carrying capacity. According to the simple mechanism for differential charge separation illustrated in Fig. 13.3, the faster-falling larger particles are fundamental to this process. The largest hailstones grown in the strongest updrafts will have the largest fall speeds, but they are unlikely to be major players in the overall charge separation because of their minuscule integrated surface area. Simple comparisons on this point for large hail and for the graupel particles believed to be of greatest importance to charge separation in both nonsevere and severe storms are shown in Table 13.3. The integrated graupel surface area is one to three orders of magnitude greater than that of the large hail.

Also shown in Table 13.3 are calculations for radar reflectivity for these two distinct populations of particles. The reflectivity calculations assume a sixth-power dependence of particle radar cross section (a result strictly valid only for long radar wavelengths in the case of large hail) and a wet surface (i.e., dielectric constant for liquid water). These results emphasize how a few large particles can completely dominate the radar response and mask the information about particle surface area. The results in Table 13.3 also point up the nonuniqueness between particle concentration and size in determining radar reflectivity, which thwarts unambiguous radar identification of hail (Burgess and Lemon 1990). Recent improvements have come from multiparameter radar data analyses (Bringi et al. 1986; Balakrishnan and Zrnić 1990; Zrnić et al. 1993).

What determines maximum updraft speed in storms? As noted earlier, in ordinary thunderstorms in barotropic environments (i.e., nearly vertical updrafts of modest width), the parcel theory predictions for vertical velocity are greatly exaggerated. In severe storms for which cloud buoyancy is the main cause for upper-level updraft and the behavior is nondilute adiabatic (Heymsfield and Musil 1982; Heymsfield and Hjelmfelt 1984; Musil et al. 1986), the prediction of parcel theory

$$W_{\max} = \sqrt{2\,\mathrm{CAPE}} \tag{13.1}$$

is more accurate (Roach 1967; Bluestein et al. 1989); Zipser, E. J., 2002: Some views on "hot towers" after 50 years of tropical field programs and two years of TRMM data. Submitted to monograph honoring Joanne Simpson) and the vertical draft speeds are substantially greater. If parcel theory is assumed valid, the penetrative cloud tops observed by Darrah (1978) can be related to vertical air speed and CAPE. (Hail was the manifestation of severe weather most sensitive to penetrative overshoot in the latter study.)

For overshoot into the isothermal atmosphere, 20 m s^{-1} of vertical motion is required for each kilometer of stratospheric penetration (Vonnegut and Moore 1958). Combining this with Eq. (13.1) leads to a connection between CAPE and tropopause overshoot, shown in Fig. 13.18. The ranges of CAPE values (0–5000 joule kg^{-1}) and tropopause overshoots (0–5 km) are in line with observations, but no pairs of points are available in the observations to test this relationship.

The updraft profile in convective storms often increases monotonically from low to midlevels and attains a maximum value high in the storm, often close to cloud top. Results from a vertically pointing Doppler radar measurement in a New Mexico (nonsevere) hailstorm are shown in Fig. 13.19. This behavior is not unexpected; large undilute air parcels are positively buoyant at midlevels and continue to accelerate upward through higher levels of the cloud, as predicted by Eq. (13.1). Because of the richer supply of liquid water at lower levels, within the traditional mixed phase region bounded by the 0° and −40°C isotherms, the particle balance level found by typical hailstorms may often be near −25°C (Nelson 1983) and not in the upper part of the storm where the updraft is stronger (note trajectory in Fig. 13.20). In severe storms mentioned earlier, cloud tops overshooting both the equilibrium level and the tropopause extend 4–8 km higher than the −40°C isotherm and exhibit active

TABLE 13.3. Physical quantities for monodisperse particle distributions.

	Formula	Cloud droplets	Graupel		Hail	
Particle radius (m)	a	10^{-5}	10^{-3}	10^{-2}	10^{-2}	6×10^{-2}
Particle concentration (m^{-3})	N	10^{8}	10^{3}	10^{0}	1.6×10^{-2}	1.6×10^{-5}
Water content ($g\ m^{-3}$)	$N\ \frac{4}{3}\pi a^3 \rho$	0.42	4.2	4.2	0.07	0.01
Specific surface area of particles ($m^2\ m^{-3}$)	$N\ 4\pi a^2$	1.3×10^{-1}	1.3×10^{-2}	1.3×10^{-3}	2.0×10^{-5}	7.2×10^{-7}
Reflectivity factor ($mm^6\ m^{-3}$)	$N\ (2a)^6$	−22 dBZ	48 dBZ	78 dBZ	60 dBZ	77 dBZ

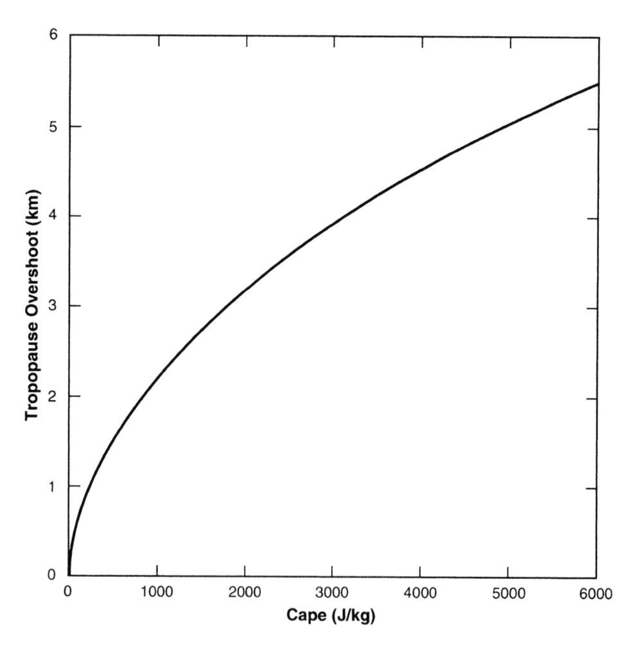

FIG. 13.18. Parcel theory calculations of CAPE vs maximum tropopause overshoot, assuming the level of neutral buoyancy is coincident with the tropopause.

flashing behavior in these very cold regions. The question about supercooled water remains an important but still unresolved issue. The importance of supercooled water pertains to both accretional growth of graupel and hail and to the process of charge separation by ice particle collisions. For reasons that may be related to ice particle growth state (Baker et al. 1987; Williams et al. 1991) but that are very poorly understood in detail, laboratory workers systematically find that the presence of supercooled water is associated with an order-of-magnitude enhancement in the charge separated when ice particles collide (Reynolds et al. 1957; Takahashi 1978; Saunders et al. 1991).

Aircraft observations of cloud-top overshoots in severe storms by Fujita (1992) provide some evidence that glaciation is taking place within the turret, thereby requiring the existence of initial liquid water. Additional evidence for the presence of liquid water at considerable height in severe convection originates from observations by Lemon (1998) of three-body scattering by hail (Zrnić 1987; Wilson and Reum 1988) whose amplitude may require the presence of liquid water. More recently, Rosenfeld and Woodley (2000) have documented supercooled water at −37.5°C in convective clouds.

In studies in the Great Plains, Fawbush and Miller (1953) found the largest values of CAPE associated with the largest hailstones (see also Williams 1995, pp. 42–44). Likewise, Seimon (1993b) associated 3 in. (7.6 cm) diameter hail with an extraordinary value for CAPE (5900 joule kg^{-1}; see also Doswell and Brooks 1993) in the Plainfield, Illinois, tornadic storm. The measured wet-bulb potential temperature of boundary layer air used in these respective computations, the single most important parameter in determining CAPE (Williams and Renno 1993), was also extraordinarily large (28.7°C). These results indicate that the wet-bulb potential temperature deserves greater consideration in the severe storms context.

The diurnal variation of hail and hailstorms showing a maximum in the late afternoon (Anthony 1994) supports a dominant role for conditional instability and CAPE, provoked by solar radiation, as the primary mechanism for updraft production and hailstone growth. Methods for forecasting maximum hailstone size based on thermodynamic soundings that originated with Fawbush and Miller (1953) often performed poorly (Doswell et al. 1982). Revised methods have been more effective (Moore and Pino 1990).

In contrast with these results, studies in Florida in a baroclinic environment in March and April have shown the occurrence of large hail with rather modest values of CAPE (Lascody and Sharp 1996). In such cases it seems likely that large updrafts were still present in the region of hail growth, but that contributions to updraft additional to cloud buoyancy were present. In strongly sheared baroclinic environments, the air entering the updraft region has already achieved a speed of a few tens of meters per second (see Fig. 13.27). If the updraft is doubled in strength on account of contributions other than buoyancy, the maximum hailstone size is quadrupled, as noted earlier. Unfortunately, in the strongly baroclinic environments in which large hail is most prevalent, the CAPE is difficult to evaluate with single thermodynamic soundings (Williams 1991b).

The melting process often provides a satisfactory explanation for the absence of large hail at the ground beneath vigorous convective storms. Figure 13.2 shows a synthesis of melting behavior with height

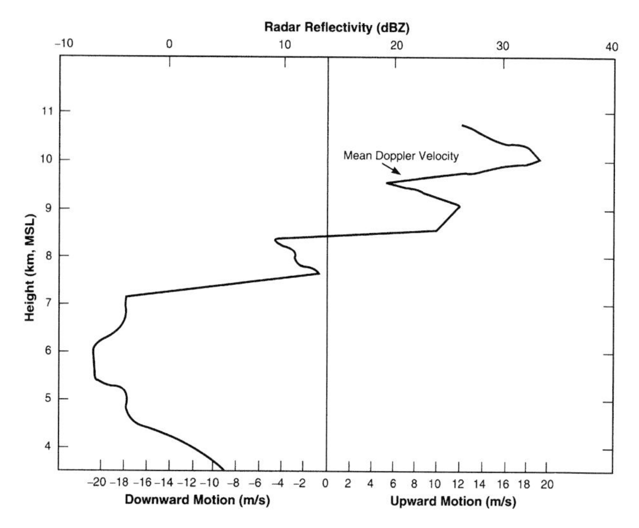

FIG. 13.19. Mean Doppler velocity vs altitude in a New Mexico hailstorm at Langmuir Laboratory. (From Williams 1981.)

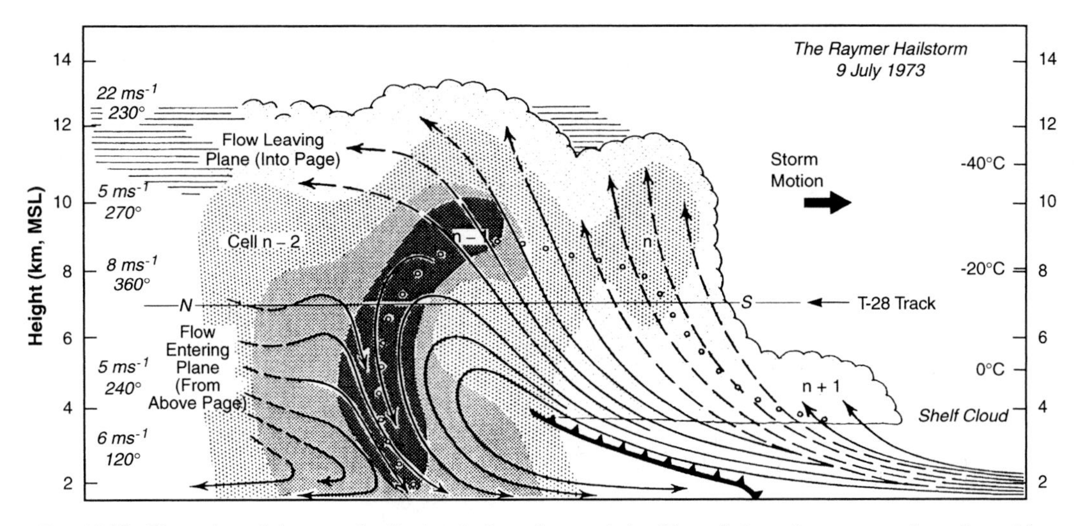

FIG. 13.20. Illustration of the strongly tilted updraft, a characteristic of baroclinic environments and one favorable to the growth of large hail. (From Browning et al. 1976.)

based on the work of Rasmussen and Heymsfield (1987). Following the earlier suggestion by Fawbush and Miller (1953), the height of the 0°C isotherm in the wet-bulb rather than the dry-bulb temperature was taken as the parameter of greatest importance in predicting the presence of large hail on the ground. For hail falling within cloudy air, however, the wet-bulb and dry-bulb melting levels are identical, and this has been assumed in the calculations summarized in Fig. 13.2. It is interesting to note that hailstones of "severe" size (1.9 cm diameter) at the melting level would only barely survive a fall of 4.5 km (typical melting level height in Florida summer thunderstorms) without complete melting, whereas stones of slightly larger diameter (i.e., "golfball" hail) would survive and thereby qualify the storm as severe. Reference to Fig. 13.1 suggests that the required minimum updraft speed in the hailstone growth region is about 30 m s^{-1}. The minimum CAPE for such a maximum velocity is 1800 joule kg^{-1}, if buoyancy is the main driver for the updraft. These calculations support the idea that otherwise severe hail is destroyed by melting in a substantial number of cases when the surface is 4 km or more from the melting level. This affords an explanation for the relatively small incidence of hail in the barotropic tropical atmosphere (Frisby and Sansom 1967; see also chapter 10 of this monograph). All other conditions being equal, in cases with elevated terrain (i.e., Colorado, United States; Kenya, Africa), hail will of course be more common.

3) TORNADOES AND VORTEX STRETCHING

Although tornadogenesis remains an area of considerable debate, there is widespread agreement that vortex stretching, or the concentration of angular momentum in the vertical, is an essential aspect. Vertical vortex stretching depends on the product of the radial gradient in tangential velocity $\partial v/\partial r$ (implying that there is initial angular momentum in the environment), and the radial convergence, which is equivalent to the vertical gradient of vertical velocity $\partial w/\partial z$. Direct measurements of these two quantities, like the updraft speed in the earlier discussion, are hard to come by, and so surrogate measurables are often used—a shear variable (like helicity) and a stability variable (like

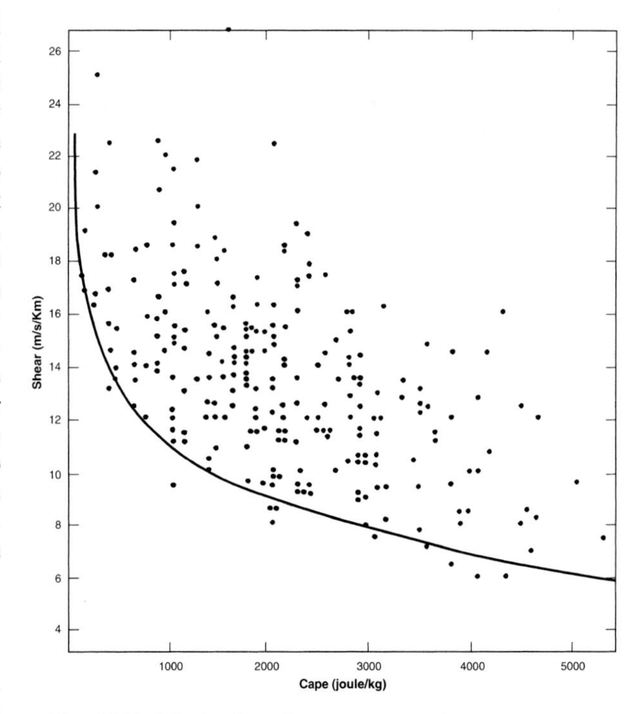

FIG. 13.21. Distribution of two parameters important to vortex stretching (shear and CAPE) for a number of observed tornadoes. (From Johns et al. 1990.)

CAPE). Some confirmation of the importance of this product of quantities is found in the work of Johns et al. (1990), in which shear is plotted against CAPE, as in Fig. 13.21. The manner in which the points representing the violent (F4–F5) tornadoes separate from the less intense tornadoes supports the idea of the product of quantities as a measure of intensity. An environment with large CAPE (and on average large $\partial w/\partial z$) requires, on average, less ambient angular momentum to match the intensity of a tornado born out of small CAPE but large angular momentum. A good example of the latter circumstance are the hurricane-related tornadoes studied extensively by McCaul (1991, 1993).

In judging the tendency for tornadogenesis based on vortex stretching, a more appropriate quantity for the instability variable from the proximity sounding is based on the parcel theory prediction for $w(z)$

$$w(z) = \sqrt{2\,\mathrm{CAPE}(z)}, \qquad (13.2)$$

where CAPE(z) is the integrated buoyancy from the level of free convection to altitude Z. Consequently, the convergence is given as

$$\frac{\partial w}{\partial z} = \frac{1}{2\sqrt{2\,\mathrm{CAPE}(z)}}\frac{\partial\,\mathrm{CAPE}(z)}{\partial z}. \qquad (13.3)$$

This appeal to the buoyancy profile, rather than the integrated buoyancy, may provide a more tightly defined threshold for violent tornadogenesis than found by Johns et al. (1990). This procedure probably deserves some attention. This "shape of the CAPE" (and the influence of the latent heat of freezing on the shape of the CAPE) may be very important in the establishment of midlevel vorticity in the mesocyclone.

4) Venn diagram for severe storm classification

The foregoing discussion of physical relationships has emphasized three aspects of severe weather: lightning, hail, and tornadoes. [The wind parameter was not examined because of the general absence of studies relating lightning and severe wind as an isolated manifestation of severe weather. One study that does pertain is Petersen et al. (1995).] With three possible states, one can imagine seven possible overall storm conditions, illustrated in the Venn diagram in Fig. 13.22. The reality of these seven conditions will now be discussed in light of the physical relationships already discussed and the available observations.

Thunderstorms with neither hail nor tornadoes are ordinary nonsevere thunderstorms, the most common condition and the subject of section 13.2. Developing in moderately unstable barotropic environments, most prevalent in tropical land zones, these storms lack the intense sloping updraft to make hail and lack the ambient angular momentum to make tornadoes.

Thunderstorms producing severe hail but no tornado are probably the most prevalent form of severe

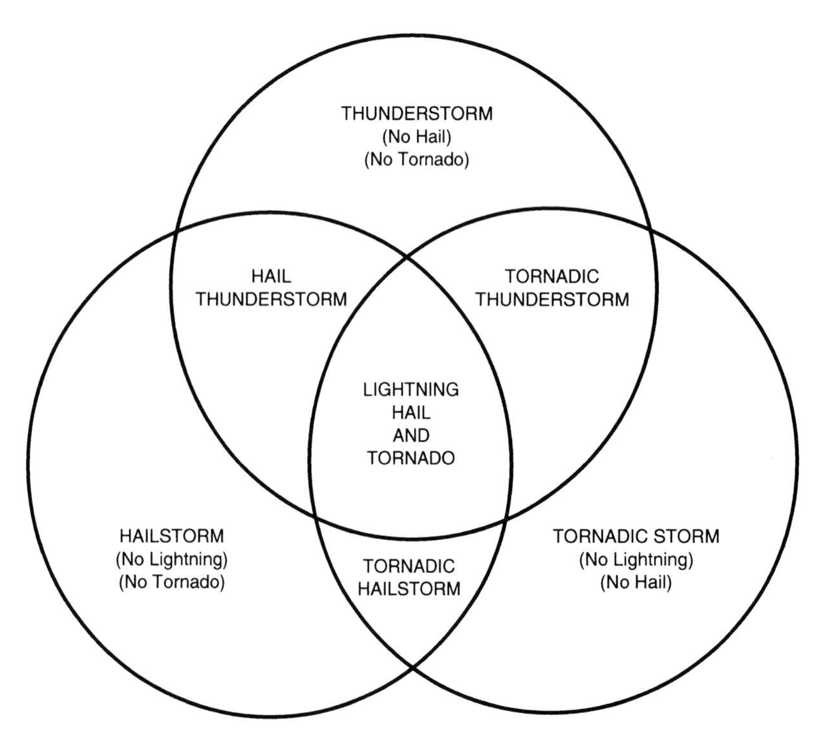

FIG. 13.22. Venn diagram of all possible storm states involving lightning, hail, and tornadoes. (Relative areas carry no significance.)

weather. A climatology of such cases for the United States can be found in Kelly et al. (1985).

Hailstorms with neither lightning nor tornadoes have been mentioned in textbooks (Byers 1944) but have not been documented in the literature as far as this author is aware. The necessity for large updraft in the mixed phase region would also seem to provide a threshold cloud size and the necessary conditions for vigorous charge separation and lightning. A far more common situation is for hailstorms to exhibit lightning and, generally, high flash rate, as revealed by results in Table 13.1 and discussed in section 13.3b.

Tornadoes in the form of waterspouts with neither lightning nor hail are likely in small clouds (Rossow 1970; Golden 1971; Wakimoto and Lew 1993). Tornadoes also appear in the vicinity of hurricanes (McCaul 1991, 1993) where ambient vorticity is abundant but CAPE is an order of magnitude less than that found in the environment of violent tornadoes (Johns et al. 1990) and the vertical development of radar reflectivity is modest. Vonnegut (1975) discusses a storm in New York State that produced a damaging tornado with little if any radio frequency emission associated with lightning. K. Knupp (1996, personal communication) has also documented mesocyclones in shallow systems without lightning.

A storm with both hail and tornado but no lightning appears to be an extremely unlikely, if not impossible, condition, given the exigencies of updraft for hail and instability for tornadoes. The strongly baroclinic systems in early spring would appear to be the only candidates.

Tornadic thunderstorms without any hail are not specifically documented in the literature (nor would they be simple to document given the sampling problems with hail observations) but would appear to be possible in conditions of moderate instability and large ambient vorticity. Such storms are likely producers of weaker tornadoes like those studied, for example, by McCaul (1991, 1993) (who did not investigate lightning). Severe storm reports for violent tornadoes noted by Darrah (1978) without accompanying hail reports are probably not an indication that hail was absent but rather an indication that all attention focused on the tornado. The severe storm reports in more recent years in which hail reports are more abundant provide substantial evidence that violent tornadic storms are consistent producers of large hail.

The Venn diagram in Fig. 13.22 with seven states is fairly complicated. If we ignore the highly improbable conditions discussed above that are not supported by observations and restrict attention to the violent tornadic storms on the Fujita F scale (i.e., F4 and F5 storms), the simpler hierarchical Venn diagram of Fig. 13.23 results. The description is now greatly simplified: All hailstorms are thunderstorms and all tornadic storms are producers of both lightning and hail. This classification is consistent with comparative observations of supercell hailstorms and ordinary hailstorms in Nelson and Young (1979). The supercell hailstorms exhibited frequent tornadoes and maximum hailstone diameters three times larger than the ordinary hailstorms that also produced no tornadoes. This observation is consistent with the tendency found by Colquhoun and Riley (1996) for CAPE to increase with F-class. This general hierarchical structure in Fig. 13.23 is also evident in Grosh (1977), who studied lightning, hail, and tornadoes (though in this study F scale values were not reported) for storms in Illinois, all in relation to radar cloud-top height. His results in Fig. 13.24 show the monotonic increase in probability for lightning, hail, and tornadoes with cloud size. General conclusion: The biggest clouds (with the high-

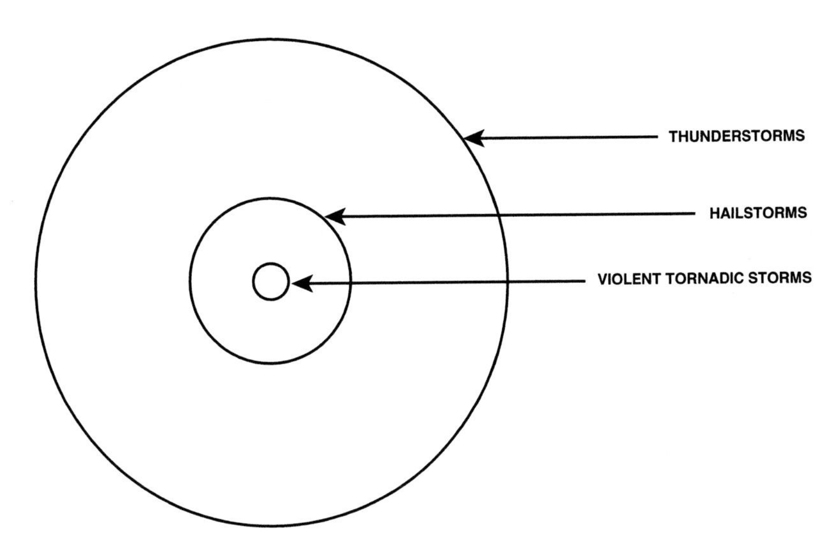

FIG. 13.23. Simplified Venn diagram when only violent (F4, F5) tornadoes are considered.

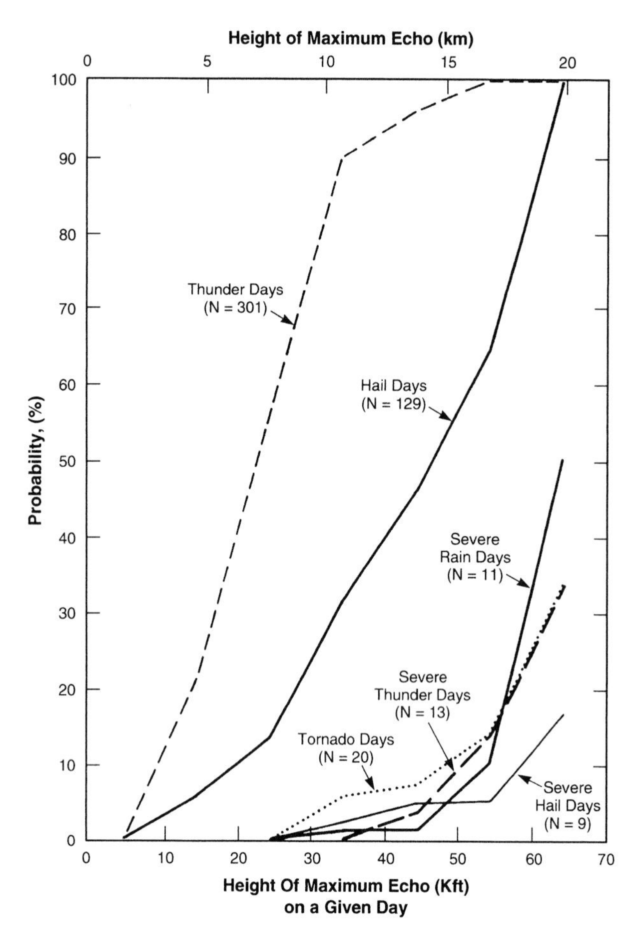

FIG. 13.24. Observed probability of lightning, hail, and tornadoes in the midwestern United States vs radar cloud-top height. (From Grosh 1977.)

est lightning rates, on average) are the most severe storms.

b. Hail and lightning

Hailstorms represent a tiny subset of the entire population of thunderstorms, and studies relating lightning activity and hail are rather sparse (Shackford 1960; Blevins and Marwitz 1968; Baughman and Fuquay 1970; Changnon 1992; Lopez and Aubagnac 1997; Carey and Rutledge 1996, 1998). The substantial sampling problem associated with narrow hailswaths is surely a factor in relating these quantities quantitatively.

Blevins and Marwitz (1968) made visual observations of lightning strokes in Colorado hailstorms. The hailstorms were defined as those thunderstorms producing hail larger than ¼ in. (6 cm) diameter, rather than the ¾ in. size defining a severe storm. The percentage of storms producing hail increased systematically with stroke rate except in the largest category ($>$70 strokes min^{-1}), for which no hail was observed (six cases), a puzzling result. The percentage of total lightning in the ground flash category diminished systematically with increasing flash rate.

Baugham and Fuquay (1970) studied hail and lightning occurrence in mountain thunderstorms in Montana. They found that hailstorms generally produce more lightning than storms without hail and that both hail and lightning activity increase with storm size. Their findings were not consistent with the finding of Blevins and Marwitz (1968) that the probability of hail diminished at extreme flash rates.

Pakiam and Maybank (1975) investigated the electrical characteristics of severe hailstorms in Alberta, Canada, using radar and mobile electric field mills. The storms exhibited bounded weak echo regions, hailstone sizes in the severe category (i.e., diameters $\geq$ ¾ in.), and radar cloud-top heights to 15 km MSL. In general, the lightning flash rate increased with both the cloud size and the maximum hailstone diameter, consistent with other results. A single outlier observation of one isolated hailstorm led the authors to a peculiar conclusion that is contradictory to other results in this review, namely, that the "charging mechanism should produce lightning of frequency 1–3 flashes per minute for each storm cell, irrespective of whether the storm is a thunderstorm, hailstorm or severe storm." The one storm on which this conclusion is based (on 9 July 1970) exhibited a radar top of 15 km, a pronounced BWER, and a stronger profile of vertical reflectivity than the storm shown in Fig. 13.10 that produced a peak flash rate more than 20 times greater than that documented by these authors. Despite their own observations that the field changes due to lightning are quite small in active stages of other storms, and despite the fact that the field mills documenting flash rate in this storm were 15–20 km distant from the storm center, the authors stand by their anomalously low flash rate of 1–3 per minute. In this author's opinion, these field mill measurements of total flash rate are in error, and so is the paper's conclusion, stated above. Similar questions have been raised (Vonnegut and Moore 1957) about the validity of field mill measurements of total flash rate reported by Gunn (1956) in a severe tornadic storm, the first entry in the severe storm section of Table 13.1.

A severe Arkansas hailstorm, with a 63 000 ft (19 km) radar top and hailstones with a diameter up to 2.75 in. (7 cm), showed a peak stroke rate on an aircraft-mounted video camera of 230 per minute (Westcott et al. 1996; O. H. Vaughan 1996, personal communication). This hailstorm also exhibited a large number of "blue jets," a new form of electrical discharge identified and named by the University of Alaska group. It is not known at present whether blue jets are preferentially associated with severe thunderstorms. Recent observations in New Mexico by P. Krehbiel (1996, personal communication) suggest that the dim blue light in blue jets is seldom seen from the ground because of Rayleigh scattering. It is therefore conceivable that blue jets are common to ordinary thunderstorms, but clearly further studies are warranted.

The most comprehensive study to date relating hail-to-ground flash activity is that of Changnon (1992). Fortuitously, the storms studied in the central United States exhibited no mesocyclones, no tornadoes, and no ground flashes of positive polarity (to be discussed in section 13.3c), and so this study isolates the hail aspect of storm severity (i.e., the middle ring in the Venn diagram in Fig. 13.23). Unfortunately, no information on hailstone size was available, so there is no guarantee that the storms studied were all severe. Nor was any information available on intracloud lightning activity. The average result for all cases studied relating ground flash rate and time of hailfall at the surface is shown in Fig. 13.25. The time of peak flash rate is within one minute of the start time for the hail. The average peak CG flash rate, 1–2 per minute, is not exceptional (see Fig. 13.28), and by itself would not provide a unique hail identifier. The CG flash rates for some cases reached 16 min^{-1}, a value large in comparison with ground flash rates in severe weather documented in other references (see Fig. 13.28). Changnon does show evidence that hailstorms with more lightning generally produced greater surface damage (Fig. 13.26). He also shows evidence for lightning centers without hail but does not compare the characteristics of the latter centers (flash rate, flash density) with those associated with hail and crop damage. Changnon considers it unlikely that hail was missed in regions exhibiting ground flashes.

The general tendency for intracloud lightning to precede ground flashes in both ordinary thunderstorms (Workman et al. 1949; Goodman et al. 1988; Williams et al. 1989b; Carey and Rutledge 1996) and in severe storms (MacGorman et al. 1989; Carey and Rutledge 1998; Weber et al. 1998; Williams et al. 1999) suggests that intracloud lightning may have preceded the hailfalls measured by Changnon by times exceeding 10 minutes (which the author alludes to at the end of his paper). The estimated range of cloud tops for these storms was 14–17 km and so would be expected to have exhibited total flash rates in excess of 10 per minute, if Fig. 13.6 is any useful guide. [Carey and Rutledge (1996) found a total flash rate of 10 min^{-1} in a nonsevere hailstorm (marble-sized hail, 1–1.5 cm) in Colorado.] Seldom did ground flashes occur in the hailfall area. This finding is consistent with the idea discussed in section 13.3a(2) that large hail is probably not a major player in the separation of electric charge in thunderstorms. Additional evidence for this assertion has been presented by Holle et al. (1994), who found a preference for hailfalls separated from ground flash activity on the south side of eastward-

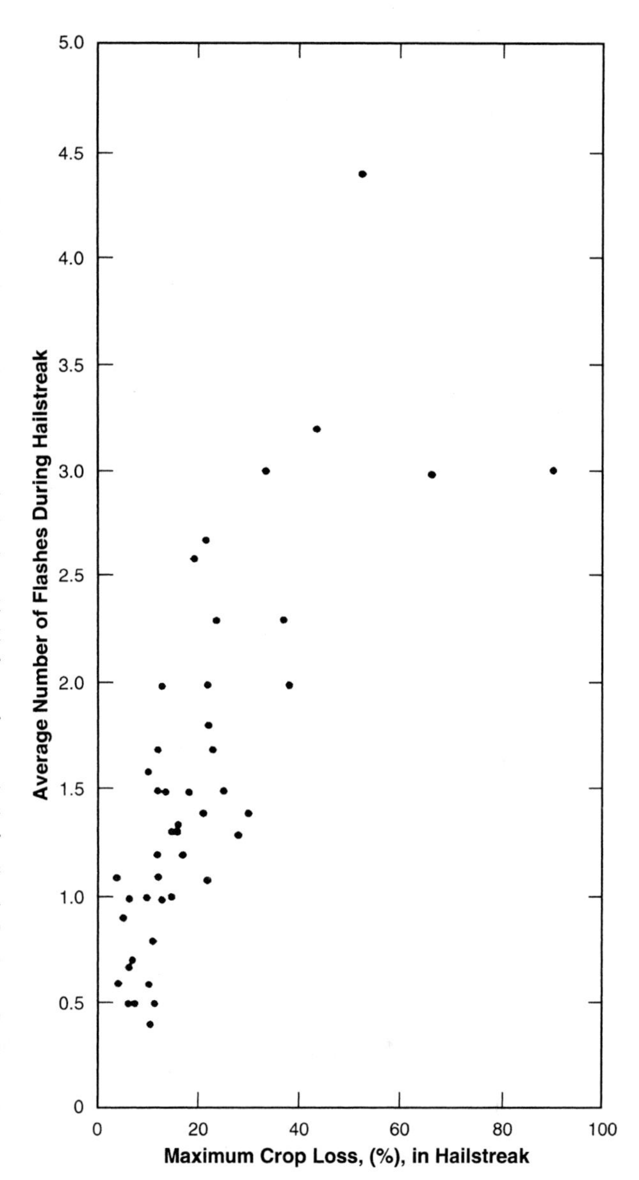

FIG. 13.26. Hail damage vs peak lightning ground flash rate. (From Changnon 1992.)

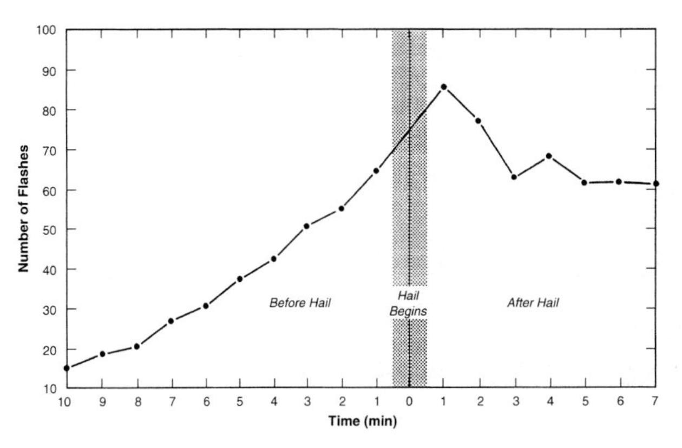

FIG. 13.25. Composite history of lightning ground flashes and hail at the ground. (From Changnon 1992.)

propagating weather, and by Carey and Rutledge (1998), who found large hail on the ground in advance of clusters of positive ground flashes.

The possible modulation of cloud-to-ground flash rate by the descent of small hail has been suggested by Lopez and Aubagnac (1997) in their analysis of a severe Colorado hailstorm. Like the nonsevere hailstorms studied by Chagnon (1992), the latter storm produced only negative ground flashes.

A violent hailstorm swept across Orlando, Florida, in March 1992 and produced some 3 in. (7.6 cm) diameter hail. RHI scans by Lincoln Laboratory's FL-2 radar (Fig. 13.27) show the characteristic bounded weak echo region and also demonstrate the strong horizontal inflow characteristic of the baroclinic environment of springtime hailstorms, which is undoubtedly a substantial contribution to the vertical motion in which the large hailstones grow. This storm illustrates another aspect of nonsummertime hailstorms: The cloud-top height may not be exceptional [in this case 48 000 ft (14.6 km)] but the height above the 0°C isotherm is usually substantial (see also the storm in Zrnić et al. 1993). Lightning Position and Ranging System (LPATS) recorded a peak ground stroke rate of 35 per minute near the time of reports of 3 in. diameter hail, though it is likely that a subset of these strokes was intracloud. No record of total flash rate is available, but the IC/CG ratio recorded in other hailstorms (Williams et al. 1999) suggests that the total flash rate easily exceeded 100 per minute.

Shafer and Carr (1990) and Kane (1991) also studied the temporal evolution of cloud-to-ground lightning and hail. Shafer and Carr found no correlation between times of large hail and the rate of flashes, and

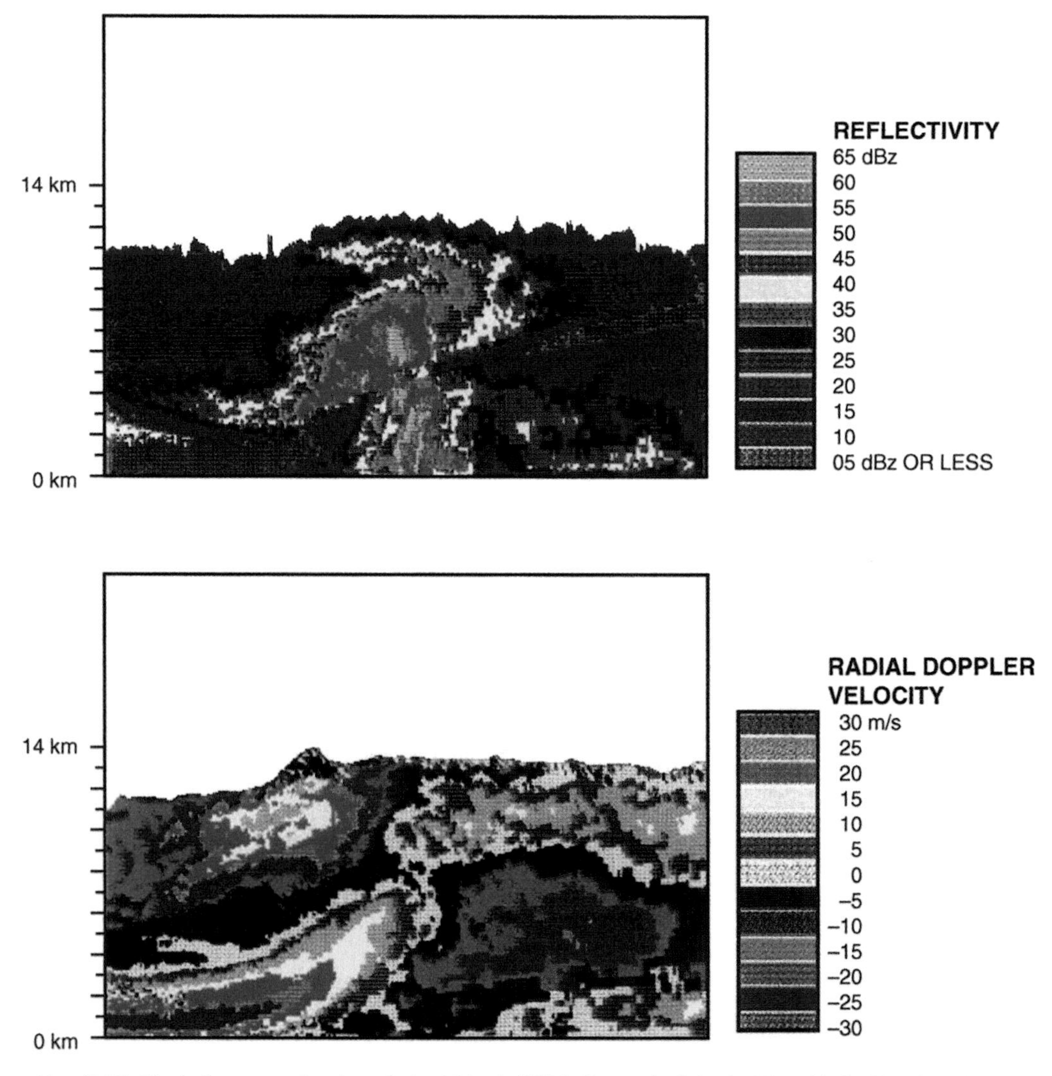

FIG. 13.27. Vertical cross section through the 6 March 1992 hailstorm in Orlando, FL, with the Lincoln Laboratory FL-2 C-band radar: (a) radar reflectivity (b) mean Doppler velocity. Evidence for a BWER, a strongly tilted updraft, and pronounced beam attenuation are clearly present. (M. Isaminger 1996, personal communication.)

Kane found peak ground flash rates prior to the hail observation time. Both these comparisons were performed on large-scale storm systems rather than on an individual cell basis, as was done by Changnon (1992), and this may account for differences in results.

Several anecdotal reports (Byers 1944; U.S. Weather Bureau 1947; D. Proctor 1996, personal communication) suggest that hail can occur in the complete absence of lightning, but no well-documented case has been presented. Such a result would be surprising in light of the evidence and discussion in section 13.3a and the recent findings in Florida in Table 13.2.

c. Tornadic storms and lightning

1) Variability in behavior

Great variability is perhaps the best characterization of the relationship between lightning and tornadic storm evolution (Keighton et al. 1991; McGovern and Nielsen 1991; MacGorman 1993; Knapp 1994; Perez et al. 1995). Such variability is understandable from at least one standpoint, given the discussion on vortex stretching in section 13.3a and the observations of pairs of parameters that control vortex stretching—one related to rotation and one related to vertical air motion. As noted by Johns et al. (1993), violent tornadoes are associated with an extremely wide range of values of shear and CAPE, though there appears to be some trade-off in these parameters, as illustrated in Fig. 13.21. Tornado-spawning storms grown in vorticity-rich environments, but with modest CAPE and modest vertical motion, and therefore modest vertical development of the ice phase, may be rather weakly electrified. Good examples might be the hurricane tornadoes studied extensively by McCaul (1991, 1993) and the shallow mesocyclone documented by K. Knupp (1996, personal communication). Tornadic storms developing in high CAPE environments with more moderate shear may exhibit extreme vertical development and extraordinary electrical activity. Good examples are the Worcester, Massachusetts, tornadic storm in June 1953 (Vonnegut and Moore 1958); the 1981 Binger storm studied by MacGorman et al. (1989) (see also MacGorman and Nielson 1991; MacGorman 1993); and the Plainfield, Illinois, supercell storm in August 1990 (Fujita 1992; Seimon 1993a; Doswell and Brooks 1993; Seimon 1993b; MacGorman and Burgess 1994). The ambiguity of the contribution to vertical air speed from buoyancy forces and from pressure gradients associated with baroclinic effects adds to the uncertainty in predicting the behavior of lightning in tornadic storms.

2) Cloud-to-ground flash rates

Cloud-to-ground lightning is now routinely recorded and archived by the National Lightning Detection Network within the United States and by numerous additional networks in other parts of the world. These systems are triggered by the rapid increase in current that occurs when lightning contacts the ground, and as a consequence, these systems selectively reject the great majority of intracloud flashes, a topic to be discussed in greater detail in the tornado context in section 13.3c(4). This extensive dataset enables a statistical comparison between ground flash rates in severe tornadic thunderstorms (Knapp 1994; Perez et al. 1995) and in ordinary nonsevere thunderstorms (Holle and Maier 1982). Figure 13.28 shows these comparisons. As with the earlier comparisons for total lightning activity in Table 13.1, there is a clear tendency in these datasets for the severe storms to produce higher peak rates of cloud-to-ground lightning than nonsevere storms. The mean rate for nonsevere Florida storms (Holle and Maier 1982) is less than one flash per minute, whereas severe storms typically produce several flashes per minute. These distribu-

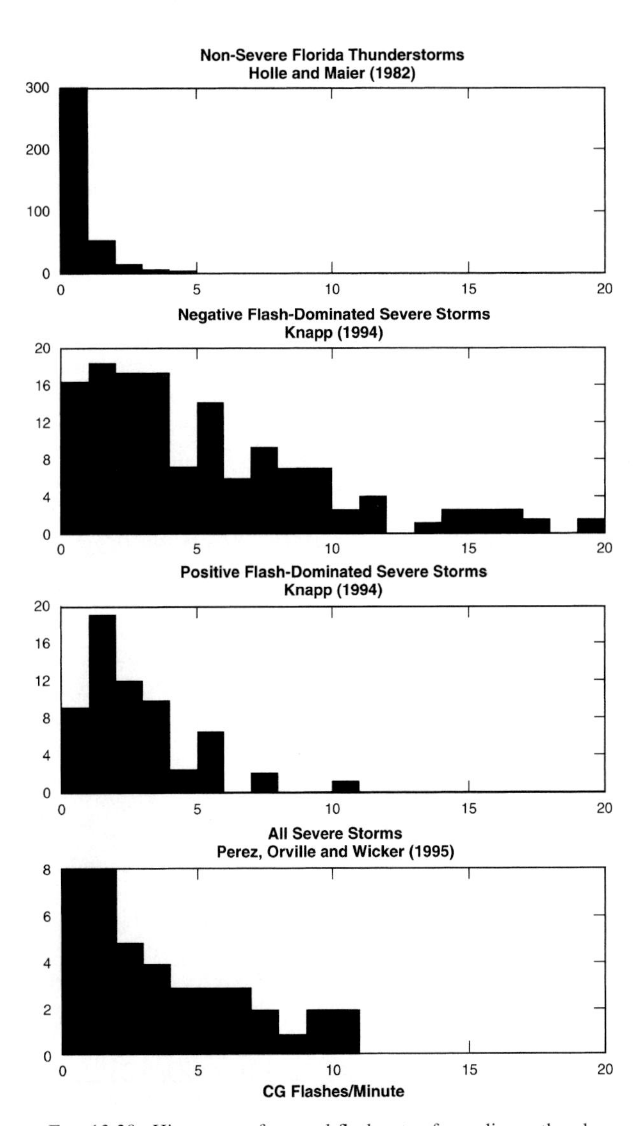

FIG. 13.28. Histogram of ground flash rates for ordinary thunderstorms (Holle and Maier 1982) and for severe storms. (From Knapp 1994; Perez et al. 1995.)

tions, however, are not completely distinct, and there appears to be no clear-cut ground flash threshold for severe weather (Mosher and Lewis 1990). Furthermore, other studies (Carey and Rutledge 1996, 1998) have shown that nearby nonsevere storms produced more ground flashes than the severe storms under investigation during some phases of severe convective weather. All ground flashes cease for periods of minutes to as long as one hour (MacGorman and Burgess 1994; Carey and Rutledge 1998).

When one examines the results of specific case studies, the variability in ground flash activity as a quantifier of severe weather is further emphasized. MacGorman (1993) has contrasted the ground flash behavior of two tornadic storms, one strong (Binger) and the other weak (Edmond). The weaker storm exhibited the higher ground flash rate. This behavior was attributed to a higher main negative charge center in the Binger storm that effectively suppressed ground flashes and enhanced the more prevalent intracloud lightning in the strong storm. These ideas are consistent with the dependence of charge structure on storm size illustrated in Fig. 13.15 and with the trends in IC/CG ratio with total flash rate (and presumably cloud size) shown in Fig. 13.8.

MacGorman et al. (1989), Goodman (1990), and Lopez and Aubagnac (1997) all noted increases in cloud-to-ground flash rate as the updraft in the mesocyclone collapses, presumably enabling charged graupel particles formed in the updraft to contribute to charge transferred to ground when these particles descend. Williams et al. (1989b), Carey and Rutledge (1996), and Petersen et al. (1996) have observed increases in ground flash activity associated with the descent of the reflectivity core in the mixed-phase region in storms not characterized by a mesocyclone.

Perez et al. (1995, 1997) studied ground flash rates for a large number of tornadic storms, but without consideration of the radar structure of the storms. They found essentially no correlation between ground flash rate and tornadogenesis, though in 74% of cases studied, the maximum ground flash rate preceded the time of tornado touchdown.

3) Positive polarity ground flashes

As noted in section 13.2 and Fig. 13.5, ground flashes of negative polarity predominate in nonsevere thunderstorms. Positive ground flashes have been identified in meteorological regimes with weak vertical draft speeds (Rutledge and MacGorman 1988; Engholm et al. 1990; Rutledge et al. 1993; Rutledge and Petersen 1994; Williams 1995b), regimes that are ordinarily also nonsevere (see also section 13.3d). The most noteworthy of these regimes is the stratiform region of mesoscale convective systems (Rutledge and MacGorman 1988; Engholm et al. 1990; Williams and Ecklund 1992; Rutledge et al. 1990; Stolzenburg et al. 1994; Williams and Boccippio 1993; Rutledge et al. 1993; Holle et al. 1994; Rutledge and Petersen 1994; Boccippio 1996; Williams 1998). In recent years, positive ground flashes have also been identified with tornadic supercell storms (Rust et al. 1981a,b; Reap and MacGorman 1989; Curran and Rust 1992; Branick and Doswell 1992; MacGorman 1993; Seimon 1993a,b; MacGorman and Burgess 1994; Stolzenburg 1994; Knapp 1994; Perez et al. 1995, 1997; Carey and Rutledge 1998; Bluestein and MacGorman 1998). This discovery has raised interest in the operational use of the positive signature as a severe storm indicator (Branick and Doswell 1992). This discovery has also raised considerable scientific interest about the cloud electrical structure that causes the clusters of positive flashes, as this behavior is at odds with the picture for ordinary nonsevere thunderstorms (Fig. 13.5). These two issues shall be addressed in turn.

In the operational context, Branick and Doswell (1992) have emphasized that the value of the positive ground flash signature for storm warning operations is "dependent on whether or not the correlation between elevated +CG rates and supercell type can be demonstrated routinely and reliably." Results in cases that may well be extremes of the extreme (Fujita 1992; Seimon 1993; MacGorman and Burgess 1994) showed periods of ground flash activity with almost exclusive positive polarity with peak rate approximately 20 minutes prior to the appearance of an F5 tornado and a complete changeover to negative flashes. The extension of these early results to a larger number of cases (MacGorman and Burgess 1994; Knapp 1994; Stolzenburg 1994; Bluestein and MacGorman 1998) does not show such "clean" behavior with an exclusive connection with F5 tornadoes, but some generalizations about average behavior can be drawn. Positive ground flash-dominated storms show, on average, higher radar tops than negative flash-dominated cases, as shown by Knapp's data in Fig. 13.29. (This result suggests higher probability for severe weather for the positive-dominated cases following results in Fig. 13.14 and higher rates of intracloud lightning following the results in Fig. 13.6.) Furthermore, the case studies in Curran and Rust (1992), Seimon (1993), MacGorman and Burgess (1994), and Carey and Rutledge (1998) show that the positive flash activity is associated with strong vertical growth of the radar cloud top and large values of the Larsen area, often to the maximum values observed over the lifetime of the storm. The majority of storms dominated by positive ground flashes also produce large hail (within 10–20 minutes of the time of the positive flash production; Carey and Rutledge 1998) consistent with larger draft speeds in these storms. Radar analysis of severe Canadian storms by S. Clodman (1996, personal communication) indicates that the positive flash activity is associated with extraordinary vertical development of radar reflectivity and the descent of ice particles that may

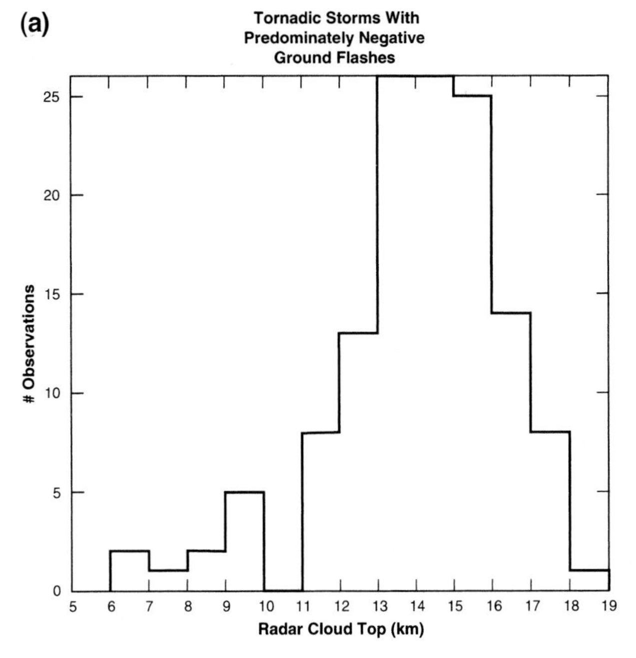

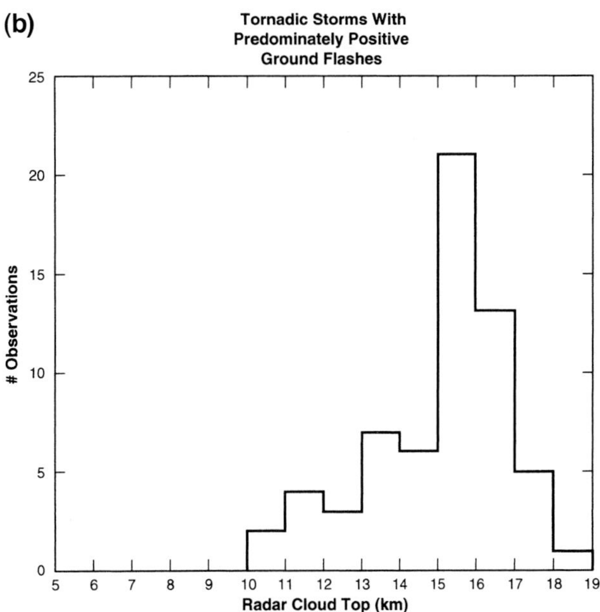

FIG. 13.29. Distributions of radar cloud-top height for negative ground flash and positive ground flash dominated severe storms. (Knapp 1994; unpublished data.)

have grown in the BWER during the strong updraft phase. The positive-dominated storms often produce tornadoes and at a time following the positive flash activity (Seimon 1993; Knapp 1994).

The mechanism for positive ground flashes in severe storms and the pronounced departure in behavior from ordinary thunderstorms is presently unknown. Various hypotheses are pictorialized in Fig. 13.30 and summarized below in the context of highly simplified overall charge structures. Some additional discussion of hypotheses can be found in MacGorman and Burgess (1994).

1. *Tilted dipole hypothesis.* The upper positive charge of the thunderstorm dipole (recall Fig. 13.3) is displaced laterally, by virtue of the sloping updraft and the upper-level wind, relative to the main negative charge. Direct transfer of positive charge to ground from great height is then possible. This hypothesis is the most conservative of those discussed below in the sense that only a large-scale distortion (tilting) of the traditional dipole structure of nonsevere storms (Fig. 13.5) is called upon. Evidence for this picture in small nonsevere winter storms in Japan is found in Brook et al. (1982). Evidence for this picture in severe storms in Oklahoma is found in Rust et al. (1981) for isolated events, and the hypothesis is further supported by observation by Curran and Rust (1992) and Branick and Doswell (1992). It would appear that extraordinary energy is required to bridge the large distance from upper cloud (10–16 km) to ground, though there is little question that this can occur (see, e.g., *Weatherwise,* August–September 1997, p. 47). Further observations are needed to show that repeated discharges of this type are responsible for the clusters of positive ground flashes that appear in NLDN observations for severe storms with large hail (Seimon 1993a; MacGorman and Burgess 1994; Stolzenburg 1994).
2. *Precipitation unshielding hypothesis.* This idea is a variant of hypothesis 1 in which the descent of negatively charged precipitation "unshields" the upper positive charge, making it more prone to initiate a positive flash to ground (Carey and Rutledge 1998). This alternative to the tilted dipole hypothesis seeks to account for the suppression of negative ground flashes in the observations, which is a long-standing weakness of the latter hypothesis. This weakness is particularly apparent when the microphysical conditions needed for negative charging [based on the laboratory results (Takahashi 1978; Saunders et al. 1991) and summarized earlier in Fig. 13.4] would appear to be present. This hypothesis is consistent with observations (Carey and Rutledge 1998) that the positive ground flash locations are not always in the precipitation cores.
3. *Inverted dipole hypothesis.* The polarity of the main thunderstorm dipole is reversed from that in ordinary thunderstorms (Seimon 1993a). Positive charge is in close proximity to ground and available for direct transfer. This hypothesis provides the simplest picture for placement of the appropriate charge reservoir in close proximity to the earth's surface, long believed a major influence in explaining the predominance of negative ground

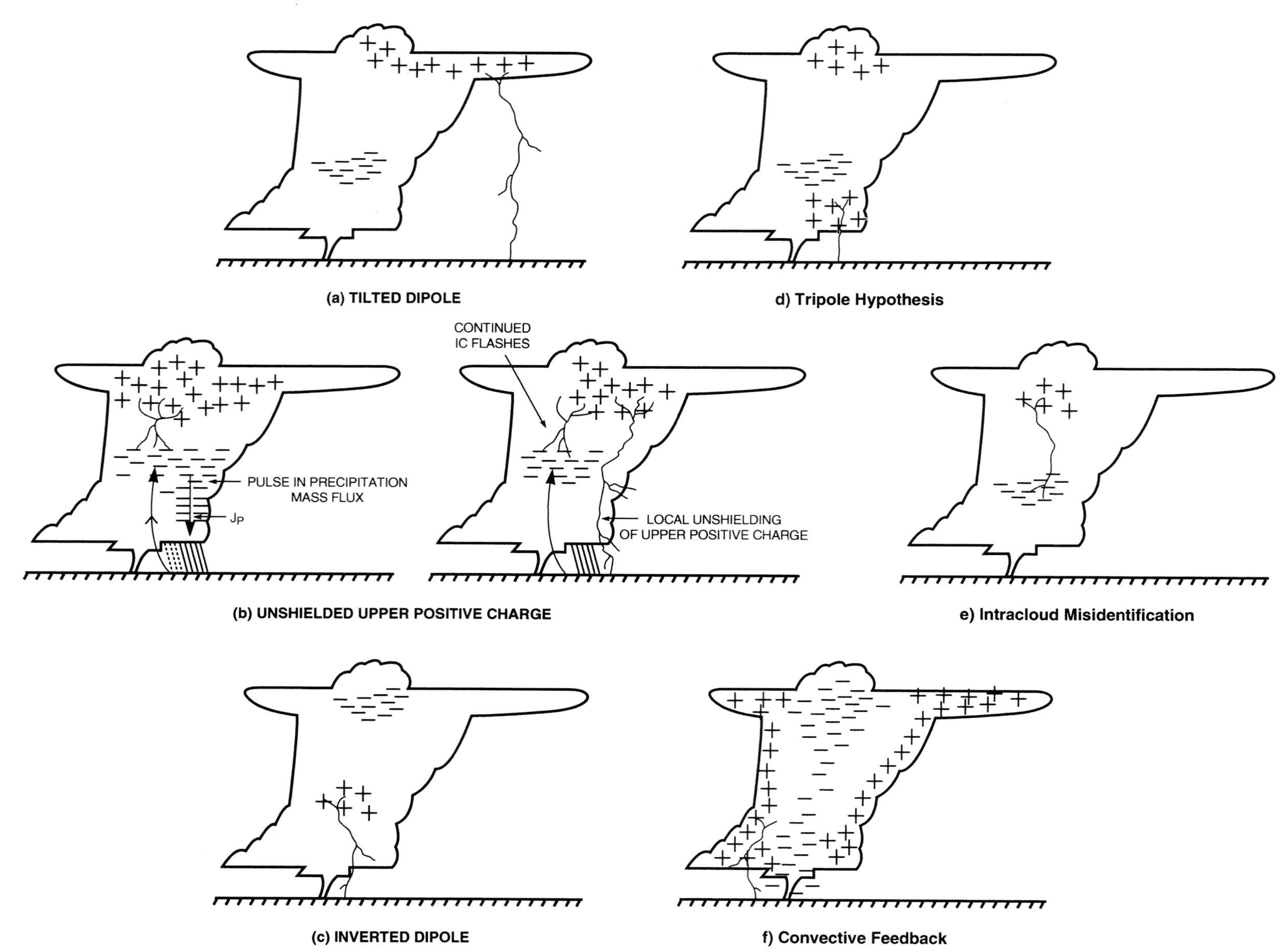

FIG. 13.30. Illustration of hypotheses for positive ground flash production in severe storms: (a) tilted dipole, (b) "unshielded" tilted dipole, (c) inverted dipole, (d) tripole, (e) intracloud misidentification, (f) convective theory.

flashes in nonsevere storms and consistent with the middle illustration in Fig. 13.5. A major problem with the inverted dipole picture is that intracloud lightning (bridging upper negative and lower positive regions) would show field changes of opposite polarity to nonsevere storms. This prediction is contrary to the behavior of active thunderstorms observed, for example, at the Kennedy Space Center.

4. *Tripole hypothesis.* The lower positive charge of the ordinary thunderstorm tripole structure (Simpson and Scrase 1937; Williams 1989b; Murphy et al. 1996) is enhanced in magnitude and plays a more dominant role in mediating ground flashes because the main negative charge region is greatly elevated (recall Fig. 13.15). The anomalous positive ground flashes are explained in this case not by the deformation (tilting) of the basic nonsevere electrical structure (Fig. 13.5) but rather by extreme conditions in the microphysical parameter space (Fig. 13.4) imposed by an extraordinary updraft. Despite notable differences in the charging behavior in the two laboratory diagrams in Fig. 13.4, they do share one common feature. For the complete range of cloud temperature explored in the laboratory, the large particles are predicted to charge positively for sufficiently large values of cloud liquid water content. This point transcends the issue of the microphysical growth state of the positively charged particles, a subject debated in the recent literature (Williams et al. 1991; Saunders and Brooks 1992; Williams 1995b). Larger values of cloud liquid water content are expected in broad undilute severe storm updrafts in which large hailstones are commonly grown. If the positive charging occurred in the elevated negative charge region above the 0°C isotherm (Fig. 13.15), then a substantial reservoir of lower positive charge would be available to explain the clusters of positive lightning in severe conditions. This picture requires that the positive ground flashes occur at least partially within the precipitation (Seimon 1993a; S. Clodman 1996, personal communication), the postulated carrier of the positive charge. As shown in Fig. 13.30, both the tripole hypothesis and the "unshielding" hypothesis above allow for sustained intracloud lightning in the upper dipole simultaneous with the occurrence of positive ground flashes. Given the common negative charge reservoir for both intracloud and negative ground flashes, the intracloud lightning may short-circuit the negative ground flashes and provide an additional explanation for their suppression in strongly severe conditions.

A second possible microphysical mechanism for positive charging of ice particles to form the postulated lower positive region in the severe storm is the process investigated by Schaefer and Cheng (1971) and Rydock and Williams (1991), based on the spontaneous ejection of negative fragments from surfaces growing by vapor deposition. This mechanism is distinct from the noninductive ice–ice process in which particle collisions are essential.

Rust and Marshall (1996) have recently advocated the abandonment of the tripole picture that originated with Simpson and Scrase (1937) on the basis that one-dimensional balloon soundings of the electric field sometimes show significant departures from this structure. In the author's opinion, based on all the evidence from the laboratory and the field, this abandonment is ill-advised.

5. *Intracloud lightning misinterpretation.* Intracloud lightning, active at the time of the deep vertical development associated with the presumed positive ground flashes, is misidentified by the National Lightning Detection Network (NLDN). The validity of this hypothesis would require that K-changes in large intracloud flashes be extraordinarily large, as measured peak currents in positive flash episodes (Seimon 1993a; M. Stolzenburg 1995, personal communication) often exceed 50 KA. Nonetheless, this possibility deserves greater scrutiny. Recent analysis of NLDN events with peak currents of the order of a few kiloamperes or less indicates that these events are mostly intracloud flashes (K. Cummins 1997, personal communication; Cummins et al. 1998; Orville and Huffines 2001). In more recent studies (Wacker and Orville 1999; Orville and Huffines 2001) a +10 KA filter has been implemented to exclude intracloud flashes.
6. *Convective feedback hypothesis.* This hypothesis follows from the convective theory for thunderstorm electrification (Vonnegut 1963) and some initial condition in the boundary layer (i.e., initial space charge) that allows ingested negative space charge to be carried aloft in the updraft. Positive charge accumulates in electrical screening layers at the cloud boundary and descends to replenish a lower positive charge region. The main cloud dipole would be inverted as in hypothesis 3. Surface observations of the Worcester, Massachusetts, tornadic storm (Vonnegut and Moore 1958) did suggest a negative cloud top, but a major weakness in this hypothesis at present is the absence of a physical feedback "switch" for a large-scale change of cloud polarity. The existence of the necessary large space charge densities at the cloud boundary (Fig. 13.30f) is not supported by lightning radar echo behavior, which shows the strongest reflections from lightning in the intense precipitation well within the cloud (Williams et al. 1989a).

It is recognized that the pictures in Fig. 13.30 and their descriptions herein are highly simplified approximations for either nonsevere or severe storms, and that the ultimate explanation for the positive ground flashes in severe weather may require a structure other than those offered here. Urgently needed are observations that constrain the location of the positive region participating in the anomalous ground flashes. High resolution rf mapping techniques and photography will be useful here in defining lightning's charge transfer.

Recent studies (Sentman 1987; Boccippio et al. 1995; Burke and Jones 1996; Huang et al. 1999) have shown that positive ground flashes have a special status in exciting the earth's Schumann resonances in the form of Q-bursts. ELF methods may prove valuable in identifying characteristics of positive ground flashes that distinguish severe from nonsevere thunderstorms.

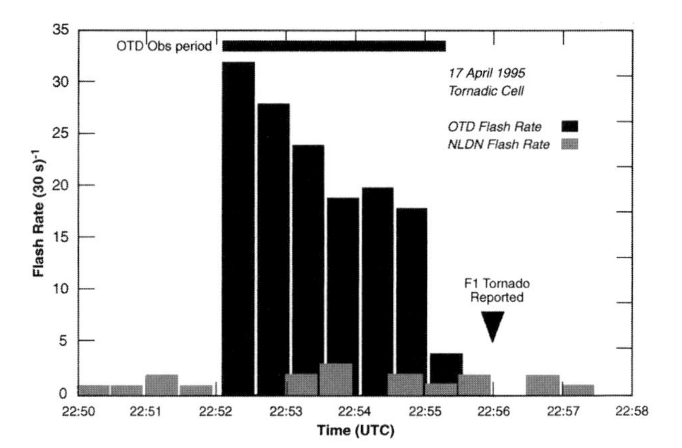

FIG. 13.31. Lightning activity observed from space by the Optical Transient Detector and from the ground-based NLDN for a tornadic storm on 17 April 1995. (From Buechler et al. 2000.) Intracloud lightning is clearly dominant.

4) INTRACLOUD BEHAVIOR AND STORM EVOLUTION

An understanding of the behavior of intracloud lightning in tornadic supercell thunderstorms benefits from the discussion on ordinary thunderstorms in section 13.2. Two key points were emphasized: 1) total flash rate increases nonlinearly with cloud size, with a fifth-power dependence best representing the average behavior (Fig. 13.6), and 2) the intracloud activity increasingly dominates the total activity at higher flash rates (Fig. 13.8) and is found at high levels in the storm (Lhermitte and Krehbiel 1979; Taylor et al. 1984; Lang et al. 2000). In light of observations that violent tornadic supercells are among the largest convective clouds in the troposphere, it is clear that intracloud lightning is a valuable quantity to study in this context, as MacGorman (1993) and Perez et al. (1995, 1997) have also emphasized. In contrast with the widespread availability of information on ground flashes in severe storms, as reviewed in an earlier section, rather little information of similar quality is available for the intracloud activity. [Exceptions are the set of results for Florida in Table 13.2 and the recent study by Williams et al. (1999).] When the latter information has been available for tornadic thunderstorms (Silberg 1965; MacGorman et al. 1989; MacGorman 1993; Williams et al. 1999; Buechler et al. 2000; Krehbiel et al. 2000), strong indications that intracloud lightning is a precursor to the tornado on the ground have been identified. Figure 13.31 shows the behavior of total lightning activity seen from space prior to a tornado touchdown (Buechler et al. 2000). Other observations to be discussed below suggest that intracloud activity will be intimately linked with tornadogenesis and may be a more robust indicator of supercell evolution than what one might believe based on relationships between tornadic storms and electrical activity in the existing literature. Recent results from Lightning Imaging Sensor Data Application Display (LISDAD) in Florida support this statement (Weber et al. 1998; Williams et al. 1999).

The literature on sferics detection of lightning activity in tornadic storms has been comprehensively reviewed in articles by MacGorman et al. (1989) and MacGorman (1993), which provide details about individual investigations. Some general observations on specific topics intended to be of a more critical nature (to emphasize where further clarification of this research area is desirable) are as follows.

- *Frequency range for detection.* As emphasized by Taylor (1972) and MacGorman et al. (1989), the detection of total lightning by electromagnetic (EM) means requires the use of frequencies of 1 MHz or higher. (The observations of severe storms in Table 13.2 were acquired at a frequency of 50 MHz.) The physical scale and level of current in intracloud lightning are not sufficient to excite strong EM emission at lower frequencies (Malan 1958). Some early sferics measurements on tornadic storms (Dickson and McConahy 1956) recorded only a small subset of the total lightning activity. Figure 13.31 from MacGorman et al. summarizes these results.
- *Meaning of sferics counts.* Sferics counts have generally not been related to flash counts, a well-defined quantity in visual/optical observations of thunderstorms, at least when flash rates are less than about one per second. For example, it is not known if the quantity "bursts per minute" (MacGorman et al. 1989) in Fig. 13.32 is directly translatable to "flashes per minute." Severe weather is often characterized by flash rates exceeding one per second (Tables 13.1 and 13.2) that begin to challenge the capability of analog recorders, including the human eye. (See also Vonnegut and Moore 1958.)

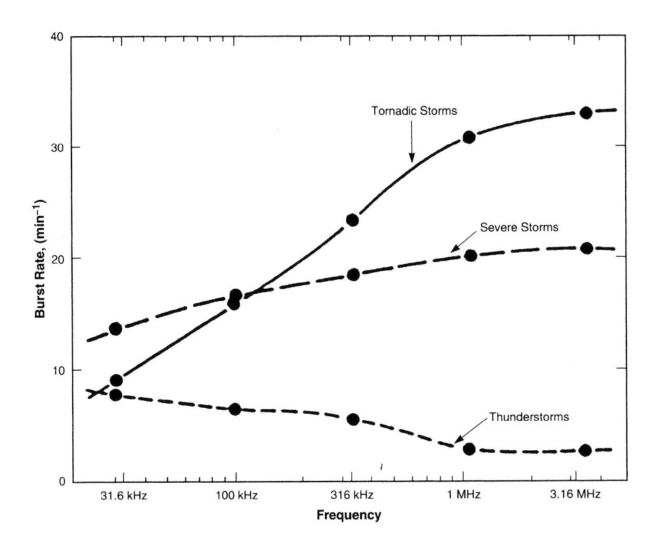

FIG. 13.32. Lightning "burst rate" vs receiver frequency for various storm classifications. (From MacGorman et al. 1989.)

- *Omnidirectional receivers.* The sferics receivers used in previous studies of tornadic thunderstorms (Dickson and McConahy 1956; Jones 1958; Scouton et al. 1972; Lind et al. 1972) have been omnidirectional and so lack localization capability. Almost invariably, when a supercell is under way, other nearby electrified convection is active and also radiating EM energy. The simultaneous reception of sferics from all of the weather may confuse the interpretation for the central region of interest. Localization capability in the case of the ground flash networks has demonstrated value in the context of tornadic storms (Seimon 1993a; Stolzenburg 1994; MacGorman and Burgess 1994). The technology for locating all (intracloud and cloud to ground) lightning in space and time has been available for more than 20 years (and is greatly facilitated with recently available GPS technology), but has not been implemented on tornadic storms until quite recently (Krehbiel et al. 2000). For example, the 23 April 1997 storm in Table 13.2 was tornadic, and the peak flash rate preceded the reported tornado time.
- *Spatial resolution of flashes.* A similar limitation in relating flash rate activity to tornadogenesis in observations from space (Turman and Tettelbach 1980) has been the lack of spatial resolution. The observed enhancements of flash activity on a synoptic scale surrounding a tornadic storm fall short of resolving the issue of precedence in the supercells of interest. The more recent study by Buechler et al. (2000) using the Optical Transient Detector (OTD) in space, with greatly improved spatial resolution, showed a well-defined result: The total lightning activity precedes the tornado touchdown (see Fig. 13.31).
- *Coordinated measurements.* The timing between the meteorological phenomenon (i.e., maximum cloud top, appearance of wall cloud, maximum mesocyclone circulation, appearance of funnel, tornado touchdown) and the electromagnetic emission (sferics) has not been carefully documented in previous studies (Lind et al. 1972; Trost and Namikos 1975; Taylor 1975; Johnson et al. 1977). To be sure, this is not an easy task. With the exception of a few enthusiasts, storm chasers have not been closely concerned with electrical activity. Given the evidence for the very sparse lightning activity at low levels in the vicinity of the wall cloud and the tornado (Davies-Jones and Golden 1975a; H. Bluestein 1996, personal communication; J. Wurman 1996, personal communication), perhaps this lack of attention is understandable.
- *Null cases.* Occasional examples of tornadic storms that lack sferics signatures have been identified (Vonnegut 1975; Johnson et al. 1977). However, little if any attention is given to the meteorological description of such cases to understand why no electromagnetic emission may have been produced. On the basis of the discussion in section 13.3a, we can anticipate a situation in which tornadogenesis without appreciable electrical activity is possible.
- *False alarms.* Taylor (1973a,b), MacGorman et al. (1989), MacGorman (1993), and Williams et al. (1999) have identified situations in which nonsevere thunderstorms and nontornadic severe storms have produced enhanced sferic activity without accompanying tornadoes. Evidently, some of these false alarm days are characterized by low values of a shear parameter in the vortex stretching term (see section 13.3a), but this point has not been studied. Little work has been done to distinguish features of EM emission (amplitude, duration, polarization, altitude of occurrence) in different storm types to reduce the probability of false alarms in an operational context.

Lightning activity is sensitively influenced by vertical air motions that are not routinely measurable by other means. Evidence is accumulating for a parallel between intracloud precursors to hazardous weather in ordinary thunderstorms and in severe storms (Williams et al. 1999). The well-established intracloud lightning precursor to microbursts in ordinary thunderstorms (Goodman et al. 1988; Williams et al. 1989b; Hoffert and Pierce 1996; Williams et al. 1995), discussed earlier in section 13.2, has its probable origin with the updraft and the growth of ice particles high in the storm. This upper-level development was often missed in operational microburst detection methods that concentrated on measurement of the differential radial velocity at the surface, thereby curtailing radar scans at upper levels. This lightning precursor also generally goes undetected if only ground flash observations are available. The ground flashes are not only infrequent but also occur later in the storm evolution (Williams et al. 1989b).

Returning now to the context of severe storms, two sets of observations suggest that violent tornadoes are preceded by strong reflectivity development aloft and by overshooting cloud tops. Early observations with radar RHI scans by Donaldson (1958, 1962) showed rapid growth of the radar top some 20 minutes before tornadoes were observed on the ground. Figure 13.33 from Lemon (1977) shows the evolution of radar-measured cloud-top height over a 2.5-h period. Upsurges in cloud top above an altitude of 16 km are seen on four occasions. The first is followed (by 15 minutes) by 5 cm diameter hail at the ground. The three subsequent cloud-top excursions are each followed by tornadoes with lags ranging from 10 to 15 minutes. Lightning was not measured in either of these studies, but the record of lightning activity in other thunderstorms in which the radar cloud top is well resolved (e.g., Fig. 13.7) strongly suggests that pronounced peaks in intracloud flash rate will accompany the cloud-top maxima and thereby precede severe weather at ground level. Recent findings with LISDAD strongly support this idea (Williams et al. 1999). It is noteworthy that the percentage variations in cloud-top height in Figs. 13.7 and 13.33 are quite modest—of the order of 10%—but the variations in lightning are factors of 2. Because the cloud is already quite deep, the level of background lightning activity may also be quite large.

The upsurges in cloud-top height prior to tornadogenesis have also been observed from satellites. Adler and Fenn (1981) and Negri (1982) have reported systematic declines in cloud-top temperature to minimum values 10–20 minutes prior to tornadoes. McCann (1983) has noted a particular signature in satellite observations associated with overshooting cloud tops with comparable lead times to tornado occurrence. All these observations are consistent with the radar observations noted earlier. A rough picture emerging from all the observations is that the mesocyclonic vorticity is stretched by the updraft with which intracloud lightning is associated, and the tornado vorticity is stretched (and carried downward to the surface) by the downdraft that lags the total lightning activity.

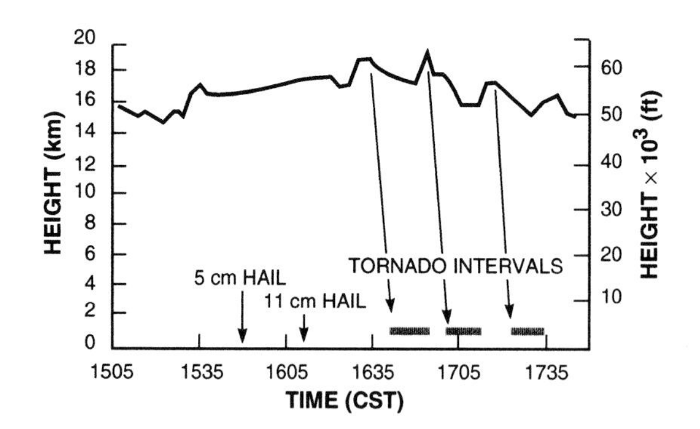

FIG. 13.33. Systematic evolution of radar cloud-top height and tornadogenesis for a tornadic supercell storm in Oklahoma on 18 June 1973. (From Lemon 1977.)

The strong evidence for a systematic evolution of the parent cloud, coupled with some improved knowledge about the large-scale electrical structure and considerably improved knowledge about the behavior of intracloud lightning, suggests that total lightning activity be reexamined with more sophisticated detectors (and in conjunction with PPI-type Doppler radar scanning) as a valuable precursor to tornadogenesis. It is important to recall the evidence that this pursuit will be most profitable for violent (F4, F5) tornadoes in supercells, and there are good reasons to believe that the bull's-eye in the simplified Venn diagram in Fig. 13.23 could be targeted. There are practical reasons for such a pursuit.

1. These storms, on account of their size and vertical velocity, are the most likely to exhibit a strong intracloud lightning precursor to both tornadoes and hail.
2. A disproportionate number of deaths are associated with violent tornadoes. [Gaffin and Smith (1995) indicate that 60% of deaths due to tornadoes are associated with F4 and F5 events, though the latter events constitute only 4% of all documented tornadoes.]
3. A disproportionate amount of property damage is also associated with F4 and F5 events. Storm data in Hales and Crowther (1991), for example, indicate that more than half of tornado property damage is associated with F4/F5 events.

5) Electrical energy for tornadoes

The physical origin of the intense wind in tornadoes has come under considerable scrutiny since the publication of the 1963 monograph on severe storms, and this issue deserves some review on both meteorological and electrical aspects. The conventional view holds that ambient angular momentum of air is concentrated by air motion caused by the release of latent heat, as discussed in an earlier section. Observations of extraordinary electrical activity in the Worcester, Massachusetts, tornadic storm in June 1953 (Vonnegut and Moore 1958) led Vonnegut (1960) to propose electrical heating as an explanation for observed tornado winds in excess of values achievable by the release of latent heat alone—the so-called *thermodynamic limit.* Ryan and Vonnegut (1971) subsequently demonstrated the formation of vigorous vortices in air caused by a high voltage discharge.

These ideas clearly influenced the electrical studies of tornadic storms in the decade to follow. Investigations of sferics activity in tornadic storms in the 1970s were clearly motivated by the idea that the electromagnetic emission originated within the funnel itself (Stanford et al. 1971; Lind et al. 1972), rather than high in the parent thunderstorm as described in sections 13.3a(1) and 13.3c(4). By 1973 Scouton et al. (1972) and Taylor (1973a) had concluded that the majority of RF noise did not emanate from the funnel. The discussion in the previous section and recent results (Williams et al. 1999) largely corroborate this finding.

Davies-Jones and Golden (1975a) criticized Vonnegut's (1960) original idea on the basis that lightning near or within the funnel was a rare event, an observation corroborated by recent storm chase teams (H. Bluestein 1996, personal communication; J. Wurman 1996, personal communication). Vonnegut (1975) countered this criticism by noting that electrical activity that may be contributing to tornado wind is obscured in daytime observations. Nighttime electrical displays, though generally unquantified in the context of electrical heating, are notably more dramatic (Vonnegut and Weyer 1966; Vaughan and Vonnegut 1976). In the latter observations, the most prevalent observation was the enhanced intracloud activity, consistent with the discussion in sections 13.2 and 13.3c(4). Unfortunately, no definitive conclusions were reached in the subsequent discussion (Colgate 1975; Davies-Jones and Golden 1975a; Vonnegut 1975; Davies-Jones and Golden 1975b) and attention to the original issue—the thermodynamic limit—was largely bypassed. It is interesting to note that Davies-Jones and Golden (1975b) did claim "little hard evidence for tornadic winds in excess of 110 m s^{-1}," and in an earlier paper, Kessler (1970) states, "We would be particularly urged to seek a fundamental role for electrical processes during genesis and maintenance of tornadoes, if tornadic winds much in excess of 200 mph (89 m s^{-1}) were proved."

In recent years, the existence of winds of this magnitude has been demonstrated by Doppler radar measurements at close range (Bluestein et al. 1993). Wind speeds of 120–125 m s^{-1} were found in one tornado, and based on a number of cases, the thermodynamic limit (Vonnegut 1960; Lilly 1969; Kessler 1970; Snow and Pawley 1984) was usually exceeded by 50%–100%. One possible explanation for the apparent violation of theory in the Doppler radar observations noted above is the invalidity of the original thermodynamic limit, a result claimed by Fiedler and Rotunno (1986). However, these authors' calculation of the most intense tornado wind speed is 110 m s^{-1}, a value still less than the observation of Bluestein et al. (1993). This rather modest discrepancy between theory and experiment may not be sufficient cause to resurrect the theory of electrical heating, though a contribution from the latter mechanism cannot be considered disproven. Clearly, additional work on the problem of tornadic winds, both theoretical and observational, is warranted. In light of the simultaneous roles of baroclinicity and conditional instability in tornadogenesis [section 13.3a(3)], it may be valuable to examine the departures from the thermodynamic limit in a diagram like Fig. 13.21 as observations of maximum wind speed improve.

d. *"Severe" lightning and sprites in nonsevere meteorological conditions*

Previous sections have emphasized that the most electrically active weather in the atmosphere is frequently severe. This observation might lead one to the conclusion that lightning deaths and lightning damage would also maximize in severe weather. Such a conclusion is probably false. Lightning injury and death are more likely when the potential victim is less aware of the hazard—near the beginning and near the end of a storm when conditions are likely nonsevere (Holle et al. 1995).

It is increasingly realized that the most energetic lightning, and by inference the most destructive to property and human life, is the positive ground flash that occurs in conditions of widespread stratiform precipitation (Krehbiel 1981; Rutledge and MacGorman 1988; Engholm et al. 1990; Williams and Boccippio 1993; Holle et al. 1994; Williams 1995b; Marshall et al. 1996; Williams 1998; Huang et al. 1999), generally resulting from the more violent convective activity that precedes it. The accompanying meteorological conditions—weak vertical air motions (tens of centimeters per second) and a laterally extensive (tens of kilometers) radar bright band (Boccippio 1996)—appear to be conducive to laterally extensive layers of electrical space charge. These layers appear to serve as large charge reservoirs for horizontally extensive "spider" lightning near the melting level. The lightning rate at this time may be only 0.1–0.2 flashes per minute, one order of magnitude less than in an ordinary thunderstorm and two to three orders of magnitude less than maximum rates in severe weather (Table 13.1). However, when the "spider" structure connects with the ground to produce a positive ground flash, hundreds and perhaps thousands of coulombs of charge can be transferred, far greater amounts than during active isolated convection (Krehbiel 1981; Brook et al. 1983). The measurement of peak current for these ground flashes by the National Lightning Detection Network may not adequately record the strength of these discharges because the long continuing current is not recorded.

It is now recognized that the large charge transfers associated with positive ground flashes are responsible (Boccippio et al. 1995; Pasko et al. 1995) for a newly discovered form of atmospheric discharge named a *sprite* (Franz et al. 1990; Sentman and Wescott 1993; Lyons 1994). The sprite is a short-lived (milliseconds), large-scale (tens of kilometers) illumination in the mesosphere (50–90 km in altitude), predominantly red in color, that often accompanies the positive ground flashes in mesoscale convective systems. The empirical area threshold of MCS necessary to produce sprites (20 000 km^2; Lyons 1996) would suggest that many sprites are preceded (by some hours) by some form of severe weather. More recently, P. Krehbiel and M. Stanley (1997, personal communication) have reported sprites over smaller Florida MCSs (2000 km^2) that probably never attained severe storm status.

13.4. Conclusions

The most basic conclusion of this review is not very profound: On average, severe weather is characterized by larger convective clouds, stronger updrafts and downdrafts, larger quantities of ice-phase precipitation, and higher lightning flash rates, both intracloud and cloud-to-ground. Recent studies (Weber et al. 1998; Williams et al. 1999) show a strong tendency for total lightning activity to lead severe weather on the ground. When individual case studies are examined, considerable variability is apparent, and relationships between lightning and other severe weather manifestations are less clear cut. For example, high flash rates (>30 flashes min^{-1}) are no guarantee for severe weather, and severe weather often occurs in the absence of clustered positive ground flash activity. Weak tornadoes can occur without any lightning. Some of this variability and apparent non-uniqueness is attributable to differences in the physical parameters that matter most to each manifestation of severe weather being compared. Other aspects of this variability are surely attributable to our still meager understanding of the detailed physical basis for cloud electrification. Laboratory studies of ice particle charging aimed at understanding the electrical behavior of ordinary thunderstorms had to be extended to a wider range of temperature and water content.

Sampling will always be a concern in investigations of severe weather that is inherently in the tail of the statistical distribution. This problem pertains to both academic and operational investigations. The greater quantitative use of lightning information, easily obtained from remote observations, may provide important improvements in this regard. The behavior of intracloud lightning activity shows the greatest promise here.

Acknowledgments. The author benefited greatly from discussions with the following individuals, many of whom have direct experience with severe weather: D. Atlas, H. Bluestein, R. Boldi, M. Brook, D. Burgess, L. Carey, S. Changnon, S. Clodman, W. Cotton, K. Cummins, J. Dessens, R. Donaldson, C. Doswell, G. Ellrod, S. Goodman, S. Hodanish, R. Holle, M. Isaminger, D. Knapp, K. Knupp, P. Krehbiel, E. P. Krider, R. Lascody, L. Lemon, C. Lennon, R. Lhermitte, W. Lyons, D. MacGorman, M. Maier, T. Marshall, A. Matlin, C. B. Moore, J. Moore, F. Mosher, S. Nelson, T. Nelson, R. Orville, W. Petersen, D. Proctor, E. Rasmussen, N. Renno, P. Richard, W. D. Rust, C. Saunders, A. Seimon, D. Sentman, D. Sharp, M. Stolzenburg, W. Taylor, O. H. Vaughan, B. Vonnegut, R. Wakimoto, M. Weber, G. Wescott, J. Wilson, J. Wurman, and D. Zrnic. David Knapp generously provided previously unpublished data on ground flashes in severe weather. I am particularly indebted to Steve Rutledge and his former students at Colorado State University (W. Petersen, L. Carey, and T. Schuur) for serving on the advisory committee for this paper.

REFERENCES

Adler, R. F., and D. D. Fenn, 1981: Satellite observed cloud top height changes in tornadic thunderstorms. *J. Appl. Meteor.*, **20,** 1369–1375.

Anthony, R. W., 1994: Severe thunderstorm climatological data for the new Jacksonville, Florida, county warning area. NOAA Tech. Memo. NWS SR-155, National Weather Service.

Atlas, D., 1966: The balance level in convective storms. *J. Atmos. Sci.*, **23,** 635–651.

Baker, B., M. B. Baker, E. R. Jayaratne, J. Latham, and C. P. R. Saunders, 1987: The influence of diffusional growth rates on the charge transfer accompanying rebounding collisions between ice crystals and soft hailstones. *Quart. J. Roy. Meteor. Soc.*, **113,** 1193–1215.

Balakrishnan, N., and D. Zrnic, 1990: Use of polarization to characterize precipitation and discriminate large hail. *J. Atmos. Sci.*, **47,** 1525–1540.

Baughman, R. G., and D. M. Fuquay, 1970: Hail and lightning occurrence in mountain thunderstorms. *J. Appl. Meteor.*, **9,** 657–660.

Blevins, L. L., and J. D. Marwitz, 1968: Visual observations of lightning in some Great Plains hailstorms. *Weather,* **23,** 192–194.

Bluestein, H. B. and M. H. Jain, 1985: Formation of mesoscale lines of precipitation: Severe squall lines in Oklahoma during the spring. *J. Atmos. Sci.*, **42,** 1711–1732.

———, and D. R. MacGorman, 1998: Evolution of cloud-to-ground lightning characteristics and storm structure in the Spearman, Texas, tornadic supercells of 31 May 1990. *Mon. Wea. Rev.*, **126,** 1451–1467.

———, E. W. McCaul Jr., G. P. Byrd, and G. R. Woodall, 1988: Mobile sounding observations of a thunderstorm near the dryline: The Canadian, Texas storm complex of 7 May 1986. *Mon. Wea. Rev.*, **116,** 1790–1804.

———, ———, ———, ———, G. Martin, S. Keighton, and L. C. Showell, 1989: Mobile sounding observations of a thunderstorm near the dryline: The Gruver, Texas storm complex of 25 May 1987. *Mon. Wea. Rev.*, **117,** 244–250.

———, W. P. Unruh, J. LaDue, H. Stein, and D. Speheger, 1993: Doppler radar wind spectra of supercell tornadoes. *Mon. Wea. Rev.*, **121,** 2200–2221.

Boccippio, D., 1996: The electrification of stratiform anvils. Ph.D. thesis, Massachusetts Institute of Technology, 234 pp.

———, 2001: Lightning scaling relations revisited. *J Atmos. Sci.*, (in press).

———, E. R. Williams, S. J. Heckman, W. A. Lyons, I. Baker, and R. Boldi, 1995: Sprites, ELF transients, and positive ground strokes. *Science,* **269,** 1088–1091.

Bonner, W. D., and J. E. Kemper, 1971: Broad scale relations between radar and severe weather reports, Preprints, *Seventh Conf. on Severe Local Storms,* Kansas City, MO, Amer. Meteor. Soc., 140–147.

Brannick, M. L., and C. A. Doswell III, 1992: An observation of the relationship between supercell structure and lightning ground-strike polarity. *Wea. Forecasting,* **7,** 143–149.

Bringi, V., J. Vivekanandan, and J. D. Tuttle, 1986: Multiparameter radar measurements in Colorado convective storms. Part II: Hail detection studies. *J. Atmos. Sci.,* **43,** 2564–2577.

Brook, M., M. Nakano, P. Krehbiel, and T. Takeuti, 1982: The electrical structure of the Hokuriku winter thunderstorms. *J. Geophys. Res.,* **87,** 1207–1215.

———, P. Krehbiel, D. MacLaughlan, T. Takeuti, and M. Nakano, 1983: Positive ground stroke observations in Japanese and Florida storms. *Proceedings in Atmospheric Electricity,* L. H. Ruhnke and J. Latham, Eds., Deepak Publishing, 365–369.

Browning, K., and G. B. Foote, 1976: Air flow and hail growth in supercell storms and some implications for hail suppression. *Quart. J. Roy. Meteor. Soc.,* **102,** 499–533.

Buechler, D. E., and S. J. Goodman, 1990: Echo size and asymmetry: Impact on NEXRAD storm identification. *J. Appl. Meteor.,* **29,** 962–969.

———, K. Driscoll, S. J. Goodman, and H. J. Christian, 2000: Lightning activity within a tornadic storm observed by the Optical Transient Detector. *Geophys. Res. Lett.,* **27,** 2253–2256.

Burgess, D. W., and L. R. Lemon, 1990: Severe thunderstorm detection by radar. *Radar Meteorology: Battan Memorial and 40th Anniversary Radar Meteorology Conference,* D. Atlas, Ed., Amer. Meteor. Soc., 619–647.

Burke, C. P., and D. L. Jones, 1996: On the polarity and continuous currents in unusually large lightning flashes deduced from ELF events. *J. Atmos. Terr. Phys.,* **58,** 531–540.

Byers, H. R., 1944: *General Meteorology.* McGraw Hill, 606 pp.

———, and R. R. Braham, 1949: *The Thunderstorm.* U.S. Government Printing Office, 282 pp.

Carey, L. D., and S. R. Rutledge, 1996: A multiparameter radar case study of the microphysical and kinematic evolution of a lightning producing storm. *Meteor. Atmos. Phys.,* **59,** 33–64.

———, and ———, 1998: Electrical and multiparameter radar observations of a severe hailstorm. *J. Geophys. Res.,* **103** (D12), 13 979–14 000.

Carlson, T. N., and F. H. Ludlam, 1968: Conditions for the occurrence of severe local storms. *Tellus,* **20,** 203–226.

Changnon, S. A., 1992: Temporal and spatial relations between hail and lightning. *J. Appl. Meteor.,* **31,** 587–604.

Cheze, J.-L., and H. Sauvageot, 1997: Area-average rainfall and lightning activity. *J. Geophys. Res.,* **102,** 1707–1716.

Colgate, S. A., 1975: Comment on "On the relation of electrical activity to tornadoes" by R. P. Davies-Jones and J. H. Golden. *J. Geophys. Res.,* **80,** 4556.

Colquhoun, J. R., and P. A. Riley, 1996: Relationship between tornado intensity and various wind and thermodynamic variables. *Wea. Forecasting,* **11,** 360–371.

Cruz, L. A., 1973: Venezuelan rainstorms as seen by radar. *J. Appl. Meteor.,* **12,** 119–126.

Cummins, K. L., M. J. Murphy, E. A. Bardo, W. L. Hiscox, R. B. Pyle, and A. E. Pifer, 1998: A combined TOA/MDF technology upgrade of the U.S. National Lightning Detection Network. *J. Geophys. Res.,* **103,** 9035–9044.

Curran, E. B., and W. D. Rust, 1992: Positive ground flashes produced by low-precipitation thunderstorms in Oklahoma on 26 April 1984. *Mon. Wea. Rev.,* **120,** 544–553.

Darrah, R. P., 1978: On the relationship of severe weather to radar tops. *Mon. Wea. Rev.,* **106,** 1332–1339.

Davies-Jones, R. P., and J. H. Golden, 1975a: On the relation of electrical activity to tornadoes. *J. Geophys. Res.,* **80,** 1614–1616.

———, and ———, 1975b: Reply. *J. Geophys. Res.,* **80,** 4561–4558.

Dickson, E. B., and R. J. McConahy, 1956: Sferics readings on windstorms and tornadoes. *Bull. Amer. Meteor. Soc.,* **37,** 410–412.

Donaldson, R. J., 1958: Analysis of severe convective storms observed by radar. *J. Meteor.,* **15,** 44–50.

———, 1962: Radar observations of a tornado thunderstorm in vertical section. National Severe Storms Project Rep. 8.

Doswell, C. A., 1985: The Operational Meteorology of Convective Weather, Vol. II: Storm Scale Analysis. NOAA Tech. Memo ERL ESG-15, 240 pp.

———, and H. E. Brooks, 1993: Comments on "Anomalous cloud-to-ground lightning in an F5 tornado-producing supercell thunderstorm on 28 August, 1990." *Bull. Amer. Meteor. Soc.,* **74,** 2213–2218.

———, and D. W. Burgess, 1993: Tornadoes and tornadic storms: A review of conceptual models. *The Tornado, Its Structure, Dynamics, Predictions, and Hazards, Geophys. Monogr.,* No. 79, Amer. Geophys. Union, 162–172.

———, J. T. Schaefer, D. W. McCann, T. W. Schlatter, and H. B. Wobus, 1982: Thermodynamic analysis procedures at the National Severe Storms Forecast Center. Preprints, *Ninth Conf. on Weather Forecasting and Analysis,* Seattle, WA, Amer. Meteor. Soc., 304–309.

———, A. R. Moller, and R. Przybylinski, 1990: A unified set of conceptual models for variations on the supercell theme. Preprints, *16th Conf. on Severe Local Storms,* Kananaskis Park, AB, Canada, Amer. Meteor. Soc., 40–45.

Dye, J. E., and Coauthors, 1986: Early electrification and precipitation development in a small, isolated Montana cumulonimbus. *J. Geophys. Res.,* **91,** 1231–1247.

Engholm, C. D., E. R. Williams, and R. M. Dole, 1990: Meteorological and electrical conditions associated with positive cloud-to-ground lightning. *Mon. Wea. Rev.,* **118,** 470–487.

Fawbush, E. J., and R. C. Miller, 1953: A method for forecasting hailstone size at the earth's surface. *Bull. Amer. Meteor. Soc.,* **34,** 235–244.

Fiedler, B. H., and R. Rotunno, 1986: A theory for the maximum windspeeds in tornado-like vortices. *J. Atmos. Sci.,* **43,** 2328–2340.

Franz, R. C., R. J. Nemzek, and J. R. Winkler, 1990: Television image of a large upward electrical discharge above a thunderstorm system. *Science,* **249,** 48–51.

Frisby, E. M., and H. W. Sansom, 1967: Hail incidence in the tropics. *J. Appl. Meteor.,* **6,** 339–354.

Frost, R., 1954: Cumulus and cumulonimbus cloud over Malaya. Meteorological Office Air Ministry, Met. Rep. 15.

Fujita, T., 1985: *The Downburst: Microburst and Macroburst.* Satellite and Mesometeorology Research Project, University of Chicago, 1–122.

———, 1992: *The Mystery of Severe Storms.* University of Chicago Press.

Fuquay, D. M., 1982: Positive cloud-to-ground lightning in summer thunderstorms. *J. Geophys. Res.,* **87,** 7131–7140.

Gaffin, D., and R. Smith, 1995: Severe weather climatology for the NW SFO Memphis County Warning Area. NOAA Tech. Memo. NWS-169.

Golden, J. H., 1971: Tornadoes and waterspouts over South Florida. *Mon. Wea. Rev.,* **99,** 146–154.

Goodman, S. J., 1990: Predicting thunderstorm evolution using ground-based lightning detection networks. NASA TM-103521, 193 pp. [Available from NTIS, Springfield, VA 22161-2171.]

———, and K. R. Knupp, 1993: Tornadogenesis via squall line and supercell interaction: The Nov. 15, 1989 Huntsville, Alabama tornado. *The Tornado: Its Structure, Dynamics, Prediction, and Hazards, Geophys. Monogr.,* No. 79, Amer. Geophys. Union, 183–189.

———, D. E. Buechler, P. D. Wright, and W. D. Rust, 1988: Lightning and precipitation history of a microburst-producing storm. *Geophys. Res. Lett.,* **15,** 1185–1188.

Grosh, R. C., 1977: Relationships between severe weather and echo tops in central Illinois. Preprints, *Tenth Conf. on Severe Local Storms.* Omaha, NE, Amer. Meteor. Soc., 231–238.

Gunn, R., 1956: Electric field intensity at the ground under active thunderstorms and tornadoes. *J. Meteor.,* **13,** 269–273.

Hales, J. E., and H. G. Crowther, 1991: Severe thunderstorm cases of July 1989 thru June 1990. NOAA Tech. Memo. NWS NSSFC-29.

Heymsfield, A. J., and D. J. Musil, 1982: Case study of a hailstorm in Colorado. Part II: Particle growth processes at mid-levels deduced from in-situ measurements. *J. Atmos. Sci.,* **39,** 2847–2866.

———, and M. R. Hjelmfelt, 1984: Processes of hydrometeor development in Oklahoma convective clouds. *J. Atmos. Sci.,* **41,** 2811–2835.

Hoffert, S. G., and M. L. Pearce, 1996: The 29 July 1994 Merritt Island, Florida microburst: A case study intercomparing Kennedy Space Center three-dimensional lightning data (LDAR) and WSR-880 radar data. Preprints, *18th Conf. on Severe Local Storms,* San Francisco, CA, Amer. Meteor. Soc., 424–427.

Holle, R. L., and M. W. Maier, 1982: Radar echo height related to cloud-ground lightning in South Florida. Preprints, *12th Conf. on Severe Local Storms,* San Antonio, TX, Amer. Meteor. Soc., 330–333.

———, A. I. Watson, R. E. Lopez, D. R. MacGorman, R. Ortiz, and W. D. Otto, 1994: The life cycle of lightning and severe weather in a 3–4 June 1985 PRE-STORM mesoscale convective system. *Mon. Wea. Rev.,* **122,** 1798–1808.

———, R. E. Lopez, K. W. Howard, J. Varrek, and J. Allsopp, 1995: Safety in the presence of lightning. *Seminars in Neurology,* **15,** 375–380.

Huang, E., E. Williams, R. Boldi, S. Heckman, W. Lyons, M. Taylor, T. Nelson, and C. Wong, 1999: Criteria for sprites and elves based on Schumann resonance methods. *J. Geophys. Res.,* **104,** 16 943–16 964.

Illingworth, A. J., 1985: Charge separation in thunderstorms: Small scale processes. *J. Geophys. Res.,* **90,** 6026–6032.

Johns, R. H., J. M. Davies, and P. W. Leftwich, 1990: An examination of the relationship of 0–2 km AGL "positive" wind shear to potential buoyant energy in strong and violent tornado situations. Preprints, *16th Conf. on Severe Local Storms,* Kananaskis Park, AB, Canada, Amer. Meteor. Soc., 593–602.

———, ———, and ———, 1993: Strong wind and instability parameters associated with strong and violent tornadoes. 2: Variations in the combinations of wind and instability parameters. *The Tornado: Its Structure, Dynamics, Predictions, and Hazards, Geophys. Monogr.,* No. 79, Amer. Geophys. Union, 583–590.

Johnson, H. L., R. D. Hart, M. A. Lind, R. E. Powell, and J. L. Stanford, 1977: Measurements of radio frequency noise from severe and nonsevere thunderstorms. *Mon. Wea. Rev.,* **105,** 734–747.

Jones, H. L., 1958: The identification of lightning discharges by sferic characteristics. *Recent Advances in Atmospheric Electricity,* L. G. Smith, Ed., Pergamon Press, 543–556.

Kane, R. J., 1991: Correlating lightning to severe local storms in the northeastern United States. *Wea. Forecasting,* **6,** 3–12.

Keighton, S. J., H. B. Bluestein, and D. R. MacGorman, 1991: The evolution of a severe mesoscale convective system: Cloud-to-ground lightning location and storm structure. *Mon. Wea. Rev.,* **119,** 1533–1556.

Kelly, D. L., J. T. Schaefer, and C. A. Doswell III, 1985: Climatology of nontornadic severe thunderstorm events in the United States. *Mon. Wea. Rev.,* **113,** 1997–2014.

Kessler, E., 1970: Tornadoes. *Bull. Amer. Meteor. Soc.,* **51,** 926–936.

Knapp, D. I., 1994: Using cloud-to-ground lightning data to identify tornadic thunderstorm signatures and nowcast severe weather. *Natl. Wea. Dig.,* **19,** 35–42.

Krehbiel, P. R., 1981: An analysis of the electric field change produced by lightning. Ph.D. thesis, University of Manchester, England, 245 pp.

———, 1986: The electrical structure of thunderstorms. *The Earth's Electrical Environment,* National Academy Press, 90–113.

———, R. J. Thomas, W. Rison, T. Hamlin, J. Hamlin, and M. Davis, 2000: GPS-based mapping system reveals lightning inside storms. *EOS, Trans. Amer. Geophys. Union,* **81,** 21.

Krider, E. P., 1989: Electric field changes and cloud electrical structure. *J. Geophys. Res.,* **94,** 13 145–13 149.

Lang, T. J., S. A. Rutledge, J. E. Dye, M. Venticinque, P. Laroche, and E. Defer, 2000: Anomalously low negative cloud-to-ground lightning flash rates in intense convective storms observed during STERAO-A. *Mon. Wea. Rev.,* **128,** 160–173.

Larson, H. R., and E. J. Stansbury, 1974: Association of lightning flashes with precipitation cores extending to height 7 km. *J. Atmos. Terr. Phys.,* **36,** 1547–1553.

Lascody, R. L., and D. W. Sharp, 1996: The record hailstorms of Florida during March, 1992. NWS Office, Melbourne, Florida.

Latham, J., 1981: The electrification of thunderstorms. *Quart. J. Roy. Meteor. Soc.,* **107,** 277–298.

Lemon, L. R., 1977: New severe thunderstorm radar identification techniques and warning criteria: A preliminary report. NOAA Tech. Memo., NWS NSSFC-1, 60 pp.

———, 1998: The radar "three-body scatter spike": An operational large-hail signature. *Wea. Forecasting,* **13,** 327–340.

Lhermitte, R. M., and P. R. Krehbiel, 1979: Doppler radar and radio observations of thunderstorms. *IEEE Trans. Geosci. Electron.,* GE-17 (4), 162–171.

———, and E. R. Williams, 1985: Thunderstorm electrification: A case study. *J. Geophys. Res.,* **90,** 6071–6078.

Lilly, D., 1969: Tornado dynamics. NCAR Manuscript 69–117, 52 pp.

Lind, M. A., J. S. Hartman, E. S. Takle, and J. L. Stanford, 1972: Radio noise studies of several severe weather events in Iowa in 1971. *J. Atmos. Sci.,* **29,** 1220–1223.

Lopez, R. E., and J.-P. Aubagnac, 1997: The lightning activity of a hailstorm as a function of changes in its microphysical characteristics inferred from polarimetric radar observations. *J. Geophys. Res.,* **102** (D4), 16 799–16 813.

Ludlam, F. H., 1963: Severe local storms: A review. *Severe Local Storms, Meteor. Monogr.,* No. 27, Amer. Meteor. Soc., 1–30.

Lyons, W. A., 1994: Characteristics of luminous structures in the stratosphere above thunderstorms as imaged by low-light video. *Geophys. Res. Lett.,* **21,** 875.

———, 1996: Sprite observations above the U.S. High Plains in relation to their parent thunderstorm systems. *J. Geophys. Res.,* **101** (D23), 29 641–29 652.

MacGorman, D. R., 1993: Lightning in tornadic storms: A review. *The Tornado: Its Structure, Dynamics, Prediction, and Hazards, Geophys. Monogr.,* No. 79, Amer. Geophys. Union, 173–182.

———, and K. E. Nielsen, 1991: Cloud-to-ground lightning in a tornadic storm on 8 May 1986. *Mon. Wea. Rev.,* **119,** 1557–1574.

———, and D. W. Burgess, 1994: Positive cloud-to-ground lightning in tornadic storms and hail storms. *Mon. Wea. Rev.,* **122,** 1671–1697.

———, ———, V. Mazur, W. D. Rust, W. L. Taylor, and B. C. Johnson, 1989: Lightning rates relative to tornadic storm evolution on 22 May, 1981. *J. Atmos. Sci.,* **46,** 221–250.

Malan, D. J., 1958: Radiation from lightning discharges and its relation to the discharge process. *Recent Advances in Atmospheric Electricity,* L. G. Smith, Ed., Pergamon Press, 557–563.

Malherbe, C., J. Pigère, P. Blanchet, C. Plessis, O. Desté, A. Bondiou, and P. LaRoche, 1992: Relation entre l'activité

electrique d'un orage et le développement du microbursts. Experience MIT/ONERA, Orlando 1991 Final Rep. 5/6154P4.
Marsh, S. J., and T. C. Marshall, 1993: Charged precipitation measurements before the first lightning flash in a thunderstorm. *J. Geophys. Res.,* 16 605–16 611.
Marshall, J. S., and S. Radhakant, 1978: Radar precipitation maps as lightning indicators. *J. Appl. Meteor.,* **17,** 206–212.
Marshall, T. C., W. D. Rust, and M. Stolzenberg, 1995: Electrical structure and updraft speeds in thunderstorms over the southern Great Plains. *J. Geophys. Res.,* **100,** 1001–1015.
———, M. Stolzenberg, and W. D. Rust, 1996: Electric field measurements above mesoscale convective systems. *J. Geophys. Res.,* **101,** 6979–6996.
McCann, D. W., 1983: The enhanced-V: A satellite observable severe storm signature. *Mon. Wea. Rev.,* **111,** 887–894.
McCaul, E. W., Jr., 1991: Buoyancy and shear characteristics of hurricane-tornado environments. *Mon. Wea. Rev.,* **119,** 1954–1978.
———, 1993: Observations and simulations of hurricane-spawned tornadic storms. *The Tornado: Its Structure, Dynamics, Predictions, and Hazards, Geophys. Monogr.,* No. 79, Amer. Geophys. Union.
Moore, J. T., and J. P. Pino, 1990: An interactive method for estimating maximum hailstone size from forecast soundings. *Wea. Forecasting,* **5,** 508–525.
Mosher, F. B., and J. S. Lewis, 1990: Use of lightning data in severe storm forecasting. Preprints, *Conf. on Atmospheric Electricity,* Tucson, AZ, Amer. Meteor. Soc., 692–697.
Murphy, M. J., E. P. Krider, and M. W. Maier, 1996: Lightning charge analyses in small convective and precipitation electrification (CAPE) experiment storms. *J. Geophys. Res.,* **101,** 29 615–29 625.
Musil, D. J., A. J. Heymsfield, and P. L. Smith, 1986: Microphysical characteristics of a well-developed weak echo region in a high plains supercell thunderstorm. *J. Climate Appl. Meteor.,* **25,** 1037–1051.
Negri, A. J., 1982: Cloud top structure of tornadic storms on 10 April 1979 from rapid scan and stereo satellite observations. *Bull. Amer. Meteor. Soc.,* **63,** 1151–1159.
Nelson, S. P., 1983: The influence of storm flow structure on hail growth. *J. Atmos. Sci.,* **40,** 1965–1983.
———, and S. K. Young, 1979: Characteristics of Oklahoma hailfalls and hailstorms. *J. Appl. Meteor.,* **18,** 339–347.
Nisbet, J. S., T. A. Barnard, G. S. Forbes, E. P. Krider, R. Lhermitte, and C. L. Lennon, 1990: A case study of the thunderstorm international project storm of July 11, 1978. 1: Analysis of the data base. *J. Geophys. Res.,* **95,** 5417–5433.
Orville, R. E., and G. R. Huffines, 2001; Cloud-to-ground lightning in the United States: HLDN results in the first decade 1989–98. *Mon. Wea. Rev.,* **129,** 1179–1193.
———, R. A. Weisman, R. B. Pyle, and R. W. Henderson, 1985: Cloud-to-ground lightning flash characteristics from June 1984 through May 1985. *J. Geophys. Res.,* **92,** 5640–5644.
Pakiam, J. E., and J. Maybank, 1975: The electrical characteristics of some severe hailstorms in Alberta, Canada. *J. Meteor. Soc. Japan,* **53,** 363–383.
Pasko, V. P., U. S. Inan, Y. N. Taranenko, and T. F. Bell, 1995: Heating, ionization and upward discharges in the mesosphere due to intense quasi-electrostatic thundercloud fields. *Geophys. Res. Lett.,* **22,** 365–368.
Perez, A. H., R. E. Orville, and L. J. Wicker, 1995: Characteristics of cloud-to-ground lightning associated with violent-tornado producing supercells. Preprints, *14th Conf. on Weather and Forecasting*, Dallas, TX, Amer. Meteor. Soc., 409–413.
———, ———, and ———, 1997: Characteristics of cloud-to-ground lightning associated with violent tornadoes. *Wea. Forecasting,* **12,** 428–437.
Petersen, W. A., G. D. Green, and S. A. Rutledge, 1995: Use of NEXRAD radar data and NLDN lightning data in the analysis of Arizona monsoon thunderstorms. Preprints, *27th Conf. on Radar Meteorology,* Vail, CO, Amer. Meteor. Soc., 642–644.
———, S. A. Rutledge, and R. E. Orville, 1996: Cloud-to-ground lightning observations from TOGA COARE: Selected results and lightning location algorithms. *Mon. Wea. Rev.,* **124,** 602–620.
Rasmussen, R. M., and A. J. Heymsfield, 1987: Melting and shedding of graupel and hail. Part II: Sensitivity study. *J. Atmos. Sci.,* **44,** 2764.
Ray, P. S., D. R. MacGorman, W. D. Rust, W. L. Taylor, and L. W. Rasmussen, 1987: Lightning location relative to storm structure in a supercell storm and a multicell storm. *J. Geophys. Res.,* **92,** 5713–5724.
Reap, R., and D. R. MacGorman, 1989: Cloud-to-ground lightning: Climatological characteristics and relationships to model fields, radar observations and severe local storms. *Mon. Wea. Rev.,* **117,** 518–535.
Reynolds, S. E., M. Brook, and M. F. Gourley, 1957: Thunderstorm charge separation. *J. Meteor.,* **14,** 426–436.
Rison, W., P. Krehbiel, L. Maier, and C. Lennon, 1996: Comparison of lightning and radar observations on a small storm at Kennedy Space Center, Florida. Preprints, *Tenth Int. Conf. on Atmospheric Electricity*, Osaka, Japan, Amer. Meteor. Soc., 196–199.
Roach, W. T., 1967: On the nature of the summit areas of severe storms in Oklahoma. *Quart. J. Roy. Meteor. Soc.,* **93,** 318–336.
Rosenfeld, D., and W. L. Woodley, 2000: Convective clouds with sustained highly supercooled liquid water down to −37.5°C. *Nature*, **405,** 440–442.
Rossow, V., 1970: Observations of water spouts and their parent clouds. NASA Tech. Note D-5854, 63 pp.
Rust, W. D., and T. C. Marshall, 1996: On abandoning the thunderstorm tripole-charge paradigm. *J. Geophys. Res.,* **101,** 23 499–23 504.
———, D. R. MacGorman, and R. T. Arnold, 1981a: Positive cloud-to-ground lightning flashes in severe storms. *Geophys. Res. Lett.,* **8,** 791–794.
———, W. L. Taylor, D. R. MacGorman, and R. T. Arnold, 1981b: Research on electrical properties of severe thunderstorms in The Great Plains. *Bull. Amer. Meteor. Soc.,* **62,** 1286–1293.
Rutledge, S. A., and D. R. MacGorman, 1988: Cloud-to-ground lightning activity in the 10–11 June 1985 mesoscale convective system observed during PRE-STORM. *Mon. Wea. Rev.,* **116,** 1393–1408.
———, and W. A. Petersen, 1994: Vertical radar reflectivity structure and cloud-to-ground lightning in the stratiform region of MCSs: Further evidence for in situ charging in the stratiform region. *Mon. Wea. Rev.,* **122,** 1760–1776.
———, C. Lu, and D. R. MacGorman, 1990: Positive cloud-to-ground lightning in mesoscale convective systems. *J. Atmos. Sci.,* **47,** 2085–2100.
———, E. R. Williams, and T. D. Keenan, 1992: The Down Under Doppler and Electricity Experiment (DUNDEE): Overview and preliminary results. *Bull. Amer. Meteor. Soc.,* **73,** 3–16.
———, ———, and W. A. Petersen, 1993: Lightning and electrical structure of mesoscale convective systems. *Atmos. Res.,* **29,** 27–53.
Ryan, R. T., and B. Vonnegut, 1971: Formation of a vortex by an elevated electrical heat source. *Nature,* **233,** 142–143.
Rydock, J., and E. Williams, 1991: Charge separation associated with frost growth. *Quart. J. Roy. Meteor. Soc.,* **117,** 409–420.
Saunders, C. P. R., 1994: Thunderstorm electrification laboratory experiments and charging mechanisms. *J. Geophys. Res.,* **99,** 10 773–10 779.
———, 1995: Thunderstorm electrification. *Handbook of Atmospheric Electrodynamics,* H. Volland, Ed., CRC Press, 61–92.
———, and I. Brooks, 1992: The effects of high liquid water content on thunderstorm charging. *J. Geophys. Res.,* **97,** 14 671–14 676.

———, W. D. Keith, and R. P. Mitzeva, 1991: The effect of liquid water content on thunderstorm charging. *J. Geophys. Res.,* **96,** 11 007–11 017.

Schaefer, V. J., and R. J. Cheng, 1971: The production of ice crystal fragments by sublimation and electrification. *J. Rech. Atmos.,* **5,** 5–10.

Scouton, D. C., D. T. Stephenson, and W. G. Biggs, 1972: A sferic rate azimuth-profile of the 1955 Blackwell, Oklahoma, tornado. *J. Atmos. Sci.,* **29,** 929–936.

Seimon, A., 1993a: Anomalous cloud-to-ground lightning in an F5 tornado-producing supercell thunderstorm on 28 August 1990. *Bull. Amer. Meteor. Soc.,* **74,** 189–203.

———, 1993b: Reply. *Bull. Amer. Meteor. Soc.,* **74,** 2218–2220.

Sentman, D. D., 1988: Polarity of ultralarge lightning strokes inferred from discrete Schumann resonance excitations. *EOS, Trans. Amer. Geophys. Union,* **69** (44), 1069.

———, and E. M. Wescott, 1993: Observations of upper atmosphere optical flashes recorded from an aircraft. *Geophys. Res. Lett.,* **20,** 2857.

Shackford, C. R., 1960: Radar indications of a precipitation-lightning relationship in New England thunderstorms. *J. Meteor.,* **17,** 15–19.

Shafer, M. A., and F. H. Carr, 1990: Cloud-to-ground lightning in relation to digitized radar data in severe storms. Preprints, *Conf. on Atmospheric Electricity,* Tucson, AZ, Amer. Meteor. Soc., 732–737.

Silberg, P. A., 1965: Passive electrical measurements from three Oklahoma tornadoes. *Proc. IEEE,* **53,** 1197–1204.

Simpson, G. C., and F. J. Scrase, 1937: The distribution of electricity in thunderclouds. *Proc. Roy. Soc. London, Ser. A,* **161,** 309–352.

Snow, J. T., and R. L. Pauley, 1984: On the thermodynamic method for estimating maximum tornado windspeeds. *J. Climate Appl. Meteor.,* **23,** 1465–1468.

Srivastava, R. C., 1987: A model of intense downdrafts driven by the melting and evaporation of precipitation. *J. Atmos. Sci.,* **44,** 1751–1773.

Stanford, J. L., M. A. Lind, and G. S. Takle, 1971: Electromagnetic noise studies of severe convective storms in Iowa: The 1970 storm season. *J. Atmos. Sci.,* **28,** 436–448.

Stanley, M., P. Krehbiel, L. Maier, and C. Lennon, 1996: Comparison of lightning observations from the KSC LDAR system with NEXRAD radar observations, Preprints, *Tenth Int. Conf. on Atmospheric Electricity,* Osaka, Japan, Amer. Meteor. Soc., 224–227.

Stolzenburg, M., 1994: Observations of high ground flash densities of positive lightning in summer thunderstorms. *Mon. Wea. Rev.,* **122,** 1740–1750.

———, 1996: An observational study of electrical structure in convective regions of mesoscale convective systems. Ph.D. thesis, University of Oklahoma, 137 pp.

———, T. C. Marshall, W. D. Rust, and B. F. Smull, 1994: Horizontal distribution of electrical and meteorological conditions across the stratiform region of a mesoscale convective system. *Mon. Wea. Rev.,* **122,** 1777–1797.

Takahashi, T., 1978: Riming electrification as a charge generation mechanism in thunderstorms. *J. Atmos. Sci.,* **35,** 1536–1548.

Taylor, W. L., 1972: Atmospherics and severe storms. *Remote Sensing of the Troposphere,* V. E. Derr, Ed., U.S. Government Printing Office, 17-1–17-17.

———, 1973a: Electromagnetic radiation from severe storms in Oklahoma during April 29–30, 1970. *J. Geophys. Res.,* **78,** 8761–8777.

———, 1973b: Evaluation of an electromagnetic tornado-detecting technique. Preprints, *Eighth Conf. on Severe Local Storms,* Denver, CO, Amer. Meteor. Soc., 165–168.

———, 1975: Detecting tornadic storms by the burst rate nature of electromagnetic signals they produce. Preprints, *Ninth Conf. on Severe Local Storms,* Norman, OK, Amer. Meteor. Soc., 311–316.

———, E. A. Brandes, W. D. Rust, and D. R. MacGorman, 1984: Lightning activity and severe storm structure. *Geophys. Res. Lett.,* **11,** 545–548.

Trost, B. N., and C. E. Nomikos, 1975: VHF radio emissions associated with tornadoes. *J. Geophys. Res.,* **80,** 4117–4118.

Turman, B. N., and R. J. Tettelbach, 1980: Synoptic scale satellite lightning observations in conjunction with tornadoes. *Mon. Wea. Rev.,* **108,** 1878–1882.

Ushio, T., S. J. Heckman, D. J. Boccippio, and H. J. Christian, 2001: A survey of thunderstorm flash rated compared to cloud top height using TRMM satellite data. *J Geophys. Res.,* in press.

U.S. Weather Bureau, 1947: Thunderstorm rainfall. Hydrometeorological Rep. 5, 155 pp.

Vaughan, O. H., Jr., and B. Vonnegut, 1976: Luminous electrical phenomena associated with nocturnal tornadoes in Huntsville, Alabama, 3 April 1974. *Bull. Amer. Meteor. Soc.,* **57,** 1220–1224.

Vonnegut, B., 1960: Electrical theory of tornadoes. *J. Geophys. Res.,* **65,** 203–212.

———, 1963: Some facts and speculations concerning the origin and role of thunderstorm electricity. *Severe Local Storms, Meteor. Monogr.,* No. 5, 224–241.

———, 1975: Comment on "On the relation of electrical activity to tornadoes" by R. P. Davies-Jones and J. H. Golden. *J. Geophys. Res.,* **80,** 4559–4560.

———, and C. B. Moore, 1957: Electrical activity associated with the Blackwell-Udall tornado. *J. Meteor.,* **14,** 284–285.

———, and ———, 1958: Giant electrical storms. *Recent Advances in Atmospheric Electricity,* L. G. Smith, Ed., Pergamon Press, 399–411.

———, and J. R. Weyer, 1966: Luminous phenomena in nocturnal tornadoes. *Science,* **153,** 1213–1220.

———, and C. B. Moore, 1989: Comments on "The electrification of thunderstorms." *Sci. Amer.,* **261,** 8.

Wacker, R. S., and R. E. Orville, 1999: Changes in measured lightning flash count and return stroke peak current after the 1994 U.S. National Lightning Detection Network upgrade. 1: Observations and 2: Theory. *J. Geophys. Res.,* **104,** 2151–2162.

Wakimoto, R. M., and J. K. Lew, 1993: Observations of a Florida water spout during CaPE. *Wea. Forecasting,* **8,** 412–423.

Watson, A. I., R. L. Holle, and R. E. Lopez, 1995: Lightning from two national detection networks related to vertically integrated liquid and echo top information from WSR-88D radar. *Wea. Forecasting,* **10,** 592–605.

Weber, M. E., E. R. Williams, M. Wolfson, and S. J. Goodman, 1998: An assessment of the operational utility of a GOES lightning mapping sensor. Project Rep. NOAA-18, MIT Lincoln Laboratory, Lexington, MA, 108 pp.

Wescott, E. M., D. D. Sentman, M. J. Heavner, D. L. Hampton, D. L. Osborne, and O. H. Vaughan Jr., 1996: Blue starters: Brief upward discharges from an intense Arkansas thunderstorm. *J. Geophys. Res.,* **23,** 2153–2150.

Williams, E. R., 1981: Thunderstorm electrification: Precipitation versus convection. Ph.D. thesis, Massachusetts Institute of Technology, 254 pp.

———, 1985: Large scale charge separation in thunderclouds. *J. Geophys. Res.,* **90,** 6013–6025.

———, 1988: The electrification of thunderstorms. *Sci. Amer.,* **259,** 88–89.

———, 1989a: Reply. *Sci. Amer.,* **261,** 8.

———, 1989b: The tripole structure of thunderstorms. *J. Geophys. Res.,* **94,** 13 151–13 167.

———, 1991a: The structure of tropical convection and its contribution to the global electrical circuit. Preprints, *25th Conf. on Radar Meteorology,* Paris, France, Amer. Meteor. Soc., 893–896.

———, 1991b: Comments on "Thunderstorms above frontal surfaces in environments without positive CAPE. Part I: A climatology." *Mon. Wea. Rev.,* **119,** 2511–2513.

———, 1995a: Meteorological aspects of thunderstorms. *CRC Handbook on Atmospheric Electrodynamics,* Vol. I., H. Volland, Ed., CRC Press, 27–60.

———, 1995b: Comment on "Thunderstorm electrification laboratory experiments and charging mechanisms," by C. P. R. Saunders. *J. Geophys. Res.,* **100,** 1503–1505.

———, 1996: Comment on "A climatological study of tropical thunderstorm clouds and lightning frequencies on the French Guyana Coast" by J. Molinié and C. A. Pontikis. *Geophys. Res. Lett.,* **23,** 1702–1703.

———, 1998: The positive charge reservoir for sprite-producing lightning. *J. Atmos. Terr. Phys.,* **60,** 689.

———, and W. Ecklund, 1992: 50 MHZ profiler observations of trailing stratiform precipitation: constraints on cloud microphysics and in situ charge separation. Preprints, *Int. Conf. on Cloud Physics and Precipitation,* Montreal, PQ, Canada.

———, and D. J. Boccippio, 1993: Dependence of cloud microphysics and electrification on mesoscale vertical air motions in stratiform precipitation. Preprints, *Conf. on Atmospheric Electricity,* St. Louis, MO, Amer. Meteor. Soc., 825–831.

———, and N. Renno, 1993: An analysis of the conditional instability of the tropical atmosphere. *Mon. Wea. Rev.,* **121,** 21–36.

———, and R. Zhang, 1993: Comment on "The effect of liquid water on thunderstorm charging," by C. P. R. Saunders. *J. Geophys. Res.,* **98,** 10 819–10 821.

———, and ———, 1996: The density of rime in laboratory simulations of thunderstorm microphysics and electrification. *J. Geophys. Res.,* **101,** 29 715–29 719.

———, S. G. Geotis, and A. B. Bhattacharya, 1989a: A radar study of the plasma and geometry of lightning. *J. Atmos. Sci.,* **46,** 1173.

———, M. E. Weber, and R. E. Orville, 1989b: The relationship between lightning type and convective state of thunderclouds. *J. Geophys. Res.,* **94,** 13 213–13 220.

———, R. Zhang, and J. Rydock, 1991: Mixed-phase microphysics and cloud electrification. *J. Atmos. Sci.,* **48,** 2195–2203.

———, S. A. Rutledge, S. G. Geotis, N. Renno, E. Rasmussen, and T. Rickenbach, 1992: A radar and electrical study of tropical "hot towers." *J. Atmos. Sci.,* **49,** 1386–1395.

———, M. E. Weber, R. Boldi, and P. Laroche, 1995: The use of lightning for the detection of microbursts and other aviation hazards. White paper prepared for Lightning Location and Protection.

———, and Coauthors, 1999: The behavior of total lightning activity in severe Florida thunderstorms. *Atmos. Res.,* **51,** 245–264.

———, K. Rothkin, D. Stevenson, and D. Boccippio, 2000: Global lightning variations caused by changes in thunderstorm flash rate and by changes in number of thunderstorms. *J. Appl. Meteor.,* **39,** 2223–2230.

Wilson, J. W., and D. Reum, 1988: The flare-echo: Reflectivity and velocity signature. *J. Atmos. Oceanic Technol.,* **5,** 197–205.

Witt, A., and S. P. Nelson, 1991: The use of single-Doppler radar for estimating maximum hailstone size. *J. Appl. Meteor.,* **30,** 425–431.

Workman, E. J., and S. E. Reynolds, 1949: Electrical activity as related to thunderstorm cell growth. *Bull. Amer. Meteor. Soc.,* **30,** 142–149.

Ziegler, C. L., and D. R. MacGorman, 1994: Observed lightning morphology relative to modelled space charge and electric field distributions in a tornadic storm. *J. Atmos. Sci.,* **51,** 833–851.

Zipser, E. J., 1994: Deep cumulonimbus cloud systems in the tropics with and without lightning. *Mon. Wea. Rev.,* **122,** 1837–1851.

Zrnić, D. S., 1987: Three-body scattering produces precipitation signature of special diagnostic value. *Radio Sci.,* **22,** 76–86.

———, V. N. Bringi, N. Balakrishnan, K. Aydin, V. Chandrasekar, and J. Hubbert, 1993: Polarimetric measurements in a severe hailstorm. *Mon. Wea. Rev.,* **121,** 2223–2238.